Mineral Processing Plant Design, Practice, and Control

PROCEEDINGS

VOLUME 2

Mineral Processing Plant Design, Practice, and Control

P R O C E E D I N G S

VOLUME 2

Edited by Andrew L. Mular, Doug N. Halbe, and Derek J. Barratt

Published by the
Society for Mining, Metallurgy, and Exploration, Inc.

SPONSORS

These organizations provided generous financial support for this publication.

Platinum Level
Canadian Process Technologies Inc.
Metso Minerals Industries Inc.

Gold Level
Newmont Mining Corporation
Norcast
OSISoft

Silver Level
Pocock Industrial, Inc.
WesTech

Cover photo courtesy of Rick Coleman of P.T. Freeport Indonesia (a subsidiary of Freeport-McMoRan Copper & Gold Inc.). The photo shows the P.T. Freeport Indonesia milling complex located in West Papua, Indonesia. The facilities currently process 245,000 tonnes per day of copper, gold, and silver ore, producing 2.8 million tonnes of concentrate annually.

Society for Mining, Metallurgy, and Exploration, Inc. (SME)
8307 Shaffer Parkway
Littleton, Colorado USA 80127
(303) 973-9550 / (800) 763-3132
www.smenet.org

SME advances the worldwide mining and minerals community through information exchange and professional development.

Copyright © 2002 Society for Mining, Metallurgy, and Exploration, Inc.

No part of this publication may be reproduced, stored in a retrieval system, or transmitted in any form or by any means, electronic, mechanical, photocopying, recording, or otherwise, without the prior written permission of the publisher.

All Rights Reserved. Printed in the United States of America

Disclaimer
The papers contained in this proceedings are published as supplied by individual authors. Any statement or views presented here are those of individual authors and are not necessarily those of the Society for Mining, Metallurgy, and Exploration, Inc. The mention of trade names for commercial products does not represent or imply the approval or endorsement of SME.

ISBN 0-87335-223-8
ISBN 978-0-87335-223-9

Preface

Nearly 25 years ago, SME published its first major symposium volume on plant design practice: *Mineral Processing Plant Design*. Two more volumes, the *Design and Installation of Comminution Circuits* (1982) and the *Design and Installation of Concentration and Dewatering Circuits* (1986), were published creating a positive worldwide affect on mining, mineral processing, and metallurgy. And although out of print, each of these publications is still in use today.

In 1998, SME, with strong encouragement from the Canadian mineral processing division of CIM, agreed to update the plant design series to provide a current overview of all facets of mineral processing plant design, control, and practice. An organizing committee and a general committee were formed to oversee this huge task.

The organizing committee, with input from some 40 members of the general committee, chose sections for the symposium as well as section co-editors. The proceedings editors worked with the section co-editors vetting the content and making the final choices for each section. In 1999, a call for papers was issued in *Mining Engineering* and various other journals. Authors were contacted by co-editors and encouraged to follow a stringent timetable to ensure that the proceedings would be published in time for distribution at the symposium.

Despite busy work schedules, our authors and co-authors managed to meet their deadlines with relatively minor delays. The section co-editors deserve significant praise for this accomplishment. They maintained close contact with corresponding authors and provided help when requested. Readers will find that the results of this hard work are most gratifying and that our authors deserve to be recognized and appreciated for years to come.

The useful information included in these two volumes will serve as an up-to-date aid for university-level professors and students in plant design courses; reference material for operators who must consider plant expansion, renovations, and new projects; and a quick reference for mineral processing plant design engineers, engineers from other disciplines, consultants in mineral processing, consultants who have minimal knowledge of mineral processing plant design, suppliers, and manufacturers.

The symposium is being held at the Fairmont Hotel Vancouver in Vancouver, B.C., Canada, from Oct. 20–24, 2002. It should be stressed that a comprehensive symposium of this type is organized once every 12 to 15 years and is international in scope. During the meeting, competition is avoided between sessions by not programming them concurrently. Everyone attends the same sessions or roundtables, thereby stimulating interaction and discussion throughout the five-day period.

The proceedings editors, section co-editors, and authors wish to thank Ms. Joette Cross, meetings manager; Ms. Tara Davis, program manager; and Ms. Jane Olivier, manager of book publishing, for their excellent assistance on this project. Despite our idiosyncrasies, their help and guidance was provided with professional courtesy and efficiency. We appreciate it.

Andrew L. Mular
Doug Halbe
Derek J. Barratt

Organizing Committee

Andrew L. Mular, *Chair*
Professor Emeritus
Dept. of Mining, University of B.C.
6350 Stores Road
Vancouver, BC V6T 1Z4
Canada

Doug Halbe, *Co-Chair*
Consultant
T.P. McNulty & Associates Inc.
P.O. Box 58667
Salt Lake City, UT 84158
USA

Leonard Harris, *Finance Chair*
Senior Consultant
9534 La Costa Lane
Lone Tree, CO 80124

Derek J. Barratt, *Committee Secretary and Technical Advisor*
Principal
DJB Consultants Ltd.
427 Silverdale Place
North Vancouver, BC V7N 2Z6
Canada

Proceedings Editors

Andrew L. Mular
Professor Emeritus
Dept. of Mining, University of B.C.
6350 Stores Road
Vancouver, BC V6T 1Z4
Canada

Doug Halbe
Consultant
T.P. McNulty & Associates Inc.
P.O. Box 58667
Salt Lake City, UT 84158
USA

Derek J. Barratt
Principal
DJB Consultants Ltd.
427 Silverdale Place
North Vancouver, BC V7N 2Z6
Canada

Section Co-Editors
Session/Roundtable Co-Chairs

1 Sampling
Graham Farquharson
Strathcona Mineral Services Ltd.
12th Floor, 20 Toronto Street
Toronto, ON M5C 2B8
Canada

Art Winckers
Arthur H. Winckers and Associates
Mineral Processing Consultant
4345 Raeburn Street
North Vancouver, BC V7G 1K1
Canada

2 Bench Scale and Pilot Plant Testwork
Jeff Austin
President
IME Inc.
13 2550 Acland Road
Kelowna, BC V1X 7L4
Canada

John Mosher
Consultant
A.R. MacPherson Consultants Ltd.
Postal Bag 4300
185 Concession Street
Lakefield, ON K0L 2H0
Canada

3 Financial and Feasibility Studies
Brian H. Johnston
Manager, Financial Analysis
Fluor Mining & Minerals
#700 1075 West Georgia Street
Vancouver, BC V6E 4M7
Canada

John W. Scott
Technical Director
Fluor Mining & Minerals
#700 1075 West Georgia Street
Vancouver, BC V6E 4M7
Canada

4 Models and Simulators for Selection Sizing, and Design
Dr. Brian Flintoff
Vice President, Technology
Metso Minerals—Minerals Processing
  Business Line
Process Technology
2281 Hunter Road
Kelowna, BC V1X 7C5
Canada

Dr. John Herbst
General Manager, Optimization
Metso Minerals—Minerals Processing
  Business Line
Optimization Services
79-7460 Mamalahoa Hwy.
M.C. Bldg. 2 #215
Kealakekua, HI 96750-2568
USA

5 Comminution (Crushing and Grinding)
R.E. McIvor
General Manager
Cliffs Technology Center
550 East Division Street
Ishpeming, MI 49849
USA

Tony Moon
Senior Consultant
Rio Tinto Technical Services
5295 S. 300 W., Suite 300
Murray, UT 84107
USA

James Vanderbeek
Manager, Mineral Processing
Phelps Dodge Mining Company
One North Central Avenue
Phoenix, AZ 85004
USA

6 Size Separation

Patrick Turner
Vice President, Mining Sales & Marketing
Krebs Engineers
5505 West Gillette Road
Tucson, AZ 85743
USA

James E. Wennen
Consulting Engineer
Consulting Services
31506 MacDougal Bay Road
Grand Rapids, MN 55774
USA

7 Solid-Solid Separation

George H. Hope
Manager, Project Development
Teck Cominco Ltd.
#600, 200 Burrard Street
Vancouver, BC V6C 3L9
Canada

Donovan F. Symonds
President
Norwest Corporation
12th Floor, 136 E. South Temple
Salt Lake City, UT 84111
USA

8 Flotation

M. Ian Callow
Manager, Technology—South America
Bechtel Chile Ltda.
Coyancura 2283, Piso 3, Providencia
Santiago
Chile

Glenn Dobby
Senior Vice President
MinnovEX Technologies Inc.
1140 Sheppard Avenue W. #6
Toronto, ON M3K 2A2
Canada

9 Solid-Liquid Separation

Frank Baczek
Region Manager, Latin America
EIMCO Process Equipment Co.
669 W. 200 S.
Salt Lake City, UT 84101
USA

Benjamin K. Pocock
General Manager
Pocock Industrial Inc.
6188 S. 300 W.
Salt Lake City, UT 84107
USA

10 Pumping, Material Transport, Drying, and Storage

Ken Boyd
President
KLB Enterprises Inc.
P.O. Box 508
Qualicum Beach, BC V9K 1A0
Canada

William E. Norquist
Manager, Metallurgy
Fluor Mining & Minerals
#700 1075 West Georgia Street
Vancouver, BC V6E 4M7
Canada

11 Pre-Oxidation

Dr. Ralph P. Hackl
Research Director
Placer Dome Inc.
P.O. Box 49330, Bentall Station
Vancouver, BC V7X 1P1
Canada

Dr. Kenneth Thomas
Managing Director
HATCH—Western Australia
Perth, WA
Australia

12 Leaching and Adsorption Circuits

Dr. Chris Fleming
Vice President & General Manager
Lakefield Research
Postal Bag 4300
Lakefield, ON K0L 2H0
Canada

Michael R. Schaffner
Process Superintendent
Coeur Rochester Inc.
P.O. Box 1057
Lovelock, NV 89419
USA

13 Extraction

Paul G. Semple
President
Penguin Automated Systems Inc.
1178 Giles Gates
Oakville, ON L6M 2S3
Canada

Ron Bradburn
Process Manager, International Operations
Newmont Mining Corporation
10101 E. Dry Creek Road
Englewood, CO 80112
USA

14 Bullion Production and Refining
Dr. Corby Anderson
Director
Center for Advanced Mineral and Metallurgical
    Processing
Montana Tech
1300 W. Park Street, Room 221 ELC Bldg.
Butte, MT 59701
USA

15 Tailings Disposal, Wastewater Disposal, and the Environment
James R. Arnold
Chief Operating Officer
Earthworks Technology
3495 Paiute Street
Winnemucca, NV 89445
USA

Dr. George W. Poling
Senior Vice President
Rescan Environmental Services
6th Floor, 1111 West Hastings Street
Vancouver, BC V6E 2J3
Canada

16 Construction Materials for Equipment and Plants
Dr. Rod McElroy
Senior Metallurgist
Fluor Mining & Minerals
#700 1075 West Georgia Street
Vancouver, BC V6E 4M7
Canada

Wesley Young
Principal Process Engineer
Fluor Mining & Minerals
#700 1075 West Georgia Street
Vancouver, BC V6E 4M7
Canada

17 Power, Water, and Support Facilities
M.N. Brodie, P. Eng.
Senior Electrical Consultant
Fluor Mining & Minerals
#700 1075 West Georgia Street
Vancouver, BC V6E 4M7
Canada

Charles R. Edwards
Chief Metallurgist
Cameco Corporation
2121 11th Street West
Saskatoon, SK S7M 1J3
Canada

18 Process Control and Instrumentation
Robert Edwards
Director, Process Engineering
Metso Minerals—Minerals Processing Business
    Line
Process Technology
2281 Hunter Road
Kelowna, BC V1X 7C5
Canada

Aundra Nix
Chief Metallurgist
ASARCO Inc.
Mission Complex
P.O. Box 111
Sahuarita, AZ 85629
USA

19 Engineering, Procurement, Construction, and Management
Roger M. Nendick
Vice President, Consulting & Technology
Fluor Mining & Minerals
#700 1075 West Georgia Street
Vancouver, BC V6E 4M7
Canada

Robert C. Schenk
Senior Consultant
509 Oak Park Drive
San Francisco, CA 94131-1028
USA

20 Start-Up, Commissioning, and Training
Ken Major
Manager, Crushing & Milling Systems
HATCH
#200 1550 Alberni Street
Vancouver, BC V6G 1A5
Canada

Mike Mular
Manager, Concentrating Operations
P.T. Freeport Indonesia
P.O. Box 61982
New Orleans, LA 70161-1982
USA

21 Case Studies
Dr. Martin C. Kuhn
Principal
Minerals Advisory Group, LLC
One W. Wetmore, Suite 207
Tucson, AZ 85705
USA

Donald C. Gale
Director Process Technology
Kvaerner—Metals E&C Division
12657 Alcosta Blvd.
San Ramon, CA 94583
USA

Contents

The Formal Basis of Design

Design Criteria: The Formal Basis of Design
J.W. Scott 3

1 Sampling

Section Co-Editors 23

Sampling a Mineral Deposit for Feasibility Studies
K.J. Ashley 25

Sampling in Mineral Processing
J.W. Merks 37

Sampling High Throughput Grinding and Flotation Circuits
J. Mosher, D. Alexander 63

Practical and Theoretical Difficulties When Sampling Gold
F.F. Pitard 77

Sampling a Mineral Deposit for Metallurgical Testing and the Design of Comminution and Mineral Separation Processes
J. Hanks, D. Barratt 99

2 Bench Scale and Pilot Plant Testwork

Section Co-Editors 117

Overview of Metallurgical Testing Procedures and Flowsheet Development
T.P. McNulty 119

Bench-Scale and Pilot Plant Tests for Comminution Circuit Design
J. Mosher, T. Bigg 123

The Selection of Flotation Reagents via Batch Flotation Tests
P. Thompson 136

Bench and Pilot Plant Programs for Flotation Circuit Design
S.R. Williams, M.O. Ounpuu, K.W. Sarbutt 145

Bench-Scale and Pilot Plant Testwork for Gravity Concentration Circuit Design
A.R. Laplante, D.E. Spiller 160

Bench Scale and Pilot Plant Tests for Magnetic Concentration Circuit Design
D.A. Norrgran, M.J. Mankosa 176

Bench-Scale and Pilot Plant Tests for Thickening and Clarification Circuit Design
B.K. Pocock, C.B. Smith, G.D. Welch 201

Bench-Scale and Pilot Plant Tests for Filtration Circuit Design
T. Kram **207**

Gold Roasting, Autoclaving, or Bio-Oxidation Process Selection Based on Bench-Scale
and Pilot Plant Test Work and Costs
J. McMullen, K.G. Thomas **211**

Bench-Scale and Pilot Plant Tests for Cyanide Leach Circuit Design
G.E. McClelland, J.S. McPartland **251**

Bench-Scale and Pilot Plant Work for Gold- and Copper-Recovery Circuit Design
D. Thompson **264**

Guiding Process Developments by Using Automated Mineralogical Analysis
D. Sutherland, Y. Gu **270**

3 Financial and Feasibility Studies

Section Co-Editors **279**

Guidelines to Feasibility Studies
J. Scott, B. Johnston **281**

Major Mineral Processing Equipment Costs and Preliminary Capital Cost Estimations
A.L. Mular **310**

Process Operating Costs with Applications in Mine Planning and Risk Analysis
D. Halbe, T.J. Smolik **326**

Financial Analysis and Economic Optimization
L.D. Smith **346**

Mining Project Finance Explained
R. Halupka **371**

4 Models and Simulators for Selection, Sizing, and Design

Section Co-Editors **381**

Mineral Processing Plant/Circuit Simulators: An Overview
J. Herbst, R.K. Rajamani, A. Mular, B. Flintoff **383**

BRUNO: Metso Minerals' Crushing Plant Simulator
D.M. Kaja **404**

PlantDesigner®: A Crushing and Screening Modeling Tool
P. Hedvall, M. Nordin **421**

JKSimMet: A Simulator for Analysis, Optimisation, and Design of Comminution Circuits
R.D. Morrison, J.M. Richardson **442**

JKSimFloat as a Practical Tool for Flotation Process Design and Optimization
M.C. Harris, K.C. Runge, W.J. Whiten, R.D. Morrison **461**

USIM PAC 3: Design and Optimization of Mineral Processing Plants from Crushing
to Refining
S. Brochot, J. Villeneuve, J.C. Guillaneau, M.V. Durance, F. Bourgeois **479**

Emergence of HFS as a Design Tool in Mineral Processing
J.A. Herbst, L.K. Nordell **495**

Reducing Maintenance Costs Using Process and Equipment Event Management
O.A. Bascur, J.P. Kennedy **507**

Enterprise Dynamic Simulation Models
D.W. Ginsberg **528**

5 Comminution (Crushing and Grinding)

Section Co-Editors **537**

Factors Which Influence the Selection of Comminution Circuits
D. Barratt, M. Sherman **539**

Types and Characteristics of Crushing Equipment and Circuit Flowsheets
K. Major **566**

Selection and Sizing of Primary Crushers
R.W. Utley **584**

In-Pit Crushing Design and Layout Considerations
K. Boyd, R.W. Utley **606**

Selection and Sizing of Secondary and Tertiary Cone Crushers
G. Beerkircher, K. O'Bryan, K. Lim **621**

Selection, Sizing, and Special Considerations for Pebble Crushers
K. O'Bryan, K. Lim **628**

Selection and Sizing of High Pressure Grinding Rolls
R. Klymowsky, N. Patzelt, J. Knecht, E. Burchardt **636**

Crushing Plant Design and Layout Considerations
K. Boyd **669**

Types and Characteristics of Grinding Equipment and Circuit Flowsheets
M.I. Callow, A.G. Moon **698**

Selection of Rod Mills, Ball Mills, and Regrind Mills
C.A. Rowland Jr. **710**

Selection and Sizing of Autogenous and Semi-Autogenous Mills
D. Barratt, M. Sherman **755**

Selection and Sizing of Ultrafine and Stirred Grinding Mills
J.K.H. Lichter, G. Davey **783**

Grinding Plant Design and Layout Considerations
M.I. Callow, D.G. Meadows **801**

Selection and Evaluation of Grinding Mill Drives
G.A. Grandy, C.D. Danecki, P.F. Thomas **819**

The Design of Grinding Mills
V. Svalbonas **840**

6 Size Separation

Section Co-Editors	**865**
Sizing and Application of Gravity Classifiers *W.M. Reed*	**867**
Hydrocyclone Selection for Plant Design *T.J. Olson, P.A. Turner*	**880**
Coarse Screening *M.A. Bothwell, A.L. Mular*	**894**
Fine Screening in Mineral Processing Operations *S.B. Valine, J.E. Wennen*	**917**
The Use of Hindered Settlers to Improve Iron Ore Gravity Concentration Circuits *S. Hearn*	**929**

7 Solid-Solid Separation

Section Co-Editors	**945**
Types and Characteristics of Gravity Separation and Flowsheets *R.O. Burt*	**947**
Types and Characteristics of Heavy-Media Separators and Flowsheets *R.A. Reeves*	**962**
Types and Characteristics of Non-Heavy Medium Separators and Flowsheets *J.K. Alderman*	**978**
The Selection and Sizing of Centrifugal Concentration Equipment: Plant Design and Layout *A.R. Laplante*	**995**
Sizing and Selection of Heavy Media Equipment: Design and Layout *D.F. Symonds, S. Malbon*	**1011**
Photometric Ore Sorting *B. Arvidson*	**1033**
Electrical Methods of Separation *A.L. Mular*	**1049**
Selection and Sizing of Magnetic Concentrating Equipment: Plant Design/Layout *D.A. Norrgran, M.J. Mankosa*	**1069**

8 Flotation

Section Co-Editors	**1095**
Overview of Flotation Technology and Plant Practice for Complex Sulphide Ores *N.W. Johnson, P.D. Munro*	**1097**
Overview of Recent Developments in Flotation Technology and Plant Practice for Copper Gold Ores *A. Winckers*	**1124**

An Overview of Recent Developments in Flotation Technology and Plant Practice for
Nickel Ores
A. Kerr **1142**

Nonsulfide Flotation Technology and Plant Practice
J. Miller, B. Tippin, R. Pruett **1159**

Design of Mechanical Flotation Machines
M.G. Nelson, F.P. Traczyk, D. Lelinski **1179**

Flotation Equipment Selection and Plant Layout
K.R. Wood **1204**

Column Flotation
G. Dobby **1239**

9 Solid-Liquid Separation

Section Co-Editors **1253**

Characterization of Process Objectives and (General) Approach to Equipment Selection
C.E. Silverblatt, J.H. Easton **1255**

Centrifugal Sedimentation and Filtration for Mineral Processing
W. Leung **1262**

Characterization of Equipment Based on Filtration Principals and Theory
G.D. Welch **1289**

Testing, Sizing, and Specifying Sedimentation Equipment
T. Laros, S. Slottee, F. Baczek **1295**

Testing, Sizing, and Specifying of Filtration Equipment
C.B. Smith, I.G. Townsend **1313**

Design Features and Types of Sedimentation Equipment
F. Schoenbrunn, T. Laros **1331**

Design Features and Types of Filtration Equipment
C. Cox, F. Traczyk **1342**

Plant Design, Layout, and Economic Considerations
M. Erickson, M. Blois **1358**

10 Pumping, Material Transport, Drying, and Storage

Section Co-Editors **1371**

Selection and Sizing of Slurry Pumps
M.J. Bootle **1373**

Selection and Sizing of Slurry Lines, Pumpboxes, and Launders
B. Abulnaga, K. Major, P. Wells **1403**

Slurry Pipeline Transportation
B.L. Ricks **1422**

The Selection and Sizing of Conveyors, Stackers, and Reclaimers
G. Barfoot, D. Bennett, M. Col **1446**

Selection and Sizing of Concentrate Drying, Handling, and Storage Equipment M.E. Prokesch, G. Graber	1463
The Selection and Sizing of Bins, Hopper Outlets, and Feeders J. Carson, T. Holmes	1478

11 Pre-Oxidation

Section Co-Editors	1491
Design of Barrick Goldstrike's Two-Stage Roaster D. Warnica, A. Cole, S. Bunk	1493
Selection of Materials and Mechanical Design of Pressure Leaching Equipment K. Lamb, J. Gulyas	1510
Barrick Gold—Autoclaving and Roasting of Refractory Ores K.G. Thomas, A. Cole, R.A. Williams	1530
Selection and Sizing of Biooxidation Equipment and Circuits C.L. Brierley, A.P. Briggs	1540

12 Leaching and Adsorption Circuits

Section Co-Editors	1569
Copper Heap Leach Design and Practice R.E. Scheffel	1571
Precious Metal Heap Leach Design and Practice D.W. Kappes	1606
Agitated Tank Leaching Selection and Design K.A. Altman, M. Schaffner, S. McTavish	1631
CIP/CIL/CIC Adsorption Circuit Process Selection C.A. Fleming	1644
CIP/CIL/CIC Adsorption Circuit Equipment Selection and Design K.A. Altman, S. McTavish	1652

13 Extraction

Section Co-Editors	1661
Zinc Cementation—The Merrill Crowe Process A.P. Hampton	1663
Selection and Design of Carbon Reactivation Circuits J. von Beckmann, P.G. Semple	1680
Selection and Sizing of Elution and Electrowinning Circuits P. Hosford, J. Wells	1694
Selection and Sizing of Copper Solvent Extraction and Electrowinning Equipment and Circuits C.G. Anderson, M.A. Giralico, T.A. Post, T.G. Robinson, O.S. Tinkler	1709

14 Bullion Production and Refining

Section Co-Editors.. **1745**

Bullion Production and Refining
C.O. Gale, T.A. Weldon **1747**

Platinum Group Metal Bullion Production and Refining
C.G. Anderson, L.C. Newman, G.K. Roset **1760**

Fundamentals of the Analysis of Gold, Silver, and Platinum Group Metals
C.G. Anderson **1778**

15 Tailings Disposal, Wastewater Disposal, and the Environment

Section Co-Editors.. **1807**

Management of Tailings Disposal on Land
B.S. Brown **1809**

Design of Tailings Dams and Impoundments
P.C. Lighthall, M.P. Davies, S. Rice, T.E. Martin **1828**

Hazardous Constituent Removal from Waste and Process Water
L. Twidwell, J. McCloskey, M. Gale-Lee **1847**

Treatment of Solutions and Slurries for Cyanide Removal
M.M. Botz, T.I. Mudder **1866**

Strategies for Minimization and Management of Acid Rock Drainage and Other Mining-Influenced Waters
R.L. Schmiermund **1886**

Environmental and Social Considerations in Facility Siting
B.A. Filas, R.W. Reisinger, C.C. Parnow **1902**

16 Construction Materials for Equipment and Plants

Section Co-Editors.. **1909**

Selection of Metallic Materials for the Mining/Metallurgical Industry
G. Coates **1911**

Elastomers in the Mineral Processing Industry
P. Schnarr, L.E. Schaeffer, H.J. Weinand **1932**

Plastics for Process Plants and Equipment
G.W. McCuaig **1953**

Commercial Acceptance and Applications of Masonry and Membrane Systems for the Process Industries
R.E. Aliasso Jr., T.E. Crandall, D.M. Malone, R.J. Storms **1962**

17 Power, Water, and Support Facilities

Section Co-Editors.. **1971**

The Development of an Electric Power Distribution System
M.N. Brodie **1973**

Selection of Motors and Drive Systems for Comminution Circuits
P.F. Thomas **1983**

Selection of Metallurgical Laboratory and Assay Equipment: Laboratory Designs and Layouts
P.F. Wells **2011**

On-Line Composition Analysis of Mineral Slurries
T.F. Braden, M. Kongas, K. Saloheimo **2020**

18 Process Control and Instrumentation

Section Co-Editors **2049**

Introduction to Process Control
B. Flintoff **2051**

Well Balanced Control Systems
T. Stuffco, K. Sunna **2066**

The Selection of Control Hardware for Mineral Processing
R.A. Medower, R.E. Cook **2077**

Basic Field Instrumentation and Control System Maintenance in Mineral Processing Circuits
J.R. Sienkiewicz **2104**

Strategies for Instrumentation and Control of Crushing Circuits
S.D. Parsons, S.J. Parker, J.W. Craven, R.P. Sloan **2114**

Strategies for the Instrumentation and Control of Grinding Circuits
R. Edwards, A. Vien, R. Perry **2130**

Strategies for the Instrumentation and Control of Solid–Solid Separation Processes
G.H. Luttrell, M.J. Mankosa **2152**

Strategies for Instrumentation and Control of Thickeners and Other Solid–Liquid Separation Circuits
F. Schoenbrunn, L. Hales, D. Bedell **2164**

Strategies for Instrumentation and Control of Flotation Circuits
H. Laurila, J. Karesvuori, O. Tiili **2174**

Pressure Oxidation Control Strategies
J. Cole, J. Rust **2196**

19 Engineering, Procurement, Construction, and Management

Section Co-Editors **2209**

Development of a Mineral Processing Flowsheet—Case History, Batu Hijau
T. de Mull, S. Saich, K. Sobel **2211**

Specification and Purchase of Equipment for Mineral Processing Plants
C. Hunker, S. Maldonado **2223**

The Management and Control of Costs of Capital Mineral Processing Plants
D.W. Stewart **2230**

Schedule Development and Schedule of Control of Mineral Processing Plants
P. Kumar **2238**

The Risks and Rewards Associated with Different Contractual Approaches
P.J. Gard **2245**

Success Strategies for Building New Mining Projects
R.J. Hickson **2250**

20 Start-Up, Commissioning, and Training

Section Co-Editors **2275**

Pre-Commissioning, Commissioning, and Training
T. Watson **2277**

Plant Ramp Up and Performance Testing
R.M. Nendick **2285**

Preparation of Effective Operating Manuals to Support Operator Training for Metallurgical Plant Start-Ups
S.R. Brown **2290**

Planning and Staffing for a Successful Project Start-Up
K.A. Brunk, L.J. Buter, K.M. Levier **2299**

Maintenance Scheduling, Management, and Training at Start-Up: A Case Study
P. Vujic **2315**

Operator Training
A. Vien **2328**

Safety and Health Considerations and Procedures During Plant Start-Up
L.A. Schack **2337**

21 Case Studies

Section Co-Editors **2343**

Sunrise Dam Gold Mine—Concept to Production
W.R. Lethlean, P.J. Banovich **2345**

A Case Study in SAG Concentrator Design and Operations at P.T. Freeport Indonesia
R. Coleman, A. Neale, P. Staples **2367**

High Pressure Grinding Roll Utilization at the Empire Mine
D.J. Rose, P.A. Korpi, E.C. Dowling, R.E. McIvor **2380**

The Raglan Concentrator—Technology Development in the Arctic
J. Holmes, D. Hyma, P. Langlois **2394**

Author Index **I-1**

Subject Index **I-3**

9
Solid–Liquid Separation

Section Co-Editors:
Frank Baczek and Benjamin K. Pocock

Characterization of Process Objectives and (General) Approach to Equipment Selection C.E. Silverblatt, J.H. Easton	1255
Centrifugal Sedimentation and Filtration for Mineral Processing W. Leung	1262
Characterization of Equipment Based on Filtration Principals and Theory G.D. Welch	1289
Testing, Sizing, and Specifying Sedimentation Equipment T. Laros, S. Slottee, F. Baczek	1295
Testing, Sizing, and Specifying of Filtration Equipment C.B. Smith, I.G. Townsend	1313
Design Features and Types of Sedimentation Equipment F. Schoenbrunn, T. Laros	1331
Design Features and Types of Filtration Equipment C. Cox, F. Traczyk	1342
Plant Design, Layout, and Economic Considerations M. Erickson, M. Blois	1358

Characterization of Process Objectives and (General) Approach to Equipment Selection

Charles E. Silverblatt[1], Jeffery H. Easton[1]

ABSTRACT

There are many processes in the Minerals Industry that require some type of solid-liquid separation. At the same time, there are many different equipment designs championed by a variety of manufacturers, each seeking to apply their own proprietary design. It is in the manufacturer's best interest to provide the most economic solution to a separation application, and every effort is made to do so. But providing the most economic design requires a cooperative effort between the process design engineer and the manufacturer. The design engineer must provide the manufacturer with a thorough understanding of the process requirements, and the manufacturer must in turn provide the design engineer with the capabilities of the separation equipment.

INTRODUCTION

The design and equipment selection process becomes much easier when the design engineer has a thorough understanding of the characteristics of the solid-liquid suspension to be separated and how these slurry characteristics affect equipment design. The simple fact is this - The **slurry** dictates the equipment design and selection, not the application engineer. The application engineer assists by providing a design that satisfies the demand of the slurry separation objectives.

This chapter provides the reader with an introduction to separation equipment options, an approach that the application engineer can use in choosing which option(s) best fits a given process, and the primary factors involved in solid-liquid separation. Later chapters will provide detailed discussions of sedimentation and filtration unit operations and the specific equipment designs that provide a suitable application for each of unit operation. Once an understanding of the various unit operations has been established, selection of the correct separation device can be confidently made. The discussions in this section are based on non-biological solids.

CRITICAL PLANT SEPARATIONS

Solid-liquid separation is an important, and many times critical, step in a mineral processing plant. Some examples of typical critical separations are as follows:

1. Operation of a seven to eight stage countercurrent decantation (CCD) washing/thickening circuit to effect high recovery of the caustic and aluminum values in an alumina plant. This must be accomplished in a closed circuit system with highly scaling slurries at starting temperatures of 100°C.
2. Operation of a vacuum filter to produce an iron concentrate cake with a moisture content required for pelletizing in order to minimize bentonite addition and maximize pellet quality. A good quality filtrate must be maintained to prevent abrasion of the internal parts of the filter.
3. Operation of a closed water system in a coal washing plant so that reclaimed water containing less than one percent suspended solids is produced for reuse to maintain beneficiation efficiency. Refuse solids must also be thickened so that they can be effectively dewatered to meet EPA requirements.

---

[1]WesTech Engineering, Inc., Salt Lake City, Utah

4. Recovery of phosphoric acid values from slurry containing gypsum and other gangue solids at an elevated temperature. This must be done while minimizing wash water usage, maximizing P_2O_5 recovery and minimizing scale problems.
5. Neutralization of acid wastes from mining or metallurgical processes to produce water suitable for discharge or recycle to the process and solids suitable for landfill.

While solid-liquid separations are important to the overall performance of the plant, they also represent a substantial capital investment and significant operating and maintenance expense. Once the equipment selection process has reduced the number of choices to a few suitable options, comparative costs become a very important part of the selection process.

AVAILABLE EQUIPMENT DESIGNS

Equipment designs intended to separate solids from liquids or concentrate slurries vary widely depending upon the specific process requirements and the characteristics of the feed slurry. There are generally two basic objectives in any solid-liquid separation process: The production of a clear liquid and a properly washed and dewatered solid. Dilute slurries frequently require some type of pretreatment before final dewatering, while concentrated slurries can usually be handled directly by the final dewatering machine. Gravity sedimentation is frequently used to clarify relatively large volumes of liquid or to concentrate dilute slurries when preparing them for dewatering.

A few of the available solid-liquid separation designs are:

Standard Clarifiers
Clarifiers With Solids Recirculation
   - Internal & External
Clarifiers With Internal Flocculation
   Chambers
Pulp Blanket Clarifiers
Standard Thickeners
Thickeners With Internal Flocculation
   Chambers
Thickeners With Self Diluting Feed Systems
Granular Media Filters - Fixed & Moving Bed

Disc Filters
Standard Drum Filters
Drum Belt Filters
Horizontal Belt Filters
Vacuum Precoat Filters
Plate & Frame Pressure Filters
Recess Plate Pressure Filters
Tower Pressure Filters
Centrifuges
Dissolved Air/Gas Flotation

How does one make a selection from such varied equipment types, particularly when there are a number of variations within each type? By determining the:

1. Objective of the separation step
2. Effect of the separation step on the overall flowsheet
3. Flow rate of the slurry
4. Type of operation: Continuous or batch. If batch, the time available to process a batch.
5. Characteristics of the slurry involved:
   a. Size and density of the solids: coarse or fine, heavy or light
   b. Concentration of the slurry: dilute or concentrated
6. Need for flocculation:
   a. Are coagulants and polymers permitted?
   b. Are the flocs fragile or tough?
7. More valuable phase: Solid or liquid
8. Required moisture and solute content of the final slurry or filter cake
9. Required clarity of the liquid
10. Need & suitability of filter aids such as DE, etc.
11. Availability of representative sample and arrangements required for testing.

This information will eliminate many possibilities, point directly to a few possibilities and help in designing a suitable testing program. The results from a test program will allow a detailed evaluation of the available equipment options. In order to best understand how the various equipment options can fit into the proposed flowsheet, one must understand the primary factors that influence equipment and process design.

PRIMARY FACTORS IN SOLID-LIQUID SEPARATION

The discussions that follow address only the more important of the many factors affecting separation.

Required Capacity

It is not surprising to find that the choice of equipment type is frequently a function of required capacity. Relatively small to medium capacity requirements may, for example, best be met by using filters, either batch or continuous, while the most economic solution for large capacity requirements frequently involves thickeners. There have been cases where the design engineer had a strong preference for filters in the washing circuit based on historical usage, but when testing showed that the filtration rate was lower than expected, a CCD circuit was found to be the more economic solution. With very large capacity requirements, thickeners, when they are applicable, are generally the more economic solution, even though they require more space than filters, as they operate with less operator attention, less maintenance, and less overall cost than filters.

Particle Size Distribution

Particle size is one of the most important factors influencing performance and cost. Information such as the percent minus 200, 325 or 400 mesh will usually provide the application engineer with an idea of the unit's capacity and expected performance. When extra fine solids are involved, specific surface measurements may be more meaningful. One thing is certain, the finer the solids, the lower the unit's capacity, the poorer it's performance and the more restrictive the equipment choices. However, fine particles with a very narrow distribution, such as may be produced by a crystallizer, are much easier to handle than a wide distribution range with the same average particle size.

Sometimes the particle size distribution or specific surface area is purposely controlled and kept within a narrow range to improve the quality of the final product. In the production of alumina trihydrate, a once through precipitation would produce a very fine product that would be difficult to dewater and wash. However, classification procedures after precipitation along with fine particle recycle and other crystal growth controls produce a relatively coarse, but narrow particle size range product that dewaters rapidly and washes efficiently.

Poor design can destroy a good size distribution. A three stage, countercurrent cyclone classification pilot plant circuit was installed to eliminate extreme fines in a plant handling soft phosphate rock. Unfortunately, greatly oversized centrifugal pumps with high shear closed impellers were employed which required excessive throttling. The resulting attrition produced a product size distribution that was essentially the same as the initial feed and product values were lost.

Particle Shape

Particle shape starts to influence performance when it varies greatly from a general spherical form. Platelets and long needles represent two extremes. Platelets act as multiple flapper valves within a filter cake and severely restrict cake formation rate, particularly at higher vacuums. Long needles can cause severe blinding of filter media by imbedding in the pores of the cloth. In this case, continuous cloth washing on a belt type filter is required to maintain cloth porosity.

Heavy (high specific gravity) particles of irregular shape, such as iron ore particles, can pack so tightly on the bottom of a thickener that they are very difficult to move. The problem is overcome by operating at low solids inventory, high rake speed and special rake blade designs.

Feed Suspended Solids Concentration

Feed solids concentration strongly influences equipment selection. Generally, dilute slurries must be concentrated, usually in a gravity thickener, before going to the final dewatering device. Since the final dewatering device almost always operates most efficiently on more concentrated slurries, it behooves the operator to control the thickener so that it consistently produces concentrated slurry.

If flocculation is required, thickener operations can be influenced in the opposite way. Dilute slurries not only flocculate more easily than concentrated slurries but yield a different floc structure that settles and concentrates more rapidly. If a relatively concentrated thickener feed must be flocculated, it should be diluted for best results. The need for thickener feed dilution was clearly demonstrated many years ago at a coal washing plant when very poor flocculation and settling resulted in a dirty overflow, a low underflow solids concentration, and a solids build up in the circulating water system. The problem was corrected by simply diluting the feed with a portion of the overflow. A rule-of-thumb developed in the coal industry is that the solids concentration of the minus 200 mesh fraction of thickener feed should be no greater than 7%.

Recent experience has shown that most required thickener feed dilution can be conveniently carried out in a self-diluting feed system.

-10 Micron Solids Content

The finest particle sizes, particularly those finer than 2 microns, have a tremendous affect on solid-liquid separation processes because of the associated very large surface areas. If fine particles are dispersed as true colloids, they exhibit Brownian movement and do not settle. Therefore, charge neutralization (coagulation), and sometimes polymer addition, are required for their removal. Even then, these finer sizes adversely affect settling rate, filtration rate, and cake moisture content. If flocculation is not employed, these finer solids can build up within the system and adversely affect the whole operation.

Flocculation

The use of long chain polymers for flocculation, frequently following charge neutralization, is an excellent and important tool in all types of sedimentation and filtration, and their use has turned many difficult separations into easy ones.

When long chain polymers are used in thickening, the operator must be very careful not to over-use or mis-use them. Over-flocculation not only leads to the creation of a gelatinous mass that is difficult or impossible to transport to the discharge point, but is an unnecessary expense. It is easy to understand how over-flocculation occurs, particularly when there is significant variation in the mass flow rate and settling ability. The easy, potentially troublesome and more costly, approach to flocculation is to set the polymer addition rate high enough to cover all circumstances, otherwise, someone must check the degree of flocculation each time there is a change in feed rate or quality. The correct approach to polymer control is to vary the polymer rate to maintain a constant settling rate, while at the same time, checking to make sure that solids do not form a gelatinous mass. The best approach to polymer control is the use of one of the computer control systems that are now becoming available.

Viscosity & Feed Temperature

Increases in fluid viscosity almost always decrease solid-liquid separation rates. This is true in sedimentation as well as in final dewatering. In the case of dewatering, the effect is usually an increase in the moisture content of the discharged solids. Thus, while production rate of the dewatering device might be maintained, moisture content of the product would increase, possibly causing downstream problems.

Viscosity is affected primarily by temperature. For example, the viscosity of water is 0.98 centipoise at 21°C, but increases to 1.7 centipoise at 2°C, a 73.5% increase. Obviously, this increase in viscosity will affect both settling rate and moisture reduction in the final dewatering step. In the filtration of fine coal, a study some years ago showed that viscosity had a substantial affect on cake moisture content, while surface tension, which is also affected by temperature, made little difference. An examination of the operating data of that same plant, which used a lot of make-up water, showed that the moisture content of the total coal plant product was a function of the temperature of the river water used in the plant.

Dissolved Solids Build Up In Feed Liquor

Overall plant economics usually dictate that a minimum of water be discharged to waste. This also means that even minor contaminates have an opportunity to build up to significant, and sometimes harmful, concentrations. While it happens infrequently, long-term solute build up can change water phase ionic strength sufficiently to require a complete re-investigation of flocculation because the original polymers and procedure failed completely. Build up of other constituents, such a chlorides, may result in unexpected corrosion.

Slurry Scaling

Whenever lime addition is involved, there is the risk of incomplete reaction such as in the precipitation of gypsum and calcium carbonate or other compounds that form supersaturated solutions. These reactions will go-to-completion and scale on equipment surfaces unless care is taken to carry out these precipitations in the presence of previously precipitated solids and with ample reaction time. Care in the design of the precipitation reaction will not only reduce scaling tendencies, but can produce larger, easier to handle particles. The various high-density sludge acid mine drainage precipitation processes and the solids recirculation units used for water softening are good examples of the application of these principles.

DETERMINATION OF SLURRY CHARACTERISTICS

Operating Experience

When considering a plant expansion, an operating plant is the ideal source for information on slurry characteristics. Even then, design engineers frequently make changes in the flow sheet that will significantly impact feed slurry characteristics. Nevertheless, data from a plant operating on a similar slurry is very valuable information that should be used to amplify laboratory or pilot plant testing results, as an operating installation will often bring to light important factors that may not be evident during a short term testing program. When these types of data are not available, a well-constructed and conducted test program will provide sound sizing and design information.

Unit Operations

There are a number of design variables in both sedimentation and filtration unit operations that must be investigated and defined in the process of selecting the proper type of equipment. Test results are usually defined in terms of these design variables. The lists that follow, although not exhaustive, contain most of the variables of concern.

Gravity Sedimentation

- **Chemical flocculation.** Current practice almost always makes use of chemical flocculation to reduce the size of the sedimentation vessel, enhance settling rate, and increase underflow concentration.
  - *Chemical flocculant selection.* A wide selection of chemical flocculants is available from a number of manufacturers. Screening tests on small slurry samples can quickly reduce the number of chemical candidates to less than three.
  - *Chemical flocculant mixing.* The method and intensity of mixing influences flocculation. Properly conducted screening tests can define the best mixing method.
- **Solids recirculation.** The minimum suspended solids concentration required for flocculation to produce a practical settling rate usually occurs within the range of 300 to 700 mg/L. Previously concentrated solids from the clarifier underflow are recirculated to the feed when the feed solids must be increased.
- **Feed dilution.** When the solids concentration is too high, flocculation is inefficient. It is almost always best to dilute the feed to less than 20 wt.% suspended solids, and frequently to as low as 10 wt.% solids. Dilution may be by external pumping or by natural internal dilution into the flocculation zone.
- **Mechanical floc growth.** Mechanical floc growth is the slow mixing of a polymer solution with the feed slurry that is required to form flocs large enough to settle at a useful rate. This mixing is usually carried out within the feedwell to prevent the floc breakdown that occurs when flocs are transferred from one vessel to another. The nominal design time is usually three times the observed bench test time.
- **Settling rate.** Settling rate is the rate at which the solids settle out of the liquid. The observed rate obtain during a bench test must be derated to account for the short circuiting that exists in every gravity sedimentation machine. One historically accepted scale-up factor is 50% of the observed settling rate.

- **Clarification time.** Clarification time is the time required for the fine particles left in suspension after the interface or bulk of the solids have settled out to agglomerate and settle out. The time observed in a bench test must be derated to account for the short-circuiting that exists in every gravity sedimentation machine. The derating factor is variable, based on the diameter to depth ratio of the clarifier.
- **Solids handling.** Time and depth are required for the solids to settle to some desired concentration. This requirement is frequently expressed as Unit Area, the area required for a ton per day of solids settle to the required concentration. This area increases at a rapidly increasing rate as the solids concentration approaches maximum.

Filtration
- **Chemical flocculation.** Flocculation immediately ahead of filter may or may not be required. If the filter follows a thickener that used flocculation, additional polymer will usually be beneficial.
  - *Chemical flocculant selection.* A wide selection of chemical flocculants is available from a number of manufacturers. Flocculation and filtration tests on small slurry samples are used to reduce the number of chemical candidates to one or two.
  - *Chemical flocculant mixing.* Casual mixing is seldom sufficient for slurries concentrated enough for efficient filtration. A power mixer is usually the most effective. The mixing system should be located adjacent to or directly above the filter.
- **Cake formation rate.** Rate at which the cake is formed. This rate is strongly influenced by feed solids concentration and by the cake thickness required for discharge. Correlating techniques are readily available in the literature.
- **Cake dewatering rate.** Rate at which residual liquid is removed from the cake during a period when only air or gas passes through the cake.
- **Cake washing rate.** Rate at which wash liquid passes through the cake.
- **Wash efficiency.** Cake wash displaces the residual liquid in the unwashed cake, and the efficiency of this displacement is the wash efficiency. Correlating techniques are available in the literature.
- **Airflow rate.** Rate at which air or gas passes through the cake during a dewatering period.
- **Steam drying.** Application of steam to a filter cake. Steam not only displaces residual liquid, but heats both the residual liquid and the cake solids. The resulting elevation in temperature of the residual liquid increases the rate of dewatering and thus reduces the final cake liquid content.

TESTING PROTOCOL

Representative Sample
Before deciding upon any type of testing program, it must be possible to obtain a representative sample, either for laboratory testing or for pilot plant testing. Laboratory tests are much less expensive than continuous pilot tests and can also be carried out in a much shorter time period. Any sample used for laboratory testing represents only a snapshot of the process, and one must be certain that a single or several snapshot samples will truly represent the process variations, or are the best representation that can be obtained. If process development is based solely on core samples, laboratory tests are the only choice, and several samples are usually required to cover the expected range of operations. Even then, allowances must be made for the fact that the commercial product will not be as clean as the product produced from core samples. Today's laboratory test scale up techniques are sufficiently dependable to use for full-scale design, particularly when the equipment supplier has experience in similar operations. The weakest link in the sizing chain is usually the sample.

Test Objective
The objective or objectives of any test program must be clearly defined, or the test program is likely to be a waste of time. A test program frequently includes several separation steps, and specific collateral data may be required to assist in sizing pumps, mixers, etc. Care must be taken to ensure that the persons carrying out the test program are familiar with the overall process and the affect of each step upon the rest of the process. Careful planning is essential.

Data Collection & Analysis

It is essential that each sample be characterized, as a bare minimum, with respect to suspended solids concentration, particle size distribution, liquid phase TDS, pH and temperature. Other important tests include a detailed chemical analysis of the liquid phase, a chemical analysis of the solids, and any other information that may be specific to the process. This information allows the results of the test program to be evaluated against the expected feed quality or the actual feed obtained after plant start up.

The data required for such standard separation steps as clarification, thickening, filtration, centrifugation, etc. are well defined by the companies supplying those items of equipment. Nevertheless, there are frequently special analytical requirements or other unusual considerations that require special planning. Details of the tests including all data sheets and all support assistance must be prearranged to ensure that all information is obtained in a timely fashion.

Equipment Selection

The results of the test program will detail equipment selection and sizing options required to meet solids-liquid separation objectives. Capital and operating costs then become the primary equipment selectors. The following chapters will detail specific equipment characteristics and the sizing tests required for each type of equipment.

REFERENCES

D.A. Dahlstrom. 19__. Chapter 6, Fundamentals of Solid-Liquid Separation. Solid-Liquid Separation, ASME _____, Edited by Harma, R. O. & Degner, V. R.

R.H. Perry & D. W. Green (Eds.). 1997. Liquid-Solid Operations and Equipment. Perry's Chemical Engineers' Handbook, 7th Ed.

D. Purchas (Ed.). 1977. Solid/Liquid Separation Equipment Scale-Up, Uplands Press, Croydon, England.

Centrifugal Sedimentation and Filtration for Mineral Processing

Wallace Leung, Bird Machine Company

ABSTRACT

Centrifugal separation has been commonly practiced in many key separation steps in mineral processing, some of which include separation, clarification, classification, degritting, dewatering, and purification. Literally thousands of sedimenting and filtering centrifuges are employed today in processes such as coal, tar sand, kaolin, potash, soda ash, calcium carbonate, and drill mud etc. In the past decade, significant advancement has been made to develop effective technology and know-how to improve the process separation and to get higher-grade product(s) with improved throughput and lower energy consumption. In this paper, the theory, types, process functions, applications, and new developments of centrifuges for mineral processing are presented.

THEORY OF CENTRIFUGAL SEPARATION

Regimes of Sedimentation

The sedimentation behavior of a suspension may be classified into four categories in accordance with the solids concentration in suspension and the degree of aggregation of solids. This is illustrated in Figure 1, which in essence is a modified Fitch diagram. For dilute concentration and low degree of solids aggregation, solid particles settle independent of each other and they follow the Stokes' law of sedimentation, which was developed for spherical particles settling under earth gravity, one g ($9.8 \, m/s^2$). As solids concentration increases, the sedimentation rate of particles is affected hydrodynamically by neighboring particles despite there is no physical contact between them. Under this condition the settling rate may be less, or even higher, than the Stokes' settling velocity.

For a given solids concentration, as the particles tend to agglomerate due to weak or negligible electric repulsion they form aggregate and settle as a large floc, which can be modeled by fractal analysis. This allows both small and large particles to settle at the same speed, also known as zone settling, without discriminating the size of individual particles. Addition of coagulant and flocculant (polymer) may further promote formation of agglomerates and flocs leading to zone settling; while introduction of dispersant extends the discrete particle settling condition well into the concentrated solids region in which hindered and zone settling normally prevail, see Figure 1. The former finds applications such as clarification of waste slurries, while the latter finds applications such as classification of valuable fine-particle slurries for coating and pigment market.

Dense thick slurry forms networking as particle concentration and the degree of aggregation both increase in a suspension. Under gravitational body force the solid network or matrix compresses downward (compaction) while liquid expresses counter-currently upward (expression). This is delineated as the region to the upper right in Figure 1.

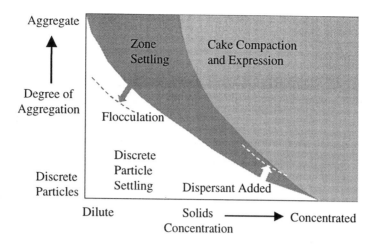

Figure 1 Different regimes of sedimentation

Stokes' Law
The sedimentation of discrete particles in a viscous fluid under centrifugal gravity, G, can be described by Stokes' law,

$$v_s = \frac{\Delta \rho G x^2}{18\mu} \lambda(\phi) \qquad (1)$$

$\Delta\rho = \rho_s - \rho_L$ is the density difference between solids and suspension, x the particle size (equivalent diameter), ϕ the solids volume fraction, λ the hindered settling function, μ the viscosity of suspension. For dilute suspension where $\phi \ll 1$, $\lambda(\phi) = 1$ and the above reduces to the Stokes' form of sedimentation for single particles if centrifugal gravity G is replaced by earth gravity g. For concentrated slurry $\lambda(\phi) < 1$, a useful correlation developed for concentrated suspension in gravity is the Richardson and Zaki's form [1] which is $\lambda(\phi) = (1-\phi)^{4.5}$. The dilute form is given by the Einstein's equation, $\lambda(\phi) = (1-2.5\phi)$. The G-acceleration can be calculated from

$$G/g = 5.591 \times 10^{-7} \Omega^2 D; \quad \Omega [RPM], D[mm]$$
$$G/g = 0.0000142 \Omega^2 D; \quad \Omega [RPM], D[in] \qquad (2a, b)$$

Boundary/Critical-Layer Model
The flow inside a centrifuge with a cylindrical clarifier is very complex especially for decanter centrifuge where the conveyor scroll is rotating at angular speed, either faster or slower, relative to the bowl. There are several models to quantify the separation capacity based on the geometry and the operating parameters including the Sigma theory, G-surface and G-volume approaches. None of these satisfactorily explains and quantifies the effect of flow in the complicated geometry and operation. The most useful model is the improved boundary-layer/critical-layer model, which is a generalization of the boundary-layer model [2]. This model is still somewhat inadequate. Nevertheless this is the only model, among all, that accounts for particle size distribution (PSD), which is vital for separation, classification and clarification. In the improved boundary layer model, a *critical* moving layer in the vicinity below the pool surface at a mean radius R_p flows across the pool. The model assumes that once a solid particle settles across the layer, it transverses a stagnant pool below and ultimately settles in the sediment or cake. A major difference between the improved model with the earlier boundary-layer model is that feed

solids of all sizes are assumed to be uniformly distributed across the entire critical layer thickness instead of concentrating at the surface of the layer.

Depending on the operating differential and the geometry of the conveyor the layer thickness can be very thin or as thick as the annular pool depth. When the critical layer is indeed thin in comparison with the radius, the actual thickness of the layer does not enter directly into the result. The G's for separation is determined from the critical radius R_p wherein $G=\Omega^2 R_p$. For some geometry with an axial-flow design, the critical radius is located at the pool surface. Whereas for other geometry and operating conditions where there is substantial mixing across the pool due to secondary flow caused by the differential speed between the conveyor and bowl, an effective radius can be used instead, such as the geometric mean of the bowl radius and the pool radius. This can account for effect due to variation in pool depth.

Solids Recovery

The solids recovery, respectively, in the cake R_s or in the centrate R_e, for the improved boundary layer model, are expressed as

$$R_s(Le) = 1 - F_f(x) + I(x_c)/x_c^2$$

$$R_e(Le) = 1 - R_s$$

$$I(x) = \int_0^x x^2 f_f(x) dx$$

$$I(x_n) \approx \frac{1}{4} \sum_{k=1}^{n} (x_{k-1} + x_k)^2 (F(x_k) - F(x_{k-1}))$$

$$f_f(x) = \frac{dF_f(x)}{dx}$$

$$x_c / x_o = \sqrt[3]{\frac{1}{\sqrt{\pi}}} Le$$

$$Le = \frac{\sqrt{Q/L_c} \sqrt{\mu / \Delta\rho \lambda(\phi)}}{\Omega R_p x_o \eta} = \frac{\sqrt{Q/L_c} \sqrt{\mu' / \Delta\rho}}{\Omega R_p x_o \eta}$$

(3a, b, c, d, e, f, g)

where

x	=	particle size
$F_f(x)$	=	% cumulative undersize x in feed slurry
$f_f(x)$	=	frequency of a given size x in feed slurry
x_c	=	cut size, i.e., maximum size of particles in the overflow
x_o	=	reference particle diameter, convenient taken as 1 μm
Q	=	flow rate
L_c	=	cylindrical clarifier length
μ	=	viscosity of slurry
Δρ	=	density difference between solid particle and suspension
Ω	=	angular rotation speed of bowl
R_p	=	critical layer radius
η	=	average feed acceleration efficiency for the feed stream in clarifier
λ	=	hindered settling function due to hydrodynamic interaction applicable for dense slurry

From a practical standpoint, it is difficult to determine the viscosity of dense slurry especially with fast settling particles and the hindered settling function. The ratio μ/λ is combined as a "viscosity factor" μ' given λ is dimensionless and is mostly less than unity. Often μ' is used as a matching parameter for a given process separation from available test data with projection to performance on other operating conditions which are unknown. For that matter, if the average clarifier acceleration efficiency, nominally 0.7–0.9, is not determined accurately, any discrepancy with the actual performance could also have been absorbed in the viscosity factor when matching the actual performance data with theoretical prediction.

Le is the dimensionless Leung number [2] governing centrifugal sedimentation. Note the solids recovery in the cake R_s, for application of separation and clarification, as well as solids recovery in the centrate R_e, for classification application are both functions of Le. As evident from Equation 3e, Le is directly proportional to the cut size in the centrate.

In fact, the cut size x_c is simply 1.69Le μm given x_o is taken conveniently as 1 μm. Suppose Le for the geometry and operating condition of the centrifuge is such that it assumes a value of 3, the maximum particle size in the centrate is thus 5.2 μm. Table 1 enlists the typical cut sizes, the corresponding operating Le, and the typical applications.

Table 1 Process functions

Cut Size		Le	Typical Applications
Tyler Mesh	Microns	[1]	
	0.1	0.06	Ultra high-G classification of valued ultrafines, waste slimes, and colloids
	0.5	0.3	
	1	0.6	Classification in coatings, pigments (e.g., kaolin, calcium carbonate, silica, mica, etc.) and drill mud with high-speed centrifuges
	2	1.2	
	5	3	
	10	6	
	25	15	Classification/degritting of oversize particles above 25 μm
325	45	27	Classification/degritting of oversize particles above 45 μm
200	75	44	Sedimenting fine particles for which continuous filtering centrifuge does not separate well
150	100	59	Relatively coarse separation with lower-speed sedimenting centrifuges
100	150	89	
150	212	125	

Note Le has taken into account, among many variables, the feed rate and speed (or G). For a given application, the feed rate and speed may be very different for different size machines, however for the same application with the intent to get the same cut size, Le has to be identical for the machines. This becomes a very important scale-up criterion for separation, classification and degritting. One can use the Le scale-up approach to design and scale-up a machine to perform with a specified feed rate and attain a specified performance such as recovery etc. Alternatively, it can be used to predict the performance of a given machine if say the feed rate or speed is changed such as in process optimization.

Most PSD has a polydispersed distribution with the population of particle size spreads out from very fine to very coarse sizes as characterized by the cumulative undersize function $F_f(x)$ or the frequency distribution function $f_f(x)$. On the other hand, monodispersed distribution results with the particle size concentrates near a relatively narrow range as with precipitated minerals (e.g., precipitated calcium carbonate) from dissolution and crystallization processes. Also particles may be populated in two seemingly distinct size ranges, i.e., bimodal. It can be found in slurry with single species or commonly found when two species are mixed (such as in weighted drill mud), each with their respective characteristic size distributions and densities. The PSD of the slurry is critical to separation, classification and clarification.

Particle Size Distribution of Overflow

The particle size x_k in the product centrate (i.e., overflow of centrifuge) from the improved boundary-layer model is given by,

$$F_e(x_k; Le) = \frac{F_f(x_k) - I(x_k)/x_c^2}{R_e(x_c)} \quad (4)$$

The centrate PSD F_e is therefore a function of the size x_k the size cut x_c which depends on Le, and the PSD of the feed F_f.

Cumulative Size Recovery

The cumulative recovery of particles with size below x_k in the product overflow from the centrifuge can be determined using the improved boundary-layer model as,

$$SR(x_k; Le) = 1 - \frac{I(x_k)}{x_c^2 F_f(x_k)} \quad (5)$$

Again SR is a function of x_k and Le through F_f and I. It is clear that the PSD of the feed slurry plays a vital role in determining the size distribution of the centrate product, size recovery and total solids recovery.

Filtering Centrifuges

The final dewatering or deliquoring of cake in a filtering centrifuge depends on the dimensionless time t_D [2] which incorporates several important operating variables in deliquoring – the G-force, time duration t, hydraulic diameter of particles x_h, cake height h, viscosity μ_L, and density of liquid ρ_L. It is defined as,

$$t_D = \frac{Gtx_h^2}{(\mu_L/\rho_L)h} \propto \frac{Gt}{\mu_L h} \quad (6)$$

Granted the particle characteristic size stays constant, the cake permeability which is proportional to x_h^2 is also constant. The final moisture of the cake depends on the G-force, retention time, inversely related to the cake height (which in turns relate to solids throughput), and also inversely related to the liquid viscosity. The latter usually decreases with increasing operating temperature. The shape of the moisture drainage curve depends on the surface properties of the cake and the kinetics of liquid film drainage from the cake surface.

PROCESS FUNCTIONS AND MACHINES

- Separation (without and with added chemicals) – In separation feed slurry is introduced into the centrifuge where a concentrated stream of solids and a liquid stream primary with low concentration of solids are desired. Polymer can be added for dewatering and separating waste stream when majority of the particles is too fine in the micron to sub-micron size range. For mineral applications, the polymer dosage is typically less than 0.1% kg polymer per kg of dry solids.

- Clarification (liquid phase product) – When liquid phase should be free of solids for reuse or discharge, solid concentration needs to be minimized and typically a tolerable limit is set on the maximum tolerable solids concentration in the centrate.

- Classification – Finer solids fraction is the desired product with coarser fraction discarded; vice versa coarser fraction is the product discarding the finer fractions, e.g., slimes with particles less than 0.5 μm.

- Degritting – This concerns removal of oversized particles, such as 25 μm, 45 μm, etc. and/or foreign particles with a different density compared with the solids in the process stream.

- Deliquoring (dewatering) – Cake moisture needs to be minimized to meet process requirements (final end product, further downstream processing such as thermal drying, purification, etc.)

- Washing – Cake is required to be washed with appropriate solvent or wash liquid with minimal contaminants when cake solid needs to meet purity requirement. Wash liquid should be minimized while achieving good cake purity.

- Deliquoring followed by reslurrying – When particles in the slurry are in microns, say 5–10 μm, and cake washing is desired to achieve high purity, the slurry is first deliquored by centrifugation followed by reslurrying. This can be done in stages, with say two stages being very common. Alternatively a basket centrifuge can also be used if feed volume is limited and storage tanks are required for silo.

TYPES OF CENTRIFUGES
There are two types of centrifuges – continuous-feed and batch basket centrifuges. Within each group there are many different designs and variations. Most applications use continuous-feed centrifuges. Solid bowl or decanter is the most versatile among all.

Solid Bowl or Decanter
A schematic of the solid bowl or decanter is shown in Figure 2. Feed slurry, after accelerated in the rotating feed compartment or accelerator, is introduced to the annular pool. Under high centrifugal force, the heavier solids migrate radially outward toward the bowl to form cake displacing the lighter liquid toward the center. Solids are compacted against the bowl wall by the centrifugal force and conveyed by the screw conveyor, rotating at differential speed relative to the bowl, to the small diameter of the conical beach for discharge. As the cake is lifted above the annular pool in the dry beach, liquid further drains back to the pool leaving a drier cake for discharge. The gear unit and/or backdrive control the differential speed between the bowl and conveyor changing the solids retention time as needed. The clarified liquid overflows the weirs located at the opposite end of the machine. The pool is controlled by the discharge diameter of the weirs. The performance of the centrifuge depends on the various operating variables such as feed rate, pool depth, rotation speed or G-force, and differential speed and should be optimized for a given process.

Feed Rate. Liquid residence (settling) time of the slurry in the bowl may directly affect the degree of centrate clarity that can be obtained. Decreasing the feed rate will increase the liquid residence time and permit more efficient settling of suspended solids.

With very dilute suspensions (solids concentration much less than 1%), gravity or cyclonic thickening upstream of the centrifuge is recommended to concentrate and reduce the total volume of feed slurry or liquid handled. Hydraulic loading affects the main-drive motor requirement from perspective of accelerating the feed stream while solids loading affect the conveyor torque load.

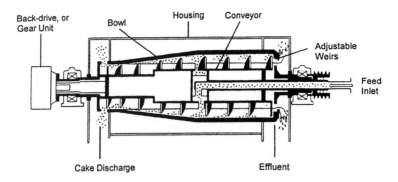

Figure 2 Solid-bowl or decanter centrifuge schematic

Pool Depth. The proper pool depth depends on the settling characteristics of the solids in the feed slurry. By reducing the pool depth, a drier cake is normally obtained because a longer dry beach is available for cake drainage before discharge. Pool level should not be lowered to a point where centrate clarity suffers or solids conveyability is hindered.

When the pool depth is increased the length of the drying beach is reduced. This generally results in higher cake moisture. A deeper pool usually improves centrate clarity, since liquid retention time is increased giving lighter and smaller particles more time to settle. Deepening the pool also eases movement of the cake due to liquid buoyancy, resulting in improved cake conveyability.

Rotation Speed and G-Force. Higher rotation speed produces higher centrifugal force acting on the solids in the cake and improved settling rate. The consequence is lower cake moisture and/or a clear centrate. However, this does not necessarily always hold. Some solids especially the finer size fraction have density very close to that of the suspension (i.e., somewhat neutrally buoyant) due to adhesion of contaminants or bubbles to the solid surfaces. They do not settle regardless of the magnitude of the centrifugal force. Also different solids tend to pack tightly (i.e., compactible cake) under high centrifugal force and some drain more readily under lower speeds where larger voids exist with higher cake permeability. For compactible cake, increasing G beyond a certain point does not warrant increase in cake dryness due to increasing cake resistance to deliquoring [2]. In general, for maximum clarity, cake dryness and least power consumption, operate at the "lowest possible" speed compatible with the material characteristics and performance requirements. It is always good practice during the initial start-up period to compare cake dryness and centrate clarity at different rotation speeds and G's. This allows selection of the optimum centrifugal forces for the specific application.

Differential Speed. By lowering the speed differential between the conveyor and the bowl, the solids residence time is increased. This causes an increase in cake depth against the bowl wall with greater compacting stress and higher cake dryness in most cases. This is often accompanied by increasing conveyance torque as well.

Lower conveyor differential provides less turbulence and less re-suspension of solids. However, low conveyor differential may have the opposite effect if the incoming feed solids rate is higher than the capability of the conveyor to remove them in which case solids un-transported buildup in the cylinder and eventually overflow at the liquid discharge ports. The latter is often accompanied with increasing conveyance torque over time. There must be a balance between solids input and solids removal to prevent plugging of the machine and loss of the centrate clarity.

Differential speed may be changed by changing the gear ratio for the same box if available or changing the gear box with a different gear ratio.

The differential speed is related to the bowl speed Ω_b and the pinion speed Ω_p by the following kinematic relationship:

$$\Delta\Omega = \frac{\Omega_b - \Omega_p}{r} \tag{7}$$

Suppose Ω_b=2600 rpm and the pinion shaft is locked stationary with Ω_p=0, with a gear ratio r=80:1 it gives a differential speed of 32.5 rpm.

An alternative is to provide an electric backdrive where the pinion is driven by a DC motor, or an AC motor, which can be controlled by a variable frequency drive (VFD). By tuning the frequency, the pinion speed is adjusted thus changing the differential while the machine is running. For example, the VFD is tuned with pinion shaft rotating at speed respectively 1000 rpm and 2440 rpm, the differential speed becomes 20 rpm and 2 rpm. The small differential speed allows longer retention time, which facilitates deliquoring cake to high dryness. Flocculant and/or coagulant are used to agglomerate fine particles improving centrate clarity. This is especially on the waste application in which polymer dissolved in liquid stream is of lesser concern [3].

Also a hydraulic pump/motor is used as backdrive for centrifuge. The pump pressure is used to control the torque while the oil flow rate in the pump is used to control the differential speed between the scroll and the bowl.

The centrifuge can be over-torqued due to plugging with unconveyed solids accumulating in the bowl. If temporary stopping of feed to the machine while continuously maintaining the differential speed between the conveyor and bowl does not clear the machine, the rotation speed and thus the counteracting centrifugal force is reduced to facilitate cake transport. Unfortunately the differential speed also reduces in lieu of Equation 7. Fortunately, a centrifuge equipped with an electric or hydraulic backdrive allows maximum differential speed to transport cake out of the machine despite the reduced bowl speed or when the bowl stops rotating.

Feed Acceleration. In a decanter the feed stream needs to be accelerated to match the tangential speed of the rotating pool to generate centrifugal force to effect separation. Also the feed should be distributed uniformly on the pool with minimal radial velocity to avoid disturbance and turbulence. Unfortunately, conventional feed-accelerator designs by-and-large poorly accelerate the feed and frequently the latter is introduced in concentrated streams jetting into the pool causing turbulence, resuspension of sediment, and wear on rotating surfaces. This is especially significant for a high volumetric rate application. Under high flow rate, feed slurry that has difficulty swallowing through the feed ports floods the feed chamber; and when serious feed may leak back along the outer diameter of the feed pipe. A comprehensive patented feed acceleration system [2, 4, 5, 6, 7] has been developed for continuous-feed centrifuges including sedimenting and filtering centrifuges. To visualize the effect of feed acceleration, two identical rotating bowls of 250-mm diameter, cantilever outward from the support, were set up side-by-side to demonstrate the visual effect of feed acceleration [7]. Water was introduced in one centrifuge bowl via an improved feed accelerator and for the other at the same flow rate with a conventional accelerator design. The surface of the pool was observed with a strobe light tuned at the frequency of the rotating bowl at 1000 rpm. A pool meter in the form of a water wheel which has a pair of paddles dipped into the rotating pool by 3 mm was set up under free wheeling condition and the pool meter was driven by the rotating pool where the feed entered. The rotational velocity of the feed at the pool based on the rotating speed of the pool meter was measured using a strobe light and compared with the bowl speed at 1000 rpm. Figure 3a is a strobe picture showing the pool region where feed was introduced via a conventional feed-accelerator design. The surface of the pool appeared like a rough sea with turbulence. A close-up of the pool in Figure 3b reveals the entire pool surface was wavy and chaotic. In fact, the pool where feed was introduced was rotating at a speed significantly less than that of the bowl. This "slip" or mismatch in tangential velocity between the incoming feed and the remaining pool is the cause of turbulence. On the other hand, Figure 3c shows the same rate of feed being introduced to the pool after flowing through an improved XL•PLUS® feed accelerator system. The pool was quiescent reflecting the image and color of the pool meter. Also the pool meter appeared stationary rotating at the same pool speed as the bowl. This fosters the "calm" condition for sedimentation where heavier particles under the driving centrifugal force settle to the bowl wall. An application of this technology is demonstrated later in Figure19. Figure 4 shows the ratio of measured pool speed to that of the bowl for different feed rates. The speed ratio, which measures the feed acceleration efficiency, stays constant at 100% independent of feed rate for the XL•PLUS® feed system while the efficiency for the conventional design drops off sharply to an disappointing level approximately 50% with high feed rate (above 7 m^3/h for the 250-mm bowl without a conveyor). This is a common experience with all conventional designs. The square of the velocity ratio, which is an important measure of the actual G-force (a measure of separation ability) versus the rated G-force – i.e., G-efficiency, decreases even more sharply. Despite this, under-accelerated feed would eventually get accelerated in a decanter by contact with the pool and the rotating surfaces but not without skidding, wear and turbulence. This is at the expense of consuming clarifier and pool volume for feed acceleration leading to lower throughput, lower solids capture and possibly feed leakage. For a decanter with a short clarifier (such as a screen bowl) this is seriously disadvantageous. Improved feed acceleration systems and related technologies have been presented in much greater detail elsewhere [2, 4, 5, 6, 7].

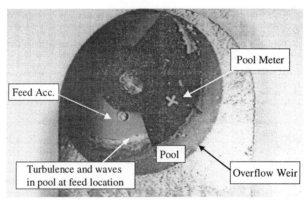

Figure 3a Strobe picture of the pool in a 250-mm rotating bowl fed with water after accelerated by conventional feed accelerator design (courtesy of Bird Machine Company)

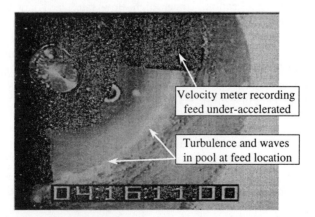

Figure 3b Close-up strobe picture of the pool for configuration of Figure 3a showing turbulence and waves on pool surface (courtesy of Bird Machine Company)

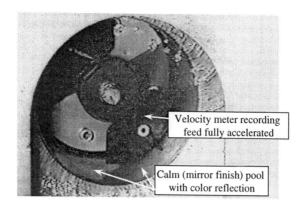

Figure 3c Strobe picture of a pool in a 250-mm bowl fed with water after accelerated by XL•PLUS® feed accelerator design (courtesy of Bird Machine Company)

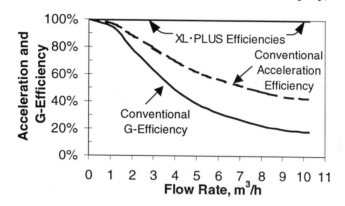

Figure 4 Acceleration efficiency (pool speed at feed entrance/bowl speed) and G-efficiency (square of acceleration efficiency) for, respectively, conventional and XL•PLUS® accelerators shown in Figures 3a-3c

Sizes. A variety of different diameters are available for decanters from as small as 150-mm (6-in) diameter to over 1400-mm (55-in) diameter. The length-to-diameter (aspect) ratio is as large as 4.2 to 4.5, and as small as 1.3. Generally, the larger machines operate at lower speed (1000g–3000g) and accept higher throughput when compared with small diameter machines operating at higher speed thus higher centrifugal gravity (3000g–4000g) processing more difficult-to-separate materials.

Materials of Construction. Duplex steel with higher yield strength is employed for high-speed decanters. For processing abrasive materials, all of the high-wear areas such as feed zone, blade tips, possibly bowl wall, and cake discharge are protected with a wear resistant coating such as ceramic, tungsten carbide tiles or spray, or other specially prepared sacrificial coatings.

Screen Bowl Centrifuge

With an added cylindrical-screen section attached to the small cake discharge diameter, a screen-bowl centrifuge can be viewed as a modified decanter centrifuge, see Figure 5. Given the distance between the two bearings are the same for a centrifuge with a given diameter, therefore the length of the larger-diameter cylindrical section of the solid bowl is reduced to accommodate for the screen length of the machine.

The cylindrical screen section has openings to accept soap-dish shaped screens, which are installed externally of the bowl held by a big band clamped on the outer diameter of the bowl, or screen panels, which are installed internally. The screen can be made of stainless steel, tungsten carbide ligaments, and ceramics. The screen is mostly profiled (such as wedge wire, see Figure 6) with the smallest opening of the cross section at the screen surface facing the cake and the largest opening at the exit outer diameter to prevent trapping solids in the screen causing blinding of the screen to flow of filtrate. A ceramic conveyor is shown in Figure 7.

Cake is washed at the initial screen region while the remaining screen section toward the cake discharge is used for deliquoring. It is important to wash the cake while it is still fully saturated with the mother liquor otherwise air bubbles trapped in the cake may block the wash liquid to the cake pores when the cake is partially desaturated.

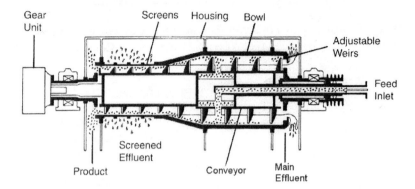

Figure 5 Screen-bowl centrifuge schematic

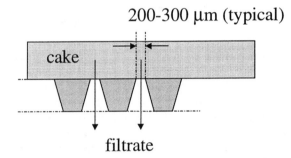

Figure 6 Self-cleaning wedge-wire profile used for screens for filtering centrifuge

Figure 7 Screen-bowl conveyor blade tips lined with ceramic tile protection (courtesy of Bird Machine Company)

Pusher Centrifuge

The pusher centrifuge is cantilevered supported and has single (single-stage pusher), or multiple cylindrical screens (multiple-stages with two- and four-stages common) with a progressively larger screen diameter. A schematic of the two-stage pusher is shown in Figure 8. A thickened feed slurry of 40%–60% after properly accelerated by the feed accelerator is introduced to the first-stage basket. An annular push plate with an outer diameter closely fitted to the inner diameter of the first-stage basket pushes the cake as the first-stage basket reciprocates along the axis. For the case of a two-stage pusher, the cake at the end of the first-stage basket drops onto the second-stage basket as the first-stage basket retracts. In the second-stage basket the cake is further deliquored under a higher G-force. For a four-stage pusher, the first two stages are equipped with an end ring (acting as a push plate) at the exit of the screen section for pushing the cake. Both the first stage as well as the third stage reciprocates. In the retrieved stroke, the cake is dropped from both the first and third stage, respectively, on the second and fourth stage. In the forward stroke, the cake is pushed from the second and fourth stage to the third stage and discharge, respectively. Deliquoring is accomplished as the cake gets thinner with capillaries between solids opened up as it spreads in successive stages at increasing diameter and higher G-force.

Cake washing is often carried out for a multiple-stage pusher at the transition between the first and second stage to remove cake impurities. The wash liquid needs to be delivered and pre-accelerated through a wash pipe equipped with a nozzle so that the velocity of the wash liquid approximately matches the tangential speed of the cake in the rotating basket to establish the same centrifugal gravity. Otherwise, the wash liquid would appear to have a "lighter" gravity and would not be able to penetrate the cake displacing the mother liquor saturated with impurities.

In applications requiring high performance, the cylindrical basket of the last stage is replaced with a conical basket. This reduces the load on the push mechanism at high solids throughput because the longitudinal component of the centrifugal gravity is assisting cake transport. There are additional enhancements on cake deliquoring with this design. As the cake is conveyed to the large diameter of the cone, the cake height is reduced with more surface area and concurrently the G-force is increased.

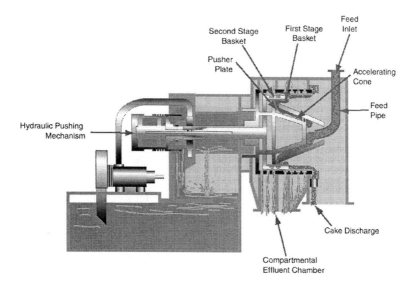

Figure 8 Two-stage pusher centrifuge schematic

Conical Screen Centrifuge

Vibrating Screen Centrifuge. The inertia generated by vibrating two eccentric rotating masses facilitates cake transport along the conical basket. Commercial sizes with 1100-mm (44-in), 1200-mm (48-in), and 1300-mm (56-in) diameter are available. The centrifugal gravity is typical under 100g.

Screen-Scroll Centrifuge. The screen-scroll centrifuge has a rotating conical basket mounted either horizontally (see Figure 9) or vertically. Feed slurry is accelerated in a conical section of the hub, which forms a feed compartment. Accelerated feed is introduced into a rotating truncated conical screen at the small diameter. After filtration, the cake is conveyed to the large diameter by a scroll with complete "wrap-around" helical blades, or a discrete number (4 or 8) of blades profiled with a small helix angle. Cake washing at the small diameter of the conical screen is possible, in which wash liquid is introduced into a separate compartment in the conveyor hub, where wash liquid is sprayed through nozzles mounted on the hub outer diameter. As the cake is conveyed to the large diameter it spreads to a thinner pile and under higher centrifugal force the cake attains low cake moisture before discharge at the large diameter of the basket. Common sizes are 250 mm (10 in), 400 mm (16 in), 700 mm (28 in), 900 mm (36 in) and 1000 mm (40 in). Vertical centrifuge operates at lower G, 230+g, while the horizontal centrifuge operates between 300-800g.

Wedge-wire profiled screen and perforated plates are commonly used. The opening of the screen is nominally 300 μm and a more open screen with 700–800 μm is employed for processing coarse materials. Laser-cut screen with uniform sieve size over the entire screen can also be used for screen-scroll centrifuge. The loss of fine solids in the filtrate can be regulated by selecting the appropriate sieve size. For example, a 60-μm sieve has an 8% opened area and provides good filtering capacity as well as screening capability. The laser-cut screens are coated with a chromium layer for wear protection.

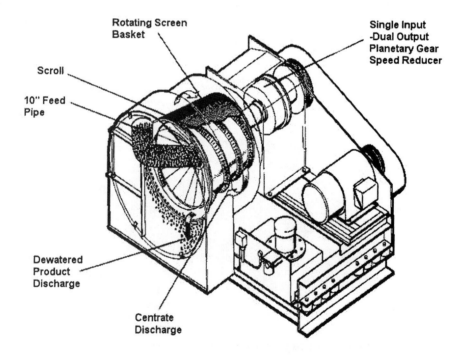

Figure 9 Screen-scroll dryer schematic

APPLICATIONS OF CENTRIFUGAL SEPARATION

Centrifuges are used in many separation processes including coal, potash, soda ash, tar sand, kaolin, calcium carbonate, drill mud, yellow cake, bauxite, and barium sulfate, etc. Only a few of these applications are discussed.

Coal

The conical screen centrifuges are commonly known as mechanical dryers in coal preparation plant. The vibrating conical screen centrifuges are used for dewatering of coarse coal from 0.5 to 50 mm top size. The cake moisture for the coarser fraction 25 mm x 13 mm coal is about 2%–3% while the moisture for the finer coal fraction 0.6 mm x 13 mm is 5%–6%. The vibrating screen centrifuge operates at about 70-80g with eccentric masses rotating at 1300–1500 rpm generating amplitude of vibration in the range of 4–5-mm for cake transport.

Screen-scroll centrifuges are used to mechanically dewater finer coal from 150 μm (100 M) to 2–3 mm top size or to 13 mm top size. The G-force is nominally 230–300g though it can go up to 500g with the horizontal design. Furthermore, the horizontal design is less susceptible to clogging with processing sticky fine materials.

Improved design of a feed accelerator [4] allows the screen scroll to process more solids throughput by instantly accelerating the feed stream to rotate at basket tangential speed thus getting full G-force for liquid drainage. Also feed is uniformly laid around the entire circumference of the basket as shown in Figure 10 utilizing the full screen area without which the cake is found to distribute non-uniformly in the screen between the one o'clock to six o'clock positions. In consequence, much higher capacity can be attained with a marginal increase in cake moisture as shown in Figure 11. In addition, significant wear is attributed to under-accelerated feed when introduced at the small end of the basket. Typical life of the basket is about 300 hours. On the other hand by reducing skidding and uniformly distributing the feed onto the basket with the improved feed accelerator technology, the life of the wedge-wire basket has been proven to increase to 500–700 hours reducing operating costs and downtime. Premium basket made of tungsten carbide has been used to withstand abrasion from coal and their life is several times that of the conventional wedge wire basket.

Figure 10 With the improved feed accelerator not only feed is accelerated instantly to speed, it is uniformly distributed around the basket, see coal from windows, respectively, at 3 o'clock, 6 o'clock, and 9 o'clock positions. In contrast with conventional feed accelerator, feed is non-uniformly distributed in the basket between the 1 o'clock and 6 o'clock positions (courtesy of Bird Machine Company)

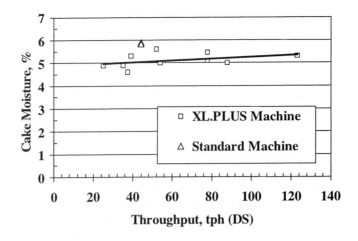

Figure 11 High performance of screen scroll with improved feed accelerator in processing coal with high throughput and low cake moisture

For clean coal 12 mm x 150 µm with slightly larger bottom size 150 µm (100 M), a horizontal 900-mm screen scroll can process 110–125 tph (dry basis) while a larger vertical 1050-mm screen scroll on the same feed can process 100+ tph.

The screen-bowl centrifuge typically takes feed from 0.5–1 mm top size down to zero size. On fine coal, a screen bowl produces exceptionally low surface moisture 5%–7% lower than the best results generally obtained from vacuum disk and drum filters yet with 20% less power to operate when compared with vacuum filters. Screen life of 15,000–20,000 hours is normal depending on the coal hardness while conveyor screws run up to 15,000 hours or more with tungsten carbide tiles. Figure 12 shows an installation of screen bowl centrifuges in a coal preparation plant for dewatering flotation coal.

For technology improvement, given the clarifier screen is much shorter compared to that of the decanter for the same size, effluent quality would benefit by getting the feed instantly accelerated to speed. The screen bowl centrifuge equipped with accelerating vane apparatus [5] has demonstrated higher solids recovery with better capture of the finer coal solids, which would have been lost in the effluent as coal refuse. On the other hand, screen bowl is operated to some degree to classify ultrafine solids, which are inorganic (ash) in nature, in the effluent stream as they do not have caloric value.

Figure 12 Eight 900-mm x 1800-mm screen bowl operating in a coal prep plant (courtesy of Bird Machine Company)

Figure 13 shows the fine coal moisture as a function of the concentration of finer materials (measured by % -45 μm) in the feed. In general, as coal gets finer in size with increasing surface area and electrical charges, the particles trap more moisture on their surfaces. The cake moisture increases from 13%–19% while the finer solids in the feed increases from 18%–29% at 665g as measured at the screen radius. At lower 413g the cake moisture is 6% higher as compared with 665g for the same fine fraction as shown in Figure 13.

Figure 14 shows the effect of presence of fines (-45 μm) in the cake on cake moisture. At 10% fines the total moisture is about 11%–13%, while at 40% fines in the cake it increases to 25%–26%. The inherent moisture of the cake in Figure 14 is 1%–1.5%.

In absence of significant dynamic effect, laboratory bucket centrifuge can simulate the cake dewatering characteristics of screen bowl or other screen typed centrifuges. Figure 15 shows test results plotted as cake moisture versus t_D, which is simplified to (G/g)t/h, for both clean coal and stock-pile coal from the same coal preparation plant. At (G/g)t/h = 5,000 s/in (197,000 s/m), the cake moisture runs at 15%–17%, which matches with the level which the plant has reported with centrifuges operating at comparable (G/g)t/h. Also at higher feed rate in which (G/g)t/h = 3,000 s/in (118,000 s/m), the cake moisture for the screen bowl is about 18% moisture which is consistent with the laboratory bucket results in Figure 15.

Solid-bowl decanter is used to process fine coal refuse with 75%–80% particles less than -45 μm (-325 M) in which cake moisture can exceed 30%. Depending on the size of the machine, solids throughput ranges between 8–30 tph on a dry basis.

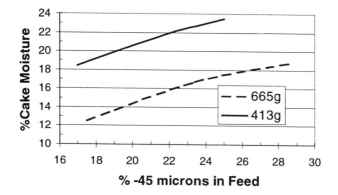

Figure 13 Cake moisture versus -45 μm in feed for two different G's

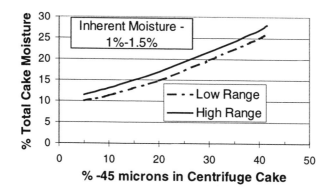

Figure 14 Cake moisture versus -45 μm in cake

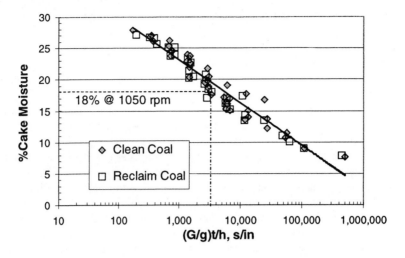

Figure 15 Coal moisture versus (G/g)-t/h for clean coal sample

Table 2 provides a summary of centrifuges as applied to coal processing.

Table 2 Typical applications of centrifuges for coal processing

Centrifuges	Size	Cake Moisture %	Solids Recovery % w/w ss	Solids Throughput tph ss
Vibrating Conical Screen 1100 mm (44 in) 1200 mm (48 in) 1300 mm (52 in)	0.6 mm x 35 mm (28 M x 1.25")	2%–6%	99%+	100–350
Screen Scroll 900 mm (36 in)	0.1 mm x 13 mm (150 M x ½")	5%–8%	95%–98%	50–100
Screen Bowl 900 mm x 1800 mm (36 in x 72 in)	0.6 mm x 0 (28 M x 0)	13%–22% (froth)	98%–99%	20–50
1100 mm x 3300 mm (44 in x 132 in)		8%–12% (cyclone)		50–100
Solid Bowl /Decanter 900 mm x1800 mm (36 in x 72 in) 900 mm x 2400 mm (36 in x 96 in) 1100 mm x 3300 mm (44 in x 132 in)	75%–80% <45 μm (-325 M)	25%–45%	93%–99%	8–30

Potash

The main use of centrifuges is for dewatering potassium salts (primarily potassium chloride) in the following processes:

1. Flotation products
   1.2 mm x 75 μm (14 M x 200 M), or 150 μm x 0 (100 M x 0)

Most flotation salt/product falls into the size category 1.2 mm x 75 µm (14 M x 200 M) and is dewatered by 1350-mm x 1750-mm (54-in x 70-in) screen bowl operating at 600 rpm, which is equivalent to 275g at the bowl diameter (cylindrical clarifier) and 240g at the smaller screen diameter. Feed rate for screen bowl ranges between 60 and 110 tph depending on the design and operation condition. Cake moisture for product with 85% +210 µm (+65 M) and less than 5% -75 µm (-200 M) is typically 3%–4% and higher moisture results when the -75 µm (-200 M) percent increases.

2. Crystallized product
   1.2 mm x 100 µm, 95% + 150 µm (14 M x 150 M, 95% + 100 M)

The feed rate to crystallized product is comparable to that of flotation product. Total moisture is between 4% and 6% depending on the inherent moisture trapped in the crystals. To remove impurities cake washing in the screen section immediate downstream of the conical beach is typical although cake washing further upstream in the conical beach has also been practiced. Regardless, it is most effective to wash the cake before it becomes unsaturated to avoid dead air pockets that cannot be penetrated by wash liquid. A 900-mm (36-in) diameter pusher can process crystallized product at 30–35 tph with moisture of 3%–4%. Improved feed accelerator in the form of conical [6], or double disk [7], provides higher capacity while maintaining acceptable cake moisture and purity.

The 1000-mm diameter screen scroll, operating at 600g, has been used to process crystallized product processing 45–50+ mtph with cake moisture at 4%–4.5%. The screen scroll is equipped with improved feed accelerator technology [4] in which feed is instantaneously accelerated to speed to effect filtration and dewatering with full centrifugal force. Solids recovery is about 85%–90%.

3. Salt Tails (sodium chloride) 1.2 mm x 0 (14 M x 0)

Salt tails are dewatered to recover the brine. Therefore, high solids recovery is essential and cake dryness is not important. Various solid-bowl decanter sizes from 1000 mm (40 in), 1100 mm (44 in) and 1350 mm (54 in) are used with top rate at 50, 60, 120 tph, respectively. The solids recovery is 95%–100% with 6%–8% cake moisture.

Soda Ash
Screen-bowl centrifuges have been used to process monohydrate, sesqui-hydrate and decahydrate. The throughput for the 1100 mm x 3300 mm (44 in x 132 in) is between 100–120 tph with 3%–8% cake moisture and 90%–97% solids recovery.

New innovation [8] has demonstrated improvement of cake washing in a screen bowl. In the screen deck, cake passes through a series of compartments, in a sequence along the cake path made up of solid-wall and screen-wall, each separated by gates attached to the conveyor. As shown in Figure 16, cake is washed in a solid-wall compartment followed by the next compartment with screen wall for drainage. Wash liquid is allowed to stay in longer contact with the cake thus promoting diffusion wash which would not have happen with normal design due to short duration. Also, as the cake passes through the rectangular opening of the gate, it is reoriented from a triangular pile against the push/driving face of the conveyor blade to a rectangular profile, see Figure 17a, b. Washing is most effective with uniform cake height and post-shear from the gate to reduce any occluded liquid trapped by pores in the cake. Figure 18 compares the results of this technology using a 450-mm x 700-mm (18-in x 28-in) screen bowl. Improved purity or less wash liquid is required to achieve the same cake purity. The reslurrying cake washing technology is directly applicable for relatively soft cake with low conveyance torque.

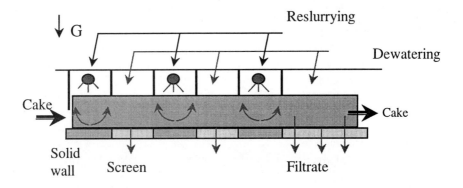

Figure 16 Reslurrying and Dewatering in sequence for CENTRIWASH™ technology (courtesy of Bird Machine Company)

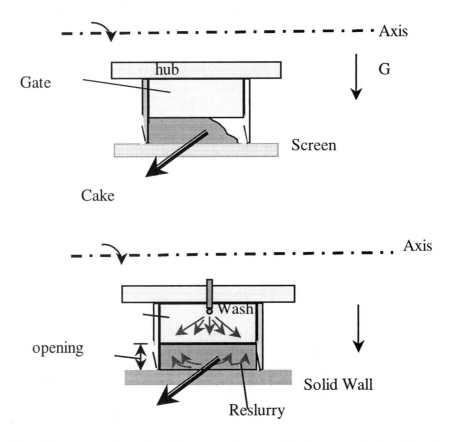

Figure 17a Cake approaches gate with non-uniform profile in screen section; (b) cake passed rectangular gate opening with uniform cake height at which cake is washed in the solid-wall section for reslurrying and/or increasing contact time to promote diffusion of solute

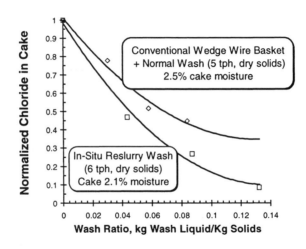

Figure 18 Wash performance with CENTRIWASH™ wash technology on soda ash using a 450-mm diameter screen-bowl centrifuge

Tar Sand

Centrifuges have been used for processing tar sand slurry from underflow of inclined plate settlers and thickener tanks. Specifically solid-bowl decanters have been used to separate solids (sand) from the liquid (hydrocarbon and water). The liquid is sent to pre-filter for further removal of solids and the filtrate is routed to high-speed disk centrifuges (single-stage and two-stage separation as necessary) in which valuable hydrocarbon is separated from water with fine residual solids.

The process of separating sand/silica from liquid is highly abrasive; as such all the mechanical components exposed to the slurry are well protected with wear resistant materials and moderate G-force is used in the separation process. The 900-mm (36-in) diameter decanter is employed to process 20–52 L/s (300–800 gpm) of feed depending on the process requirements. The median size of the feed solids can range from less than 20 μm to as large as 100 μm. The finer is the feed size the lower is the solids recovery, alternatively the feed rate to the centrifuge can be reduced to maintain acceptable solids recovery.

Kaolin

Solid-bowl decanters have been used to classify kaolin slurries. Mined kaolin in the form of slurry has solids concentration ranging between 30% w/w (norm) and 50% w/w (high-density). For example, the feed to the centrifuge may contain 70%–80% -2 μm and 50%–60% -1 μm. After classification by centrifuge, the centrate (product) may have 90–99% -2 μm, and 80–90% -1 μm. It is essential to maintain a high total solids recovery of all sizes in the overflow between 60–70+%. It is further desirable to maintain a high cumulative size recovery of -2 μm particles above 85%. The narrow particle size distribution together with complete removal of heavy metals in the centrifuge centrate assures the brightness of the product to be within specification. Obviously, a higher size cut can be readily achieved when the feed grade is finer as compared with feed slurry that is a relatively lower grade, i.e., coarser in size distribution.

Various decanter sizes have been used including 600 mm (24 in), 750 mm (30 in), 900 mm (36 in), and 1000 mm (40 in). Some modern decanters have aspect ratio, length-to-diameter between 3 and 4.4. The G-force varies between 1000g–2000+g depending on the process requirements. To produce finer quality product, high G-force is required and it is advantageous to use improved feed accelerator design [5] so that the feed attains the tangential speed of the pool as it is laid on the pool surface to effect classification. This reduces turbulence, skidding, wear, and undesirable resuspension of the sediment. Figure 19 shows the 1-μm product obtained from the decanter with an improved feed accelerator system is much greater compared with the other with a conventional feed accelerator. Another way to interpret this is that for the same quality product, the improved feed accelerator decanter produces much higher capacity compared with a standard machine.

The rheology of the rejected cake is non-Newtonian and it generally exhibits shear-thinning behavior. It may be difficult to convey fluid-like cake that flows back to the conical beach toward the pool especially under G-force. A cake weir can be installed in the conical beach whereby it allows a differential hydrostatic head across the weir, with the liquid level higher in the clarifier and lower in the conical beach. This differential hydrostatic head facilitates cake transport in additional to the differential speed of the conveyor. It is critical to provide the proper driving hydrostatic head. Too large a hydrostatic head would result in "wash-out" of the cake, i.e., pool spilling at the conical beach, while too little would undermine the cake transport resulting in intermittent cake discharge. An adjustable cake baffle [9] is preferred, wherein the "resistance" or gap through which the cake has to be transported across is adjustable thus compensating for the hydrostatic liquid head, see schematic in Figure 20. The solids throughput can increase by 20+% on production-size 1000-mm (40-in) decanter in kaolin processing.

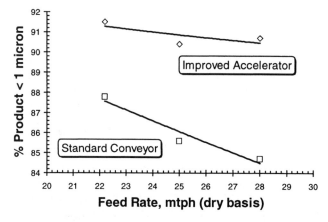

Figure 19 Comparing %<-1 μm in product between improved feed accelerator and conventional accelerator

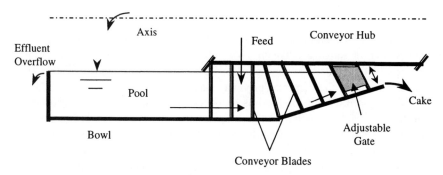

Figure 20 VARIGATE™ centrifuge schematic (courtesy of Bird Machine Company)

A new style nozzle decanter has been used to classify kaolin slurry, see Figure 21. The nozzles are located at the conical beach-cylinder junction in the machine for discharging the underflow instead of conveying cake up the conical beach. The feed and the centrate product both flow concurrently, and given the centrate product overflows at the discharge diameter of the conical beach, this design takes advantage of the full length of the machine for classification of fine particles.

Figures 22a, b, and c show an example of classification results from a CENTRIZISER® centrifuge with 650-mm (25.6-in) diameter. The size cut and total solids recovery measurements in the centrate as given, respectively, by Figures 22a and 22b, are in good agreement with the theoretical prediction of the improved boundary-layer model, Equations 3 a-f, 4, and 5. Figure 22c shows a reasonable comparison of the measurements with the theoretical "S" shaped curve for cumulative size recovery. As can be seen in Figures 22a, b, and c, increasing Le (lower G/speed, higher rate) increases total solids recovery and size recovery in the product but this is at the compromise of coarser product, and vice versa. The model prediction together with the limited measurements, which provide validation, allow optimization of the process for existing installation as well as sizing of new machines for the same process for different rates and operating conditions.

Another comparison between prediction and production measurement was made on a conventional 1000-mm (40-in) decanter in a different kaolin installation and the results expressed as %undersize in product versus Le are displayed in Figure 22d. The agreement on measured different particle sizes in the centrate product with the theoretical prediction is quite remarkable.

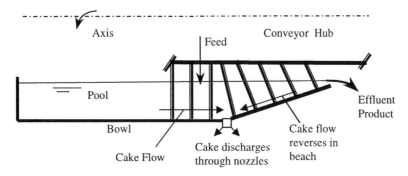

Figure 21 Schematic of a CENTRISIZER® nozzle decanter (courtesy of Bird Machine Company)

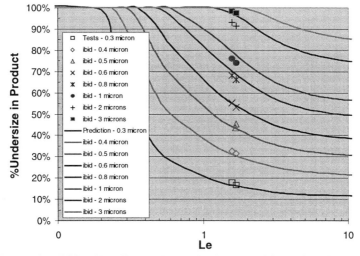

Figure 22a Comparing PSD of kaolin product with improved boundary layer model prediction expressed as function of Le for 630-mm nozzle decanter in Plant A

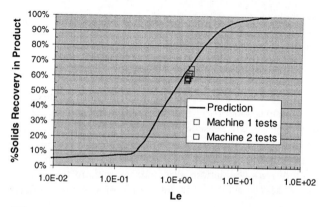

Figure 22b Total solids recovery of kaolin as function of Le for 630-mm nozzle decanter in Plant A

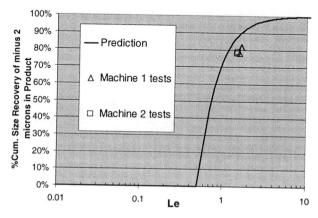

Figure 22c Cumulative size recovery of -2 µm of kaolin product for a 630-mm diameter nozzle decanter in Plant A

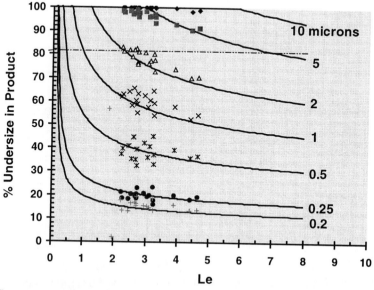

Figure 22d Comparing PSD of kaolin product with prediction for a 1000-mm diameter decanter centrifuge in Plant B

In some installations, nozzle-disk centrifuges have also been used for high-G (3000g+) classification. It is typically used as the second of the two-step classification process using higher G-force to get the size fraction. The solid concentration and loading in the feed is somewhat limited.

Drill Mud

There are several key applications of centrifuges in drill mud applications. The first is the traditional solids control where the solid-bowl decanter is used to strip the undesirable solids, recover the barite for weighted mud, and restore the fluid properties (viscosity and density) of the original mud so that it can be reused for drilling.

For a spent weighted mud 12–16 lb/gal, the centrifuge operates at low speed and G-force about 400–700g to sediment the barite in cake leaving the drill solids and the valuable fine solids in the effluent. The effluent with density 9–10 lb/gal is forward to a high-speed centrifuge operating at 1500–2500g to remove the unwanted solids (drill solids, silica, calcium carbonate, ultrafine barite solids, etc.) keeping the dissolved chemicals and bentonite in the effluent. The effluent is mixed with the previously separated barite to obtain the proper mud density, with addition of make-up mud and chemicals before reuse. In shallow drilling, unweighted mud is used and spent unweighted mud with density 9.5–10 lb/gal is sent to a high-speed centrifuge to strip out the drill solids similar to the second stage of a two-step process for the weighted mud.

Density is monitored and controlled for the feed and effluent to ensure quality from separation. The feed rate for a given centrifuge varies depending on the density of the feed. The feed rate is reduced for high density and high viscosity feed.

More modern machines use high aspect ratio (length-to-diameter), 3:1 or 4:1, to get the benefit of a longer cylindrical clarifier for classification although some low aspect ratio machines are still in use from the past such as 450 mm x 700 mm (18 in x 28 in), 600 mm x 950 mm (24 in x 38 in), and 600 mm x 1125 mm (24 in x 45 in), etc. In some applications, it is desirable for the centrifuge to be equipped with a VFD main drive so that the same machine can be used for either low-speed, high-gravity solids (barite) recovery, or high-speed classification of low gravity solids. VFD or hydraulic backdrive is also employed to control the differential speed and to provide adequate cake transport during process, cleaning and shut down.

Calcium Carbonate

There are several applications for centrifuges in processing calcium-carbonate slurries.

For processing ground calcium carbonate (GCC) slurries, centrifuges are used in conjunction with the ball mill to produce finer particles with median size 5–10 μm with zero oversize particles greater than 45 μm. High solids recovery of the fine particles less than 45 μm in the overflow or centrate is desirable as the cake, which may contain the finer particles, is recirculated back to the mill.

The second processing step is to dewater the fine-particle slurry with median size 5–8 μm to high cake dryness in excess of 70% w/w and high solids capture is essential. The cake is reslurried in dispersant and sent for downstream mill to produce further finer particles. Here the objective of the centrifuge in cooperation with the mill is to remove or degrit oversized particles in excess of 25 μm, or any foreign particles such as grinding media from the mill, from dense, high-viscosity, suspension containing 70+% solids w/w.

For precipitated calcium carbonate (PCC) process, after the slurry leaves the reactor it is sent to the centrifuge for dewatering. Particles in suspension are typical approximately mono-dispersed with mean size between 1–3 μm. Increasing feed rate beyond a critical level might result in a large drop in the solids recovery due to the feed PSD [2]. The cake is mixed with dispersant and sent to a second-stage centrifuge for high-density degritting similar to that of GCC.

Some common size centrifuges are 450-mm (18-in), 750-mm (30-in), 1100-mm (40-in) and 1100-mm (44-in) diameter deployed for use in the calcium-carbonate separation processes.

Dewatering of the slurry is usually limited by the solids recovery and at other times the cake dryness may be the limiting factor. In some applications, backdrive is used to provide good control on differential speed for process and machine cleaning.

The G-force can range between 100–1200g for different classification steps to 1000–3000g for dewatering depending on the process (feed rate, PSD) and the size of the machine. The feed rate also varies depending on the process and the size of the machine.

A new compound-beach design [10] has been developed to dewater fine-particle slurry such as PCC and GCC where cake solids are 5%–10% above and beyond which can be obtained from conventional design. This new high-G design, as illustrated in Figure 23, has a compound beach profile with a steep conical beach followed by a zero-degree cylindrical beach. An adjustable gate is located near the exit of the machine to meter the cake to be discharged. Across the entire cake layer only the driest cake layer near the bowl wall is skimmed for discharge while the wet cake near the cake surface is recycled back upstream. This increases the solids retention time in the machine. With the compound beach geometry, increasing solids capacity does not compromise cake dryness and vice versa [10] as with conventional decanter. Figure 24 compares the result between two 450-mm (18-in) diameter machines, a conventional design operating at 3000g and the compound-beach design with an adjustable cake-flow control. With conventional design, as feed rate increases the cake layer also increases leading to slow expression of water out of the thick cake and wetter cake results with higher throughput. This is distinctly different for the new design for reasons as discussed. As throughput increases, cake dryness actually increases instead of decreases as with conventional design. In fact, a significant increase in cake solids of 3%–5% is obtained for dewatering of fine PCC slurry with 1-µm mean particle size at 3000g, see Figure 24. As the G-forces further increase to 3850g, cake solids concentration increases with increasing throughput and G-force for the new design. The compound beach with cake flow control can be applied to dewater fine-particle slurries, including kaolin that has been traditionally carried out with low-speed, large-diameter (i.e., low specific throughput) vacuum filter.

Mineral Tails

Dewatering of mineral slurries can be done similar to that of municipal sewage using decanter centrifuges [3]. A drier cake product with lower polymer dose can be obtained with an adjustable gate design [11]. Higher dryness usually accompanies with high torque [12].

CONCLUSIONS

Centrifuges have been used quite extensively for various important separation steps in mineral processing. In the last decade, more effective and new centrifugal separators together with process know-how have been developed to address the needs from this process industry.

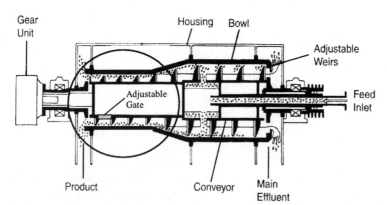

Figure 23 ULTRADRY™ decanter with compound-beach decanter and adjustable exit gate for fine-particle dewatering (courtesy of Bird Machine Company)

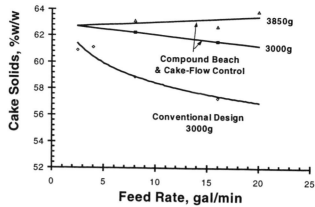

Figure 24 Compound-beach with cake-flow control versus conventional design on dewatering fine-particle (1-2 μm) slurry

LEGEND

D	bowl diameter, mm
G	centrifugal gravity, m/s²
g	earth gravity, 9.8 m/s²
F_e	centrate PSD
F_f	feed PSD
f_f	feed, frequency of a given size
h	cake height, m
I	integral, Equation 3c
L_c	clarifier length, m
Le	Leung number, [1]
Q	feed rate, m³/s
SR	cumulative size recovery, %
t	time, s
t_D	dimensionless time, Equation 6, [1]
R	gear ratio
R_e	solids recovery in centrate, %
R_p	critical pool radius, m
Rs	solids recovery in cake, %
r	gear ratio
v_s	settling velocity, m/s
x	particle size, μm
xc	cut size, maximum size in centrate, μm
x_o	particle reference size, μm
x_h	hydraulic diameter of particles in cake, m

Greek

η	clarifier acceleration efficiency, [1]
Δρ	density difference, kg/m3
$ρ_L$	liquid density, kg/m3
μ	suspension viscosity, cp
μ'	viscosity factor, cp
$μ_L$	liquid viscosity, cp
μm	micron

ϕ   solid volume fraction, [1]
Ω   rotation speed, rpm
ΔΩ differential speed, rpm
λ   hindered settling function

REFERENCES

[1] Richardson J. and Zaki W. "The sedimentation of a suspension of uniform spheres under conditions of viscous flow," Chem. Eng. Sci., 3, 1954, pp. 65–73.

[2] Leung, W. Industrial Centrifugation Technology, McGraw-Hill, 1998.

[3] Leung, W. and Havrin R. "High-Solids Decanter Centrifuge," Fluid-Particle Separation Journal, Vol. 5, No. 1, pp. 44–48, March 1992.

[4] Leung, W. et al. "Feed Accelerator Improves Coal Dewatering Centrifuge Capacity and Basket Life with Improved Cakes and Recovery," 10th Pittsburgh Coal Conference, Sept. 20–24, 1993.

[5] Leung, W. and Shapiro, A. "An Accelerating Vane Apparatus for Improved Clarification and Classification in Decanter Centrifuges," Transactions of Filtration Society, 1 (3), pp. 61–67, 2001.

[6] Leung W. and Shapiro, A. "Improved Design of Conical Accelerators for Decanter and Pusher Centrifuges," Filtration and Separation, Sept. 96, pp 735–738.

[7] Leung W. and Shapiro, A. "Efficient Double-Disk Accelerator for Continuous-Feed Centrifuge," Filtration and Separation, Oct. 96, pp 819–823.

[8] Leung, W., Yarnell, R., and Quinn, T. "In-Situ Reslurrying and Dewatering in Screenbowl Centrifuges," Filtration and Separation, Vol. 37, #7, pp. 42–49, Sept. 2000.

[9] Leung, W., Shapiro, A., and Yarnell, R. "Classification of Fine-Particle Slurries Using a Decanter with Adjustable Cake-Flow Control and Improved Feed Accelerator," Filtration and Separation, pp. 32–37; Vol. 36, No. 9, Nov. 1999.

[10] Leung, W. and Shapiro, A. "Dewatering of fine-particle slurries using a compound-beach decanter with cake-flow control," Minerals and Metallurgical Processing, Vol. 19, No. 1, February 2002.

[11] Leung, W. "Dewatering Biosolids Sludge with VARIGATE Decanter Centrifuge," Transaction Filtration Society, Vol. 1 (2), April 2001.

[12] Leung, W. "Torque Requirement for High-Solids Centrifugal Sludge Dewatering," Filtration and Separation, Nov. 1998, pp 882–887.

Characterization of Equipment Based on Filtration Principals and Theory

Glenn D. Welch [1]

ABSTRACT

The characteristics of the fluid/particle mixture to be filtered, its intrinsic filtration properties, and the objective of the solid-liquid separation dictate the equipment selection that would be technically and economically feasible for a prospective application.

Definitions of the physical and chemical characteristics of the slurry to be filtered and correlations obtained from testing representative feed samples by applying principals of filtration theory are used as a basis for filter design.

A systematic approach is used initially to reduce the number of options considered feasible for a given filtration application. A brief discussion of important considerations relating to equipment design is provided as a further guide in the selection process. Testing to develop design criteria is the final requirement to complete the selection process.

INTRODUCTION

The selection of filtration equipment is an integral part of overall process design. The main focus of this paper is to provide a general guide for the engineer to reduce filtration options for a specific application and to develop sufficient information to properly select the optimum design.

The economic selection of filtration equipment begins with choosing a design basis, and narrowing down filtration equipment options through systematic consideration of solid-liquid separation system operations prior to the filter, post-treatment requirements, overall process goals, and characterization and nature of the feed material. A test program based on the filtration options selected is then the final step necessary for further evaluation. The test program is focused on the goals of the specific application and serves to define the important aspects of the filtration cycle time (i.e. form time, wash time, and dry time) required for filter design. Once the data has been obtained and correlated, final selection by evaluation of engineering requirements and economic analysis can commence.

Evaluation of the rate of cake formation and/or the rate of filtrate production achieved for an applied pressure drop (driving force) dictates the type and size of the filter that will be required based on the tonnage to be processed.

The important properties of the feed material affecting filtration equipment design can be accurately defined by performing tests on representative samples in a suitable laboratory scale testing apparatus through correlation of test data by applying standard filtration principals. A complete discussion of correlation methods for filtration data can be found in the cited references (Smith and Townsend 2002).

CHOOSING A BASIS FOR FILTRATION DESIGN

The design of filtration systems will either be based on:

- Cake formation (removal of bulk solids from liquid), or
- Clarification (removal of solids from bulk liquid).

---

[1] Pocock Industrial, Inc., Salt Lake City, Utah

Which of these general categories an application falls into has an affect on the test work to be done, and the equipment type and choices to consider.

Cake Formation
Cake formation processes involve the removal of solids that are present in bulk from the liquid phase. Either the solids or the liquid may be the important recovered product. Definition of whether the solids or the liquid is the important phase has bearing on the method of treatment and the filtration equipment selection. In either case, filtrate clarity, washing of the filter cake, and final cake moisture content are important design considerations. Common types of filtration equipment used for cake formation applications include: Continuous vacuum filters (horizontal belt, disc, and drum), pressure filters (recessed plate, plate and frame, and tower presses), belt presses, and filtering type centrifuges. Gravity thickening is often used to increase the solids concentration of the filter feed.

Clarification
Clarification processes involve the removal of low concentrations of solids from the liquid phase present in bulk. The liquid phase is usually the important recoverable quantity. Solids are removed to meet specifications for downstream liquid processing, and/or to meet environmental regulations. Clarification filtration equipment may include: Precoat drum filters, pressure filters (recessed plate, plate and frame, and tower presses), deep-bed granular filters, cartridges, pressure leaf or bag type filters, and filtering type centrifuges. For many applications involving vacuum and pressure filters, some type of filteraid is used to improve filtration rates and filtrate quality.

SOLID-LIQUID SEPARATION SYSTEM OPERATIONS
In the design of any filtration system, there are four SLS system operations to consider:

- Pretreatment
- Solids Concentration
- Solids Separation
- Post Treatment

Pretreatment
Some method of chemical or physical pretreatment may be required to improve the properties of the stream for SLS, or to make it amenable to separation. Pretreatment of a stream generally involves increasing the particle size of the solids present, and/or reducing the viscosity of the liquid. Hence, pretreatment of the feed makes the SLS separation easier by improving the sedimentation, and/or filtration properties. Some common forms of chemical pretreatment include the addition of coagulants or flocculants, extenders such as bentonite clay, and pH adjustment. Common forms of physical pretreatment may include temperature changes, crystallization, aging, freezing, and the addition of admix materials such as diatomaceous earth to modify slurry properties.

Concentrating Solids
Gravity thickeners or clarifiers are the most common types of SLS equipment used to concentrate solids in a dilute stream prior to filtration. Gravity sedimentation can be used alone, or in combination with a variety of filtration options depending on the extent of liquid removal required for the application. Control of solids concentration fed to filters will improve separation rates and process operation as well as reduce the size of the equipment required.

Washing in counter-current thickening or clarification circuits (CCD) provides an efficient means to remove/recover soluble components. A CCD circuit may also be used in combination with a filter or centrifuge to provide a more efficient and economical means of meeting process goals when large tonnages are involved. To achieve the same wash efficiency as a CCD circuit by using vacuum or pressure filtration alone, staged filtration steps with intermittent repulps would be required.

Cross-flow filters may also be used to concentrate solids in some applications.

Solids Separation

Selection of the type of equipment used for the final separation step depends on the characterization of the feed, the process goals, and the post treatment required.

When filtration is to be considered as a final SLS step, the most important aspect related to equipment selection is the rate of cake formation. The rate of cake formation can be used as an initial guide in selecting filtration equipment. The following list shows a breakdown of equipment that may be considered based on cake formation rates achieved (Tiller 1974):

- Rapidly filtering materials (>5cm/s): Gravity pans, screens, horizontal belt or top-feed drum filters, or filtering centrifuges.
- Medium filtering materials (0.05 to 5 cm/s): Vacuum drum, disc, horizontal belt or pan filters.
- Slow filtering materials (cm/hr): Pressure filters, or disk and tubular centrifuges.
- Clarification (negligible cake): Cartridges filters, deep granular bed filters, precoat drums, and admix filters.

Post-treatment

Frequent review of the post-treatment goals to be accomplished by the separation is an important step in the selection of filtration equipment. Important factors affecting the design of filtration operations include:

- Filtrate clarity
- Soluble material in the filter cake
- Liquor or moisture content of the filter cake
- End use of solids

The type, size, and cost of filtration equipment depends upon process requirements. Equipment required to produce a small change in any listed item could significantly increase the cost of equipment. Economics demand that only minimum requirements are met. Specification of equipment to yield results in excess of the basic requirements could be very costly and of little value to overall process goals.

Filtrate clarity. Post-treatment of filtrate produced from a filter may be somewhat relaxed if clear filtrate is not a requirement and a thickener or clarifier is provided upstream of the filter as solids can be recovered by recycling the filtrate to the thickener feed.

If a thickener or clarifier is not included in the flowsheet, the filtrate can be collected and a polishing filter used to clarify the primary filtrate. This will eliminate the necessity of using the primary filter as a clarifier. Specific equipment types of gravity media, pressure, and vacuum filters are used to clarify streams containing low solids concentrations.

Soluble material in the filter cake. To remove/recover soluble components in a filter cake washing is required. The amount of wash solution needed to achieve a specific goal depends on the efficiency of the washing method, and the characteristics of the material in the filter cake. For filtration systems wash volume is defined by the number of displacements of liquor in the cake with wash solution (wash ratio). Wash efficiency is determined during testing, and is usually displayed as a function of wash ratio.

The efficiency of washing on filters depends on the properties of the solids, the porosity of the filter cake (porosity is a function of particle size), and the method of washing used. There are two basic methods of washing on a filter, flood washing and spray washing. In flood washing, the wash solution is applied and forms a pool on top of the cake until forced through by the driving force of the filter. Flood washing is the most efficient method of washing. Pressure filters all use some form of flood washing to wash filter cakes. In spray washing, the wash solution is sprayed on the cake at a rate that equals the rate of passage through the cake. This type of washing is used for rotary drum applications.

Filter cake washing can be accomplished in vacuum and pressure filtration equipment. The extent of washing required dictates equipment choices. For vacuum filtration equipment involving washing, horizontal belt, pan, and rotary drum are the best equipment options. These filter types allow for separate collection of strong and weak filtrates and facilitate counter-current washing. Most pressure filters are well suited for cake washing but counter-current washing is more complex. Counter-current washing can be accomplished in a pressure filter circuit by using cycled wash solutions from surge tanks or by successive re-pulping and re-filtering. For processes where extensive washing is required, inclusion of a CCD circuit prior to the pressure filter is common practice.

Cake moisture content. Cake moisture content is a function of particle size, shape, and composition, and is also affected by temperature, liquid viscosity, and cake porosity. Minimum cake moisture is obtained by maximizing dewatering time and driving forces that control the rate of airflow through thin filter cake. Steam and heated air are also used to reduce cake moisture content. Mechanical expression or squeezing cake to remove excess moisture is also used in some types of recessed plate and tower (vertical) presses. Whether a cake can be successfully dewatered by air or mechanical means depends on capillary backpressure, which is a function of pore size. For air blowing alone to be effective, solids particles are usually larger than 10 microns diameter.

Pressure filtration is used for smaller particle sizes, and for slower filtering materials when low cake moisture content is required. Vacuum filtration is limited by low driving forces and airflow rates during cake dewatering.

End use of solids. Definition of filter cake end use impacts filter selections. Cake handling characteristics and downstream process requirements will often dictate the type of filter utilized. For example, if the filter is required to produce a constant supply of solids to downstream processes, continuous vacuum filtration may be applicable. Pressure filtration cakes are produced batch-wise, but could be stored for supplying continuous systems. Pressure filter cakes are also typically drier, and may require breaking or pulverizing to meet further processing demands.

FILTRATION EQUIPMENT DESIGN CONSIDERATIONS

Overall Process Goals
Process and logistic factors affecting the overall flowsheet require consideration throughout the equipment selection process.

There are many types of filtration equipment available to accomplish a particular separation goal, some types may be more effective than others, or involve different techniques or mechanisms. The selection process should focus on choosing a generic equipment type that can provide required performance reliably and economically.

Characterization of the Feed
Characterization of the feed slurry is an important initial step. Aspects of the feed to be defined include:

- Particle Size Distribution
- Suspended solids concentration
- Dissolved solids concentration
- pH requirements
- Temperature and Volatility

- Tonnage to be processed
- Special properties and nature of the solids or liquid present

Particle size distribution. Has a significant impact on cake formation rates and subsequent filtration equipment selection. The amount of material present below 44 microns can have an especially significant impact on cake formation rates, and be a deciding factor in equipment selection. For this reason, samples used for testing must be prepared very carefully or they will bias results for equipment design.

Over-grinding or mixing during preparation of a sample may also severely bias test results. Softer materials tend to slime depending on the material(s) present, posing a significant problem. For harder materials or precipitates sliming may not be an issue. However, the impact of attrition on the sample to be tested should be carefully considered.

Suspended solids concentration. Plays a key role in equipment selection. Feeds containing high solids concentrations (usually 5% or greater) are typically concerned with cake filtration (removing bulk solids from the liquid). Cake accumulation rates for high solids feeds usually vary from several seconds (cm/s) to several minutes (cm/min).

Feeds containing low solids concentrations (usually less than 5%) involve the removal of small amounts of solids from the bulk liquid (clarification). Selecting equipment for clarification processes often involves equipment combinations. By their nature, clarification processes typically involve removal of ultra fine colloidal solids particles that are difficult to filter unless pre-concentration is practiced.

Dissolved solids concentration. Impacts solute viscosity and is a function of the degree of saturation for the components present. Dissolved solids can potentially cause scaling in the internals of the equipment or blind the filter cloth due to precipitation. The latter can be initiated by pressure drop or temperature changes during the separation process. This may be an important consideration depending on the degree of solute saturation, and nature of the dissolved solids present in the treated stream.

The pH of the feed stream. The initial pH of the feed stream, and the flexibility of making changes to the pH in the process has potential impact on filtration performance. Changes in pH can provide some flexibility in pretreatment of the slurry as a result of its affect on the behavior of particle surface chemistry. Use of coagulants and flocculants are affected by pH.

Construction materials and their selection is impacted by pH and solute corrosion activity. Corrosivity may limit the selection of certain types of equipment.

Temperature and volatility. Of the feed liquid present have impact on equipment design and selection. For operating temperatures above 50°C, and/or for volatile liquids, vacuum filtration equipment will need to provide for the increased vapor load to be treated as flashing occurs. Vapors produced could also present various explosion or health hazards, or may cause damage to the vacuum pump.

For high temperature, and/or for volatile liquids, pressure filtration options should be considered.

Process tonnage. The tonnage to be processed has an impact on the economics of equipment selection. Continuous filtration equipment may not be economically justified for low tonnages. For processes involving low tonnages and/or solids concentrations, there may be a definite economic advantage in manual or semi-automated batch operations, such as pressure filtration.

Some pressure filters are fully automated and specifically designed to handle large tonnages. Pressure filters of this type approach the economics of continuous vacuum filtration. The economics of continuous versus batch filtration equipment is governed by the filtration rates achieved, tonnage processed, and cake moisture attained.

Special properties and nature of the solids and liquid. May impact filtration equipment selection. Problems such as scale formation and cloth blinding can usually be overcome or prevented if recognized early in the design process.

SLS TESTING: THE FINAL PHASE IN EQUIPMENT SELECTION

After the above aspects of the type of process and separation have been examined, the engineer should now be at a point where some basic equipment type(s) can be selected to achieve the specific requirements for the application at hand.

Testing is the next important phase required to complete the process of equipment selection by providing the needed design criteria. The equipment choices should reflect the complexity of the separation and the importance of the specific process requirements to be met.

To make the selection choice easier, testing relevant to all the types of equipment chosen may be conducted. Once testing is complete and design criteria provided, an economic analysis can be completed, and final judgments for equipment selection can be made. Final equipment selection must also take into account process reliability and robustness, which are significant considerations in the overall economic picture.

A typical SLS testing program will provide the following information:

- An in-depth analysis of pretreatment options, including dosage requirements for all chemicals and filteraid added to the system, and details of any necessary physical treatment.
- Settling rates and sizing basis required for thickener and/or clarifier design.
- Suspended solids concentrations achievable for the type of clarification method desired.
- Selection of suitable filter media and dischargeable cake thickness.
- Filtration cycle time needed for filter design (form time, wash time, and dry time).
- Production rate, washing efficiency, cake moisture, and air flow data required for the design of continuous vacuum filtration equipment (horizontal belt, disc, or drum type vacuum filters).
- Area or volume requirement, washing efficiency, and cake moisture data required for the design of batch pressure filtration equipment (manual, semi-automatic, or fully automatic recess plate, plate and frame, leaf or bag type, horizontal and/or vertical pressure filters).
- Filtration test results may also include; steam, ambient, or heated air drying data, compression drying data, data indicating scaling or cloth blinding tendencies, and cake discharge recommendations and properties.
- Viscosity data needed for slurry pump and pipeline design.

REFERENCES

R. C. Emmet, and C. E. Silverblatt. 1974. When to Use Continuous Filtration. Chemical Engineering Progress (Vol. 70, No. 12).

B. Fitch. Choosing a Separation Technique. Chemical Engineering Progress (Vol. 70, No. 12).

C. E. Silverblatt, Hemant Risbud, and Frank M. Tiller. Batch, Continuous Processes For Cake Filtration. Chemical Engineering (April 29).

Cory B. Smith, and Ian G. Townsend. 2002. Testing, Sizing and Specification of Filtration Equipment. Article is included with this SME publication.

Frank M. Tiller, and Joseph Wilensky. 1974. Pretreatment of Slurries. Chemical Engineering (April 29).

Frank M. Tiller. 1974. Bench-Scale Design Of SLS Systems. Chemical Engineering (April 29).

Testing, Sizing, and Specifying Sedimentation Equipment

Tim Laros[1], Steve Slottee[1], and Frank Baczek[1]

ABSTRACT
Various methods of testing have been developed over the years to study the sedimentation characteristics of solids for the purpose of sizing thickening or clarification equipment. Proven testing and data correlation procedures are presented for developing design specifications for the most common types of sedimentation equipment. The paper covers a brief review of sedimentation theory, feed sample characterization, flocculant selection, flocculation conditions, solids flux optimization, thickening rate, detention time, and thickened solids rheology. The paper reviews data evaluation methods to specify equipment features and predict performance.

INTRODUCTION
Sedimentation is the partial separation of suspended solid particles from a liquid by gravity settling. There are two primary sedimentation operations: thickening and clarification. Thickeners increase the concentration of solids in a stream and maximize liquid removal. Clarifiers remove relatively small quantities of suspended particles to produce a clear effluent. Sedimentation equipment designed for either function typically looks the same, having similar features; however, the approach to design is different and feed conditioning features are typically based on achieving either underflow slurry density or overflow liquor clarity.

The basic principles and testing procedures for determining the size, performance, and basic design guidelines of thickeners and clarifiers are the subjects of this chapter. In recent years, new designs of high efficiency ultra high rate and high density underflow slurry thickeners have been introduced to the market. Testing techniques for sizing these newer technology units is in the hands of the specific equipment vendors as proprietary information. However, the testing techniques discussed in the following sections apply in principle to the majority of proven sedimentation equipment designs.

A BRIEF REVIEW OF SEDIMENTATION THEORY
The primary forces present in sedimentation separations are gravity, buoyancy, and friction. The factors influencing these forces are:

- Liquid density
- Particle density
- Particle size and shape
- Liquid viscosity
- Temperature

- Particle flocculation
- Particle concentration
- Thickened slurry rheology
- Distance of settling
- Horizontal and vertical motion

The solids settling and rate of separation can be theoretically related to many of these factors, and therefore, influence thickener and clarifier design within the specified process requirements. While it is important for the designer of a thickener or clarifier to understand this, there is a point where theory becomes incomplete and empirical testing begins. The sizing of a thickener and clarifier is a combination of applied theory and testing.

---
1 EIMCO Process Equipment Company, Salt Lake City, Utah.

The relative settling characteristics of particles can be separated into three basic regimes:

- Free independent particle settling systems – settling rates are independent of particle concentration.
- Hindered settling – settling rate steadily decreases as the particle concentration increases.
- Compression – particles settling rate is restricted by the mechanical support of particles below, causing compaction and deformation.

All three regimes normally exist in each sedimentation operation, with one typically having the primary influence on the size and design of the equipment. Each regime is a function of solids concentration and how "flocculent" a particle is. "Flocculent" refers to the tendency of particles to cohere or stick together. Flocculent particles will generally be less than 20 microns in diameter, metal hydroxides, chemical precipitates, and most organic particles. Free, independent particles are discrete and include many mineral solids, salt crystals, and particles will little tendency to cohere. Figure 1 illustrates how the settling regimes are related to concentration and flocculent nature. Understanding the particle settling regimes is required to determine which tests to conduct for a specific thickening or clarification application.

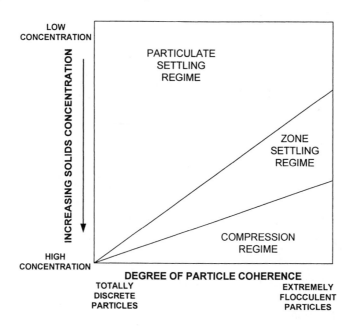

Figure 1 Effect of particle coherence and solids concentration on the settling characteristics of a suspension.

Sizing Nomenclature

Thickener or clarifier area must be sufficient to allow the slowest settling particle to reach the bed of compressing solids or bottom of the tank before its associated liquor overflows. There is a critical concentration for which the settling rate, or upward liquor flow rate will limit the solids throughput rate. The area requirements for thickeners are frequently based on the solids settling rates measured in the hindered or zone settling regimes, where the settling rate steadily decreases as the particle concentration increases. Since the surface area of the thickener is the main variable, the mass flow through the system is considered a flux (the product of concentration and velocity), such as tonnes solids/m^2/day. Frequently the inverse of this flux is used, called "Unit Area", m^2/tonnes solids/day. This allows system properties to be described in a general way for any size unit. The same nomenclature applies to clarifiers, where "rise rate", usually expressed as m/min or m^3/min/m^2, limits or sets the criteria for overflow clarity.

WHEN IS TESTING RECOMMENDED?

Data from full-scale sedimentation equipment, operating in the application under consideration, is always a first choice for sizing new equipment. However, care must be taken in evaluating the applicability of these data. The characteristics of the feed stream for the new application (i.e., mineral characteristics, particle size, viscosities, pH, etc.) must be identical to the existing application. It is also necessary to know whether the existing equipment is operating at its "capacity", and what factors might be influencing how close to "capacity" it is. To the extent that the characteristics and operating conditions are different, bench or pilot scale testing may be required. Also, various sedimentation equipment designs may be offered by equipment vendors, with the potential of more efficient operation, and the specific design or features may require special testing to verify sizing or benefits. Any significant difference between the existing and new feed streams, or new equipment design features, will most likely require testing.

OVERALL APPROACH TO TESTING SEDIMENTATION EQUIPMENT

The following data is required to design either a thickener or clarifier, and testing must be designed to produce this information:

- Feed stream characteristics
- Feed flow rate
- Expected underflow slurry density and/or overflow liquor clarity
- Flocculant type, solution concentration, and dose (if used)
- Conditions for flocculation (solids concentration, mixing time and energy)
- Vessel area and depth
- Settled solids rheology (for raking mechanism design and drive torque specification)
- Site-specific requirements: seismic zone, weather related specifications, local mechanical design codes, and the user's preferred design specifications.
- Local operating practices

There are three basic approaches to testing for sedimentation equipment:

- **Continuous piloting** – a small diameter thickener or clarifier of same configuration as the full-scale equipment.
- **Semi-continuous bench scale tests** - using laboratory pumps which pump feed slurry and flocculant solution into settling cylinders from which overflow liquor and underflow slurry are continuously collected.
- **Batch bench scale settling tests** – the conventional procedure requiring a relatively small amount of sample and graduated cylinders in which a sample is placed and allowed to settle.

Generally, batch bench testing is sufficient for scale up if the testing engineer has previous experience in the application being evaluated, and proven scale-up methods are used.

Larger scale testing is dictated by the following factors:

- Sample size – small amounts of sample will limit testing to batch bench scale
- Test costs – semi-continuous bench scale tests are more expensive than batch tests and pilot tests are significantly more expensive than both.
- Accuracy – although bench scale testing is relatively well developed and accurate, scale-up factors are still required. Pilot scale testing may be required for certain applications where the scale-up factor is uncertain or function of the equipment is otherwise uncertain (i.e. Will a thickener be able to discharge underflow with the rheology produced in the batch test? Are the feed characteristics expected to vary frequently, and will batch testing be able to quantify the effects on performance?).

- Demonstration – observing a clear overflow or a thick underflow from pilot equipment of the same design as the full-scale equipment significantly reduces uncertainty.

Sample Characterization

Whether clarification or thickening tests are conducted, sample characterization is common and necessary for both. Any sizing from testing is based on the characteristics of the sample tested. Without this data included in the basis of design, the sizing and predicted performance cannot be validated for the specified feed stream. Characterization requires the following measurements as a minimum:

- Particle size distribution – include coarse (+100 micron) and fine (minus 20 micron) particle diameters
- Particle Surface Area
- Particle specific gravity
- Liquid specific gravity
- Dissolved materials, if any
- Temperature
- pH
- Feed solids concentration

Coagulant and Flocculant Screening

Coagulants and flocculants are widely used to enhance settling rate which reduces thickener and clarifier diameter and improves overflow clarity and/or underflow slurry density. The terms "coagulation" and "flocculation" are sometimes used interchangeably, however, each term describes separate functions in the agglomeration process.

Coagulation is a preconditioning step that may be required to destabilize the solids suspension to allow complete flocculation to occur in clarification applications. Flocculation is the bridging and binding of destabilized solids into larger particles. As particle size increases, settling rate generally increases. The science of flocculation is not discussed here but can be found in numerous texts and literature which is readily available from flocculant vendors.

Both coagulation and flocculation are typically considered in designing clarifiers, whereas, flocculation is normally the only step in designing thickeners.

Coagulants may be either organic such as polyelectrolytes or inorganic such as alum. Coagulants can be used alone or in conjunction with flocculants to improve the performance of the flocculant or reduce the quantity of the flocculant required. In some systems, where a flocculant has been used in an upstream process, a coagulant may be needed to allow additional flocculant to be effective.

There are two primary types of flocculants:

- Natural flocculants - Starch, guar, and other natural materials have historically been used for sedimentation flocculation, but have been replaced by more effective synthetic polymers.
- Synthetic polymeric flocculants – There are hundreds of synthetic polymers available developed for specific mineral types and applications

Because of the many available flocculants, a screening program is necessary to choose an effective flocculant. Testing time and expense usually limits the number of flocculants screened to less than ten. The choice of flocculant can be narrowed by considering the following:

- Prior experience with flocculants on the feed stream under evaluation is always a good source of data.
- Experience of the test engineer or end user with a particular type of flocculant for the application (e.g., copper tailings, alumina red mud, etc)
- Test one each of the major types of flocculant charge: anionic, nonionic, and cationic.

- Test one each of the synthetic polymer length: long chain, short chain.

The purpose of the screening tests is to select a coagulant or flocculant whose generic type will most likely be effective in plant operation, and therefore, suitable for clarifier or thickener testing. Although a thickener or clarifier may be started up on the flocculant selected in the testing, it is very common to conduct further tests on the full-scale machine to further optimize dosage or flocculant type. The flocculant manufacturer can be a source of great assistance both in the testing and the full-scale optimization of flocculant use.

Equipment: Several 100 - 250 ml beakers for testing slurries for thickening, or,
laboratory gang stirrer apparatus for testing liquors for clarification
Several flocculants of different charge, polymer length
A stirrer or hand-held laboratory spatula
Syringe or burette for measuring the dosage

Procedure: Coagulant or flocculant solutions should be made up according to the manufactures instructions and used within the shelf life recommended. The solution concentration recommended for testing is typically more dilute than the recommended "neat" concentration so that the viscosity is lower to make dispersion more rapid during testing.

In the screen tests, each coagulant or flocculant is added to the beaker samples of representative of slurry or liquor in a drop-wise fashion, while the sample is mixed with a spatula or stirrer. The amount of coagulant or flocculant required to initiate floc particle formation is noted along with relevant notes as to the size of the floc, capture of fines, resultant liquor clarity, and stability of the floc structure. The dosage is typically noted in g/tonne solids it the sample is primarily solids (thickener design), or in mg/l liquor if the sample is primarily for clarification and the solids concentration is low.

Once a coagulant and/or flocculant is chosen, and a dosage estimated, optimization testing is conducted to further quantify the conditions for the larger volume design tests.

Optimization of Thickening Test Conditions

After a flocculant type is selected, the next step is to conduct a range of tests on larger sample volumes, using the selected flocculant, to gather data on the effects of feed slurry solids concentrations on flocculant dosage and settling rate. There is a range of feed solids for which flocculation effectiveness is maximized, resulting in improved settling characteristics. Operating within this feed solids range, results in smaller diameters, higher underflow slurry densities, better overflow liquor clarity, and lower flocculant dosages.

Equipment: 500 ml beaker
Laboratory spatula
Syringe or burette for measuring flocculant solution
Stop watch

Procedure: Prepare a series of equal volume feed slurry samples at different solids concentrations. A good method is to start with the design feed slurry solids concentration, and then prepare a series of samples, decreasing in increments of 5% solids. For some very fine solids samples (e.g., alumina red mud, clays, leached nickel laterites, etc), it is recommended to also check a sample diluted to 2-3% solids. The final volumes if each sample should be around 350-400 ml so there is room in the beaker for adding the flocculant and mixing.

Begin adding the flocculant solution drop-wise and make notes on the dosage at which flocculation begins, and the settling velocity. Continue adding flocculant incrementally and noting the floc structure, fines capture, liquor clarity, and settling velocity. Once the settling velocity remains constant for a few tests, sample testing can be stopped, and then move on to the next sample. From the above tests, the plot shown in Figure 2 can be drawn and the results used to set conditions for the larger and final tests for sizing the thickening equipment. The test procedure for the design tests should be structured to span both the optimum solids concentration and concentration points higher and lower. The flocculant dosage should be checked at the optimum and at dosages slightly higher and slightly lower than that determined in the above tests.

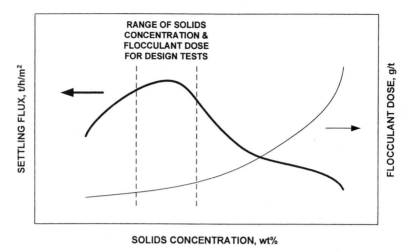

Figure 2 Effect of solids concentration on flocculant dose and settling flux.

Optimization of Clarification Test Conditions

As discussed later in this chapter, clarification is typically controlled by the time necessary for the dilute suspension of solids to coagulate, flocculate, and then settle. In many instances, the rate of clarification is enhanced by increasing the solids concentration in the flocculation zone of the clarifier. This is done in a full scale operation, by internally or externally recycling previously settled solids into the flocculation zone where they are mixed with fresh, coagulated feed. The higher population of solids improves the flocculation efficiency and settling rate.

To conduct these tests, a large sample of feed liquor (typically enough to contain ten grams of suspended solids) is first coagulated at the dosage and mixing intensity determined in the screening tests and flocculated according to the screening test. The solids are allowed to settle and the supernatant carefully decanted. The settled solids are then transferred into a smaller beaker and saved for dosing samples of freshly coagulated feed samples.

The final, optimizing tests are conducted by preparing approximately 750 ml samples and coagulating as previously determined. Recycle solids are added, as a slurry, to give suspended solids concentrations of 1, 2, 3, and 5 g/l. Flocculant is then added and the mixing stopped. Settling rate is measured and clarity observed. The recycle solids concentration that gives the best clarity is selected for the final, larger design tests. Settled solids from these tests should be saved for the final tests since they are representative of recycle stream solids.

In some leach solutions, very fine colloidal solids are present and are very difficult to coagulate. In these cases, it is typically necessary to test for mixing intensity and mixing time to obtain coagulated solids that are more amenable to flocculation.

TESTING STRUCTURE AND EQUIPMENT TO SIZE THICKENERS

Thickener sizing requires determination of two values: the area necessary to prevent the formation of a critical concentration zone of solids which can rise to completely fill the thickener (unit area), and the bed depth needed to attain the desired underflow concentration. Sizing of thickeners normally requires evaluating settling data to identify the critical solids settling flux to establish the minimum unit area for liquor release and separation from settling solids. For applications of slow thickening suspensions, such as clays, or where a high underflow density approaches paste consistency, the testing should also study the effect of the compression zone to determine the settled mud bed unit volume for thickening. Which ever is the limiting condition will have an effect on sizing and design of the thickener. In some instances, the calculated compression depth may be too great to be practical, and the area must be increased as necessary to provide the needed retention time with a lesser depth.

Method of Coe Clevenger for Non-Flocculated Pulps

In 1916, Coe and Clevenger proposed a method for sizing thickeners using a zone sedimentation or compression subsidence model, which has been proven to be valid in applications where the slurries do not require flocculation, or use only reagents such as lime which cause little variation in the floc size or dewatering characteristics. The basis for this sizing approach is that the settling rate of the pulp is a function only of the solids concentration, as represented by the zone settling regime of Figure 1. By definition, the critical point will be that solids concentration just before, to the beginning of, the compression zone. To test the initial settling rate, of a number of pulp samples ranging in solids concentration from feed slurry to underflow slurry is measured using conventional laboratory settling tests. For each concentration, an area value can be calculated from the observed settling rate and the volume of liquor which would report to the overflow, based on a pre-selected underflow concentration. A limiting maximum area per unit weight of solids per day can be determined.

The equation used is:

$$\text{Unit Area}, \, m^2/tpd = (1/C - 1/C_u) / v \qquad (1)$$

Where, C = test solids concentration, kg/L
C_u = underflow solids concentration, kg/L
v = initial settling rate at test conditions, m/d

At the end of the series of tests, the solids are dried and weighed and values substituted in Equation (1) to calculate the corresponding unit area. A maximum value will be obtained, representing the limiting or size-determining conditions, as illustrated in Figure 3.

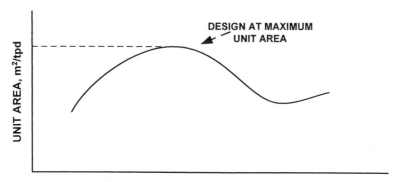

Figure 3 Coe Clevenger method of determining thickener unit area.

Methods Derived from Kynch Theory for Flocculated Pulps

When polymers are used to flocculate a suspension, dilute suspensions will produce much larger flocs, compared to higher concentrations as discussed earlier, and it is preferable to run a single test at the expected feed solids concentration. Procedures derived form Kynch theory can be used to determine the unit area, m^2/tpd. One of these methods is the Talmage and Fitch method which uses Equation (2):

$$\text{Unit Area}, \, m^2/tpd = t_u / C_o H_o \qquad (2)$$

Where, t_u = settling time, days
C_o = test of feed solids concentration, kg/L
H_o = initial height of pulp in the test, m

The value of t_u is determined from the settling curve by any of various methods, the selection of which depends on the particular testing organization and its experience in scaleup from this approach and organization's proprietary methods proven in actual practice. One commonly employed system which produces conservative results is illustrated in Figure 4, and is based on the bisection of the angle formed by two tangents to the straight line portions of the curve, the intersection of the bisection and the settling curve defining the critical point. A tangent to the curve drawn at the critical point intersects a line representing the height of the pulp at underflow concentration, giving the value of t_u.

Another approach, called the Oltmann method employs a straight line drawn from the start of the settling curve to a discernible point on the curve at which compression is believed to begin. The extension of this line to the underflow line, as also shown in Figure 4, gives the value of t_u to be used in the equation.

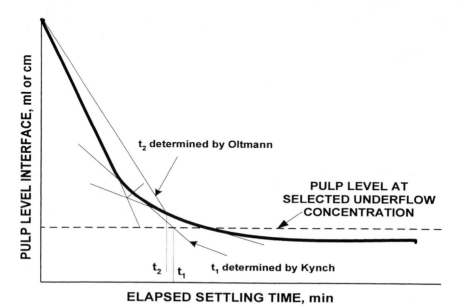

Figure 4 Sizing a thickener from batch data using Talmage and Fitch equation, and selecting time, t, by Kynch or Oltmann methods

Both methods described above usually apply a scale up factor which accounts for the mud bed depth in an actual thickener.

Various other sizing methods have been developed over the years. The Warren Spring Laboratory published an extensive review on 1977. one such "flux" type model is the Wilhelm-Naide model which is an adaptation of the Yosioka & Hassett models of the 1950's and 60's. The flux type models were developed to overcome some of the over and under sizing problems that were found to occur with Kynch and Talmage & Fitch, especially with modern flocculants.

The Wilhelm-Naide model uses the total thickener flux, settling flux in addition to withdrawal flux, to define unit area as:

$$\text{Unit Area} = \frac{[(b-1)/b]^{b-1}}{a\,b} \; C_u^{b-1} \qquad (3)$$

Where,   a and b = coefficients developed by the equation $v_i = aC_i^{-b}$ where C_i is measured as weight of dry solids per unit volume, and v_i is settling velocity of a layer with suspended solids concentration C_i, length/time.

C_u = underflow solids concentration, in weight dry solids per unit volume.

As proved by Kynch, tangents to the test settling curve extrapolated to the vertical axis permits determination of the solids concentration or C_i. The slope of the tangent line yields the settling rate. Thus, by plotting velocity as a function of solids concentration on log-log paper, the values of a and b are determined. It is interesting to note that normally 3 straight line relationships are obtained and at least two in any case. Accordingly, the appropriate values of a and b are used in Equation 3 to calculate the unit area value at the desired underflow solids. The reader is referred to the literature for the complete development of the method.

Determining Effects of Compression on Sizing

As pulp enters the compression regime, different factors come into play. Compression bed depth will typically have an effect on the overall thickening rate and higher bed depths will reduce the unit area. Predicting the effect of increased bed depth is not possible from theoretical considerations alone, as too many other factors have an influence, including the metallurgical nature, particle size distribution, and particle of the solids. The flocculant dosage and the characteristic floc structure affects the unit area, as well as the mechanical action of the rake and the particular rake design.

Ideally it would be desirable to carry out this test in a cylinder in which the pulp could be maintained at a depth approaching that of a full-scale thickener. Usually, this is impractical and appropriate scale-up methods are needed.

If compression is expected to be a critical factor, additional tests should be carried out in deep cylinders, and the values plotted so as to permit extrapolation to grater depths, using the log-log relationship shown in Figure 5. Generally, this extrapolation should not be extended beyond a pulp depth in compression of 1m unless full-scale data indicate the benefit of greater depths.

The mechanical action in the compression zone and the depth of the pulp in actual compression will influence the rate of thickening. In evaluating the compression zone requirements from a batch test, one can calculate the unit volume (cubic meters of compression zone volume per tonne per day) from tests conducted under identical conditions except in cylinders of deeper pulp depths. It will be found that the calculated volume increases as the cylinder depth increases. Using the unit volume calculated from tests at a normal cylinder depth, as a design basis, can result in an undersized unit if the compression zone determined the size of the thickener and if only the detention time were considered. This is generally manifested as a lower unit density than expected or the need for a greater amount of flocculant than indicated in the original test work.

In determining the compression zone unit volume, the following equation can be applied:

$$\text{Unit volume, V, m}^3/\text{tpd} = t_c (\rho_s - \rho_l) / [\rho_s(\rho_p - \rho_l)] \tag{4}$$

Where, t_c = compression time required to reach a particular underflow concentration, days.

ρ_s, ρ_l, and ρ_p = densities of solids, liquid, and pulp (average), respectively, tonne/m^3.

The log-log plot of unit volume versus average pulp depth (Figure 5) will usually produce a straight line having a slope less than 1, decreasing in value as underflow concentration increases.

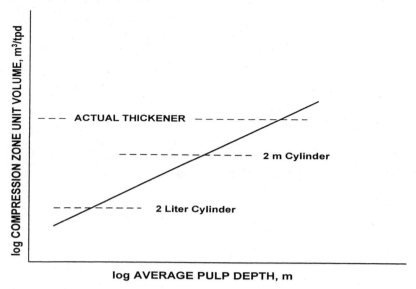

Figure 5 Log-log plot of compression zone unit volume vs average pulp depth

Batch Settling Test

Tests are conducted using the apparatus illustrated in Figure 6. Figure 7 shows a typical batch test data sheet.

> **Equipment:** 2-liter cylinders with markings to measure height, settling rate, and volumes
> Picket rakes with motors (rakes should turn at 6 rph)
> Flocculant addition and mixing apparatus
> Stop watch, or suitable timing device
> Balance for weighing cylinders and samples
> Apparatus to filter, wash (if required), and dry solids
> Flocculant solutions

Procedure: Record the tare weight of all cylinders and beakers used for gathering sample weights, and prepare flocculant solutions. Add the slurry sample to the cylinder in a manner so that the slurry sample is representative of the characterized feed sample, and prepare the dosage of flocculant determined for the test from previous screening tests and the planned testing program. Measure the weights and volumes as required by the data sheet.

When ready to begin the test, mix the cylinder of slurry using a suitable apparatus and add the flocculant solution while mixing. Figure 6 shows a plunger type mixer made with a rubber stopper mounted to a hollow tube with a syringe attached to the top for delivering a pre-measured amount of flocculant solution while mixing.

Once the flocculant is mixed into the slurry, the mixing apparatus is withdrawn, rakes are quickly inserted, and the settling rate of the slurry interface is recorded. The time for each test depends on how long it will take for the settled slurry to reach its final level and density. When the test sample has reached terminal density, the rakes are carefully removed from the cylinder in a manner to minimize re-suspension of the settled slurry. Allow the cylinder and sample to set for a while after the rakes are removed so any disturbed slurry resettles, and take the following data:

- Total volume of sample before decanting
- Total weight, plus tare, of the cylinder and sample
- Total slurry volume after decanting supernatant
- Settled slurry weight after decanting
- Settled slurry height

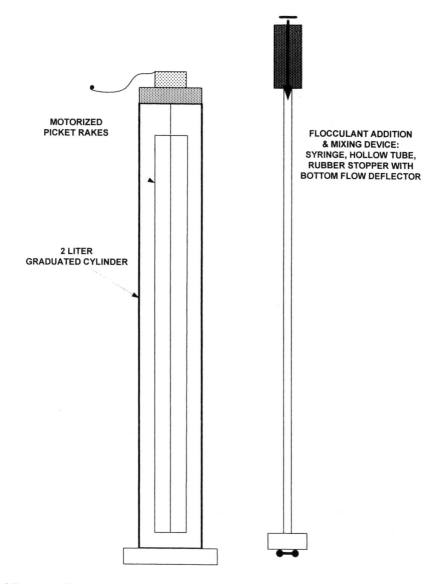

Figure 6 Batch settling test apparatus.

After the supernatant has been decanted and the above measurements made, repulp the settled slurry with the mixing apparatus and pour off as much as will flow into a beaker. With a tarred syringe, sample 25mls of the slurry and weight the sample to obtain the density, and then dry the sample to obtain the per cent solids concentration by weight. If dissolved solids are present in the liquor, the dried sample weight must be corrected for the dissolved salt concentration.

Complete the data sheet and correlate the data using one of the methods described earlier. Most testing laboratories that conduct thickening tests on a regular basis, typically have a computer programmed that will plot the settling curve and correlate the data to determine a unit area for a specified underflow slurry density.

THICKENER TEST DATA SHEET

COMPANY: ____________________ TEST No.: __________

____________________ DATE: ______ BY: ______

CONTACT: ____________________ LOCATION: __________

MATERIAL: ______________________________

OBJECT OF TEST: ______________________________

TEST I.D.: __________

THICKENER TYPE: __________

PICKET RAKE SPEED, RPH: __________

FLOCCULANT MIX INTENSITY: __________

FLOCCULANT: __________

CONCENTRATION, g/l: __________

VOLUME ADDED, ml: __________

DOSAGE, g/t: __________

FEED SLURRY INFO:

TEMPERATURE, °C: __________

pH: __________

DESCRIPTION: __________

SETTLING VESSEL SIZE, ml/ft: __________

UNDECANTED SLURRY VOLUME, ml: ____

SLURRY WEIGHT, gm: ______ TARE: ___

DECANTED SLURRY VOLUME, ml: ______

SLURRY WEIGHT, gm: __________

DRY SOLIDS WEIGHT, gm: ____ TARE: ___

ULTIMATE INTERFACE HEIGHT, ml: ______

PARTICLE SIZE DISTRIBUTION--------

Mesh	wt%	Mesh	wt%
___	___	___	___
___	___	___	___
___	___	___	___
___	___	___	___

SUPERNATANT DESCRIPTION--------

THICKENED PULP DESCRIPTION----

SP. GRAVITY LIQUOR: __________

SP. GRAVITY SOLIDS: __________

INITIAL % SOLIDS: __________

FINAL % SOLIDS: __________

REMARKS:

	TIME, min.:	INTERFACE HEIGHT, ml:
1.		
2.		
3.		
4.		
5.		
6.		
7.		
8.		
9.		
10.		
11.		
12.		
13.		
14.		
15.		
16.		
17.		
18.		
19.		
20.		
21.		
22.		
23.		
24.		
25.		
26.		

Figure 7 Batch thickening test data sheet.

Semi-Continuous Settling Tests

A semi-continuous settling test can be effective is determining the initial settling velocity for various feed solids concentrations and flocculant dosages if enough sample is available for conducting the test through all the variables desired. The test is usually representative of the initial slurry free settling zone seen on the batch test settling curve (Figure 4). To obtain compression zone settling data, the flocculated sample is withdrawn from the bottom of the test apparatus and the settling rate is measured in a separate 2-liter cylinder fitted with picket rakes. Compression zone data is collected in the same manner as that described above for the batch settling tests. Some thickener vendors use a semi-continuous testing apparatus with a deep cylinder that is raked so that the test can be interrupted and the flocculated slurry allowed to settle to obtain the compression zone settling data. The same data is typically collected as in the batch tests.

Figure 8 shows one type of semi-continuous testing apparatus. A well mixed, homogeneous, feed slurry sample is pumped into the test apparatus. Dilution water can also be pumped into, and mixed with the feed slurry, prior to entering the apparatus to adjust the feed slurry to the different feed solids concentrations determined for the test. Flocculant solution is then pumped and mixed with the feed slurry in a mixing device that is integral with the feedwell of the test apparatus. In addition to volumetric flow rates of the different feed streams, the volumetric flow rates and density of the overflow liquor and underflow slurry must be measured to obtain a mass balance for the particular test conditions.

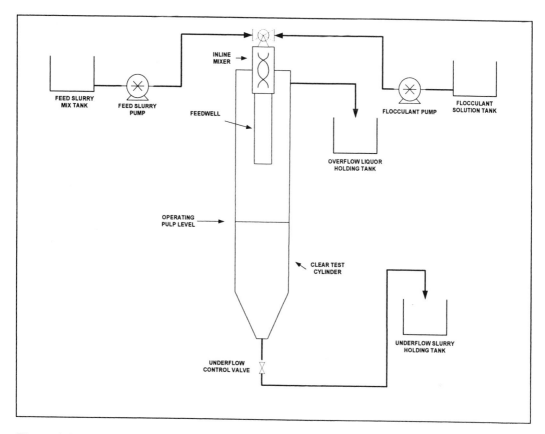

Figure 8 Semicontinuous test apparatus

Rheology Tests and Drive Torque Selection
The basic test methods described above will allow the design engineer to select a unit area for design of a thickener for a particular application. They will also set the feed solids concentration and flocculant type and dosage required for the particular unit area. Selection of a rake drive torque can be done from historical data of operating units in the same or similar applications. A table presented in the chapter "Design Features and Types of Sedimentation Equipment" gives guidelines on the use of drive torque K factors, based on experience across a range of applications, which are applied in Equation 5 for the selection of drive torque:

$$\text{Torque, Nm} = K D^2 \quad (5)$$

Where, K = torque factor
D = thickener diameter, m or ft

The information given in the table relates to thickeners operating at underflow slurry densities with yield stress values in the range of 10 – 70 Pa. In recent years, thickening and thickener design technology has advanced to allow the production of much higher underflow slurry densities that exhibit much higher yield stresses and approach the limit of flowability. In these particular applications and designs, the rheology and yield stress of the underflow slurry becomes a critical design guideline for rake mechanism design and drive torque selection.

For paste and high density applications, the K factor is related to the yield stress of the thickened slurry. When a thickener is designed to produce a very high underflow slurry density, or paste, the yield stress of the underflow increases rapidly as the solids concentration increases and approaches the consistency if a formed, but un-dewatered, filter cake. Since this is a relatively new field, most vendors of thickener drives have developed company proprietary information on the capability of their respective drives.

Figure 9 shows a typical curve of yield stress versus slurry solids concentration with the corresponding effect on the required torque capability of the thickener drive. The particular curve must be developed independently for each application since the rheology characteristics will vary depending on the particle size distribution, particle shape and surface area, slurry solids concentration and mineral characteristics.

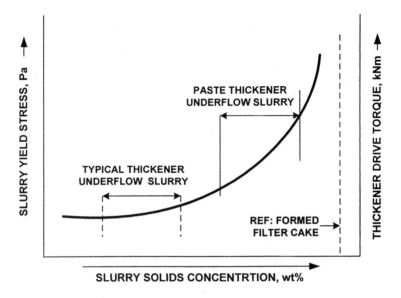

Figure 9 Typical curve showing how required drive torque increases relative to slurry yield stress.

Sizing Rakeless Thickeners
The new ultra high rate rakeless thickeners are typically sized from pilot plant operating data obtained with a pilot scale unit of the rakeless thickener. The unit's design incorporates feed dilution to optimize flocculation and maximize settling rate, and also internal dewatering cones that enhance liquor removal and the rate thickening. The particular effects of these internal features are difficult to simulate on the bench, and the most accurate sizing is obtained from the pilot plant testing.

The unit design incorporates a 60 degree cone which facilitates thick underflow slurry movement to the discharge cone without rake action.

TESTING STRUCTURE AND EQUIPMENT TO SIZE CLARIFIERS
There are a variety of clarifier designs applied in minerals processing, and most are designed to use coagulants and flocculants to improve the efficiency and rate of clarification. These are covered in the chapter "Design Features and Types of Sedimentation Equipment" of this manual. The particular stream to be clarified, the concentration and characteristics of the solids to be removed, and the degree of clarification required will usually dictate the feed conditioning features of the clarifier. These feed conditioning features typically are incorporated into the feedwell system, where the flocculant and recycle solids are combined with the fresh feed, and the mixing intensity and retention time designed to provide floc formation and solids capture. The current practice and equipment designs used in similar applications, and which are performing well, can be used to narrow the options, and a testing program can then be designed to optimize the feed preparation conditions, special feed conditioning features and equipment size.

Clarifier testing involves an understanding of the free particle settling regime and how it is affected by the flocculent mature of particles and the imperfections in full-scale clarification resulting from thermal and density velocity gradients. If the process allows, the use of coagulants and flocculants greatly improves the efficiency and rate of clarification, as well as the size of the clarifier.

Clarifiers are sized on a combination of the area requirements which are based on the free settling of particles, or overflow liquor rise rate, and the detention time required to achieve the desired overflow liquor clarity. These values define the clarifier diameter and depth.

The area requirements are based on the free settling rate of the slurry in m/hr divided by an accepted scale up factor, either at 0.5 or 0.75, to give a rise rate. The rate of liquor overflowing the clarifier must be less than the free settling rate of the particles, hence the 0.5 or 0.75 scale up factor depending on the application. The quantity of liquor overflow divided by the rise rate determines the required clarifier area and hence diameter.

The detention time requirements are based on a second order detention curve relating liquor clarity to settling time. The laboratory detention time is usually multiplied by 4 to give a required dynamic detention time. This dynamic detention time then gives the volume of liquor that needs to be detained or stored in the clarifier, generally below the bottom of the feedwell. Together with the diameter, and feedwell dimensions, the clarifier height is then determined.

Jar Tests

For clarification, the polymer, type of addition, and mixing is generally critical. Determination of the type of polymer addition (staged or single application) and the degree of mixing (flash or gentle) is usually accomplished with jar tests. These tests are carried our in a gang stirrer with variable speed, flat paddle stirrers in a standard testing apparatus as manufactured by the Phipps Bird Company. From these tests, polymer type, dose, and the degree of mixing and mix time are determined for the best overflow liquor clarity.

Settling Tests

Once the polymer and mixing requirements are determined, a standard two liter settling test is conducted, in which interface height versus time is recorded. Generally, picket rakes are not used in this test since the objective is not to define a thickener size, but a clarifier size. This test yields the free settling rate.

Detention Test

The detention test is carried out preferably in a relatively large vessel, 2 to 4 liters in volume, with a diameter-to-depth ratio of 0.5:1, thermally insulated and covered if at an elevated temperature. The sample is placed in the vessel, flocculated as determined in the screening tests, and allowed to settle for a length of time which will yield the desired clarity. Samples are withdrawn from a point near the middle of the vessel and analyzed by any suitable means. It will be observed that the clarity in this test vessel generally is about the same in a zone from 1 to 2 cm below the surface to just above the settled solids; thus, the depth of sampling is not important, provided it is near the center of this zone and care is taken not to stir up the settled material during sampling. Normally, five or six samples, collected at 5, 10, 20, 30, and 60 minutes, will be adequate to define the settling characteristics of the suspension. Data from the test are plotted as a log-log plot of clarity (suspended solids concentration) versus time as shown in Figure 10.

Surface area is then selected to make sure the rise rate is not excessive and in a safe region. If interface settling occurs in the batch test, the interface settling or bulk settling rate can be used to determine a minimum area. The settling velocity or bulk steeling rate is measured and converted to l/min m^2 to select the minimum area, and a scaleup factor applied based on experience in the application.

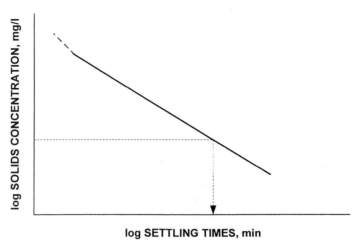

Figure 10 Log-log plot of suspended solids concentration vs time

Inclined Plate Clarifiers

Inclined plate clarifiers, commonly known as lamella clarifiers, are shallow depth sedimentation devices in which the inclined plates serve to increase the total settling area of the clarification zone; essentially creating a series of small clarifiers in parallel. The settling area of an inclined plate clarifier is the horizontal projected area of each plate multiplied by the number of plates.

Sizing inclined plate clarifiers is similar to sizing conventional clarifiers. First, a jar test is conducted to determine polymer type, dose, and mixing conditions to achieve the desired clarity. A detention time test is then conducted to generate the second order detention curve. This test is slightly different from the conventional detention time test in that the sample of clarified liquor collected a prescribed distance below the surface of the liquor. This distance is normally the spacing of the plates in the clarifier.

From the detention curve, a required time to achieve a desired liquor clarity is determined. From this time, a settling velocity is calculated according to Equation (6):

$$Vs = (Sp - Suf) / Tr \qquad (6)$$

Where, Vs = Settling velocity, m/sec
Sp = Plate spacing, m
Suf = Scale up factor, usually 0.5
Tr = Required time, sec

The Projected Plate Area, PPA, in m^2 is then determined from Equation (7):

$$PPA = Q / Vs (3600) \qquad (7)$$

Where, Q = Overflow from the clarifier, m^3/hr
Vs = Settling velocity, m/sec

The appropriate manufacturer's model having the required projected plate area can then be selected.

Counter Current Decantation (CCD) Design Methods

Washing in a thickener consists of mixing solids and associated solution with wash water, settling the solids, decanting the clarified solution and then repeating the process as required until a targeted removal of dissolved material, which was present in the original slurry liquor, is achieved. The variables in the design of a CCD circuit to achieve a desired dissolved solids removal are:

- The number of washing stages.
- The ratio of overflow volume to the volume of liquor in the underflow, which is often referred to as "wash ratio."
- The "efficiency" of each stage which refers to the completeness of mixing of the underflow and overflow slurries and liquors which enter each washing stage.

The objective of CCD testing is to quantify each of these design parameters. This is usually done by using data collected from bench settling tests to determine the expected underflow slurry density of a generic stage. A mass balance around the CCD circuit, using the "LDC" method, is used to determine the number of stages required at varying design variables of wash ratio and inter-stage mixing efficiencies.

Figure 11 shows the "LDC" method of calculating a mass balance around the CCD washing circuit to determine the overall circuit washing or recovery efficiency. In the method,

L = kg liquor / kg solids,
D = kg solute / kg solids,
C = kg solute / kg liquor,
$C = D / L$,

and the calculations begin at the last stage and work towards the first stage. To begin the calculation, set $D=1$ at the last stage to represent an unknown value. Once the mass balance is complete, the efficiency, or recovery of solute, is determined by dividing the value D for the last stage underflow ($D=1$) by the value of D for the feed and subtracting from 1, then converting the decimal result to a percent as shown in Equation 8:

$$\text{CCD Circuit Recovery, \%} = (1-(1/x))100 \qquad (8)$$

Where, x = feed solute value of D, kg solute / kg solids, calculated from the mass balance.

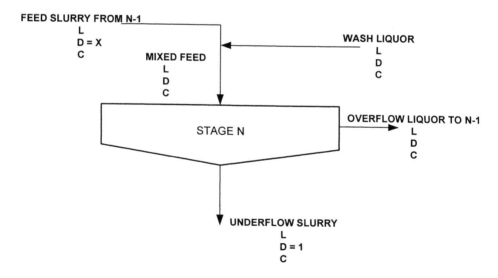

Figure 11 Example of mass balance approach to determining CCD recovery efficiency.

CONCLUSION

The overview of testing and sizing sedimentation equipment should give the reader an understanding of the general approach to designing and defining a test program for sizing sedimentation equipment for different applications and desired results.

Although testing techniques will vary depending upon the engineer's experience and between vendors of sedimentation equipment, the same basic principles apply to all.

REFERENCES

Dahlstrom, D. A. 1986. Selection of Solid-Liquid Separation Equipment. In *Advances in Solid-Liquid Separation*, ed. H. S. Muralidihara, Chapter 9. Columbus: Battelle Press.

Baczek, F. A., R. C. Emmett, E. G. Kominek. 1988. Sedimentation. In *Handbook of Separation Techniques for Chemical Engineers*, ed. P. A. Schweitzer, 2nd ed., Section 4.8. New York: McGraw Hill Book Company.

Pearce, M. J. 1977. *Gravity Thickening Theories-A Review*. Warren Spring Laboratory, Department of Industry.

Baczek, F. A. 1985. Equipment Design and Process Consideration in Flocculation and Sedimentation. *Proceedings of the Engineering Foundation Conference.* 261-272.

Testing, Sizing and Specifying of Filtration Equipment

Cory B. Smith[1], Ian G. Townsend[2]

ABSTRACT

Specification of filtration equipment to meet process performance and economic objectives can be reliably determined by bench scale testing when representative samples are available. Through development of a series of generalized vacuum and pressure filtration relationships, filter selection and sizing can be obtained for specified feed samples. Impact of process variables on filtration performance can be assessed as an integral part of economic evaluation. Complex filtration systems that include counter-current filtration and various washing schemes can also be reliably evaluated using bench-scale techniques.

Test sample representation is the key to obtaining results that can be confidently used to select and size filtration equipment. Pilot plant scale testing may be required if samples are affected by aging, particle size variation, temperature effects, and other process variables that would change filtration characteristics of the feed sample. Pilot scale operation is also required when bulk samples of filter cake or filtrate are required for downstream process testing. Pre-thickening of the sample and use of flocculants to enhance filtration rate requires careful evaluation to obtain reliable results.

INTRODUCTION

The purpose of this paper is to give an overview of the steps in sizing and specifying filtration equipment. These steps include determining the ultimate objectives, obtaining representative samples of the materials to be filtered, conceiving and executing a testing program, and finally, interpreting test data to determine filter sizing.

The first step should always be to determine what objectives must be accomplished. Selection of filtration equipment requires choosing the optimal type of equipment. For example, horizontal belt, rotary drum and disc types of vacuum filters, as well as horizontal or vertical filter presses may all be candidates for filtration of many types of metallurgical solids. Each type, however, requires a different testing approach. Furthermore, process requirements must be understood prior to initiating laboratory work. Will the process require cake washing to reduce or remove the liquor solute content of the cake? What moisture targets are there for the cake produced? These types of questions should be answered prior to starting a testing campaign to insure all data required to meet the process objectives are collected during testing.

Once objectives are understood, a test program can be conceived. Vacuum and/or pressure filtration tests can be planned as needed to determine unit production rates for each relevant piece of equipment. Planning should include collection of all necessary data to describe various process options and their associated operational trade-offs. Cake washing rates and efficiency data should be collected if the filter is to be a washing application. Similarly, drying cycle data should adequately describe the trade-off between cake dryness and production rate. The effects of expected variations in feed conditions (such as feed slurry solids content, temperature, applied vacuum, mineralogy, precipitation conditions etc.) should be considered in the testing scope as well.

Once a filtration testing program is properly conceived and outlined, the execution of the testing can be confidently undertaken. Data interpretation and correlation are relatively straightforward with an understanding of filtration theory basics.

---

1 Pocock Industrial, Inc., Salt Lake City, Utah.
2 Larox, Buckinghamshire, UK.

BRIEF REVIEW OF FILTRATION THEORY
Classical treatment of filtration theory nearly always begins with Equation (1) Poiseuille's equation (Carman, 1938; McCabe and Smith, 1976). This form of the equation contains a couple of cumbersome terms, V for the volume of filtrate, and c for solids concentration expressed as the weight of solids per unit volume of filtrate.

$$\text{Eqn (1):} \quad \frac{dV}{A d\theta_f} = \frac{\Delta p}{\mu \left[\frac{\alpha c V}{A}\right]}$$

For more practical use, this can be converted to:

$$\text{Eqn (20):} \quad W = \sqrt{\frac{\theta_f \, 2\Delta p \rho}{\mu \alpha} \left[\frac{S}{1 - \frac{S}{S_c}}\right]}$$

By solving the expression, and then plotting W as a function of θ_f on log-log paper, a straight line of 0.5 slope should result (and it generally does). This equation will be the basis of the cake form rate correlation, which is explained later in this paper and is used for filter sizing. A graphical representation of this correlation is shown in Figure 2 later in this paper. The full mathematical derivation of equation (20) is shown for reference in the Appendix to this paper.

The notation used above in equations (1) and (2), and hereinafter, is as follows:

θ_f = Cake formation time.
V = Volume of filtrate.
μ = Liquid viscosity.
A = Filtration area.
Δp = Pressure drop across the filter cake.
α = Average specific (filtration) resistance (units of length/mass).
c = feed slurry solids concentration in terms of mass of solids per unit volume of filtrate.
R_m = Resistance of the filter medium (units of length^{-1}).
W = Mass of dry solids in the filter cake per unit area.
ρ = Liquid density.
S = Weight fraction of solids in the filter feed slurry.
S_c = Weight fraction of solids in a formed, but undewatered filter cake.

FILTRATION TEST PLANNING
Define Objectives
Process needs should drive the testing objectives; hence, it is crucial to understand these process requirements prior to testing. Examples of possible process requirements for filtration equipment include, but are not limited to:
- Feed slurry dewatering.
- Cake washing.
- Filtrate clarity requirements.
- Cake moisture content targets.
- Cake transportability.
- Filter production rate.

All process requirements must be envisioned prior to planning the test strategy.

Sample Selection, Characterization, and Preparation

Selection of a representative samples is critical if test results are to apply to the actual processes. Sample availability may be plentiful when testing materials from an existing operation; however, sometimes sample quantity may be extremely scarce, such as during pre-feasibility research programs. In any case, the sample should be matched as closely as possible to the actual expected stream and conditions for which the filtration equipment is to be specified.

Sample characteristics that should be considered when determining if a sample is representative include several physical and chemical parameters. Solids concentration, specific gravity, size distribution, slurry pH, ORP, and chemical dosage (such as flotation reagents or flocculants) should be matched as closely as possible. Variations in ore type or mineralogy in a mine can significantly affect filterability of flotation concentrates and should therefore be taken into account. Sample age can also be extremely important. Many metallurgical solids are reactive and change with time significantly enough to alter filtration characteristics. An example would be sulfide flotation concentrates, which will oxidize over time. For this reason, samples for filtration testing should be as fresh as possible. Ideally, testing should be conducted immediately after the sample material is produced. This could mean that testing must be conducted at a mine or plant site, or at a pilot plant where flow-sheet development work is underway. Samples should not be allowed to freeze, dry, or be dewatered and then re-pulped prior to testing, since the resulting sample will probably not behave the same as it would have originally.

Finally, expected process parameters should be closely matched in preparation of the sample for filtration testing. Feed solids concentration, temperature, pH, etc. should be carefully matched to the process conditions. When testing metallurgical slurries, a fall in temperature can cause crystallization of salts, thereby changing cake-washing process to re-dissolution, with very different kinetics. Temperature changes also influence filtrate viscosity, and cloth blinding through salt deposition. Feed solids concentration is especially import to consider, since any de-watering process will be greatly affected by the quantity of water present. Often processes produce samples at solids concentrations that are not amenable to filtration. In these instances it is generally recommended to pre-thicken filtration feed. An example would be flotation concentrates, which generally are thickened prior to filtration. If thickening of the slurry is to be conducted, it should be carefully done so that the resulting filter feed is at a reasonable pulp density for thickener underflow, and contains a proper dose of the correct flocculant. It is often advantageous to conduct all solids/liquid separation testing in concert, since (for example) thickening test underflow material could be used for rheology and filtration testing feeds.

Equipment Needed

Vacuum filtration testing. The primary equipment required for vacuum filtration test work consists of a grid of known area covered with an appropriate filter cloth and surrounded by a metal or plastic shim to contain the pulp sample. This drainage grid or filter leaf is supported vertically on a vacuum flask. Alternately, a similar vacuum leaf can be attached to a flexible tube that connects to a vacuum flask. This alternate arrangement would allow for the leaf to be inverted and immersed in a container of filter feed slurry to simulate drum type vacuum filter operation. The differential pressure is translated from the vacuum pump to the filter leaf surface through large bore tubing and fittings. The vacuum pump should be equipped with an internal bypass system to control the vacuum level without the introduction of bleed air. More information on testing equipment, as well as diagrams of possible testing apparatus can be found in Perry's Chemical Engineering Handbook (Perry and Green, 1991).

Pressure filtration testing. Bench-scale pressure filtration test work can be performed using a pressure bomb device. The apparatus consists of a 250 mm section of nominal 50 mm pipe, capped with two flanges. The upper flange contains fittings for air pressure connection and the sample feed port. The lower flange contains an integral drainage grid, which supports the filter media. The filtrate port is centered in the bottom flange, below the filter media. . More information on testing equipment, as well as diagrams of possible testing apparatus can be found in Perry's Chemical Engineering Handbook (Perry and Green, 1991).

DATA COLLECTION

Vacuum filtration and washing tests should be conducted to collect a general set of filtration and wash efficiency data to design and size vacuum filters. Testing should (as applicable) examine the effect of applied vacuum level, feed solids concentrations, cake thickness, dry time, filter aid (flocculant or D.E. etc.) and the volume of applied wash solution on production rate, filter cake moisture and wash efficiency. Similarly, pressure filtration and washing tests should be conducted to collect a general set of filtration and wash efficiency data to design and size pressure filters. Tests should examine the effect of applied pressure, feed solids concentration, cake thickness, air blow time and the volume of applied wash solution on production rate, filter cake moisture and wash efficiency.

General Procedures

Vacuum filtration testing. To produce a test filter cake from a slurry sample, a given weight of pulp at the proper temperature and known to yield an approximate cake thickness should be poured onto the upturned test leaf while the ball valve connecting the leaf to the vacuum flask is opened to apply the differential pressure. As the last of the liquid phase disappears through the surface of the formed cake, the form time ends and is noted, and a known volume of wash water is poured onto the surface of the newly formed cake (only applicable if wash is to be tested). Once the last of the liquid wash solution disappears through the surface of the cake, the wash time is ended and noted and the subsequent dry time begins.

At the end of the dry time, the filter cake is discharged from the leaf, and the wet weight and cake thickness are determined and recorded. After drying, the dry cake weight is determined and recorded for cake moisture calculations.

Additional cakes should be collected in the same manner, each at different test conditions, until the range of test variables selected has been adequately covered.

Pressure filtration testing. To produce a test pressure filter cake from sample slurry, a given weight of pulp at the proper temperature and known to yield an approximate cake thickness is poured into the pressure chamber. The sample port is closed and air pressure applied above the feed slurry to facilitate initial cake formation and dewatering. As the last of the filtrate is produced, shown by rapid air flow through the drainage grid, the form time is noted and recorded.

Immediately following the form time the filter is depressurized, and a known volume of wash solution is poured onto the surface of the newly formed cake (for washing cases). The sample port is then closed and air pressure applied to facilitate cake washing and dewatering. Once the last of the liquid wash solution is produced, the wash time is noted and the subsequent dry time begins (only applicable if wash was applied).

At the end of the timed air blow, the filter cake is discharged from the filter, and the wet weight and cake thickness are determined and recorded. After drying, the dry cake weight is determined and recorded for cake moisture calculations.

Additional cakes should be collected in the same manner, each at different test conditions, until the range of test variables selected has been adequately covered.

Wash procedure discussion. Experience has shown that the construction of a washing curve from data is often made more difficult than necessary by the choice of methods employed to collect these data. Hence, when washing effectiveness data are collected, the method of testing should be varied as described in the following section.

A washing stage in vacuum and pressure filtration consists of cake liquor displacement by both air and the applied wash fluid. Remaining cake liquor content will be a function of time of air displacement, cake thickness, pressure drop across the cake, particle characteristics, liquor viscosity and other factors. The wash stage begins the instant wash fluid is applied, continues as the fluid passes through the cake, and ends with a period of liquor displacement by air. The end of a wash stage may precede cake discharge from the filter, or the application of a subsequent wash (in cases with multiple wash stages).

A correlation of washing data should fall on a single curve and should illustrate the relationship between the volume of wash applied and the degree of solute removal without being affected by variations in the liquor concentration throughout the washing stage. It is possible to achieve this type of correlation using a parameter R, which is plotted as a function of a second parameter N, where N and R are defined as follows:

N is defined as the volume of wash applied divided by the volume of liquor in the cake at the end of the wash stage, which includes some liquor displacement by air. If L_w is defined as volume of applied wash and L_2 as volume of liquor in the cake at the end of the wash stage, then N is defined:

$$N = \frac{L_w}{L_2}$$

In situations where the wash fluid applied contains no soluble value, R is defined as the concentration of solute in the cake liquor at the end of the wash stage divided by the concentration of solute in the cake liquor at the beginning of the wash stage. In cases where the wash fluid contains soluble value, this concentration must be accounted for. Theoretically, with infinite wash applied, the concentration of the solute in the cake liquor would equal the concentration of solute in the wash liquor. Therefore if C_2 is defined as the concentration of solute in the cake liquor at the end of the wash stage, C_1 is defined as the concentration of solute in the cake liquor at the beginning of the wash stage, and C_w is defined as the concentration of solute in the applied wash fluid then R is defined:

$$R = \frac{C_2 - C_w}{C_1 - C_w}$$

The method of determining cake liquor concentration at the end of a washing stage has been found to be critical in terms of the quality of the correlation achieved. Unsatisfactory results are obtained if either the wash filtrates or dry (washed) cake residues are analyzed for residual soluble value content.

The suggested approach consists of repulping each filter cake of a wash series in a known volume of pH adjusted de-ionized water. After repulping, the resultant slurry is filtered using a new filter mat and clean and dry buchner funnel and vacuum flask for each cake. Filtrate reporting to the vacuum flask is submitted for assay while filter cake solids are dried to obtain dry suspended solids. Complete accountability for all cake solids must be maintained throughout this repulp procedure.

When the analytical results are obtained from the cake repulp liquor, the actual concentration of soluble value in the liquor associated with the cake upon discharge from the filter can be calculated by solving the following expression:

$$C_1 = \frac{\frac{W_1}{\rho} + V_r}{\frac{W_1}{\rho}} C_{rl}$$

Where:
C_1 = Cake liquor concentration at discharge (g/ml).
W_1 = Weight of cake liquor (grams).
ρ = Density of cake liquor (g/ml).
V_r = Volume of repulp fluid used (ml).
C_{rl} = Concentration of solute in repulp liquor (analytical result).

GENERAL FILTRATION TESTING DATA INTERPRETATION AND CORRELATION

Data collected during testing can be correlated to show several general relationships, which can be subsequently used to predict filter performance for given conditions (Dahlstrom, 1957; Silverblatt, 1974). The key correlations include:

- Cake weight versus cake thickness.
- Cake formation rate verses cake weight (or thickness).
- Cake moisture content verses relative dry cycle length.
- Cake wash penetration verses cake thickness and applied wash volume.
- Solute removal verses applied wash volume.

These correlations can each be used to describe filter operation during each portion of a complete cycle, and when used together will describe the total filter performance and cycle time for the conditions tested. An example data set is used below to illustrate the correlations described above. An example of each correlation will be shown in Figures 1 through 5.

During testing, several cakes of various thickness should be produced so that data for the first two correlations can be collected. The first correlation shown in Figure 1 demonstrates the relationship between wet filter cake thickness (in millimeters) and dry filter cake weight W (with units of dry kg/m^2). For the purpose of design, a possible cake thickness of 15 mm could be used for horizontal belt filters. According to Figure 1 the unit weight of a 15 mm cake would then be 2.9 dry kg/m^2 when flocculant is added, while a 15 mm cake produced without flocculant would weight 3.9 dry kg/m^2

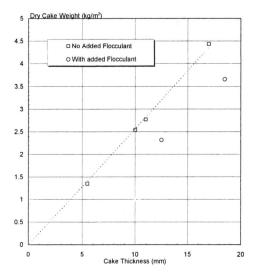 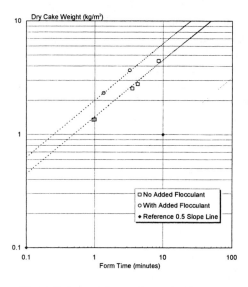

Fig. 1 Cake weight vs. cake thickness   Fig. 2 Cake weight vs. form time

Figure 2 displays the logarithmic relationship of dry cake weight, W, with units of dry kg/m^2, as a function of cake formation time, in minutes. As predicted by theory, the slope of each curve is ½ (Silverblatt, 1974). The correlation shown in Figure 2 indicates that, for the case where flocculant is used, a 15 mm cake, which weighs 2.9 dry kg/m^2 will form in approximately 2.1 minutes, while when flocculant is not used, a 15 mm cake, which weighs 3.9 dry kg/m^2 will form in approximately 7.2 minutes.

The relationship between filter cake moisture at discharge and the dry time factor (θ_d/W, with units of min•m^2/kg) is shown in Figure 3. The dry time factor is the dry time (θ_d, in minutes) divided by the dry cake weight per unit area (W, with units of dry kg/m^2). The dry time factor permits a correlation between cake moisture and dry time for all variations in cake thickness, by normalizing the dry time for cake weight, which depends on cake thickness. The correlation indicates that, when flocculant is employed, approximately a 2.9-minute dry time (θ_d/W =1.0 min•m^2/kg and W = 2.90 dry kg/m^2) following the cake wash will yield filter cake with 79% moisture. Similarly, when flocculant is not used, approximately a 4.9-minute dry time (θ_d/W =1.27 min•m^2/kg and W = 3.9 dry kg/m^2) following the cake wash will yield filter cake with 75% moisture.

Figure 4 shows the correlation of the wash time (θ_w) and the wash time factor (W•Vw), plotted as θ_w in (minutes) vs. W•Vw in (kg•liter/m^4). Similar to the dry time factor, the wash time factor permits a correlation between wash time and specific wash volume Vw (liter/m^2) for all variations of cake thickness. From this plot it can be seen that for a W•Vw value of 27.82 kg•liter/m^4 (corresponding to a wash ratio of N = 0.5) a wash time of 8.98 minutes is required when no flocculant is added, similarly, when flocculant is used, a W•Vw value of 19.89 kg•liter/m^4 (corresponding to a wash ratio of N = 0.5) a wash time of 1.37 minutes is required.

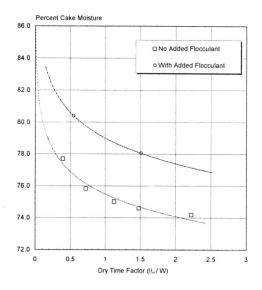

Fig. 3 Cake moisture vs. dry time factor

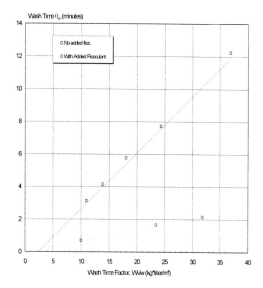

Fig. 4 Wash time vs. wash time factor

Figure 5 shows the relationship of fraction of solute remaining (R), determined from assay values on the repulped filter cakes, with respect to wash ratio (N). The wash ratio "N" is defined as the number of cake liquor displacements with wash solution when the filter cake is at the discharge moisture content.

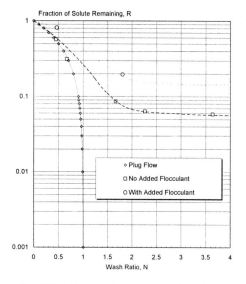

Fig. 5 Solute remaining vs. wash ratio

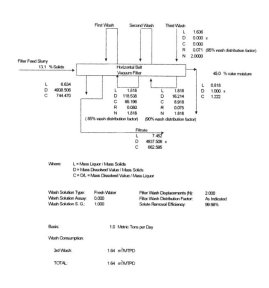

Fig. 6 Cake washing material balance

The recovery data correlation displayed in the in Figure 5 (often referred to as a "wash curve") allows a quick estimate of where increases in wash volume produce little effect, so that the optimal level of wash can be selected. The predicted "R" values also allow for construction of accurate material balances for a variety of washing filters. An example of a material balance for a three-stage counter-current washing belt filter is shown in Figure 6. While this is a fairly complicated filter arrangement, data from the wash curve allows recovery prediction.

VACUUM FILTER CYCLE TIMES AND PRODUCTION RATE CALCULATIONS

Horizontal Belt Vacuum Filters.

Cycle time for horizontal belt filters is the summation of the various filter function times required, that is, the sum of cake formation time, cake washing time, and cake dry time is the cycle time. Production rate (with units of weight per filter unit area per time) for a horizontal belt vacuum filter is calculated as follows:

$$\text{Production Rate} = \frac{C \times W}{\text{CycleTime}}$$

An empirical scale-up factor is included in the C term as well as the required constants for engineering unit conversions. Minimum cake thickness is considered during selection of cycle times used to calculate production rates. Minimum design cake thickness for horizontal belt vacuum filters range from 1/8" to 3/16" (3.2 – 4.8 mm) depending upon material characteristics.

Rotary Drum Vacuum Filter Configuration, Cycle Time, and Production Rate.

Standard data, such as that discussed for Horizontal Belt Vacuum Filters above, are utilized in the same manner for Rotary Vacuum Drum and Disc filters with a few modifications that account for the configuration of a rotary vacuum filter. Form time, wash time (when applicable) and dry time and thus total cycle time and production rate are generally based on vacuum drum filters apportioned as shown in Table 1.

Table 1 Vacuum drum filter area apportionment

Drum Filter Zone	Drum Filter w/o Wash		Drum Filter with Wash	
	Arc Length	% of cycle	Arc Length	% of cycle
Form Zone	108°	30%	108°	30%
Wash Zone	NA	NA	90°	25%
Dry Zone	180°	50%	65°	18%
Discharge/Pre-wash/Dead	72°	20%	97°	27%
Total:	360°	100%	360°	100%

These fractions of the total cycle time result from the geometry of the drum filter itself. Refer to Figure 7 as an example of how a vacuum drum filter might be proportioned.

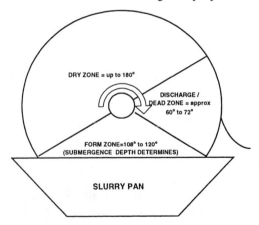

Fig. 7 Vacuum drum filter zone configuration without wash

Hence, for the general case, cycle time is given by the following equation:

$$\text{Cycle Time} = \frac{\text{Form Time}}{\text{Form Zone Fraction}} = \frac{\text{Dry Time}}{\text{Dry Zone Fraction}} = \frac{\text{Wash Time}}{\text{Wash Zone Fraction}}$$

For a specific example, the cycle time for a washing drum filter is given by:

$$\text{Cycle Time} = \frac{\text{Form Time}}{0.30} = \frac{\text{Dry Time}}{0.18} = \frac{\text{Wash Time}}{0.25}$$

Filter operation can be considered to be form time limited, wash time limited, or dry time limited based on which portion of the cycle is shown to be limiting during testing. The cycle time used is always the largest of the fractions shown above, since this slowest portion of the cycle limits the overall filter operation. A form, wash, or dry time limited operation is possible within limits imposed by the maximum and minimum rotational speed of the drum. By determining the times needed for cake formation, washing, and drying for a given cake thickness in consideration of washing and moisture targets, it is possible to calculate the total required cycle time.

It is possible for the form cycle time to be limited by the minimum cake thickness required for discharging. The minimum design cake thickness for various types of vacuum drum filter discharge methods are shown in Table 2.

Table 2 Vacuum drum filter minimum design cake thickness.

Discharge Method	Cake Thickness	
	Inches	mm
Belt	1/8 – 3/16	3.2 – 4.8
Roll	1/32	0.8
Scraper	1/4	6.4
Coil	1/8 – 3/16	3.2 – 4.8
String	1/4	6.4

The limiting zone of the drum governs the cycle time for a complete circular rotation. Non-limiting zones may not require the full available zone in order to meet individually required targets. The form zone may be limited to the specific required form time by placing rotary valve bridge block settings within this zone.

The production rate (with units of weight per filter unit area per time) for a rotary drum vacuum filter is calculated using the following equation where W is the dry cake weight per unit filter area and C is a constant including an empirical scale-up factor as well as required constants for engineering unit conversions:

$$\text{Production Rate} = \frac{C \times W}{\text{Cycle Time}}$$

Rotary Disc Vacuum Filter Configuration, Cycle Time, and Production Rate.
Form time and dry time, and thus total cycle time and production rate, are generally based on vacuum disc filters apportioned as shown in Table 3.

Table 3 Vacuum disk filter area apportionment

Disc Filter Zone	Disc Filter	
	Arc Length	% of cycle
Form Zone	108°	30%
Dry Zone	135°	37.5%
Discharge/Pre-wash/Dead	117°	32.5%
Total:	360°	100%

Hence, for the general case, cycle time is given by the following equation:

$$\text{Cycle Time} = \frac{\text{Form Time}}{\text{Form Zone Fraction}} = \frac{\text{Dry Time}}{\text{Dry Zone Fraction}}$$

Specifically, the cycle time for a rotary disc vacuum filter is given by:

$$\text{Cycle Time} = \frac{\text{Form Time}}{0.30} = \frac{\text{Dry Time}}{0.375}$$

Filter operation can be considered to be form time limited or dry time limited based on which portion of the cycle is shown to be limiting during testing. The cycle time used is always the largest of the fractions shown above, since this slowest portion of the cycle limits the overall filter operation. A form or dry time limited operation is possible within limits imposed by the maximum and minimum rotational speed of the disc. By determining the times needed for cake formation and drying for a given cake thickness and moisture target, it is possible to determine the total required cycle time.

It is possible for the form cycle time to be limited by the minimum cake thickness required for discharging. The minimum design cake thickness for a rotary disc vacuum filter ranges from 3/8" to 1/2" (9.5 – 12.7 mm) depending upon material characteristics.

The limiting zone of the disc governs the cycle time for a complete circular rotation. The non-limiting zone may not require the full available zone in order to meet the design target. The form zone may be limited to the specific required form time by placing a rotary valve bridge block setting within this zone.

The production rate (with units of weight per filter unit area per time) for a rotary disc vacuum filter is calculated using the following equation; where W is the dry cake weight per unit filter area and C is a constant including an empirical scale-up factor as well as required constants for engineering unit conversions:

$$\text{Production Rate} = \frac{C \times W}{\text{Cycle Time}}$$

Pressure Filter Cycle Time and Filter Press Sizing
Pressure filter cycle time will simply be the sum of the individual portions of the cycle time for each operation, that is: cake formation time, cake washing time, dry time (including diaphragm squeeze time, if applicable), and miscellaneous time required to open and close the press, discharge the cake, and clean the cloth, etc. Therefore, the total cycle time is given by:

$$\text{Cycle Time} = \text{Form Time} + \text{Dry Time} + \text{Wash Time} + \text{Misc. Time}$$

Once a cake thickness is selected (based on filter press chamber thickness), the components of the cycle time can each be predicted by the correlations shown in Figures 1 through 4. Note that within a vertical recessed plate or plate and frame pressure filter chamber, two (2) filtration surfaces exist; hence, a half-cake forms from each filtration surface simultaneously. This half-cake thickness must be used for form time prediction from the correlation in Figure 2. In pressure filters with horizontal plates filtration is single-sided, so full cake thickness is used for form time calculation. Total cake thickness (chamber thickness) should be used for all other calculations. The correlation in Figure 1 relates cake thickness to cake weight per unit area (W), and the correlation in Figure 2 relates form time to cake weight W. The correlations in Figures 3 and 4 similarly predict washing and dry times based on selected wash volume and cake moisture. The miscellaneous time will be equipment specific, and should be supplied by the vendor of the selected equipment

Once the pressure filter cycle time is known, the filter sizing can be undertaken. This is not as simple as determining the production rate based on test data. For example, if filtration rate alone determined pressure filter sizing, the production rate (with units of weight per filter unit area per time) would be given by:

$$\text{Theoretical Production Rate} = \frac{W}{\text{Cycle Time}}$$

This equation is generally only valid for hydraulically limited filters, for which the production rate is relatively small. Most pressure filters are actually limited by the volume capacity of the filter, which limits the amount of cake that can be produced in a single cycle.

Since pressure filters are closed chambered devices, the fixed volume of the chambers limits the volume of cake that can be produced in a single cycle. Hence the dry bulk cake density (dry kg/m³) is an extremely important experimental value. This value can be inferred from the correlation in Figure 1, which describes the dry cake weight per unit area per unit of cake thickness. With the dry bulk density known, the desired tonnage will set the total volume of cake to be produced in a given day. The cycle time can now be used to set filter sizing, since the total hours of filter operation per day will limit the number of cycles that can be run by the filter.

$$\text{Cycles Per Day} = \frac{\text{Hours Per Day Worked}}{\text{Cycle Time (hours)}}$$

Therefore the filter press required cake volume (in cubic meters) must be:

$$\text{Filter Press Volume} = \frac{C \times 1000 \times \text{Daily Tonnage}}{\text{Cake Dry Bulk Density} \times \text{Cycles Per Day}}$$

Where C is an empirical scale-up factor, tonnage is in metric tons, and cake dry bulk density has units of kg/m³. This volume specification can be considered to be the required filter sizing

The effective production rate for a filter sized previously described would then be:

$$\text{Effective Production Rate} = \frac{\text{Cake Weight Per Cycle}}{\text{Filter Press Area} \times \text{Cycle Time}}$$

If this effective production rate (for volume limited filter presses) is less than the theoretical production rate previously described, then the filter design can be considered to be volume limited. Since this happens to be the case in many instances, this shows the danger inherent in sizing a pressure filter based on production rate alone.

PILOT SCALE TESTING

Laboratory testing described in the previous sections will generate filter sizing data using standard, bench scale equipment. It is particularly useful when only small quantities of representative sample are available. The test data can be used to evaluate many vacuum and pressure filtration options as part of the flowsheet and equipment selection process. A short-list of preferred filter types can then be produced. While laboratory scale testing focuses on the filtration process, the scope of pilot scale testing is larger, and includes data collection for full-scale plant design including ancillaries.

If sufficient representative sample is available, for example at a production plant considering an upgrade, testing can proceed to pilot scale. Pilot scale testing may not always be necessary, but is conducted when factors including the following are important:

- To evaluate a specific type of filter, and to demonstrate that it will work as predicted at laboratory scale.
- To check filter operating characteristics that cannot be demonstrated at laboratory scale, such as cake discharge and cloth blinding.
- To produce large samples of filter cake or filtrate to develop downstream unit operations, or for marketing purposes.
- To collect engineering data for filter plant design and ancillary equipment selection.
- To compare existing and proposed filters in parallel over time, and over a wide range of operating conditions.

- To gain operating and maintenance experience on a filter as part of the selection process, or to train staff before the production unit is installed.

Whereas standard laboratory equipment can be used to evaluate almost all types of vacuum and pressure filters, pilot scale units tend to be more specific. Pilot scale filters are often supplied by manufacturers, and are designed to replicate the operation of that company's equipment. This should be clearly understood when analyzing test data and trying to apply it generally.

"Pilot Scale" operation is not rigidly defined, and test units can have filtration areas ranging from 0.1 to $5m^2$. Pilot plants may be located permanently, for example at in a fixed location such as a research center. However, they are commonly mobile units that can be transported and operated wherever needed. Testing on-site often ensures that samples are both representative and fresh. Smaller pilot scale units are operated "off line" in batch tests, whereas larger units can be integrated into an existing process for continuous operation. In both cases, preparations for pilot scale testing must be made well in advance. Although some transportable pilot filters may be delivered with their own ancillaries such as feed pumps and compressed air supply, others may need these to be provided at the test site. In all cases, plans must be made to provide electricity and water, analytical services, and assistance with installation and maintenance. A single technician can conduct laboratory scale testing, but pilot scale operation is significantly more demanding on resources.

Figure 8 shows a test pressure filter with a filtration area of $0.1m^2$. This unit is manually operated and simulates a single cycle of a membrane pressure filter including cake formation, cake compression, cake washing, second cake compression, and air-drying. This type of test filter can be pneumatically powered, although electricity may be required for slurry heating and agitation. The test unit includes all necessary ancillary equipment such as feed tank, feed pump, pressing pump and cake wash liquid tank. The feed tank holds 100 liters slurry, and filter cakes of several kilograms can be produced. Different chamber depths can be used, and flow rates and pressures can be adjusted. The test unit illustrated is 2 meters long, 1.5 meters high and 0.6 meters wide and weighs 400 kilograms. Such units can be easily moved through the plant to the test location on wheels, with a forklift truck, or by crane.

Fig. 8 Pressure filter ($0.1m^2$)(courtesy of Larox)

Similar test filters can be fully automated, with continuous recording of pressures, flowrates and filtrate volume direct to computer. Figure 9 shows a $1.6m^2$ fully automatic pilot scale pressure filter built into a standard shipping container. This pilot unit includes all necessary ancillaries, and can be integrated into a continuous process.

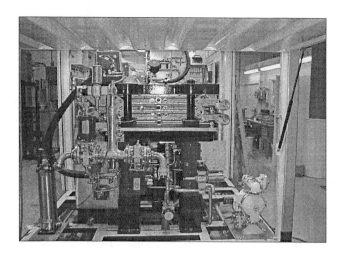

Fig. 9 Automatic pilot scale pressure filter (1.6m^2)(courtesy of Larox)

Data collection and analysis methods are similar to those used in bench-scale testing. Of course larger sample volumes are involved, and sample handling logistics more demanding. Power, air and water consumption are monitored during pilot scale testing. The aim is to optimize the filtration cycle for energy consumption, as well as process results, and also to avoid incorrect sizing of ancillaries.

Pilot scale testing will generate large samples of filtrate and filter cake. Valuable information on materials handling can be obtained by observing ease of cake discharge, and transport characteristics. For example, filter cake samples can be used for angle of repose tests for downstream surge bin design. Extended pilot plant trials give an opportunity to test different materials of construction, especially in aggressive, high temperature and corrosive conditions. Above all, pilot-scale testing provides tangible evidence to decision makers that the filter provisionally selected during laboratory scale testing will, in fact, perform as expected at production scale.

APPENDIX
DERIVATION OF CAKE FORMATION RATE EQUATIONS

Classical treatment of filtration theory nearly always begins with Equation (1) Poiseuille's equation (Carman, 1938; McCabe and Smith, 1976). This form of the equation contains a couple of cumbersome terms, V for the volume of filtrate, and c for solids concentration expressed as the weight of solids per unit volume of filtrate.

More succinctly, most applications we encounter in either vacuum or pressure filtration deal with the solids filtration rate rather than the liquid filtration rate. It follows that an expression for the rate of cake formation that avoids the V and c terms will be more useful in practical applications.

The choice of an equation from which to start a revised derivation for the purpose stated above depends on the reference source selected. The same is generally true for the notation that is used. The starting equation is usually derived from Poiseuille's equation and is typically in the following form which shows the instantaneous rate of filtrate flow as the ratio of the driving force to the product of the viscosity and the sum of the filter cake and filter media resistances:

Eqn (1):
$$\frac{dV}{A\, d\theta_f} = \frac{\Delta p}{\mu\left[\dfrac{\alpha c V}{A}\right]}$$

This can be converted to:

Eqn (2):
$$\frac{d\theta_f}{dV} = \frac{\mu}{A\Delta p}\left[\frac{\alpha c V}{A} + R_m\right]$$

The notation used above in equations (1) and (2), and hereinafter, is as follows:

- θ_f = Cake formation time.
- V = Volume of filtrate.
- μ = Liquid viscosity.
- A = Filtration area.
- Δp = Pressure drop across the filter cake.
- α = Average specific (filtration) resistance (units of length/mass).
- c = feed slurry solids concentration in terms of mass of solids per unit volume of filtrate.
- R_m = Resistance of the filter medium (units of length^{-1}).
- W = Mass of dry solids in the filter cake per unit area.
- ρ = Liquid density.
- S = Weight fraction of solids in the filter feed slurry.
- S_c = Weight fraction of solids in a formed, but undewatered filter cake.

Integrating equation (2):

Eqn (3):
$$\int_0^{\theta_f} d\theta_f = \frac{\mu}{A\Delta p}\left[\frac{c\alpha}{A}\int_0^V V\,dV + R_m \int_0^V dV\right]$$

Eqn (4):
$$\theta_f = \frac{\mu}{\Delta p}\left[\frac{c\alpha}{2}\left(\frac{V}{A}\right)^2 + R_m\frac{V}{A}\right]$$

However, where:

Eqn (5): $$W = \frac{cV}{A}$$

Eqn (6): $$\theta_f = \frac{\mu}{\Delta pc}\left[\frac{(W)^2 \alpha}{2} + WR_m\right]$$

and, it follows:

Eqn (7): $$\theta_f = \frac{\mu W}{\Delta pc}\left[\frac{W\alpha}{2} + R_m\right]$$

It should be clear that the rate of cake formation, as we usually express it, (in terms of weight of dry cake solids per unit area and per unit time), is equal to W/θ_f. Rearranging equation (7) yields:

Eqn (8): $$\frac{W}{\theta_f} = \frac{\Delta pc}{\mu}\left[\frac{1}{\frac{W\alpha}{2} + R_m}\right]$$

This is a rather simple equation, and it avoids the use of the V term. However, the c term is one that has an unusual definition, and this definition is probably responsible for more misunderstanding regarding filtration theory than any other factor. Hence, the term is ripe for change, and must be replaced with another term for feed slurry concentration that can readily be understood, i.e., weight % solids. Actually, the term that will be used is S, the weight fraction of solids in the feed slurry, which is almost the same thing and easier to manipulate.

The definition of c is the weight of dry solids in the feed slurry per unit volume of filtrate. The liquid remaining in the formed (but yet undewatered) cake is, of course, not part of the filtrate, but it certainly was part of the feed slurry and it must be taken into account. Note the reference to the formed but undewatered filter cake. The mechanism of dewatering or draining of a filter cake by displacing the entrained liquid with air (as occurs in the cake drying portion of a pressure or vacuum filter) has nothing to do with the mechanism of cake formation, which is described by equation (8). Therefore, because there are two mechanisms involved, another term, Sc, denoting the weight fraction of solids in a formed but undewatered cake, must also be used.

The definition of c could be expressed as:

Eqn (9): $$c = \frac{\text{Weight of Solids}}{\text{Volume of Filtrate}}$$

By dividing c by ρ, the density of the liquid, it is obvious that:

Eqn (10): $$\frac{c}{\rho} = \frac{\text{Weight of Solids}}{\text{Weight of Filtrate}}$$

Also, S_c can be defined as:

Eqn (11): $$S_c = \frac{\text{Weight of Solids}}{\text{Weight of Formed (but undewatered) Cake}}$$

Or, also as:

Eqn (12): $$S_c = \frac{\text{Weight of Solids}}{\text{Weight of Slurry} - \text{Weight of Filtrate}}$$

Therefore,

Eqn (13): $$\text{Weight of Filtrate} = \text{Weight of Slurry} - \frac{\text{Weight of Solids}}{S_c}$$

and,

Eqn (14): $$\frac{c}{\rho} = \frac{\text{Weight of Solids}}{\text{Weight of Slurry} - \frac{\text{Weight of Solids}}{S_c}}$$

By multiplying the term containing Sc by (weight of slurry/weight of slurry),

Eqn (15): $$\frac{c}{\rho} = \frac{S}{1 - \frac{S}{S_c}}$$

and,

Eqn (16): $$c = \frac{\rho S}{1 - \frac{S}{S_c}}$$

Substituting equation (16) into equation (8),

Eqn (17): $$\frac{W}{\theta_f} = \frac{\Delta p \rho}{\mu} \left[\frac{S}{1 - \frac{S}{S_c}} \right] \left[\frac{1}{\frac{W\alpha}{2} + R_m} \right]$$

Note that the steps between equation (1) and equation (17) were made using only mathematics, with no assumptions at all, so if the validity of equation (1) is accepted, then equation (17) is rigorous. Equation (17) can be particularly useful in ascertaining what the effect of changing a variable on the right side of the equation has on the left side (the "form filtration rate"). For instance:

If Δp is doubled, the form rate will double. Note that doubling the form rate can only be due to θf being halved as W was not changed on the on the right side, W cannot be changed on the left side (since they are the same). This illustration is filtration at constant cake thickness. On an actual vacuum filter, if the operator doubled Δp, he would have to increase the filter speed by a factor of 2, to get the same cake thickness (he had before Δp was doubled). On a pressure filter, if slurry feed or fill time to the chamber was held constant, but air-fill and diaphragm pressures were doubled, the form rate would be halved.

If $\dfrac{S}{1 - \dfrac{S}{S_c}}$ (the newly defined feed slurry concentration) were doubled, the form rate would double. Once again, this is true only if the cake thickness were held constant by changing the filter cycle time (vacuum filter), or the fill time (feed time) were halved (pressure filter).

A seemingly logical question follows: What is the effect of changing a variable if the appropriate filter function, such as cycle time or fill time, were not changed? In other words, consider the effect on form filtration rate, now at constant cycle or function time.

To evaluate this possibility and at the same time avoid the use of a complicated equation, a simplifying assumption must be made, that is, the resistance due to the filter medium, R_m, is small compared to the resistance due to the filter cake, $W\alpha/2$, which is usually the case in most vacuum and pressure filter applications. After setting R_m equal to zero, we obtain:

Eqn (18): $\qquad \dfrac{W}{\theta_f} = \dfrac{2\Delta p \rho}{\mu W \alpha} \left[\dfrac{S}{1 - \dfrac{S}{S_c}} \right]$

Rearranging:

Eqn (19): $\qquad (W)^2 = \theta_f \dfrac{2\Delta p \rho}{\mu \alpha} \left[\dfrac{S}{1 - \dfrac{S}{S_c}} \right]$

and further,

Eqn (20): $\qquad W = \sqrt{\dfrac{\theta_f \, 2\Delta p \rho}{\mu \alpha} \left[\dfrac{S}{1 - \dfrac{S}{S_c}} \right]}$

By solving the expression, and then plotting W as a function of θ_f on log-log paper, a straight line of 0.5 slope should result (and it generally does).

By dividing W by θ_f to calculate the form filtration rate, we obtain:

Eqn (21): $\qquad \dfrac{W}{\theta_f} = \sqrt{\dfrac{2\Delta p \rho}{\mu \alpha \theta_f} \left[\dfrac{S}{1 - \dfrac{S}{S_c}} \right]}$

By solving the expression, and then plotting the form filtration rate, W/θ_f, as a function of form time, θ_f, on log-log paper, a straight line of minus 0.5 slope should result (and it generally does).

REFERENCES

Carman, 1938. *Trans. of the Inst. Chem. Eng. (London)* 16: 174.

McCabe and Smith, 1976. *Unit Operations of Chemical Engineering, 3rd Edition.* New York: McGraw Hill. p937.

Dahlstrom D.A., and Nelson, P.A., 1957. Moisture-Content Correlation of Rotary Vacuum Filter Cakes. *Chemical Engineering Progress.* 7: 320-327.

Silverblatt, C.E., Hemant Risbud, and Tiller, F. M. 1974. Batch, Continuous Processes for Cake Filtration. *Chemical Engineering.* 4: 127-136.

Perry, R.N., and Green, D.W., eds.1991. *Chemical Engineering Handbook. 6Th Edition.* New York: McGraw Hill.

Design Features and Types of Sedimentation Equipment

Fred Schoenbrunn[1] *and Tim Laros*[1]

ABSTRACT

Sedimentation equipment can be specialized to cover a wide range of process goals. Various equipment designs are available that have been optimized for different processes and process objectives. Some of the common designs used in the minerals industry include clarifiers, solids contact clarifiers, inclined plate clarifiers, conventional thickeners, high rate thickeners, high rate rakeless thickeners, high density thickeners, and Alcan deep thickeners. Within these designs, options include bridge and center column mounting, as well as a variety of tank construction options. Selection of the optimal design for a project depends on the process objectives and can be a function of capital versus operating costs, with project life, maintenance, materials selection, and site layout considerations also being factors.

INTRODUCTION

There are a variety of sedimentation equipment designs available to the minerals industry. Selection and sizing of the proper piece of equipment depends on the process and objectives. Details of the design of the equipment can vary not only with process objectives, but also with site topography, material specific gravity and particle size, and availability of local building materials, as well as plant operating philosophy.

The development of modern flocculants has lead to high rate and high capacity designs that are optimized for their use. Modern flocculants can decrease the required sizing for sedimentation equipment by an order of magnitude or more. The cost of flocculant is a significant operating cost and should be considered in an economic evaluation. There is usually a trade off between equipment size and required dosage, with smaller units requiring a higher flocculant dosage up to a point. It used to be that "conventional" meant that flocculant was not being used. Now it implies a mechanism that is not specifically designed for the optimal use of flocculant, and is usually used as a reference point regarding the amount of risk in a design. However, it should be noted that a well-designed high capacity thickener could operate more consistently and with lower risk than a larger well-designed conventional thickener using flocculant.

In the minerals industry, sedimentation equipment is ubiquitous. The most common applications in base metals involve thickening of tailings and concentrates. In alumina, nickel laterite, uranium, gold, and some of the other leach circuits, counter current decantation (CCD) is a fundamental part of the process, using as many as 8 thickeners in series. Other applications include clarification of plant discharge water, process water treatment, leach liquor clarification, removal of precipitates, pre-leach, and grind thickening.

There are three general types of thickener rake drive mechanisms; bridge mounted, center column mounted, and traction drives. Bridge mounted drives are centered on a bridge that spans the thickener, with a shaft attached to the rakes. Because of the bridge, there is an upper limit on the size of machine this design can be economically applied to, generally around 40 - 50 m diameter. Center column drives are used on larger thickeners (or small ones without lifts) and are mounted on a center column that typically also supports an access bridge that spans one half of the tank. The shorter bridge compensates for the addition of the column and the use of a cage around

---

1 EIMCO Process Equipment Company, Salt Lake City, Utah.

the column to drive the rakes. These are economical starting at about 30 m diameter and are currently available up to a diameter of 130 m. Traction thickeners use a peripheral drive mounted on one (or two) of the rake arms. These units can develop very high torques, but do not have lift capabilities and are sensitive to environmental conditions regarding contact between the drive wheel and the rail or traction surface. They are usually only considered for large diameter applications.

Conventional thickener design uses center underflow outlets with the floor sloped at 1:12 – 2:12 towards the center. Lighter duty applications such as clarifiers usually use a 1:12 slope. Large diameter machines often use a dual slope design, with an inner slope of 2:12 and an outer slope of ½:12 or 1:12, to avoid making the machine excessively deep. Some applications such as uranium yellow cake and magnetite use steeper slopes such as 3:12. The tank sidewall depth is usually about 3 m and is determined by both process and mechanical considerations. A freeboard of 150 – 300 mm is usually used on both the tank and feedwell. The center outlet is typically a 45° cone on bridge type thickeners and a trench on center column thickeners. Either requires a scraper assembly attached to the rake structure to keep the material moving. A generalized thickener is shown in Figure 1. Figures 2 and 3 are photos of typical center column and bridge mounted thickeners respectively.

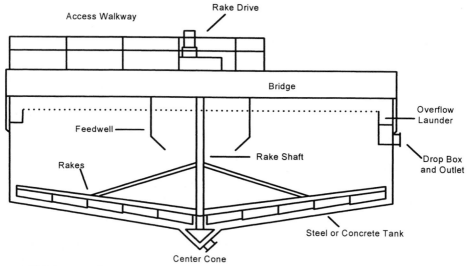

Figure 1 Thickener schematic

DESCRIPTION OF THICKENER COMPONENTS

Feedwell
Thickeners and clarifiers are typically fed in the top center and use a feedwell to still the feed stream. The feedwell is also the most common point for flocculant addition. Properly designing the feedwell and flocculant addition location to maximize the effectiveness of the flocculation can have great impact on the operation.

The feed slurry concentration for optimum flocculation and most economic thickener size is not necessarily the slurry concentration reporting to the thickener from the upstream process. Quite often, the feed slurry requires dilution prior to addition of flocculant to achieve best flocculation and thickening performance. This effect is presented in Figure 4. This figure shows a maximum in the settling flux at a relatively low solids concentration. The settling flux is the amount of solids settling through a given area, which is the product of the solids settling rate times the concentration. The optimal concentration depends on the characteristics of the feed solids. Very fine solids may need to be diluted to 5 wt% where the maximum flux may occur at 15 wt% for a coarse grind tails. For clarification, the feed slurry may be too dilute and may benefit from an increase in concentration by recycling underflow slurry back to the feed slurry. This will be discussed under Solids Contact Clarifiers.

Figure 2 Center column mounted thickener. Note the concrete on-ground tank, truss type rake arms, and feed launder mounted under the bridge. Photo courtesy of EIMCO Process Equipment Company

Figure 3 Bridge mounted thickener. The unit uses an elevated steel tank. Photo courtesy of EIMCO Process Equipment Company

Many methods of feed dilution have been tried since the advent of synthetic flocculants. Pumping of overflow liquor into the feed slurry from the overflow launder or directly from the top of the thickener with submersible pumps is quite common. This method requires a pump, which can be quite large and expensive in cases where large quantities of dilution are required. Early dilution methods without external pumping used slots or holes cut into the feedwell to draw dilution liquor into the feed slurry due to the density differential between the feed stream and the clarified liquor. These types of feedwells are often referred to as Cross type feedwells, as Harry

Cross developed one of the first units. The Outokumpo Floc-Miser is an example of this type of feedwell.

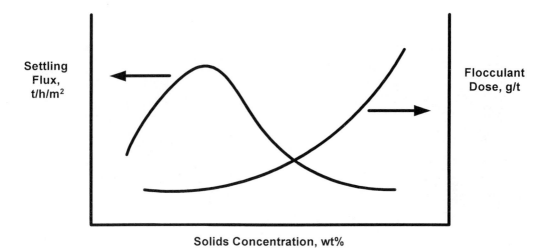

Figure 4 Solids concentration versus settling flux and flocculant dosage

An alternative method of dilution is to place an eductor or jet pump in the feed line to dilute, flocculate, and mix the slurry prior to entering the feedwell. The degree of dilution can be designed into the eductor through the geometry of the eductor and the driving head of the feed slurry through the eductor. This type of feedwell is distinguished by the EIMCO E-Duc® Self-Diluting Feedwell. An example of this is shown in Figure 5.

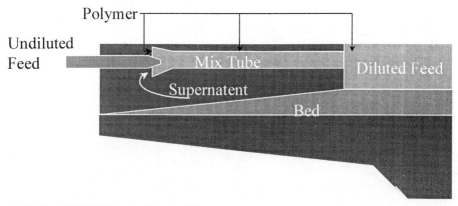

Figure 5 E-Duc® Self-Diluting Feedwell

Rakes
Most thickeners use a set of arms that move through the pulp to help thicken the pulp and move the thickened material to the underflow outlet. These typically have blades set at 30-45 degrees to the tangent of motion to push the solids towards the outlet. The rake arms must be strong enough to transmit the torque needed to push the solids towards the thickener discharge.

Rakes are usually designed for specific process applications. Processes that produce a heavy scale build up on the rakes such as alumina refining require a rake containing a minimum of steel surface area. Rakes for magnetite thickening usually have spikes attached to the blades such that heavily thickened magnetite can be resuspended. Some rakes are streamlined to help reduce the torque on the rake structure. Rakes may have pickets or Thixoposts (posts to distance the blades from the rake arms) attached to them for processes in sticky, viscous materials. In general, rake design is process dependant.

Rake Drives
The rake drive provides driving force (torque) to move the rake arms and blades against the resistance of the thickened solids. The drive also provides the support for the rotating elements. Bridge type thickeners can use spur gear, worm gear, or planetary or other commercial reducer drives. Column mounted thickeners almost always use spur gear drives. Since the rake steel is usually designed based on the 100% torque rating of the drive, two or more levels of overload protection are often designed into the drive to prevent torque overloads and associated rake damage. Reliability is a critical issue with the drive, since failure frequently means digging out the thickener. Key drive elements are hardened steel gears, large precision bearings, oil bath lubrication, accurate torque measurement, and strong housings.

There are a variety of torque descriptions such as; peak torque, design torque, normal operating torque, duty rated torque, AGMA 20 yr torque, cutout torque, etc. This is partly due to the relationship of torque to gear life, where gears last a very long time at low torque, but exponentially shorter as the torque increases.

There are two methods of motive power for drives, either electric motors or hydraulic power packs. Using hydraulics offers features such as soft starts, variable speed, torque indication by hydraulic pressure, low speed hydraulic motors, excellent torque sharing on multiple pinion drives, and pressure relief as an overload protection. The downsides are low efficiency, higher cost, maintenance, and the added complexity of another system. Similar features are now available for electric drives using electric VFDs. Electric drive motors are relatively simple and can use mechanical, load cell type, or electronic load sensing torque measurement and protection.

Rake drive sizing is dependant on the application with variables such as particle size distribution, flocculant use, solids loading, rake design, and design underflow concentration affecting the selection. Since the torque is related to the diameter by a square power function, a "K" factor is usually used to refer to a drive size independent of diameter, where Torque = K x Diameter^2. A table of typical values for standard duties is shown below, covering clarifiers, conventional, and high rate thickeners. However, the selection of an appropriate K factor should also consider the variables listed above, and may need to be significantly different from those listed. For example, very high underflow densities may dictate a K factor an order of magnitude higher.

Duty	Examples	K Factor (N/m)	K Factor (lbs/ft)
Light	River, or lake water clarification, Metal hydroxides, Brine clarification	15-60	1-4
Standard	Magnesium oxide, Lime softening, Brine softening	70-130	5-9
Heavy	Copper Tails, Iron Tails, Coal refuse tank, Coal, Zinc or lead concentrates, Clay, Titanium oxide, and phosphate tails	150-290	20-40
Extra Heavy	Uranium Counter Current Decantation (CCD), Iron Ore concentrate, Iron Pellet feed, Titanium Ilmenite	290+	40+

Lifts

Rake lifts are used to protect the drive from high torque. These can be used with either bridge or center column designs. Traction type drives are generally not compatible with rake lifts, although some have been supplied using the cable-type design described below. Lifts are typically powered by a separate motor from the drive and must be designed for precise control. Lifts can prevent shutdown of the machine in plant-upset conditions when large amounts of coarse material are encountered. Their use in minerals applications is fairly ubiquitous, although there are many applications that do not need a lift, or where extra torque would make a lift unnecessary.

Lifts have been used to aid in storing material, especially mineral concentrates, in a thickener. The rake is lifted allowing thickened concentrate to accumulate below the rake. The rake is then slowly driven into the dense concentrate to evacuate the thickener. This technique is not commonly used. However, it illustrates the need for the rakes to be designed to accommodate a downward force from the lift.

Cable-type designs use an arm hinged at the center connection and are supported and towed by cables, so that the arms can pivot upwards when an obstruction or high torque is encountered. The advantages of this design include a streamlined arm design and low cost. The main disadvantages are the lack of positive control of the arm position and the inability of the rakes to lift significantly in the center, where the heaviest accumulations are usually found.

Effluent Launders

Most thickeners use a peripheral effluent launder to collect the clarified overflow and bring it to a single or double discharge point. The effluent should flow into the launders uniformly around the periphery of the tank, and should not be back-flooded into the thickener. V-notch weirs are often provided to assist in distributing the effluent around the periphery of the tank. The weirs can be built into the tank launder or can be a separate, adjustable element. Froth baffles can be located at the liquid level just inboard of the launder and are used to prevent floating material from getting to the launder. Flotation concentrate thickeners are almost always provided with froth baffles and often with some method of froth management such as water sprays.

Other methods of handling effluent include radial launders, single point discharge, and bustle pipes. Radial launders are often used on solids contact clarifiers to help distribute the effluent removal evenly over the surface of the clarifier. Conversely, some applications can use a simple single point discharge nozzle and still have acceptable overflow clarity. Bustle pipes are often used where liquor storage is desired at the top of the tank or the tank liquor level is variable, using submerged pipes with orifices to distribute the effluent discharge.

The size, slope, and number of discharge points of a launder are determined using hydraulic flow equations developed by the thickener manufacturers.

Tank Design

There are a number of possible tank styles that can be used, with attendant tradeoffs. Most mineral applications require good access to the underflow piping, and it is usually preferable to locate underflow pumps in close proximity to the thickener underflow outlet. This leads to the need for either elevated tanks or underflow access tunnels. Due to the complexity, cost and safety issues with underflow tunnels, elevated tanks are generally preferred for small to medium size thickeners, up to about 40 m diameter. Other benefits of elevated tanks are storage space underneath, unhindered pump access, and ease of leak detection and repair. On-ground tanks do not require the structural steel needed for elevated tanks, and so they are generally preferred for applications where the underflow pipe access is not critical as well as for large diameter thickeners and clarifiers, generally 50 m diameter and beyond. If the process allows the underflow pipe to be buried to the edge of the thickener, an on-ground tank is significantly less expensive. As a result, on-ground tanks are fairly common in clarification and water treatment applications.

Within the realm of on-ground tank design, there are number of construction methods available using steel or concrete. Anchor channel construction refers to a steel channel embedded in a concrete footing at the wall, with the steel shell welded to the anchor channel. With this construction, the floor can be concrete, membrane or fill. The other method of construction for a steel wall tank is to use a steel floor on a compacted foundation. For concrete wall tanks, the floor can be concrete, membrane or fill. If fill material is used for a floor, it should be engineered, properly placed, and compacted as part of the overall tank construction. Various linings, covers and insulation can be applied to suit the process.

SEDIMENTATION EQUIPMENT DESIGNS

Clarifiers

Standard clarifiers are often used for water and waste water applications. These units usually have relatively light mechanisms since the amount of raking, particle size and solids density are usually on the light end of the spectrum, and rake lifts are frequently not needed. In general, clarifiers are very similar in appearance to thickeners. The feedwell is generally larger in diameter and the tank depth higher. This is done to provide more time for feed slurry flocculation in the feedwell, slower velocities exiting the feedwell, and a longer liquor detention time in the clarifier. These features are necessary to achieve optimum overflow liquor clarity

Classic clarifier hydraulic loadings start at about $1 \text{ m}^3/\text{h}/\text{m}^2$ (0.4 gpm/ft^2) and goes up as high as $6 \text{ m}^3/\text{h}/\text{m}^2$ (2.5 gpm/ft^2) for some solids contact clarifiers with optimal solids. Nominal sizing at $2.4 \text{ m}^3/\text{h}/\text{m}^2$ (1 gpm/ft^2) is a good starting point for many applications. These numbers consider using coagulants or flocculants, which greatly aid particle settling rate. Chemical addition is almost always used in clarification applications due to its effectiveness at increasing the effluent clarity.

Solids Contact Clarifiers

This equipment is designed to internally or externally recirculate solids and to promote particle contact and flocculation in the reaction well or to enhance solids precipitation and hardness reduction. This can be done externally by pumping a portion of the underflow back to the feedwell. The same thing can be accomplished internally using a draft tube and a turbine.

The design philosophy for solids contact clarifiers is to maximize the size and concentration of particles in the reaction zone by having the pumping capability to suspend a high concentration of particles. By recirculating centrally and directly above the tank floor, the heavier particles necessary for improved settling velocities are mixed with the incoming feed to allow particle contact and growth. Flocculation and recirculation are accomplished symmetrically within the reaction well for the most efficient use of reactor volume and turbine energy.

A large feedwell provides the required detention time, and allows precipitation to take place prior to the slurry entering the clarification zone. Chemical addition is introduced into the recirculation drum in the presence of previously formed precipitated solids prior to passing through the turbine for optimum mixing and flocculation.

Typical hydraulic loadings for solids contact clarifiers are $2.4-4.8 \text{ m}^3/\text{h}/\text{m}^2$ (1-2 gpm/ft^2), upwards to $12-24 \text{ m}^3/\text{h}/\text{m}^2$ (5-10 gpm/ft^2) in some steel mill wastewater clarification applications.

Inclined Plate Clarifiers

Inclined plate clarifiers use plates to increase the effective settling area in a small tank. The plates are set at a 45°-60° angle to allow settled solids to slough off. The plates are typically stacked with 50 mm spacing, although this can be varied for the application. The effective area is the sum of the horizontal projections of the plate areas. They can be supplied either with rakes or with steep cone bottoms, and are often available as packaged units. These units are very effective putting a lot of area in a small tank, but are susceptible to solids buildup and plugging.

Conventional Thickeners

There are many applications for which polymer is either not needed or may adversely affect downstream processing, and hence is not used. For these situations, a conventional thickener is all that is required. However, there are many conventional thickeners that use flocculant, to improve overflow clarity, handle a higher tonnage, or aid in achieving the desired underflow density. These units are often fairly simple, although relatively sophisticated mechanisms are used for particular applications. Due to the relatively large size, they are somewhat forgiving in operation and can have the storage capacity to absorb some plant upsets without affecting downstream operations.

Typical features include a drive and rakes, a relatively shallow feedwell, and a bridge to support the feed pipe or launder and allow center access. The drive size is dependant on the application. Conventional thickeners can be either bridge, center column or traction design. Early thickeners mostly fall into this category. A table of conventional thickener sizing in included in the Appendix.

High Rate Thickeners

With the advent of synthetic flocculant, the terms High-Rate and High-Capacity emerged as a type of thickener, as the throughput rates for the now flocculated feed slurries were considerably higher than for unflocculated slurries. These terms now generally refer to designs optimized for use with flocculants, although further improvements on earlier optimized designs are frequently possible, as the technology has continued to evolve and improve. Thickener size or throughput is directly dependant on flocculant dose and feed slurry concentration, again as presented in Figure 4. Because of this, most high rate thickeners use feed dilution systems. The optimum size of these thickeners is governed by capital and the primary operating cost of flocculant. A smaller thickener may be less expensive to install, but more costly in the long run due to a higher overall flocculant cost, and vice versa.

High Rate thickeners are generally small to medium sized bridge type thickeners, although large center column thickeners processing very high tonnage can also fall into this category. Flocculation is required and feed slurry dilution systems are often needed for optimal performance. These are the most common type of thickener in the minerals industry today. Grind, tails, leach, preleach, CCD, and concentrate thickeners often fall into this category.

Common features include a deep self-diluting feedwell, heavy duty drive, streamlined rake arms, and large effluent launders and underflow outlets.

High Rate Rakeless Thickeners

This new class of sedimentation equipment utilizes a deep tank and steep bottom cone to maximize the underflow density while eliminating the rake and rake drive. The design is based on maximizing throughput rate in a small diameter while achieving good underflow density and overflow clarity. Flocculant is always used and feed dilution is usually built into the design.

Since these units have a very low residence time, startup and shutdown are quick, typically requiring only about 30 minutes to reach steady state operation. They are operated as continuous process equipment, and cannot be used for storage. The lack of a rake mechanism makes these units very simple to operate. Current examples are the EIMCO E-CAT^{TM} Clarifier/Thickener and the Bateman Ultrasep Ultra High Rate Thickener. A diagram of the internal flow pattern of an E-CAT^{TM} Clarifier/Thickener is shown in Figure 6.

High Density Thickeners

This technology is an extension of high rate thickening, utilizing a deeper mud bed to augment the thickening capacity. Also known as high compression thickeners, these machines usually add depth to a high rate design to aid in increasing the underflow density. Deeper mud beds increase the mud compressive force, reducing the time required for thickening and increasing the underflow density. These machines may require significantly more torque (2-5x) than high rate machines due to the increased mud viscosity.

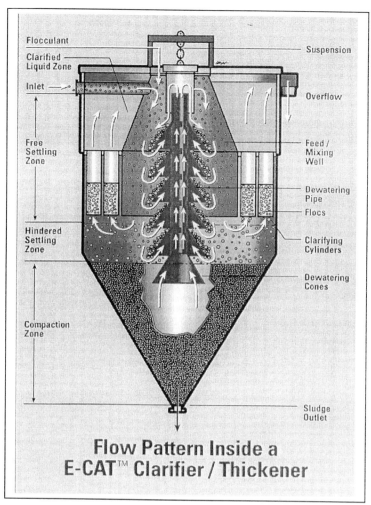

Figure 6 E-Cat™ Clarifier/Thickener showing the internal flow patterns.

Alcan Deep Thickeners

Alcan Deep Thickeners originated in work done by the British Coal Board in the 1960's, utilizing steep bottom tanks without rakes to produce underflows with very high solids contents. The goal of the work was to produce a material that could be put on a conveyor, and while this was not achieved consistently, very thick material could be produced. More recent development in the alumina industry eventually led to commercialization of this technology outside alumina with applications ranging from high efficiency CCD washers to underground paste disposal and wet stacking of surface tailings. The EIMCO Deep Cone™ Paste Thickener and EIMCO Hi-Tonnage Paste Thickener are examples of this design. It is possible to produce material at the limits of pumpabilty with these units.

In some applications, underflow with the consistency of paste can be achieved by high rate, high rate rakeless, or high density machines. However, deep thickeners are currently the best technology for achieving maximum underflow densities utilizing sedimentation equipment alone. These units typically utilize very deep mud beds in order to take maximum advantage of mud compressive forces for dewatering and provide sufficient time for the mud to dewater to a paste consistency. The tank height to diameter ratio is frequently 1:1 or higher. Due to the high underflow viscosities, mechanism torques can be 5-10 times higher than high rate machines on similar materials. Applications include surface tailings disposal by wet stacking, underground paste backfill, and countercurrent decantation.

Major Factors Influencing Thickener Design

- Process requirements for the overflow liquor quality and underflow slurry density. These determine the mechanism design.
- The quality of solids to be handled. Usually expressed as area per unit weight of dry solids per day. High rates usually require a combination of a stronger thickener mechanism and a lifting device.
- The amount of material larger than 250 micron (+60 mesh) in the feed. This affects tank bottom slope, drive and strength of mechanism. It may also require a rake lifting device.
- Specific gravity of the solids. The greater the specific gravity the more likely a stronger drive and mechanism will be required.
- Feed, overflow, and underflow systems capable of handling additional material when other thickeners are out of service.
- Feed and underflow material settling characteristics that may require special rake construction such as blades located a distance below the rake arms on posts or spikes on the blades to cut into packed solids.
- Scale build up tendency of feed slurry may require special arms and drive.
- An operating requirement to accumulate solids for defined periods of time will require a special mechanism design, as it is not a normal operating procedure.
- Froth control or removal may require sprays, froth baffles or skimmers.
- Slurry temperature, vapors, gases, etc. may require covered and/or insulated tanks with attendant seals.
- Soil conditions and ground water elevation affect foundation design and may determine tank and mechanism type.
- Climatic conditions may require special considerations, such as enclosures around the drive and instrumentation.

CONCLUSIONS

Different sedimentation equipment designs are available for various applications and process objectives. Utilizing the proper design for an application can make the difference between a smooth operating process step and continual problems that prevent a plant from realizing its potential. Taking the time in the project planning stage to make sure the correct design is being used can make for a successful project.

REFERENCES

H. Cross, "A New Approach To The Design And Operation Of Thickeners", Journal Of The South African Institute Of Mining And Metallurgy, (February 1963)

F. M. Tiller, D. Tarng, "Try Deep Thickeners and Clarifiers", Chemical Engineering Progress, pp. 75-80, (March 1995).

R. Klepper, T. Laros and F. Schoenbrunn, "Deep Paste Thickening Systems", Proceedings of Minefill '98, Brisbane, Australia, (1998).

R.C. Emmett, T.J. Laros, K.A. Paulson, "Recent Developments in Solid/Liquid Separation Technology in the Alumina Industry", Light Metals, 1992, edited by E.R. Cutshall, pp 87-90.

E.S. Hsia, F.W. Reinmiller, "How to Design and Construct Earth Bottom Thickeners", Mining Engineering, August 1977.

APPENDIX

Conventional Thickener Sizing

APPLICATION	% SOLIDS		UNIT AREA FT^2/TON/DAY	OVERFLOW RATE GPM/FT^2	K* FACTOR FT-LBS/FT^2
	FEED	UNDERFLOW			
Alumina, Bayer Process					
Red Mud					
Primary	3-4	25-30	20-50	---	35-40 (1)
Washers	6-8	30-35	10-40	---	35-40
Final	6-8	35-40	10-30	---	35-40
Hydrate	2-8	30-50	12-30	---	25-30
Brine Purification	0.1-2	8-15	---	0.2-0.5	10-25
Coal					
Refuse	0.5-6	20-40	100-120	0.3-1.0	20-30
Clean Coal Fines	---	20-50	---	0.4-1.0	15-25
Heavy Media (Magnetite)	20-30	60-70	---	3.0	35-40
Flue Dust					
Blast Furnace	0.2-2	40-60	---	0.4-1.5	20-30
BOF	0.2-2	30-70	---	0.4-1.5	15-25
Magnesium Hydroxide from Brine	8-10	25-50	60-100	---	10-25
Magnesium Hydroxide from Sea Water					
Primary	2-3	15-20	100-250	---	10-15
Washers	5-10	20-30	100-150	---	10-15
Copper Concentrates	15-30	50-75	2-6	---	35-40
Copper Tailings	10-30	45-65	4-10	---	25-30
Iron Ore					
Fine Concentrates (65- 90%- 325 m)	20-35	60-70	0.4-0.8	1.2	20-25
Coarse Concentrates (45- 65%- 325 m)	25-50	65-80	0.2-0.5	---	20-25
Tailings	1-10	40-60	4-10	---	15-20
Gold CCD	10-15	40-55	1-4		20-35
Molybdenum Concentrate	8-10	25-35	10-15		15-25
Zinc Concentrate	10-20	60-70	3-7		25-35
Uranium					
Acid Leached Ore	10-30	45-65	1.5-6	---	20-25
Alkaline Leached Ore	20	60	10	---	20-25
Uranium Precipate	1-2	10-25	50-125	---	20-25

*$K = T/D^2$ also, Torque = KD^2

(1) Torque based on scale load

Design Features and Types of Filtration Equipment

C.Cox[1] and F.Traczyk[2]

ABSTRACT
The chapter covers the various types of vacuum and pressure filtration equipment used in minerals processing, in addition to a few special designs for very fine particle applications. The basic design features of each type are presented, and the typical applications where the various types are used are discussed. Features within the overall equipment type designs which add to functionality within a specific application are also reviewed.

INTRODUCTION
The mechanical separation of solids from liquids is often the final major step in a mineral production process. Typically, all of a plants valuable product is handled by this stage; thus selection of the correct equipment is critical. Consideration must be given to feed characteristics, final product specification; i.e. cake percent moisture and/or filtrate clarity, and production and availability requirements.

This section will discuss the major types and respective design features of today's filtration equipment. Category of equipment to include vacuum and pressure, with discussion of filter media. The typical range of filter application with respect to particle size and final product moisture envelopes will also be presented.

MECHANICAL DEWATERING BY FILTRATION
Liquid-solid separation by filtration requires a differential pressure, ΔP, across a cake of solids. The ΔP required can be defined in a general way via Kelvin's Law, which describes capillary forces within the interstitial pores of a cake of solids and solution [1]:

$$\Delta P = \frac{4 \cdot T \cdot \cos \theta}{D}$$

Where T = surface tension
θ = contact $\angle$
D = pore diameter

Thus the smaller the pore size the greater the force or ΔP required to overcome the capillary forces and displace the interstitial solution and achieve a desired final cake moisture. Pore size has a direct relationship to the p80 and p10 size of the material to be dewatered.

For mineral dewatering applications requiring less than 1 Bar of ΔP, vacuum filtration methods are generally employed. At application requirements greater than 1 Bar ΔP, pressure filtration in its various forms is then utilized. The relationship of feed particle size to achievable,

---

1 Metso Minerals Inc., Colorado Springs, Colorado.
2 EIMCO Process Equipment Company, Salt Lake City, Utah.

residual cake moisture by the typical range of vacuum and pressure filtration equipment is presented in Figure 1. In Figure 1, the region 1 represents the range of typical vacuum filtration equipment where as regions 2 and 3 indicate where pressure type filters are applied.

It is important to note in Figure 1, that in today's mineral processing world, due to the fact the mineral concentrate and tailings products generally have a p80 of 40μ or less, pressure filtration is playing an ever-increasing role. As a general rule however, the greater the pressure drop required the greater will be the corresponding capital and operating cost per ton of product.

In the subsequent sections, the various types and features of vacuum filters and pressure filters will be discussed. Specialty filtration devices, which also find application in plant design, such as the tube press and belt press, will also be reviewed.

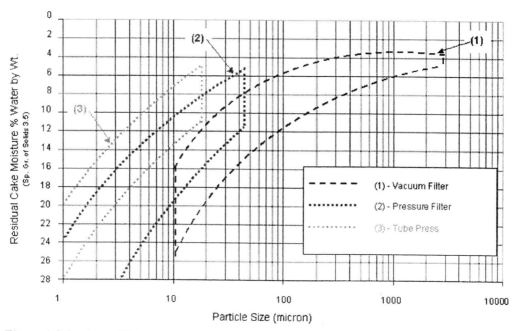

Figure 1 Selection of filters

VACUUM FILTRATION

Vacuum filtration is a well-established technique used in industrial dewatering. All vacuum filters operate on similar principle. Within a slurry tank a pressure differential between the filter medium surface and the inside of the drum, disk or belt is applied by means of vacuum. This pressure differential causes transport of liquid through the filtration surface while the filter medium arrests solid particles thus forming a cake. As the unit rotates, the cake rises above the slurry level and air is drawn through the cake, forcing out liquid. The liquid (filtrate) exits the filter through the internal piping and the vacuum head. A typical vacuum filtration installation is illustrated in Figure 2.

The vacuum filter is fitted with a filter cloth or screen where the solids are deposited. A grid or drainage section supports the cloth where the vacuum is applied.

Vacuum receivers are used to collect the filtrate. Filtrate pumps can either be mounted separately or attached directly to the side of the vacuum receivers. Multiple receivers are also used for horizontal belt filters and where individual filtrates may be separated or washing is used.

Moisture traps prevent liquids from being pulled over into the vacuum pump. This is extremely important in the case of corrosive liquids or dry vacuum pumps. This can sometimes be eliminated when the solids or liquids contained in the filtrate would not be harmful and a liquid seal pump, such as a Nash, is used.

The vacuum applied to the filters is the atmospheric pressure less approximately 0.2 bars for line losses. In cases where filter cake cracking occurs, the vacuum pressure can be reduced to prevent the air from short-circuiting. Note special mediums such as ceramic can eliminate air bypass by utilizing small enough openings to take advantage of capillary forces, which are greater than that which can be generated by a vacuum.

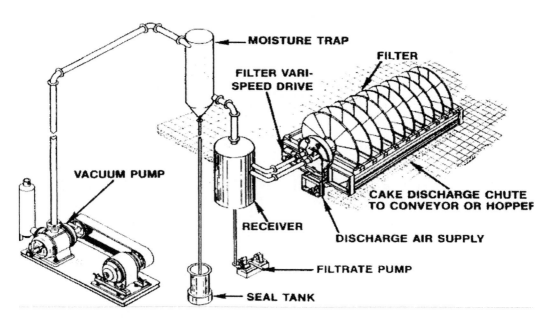

Figure 2 Typical vacuum filtration installation

Disk Filters

Vacuum disk filters are particularly suitable for relatively simple dewatering applications where high capacity is the principal requirement. The disk design permits a greater filtration area per unit of floor space compared to the drum or belt type designs.

The disk filter consists of individual sectors placed together to form a circular disk. Vacuum filtration occurs on both sides of the disk and the filtrate is collected internally and fed through a collection pipe or center barrel. Vacuum is applied through a rotary distribution valve fitted to one or both ends of the unit. Individual sectors are covered with sector bags of cloth or screen. The life of the various mediums can vary from days to months depending on the type of material being filtered.

The sectors are rotated in a tank that is mechanically agitated to prevent sanding and provide a consistent mixture of solids throughout the slurry. A variable speed drive is used to rotate the sectors at a normal speed of 1 to 10 minutes per revolution. Disk submergence can also be used along with cycle speed to adjust pick up and dewater times.

Vacuum is applied in the submerged area or pickup zone on the disk for forming the cake. As the disk rotates out of the slurry cake dewatering continues until the cake discharge portion of the cycle is reached. For filters with conventional cloth media, the discharge area is separated by means of bridge blocks in the distribution valve and an airblow is applied to inflate the sector bag to discharge the cake. High pressure blows, or snap blow, is often used for difficult to discharge material.

The advantages of the disk filter are the large filtration area, low initial investment cost, and minimum floor space requirements. The disadvantages are that the washing of the filter medium for cloth type systems is difficult and effective cake washing is not possible. Recent advancements have been made in hydraulic engineering to increase filtration capacity.

Ceramac Disk Filter. A variation of the vacuum disk filter is the Outokumpu Ceramac Disk, Figure 3 (courtesy of Outokumpu Technologies Inc). In the case of the Outokumpu filter, the sectors are a complete one-piece ceramic design (see filter medium section). The primary driving force of the fluid is done by capillary action. The filter requires less power as the vacuum pump acts primarily as a filtrate pump. The capillary action can produce very low moistures that are generally less than conventional vacuum filtration and approach pressure filtration. The hydraulic capacity is somewhat limited, so high feed solids concentrations can produce the best results. The main applications are mineral concentrates.

The Outokumpu Ceramic Disk provides the ability to maintain the sectors via ultrasonic cleaning and internal acid washing.

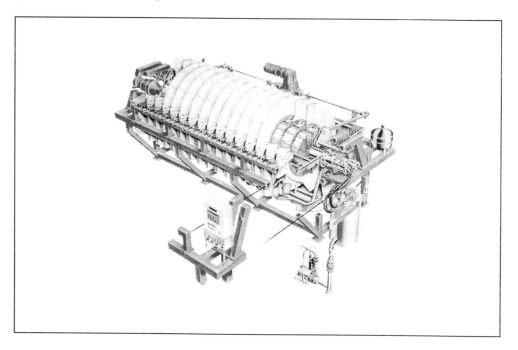

Figure 3 Outokumpu ceramic disk filter

Drum Filters

Vacuum drum filters have an extended range of application and are considered to have lower maintenance costs compared to disk units. Drum filters are preferred in applications that require lower moisture and or where effective cake washing is required. (Figure 4 courtesy of WesTech)

Like a disk filter, the drum or shell is rotated in an agitated slurry tank. Drainage grids are mounted on the surface of the drum, which are then covered by filter media. The drainage grids are made in sections and each section contains a pipe or pipes that apply vacuum to that section. The pipes are collected in one or both ends of the drum to a rotary distribution valve.

By having separate individual sections on the surface of the drum, the distribution valve can be used to change the proportion of cake form, wash, dewater and discharge zones. Also similar to the disk filter the submergence of the drum can also be varied.

A displacement wash can be effectively applied to the drum by dripping or spraying liquid across the surface of the cake. When high wash efficiencies are required, two or more drum filters can be used in series. Repulp stages are used between filters to increase washing efficiency. More that three drum filters in a series are seldom used if flocculation is required as the cake usually becomes progressively harder to handle and filter.

Figure 4 Drum Belt filters used for barite processing

Drum filters offer the flexibility to handle a wide range of material types and p80 sizes by modifying feeding point and by selection of the type of discharge method.

For applications with coarse particle slurries a top feed drum filter should be considered. The top feed principle promotes segregation of the coarser particles in the feed box, which are then deposited first on the media forming a layer with large porosity, which increases filtration rate.

Selection of the discharge method will depend upon material characteristic such as, tendency to stick to or blind filter media, fibrous content, and filtrate clarity requirements. Each type of discharge type has its own characteristics and is discussed separately. The most common discharge types are shown in Figure 5 (courtesy of WesTech).

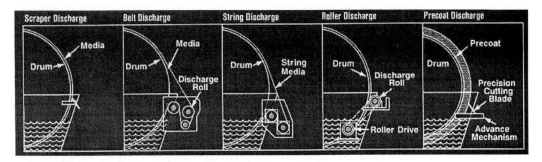

Figure 5 Common types of drum filter discharge mechanisms

Scraper Discharge. Scraper discharges consist of a scraper blade near or on the surface of the drum. During the discharge portion of the cycle, low-pressure air blow is used to expand the filter media and aid in cake release. Scraper discharge can be used with solids which do not blind the filter media, and the cake is readily released. For applications requiring high discharge blow pressures, the surface of the drum can be wound with wire to prevent damage to the cloth.

Belt Discharge. In applications where cloth blinding is a problem, the Belt Discharge method is used. With a belt type drum, the cloth is removed from the drum after passing through the dewatering zone. The cake is discharged and the filter belt is the washed over its full width typically by high-pressure sprays, one above and one below the cloth. Wash water is confined and collected for separate discharge through a launder. The wash cloth is then returned to the drum for the next cycle. The belt discharge method is more involved as a cloth tracking system is required.

Roll Discharge. Roll discharges are commonly used in the clay industry where material is sticky and difficult to release. The cake from the drum adheres to a roll which is in contact with the drum and which is driven at a slightly higher rotational speed. A scraper blade is set to remove or peel the cake leaving a heel in place on the roll. This discharge method is well suited for sticky materials which has a tendency to adhere to itself rather that the filter cloth.

String Discharge. String discharge consists of a rotary drum similar to the scraper discharge filters. Strings are used across the face of the drum to remove cake or pulp from the surface and release it over a set of rolls. This is an effective discharge method when fibrous pulps or sufficiently strong cakes are produced. This method of discharge gives the advantage of a clean drum surface that can be washed intermittently from the outside. A draw back is the obvious maintenance of the string media.

Precoat. Precoat filters are used either for clarification or for filtering of solids that are very slimy and do not produce a thick enough cake for discharge on other types of discharge mechanisms. The filter is first coated with a precoat material such as diatomaceous earth after which the slurry is introduced to the filter. Vacuum is maintained throughout the complete rotation cycle to ensure adhesion of the precoat. A doctor blade or knife is used to continuously cut a thin surface of the precoat to discharge the solids that have been deposited. This also presents a cleaned surface to the slurry for the next cycle. Precoat filters can produce a filtrate with a very low solids content thus they are often used for polishing applications.

Horizontal Belt Filters

Horizontal vacuum belt filters are generally used for handling coarse solids and, or where high washing efficiency is required. They also can achieve lower cake moistures compared to vacuum disk or drum types. Belt filter equipment size ranges from 1 square meter up to as large as 154 square meters.

A typical belt filter is shown in Figure 6. The units consist of a rubber drainage belt supported by pulleys on both ends and travels over a vacuum box. Vacuum is pulled through holes in the center of the drainage belt, which is grooved to allow the liquid to flow to the center. The stationary vacuum box is joined to the moving drainage belt either by a series of traveling wear belts or lubricated wear strips. The vacuum box can normally be automatically lowered for easy access and maintenance.

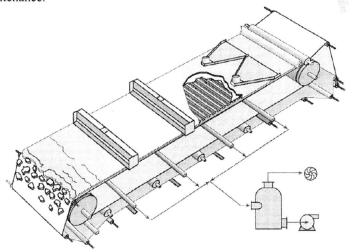

Figure 6 Eimco Extractor horizontal belt filter

A separate filter media travels along the surface of the drainage belt in the vacuum area and is removed in the discharge and washed before returning to the filter. The drainage belt is supported on a deck with air or water lubrication. Filter fabric media has been developed specifically for belt filters to provide single width heavy fabrics to eliminate tracking problems. The vacuum pan can be divided into multiple sections, and each section can have its own vacuum receiver. This type of filter is very sell suited for co-current or counter-current washing applications such as uranium or gold. When co-current or counter-current washing applications are used, the wash rations can be at any level desired since the equipment is not constricted geometrically. In addition, the pulp is only flocculated once and does not become more difficult to handle as in a series of drum filters. Thus compared to disk or drum, belt filters offer the most efficient washing capabilities.

Major recent innovations include improved maintenance around the vacuum pan, support deck, and wear parts. Improved drainage belts with high curbing and 4.2 meter wide belts are becoming standard.

Steam hoods can be used to decrease moisture content. Figure 7 shows a steam hood on an Eimco Extractor that utilizes a proprietary control system to prevent overheating of the drainage belt.

The horizontal belt filters are used on a wide variety of minerals including concentrates, coal, industrial minerals, tailings, and washing applications.

Figure 7 Eimco steam hood

PRESSURE FILTRATION

The mineralogy of today's ore bodies is such that economic liberation size of valuable components is becoming finer and finer for mineral concentrates. Final product p80 size of 30 microns or less is now common in Cu, Pb, and Zn processing as well as auriferous pyrites. Also shipping moistures and smelter schedules are requiring these fine size concentrates to contain moisture contents in the range of 8 to 10 weight percent. As a result, a greater pressure drop across the cake, than can be achieved by vacuum, is required. Thus the trend in new plant design and plant modernization is pressure filtration.

Pressure filtration is an old concept. The challenge in the past has been to accomplish this with a cost effective, reliable method. The equipment industry has responded by developing ever larger machines and by taking advantage of state of the art manufacturing methods and incorporating up to date instrumentation and computer controls along with corresponding advancements in filter cloth technology. Today pressure filters handle the entire output of a large concentrator in one or two machines with online availability of greater than 95%. Operator supervision is minimal with supervision in some cases being managed by remote communications.

Today's filters have evolved from two basic roots: actuation either horizontally or vertically. Such filters have been adapted to address the specific difficult requirements of the minerals industry and are discussed in specific below.

Horizontal Pressure Filters

A picture of a typical Horizontal Pressure Filter is presented in Figure 8 (courtesy of Metso Minerals Inc.).

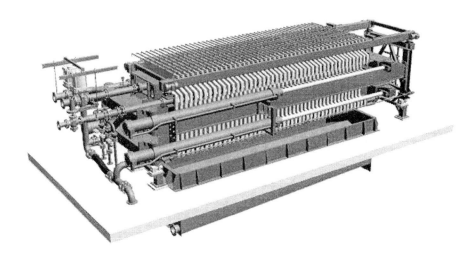

Figure 8 Horizontal pressure filter

In this configuration filter plates, generally constructed from lightweight polymer are suspended upon a steel frame. The plates are linked together and are opened and closed by hydraulic cylinders. The plates have recessed chambers and between plates is hung individual filter cloths. The two cloths per chamber define the filter volume see Figure 9.

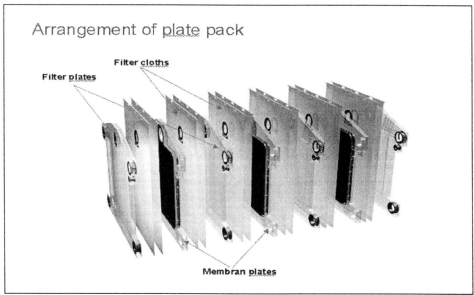

Figure 9 Filter plate pack

Feed slurry is pumped to the filter and is distributed to completely fill each chamber. During the feed cycle, the pump output pressure is ramped up to typically 6 bar. Dewatering commences as soon as feed is introduced with filtrate exiting through both sides of the chamber. When the chambers are full of materials, select filters have membranes in one side of the filter chamber which are pressurized, typically by air, to hold the cake in place to prevent cracking and air by pass during the air dewatering cycle.

A common misconception is that pressure filtration is accomplished by a mechanical squeeze of the cake. Since most concentrates are non-compressible the primary dewatering method is by pressurized air being forced through the cake. Such "air through blow" is performed at 5 to 8 bar effectively displacing liquid as it passes through the cake. The relative effect on dewatering by the feed, membrane squeeze, and air-drying is depicted in Figure 10 (dewatering curve). The curve was generated by a machine, which incorporates load cells to accurately determine the net cake weight throughout the entire filtration cycle. The use of a load cell ensures that the filter is properly filled with solids and that all solids have discharged at the end of the cycle. Also a load cell machine can be run based on timed cycles or weighed cycles. The curves show that the majority of the dewatering takes place during the air blow cycle. A wash cycle can be incorporated if required. During wash, a liquid takes the place of the air blow and is passed through the cake. After wash, air blow is restarted to achieve the final moisture.

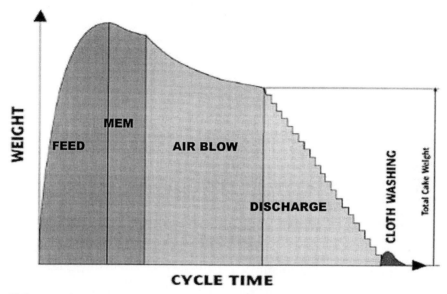

Figure 10 Dewatering curve showing cake weight through a typical cycle time

At completion of the air blow cycle the chambers are opened and the cake falls by gravity through a chute onto a load out conveyor. Chambers are open in sequence thus evenly distributing the discharge weight.

Following discharge, the filter cloth is cleaned by a combination of vibration and spray wash. The wash down material is directed back to the filter feed tank for recovery. The chambers are then closed and the cycle repeats.

The operation is batch, however with short cycle times of 8 to 10 minutes, and with the use of a filter feed tank with adequate storage, the filtration appears continuous to the overall operation.

To match the demands of the industry, horizontal filters now include units with 2 meter by two-meter plates with up to 60 chambers, Figure 11. Such units can now economically handle applications such as dewatering mineral concentrates, tailings and even fine coal.

Vertical Pressure Filters

Vertical pressure filters differ from the horizontal units in that the individual chambers are stacked one on top of the other. The chambers are linked and are open and closed by vertically operating hydraulic cylinders as depicted in Figure 12 (courtesy of Larox Oy).

Figure 11 Filter frame for two x two meter plates

Figure 12 Vertical Pressure Filter

Plate and chamber design incorporates membranes or diaphragms to provide high-pressure squeeze on the cakes. Unlike the horizontal units with the individual filter cloths hung between each chamber, the vertical units operate with a continuous cloth.

The filtration cycle is similar to that described above for the horizontal type design. At the beginning of each cycle, all plates are closed simultaneously and slurry is fed to all chambers simultaneously. Filtration begins immediately as the cakes are being formed. When the chambers are full, high-pressure water is pumped behind the diaphragms to mechanically force additional liquid from the cake (air is used for Larox filters with filtration areas > $30m^2$). If washing is required, the diaphragms are drained of high-pressure water and wash water is pumped in on top of the cake. The diaphragms are again pressurized and the wash water is forced through the cake. For final dewatering, the diaphragms are again depressurized and high-pressure air is blown through each cake. Upon completion of air blow, all filter plates are opened and the cloth is advanced through the unit chambers achieving cake discharge followed by cloth washing.

Recent improvements include automatic monitoring of the filter cloth condition, seam position and tracking. This eases maintenance and improves cloth life. Other developments include use of sensors on the filter to monitor and control cake thickness, cake pressing (squeezing) time, and air blow flow rate and time. The objective is to provide full process control to achieve consistent results during time of process upsets or mineralogy changes.

Tube Press
Select dewatering requirements of ultra fine (<10 micron) materials required specialty equipment. The extremely powerful capillary forces demand greater pressure drops than can be achieved with the horizontal or vertical type units described above. For such ultra fines, pressure requirement to achieve target moistures exceed 100 bar. Thus filtration takes place in a pipe or tube to handle the mechanical forces of the high pressure.

Tube presses were initially developed to dewater fine Kaolin. It has since been applied to a variety of difficult filtration operations. A cross section of the Tube press unit is shown in Figure 13 (courtesy of Metso Minerals Inc.). In summary the assembly consists of an outer casing which has a flexible membrane (bladder) fastened at each end. An inner "candle" provides the filtration service through surrounding layers of backing mesh, backing felt and cloth media. The candle is drilled with holes to provide drainage for filtrate.

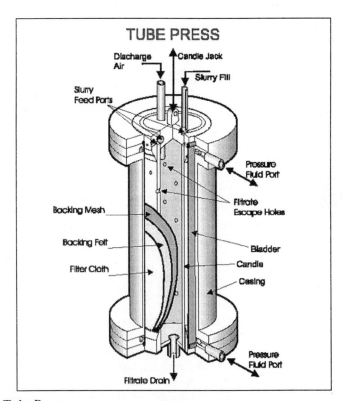

Figure 13 Tube Press

The filtration operations follow essentially the same sequence described previously for pressure units. Feed is introduced in the tube press under pressure. When a cake is formed, dewatering pressure is applied via the membrane and air blow. If a wash is required the membrane is relaxed and the wash liquor is supplied and then forced through the cake by resqueezing the membrane. A subsequent air blow can then be applied.

When the filtration, washing and dewatering are complete, a hydraulic vacuum is applied to retract the membrane, and the candle is lowered to discharge the cake. Air can be blown in behind the filter media to aid in discharge. The candle is then reinserted into the casing for the next cycle.

Plate and Frame
The earliest types of filter presses date back to the Dark Ages when monasteries used filter presses for production of grape juice. The early modern day pressure filters were Plate and Frame filters. The filters consist of alternate plates and frames that are supported overhead or on side bars (Figure 14). The plates and frames are pressed together to form chambers that are filled under pressure with pumps. The plates are covered with filter media or can also use paper where slimes are present that would blind media.

Plate and Frame filters are limited to about 7-bar pressure since the media extends outside the plates. Plate and Frame filters are low cost but still are labor intensive. They are widely used for low volume or long cycle time applications or where paper media may be required.

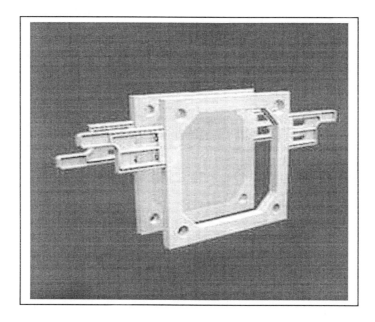

Figure 14 Plate and frame – side bar design

Recessed Plate Pressure Filters
Recessed Plate filters consist of single plates that are recessed on both sides . When the plates are pressed together, a chamber is formed. The plates can also be gasketed for applications where spillage of the filtrate is important. Recessed plate filters can operate at a higher pressure than Plate and Frame filters with normal operating pressures of 2-15 bars. Recessed Plate filters can also be fitted with membranes to help in cake washing and provide lower moistures. Figure 15 shows an overhead design recessed plate automatic filter (courtesty of Eimco Process Equipment Co.).

Figure 15 Overhead design recessed plate filter

The recent advancements in the Plate and Frame and Recessed filters include automatic plate shifting and high-pressure plate washing that reduces cycle time. Recessed filters have higher volume plates to reduce filter size requirements. Filter plate material of construction has also greatly improved with higher operating pressures and temperatures. Polypropylene plates can operate at temperatures of 95 degrees centigrade.

Belt Filter Press
Selected dewatering applications involve producing a material that can be handled, transported and stored as a solid. Reaching ultimate low moisture is secondary. Examples include aggregate or milling operations, which have minimal land available for tailing storage, and environmental considerations restrict the use of storage ponds. Such tailings typically contain a high percentage of minus 200 mesh material and as such are difficult to economically dewater to a state that can transported by conveyor or truck.

Belt filter presses have been field proven to address the above challenge by dewatering slurries containing 40 to 50 percent solids to a solid phase material at 70 percent solids and higher.

A diagram of a belt filter press cross section is provided in Figure 16 (courtesy of Phoenix Process Equipment Co.). The feed to the belt filter (1) press is dosed with flocculent, passes through an in-line-mixing device and is distributed evenly into the gravity drainage zone (2). The feed is contained within the gravity drainage zone by a frame-mounted feed containment box. The gravity drainage section (3) is inclined to facilitate the drainage of free water through the lower filter belt. A frame mounted, wear resistant dewatering grid supports the lower filter belt in the gravity drainage zone. Filtrate is collected in a gravity drainage collection pan. The drained solids leaving the gravity drainage zone are feed into the adjustable compression zone where the upper filter belt converges (4) to gently apply compression. Liquid that has been pressed through the screen is collected in the wedge zone collection pan.

Pressure and dewatering efficiency are increased by entry into the large roll compression/shear zone. As the belt and cake progress through the roll train (5) the diameter is progressively decreased to form the high-pressure high-shear zone. Discharge (6) of the dewatered cake from the press is accomplished by the use of plastic doctor blades, which peel the low moisture solids away from the filter belts. On the return to the feed section of the press, each filter belt passes through the continuous support functions of belt washing, belt alignment and belt tensioning. [2]

Belt filter presses are available in widths up to 3.0 meters and can handle up to 30 st/hr of dry solids. The main operating costs include belt replacement and flocculent consumption, which can range from 50 g/t to 500 g/t.

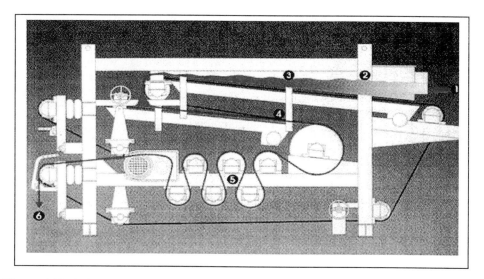

Figure 16 Belt filter press diagram

FILTER MEDIA

Vacuum and pressure filters all employ filter cloth or paper as the filtration medium. Experience has shown that specification of cloth type is critical to the success of a filter application, both in terms of operating cost and filtrate clarity.

Media or cloth consumption is a major component of a filter's operating cost per ton. Initial filter economics will include an assumption of the number of cycles or life expectancy of the filter cloth. Such assumptions will be based on experience with similar or like materials. However, often, the material encountered during operation can have different characteristics with regard to the degree of fines, clay content, or abrasiveness. As such the cloth pre-selected can blind or wear out prematurely. Assumed life cycles of for example 6000 can degrade to less than 1000 with an attendant major impact on cloth consumption cost, maintenance man hours, and filter availability.

Therefore it is important to define as well as possible the cloth specifications for a given application. Following start up, cloth performance will be reviewed and monitored. Working with the filter supplier and their cloth suppliers is recommended to arrive at the best balance of cloth permeability, resistance to blinding and abrasion, chemical suitability, mechanical requirements, and cost options. Cloth technology is constantly evolving thus a good partnership with suppliers will ensure that new innovations can be incorporated and performance continually upgraded.

The following filter cloth basics and a summary of filter cloth terms presented in Table 1 are provided by Crosible Filtration, Inc.

There are two types of cloth, woven and non-woven. For the most part it has been determined that woven cloth offer the best combination of filtration efficiency and low blinding tendencies.

Woven cloths have three basic yarn compositions: spun, multi-filament and monofilaments. A cloth can be made from 100 percent of one of these yarns or 50 percent of two of them; that is a different warp and filling.

An all-spun fabric will have the highest particle capture rate for a given porosity; followed by multi-filament and then monofilament which has the lowest rating. As a trade off however the reverse order is the case with regard to the cloth's susceptibility to blinding. Thus to meet a particular filter application requirements today's cloths tend to be blend of multi/spun or mono/multi or even mono/spun to take advantage of each yarn's strengths.

Yarns can be produced from several materials. The most popular is polypropylene as it is the least expensive, most chemically resistant and is easiest with which to work. Nylon and polyester are also considered when temperature and or chemical incompatibility issues preclude the use of polypropylene. **Specialty fabrics** like Nomex, Ryton and PTFE, which are very expensive and are employed in the harshest of chemical and temperature environments.

The type of weave is also important in the performance of cloth. A plain weave is the tightest also most apt to blind. The twill weave, usually a 2 x 2, was a standard as a way to offset blinding and give a thicker cloth for filtration. Current design favors the sateen weave where a yarn will travel over several perpendicular yarns then go under one then repeat the sequence. This produces a very smooth surface, which resists blinding and easily cleaned. It is a matter of determining the proper yarn or yarns and type of weave for optimum filtration.

In addition to conventional cloth technology, Outokumpu Mintec Oy has developed and commercialized ceramic media termed CERAMEC for vacuum disk application. A cross section of the CERAMEC disk is present in Figure 17.

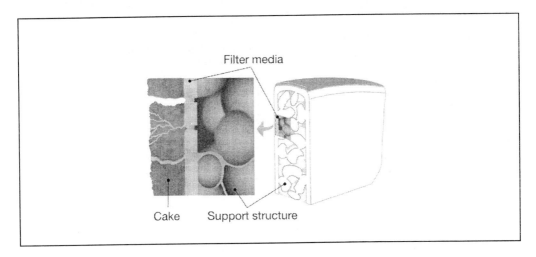

Figure 17 CERAMEC disk cross section

CERAMEC filter discs are patented sintered alumina membranes with uniform micropores to create capillary action. This microporous filter media allows only liquid to flow through. Even at a perfect vacuum there is insufficient pressure differential to force air through the media openings. The capillary phenomenon is based on the Young-Laplace law, which states that the pores of a certain diameter cause a capillary effect due to surface tension and the contact angle of the liquid. As the discs are immersed into the slurry basin, a pressure difference maintained with a small vacuum pump causes cake formation on the surface of the discs and dewatering takes place as long as free liquid is present.

CONCLUSIONS
Filtration equipment is available to meet dewatering requirements for a wide range of material characteristics taking into account feed particle size distribution, feed percent solids, desired product moisture and product handling specifications. Equipment capacities have expanded to address current large tonnage requirements while incorporating the latest in computer control and sensor design to allow automatic independent operation with self-diagnostic capabilities. In support of filtration equipment, media technology has progressed to offer a broad scope of materials and construction to provide the necessary cycle life and performance to ensure economic operation.

REFERENCES
Mapes, Christopher, Phelps Dodge Corp., "Development and Application of Ceramic Media Disk Filter Technology at Phelps Dodge Morenci, Inc.

Petrey, Pete, Phoenix Process Equipment, " Design Considerations in the Elimination of Slurry Ponds" Aggregates Manager June 1998

Table 1
FILTER CLOTH TERMS

Blinding
Plugging up of a filter fabric resulting in reduced flow rates and filtration efficiency.

Cake Release
Ability of a filter cloth to completely discharge the cake from the cloth.

Monofilament-
Single, long continuous strand of a synthetic fiber extruded in fairly coarse diameter.

Multifilament-
Smooth yarn consisting of two or more monofilaments twisted tightly together.

Needled Felt-
A fabric (felt) made by the mechanical interlocking of individual fibers in a random orientation.

Nylon-
A synthetic fiber manufactured from, water and air. Has good resistance to alkalis, but is affected by strong acids and solutions in pH range 1-6. Max. Operating temperature is 225F.

Oxford Weave-
Modified plain weave where both warp and weft yarns are multiple yarns but not of equal number.

Plain Weave-
The simplest and most common weave produced by passing the weft thread over and under each successive warp thread. Offers low permeability and excellent particle retention, however is susceptible to blinding.

Polyester-
A synthetic fiber manufactured from Terephthalic acid and Ethylene Glycol, the resulting polymer has excellent resistance to most mineral acids and solutions in the H range 1-8. Max. Operating temperature = 285F.

Polypropylene-
A synthetic fiber manufactured from a petroleum industry by-product. Excellent resistance to full pH range and has max. Operating temp. of 190F. Is affected by oxidizing agents.

Porosity (Permeability)-
The rate of flow of air under differential pressure through a cloth. Generally measured in ft^3/min (at ½" water pressure).

Satin (or Sateen) Weave-
Weave in which warp yarns are carried uninterruptedly over many weft yarns to produce a smooth-faced fabric. Offers superior cake release and excellent resistance to binding.

Spun (Staple)-
Yarns made from filaments that have been cut into short lengths, then twisted together. Staple fibers offer good particle retention; however, cake release may suffer due to the hairiness of the yarns.

3 X 1 Double Weave-
Special weave in which both sides of the cloth show a smooth 3 X 1 broken twill surface. Offer excellent stability, retention and cake release properties.

Twill Weave-
Weave in which the weft threads pass over one and under two or more warp threads to give the look of diagonal lines. Offers medium retention and blinding properties with high abrasion resistance and good flow rates.

Warp-
The yarn that runs lengthwise in cloth as it is woven on a loom.

Weft-
The yarns that run widthwise in cloth as it woven on a loom. Also known as filling yarns.

Plant Design, Layout and Economic Considerations

Mark Erickson[*] & Mike Blois[*]

ABSTRACT

In the context of this handbook, solid/liquids separation predominantly refers to in-plant and tailings thickening (and tailings dam management to some extent), concentrate thickening, concentrate filtration and clarification.

The incorporation of solids/liquids separation steps within a process flowsheet affects the plant design and layout in two broad aspects:
- MACRO LEVEL – Impact of major plant sections on the overall block plan of the processing facility
- MICRO LEVEL – Impact of individual systems on the general arrangement / plant layout within one of the major plant sections

Because of the large equipment sizes that are typically used, decisions regarding sedimentation equipment, e.g. thickener and clarifiers, generally have a greater impact on the macro level of the overall block plan. Environmental factors, such as weather and geotechnical aspects, are also influential and are presented in this chapter.

The selection of filtration equipment usually impacts the plant design on the micro scale, such as the arrangement of filters within a building. At this level, an understanding of the system that includes the solids/liquids separation stage is important. This chapter also presents issues such as design and layout implications for both upstream and downstream equipment.

INTRODUCTION

The decision-making process for the choice of a plant siting and the development of a plant layout has been described as an "art rather than an exact science because the factual demands of process design must be combined with experience, e.g. to anticipate unsolved mechanical design problems, and to provide for the human element in operation and maintenance".[1]

The literature goes on to say "The key to economical construction and efficient operation is a carefully planned, functional arrangement of equipment, piping and buildings. Furthermore, an accessible and aesthetically pleasing plot plan can make major contributions to safety, employee satisfaction and sound community relations.

In fact, aside from the process design aspects, no single factor in a process plant project is as important as the physical layout of the equipment itself. A modern process unit erected today is likely to remain in use for 20 or more years. Any errors in the beginning will be costly to rectify later".

Although these opening thoughts were written for the chemical engineering industry, they are just as applicable to the minerals processing and metallurgical industries. Furthermore, these words, originally written over 30 years ago, remain just as valid in the "New Millennium".

To identify all the engineering details that are necessary to solve plant layout problems related to solids/liquids separation is a task far beyond the scope of this chapter. This chapter illustrates, by means of examples, the range of factors that need to be reviewed when considering the plant design and layout of solids/liquid separation facilities. The factors in this chapter can be used as a checklist to ensure that none are overlooked in the study and design stages.

---

[*] Bechtel Corporation, Denver, Colorado

The concept of designing a plant layout is divided into two aspects of influence: macro and micro:
- MACRO: The block plan (arrangement, and inter-relationship, of buildings or plant sections that make up the whole plant)
- MICRO: The general arrangement (the layout within a building or plant section).[2] This chapter emphasizes the importance of understanding the inter-relationship between individual systems within a solids/liquids separation stage.

MACRO FACTORS INFLUENCING SOLIDS/LIQUIDS SEPARATION FACILITIES

Overview of Macro Factors

The macro factors that influence the layout of a solids/liquids separation facility can be divided as shown in Table 1.

Table 1 Macro Factors

SITE CONDITIONS	Topographical
	Available Space
	Geotechnical
	Water Availability
	Environmental Issues
	Climatic Conditions
PROJECT CRITERIA	Expected Life of Operation
PLANT EXPANSION	Expansion Possibilities
PROCESS SYSTEMS	Continuous Operation
	Batch Operation

Site Conditions

Topographical. "The topography of a proposed plant site can be a factor of considerable importance. The ideal site is a level or gently sloping terrain, permitting optional arrangements of the mill and ancillary facilities (and optimal use of gravitational fluid flows). In rugged mountainous terrain, the selection of a plant site can be difficult. It is sometimes necessary to consider a longer ore haulage distance to a more acceptable site. Construction of plants in rugged terrain generally requires a large volume of excavation, extensive retaining walls, deep foundations or piling in filled areas and crowding of buildings because of the space limitation. The choice of plant layouts is restricted. It is usually necessary to locate the plant and auxiliary buildings at different elevations to obtain sufficient area for the complete installation"[3]

The above paragraph was written with the whole of the metallurgical plant in mind but it is particularly true of the thickening stage in most solids/liquids separation facilities.

Available Space. For a green fields project, abundant space may be available; however plants that are excessively spread out can be problematical from a plant operator's perspective and contribute to operating inefficiencies.

Conversely, sites with little available space present other challenges to the layout engineer. An example of one of the most constrained sites for minerals processing plants would be the diamond recovery vessels operated by De Beers Marine Namibia and others. Vessels, such as the "Debmar

Atlantic", have the diamond separation plants located on board where space is obviously at a premium. Although solids/liquids separation is not used in the traditional sense, the heavy medium recovery circuits employ gravitational separators.

As mentioned above, the nature of the topography is likely to be a major determinant of the available space. Thickening systems are more likely to be affected simply because of the space requirements of large diameter thickener tanks.

Geotechnical. The large areas covered by many modern thickeners make the use and understanding of applicable geotechnical information a critical issue. This is especially true of the thickeners installed at the large copper concentrators located in earthquake prone regions, such as the Andean region. The results from the geotechnical analyses could influence the economics of the method of construction of the thickener tanks, i.e. steel construction compared with concrete wall construction or in-situ configurations using membrane liners and earthen berm walls.

As an example of the geotechnical influence on the design of a solid/liquid separation facility, the La Coipa operation in Chile utilizes tailings filtration to produce a filter cake that could be stacked as a method of tailings management. The high seismic potential of the area resulted in the following considerations:

- The instability of conventional tailings deposition was considered to be too risky in the immediate vicinity of the mine.
- Pumping of the tailings to a more stable deposition site was too expensive.
- The use of tailings filtration minimized the percolation of cyanide bearing solutions into the ground water.
- The use of tailings filtration also provided capabilities for increased recovery of both water, in an arid area of high evaporation rates and cyanide reagent for reuse and recirculation.

Plant operating experience at La Coipa indicates that the "recovery of cyanide pays for the filter plant operation".[6] The La Coipa operation was originally designed to handle 15 000 t/d and the current throughput is approximately 18 000 t/d. The tailings filtration is carried out using twelve 100 m^2 Delkor horizontal belt filters.

Water Availability. The supply of water is a fundamental requirement for most mineral processing operations. In areas where water is in short supply, the focus of a solids/liquids separation stage may be for water recovery rather than separation of the valuable constituents. Water recovery process often takes the form of tailings thickening, the choice of which may also be impacted upon by the method of tailings management.

The 10,000 t/d Mantos Blancos operation, in Chile, is an example where the cost and availability of water are such that tailings filtration for water reclamation is economical. "Tailings from flotation are cycloned, with the coarse plus a portion of thickened fines being filtered. Originally, disc filters were used, but were subsequently replaced with three 100 m^2 Delkor horizontal belt filters.

In addition to the economic benefits of the water reclamation, the tailings deposition area had a high subsurface salt content. Deposition of the wet or unfiltered tailings could have resulted in serious salt dissolution problems resulting in soil instability problems under the dam site."[6]

Environmental Issues. Although many large capacity concentrators now discharge flotation tailings directly to the tailings management facility, the use of tailings thickeners is also common. The diameter of these large thickeners can exceed 120 metres and therefore can present a significant visual impact.

The environmental issues associated with the storage and management of concentrator tailings are beyond the scope of this chapter; however prevention of breaching and containment of seepage are primary considerations.

The use of vacuum as a method of filtration requires the use of vacuum pumps. These units tend to be noisy and specific enclosures, e.g. brick buildings and sound proofing, are usually required to ensure that the vacuum pumps operate within the noise regulations.

Climatic Conditions. The drive to higher tonnage operations and single line facilities has resulted in an increase in the size of conventional and high rate thickeners. Most of these large thickeners are installed outside of buildings. However mining operations in cold climates such as Northern Canada require that thickeners need to be installed inside the concentrator building, both for operator protection and to prevent freezing of the process streams. An example of a cold weather installation is the EKATI™ diamond mine in Canada. The EKATI™ mine, which commenced operations in October 1998, is located in the Northwest Territories approximately 300 kilometres NNE of Yellowknife in Canada. Fine tailings, consisting of minus 0.65 millimetre material, is thickened in deep-bed compression ultra-high rate thickeners, supplied by Wren Technologies (Pty.) Ltd.[†]

"The advantage of this unit lies in its small footprint, high capacity per unit area and the fact that there is no raking mechanism and therefore no moving parts."

"Below is a list of the advantages and disadvantages of the Wren unit (E-Cat thickener) that were considered, by BHP Diamonds Inc., as most important for this application, as compared with those of other thickening and de-watering devices."

Potential advantages of the Wren unit (E-CAT) include:
- "high throughput rates
- low capital cost, mainly due to the simplicity of construction
- small floor area required

Potential disadvantages of the Wren unit (E-Cat) include:
- "considerable pumping capacity required to overcome the large static head, because of the large height to diameter ratio
- limited solids surge capacity; therefore, any problem at the thickeners or downstream requires a rapid shutdown of the solids feed to the thickener

A Wren pilot scale CAT was used at BHP's Koala bulk sampling plant, prior to the design and construction of the main EKATI™ process plant. Extensive testing of this pilot unit proved its suitability for the application. In this particular instance, the deep bed compression type thickener was the only feasible option due to the limited space."[4] Figure 1 below shows an indoor installation.

Figure 1 E-Cat™ Clarifier/Thickeners installed indoors for cold climate operation

---

[†] Wren Technologies (Pty.) Ltd. was acquired by Eimco Process Equipment Company. These deep-bed compression ultra-high rate thickeners are now called the "Eimco E-Cat thickeners."

Project Criteria

The overall project criteria will define the expected life of a potential mining project.

The life of a mining operation / mineral processing plant typically depends on the size of the deposit and the rate of mining, and the design life will affect the selection of solid/liquid separation equipment and facility arrangement.

Plant Expansion

"The history of the mining and metallurgical industry abounds with expansions of operations beyond the original capacity. This usually results from changes to the original design criteria such as "the mining plan" and "the final pit limits."[1] The space requirements of thickeners, particularly tailings thickeners, are often substantial. A clear understanding of the plant expansion possibilities is needed and this understanding should be incorporated into the Plant or Process Design Criteria so that potential future misunderstanding can be minimized.

Unless it has been defined as a prerequisite, the plant designer of the original plant cannot be expected to anticipate future expansion needs or provisions. However he should investigate the operator's vision, and strive for acceptable practicable means to allow for future flexibility.

"Generally speaking, the expansion of a metallurgical plant embraces two categories of equipment:
- that which can be increased in capacity by simply adding more units within existing space limitations
- that which is one of a kind and simply has to be enlarged or replaced by a larger machine with attendant waste of investment capital and loss of production"[3]

Process Systems

Another of the more important aspects of designing a plant layout is an understanding of the type of process system and the grouping of these process units. There are three types of process streams that are commonly found in the mineral processing industry

Batch Systems. Batch systems in the mineral processing industry tend to be confined to relatively small capacity, high value products. An example would be the use of plate and frame type filter presses for copper and other concentrates; these units can operate in a "fill, pressure squeeze, air blow and cake discharge" cycle. During this cycle, the filling of the filter may only occupy three minutes out of a twenty-minute cycle, i.e. filling only takes place for 15 percent of the time. As a result, sufficient storage of both the filter feed slurry and the product filter-cake must be provided if the filter is part of a continuous system.

Batch systems tend to require more frequent and a greater degree of sophisticated attention by the plant operator. Further these systems tend to have a greater degree of instrumentation associated with them.

Continuous Process Stream System. In this system, the layout follows the process flowsheet. Many mineral processing plants are designed around this concept. The pipe routes tend to be minimized with this layout.

Functional Grouping System. In this system, similar process functions are grouped together. This system is common in the chemical engineering industry.

An example of this continuous functional grouping system would be the grouping of filters with different duties within a single filtration building. The piping routes tend to be longer in this system but the ancillary services such as the reticulation for vacuum and compressed air systems would be more localized.

MICRO FACTORS AFFECTING SOLIDS/LIQUIDS SEPARATION SYSTEMS

Overview of Micro Factors
The micro factors that influence the design of a solids/liquids separation facility can be divided as shown in Table 2.

Table 2 Micro Factors

FEED	Feed Streams
DISCHARGE	Solids Discharge
	Liquid Discharge
SAFETY ISSUES	Safety in Operation
	Safety in Construction
OTHER ISSUES	Process Considerations
	Plant Height
	Operability
	Maintainability
	Construction Considerations

MICRO FACTORS AFFECTING THICKENING SYSTEMS

Introduction
The two key aspects of the micro factors that affect solids/liquids separation facilities are:
- Input or feed streams
- Output streams covering both the solids and the liquids discharge

These three streams, for the thickening stage of solids separation are usually in the form of slurries.

Feed Streams
The feed stream to a thickening stage is either by gravity flow or from the discharge of a pumping stage. Thickeners are often fed by gravity when the feed stream is the overflow from a cycloning stage in a milling circuit as the cyclones are usually positioned at a height sufficient to permit gravity flow. Similarly, the tailings stream from a flotation plant often utilizes gravity flow to the tailings thickeners. As an operating precaution, the gravity flow from the flotation cells allows the flotation cells to be partially drained in the event of a loss of power or control. The cell discharge valves would operate in the "fail–open" mode. This arrangement allows some of the slurry volume, held within the cells, to report to the tailings thickener and thus preventing the cells from sanding up. In the case of the very large copper concentrators, a modified control system could be adopted to reduce the loss of un-floated valuable constituent to tailings and resulting loss of revenue.

In the case of the ultra high rate thickeners, where the height to diameter aspect ratio is high, the thickeners usually need to be pump fed. This additional pumping stage would add to both the capital and the operating costs.

Whatever the method of feeding, the incoming slurry to a thickener should enter the feed well in a manner which minimizes turbulence and air entrainment. The Fitch type of feed well with its opposing "race-ways is an example of a split feed launder design that has been used for conventional thickeners.

If the slurry is being pumped, the problem of air entrainment can be greatly reduced by pumping from a level-controlled sump so that the pump does not aspirate air when the slurry level in the sump drops to the level of the pump suction.

Discharge Streams

"Underflow pumping has four basic arrangements: (1) The underflow pump at the tank perimeter with buried piping from the discharge cone, (2) the underflow pump at the tank perimeter or directly under the discharge cone in a tunnel or elevated tank, (3) the underflow pump at the tank perimeter with a peripheral discharge from the tank sidewall and (4) the underflow pump in the center of the tank or at the perimeter with center column piping"[5]

"A high percentage, if not the majority, of thickener operating problems can be accounted for by deficiencies in the underflow removal system. As a general rule, the closer the pump or throttling system, such as an orifice or control valve, is to the bottom outlet of the thickener or clarifier, the better. Exceptions can be made for underflow slurries that are relatively fluid and have little tendency for sand separation such as magnesium hydroxide and waste treatment suspensions. Also some fine-size mineral applications in which very low flocculant dosages are utilized; these dosages would be insufficient to affect the slurry viscosity."[7]

For the majority of mineral applications, speed regulated centrifugal pumps provide adequate control for the underflow pumping system. The high suction head on the underflow pump means that gland seal water is usually needed on a centrifugal pump. To prevent excessive dilution of concentrate slurries that are pumped to filtration stages, the use of variable speed peristaltic pumps, such as the Bredel™ pump, have found increasing acceptance; further they are able to handle dense slurries. The use of pulsation dampers on the discharge piping is required with this type of pump and should be regarded as a critical design issue.

The use of gravity discharge for thickener underflows is not uncommon in the mineral processing industry. Those systems, which involve large tonnages, generally work best since larger orifices can be used and plugging from tramp oversize is less likely. This gravity discharge approach can be used where the disposal of the slurry, usually tailings, is downhill from the thickener. Southern Peru Copper Company's (SPCC) Toquepala concentrator has three peripheral/traction driven 100 metre diameter tailings thickeners and the underflow from these gravitates to the disposal area.

Safety Issues

There is always the possibility of someone falling into a thickener and drowning (or poisoning in the case of cyanide solutions). Life buoys should be located along the length of the thickener bridge. If there is operator access to the thickener overflow weir, life buoys should also be placed here. Although not strictly a personnel safety issue but rather more of an equipment interruption issue, hard hats worn by the operators have a tendency of falling into the thickener and eventually causing plugging problems in the thickener underflow system. In order to reduce the potential of the hard hats falling into thickeners, operators and maintenance staff working on or near thickeners are typically required to remove their hard hats and place them in a box located near the edge of the thickener; as an alternative hard hats can be equipped with chin straps.

As part of a HAZOP analysis of the thickener installation, the possibility of a failure in the underflow piping must to be considered. The risk of failure in the underflow pumping system is greater if the pumping system is located under the center of the tank compared with the pumps located at the periphery. However, the pump suction piping may be more prone to plugging if the pumps are located at the periphery. With an elevated thickener, the spillage from the underflow system is likely to be collected within a bunded area. However, if the thickener is installed at ground level and the underflow system is installed beneath the thickener, the possibility exists of flooding the underflow pumping area and the access tunnel. The tunnel could be sloped towards the center so that personnel move to shallower ground on leaving the tunnel; obviously there would have to be a significant spillage handling system located at the center. Alternatively, the tunnel could be sloped away from the center, in which case the spillage system would be at the periphery. In either case, and especially for large thickeners a personnel emergency escape tunnel should be considered.

Other Issues

Process Issues. The use of Counter Current Decantation (CCD) circuits is not confined to the recovery of dissolved valuable components. Newmont's Batu Hijau operation in Indonesia uses seawater as the flotation medium. "The final product, copper-gold concentrate containing 30 percent to 33 percent copper, is washed in a counter-current decantation system to remove chloride before being pumped as a slurry to the filter plant"[8]

The chemical composition of the liquid phase must be fully understood. In acid base metal recovery process, the liquid phase contains varying concentrations of sulfuric acid. In the potash industry, the liquid phase may contain varying concentrations of salts and especially the chloride ion. Obviously the chemical composition of the liquid phase will impact the selection of materials of construction with which the fluid comes in direct contact. However, care must be taken to understand the impacts on the materials with which there is no direct and intentional contact. For example, the acid mist generated in the filtration sections of some base metal refineries requires that the building structural steel and cladding be suitably protected. In addition to structural steel protection, the chloride environment requires that consideration be given to the protection of the outside of piping, electric motors and concrete.

Construction Considerations. A number of recent thickeners have been built using the compacted bed approach with either a concrete thickener floor or a plastic membrane covered with sand being placed upon the compacted bed. There have been a number of reported failures of the compacted beds; these include the Hartley Platinum Project in Zimbabwe, and the Murrin Murrin nickel laterite project in Australia. Whilst the specific reasons for each failure are not known, these failures would suggest insufficient quality control in either or both the design and construction stages.

MICRO FACTORS AFFECTING FILTRATION SYSTEMS

Introduction

The two key aspects of the micro factors that affect solids/liquids separation facilities are:
- Input or feed streams
- Output streams covering both the solids and the liquids discharge

The solids discharge stream for the filtration stage of solids/liquids separation is usually in the form of a filter cake containing 5 percent to 20 percent moisture depending on the characteristics of the material being filtered.

"Although modern automatic pressure filters have short cycle times, they are nevertheless batch units. Filter feed and cake discharge is intermittent. This must be taken into account when placing pressure filters into a flowsheet for a continuous process such as a flotation plant. Adequate upstream and downstream surge capacity must be provided. It is critically important to understand the stages of the pressure filtration cycle, and the relationship between average and instantaneous flows."[9]

Feed Streams

A filter feed surge tank is needed to provide buffer capacity because the thickener underflow is a continuous stream but the feed to the filter is intermittent. The size of the buffer storage depends on the type of filter selected and the number of operating filters; however good practice is to have a minimum of one hour's storage at maximum production rates.

Manufacturers of filtration equipment are recommending that trash screens be installed prior to the filter feed surge tanks. "All process streams contain trash, which can block pipelines and filter plate feed ports. The source of trash can be a mystery, and the type of trash varies from stones and hardened slurry to rubber gloves and cable ties. Trash removal is essential for good operation and to prevent blocked pipelines or filter feed ports that could damage equipment. A self-cleaning screen discharging trash to a container outside the bund is more reliable than a manually emptied screen basket."[9] Delkor linear screens are becoming increasingly accepted as trash screens ahead of both thickener and filter facilities.

The design of the filter feed system is critical to the successful operation of a filter. Some filter suppliers prefer to be responsible for the filter feed system and therefore they can include a smaller intermediate feed tank and the feed pumps within their supply.

"For most mining and metallurgical applications, the filter feed pump runs only during filter feeding, and is started and stopped under control from the filter's PLC. In special cases, the feed pump runs continuously, with a recycle (at lower pump speed) to the feed tank during other phases of the filtration cycle. A pressure filter is a batch unit and the filling or filtration stage usually takes only 20 percent of the total cycle time. As a consequence, the instantaneous flow rates during filtration are quite high. Centrifugal slurry pumps with gland service water are the standard choice. Gland seal arrangements are best determined by the pump manufacturer, but they need to account for the start/stop operation of the pump and the wide range of operating conditions. Where full flow water flushing of the gland is provided it can cause considerable slurry dilution between filling cycles, and solenoid control of gland water, or reduced flow sealing systems should be considered.

"The pump runs against increasing head as the filter cake builds (Figure 2), and an understanding of the system curve is important for correct pump sizing. Once the chamber is about 50 percent full, the pressure will rise as the cake resistance increases, causing the operating point to move back up the pump curve. At the end of the filtration stage, the pump will be operating at close to full pressure and the flow will have decreased to 10 to 40 percent of the initial flow rate. Where slurries are known to have a highly abrasive nature, the maximum flow rate may have to be lower. Where pump selection shows that flows greater than the recommended maximum are possible, a variable speed drive pump should be used. Pump speed may then be increased as system head increases. To calculate the flow, the frictional losses for the pipeline and static head need to be known. The resultant curve should then be drawn onto the pump curve for the proposed pump. The intersection of the system curve with the pump curve at the proposed speed will then determine the maximum feed flow rate. Where large feed tanks are used, the flow rate should be checked with full and empty tank levels. With flat system curves there can be a significant change in flow with changes in tank level. Similarly, where there are significant changes in feed density, the flow with minimum and maximum densities should be checked. Once the maximum flow rate has been determined the Net Positive Suction Head (NPSH) required should be checked against the NPSH available with the proposed suction pipe and tank design."[9]

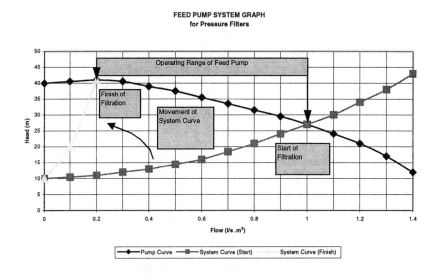

Figure 2 Filter feed pump and system curves (modified from Townsend)

Discharge Streams

Solids Discharge Streams. An important criterion for the design of a filter plant is the arrangement for handling of the solids product (filter cake) from a filtration stage. The height of installation of the filters is often determined by the handling/transfer arrangement of the filter cake. Filters can be located directly above stockpiles and thus the discharge would be by gravity directly onto the stockpile. Although this would minimize the initial handling of the filter cake, there would be two distinct disadvantages. Firstly, the elevation of the filters would likely be higher than if the product was transferred away from the filter; as a result both the capital and operating costs of the installation would probably be higher. Secondly the control of spillage and hose-down water would be difficult; it is not good practice to allow the spillage to fall down onto the stockpile. In general, the filter cake should be transported from the filter and the surrounding wet area into a dry area.

The method of discharge of the filter must be understood prior to the design of the transport system; the instantaneous discharge rate can be significantly higher than the average hourly rate. A horizontal belt filter discharges the cake continuously and therefore at a rate very close to the average hourly rate. A plate and frame filter generally drops the filter cakes one at a time throughout the discharge period; therefore the instantaneous discharge rate is not as great as for a Larox type filter whose discharge period is usually between 10 and 30 seconds per cycle.

Filtrate Streams. As with the solids product, the type of filter determines the approach to the collection of the filtrate and especially whether the filtrate flow is continuous or batch. The filtrate is usually at atmospheric conditions for pressure filters and can be gravitated; however for rotary drum and horizontal belt filters, the filtrate needs to be separated from the vacuum system.

Vacuum systems require the use of filtrate receivers, barometric legs and moisture traps. These units should be installed in a manner that minimizes the pressure loss between the filter and the vacuum pumps. Excessive pressure losses result in energy wasted in overcoming these losses. The minimization of pressure losses, in the form of flow restrictions, between the filter and the receivers, will reduce wear caused by filtrate containing abrasive solids. "The line to the vacuum pump should also be designed with the same objective and if a velocity of approximately 70 feet per second is used in selecting the pipe size, the pressure drop will be minimized in the line. With dry vacuum pump installations it is essential to include a moisture trap between the receiver and the pump; this is possibly followed by a scrubber if a corrosive solution is being filtered. The height of the moisture trap is not important as long as the discharge line, which will act as the barometric leg, is of a length greater than needed to sustain a column of water under full vacuum, typically a height close to 35 feet. The barometric leg should be sized to handle the total filtrate volume at a line velocity not greater than 5 feet per second. Water sealed vacuum pumps generally will not require moisture traps unless the solution is corrosive or valuable, or contains free lime. Filtrate pumps must be selected with NPSH characteristics that match the expected operating conditions."[7]

The Krogh pump is a good example of a filtrate pump that mounts directly onto the filtrate receiver, thus doing away with suction piping and balance legs.

The batch nature of pressure filters means that, like the solids, the filtrate flow will vary from a maximum at the start of slurry feeding to almost zero at the end of the feed cycle. If an air blow step is included, there is likely to be very little filtrate entrained in the exhaust air stream. Thus the filtrate collection system needs to be sized for a range of conditions and needs to recover the fluid from what can be a very high airflow. At the start of the feed cycle, the filtrate can contain a significant quantity of suspended solids; as a result the filtrate is usually returned to a thickening step for the recovery of the solids. If the thickening stage is omitted, the implications and quantity of the suspended solids must be understood prior to selecting the discharge point of the filtrate.

Safety Issues

"As a result of the new labor reducing designs and new types of filter cloth developed by the cloth manufacturers, the pressure recessed plate filters are being used in applications that have traditionally been processed on vacuum disc or drum filters."[10] The improvements in the design of both the filter cloth and especially the recessed plates has lead to increases in both the pressure used for closing the filters and the slurry inlet feed pressures. The filter manufacturers have placed an increased emphasis on the safety aspects of the filter operation. The use of light curtains, which stop the closing of the presses, have become common. Filter cake removal systems and cloth washing, in both the vertical type of filter and the horizontal suspended recessed plate filter, have ensured that good cloth seals are obtained upon filter closing. Thus the chances of pressurized slurry streams squirting from between the plates have been reduced.

Other Issues

Process Considerations. The selection of filtration for a solids/liquids separation stage can also be influenced by process considerations. Twenty-six 83 m^2 Delkor horizontal belt filters were installed at the Nchanga Copper Tailings Leach Plant, in Zambia. "The major process consideration was whether to recover the copper solutions produced in the acid leaching process with either:
- conventional counter current decantation (CCD) thickeners, or
- large horizontal belt vacuum filters recently applied to other high-tonnage mining applications.

The principal advantage of the filters was that they recover valuable solutions from unwanted solids in one step at greater efficiency and at a lower capital cost than a train of say five thickeners, each 76 metres in diameter."[11]

Plant Height. The discharge arrangement for the solid product, as mentioned above, is often the determining factor for the height at which filters can be installed. Both capital and operating costs are increased with increasing height of installation.

Operability. " Pressure recessed plate type filters have undergone extensive redesign to minimize operator attention. Almost all of these pressure filters have automatic opening and closing mechanisms to either open all of the plates at the same time, or to index and open 1 to 6 plates at the same time."[10] The increased use of reliable automation has improved the operability of filtration equipment, especially pressure filters.

The increased the use of automation has enabled filter systems to operate unattended. However, it is good practice to locate them in an accessible and frequently visited part of the plant to ensure that regular checks and routine maintenance are carried out.

Maintainability. In order to provide an efficient maintenance environment, there should be adequate clear space around the filter and the building should allow for the safe lifting of components from the ground floor to the filter. In most installations, an overhead crane is required above the filter. The crane's access to the filter should not be impeded by filter feed pipes or other process connections.

LAYOUT METHODOLOGY

"There is no single technique leading to the best arrangement in any layout problem; several stages may be required with different techniques appropriate to each.

The development of a plant layout is an interactive process between all factors both macro and micro. The fact that plant layout problems are 3-dimensional in nature is initially the main reason for the interactive process. During the detailed design phase, additional information now available may require re-evaluation of some earlier constraints. As a result, re-examination of earlier alternatives may be required."[2]

"There are three basic principles of layout planning:
- <u>Plan the whole then the detail</u>: Individual aspects must be subservient to the whole and sub-optimization avoided.

- Plan the ideal and from it the practical: The ideal is free from restrictions and gives a datum, the cost of departing from which can be set against the advantages to be gained.
- Plan more than one layout: It is seldom that a single layout is "best" for each criterion. Planning more than one permits comparisons and leads to greater confidence in making the final selection."[12]

CONCLUSIONS

This chapter has, by means of a variety of examples, illustrated the range of factors that need to be reviewed when considering the plant design and layout of solids/liquid separation facilities.

The relative importance of each of these factors is very much situation dependent. The diversity of solids/liquid separation situations faced by mineral processing and design engineers is large and therefore this chapter has deliberately not focused on one particular aspect.

The factors highlighted in this chapter can therefore be used as a checklist to ensure that none are overlooked in the study and subsequent design stages.

REFERENCES

1 House, F.F. "An Engineer's Guide to Process-Plant Layout" Chemical Engineering 120 (July 28, 1969)
2 Blois, M.D.S. and Talocchino, L. "Plant Siting and Layout – A Metallurgist's Perspective". South African Institute of Mining and Metallurgy, School of Metallurgical Process Design in the 90's. 1993
3 Weiss, N.L. "SME Mineral Processing Handbook". SME of the American Institute of Mining, Metallurgical and Petroleum Engineers Inc. New York (1985)
4 Mohns, C.A. and Paradis, T.G. "Deep Bed Thickener Operation at the EKATI™ Diamond Mine", Paste Technology for Thickened Tailings Symposium November 1999. School of Mining and Petroleum Engineering, University of Alberta (1999).
5 King, D.L. and Baczek, F.A. "Characteristics of Sedimentation-Based Equipment" Design and Installation of Concentration and Dewatering Circuits SME of the American Institute of Mining, Metallurgical and Petroleum Engineers Inc. New York (1986)
6 Minson, D.N. and Williams, C.E. "Filtering Systems for Dry Tailings Deposition" Canadian Institute of Metallurgy, 38th Annual Conference of Metallurgists. Quebec (1999)
7 Emmett, R. "Private Communication" September 7th, 2001
8 DeMull, T.J., Spenceley, J. and Hickey, P. "Planning and Teamwork Lead to Successful Start-Up at Batu Hijau" SME Annual Meeting 2001. SME of the American Institute of Mining, Metallurgical and Petroleum Engineers Inc New York (2001)
9 Townsend, I. "Plant Design Considerations for Automatic Pressure Filtration of Flotation Concentrates" XXV Convencion de Ingenieros de Minas del Peru. Arequipa, Peru (2001)
10 Moos, S.M. and Klepper, R.P. "Selection and Sizing of Non-Sedimentation Equipment" Design and Installation of Concentration and Dewatering Circuits SME of the American Institute of Mining, Metallurgical and Petroleum Engineers Inc. New York (1986)
11 Hampsheir, P.R. "Design and Construction of the Nchanga Copper Tailings Leach Plant Stage 3", The Institution of Mechanical Engineers Volume 200 No. 131 (1986)
12 Mecklenburgh, J.C. Plant Layout "A Guide to the Layout of Process Plant and Sites." The Institution of Chemical Engineers, London (1973)

10

Pumping, Material Transport, Drying, and Storage

Section Co-Editors:
Ken Boyd and William E. Norquist

Selection and Sizing of Slurry Pumps
M.J. Bootle **1373**

Selection and Sizing of Slurry Lines, Pumpboxes, and Launders
B. Abulnaga, K. Major, P. Wells **1403**

Slurry Pipeline Transportation
B.L. Ricks **1422**

The Selection and Sizing of Conveyors, Stackers, and Reclaimers
G. Barfoot, D. Bennett, M. Col **1446**

Selection and Sizing of Concentrate Drying, Handling, and Storage Equipment
M.E. Prokesch, G. Graber **1463**

The Selection and Sizing of Bins, Hopper Outlets, and Feeders
J. Carson, T. Holmes **1478**

Selection and Sizing of Slurry Pumps

Michael J. Bootle[1]

ABSTRACT

The mineral processing industry utilizes centrifugal slurry pumps in a wide range of key applications, including hydrotransport, grinding circuits, flotation circuits, thickening, and tailings disposal. Optimum system operation is dependent upon a large number of critical factors. Foremost, a detailed knowledge of the solid and slurry properties is required for sump and piping design, in order to prevent settling of the solids and to minimize air entrainment, as well as, to determine the flow, friction losses, head and pressure requirements of the pump. This same knowledge of slurry properties is also used to properly select a pump with the appropriate combination of geometry and materials to yield the best balance of the often conflicting pump priorities of stable operation, maximum wear life, and minimal energy consumption. These factors and choices will be discussed along with methods of applying the appropriate slurry corrections to the final pump selection.

INTRODUCTION/OBJECTIVE

The objective of this chapter is to provide the reader with a basic introduction to the critical factors, which need to be considered to create a successful solids handling pumping system. It is hoped that upon completion of reading this chapter, the reader will have a better understanding of the following:

- The basic construction features of a centrifugal slurry pump.
- The advantages of various impeller and casing designs.
- The benefits and limitations of elastomers and metals for wear components.
- A basic understanding of pump performance curves, system curves and their interdependent relationship with respect to point of operation, i.e. flow and head.
- The difference between settling and non-settling slurries, viscous Newtonian and viscous non-Newtonian slurries.
- The effect of each of the above slurries on pump performance and how to apply the appropriate corrections to account for these effects.
- The importance of determining the settling velocity of the solids in the slurry and a method of calculating settling velocity for a heterogeneous slurry.
- Which of the above factors will most likely be involved in the various applications within the mineral processing industry and what pump type and appropriate correction factors apply.

In an attempt to accomplish all of the above in one short chapter, it is obvious that for most of the above topics only the basic details will be discussed. It is the writer's deliberate intent to give preference to an overview of the system design process and very basic, but important potential pitfalls to be avoided at the expense of excessive detail. When appropriate, references will be given to allow for additional research, if and when required.

---

[1]Weir Slurry Group, Inc., Madison, Wisconsin.

Copyright © Weir Slurry Group, Inc. 2002

PUMP CONSTRUCTION

General

The design of a horizontal centrifugal slurry pump is a balance of design considerations to best meet the requirements of a particular slurry duty. These requirements may include one or more of the following:

- The ability to pump high density abrasive slurries with adequate wear life.
- The ability to pass large diameter solids.
- The ability to handle air entrained and/or viscous fluids with reliability and minimal performance corrections.

When compared with clear liquid pumps, the above requirements often result in the slurry pump being larger than its clear liquid counterpart and sacrificing maximum efficiency in exchange for the ability to achieve the above goals.

Within a slurry pump it is expected that the components which come into contact with the abrasive slurry will wear. It will be shown that minimizing wear is done through appropriate pump design, proper material selection and proper pump application.

Figure 1a illustrates a typical unlined horizontal centrifugal slurry pump and Figure 1b illustrates a fully lined version of the same pump. The corresponding wear components for each have been appropriately marked. To maximize wear life, thick casting sections are provided on the impeller, the casing of the unlined pump, and the wear liners of the fully lined pump. Often, the unlined casing casting thickness is greater than fully lined version casing liners. This additional thickness requirement is due to a greater need for the unsupported, unlined casing to safely handle the internal pressure of the pump with an adequate factor of safety to account for wear. The lined pump also allows for the use of a wider variety of materials, such as elastomer liners, which often outperform metal in fine particle and corrosive applications. Therefore, while the unlined pump may offer the lower initial capital cost, the lined version allows for a greater number of material choices, which may have longer wear life and lower replacement spares cost. The lined design is also inherently safer from a pressure containment standpoint.

Large clearances are provided within the impeller and casing to allow for the passage of large diameter solids, while also reducing internal velocities and corresponding wear.

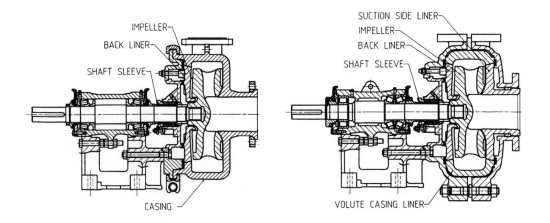

Figure 1a: Wear components on an Unlined Centrifugal Slurry Pump.

Figure 1b: Wear Components on a Fully Lined Centrifugal Slurry Pump.

Slurry pump impellers tend to be larger than their clear liquid counterparts. To minimize speed and maximize wear life of both the impeller and suction side liner, rarely are slurry pump impellers trimmed in diameter to meet the duty point. On low horsepower applications (below 300 kW, 400 hp), belt drives are the most popular means of achieving the required speed for the duty point. Belt drives are inherently quiet and also allow for relatively easy speed changes if required. For higher horsepower applications, gearboxes are commonly used to meet the desired speed and duty point. On applications where variable flow is required, variable frequency drives are used to provide the necessary continual speed changes. Most often these VFDs operate with other speed reduction devices, such as belt drives and gearboxes, to allow for the use of higher speed, lower cost motors.

Bearing Assembly
The bearing assembly of a heavy-duty slurry pump should incorporate larger diameter shafting and bearings than those found on a clear liquid pump. Often, roller bearings are used to provide further additional load carrying capacity to handle the added forces associated with the greater specific gravity liquid, the inherent imbalance due to wear, the shock loading from large particles and the hydraulic imbalance from air entrained liquids. On belt drive applications, the drive end bearing handles the majority of the side load due to belt pull, which can be substantial. Most properly sized and properly tensioned v-drives exert approximately 9 KN of belt pull for every 100 kW of motor power (1500 lbf for every 100 Bhp).

Figure 1a illustrates a bearing assembly with identical low angle, angular contact bearings on both the pump end and drive end. The low angle bearing is well suited to the radial loads from the impeller or belt pull from the drive (if so equipped).

Figure 1b illustrates a higher capacity bearing assembly, which fits within the same packaging dimensions. This assembly features a duplex angular contact roller bearing on the pump end and a cylindrical roller bearing on the drive end. The duplex angular contact bearing has a higher bearing angle than the bearings shown in Figure 1a, making it more suitable to handle higher axial loads, such as those seen with smooth backed impellers, open faced impellers and series pumping applications. In this configuration, the cylindrical roller bearing on the drive end handles none of the axial load, but has tremendous radial load capability, making it well suited to belt drive applications.

Grease or oil are both suitable means of providing lubrication to the bearings. Grease has the advantage of providing greater contamination protection. Oil has the advantage of higher speed capability and is easier to change if it becomes contaminated. Oil lubrication also provides the opportunity to install a bearing cooling system should it be judged necessary. With remote installations, care must be taken with oil lubrication to insure level mounting.

When the shaft seal design allows it, shorter shafts with reduced impeller overhangs result in reduced bearing loads, shaft stress and deflection through the seal area. When packing is used as a shaft seal, a hardened and/or ceramic coated shaft sleeve is recommended to prevent shaft wear.

Impeller Design
As mentioned above, slurry pumps typically have impellers that are larger than their clear liquid counterparts. This is to lower the impeller speed required to achieve a given head and to provide more material for wear purposes. For high wear applications closed impellers are preferred. In coarse particle applications, expelling vanes are recommended on the face of the front shroud. The purpose of these expelling vanes is to prevent large particles from becoming trapped between the impeller and suction side liner and to minimize recirculation. The benefit is a reduction in gouging and recirculation wear, at the expense of two to three percentage points of efficiency.

Expelling vanes are also often used on the back shroud of the impeller in coarse particle applications to prevent the trapping of large particles between the impeller and back liner. In this location they also serve to reduce the forward axial load (improving bearing life) by lowering the pressure acting on the back shroud and beneficially reducing the pressure at the hub and packing. This reduces the pressure differential at the shaft seal and reduces the tendency for slurry leakage from the pump. As with expelling vanes on the front shroud, "backvanes" usually absorb two to three percentage points of efficiency.

To combat wear and allow for the passage of large diameter solids, heavy-duty slurry pump impellers feature thicker main pumping vanes and fewer of them. Both of these factors further contribute to a reduction in efficiency when compared with a clear liquid counterpart. While a clear liquid impeller usually has five to nine vanes, most slurry pump impellers have two to five, with four and five vane designs being the most common. Two and three vane designs are usually reserved for very large particle passing, as required in dredging applications.

Figure 2 illustrates the difference between a four vane heavy duty slurry design with front and back expelling vanes and a six vane high efficiency slurry design with smooth front and back shrouds. Note the short "blocky" vanes on the heavy-duty design and the thin long length, long wrap vanes on the high efficiency design.

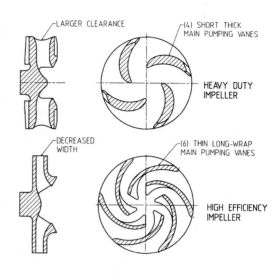

Figure 2: Differences between heavy duty and high efficiency style impellers.

Work has been done which shows that on large particle applications, the large solids follow a different path from the fluid due to the inertial effect of the heavy solids (reference 1). This is illustrated in Figure 3. This results in gouging wear at any location where the fluid is required to make an abrupt change in direction. For this reason, on large particle applications, heavy duty designs with blunt leading edges, wide between shroud spacing, and thick back shrouds are recommended to combat the impact of the large particles. Short main pumping vane lengths with minimal vane overlap allow the large particles to pass unimpeded.

This heavy-duty geometry results in a head versus capacity curve that is flatter than the typical clear liquid pump. Figure 4 shows a typical performance curve of a four vane heavy-duty impeller for a 10 inch suction, 8 inch discharge horizontal slurry pump. The total combination of fewer, thick, short main pumping vanes, combined with expelling vanes on the front and back shroud can result in slurry pump efficiencies, which are as much as ten percentage points lower than a comparable clear liquid impeller. These differences are minimized on larger pumps.

On fine particle service ($d_{85} < 100$ microns) it has been shown that the particles follow the fluid path (Figure 3). In these instances, high efficiency designs, which minimize the presence of vortices, have been shown to not only improve efficiency, but also improve impeller life. In these fine solid particle applications, the lack of expelling vanes on the front and back shroud has also been shown to improve side liner wear life.

Figure 5 illustrates the performance for a four vane high efficiency design, similar to the six vane design illustrated in Figure 2. This impeller is an alternative impeller for the 10/8 slurry pump with heavy-duty 4 vane performance illustrated in Figure 4. Note the steeper head versus capacity curve for the high efficiency design with longer wrap vanes and lower impeller vane exit angle. Also note higher efficiency (81% versus 75%).

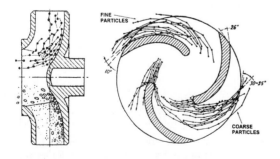

Figure 3: Fine and coarse particle trajectories in an impeller.

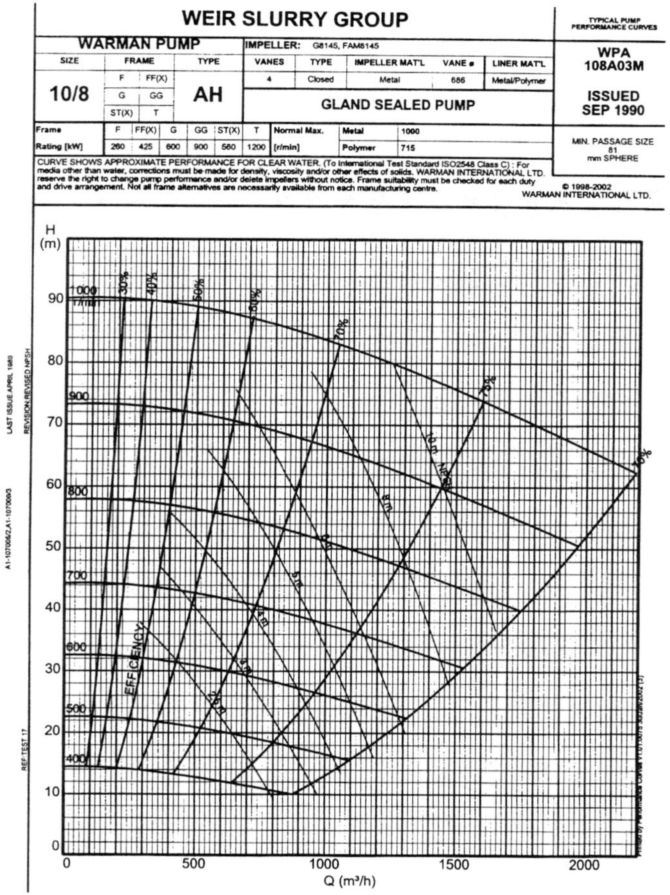

Figure 4: Performance curve for 10/8 slurry pump with heavy-duty 4 vane impeller.

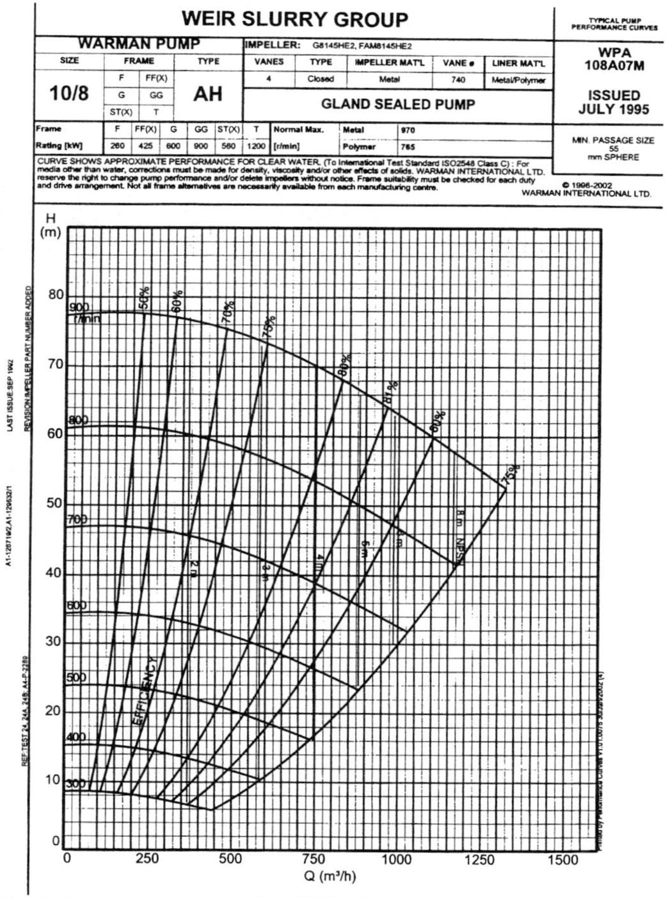

Figure 5: Performance curve for 10/8 slurry pump with high efficiency 4 vane impeller.

Specific Speed

Specific speed is a dimensionless number which defines the relative proportions of the impeller, and pump in general, in terms of its hydraulic performance at the best efficiency point (BEP) for any given speed. The equation for specific speed, N_s, is:

$$N_s = \frac{N\sqrt{Q}}{H^{3/4}}$$

where: N = pump speed in rpm,
Q = capacity in m3/sec at BEP,
H = total head per stage in meters at BEP.

For the heavy duty impeller with performance illustrated on curve WPA108A03M in figure 4, the N_s is calculated as:

$$N_s = \frac{1000 rpm \sqrt{(1600 m^3/hr)/(3600 \sec/hr)}}{(74m)^{3/4}} = 26.4$$

Note: 1000 rpm was arbitrarily chosen as the reference speed, but similar values would have been obtained using BEP performance at other speeds.

Figure 6 (reference 2) illustrates the relationship between specific speed and impeller geometry. Slurry pumps typically have specific speeds in the range of 15 to 40 for N_s derived from terms of m3/sec, m head and rpm.

As of the time of this writing, in North America, specific speed is still mainly derived from units of USGPM flow, feet of head and rpm, resulting in slurry pumps with specific speeds in the range of 750 to 2000 N_s, using these terms. The conversion from N_s in terms of m3/sec and m head to N_s in terms of gpm and ft of head is a multiplier of 51.67.

As the equation for N_s incorporates BEP flow and head conditions (in the numerator and denominator, respectively), it is clear that high specific speed pumps are better suited to high flow, low head applications, while low specific speed pumps are better suited to low flow, high head applications. The impeller geometry illustrated in Figure 6 supports this idea, as flow is proportional to inlet area and head is proportional to impeller diameter (squared) for any given speed. Thus, as illustrated, the low specific speed pumps have larger outlet diameter to inlet diameter ratios.

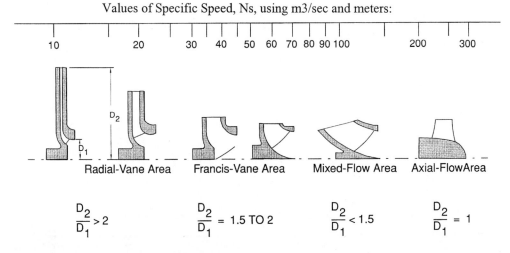

Figure 6: Impeller geometry and its effect on specific speed (courtesy of Hydraulic Institute).

Maximum efficiency occurs at N_s of 42. Nevertheless, from a wear standpoint it may not be desirable select a pump with this high a specific speed, as higher specific speed pumps (with smaller diameter impellers) are required to operate at higher speeds to achieve any given head. Similarly, low specific speed pumps are not always desirable due to their lower efficiency and poor ability to pass large diameter solids.

Pumps with N_s of 20 to 25 are recommended for severe duty applications (reference 3). For medium to heavy-duty applications an N_s of 25 to 33 is recommended. Experience has shown pumps with Ns of 42 can be safely used on medium duty applications with heads of 30 meters or less.

Casing Design
The three basic casing design types used in a centrifugal slurry pump are the true volute, near or semi-volute, and the circular volute. These three volute types are illustrated in Figure 7a, 7b, and 7c, respectively (reference 4):

The true volute, Figure 7a, has the highest efficiency due to its tight impeller to cutwater clearance. The cutwater is the "V" shaped diverter/flow splitter between volute and discharge. This tight clearance increases efficiency by minimizing recirculating flow at the best efficiency point. At flows below the best efficiency point, however, the increased recirculating flow has a high velocity through the small impeller to cutwater area. This results in high wear at low flows for this design in the area at and just beyond the cutwater. This design also has highest hydraulic radial loading at flows other than the BEP.

The near volute, Figure 7b, is similar to the true volute, only the impeller to cutwater clearance has been increased to reduce velocity at the cutwater area at lower flows. This design results in significantly reduced wear at this area at low flows. This design also has significantly lower hydraulic radial force at conditions away from the BEP. For most slurry applications, this design offers a good balance of wear life and efficiency.

The circular volute, Figure 7c, has uniform area between the impeller and casing all around the volute. This volute design has the lowest hydraulic radial force at conditions off the BEP. This makes it ideal for high head, low flow applications.

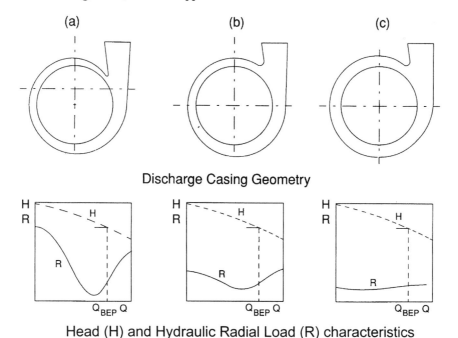

Figure 7: Slurry Pump Casing Types.

Normal recommended operating conditions for the three volute types are listed below:

- True volute: 80 to 120 percent of BEP flow
- Near volute: 60 to 110 percent of BEP flow
- Circular volute: 40 to 100 percent of BEP flow

For severe duty slurries, one would like to select a pump with an operating point in the middle of the above ranges to minimize volute wear. For relatively light slurries, the above ranges can be widened.

MATERIALS

General

The primary materials used for wear components in centrifugal slurry pumps are hard metals, elastomers, and to a lesser extent ceramics. In very simple layman's terms, hard metals and ceramics combat erosion due to their high hardness values. Elastomers combat erosion by their ability to absorb the energy of the impacting particle(s) due to their resilience and tear resistance.

Elastomers generally outperform hard metals, in terms of erosion resistance, in those applications where particle size is smaller than 250 microns, impeller tip speed is within the limits of the elastomer, and there is no risk of large particle "tramp" damage.

Metals

The three basic types of metals used to combat erosion in centrifugal pumps fall under ASTM A532 (class I, II, and III). These are the Martensitic White Irons (class I), the Chromium-Molybdenum White Irons (class II), and the High Chrome Irons (class III). These materials consist of hard carbides within a supporting ferrous matrix. The two types of carbides found within these materials and their approximate Vickers hardness range are listed below:

- Iron Carbide   850 to 1000 HV
- Chromium Carbide (Eutectic $(Fe,Cr)_7C_3$)   1200 to 1500 HV

The matrix types and corresponding hardness ranges are listed below:

- Ferrite   150 to 250 HV
- Austenite   300 to 500 HV
- Martensite   500 to 1000 HV

The bulk hardness of the material is dependent not only upon the carbide and matrix type, but also upon the volume of the carbides within the matrix. For medium to large particle applications the bulk (combined) material hardness is of primary importance. For small particle applications, a fine microstructure, with smaller intercarbide spacing, is more important to minimize erosion of the softer matrix. For very large particle applications (greater than 4 inch diameter) fracture toughness of the matrix is most important.

Ni-Hard 1 and Ni-Hard 4 are martensitic white irons, which fall under ASTM A532, class I. They consist of iron carbides in a martensite matrix with some retained austenite. In the as cast form, the material typically has a hardness in the range of 500 to 550 Brinell (540 to 600 HV). Heat-treatment is used to reduce the retained austenite and increase the matrix hardness and, therefore, bulk hardness, however the erosion resistance is not as good as most of the following metals. The low chromium content (approximately 3%) also provides little corrosion resistance. For these reasons, the martensitic white irons have been largely superceded by the chromium-molybdenum white irons and high chrome irons in heavy duty slurry applications.

The chromium-molybdenum white irons (ASTM A532, class II) and the high chrome irons (ASTM A532, class III) consist of extremely hard chromium carbides (approximately 20 to 30% by volume) in a martensitic matrix with retained austenite. Heat-treatment of these materials not only increases hardness by reducing the retained austenite through conversion to martensite, but also initiates the precipitation/formation of secondary (fine) chromium carbides in the matrix. As the chromium carbides are refractory materials, they are generally inert to corrosion from most slurries and the corrosion resistance of these material is generally dependent upon the chromium content and corrosion resistance of the matrix. 15-3 Alloy is a 15% chromium-3% molybdenum white iron that falls under class II, which can be fully hardened to 750 Brinell. It suitable for use in slurry pumps where high erosion resistance is required and where impact loading conditions and corrosion rates are minor to moderate. 27% high chrome iron falls under class III and can be fully hardened to 650 Brinell. It is suitable for use in slurry pumps where erosion resistance, corrosion resistance and fracture toughness requirements are moderate to high.

Many propriety variations of the high chrome iron are available:

- Lowering the carbon, and raising the chromium results in less of the chromium being removed from the matrix to produce chromium carbides. This lowers the percent by volume of carbides and correspondingly the hardness, but retains additional chromium in the matrix for added corrosion resistance. Increasing the chromium content provides further gains in corrosion resistance. Currently, there are proprietary alloys with 450 Brinell hardness and a duplex (austenite-ferrite) matrix with the corrosion resistance of CD4MCu.
- Advanced high chrome irons are available with 50 to 75% by volume carbides and hardness above 700 Brinell. Through proprietary techniques, these materials are able to be produced with a relatively fine microstructure, which when combined with the increased volume of carbide, reduces the intercarbide spacing and have shown the ability to achieve two to three times the life of standard high chrome iron.
- Tougher high chrome irons with austenitic matrix and high impact resistance are available for extremely large particle dredging applications.

Elastomers

Typically are used in applications with particle diameters not greater than 10 mm.

Elastomers can be broadly broken into two categories: natural rubber and synthetic elastomers. Based solely on erosion, natural rubber is the clear winner due to its significantly greater resilience and tear resistance. Resilience is a measure of how high a ball of the material will bounce measured as a percentage of the initial drop height. Typically, this value ranges from 65 percent to 90 percent dependent on the rubber blend. Tear initiation resistance for natural rubber is typically in the range of 30 to 110 N/mm, dependent upon blend. For natural rubber, tear resistance tends to improve with increasing hardness, while resilience tends to decrease with increasing hardness. For fine particles applications (less than 100 microns) resilience has been shown to be more important in combating wear. For larger particles (greater than 500 micron) tear strength (resistance) is more important. As mineral processing slurries have a mixture of particle size, the best performing natural rubber will be one with the optimum combination of resilience for fine particle wear resistance and tear resistance to prevent larger particle damage.

Synthetic elastomers are used in small particle applications where natural rubber would be subject to chemical attack, causing swelling, hardening or reversion of the natural rubber. The most commonly used synthetic elastomers for wear materials and some of their typical applications are listed below:

- Nitrile: Generally used in fats, oils and waxes. Moderate erosion resistance. Limited resistance to acids and alkali environments.

- Butyl: Hydrochloric acid, phosphoric acid and sodium hydroxide. Sulphuric acid causes degradation and chlorinated hydrocarbons cause swelling.
- Hypalon: Primary use in acid conditions with some resistance to vegetable and mineral oils. Not recommended for use in ketones or chlorinated solvents.
- Neoprene: Moderate resistance to oils, fats, grease and some hydrocarbons. Can also be used in some mild oxidizing acids.

Synthetic elastomers also have higher temperature limits than natural rubber. While natural rubber is limited to 75 to 85 degrees C, dependent upon blend, the above synthetic elastomers have temperature limits ranging from 95 degrees C for Nitrile to 110 degrees C for Hypalon.

Further comparisons can be made with respect to mechanical properties. The tear resistance for the above synthetic elastomers range from 30 N/mm for Nitrile to 50 N/mm for Neoprene (compared to 30 to 110 N/mm for natural rubber). Neoprene has the highest resilience of the synthetic elastomers at 58 percent (compared to 65 to 90 percent for natural rubber). The superior mechanical properties of natural rubber indicate it should be used in preference to synthetic elastomers, unless temperature and/or chemical resistance are overriding factors.

Polyurethane is an elastomer that is used in applications where there is a good chance of large particle "tramp" damage, which would otherwise cut natural rubber and synthetic elastomers. It excels at its cut resistance due to its high tear strength (50 to 100 N/mm). Polyurethane is generally limited in temperature to 70 degrees C, due to swelling from hydrolysis attack, although polyurethanes are currently being produced, which reportedly can operate at 110 degrees C with no degradation. Generally speaking, in most small particle applications, where "tramp damage" is not a problem, natural rubber will outperform polyurethane.

Tip speed limit

Another important factor to consider in the selection of materials is the impeller tip speed limit. For elastomers, the concern is vibration or fibrillation of the material due to the relative motion of the impeller with respect to the side liners. This vibration can lead to heat generation within the elastomer and a thermal breakdown of the material. For natural rubber this often results in the elastomer reverting back to its natural "gummy" state. Tip speed problems on elastomers are easy to diagnose, as the damage is always most severe at areas with highest relative speed. For elastomer impellers this would be the periphery of the impeller and for elastomer side liners, this would be the area adjacent to the periphery of the impeller.

Generally speaking, the tip speed limit is a function of the hardness of the elastomer and its ability to dissipate heat. Typically, natural rubber has a tip speed limit of approximately 27.5 m/sec. Highly wear resistant soft natural rubber can have a tip speed limit as low as 25 m/sec, while proprietary blends with improved thermal conductivity can operate at 30 m/sec. For a centrifugal slurry pump this tip speed limit is important, as the head generated (meters) is approximately equal to the quantity (0.5 times the square of tip speed in meters per second) divided by the acceleration due to gravity (9.8 m/sec^2). Approximate speed limits and corresponding approximate BEP head limits for various materials are listed below:

- Highly wear resistant soft natural rubber   25.0 m/sec   32 meters head
- Typical natural rubber   27.5 m/sec   39 meters head
- Anti-thermal breakdown rubber   30.0 m/sec   46 meters head

- Nitrile   27.0 m/sec   37 meters head
- Butyl   30.0 m/sec   46 meters head
- Hypalon   30.0 m/sec   46 meters head
- Neoprene   27.5 m/sec   39 meters head

- Polyurethane   30.0 m/sec   46 meters head

- Hard metal (impellers)   38.0 m/sec   74 meters head

For the hard metal impellers, the 38 m/sec (7500 ft/min) tip speed limit is based on the limited ductility of the material and not thermal breakdown.

PUMP WEAR

In general, wear in a centrifugal pump will consist of various modes of abrasion and erosion.

Abrasion, which is the forcing of hard particles against a (wear) surface, only takes place within a slurry pump on the shaft sleeve and between the tight tolerance wear ring section of the impeller and the suction side liner.

Erosion is more commonly used to describe the progressive wear loss from the interaction or impingement of the fluid and particles against the wear components. The three primary modes of erosion and their wear locations within a centrifugal slurry pump are described below:

- Deformation wear: Direct impact to the leading edge of the impeller vanes, the back shroud of the impeller and the "protruding" cutwater within the volute.

- Random impingement: Random impacts to the impeller shroud and trailing edge of the main pumping vanes.

- Low angle impact: Wear from the tangential or near tangential movement of particles against the volute casing or vane surface.

Of these modes of erosion, deformation wear (direct impact) is the most severe and low angle impact is the least severe. The degree of wear is dependent upon the following:

- The kinetic energy of the particle: particle mass (specific gravity) and velocity.
- The particle shape: sharp particles have small contact area and high local stress, so wear is more severe than with rounded particles.
- The slurry concentration: higher percentages of solids result in more impacts for a given flow.

Based on the above, slurries can be classified according to the following criteria (reference):

- Heavy duty: C_w>35%, d_{85}>400 µm, SG_s>2.0, sharp particles.
- Medium duty: 20%<C_w<50%, 150 µm<d_{85}<400µm, SG_s>1.4, angular particles.
- Light duty: C_w<20%, d_{85}<150 µm, SG_s>1.4, rounded particles.

In Addition to the general specific speed limits and the material tip speed limits discussed previously, to maximize wear life, the following general impeller tip speed limits can be applied based on the severity of the duty:

- Heavy duty: 25 m/sec max. 32 meters head @ BEP
- Medium duty: 32 m/sec max. 52 meters head @ BEP
- Light duty: 38 m/sec max. 74 meters head @ BEP

Further recommendations can be made with respect to impeller type and flow range:

- Heavy duty: heavy duty impeller @ 0.60 to .80 BEP
- Medium duty: heavy duty impeller @ .70 to .90 BEP
- Light duty: high efficiency impeller @ .80 to 1.1 BEP

CAVITATION:

Cavitation is another cause of wear, which should be briefly mentioned. It is both system and pump design dependent. It is due to the local velocities at the impeller inlet reducing the suction pressure, which is system dependent, to a level below the vapor pressure of the liquid at the temperature being pumped and causing the fluid to boil. The increase in pressure as the fluid flows through the impeller results in a collapse and implosion of the vapor bubble(s) resulting in localized wear.

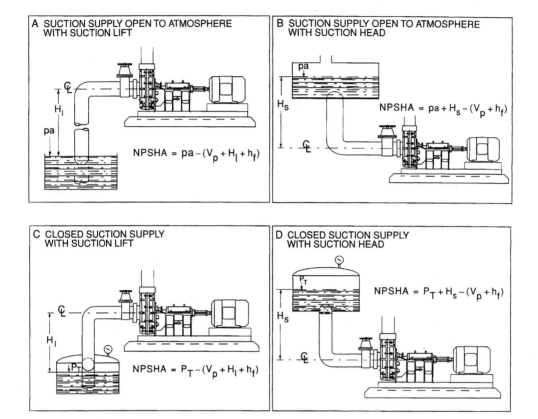

pa = Barometric pressure in meters absolute.

V_P = Vapor pressure of the liquid at the pumping temperature in meters absolute.

P_T = Pressure on the liquid surface inside a closed tank in meters absolute.

H_l, H_s = Static suction lift and head in meters absolute.

h_f = Friction loss in suction pipe in meters.

Figure 8: Calculation of NPSHA for various suction conditions.

Depending on the severity of the cavitation, high noise and vibration may be present and a reduced head output, lower flow, or complete loss of prime also may occur.

Whether or not cavitation occurs is dependent upon the suction characteristics of the pump and the system in which it is placed. The suction characteristics of a pump are expressed using the term net positive suction head required, or NPSHR. The suction characteristics of the system are expressed using the term net positive suction head available, or NPSHA. Even though cavitation is a pressure phenomenon, as friction in the pipeline and head output from the pump are expressed in terms of meters of head, so are NPSHR and NPSHA. For cavitation not to occur the NPSHA should exceed the NPSHR by 15 to 30 percent.

Values for NPSHR for a particular pump are a function of the pump design and are read off the manufacturer's performance curve at the duty point. Generally, NPSHR increases with flow and pump speed, although NPSHR can often rise at extremely low flows near pump shut off.

Values of NPSHA for the system are not simply the height of liquid above the centerline of the pump. Instead, NPSHA is a function of the local atmospheric pressure, the height of the liquid relative to the pump centerline, the friction losses in the suction pipe and the vapor pressure of the liquid at the pumping temperature. Means of calculating NPSHA for various suction conditions are illustrated in Figure 8 on the previous page (reference 4).

Under suction lift conditions as shown in Figure 8a, it is recommended the atmospheric pressure (pa) be reduced by dividing by the specific gravity of the liquid. This is to account for the weight of the liquid in the column and the negative effect this has on the suction pressure at the pump. It is not recommended that any positive correction be applied under flooded conditions. The reader is encouraged to review references (2), (4) (5) and (6) for more information on NPSHR and NPSHA.

Cavitation can easily be prevented during the system design process by insuring there is adequate suction height, minimal suction friction and proper pump selection. After system construction, these factors are far more difficult to change.

HYDRAULICS:

Because the success or failure of a centrifugal slurry pump is highly dependent upon its operating point, it is important to have a clear understanding of the relationship between the pump performance curve and the system curve.

As seen in Figure 4 and Figure 5, centrifugal slurry pumps typically have slowly drooping head versus capacity curves, with the highest head for a given speed being produced at the lowest flow. Changes in speed result in the generation of a new head versus capacity curve, which follows a set of rules called the affinity laws. These rules indicate that the relative location of points of equal efficiency will occur at a capacity equal to the original capacity multiplied by the ratio of speed change and at a head equal to the original head multiplied by the ratio of the speed change squared. Within reasonable speed changes, this useful tool allows one to plot the characteristic pump performance curve at a multitude of speeds given a curve for one reference speed. Nowadays, most centrifugal slurry pump manufacturers plot performance at a wide range of speeds (as shown in Figure 4 and figure5), so it is very easy to estimate the characteristic pump performance curve at a given speed without the use of these rules.

It is important to note that the point of operation (flow and head), is system dependent and does not necessarily follow the points of equal efficiency predicted by the affinity laws, as will be shown.

The system curve is a graphical representation of the piping systems resistance to flow. It consists of a fixed component called the static head and a variable component called the friction head. In systems that are open to atmosphere, the static head is the elevation of the discharge pipe (or discharge liquid level for submerged discharge), minus the elevation of the liquid in the suction tank. In systems where the suction or discharge are pressurized or under vacuum these conditions must be accounted for after being converted to appropriate head values. The friction head is the resistance to flow through the piping and fittings, which varies with the flow rate. The reader is encouraged to refer to (2), (4), (5) and (6) for more information on system curves.

The point of operation for a centrifugal pump is the intersection of its head versus capacity curve with the system curve. Figure 9 illustrates a typical system curve with 10 m of static head and friction that increases exponentially with flow. Also shown are pump head versus capacity curves for the heavy duty impeller illustrated in Figure 4 at 600 and 800 rpm. For each speed, the operating point occurs at a flow where the head output from the pump matches the total resistance in the system at the same flow. As stated above, this is the intersection of the head versus capacity curve and the system curve. For this pump in this hypothetical system, the expected operating points are 1000 m3/hr @ 26 m head at 600 rpm and 1400 m3/hr @ 44.5 m head at 800 rpm.

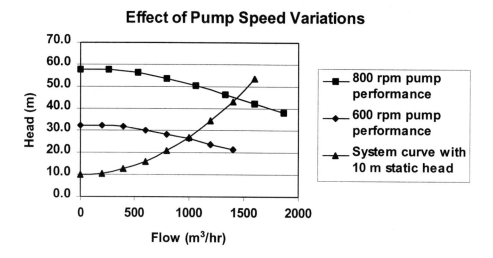

Figure 9: Effect of speed on point of operation.

Figure 10 illustrates the effect of a change in static head for the same system. In this instance we have increased the static head from 10 m to 25 m. Note the system curve shifts directly up by the increase in static head (15 m). The result is the intersection of the system curve with the 600 rpm and 800 rpm performance curves occurs at lower flows, reducing pump output. This effect is observed when there is a drop in suction level or an increase in elevation, such as an increasingly tall tailings dam.

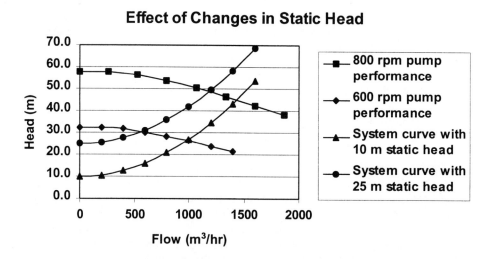

Figure 10: Effect of changes in static head on point of operation.

Figure 11 illustrates the effect of a change in friction. In this case we have increased the friction, which results in a steeper system curve and correspondingly lower output from the pump at both 600 and 800 rpm. Examples of this effect are the partial closing of discharge valve (not recommended in slurries), a change to smaller diameter pipe, a lengthening of the discharge pipe or increase in friction due to an increase in concentration of the slurry. The opposite effect of that shown in figure 11 is a decrease in friction, which results in a shallower system curve and increased pump output. This condition usually occurs when an engineer, with good intentions, but poor judgement, has added a factor of safety to his friction calculations.

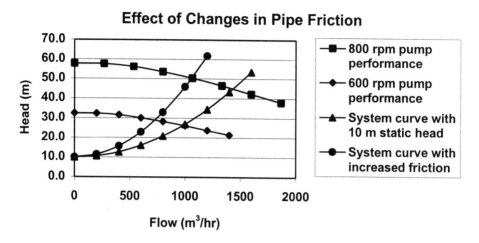

Figure 11: Effect of changes in pipe friction on point of operation.

Figure 12 illustrates the effect of having two identical pumps at the same speed (in this case 600 rpm) operating in series. The net effect is a summation of the individual heads of each pump at a given flow. This results in a steeper combined pump head versus capacity curve and increased head at any given flow. This, of course, results in an intersection with the system curve at a higher flow and higher flow output. Typically, the use of two or more pumps in series is not to increase flow output per se, but instead to overcome a large amount of friction, such as in a long tailings line or to improve wear life by lowering pump speed.

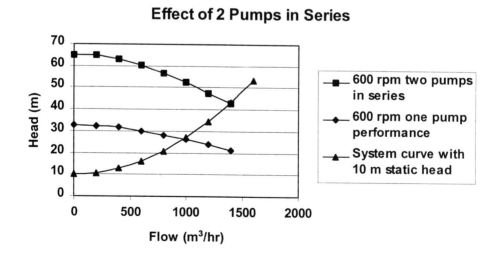

Figure 12: Effect of 2 pumps in series on point of operation.

Figure 13 illustrates the effect of having two pumps in parallel in the same system. Here the combined performance curve is the horizontal summation of the capacities at the same pump head. This results in a shallower combined head versus capacity curve and slightly greater flow output. <u>Note that the use of two pumps in parallel does not result in a doubling of the flow.</u> Only when the system curve is very shallow (due to low friction) do we see an appreciable increase in flow with parallel operation.

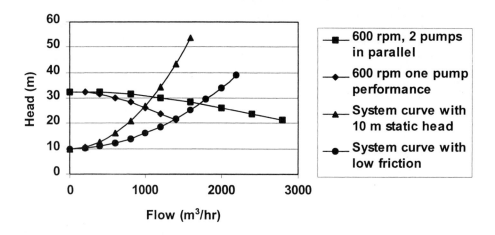

Figure 13: Effect of two pumps in parallel on point of operation.

Parallel operation is not recommended in slurry handling systems for the following reasons:

- The shallow pump performance curves and slight variations in head output from each pump, due to wear and manufacturing tolerances, can result in substantially different flows through each pump.
- This in turn can accelerate wear in one pump relative to the other, accelerating the problem.
- There is increased chance of sanding one of the legs of the system if the other pump is producing substantially more head.

SETTLING VELOCITY:
From the above it would seem that the ideal system would be one with large diameter pipe to minimize the friction and allow for a lower pump speed, lower pump head output, lower wear and lower horsepower draw. However, when handling slurries, one must first determine if the slurry is heterogeneous (settling) or homogeneous (non-settling). If the slurry is heterogeneous, then care must be taken to operate above the settling velocity of the solids, which dictates a maximum pipe diameter for a given required flow.

Figure 14 (reference 7) is a chart, which determines if the solids in a water based slurry are considered settling or non settling on the basis of average particle size and solids specific gravity. If the slurry is settling, then a determination of the settling velocity must be made to insure operation above it to prevent solids deposition and possible pipe plugging.

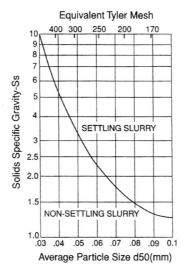

Figure 14: Determining if slurry is settling (reproduced with permission of the Turbomachinery Lab, (reference 7))

Considerable work has been done on the effect of particle size and density on the settling velocity of closely graded particle, heterogeneous slurries (references 8, 9, and 10). Durand's much published work estimates settling velocity (V_L) using the following formula:

$$V_L = F_L \sqrt{\frac{2gD(S - S_l)}{S_l}}$$

where: F_L is a settling velocity parameter dependent upon particle sizing and solids concentration,
g is the acceleration due to gravity (9.81 m/sec^2),
D is the pipe diameter in meters,
S is the specific gravity of the solids,
S_l is the specific gravity of the liquid.

For closely graded particle sizing, F_L is obtained from Figure 15 (reference 9). For the purpose of this paper, closely graded slurries are ones where the ratio of particle sizes does not exceed 2:1 for at least 90% of the weight of the solids in the sample.

Figure 15 has been found to be conservative for slurries with more coarsely graded solids and significant portions of particles finer than 100 micron. For more widely graded particle sizing, Figure 16 should be used to estimate V_L (reference 10).

For a heterogeneous slurry, accurately predicting the settling velocity is perhaps the first most critical factor in the design of a slurry pumping system. The settling velocity determines the required pipe diameter for a given desired flow, which is then used to determine the pipe friction. Only then is this information used to make the appropriate pump selection, determine the proper operating speed, determine the appropriate materials of construction, and determine the expected horsepower draw.

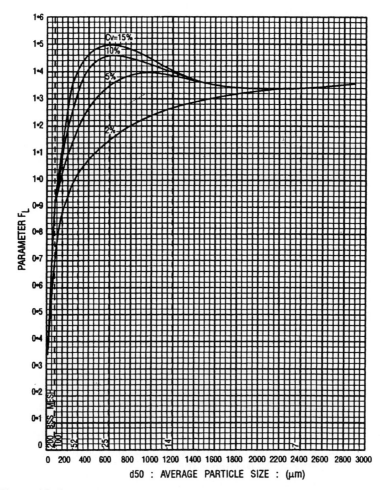

Figure 15: Durand's limited settling velocity parameter (for closely graded particles).

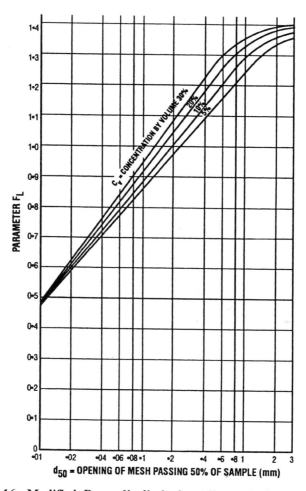

Figure 16: Modified Durand's limited settling velocity parameter (for particles of widely graded particle size).

PERFORMANCE DERATING:

The operating point for a centrifugal pump is dependent not only upon the pump speed and the system in which it is placed, but also upon the material being pumped. Just as pipe friction can vary with the concentration of solids and the viscosity of the fluid being pumped, pump head output and efficiency can also deviate from expected water performance. It is desirable to be able to accurately predict these deviations, or deratings, in order to be able to determine if a centrifugal pump is the appropriate pump choice and to be able to compensate for these deratings by making the appropriate speed changes and providing adequate motor horsepower.

Slurry Classification:

Prior to calculating the appropriate head and efficiency derating, it is necessary to classify the slurry in order to apply the appropriate corrections. The vast majority of slurries seen by a centrifugal pump typically fall into one of the following three categories:

- Heterogeneous slurry: The bulk of the solids handling slurries. A settling slurry. Typically water based with an overwhelming percentage of solids being greater than 100 microns in size. A relatively low content of fines (solids less than 100 microns) results in the carrier fluid (water plus fines) being essentially similar to water.
- Viscous Newtonian: Technically, this refers to any slurry or fluid, which has no yield stress and constant viscosity (greater than, equal to, or less than water). For the purpose of discussion, we will consider a viscous Newtonian slurry to be one where the combined carrier fluid and soluble particles has a viscosity greater than water, but is still free flowing, albeit at a slower rate. Oil is perhaps the most common viscous fluid.
- Bingham plastic: A slurry where the carrier fluid contains sufficient fines content to provide a yield stress. Bingham plastic fluids are not free flowing unless there is sufficient force to overcome the shear stress. Ketchup, red mud and cement slurries are examples of Bingham plastics.

Figure 17 illustrates the rheological differences between the three slurry types (reference 11). For sake of discussion, we assume the heterogeneous slurry to have a stress versus strain plot similar to water. For reference, these plots of shear stress versus shear strain are called rheograms.

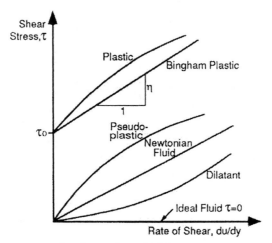

Figure 17: Rheological properties for various fluids.

The lab analysis of a small sample of representative slurry for rheological properties and solids composition can provide the most useful/helpful information for the system designer and pump application engineer.

While we will be focusing primarily upon pump derating, is equally important to remember that the properties of the slurry also effect the system and also must be accounted for in the frictional data given to the pump supplier.

Heterogeneous Slurry

The two principal reasons for deration of centrifugal pump performance when handling solids are:

- Slip between the water and the solid particles during acceleration and deceleration of the slurry (as it passes through the impeller). This results in energy losses.
- Increased friction losses. These losses increase with increased solids concentration of the slurry.
- Inability of the suspended particles to store or transmit pressure energy.
- Mechanical friction changes in the gap between the impeller and side walls, which affect energy consumption (efficiency).

When discussing centrifugal pump deratings for heterogeneous slurries, we refer to the terms, head ratio and efficiency ratio (identified by HR and ER, respectively), where:

$$HR = \frac{Head_{slurry}}{Head_{water}}$$

$$ER = \frac{Efficiency_{slurry}}{Efficiency_{water}}$$

Extensive work (reference 12) showed the degree of derating is dependent upon the following factors:

- Solids specific gravity.
- Percent concentration by volume.
- Relative particle size (d_{50} size of particle/impeller diameter).

The results of this work are summarized in the nomogram in Figure 18, which is self explanatory.

Using the derating example given in Figure 18 (30% by volume slurry with 2.65 specific gravity dry solids, 1500 kg/m3 slurry and 350 micron d_{50} average particle size), we have plotted the derating of a 6/4 pump at 1200 rpm with same 365 mm diameter impeller on a hypothetical system curve and plotted it in Figure 19. In this example the intersection of the 1200 rpm water performance curve and the system curve occurs at 300 m3/hr at 28 m head and 70% pump efficiency. With slurry correction (.84 HR and .80 ER), the intersection of the derated pump performance curve and the system curve occurs at only 275 m3/hr at 24.5 m head and the pumps efficiency has been derated to 56%. Power draw at the slurry condition of 275 m³/hr is expected to be:

$$kW_{draw} = \frac{(275 m^3/hr)(1 hr/3600 \sec)(24.5 m)(1500 kg/m^3)(9.8 m/\sec^2)}{(56 percent/100)(1000 W/kW)} = 49.1 kW$$

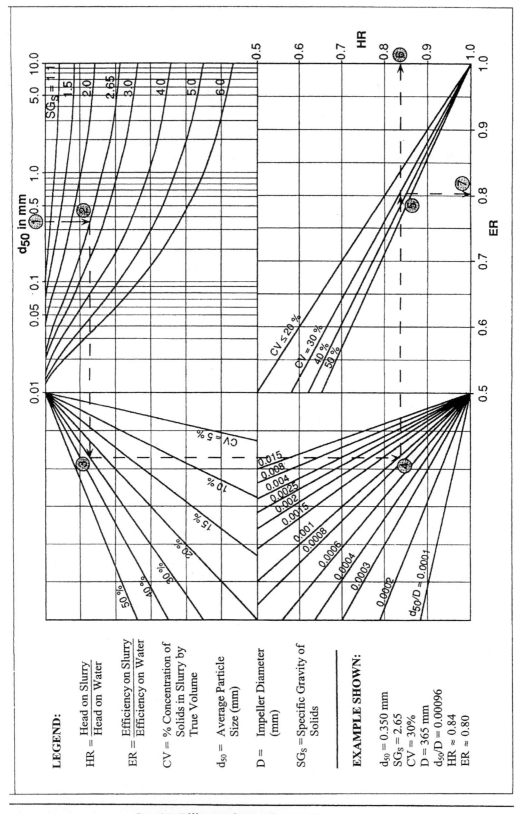

Figure 18: Pump head and efficiency deratings for settling slurries.

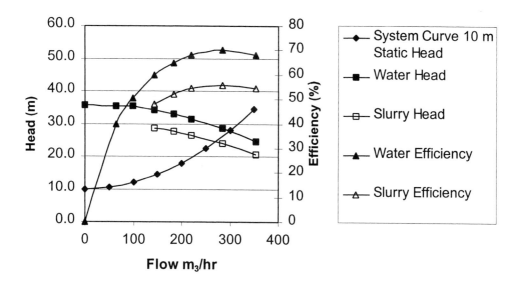

Figure 19: Effect of slurry correction on pump performance

If it were desired to operate at 300 m³/hr in this system (instead of the derated 275m³/hr shown above) then the required pump water head would need to be greater than the 28 meter system head at 300 m³/hr by the inverse of the head ratio. This means the pump would be required to deliver 28 meter/0.84 head ratio = 33.3 meters head on water to achieve 28 meters head on slurry and 300m³/hr in this system. Using the affinity laws or referring to the manufacturer's characteristic curve indicates this occurs at approximately 1290 rpm with a water efficiency of approximately 70%. Derating the water efficiency with .80 efficiency ratio yields 56% slurry efficiency and the following power draw at 300 m³/hr:

$$kW_{draw} = \frac{(300 m^3/hr)(1 hr/3600 \sec)(28 m)(1500 kg/m^3)(9.8 m/\sec^2)}{(56 percent/100)(1000 W/kW)} = 61.3 kW$$

Newtonian Viscous Slurry:
Purely viscous Newtonian fluids are not very common in slurry pumping. In most instances, the carrier fluid is water with a small percentage of fines in suspension and has a viscosity similar to water. In these instances, when the majority of particles are of a size greater than 100 microns, the pump deratings come from the slip and friction caused by the relatively large particles and the heterogeneous slurry correction discussed above is recommended.

Recall from Figure 14, when there are particles less than 100 microns with the appropriate specific gravity, these particles are considered non-settling and remain in suspension. These fines effectively increase the density of the carrier fluid. If there is sufficient quantity of these fines in suspension then carrier fluid viscosity increases and at high concentrations the carrier fluid exhibits a yield stress (i.e., is not free flowing).

Only in those instances when there is no yield stress present and when the slurry has a viscosity different from water should the traditional viscosity correction charts be used. This indicates the benefit to testing a sample of the slurry for its rheological properties to determine its slurry type. For reference, Figure 20 shows the effect of a Newtonian (zero yield stress, constant viscosity) "traditional" viscous fluid on our 6/4 pump example, for 200 and 400 centistoke slurries.

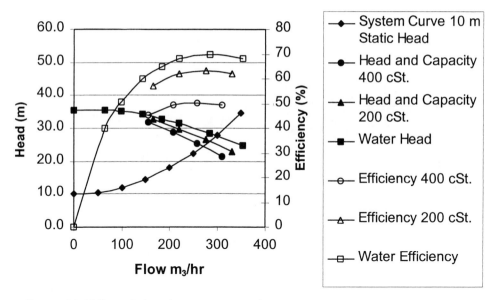

Figure 20: Effect of viscosity on pump performance

Note the effect of increased viscous correction, with respect to head, at higher flows. This is illustrated in the corrections for the above example, which are listed below:

		200 cSt	400cSt
Capacity correction:	C_Q:	0.97	0.91
Head correction:	C_H: @ 0.6 x BEP:	0.98	0.96
	@ 0.8 x BEP:	0.96	0.925
	@ 1.0 x BEP:	0.935	0.89
	@ 1.2 x BEP:	0.915	0.86
Efficiency correction:	C_E:	0.68	0.54

Also note the correction for capacity. These corrections are taken from the Hydraulic Institute viscosity correction chart which provides corrections based on flow (an indication of pump size), head (an indication of pump speed) and viscosity. These corrections do not apply for other size pumps at other conditions. The reader is encouraged to refer to the Hydraulic Institute Standards (reference 2) for further clarification on Newtonian viscosity correction.

Bingham Plastic Fluids:
A Bingham plastic fluid or slurry is a material that has a yield stress. This means it is not free flowing unless there is sufficient stress to overcome the yield stress. In simple terms, Bingham plastic fluids are "pastes". As mentioned above, when non-settling fines are present in a slurry, they have the effect of increasing the density of the carrier fluid. At a certain concentration level (typically in the range of 30 to 50 percent by weight, dependent upon material), the slurry starts to exhibit a yield stress. Within the mineral processing industry, the addition of flocculants and coagulants can also be used to develop a yield stress. Figure 21 (next page) illustrates a rheogram of a red mud slurry with Bingham fluid characteristics.

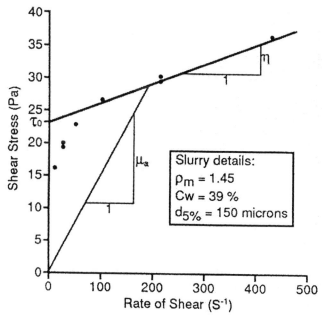

Figure 21: Rheological data for red mud slurry with 95 % of solids less than 150 microns.

Important parameters for a Bingham plastic fluid are the yield stress (τ_0) and the coefficient of rigidity (η), which is also commonly referred to as the plastic viscosity (μ_P). The plastic viscosity is the slope of the shear stress versus shear rate line at high shear rates. The yield stress is the point along the y axis where this straight line intersects the y axis. For the example above, the yield stress is approximately 23 Pascal. For this example, the plastic viscosity is calculated below:

$$\mu_P = \frac{(38 Pa - 23 Pa)}{(500 \sec^{-1} - 0 \sec^{-1})} = .03 Pa \cdot \sec = 30 \, centipoise$$

It is important to differentiate between apparent viscosity (μ_a) and plastic viscosity (μ_P). Recall that for a Newtonian fluid, there is no yield stress and the viscosity is constant. As such, its rheological plot is a straight line starting at the origin with slope equal to the viscosity (refer to figure 17). A similar straight line can be drawn from the origin to any point on rheological plot of a non-Newtonian, Bingham plastic fluid, yielding a line with "apparent" viscosity slope for the particular shear rate and shear stress chosen. Note that the apparent viscosity is dependent upon the chosen point. At other shear stresses and rates of shear, the "apparent" viscosity will differ. As an example, for our red mud slurry:

$$\mu_a @ 200 \sec^{-1} = \frac{(29 Pa - 0 Pa)}{(200 \sec^{-1} - 0 \sec^{-1})} = .145 Pa \cdot \sec = 145 \, centipoise,$$

$$\mu_a @ 500 \sec^{-1} = \frac{(38 Pa - 0 Pa)}{(500 \sec^{-1} - 0 \sec^{-1})} = .076 Pa \cdot \sec = 76 \, centipoise.$$

Because the "apparent" viscosity decreases with increasing shear rate, this material is shear thinning. At infinite shear rate the apparent viscosity and plastic viscosity converge.

In 1984 Walker and Goulas (reference 13) performed a series of 3 inch pump tests using kaolin clay and coal slurries of varying yield stress and plastic viscosity. One pump had a closed 4 vane impeller of diameter .352 m and a low N_s of 14.3 (738 in US units). The second pump was trialed with a five vane closed impeller of .244m diameter with Ns of 27 (1394 in US units) and an open 3 vane impeller of .240 m and Ns of 29.6 (1531 in US units).

On high yield stress material (approximately 10 to 20 Pa) they found a severe drop in head at low flow. This general effect is illustrated in Figure 22.

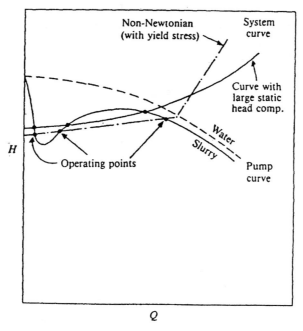

Figure 22: Effect of high yield stress on pump performance.

The significance of this drop in head is that the pump performance curve crosses over the system curve at three points. A typical "turbulent flow" exponentially rising system curve with high static head and a high yield stress non-Newtonian system curve with laminar (low flow) and turbulent (high flow) regions are plotted. The intersection of the pump performance curve with the system curve(s) at more than one location means the possibility of more than one operating point and the pump swinging between high and low flows. The writer has observed this phenomenon on high yield stress cement slurries and thickener underflow.

Walker predicted the observed performance derating in terms of a modified pump Reynolds Number (Re_p), where:

$$Re_p = \omega \cdot D_i^2 \cdot \rho_m / \eta$$

where: ω is the pump rotational speed in radians/second,
D_i is the impeller diameter in meters (then squared),
ρ_m is the slurry density in kg/m^3,

and η is: 1) the plastic viscosity for determining derating at the BEP in Pa·sec,
2) the apparent viscosity at a shear rate of 2ω for determining derating at 10 % of BEP in Pa·sec.

Figure 23 (reference 11) illustrated on the next page shows the effect of Modified Pump Reynolds number on pump derating.

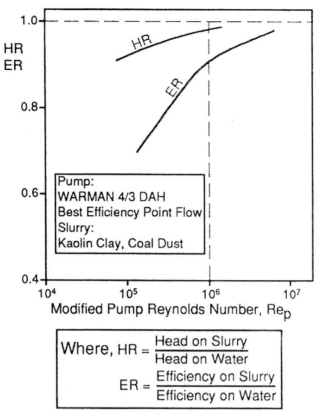

Figure 23: Typical pump performance derating when handling Bingham plastic slurry.

The writer's review of the data showed a bit of spread to the data points with the low (738) N_s pump performing worse than the medium (1394-1531) N_s pump and, more importantly, a severe drop off in performance at Re_p below 1×10^5. Note that lower values of Re_p indicate higher yield stress and/or viscosity. Nevertheless, the above chart is a very good first guide to the feasibility of a standard centrifugal pump on high yield stress material.

Recent testwork contracted by Shell, performed by the Saskatchewan Research Council and monitored by the writer showed the ability of a proprietary, modified version of the above medium N_s pump to exceed the above performance. A kaolin clay slurry with 200 Pa yield stress and Re_p of less than 1.2×10^4 was pumped no head derating. Additional testwork is planned to confirm the ability to duplicate these results and potential limitations of the suction piping on high yield stress fluids.

APPLICATIONS:
It is hoped that the preceding work will provide insight and a basis for the brief recommendations for the following applications.

Grinding Circuits:
Grinding circuits are typically the most severely erosive of pump applications. Typically, the solids size is relatively large, the density tends to be high and the particles being freshly ground have sharp edges. Fortunately most grinding circuits are low head applications allowing for relatively low pump speed.

Because of the large particles, heavy duty impellers with low specific speed operating at 60 to 80% of BEP flow are recommended. Hard metal impellers are required to handle the direct impact of the solids on the leading edge of the main pumping vanes. High tear strength natural rubber is an option on the casing liners, which primarily see low angle impact wear, if the head, and correspondingly, tip speed and particle kinetic energy are low. While 10 mm is typically considered the general limit for elastomers, the writer has seen high tear strength rubber casing liners outperform metal on wet-crusher service with 20 mm solids. On higher head grinding applications and applications where there is a chance of large particle tramp damage, metal volute liners or casings are recommended.

Proper sump design is key to promoting stable operation and minimizing wear. Inadequately sized sumps increase the chance of air entrainment and reduced head and flow output. This in turn results in higher pump speed and increased wear. Sump volume should be a minimum of 1 minute retention time and preferably two minute retention time. The sides and back of the sump should be sloped at an angle above the angle of repose of the solids and a dead zone or dead box should be provided at the bottom of the sump to contain any oversized solids. Reference (4) provides additional information on good sump design.

Hydrotransport and Tailings Circuits:
The high flow and high head of these applications dictates the need for a balance of wear and efficiency in these relatively high horsepower applications. Typically pumps with N_s in the range of approximately 27 to 33 (1400 to 1700 in US units) provide the optimum balance of these two factors. To further optimize efficiency, operation closer to the best efficiency point (70 to 100% of BEP) and impellers with thinner and longer wrap vanes than the heavy duty design used in mill circuits are recommended.

Hard metal is typically used on impellers and side liners, due to the high tip speeds, although the writer has seen elastomers outperform metal on suction side liners at heads above 50 meters. Casing liners can be elastomer or metal depending upon particle size and the possibility of large particle "tramp" damage.

For series pumping applications high pressure designs are required. On unlined pumps, the volute, which is naturally a sacrificial wear component, is also the casing/pressure retaining vessel, so additional material and ribbing are required to obtain the support the operating pressure, while also allowing adequate allowance for wear. On fully lined pumps, the non-wearing outer shell is designed to support the operating pressure, so there is inherently greater safety with this double walled design. Figure 24 illustrates a high pressure fully lined design. On larger sized pumps a separate suction cover (as illustrated) allows for removal and replacement of the impeller and suction side liner without disturbing the discharge piping.

Thickeners:
Metal or elastomer wear components are used depending upon the solids size handled by the thickener. In high yield stress applications there appears to be a benefit to staying away from low Ns designs. Low NPSHR designs have the ability to better handle low suction pressure conditions. On high yield stress applications open impellers with flow inducing vanes are more likely to provide stable operation.

Suction piping should be kept to minimum length to minimize friction losses associated with high yield stress, high viscosity slurries. Generous suction levels above pump centerline are recommended to provide maximum suction pressure to feed the impeller.

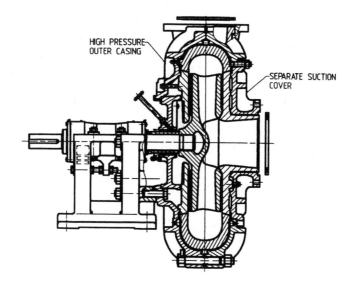

Figure 24: High pressure pump for hydrotransport and tailings.

Froth:

An additional chapter could be written on froth handling. Past recommendations have resulted in froth factors as great as six being applied to the desired volume flow. This oversizing was thought to provide the benefits of an increased eye area to handle larger air volumes without losing prime, a larger diameter impeller to keep speeds and inlet velocities low to minimize separation of gas and fluid and a lower NPSHR. These recommendations resulted in the application of "typical" slurry pumps at least one size larger than normally required for an air free slurry.

Recent designs specifically developed for froth handling have been produced with separate inducers or oversized inlets and flow inducer vanes to lower NPSHR and provide positive displacement characteristics to feed the main pumping vanes. These designs show the promise of providing pumps of lower initial capital cost due to reduced size while also offering improved efficiency.

From a system design standpoint, oversized sumps with baffles and additional retention time to assist in venting of the gas prior to it entering the pump are recommended. Figure 25 illustrates an appropriate baffle arrangement. Suction pipe and impeller eye vents are often used to vent the suction side of the pump, while rotation of the casing to top 45 degree discharge or top horizontal discharge is used to prevent the cutwater from trapping air and help vent the discharge side of the casing.

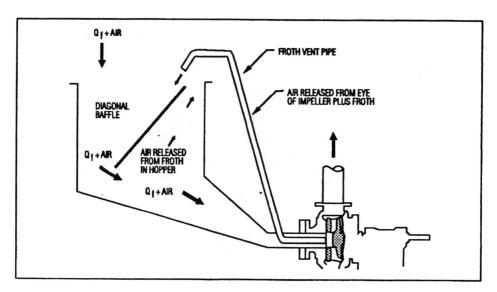

Figure 25: A sump design to provide maximum air/gas venting.

Conclusion:
The successful application of a centrifugal slurry pump requires not only an intimate knowledge of the properties of the slurry being pumped, but also a knowledge of centrifugal slurry pump design and the effect of the system design decisions on pump operation. The system designer requires this knowledge to determine the appropriate pipeline velocities, pipeline friction, and sump design, while the pump application engineer requires this knowledge to determine the appropriate pump selection, materials selection, slurry deratings, speed correction and horsepower requirements.

REFERENCES:
(1) V. K. Suprun. 1972. Abrasive Wear of Slurry Pumps and Means of Combating It. Moscow, Mashinostroyenie, 104 pages.
(2) Hydraulic Institute Pump Standards. 1994. Hydraulic Institute, Parsippany, NJ. www.pumps.org.
(3) C. Walker and G. Bodkin. 1991. Warman Group Development Technical Bulletin #24: Selecting Slurry Pumps to Minimize Wear. Warman International, Sydney, Australia.
(4) A. Roudnev and T Angle. 1999. Slurry Pump Manual. Envirotech Pumpsystems, Salt Lake City, Utah.
(5) Cameron Hydraulic Data. 1994. C.C. Heald. Ingersoll Dresser Pumps, Liberty Corner, NJ.
(6) Warman Slurry Pumping Handbook. 1999. Warman International, Inc., Madison, WI.
(7) G. Wilson. 1987. The Effects of Slurries on Centrifugal Pump Performance. Proceedings of the Fourth International Pump Symposium. Turbomachinery Laboratory, Texas A&M University, College Station, Texas. pp. 19-25.
(8) K.C. Wilson. 1979. Deposition-Limit Nomogram for Particles of Various Densities in Pipeline Flow. Sixth Annual Conference on the Hydraulic Transport of Solids in Pipes. Held at University of Kent, Canterbury, U.K., September 26th-28th, 1979.
(9) R. Durand. 1952. Hydraulic Transportation of Coal and Solid Material in Pipes. London Colloquium of the National Coal Board.
(10) R. McElvain and I. Cave. 1972. Tailings Transportation. World Mining Tailings Symposium, November 2, 1972.
(11) P. Wells. 1991. Warman Group Development Technical Bulletin #14: Pumping Non-Newtonian Slurries.
(12) C. Walker and P. Wells. Warman Group Development Technical Bulletin #24: Influence of Slurry on Pump Performance. Warman International, Sydney, Australia.
(13) C. Walker and A. Goulas. 1984. Performance Characteristics of Centrifugal Pumps When Handling Non-Newtonian Homogeneous Slurries. Proc I Mech E Vol 198a, No1, pp. 41-49.

Selection and Sizing of Slurry Lines, Pumpboxes and Launders

Baha Abulnaga, Mazdak International Inc., Sumas, Washington, USA

Ken Major and Peter Wells, HATCH Associates Ltd., Vancouver, BC, Canada

ABSTRACT
A significant component in the design of a mineral processing plant to ensure an efficient transition from concept (engineering) to reality (plant operation) is materials handling, the moving of ore through the different unit operations. The design of slurry systems using pipelines and launders needs to consider a number of different variables including slurry rheology, density, viscosity and particle size. In this paper the different regimes of slurry flows are reviewed outlining methodology for sizing full flow pipes, launders and upcomers. With the advent of modern personal computers, it is possible to size up slurry lines more precisely using specialty programs.

INTRODUCTION
Mineral processing is the combination of the many unit operations that are required to produce a marketable product from an ore. Although the coarse size reduction steps (crushing) are dry, in most project flowsheets the recovery steps are in slurry form. In mineral processing, slurries are a mixture of water and/or chemical solution with ground rock. A simplified process flowsheet will have four or five unit operations. The design of the slurry handling system is very important to the efficient operation of the process because it provides the links between these unit operations. A problem with surging flow in the first unit process, typically grinding, has the potential to impact the behaviour of the downstream processes.

SIZING OF PIPING SYSTEMS
A mineral processing plant typically involves a large number of unit operations. Each of these different operations may have specific slurry system requirements depending on the slurry characteristics. The piping system must be designed to ensure that the opportunity for sanding and plugging of the pipeline is minimized while at the same time ensuring that the line velocities to prevent this from happening do not lead to excessive wear conditions of the pipeline. To maintain suspension of a particle in the slurry the velocity of the slurry in the pipeline must be greater than the critical velocity. The main design factors to consider when selecting the pipe size are:

- Particle size
- Slurry density
- Viscosity
- Flowrate
- Friction losses

Particle size in a mineral processing plant will vary based on the unit operation and process criteria. For example, grinding circuits are typically characterized with large particles (+19 mm SAG grinding, +2mm rod mill, ball mill) and high densities (45% solids to 60% solids). A flotation concentrate stream after multiple regrind stages may have a P_{80} of 10 microns.

As the slurry density increases and/or the particles become finer the hindered settling velocity in the slurry will increase making it easier to keep the larger particles suspended. The finer particle size will also increase the apparent viscosity of the slurry. This is observed in many of the Nevada gold mines with high clay contents in the ore.

For all the accuracy that can be applied to pipeline sizing through calculations to determine settling rates, slurry velocity and pipe diameters the process becomes flawed when operation needs are defined. Many of the ore deposits have multiple ore types with varying processing requirements. In low grade copper and gold mills economics are predominated by the need to maximize throughput. With a change in ore from hard to soft, relative to SAG grinding conditions, throughputs can change radically. Doubling the feed grade in a flotation plant will result in doubling the concentrate production.

Figure 1: Computer representation of a SAG mill circuit showing both pumped and open channel flows.

A typical plant layout in a SAG mill circuit, Figure 1, may involve pumped flow such as from the discharge of the SAG mill to the hydrocyclones or gravity flow in launders from the underflow of the cyclones to ball mills. The location of hydrocyclones high in a plant, may also involve gravity flow to downstream processes. It is important to appreciate these two different kinds of flows.

SLURRY PIPE FLOWS
Slurry flows are classified as:

- heterogeneous (settling flows).
- homogeneous (non-settling) flows.

Heterogeneous flows with typical particle size, $d_{50}>44$ microns constitute the majority of circuits encountered in plants. The flow is characteristically Newtonian. There are however instances where clays, very fine grinding of the ore leads to non-Newtonian flows particularly in certain bauxite (red mud, or kaolin), and even gold-copper associated with clays at weight concentration in excess of 40%

There are essentially four main regimes of Newtonian flow in a horizontal pipe:

- Flow with a stationary bed.
- Flow with a moving bed and saltation (with or without suspension).
- Heterogeneous mixture with all solids in suspension.
- Pseudo-homogeneous or homogeneous mixtures with all solids in suspension.

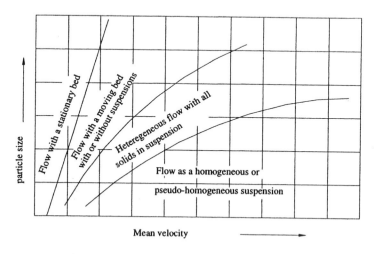

Figure 2: Regimes of Newtonian flows for slurry mixtures in a horizontal pipe[1]

Transitional Velocities

These four regimes of flow can be represented by a plot of the pressure gradient versus the average speed of the mixture as illustrated in Figure 2.

The transitional velocities are defined as:

- V_1 = the velocity at or above which the bed in the lower half of the pipe is stationary. In the upper half of the pipe some solids may move by saltation or suspension.
- V_2 = velocity at or above which the mixture flows as an asymmetric mixture with the coarser particles forming a moving bed.
- V_3 or V_D = velocity at or above which all particles move as an asymmetric suspension and below which the solids start to settle and form a moving bed.
- V_4 = velocity at or above which all solids move as a symmetric suspension.

---

[1] Abulnaga B.E. 2002 – *Slurry Systems Handbook* – McGraw-Hill

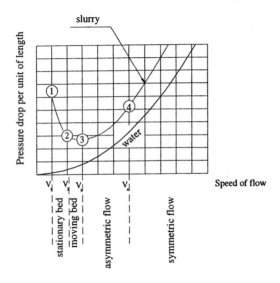

Figure 3: Regimes of velocity for Newtonian flows of settling slurries in horizontal pipes.[1]

The deposition velocity V_D or V_3 is usually established by the Durand equation

$$V_D = V_3 = F_L \{2 * g * D_i[(\rho_s - \rho_L)/\rho_L]\}^{1/2} \qquad (EQ\ 1)$$

F_L = is the Durand factor based on grain size and volume concentration
V_3 = the critical transition velocity between flow with a stationary bed and a heterogeneous flow.
D_i = pipe inner diameter (m)
g = acceleration due to gravity (9.81 m/s)
ρ_s = density of solids in a mixture (kg/m³)
ρ_L = density of liquid carrier (kg/m³)

The Durand factor F_D is typically represented in a graph for single or narrow graded particles after the work of Durand (1953). However, since most slurries are a mixtures of particles of different sizes, this plot is considered to be too conservative. The Durand velocity factor has been refined by a number of authors. Schiller (1991) proposed the following equation for the Durand velocity factor based on the d_{50} of the particles

$$F_L = \{(1.3 \times C_v^{0.125})(1 - \exp(-6.9\ d_{50}))\} \qquad (EQ\ 2)$$

Equation 2 can be solved using a simple computer program. The Durand factor F_L is related to the Froude Number by the following equation:

$Fr = F_L * \sqrt{2}$.

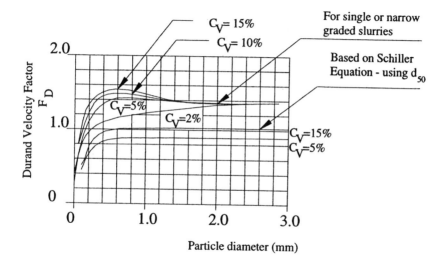

Figure 4: Comparison of the conventional Durand factor for single graded slurries and the factor using Schiller equation.[1]

The Schiller's equation is valid for viscosities of the order of 1 cP. For settling mixtures with higher viscosities. To estimate the deposition velocity V_3, Gilles et al (1999) developed an equation for the Froude Number based on the Archimedean number

$$Fr = a\, Ar^b \qquad (EQ\ 3)$$

$$Ar = \frac{4}{3\mu_L^2} d_p^3 \rho_L (\rho_s - \rho_L) g \qquad (EQ\ 4)$$

μ_L = viscosity in Pa-s
d_p = particle diameter in meters
ρ_L = density of liquid in kg/m³
ρ_s = density of solids in kg/m³
g = acceleration due to gravity in m/s²
Fr is non-dimensional

For Ar>540, a= 1.78, b= -0.019

For 160<Ar<540, a= 1.19, b= 0.045

For 80<Ar<160, a=0.197, b=0.4

For Ar<80, the Wilson and Judge (1976) can be used, which expressed the Froude Number as:

$$Fr = (\sqrt{2}) * \left\{ 2.0 + 0.30 \log_{10} \left(\frac{d_p}{D_i C_D} \right) \right\} \qquad (EQ\ 5)$$

This correlation is useful in the range of:

$$10^{-5} < \left(\frac{d_p}{D_i C_D} \right) < 10^{-3}$$

The Wislon-Judge method requires a computer program due to the various ranges of Archimedean numbers.

To determine the drag coefficient the actual density of the liquid should be used while the viscosity should be corrected for the presence of fines and for volumetric concentration (Figure 5).

For volume correction to the viscosity, Thomas (1965) proposed the following equation with an exponential function:

$$\frac{\mu_m}{\mu_L} = 1 + K_1 C_v + K_2 C_v^2 + A \exp\{B C_v\} \tag{EQ 6}$$

K_1 is the Einstein constant of 2.5
K_2 = 10.05
A = 0.00273
B = 16.6
C_v = volumetric concentration

The magnitude of K_2, A and B, may however changer for very fine particles:

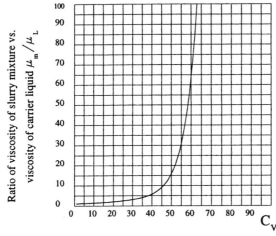

Figure 5: effect of volumetric concentration on viscosity of a slurry mixture (after Thomas 1965)

From the calculated deposition velocity, the pipe size can be determined based on the flow rate. With the short piping runs generally found in a mill operation there are some simplified guidelines that can be adopted for selecting pipe diameters:

- Coarse particle applications   SAG mill discharge   3.5 to 5 m/s (12 to 16 ft/sec)
- Medium particle sizing   Leach / Flotation Feed   2.5 to 3.5 m/s (8 to 12 ft/sec)
- Fine particle sizing   Concentrates   1.5 to 2.5 m/s (4 to 8 ft/sec)

The selection of the pipeline diameter should also consider the slurry density, with higher velocities used at the lower densities, and slurry viscosity.

The selection of a pipe diameter that results in a line velocity higher than required will result in an increased wear rate in the pipe and an increase in operating costs. The smaller pipe diameter will also result in a higher friction loss per unit length that will increase the power requirements for the pump and the unit operating costs.

The selection of a pipe that is too large for the projected flow regime will result in settling in the line and the creation of a dead zone. The dead zone will continue to form a deeper bed in a horizontal line until the apparent line velocity is sufficient to keep the particles suspended in the slurry or moving through the pipe by saltation. The formation of the dead bed will create an irregular shaped pipe with a very rough surface factor increasing the unit friction loss and power requirements for the pump. In addition the formation of a bed will result in two contact lines along the wall of the pipe that will be exposed to an increased wear rate from the sliding bed.

An oversized pipe on a vertical pipe will create a different set of issues. Coarse particles that cannot be suspended in the nominal slurry conditions will be in transition between settling and suspension until the apparent density is sufficient that the solids are carried as a high density "slug" to the next unit operation or until enough mass builds up in the horizontal to vertical transition to plug the line.

A typical example would be a SAG mill discharge application. The size of the coarsest particle in the SAG mill discharge pumping system (combined SAG mill / ball mill cyclone feed) will be dependent on the mill grates and the discharge screen openings. Normal maximum particle size is 19 mm (3/4") with coarser particles experienced with grate and/or screen failures. The slurry stream will also contain some loading of steel chips from the grinding media. These ball chips will have a higher specific gravity and irregular shapes and have a greater tendency to sink rather than be carried by the slurry. A low velocity line that plugs requires a significant effort and mill downtime to clean out. A low velocity line that tends to "slug" can result in plugging problems at the cyclones and/or create surging in downstream operations.

COMPOUND MIXTURES

It is now an accepted fact that mineral slurry typically consists of a mixture of coarse and fine solids. The coarse flow at the lower bottom of a horizontal pipe while the finer particles flow above the bed. Wasp et al (1977) proposed therefore that the pressure loss for each layer be computed. The concentration of particles of a certain size are established in reference to a layer "a" which is typically 8% of the pipe diameter. The Wasp method was derived from extensive research by various authors on coal slurries. The Wasp method requires repetitive iterations. A suitable computer program for such a method is presented by Abulnaga (2002).

The Wasp method is limited to slurries with $d_{50}>44$ microns. For finer slurries, non-Newtonian models should be used.

Various models are however based on a bi-modal distribution, meaning fine and coarse. A review of these models was presented by A.R Khan and J.F. Richardson 1996. Equations are then developed for what is essentially two layers, a bottom layer of coarse particles and an upper layer of fine particles. The two-layer model has gained acceptance in the oil sand industry and computer models are available from appropriate research labs such as the Saskatchewan Research Institute in Canada.

The two-layer method is limited to slurries with $d_{50}>74$ microns. Certain grades of oil sands have been found to yield $d_{50}<74$ microns.

The plant design engineer should be aware of the limitations of both methods. In many instances the routing of the pipe is short and involves essentially static rise or drop, with minimal friction losses when the pipe is properly sized. Engineers have been able to use simple formulae based on a Hazen-Williams factor or a Zandi factor. For long in-plant piping or pipelines more correct methods are required.

Models for non-Newtonian flows have been developed by various authors such as Darby, Heywood, Torrance, Wilson and Thomas, Slatter.

LAUNDER SIZING

In process plants where the lay out is conducive to utilizing gravity transportation of slurries open launders and gravity pipes are often used. The design of open launders and gravity pipes has traditionally been based on empirical formulas. The well-known Manning formula is usually used for designing launder systems however it determines the slope and configuration as if it were transporting water. Recently the Graf-Acaroglu relationship to size open launders as a function of density, particle size, hydraulic radius and volumetric concentrate has been used.

Open launders are usually utilized where access to the slurry is require ie for visual inspection where reagents are added. The slope of the launder is critical as too much slope increases the equipment elevation unnecessarily and too little slope results in spillage. Launders and pipes should be sized to run half full to alleviate the problems associated with surges in plant throughput.

By definition an open channel/launder is not full. The hydraulic diameter is the defined as the equivalent diameter of flow for an open channel. The hydraulic radius is defined as the ratio of the area of the flow by the wetted perimeter. It is also called in certain European books the hydraulic mean depth.

$$R_H = \frac{A}{P} \tag{EQ 7}$$

A = area in m^2
P = perimeter in meter
R_H = hydraulic radius in meter

The Manning number is correlated to the hydraulic radius R_H, and to the Fanning friction factor for flow in a launder by the following equation.

$$n = R_H^{1/6} / \sqrt{\frac{2g}{f_N}} \tag{EQ 8a}$$

n = manning roughness number
g = acceleration due to gravity or 9.8 m/s^2
f_N = fanning friction factor

if the Darcy friction f_D factor is used instead of the fanning friction factor, the Manning roughness number is expressed as

$$n = R_H^{1/6} / \sqrt{\frac{8g}{f_D}} \tag{EQ 8b}$$

For a fully developed and uniform flow, the slope or energy gradient of an open launder is established in terms of the head loss per unit of length (Henderson 1990).

$$S = \frac{H}{L} = \frac{f_N U^2}{2gR_H} \tag{EQ 9}$$

The slope S is expressed in decimals.

Figure (6) shows typical values of the hydraulic radius.

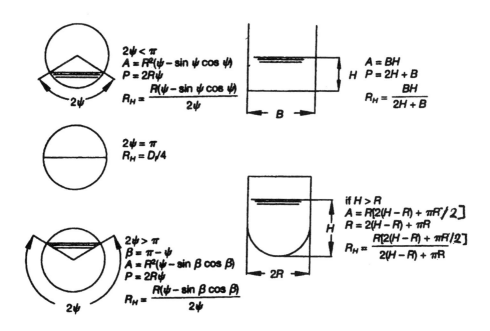

Figure 6 - Hydraulic Radius of open launders

Table 1: Typical values for the Manning's Number "n" for water flows (do not use for slurries)

Channel Surface	Manning factor "n" in ft$^{-1/3}$ s^{-1}	Manning factor "n" in m$^{-1/3}$ s^{-1}
Glass, plastic, machined metal surface	0.011	0.016
Smooth steel surface	0.008	0.012
Sawn timber, joints uneven	0.014	0.021
Corrugated metal	0.016	0.024
Smooth concrete	0.0074	0.011
Cement plaster	0.011	0.016
Concrete culvert (with connection)	0.009	0.013
Glazed brick	0.009	0.013
Concrete, timber forms, unfinished	0.014	0.0208
Untreated gunite	0.015 – 0.017	0.022 – 0.0252
Brickwork or dressed masonry	0.014	0.0208
Rubble set in cement	0.017	0.0252
Earth excavation, clean, no weeds	0.020	0.022
Earth, some stones and weeds	0.025	0.037
Natural stream bed, clean and straight	0.020	0.030
Smooth rock cuts	0.024	0.035
Channels not maintained	0.034 – 0.067	0.050 – 0.1
Winding natural channels with pools and shoals	0.033 – 0.040	0.049 – 0.059
Very weedy, winding and overgrown natural rivers	0.075 – 0.150	0.111 – 0.223
Clean alluvial channels with sediments	0.031 $(d_{75})^{1/6}$ using d_{75} size in feet	0.0561 $(d_{75})^{1/6}$ using d_{75} size in m

After Manning R(1895) and Henderson (1990)

Designing Launders for Slurry

The presence of solids accentuate the slope of the launder as determined by the size of the particle. Tournier and Judd (1945) reported that the specific gravity of the ore is an important factor to consider. Heavier ores require more slope to be transported in an open channel; as shown in Figure 7.

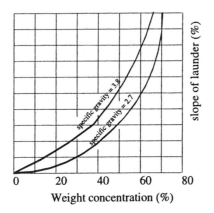

Figure 7: Launder Slope as a function of specific gravity and solids concentration.[1]

Tournier and Judd (1945) reported that the size of the particles play an important role, and larger particles require more slope as shown in Figure 8.

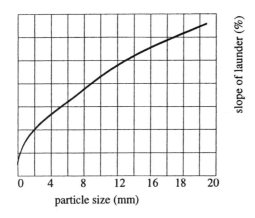

Figure 8: Launder slope as a function of weight

The first step when designing a launder for slurry is to determine the deposition velocity for solids. Dominguez et al. (1996) published an equation based on experimental data measured at Codelco and at the Chilean Research Center of Mining and Metallurgy. For cases where the viscosity effects are negligible:

$$V_D = 1.833 \, [8gR_H (\rho_S - \rho_m)/\rho_m]^{\frac{1}{2}} (d_{85}/R_H)^{0.158} \tag{EQ 10}$$

V_D = deposition velocity in m/s
g = acceleration due to gravity or 9.8 m/s^2
ρ_S = density of solids (kg/m^3)
ρ_m = density of mixture (kg/m^3)
d_{85} = 85% passage diameter of solids in m
R_H = hydraulic radius in m

However, in cases where the viscosity of the carrier liquid is instrumental, such as with alkaline water, Dominguez et al. (1996) derived the following equation:

$$V_D = 1.833 \, [8gR_H(\rho_S - \rho_m)/\rho_m]^{1/2} (d_{85}/R_H)^{0.158} \, 1.2^{(3,100/J)} \qquad (EQ\ 11)$$

$J = R_H(gR_H)^{1/2}/\mu_m$
V_D = deposition velocity in m/s
g = acceleration due to gravity or 9.8 m/s^2
ρ_S = density of solids (kg/m^3)
ρ_m = density of mixture (kg/m^3)
d_{85} = 85% passage diameter of solids in m
R_H = hydraulic radius in m
μ_m = the absolute viscosity of the mixture in Pa-s

These two equations clearly indicate that the deposition velocity is a function of the hydraulic radius, the density as well as the particle diameter. It is a far cry from the Manning based equations.

The characteristics of the slurry flow in open launders depends on the Froude Number defined as:

$$Fr = V / \sqrt{(gy_m)} \qquad (EQ\ 12)$$

At low Froude number (Fr<1), the flow is called subcritical by Civil engineers, a term that would tend to confuse slurry experts. Bed forms in the shape of ripples and anti-dunes, while at Fr>1, dunes form. The formation of these bedforms tend to throw away the concept of Manning roughness which is often replaced by a concept of roughness based on the particle size. Numerous equations for friction losses which take in account the particle size have been discussed by Abulnaga (2002), Graf (1971) and Yalin(1977). Green et al (1978) recommended that the launder be designed for a Froude Number in excess of 1.5 to minimize the formation of bedforms. This is not always possible when the topography may dictate the slope of the launder.

Recognizing the presence of bedforms in many launders, and considering that the effective roughness is equivalent to the average particle size, Graf and Acaroglu (1968) developed a method to compute the slope based on the particle size, the volumetric concentration, the hydraulic radius and the slope of a launder.

Velocity is expressed as:

$$U_f = \sqrt{\left(\frac{\tau_w}{\rho}\right)} = \sqrt{(R_H S g)} \qquad (EQ\ 13)$$

U_f = friction velocity
τ_w = shear stress at the wall in Pa
ρ = density of liquid in kg/m^3
S = slope of the launder in decimals
g = acceleration due to gravity in m/s^2

By assuming that the absolute roughness of the bed is equal to the particle diameter, Acaroglu and Graf (1968) proceeded to define the shear intensity parameter as:

$$\Psi_A = \frac{(\rho_s - \rho_L)d_\rho}{\rho_L S R_H} \qquad (EQ\ 14)$$

The power consumed with friction or head losses in the open channel is expressed in terms of the energy slope (head loss per unit length) and a non-dimensional transport parameter is derived as:

$$\varphi_A = \frac{C_V U_{av} R_H}{\sqrt{[(\rho_s/\rho_L - 1)g d_p^3]}} \quad \text{(EQ 15)}$$

By examining data from various authors and by regression analysis, Graf and Acaroglu extrapolated the following relationship:

$$\varphi_A = 10.39 \left(\Psi_A \right)^{-2.52} \quad \text{(EQ 16)}$$

or

$$\frac{C_V U_{av} R_H}{\sqrt{[(\rho_s/\rho_L - 1)g d_p^3]}} = 10.39 \left[\frac{(\rho_s - \rho_L) d_p}{\rho_L S R_H} \right]^{-2.52} \quad \text{(EQ 17)}$$

C_V = volumetric concentration of solids in decimals
U_{av} = average velocity in the launder in m/s
R_H = hydraulic radius in m
S = slope in decimals
d_p = average particle diameter in m
g = acceleration due to gravity or 9.8 m/s^2
ρ_S = density of solids (kg/m^3)
ρ_L = density of liquid (kg/m^3)

This equation was obtained for finely graded sand with a particle diameter between 0.091 mm and 2.70 mm. (0.0036 - 0.1063 in) and was studied in rivers and open channel flumes. This equation applied to both closed conduits and open channels as Graf (1971) explained. It is particularly well suited for saltation flow in closed channels.

$$\Psi_A = \frac{(\rho_s - \rho_L) d_p}{\rho_L S R_H}$$

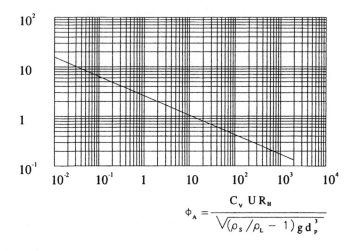

Figure 9: The Graf-Acaroglu relationship to size open launders as a function of density, particle size, hydraulic radius, volumetric concentration – adapted from Graf and Acaroglu (1968)

Non-Newtonian gravity flows are encountered with concentrate pipelines and with thickener underflow. Abulnaga (2002) discussed the methodology for design of such launders.

GRAVITY FLOW PIPELINES

During development of the plant equipment layouts it is important for designers to look for opportunities to use gravity to transport slurry between unit operations or between stages. The ability to use gravity eliminates unnecessary pump installations and reduces power requirements. Slurry transportation using gravity will typically use launders (flotation concentrates) or pipes (tank interconnections for leach and CIL tank trains).

Sizing of gravity flow pipelines for slurry follows a similar approach to that for pressurized lines. Determining and maintaining a slurry velocity is important to prevent plugging of the pipeline and some design allowance must be provided for varying flowrates. With the design of gravity flow pipelines it is important to remember that for a given flow and pipeline slope the line velocity in a ½ full pipeline is the same as for a full pipeline as a result of friction losses based on the slurry volume flow rate and wetted perimeter. For a given flow maximum velocity is through a ¾ full pipeline. In selecting a pipeline the diameter can be selected to allow capacity to increase by about 100%.

From the Manning equation for open channel flows of water:

$$Q_f = 0.463 * d^{8/3} * S^{1/2} / n \tag{EQ 18}$$

Q_f = full flow volume, cubic feet per second
d = pipe diameter in feet
S = energy loss, ft per ft of conduit length, approximately the slope of the conduit invert
n = pipe roughness (Manning number)

$$V_f = 0.590 * d^{2/3} * S^{1/2} / n \tag{EQ 19}$$

V_f = full flow volume, feet per second

From the process design criteria; nominal plant throughput, slurry density and the full volume flow rate is known. Utilizing the target pipeline velocities discussed previously it is possible to solve EQ (18) and EQ (19) to provide a preliminary sizing for pipe diameter, d and the slope, S.

UPCOMER DESIGN

The development of leaching to recover gold and carbon-in-pulp (CIP) and carbon-in-leach circuits (CIL) the design of gravity flow systems between tanks became very important. The movement of slurry in and out of the tanks to ensure good mixing conditions and to minimize short circuiting was also identified as a key area of piping design.

Upcomers are provided to allow the slurry to be collected near the bottom of the tank, drawn to the top for feeding to the top of the next successive tank, typical of a leach tank operation. For CIP and CIL circuits the use of an upcomer or downcomer is dependent on the type of in-tank carbon screen and the plant layout. The screens are mounted at the surface of the tanks and it is desirable to direct the slurry discharging through the screen near the agitator blades in the subsequent tank to effect the best mixing. Downcomer design must ensure that there is no restriction to the flow, as this can create a back up in the tanks preventing maximum throughput opportunities. Oversizing a downcomer is not critical. The downcomer shouldn't extend below the bottom agitator blades. Critical to downcomer operation is maintaining a consistent slurry density. A lower density slurry discharging to the top of an agitated tank will be distributed in to the tank slurry quickly. A lower density slurry delivered to a tank through the downcomer must have sufficient head to overcome the difference in density or a back up will occur in the tank train with the possibility of overflowing tanks.

Upcomer design considerations are more specific. It is desirable not to oversize the upcomer because the upflow velocity will be too low to carry the larger particles. Underdesigning the upcomer could lead to flow restrictions at higher throughputs. Recommendations from agitator suppliers suggested an upcomer flow velocity of 6 times the hindered settling velocity. The hindered settling velocity, as before, is a function of particle size and density. Upcomer design should be evaluated for nominal conditions. A check can be easily made for upset conditions that will be a result of low density, coarser grind and reduced throughput.

Determining the terminal settling velocity[2]:

$$V = \frac{(2gr^2)(d_1-d_2)}{9\mu} \qquad (EQ\ 20)$$

g = acceleration of gravity cm/sec^2
r = radius of particle cm
d_1 = density of particle g/cm^3
d_2 = density of medium g/cm^3
μ = viscosity of medium dyne sec/cm^2

Hindered settling velocity:

$$U_{ts} = V(1-Cv)^n \qquad (EQ\ 21)$$

Cv = Volume percentage of solids
n = 4.65

Upcomer velocity:

$$V_u = U_{ts} \times 6 \qquad (EQ\ 22)$$

PUMPBOX DESIGN

The pumpbox is an integral component of the pumping system and its design is critical for the successful pumping of slurries.

In determining a pumpbox size it is normal practice to keep the pumpbox height to a minimum whilst allowing sufficient volume for fluctuations in flow and sufficient retention time. The retention time is normally set at one minute to allow sufficient time for entrained air to escape. Plant layouts sometimes make it difficult to achieve this especially when sloped pumpboxes are used. Plant capacity will also limit retention time. It may not be practical to install a pumpbox with sufficient capacity to achieve the desired retention time. Pumpbox level sensors controlling pump speed and/or water addition will be important for minimizing process flow surges and reducing spills.

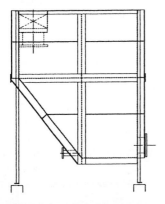

Figure 10: Typical Rectangular Pumpbox with a sloped bottoms

---

[2] Chemical Engineers' Handbook, Perry Chilton

A typical rectangular pumpbox with a sloped bottom is shown in Figure 10. The sloping of the pumpbox forces the solids into the pump before they can settle out. If the solids are allowed to settle periodic sliding of the settled material into the pump suction can occur. When this type of sliding of coarse material into the pump suction occurs the pump will operate erratically and can choke.

The design of pumpboxes for froth applications requires particular attention due to the blinding effect of froth on the pump and subsequent loss of pumping action. Standpipes are commonly used where the plant layout is suitable this allows a crushing effect on the froth and prevents it from being drawn into the pump suction. Froth factors are usually plant specific and normally range from:

1.1 to 1.5 – Brittle froth
1.5 to 2.0 – Tenacious froth

In certain plants froth factors of up to five have been used.

A common error in pumpbox design is to have the pipe delivering the slurry discharge to close to the suction of the operating pump. This will result in air being drawn into the pump suction adversely affecting the pump performance.

When a pumpbox has a standby pump consideration must be given to prevent or minimize a build up of solids in front of the standby pump when it is not in operation. A quick release dump valve on the pump suction and/or pumpbox should also be installed.

Where varying flows are expected the pumps should be fitted with VFD's unless the pumpbox has sufficient volume to cope with flow variations.

Where abrasive materials are encountered pumpboxes should be rubberlined. It is difficult to effect repairs to a rubberlined pumpbox. The main wear area will be around the discharge nozzle. This should be designed using sacrificial inserts.

PLANT PIPING LAYOUT

The routing of slurry piping will have a direct influence on the successful operation of a slurry transportation system. Pipe routing should be kept as straight as possible as every bend, elbow and "T" piece results in a potential area of wear and subsequent failure.

Methods of reducing pipe wear is to rotate straight lengths of pipe, utilize long radius bends and hoses. Short radius bends, tees and elbows should be avoided where possible. In high tonnage (high flowrate) plants the application of 3D or greater bends is often impractical because of support locations or space limitations. The use of "T's" or fabricated wear boxes provide effective solutions.

The routing and layout of large diameter (+150 mm) process plant piping should be taken into consideration in the early stages of design. Priority routing should be provided to the main process slurry lines because of the potential production impacts resulting from sanded lines or line failures. Wherever possible pipe runs should be designed in the horizontal and vertical planes. This facilitates design of the pipe supports. In addition, pipelines installed at an angle will wear out faster as a result of the sliding bed that will form on the bottom. Horizontal pipelines should be installed with a 2° to 3° slope to promote drainage from the line during shutdown. This slope is not expected to drain the solids. Some of the solids will settle to the bottom of the line but will be picked back up once the system is charged and pumping slurry again. The pipelines should be sloped back to the pump and/or to the discharge point, if opened to the atmosphere. Drains, or break connections, should be installed near potential plug points, for example, the horizontal to vertical transition on a cyclone feed line.

Plant piping design should minimize valves in slurry applications. Also pump suction lengths should be kept as short as practical. Long suction lines or obstructions (screen, valves) on the line can have significant negative impacts on the operation of the pumping and piping system. The use of 3D models assists greatly in preventing clearance, maintenance and pipe support problems.

For gold plants incorporating carbon absorption technology, carbon transportation will be either carbon/water or carbon/process slurry. Piping systems must be designed to minimize carbon degradation as this will impact gold recovery. Where practical, operations have achieved success field running HDPE pipe to the natural bending radius.

PIPING MATERIAL SELECTION

For slurry operations, steel pipe is rubber lined with natural rubber with a shore hardness of 40.

Polyurethane is an acceptable substitute to rubber particularly for grooved pipes connected by Victaulic and alternative grooved couplings. The maximum length of rubber-lined pipes is usually 12 m or 18 m depending on the capabilities of the fabricator.

High Density Polyethylene (HDPE) has been accepted as a substitute to rubber for fine particles of mild abrasivity up to 3.3 m/s but has failed on taconites and certain nickel and laterite ores. However HDPE can be used as a lining for pipes up to a length of 1 km, as was done on the Collahusi copper concentrate pipeline. The maximum rating of HDPE piping is typically 1.4 Mpa in Northern climates. HDPE is particularly suited for granular carbon pumping systems, sulfuric acid processes, and cyanide leaching.

For slurry with coarse particles, larger than 6.4 mm (¼-in), hardened steel pipe will be an alternative to rubber-lined steel. Some fittings are cast in ASTM A532 grades (Ni-hard, 16% chrome to 28% chrome white irons). Steel pipes can be manufactured for very high pressures up to 17 Mpa at 232°C. Mineral processing applications this would apply to include autoclave feed, paste-filling, long distance concentrate and tailings pipelines.

To minimize layout issues with piping material conducting hose has been used successfully for slurry applications. It is not common to find material conducting hose used for the full length of the line because the line needs to be completely supported along the horizontal. The main applications for hose have been to substitute for elbows (providing long radius bends for cyclone feed line applications, etc.) or for providing alignment and connection between sections of a pipeline.

Phosphate and phosphoric-acid based slurries, such as those used to pump phosphate matrix attack steel pipe, this creating a mechanism of erosion-corrosion. Some manufacturers offer special alloys, which are forms of stainless-steel reinforced with carbides.

The selection between a plastic (HDPE) and a steel pipe should be based on more than just the cost of the pipe. The plastic pipe will require more supports and on the rate of expansion. HDPE in warm South American or Middle Eastern and Australian environment the plastic pipe will tend to expand at higher rates than steel.

VALVE SELECTIONS FOR SLURRY APPLICATIONS

A number of specific types of valves are used in a typical mineral beneficiation plant and these each have specific uses.

The vales normally utilized in process plants slurry applications are:

1. Knife Gate Valves
2. Pinch Valves
3. Diaphragm Valves
4. Tech Taylor
5. Ceramic ball valves

Knife Gate Valves

- Knife gate valves have in recent years been extensively used in slurry pumping applications.
- Knife gate valves provide relatively trouble free on/off control for slurry applications.
- The knife gate utilizes a thick elastomer sleeve which has replaced troublesome hard seats, guides and packing. Various elastomer and knife materials are available to suit the required application.
- The knife gate valve can be operated manually or with actuators.
- One of the major advantages of the wafer style knife gate valve is that it takes very little room to install.

- The knife gate valve should only be used in on/off conditions and should not be used to control flow.

Pinch Valve

- Pinch valves offer a cost effective method of on/off and flow control for slurry applications.
- This type of valve has the advantage that the slurry only comes in contact with the rubber sleeve and the mechanism air or mechanical operates outside of the slurry.
- A disadvantage of this type of valve is that its bulky in construction especially where higher pressures are required. The overall length of the valve can also create problems with the piping layout.
- When using the valve as a control valve it offers durability and relatively accurate control.
- Pinch valves come in a variety of materials such as rubber, neoprene and urethane and this allows for the selection of the material to suit the application.

Diaphragm Valve

- Diaphragm valves are similar to a pinch valve however the diaphragm closes against a valve seat which is in contact with the slurry.
- The diaphragm valve is usually used for flow control. As the slurry passes between a valve seat and the diaphragm high maintenance costs result due to high failure rates.

Tech Taylor Valves

- The Tech Taylor type of valve is used where a standby pump delivers into a common line with the operating pump. The pressure from the slurry in the operating line keeps the valve ball in place over the stand-by pump discharge line. When the stand-by pump is engaged and the operating pump shutdown the ball is relocated by the slurry flow. The advantage of the Tech Taylor valve is that it operates automatically and allows either or both pumps to operate without any external operator.
- The Tech Taylor valve also has the advantage that it reduces the costly high maintenance that is required for this type of pump layout.
- Plug, gate and butterfly valves are normally not suitable for use in an abrasive slurry application.

Ceramic Ball Valves

- Ceramic ball valves are available for tailings and concentrate lines up to a diameter of 150 mm (6"). Because they are very expensive their use is limited to very high pressures, as those associated with positive displacement pumps.

INSTRUMENTATION AND MONITORING

Instrumentation and monitoring of the pipelines will be related to the process requirements and are discussed in detail elsewhere in the text. The main instruments related to piping for plant design will typically include:

- magnetic flow meter
- radiation gauges
- ultrasonic flow meters
- diaphragm-sealed pressure gauges
- float switches and level indicators
- ultrasonic level sensors

- bubble tubes
- differential pressure (dp) sensors

An important aspect of the instrumentation is to control the speed of flow in pipes and to maintain it above deposition speed. A magnetic flow meter, a nuclear radiation gauge or ultrasonic meter is non-intrusive. Various pieces of equipment from a pump, to a tumble mill must operate at the correct density and volume of slurry and need to be adjusted through instrumentation.

A level switch, float switch on a pump box is often used to adjust the flow of makeup water or the speed of a pump. A float switch on a floor sump is used to activate and shutdown a vertical slurry pump.

CONCLUSION

The last thirty years have seen considerable advancement in the science of slurry pumping. Process engineers have gradually increased the concentration of slurry, and the technology of grinding permits to produce much finer slurries for better recovery of ores. It is now accepted that the slurry in a plant is far from single-graded but consists of mixtures of fines and coarse or particles of different sizes. The new models for the deposition velocity take in account the d_{50} particle size. Friction calculations are also based on the characteristics of the layers of fines and coarse whether a multi-layer model or a two layer model is used. For fine mixture with $d_{50}<44$ microns, it is recommended to use non-Newtonian models.

A better understanding of open-channel flow is now achievable. The conventional Manning equations are now limited to very dilute slurry, as they do not account for particle size, volumetric concentration of solids and bedforms. Recent work from Dominguez et al (1996) based on extensive tests in Chile has yielded an acceptable equation for deposition speed in launders. The Graf-Acaroglu equation establishes a correlation between the volumetric concentration, the hydraulic radius, the d_{85} size, and the specific gravity of the ore.

More research is needed to understand non-Newtonian gravity flows as they are being observed more frequently in tailings disposal, thickener underflow and concentrate circuits.

REFERENCES

1. Abulnaga B.E. 2002 – *Slurry Systems Handbook* – McGraw-Hill
2. Dominguez, B., R. Souyris, and A. Nazer. 1996. Deposit velocity of slurry flow in open channels. Paper read at the symposium, *Slurry Handling and Pipeline Transport. Thirteenth annual International Conference* of the British Hydromechanic Research Association, Johannesburg, South Africa.
3. Durand, R., and E. Condolios. 1952. Experimental investigation of the transport of solids in pipes. Paper presented at Deuxieme Journee de l'hydraulique, Societe Hydrotechnique de France.
4. Gillies, R. G. J. Schaan, R. J. Sumner, M. J. McKibben, and C. A. Shook. 1999. Deposition Velocities for Newtonian Slurries in Turbulent Flows. Paper presented at the Eng. Foundation Conference, Oahu.
5. Graf, W. H. 1971. *Hydraulics of sediment transport.* New York: McGraw-Hill.
6. Graf, W. H., and E. R. Acaroglu. 1968. Sediment transport in conveyance systems. Part I. *Bulletin. Intern. Association of Sci. Hydr., XIIIe Annee,* no. 2
7. Green, H. R., D. H. Lamb, and A. D. Tylor. 1978. A new launder design procedure. Paper read at the Annual Meeting of the Society of Mining Engineers, March, in Denver, Colorado.
8. Hanney K.E.N. 1982. Selection and sizing of slurry lines, pump boxes and launders. Design and Installation of Comminution Circuits. SME.
9. Henderson, F. M. 1990. *Open channel flow.* New York: Macmillan Publishing Co., Inc.
10. Manning R.1895. On the flow of Open Channels and Pipes. Transactions, Institution of Civil Engineers of Ireland, Vol 10, pp 161-207;14
11. Khan, A. R., and J. F. Richardson. 1996. Comparison of coarse slurry pipeline models. Paper presented at Hydrotransport 13 pp 259 –281 published by the BHR Group, Cranfield, UK
12. Schiller, R. E., and P. E. Herbich. 1991. Sediment transport in pipes. Published in *Handbook of Dredging.* Edited by P. E. Herbich. New York: McGraw-Hill, Inc

13. John C. Loretto and ET Laker. Process Piping & Slurry Transportation. 1978. Mineral Processing Plant Design.
14. Thomas D. G. 1965. Transient characteristics of suspensions: part VIII. A note on the viscosity of Newtonian suspensions of uniform spherical particles. *J. Colloid Science* 20:267.
15. Tournier E.J. and E.K.Judd.1945. Storage and Mill Transport – Section 18 – Handbook of Mineral Dressing – John Wiley & Sons.
16. Wasp, E. J., J. P. Kenny, and R. L. Gandhi. 1977. *Solid-Liquid Flow – Slurry Pipeline Transportation.* Aedermannsdorf, Switzerland: Trans-Tech Publications
17. Yalin, M. S. 1977. *Mechanics of sediment transport.* 2nd Edition. Toronto: Pergamon Press.

Slurry Pipeline Transportation

Brad L. Ricks[1]

ABSTRACT

Before the 1950's slurry pipelines were used mainly for waste (tailings) disposal where the volumetric flow was high and the distances were relatively short. Solids transport was achieved via gravity systems with high line velocities. However, over the last few decades, slurry technology has made sufficient advances to become a very reliable long distance transportation[2] method for many useful products such as steam coal, mineral concentrates (iron and copper in particular), phosphate, and other minerals.[3] Also, in order to address environmental and public safety issues, slurry technology advances have been utilized for long term and long-distance tailings disposal systems. This sub-section (D-G-4) deals with the current technical status of long-distance slurry pipeline systems.

INTRODUCTION

The idea of hydraulic transport of solids is not new and in fact demonstration models and patents were issued prior to the turn of the century. However, the development of the modern slurry pipeline industry really began in the early 1950's with the systematic experimentation of coal slurries by Edward J. Wasp, then of Consolidation Coal. Wasp synergistically combined the works of several scientists and engineers to develop his model that relates the homogeneous (uniform solid's distribution throughout the pipe or conduit's cross section) and the heterogeneous (non-uniform solid's distribution) flow characteristics of slurry for the design of long distance slurry pipelines. (Thompson 1981.)

During the course of the development effort, Wasp realized that the key to designing reliable slurry pipeline systems was an understanding and control of the slurry properties rather than in the development and selection of exotic materials or special equipment. (Wasp 1977.) This philosophy was instrumental in making the industry viable. Long distance slurry pipelines were then developed using conventional equipment and construction practices developed by the oil and gas industry in prior decades. The fundamental problems were and remain:

      1. Stable flow conditions throughout the pipeline and
      2. Control or elimination of the pipe's internal corrosion and/or erosion to the extent that design lives up to 40 years are possible.

Based upon these concepts the industry has built several pipelines. In fact, known to this author, over 4,000 km of long distance slurry pipelines have been constructed with over 125 million tones per year combined capacity.

Nevertheless, in recent years, special equipment and materials have been developed that are utilized to enhance the economics and reliability of long distance slurry pipelines. Also, special construction techniques and procedures have been utilized.

---

[1] Brad Ricks Advanced Slurry Systems {BRASS}, San Ramon, California {www.brassengineering.com}
[2] The definition of 'long-distance" systems is somewhat arbitrary. Some systems lengths are only a few kilometers while other are hundreds of kilometers.
[3] Other minerals include, zinc concentrates, lead concentrates, mineral ores, limestone, Gilsonite, and kaolin.

And lastly, slurry preparation techniques have been developed for not only hydraulic transport of the solids, but also to better meet the needs of the end user.

The following article will mention many of these advances.

TRENDS

Not only are slurry pipelines economic but, with concerns for the environment, long distance slurry pipelines have become more attractive as a means of bulk solid's transportation, mainly because they are: 1/ buried and out of sight, and 2/ the route is quickly reclaimed by the native environment.

In the case of tailings disposal, impoundment sites are selected more for environmental reasons than for convenient locations, thereby necessitating longer pipelines. By necessity, modern materials, equipment, slurry preparation and designs are employed.

Whether, the slurry pipeline transports mineral products or tailings, in most cases the slurry pipeline is the most pragmatic choice. This is because they must transverse rough terrain were other haulage methods can not. Shown in Figure 1 is a bridge that spans a gorge in a mountainous sector.

Figure 1: Alumbrera Copper Concentrate slurry pipeline's bridge spanning the Arroyo Cangrejillo Gorge. Photo was provided courtesy of Minera Alumbrera Ltda., Catamarca, Argentina (Ricks, Connelly, and Moreiko, 1998).

In many cases beneficiation processes produce mineral concentrates that are suitable for long distance slurry pipeline transport without any further processing. Common examples are iron, copper, zinc, phosphate, limestone and kaolin. In other cases, such as steam coal, the slurry is prepared to meet specified characteristics. Nevertheless, in order to optimize the development of a mineral deposit, one must consider the ore dressing process, the slurry transportation requirements, and utilization requirements of the overall system.

Slurries can be classified into one of three categories:

1. Homogeneous (due to the force of gravity, it is recognized that truly homogeneous slurries are nonexistent but rather are pseudo-homogeneous)
2. Heterogeneous where the concentration of the slurry particles varies greatly from top to bottom of the pipe's cross section
3. Complex (not completely homogeneous nor heterogeneous)

Heterogeneous slurries are typically composed of coarse particles which are not uniformly suspended in the pipeline and tend erode the pipe bottom. Associated with heterogeneous slurries are high line velocities and high-energy consumption. Because of these drawbacks, heterogeneous long distance pipelines are generally not feasible.

Most slurries are either homogeneous or complex and behave in a non-Newtonian manner. Typically, with the exception of coal, the solid's volume fraction is less than 0.40. These slurries are considered to be "conventional".

In general, for long distance pipelines the slurry properties are such that during transport, the slurry is uniform from top to bottom in the pipe. That is to say, the slurry should be a homogeneous fluid having the solids uniformly suspended through the pipeline's cross section. In reality complete homogeneity is approached but not achieved (pseudo-homogeneous). Usually the slurry specifications require a somewhat "fine" particle size distribution (PSD) and sufficient concentration to produce a "suspending" viscosity. The line velocities are such that friction losses are reasonably low in order to be economic and erosion of the pipe wall does not occur. Yet at the same time, the line velocities must be sufficient to generate enough turbulence to uniformly suspend the slurry particles. Normally, the long distance pipeline transports conventional slurry

For short distance pipelines, the degree of homogeneity is not as critical. Therefore, the design approach can be quite different than for a long distance system. For example, the slurry PSD may be relatively coarser for the short distance pipeline resulting in a fluid with non-uniform characteristics across the pipe's cross section (heterogeneous flow). Usually the line velocities are higher, relative to long distance systems, and the system design allows for some erosion of the pipe's wall. These pipelines are designed to be periodically rotated, or the pipe material may be erosion resistant, or it could be steel pipe with erosion resistant liners. The most common erosion resistant materials are rubber and polyurethane using a pipe joining method that employs a suitable flange or coupling.

Alternately, "non-conventional" slurry transport systems are emerging. The non-conventional slurries have high solids concentrations and are highly viscous. Associated with these systems are high friction losses and relatively low line velocities. The main advantage is that carrier fluid (usually water) quantities are reduced and any end "de-watering" process is eliminated.

RELIABILITY

Slurry pipelines are now regarded as a very reliable transportation mode For example, the Black Mesa Pipeline in the USA has historically been over 99.5 percent reliable (Montfort 1981) in its 32 years of operation. The Samarco Pipeline in Brazil has a similar reliability (Ricks 2000). In one case, a grassroots slurry pipeline system was constructed and put in operation in spite of the fact that it paralleled an existing railroad for much of its route. In that case the pipeline was selected not only for its economic benefits but also because of high reliability. Shown in Figure 2 is "Availability versus Probability" plot of a long distance slurry pipeline system derived from a detailed Monte Carlo Simulation. The figure considers natural events, corrosion, third party incidents, major component failure, and system component maintenance and repairs.

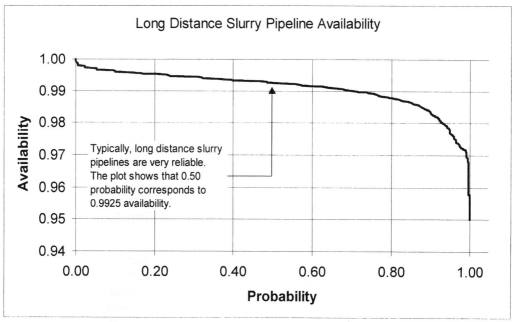

Figure 2: Long distance slurry pipeline availability versus probability derived from a detailed Monte Carlo Simulation. Note that the average result (0.50 probability) is an availability of 0.9925.

SLURRY PIPELINE DESIGN

In every case, the decision to develop a project is political. We use technical and economic projections to persuade the involved parties to proceed. Historically, slurry pipelines have been technically and economically persuasive. They enjoy the same type of economy of scale as gas and petroleum pipelines. In general this is true for relatively high volume and moderate distances. However, depending upon the slurry and terrain, it would be remiss not study the feasibility of slurry pipelines for almost every bulk material transportation project. For example for coal slurry pipelines, distances over 80 km and mass rates of 4 or 5 million tones per year are viable. For copper concentrates, distances over 15 kilometers and volumes over 0.75 million tones per year are viable. For a typical slurry pipeline, the major contributions to unit transportation costs are capital related fixed costs that are not subject to inflation.

The major elements of a slurry pipeline system are:
- the slurry preparation facilities such as a mineral concentrator;
- carrier liquid supply (e.g. water)
- slurry surge storage at the preparation facility;
- the mainline with pump stations, valve stations, and in some cases energy dissipation stations
- communications and control system
- intermediate slurry storage facilities (if required)
- Terminal slurry storage facilities.

Technical Design Bases

Although cost factors are normally considered an important element in system design, they are generally very specific to each project and are not discussed in this article. So the purpose of this discussion is held to technical issues. The technical design bases for slurry pipelines fall under three general categories:

1. Slurry Properties
2. Capacity
3. Terrain

Slurry Properties. Slurry characteristics are determined by the carrier fluid (usually water), the solids particles, and interaction between the carrier fluid and solids in the mixture.

Liquid. Several carrier fluids have been proposed and studied, such fuels, oil, alcohol, and liquid carbon dioxide (Ricks 1988). However in nearly all the operating systems the carrier fluid has been water at or near ambient temperatures. The relevant physical properties for the carrier fluid are 1/ specific gravity, 2/ bulk modulus, 3/ viscosity, and 4/ vapor pressure. These properties are well understood as a function of temperature. Notwithstanding, for most slurries with volume concentrations greater than 30 percent, the temperature of the carrier fluid has a diminished effect upon the overall slurry properties.

Solids. The overall slurry properties are affected by the following solid's properties: 1/ specific gravity (SG), 2/ particle shape, 3/ particle size distribution (PSD), and 4/ any surfactants acquired during the beneficiation process.

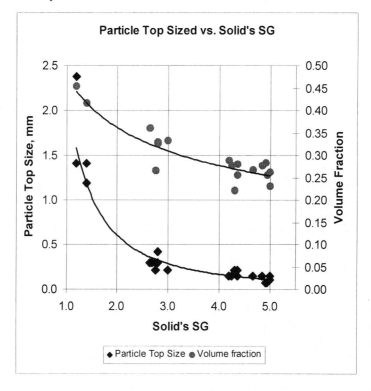

Figure 3: The relationship between solids specific gravity and "conventional" slurry top particle size and solid's volume fraction. The volume fraction relationship assumes that the carrier fluid is water at ambient temperatures. The volume fraction represents the higher values. Many tailings systems operate with lower volume fraction values and are not shown in the Figure.

Although it is possible to pump virtually any combination of solids and carrier fluid, strict restriction on solids PSD and concentration are required to ensure a slurry: 1/ will maintain predictable and stable flow conditions, 2/ that will it not wear the pipe bottom, 3/ and it can be shutdown and restarted. Inter-related is the solids SG with both concentration and PSD. Shown in Figure 3 is a rough relationship between a conventional slurry particle's "top size", volume concentration and the solid's SG.

Mixture. The solid's concentration, particle shape, and acquired surfactants effect the particle's settling characteristics and the slurry rheology. In general, most slurries are characterized as non-Newtonian. A comparison of selected rheological types is shown in Figure 4 showing rheogram curves for each type.

Newtonian	$T = \mu \gamma$	
Pseudoplastic	$T = K\gamma^n$	$n < 1$
Dilatant	$T = K\gamma^n$	$1 < n$
Bingham Plastic	$T = T_o + \eta\gamma$	
Yield-Power Law	$T = T_o + \eta\gamma^n$	

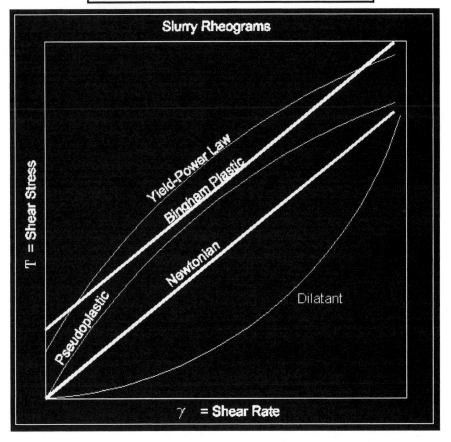

Figure 4: Rheogram comparison of selected constitutive equations for rheological models associated with slurries. The most common rheological models used for conventional slurries are the Newtonian model and Bingham-Plastic. Technically the nomenclature for the above is: "μ" is defined as the Newtonian viscosity; "K" is defined as the power-law consistency index; "n" is the power-law flow behavior index; and "η" is defined as the Bingham-plastic coefficient of rigidity.

Establishing the Slurry Concentration Range. Holding the PSD, solid's SG, and process reagents (such as pH and process surfactants) constant for a given slurry, the rheological properties are a strong function of concentration. For most conventional slurries, a solid's volumetric concentration up to 40 percent is readily handled per Figure 3. For these slurries, the solid's surface area to volume ratio is relatively low. However for some slurries such as sludge and minerals with fine clays, the solid's surface to volume ratio is quite high. In these cases, the solid's volumetric concentrations of 15 percent or less is often the practical limit. For these slurries, knowledge of the rheological characteristics is important.

As a rule of thumb, usually a comfortable concentration is 10 to 15 volume percent less than the final settled concentration (Aude and Thompson 1985.) The settled concentration can be determined by allowing the slurry to settle and then calculating its settled concentration by comparing the column height of the fluid to the column height of the liquid-slurry interface.

Shown in Figure 5 is a typical plot of slurry viscosity[4] versus the slurry concentration. The concentration range is from 59 to 64 percent. This range produces viscosities that provide sufficient suspending characteristics yet the slurry is not extremely sensitive to concentration changes.

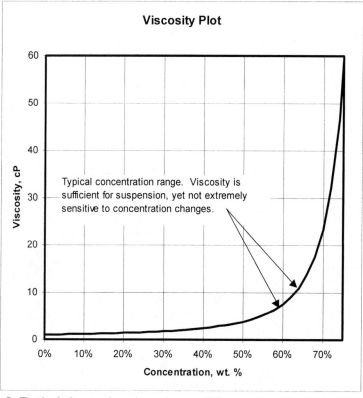

Figure 5: Typical slurry viscosity versus solid's concentration plot for conventional slurry. Note the pumping range is from 59 to 64 percent

---

[4] In this case the reference to viscosity is loosely applied. It could refer to viscosity for a Newtonian fluid or the coefficient of rigidity for a Bingham plastic fluid.

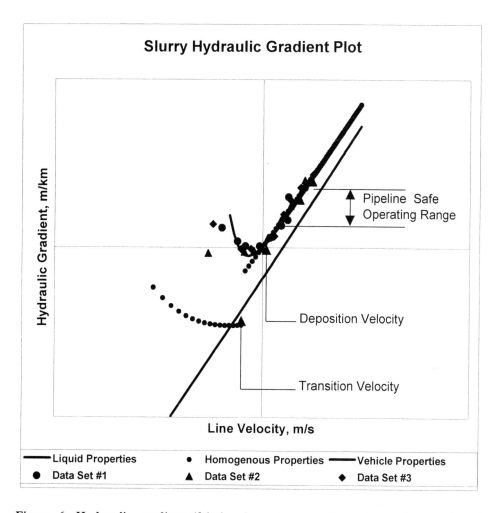

Figure 6: Hydraulic gradient (friction losses) versus line velocity for a selected slurry test. A Bingham plastic rheological model is used. A comparison is shown for the carrier fluid (water a Newtonian fluid) and calculated losses based upon slurry homogeneous properties (homogeneous fluid) and vehicle properties (complex fluid). Note as the line velocity increases, the homogeneous and vehicle calculations become indistinguishable. For buried long distance systems, such as concentrate pipelines, the "safe" operating range is well above the deposition velocity. For short systems such as tailings disposal, the normal operating range need not be as conservative.

Predicting Friction Losses. The mainstay of the slurry design calculation is the Wasp model. At given flow conditions in a pipe, the model determines the degree of heterogeneity (or conversely, the degree of homogeneity) of the solid's particles. It then determines the friction losses contributions of the vehicle (the supporting pseudo-homogeneous slurry) which suspends the heterogeneous solids (the bed). The total friction loss is calculated by summing losses due each. However, the model assumes Newtonian behavior of the vehicle. Most conventional slurries fall within the experience represented by the model, particularly in the turbulent flow regime. In this case the degree of "non-Newtonianess" is not significantly large to affect its predictive accuracy. Refer to Figure 6 showing a plot of friction losses in terms of a hydraulic gradient versus line velocity for selected slurry tests. Note the plots compares the calculated friction losses for the carrier liquid to that of the slurry based upon "homogeneous" properties and "vehicle" properties. Note also, that at higher velocities, where the fluid turbulence is sufficient, the vehicle and homogeneous friction losses correspond. This is because, at these velocities, the turbulence is sufficient to produce a pseudo-homogeneous suspension throughout the pipe or conduit's cross-section. In this case the vehicle and homogeneous conditions are indistinguishable.

Referring to Figure 6. If the line velocity is too low, say below the deposition velocity, then a bed of slurry forms and it is dragged along the bottom of the pipe. Pipe bottom wear results. If the velocity is too high, then high friction losses are encountered. If the line velocities become too high, then abrasive conditions develop. The preferred operating conditions are in the sector denoted by the "safe operating range." The velocities are sufficient to suspend the solids yet sufficiently low to keep friction losses within economic levels. At the same time the velocities are low enough that abrasion does not occur. Refer to Figure 7 showing the calculated solids distribution across the pipe cross-section. It shows the concentration ratio of a particular particle size at the top of the pipe compared to the bulk concentration.

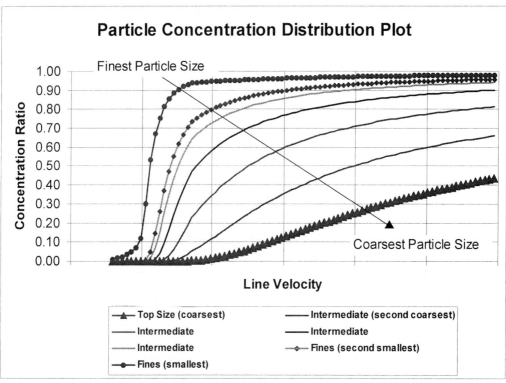

Figure 7: Particle concentration distribution plot showing the ratio of the concentration of a particular particle size at the top of the pipe compared to the bulk concentration. As the line velocity increases, turbulence becomes sufficient to suspend the particles more uniformly. Note the larger particles are not as readily suspended as the fine particles. Also, at low velocities the slurry is mainly a heterogeneous fluid, but as the particles for each size are suspended, the slurry becomes a complex mixture, and finally for the fine particles, the slurry becomes nearly homogeneous.

Conventional Mineral Slurries. To date mineral slurries have been conventional slurries and they are usually prepared as a natural consequence of their beneficiation process. These process usually involve crushing and grinding, and separation of refuse material through washings, flotation, or other separating techniques. The mineral concentrates are usually thickened and held in agitated slurry storage tanks prior to being committed to the pipeline. In these cases, the slurry pipeline designer does not have much control over the PSD but has some latitude over the concentration and line velocities. The majority of mineral concentrates are iron, copper, phosphate, and limestone.

Conventional Tailings Slurries. Due to environmental reasons, the length of tailings disposal pipelines have increased. Gone are the days of dumping the tailings at the most convenient downhill slope. Typical of many tailings systems, the beginning is at high elevations and the impoundment is at lower elevations. Conventional tailings slurries lack the high degree of concentration and particle size control normally associated with concentrate slurry pipelines. Therefore, the tailings pipeline designer prepares for many scenarios using heterogeneous and homogeneous models in order to account for all possibilities. Sometimes systems must dissipate excess energy in order to control the pipeline velocities. Choking is evolving as preferred methods of energy dissipation though cascades and drop boxes are used.

Non-conventional Slurries. Non-conventional slurries are highly concentrated and are non-Newtonian, with most being significantly non-Newtonian. Consequently, other rheological models have been used to describe their behavior. Refer to Figure 3 for Bingham plastic, pseudo-plastics, dilatant, and yield-power law fluids. These slurries have a high degree of homogeneity and are treated accordingly. It is argued that if a slurry is modeled for laminar flow (rheogram), then the same momentum transfer will also be carried into the turbulent regime. For example, if a slurry is described by a Bingham plastic model, then terms such as the yield stress must be accounted for in the turbulent flow calculation. These models have been worked out and are now being used (Hanks 1978; Darby 2001; Hanks and Ricks 1975; Hanks and Dadia 1971.)

Thickened Tailings Disposal (Non-conventional). In recent years, the concept of thickened tailings disposal (TTD) is gaining recognition (Robinsky 1999). It has the advantage of disposing of highly concentrated tailings and therefore reduces the storage volume required. Also, it eliminates the need to install water reclaim systems. To put the difference in perspective, conventional tailings slurries usually are transported between 15 and 35 percent concentration by weight while TTD concentrations are around 65 to 70 percent. Because the tailings are waste material, a wide range of solid's particles and tramp material is introduced into the disposal line. A TTD slurry is viscous and homogeneous throughout the flow range. For conventional slurries the velocities are usually from 2.5 to 3 meters per second in order to maintain the slurry solids in suspension. In contrast, TTD transports with line velocities below one meter per second.

Non-conventional Steam Coal Slurries. Non-conventional steam coal slurries have been proposed that avoid the de-watering process prior to being fed to the boiler. Slurries such as coal-oil-mixtures (COM), and coal-water-mixtures (CWM), slurries using liquid carbon dioxide, and other fuel mixtures are being developed. Of the above, the most advanced to date is CWM. CWM consists of a highly concentrated slurry that can be directly fired in a boiler. The slurry is homogeneous and the line velocities are low (laminar flow). Pilot systems have been built and tested. A commercial system has been built in Russia. The results are not well known, it but appears that there is more development required.

Determining Minimum Velocities. Referring to Figure 6, note that two velocities are specifically annotated, namely "transition velocity" and "deposition velocity." The transition velocity corresponds to the critical velocity where laminar flow transitions to turbulent flow. For a Newtonian fluid this would correspond to the laminar-turbulent transition for the classic Reynold's number of 2100. The deposition velocity corresponds to when the slurry particles begin to saltate. Usually the large particles require higher velocities for suspension than do the finer particles.

Transition Velocity. The transition velocity is determined for homogeneous fluids. For a Newtonian fluid, that velocity would correspond to the well know critical Reynold's number value of 2100. For non-Newtonian fluids, the "critical Reynold's number" value varies depending upon the degree of non-Newtonian properties of the fluid. Since a Bingham plastic model is used for most conventional slurries, it is used to illustrate the transition velocity determination. The degree of non-Newtonian behavior can be described by a dimensionless group of factors known as the

Hedstrom number (N_{He}). The larger the Hedstrom number for a given slurry the more non-Newtonian its behavior. It has been shown that there is a relationship between the critical Reynold's number (N_{ReC}) and the Hedstrom (Hanks and Dadia 1971):

$$N_{ReC} = N_{He} \frac{\left\{1 - \frac{4}{3}\xi_{oC} + \frac{1}{3}\xi_{oC}^4\right\}}{\xi_{oC}}$$

Where

$$\frac{\xi_{oC}}{(1-\xi_{oC})} = \frac{N_{He}}{16,800}$$

Similar relationships have been developed for pseudo-plastics and yield-power law models (Hanks and Ricks 1974, 1975.)

Deposition Velocity. The deposition velocity is related to complex slurries and heterogeneous flow. Durand did the classic work where he proposed that the deposition velocity was proportional to the square root of the pipe's diameter multiplied by a density factor (Durand 1952):

$$\text{Deposition Velocity} \sim \sqrt{D * (S - 1)}$$

$$S = \text{Solids SG / Liquid SG}$$

Wasp added to the above an additional factor to account for the solid's particle size (Wasp 1977):

$$(d / D)^{1/6}$$

However, according to experience the slurry concentration has a significant role in the deposition phenomena. It has been observed that for conventional mineral concentrates the degree of heterogeneity or complexity of slurry decreases as the concentration increases. Therefore, the deposition velocity decreases with concentration up to the point where the viscosity (concentration) is sufficient to produce a homogeneous slurry. Thereafter, the deposition velocity and the transition velocity become virtually identical. To illustrate, refer to Figure 8 showing the relationship between deposition and transition velocities for a selected concentrate slurry. A much more sophisticated calculation is used to determine this relationship.

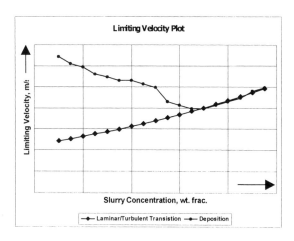

Figure 8: Limiting velocity plot of a selected conventional slurry showing transition velocity and deposition velocity relationship.

Capacity and Diameter Selection. Determining the pipeline diameter is an iterative process. Friction losses, limiting velocities, concentration ranges, power consumption, energy dissipation, material costs, and installation costs are all balanced in order to design the most cost-effective system. Furthermore, a overall approach to the entire project may be conducted in order to consider the mine, ore beneficiation, transportation, and utilization of the mineral in order to arrive at an overall system optimization. Usually the concentration range is established and then to start, for the concentration range, candidate diameters are determined based upon the capacity requirements of the system. Normally, the velocities will fall in the range of 1.25 to 2.1 meters per second, with the most likely bulk velocities falling inside 1.65 to 1.85 meters per second. In the end, the selected pipeline diameter must result in bulk velocities above the minimum operating velocity yet still low enough to keep friction losses within reason in order to provide economic transportation. Turndowns are generally controlled via flow and concentration control and are of the order of 0.75. to 0.80.

Terrain. In many cases slurry pipelines are selected because of the terrain. Usually the route is in difficult and remote locations. The slurry pipeline lends itself to such situations for the following reasons:
- Pipelines are easier to construct than railroads and truck haul roads because the pipeline grade (slope) requirements are less restrictive
- Construction rates are faster using pipeline construction techniques developed over many years
- The land is naturally reclaimed in a relatively short time period
- The pipeline is underground and less susceptible to third party incidents
- The system is remotely controlled
- Major maintenance is usually at pre-determined locations such as pump stations, valve stations and terminal facilities. Maintenance at intermediate points along the route is infrequent and usually does not affect production while being done.

Usually candidate route corridors are identified and studied regarding technical and economic issues using topographic maps or preliminary surveys. Once the corridor is determined, the route is "staked" based upon length, slope criteria, construction criteria and property issues. Right-of-way and easements are finalized and detailed surveys are performed. Finalizing a route can be iterative. As the route is finalized, the system design is refined and finalized.

Computer Simulations. Due to the development of computer technology over the last couple of decades, dramatic changes in design methods and analysis have transpired (Ricks and Aude 1992.)

Over the last ten years, computer simulations have been employed from a project's beginning through to commercial operation and control: namely conceptual engineering, basic engineering, detailed engineering, commissioning and startup, and into operation and control. Shown in Figure 9 is a graphic display of a simulation program. The ability to view results of steady state and dynamic graphic displays has given the engineer immediate feedback on cases under analysis. Also, from a practical standpoint, the input data and design assumptions are checked and verified.

The benefits of computer simulations have resulted in:
- Reduced engineering labor costs
- More cost effective design
- Better designer understanding of operations and better development of operating procedures
- Better operator understanding of system operations
- And finally designs that are rigorously proven to be within safety codes.

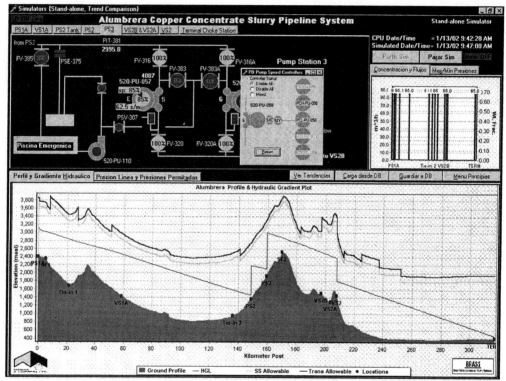

Figure 9: Graphic display of a computer simulation program used for system design, operating procedure development and system monitoring. Provided courtesy of Minera Alumbrera Ltda.

Codes

Since 1986, slurry pipelines used the American National Standard (ANSI), American Society of Mechanical Engineers (ASME) B31.11 as the design code for slurry systems. Its origin stems from the initial ASME code for pressure piping, B31, and in particular it is a modification of ASME B31.4, "Liquid Petroleum Transportation Piping Systems." Prior to 1986, B31.4 was used for cross-country slurry pipeline design. B31.11 is basically the same as B31.4 with the following exceptions:

- For non-hazardous water based slurries, provisions in the code for flammable and explosive petroleum products are dropped or not as stringent
- Internal corrosion and abrasion aspects are treated more stringently
- The design factor has been increased from 0.72 to 0.80

An update is issued approximately every three years.

System Components

Support components such water supply, slurry preparation, and dewatering facilities are not discussed herein. Also, not discussed are communications and controls components. Those component covered here are:

- slurry surge storage tanks
- high pressure mainline pumps
- slurry flow control valves
- mainline pipe materials
- energy dissipation

Slurry Storage. It is common practice to install surge storage at facility interface locations. The most common surge storage location is between the slurry pipeline or its first pump station (if required) and its preparation facility. For most long distance mineral slurry pipelines, agitated storage tanks are used. For many systems, the surge storage is a sump. The surge capacity may range from a few minutes of pipeline flow to several hours, or weeks, or months. The storage can be classified as active or inactive. Active storage refers to slurry that is kept homogenized (solids in suspension) and is ready for transportation. Inactive storage refers to long term storage where the solids are allowed to settle. The purpose and location of slurry storage considers the following:

- at pipeline feed locations, such as concentrators or preparation plants, storage is used to separate different types of slurry such as copper concentrate and zinc concentrate, or caps and tails for coal systems
- at intermediate pump stations to facilitate for re-homogenization of slurries restarted after shutdowns, or slurries with density waves, or to facilitate batch operations
- at pipeline terminals as surge storage between the pipeline and terminal processing facilities
- at pipeline feeder junctions or distribution junctions
- emergency dump storage (inactive storage)

Each storage facility is selected based upon logical criteria appropriate to any specific project.

Agitated Slurry Storage Tanks. Agitated slurry storage tanks are designed using API Standard 12D, "Specification for Field Welded Tanks for Storage of Production Liquids", or most commonly, API Standard 650 "Welded Steel Tanks for Oil Storage" for guidance. Tank tops are not required and the tank design must be structurally modified to mount an agitator on a bridge and account for the forces generated by the agitator. The tanks are cylindrical, baffled and having a "flat" bottom[5] with a wear plate and bump ring installed directly below the agitator's impeller/s. Sizing criteria are usually based upon the number hours pipeline capacity considered appropriate for the project under consideration. These values can range from a few minutes to several hours, but are normally from 8 to 24 hours of pipeline capacity. Since vessel materials are minimized for a fixed volume at about 1.05 height-to-diameter ratio (Z/D) it has been common practice to select

---

[5] The tank bottom has a slight slope from the center to rim.

Z/D ratios between one and 1.05. However, one study has shown that this ratio is not necessarily the most economic. Depending upon the circumstances, the optimized Z/D ratio can vary from 0.63 to 1.05 (Von Essen and Ricks 1999).

Two types of axial flow impellers are used: 1/ hydrofoil, and 2/ pitched-blade turbine. The hydrofoil type has the advantage of high pumping efficiency, meaning low power consumption. It has the disadvantage of a low live volume to total volume ratio because the impeller must be mounted about to 25 to 30 percent of the tank height above the floor. Recent installations have employed the both impeller types, using hydrofoil main blade and a smaller pitched blade turbine near the tank floor. This is done to utilize as much of the total volume as possible. In Figure 10 is a sketch showing typical dimensions for an agitated mineral-concentrate slurry storage tank.

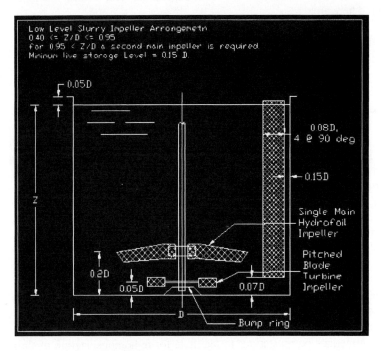

Figure 10: Typical Agitated Slurry Tank. Data for the sketch were provided courtesy of Philadelphia Mixers, Palmyra, Pennsylvania.

Mainline Pumps.
Depending upon the pipeline's route and pipeline size, pump stations may or may not be required. If pump stations are required then is becomes a matter of whether to use centrifugal pumps or positive displacement pumps (PD pumps).

Centrifugal Pumps. Centrifugal pumps are used for relatively low discharge pressures and large flows. Centrifugal pumps are not as sensitive as PD pumps to solid particle size or tramp material in the slurry. For PD pumps the solid's particle size is usually limited to 6 or 7 millimeters in order to protect their valves. However, centrifugal pumps operate and lower efficiency than the PD pump. Usual efficiency values are around 0.60 to 0.65. For slurries with coarse particles, the head ratio (ratio of "tdh" of slurry to that of water) is usually less than one (1.00). For most concentrate slurries, the head ratio value can be considered as unity. For low head systems, the centrifugal pump trim is usually rubber-lined or lined with other polymers. For higher head systems, hard metal trim is used. The initial cost of centrifugal pumps is much less than PD pumps and since they are commonly used in most process facilities, maintenance and operating personal need not be specifically trained. For PD pumps personnel require special training. If impeller tip speeds can be keep low (low head requirement system), then pump durability is good.

On the other hand, if head requirements are high requiring high pump speed, then wear parts are consumed at a high rate.

Some "grinding" of the slurry particles occurs as the slurry passes through a centrifugal pump. This is due to the high shear conditions inside the pumps. Consequently, the slurry properties are affected, in particular the PSD and rheological properties. These changes can not only affect the pipeline transportation process, but also, processes downstream of the pipeline system. Therefore, for centrifugal systems, inordinate numbers of stages or numbers of pumps (and pump stations) in series are avoided. Furthermore for multiple pumps, dilution from seal water can adversely affect the delivered slurry.

PD Pumps. Normally for high-pressure systems positive displacement pumps are used. The genesis of these pumps was in the "oil patch." Early pipelines were driven by oil field pumps that were modified for heavy continuous duty with a design life of 25 years or more. The largest pumps in commercial operation have a frame size of the order of 1600 or 1700 horsepower.[6] For high volume pipelines, such as would be required for coal slurry pipelines, larger pumps have been designed and tested up to a frame size of 3600 horsepower. They have been designed in modular form from the power-end to the fluid-end so that singleplex, duplex, or triplex machines can be fabricated from the modular components. None of these are in commercial operation.[7]

PD pumps can generate high discharge pressures (currently over 21,000 kPa), but their flow capacity is limited. The following table shows the PD pump types that have been used:

PD PUMP TYPE	SLURRY TYPE
Piston	Non abrasive
Plunger	Abrasive
Piston-diaphragm	Abrasive

Abrasivity. To select the PD pump type, an "abrasivity" test is performed. Refer to ASTM G75-89 (Miller and Miller 1989). A Miller number is an index of how abrasive slurry may be in contact with moving pump parts. For slurries having a Miller Number less than 60, a piston pump could be used. For slurries with Miller Numbers greater than 100 a plunger or piston-diaphragm pump should be used. As in all cases, these values are not considered rigid criteria for pump selection. The designer's experience is very important. For a given project, the designer may wish to determine a slurry's SAR number. SAR means Slurry Abrasion Resistance. Rather than use standard test specimens, "non-standard" specimens are used. The non-standard specimen material is usually specified in order to be pertinent to a specific each project.

Piston Pumps. Of the above listed PD pumps, the piston pump is probably the simplest and least expensive. The origin of slurry pipeline pumps evolve from mud pumps used for oil well drilling that were "beefed-up" for use with slurry pipelines. One of the early uses for piston pumps was the Black Mesa pipeline. This pipeline has been operation for more than 30 years. With a piston pump the slurry comes in direct contact with the pump's seals, piston, liner, and valves. Refer to Figure 11. The wear parts are:

- piston seals

---

[6] The frame size is not indicative of the motor horsepower. The manufacture usually determines the frame size based upon the anticipated push-rod and bearing loads.

[7] Actually there is a large pump with a frame size of 2000 horsepower in operation. It is installed on a large copper concentrate pipeline. Its flow capacity is over 300 cubic meters per hour. Another large pump that is approximately the same frame size should be commercially operating by the time this article is published. It is installed on a magnetite concentrate pipeline. Its flow capacity is approximately 200 cubic meters per hour.

- piston rubbers
- piston liners
- Valve sets.

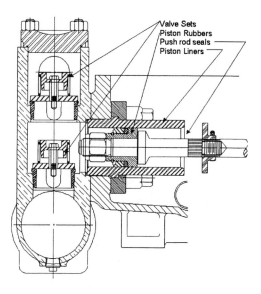

Figure 11: Piston pump fluid end view provided courtesy of National-Oilwell, Houston, Texas.

Plunger Pumps. Plunger pumps are used for "abrasive" slurries. The pumping action occurs as the plunger displaces slurry in the fluid end. The abrasive action of the slurry is mitigated by "flush water" that is injected around the plunger in order to avoid direct slurry contact between the plunger and stuffing box components. The stuffing box components include lantern rings for even distribution of flush water around the plunger, and packing sets that are held between bushings for sealing. Refer to Figure 12. The wear parts are:

- plunger barrel
- throat bushings and lantern rings
- packing sets
- And valve sets.

The purpose of the plunger flush system is to increase the wear parts utility life; however, it introduces more complexity for maintenance and operations. If the slurry system requires multiple pump stations, the design must account for slurry dilution due to flush water injection. The usual flush water volume is 2.5 to 4 percent of the total pump flow.

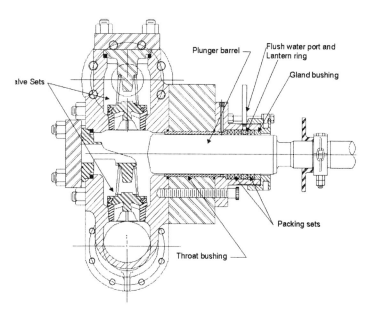

Figure 12: Plunger pump fluid end view provided courtesy of National-Oilwell, Houston, Texas.[8]

Piston-Diaphragm Pumps. Piston-diaphragm pumps incorporate a design that avoids contact of the stuffing box components with abrasive slurry. A fluid such a lubricated-water or oil is used in the stuffing box to drive a flexing diaphragm. The diaphragm action draws and displaces the pumped volume through the valves. Refer to Figure 13 showing a typical pump power end, propelling fluid chambers, and fluid end where the slurry is pumped. Since most of the wear parts are in contact with the propelling fluid, their utility lives are relatively long. The wear parts in direct contact with the slurry are limited to the:

- Diaphragm
- And valve sets

---

[8] In Figures 11 and 12 the suction-discharge valve arrangement is known as an "over/under" fluid end block because the suction valve is directly under the discharge valve. In many installations, the fluid end valve arrangement is modified into a "L" shaped fluid end block in order to facilitate suction valve maintenance.

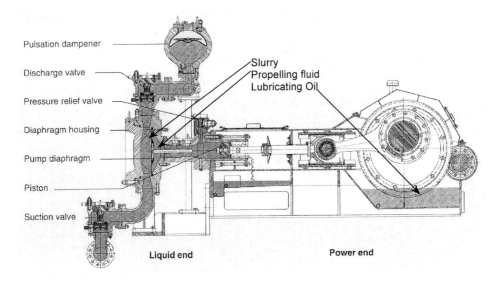

Figure 13: Piston-diaphragm pump sketch provided courtesy of EnviroTech Pumpsystems (Geho), a Weir Company, AE Venlo, The Netherlands. Note that the slurry is on the left-hand side of the diaphragm and propelling fluid is on the right-hand side.

These pumps are more sophisticated and somewhat more expensive than piston pump and plunger pumps but they reduce and simplify routine maintenance.

Slurry Flow Control Valves In a number of recently designed systems, the routine operating procedures allow for pipeline shutdowns while full of slurry. With this operating plan in mind, the importance of critical valves has increased significantly and could be considered equal to mainline pumps. The value duty may, in some cases, be considered even greater than that of the pumps, since there is usually an installed spare pump. Installation of installed spare flow paths and valves is not common practice. The valves are used in an "open/shut" application and not used for throttling flow.

The severity of the slurry valve duty can be put into perspective by the following example. It is commonly believed that abrasive wear for given slurry increases by the velocity to a power of somewhere between 2 and 3.5 (Wiedenroth 1984; Rao and Buckly 1984). The maximum velocity through a valve during the opening/closing cycle can be estimated using its characteristics and differential pressure. If a moderate differential pressure across a valve of 10,000 kPa is assumed, as the port opens or closes, velocities up to 90 meters per second can occur. Therefore, if the slurry is mildly abrasive at a velocity of three meters per second, then it will be at least 900 times more abrasive at 90 meters per second (Ricks and Aude 1989).

Low and Moderate Pressure Slurry Valves. There are variety of low pressure slurry valves, such knife gate, pinch, ball and plug. Low pressure meaning ANSI Class 150 and below and moderate pressure meaning up to ANSI Class 300 rating. The most commonly used low-pressure valves are knife gates. Typically the maximum pressure across these valves are less than 500 kPa. Shown in Figure 14 is a typical knife gate.

Figure 14: Slurry knife gate valve. Cut-away view of AK Knife Gate Slurry Valve provided courtesy of the Weir Slurry Group, Madison, WI.

For moderate pressures, such as branch line isolation or choke line isolation, plug valves or hard metal ball valves have been used. In recent years, valves whose trim is lined with abrasion resistance rubbers or polymers are being developed.

High Pressure Slurry Valves. Most high-pressure applications are found at pipeline terminals and intermediate valve stations. These valves are used to shut-off or start slurry flow beginning with high differential pressures. Because of the specific gravity of typical concentrate slurries and the attendant mountainous terrain, these valves must be rated up to ANSI Class 1500 (26,000 kPa), the most common rating being ANSI Class 900 (16,000 kPa.) However, due to technological advances, higher pressures are becoming the norm rather than the exception. The most common valves for this duty have been lubricated plug valves and hard metal ball valves.

Lubricated plug valves are typically used in "open/shut" applications where a full round opening is not required. The plug and body cavity must have hard, wear-resistant surfaces. Conscientious application of lubricant/sealant is very important to tight shut-off and long life. The sealant lubricates the plug and at the same time purges slurry solids from between the plug and valve body. Shown Figure 15 is a typical lubricated plug valve.

Hard metal ball valves are used as station isolation valves where full port openings are required in order to pass a pipeline "pig". They also used for open/shut applications and as isolation valves for high-pressure mainline pumps. As with lubricated plug valves, preventing accumulation of solids between the moving surfaces in the body cavity is essential. Also, the slurry must be prevented from restricting movement in the stem's seal rings. Show in Figure 16 is a typical hard metal slurry ball valve.

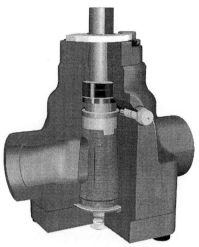

Figure 15: Lubricated slurry plug valve. Cut-away view of a lubricated plug valve provided courtesy of the Nordstrom Valve Incorporated, Sulfur Springs, Texas.

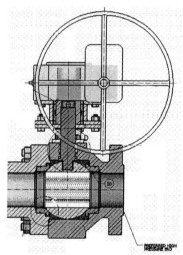

Figure 16: View of a hard-metal slurry ball valve courtesy of Mogas Industries, Inc., Houston, Texas.

Mainline Pipe Materials. Due to high pressure usually required in slurry pipelines, high test carbon steel continues to be the main material used. For short distance lines or low pressure lines other material may be considered.

Corrosion/Erosion Protection. For many long distance systems, the design life is from 20 to 40 years. The pipe wall thickness must be sufficient to contain the internal pressures after allowing for losses due to internal corrosion and/or abrasion. Therefore control of internal corrosion and erosion is important. More often than not, the slurry particles increase the corrosion rate because they tend to scour protective corrosion products from the pipe wall leaving a clean new surface ready for further attack.

To control corrosion in pipelines using ferrous pipe material (usually steel), the slurry pH is adjusted, if necessary, to values between 9 and 11 in order to chemically inhibit the steel from corrosive attack. If necessary, the dissolved oxygen content is reduced by using oxygen scavenger reagents.

Inserting corrosion resistant liners into the pipe can eliminate corrosion. For new pipelines, the costs of the liner and its installation are weighed against the cost of extra steel (corrosion allowance), corrosion inhibitor systems, and inhibitor reagents and consumable supplies. Since the liner is a passive protection form, it is usually preferred over corrosion inhibitor systems that require constant maintenance and process attention.

Steel. Carbon steel continues to be the principal pipe material used. It is relatively inexpensive and easily obtained. The pipeline design can choose from many grades of steel. Usually for low and moderate pressures Grade B material is used. For high-pressure applications, high test line pipe material is used. These grades can vary from API 5L X42 to X70. The most common grades are X60 and X65. The grade and wall thickness is interrelated depending upon the corrosion allowance, availability, and construction criteria under consideration. Long distance slurry pipelines are designed and operated under flow conditions such that erosion of the pipe wall does not occur. However, corrosion control is required either by chemical inhibitors or corrosion resistant liners. For lines where wear can occur (usually bottom wear in tailings systems or similar), steel lines are designed so that they can be periodically rotated in order to maintain uniform wear around the pipe or abrasion resistant liners such as rubber, polyurethane, or concrete are used.

Rubber. Many abrasion resistant materials have been developed such as basalt lined pipe, ceramic lined pipe, hardened metals, and many polymers. However, rubber is still the most commonly used in pipes, pumps, and other process equipment. It has provided good service for slurries that are uniformly abrasive.

High Density Polyethylene (HDPE). HDPE has seen substantial increase in service over the last ten years. It is relatively inexpensive and its applications are quite flexible. Improvised pipelines have been constructed in very short time periods by using a field joint fusing process that is quick and effective. HDPE appears to be effective for slurries that are uniformly abrasive. For slurries with large and/or jagged particles that may cut the HDPE, it is not as effective against wear.

HDPE is a good corrosion resistant material. It can be used on its own as the primary pipe when internal pressures are less than 1500 kPa. For high pressure applications it is used as an effective liner inside steel pipes. In this manner the corrosion resistance properties, and somewhat abrasion resistance properties, are combined with the strength of steel. Pipelines with pressures well over 21,000 kPa are using HDPE liners.

For long distance pipelines, HDPE liner is pulled through existing steel pipelines at about one kilometer intervals. This technique is used to rehabilitate old pipelines or depending on project economics, it is installed in new pipelines.

Polyurethane. Polyurethane has gained a reputation for toughness against abrasion. Even for abrasive conditions such as for cutting and gouging, polyurethane performs well. Generally it is more expensive than HDPE or rubber but it can be shown to be cost effective for many applications. It finds use as pipeline liners and also it is used for coatings and liners for variety equipment, valves, and components. Normally a joint is made by using a flange or coupling. For Long distance pipelines these methods can be expensive in terms of materials and installation labor. In order to reduce costs techniques are being developed that employ joint welding similar to steel pipeline construction.

Energy Dissipation Stations. In many cases, slurry pipelines transport material from high mountainous locations to lowland facilities. In order to avoid abrasive line velocities and maintain control of the flow energy dissipation is required. Several mean are available such as drop boxes, cascades, and choke stations. For most long distance mineral concentrates, choking is used to control the flowrate. The choke material is composed of abrasion resistant ceramics or composite materials.

Figure 17: View of slurry pipeline choke bean and holder assembly.

SUMMARY AND CONCLUSIONS

Over the last several years technical advances in terms of slurry fluid mechanics, equipment and materials have made slurry pipelines (minerals, concentrates, and tailings) a very reliable and dependable transportation method. It is expected that advances will continue.

REFERENCES

Aude, T. C., Thompson, T. L., (Edited by Kulwiec, R. A.), Materials Handling Handbook, 2nd Edition, John Wiley & Sons, Inc., 1985.

Darby, R., "Take the Mystery out of Non-Newtonian Fluids," Chemical Engineering, March 2001.

Durand, R., "The hydraulic transportation of coal and other materials in pipes," Colloq. Of National Coal Board, London, U. K., November 1952.

Hanks, R. W., "Low Reynolds Number Turbulent Pipeline Flow of Psuedohomogeneous Slurries", Fifth International Conference of the Hydraulic Transport of Solids in Pipes, May 1978,

Hanks, R. W., Ricks, B. L., "Laminar-turbulent transition in flow of psuedoplastic fluids with yield stresses", Journal of Hydronautics, October 1974.

Hanks, R. W., Ricks, B. L., "Transitional and Turbulent Pipe flow of Pseudoplastic Fluids", Journal of Hydronautics, January 1975.

Hanks, R. W., Dadia, B. H., "Theoretical Analysis of the turbulent flow of non-Newtonian slurries in pipes", AIChe Journal, May 1971.

Miller, J. E., Miller, J. D., "Slurry Abrasion Testing – The Miller Number and The SAR Number", White Rock Engineering, Inc., Dallas, Texas

Montfort, J. C., "Operating experience is described for Black Mesa coal- slurry pipeline," Oil and Gas Journal, 27 July 1981.

Rao, P. V., Buckley, D. H., "Solid Impingement Erosion Mechanisms and Characterization of Erosion Resistance of Ductile Metals," Journal of Pipelines, Elsevier Science Publishers B. V., Amsterdam, 1984.

Ricks, B. L., Yearbook of Science and Technology, "Slurry Pipeline Transportation", McGraw-Hill (Parker, S. P., Publisher), 1989.

Ricks, B. L., private communication, June 2000.

Ricks, B. L., Aude, T. C., "Slurry Pipeline Valves", Proceedings of International Freight Pipeline Conference, May 1989.

Ricks, B. L., Aude, T. C., "90's Pipeline Design Methodology – Dynamic Modeling", 17th International Conference on Coal and Slurry Technology, CSTA, Washington, D.C., April 1992.

Ricks, B. L., Connelly, M., Moreiko, F., "The Alumbrera Pipeline, The World's Longest Copper Ore Concentrate Slurry Pipeline," 1998 SME Annual Meeting, Orlando, Florida, March 1998.

Robinsky, E. I., <u>Thickened Tailings Disposal in the Mining Industry,</u> E. I. Robinsky Associates Limited, Toronto, Canada, November 1999.

Thompson, T. L., "From Reynolds to Wasp – Development of an Industry," presented at the Sixth International Technical Conference of the Slurry Transport Association, Las Vegas, Nevada, March 1981.

Von Essen, J. A., Ricks, B. L., "Optimum Design of Pipeline Slurry Storage Tanks and Agitators," Chemical Engineering Progress, November 1999. Also presented at Hydro Transport in The Netherlands, September 1999.

Wasp, E. J., Kenny, J. P., Gandhi, R. L., <u>Solid-Liquid Flow Slurry Pipeline Transportation,</u> TransTech Publications, Clausthal, Germany, 1977.

Wiedenroth, W., "An Experimental Study of the Wear of Centrifugal Pumps and Pipeline Components," <u>Journal of Pipelines,</u> Elsevier Science Publishers B. V., Amsterdam, 1984.

The Selection and Sizing of Conveyors, Stackers and Reclaimers

Greg Barfoot[1], Dave Bennett[2] and Martin Col[3]

ABSTRACT

The selection of the conveying, stacking and reclaiming equipment requires not only determination of the width, speed and size requirements, but also the optimization of the layout and equipment interactions. This paper outlines the traditional methods of selection and sizing of conveying, stacking and reclaiming equipment as well as reviewing computer simulation techniques that are valuable tools in the detailed design of these systems. Selection and sizing equipment as a system produces a plant that minimizes capital and operating cost and maximizes availability.

INTRODUCTION

In recent years the tools available to the material handling engineer for the design of conveying, stacking and reclaim systems have improved dramatically. This improvement has been in two main areas. The capacity and availability of computer design tools, and the ability to measure and document the performance of existing systems. This paper will focus on the selection process. Determining what type of system is appropriate and developing the information that can be used for more detailed calculations. It will also assist the reader to interpret some of the results generated by advanced design tools.

BELT CONVEYOR SYSTEM TYPES AND SELECTION CRITERIA

This section will briefly outline the basic types of conveyor systems available and the key features for each system that influence the decision making process.

Conventional Troughed Conveyor

The most common type of conveyor system. Very versatile catering to a wide range of tonnages and applications. Components for this type of system are freely available from multiple sources all over the world. Capable of throughputs up to 10,000 mtph, this type of system can operate at speeds up to 10 m/s. Conventional belts can typically negotiate inclines and declines up to 18 degrees and have been used in horizontally curved applications with radii of a few thousand meters depending on the belt strength and tensions.

Key Features:

1. High tonnages, large lump size
2. Readily available components
3. High speeds
4. Up to 18 degree inclines
5. Horizontal curves.

---

[1] Fluor Daniel, Denver, Colorado
[2] Fluor Daniel, Vancouver, Canada
[3] RAHCO International, Spokane, Washington

Pipe Conveyor

Pipe conveyors are similar to conventional troughed conveyors in terms of the components used but rather than forming a trough the belt is formed into a pipe. This type of system has the advantage of enclosing the transported material. This has benefits where the material needs to be protected or where dust from the material may be a problem. The disadvantages are lower tonnages and smaller lump size.

Key Features:

1. Lower tonnages than conventional troughed systems
2. Smaller lump size, may require extra crushing
3. Readily available components (uses most of the same components as conventional belts
4. Capable of higher inclines and tighter horizontal curves than conventional belts.

Cable Belt

Unlike conventional and pipe conveyors, cable belt systems utilize two steel cables to provide the driving force of the system. The belt itself rests on top of these cables. These systems have some particular advantages for long distance conveying systems. Because the driving force is delivered through the cables intermediate drives can be used without the need for transferring material. The system can negotiate tighter horizontal curves than conventional systems.

Key features:

1. Lower tonnage and smaller lump size than conventional systems
2. Specialized components
3. Capable of tighter horizontal curves than conventional systems
4. Some technical advantages for long distance systems.

Steep Angle Conveyor Systems

For applications where inclines exceed 18 degrees, even to vertical applications, there are a number of systems, which are capable of conveying material. Conveyors where the material is kept in place by buckets and may also include walls as part of the design. Sandwich type systems where the material is kept in place by a second belt pushing down on top of the material.

Key features:

1. Capable of high inclines, up to vertical
2. Specialized components
3. Lower tonnages and smaller lump size than conventional systems.

BELT CONVEYOR SYSTEM COMPONENTS

Regardless of the type of conveyor used the basic components of a conveyor system utilize similar names. Although the following is based on a conventional troughed conveyor the basic components can be applied to most types of conveyor system. Figure 1 shows the basic components that make up a belt conveyor system.

Feed Chute

Loads the material onto the conveyor system. It is designed to minimize impact on the belt and provide some acceleration of the material in the direction of travel. This area can employ rock boxes or specially designed curved chutes to minimize wear and are typically lined with wear resistant material.

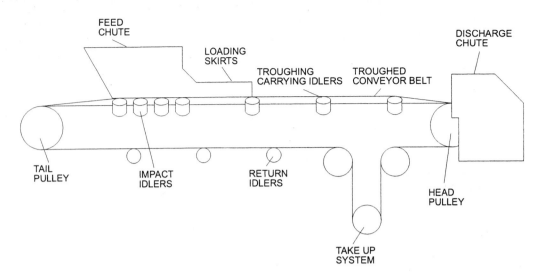

Figure 1 Basic conveyor system components

Loading Skirts
This area prevents spillage of the material until the load has stabilized after the loading area. It also ensures that the material is centrally loaded on the conveyor. Usually employs a sealing mechanism to prevent spillage and to aid in dust collection.

Troughing Carrying Idlers
The carrying idlers form the belt into a trough and carry the material load. Typically forming the trough into three sections, the outside sections are typically at 20, 35 or 45 degrees to the center. Higher capacity systems will utilize the higher toughing angles.

Troughing Conveyor Belt
Usually manufactured from rubber around a reinforcing carcass of fabric layers (lower strength applications) or steel cables (high strength applications). Other materials such as PVC are utilized for special applications such as high temperature or flameproof applications.

Discharge Chute
Takes the material stream from the conveyor and directs it to the next component in the system.

Head Pulley
The belt flattens out as it passes over the head pulley to be returned to the loading point along the return strand of the system. This is usually the high-tension area of the system and as such, the head pulley is often the drive pulley (the place where the motor drives the system). The power from the motor is transmitted to the belt by the friction between the pulley and the belt surface.

Take-Up System
This system maintains the tension in the system. The tension must be high enough to prevent too much sag between idlers, particularly on the carrying side, and to prevent slip at the drive pulley due to the power being imparted into the belt. Tension is adjusted by physically adding or removing belt from the system using the pulleys. The movement can be created by gravity type systems, which use a large weight, or by a winch arrangement.

Return Idlers
These rolls carry the empty belt back to the loading point. Because there is no material loading these a spaced at longer intervals than the carrying idlers and can be flat or on longer systems form a small 'v' trough to keep the belt running in the center of the structure.

Impact Idlers
At the loading point the vertical forces produced by the material falling onto the belt must be absorbed by the impact idlers. To accommodate this, these idlers are spaced very close together and are usually covered in impact absorbing material such as rubber.

Tail Pulley
Re-directs the belt back to the loading point from the return strand of the system.

BASIC BELT CONVEYOR SYSTEM CALCULATIONS
As with the previous section, these calculations are based on the conventional troughed conveyor system although they can frequently be applied to other types of systems as well.

Conveyor System Power and Belt Strength
There are a number of methods and standards for calculating the power and belt tension (required belt strength) for a conveyor system. They are all, however, based on the same principle. The conveyor system is modeled as a mass being dragged over a surface as shown in Figure 2. The friction between the mass and the surface is used to represent the losses in the idlers, indentation of the belting itself and the flexing of the material as it passes over the carrying idlers. The forces required to lift or lower the material itself is a matter of simple physics. The calculation outlined below is a combination of a number of methods and forms a very simple calculation for tension and power. This calculation is designed to give basic results suitable for conceptual layouts, cross checking of other results and for basic technical viability decisions. It is strongly recommended that more detailed analysis be undertaken for the detailed design of a conveyor system.

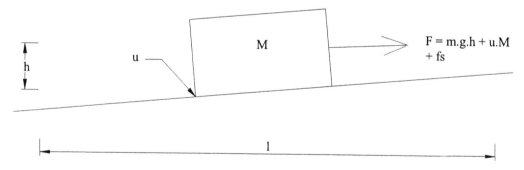

Figure 2 Conveyor System Model

Material Loading
In order to proceed with a basic conveyor calculation we need to establish some design criteria. The first being material throughput. Based on the required tonnes per day, and an estimate for conveyor availability, a design throughput can be established. For example: If a system is required to deliver 36,000 tonnes per day with an expected availability of 18 hours per day, the required design throughput for the conveyor is 2000 tonnes per hour.

The next criteria to select are belt speed. Standard conveyor speeds vary from 0.5 m/s up to 6.5 m/s. The lower speeds are suitable for low capacity systems, a few hundred tonnes per hour, or where the properties of the material may require low velocity. Such as very light material easily moved by air currents. Typical speeds for most applications will fall between 1.5 m/s and 3.5 m/s.

For high capacity systems or long conveying lengths, greater than 2 km, speeds up to 6.5 m/s can be used. In order to determine the material loading required in following calculations we use Equation (1).

$$m = \frac{1000\,T}{3600\,v} \tag{1}$$

Where
m = mass of material [in kg/m]
T = design throughput [in tonnes/hour]
v = belt speed [in m/s]

Belt Width Selection
The bulk density of the material is the next criteria required. Standard values for the bulk density of most common materials can be found in many design standards. This value combined with the result of Equation (1) allows us to calculate the cross sectional area of the material stream using Equation (2).

$$A = \frac{m}{\rho} \tag{2}$$

Where
A = cross sectional area of material [in m^2]
m = mass of material [in kg/m]
ρ = material bulk density [in kg/m^3]

Once the required cross sectional area of the material stream is known, the profile of the conveyor belt and material can be estimated. This can be done graphically or by using one of the methods described in conveyor design standards. The result of this exercise will be the width of the belt that is actually in contact with the material.

A fully loaded conveyor belt requires a certain amount of clearance at each edge. This clearance prevents spillage of material. As the maximum size of material lump increases the required edge clearance to prevent spillage will also increase. The calculation of usable belt width (or the amount of belt actually in contact with the material), for a standard three idler configuration where material lump size is not a factor, is given in Equation (3).

$$\begin{aligned} b &= 0.9\,B - 0.05 \quad (\text{for } B < 2m) \\ b &= B - 0.25 \quad (\text{for } B > 2m) \end{aligned} \tag{3}$$

where
b = usable belt width
B = actual belt width

Conveyor system calculations are an iterative process. At this point a mass for the conveyor belting should be selected and used for the following calculations. Once we have our first result, the conveyor belt rating is known and the selected value for belt mass can be checked. It may be necessary to repeat the process several times to obtain the final result.

Idler Loading
An estimate of idler mass is required to proceed with the calculations. This can be determined from manufacturers data based on the idler vertical loading that is calculated using Equation (4).

$$L_i = \frac{m + m_b}{a} \tag{4}$$

Where
L_i = idler load [in kg]
m = material loading [in kg/m]
m_b = belt mass [in kg/m]
a = idler spacing [in m]

Slope Resistance

The first part of the driving force for the conveyor, the force to lift or lower the material, is called the Slope Resistance and is described by Equation (5).

$$F_s = m.g.h \tag{5}$$

Where
F_s = force to lift (or lower) material [in N]
m = mass of material [in kg/m]
g = acceleration due to gravity
h = change in elevation from loading to discharge [in m]

Main Resistance

The Main Resistance of the conveyor is the force required to move the belt over the idlers overcoming rolling (idler bearings and seals), indentation (of the rubber belt) and flexure (of the material) resistance. The Main Resistance is calculated using the formula in Equation (6).

$$F_M = f L g [m_R + m_C + (2m_B + m_M)\cos\delta] \tag{6}$$

Where
F_M = Force to move the belt [in N]
f = Friction coefficient
L = Length of conveyor [in m]
g = Acceleration due to gravity
m_R = Mass of return idler rotating parts [in kg/m]
m_C = Mass of carry idler rotating parts [in kg/m]
m_B = Mass of belt [in kg/m]
m_M = Mass of material [in kg/m]
δ = Angle of Incline [in degrees]

The mass component of the basic friction force equation shown in Figure 2 is replaced with the mass of the conveyor system calculated from the length of the conveyor and the weight of the components in mass per unit length. This illustrates the basic principle of how a conveyor system is modeled and the forces required to drive the system calculated.

The selection of the friction factor f is crucial to the calculation outlined in Equation (6). A typical value for this friction factor is 0.02, but conditions can exist which vary the value between 0.013 and 0.03. Conditions that require higher values of friction factor include short conveyor systems, poor alignment or difficult conditions, low tensions and high sag values and large spans between idler sets. Conversely, some conditions that require lower values of friction factor are long conveyor systems, well aligned low resistance systems, and low energy loss designs for belting and components.

Secondary Resistance

The secondary resistance is a component in the calculation which accounts for the forces created by belt cleaning devices, losses in pulley systems, drag forces from skirting and other miscellaneous items not included in the main calculation. Generally, these items represent a small fraction of the total forces but for short length conveyors, less than one kilometer, they can be significant. A reasonable approximation of secondary resistance can be made by increasing the main resistance value by 5% for conveyors between one kilometer and five kilometers in length.

Longer than five kilometers the value of secondary resistance is negligible for this type of calculation. For conveyors less than one kilometer in length the secondary resistances are significant and worthy of individual calculation. Formula for calculating these are available in most of the recognized conveyor design standards.

Once these basic calculations are complete there is sufficient information to make an estimate of the size of the conveyor components required for the system. Some iterative calculation will be necessary to make an accurate selection of belting and idlers for the application. In most cases, further calculations will be necessary to develop the detailed design of the system. These further calculations are beyond the scope of this paper and would generally be done by a knowledgeable material handling specialist.

ADVANCED BELT CONVEYOR SYSTEM DESIGN TOOLS

The calculation outlined in the preceding section is designed to allow a rough calculation of conveyor forces for the purpose of estimates, feasibility studies or as a cross check of information supplied by others. There are of course, many detailed calculations related to conveyor system design that are outside of the scope of this paper. However, the reader should be aware of these and what purpose they serve in the development of a conveyor system design. It is recommended that once the project has progressed passed the basic conceptual stage a material handling engineer be employed to complete the detailed calculations and design of the system.

Drive Calculations

Once the tensions and driving force for the conveyor system has been calculated, the layout of the drive system can begin. The detailed calculation will include the required wrap angle for the drive. The wrap angle is simply the amount of belt in contact with the drive pulley surface; the more contact area, the more force can be transmitted to the belt by that drive pulley. In a high capacity conveyor, it is usual to have multiple drive pulleys to effectively increase the contact area between the belt and pulleys and allow transference of sufficient power to drive the system. This calculation will include variables such as angle of wrap, friction coefficient (between the pulley and belt) and belt tension before and after the drive.

Take-Up Calculations

The size, type and location of the take-up system will vary from system to system. Take-up location is a function of the drive location and the belt tensions. For example, in a high lift conveyor system the drive will most likely be located at the top, or head end, of the conveyor. This is the high tension area of the system and locating the drive at this location means there are higher forces between the belt and the pulley surface making the chance of the drive pulley slipping very low. As one of the functions of the take-up is to maintain tension at the drive during start up of the system it is often beneficial to locate the take-up immediately after the drive, however, locating a take-up in a high tension area means the take-up itself needs to be very large. It may make more sense to locate the take-up at the tail end of the system thus reducing the size of the take-up required.

Vertical Curve Calculations

This involves selecting the radius for vertical curves in the system (where the incline of the conveyor changes). The selected radius needs to ensure that the belting will not lift out of the idlers, for a concave curve, or exceed the rating of the belting and/or idlers, for convex curves. The calculation takes into account the mass of the belting, how the belt tensions vary with material loading, particularly for partially loaded conditions, and how the running forces of the system may vary with temperature and operating conditions.

Horizontal Curve Calculations

Horizontally curved conveyors are becoming more common. The ability to have a conveyor system negotiate curves can eliminate transfer points and optimize the design of the system. Calculating the minimum radius for a horizontal curve is a matter of balancing the gravitational forces, the weight of the belt and material, with the tension forces trying to pull the belt into the curve. To balance these forces the idlers for the system are angled opposite to the direction of the curve, i.e. opposite to a camber used on a highway. As with vertical curves this calculation needs to account for varying load conditions and the effect of temperature and operating conditions.

Dynamic Calculations
Some of the more advanced computer models for conveyor systems are capable of simulating the changing tensions and velocities that occur during starting, stopping and load variations of the system. These calculations are essential in designing long, complicated belt conveyors and can identify major problems with operation at the design stage. Typically, however, such levels of calculation are not required for standard, straight belt conveyors. For a given conveyor system these programs can give good estimates for drive torque requirements, test starting control options, identify peak tensions for vertical and horizontal curve calculations, evaluate braking and stopping functions, determine take-up requirements and identify potential tension problems arising during starting and stopping. Figure 3 and Figure 4 show some examples of the results generated by dynamic calculations.

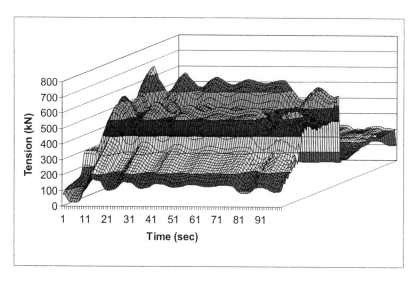

Figure 3 Dynamic Calculation of Belt Tension

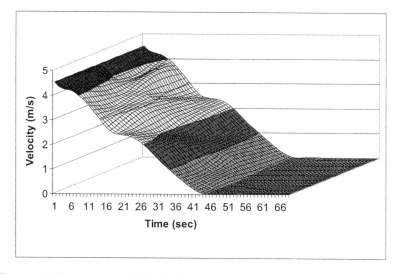

Figure 4 Dynamic Calculation of Belt Velocity

STOCKPILE SYSTEM TYPES AND SELECTION CRITERIA

Stockpiles in bulk solids handling system generally provide a buffer between two processes or modes of transportation. At a transshipment facility, stockpiles absorb the difference in the inter-arrival time of incoming and outgoing product. At a mine or quarry stockpiles are commonly used to provide surge capacity up stream of a mill or crushing and screening circuit to ensure a uniform feed is available to the process. In addition to providing surge storage, stockpiles at iron and steel plants and other process facilities are often used to blend different materials before being reclaimed and delivered to the downstream process. Coal fired power plants use stockpiles to ensure both short term and long-term coal stocks are available. One additional and unique "stockpile" is the stacking of and sometimes reclaiming of heap leach material (most commonly found in the gold and copper industries).

The type, size and purpose of the stockpile have a direct impact on the methods of stacking and reclaiming to be used. Before a specific method of stacking and reclaiming can be considered it is important to established some basic criteria, the following are some of the main factor that must be considered:

1. Does the pile have to be covered?
2. Does material have to be stacked and reclaimed simultaneously?
3. Is a first in, first out (FIFO) stacking and reclaim system required?
4. How many products have to be stockpiled?
5. Is cross contamination of products a concern?
6. What are the material characteristics? Bulk density; Moisture content; Size distribution; Friability; Hardness, Susceptibility to spontaneous combustion or oxidation, Stickiness, Abrasiveness, etc.
7. What is the method of product delivery to the stockpile facility and how uniform is it?
8. What is the method of product delivery from the stockpile facility and how uniform is it?
9. How much storage is required?
10. What proportion of live to dead storage is required?
11. For a blending bed or stockpile, the proportions and availability of the constituent elements have to be defined.
12. In heap leaching applications, is it a multiple lift (permanent pad) or an on/off (reusable pad)?

This paper cannot cover all of the potential applications and materials that require stockpile facilities; consequently, it will deal with the typical requirements at coal and hard rock mines and transshipment facilities.

Once the basic criteria have been established the next step in the design of a stockpile/reclaim system is to select the basic layout. This section reviews the basic conceptual layouts and some of the advantages and disadvantages of the commonly used systems:

Stacking systems
1. Fixed stacker – conical pile
2. Overhead tripper or shuttle – longitudinal pile
3. Radial stacker – circular or kidney pile
4. Cascading conveyor system with radial stacker – rectangular or odd shaped pile
5. Traveling stacker or stacker /reclaimer – longitudinal pile
6. Mobile conveyor with traveling tripper – longitudinal and/or circular piles
7. Silo storage – fed by a conveyor.

Reclaim systems
1. Front end loaders to a loading hopper
2. Gravity to a feeder system below the pile or silo
3. Bucket wheel reclaimer
4. Scraper reclaimer
5. Drum reclaimers.

STACKING SYSTEMS

Fixed Stacker

A fixed stacking system typically consists of a cantilevered conveyor, which creates a conical stockpile generally up to 50m in height, although larger systems are possible. The reclaim from such a pile can be by front-end loader for intermittent, low through put facilities. For facilities requiring an automated uniform reclaim rate, vibrating, apron or belt feeders are normally located in a tunnel beneath the pile. The feeders draw material from the pile and discharge it onto a belt conveyor for transportation to the down stream process. The number, type and size of the feeders selected are normally based on the material flow characteristics, feed rate required and the live to dead ratio required in the pile. For most situations, this is lowest cost stacking/reclaim system. However, it is not efficient in terms of the live to dead ratio. With a single line of feeders the live capacity often less than 20% of the total even with relatively free flowing material and with only a single feeder with less free flowing material the live capacity may be less than 10%. However, this is frequently a good option where space is readily available and the dead portion of the pile can be utilized to provide feed to the plant during infrequent interruptions in the stacking system. When stacking and gravity reclaiming from the live portions of the pile a FIFO (first in first out) regime exists and size segregation is not normally a problem. However, during initial pile building, natural size segregation occurs which results in coarser material collecting at the edges of the pile with finer products at the center. To reduce potential dust problems, a telescoping chute can be fitted to the head of the stacking conveyor but these units can be a major wear and maintenance item in situation where abrasive coarse ore is being handled. If necessary, the system can easily be covered to improve dust control and protect the product from the elements.

Overhead Tripper or Shuttle

A fixed conveyor with a traveling tripper or a reversible shuttle conveyor is supported on a structural frame over the stockpile. An example of this type of system is shown in Figure 5. Material is discharged from the tripper or shuttle directly into the stockpile or storage bin system below. This stacking system allows materials to be stacked in different piles or to be blended in a single pile. The other main features associated with the fixed stacker system are also applicable to the overhead tripper/shuttle, with the exception that a telescoping discharge is not practical on a tripper. The most significant differences between the overhead tripper/shuttle and the fixed stacker are the size of stockpile and the possibility to use automated reclaimers, as an alternative to gravity reclaim. The overhead tripper/shuttle system is also considerably more expensive than a simple fixed stacker.

Reclaiming from a longitudinal pile formed by an overhead tripper/shuttle can be by gravity to feeders beneath the stockpile, as described above for fixed stackers, or by automated mechanical reclaimers such as bridge mounted bucket wheels, drum reclaimers, portal or full face scraper reclaimers. These automated mechanical reclaimers generally have limitations on the lump size they can handle and depending on the product, they can be a source of high wear and maintenance. Additional aspects of these reclaimers are discussed below in the section on traveling stackers.

When under pile reclaim feeders are used they can be arranged along the length of the pile, as is commonly used at gravel plants, or across the width of the pile to feed different process lines as often used at base metal concentrators.

Figure 5 LAXT Coal Terminal (Courtesy KRUPP)

Radial Stacker

There are several configurations for this type of stacker. The most common are those that comprise an inclined belt conveyor supported on a pivot at the loading point and a structural leg approximately two thirds along the conveyor. The structural leg can be mounted on pneumatic tires or steel wheels and a curved rail. The stacker slews about the pivot point to form a kidney shaped pile. The gravity reclaim systems used with this type of stacker are similar to those described above for both fixed stacker and overhead tripper/shuttle type systems. The arc described by the inclined radial stacker can be 270 degrees or more. However the most efficient live to dead ratio that can be achieved using a single reclaim tunnel occurs when the center line of the reclaim tunnel forms the cord of an arc described by a 60 degree slew of the stacker. Automated mechanical reclaimers are not normally used on kidney shaped stockpiles and it is difficult to efficiently cover this type of stockpile. This type of stockpile is commonly used in quarries and gravel plants.

Another type of radial stacker that is used in automated systems for blending, stacking and reclaiming comprises a pedestal mounted radial conveyor boom. Product is discharged from a feed conveyor to the radial stacking boom at the central pedestal. The stacking boom can luff and slew as it slowly rotates around the central pedestal, thus allowing product to be discharged in chevron layers or as cone shells to form a continually rotating "kidney shaped" stockpile. The open part of the "kidney" is where the reclaimer operates, it follows in the same direction as stacker boom, reclaiming material from the pile and discharging it at the central pedestal, onto an under ground transfer conveyor. Reclaimers in this type of system are normally of the bridge mounted bucket wheel or scraper reclaimer type. They are supported at one end by the central pedestal and the other on a circular rail around the perimeter of the pile. This type of system can be fully automated and is ideal for stacking and blending relatively fine material (minus 25 mm). It is not suitable for a coarse ore storage system. The maximum economically practical Reclaimer Bridge defines the limit on size for this system. Although 40m span reclaimers have been built, 35 m spans may prove to be a more economic length, this would result in a system diameter of over 75 m and if required could be fully enclosed. A key feature of these mechanical reclaim systems is that the entire stockpile is live and they are better at handling less free flowing materials that may be very difficult to reclaim with a gravity based system.

Cascading Conveyor System

The cascading or commonly called "grasshopper" system is often used to build piles of ore for heap leaching. This system incorporates a series of portable conveyors, which feed another in a "cascading" fashion. The last portable conveyor in the line discharges material onto a "bridge" conveyor, which is, positioned 90° to the portable system. The bridge conveyor, a horizontal conveyor mounted on either rubber tires or crawler tracks or a combination of the two, is generally 20-110% longer than the portable cascading conveyors to allow flexibility when removing one (or two) portable conveyor(s). In turn, the bridge conveyor feeds a radial stacker, which can be either rubber tire or crawler track mounted. The heaps are constructed in a retreat fashion. An example of this type of system is shown in Figure 6.

Figure 6 Cascading or "Grasshopper" System (Courtesy RAHCO)

Traveling Stacker or Stacker/Reclaimer
These machines are commonly used on longitudinal stockpile systems at transshipment terminals and process plants. Although traveling stackers can be used with gravity reclaim systems they are usually combined with mechanical reclaimers and are often built as combined stacker/reclaimers. The stackers run on rail tracks along the side of the stockyard and are equipped with a tripper that raises the feed conveyor belt to transfer the product onto the stacker's boom conveyor, which in turn discharges it into the stockpile. There are several different configurations of stacker boom such as: fixed or luffing, single or double booms and single boom machines that luff and slew. The selection of stacker type must take into consideration the reclaim system to be used as well as the stockyard layout to be adopted. The length of the stacker boom is directly related to the width of stockpile required and may be influenced by the method of stacking to be adopted. Although not commonly used today windrow stacking was used as a method of reducing the effects of size segregation that occurs with chevron stacking. The cost of the longer boom required to build windrows is difficult to justify in most applications so that luffing/slewing stackers that reach the center of the stockpile and build chevron or cone shell piles have become more common.

Boom mounted bucket wheel reclaimers also run on rail tracks along side of the stockpile and varies in boom length and capacity. Boom lengths of up to 60 m are common at coal and iron ore terminals with reclaim rates in the order of 6,000 t/h to 10,000 t/h. Many terminals prefer to use multiple combined stacker/reclaimers that provide additional versatility when managing a large number of grade or types of material in a stockyard.

Traveling stackers are also used to stockpile material in blending beds and in stockpiles where portal scraper reclaimers are used. To build an efficient blending bed the entire constituent parts of the blend must be laid down in the pile so that they will occur consistently in a predetermined mix when the pile is reclaimed. The best blending bed reclaimers are therefore machines that take a full slice from the exposed face of the stockpile. Typical machines that fall into this category are drum reclaimers, bridge mounted bucket wheels and scraper reclaimers. The maximum width of this type of stockpile is set by the economic span of the reclaimer that is in the order of 40 m. Where precise blending is not required larger wider stockpiles can be reclaimed with a portal type scraper reclaimer that can span pile widths of 60 m.

The maximum economic capacity of reclaimers for stockpile use varies depending on the type of machine and the materials to be handled. Bucket wheel and drum reclaimers with capacities of 6,000 t/h are relatively common in coal terminals and machines with capacities up to 10,000 t/h are not uncommon at iron ore terminals.

The main advantages of the rail mounted stacker and reclaimer systems are their flexibility to handle multiple products with varying flow characteristics with little or no dead storage. While these systems can be designed to handle lump sizes of 150 mm, they are better suited to handle material of less than 75 mm. For most facilities using this equipment, it is not practical to cover the stockpile system and alternative methods of dust control have to be adopted. Figure 7 shows an aerial view of a traveling stacker system. Figure 8 shows a closer view of this type of stacker and reclaimer.

Figure 7 Traveling Stacker and Reclaim Systems (Courtesy KRUPP)

Figure 8 Blending Yard Stacker and Portal Style Reclaimer (Courtesy FAM)

Heap Leach Stacking and Reclaiming System

The reusable or on/off leach pad has become very common particularly in the copper industry. One of the main differences between the standard blending yard or port facility and the reusable leach pad is the strict "first-in, first-out" material flow. Typically, a pair of rectangular leach pads are placed side-by-side with feed and discharge conveyors in the corridor between the pads. A mobile tripper (usually rail-mounted) transfers material from the feed conveyor to the Mobile Stacking Conveyor (MSC) as shown in Figure 9. The MSC is mounted on crawler tracks and has a rail-mounted tripper, which travels the length of the MSC as it discharges material to form the heap for leaching. Once leached, the material is reclaimed using a Bucketwheel Reclaimer (BWR) and fed into a rail-mounted hopper, which travels the length of a second crawler-track mounted mobile conveyor, the Mobile Reclaim Conveyor (MRC) as shown in Figure 10. Once the BWR has completed one pass along the face of the heap (parallel to the MRC), the BWR and MRC both take a step toward the pad and repeat the process Approximately 1.0-1.5 m of material is removed with each pass. The reclaimed spent ore is then transferred to a mobile hopper, placed on the spent ore conveyor and stacked on a permanent spent ore pile. A shiftable conveyor and a large crawler mounted spreader or a third mobile conveyor typically stack the spent ore.

Figure 9 Typical Reusable Heap Leach Pad (Courtesy RAHCO)

Figure 10 Mobile Stacking Conveyor (Right) and Mobile Reclaim Conveyor with Bucketwheel Reclaimer (Courtesy RAHCO)

Silo storage

This type of storage system is used in special circumstances where relatively small quantities of product are to be stored, or some process or product requirements favor such an enclosed environment. The main use of these silos in a mining related environment is at high-speed rail load out facilities for coal. A typical 10,000 t capacity silo is filled by a conveyor directly from the coal preparation plant and discharges directly into the moving train as it passes through the silo. The silos are normally slip or jump formed concrete structures with multiple loading gates at the bottom to control the flow of coal into the railcars, refer to Figure 11.

Silos are also used for fine ore storage at some hard rock mines as surge storage ahead of agglomerators at heap leach operations where typical capacities would be in the order of 6,000 t.

Figure 11 Silo Storage (Courtesy FAM)

Table 1 Common Repose Angles, CEMA Belt Conveyors for Bulk Material, Fifth Edition, Pages 29-43

Material	Angle of Repose
Ash, coal	45°
Ash, fly	42°
Bauxite, mine run	31°
Bauxite, -75 mm	30-44°
Borate of Lime	30-44°
Cement, Portland	30-44°
Clay, dry, fines	35°
Coal	27-40°
Coke	30-44°
Copper Ore	30-44°
Diatomaceous earth	30-44°
Earth, dry as excavated	35°
Earth, wet containing clay	45°
Gravel, pebbles	30°
Gypsum –12 mm	40°
Gypsum, 30-60 mm	30°
Ilmenite ore	30-44°
Iron ore	35°
Iron ore pellets	30-44°
Kaolin Clay –75 mm	35°
Lignite, air dried	30-44°
Limestone, crushed	38°
Phosphate rock, broken, dry	25-29°
Phosphate rock, pulverized	40°
Potash (muriate), dry	20-29°
Potash (muriate), mine run	30-44°
Potassium nitrate	20-29°
Potassium, sulfate	45°
Quartz	20-29°
Salt, common dry, fine	25°
Salt cake, dry, coarse	38°
Sand, bank, damp	45°
Sand, bank, dry	35°
Sandstone, broken	30-44°
Soda ash, briquettes	22
Soda ash, heavy	35°
Soda ash, light	37°
Sodium phophate	37°
Taconite, pellets	30-44°
Zinc ore	38°

BASIC STOCKPILE SYSTEM CALCULATIONS

Calculating Live Storage

Calculating the live storage of a stockpile is pure geometry. Using the estimated reposes and draw down angles of the material, refer to Table 1, a geometrical shape of the full and empty (fully reclaimed) stockpile is generated. Many three dimensional CAD packages are extremely useful for creating these shapes. The live storage of the stockpile is the volume of the completely reclaimed stockpile minus the volume of the full stockpile.

CONCLUSION

The handling of bulk material via conveyors is quite common and has become much more flexible in the last decade. A wide variety of materials are conveyed, stacked and reclaimed using a broad range of equipment in many configurations. This paper has attempted to introduce the selection and sizing of this equipment and will provide a base from which a conveying, stacking and reclaiming system could be developed. As many a user of such equipment has said, "the only limit is our own imagination."

ACKNOWLEDGMENTS

The authors would like to thank the equipment manufacturers noted in this paper for the use of information and photographs of material handling systems.

REFERENCES

Belt Conveyors for Bulk Material, Fifth Edition, Conveyor Equipment Manufacturers Association.
ISO 5048 – Continuous Mechanical Handling Equipment, Belt Conveyors with Carrying Idlers, Calculation of Operating Power and Tensile Forces. 1989

Selection and Sizing of Concentrate Drying, Handling and Storage Equipment

Michael E. Prokesch, P.E., FFE Minerals Inc., Grant Graber, P.Eng., AMEC

ABSTRACT
The selection of appropriate drying equipment is a function of a material's physical properties and drying rate, and desired product characteristics. Proper material characterization and analysis must be followed by equipment selection based on established criteria. This paper covers material characterization, reviews available drying systems with a focus on rotary, fluid bed and flash systems, and outlines basic selection criteria given the advantages and disadvantages of each. Concentrate handling and storage equipment and systems are also reviewed, taking numerous factors into account including moisture content, bulk material properties, climatic conditions, transportation logistics, and market conditions.

DRYING
Drying is the use of thermal energy to reduce the moisture content of a material through vaporization. Typically, free moisture is removed, as opposed to bound or crystalline water that often requires higher temperatures and specific energy input. Since mechanical dewatering generally cannot reduce concentrate moisture content below 10-20%, drying is frequently required to attain moisture levels that maximize the thermal efficiency of a downstream process, provide the desired consistency for handling and blending, satisfy end-use specifications, etc.

Available drying systems handle materials ranging from slurries to filter cakes to free-flowing crushed ores. Many of these systems also dry pyrophoric ores, using controlled gas composition and temperature to limit the potential for oxidation. Proper system selection must be based on a thorough understanding of material properties and product requirements.

DRYER SELECTION CRITERIA
Material Properties. The selection of a dryer is highly dependent on the material's physical and specific drying characteristics. Drying is a three-step process (see Figure 1): first, sensible heat brings the material to the vaporization temperature. The surface then remains at a constant temperature until the critical moisture point, at which all its moisture has been removed. Final drying occurs at a decreasing rate, as surface temperature increases and internal moisture is driven to the surface and volatilized. During this stage, materials with high levels of internal moisture may be subject to high stresses, resulting in structural degradation.

The drying rate curve characterizes a material's basic drying properties. This is developed by drying a material at a constant gas temperature and determining the rate of weight loss. Figure 2 shows drying curves for oil shale, clay and garnierite ore. The curve does not provide an absolute time/temperature relationship for dryer design, but gives a relative comparison to other materials, so that required dryer operating conditions can be calculated from experience. The drying rate curve is used to select dryer operating parameters, rather than dryer type.

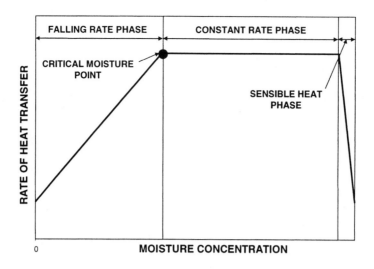

Figure 1 Drying cycle

The physical properties of the dryer feed stream have the largest impact on dryer selection and design. These include moisture content, disposition of moisture (surface and internal), particle size distribution, density, particle integrity (i.e., friable), temperature sensitivity, abrasiveness and corrosivity. Of these, particle size distribution and consistency are used to define the category of dryers suitable to an application.

Dryer Classification. Dryers may be classified as adiabatic (direct heat transfer) and non-adiabatic (indirect), or by continuous or batch operational mode. Some designs incorporate both direct and indirect heat exchange, but their application is limited. Most concentrate drying applications require large continuous throughputs and maximum thermal efficiency, so this paper will describe only continuous adiabatic systems. For the limited concentrate applications requiring non-adiabatic and/or batch operations, an expert should review the many equipment options available. These applications often include materials that cannot be exposed to combustion gases or an oxidizing atmosphere during the drying process.

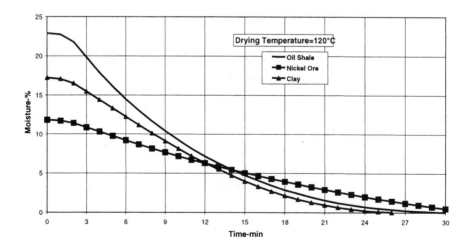

Figure 2 Drying rate curves

Dryer Selection. Table 1 provides general guidelines for the initial selection of an adiabatic, continuous feed dryer system based on feed consistency and particle size. Other adiabatic, continuous feed designs are available, but these are typically modifications of the basic systems shown in Table 1, and have been excluded to simplify the selection process.

Table 1 Dryer selection based on feed properties

Slurries		Pumpable Sludge	
Spray dryer		Spray dryer	Rotary dryer
		Rotary tray dryer	Hammermill dryer w/recycle

Filter Cakes/Thick Sludges:			
Fine (under 6 mm)		**Coarse**	
Flash dryer w/recycle	Rotary dryer	Rotary dryer	Rotary tray dryer
Fluid bed dryer w/inert bed	Rotary tray dryer	Hammermill dryer	Band dryer
Band dryer w/performing	Hammermill dryer w/recycle		

Free-Flowing:			
Fine (under 6 mm)		**Coarse**	
Flash dryer	Rotary dryer	Rotary dryer	Rotary tray dryer
Fluid bed dryer	Rotary tray dryer	Hammermill dryer	Band dryers
Hammermill dryer			

The term "recycle" refers to conditioning the wet feed stream with a portion of the dry product stream to obtain flow properties compatible with the dryer system. For example, a sticky sludge containing 50% moisture may be mixed with dry product to reduce its moisture content to 20-25% and produce small free-flowing agglomerates that may be properly entrained in a flash flyer's gas stream. This mixing may be performed using a simple pug mill device.

The inert bed for filter cake/sludge processing in a fluid bed usually comprises a coarse stone (i.e., 13 mm crushed limestone). This stone is maintained in a highly active state of fluidization, which enables it to break up sticky agglomerates and disperse the material through the bed for rapid drying. The inert material is replenished as it slowly degrades in the bed.

Following selection of an appropriate system, additional feed properties should be considered to narrow the system options. Flash dryer, fluid bed dryer with inert bed and hammermill dryer systems are not recommended for friable materials for which fines generation must be minimized. This must also be considered when designing a rotary dryer with internals, due to the increased potential for particle degradation.

Temperature-sensitive materials are not well-suited to gas suspension flash drying, because during the falling rate phase, the system's short retention time (1-3 sec) requires a high temperature differential between the particle surface and core. Materials at risk for high-temperature degradation or reactions (i.e., sulfide ores) are better suited to rotary, fluid bed, rotary tray and band dryer systems, which have longer material retention times.

Drying capacity, spatial requirements and capital cost may further narrow the options. Rotary tray and band dryers are high-cost and limited to several tons per hour, giving them a high cost per per unit of material processed. Rotary dryers require a larger footprint than other systems. Flash dryers have the greatest vertical clearance requirement (10-15 m).

Design summary. A number of items must be considered in dryer system selection (although not are necessary for initial selection and sizing):

- Material drying properties – drying curve
- Free moisture level – loss at 105 °C

- Particle size distribution
- Wet and dry bulk density
- Specific gravity
- Feed flow properties: slurry, sticky sludge, free-flowing, etc.
- Particle degradation issues
- Temperature sensitivity
- Corrosiveness
- Potential emissions requiring control
- Capacity in wet or dry mtph
- Mode of operation – 24 h/d or campaigned
- Product moisture specification
- Fuel type
- Installation area description / limitations
- Operating environment.

DRYING EQUIPMENT

Rotary Dryers. Rotary dryers are the commonest system for mineral processing, since they offer the greatest flexibility in terms of capacity, retention time, operating temperatures, ease of operation, operational availability, the ability to process different feeds, handling variations in feed properties, and operating at reduced throughputs. A single unit can process several thousand tons per day of wet feed.

A typical rotary dryer system (Figure 3) includes an unlined carbon steel shell with supports and a drive mechanism. Product purity requirements or corrosive applications may dictate more robust materials of construction. Hoods at both ends of the tube accommodate the transfer of solids and gas. Fuel is combusted in a refractory lined external combustion chamber, and the resulting hot gas enters the tube and flows parallel with the material stream. Material drops off the far end of the tube into a material handling circuit, and the dryer offgas is filtered, scrubbed if necessary and then emitted to atmosphere. An induced draft fan downstream of the offgas filter controls the static pressure (slightly negative in the discharge hood) inside of the dryer.

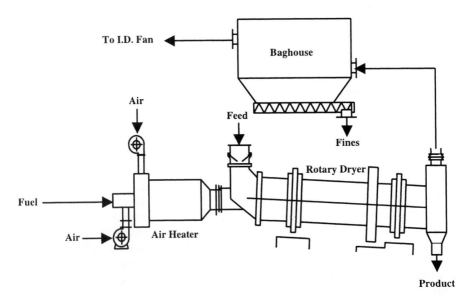

Figure 3 Rotary dryer system

The earliest application of the rotary dryer used a straight tube. Since then, efficiency has been improved through the use of internals such as chains, lifters and dams. Figure 4 includes examples of several common lifter configurations. The purpose of these devices is as follows:

- Chains break apart agglomerates for increased surface area and absorb heat from the gas stream for transfer to the material load.
- Lifters shower the material through the gas stream. Lifter geometry is varied to adjust the angle at which the material is showered through the gas stream. Lower angles are used for fine or friable materials to minimize entrainment and degradation, or to accommodate sticky materials. Lifter sets are staggered to maximize effectiveness.
- Dams increase the material loading and retention time in the tube. A typical loading rate is 10-15%.

Material retention time in a rotary dryer typically ranges from 15-30 minutes. Most dryers include a variable speed drive to permit changes in retention to maximize capacity. Retention time is a function of material angle of repose, shell diameter, shell length, shell slope, shell speed, gas velocity and the design of internals. Retention time changes linearly with shell speed changes.

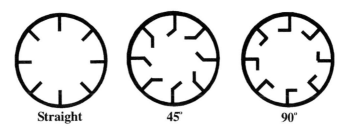

Figure 4 Rotary dryer lifter configurations

Typical design ranges for rotary dryer systems are as follows:

- Diameter: 6 m maximum
- Length: equivalent to 5-10 times the diameter
- Retention time: 15-30 minutes
- Exit gas velocity: 3-5 mps
- Loading: 10% of total shell volume
- Shell slope: 2.4°
- Shell peripheral speed: 0.1-0.5 mps
- Inlet temperature: approximately 900°C maximum
- Outlet temperature: 125-150°C
- Evaporation load: 30-110 kg/h/m^3, depending on material properties.

The rotary can operate in counter-flow or parallel-flow mode. Most operate in parallel mode, which permits operation with a high inlet gas temperature to maximize dryer capacity and efficiency without overheating the solids. The higher differential gas temperature in the parallel mode equates to lower specific energy consumption per unit of dry product than counter-flow.

Parallel-flow may be used to dry pyrophoric ores. This requires limiting the oxygen content in the system to a maximum of 10% through gas injection or flue gas re-circulation. The rate of pyritic sulfur oxidation at this concentration is significantly retarded. In addition, the particle surface temperature must be limited to <300°C during the falling-rate drying period.

The rotary dryer's disadvantages include high price, space requirements and maintenance costs. These should be weighed against their flexibility and ease of operation.

Fluid Bed Dryers. Fluid bed dryers use fluidization or suspension of solids in a gas stream to transfer heat between gas and solids. Close contact is maintained between gas and solids through vigorous mixing, promoting higher product uniformity than a rotary dryer. These mixing and heat transfer properties, coupled with a relatively long retention time (several minutes), enable a fluid bed to easily meet low-moisture specifications. They can also functioning as classifiers to segregate coarse and fine fractions.

A conventional fluid bed dryer system (Figure 5) includes a cylindrical vessel with an air distribution plate in its lower portion to form a plenum below the plate, and bed/freeboard zones above the plate. Hot air from a heater passes through the plate and enters the bed. Wet feed is added through the side of the vessel at the top of the bed or through the top of the unit. Dry product either overflows from the top of the bed across from the feed inlet, or is discharged through an underflow at the level of the distribution plate. Gas containing entrained fines passes through the expanded freeboard zone, followed by a cyclone and/or baghouse for particulate removal. These fines may be kept separate from the coarse product or recombined. An ID fan downstream of the baghouse maintains a neutral static pressure in the fluid bed freeboard zone.

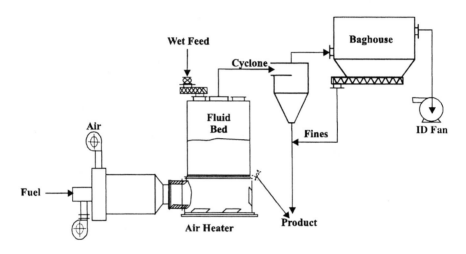

Figure 5 Fluid bed dryer system

The air distribution plate generates a minimum pressure drop of ~3.4 kPa, to ensure even gas distribution across the bed cross-section, essential for proper fluidization. Uneven distribution would allow the gas to channel through and over-fluidize portions of the bed; with sulfide ores, the remaining static portions would overheat and begin to oxidize. Figure 6 illustrates several modes of fluidization; active fluidization, or "bubbling bed" mode, is preferred.

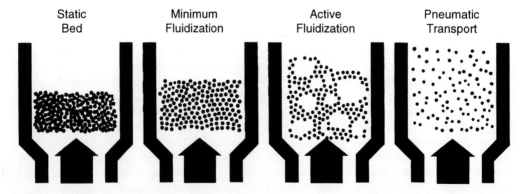

Figure 6 Modes of fluidization

A material's fluidization properties are determined in the laboratory by a cold fluidization evaluation. This enables the designer to observe the fluidization behavior of the material, determine the velocity required for minimum fluidization, determine the velocity range for active fluidization, and measure the quantity and particle sizing of the fines elutriated from the bed during active fluidization. Improper fluidization leads to poor drying efficiency, a reduction in capacity and overheating of the material bed.

Round orifices, bubble caps or nozzles usually distribute the air. Plate design is influenced by material particle size distribution, abrasiveness, fluidization velocity, target pressure drop and operating temperature. Most fluid bed dryers operate with a maximum inlet temperature of 800°C, employing a stainless steel plate. Higher temperatures are possible with exotic alloys, ceramics or refractory materials.

The key to the success of fluid bed technology is a feed material with acceptable flow characteristics and consistent size. The incoming feed must break up and disperse in agglomerate sizes comparable to or less than the top size of the fluidized bed of particles. Otherwise, the large agglomerates will sink to the air distribution plate, remain semi-static and become overheated. This accumulation eventually disturbs the fluidization of the entire bed.

Sticky sludges or hard filter cakes are usually processed by simply mixing dry product into the wet feed to improve its flow properties. A second method uses an inert bed of coarse particles such as limestone, slag or other hard material that will not contaminate the dry product as it slowly degrades in the bed. The inert particles are fluidized at high velocities (10-15 fps), which serves to break apart and disperse the incoming feed throughout the bed. As the feed breaks apart and dries, it is swept out of the fluid bed to a cyclone for collection. No bed overflow or underflow is used. The pressure drop across the bed is monitored to track the inert material inventory. As the material degrades and the pressure drop decreases, fresh inert material is added to maintain the inventory.

The inert fluid bed system has been successfully used to dry pyrophoric concentrates, which are broken down in the inert bed zone and quickly quench the incoming hot gas stream to well below the reaction temperature. The fine concentrate particles are entrained and then carried through the relatively low-temperature freeboard zone. To further minimize the potential for oxidation reactions, oxygen concentration in the process gas stream is reduced to below 10% through nitrogen injection or gas recirculation.

The inventory of material in the fluid bed zone enables the unit to handle minor fluctuations in feed moisture content or flow rate. The unit is also very simple to operate once the fluidized bed has been established. However, it cannot accommodate wide variations in feed particle size or moisture content without jeopardizing system stability.

Typical ranges for fluid bed dryer design and operation are as follows:

- Diameter: 5 m maximum
- Height: 2m maximum
- Retention time: 5+ minutes
- Bed gas velocity-conventional: 0.15-2 mps
- Bed gas velocity-inert bed: 3-5 mps
- Distributor plate pressure drop: 3.4 kpa minimum
- Plenum temperature: 800°C maximum
- Bed outlet/freeboard temperature: 125-150°C.

Fluid bed drying systems have relatively low capital/maintenance costs, and a small footprint (except for fine materials that require very low velocities, which have a much smaller capacity per unit bed). Their specific power consumption costs are higher, due to the pressure drop across the air distribution plate and suspended bed.

Flash dryers. Flash dryers are becoming more popular due to their low cost, small footprint, efficiency, low maintenance and ease of operation. Material is introduced to a hot gas stream at a velocity sufficient to entrain all feed particles and carry them vertically up the flash tube. The excellent gas/solids contact enables most materials to be completely dried in 1-3 seconds. The gas/solids mixture passes through a cyclone collector for solids removal, followed by a baghouse for fine particulate capture. Figure 7 illustrates a typical gas suspension flash drying system, with provisions for dry product recycle to handle feed materials that are not free-flowing.

Adequate velocity must be maintained to ensure that all feed particles and agglomerates are entrained and carried through the unit. In addition, particularly with large diameter flash tubes, the material must be dispersed uniformly across the tube. This can be accomplished using multiple feed locations and by using splash plates at the feed inlets. A velocity of 3-4 mps is needed to convey particles up to 850 µ with specific gravity <3. Units may be designed for velocities >18 mps to convey 5 mm dense particles (SG up to 4.5). However, as gas/material retention time is a function of flash tube length, cost becomes prohibitive at higher velocities due to the height of the tube and support structures. Tube wear also becomes an issue with coarse abrasive materials, even with the use of wear resistant refractory materials. Therefore, most flash dryers are designed for particles >2.5 mm with a maximum velocity of 10 mps and a retention time of 1-2 seconds.

When a fine feed is unsuitable for a conventional flash dryer due to the presence of a small percentage of +2.5 mm particles, a hybrid system can be considered. This hybrid system combines the features of fluidized bed and flash drying. A shallow, low-pressure drop fluid bed zone is developed in the bottom of the tube using an air distribution plate or high-velocity throat. This zone operates at 4 mps to support active fluidization of a dilute bed of coarse particles. The freeboard zone above the bed is extended to provide 1-2 seconds of retention time at 4 mps, to serve as the flash drying zone for the fine fractions. The dried coarse fraction from the bed overflow and the dry fine fraction from the collection cyclone are then combined outside of the dryer. This allows the dryer system to accommodate much wider variations in particle size.

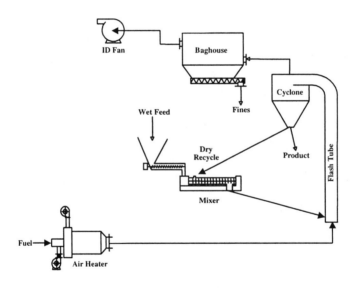

Figure 7 Flash dryer system

Typical flash dryer design and operating parameters are as follows:

- Diameter: 5 m maximum
- Height: 8-30 m
- Retention time: 2-3 seconds
- Gas velocity: 4-10 mps
- Feed top size: 2-2.5 mm
- Inlet temperature: up to 1000°C with refractory inlet
- Outlet temperature: 150-200°C.

Air-Swept Hammermill Dryer. This dryer has the ability to mill and dry materials with a wide range of particle sizes and moisture levels. If size reduction is desired or is not an issue, then this system offers the best combination of performance, flexibility and ease of operation.

The circuit (Figure 8) resembles a flash dryer with a hammermill between the air heater and flash tube. Material is introduced to the mill with the hot gas stream. As the material passes through the high-speed rotor, it is reduced in size, entrained in the gas flow and carried into the flash drying zone. This zone provides 1-2 seconds of retention time to complete the drying process. In most cases, a static or dynamic separator in the upper portion of the flash drying zone separates and returns coarse particles to the mill for further reduction. The gas and material passing through the separator are directed to a cyclone and baghouse for product collection. If the feed material is too sticky to flow through the mill housing, dry product recycle is used to condition the feed. The system can produce a fine, dry product from coarse feedstocks with moisture levels exceeding 50%. For limestone or software materials, top size can be reduced to 80-100% passing 150 μ. While footprint and ease of operation are similar to a flash dryer, a hammermill dryer is costly, consumes much power and is not easy to maintain. A rotary dryer and crushing/milling circuit combination may be equally worthy of consideration.

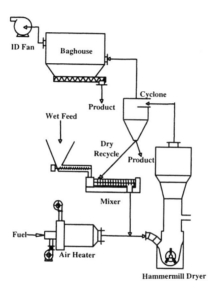

Figure 8 Hammermill dryer

Typical ranges for hammermill dryer design and operating parameters are as follows:

- Capacity: >200 mtph dry product
- Rotor width: 4 m maximum
- Rotor diameter: 4 m maximum
- Heat input: 100 MM Btu/hr maximum
- Inlet temperature: 870°C maximum
- Outlet gas temperature: 80-95°C
- Gas retention time: 1-2 seconds.

Rotary Tray Dryer. The rotary tray dryer, also known as the turbo tray dryer, can handle feeds ranging from viscous slurries to free flowing solids, and can provide long residence times. It provides gentle handling and heating of friable and temperature-sensitive materials. Its capacity is lower, and its capital cost per unit output higher, than most adiabatic drying systems.

The dryer (Figure 9) comprises a series of round trays stacked in a vertical plane. Feed is dropped onto the top rotating tray where it is leveled. Following approximately one revolution, the material is conveyed onto the tray below using a stationary wiper. Additional fixed devices mix and level the material on this tray. This mode of transport and mixing continues until the material is discharged from the bottom tray. Retention time is controlled by tray speed. Hot gas enters the bottom of the dryer and is circulated around the trays in a counter-flow orientation using central fans. The gases exit the top of the unit and are directed to a dust control circuit.

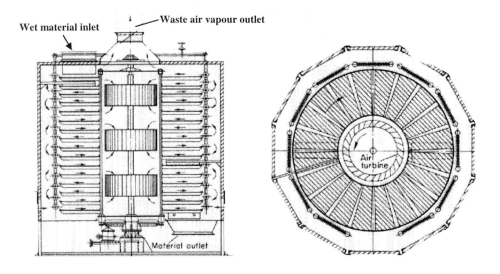

Figure 9   Rotary tray dryer

Typical rotary tray dryer design and operating parameters are as follows:

- Diameter: 10 m maximum
- Height: 20 m maximum
- Capacity: 15 mtph product maximum
- Evaporative load: 1,100 kg/hr maximum
- Material travel on tray: 80-85% of circumference
- Tray speed: 0.1-1.0 rpm
- Material retention time: 30-60 minutes.

Band Dryer. The band dryer, or conveyor dryer, dries a permeable bed of coarse particles or fine particles that have been preformed into large agglomerates such as briquettes or extrusions. It is applied before end use or high-temperature processing. The feed lies on a perforated belt that travels in a horizontal plane similar to that of a conventional belt conveyor. The belt sits inside a chamber supplied with hot gas that circulates around and through the bed of static material. The drying chamber is usually divided into cells, with gas recirculation to control the drying temperature profile. The last cells are often used for ambient air cooling. The most common configuration includes a single conveyor dryer. Multiple conveyors (in a multi-stage dryer) may be used in a vertical configuration when the material is very heat-sensitive, to improve efficiency and/or capacity. A single-stage band dryer is shown in Figure 10.

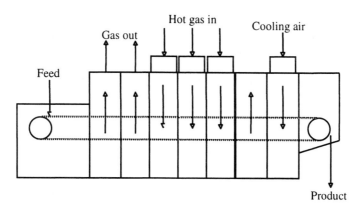

Figure 10 Band dryer

The dryer's gentle handling virtually eliminates degradation of even the most friable materials. The most critical part of the operation is the delivery of material to the belt in a manner that establishes a uniform bed depth to promote even air distribution and drying. Breaking devices may be employed at one or more locations along the belt to provide some degree of material agitation and mixing.

The typical ranges for band dryer design and operating parameters are as follows:

- Belt width: 0.5-3.0 m
- Belt length: no typical range
- Belt velocity: 0.001-0.02 mps
- Bed depth on belt: 20-150 mm (typically <40-60 mm to limit dP)
- Superficial velocity through bed: 0.5-2.0 mps
- Drying temperature: 100-200°C.

Spray Dryer. The spray dryer is the most effective for drying pumpable slurries. The feed is atomized to produce fine droplets of uniform size characterized by a high surface area (35,000-310,000 m^2/m^3 slurry), resulting in almost instantaneous drying that achieves very low moisture levels without particle overheating due to the high rate of evaporation. For slurries with multiple components, of spray drying is effective at fixing particles that are homogeneous in composition.

The primary components include an atomizer, drying chamber and product collection circuit. The most common configuration (Figure11) introduces the atomized slurry into the top of the vessel. Hot gas also enters the top of the vessel, flows down and sweeps the dried particles out the bottom to the product collection device. This figure illustrates concurrent flow, but flow may also be countercurrent or mixed.

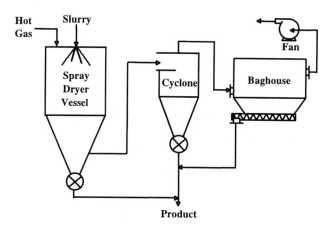

Figure 11 Spray dryer system

Three methods of slurry atomization are utilized: single fluid nozzle, pneumatic nozzle and centrifugal discs. The single fluid nozzle is generally used to produce fine droplets (120-250 μ), with the droplet size being a function of pressure drop across the nozzle. The pneumatic nozzle offers more flexibility in terms of droplet size, with size being a function of nozzle velocity and slurry properties. This design can also produce the smallest droplet size of the three nozzle options (10-20 μ). The centrifugal disc is the most widely applied method of atomization due to its simplicity, ease of operation, ability to handle abrasive slurries, rare occurrence of blockages, and atomization is not as sensitive to slurry properties. Slurry is dropped onto a disc rotating with a linear speed of 90-210 mps, and droplet size (30-120 μ) may be controlled by adjusting disc speed. A larger diameter drying vessel is required when the centrifugal disc is utilized in order to accommodate the wider spray pattern. Typical spray dryer design and operating parameters include:

- Evaporative loading: up to 7,000 kg/h
- Inlet gas temperature: 93-760°C
- Fluid pressures:
- Single fluid nozzle: 2,000-27,000 kPa
- Pneumatic nozzle: 0-415 kPa
- Centrifugal disc: 0
- Retention time: 3-30 seconds.

CONCENTRATE HANDLING

Bulk mineral concentrates are handled by an array of equipment including trucks, railcars, ocean vessels, front-end loaders, and conveyors. (Slurry pipelines are also used to transport concentrates, but these are addressed separately.) Concentrates are typically stored in bins, buildings or open stockpiles at one or more points between production and delivery to the end user, which may be a smelter, transport contractor or other customer.

Flow Properties. Flow properties of concentrates vary significantly not only with mineral type, but also within a particular concentrate due to variations in moisture content, grain size, climate conditions, and consolidation (i.e., due to storage time). The need to understand the flow properties under all conditions cannot be overemphasized when designing and selecting handling equipment. For example, chute and hopper valley angles of 45° may be sufficient for reliable flow for one particular concentrate, but not another. In fact, some concentrates require slopes of at least 75°. Analysis of concentrate flow properties is commonly omitted to save money, but the resulting design often proves problematic, with operating and maintenance cost increases that more than offset the front-end savings. A simple flow test program provides useful design information such as minimum chute valley angles and hopper opening sizes, eliminating the risk of material "hang-up" in chutes and hoppers, and bridging and ratholing in bins.

Moister concentrate is usually harder to handle, primarily because it tends to adhere to conveyors as well as surfaces such as the corners of truck and railcar bodies. Most concentrates are produced with moisture content between 4 and 8%, though higher content is not uncommon. For every concentrate and concentrator operation, there is a point at which the incremental cost of reducing moisture is no longer offset by the expense of transporting water to the end user.

Long-term storage (e.g., several weeks or months) can introduce storage and handling problems if consolidation effects are not addressed in the handling equipment design. For example, as moisture migrates out of the pile over time due to consolidation and/or vibration, the concentrate can become further compacted, which may introduce reclaim and handling problems. The exposed exterior of the pile will effectively dry out, which can lead to fugitive dust problems. Extended exposure to cold can form frozen lumps in stockpiles, which may not break up during reclaim operations.

Handling Equipment. The most common concentrate handling equipment items are front-end loaders and belt conveyors, with the latter including belt feeders, overland conveyors, stackers, and tripper conveyors. Mechanical conveying devices such as screw, drag, and pneumatic conveyors are also used, but their application is usually limited to low flow rates of concentrates with moisture content below 4%.

Belt conveyors provide economic, reliable handling, provided that transfer chutes are properly designed and belt cleaning devices are properly specified and maintained (especially for moist concentrates). Chute valley angles must be steep enough to promote reliable flow under all conditions. Chute liners, such as ultra-high-molecular-weight (UHMW) polyethylene and stainless steel, are a relatively low-cost means of preventing hang-up, and can be installed new or as retrofits. Properly specified and maintained belt-cleaning devices will minimize the carry-back of concentrates on the belt, reducing cleanup and maintenance costs.

In the case of outdoor belt conveyors, wind and precipitation can result in material loss, fugitive dust generation, and product degradation. Conveyors can be provided with partial or full belt covers, or fully enclosed in a steel structure, to minimize or eliminate such problems..

Belt conveyors can usually be inclined $15°$ or higher, depending on the concentrate characteristics. Finer grain size and/or higher moisture content usually permit steeper inclines due to greater internal cohesion.

Concentrates usually have higher bulk density than coal or crushed ore. Over time, accumulation and subsequent compaction of accumulated dust and/or spillage can result in relatively high loads on conveyor support and access structures. It is therefore critical to consider such loads in the design of these structures.

Some specialized belt conveyor systems are seeing increased application in concentrate handling. Sandwich-type high-angle conveyors, Pocketlift™ and pipe conveyors offer several advantages over conventional belt conveyors. High-angle and Pocketlift™ conveyors allow inclines of up to $90°$, while pipe conveyors can provide effective dust containment without the need for enclosures. These conveyors are usually more sophisticated than conventional belt conveyors in terms of design and componentry, and equipment manufacturers should be consulted when considering their application.

Relatively free-flowing, low-moisture concentrate may permit the use of mechanical conveying systems such as drag, screw, and pneumatic conveyors. However, the abrasiveness and particle size of the concentrate must be considered.

Bottom-dump, end-dump and side-dump trucks are commonly used for concentrate transportation, along with railcars (usually bottom-dump type). Concentrate flow properties must be considered when selecting a truck body type. Bottom-dump hoppers must be designed for reliable discharge of consolidated concentrates, while the tipping angle of side- and end-dump trucks must be sufficient to promote sliding discharge without the need for manual intervention. UHMW or stainless-steel liners in truck bodies are often helpful.

Weighing and Sampling. Concentrate weighing and sampling are usually performed at one or more points in the handling process. Conveyor belt weigh scales or weightometers are commonly used for process feedback control by providing instantaneous mass flow rate information, and for load measurement. For example, integration of the instantaneous flow rate data over a discrete duration can provide an estimate of concentrate discharged into a bin, railcar, or truck. When installed on a conveyor forming part of a truck, train or shiploader, higher-accuracy "certified for trade" scales can also be used for accounting purposes. Concentrates can also be weighed indirectly via truck and railcar scales, draft survey or in batch weighing systems under loadout bins.

Sampling is normally performed at one or more points for analysis of moisture content, grade, bulk density, etc. This is normally done by manual or automated grab sampling, or with automated sweep or cross-cut type systems installed over a belt conveyor. The method is dictated by accuracy and precision requirements as well as concentrate flow properties.

Dust Suppression, Safety, and Environmental. Concentrates inevitably generate fugitive dust when handled or exposed to wind. As most contain various metals and other hazardous elements, dust emissions can present significant safety and environmental hazards if not mitigated. The amount of dust generated depends primarily on the handling method. concentrate moisture content and particle size. Conveyor transfers, front-end loader operation, and loading of railcars, trucks and ocean vessels will generate dust due simply to the displacement of air; clearly, dust generation can be reduced by minimizing the number of times that concentrates are transferred from one point to another and the heights through which they are dropped.

Design of storage facilities should include effective sealing of building panels and doors. Conveyor system transfer chutes should be designed with minimal opening sizes. For relatively free-flowing and not overly abrasive concentrates, chutes can be designed such that the material slides along the inner chute surfaces, rather than hitting them. Cartridge or baghouse dust collection systems are also effective in controlling fugitive dust, provided that they are properly designed for the concentrates' flow rate, particle size, specific gravity, and losses from chute openings and system inefficiencies. Water spray dust suppression systems can be useful, although attention must be given to runoff collection and their effect on concentrate moisture content.

In enclosed storage buildings where concentrates are reclaimed by front-end loaders, emissions of both concentrate dust and equipment exhaust can pose potential health and safety hazards. Adequate positive building ventilation is typically impractical for large storage buildings due to the prohibitive costs of fans and power. Operators should wear dust masks, and emissions control devices (such as scrubbers and diesel particulate traps) should be installed on the equipment.

STORAGE

Capacity. Concentrates are usually stored for some period of time upon discharge from the concentrator, for periods ranging from a few days to several months.

Short-term storage facilities are typically sized to effectively decouple concentrate production and its subsequent handling operation; therefore, production rate and availabilities of the mill and downstream transportation are key criteria. A properly sized facility eliminates the need to reduce or stop production when, e.g., a bulk carrier is unavailable for loading or a railcar loading facility is down for maintenance. Conversely, as a fundamental part of planning mill shutdowns, the storage facility can be "topped up" to allow production to stop without interrupting concentrate loading and transportation.

Sizing of long-term storage facilities is sometimes mandated by system availabilities or market conditions. Supply contracts are typically be linked with storage capacity. Mills that produce more than one type or grade of concentrate require storage facilities that allow different volumes of different products to be stored depending on market conditions. Many operations in Arctic regions require facilities that store up to one year's production, since the ocean transport of product is limited to short ice-free shipping seasons.

Storage Facility Design. Concentrates are typically delivered to a storage facility by truck, rail, front-end loader or conveyor, which may comprise a fixed conveyor feeding a stockpile or bin, radial stackers or overhead tripper conveyor.

While concentrates can be stockpiled in the open, they are usually stored in bins, silos, or buildings for protection from the weather and to comply with environmental regulations. Without protection, wind can create dust problems and erode valuable product. Precipitation can increase concentrate moisture content, adversely affecting its flow properties and increasing water weight. In some short-term storage facilities or in dry areas, covered storage may not be economically justified; in such cases, water spray systems may be used to mitigate fugitive dust emissions.

Stockpiles may or may not require underlying concrete slabs, depending on geotechnical conditions, product or ground contamination, and reclaim method. In many cases, a compacted fill floor is adequate and less expensive; alternatively, a compacted fill floor can be initially covered with a thin (e.g., 0.3 to 1.0 m) layer of compacted low-grade concentrate to prevent soil from being excavated with the concentrate during reclaim operations.

Multiple types and/or grades of concentrate can be stored in the same area, provided piles are separated. Cast concrete or concrete block bulkheads are a useful means of separation.

Bins are common for relatively short-term storage requirements, such as smelter feed systems or surge bins. In designing bins, it is essential to understand (and preferably test) the concentrate bulk material properties under all potential conditions, so that hopper geometry and reclaim equipment are properly designed to ensure reliable reclaim operations. The effects of ambient conditions, moisture content, and consolidation on properties such as internal and wall friction angles, and bulk density can have highly detrimental results on reliability.

Reclaim. Reclaiming of concentrates from open or covered stockpiles is typically performed either by front-end loaders (which transfer concentrates directly to trucks, railcars or to a hopper feeding a belt conveyor), or by an automated reclaim system. Vibratory feeders draw concentrates from under a stockpile through one or more draw-down openings and onto a belt conveyor; a system with multiple draw-down points and vibratory feeders can permit blending of product from more than one stockpile, but vibratory feeders are better suited to relatively free-flowing concentrates. Bucket-wheel or scraper reclaimers are relatively expensive, but offer relatively low operating costs when handling high throughputs on a continuous basis. In specifying such equipment, it is important to consider the worst-case scenario of consolidated or frozen concentrate in the stockpile, since this can pose a hazard to downstream handling equipment. If lumps are relatively rare, they can usually be broken up using mobile equipment. Otherwise, a feeder-breaker unit can be installed.

CONCLUSIONS AND RECOMMENDATIONS

The wide array of available drying systems may make the selection process seem difficult, but each has distinct advantages and disadvantages depending on material feed properties, capacity and installation requirements, and other factors. Options can usually be narrowed down to one basic system type, after which plant personnel may contact suppliers to determine the system configuration that best suits their needs. The design and operation of storage and handling systems requires attention to many factors including economics, moisture content, flow properties, ambient climate conditions, production rate, storage duration, transport method, and customer requirements. Concentrate flow properties should be analyzed prior to design and selection of handling equipment to ensure reliable, cost-effective operation.

REFERENCES

Williams-Gardner, A. 1971. Industrial Drying. Leonard Hill. London, England.

Van't Land, C. M. 1991. Industrial Drying Equipment-Selection and Application. Marcel Dekker Inc. New York, NY.

Kolthammer, K.W. 1980. Concentrate Drying, Handling and Storage. Mineral Processing Plant Design, 2nd Edition, Society of Mining Engineers, USA.

The Selection and Sizing of Bins, Hopper Outlets, and Feeders

Dr. John Carson and Tracy Holmes, Jenike & Johanson, Inc.

ABSTRACT
This paper provides practical guidelines for the selection of bins, feeders, hopper outlets and gates, outlines the principles in their selection, and touches on the basic need to know the bulk flow properties of the material being handled.

INTRODUCTION
To cover this very broad topic in a manner that is easy to follow, we have divided this paper into three individual sections:

- Bin selection
- Hopper outlet sizing
- Feeder selection.

BIN SELECTION
A bin (silo, bunker) generally consists of a vertical cylinder and a sloping converging hopper. Based on the flow pattern that develops, there are three types of bins: Mass Flow, Funnel Flow and Expanded Flow, all of which will be discussed in more detail below. Whatever type of bin is selected, it needs to have the desired capacity, be capable of discharging its contents reliably on demand, and be safely constructed. Here are some of the important things that need to be done in order to ensure that a bin will perform these functions adequately:

Step 1. Define your storage requirements
Identify the operating requirements and conditions. Some of the more important include:

- Capacity. This will vary with your plant's operating philosophy, and where the bin is to be located (e.g., start of your process, at an intermediate process step, or at the end).
- Discharge rate. Consider average and instantaneous rates, minimum and maximum rates, and whether the rate is based on volume or mass.
- Discharge frequency. How long will your material remain in the bin without movement?
- Mixture and material uniformity. Is particle segregation a concern in terms of its effects on material discharge or, more importantly, downstream processes?
- Pressure and temperature. Consider differences between the bin and upstream and downstream equipment.
- Environmental. Are there explosion risks, human exposure concerns, etc.?
- Construction materials. Abrasion and corrosion concerns may limit the types of materials you can use to construct your bin.

Step 2. Calculate approximate size of your bin
Initially ignore the hopper section. Use the following formula to estimate the approximate height of the cylinder section that is required to store the desired capacity:

$$H = (C / \gamma_{avg} A) \tag{1}$$

where H = cylinder height, ft
C = bin capacity, ft^3
γ_{avg} = average bulk density, lb/ft^3
A = cross-sectional area of cylinder section, ft^2

The actual cylinder height will have to be adjusted to account for volume lost at the top due to the material's angle of repose as well as for the volume of material in the hopper section.

In general, the height of a circular or square cylinder should be between about 1.5 and 4 times the cylinder's diameter or width. Values outside this range often result in designs which are uneconomical or have undesirable flow characteristics.

It is important to recognize that a bin's storage volume and its active (live, useable) volume are not necessarily the same. With a funnel flow or expanded flow pattern (described below), significant dead (stagnant) volume may need to be taken onto account.

Step 3. Determine your material's flow properties

The flow characteristics of a bulk solid must be known in order to predict or control how it will behave in a bin or hopper. These characteristics can be measured in a solids flow testing laboratory under conditions that accurately simulate how the solid is handled in your plant. Tests should be conducted on-site if your solid's properties change rapidly with time or if special precautions must be taken.

The most important bulk solids handling properties that are relevant to predicting flow behavior in bins and hoppers are listed in Table 1. Each of these parameters can vary with changes in the following:

- Moisture
- Particle size, shape, hardness and elasticity
- Temperature
- Time of storage at rest
- Chemical additives
- Pressure
- Wall surface.

The appropriateness of these bin design parameters has been proven over the last 40 years in thousands of installations handling materials as diverse as fine chemical powders, cereal flakes, plastic granules and mined ores.

Table 1 Important flow properties

Parameter	Measured by	Useful for calculations of
Cohesive strength	Shear tester	Outlet sizes to prevent arching and ratholing
Frictional properties	Shear tester	Hopper angles for mass flow, internal friction
Sliding at impact points	Chute tester	Minimum angle of chute at impact points
Compressibility	Compressibility tester	Pressure calculations, bin loads, feeder design
Permeability	Permeability tester	Discharge rate calculations, settlement time
Segregation tendency	Segregation tester	To predict whether or not segregation will occur
Abrasiveness	Abrasive wear tester	To predict the life of a liner
Friability	Annular shear tester	Maximum bin size, effect of flow pattern on particle breakage

Step 4. Understand the importance of flow patterns

Although it is natural to assume that a bulk solid will flow through storage or conditioning vessels in a first-in/first-out sequence, this is not necessarily the case. Most bins, hoppers, silos and conditioning vessels move solids in a funnel flow pattern.

With funnel flow, some of the material moves while the rest remain stationary. This first-in/last-out sequence is acceptable if the bulk solid is relatively coarse, free flowing, non-degradable, and if segregation is not important. If the bulk material and application meet all four of these criteria, a funnel flow bin is the most economical storage device.

Unfortunately, funnel flow can create serious problems with product quality or process reliability. Arches and ratholes may form, and flow may be erratic. Fluidized powders often have no chance to de-aerate. Therefore, they remain fluidized in the flow channel and flood when exiting the bin. Some materials cake, segregate or spoil. In extreme cases, unexpected structural loading can result in equipment failure.

These problems can be prevented with storage and conditioning vessels specifically designed to move materials in a mass flow pattern. With mass flow, all material moves whenever any is withdrawn. Flow is uniform and reliable; feed density is independent of head of solids in the bin; there are no stagnant regions, so material will not cake or spoil and low-level indicators work reliably; sifting segregation of the discharge stream is minimized by a first-in/first-out flow sequence; and residence time is uniform, so fine powders are able to de-aerate. Mass flow bins are suitable for cohesive materials, powders, materials that degrade with time, and whenever sifting segregation must be minimized.

A third type of flow pattern is called expanded flow. In this, the lower part of a bin operates with flow along the hopper walls as in mass flow, while the upper part operates in funnel flow. An expanded flow bin combines the best aspects of mass and funnel flow. For example, a mass flow outlet usually requires a smaller feeder than would be the case for funnel flow. This flow pattern is suitable for storage of large quantities of non-degrading solids. It can also be used with multiple outlets to cause a combined flow channel larger than the critical rathole diameter.

Step 5. Follow these detailed design procedures

Step 5A. Mass flow. In order to achieve a mass flow pattern, it is essential that the converging hopper section be sufficiently steep and have low enough friction to cause flow of all the solids without stagnant regions, whenever any solids are withdrawn. In addition, the outlet must be large enough to prevent arching and to achieve the required discharge rate.

Typical design charts showing the limits of mass flow for conical- and wedge-shaped hoppers are given in Figure 1. Hopper angle (measured from vertical) is on the abscissa, and wall friction angle is on the ordinate. For example, mass flow will occur in a conical hopper which has an angle of 20° and is constructed from or lined with a wall material which provides a wall friction angle of 23° or less with the stored bulk solid. Making the hopper walls less steep by 4° or more could result in funnel flow. Alternatively keeping the wall angle at 20° but increasing the wall friction angle to 28° or more would also result in funnel flow.

Calculating the outlet size needed to overcome arching is more difficult. It involves measuring the cohesive strength and internal friction of the bulk solid, then following the design procedure outlined in Reference 1.

Sizing the outlet for discharge rate is covered in this article.

Step 5B. Funnel flow. The key requirements for designing a funnel flow bin are to size the hopper outlet large enough to overcome arching and ratholing, and to make the hopper slope steep enough to be self-cleaning.

Minimum dimensions to overcome arching and ratholing require knowledge of your material's cohesive strength and internal friction. Design procedures for funnel flow are also given in Reference 1

The requirement for self-cleaning can usually be met by making the hopper slope 10° to 15° steeper than the wall friction angle.

Step 5C. Expanded flow. Consideration must be given to both the mass flow and funnel flow sections. In the lower mass flow section, the procedure outlined above for a mass flow hopper should be followed. In addition, the flow channel must be expanded to a diagonal or diameter equal to or greater than the material's critical rathole diameter, which can be calculated using the procedure in Reference 1.

Here too, the hopper slope in the funnel flow portion should be steep enough for self-cleaning.

Step 6. Consider the bin's shape.

At first glance, it might appear that a square or rectangular straight-sided section at the top of a bin is preferable to a circular cross-section. Such cylinders are easier to fabricate and have greater cross-sectional area per unit of height. However, these advantages are usually overcome by structural and flow considerations.

A circular cylinder is able to resist internal pressure through hoop tension, whereas flat walls are subjected to bending. Thus, thinner walls and less external reinforcement are required with circular cross-sections. In addition, there are no corners in which material can build up. This is particularly important when interfacing with a hopper at the bottom.

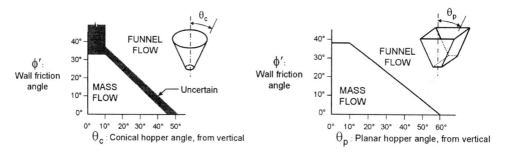

Figure 1 Typical chart determining mass flow wall angles

Several factors to consider when choosing hopper geometry are listed below:

- Sharp versus rounded corners. Pyramidal hoppers usually cause a funnel flow pattern to develop because of their inward-flowing valleys that are less steep than adjacent sidewalls. Conical, transition and chisel shapes are more likely to provide mass flow because they have no corners. See Figure 2.
- Headroom. Typically, a wedge-shaped hopper (e.g., transition or chisel) can be 10° to 12° less steep than a conical hopper and still promote mass flow. This can provide significant savings in hopper height and cost, which is particularly important when retrofitting existing equipment in an area of limited headroom. In addition, a wedge-shaped hopper design is more forgiving than a cone in terms of limiting hopper angles and wall friction.
- Outlet sizes. In order to overcome a cohesive or interlocking arch, a conical hopper has to have an outlet diameter that is roughly twice the outlet width of a wedge-shaped hopper (provided the outlet length is at least three times its width). Thus, cones generally require larger feeders.
- Discharge rates. Because of the increased cross-sectional area of a slotted outlet, the maximum flow rate is much greater than that of a conical hopper.
- Capital cost. Each application must be looked at individually. While a wedge-shaped hopper requires less headroom or a less expensive liner than a cone, the feeder and gate (if necessary) may be more expensive.
- Discharge point. In many applications, it is important to discharge material along the centerline of the bin in order to interface with downstream equipment. In addition, having a single inlet point and single outlet, both located on the bin's centerline, minimizes flow and structural problems. Generally, conical hoppers are better for these situations, particularly if only a gate is used to stop and start flow.

- Mating with a standpipe. If material is being fed into a pressurized environment, a circular standpipe is often preferred to take the pressure drop.

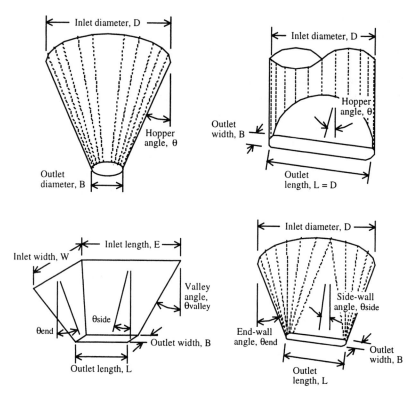

Figure 2 Hopper geometries (conical-top left; chisel-top right; pyramid-bottom left; transition-bottom right)

Step 7. Consider other important factors
Some additional considerations include:

- Gate. A slide gate at the outlet of a bin must generally only be used for maintenance purposes, not to control or modulate the flow rate. Therefore, it should only be operated in a full-open or full-closed position.
- Feeder. The feeder's design is as important as that of the bin above it. The feeder must uniformly draw material through the entire cross-section of the bin's discharge outlet to be effective. The section below on Feeder Selection covers this in more detail.
- Mating flanges. The inside dimensions of the lower of two mating flanges must be oversized to prevent any protrusions into the flowing solid. The amount of oversize depends on the accuracy of the construction and erection. Usually 1 inch overall is sufficient. If gaskets or seals are used, care must be taken to ensure that these too, do not protrude into the flow channel. All flanges should be attached to the outside of the hopper with the hopper wall material being the surface in contact with the flowing solids. This ensures that the flange and gasket do not protrude into the flowing solids.
- Interior surface finish. Whenever possible, welding should be done on the outside of the hopper. If interior welding is necessary, all welds on sloping surfaces must be ground flush and power brushed to retain a smooth surface. After welding, all sloping surfaces must be clean and free of weld spatter. The surface finish is most critical in the region of the hopper outlet. Therefore, any blisters in this area from exterior welding must be brushed smooth. Horizontal or diagonal welded connections should preferably be lapped

with the upper section on the inside so the resulting ledge does not impede flow. If horizontal butt welds are used, care must be taken to avoid any protrusion into the flowing solid.
- Liner attachment. Inside liners, such as stainless sheet or ultra-high molecular weight (UHMW) polyethylene, must be placed on sloping surfaces with horizontal or diagonal seams lapped with the upper liner on top in shingle fashion. Vertical seams may be either lapped or butted.
- Abrasive wear considerations. In mass flow, a bulk solid flows against the hopper and cylinder walls. Handling an abrasive bulk solid may result in significant abrasive wear of the wall material including coatings and liners. Therefore, when designing a mass flow hopper, it is important to assess the potential for abrasive wear. Generally, a hopper surface becomes smoother with wear. However, occasionally a wall becomes rougher, which may upset mass flow. The life of a given wall material can be estimated by conducting wear tests.
- Access doors and poke holes. In general, poke holes are not recommended in mass flow bin designs as they have a tendency to prevent flow along the walls, thus creating a problem that mass flow bins are intended to solve. Access doors are also a frequent cause of problems. If they are essential, it is better to locate them in the cylinder, rather than in the hopper section.
- Structural design issues. It is important that the bin be designed to resist the loads applied to it by both the bulk solid and external forces. This is particularly important when designing, or converting, an existing bin to mass flow because unusually high localized loads may develop at the transition between the vertical section and the mass flow hopper.
- Bulk materials of inferior flowability (*e.g.*, more cohesive with larger critical arching and ratholing dimensions than the material upon which the design was based, or more frictional requiring steeper wall angles) should not be placed in the bin because flow obstructions are then likely to occur. Such obstructions may lead to the development of voids within the bin and impose dynamic loads when material collapses into the voids. Bin failures have occurred under such conditions.
- Prefabrication drawing review. Before fabrication of the bin and feeder, an engineer trained in solids flow technology should review all detailed design drawings. This review is necessary to ensure that the design follows the recommendations and that any design details or changes are consistent with reliable bulk solids flow.

HOPPER OUTLET SIZING

In most applications, a feeder is used to control discharge from a bin or hopper (see section below on Feeder Selection). For such applications, the maximum achievable flow rate through the hopper outlet must exceed the maximum expected operating rate of the feeder. This ensures that the feeder will not become starved. This is particularly important when handling fine powders, since their maximum rate of flow through an opening is significantly less than that of coarser particle bulk solids whenever a mass flow pattern is used. In addition, any gates must not interfere with material discharge.

The following step-by-step procedure will assist in properly sizing a hopper outlet.

Step 1. Calculate the ratio of outlet width or diameter to particle size

Flow stoppages due to particle interlocking are likely if the diameter of an outlet is less than about six times the particle size. With an elongated outlet, problems are likely if the ratio of outlet width to particle size is less than about 3:1.

Step 2. Determine into which of the following categories the material fits: Coarse, Easy Flowing, and/or Fine Powder

For purposes of flow rate calculations, a bulk solid is often considered coarse if no more than 20% will pass through a 1/4 in. screen.

Whether or not a material can be considered easy flowing depends upon the cohesiveness of the bulk solid, the dimensions of the container in which it is stored, and whether or not any excess pressures are applied to the material (e.g., the container is vibrated after being filled). If the combination of these factors results in no flow stoppages at the vessel's outlet (due, for example, to arching or ratholing), the material can be considered easy flowing in that application.

A fine powder is a bulk solid whose flow behavior is affected by interstitial gas. Common household examples of fine powders include flour and confectioners (icing) sugar.

Step 3. Determine maximum achievable flow rates

Step 3A. The bulk material is coarse but not easy flowing. Either the outlet size must be increased or the material's cohesive strength must be decreased to allow the material to flow.

Step 3B. The bulk material is coarse and easy flowing. If the ratio of outlet size to particle size is sufficiently large to prevent particle interlocking, then the maximum achievable rate through an orifice of a coarse, easy flowing bulk solid such as plastic pellets is given by the following equation:

$$Q = 3600 \gamma A \left[Bg/(2(1 + m) \tan \theta) \right]^{1/2} \quad (2)$$

Where

Q = flow rate, lb/hr
γ = bulk density, lb/ft^3
A = outlet area, ft^2
B = diameter of circular outlet or width of a slotted outlet, ft
g = gravitational constant, 32 ft/sec^2
m = 0 for long slotted outlet, 1 for conical hopper
θ = flow channel angle (measured from vertical), deg

A modification can be made to Equation (2) to consider particle size. This modification is only important if the particle size is a significant fraction of the outlet size.

Note that, with a mass flow hopper (see section on Bin Selection), the flow channel coincides with the hopper wall; hence θ is the hopper angle. On the other hand, with a funnel flow hopper, the flow channel forms within stagnant material and, while it is steeper than in mass flow, it is variable. Thus, the maximum flow rate from a funnel flow hopper is generally higher but less predictable than the flow rate from a mass flow hopper having the same outlet size.

Step 3C. The bulk material is a fine powder. Fine powders are often mishandled in funnel flow bins. As noted in the section on Bin Selection, fine powders have little or no chance to de-aerate in such bins. Instead, they often remain fluidized in the flow channel and flood uncontrollably when exiting the bin. A funnel flow pattern should therefore be avoided when handling fine materials.

Mass flow bins, on the other hand, can provide predictable and controlled rates of discharge of fine powders as well as other bulk solids. Unfortunately, a fine powder's maximum rate of discharge through a mass flow hopper outlet is often several orders of magnitude lower than the limiting rate of a coarse particle material having the same bulk density. This severe flow rate limitation is the result of the upward flow of air through the hopper outlet caused by the slight vacuum condition, which naturally forms in the lower portion of a mass flow hopper as material flows through it.

The limiting rate of material flow through a mass flow hopper outlet can be calculated once the powder's permeability and compressibility have been measured.

If the limiting discharge rate is too low, there are several possible ways to increase it:

- Increase the outlet size. Since the limiting rate is approximately proportional to the cross-sectional area of the outlet, doubling the diameter of a circular outlet increases the maximum discharge rate by roughly a factor of four.
- Decrease the level of material in the bin. Fine powders do not behave like fluids. Thus, lower heads result in higher discharge rates, although the effect is generally not too pronounced.
- Provide an air permeation system as shown in Figure 3. This has the effect of partially satisfying the vacuum condition that naturally develops. As a result, there is less need for air to be pulled up through the outlet counter to the flow of particles. With such a system, the maximum flow rate can often be increased by a factor of between 2 and 5.

If none of these will allow the desired discharge rate, fluidization, as discussed in the next step, can be considered.

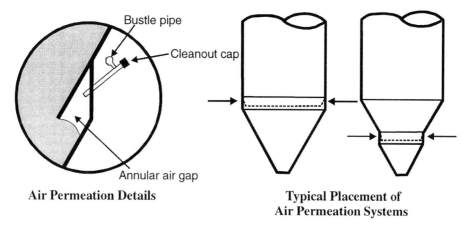

Figure 3 Air permeation systems

Step 3D. The bulk material is a fine powder and the required discharge rate is high. If the limiting flow rate from a mass flow hopper is still too low, consideration should be given to fluidizing the fine powder and handling it like a liquid. For this technique to be successful, it is generally necessary that the powder have low cohesion and low permeability. Low cohesion allows the material to fluidize uniformly, so the air does not channel around large lumps. With a low permeability material, significant pressure gradients can be established, and the material takes a long time to de-aerate.

The Geldart chart, shown in Figure 4, provides a rough indication of whether or not a particular material is a good candidate for fluidization. Powders falling within classifications A and B are generally considered good candidates, while category C materials are difficult to fluidize. Category D materials are acceptable for fluidization, but the bed settles quickly and high gas rates are required.

If the bin is small, it may be practical to fluidize the entire contents. With larger bins, this is neither practical nor necessary. However, if only localized regions are fluidized, consideration must be given to the potential for arching and ratholing in non-fluidized regions.

In some cases, it is necessary to only fluidize intermittently, while in other cases, continuous fluidization is required during discharge. Whether batch or continuous discharge is required will influence this, but there are other factors to consider as well.

When considering the fluidization option, several operational requirements must be evaluated:

- The bulk density of the discharging material will be lower than if the material were not fluidized. Therefore, a given mass will occupy more volume, which could result in downstream equipment (such as a bulk bag, hopper or railcar) receiving less than the desired mass even though it is full.
- The material's bulk density will also vary with time depending on the degree of fluidization of the discharging material. This can present process problems downstream.
- Some materials are hygroscopic, while others are explosive. In such cases, dry air or inert gas may be required for fluidization.
- Higher energy and gas consumption rates must be taken into account as an additional operating cost.
- The feeder controlling the discharge must be capable of metering fluid-like materials.
- This technique should be avoided if particle segregation is a concern.

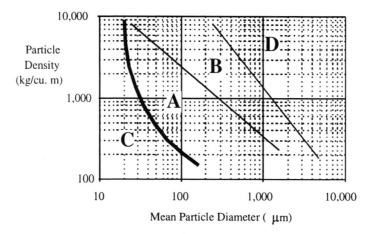

Figure 4 Geldart's fluidization classification

FEEDER SELECTION

A feeder is used whenever there is a need to control the solid's flow rate from a bin or hopper. Conveyors are incapable of performing this function, because they only transport material and do not modulate the rate of flow.

Dischargers are sometimes used to encourage material to flow from a bin, but by themselves cannot control the rate at which material flows. Thus, they are not a feeder, either. Consequently, when a discharger is used, a feeder is also required to control the flow rate from a storage system.

Criteria For Feeder Selection

Regardless of what type of feeder is used, it should provide the following:

- Reliable and uninterrupted flow of material from some upstream device (typically a bin or hopper).
- The desired degree of discharge flow rate control over the necessary range.
- Uniform withdrawal of material from the outlet of the upstream device. This is particularly important if a mass flow pattern is desired, perhaps in order to control segregation, provide uniform residence time, or to minimize caking or spoilage in dead regions.
- Minimal loads acting on the feeder from the upstream device. This minimizes the power required to operate the feeder, as well as minimizes particle attrition and abrasive wear of the feeder components.

Plant personnel often prefer a certain type of feeder because of past experience, availability of spare parts, or to maintain uniformity to make maintenance easier throughout the plant. Such personal preferences can usually be accommodated because, in general, several types of feeders can be used in most applications, if they are designed properly.

Volumetric or Gravimetric
Feeders can be divided into two basic types - volumetric and gravimetric. A volumetric feeder modulates and controls the volumetric rate of discharge from a bin (e.g., cu ft/hr). Many types of volumetric feeders are available on the market. The four most common types of such feeders are screw, belt, rotary valve and vibrating pan. Each has inherent benefits and limitations, many of which are spelled out in this guide. Special designs can often overcome many of the weaknesses stated.

In contrast to a volumetric feeder, a gravimetric feeder modulates the mass flow rate. This can be done either on a continuous basis, in which the feeder modulates the mass of material discharged per unit time; or on a batch basis, in which a certain mass of material is discharged and then the feeder shuts off.

Follow this step-by-step procedure for selecting a helical screw, belt, rotary valve (sometimes referred to as a rotary air lock), or vibrating pan for your application.

Step 1. Determine maximum particle size of your material
If it's less than about 1/2 in., almost any type of feeder can be used. If over about 6 in., the choices are limited. In most industrial plants, this generally means either a belt or vibrating pan feeder.

Step 2. Establish whether particle attrition is a concern
If particle attrition were a concern, a vibrating pan would be a good choice. Feeders that have pinch points (screws, rotary valves) should be avoided.

Step 3. Evaluate likelihood and frequency that material will drop directly onto the feeder, such as when the bin is empty
A vibrating pan is a good choice for this application, since it is more rugged than a belt yet it has a smooth surface, which limits buildup such as can occur with screws or rotary valves.

Step 4. Identify outlet configuration of hopper to which feeder will be attached
Square and round outlets present no restrictions in choice of feeder. Elongated outlets, on the other hand, generally require either a screw or belt. An elongated rotary valve, called a star feeder, can be used to feed across the narrow dimension of a slotted outlet. A vibrating feeder can also be oriented to feed across this dimension. This kind of feeder may require several drives to accommodate extreme width, although the drives will be small because of the feeder's short length.

Step 5. Decide whether volumetric or gravimetric control is required
Screws, rotary valves, and vibrating pans can only control flow on a volumetric basis. Belts, on the other hand, can be used for either application.

Step 6. Determine maximum operating temperature
A belt is generally limited to about 450°F, unless special materials of construction are used. Screws, rotary valves, and vibrating pans can be used with temperatures in excess of 1000°F.

Step 7. Determine design throughput
   Step 7A. Maximum. The highest throughput can generally be achieved with a belt feeder, followed by a screw, and then a rotary valve or vibrating pan. For example, with material having a bulk density of 100 pcf, the maximum capacity of a typical belt feeder is about 3,000 ton/hr. The maximum capacity of a rotary valve or vibrating pan is about one-sixth this value. As discussed

above, as well as the mechanical limits of the feeder, consideration must also be given to the maximum achievable discharge rate through the hopper outlet to which it is attached.

Step 7B. Minimum. The minimum rate of throughput is required in order to determine the turndown required for the feeder. Most feeders can easily achieve a 10:1 turndown. If significantly higher turndown were required, a good choice would be a vibrating pan.

Step 8. Look at other specific operational requirements

If the bulk material is to be fed into a pressurized environment (e.g., positive pressure pneumatic conveying line) a rotary valve is an excellent choice. A screw feeder can be used if it is designed with a moving plug at the discharge end.

If return spillage is a concern, a belt feeder should be avoided. Any of the other types of feeders being considered would not have this problem.

Step 9. Determine material characteristics that might affect feeder choice

With fine dry material, flooding and dust generation are likely to be a concern. Therefore, having a feeder that seals the outlet and/or is totally enclosed is important. A rotary valve is an excellent choice as can be a screw, if it designed properly. If the only concern is dust generation and not flooding, then either of these two types of feeders or an enclosed vibrating pan feeder is a good choice.

If the bulk material is pressure sensitive, avoiding pinch points and minimizing excessive compaction are important considerations. A belt feeder is often the best choice for these types of materials.

When handling materials that degrade easily, belts and vibrating pans are good choices. They are easy to clean and by their nature do not have stagnant zones within the feeder itself. Screws and rotary valves are not as good in this application as they do have stagnant zones; however, they can be designed for quick disassembly for cleaning.

If the bulk solid is expected to contain tramp material, belts and pan feeders are good choices whereas a screw is only a fair choice and a rotary valve is a poor choice.

Screw Feeders

The key to proper screw feeder design is to provide an increase in capacity in the feed direction. This is particularly important when the screw is used under a hopper with an elongated outlet. One common way to accomplish this is by using a design as shown in Figure 5.

Figure 5 Mass flow screw feeders draw uniformly from the entire outlet

Uniform discharge across the entire hopper outlet opening is accomplished through a combination of increasing pitch and decreasing diameter of the conical shaft.

Unfortunately, normal tolerances of fabrication are such that extending the length under the hopper outlet to greater than about six to eight times the screw diameter often results in a poorly performing screw. This length can be extended with special design and fabrication techniques.

Through special design techniques, a moving plug can be formed at the discharge end of the screw, allowing material to be fed into a higher pressure environment while minimizing leakage back into the feed bin, that can potentially create arching and ratholing problems.

Belt Feeders

As with screw feeders, the key to proper belt feeder design is to provide increasing capacity (draw) along the length of the bin outlet.

Without this, material will channel at one end of the hopper and disrupt mass flow, potentially creating arching and ratholing problems.

An effective way to increase capacity is to cut a converging wedge-shaped hopper in such a way that it is closer to the feeder at the back of the outlet than at the front. This provides expansion in both plan and elevation as is shown in Figure 6.

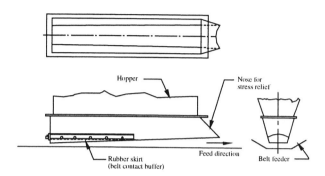

Figure 6 Typical mass flow belt feeder interface

It is important that the bed depth of material at the front of the outlet be at least 1.5 to 2 times the largest particle size to prevent blockage.

Vibratory Feeders

Vibratory feeders are excellent in providing a nearly continuous curtain of material discharge. Electromagnetic vibratory feeders are extremely rugged and simple in construction; thus, they are well suited to being used in hostile and dirty environments.

Like screw feeders and rotary valves, they can be enclosed to eliminate dusting and product contamination. They are, however, limited for the most part to feeding from round, square, or slightly elongated openings.

Rotary Valves

Rotary valves are generally limited to being used with hoppers having circular or square outlets. Thus, they are not as useful when handling cohesive bulk solids as, for example, a screw or belt feeder. A rotary valve can also be used as an air lock when feeding into a higher or lower pressure environment, such as a pneumatic conveying line.

Gates

To make it easier to perform maintenance on a feeder, various types of gates, such as clamshell or slide gates, are used to isolate the feeder from an upstream bin.

If the bin is designed for mass flow, it is important that there be no protrusions into the flow channel when the gate is open. Thus, the inside dimensions of the gate must exceed those of the bin outlet.

In addition, gates must generally be operated only in a full-open or full-closed position, and not to modulate the rate of solids flow. This is the job of the feeder. A partially opened gate will allow stagnant regions to form above it, resulting in a funnel flow pattern.

The height of the gate should be minimized to reduce the additional head pressure on the feeder.

REFERENCES

Jenike, A.W. Storage and Flow of Solids, University of Utah Engineering Experiment Station, Bulletin No. 123, Nov. 1964.

Carson, J.W. and Marinelli, J. Characterize Bulk Solids to Ensure Smooth Flow, Chemical Engineering, Vol. 101, No. 4, April 1994, pp. 78-90.

11

Pre-Oxidation

Section Co-Editors:
Dr. Ralph P. Hackl and Dr. Kenneth Thomas

Design of Barrick Goldstrike's Two-Stage Roaster
D. Warnica, A. Cole, S. Bunk **1493**

Selection of Materials and Mechanical Design of Pressure Leaching Equipment
K. Lamb, J. Gulyas **1510**

Barrick Gold—Autoclaving and Roasting of Refractory Ores
K.G. Thomas, A. Cole, R.A. Williams **1530**

Selection and Sizing of Biooxidation Equipment and Circuits
C.L. Brierley, A.P. Briggs **1540**

Design of Barrick Goldstrike's Two-Stage Roaster

David Warnica[1], Andy Cole[2], and Stanley Bunk[3]

ABSTRACT

Two fluid-bed roasters were constructed in 1999 to pretreat carbonaceous refractory ores at Barrick Goldstrike Mines. These are the largest gold-ore roasters in the world, which use almost pure oxygen to remove contained organic carbon and sulfide sulfur prior to a conventional carbon-in-leach process for gold extraction. This article describes the overall roasting process and design features that enable these roasters to exceed their design objectives.

INTRODUCTION

Indications in the early 1990's that ore reserves were becoming increasingly carbonaceous at Barrick Goldstrike Mines led to reviews of alternate pretreatment technologies. These carbonaceous ores are called double refractory, because the gold is locked within sulfide minerals and the carbonaceous matter adsorbs gold during conventional cyanide leaching (Chryssoulis and Cabri, 1990; Afenya, 1991). The existing pretreatment using pressure oxidation (autoclave leaching) would not overcome the preg-robbing characteristic of carbonaceous material, which would significantly hurt gold recoveries. The carbonaceous ores were to be mined starting in late 1999, and comprise a reserve of approximately 80 million tons with an average grade of 0.177 troy ounces per ton. An alternate pretreatment had to be determined in the meantime to maintain annual gold production at about two million ounces.

Two metallurgical processes for pretreatment were considered:

- Acid or alkaline autoclaving followed by ammonium thiosulfate (ATS) leaching
- Whole ore roasting

Both pretreatment options were the subject of bench-scale metallurgical testing and pre-feasibility engineering studies. These studies concluded that whole ore roasting was the most proven technology at the time. Other pretreatment options were discounted if they were inherently unsuitable to the Goldstrike ores or had not received significant commercial applications. Different roasting options were therefore considered, including Lurgi's circulating fluid bed (Folland and Peinemann, 1989) and Freeport-McMoRan's oxygen-enriched roasting (Smith, McCord, and O'Neil, 1990). Based on extensive testwork, oxygen-enriched roasting was selected for pretreatment due to its ability to achieve high gold recoveries from the Goldstrike ores. This technology was also well proven in commercial operation (Brittan, 1995), with two plants operating since 1989 at Jerritt Canyon and Big Springs (now decommissioned).

By mid-1997, a preliminary process flowsheet was defined with three roaster circuits. Subtle changes in the mine plan and further engineering studies enabled the use of only two roasters, with

---

[1] Hatch Associates Ltd., Mississauga, ON.
[2] Barrick Goldstrike Mines Inc., Elko, NV.
[3] Technip-Coflexip, San Dimas, CA.

corresponding refinements in the oxygen plant and the gas cleaning system. One of these roasters is shown in Figure 1, with parts of the ore feed and calcine quench systems.

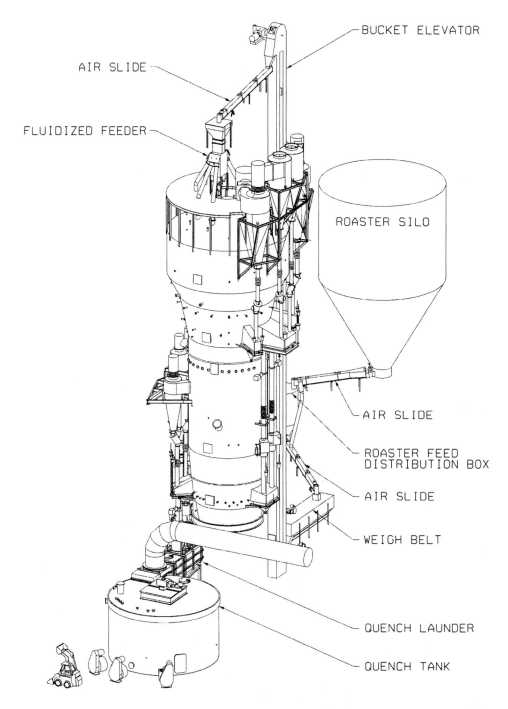

Figure 1: Barrick Goldstrike Roaster
(showing parts of the ore feed and calcine quench systems)

Detail engineering was based on the following unit operations for pretreatment of carbonaceous ore reserves at an average rate of 12,000 short tons per day (STPD) (Cole et al., 1999 and 2001):

- Primary and secondary crushing run-of-mine ore to obtain 80% passing 1.9 cm (0.75 in)
- Dry grinding of crushed ore using two double-rotator mills to obtain 80% passing 74 microns
- Whole ore roasting of ground ore using two 2-stage, oxygen roasters to remove contained organic carbon and sulfide sulfur by oxidation
- Two gas quench and dust scrubbing systems, one for each roaster, including off-gas condensers and mist eliminators
- One common gas cleaning train, including electrostatic precipitation and removal of mercury, SO_2, CO, and NO_x
- Conventional carbon-in-leach (CIL) process for gold extraction from the roasted ore.

The existing autoclave plant was to continue processing non-carbonaceous ores, and gold desorption and recovery were to be carried out using existing facilities at the Goldstrike plant.

PROCESS DESCRIPTION

Ore Mineralogy

The majority of the carbonaceous ore reserves at the Barrick Goldstrike property are in distinct zones within the footprint of the Betze-Post pit. The remainder of this ore comes from underground, primarily the Rodeo and Griffen deposits and to a lesser extent Barrick's Meikle mine.

The mineralization is generally within a Devonian Popovich formation. The host rock is typically decalcified muddy limestone and silicified sedimentary breccias. The gold is mainly present as colloidal gold occluded in the arsenian pyrite and marcasite, so the gold concentration generally increases in the fine-grained sulfide minerals. There are also trace amounts of orpiment, realgar, stibnite, arsenopyrite, and cinnabar present in the deposit. Sulfide sulfur concentrations vary from 0.5%–3.5% throughout the reserves. An average sulfide grade of 1.9% is expected initially but is predicted to drop slightly in the later years of processing.

There is significant variation in the carbonate content throughout the reserves. For the first five years of processing, the carbonate concentration is expected to be about 5%, which is near the historical Betze-Post level. A significant increase is projected as mining progresses to the west of the Betze-Post pit, where carbonate values are anticipated in the range from 15%–20%.

Organic carbon content in the ore ranges from 0.5%–4%. Tests conducted on the carbonaceous material indicate the ore is strongly preg-robbing as defined by the standard preg-rob and Barrick Goldstrike Mines (BGMI) bleach leach procedures. Further mineralogical evaluations find that the strongly carbonaceous matter is highly amorphous and generally has a small crystalline structure. All the carbonaceous materials analyzed from Goldstrike ores are comparable to anthracite or higher-grade coal.

Roaster Chemistry

Various chemical reactions occur in the roaster, including:

- Combustion of organic carbon
- Combustion of carbon monoxide
- Combustion of sulfide sulfur
- Fixation of sulfur dioxide with ore, lime, and hematite
- Reaction of lime with carbon dioxide
- Dehydration of ore

- Vaporization of mercury
- Oxidation of nitrogen.

These reactions are described in more detail below. Based on chemical analyses of the reactants and products in pilot and commercial-scale tests, this process chemistry provides an adequate model for mass and energy balances of the roasting process.

Combustion of Organic Carbon. Carbon from all sources (ore, coal, and oil) oxidizes within the roaster fluid beds to carbon monoxide and carbon dioxide with a resulting split between CO and CO_2 estimated at 4% and 96% (percent carbon by weight). Carbon monoxide may oxide further in the roaster freeboard to reduce CO levels in the off-gas. The carbon reactions occur as follows:

$$2\ C(s) + O_2(g) \rightarrow 2\ CO(g)$$
$$2\ CO(g) + O_2(g) \rightarrow 2\ CO_2(g)$$

The overall extent of organic carbon oxidation in the ore is estimated to range from 81% to 89%, depending on the ore mineralogy. The extents of coal and diesel-oil oxidation are approximately 99.5% and 100% overall, if applicable. Carbon monoxide remaining in the roaster off-gas is removed by a CO incinerator.

Combustion of Sulfide Sulfur. Oxidation of orpiment (As_2S_3), realgar (As_2S_2), and arsenopyrite (FeAsS) proceed simultaneously until they are fully reacted. Due to the highly oxidizing environment, essentially all of the arsenic is converted to solid ferric arsenate ($FeAsO_4$) or arsenic pentoxide (As_2O_5), which report to the calcine:

$$FeAsS(s) + 3\ O_2(g) \rightarrow FeAsO_4(s) + SO_2(g)$$
$$2\ As_2S_3(s) + 11\ O_2(g) \rightarrow 2\ As_2O_5(s) + 6\ SO_2(g)$$
$$2\ As_2S_2(s) + 9\ O_2(g) \rightarrow 2\ As_2O_5(s) + 4\ SO_2(g)$$
$$4\ FeS_2(s) + 2\ As_2O_5(s) + 11\ O_2(g) \rightarrow 4\ FeAsO_4(s) + 8\ SO_2(g)$$

Decomposition and partial oxidation of pyrite (FeS_2) proceeds to pyrrhotite (Fe_7S_8), most of which is further oxidized to hematite (Fe_2O_3):

$$7\ FeS_2(s) + 6\ O_2(g) \rightarrow Fe_7S_8(s) + 6\ SO_2(g)$$
$$4\ Fe_7S_8(s) + 53\ O_2(g) \rightarrow 14\ Fe_2O_3(s) + 32\ SO_2(g)$$

The overall extent of sulfide combustion is estimated at 99%, with 97.5% reacting in the first-stage bed and 1.5% in the second-stage bed.

Fixation of Sulfur Dioxide. Sulfur dioxide is fixed by reactions with minerals in the ore (carbonates and hematite) and with the lime added to the ore prior to dry grinding. Based on pilot test results, sulfur dioxide fixation ranges from 54.5%–89.5% by reaction with the ore minerals. The lime addition rate is controlled at 50% of the stoichiometric requirement for the remaining sulfur dioxide, with a lime utilization of 60%. Therefore, the lime fixes 30% of the sulfur dioxide remaining after reaction with ore minerals. The total sulfur dioxide fixation in the roaster is estimated to range from 68.1%–92.7%, depending on the carbonate content of the ore.

Sulfur dioxide reacts with carbonate and hematite in the ore:

$$2\ CaCO_3(s) + 2\ SO_2(g) + O_2(g) \rightarrow 2\ CaSO_4(s) + 2\ CO_2(s)$$
$$2\ Fe_2O_3(s) + 6\ SO_2(g) + 3\ O_2(g) \rightarrow 2\ Fe_2(SO_4)_3(s)$$

Additional SO_2 fixation is by reaction with lime added to the ore:

$$2\ CaO(s) + 2\ SO_2(g) + O_2(g) \rightarrow 2\ CaSO_4(s)$$

Sulfur dioxide remaining in the roaster off gas is removed by the gas-cleaning system and SO_2 scrubber, which has an overall removal efficiency of 99.95%.

Reaction of Lime with Carbon Dioxide. Ten percent of the lime is estimated to react with carbon dioxide:

$$CaO(s) + CO_2(g) \rightarrow CaCO_3(s)$$

Dehydration of Ore. The dehydration of clays or other hydrated minerals occurs predominantly in the first-stage bed as evaporation of water to superheated vapor. Moisture in the roaster feed ore and water vapor in air additions are also superheated.

Vaporization of Mercury. All of the mercury in the feed ore vaporizes to elemental mercury in the roaster off-gas, which is removed by the gas cleaning system..

Oxidation of Nitrogen. Nitrogen is thought to be oxidized in the roaster with a 9 to 1 volumetric ratio of NO to NO_2:

$$N_2(g) + O_2(g) \rightarrow 2\ NO(g)$$
$$2\ NO(g) + O_2(g) \rightarrow 2\ NO_2(g)$$

Although there is usually little conversion of atmospheric nitrogen to NO_x at temperatures below 980°C (1,800°F), the presence of NO_x in the off-gas may be due to feed nitrogen or the high partial pressure of oxygen.

Roaster Operation

The roasting process is accomplished in two parallel circuits to heat carbonaceous refractory gold ores and oxidize the contained organic carbon and sulfide sulfur. As shown by Figure 2, each roaster is comprised of two bubbling, fluid-bed reactors in a single vessel with an average design capacity of 6,000 STPD. This equipment includes a fluidized feed system, first and second-stage cyclone systems, and ancillary systems. The following description refers to one of the roaster circuits. Key process parameters are established from extensive testwork on Goldstrike ores.

As shown by the process flow diagram in Figure 3, the fluidized feeder distributes ore continuously from its hopper to the first-stage (upper) bed of the roaster. The feeder is fluidized with air and overflows into the roaster through standpipes extending into the first-stage bed. The roaster feed ore is 80% passing 74 microns. Up to 30% of the ground ore is smaller than 10 microns, depending on the extent of particle agglomeration.

The upper bed is maintained at constant temperature in the range from 524°–593°C (975°–1,100°F). The heat source for the first stage is provided by the ore's net heat of reaction. Coal may be added to the ground ore to provide additional heating for low fuel-value ores, or fresh water may be injected to cool the first-stage bed for high fuel-value ores. Water may also be sprayed into the first-stage freeboard to cool the roaster off-gas, if excessive freeboard combustion occurs. The ore is oxidized predominantly in the first stage.

Solids discharge continuously from the first stage through the inter-stage solids transfer system to the second-stage (lower) bed, which is maintained at a constant temperature in the range from 524°–621°C (975°–1,150°F). If heating is required, diesel oil is injected through oil guns around the bed circumference. Water may be added with the diesel oil to reduce flame temperatures and sintering, in amounts up to 75 percent relative volume. Oxidation is essentially completed in the second stage. The overall oxidation is typically 99 percent for sulfide sulfur and 88.5 percent for organic carbon.

Solid product discharges by gravity from the second-stage bed to the calcine quench system.

Low-pressure, high-purity oxygen (99.5% by volume) is introduced as the fluidizing medium through the cold windbox to the second-stage bed and the solids transfer boxes. Hot exhaust gases from the second-stage reactor are conveyed through the inter-stage gas transfer system and the hot windbox to fluidize the first-stage bed. The hot, oxygen-rich gas from the second stage promotes

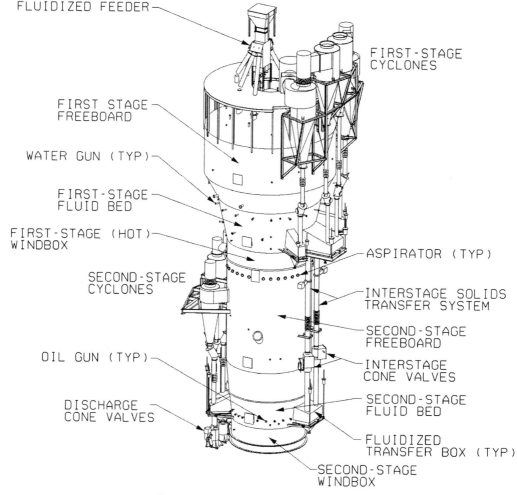

Figure 2: Barrick Goldstrike Roaster
(showing equipment details)

rapid oxidation of the organic carbon, sulfur, and fuel in the first-stage bed. The first-stage exhaust gases are de-dusted in the gas discharge system before passing to the gas quench and cleaning train.

The exhaust gas from each roaster stage is de-dusted in a set of high-efficiency, primary and secondary cyclones. In each stage of the roaster, a coarse fraction of elutriated solids is recovered by two primary cyclones and returned to the bed. A fine fraction is recovered by two secondary cyclones and fed forward through the reactors, to reduce accumulation of fines within the roaster. Fines from the first stage are therefore fed forward to the second stage, and fines from the second stage are fed forward to the calcine product.

Dust is carried over to the roaster off-gas cleaning circuit from the first-stage cyclone separators. The off-gas system is designed to process dust in amounts up to 4.5% of the dry ore feed, but the actual amount is much less in practice and depends on the off-gas rate. Dust collected within the roaster area is recovered in the roaster dust collection baghouse and returned to the roaster feed distribution box. Cleaned air is emitted to the atmosphere through the roaster baghouse stack.

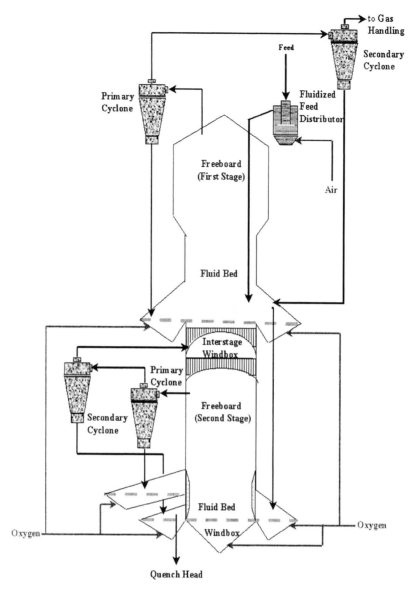

Figure 3: Process Flow Diagram

Typical operating curves are shown in Figure 4, which describe many aspects of the Goldstrike roasters. These curves reveal relationships between different equipment capacities and summarize roaster performance over a range of ore composition. At first glance, these curves appear complicated, but they provide an invaluable tool for understanding the roaster operation. Figure 4 shows three graphs with respect to the ore fuel value along a common abscissa. The ore fuel value is correlated with the contained organic carbon (%TCM) and sulfide sulfur (%S) as follows:

Ore Fuel Value [BTU/lb] = $120 \times \%TCM + 72.6 \times \%S$

where units of composition are percent of dry weight.

The bottom graph shows how the maximum instantaneous ore rate can vary with the fuel value. The total oxygen plant supply is limited to 1,100 STPD for two roasters, so the maximum ore rate decreases with increasing fuel value and hence oxidant requirement. This theoretical ore rate tends to be low during early years of the mine plan, due to the high carbon and sulfur contents, which decrease during later years when more ore can be processed with the same oxygen supply. The total ore rate for two roasters is an average rate assuming 90% operating availability. The grinding limit of 12,000 STPD for two mills therefore corresponds to an instantaneous rate of 278 short tons per hour (STPH) per roaster.

The retention limit (36 minutes) corresponds to a bed inventory of 208 tons in one roaster, which is the combined inventory of both stages. This minimum retention time is established by testwork to yield the desired gold recovery.

The coal rate shows that first-stage heating is only required in a few years with low ore fuel values. Water is injected for bed cooling in the other years with higher fuel-value ores. The autogenous limit (225 BTU/lb) is the vertical line that separates these operating regimes.

The bottom graph of Figure 4 therefore shows that the roasters can process a wide range of ore compositions at the desired average rate (12,000 STPD), corresponding to the dry-grinding capacity. The oxygen plant has some excess capacity for most of the ore feeds, except very high fuel-value ores that may require additional bulk oxygen (LOX) to supplement the plant oxygen supply.

The top two graphs represent the operating characteristics of each stage, showing gas volumes and the corresponding superficial gas velocities. In each of these graphs, a single curve is able to represent the gas volume, and bed-top and freeboard velocities, due to inherent relationships between these parameters. A different curve represents the bed bottom velocity. A common ordinate is used to read both bed velocities (top and bottom). Gas velocities differ between the bed top and bottom, due to differences in vessel diameter and the possible presence of first-stage water injection.

These graphs are prepared for nominal design temperatures where the first-stage (upper) reactor is maintained at 552°C (1025°F) by either coal addition or water injection. The water rate increases with the ore fuel value so there are corresponding increases in gas volumes and velocities. Operating conditions of the second-stage (lower) reactor are relatively uniform, regardless of fuel value. The temperature of lower reactor is nominally increased to 566°C (1050°F) by oil injection, but may be allowed to float. The actual operating temperatures of both stages depend on characteristics of the feed ore and its amenability to cyanide leaching.

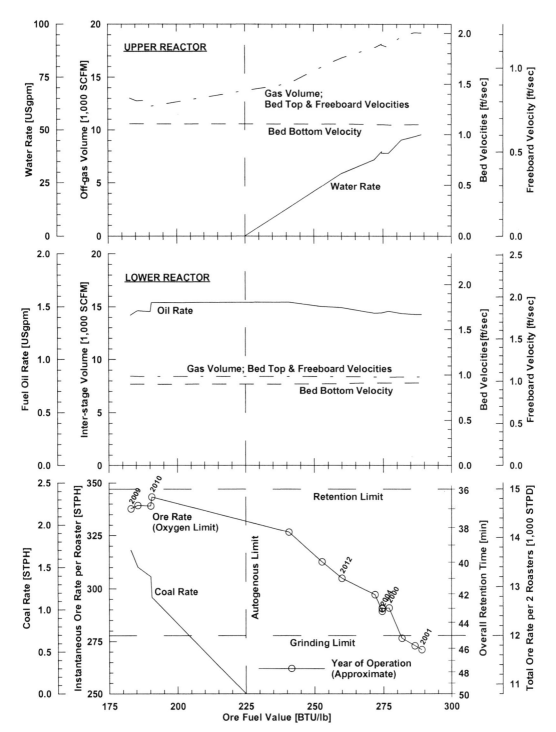

Figure 4: Barrick Roaster Operation

EQUIPMENT SELECTION AND SIZING

Project Design Criteria

In general, the project philosophy was to use proven technology and equipment where possible to avoid custom engineering and unpleasant surprises during start up. It was also a project standard to install spare mechanical equipment and avoid unnecessary roaster shutdowns for routine maintenance. These criteria had implications for equipment selection and flowsheet development. For example, unsupported refractory domes were limited to the typical maximum diameter that has demonstrated operational experience and long-term campaigns (6.71 m [22 ft]). Also, installed spare equipment was supplied for most blowers, and all pumps and solids transfer systems. The only exceptions were the start-up and preheat air blowers, which are used infrequently and can be tested off line.

The roasters themselves are custom designed by virtue of the specific process objectives and their size. Key features of the process are:

- Fluid bed roasting using almost pure oxygen as the process gas
- Two roasting stages in a single vessel to ensure a high level of both organic carbon and sulfide sulfur oxidation.

The Goldstrike roasters are the largest of their kind, with freestanding vessels about 33.5-m (110-ft) tall. Their shear size is evidence of the level of custom engineering that was required for process, mechanical, and structural engineering.

Process Design Basis

The process design basis incorporates many aspects, but only a few are presented to reveal some of the main considerations for equipment selection and sizing.

The base-case ore composition was assumed to have

- 1.9% sulfides as S
- 1.2% organic carbon as C
- 5.5% carbonates as CO_3.

with a fuel value of 282 BTU/lb. This composition represented an average feed for the first five years of operation. The corresponding reactions were modelled using the roaster chemistry to yield a mass and energy balance of the system, which was generally used for sizing the vessel and other equipment. The oxygen plant supply was to be 1,100 STPD for two roasters, with a purity of 99.5% by volume. The process was to be controlled so the concentration of oxygen in the off gas is not less than 10% (wet basis).

A special design case was also used to represent the maximum velocities and temperatures expected in the first-stage freeboard, for sizing this part of the vessel and the off-gas system. The maximum velocities were assumed to occur for a theoretical ore composition with the maximum concentrations of organic carbon and sulfide sulfur, not necessarily for the same year of the mine plan. The corresponding ore fuel-value was 302 BTU/lb. The maximum temperature was selected at 760°C (1,400°F) to account for possible combustion of carbon monoxide in the first-stage freeboard.

Number of Roasters

Three roasters were initially thought to be required, but changes in the mine plan and consideration of equipment capacities made it possible to use only two roasters. The two-roaster configuration was able to process the same amount of ore as three roasters (12,000 STPD), but used less oxygen (1,100 STPD instead of 1,350 STPD). This equipment reduction was made possible by a revised mine plan with lower ore fuel values than what were initially projected. The lower oxygen consumption resulted in lower gas volumes and therefore reduced the required cross-sectional areas for gas flows.

Roaster Sizing

One of the main constraints on roaster size is the diameter of the self-supported refractory domes that form the top and bottom of the hot windbox. These domes are subject to a combination of loads due to weight, differential pressure, and thermal expansion. The inside diameter of 6.71 m (22 ft) is based on inherent mechanical limitations and operational experience from other processes. Although larger domes have been constructed, the selected size has demonstrated long-term reliability. This diameter determines the size at the bottom of the first-stage bed and the second-stage freeboard. Gas velocities in the fluid beds and freeboards determine other diameters.

A maximum superficial velocity of 0.46 m/s (1.5 ft/sec) determines the diameter of the first-stage freeboard, using the special design case with a high fuel-value ore for gas volumes. This maximum velocity keeps dust carry over within acceptable levels for the off-gas system. The resulting freeboard diameter is 10.8 m (35.5 ft). Two conical sections provide transitions from the hot windbox. One of these transitions is gradual over the bed height of 4.0 m (13 ft), and the other is more abrupt in the lower freeboard.

A minimum fluidization velocity of 0.21 m/s (0.7 ft/sec) or mechanical limits of the refractory domes determine the diameter at the bottom of the beds. Bed velocities may be somewhat higher than the minimum fluidization value, depending on the ore composition, but they remain within accepted limits of bubbling fluid beds. The diameter at the bottom of the second-stage bed is 5.8 m (19 ft), and one conical section provides a gradual transition over the bed height of 5.2 m (17 ft).

As shown by the operating curves in Figure 4, the bed and freeboard velocities satisfy these design criteria over the entire range of ore fuel values in both the first and second-stage reactors.

The internal dimensions of the roasters are summarized in Table 1, where freeboard heights are determined by the transport disengaging height of elutriated solids. External dimensions depend on the total refractory and shell thickness that is approximately 36 cm (14 in).

Table 1: Roaster Internal Dimensions

Vessel Location		Internal Dimension	
First-stage Diameters	- Freeboard	10.82 m	(35.5 ft)
	- Bed Top	8.38 m	(27.5 ft)
	- Bed Bottom	6.71 m	(22 ft)
Second-stage Diameters	- Freeboard	6.71 m	(22 ft)
	- Bed Top	6.71 m	(22 ft)
	- Bed Bottom	5.79 m	(19 ft)
First-stage Heights	- Freeboard	6.40 m	(21 ft)
	- Bed	3.96 m	(13 ft)
Second-stage Heights	- Freeboard	6.17 m	(20.25 ft)
	- Bed	5.18 m	(17 ft)

Bed Temperature Control

Due to inherent trends of the mine plan, the roaster design had to be sufficiently flexible to accommodate wide variations in the feed ore composition. Depending on the organic carbon and sulfide sulfur content, which affect the ore fuel value, the overall system energy requirement ranges from slightly endothermic to highly exothermic in the first-stage reactor. Laboratory tests reveal that subtle temperatures of the first and second-stage reactors would affect gold recovery. A critical relationship exists between roasting temperature, residence times, and the subsequent amenability of the ore to cyanide leaching. Precise control of these parameters is essential, so independent temperature control is required for each stage.

If the organic carbon and sulfide sulfur content of the feed ore is low, coal can be added to the feed streams upstream of each roaster. If the feed ore contains a large amount of organic carbon or sulfide sulfur, an excess amount of heat can be generated in the first stage, which must be removed. Two options were considered for cooling the first-stage bed: direct quenching by water injection and indirect cooling by bayonet heat exchangers. Water injection was selected instead of indirect cooling, due to its lower capital cost, maintenance requirements, and engineering complexity. Using this system, the first-stage bed temperature can be controlled precisely over the desired range from 524°–593°C (975°–1,100°F).

The feed ore is oxidized predominantly in the first stage, and the remaining fuel value is only sufficient to offset some heat losses from the second-stage and the sensible heating of the fluidizing oxygen. Since gold recovery may be enhanced by heating the second stage to a higher temperature than the first, instead of letting it float, two options were considered to heat the second-stage bed: preheating the fluidizing oxygen and direct fuel injection. Oxygen preheating was dismissed to avoid the increased capital cost and engineering complexity of adding heat exchangers and replacing steel gas distributors with refractory domes, because oxygen temperatures greater than 315°C (600°F) would have been required. Direct fuel injection was therefore selected, but special considerations were given to reduce flame temperatures and avoid sintering the bed, which is thought to occur at bed temperatures above 650°C (1,200°F). Provisions were made to inject an emulsion of diesel oil and water, which can control second-stage bed temperatures over the desired range from 524°–621°C (975°–1,150°F) and reduce the tendency for sintering.

Roaster Off-Gas Cleaning

The roaster off-gas system quenches the hot off-gas and then processes it for removal of particulate solids, mercury, SO_2, CO, and NO_x. Independent off-gas systems were considered initially for each of three roasters, but this system was reviewed when the number of roasters was reduced. The outcome of this review was an independent gas quench and dust scrubbing system for each of two roasters, and a common gas cleaning train for removal of mercury, SO_2, CO, and NO_x. Each gas quench system included an off-gas condenser and mist eliminator, and the common gas cleaning train included electrostatic precipitation.

Oxygen Plant

The basis for design of the oxygen plant is a cold box capacity of 1,100 STPD with a total production capability of 1,200 STPD by periodic receipt of up to 100 STPD of external LOX. This significant reduction in the required capacity is due to changes in the mine plan.

Oxygen is supplied to the roasters at two pressures. Low-pressure oxygen up to 100 kPa (15 psig) is used predominantly for fluidization. Aspirators that purge solids from the hot windbox use medium pressure oxygen at 550 kPa (80 psig). All oxygen content is 99.5% by volume.

The use of high-purity oxygen at elevated temperatures in the roasters required careful attention to compatibility with materials of construction. Mild steel is used generally where gas velocities are low and there is no impingement. Stainless steels or nickel alloys are required in other critical locations such as orifice plates and aspirators. Thread sealants, gaskets, and packing materials also required careful selection.

DESIGN FEATURES

Solids Feed and Distribution System

Based on other plant experience, multiple feed chutes are used to improve operations and product quality. Splitting a high solids feed rate into several equal streams presents a challenge. The fluoseal feed distributor aerates the solids, and overflow weirs yield uniform distribution to each of six feed chutes. This device has been used successfully for many calcining and roasting applications. The feed chutes extend into the first-stage (upper) fluid bed, to maintain a pressure seal and avoid venting hot roaster gases to the dust collection system of the feed distributor. Feed chutes are vented to the roaster freeboard to avoid solids slugging and maintain constant flow. The internal support arrangement of the feed legs allows thermal expansion in a dust-laden environment, using lateral braces with guides on the vessel shell.

Solids Transfer Systems

In general, the solids flow easily when aerated, but not after being stagnant during shutdown periods. An almost vertical angle of repose is evidence of this undesirable behaviour of stagnant solids. All solids transfer chutes are therefore vertical, except very short runs from the fluoseal feeder to the feed legs. These angled runs do not pose transfer problems, however, because solids entering these pipes are generally well aerated and the pipes are never filled with stagnant solids. Purge air is introduced at critical locations such as the cone valves which are used to control bed levels.

Fluidized transfer boxes are used exclusively to convey solids between chutes and fluid beds. Transfer boxes on the Goldstrike roasters are the largest ever built, due to the large difference between freeboard and bed diameters. All transfer boxes are hung from spring hangers for additional support that accommodates thermal expansion of the vessel. Thorough fluidization of the transfer boxes is an important aspect to aerate the solids and maintain flow, especially in regions below the incoming chutes.

First-stage Gas Distribution and Hot Windbox

Hot exhaust gases from the second-stage freeboard are conveyed to the hot windbox that separates the first and second-stage reactors. The hot windbox is formed by the space between the two self-supporting refractory domes. Gas enters the side of this windbox and flows upwards through 701 tuyeres to fluidize the first-stage bed. Although the gas is cleaned to some extent by primary and secondary cyclones, fine particulate solids are unavoidable, which pose long-term maintenance requirements of either solids build-up (scaling) or erosion of the tuyere holes. Tuyeres are designed to minimize these possibilities, by using relatively large vertical holes to reduce gas velocity and pressure drop. Based on test experience, the hole size is selected to allow solids bridging and avoid draining the bed during shutdowns. During normal operation, the tuyere holes gradually plug due to scaling, which is removed during periodic maintenance shutdowns. Tuyere plugging is detected by increased pressure drop.

Aspirators

Particulate solids that enter the hot windbox with the gas stream and by sifting through the first-stage tuyeres are removed continuously during normal operation by aspirators. Twenty-four holes are distributed over the bottom refractory dome of the hot windbox, between the hot windbox and the second-stage reactor. Using principles of fluid induction, medium-pressure oxygen is injected through each hole by an aspirator to induce gas flow and purge solids from the windbox. The design of these aspirators is based on experience with phosphate calciners. The removal of solids from the hot windbox significantly reduces the maintenance requirements due to tuyere scaling and erosion.

First-stage Water Injection

Water guns are located around the bed circumference at two elevations. Twelve guns at the higher elevation are generally used during normal operation with full bed levels. Six guns at the lower elevation are used only during start up when the bed level is low. Water is injected typically into the splash zone near the top of a fluid bed, where there is intimate contact with solids circulating between the bed and freeboard but water vapour doesn't increase bed velocities.

Second-stage Oil Injection

Depending on the desired temperature, the second-stage bed may be heated by injecting an emulsion of diesel oil and water through oil guns located around the circumference. Diesel oil is supplied evenly to each of 16 oil guns by positive displacement pumps with multiple heads. Water is then mixed with the oil by an in-line mixer located just upstream of each oil gun. Water can be added in amounts up to 75% by volume to reduce flame temperatures as required, depending on operational experience and gold recovery. Air from the purge air blower conveys the oil and water emulsion into the bed, and keeps oil guns clear of solids when not being used. Air is also used to cool the oil guns.

Freeboard Water Sprays

Six atomisation nozzles are installed in the roof of the first-stage freeboard, to spray water and protect the roaster and off-gas system from high temperatures due to possible combustion of carbon monoxide. Fresh water is delivered by a high-pressure turbine pump, as required at rates up to 1.4 m^3/hr (6 USgpm). Different numbers of nozzles admit water, depending on the freeboard temperature. This type of system has been used for numerous fluid-bed reactors, primarily for incineration applications.

Start-up Air System

The roasters are normally operated using high-purity oxygen, but started up using air. One multistage centrifugal blower is connected to both roasters, and isolated as required to start one roaster at a time. There is a smooth transition from start-up air to oxygen as the desired bed temperatures and levels are achieved.

Preheat Burner System

Two preheat burners are mounted on the second-stage freeboard to heat a shallow first-stage bed above its autoignition temperature. Preheat equipment is used only for cold starts or after extended shutdowns. Ore is introduced and bed levels are increased after the desired temperature is achieved. Each burner is rated at 2,930 kW (10 million BTU/hr). One multistage centrifugal blower is connected to both roasters, and isolated as required to start one roaster at a time. The preheat burner blower supplies 100% excess air to two burners.

Purge Air Blower System

Air is injected into the roasters at various locations to

- Fluidize the feed system
- Provide the transport medium for diesel oil and water injection
- Cool oil guns
- Purge solids from instruments and other equipment.

Three positive-displacement, purge air blowers are used for this air supply. One blower provides the air requirements of each roaster, and an additional blower is connected as an installed spare to provide continuous operation during blower maintenance. Each blower has a capacity of 1,200 Nm^3/hr (700 ft^3/min), which is small in comparison with the oxygen supply.

OPERATING EXPERIENCE

The commissioning and start-up of the roasters proceeded smoothing. Start-up of the roasters was separated by one month for both construction and ease of commissioning reasons. This proved to be an advantage for the start-up of the second train in that design deficiencies were corrected prior to going hot. Design capacity for the first train was reached in about two weeks after start-up while the second train reached capacity within hours of start-up.

There were very few modification required during the start-up phase and those problems needing correction were minor. The most significant issue was associated with solids feed and distribution system. The feed system turned out to be over-sized, causing the feed to slug flow and break seal, which ended up producing fluctuation in the roaster's freeboard pressure control. A very simple fix of blinding two out of the six feed legs was all that was required to stabilized the feed rate.

Dust leaks were probably the most troublesome struggle during the first few months of operation. Frequent cycling of the roasters due to upstream and downstream problems caused the standard stainless bolts installed during construction to stretch and lose strength. Those bolts were eventually replaced with bolts made of alloy steel having a greater tensile strength and increased ductility. Another problem was erosion of vent lines providing equalization between the cyclones and fluoseal and the roaster. This would not have been as serious of a problem if isolation valves had been provided on both ends of the vent line. Eventually valves were retrofitted on every connection coming to and from the roaster.

Probably the most significant problem has been with the second stage of the roaster. Upon defluidization pressure is retained in the bed causing severe back-sifting of calcine into the cold windbox. The result has been accelerated erosion of the club head tuyeres in the second stage. The cause of the problems is associated with the deeper second stage and the design of the discharge from the underflow of the secondary cyclones. Previous designs had the cyclone discharging to a quench head rather than a common transfer box. It is thought that with a quench head the pressure is more easily relieved. The solution has been a two-fold approach. First, a pressure relief system has been designed to relieve pressure from the freeboard of the second stage to the freeboard of the first stage. When defluidization occurs, valves open to release the pressure. Ceramic tuyeres are also being tested in hopes of extending tuyere life. The back sifting problem has necessitated the replacement of all second stage tuyeres with less than four years of service.

On the positive side, one of the most significant and pleasurable surprises has been the success of the aspiration system. Time between cold shut downs for most roasters is dictated by the accumulation of solids and the ultimate build of pressure across the hot windbox. The staggering of the aspirators and the use of pure oxygen has made the system reliable and effective. Based on the first two years of experience solid/pressure build-up across the hot windbox will not be an issue with these roasters.

The performance of the roasters during the first two years of operation has exceeded expectations with respect to both throughput and operational availability. As shown in Figure 5 and Figure 6, the performance and reliability have been exceptional.

The increase in production is the result of a trade-off between fuel value (BTU/lb ore) and processing rate. The roasters were designed to treat a base case heat value (282 BTU/lb ore at 278 STPH) for a set amount of oxygen. Instantaneous processing rates have increased by leveraging the roasters' capacity to treat a given BTU per unit of time.

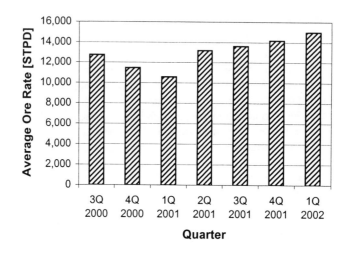

Figure 5: Roaster Throughput by Quarter

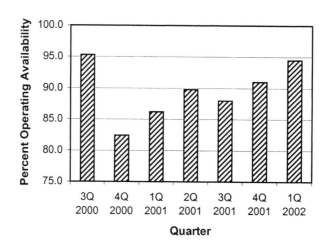

Figure 6: Roaster Availability by Quarter

ACKNOWLEDGEMENTS

The authors gratefully acknowledge contributions of Barrick Gold, Barrick Goldstrike Mines, Hatch, SNC-Lavalin, Crescent Technology, Technip USA, Air Products, and Freeport-McMoRan, and express gratitude to the dedicated people who contributed to the success of this project.

REFERENCES

Afenya, P. M. 1991. Treatment of carbonaceous refractory gold ores. *Mining Engineering.* 4:7-11:1043-1055.

Brittan, M. 1995. Oxygen roasting of refractory gold ores. *Mining Engineering.* 47:2:145-148.

Chryssoulis, S. L. and L. J. Cabri. 1990. Significance of gold mineralogical balances in mineral processing. *Transactions of the Institution of Mining & Metallurgy, Section C.* 99: 1-10.

Cole, A., S. Bunk, S. Dunn, and T. McCord. 1999. Refractory gold ore treatment by fluidized bed roasting for Barrick Goldstrike. *Proceedings Randol Gold & Silver Forum '99.* 79-84.

Cole, A., J. McMullen, K. Thomas, and S. Dunn. 2001. Barrick Goldstrike roaster facility: Roasting and gas handling. *Proceedings 33rd Canadian Mineral Processors.*

Folland, G. and B. Peinemann. 1989. Lurgi's circulating fluid bed applied to gold roasting. *Engineering & Mining Journal.* October: 28-30.

Smith, J. C., T. H. McCord, and G. R. O'Neil. Treating refractory gold ores via oxygen-enriched roasting. U.S. Patent No. 4,919,715 (24 April 1990).

Selection of Materials and Mechanical Design of Pressure Leaching Equipment

Ken Lamb, AMEC Mining & Metals, James Gulyas, SNC-Lavalin

ABSTRACT
Numerous pressure oxidation and pressure leaching autoclave plants have been installed throughout the world over the past 50 years. The technology is reaching maturity, as evidenced by the successful design and construction of plants to treat a wide range of feedstocks. Important aspects of the design of such plants include selection of materials, slurry pumping, slurry heating and heat recovery, pressure control and let-down, vessel design, agitator design, safety, and ancillary systems.

INTRODUCTION
The technology involved in pressurized hydrometallurgical processes was first developed in the early 1900s, when the Bayer process was introduced for the production of alumina. The development of pressure hydrometallurgy for the base metals industry started in the 1950s, with pressure acid leaching of nickel laterites and ammonia leaching of nickel sulfides, and has continued through the subsequent decades. Autoclave circuits have been used or proposed to process a variety of metals, including copper, gold, molybdenum, nickel, titanium dioxide, uranium, and zinc. The circuits may require different autoclave operating conditions and reagent additions, but can generally be grouped as either pressure oxidation, requiring oxygen addition in some form, or pressure leach, operating in alkaline or acid conditions.

This paper is intended to provide the reader with an overview of the important factors for mechanical design and operating practice related to autoclave circuits.

OVERVIEW OF CIRCUIT DESIGN
Typical unit operations in a pressure leach or oxidation flowsheet are indicated in Figure 1. The operations include:

- Slurry preheat
- Autoclave processing
- Flashing
- Vent scrubbing.

The mechanical design of these circuits must address:

- Materials of construction
- Vessel design
- Piping and valve design
- Preheat circuit design
- Pump design
- Pressure relief
- Pressure let-down system design
- Agitation
- Seals and seal water system.

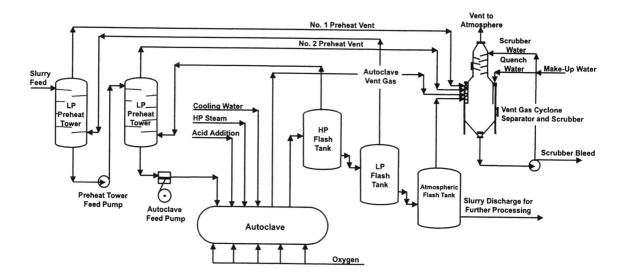

Figure 1 Typical unit operations in a pressure leach leach or oxidation flowsheet

Materials of Construction

In common with most industries, the selection of the most economic materials for specific applications is one of the first tasks facing the design engineer. Pressure oxidation and leach circuits deal with a variety of fluids, many of which are highly corrosive and/or abrasive, and operate at high temperatures.

The basic input data required for material selection include the design flows, pressures, temperatures, and fluid compositions that are usually given in the process heat and mass balances. However, the mechanical designer must be aware of potential excursion limits in determining the final design parameters.

Factors that must be considered in the selection of metallic alloys are corrosion rate, material cost, fabrication costs, availability of the alloy in all component forms (plate, bar, pipe), and anticipated maintenance costs over the life of the project. The use of high-purity oxygen in pressure oxidation circuits further complicates material selection owing to safety issues related to the potential for spontaneous ignition.

Consideration must also be given to the non-metallic materials that are commonly used for membranes and brick linings in autoclaves and flash vessels, and to other materials suitable for the relatively low temperature sections of the circuit.

Material Selection. Equipment and material suppliers can usually assist in material selection for equipment that handles the more common fluids encountered in the industry, such as reagents, including sulfuric acid.

For autoclave applications, however, the range of suitable materials is relatively limited; typically, high-grade alloys or refractory metals are required. Non-metallic materials have been used as autoclave membranes and lining materials, and polymers have been developed for valve seat materials to withstand 260°C.

Some vendors of specialty alloys and industry organizations such as the Nickel Development Institute and NACE (National Association of Corrosion Engineers) can provide technical assistance in materials selection through their substantial databases. Unfortunately, the databases cannot cover the almost infinite variations in slurry composition, impurities, and temperatures encountered in autoclave circuits, and designers have to rely on predictions of the anticipated alloy performance.

Key parameters in the selection of materials are operating temperature, acid concentration (typically sulfuric) or pH, chloride content (particularly important for stainless steels and titanium), and concentration of metallic ions such as nickel (Ni^{2+}), copper (Cu^{2+}), and iron (Fe^{3+}). In addition, fluoride ion can have a detrimental impact on titanium and, to some extent, on stainless steels and brick linings.

In ascending order of corrosion resistance, and usually in cost, materials used in autoclave circuits include the 300 stainless steels, super austenitic and super duplex steels, nickel alloys, and titanium. It should be noted, however, that there are some anomalies in this ranking; for example, Inconel Alloy 625 has been used in the construction of autoclave internals in gold circuits and has experienced very high corrosion rates. For this application it has therefore been replaced with duplex stainless steel alloys such as Ferralium Alloy 255, which has proved to be much more corrosion resistant. However, Inconel Alloy 625 weld overlay, applied to the carbon steel autoclave nozzles and protected from the direct process environment by brick and mortar, has been relatively successful. Titanium is the only material that has been found to withstand the autoclave environment without significant corrosion, but there is reluctance to use it in oxygen dip pipes because of the potential for spontaneous ignition in the presence of high-purity oxygen. The development of an titanium-niobium alloy has alleviated this concern to some extent, although it could still ignite (see Figure 2).

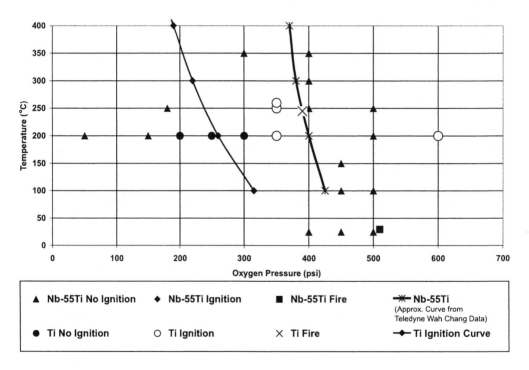

Figure 2 Effect of temperature on spontaneous ignition of ruptured unalloyed titanium & niobium-titanium in oxygen

In North America, materials are identified by UNS (Unified Numbering System) numbers, which specify an allowable chemical composition range for the alloy. They are not material supply specifications, however, and as such must be used with caution. Some vendors may supply material at the lower end of the chemical analysis which would reduce costs, but would also result in poorer corrosion resistance and mechanical properties. A more-restrictive chemical analysis, and corresponding new UNS number, has been applied to some alloys to overcome these problems. For example, the more general chemical composition wrought Duplex Alloy 2205, UNS S31803, has been supplemented with a tighter chemical specification UNS 32205.

Nickel alloys, stainless steels, and titanium are supplied to an ASME (American Society of Mechanical Engineers) or ASTM (American Society for Testing Materials) specification that includes requirements for ordering, manufacturing, testing (including tensile tests and minimum tensile requirements), inspection and marking, and allowable dimension variations, as well as for chemical composition. At present, titanium alloys are the only ones not referred to by their UNS numbers in these specifications; however, this is expected to change in future editions.

Current metallurgical trends are towards more highly alloyed austenitic and duplex stainless steels, which may be used for equipment and piping outside the high-temperature autoclave environment. Corrosion resistance generally increases with increasing chromium, molybdenum, and nitrogen content in stainless steel alloys. More super austenitic and super duplex alloys offering higher pitting resistance equivalent numbers (PRE, or PREN when nitrogen is included) are becoming available, which offer improved localized pitting and crevice corrosion resistance in aqueous solutions containing halides (chlorides and fluorides) at low pH. Halides increase acid corrosion by making it difficult to maintain the passive surface layer, and high chloride concentrations and low pH, as well as high temperatures and stagnant conditions, increase the probability of pitting and crevice corrosion. The PREN numbers establish a measure of ranking for pitting and crevice corrosion resistance by calculating the sum of the most important alloying elements in a weighted form. The higher the PREN number, the better the resistance to crevice corrosion. Table 1 shows typical PREN numbers.

Table 1 Typical PREN numbers

UNS	Material	Type	Form	%Ni	%Cr	%Mo	%W	%N	Other	PREN
S 31603	316L	SS	Wrt	12.5	17	2.5		0.13		27.3
J92800	CF3M	SS	Cast	11	19	2.5				27.3
N08020	Alloy 20Cb3	SS	Wrt	35	20	2.5				27.3
J93370	Cd4MCu	Dup	Cast	5.5	25.5	2			3.0 Cu	32.1
S31803	2205	Dup	Wrt	5.5	22	3		0.15		34.3
N/A	Cd4MCuN	Dup	Cast	5.5	25.5	2		0.2	3.0 Cu	35.3
N08904	904L	SS	W/C	25.5	21	4.5			1.5 Cu	35.9
S32550	Ferralium 255	Dup	W/C	5.5	25.5	3		0.18	2.0 Cu	38.3
N08028	Sanicro 28	SS	Wrt	32	27	3.5			1.0 Cu	38.6
S32750	2507	Dup	Wrt	7	25	4		0.25		42.2
S32760	Zeron 100	Dup	Wrt	6.5	25	3.7	0.7	0.25	0.7 Cu	42.4
S31254	254SMo	SS	Wrt	18	20	6.25		0.2	0.75 Cu	43.8
N08367	Al6XN	SS	Wrt	24.5	21	6.5		0.2		45.7
N06985	Hast G3	Ni	Wrt	Base	22.5	7		1	3.4 Co	47.3
N06625	Alloy 625	Ni	Wrt	Base	21.5	9				51.2
S32654	654 SMo	SS	Wrt	22	24	7.3		0.5	0.5 Cu	56.1
N06455	Alloy C4	Ni	Wrt	Base	16	15.5			3.0 Fe Mx	67.2
N06022	Alloy C22	Ni	Wrt	Base	22	13.5	3		3.0 Fe	71.5
N10276	Alloy 276	Ni	Wrt	Base	15.5	16	3.75		5.5 Fe	74.5

PREN numbers calculated from: %Cr + 3.3 x (%Mo + 0.5%W) + 16 x (%N)
Wrt - wrought   W/C - wrought or cast

Ignition of titanium. The application of titanium in oxygen environments has received considerable attention in pressure oxidation applications because a freshly exposed surface of titanium alloy may spontaneously ignite in oxygen concentrations greater than 35% by volume. The oxygen content in the vapor space of typical pressure oxidation autoclaves is usually less than 35% during normal operation, with only the oxygen dip pipes subject to high oxygen partial pressures. Designs and materials have been developed to mitigate the damage from oxygen usage. One example is the use of heavy-weld neck flanges on the oxygen dip pipe inside the vessel to act as a heat sink and limit flame propagation if it should ignite. Also, titanium should not be used in areas where a high-velocity, oxygen-enriched gas jet may occur, such as at a seal between autoclave pressure and the atmosphere.

Alloy Ti-45Nb has a wider operating regime than the more-common Grade 2 titanium in oxygen environments with respect to reduced potential for spontaneous ignition, and has been used where gas enriched with oxygen can be anticipated.

Figure 2 shows the ignition limit for titanium and titanium-niobium.

A number of titanium fires have occurred at autoclave installations over the years, causing considerable equipment damage and in some cases personal injury. The safety aspects of material selection cannot be taken lightly. A full HAZOPS analysis, together with careful design by experienced personnel, will reduce the likelihood of such incidents.

Vessel Design

General. The most common codes used in the design of pressure vessels for pressure hydrometallurgy are ASME Div 1(North American), AS-1210 (Australian), and BS-5500 (British). While the codes are similar, they have some variances in allowable stresses and design requirements (e.g., for masonry lining) that will result in small differences in shell thickness.

This section outlines design considerations based on the ASME code.

In North America unfired pressure vessels are designed to ASME Section VIII, Division 1 or Division 2. Division 1 uses approximate calculation methods that are adequate for most services. Division 2 provides an alternative to the minimum construction requirements of Division 1. Division 2 rules are based on more-precise calculation methods and are more restrictive. They do not permit the use of some materials allowed by Division 1, they prohibit some common design details, and they specify which fabrication procedures may be used. Engineering and manufacturing costs for materials meeting Division 2 criteria are therefore higher than those meeting Division 1 criteria.

Where the stress intensity is controlled by ultimate or yield strength, Division 2 permits the use of higher design allowable stress values in the range of temperatures covered. Hence, for autoclave applications, Division 2 may be considered where savings in materials and labor justify the costs of the necessary engineering analysis and more rigorous construction requirements. Where expensive materials or large vessels are being used, this may be the most cost-effective approach.

In some cases, material savings accompanying the Division 2 criteria for carbon steel are not substantial until the shell thickness approaches 100 mm (4 inches).

ASME Section VIII pressure vessel code changes. Various criteria are used in establishing maximum allowable stress values for pressure vessel codes.

It is important to note that the Division 1 safety factor was recently changed from 4:1 (in use since 1944) to 3.5:1 and brings the stresses closer to those of most European pressure vessel codes. This change was issued on July 1, 1999.

The temperature at which allowable stress values are affected has also been changed. Previously, a design temperature of 650°F or less did not affect the allowable stress value of SA-516-70, a carbon steel plate material in the pressure vessel code commonly used in vessel fabrication. That temperature value has been changed to 500°F. However, the allowable stress value for SA-106-B, a carbon steel seamless piping material also used in vessel fabrication, does not change until 650°F. Each item needs to be reviewed to determine the allowable stress at the vessel design temperature. It should also be noted that the design temperature always affects external pressure calculations and flange ratings.

Another code change has been made in the area of required minimum hydro-test pressure; this has changed from 1.5 times MAWP (maximum allowable working pressure) to 1.3 times MAWP.

Vessels designed to the new Division 1 criteria will be thinner, lighter, and cost less than vessels installed under the earlier code criteria. The material cost savings between Division 1 and 2 vessels will not be as significant under the new code rules.

The increase in allowable stresses could lead to existing vessels being re-rated for higher design pressures, which could benefit some processes. A National Inspection Board Code interpretation indicates that re-rating due to code changes in allowable stresses is acceptable as long as certain criteria are fulfilled; however, some jurisdictions will still not allow it.

Autoclaves. Currently accepted construction options for autoclave vessels are titanium clad steel or a carbon steel shell protected by membranes of lead or lead with fiberglass, and acid-resistant brick.

Brick-lined vessels. Historically, the most common design for pressure oxidation applications has been horizontal vessels with a carbon steel shell, lead and/or vinyl ester membrane, and two layers of acid-resistant brick lining.

In the case of brick-lined vessels, the steel shells, internals (compartment walls, nozzle inserts, dip pipes and baffles, typically made from non-ferrous materials), and brick lining have usually been specified and purchased separately. Where a complete unit was supplied, the lining manufacturer took the coordinating role and subcontracted the other work. This purchasing concept – where the vessel design is done in coordination with the lining design – reduces the work done by the owner's agent, i.e., the engineering consultant, but does restrict the number of companies that are willing to bid.

Other than accounting for the lining weight, the ASME pressure vessel codes do not specifically consider brick linings. However, the codes do state that the additional loads placed on the shell from the lining must be determined. The British code, BC-5500 contains design guidelines for lined vessels.

A primary objective of lining design is to balance the stresses between the steel shell and the lining. Since the tensile strengths of the acid brick and mortars are relatively very low, they must be maintained under compression or, at worst, with a small amount of tension. On the other hand, the compression must always be below the compressive strength of the brick to prevent spalling or crushing and not impose severe additional tension on the steel vessel wall.

A number of variables contribute to lining stresses, such as:

- Selection of materials for acid resistance and erosion resistance (includes brick mortar and membrane). The fact that bricks irreversibly swell in wet acidic conditions must be considered in selecting materials for the bricks.
- Thermal properties of the brick and mortar (used to calculate thermal gradient across lining system and hence the temperature of each component and its corresponding expansion)
- Number of brick layers and thickness
- Operating pressure, temperature, vessel diameter and material, and corresponding shell thickness.

In lined pressure vessels, stresses are purposely introduced into the lining system before operation. A newly lined vessel is cured in an acid solution at a temperature around the boiling point. During this curing time, a chemical reaction occurs in the brick that causes it to swell irreversibly, thereby increasing the stresses and "tightness" of the lining system. Bricks have been known to swell during an initial period of a few weeks, stop, and continue months later.

The lining system will be subject to various stresses during start-up, normal operation, and shutdown and must be designed accordingly. In addition, procedures must be developed to minimize stresses that could damage the lining during heat-up and cool-down. The procedures include rates for pressurization and depressurization, temperature increase and decrease, and time for soaking.

Additional loads can be imposed on the lining system by the vessel itself. For example, if the vessel is out of roundness, or the vessel weight and support locations introduce unacceptable deflections at the vessel centreline. Tolerance criteria for out of roundness and deflection have been developed to minimize the transfer of these additional loads to the brickwork.

Three major companies design and install pressure vessel linings in North America: Stebbins, Didier, and Koch. Each company has recommendations about materials for membranes, bricks, and mortars, as well as installation details and techniques. For example, Koch prefers to supply a sulfur-enhanced elastomer (Pyroflex) membrane, which, owing to its flexibility, will take up some of the brick expansion. Stebbins use its vinyl ester (AR 500) membrane and Hydromet (lead based) mortar for difficult vapor zone applications in pressure oxidation applications. Didier has developed a potassium silicate (Stellakit A) mortar for the vapor zones. The companies have bricks that differ in swell properties and resistance to vapor zone conditions of steam condensation, where the bricks and mortar tend to soften.

Lead has been the traditional membrane lining for pressure oxidation service. Given the increasing awareness of the health hazards of lead, the development of materials to replace it has become a priority for vendors. The installation of a lead liner requires substantial safety precautions, is expensive and time consuming, and may still incur corrosion in areas around the vapor space nozzles where operating temperatures are highest.

One of the brick lining companies is testing other types of polymer membranes to replace lead.

An experienced pressure vessel lining company, with a history of successful design, should be selected. Such companies are most familiar with the performance and limitations of their products, such as the thermal expansion and swell properties, and are therefore in the best position to determine vessel design loads, roundness, and deflection tolerances and to pass this information along to the vessel fabricator.

Titanium-clad vessels. These vessels are fabricated from carbon steel with an explosively bonded titanium cladding (see Figure 3). Titanium and iron are not metallurgically compatible at high temperatures, and under the conditions normally used for weld overlay or hot roll bonding, they instantly react to form brittle compounds. Consequently, explosion cladding is the preferred process for the manufacture for titanium-clad steel. This is a solid-state metal-joining process that uses explosive force to create an electron sharing metallurgical bond between two metal components. The titanium cladding is used as a corrosion barrier only; its strength is not taken into account when designing the shell wall thickness. Optimum bond mechanical properties and plate sizes are produced when the yield strength of the titanium and base metal is below 345 MPa. Therefore, the optimum bond strength and toughness of titanium cladding results from a combination of titanium-clad steel to a moderate-strength, pressure vessel steel such as SA 516 Gr 70. Titanium grades 1, 11, and 17 exhibit similar yield strength and bond performance and have been used to clad autoclaves used for nickel processing.

Special considerations must be taken in design, fabrication, welding, and testing to ensure a reliable product. Typically, a batten strap technique is used for cladding fabrication, as shown in Figure 4.

The cladding is applied to a flat sheet of carbon steel thick enough for the pressure vessel design plus an additional minor allowance for compression of the steel during cladding. The plate is formed into heads and rolled into sections. Joining two sections together usually requires the removal of approximately 12 mm (0.5 inches) of titanium from each side of the weld preparation edge. The steel is welded conventionally and the vessel is then cleaned and prepared for titanium welding. In the batten strap technique, a filler (metal strip of copper or titanium) is used to fill the space where the titanium has been removed, a titanium batten strip is applied over the plate, and the edges are fillet-welded to the clad titanium. The wider the joint, the higher the stresses in the fillet-weld during operation. Large-diameter nozzles are frequently fabricated from clad plate using the same procedures. Small nozzles are typically lined with titanium sleeves and seal-welded in the shell.

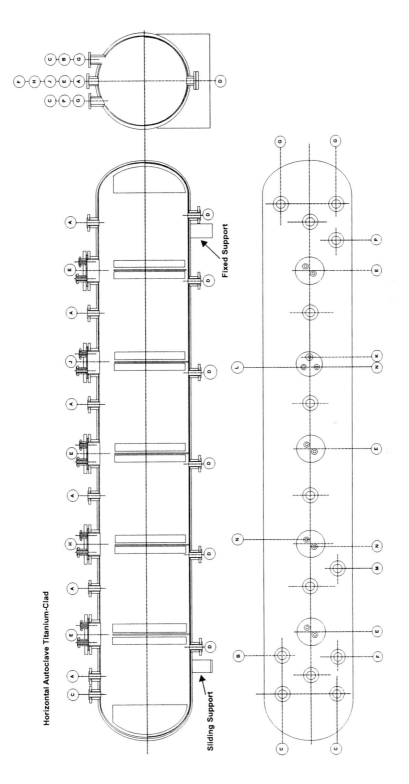

Figure 3 Explosively bonded titanium cladding

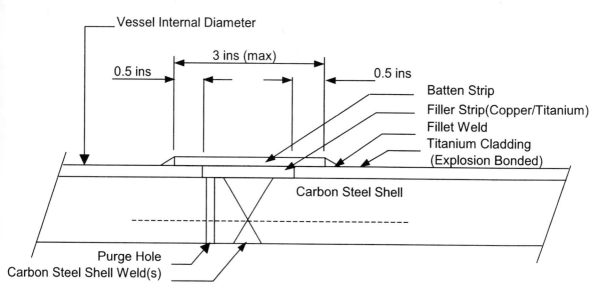

Figure 4 Batten strap technique

More than one of the nickel autoclave plants currently in operation have had problems related to titanium material selection, principally due to chloride stress cracking. The batten strips on the vessels have cracked due to thermal expansion. Batten strip design and test procedures, such as hot cycle testing, have been developed to minimize failures during start-up and operation.

Vessel heads. Some options are available for the heads of vessels in a pressure hydrometallurgical circuit. The selection is generally based on cost and delivery. For higher-pressure systems, the head selection is between hemispherical and ASME 2:1 elliptical. While a 2:1 head may be preferable because the agitator is placed in the "volumetric" center of the compartment, the difference that a hemispherical head brings to bear on this criterion does not effect operations.

Both types of head can be brick lined or explosion clad.

Flash and preheat vessels have been supplied variously with 2:1 elliptical heads, conical bottom heads, and, for lower-pressure systems, ASME Flanged and Dished heads. The conical heads were used ostensibly to aid in slurry flow, but they are more expensive, and the other head profiles have been just as successful.

Nozzles. Historically, the most troublesome issue in the maintenance of brick-lined autoclaves concerns the vessel nozzles – particularly the nozzles located in the vapor space. An individual vessel can have up to 30 nozzles. To minimize penetration in the brick linings, nozzles are grouped together (usually two to four nozzles) into a "multiple nozzle," with one lining penetration. In the initial autoclave designs, the lead membrane in the vessel shell extended through the nozzle and over the face of the flange. One of the principal problems was keeping the lead at a low enough temperature (<85°C) to prevent creep and corrosion. The annular space provided in the nozzle was insufficient to allow for the temperature drop from the operating temperature to an acceptable membrane temperature. Insulating bricks, polyethylene blocks, and insulating rope have been tried with limited success. Designs with fiberglass over the lead, or replacing the lead with a sulfur-enhanced elastomer, have also been attempted.

The most successful design has used an Inconel 625 overlay through the length of the nozzle and into a "bull's eye" in the vessel shell, with the overlay protected by brick and/or mortar and an insert. The temperature is not critical in this design, and the Inconel has withstood the environment that exists behind the brick lining. The insert is used to protect the bricks in the nozzle from mechanical damage and limits the exposure to condensate that would otherwise form and flow down the bricks and mortar, causing premature failure.

The terminations of the membrane(s), the nozzle insert, and the sealing surface are all at the face of the nozzle flange. This requires the fabrication of a custom-designed Inconel or other high-alloy seal ring to be fitted onto the face of the flange. The rings are designed to prevent the autoclave environment from reaching the steel shell by sealing the ends of the membrane(s) and across the nozzle insert, and include machined sections for the gasket(s). Spiral-wound gaskets and non-metallic gaskets have both been used successfully. Because spiral-wound gaskets have very high seating stress requirements, the gasket width should be carefully selected so that the bolting and flange thickness do not have to be custom designed. A typical sealing face arrangement for brick-lined and titanium-clad vessels is shown in Figure 5.

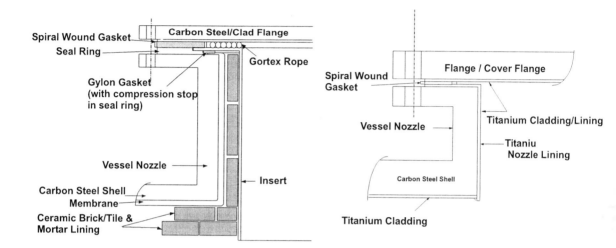

Figure 5 Typical sealing face arrangements for brick-lined and titanium-clad vessels

Nozzles on pressure vessels may be supplied as integrally reinforced components or may be fabricated from pipe and flanges with reinforcing pads. The Pressure Vessel Code requires that the area of the shell removed by the opening must be replaced with reinforcement, within dimensional limits set by the code.

Some installations have extremely large agitator nozzles that allow the agitator to be withdrawn in its entirety to reduce maintenance downtime. These nozzles require significant reinforcement and usually custom design of the flange and bolting. The design also results in additional stresses within the lining system. When a large section of the lining is removed, additional loads are transferred to the remaining sections, further increasing the compressive loads on the bricks. Other design considerations are the need to reinforce the nozzle insert to handle the increased lining stresses and possible compromise of the self-supporting nature of the lining system when such a large section of the arch is removed. This design is costly, not only because of the large nozzle and the associated blind flange, but because of the need to increase crane capacity and building height and to provide a structure to house the agitator assemblies. The maintenance savings must therefore be judged in light of the capital costs and lining issues.

Piping and Valve Design

Piping design and material selection are critical to the safe and prolonged operation of autoclave circuits. The design must accommodate the hot acidic slurries and high operating temperatures and pressures, and the piping layout must allow for ease of operation, maintenance access, and egress in the event of an emergency. Some design considerations are:

- The use of lower alloy or carbon steel back-up flanges for high alloy and refractory pipe to reduce cost
- Double check valves on oxygen, steam, and high-pressure water
- The use of "blowback traps" for steam and oxygen as backup to check valves
- Low velocities in autoclave vent systems
- Correct design and specifications for oxygen piping systems, i.e., material/velocity and cleanliness
- Proper selection of valves, mainly for seat materials for the temperatures and fluids handled
- Double block and bleed for oxygen, steam, and water into the autoclave, along with proper purge and flush connections
- Consideration of drainage or pumping out of the autoclave
- Proper pressure relief design
- Provision of control valve bypass systems
- Adequate pipe stress analysis during design
- Adequate inspection during piping fabrication and installation
- Consideration of dual feed lines.

Piping materials will vary for the different circuits and materials handled. A selection of piping materials for a typical refractory gold installation are outlined in Table 2.

Table 2 Piping materials for a refractory gold pressure oxidation circuit

Fluid	Material
Demineralized water, seal water	Stainless steel TP 304L or TP 316L
Oxygen, air, steam, quench water (autoclave to trap)	Stainless steel TP 316L
Oxygen, air, quench water (before trap)	Carbon steel A-53
Steam	Carbon steel A-106
Process slurry	Ferralium 255
Process slurry (high temperature)	Titanium Ti Gr 2
Autoclave vent	Titanium – Niobium Ti-Nb 45
Flashed slurry	Duplex stainless steel 2507 or Zeron 100

The high-temperature autoclave piping is usually designed to ASTM B31.3.

Flange Ratings. In designing the piping systems, care must be taken in understanding and applying the flange designations used in various places in the world. The class designation system is used in the United States, Canada and some other countries, whereas the nominal pressure (NP) designation system is used in Europe and most other parts of the world. In both cases, the numerical designation offers a convenient round number for reference purposes; however, the PN designation is nominally the cold working pressure in bar.

A flange rating for a specific material and class will give the allowable working pressure at various temperatures. Ratings are calculated using the material allowable stress and yield values taken from the Pressure Vessel Code. As indicated earlier, stress values have recently been increased for some materials. It could be expected that allowable working pressures for flanges would also increase, but this is not the case. The original allowable pressures, which were developed over 70 years ago, were generally based on asbestos and elastomer gasket materials that have generally been replaced with new non-metallic and spiral-wound gaskets. These new gaskets, particularly the spiral-wound type, require high compressive loads to become sealed, resulting in stresses that are marginal for some flange ratings and sizes used in current designs. Hence, the increases in the Pressure Vessel Code allowable stresses will have no major impact on the current allowable working pressures for flanges. This must be taken into consideration when existing pressure vessels are evaluated for re-rating.

Table 3 shows the pressure ratings, in bar, for flanges supplied in various materials taken from B16.5, 1996, or calculated using the procedures given in B16.5 and converted to metric units.

Table 3 Maximum allowable non-shock pressure, bar

Forged Flanges Flange Rating	Allowable Pressure, bar Temperature, °C					
Class 150	37.8	93.3	148.9	204.4	260.0	315.6
A 105 (Carbon Steel)	19.7	17.9	15.9	13.8	11.7	9.7
316L (Stainless)	15.9	13.4	12.1	11.0	10.0	9.7
2507 (Duplex Stainless)	20.0	17.9	15.9	13.8	11.7	9.7
Inconel 625	20.0	17.9	15.9	13.8	11.7	9.7
Titanium Gr 2	16.2	14.1	11.7	9.7	8.6	7.2

Forged Flanges Flange Rating	Allowable Pressure, bar Temperature, °C					
Class 300	37.8	93.3	148.9	204.4	260.0	315.6
A 105	51.0	46.6	45.2	43.8	41.4	37.9
316L	41.4	34.8	31.4	28.6	26.2	24.8
2507	51.7	51.7	50.3	48.6	45.9	41.7
Inconel 625	51.7	51.7	50.3	48.6	45.9	41.7
Titanium Gr 2	42.4	36.6	30.3	25.2	22.4	19.3

Forged Flanges Flange Rating	Allowable Pressure, bar Temperature, °C					
Class 600	37.8	93.3	148.9	204.4	260.0	315.6
A 105	102.1	93.1	90.7	87.6	82.8	75.5
316L	82.8	70.0	62.8	56.9	52.8	49.7
2507	103.4	103.4	100.3	97.2	91.7	83.4
Inconel 625	103.4	103.4	100.3	97.2	91.7	83.4
Titanium Gr 2	84.5	73.4	61.0	50.3	44.8	38.3

Valving Materials. A significant component of the capital cost associated with the installation of a conventional autoclave circuit is the valves, which are typically fabricated in titanium or duplex stainless steel. Any scaling problems must also be addressed in the selection of valves.

The major types of valves and their materials of construction are listed in Table 4.

Table 4 Summary of typical valves in a gold pressure oxidation circuit

Description	Fluids	Body Mat'l	Trim Mat'l	Seats Mat'l	Class
Ball Valve Socket Weld <50 mm	Demin. Water, Seal Water, Quench Water	ASTM-A182-GrF3161-	AISI 316SS Ball/Stem	PTFE	CL800
Ball Valve Flanged	A/C Oxygen Feed	Fe255	Fe255 c/w Hardcoat	Fe255 c/w Hardcoat	CL300
Ball Valve Flanged	A/C Slurry Drain Ports	ASTM-13348-Gr5	Ti Gr 2/3 c/w Hardcoatoat	Ti Gr 2/3 c/w Hardcoat	CL300
Ball Valve Flanged	A/C Oxygen Feed	Fe255	Fe255 c/w Hardcoat	Fe255 c/w Hardcoat	CL300
Ball Valve Flanged	Autoclave Vent Quench Water	Fe255	Fe255 c/w Hardcoat	Fe255 c/w Hardcoat	CL300
Ball Valve Flanged	Quench Water Process Slurry	ASTM-A351-GrCF8M	AlSl 316SS Hardcoat	AISI 316SS Hardcoat	CL300
Ball Valve Flanged	Oxygen, Plant Air	ASTM-A351-GrCF8M	SS316 Ball/Stem	Reinforced PTFE	CL300
Ball Valve Butt Weld	Oxygen, Plant Air	ASTM-Al82-GrF316	SS316 Ball/Stem	Reinforced PTFE	CL800
Ball Valve Flanged Dn20-250	A/C Slurry	Fe255	Fe255 c/w Securacoat	Fe255 c/w Securacoat	CL150
Ball Valve Butt Weld	Quench Water (A/C Bleed)	ASTM-A182-GrF3116	SS316 BIS Hardcoat	SS316 Hardcoat	CL800
Ball Valve Butt Weld	Agitator Seal Water	ASTM-A182-GrF316	SS316 Ball/Stem	Reinforced PTFE	CL800
Ball Valve Flanged	Quench Water Autoclave	Fe255	Fe255 c1w Hardcoat	Fe255 c/w Hardcoat	CL300
Ball Valve Flanged	Plant Air Autoclave	Fe255	Fe255 c/w Hardcoat	Fe255 c/w Hardcoat	CL300
Ball Valve Flanged	Process Slurry (Drain Ports)	ASTM-A351-GrCF8M	SS316 Ball/Stem	Reinforced PTFE	CL300
Ball Valve Flanged	Autoclave Vent	Ti-45Nb	Ti-45Nb cfw Hardcoat	Ti-45Nb c/w Hardcoat	CL300
Check Valve Flanged	Autoclave Oxygen Feed	ASTM-A351-GrCF8M	SS616 c/w Stellite	SS316	CL300

Valve companies have developed non-metallic seat material for operations up to 260°C. This type of valve usually costs less than ball valves and so is becoming more popular.

Preheat Circuit Design
When leaching laterites or oxidizing ore of low sulfur grade, there is an economic advantage to recovering heat from the slurry discharged from the autoclave and using it to preheat the feed. In most of the recent gold and base metal installations, this has been accomplished by recovering steam that flashes from the discharge slurry as the pressure is reduced and contacting it directly with the incoming feed slurry. In zinc pressure leach plants, the flashed steam is used to indirectly heat the recycled leach solution in a shell-and-tube exchanger while the incoming slurry is left unheated. This latter technique is also used in the alumina industry.

Although indirect heating offers a number of advantages, it has been avoided, particularly where the flashed steam may be dirty and would increase tube fouling rates. Extensive pilot testing would be required to establish relevant design information such as data on heat transfer coefficients and fouling factors. Economic advantages that could be realized include:

- Less steam addition to the autoclave (hotter feed)
- Smaller autoclave (less dilution of the feed slurry from condensed steam)
- Lower temperature feed pumping to the autoclave (pump before, not after preheating)
- Less acid addition (where a discharge acid concentration is being maintained)
- Smaller downstream equipment (for base metal leaching).

However, if heat transfer coefficients are low, which may be the case with a viscous slurry not amenable to achieving a turbulent flow, or if the heat transfer surfaces have a propensity to scaling, the capital and ongoing maintenance costs of heat exchangers may be excessive. Thus if pilot testing is not undertaken and a proper design applied, any potential economic advantage may well be lost. In recent years, considerable pilot testing has been done with shell-and-tube exchangers for this service; one nickel laterite facility currently in design is using indirect heating for a number of stages.

In most recent installations requiring preheat, flash steam is contacted directly with incoming slurry in a series of columns, each with a number of segmental baffle trays that provide a curtain of slurry for steam contact (Figure 6). The baffle trays extend to at least half the diameter of the column and are slightly sloped to avoid the accumulation of solids. Five or six baffle trays are usually adequate for refractory gold installations, while 40% to 50% more are needed for more-viscous feeds such as nickel laterites. For large-diameter vessels, a disc-and-doughnut design (see Figure 7) with sloped trays may be used instead of the segmental baffle design. These designs, though possibly not the most efficient, have proven effective for slurry duty. Other designs, including bubble-cap and packed towers, would be more prone to fouling with deposited solids or scale and are more difficult to clean of such accretions.

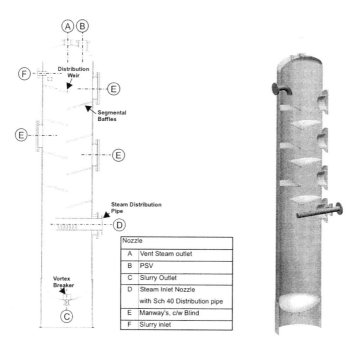

Figure 6 Alloy preheat vessel segmental baffle design

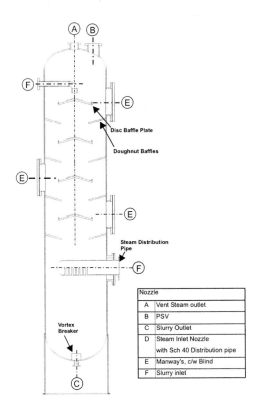

Figure 7 Alloy preheat vessel disc and doughnut design

Mechanical design of the preheat vessels follows the same principles as those outlined for autoclaves, taking into account the lower temperatures and pressures. Preheater vessels can be free-standing on skirts or supported in steel structures, depending on layout and economics.

The number of preheat stages is determined from an overall heat balance for the entire autoclave system that includes temperature restrictions on feed pumps and autoclave operation.

Pump Design
The operating temperatures of the slurry transfer pumps between preheat tower stages and the autoclave will depend on the approach temperature, i.e., difference in temperature between the incoming steam and discharging slurry. Since the steam condensing coefficient is high, approach temperatures can be close and have been measured as low as 2C°. A recommended approach for design is 5C°.

Consideration must be given in the circuit layout to pump NSPH (net positive suction head) requirements, i.e., the suction head required to prevent cavitation and flow loss in the pump. For acceptable operation and to allow for variable conditions, the available NPSH should exceed the NPSH required by 2 m to 3 m. This head is usually provided by elevating the preheat vessels or, in the case of the autoclave, by adding a centrifugal charge pump ahead of the feed pump.

The mode of operation of the preheat vessels must be taken into consideration when selecting the slurry feed pumps and control system. If there is a substantial reduction in heating steam, for instance, and the slurry feed remains constant, the vessel pressure will drop because all the steam will have condensed. Once the head that the pump is pumping against has dropped, the flow from the pump will inherently increase, further reducing the vessel pressure. Typically, a variable-speed drive in combination with mass flow measurement is used to control the slurry feed flow rate; a steep pump performance curve is required for improved control.

The feed pump is normally a positive displacement pump, which incurs an additional energy loss. This is the energy required to accelerate the fluid within the piping, which is a function of the length of suction piping (mass of fluid) and type of pump (duplex single acting, duplex double acting, triplex single acting). Installing a pulsation dampener at the inlet of the pump reduces this loss, making it a function of the piping between the pump and the dampener only, which is relatively low. Discharge dampeners are fitted to reduce pressure pulsations in the discharge piping.

The most common type of positive displacement pump used in this service is the piston diaphragm pump. In simple terms, the piston diaphram pump consists of a hydraulically operated diaphragm and two check valves. Unfortunately, elastomeric materials suitable for high temperatures are not available, and so a special design had to be developed for circuits with preheat vessels. The first design for high temperatures was installed at Barrick Mercur in Utah, which had a vertical water-cooled "leg" between the check valves and the diaphragm. More recent designs have incorporated horizontal cooling legs and modified cooling systems. Since check valves include elastomer sealing components that are subject to high wear, the circuit design must allow for maintenance of these items. The current maximum operating temperature for these piston diaphragm is around 200°C.

Pressure Relief

Pressure relief systems are a critical safety aspect in the design of pressure hydrometallurgy systems.

The ASME Code specifies safe practices in design, construction, inspection, and repair of unfired pressure vessels. Pertinent requirements of the code are:

- The maximum allowable working pressure is the internal pressure at which the weakest element of the vessel is loaded to the ultimate permissible point, when the vessel is assumed to be:
  a) in corroded condition
  b) under the effect of a designated temperature
  c) under the effect of other loadings (e.g., hydrostatic pressure) which are additive to the internal pressure
- Size of the outlet pipe is to be such that any pressure existing or developing in the discharge line does not reduce the relieving capacity of the protective devices below the requirements for adequate prevention of overpressure.

The design pressure is that at the top of the vessel, and it usually coincides with the relief valve setting unless a rupture disc is fitted in combination with the relief valve. The pressure relief system for early pressure oxidation autoclaves consisted of combined rupture discs and relief valves, with the rupture disc preventing the valve from being continuously exposed to the corrosive autoclave environment. Cheaper materials could therefore be used for the valve. This approach required a higher design pressure for the vessel to provide sufficient margin between the operating pressure and the rupture disc bursting pressure, thereby preventing premature failure of the rupture disc due to fatigue or creep.

Unfortunately, the failure of rupture discs has been a frequent problem. Various materials have been tried without success, with one plant even considering gold foil over the disc. Titanium could withstand the high-temperature, corrosive environment but is not suitable for a safety device in the presence of oxygen because of its potential for spontaneous ignition. One approach originally developed in the nuclear industry uses a water-filled U tube fitted to the relief nozzle on the autoclave, with the safety valve at the other end protected from the autoclave environment by the water leg. Although this design has been successful, care must be taken in the design and support of the discharge piping to handle a potential slug of water passing through the system. In addition, the inlet piping to the valve must be sized so that the maximum inlet pressure loss (inlet loss and line loss a maximum flow) to the relief valve is a maximum of 3% of the set pressure.

This is an ASME code requirement. One publication suggests that, as a conservative guideline, the equivalent length-to-diameter ratio (L/D) of the inlet piping to the relief valve size should be kept at 5 or less.

A set of equipment interconnected by a piping system that does not include valves for isolating individual items can be considered a single unit, requiring a single relief valve(s), but the set pressure should be that of the lowest pressure piece of equipment. Hence, a flash vessel coupled to a preheat vessel without any isolation valves in the interconnecting piping will require only one relief valve or rupture disc. This is usually installed on the flash vessel, which is the source of pressure to the system, but it could be on the piping as long as the set pressure takes into account any losses in the piping between the vessel being protected and the location of the valve.

Pressure Let-Down System Design

In all pressure oxidation or leach processes, hot acidic slurry exits the autoclave under pressure and must be let down to atmospheric conditions in a controlled manner. This may be accomplished in one or more stages, depending on whether or not there is a use for the steam evolved. Considerable energy is released in the process, and care must be taken in design to avoid severe erosion of valves and piping.

An important feature of the first stage of pressure let down is the ability to control the liquid level in the last compartment of the autoclave. The flow control "equipment" (valve, choke, or combination valve and choke) will sustain a pressure drop, and when the pressure drops close to saturation, flashing will occur.

There are a variety of designs for the flash vessel, all aimed at minimizing erosion from the high-velocity, two-phase stream exiting the choke.

Most installations control the flow in one stage consisting of a valve located at the top of the vessel, with the high-velocity jet discharging down into the vessel through a large-diameter, ceramic-lined "blast" tube and into the slurry. Energy is dissipated through turbulence of the slurry. The slurry depth is maintained by discharging through a nozzle in the side of the vessel. One of the principal functions of the flash vessel is to remove particulate and liquid carryover from the discharging steam. This is achieved through gravity separation, which requires a quiescent zone for the separation to occur. To provide this, the blast tube is fitted around the jet exiting the valve. The annular area between the blast tube and vessel shell serves as the liquid-vapor separation zone.

In some installations, the slurry jet has passes through the slurry and wears out the bottom of the vessel. Typically, a ceramic impact plate is installed in the bottom of the vessel to prevent this. Blast tubes have also experienced high wear.

A design developed in the geothermal brine energy recovery industry that utilizes one-stage flow control has the control valve located at the bottom of the vessel with the jet projecting vertically into the slurry, causing the slurry to circulate, hence, dissipate the jet energy. This arrangement has been used successfully in the low pressure flash vessel at one of the Western Australian laterite operations.

Another design uses a valve and ceramic choke in series, with most of the pressure drop occuring in the choke. The valve allows for adjustment of the upstream pressure to the choke and can thus be used for flow control. Under normal operation the pressure drop across the valve is less than that by which the autoclave operating pressure exceeds the steam vapor pressure. Thus there is no flashing of steam between the valve and choke. If this were not the case, severe erosion of the interconnecting pipe would occur.

In this two-stage flow control system, the control valve can be installed at either the top or the side of the vessel, with the choke located inside the vessel at approximately mid height (see Figures 8 and 9). The choke is sized to provide a backpressure on the valve, thereby preventing flashing in the line between the choke and the valve inside the vessel. The valve trim must be selected so that pressure drop across it, over the normal operating range, is less than the available overpressure at the valve. In other words, this will be the total autoclave pressure minus the steam partial pressure and static head difference between the autoclave and valve location. The overpressure is provided by oxygen and inerts in the case of pressure oxidation systems and by the addition of air in the case of straight leaching systems. The valve is used to modify the inlet pressure on the choke and thus control flow, with most of the pressure drop being taken across the ceramic choke. With a relatively short and properly designed choke, there is a fan effect rather than a jet, minimizing wear on the bottom of the vessel.

A third system, often called enthalpy control, adds coolant to the autoclave discharge slurry before it passes through a choke to control flow. This is a thermodynamic process, not simply one of substituting coolant flow for autoclave discharge flow; its effect can be much greater. The coolant addition modifies the flash point in the choke, resulting in an increase in mass flux (flow) in the choke. This control system may be used in conjunction with the other options to extend the flow range by "retarding" flashing in the valve. The addition of coolant will reduce the amount flashed steam produced in the flash vessels.

Figure 8 Flash vessels

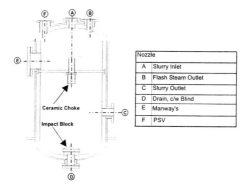

Figure 9 Carbon steel flash vessel with brick lining

Agitation
Agitator impeller types generally used in pressure hydrometallurgy are the multi-blade, radial disk turbine for gas dispersion and the pitched-blade or hydrofoil type for blending and suspension

applications. Impeller tip speeds range from 4 to 6 m/s and are selected by the process designers. The materials of the impellers are the same as or similar to the materials of the internals. The impellers are subject to wear and cavitation, which can be minimized by the proper design of sparging systems for oxygen, air, or steam, and by providing proper baffling to minimize vortexing. Sparge system design for gases is critical, not only from the cavitation standpoint, but to maximize the dispersion of the gas. A minimum gas velocity is required on exiting the sparger; if the velocity is too low, a slug flow of gas is produced and dispersion by the impeller is poor.

Mechanical considerations and practices in the specification of agitators consist of:

- A drive arrangement with shaft-mounted gearboxes (the most common, as it represents a practical and economical approach)
- AGMA and DIN standards provide the guidelines for the gear and gearbox designs
- Bolted shaft couplings are the most common
- Balanced double mechanical cartridge seals are the standard for autoclave agitators
- Keyed, clamped hubs are commonly provided so that the disks and/or blades can be readily replaced.

Seals and Seal Water System

The choice of materials for the double mechanical seals is critical, and advances have been made over the years that now yield a relatively long seal life. The major components of the seals are the stationary and rotating seal faces. Earlier seal components had to be of different materials because of wear and spalling if the two faces were the same. Typically carbon and chrome oxide were the materials of choice in the autoclave atmosphere. Today both silicon carbide and tungsten carbide seal face materials are used. The silicon carbide materials are brittle and must be handled with care. The tungsten carbide seals require nickel or cobalt binders that leach when exposed to the autoclave environment, however, development and testing of a resistant binder is in progress. Other materials in the seal consist of fluoropolymers for O-rings and stainless steels for the springs and other hardware. One of the advances in seal design is the addition of a loose seal inside the vessel, supplied with its own water source. This flushes away slurry or foam that could enter the seal cavity and thus prolongs seal life.

The most critical aspect of seal reliability and life is cooling and lubrication of the seal via the seal water system. The philosophy behind a good seal water system is that it must not only provide the cooling and lubrication, but also contain enough fluid that the autoclave can be safely shut down or continue to operate with one or two damaged seals, and that pressurized water is available in the event of a power loss. Providing only enough fluid to maintain cooling and leakage is an inadequate system for safe autoclave operation.

There are two approaches to seal system design. One is to provide individual reservoirs with their own circulation systems fed by a common supply. This is a costly approach, however. A simpler system is to provide all the seal water from one control pumping system, with adequate flow and safety backup. To follow the philosophy of seal operation as outlined above, the seal water system typically consists of a surge/make-up tank, seal water feed pumps (with redundancy and sometimes a separate drive method), filters, an accumulator, and a seal water cooler on the return side of the system. The accumulator is connected to a high-pressure gas source (nitrogen or air), which, upon power failure, provides the requisite pressure on the seals. All materials are stainless steel, and the water supplied must be reverse osmosis quality, or preferably demineralized water, at a maximum temperature of <60°C at the seal discharge. Some installations add glycol (around 10%) to aid in seal lubrication, which considerably increases the cost of the operation.

CONCLUSION

Various pressure hydrometallurgical processes are now widely used on a range of ores and concentrates, and equipment has been developed and commercially proven to handle these conditions. While some materials issues are outstanding and other improvements could be made, the designs of these systems, and their safety and reliability, have matured to the point that they can be installed with confidence.

As more complex ores continue to be exploited and environmental constraints make other treatment routes less attractive, medium- and high-temperature processing is bound to become more prevalent.

REFERENCES

ASME Boiler & Pressure Vessel Code, Section VIII Div 1, ASME/ANSI B16.5 Pipe Flanges and Flanged Fittings.

Seminar by Goulds Pumps Canada Inc. Current Trends in Pump Technology.

C.C. Smith, D.G. Dixon, M.R. Luque, J.C. Robison, S.R. Chipman. Analysis and Design of Flashtubes for Pressure Letdown in Autoclave Mining Operations.

J.G. Banker, J.P. Winski, Titanium/Steel Explosion Bonded Clad for Autoclaves and Vessels.

Barrick Gold – Autoclaving & Roasting of Refractory Ores

K.G. Thomas[1], A. Cole[2] and R.A.Williams[3]

INTRODUCTION

Over the last two decades several facilities, utilizing autoclaves and oxygenated roasters have been installed for processing refractory gold ores. Two such facilities have been installed within Barrick Goldstrike and this paper reviews the processes, inclusive of associated capital and operating costs.

In this paper the two Barrick Goldstrike refractory processes are described being as follows:

- Acid autoclaving for sulphide ore
- Oxygenated roasting for double refractory ore; sulphide and carbonaceous

In summary there is a place for both autoclaving and roasting of gold ores, the latter requiring appropriate but significant investment in environmental abatement equipment.

BARRICK GOLDSTRIKE MINE[1]

The Goldstrike property is located in the Tuscarora Mountains, Elko and Eureka counties in north-central Nevada, USA, on the Carlin geological trend. The deeper deposits at Goldstrike occur in silty limestones, which have been silicified and argillized. Gold in the sulphide ore occurs mainly as inclusions in fine-grained pyrite and marcasite. Oxidized ore lies in the upper parts of the deposit. Over 75% of the reserves are in the Betze-Post open pit deposit and the remainder in the high grade Meikle mine and Rodeo/Griffin underground deposits. Presently gold reserves are in refractory sulphide or refractory carbonaceous/sulphide ores. This makes autoclaving particularly important to Barrick Goldstrike. As the reserves continued to grow, so did the quantity of carbonaceous/sulphide ore. Half of the ounces that are contained in the current reserves are carbonaceous/sulphide. To treat this growing quantity of double refractory material an oxygenated roaster, at 12,000 stpd capacity, was commissioned in the second quarter 2000.

Besides the 24 million ounces in reserves, over 20 million ounces of gold have been produced from the Goldstrike property since acquisition in 1987. In 2000 total cash costs were US$170 per ounce and total production costs US$220.

---

[1] Managing Director, Mining & Mineral Processing, Hatch (formerly Senior Vice President, Technical Services Barrick Gold Corporation)
[2] Roaster Superintendent, Barrick Goldstrike Mines Inc.,Nevada, USA
[3] R. A. Williams, Manager Process, Pascua Project (formerly Autoclave Superintendent Goldstrike) Barrick Gold Corporation, Toronto, Ontario.

BARRICK GOLDSTRIKE ACIDIC AUTOCLAVE

Figure 1: Detailed representation of the Autoclave Circuit

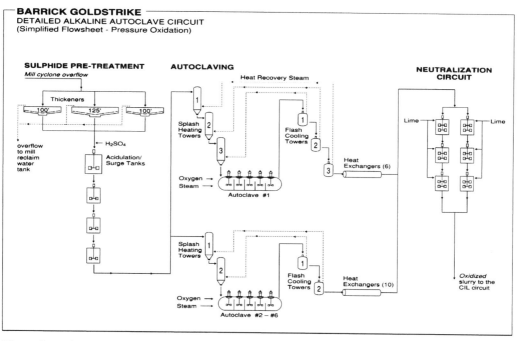

Slurry from the conventional wet SAG/ball grinding circuit, at approximately 35% solids and 80% passing 135 microns is pumped to three preoxidation thickeners. Thickener underflow, at approximately 54% solids, is pumped to a train of four acidulation tanks. Sulfuric acid is added to the slurry to destroy sufficient carbonate (CO_3) prior to entering the autoclave circuit. Process air is also injected into the acidulation tanks to aid in carbon dioxide removal. Carbonate levels are typically reduced to less than 2% in the acidulation tanks. As a rule of thumb, 1% sulphide sulphur ($S^=$) in the autoclave feed destroys 0.9% CO_3 and typically the autoclave feed ranges from 2.0% to 2.5% $S^=$.

The discharge from the high pressure splash vessel is pumped by two positive displacement pumps into the autoclave. The pumps are operated in parallel with individual suction and discharge lines. Each pump can deliver approximately 60% of the required feed rate to the autoclave, the parallel configuration enhances on line availability to 92%. Scheduled maintenance, per autoclave, is a shutdown taken every 12 months, hence the high availability.

All six autoclaves have an outside diameter of 15 feet, autoclave No. 1 is 75 feet in total length, and autoclaves No. 2 through No. 6 are 82 feet in total length. Each autoclave is divided into five compartments by brick walls, with each compartment containing an agitator and injection pipes for oxygen, steam and water. Autoclave retention time ranges from 40 to 60 minutes and the vessel operates at approximately 420 psig and 420 to 430°F. The lining of the autoclave vessels is 5/16 inch lead on the carbon steel shell, followed by 1/8 inch fibrefrax paper and 9 inch thick acid brick, 2 to 3 layers, exposed to the slurry. Another key factor for the high on line availability is the correct selection of materials of construction for various parts of the circuit. This is detailed in Table 1[2]. The selection of materials of construction is related to temperature, pressure and process chemistry, especially pH (freeacid). This can be seen by comparing the Barrick Mercur alkaline autoclave materials versus the Goldstrike acid autoclave materials.

Table 1: Materials of Construction for Autoclave Facilities (Wetted Surfaces)[2]

Item	Barrick Goldstrike	Barrick Mercur
Medium	Acid	Alkaline
Autoclave Discharge pH	1.2 to 2.0	7.5 to 8.5
Splash/Flash/Autoclave	Acid-Proof Brick	Acid-Proof Brick/Semag
Mortar in Autoclave	WiPiSe/Hydromet 50	SEMAG
Inter-stage Pumps (Splash)	High Cr Iron	High Cr Iron
Geho-Housing	CD4MCu	Nodular Cast Iron
Geho-Valves	CD4MCu	Stellite Coated Steel
Geho-Diaphragm	EPDM	EPDM
Injection Tubes	Ferralium 255	Inconel 625
Autoclave Agitator	Titanium	316 L Stainless Steel
Flash Valve Body	Titanium	Hastaloy C
Flash Choke	Ceramic Hexoloy SA	Ceramic Hexoloy SA
High Pressure Steam Duct	Titanium	316L Stainless
Low Pressure Steam Duct	316 L Stainless Steel	316 L Stainless Steel
Cooler Feed Pumps	Chlorobutyl	High Chrome Iron
Cooler Shell	Carbon Steel	Carbon Steel
Cooler Tubes	316 L Stainless Steel	316 L Stainless Steel

A typical cut away view of the splash, autoclave and flash circuit is shown in Figure 2.

Figure 2: Cut Away of Autoclave Circuit

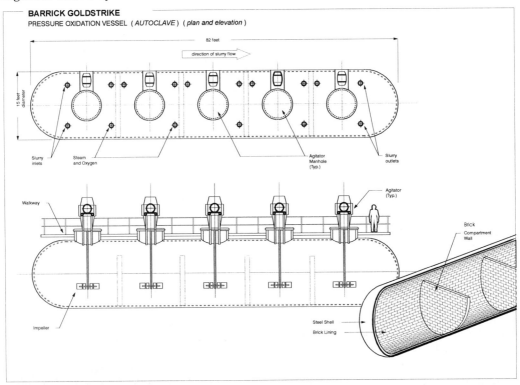

Sulphide sulphur oxidation in the autoclave is typically in the order of 90 to 92%. Residual $S^=$ exiting the autoclave is targeted at less than 0.2%. Values greater than 0.25% $S^=$ are reflected in lower gold recoveries in the carbon-in-leach (CIL) circuit. Free acid levels in the autoclave discharge are a function of $S^=$ and CO_3 in the autoclave feed but normally range from 10 to 25 g/l.

After the slurry passes through the heat exchange slurry coolers it is pumped to two parallel trains of neutralization tanks where the pH is elevated from approximately 1.5 to about 10.5.
Neutralized slurry from the pressure oxidation circuit is pumped to two parallel streams of CIL tanks for gold extraction and subsequent recovery of gold in a conventional Zadra circuit.

Suitable ancillary facilities are installed to supply flocculant, sulphuric acid, oxygen, lime, steam and compressed air to the autoclave facility. The control system is a Bailey distributed control system (DCS).

The objective of the pressure oxidation/autoclaving process is the destruction of the sulphides, pyrite, marcasite and/or arsenopyrite, thereby liberating the occluded gold. The gold is then amenable to recovery by the cyanidation process.

For an acid autoclave operating at temperatures greater than 350°F and a pH below 2, as at Barrick Goldstrike, the reactions[3] can be presented by:

$$2FeS_2 + 7O_2 + 2H_2O = 2FeSO_4 + 2H_2SO_4 \quad (1)$$

$$2FeSO_4 + H_2SO_4 + 1/2O_2 = Fe_2(SO_4)_3 + H_2O \quad (2)$$

$$Fe_s(SO_4)_3 + 3H_2O = Fe_2O_3 \downarrow + 3H_2SO_4 \quad (3)$$

The highly oxidizing conditions within the acid autoclave is important as ferrous sulphate is converted to ferric sulphate (reaction 2). This reaction is beneficial as ferrous sulphate consumes cyanide, in the cyanidation step of the CIL circuit, which adversely increases operating costs.
In acid autoclaving silver is precipitated as a jarosite but can be recovered by a lime boil pre-treatment process but is generally not practiced as it is not economical.

The key to successful autoclaving to minimize maintenance and operating costs is to layout the plant for ease of equipment access, by crane, as shown in Figure 3[4].

Figure 3: Autoclave Plant Plan & Section at High Pressure Flash Tank

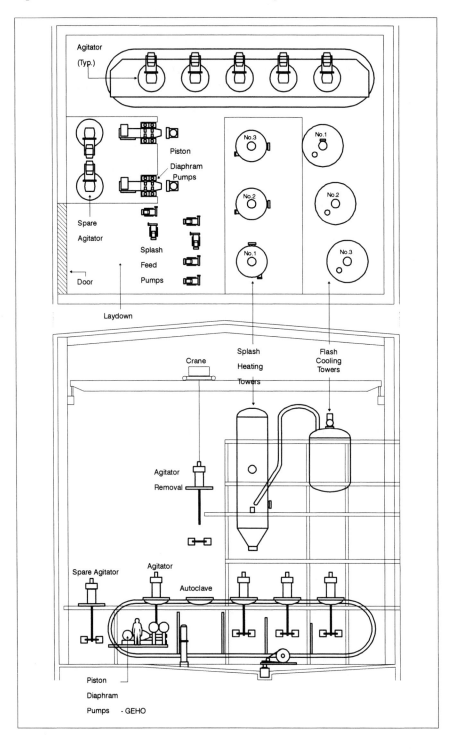

BARRICK GOLDSTRIKE OXYGENATED ROASTER[5]

A 12,000 stpd oxygenated roaster facility was commissioned at Barrick Goldstrike in the second quarter of 2000. The roaster treats double refractory ore, carbonaceous and sulphidic, at a temperature of 1000°F using 99.5% pure oxygen in a double "two-stage vertical" roaster. A flowsheet of the process is shown in Figure 4.

Figure 4: Barrick Goldstrike 12,000 TPD Oxygenated Roaster

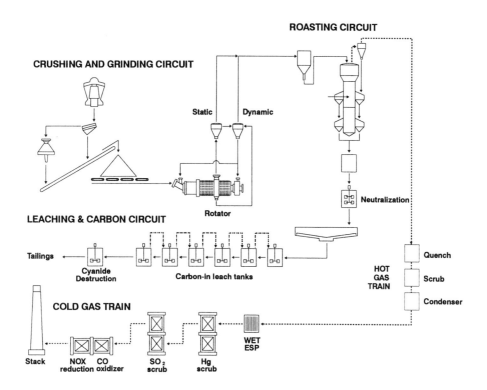

As the drilling of the Goldstrike property progressed from 1988 to the late 1990's significant quantities of double refractory ore were added to the reserves, over the original ores classified as a single refractory, that is sulphidic, which are treated with ease in autoclaves. Of the 24 million ounces in reserves at the end of 2000, approximately 9 million ounces were double refractory, requiring treatment by another process. Oxygenated roasters were selected.

The roaster circuit at Goldstrike incorporates unique equipment, that is, dry grinding using Krupp Polysius cement technology double rotators as shown in Figure 5, and Freeport McMoRan two stage vertical oxygenated roasters as shown in Figure 6. Crushed ore is fed into the drying chamber of each of two parallel double rotators, followed by coarse and fine grinding, in two separate chambers within the same mill. Ore is reduced from 40 mm to 74 micron, with the aid of static and dynamic classifiers, and the product captured in a series of baghouses before advancement to the roasting facility.

Figure 5: Double Rotator Grinding Mill by Krupp Polysius

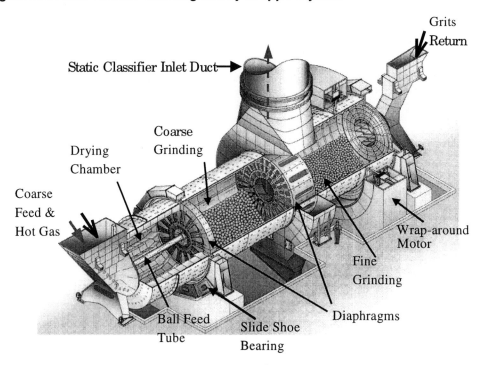

Figure 6: Two-stage Fluid Bed Roaster Freeport McMoRan Type

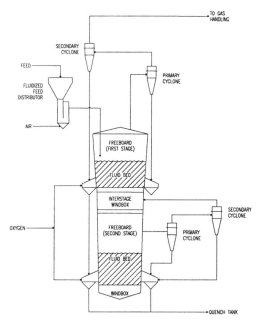

Two parallel roaster circuits are utilized to oxidize the carbonaceous matter and sulphide to release the gold for subsequent gold cyanidation. Low pressure, high purity oxygen oxidizes the ore, with the majority of the oxidation taking place in the upper bed. The lower bed completes the oxidation of the carbonaceous matter to 80%; sulphide sulphur oxidation is 99%, mainly in the first stage. The process is autogenous but can be controlled by water quenching or coal addition, depending of the calorific value of the ore. The exhaust gas is extensively cleaned, as shown in Figure 4, in a series of process equipment:

- Quenching and de-dusting
- Wet gas condenser
- Mist eliminator
- Wet electrostatic precipitator
- Mercury removal
- SO_2 scrubbing
- CO incinerator
- NO_x removal

Environmental abatement equipment at the roaster was US$40 million out of a total capital cost of US$330 million. As in the autoclave circuit suitable ancillary equipment are installed to supply reagents. Also as in the autoclave facility Bailey distributed control system is installed as a control system.

The chemistry of the double refractory ore oxidation can be represented as follows:

$$2C_{(s)} + O_{2(g)} = 2CO_{(g)} \uparrow \qquad (4)$$

$$2Co_{(g)} + O_2 = 2CO_{2(g)} \uparrow \qquad (5)$$

$$14FeS_2 + 38\tfrac{1}{2}O_2 = 7Fe_2O_3 \downarrow + 28SO_2 \qquad (6)$$

An interesting feature of the roaster chemistry is the fixation of the SO_2 by the carbonates contained within the ore, forming gypsum. Between 40% to 70% of the SO_2 is fixed by the naturally occurring calcium carbonate, which reduces subsequent processing costs. Fixation is enhanced with the addition of lime to the grinding circuit.

A key piece of equipment in the double rotator grinding circuit is the dynamic classifier. A very accurate classifier using a variable speed drive, it cuts the product, with consistency, to 80% passing 74 micron. A cut away view of the dynamic classifier is shown in Figure 7.

Figure 7: Dynamic Classifier:

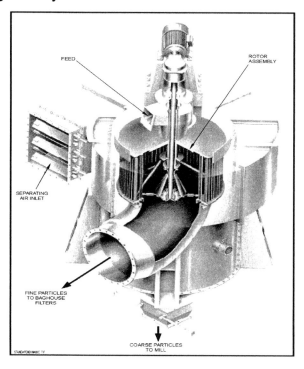

Following oxidation the ore is slurried, neutralized and advanced to CIL for gold extraction. Gold bearing carbon is transported to the Zadra circuit in the existing autoclave facility for gold recovery.

OPERATING COST COMPARISON
The two processes display a marked difference in operating costs as detailed in Table 2 below:

Table 2: Operating cost comparison US$/short ton

Cost Item	Acid Autoclave (adjusted)*	Roaster
Crush	1.05	1.05
Grind	4.95	5.25
Roast	-	5.35 **
Autoclave	9.40 **	-
CIL	1.90	1.85
Strip & Electrowin	0.25	0.30
TOTAL	17.55	13.80

* Adjusted refers to grinding to 74 micron; normally 90 microns
** Assumes oxygen plants purchased

The difference between the acid autoclave and oxygenated roaster costs is explained as follows:

- The grinding costs of wet and dry grinding plants are very similar, although the processes are markedly different. This is due to the advancements pioneered by Krupp Polysius in adopting double rotator cement technology to the mineral industry.[6]

- The two stage oxygenated roasters, 99.5% purity oxygen, pioneered by Freeport McMoRan, New Orleans, U.S.A., has allowed an inexpensive alternate to acidic autoclaving to be developed. In the roaster facility acid is not required to remove CO_3 from high carbonate ores as in the autoclave circuit. Thus saves up to $3 per ton ore, on acid costs for the Barrick Goldstrike relatively high carbonate ore (5% CO_3).

CAPITAL COST COMPARISONS
For a 12,000 stpd oxygenated roasting or acid autoclaving facility, inclusive of comminution and gold recovery, capital costs are similar at US$330 million. As mentioned earlier the oxygenated roasting facility has significant environmental abatement equipment amounting to US$40 million, which is not required to the same extent in an autoclave circuit.

PROCESS SELECTION
For a double refractory ore, carbonaceous and sulphidic, the process of choice is oxygenated roasting. This may change in the not too distant future, if a thiosulphate process can be developed for treating carbonaceous ore. A thiosulphate lixiviant does not suffer from pre-robbing as with cyanide and accordingly could be part of a gold recovery circuit following autoclaving.

Whether an oxygenated roaster or acid autoclaving process is selected for a sulphidic ore depends on the mineralogy and the gold price. Assume a 50 million ton ore body, 0.25 ounces per ton, US$315 per ounce gold and a US$2 per ton operating cost lower in favour of oxygenated roasting. This example assumes a low carbonate level in the ore body, that is below 2% CO_3, that is, reducing acidulation costs by US$ 2 to 4 per ton depending on carbonate levels in the ore. Over the life of mine this saves US$100 million operating costs. But oxygenated roasting, generally has on average 2.5% less recovery. This gives a break-even project between oxygenated roasting and acid autoclaving owing to $100 million less revenue. This simple example illustrates the influence that ore grade, gold price, mineralogy and differential operating cost and recovery has in process selection. If the carbonates are above 2% CO_3 then the economics are in favour of the roaster owing to the acid costs required for acidification prior to autoclaving.

Since the 1980's acid autoclaving has generally been the preferred process in treating sulphidic refractory gold ores and concentrates. This is especially the case with high arsenic ores, as the arsenic is fired to environmentally stable ferric arsenate during autoclave chemical reactions. Oxygenated roasting, developed in the late 1980's has in the last 5 years been developed to a satisfactory level, especially with the utilization of dry grinding double rotators from Krupp Polysius (refer to Figure 5). Accordingly there is now a choice between autoclaving and oxygenated roasting for treating sulphidic ores.

CONCLUSION

Oxygenated roasting is the process of choice for treating double refractory ore. There is now a choice between acid autoclaving and oxygenated roasting for treating sulphidic ores, using the whole ore process, and process selection depends on several factors, such as gold price, mineralogy and recovery.

ACKNOWLEDGEMENTS

The authors would like to thank Barrick Gold Corporation for permission to publish this paper and associates for their assistance over the years.

---

1. H.J.H. Pieterse and K.G. Thomas; "Pre-treatment Methods, Autoclaving;" Prepared for the Second International Gold Symposium; Lima, Peru; May, 1996
2. K.G. Thomas & R.A. Williams; "Alkaline & Autoclaves at Barrick Gold", EPD Congress 200, TMS, Nashville, Tennessee, U.S.A., March, 2000.
3. K.G. Thomas; Research, Engineering Design and Operation of a Pressure Hydrometallurgy Facility for Gold Extraction; Technical University of Delft, The Netherlands; May, 1994.
4. K.G. Thomas, "Barrick Gold Autoclaving Process", World Gold '91, Cairns, Australia, April 1991.
5. A. Cole, J. McMullen, K.G. Thomas & S. Dunn; "Barrick Goldstrike Roaster Facility Roasting & Gas Handling; Canadian Mineral Processors, 2001, Ottawa, Ontario, Canada; January, 2001.
6. N. Patzelt; Private Communications; Krupp Polysius; Germany; 1998 to 2000.

Selection and Sizing of Biooxidation Equipment and Circuits

Corale L. Brierley[1] and Andrew P. Briggs[2]

ABSTRACT

The first commercial, stirred-tank bioleach plant was commissioned in 1986 to pretreat a sulfidic-refractory gold concentrate, thus enhancing gold recovery. There are now 11 full-scale, stirred-tank bioleaching plants employing three different technologies to process sulfidic-refractory precious metal concentrates and concentrates of pyritic cobalt and chalcopyrite. Bioleaching is also used at commercial scale to heap leach secondary copper sulfide ores and to pretreat sulfidic-refractory gold ores in heaps. Heap leaching of chalcopyrite ore has been pilot tested and commercial development is anticipated in the near future. Bioleaching of concentrates in heaps is an emerging technology. This chapter describes the design and sizing of aerated stirred-tank and heap operations and discusses the selection of technology and equipment for these plants. Bioleaching, also called biooxidation, employs microorganisms as catalysts to generate ferric iron, an oxidizing agent that degrades sulfide minerals. The microorganisms also oxidize chemically reduced sulfur compounds producing acid. Microorganisms, commercially exploited in bioleaching, are divided into three groups based on the temperature range at which they grow and function. Bioleaching can be accomplished at temperatures ranging from about 20° to 95°C; organisms that function at the higher temperature ranges continue to catalyze reactions despite the large amounts of heat generated from the rapid oxidation of sulfide minerals in tank and heap reactors. These thermophilic microorganisms are also necessary for effective oxidation of chalcopyrite. The fundamental principles of biooxidation are included in this chapter as well as a description of the important microorganisms and factors that affect their performance.

INTRODUCTION

Mineral biooxidation, also called bioleaching, uses certain microorganisms to oxidize sulfide minerals present in ores or concentrates. In this process base metals are released into a dilute sulfuric acid solution for recovery by conventional methods. Precious metals occluded in sulfide minerals are exposed for enhanced extraction by cyanide or other lixiviants. Biooxidation has undoubtedly been employed unwittingly for centuries for the extraction of copper. However, the underpinnings of the unit process commercially employed today had its genesis some 50 years ago when copper was scavenged from run-of-mine, submarginal grade materials in large dump leach operations. In the last 20 years process design and equipment availability for biooxidation have achieved a higher degree of sophistication, and bioleaching is now applied in heap leaching of crushed, secondary copper ores and sulfidic-refractory gold ores and in tank leaching of base and precious metal concentrates.

---

[1] Brierley Consultancy LLC, Highlands Ranch, Colorado
[2] Fluor Mining and Minerals, Denver, Colorado

The aim of this chapter is to describe the process design and equipment selection and sizing for biooxidation circuits. Specifically considered are:

- The principles of biooxidation and incentives for commercial use of the technology,
- Process design of crushed, sulfide ore biooxidation circuits with emphasis on sulfidic-refractory gold ores,
- The design and equipment selection and sizing of aerated, stirred-tank biooxidation circuits for concentrates and high value sulfide materials,
- Current commercial biooxidation operations and their performance and problems, and
- New developments in biooxidation processing.

PRINCIPLES OF MICROBIAL OXIDATION OF MINERALS

Bioleaching is only applicable to sulfide minerals be they base metal sulfides or sulfide minerals such as pyrite and arsenopyrite that host precious metals. This is because the microorganisms, employed in bioleaching technology, derive energy for their existence and reproduction from the oxidation of ferrous iron and chemically reduced sulfur compounds much like humans obtain energy from the foods we eat. Chemical reactions, catalyzed by the microorganisms associated with the bioleach process, are responsible for releasing base metals into a dilute sulfuric acid solution and making precious metals available for dissolution by cyanide and other chemical lixiviants. Bioleaching is not a process involving a single type of microorganism, but rather involves groups of microorganisms that catalyze different reactions and thrive under different pH, chemical and temperature regimes (Norris et al. 2000). Bioleaching is not a stand-alone technology; rather it is the amalgamation of three separate disciplines – biotechnology, chemistry and metallurgy – whose requirements must converge to achieve a seamless process. This section briefly clarifies the fundamentals of biotechnology, chemistry and metallurgy that must interrelate in the design and performance of the mineral biooxidation process.

Microorganisms Involved in Bioleaching

In bioleaching technology microorganisms are divided into three groups based on the temperature ranges at which they grow:

1. Mesophilic bacteria, which actively function in the 15° to 45°C temperature range,
2. Moderately thermophilic ("heat loving") bacteria that oxidize sulfur and iron compounds in the range of 40° to about 65°C, and
3. Extremely thermophilic microorganisms, which are not bacteria but rather are *Archaea*, a unique evolutionary grouping of organisms that evolved from ancient life forms on Earth, and flourish at temperatures from about 60°C to nearly 95°C.

Although these three groupings of organisms proliferate at different temperature ranges, they all have some features in common. They all derive energy from the oxidation of ferrous iron, certain chemically reduced sulfur compounds or both:

$$4\ Fe^{2+} + O_2 + 4\ H^+ \rightarrow 4\ Fe^{3+} + 2\ H_2O \qquad \text{(Reaction 1)}$$

$$S^\circ + 2\ H_2O + O_2 \rightarrow 4\ H^+ + SO_4^{2-} \qquad \text{(Reaction 2)}$$

All organisms involved in bioleaching need oxygen (O_2) and carbon dioxide (CO_2), both supplied in the gaseous forms; O_2 is the electron acceptor for the redox reactions the organisms carry out for their metabolic energy requirements, and CO_2 is the carbon source from which they construct their cellular components, such as proteins, DNA, carbohydrates, etc. They all require ammonium (NH_4^+) and phosphate (PO_4^{3-}) ions and certain trace elements as building blocks for amino acids, DNA and other constituents. The trace elements required (Mg^{2+}, K^+, etc) are often abundant in the ore or concentrate feedstock. The bioleaching microorganisms are acidophilic, that is "acid-loving", and require a pH range of less than 2.5 and preferably greater than pH 1.0. This pH range assures that Fe^{2+} is readily available in solution for oxidation by the microorganisms.

Mesophilic bacteria. The most common mesophilic bacteria present in sulfide leaching operations are *Acidithiobacillus ferrooxidans*, *Acidithiobacillus thiooxidans*, and *Leptospirillum ferrooxidans*[3]. *Thiobacillus caldus*, technically a moderately thermophilic bacterium because of its optimum growth at 45°C, will also grow at mesophilic temperatures. These microorganisms are the workhorses in heap leaching of secondary copper sulfide minerals and during the early phase of sulfidic-refractory gold ore heap leach operations before heap temperatures exceed the tolerance level of these microorganisms. Many tank leach plants for precious and base metals rely on mesophilic bacteria. Because the mesophilic bacteria are ubiquitous in the environment, they develop naturally within the heap leach environment when the ore is moist and properly acidified, O_2 is available and the temperature is between about 15° and 40°C. When heap conditions are optimum these small (approximately 0.5 μm in diameter by 1.0 μm in length), rod-shaped organisms proliferate until they number 10^6 to 10^7 per gram of ore. When concentrate is bioleached in aerated, stirred-tank reactors at mesophilic temperatures, the reactors are inoculated with strains of mesophilic bacteria that have been pre-adapted to the feed and the high metal concentrations anticipated during leaching. In these reactors the bacterial population numbers 10^9 to 10^{10} per ml of solution. Rawlings (1997) has written a more comprehensive treatise on these organisms.

Moderately thermophilic bacteria. The moderately thermophilic bacteria are easy to find and culture from volcanic areas, acidic thermal pools, warm, acidic mine waters, and sulfidic stock and waste piles where temperatures and conditions support their proliferation. The moderately thermophilic bacteria have not been as well studied as the mesophilic leaching bacteria and therefore the taxonomy (identification and naming of the organisms) is not as well developed. Moderate thermophiles, common to bioleaching operations, include *Sulfobacillus thermosulfidooxidans*, *Sulfobacillus acidophilus*, *Acidophilus ferrooxidans*, and *Thiobacillus caldus* (Norris 1997). These rod-shaped bacteria, somewhat larger in size (1 μm in diameter by 2-3 μm long) than the mesophiles, appear naturally (without deliberate introduction) in sulfidic-refractory gold heaps and pyrite-rich copper dump leach operations, as temperatures rise to 40°C from the exothermic oxidation of sulfide minerals. When the maximum temperature limit (45°C) of the mesophilic bacteria is exceeded they die, because their protein structures destabilize, and the moderate thermophiles then dominate. The moderate thermophiles perform the same oxidation reactions with the same degree of efficiency as the mesophiles. The moderate thermophiles have also been harnessed for industrial application in aerated, stirred-tank reactors. Chalcopyrite bioleaching, a developing commercial process, is accomplished using moderate or extreme thermophiles, because improved copper leach rates and recoveries are observed with these two groups of organisms.

Extremely thermophilic microorganisms. The extreme thermophiles are not bacteria, but *Archaea*, a distinct branch of life evolved from ancient life forms on Earth. These organisms, whose existence was discovered in the late 1960s, are very unlike the mesophilic and moderately thermophilic bioleaching organisms in that they are 1 μm spheres lacking a rigid wall surrounding the cells. These "extremophiles", as they are referred to, proliferate in acid environments rich in chemically reduced sulfur and iron compounds only when temperatures approach 60° to 65°C. At these temperatures the moderately thermophilic bacteria reach their maximum heat tolerance and die. The *Archaea*, at home in volcanic regions and hot acid springs, not only tolerate but reproduce at temperatures approaching 95°C (Norris 1997). Although these organisms have been cultivated from hot, sulfidic coal waste piles, they have not been found to naturally occur in hot copper leach dumps or hot sulfidic-refractory precious metal heap leach plants. This does not mean, however, that they are not there; it may simply be that nobody has extensively looked for them using the

---

[3] Certain species of the genus *Thiobacillus* were recently reclassified as *Acidithiobacillus*

right tools. The best studied of the extreme thermophiles are *Sulfolobus acidocaldarius*, *Sulfolobus metallicus*, and *Acidianus brierleyi*. The extremely thermophilic *Archaea* are intentionally added to heaps used for pretreatment of sulfidic-refractory gold ores and to aerated, stirred-tank reactors for leaching of chalcopyrite concentrates. The advantages of using *Archaea* in these applications will be discussed in more detail later.

Other microorganisms. In commercial-scale heap and dump leach operations there are a wide diversity of microorganisms present. Many of these are microorganisms that use organic matter for their metabolism and reproduction. These "heterotrophic organisms" include bacteria, fungi and certain members of the high temperature *Archaea*. These acid-loving microorganisms scavenge and oxidize organic matter present in the leach system (Johnson and Roberto 1997). The organic compounds come from the degradation of organic matter in the ore, residual organics from the solvent extraction process and organic compounds excreted by the leaching organisms, and dead biomass. The exact role of the heterotrophic organisms in leaching is not known, but their presence does influence both the design and performance of the plant. Because these organisms require O_2, their consumption must be considered when designing the aeration system. Heterotrophic organisms, particularly fungi, can be a nuisance as they proliferate to the extent of plugging pipes, disrupting solution flow by growing on the picket fences in the solvent extraction circuit, and forming a crud layer between the organic solvent and aqueous layer in the SX plant.

Chemistry of Bioleaching

The exact mechanism employed by the mesophilic, moderately thermophilic and extremely thermophilic microorganisms to leach sulfide minerals has been studied and debated for decades. It is well known that many microorganisms present in a bioleach system are firmly attached to mineral surfaces while other microorganisms are suspended in the aqueous phase. Sand and his colleagues (1995) argue that bioleaching of metal sulfides is achieved by way of microbial Fe^{3+} generation and H_2SO_4. These researchers (Sand et al. 1995; Schippers and Sand 1999) propose that disulfides (e.g. FeS_2 MoS_2, WS_2) are oxidized by Fe^{3+} with thiosulfate ($S_2O_3^{2-}$) formed as an intermediate product. In contrast, metal sulfides (e.g. ZnS and $CuFeS_2$) are degraded with a combination of Fe^{3+} and H^+ with the main intermediates being polysulfides (S_8) and $S°$. The degradation mechanisms between disulfides and metal sulfides differ based on differences in electronic structures and solubilities. Sand and his colleagues (1999) further propose that microorganisms, adhering to the surfaces of minerals, are encased by a polymeric substance (biofilm). The organisms attach to minerals via electrostatic interaction between the biofilm and the mineral surface. The polymeric biofilm complexes Fe^{3+} iron, which Sand et al. (1999) assert can concentrate to 53 g/L within the biofilm. They propose that the interface between the microbe's polymeric layer and the mineral surface is the reaction zone where metal dissolution from the sulfide mineral takes place. This bioleaching mechanism, which was developed using *A. ferrooxidans* as the test organism, is illustrated in Reaction 3 with oxidation of the disulfide, FeS_2 by ferric hexahydrate molecules (Schippers et al. 1996):

$$FeS_2 + 6\ Fe(H_2O)_6^{3+} + 3\ H_2O \rightarrow Fe^{2+} + S_2O_3^{2-} + 6\ Fe(H_2O)_6^{2+} + 6\ H^+ \quad \text{(Reaction 3)}$$

Fe^{2+} is rapidly re-oxidized to Fe^{3+} (Reaction 1) by iron-oxidizing microorganisms, such as *A. ferrooxidans*, *L. ferrooxidans*, and a variety of moderate and extreme thermophiles that are attached to the mineral surfaces as well as suspended in the aqueous phase. At the low pH of leach systems the unstable $S_2O_3^{2-}$ is converted to polythionates (e.g. $S_4O_6^{2-}$) and elemental sulfur ($S°$); these are oxidized to sulfate (SO_4^{2-}) by attached and unattached, sulfur-oxidizing microorganisms, such as *A. thiooxidans*, *T. caldus* and other thermophiles in aerated heap and stirred-tank operations.

Regardless of the exact mechanism of the bioleaching microorganisms, the crux of bioleaching is redox potential. The microbial oxidation of Fe^{2+} increases the Fe^{3+} to Fe^{2+} ratio, increasing the redox potential. Sufficient soluble iron must be present that, when biologically oxidized, the redox potential is high enough to oxidize the target minerals. Oxidation of the target sulfide minerals consumes Fe^{3+}, increasing the Fe^{2+} concentration and decreasing the redox potential. To avoid a decrease in the redox potential, plant conditions must be favorable for the immediate re-oxidation of Fe^{2+} by the microorganisms. In stirred-tank reactors, the key is maintaining a very high redox potential at all times so all sulfide minerals oxidize rapidly. The only way to maintain the high redox potential is to ensure that everything the organisms require (O_2, CO_2, acidic conditions, nutrients, optimum temperature conditions, etc) is optimized. In heaps the same is true, but achieving optimum conditions in a heap for the organisms is challenging.

Factors Affecting Microorganisms

Temperature. Temperature, as discussed above, clearly impacts bioleaching by selecting for the group of microorganisms that will predominate at a specific temperature range.

Chemical reactions generally proceed more rapidly at higher temperatures; theoretically there is a doubling of the reaction rate for a 10°C rise in temperature, so biooxidation circuits tend to be operated at the higher temperature limit for each microbial group. Higher operating temperatures also provide some benefits in terms of decreased cooling requirements in stirred tank reactors. Chalcopyrite leaching in stirred tanks must be performed at higher temperatures to effectively leach the mineral.

There is increasing evidence that operating sulfidic-refractory precious metal heap leach operations at higher temperatures improves gold recoveries without increasing lime or cyanide consumption. As discussed later, high temperature heap leaching is a developing technology for extraction of copper from chalcopyrite ores.

pH. All microorganisms important in bioleaching are acidophilic and perform optimally when the pH is between 1.2 and 2.3. Above pH 2.5 soluble iron hydrolyzes and precipitates from solution. What this means in an operating plant is that the key microbial energy source (Fe^{2+}) and the product of the microbial oxidation and sulfide mineral oxidant (Fe^{3+}) become limited. The higher pH is also not favorable for the solubilization of the product metal cations. The consequence of a pH higher than optimum is a decline in PLS tenor for base metals and lower than anticipated extraction of precious metals.

As pH in an operating plant declines below about 1.2 and acidity increases, escalating selective pressures are placed on the resident microbial population (Norris et al. 2000). The higher the hydrogen ion (H^+) concentration in the surrounding solution, the greater the difficulty some leaching microorganisms have in rejecting the transport of H^+ across the cell membrane. When H^+ is transported across the cell membrane, the pH of the cell's cytoplasm plummets and the organism dies. Should pH of an operating leach plant decline precipitously, the harsh conditions cause natural selection for those types of organisms able to tolerate the adverse situation. The more adverse the condition the fewer types of organisms survive, limiting metabolic diversity; this in turn decreases metal production. An example is inadvertent selection of the acid tolerant, *Acidithiobacillus thiooxidans* by operating a plant at low pH. *A. thiooxidans* can only oxidize chemically reduced sulfur compounds; which means that, in the absence of any iron-oxidizing microorganisms, ferrous iron concentrations increase, redox potential declines and mineral sulfide oxidation slows and then ceases.

Declining pH will also select for *Leptospirillum ferrooxidans* over *Acidithiobacillus ferrooxidans*, because *L. ferrooxidans* is more acid tolerant (Norris 1983). However, because both of these organisms oxidize Fe^{2+}, this selection has less impact on the performance of a commercial bioleach plant.

Microorganisms intensely dislike abrupt changes and this includes abrupt changes in pH and acidity. Microorganisms are remarkably adaptable, and slow changes in acidity and other operating parameters allow time for microbial populations to adapt to a range of adverse conditions without the loss of metabolically important members of the resident population.

Redox potential. The ratio of ferrous iron to ferric iron also selects for microorganisms in an operating plant (Rawlings et al. 1999). When the redox potential is low and more Fe^{2+} is in solution, *Acidithiobacillus ferrooxidans* will predominate, because this organism has a faster growth rate and will build up a larger number of cells in the system. However, as the redox potential increases due to a higher $Fe^{3+}:Fe^{2+}$ ratio, *Leptospirillum ferrooxidans* will predominate, because these organisms have a higher affinity for Fe^{2+} than does *A. ferrooxidans*. *A. ferrooxidans* is also more sensitive to inhibition from high concentrations of Fe^{3+} in solution. Therefore, in a stirred-tank reactor, in which the redox potential remains relatively constant and is high, *L. ferrooxidans* is the primary iron oxidizer.

Oxygen. Bioleaching microorganisms require O_2. O_2 accepts the electrons in the redox reactions catalyzed by the microorganisms. The surest way to cause a production problem in a bioleaching operation is to limit O_2. Getting air into the circuits and distributing it efficiently are significant engineering challenges in the design of bioleaching plants.

Oxygen utilization efficiencies (oxygen used in reactions as a percentage of that provided) are in the order of 30 to 40% for stirred tanks and 20 to 30% for heaps. Higher figures have been reported, but these could be an indication of insufficient addition. In stirred tanks, the dissolved oxygen level is about 2 ppm in the solution at the top of the tank; this number is higher at the tank bottom due to hydrostatic pressure. Solutions leaving bioheaps should also contain about 2 ppm oxygen. Dissolved oxygen concentrations lower than 2 ppm indicate a shortage of air addition and this will slow the oxidation rate; concentrations higher than about 4 ppm indicate sufficient or excess air is being added or that microbial activity is minimal for some reason (for example, the heap is depleted in sulfide, and, therefore, the O_2 is not being used).

Nutrients. The leaching microorganisms have few nutritional requirements: PO_4^{3-}, NH_4^+ and a few trace elements. Trace elements, such as Mg^{2+} and K^+, are generally present in sufficient quantities from the degradation of rock in the acid leach. PO_4^{3-} and NH_4^+ are added to stirred-tank bioleach operations, and NH_4^+ is occasionally added to heap leach operations. K^+ is added in many tank bioleach plants as the hydroxide, sulfate or occasionally as the phosphate.

Carbon dioxide. The microorganisms require carbon for synthesis of cellular components. They obtain this carbon by reduction of CO_2 in a complex metabolic pathway. Microorganisms expend considerable energy in "fixing" this carbon. CO_2 is generally available from the air added for oxidation or from the acid neutralization of limestone added for pH control in tank reactors.

Energy (food) source. The bioleaching microbes require a food source and that food source is ferrous iron (Fe^{2+}) for the iron-oxidizing microbes and chemically reduced sulfur compounds such as $S°$ for the sulfur-oxidizing microorganisms. Microorganisms obey the laws of thermodynamics; they do not perform any oxidation reactions that are not thermodynamically possible. Microbes are referred to as "catalysts" because they speed up certain reactions. For example, the oxidation of Fe^{2+} to Fe^{3+} in an acid solution is extremely slow chemically; microorganisms increase the rate of this oxidation by some 500,000 times (Lacey and Lawson 1970). The reason the organisms are so good at iron oxidation is because they must oxidize a lot of it to obtain enough energy to fix CO_2 and synthesize complex proteins, carbohydrates, DNA, etc.

Salinity. The microorganisms involved in bioleaching are relatively intolerant to the chloride ion (Cl^-); the kinetics of Fe^{2+} oxidation by *A. ferrooxidans* are significantly slowed by 5 g Cl^-/L. Attempts to adapt the organisms to higher Cl^- concentrations have been unsuccessful (Lawson et al. 1995) due to the unfettered transport of the ion across the organism's cell membrane. There are efforts underway in laboratories around the world to find iron- and sulfur-oxidizing microorganisms that are naturally tolerant to Cl^- yet exhibit the same rates of iron oxidation that current leaching microorganisms possess. Finding such organisms is very important from a commercial standpoint, because in arid climates, saline and brackish waters are often the only water available for processing. Promising test results were declared by an Australian mining company that evaluated oxidation rates and metal recoveries from nickel, copper and zinc bearing ores using microbial strains capable of tolerating salt concentrations nearly six times that of seawater (about 125 g Cl^-/L) (Titan 2002). No test details have been published.

Soluble cation and anion metal/metalloid concentrations. Leaching microorganisms are tolerant to high concentrations of most heavy metal cations and can be readily adapted to even higher concentrations. In stirred-tank bioleach plants where heavy metal concentrations can easily exceed 20 or 30 g/L of Cu^{2+}, adaptation of the microbial culture to anticipated metal concentrations is an important design step. There are some cationic metals/metalloids, which can be toxic to the organisms. For these substances to be toxic, they must be soluble. Mercury and silver, though toxic, are usually not serious problems, because silver has a low solubility in acidic leach solutions and mercury adsorbs to rock, mitigating its toxic effect. As^{5+} is not toxic, but As^{3+} is. It is important, particularly in heap leach operations, that the redox potential is sufficiently high to ensure that, when arsenic-bearing minerals such as realgar (AsS), orpiment (As_2S_3) and arsenopyrite (FeAsS) are degraded, that As^{3+} is oxidized to As^{5+}. High concentrations of Al^{3+} have been implicated in toxicity; however, this toxicity may largely be attributable to high TDS (total dissolved solids) in which several potentially toxic cations and anions are collectively at levels that may slow microbial iron oxidation.

The nitrate anion (NO_3^-) presents toxicity issues, and concentrations in excess of 200 mg/L slow the rate of Fe^{2+} oxidation. Like other anions, such as Cl^-, the mechanism of toxicity is likely to be disruption of the cell membrane and uncontrolled transport of NO_3^- into the cell, which suggests that adaptation of the microbes to NO_3^- may not be effective. Discovering leaching microorganisms with innate NO_3^- tolerance may be the answer to this toxicity problem.

Process reagents and materials. An important consideration in the design of bioleach plants is assurance that process reagents are not toxic to the leaching organisms. For example, flotation reagents that carry over in the feed to the first stage reactor in a stirred-tank bioleach plant must be evaluated to make sure that they do not inhibit the microbes. All materials, such as rubber linings in tanks, leach pad liners, and all materials that microbes come in contact with in the process, must be evaluated in lab tests to ensure that there are no inhibitory effects.

Tailings waters containing traces of cyanide (CN^-), thiocyanate (SCN^-), or cyanate (CNO^-) must not be used as process water or make-up water to bioleach circuits. These agents are respiratory inhibitors that deactivate microbial enzymes, and, if they enter the circuit, the result is significant loss in plant performance at best (Bell and Quan 1997) and a total loss of microbial activity at worst. Obviously, the toxicity of cyanide has implications in the treatment of concentrates that have been previously cyanide leached.

Oils, greases, hydraulic fluids, water treatment chemicals, dust suppressors, etc., all substances common to metallurgical plants, are potential inhibitors to the leaching microorganisms. Some of these agents are surfactants, which damage the organism's cell membrane causing the membrane to break open. Little quantitative data are available on the exact concentrations that induce problems. Good housekeeping in metallurgical plants is necessary to avoid contaminating anything in which the microorganisms come in contact with.

Biocides are used in cooling circuits to eliminate microbial contamination. For obvious reasons biocides must never be allowed in any part of the circuit in which the leaching microorganisms are employed.

Physical Characteristics Affecting Mineral Biooxidation

Crystal imperfections and grain boundaries. Microorganisms tend to adhere to locations on mineral surfaces where there are visible imperfections (Sand et al. 1999). These sites are often associated with crystal defects due to dislocations (grain boundaries) and inclusions. Lattice strain in a mineral crystal can be caused by the presence of impurity atoms, such as gold in pyrite and arsenopyrite, cobalt in pyrite, and indium in sphalerite, and these impurities increase the possibility of chemical/biological attack at that site. Thus, the more fractured and impure the sulfide mineral, the faster the rate of oxidation is likely to be.

For some sulfidic-refractory ores, the gold particles are often concentrated at pyrite and arsenopyrite grain boundaries. In these situations, not only is the oxidation rate enhanced, but the amount of sulfide that has to be oxidized to attain good gold recovery is diminished, because the majority of the gold is released with oxidation of the sulfide adjacent to the gold grain.

Particle size effects. Generally, the smaller the particle size, the more rapid the oxidation. It makes sense that the smaller the particle diameter and thickness, the greater is the overall surface area. However, small particles in a stirred tank system may increase the apparent viscosity of the slurry and decrease oxygen transfer rates. A particle size of 80% passing 53 to 75 μm is usually considered ideal for stirred tank reactors. In crushed ore heaps the size of the sulfide particles cannot be influenced by the ore crushing circuit. Therefore, in heap processes it is important that the crush size is sufficient to liberate the sulfide particle, thus allowing it to be oxidized.

Morphology of sulfide particles. Microorganisms and reagents (Fe^{3+}, H_2SO_4 and nutrients) must reach sulfide surfaces for oxidation to commence. This may not be possible if there are oxidized surface coatings or flocculant layers. To minimize surface coating effects, a regrind circuit is often employed for flotation concentrates and partially oxidized concentrates. For whole ore heaps the issue does not arise; the material must be oxidized, as is, and this is one reason why oxidation on heaps takes so much longer than in stirred tanks.

It should be noted here that during the oxidation of some minerals, surface coatings form which then leads to a rapid decrease in oxidation rate. Galena is an example; initially rapid oxidation quickly ceases, due the formation of an insoluble $PbSO_4$ layer.

Slurry solids concentration. In stirred tank processes, the solids slurry concentration affects the oxygen transfer rate from the gaseous to the aqueous phase. To avoid limitation in O_2 and CO_2 transfer rates, most stirred tank operations run at about 20% solids or less.

AERATED, STIRRED-TANK BIOLEACHING

The first aerated, stirred-tank bioleach pilot plant was established at the Fairview Mine in South Africa in 1984 to pretreat a sulfidic-refractory gold flotation concentrate. The bioleach technology, called BIOX®, was pioneered by GENCOR S.A. Ltd. After two years of successful operation, the commercial-scale plant, which treated 40% of the mine's production, was commissioned. In 1991 the Fairview plant was expanded to 40 tonnes per day of flotation concentrate (Dew et al. 1997). Although Fairview is no longer owned or operated by GENCOR and its successor companies, the bioleach plant continues operation today. The success of BIOX® spawned bioleach technologies developed by BacTech Environment Corporation and BRGM for treating sulfidic-refractory gold and base metal concentrates. Collectively the three processes have resulted in 11 tank-bioleach, commercial plants around the world (Table 1). All of the plants listed in Table 1 treat sulfidic-refractory gold concentrates except for Kasese, which oxidizes a cobalt-containing pyrite, and Pering, which is the first industrial-scale chalcopyrite concentrate bioleach plant.

The aim of this section is to examine the similarities and differences among the three tank, bioleach technologies, consider the design of a tank bioleach plant and present case histories. Emphasis is placed on tank biooxidation as a pretreatment technology for sulfidic-refractory precious metal concentrates, however, this section will conclude with a discussion on bioleaching of chalcopyrite concentrates in tanks.

Proprietary Tank-Bioleaching Technologies

Biooxidation of sulfidic-refractory precious metal and base metal concentrates in aerated tanks entails essentially the same process flowsheet, regardless of whose technology is used. A typical flowsheet is shown in Figure 1. The ore is crushed, milled and subject to flotation; the tails are dewatered and discharged or may be further treated. The concentrate feed is dewatered, and in some cases the concentrate is reground to improve metal recovery. The concentrate is fed to the biooxidation circuit, which is detailed in the bottom portion of Figure 1. The biooxidation circuit consists of up to four stages, each stage composed of one or more tanks in parallel. The first stage typically has the larger number (usually three) of equally sized tanks in parallel, because the retention time of the feed in the first stage is longer. The purposes of the longer holding time are to establish the microbial population and to allow attachment of the microorganisms to the mineral feed, preventing "washout" (loss) of the organisms from the circuit. More sulfide is oxidized -- up to 60% of the total amount requiring oxidation -- in the first stage than in subsequent stages. Blowers are used to supply air to each tank. Large amounts of heat are generated from the oxidation of sulfides, and this heat must be removed to maintain the temperature in the range for the types of microorganisms used in the biooxidation circuit. A cooling system is integral to the

Table 1 Commercial, aerated, stirred-tank, bioleach plants

Plant	Design (t concentrate/day)	Technology Used	Operating Years
Fairview, South Africa	Initially 10 Expanded to 35 (1991) Expanded to 40 (1994)	BIOX®	1986 - Present
Sao Bento, Brazil	Originally 150 Expanded in 1994 & 1997	BIOX® Expansion-Eldorado	1990 – Present (BIOX shut down - energy saving)
Harbour Lights	40	BIOX®	1992 - 1994 (Ore depleted)
Wiluna, Western Australia	115 Expanded to 158 in 1995-1996	BIOX®	1993 - Present
Sansu, Ghana	720 Expanded to 960 in 1995	BIOX®	1994 – Present
Youanmi, Western Australia	120	BacTech	1994 – 1998 (Ceased operation; high mining cost)
Tamboraque, Peru	60	BIOX®	1999 – Present
Kasese, Uganda	250	BRGM	1999 - Present
Beaconsfield, Tasmania, Australia	~70	BacTech	2000 - Present
Laizhou, Shandong Province, China	~100	BacTech	2001 – Present
Pering, South Africa	(300 m³ tanks)	BioCOP®	2001 - Present

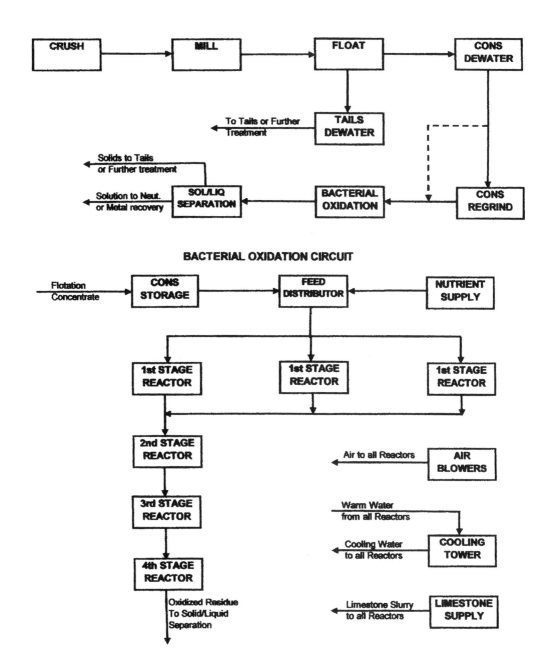

Figure 1 Flowsheet for biooxidation of concentrates in stirred tanks

biooxidation plant. Provisions are made to supply limestone for neutralization of the solution in the tanks to maintain the proper pH for the organisms. After the final stage of biooxidation, the contents of the reactor(s) are subject to solid/liquid separation. If the plant is used to oxidize sulfidic-refractory precious metal concentrates, the solids are neutralized with lime and subject to CIP/CIL for gold recovery. The solution from the solid/liquid separation is neutralized with limestone followed by lime and discharged to the tailings dam. However, if the biooxidation plant is used for recovery of base metals, the product of value is in the solution. Therefore, following solid/liquid separation of the contents of the final biooxidation stage, the solids are neutralized and discarded to tailings, and the solution is processed to recover the metal value, for example by solvent extraction to recovery copper.

BIOX® process. The BIOX® technology developed by GENCOR, first tested at Fairview and later applied in other plants listed in Table 1, utilizes a mixed culture of mesophilic bacteria and biooxidation plants are operated in the 40° to 45°C range. In the reorganization of GENCOR, the BIOX® technology relating to precious metals was transferred to Gold Fields Ltd and the base metal biotechnology shifted to Billiton Ltd. Using the BIOX® technology as a foundation, Billiton then developed similar biotechnologies for recovery of copper (BioCOP®), nickel (BioNIC®), and zinc (BioZINC®). Billiton subsequently made advances in the bioleaching of chalcopyrite using thermophilic microorganisms. The technology was piloted in cooperation with CODELCO; a full-scale test reactor was commissioned at the Pering Mine in South Africa to prove the viability of the process, and the technology is now being commercialized via a Billiton-CODELCO joint venture called Alliance Copper (Craven and Morales 2000). Billiton's interests in biotechnology are now merged with BHP (Billiton 2002).

BacTech process. BacTech (Australia) Pty, founded in the early 1980s, developed moderately thermophilic bioleaching technology researched at Kings College, London. BacTech's first commercial plant was Youanmi, a sulfidic-refractory gold concentrate bioleach plant in Western Australia (Table 1). This plant operated at about 50°C. In 1997 BacTech formed a partnership with Mintek (Johannesburg, South Africa) to pool bioleaching technology. Subsequent commercial plants at Beaconsfield (Tasmania) and Laizhou (Shandong Province, China) employ BacTech/Mintek mesophilic bacterial processes. BacTech Enviromet Corporation is a public company with headquarters in Toronto. BacTech/Mintek, too, are focusing on tank bioleaching of chalcopyrite at thermophilic temperatures. This effort is via a joint venture company, Procesos Biometalurgicos SA de CV (PMB), with Penoles SA de CV of Mexico (BacTech 2002).

BRGM process. In 1989 BRGM, France, initiated a study on bioleaching cobalt-containing pyrite tailings from the Kilembe Mine in Uganda (d'Hughes et al. 1997 and 1999). This study culminated in the operation of a 65 m^3 pilot reactor onsite in 1993. A full-scale plant design to bioleach some one million tonnes of pyritic tailings for cobalt recovery was completed and the plant was commissioned in 1999 for Kasese Cobalt Company, Uganda (Briggs and Millard 1997). The biotechnology entails leaching the pyrite with mesophilic bacteria. BRGM, also, is focusing on development of, tank bioleaching of chalcopyrite concentrates using a process the company calls HIOX® (d'Hughes et al. 2001).

Selection of the Tank Bioleach Process

Biooxidation is but one process option for the pretreatment of ores and concentrates in which the precious metal values are locked in a sulfide matrix; the other options are pressure oxidation and roasting. Fine grinding followed by cyanide leaching may be suitable for those materials in which the metal is in the metallic form or some other readily cyanide-soluble form. In some cases, simply making a concentrate and selling that concentrate is an option. All options must be considered when examining a process route for sulfidic-refractory materials. Tank biooxidation has certain advantages over the competing processes and is selected for these reasons:

- There are lower capital costs for small to medium sized plants,

- The process is flexible, easily controlled and can be managed for only partial oxidation or oxidation of specific minerals,
- No sophisticated equipment is required,
- The process can be operated using relatively unskilled labor,
- There is the potential for higher metal recoveries, and
- The process is operated at low temperature and atmospheric pressure; therefore, biooxidation is generally considered safer and a "greener" technology as there are no gaseous emissions.

The selection of tank biooxidation over a heap must take several questions into consideration. A few of the questions that must be asked are:

- Is the precious metal grade high enough to warrant the cost of upgrading the material to a concentrate? If so, tank biooxidation is an option.
- Is the ore amenable to concentration? If not, tank biooxidation may not be an option unless the ore grade is high.
- What is the mineable tonnage and life-of-mine? Is there sufficient value present to warrant the cost of making a concentrate and biooxidizing it in aerated, stirred-tanks?

Design of an Aerated, Stirred-Tank Biooxidation Circuit
Five fundamental criteria are used to design a tank biooxidation circuit: (1) oxygen requirements for air/O_2 addition, (2) heat balance for temperature control, (3) acid balance for pH control, (4) nutrient requirements, and (5) process control (Slabbert et al. 1992; Dew et al. 1993; van Aswegen 1993; Nicholson et al. 1994; Dew et al. 1996; Pinches et al. 2000; do Carmo et al. 2001). Oxygen requirements, heat balance and acid balance are closely related, because the oxidation of sulfide minerals consumes oxygen, generates heat, and either requires acid or produces acid (Dew et al. 1997a).

Testing of tank biooxidation usually entails several phases. The first phase involves small bench-scale, stirred (or shaken), batch reactors providing information on (1) amenability of the feed to biooxidation, (2) potential toxicity problems with the feed and site water, (3) ultimate recovery estimates of the precious or base metals, and (4) performance of the different suites of microorganisms. The next test phase is also conducted in the laboratory and involves several stages of continuous stirred-tank bioreactors to simulate the full-scale operation. This phase of testing provides information on the (1) rates of oxidation for various sulfide minerals present in the feed, (2) pH, acid and redox conditions under which the sulfides are oxidized, (3) acid balance data, and (4) the degree of sulfide oxidation achieved in each stage. If the continuous test was performed at a scale larger than "mini-pilot plant" scale, say 20 to 30 kg/day of feed, that would be the final testing before design, construction and commissioning of the full-scale plant. For sulfidic-refractory precious metal ores there is a sufficient knowledge base derived from existing operations that minimal laboratory and pilot-scale testing is needed for new operations. This is not the case for base metals; on-site piloting was done at Kasese to evaluate the bioleaching of the cobaltous pyrite feed. However, this is often required to generate sufficient sample for downstream process test-work for metal recovery. Large-scale demonstration plants have also been operated for extended time to evaluate the bioleaching of chalcopyrite concentrates.

Oxygen requirements. The requirement for O_2 by microorganisms and the reactions they catalyze were detailed earlier in this chapter. From a metallurgical point of view, it doesn't matter which reactions require the presence of microorganisms or occur chemically, because the quantity of oxygen required remains the same. Therefore, oxygen requirements are calculated from the oxidation reactions (Dew et al. 1997; Miller et al. 1999), the most common being:

Pyrite: $4\ FeS_2 + 15\ O_2 + 2\ H_2O \rightarrow 2\ Fe_2(SO_4)_3 + 2\ H_2SO_4$ (Reaction 4)

Arsenopyrite: $2\ FeAsS + 7O_2 + 2H_2O + H_2SO_4 \rightarrow 2\ H_3AsO_4 + Fe_2(SO_4)_3$ (Reaction 5)

Pyrrhotite: $4\ FeS + 9\ O_2 + 2\ H_2SO_4 \rightarrow 2\ Fe_2(SO_4)_3 + 2\ H_2O$ (Reaction 6)

Chalcopyrite: $4\ CuFeS_2 + 2\ H_2SO_4 + 17\ O_2 \rightarrow 4\ CuSO_4 + 2\ Fe_2(SO_4)_3 + 2\ H_2O$ (Reaction 7)

Chalcocite: $Cu_2S + 5/2\ O_2 + H_2SO_4 \rightarrow 2\ CuSO_4 + H_2O$ (Reaction 8)

Covellite: $CuS + 2O_2 \rightarrow CuSO_4$ (Reaction 9)

Bornite: $Cu_5FeS_4 + 5/2\ H_2SO_4 + 37/4\ O_2 \rightarrow 5CuSO_4 + \frac{1}{2}\ Fe_2(SO_4)_3 + 5/2\ H_2O$ (Reaction 10)

Pentlandite: $2Ni_9S_8 + 33\ O_2 + 2\ H_2SO_4 \rightarrow 18\ NiSO_4 + 2\ H_2O$ (Reaction 11)

Sphalerite: $ZnS + 2\ O_2 + \rightarrow ZnSO_4$ (Reaction 12)

Many other reactions and sub-reactions take place during oxidation of these species depending on which microorganisms are present. These side reactions can affect not only the oxygen requirements, but acid and heat balances as well. To avoid confusing side reactions, the stoichiometric oxygen requirements for plant design of a sulfidic-refractory gold bioleach plant are based on data presented in Table 2 (Dew et al. 1997).

Table 2 Process data for oxidation of sulfide minerals associated with sulfidic refractory gold plants (Dew et al. 1997)

Mineral	Heat of Reaction		Oxygen Requirement		H_2SO_4 Demand (kg/kg mineral)	Pyritic S Content (% by weight)
	Mineral (kJ/kg)	Sulfide (kJ/kg sulfide)	mole O_2/mole mineral	kg O_2/kg sulfide		
Pyrrhotite FeS	-11,373	-31,245	2.25	2.25	0.557	36.4
Arsenopyrite FeAsS	-9,415	-48,036	3.5	3.5	0.301	19.7
Pyrite FeS_2	-12,884	-24,173	3.75	1.88	-0.408	53.3

Note that the O_2 requirements for the plant are typically not determined from the lab-scale and pilot plant test work, but rather from theoretical calculations. The test work does provide information on what reactions, sub-reactions and secondary reactions are taking place. For calculating process air demand and pressure requirements for stirred tanks, these factors are considered:

- O_2 utilization efficiency for stirred tanks   30-40% (depending on agitator type)
- O_2 content of dry air   23.15% by weight
- Air SG (specific gravity)   1.293 kg/Nm^3/h
- Normal (N) temperature   0°C
- Normal (N) pressure   1 atmosphere

- Pressure requirement Hydrostatic head on sparge ring below impeller + 15 kPa line loss

Blower power for aerated, stirred tanks is calculated using the Perry and Chilton (1979) formula, adapted to metric units:

$$kW_{(AD)} = 1.0195 \, WT[(P_2/P_1)^{0.283} - 1]$$

where, $k = 1.395$
W = air flow in kg/s
T = inlet air temp in K
P_1 = inlet pressure
P_2 = delivery pressure
(P_1 & P_2 are in the same set of units and are absolute pressure <u>not</u> gauge pressure)

Blower efficiency is taken to be 74 to 78 %. The agitator in stirred tank reactors is mainly for air dispersion, and a first pass estimate of the power required is 27 W/Nm³/h air added. However, this depends on the agitator design.

Heat balance: In stirred-tank biooxidation heat is gained from reactions involving the oxidation of sulfide minerals (Table 2), from agitation energy, and from air that is added. Heat losses comprise (1) evaporation as the air, saturated with moisture, leaves the reactors, (2) convection through tank walls, (3) heating the incoming slurry, and (4) air expansion (Joule/Thompson effect). Heat balance is determined theoretically, because the lab-scale and pilot-scale tests are so small and the reactors provide such a large surface area to volume ratio that heat generation is not detected. Data from tests showing those sulfide minerals that are leached and the rates of leaching are used in the heat balance determination.

Cooling water requirements for stirred tank reactors are derived from heat balance equations and calculated accordingly:

$$\text{Flow rate (m}^3\text{/h)} = \frac{\text{Net Heat Load (kW)} \times 3.6}{\Delta T \times 4.186}$$

ΔT = Temperature difference (°C) between inlet and outlet water temperature

Note that cooling water leaving the reactor will be about 4°C cooler than the reactor temperature.

Acid balance: Some oxidation reactions generate acid (Table 2) and increase the sulfate concentration in solution. There are also secondary reactions, which are dependent on temperature, pH, ionic strength of the solution and other species in the system; these secondary reactions can also affect the acid balance. Several of these secondary reactions are (Dew et al. 1997):

Ferric arsenate precipitation (generates acid):
$2 \, H_3AsO_4 + Fe_2(SO_4)_3 \rightarrow 2 \, FeAsO_4\downarrow + 3 \, H_2SO_4$ (Reaction 13)

Acid dissolution of carbonates (consumes acid):
$CaMg(CO_3)_2 + 2 \, H_2SO_4 \rightarrow CaSO_4\downarrow + MgSO_4\downarrow + 2 \, CO_2\uparrow + 2 \, H_2O$ (Reaction 14)

Precipitation of jarosite (generates acid):
$3 \, Fe_2(SO_4)_3 + 12 \, H_2O + M_2SO_4 \rightarrow 2 \, MFe_3(SO_4)_2(OH)_6\downarrow + 6 \, H_2SO_4$ (Reaction 15)
where $M = K^+, Na^+, NH_4^+$ or H_3O^+

Lab and pilot plant tests do provide information on the acid balance. These data are coupled with theoretical calculations to arrive at the acid balance.

In some circuits the solution pH is controlled during biooxidation by the addition of lime, $Ca(OH)_2$, or limestone, $CaCO_3$. The addition of limestone to biooxidation reactors for pH control also adds CO_2, required by the microorganisms. Reactor waste products are neutralized with limestone and lime before final disposal.

Nutrient additions: Microorganisms involved in biooxidation require nitrogen, potassium and phosphorous, and these nutrients must be added to stirred-tank biooxidation circuits. These are often supplied in the form of $(NH_4)_2SO_4$, KOH and H_3PO_4. Solid $(NH_4)_2SO_4$ and KOH are mixed with H_3PO_4 and water, and these nutrients are pumped to the feed distributor and into the primary reactors with the sulfide feed concentrate. Because of the difficulty in handling these reagents, their price and their availability, different chemicals, including agricultural grade nutrients, are considered as alternatives. Whichever chemicals are used, it is imperative to test them in the laboratory to ensure that they provide a high level of microbial activity and that they do not in any way inhibit the microorganisms. Sometimes there are impurities in the reagents that cause problems.

The amounts of NH_4^+, K^+ and PO_4^{3-} added differ somewhat depending on which company's technology is being applied (Table 1), but is by and large based on the chemical composition of the microorganisms, $CH_{1.67}N_{0.2}P_{0.014}O_{0.27}$, and the number of cells per ml that develop in the solution of the reactor (Pinches et al. 1994). This number is in the 10^9 to 10^{10} per ml range. The nutrient concentration is then calculated and reported as kilograms of N, P and K per tonne of FeS_2 oxidized or tonne of concentrate fed to the circuit. Table 3 lists examples of nutrient concentrations that have been used and also lists the stoichiometric requirements.

Table 3 Example nutrient concentrations for tank bioleach operations

Company	CO_2 (kg/t FeS_2 oxid.)	N (kg/t FeS_2 oxid.)	P (kg/t FeS_2 oxid.)	K (kg/t FeS_2 oxid.)
Mintek (Pinches et al. 1994)	50	3.177	0.447	
GENCOR		3.4	0.6	1.8
BRGM (Kasese)		9.3	2.6	3.7
Stoichiometric		0.22	0.32	

The disparity observed in nutrient concentrations (Table 3) likely arises because (1) nutrient limitation adversely affects production, so excess nutrients are added to ensure that no limitation occurs, and (2) the exact number of microorganisms in the biooxidation circuit is not known. The microorganisms in solution are relatively easy to count, but the number of microorganisms adhering to the mineral surfaces is not so easily determined. Therefore, to guarantee that sufficient nutrients are available, excess is often added. In stirred-tank biooxidation, adding excess nutrients is not a serious problem from an operating standpoint, but it does increase the operating cost. Excess nutrients added to the system may simply precipitate as $FePO_4$ and ammonium and potassium jarosites and may be wasted.

CO_2 is supplied by limestone added to the biooxidation circuit for pH control and occasionally by providing gaseous CO_2. Sometimes sufficient CO_2 is provided by the feed, which may contain some carbonate minerals that produce CO_2 during acid dissolution.

Circuit conditions and process controls. The typical circuit conditions for the pretreatment of a sulfidic-refractory precious metal concentrate in an aerated, stirred-tank biooxidation circuit are listed in Table 4.

Table 4 Conditions for sulfidic-refractory precious metal concentrate tank bioleach

Element	Control
Slurry density	15 – 20% solids
Number of reactors	3 primary in parallel 3 secondary in series (some exceptions)
Residence time of slurry in circuit	4 to 6 days (2 to 2.5 days in 1st stage)
Temperature	40 - 45°C for mesophiles 45 - 50°C for moderate thermophiles
pH	Approximately 1.2 to 1.6
Dissolved O_2	2 ppm

Flotation concentrates are generally dewatered after flotation for density control and removal of flotation reagents that can adversely affect biooxidation rates. Some flotation reagents are toxic to the organisms and may also interfere with the attachment of the organisms to the mineral surface (Huerta et al. 1995).

At plant sites where process water quality is a problem, separation of water circuits is necessary. "Medium" quality water is used for flotation and other mineral processing steps. The highest quality water is used for the bioleach circuit or, if limited, is used to dilute saline or brackish waters. When water quality is a concern, good lab testing is essential to determine what site water is acceptable. Tailings return water from gold recovery circuits must not be used as make-up water in the biooxidation circuit as cyanide species are toxic to the organisms.

The residence time of the slurry in the first stage reactors is based on the time required for the microbial population to stabilize. Microorganisms multiply by dividing and in continuous reactors, when conditions are optimum, the growth rate (also called doubling time) is exponential. Different microorganisms have different doubling times. For example, *Acidithiobacillus ferrooxidans* may double every 6 to 12 hours, depending on conditions. Therefore, the residence time in the first stage reactor should be sufficiently long to ensure that the microbial growth rate is greater than the number of microbes leaving the first stage with the slurry flow. From this discussion it is clear why a condition in the first stage reactor that slows the growth rate of the microorganisms, such as insufficient O_2, a nutrient limitation or a toxin, results in "wash-out" of the microbial population; when microbial growth rate is slowed, the passage of slurry through the circuit is faster than the doubling time of the microbes, and the organisms simply wash out of the circuit. Carrying this concept one step further, it is also possible to cause washout of a single type of microorganism. For example, if the pH should decline in the first stage reactors, that population of microorganisms that is inhibited by low pH will washout, leaving other microbes more tolerant to low pH. The remaining microbial population may not oxidize Fe^{2+} in which case production will decline precipitously.

In biooxidation circuits recycle of spillage and wash-down solutions is to be avoided. These solutions may contain constituents that are inhibitory to the microorganisms in the biooxidation circuit and result in washout of the microbial population. Excess spillage could lead to washout simply by increasing the volumetric flow-rate through the circuit, thereby reducing slurry residence times. Because oils and greases can also be inhibitory, agitator gearboxes should be equipped with oil spillage trays and spills or drainage from the gearboxes should be directed outside the bounded area of the biooxidation circuit.

The control system employed at biooxidation plants is PLC based. Operator communication with the PLC is maintained by PCs equipped with software allowing continuous data acquisition and control instrumentation scanning. Control loops from the biooxidation plant include temperature and airflow to each reactor, concentrate, feed and dilution control, nutrient proportional additions, and on the downstream processing, CCD discharge rates, neutralization pH and limestone addition rates. Much data required for indirectly monitoring microbial activity are obtained by routine sampling and analysis for dissolved O_2, redox potential, total iron, Fe^{2+}:Fe^{3+} ratios, arsenic in solution, total dissolved solids, and other metals of importance in the feed.

Equipment for Aerated, Stirred-Tank Biooxidation Circuits

Materials used in the construction of biooxidation circuits for sulfidic-refractory precious metals and base metals must be able to withstand highly oxidizing and acidic slurries at temperatures of 50°C or higher. The selection criteria for materials of construction are:

- Suitability of the material for both the mine site environment and the microbial environment,
- Chloride levels in the slurry,
- Non-toxicity of the materials to the microorganisms,
- Location of the source of the material,
- Delivery time, including shipping,
- Cost of the materials,
- Availability of skilled labor to fabricate and install the selected material, and
- Ability to repair fabricated equipment and the cost of the repair.

Tanks and surrounding area. Most biooxidation tanks used in operations up to and slightly exceeding 50°C are constructed of stainless steel (SAF 2205). At the Wiluna plant Linatrite N50 rubber (Linatex) was selected to line the leach tanks and agitators. Due to commissioning difficulties, cost and maintenance issues, this liner has not been used in subsequent operations.

Owing to the corrosive nature of the bioleach solutions, areas surrounding the circuit must be free of galvanized materials including cable trays, pipe, pipefittings and light fittings. Any contact of the acidic slurry containing soluble arsenic with reducing metals, such as brass, copper, zinc, and aluminum, will generate arsine gas, which is considerably more toxic than hydrogen cyanide.

Cooling circuit. In designing the cooling water circuit for the biooxidation plant, criteria that must be considered are:

- The heat load at design capacity,
- The wide variation in heat load due to treatment of transitional ores and high-sulfide ores,
- The airflow and water flow through the cooling towers, and
- Environmental factors, including dust load, insect control, scale and microbiological control.

Some biooxidation plants are constructed using stainless steel cooling coils installed inside of the reactor tanks. Proper anti-scalents must be selected to ensure non-toxicity to the leaching microorganisms in the event of a leak in the cooling system. The design of the coils must allow for easy removal and replacement in the event of circuit failure. A water jacket around the bottom half of the bioreactor tanks is an option, if the amount of heat generated is limited. A water jacket design greatly simplifies the piping.

Air supply. The air supply is a critical component of the biooxidation circuit. Because of the wide variation of sulfide contents experienced in the life of an operation, the system must have a large turndown ratio. At existing operations the air is injected into the high shear zone below the agitator impeller via an air sparge ring. The blowers are usually high speed, turbine types with variable inlet and outlet vanes, which give a turndown capability to 40% of rated maximum capacity of each blower. On this basis air output can be varied over a significant range with the installation of multiple units. Each blower in the circuit is fitted with an aftercooler to reduce air temperature to 50°C to protect the microorganisms from high temperature zones in the reactors. The materials of construction of the aftercoolers are important, because acidic slurries can be siphoned into the system under some conditions.

The reliability of the air is of paramount importance as loss of air for a period of more than 1.5 hours can have serious consequences for the circuit. The loss of air at Wiluna (Table 1) for about 50 minutes in late 1993 resulted in the loss of microbial activity in the primary reactors. It required some seven to eight days to restore microbial activity and 10 days before production levels returned to normal. To avoid this type of problem, at least one standby blower is required which should be operated through a separate power supply and control circuit.

Agitators. The agitator design must allow for:

- Provision of sufficient power to prevent "flooding" of the impeller by the air flow,
- An oxygen mass transfer rate that exceeds that required for the oxidation reactions at the reactor temperature used, and
- An agitator pumping rate that will maintain uniform solids suspension at high solid's specific gravity in dilute slurries.

More details of these design criteria are presented in a paper by Batty and Post (1999).

A comparison of bioleach tank sizes and installed power in the early plants is presented in Table 5. The considerable power addition to the reactors means that all internal structures must be properly secured.

Table 5 Installed power and air addition at several biooxidation plants

Plant	Tank Volume (kW/m^3) (primary/secondary)	Air Addition (kW/1,000 m^3/h) (primary/secondary)
Sao Bento	0.58	31.7
Wiluna	0.39/0.19	27.2/30.7
Youanmi	0.15/0.11	27.7/50.0
Typical CIL tank	0.03 – 0.05	--

Sulfidic-Refractory Gold Biooxidation Plant Case Studies

Two operations are briefly described to illustrate the process design of sulfidic-refractory gold biooxidation plants. Wiluna and Youanmi are selected (Table 1), because they have similar concentrate throughputs, and hence similar sized oxidation reactors. However, their sulfide oxidation levels are quite different and hence air addition and heat removal requirements are therefore different. The oxidation circuits as described for these plants apply to all the concentrate oxidation circuits for gold recovery thus far constructed, and also to Kasese where pyrite is oxidized for cobalt recovery. Only after oxidation do base metal recovery circuits differ from gold circuits. The BioCOP® process, developed by Billiton for copper recovery from chalcopyrite, is also similar in the oxidation circuit, though operating temperatures and hence materials of construction differ significantly.

Wiluna Sulphides, Western Australia. When built in 1993, Wiluna comprised the crushing, grinding and flotation of 400,000 tpa of ore and the oxidation of 115 tpd (42,000 tpa) of concentrate. Oxidation residues and flotation tailings were recombined for gold leaching. The biooxidation design criteria for the BIOX® process were provided by Gold Fields (formerly GENCOR) (Brown et al. 1994; van Aswegen and Marais 1999).

For the original design at Wiluna, the ore grades were 6.0 g/t Au, 1.85% S and 0.7% As, and the concentrate grades were 92.9 g/t Au, 24.0% S and 10.0% As. The milling circuit was standard and will not be described. Flotation at Wiluna is comprised of a rougher scavenger circuit with cleaning and re-cleaning of concentrates. For biooxidation, concentrates are not required to contain a minimum sulfur grade, as needed for auto-thermal roasting, and thus flotation can be adjusted for maximum gold recovery into the concentrates. However, it is necessary to control the ratio of acid consuming gangue components to sulfide grade to ensure that a positive acid balance is maintained in the oxidation circuit. This justifies the cleaner and re-cleaner circuit.

The plant provides a five-day residence time in biooxidation at a slurry density of 20% solids, allowing for sulfide oxidation in excess of 90%. The plant operates at 40-42° C, and the waste heat is removed by cooling water through cooling coils. The pH in the reactors remains in the range of 1.2 to 1.6. The oxidation reactors are fabricated from rubber-lined mild steel.

The quantity of air supplied is based on the sulfide oxidation required; an estimated utilization efficiency of the oxygen added is 25%. This leads to a total addition rate of 31,300 Nm^3/h of air (40.5 tph) or about 8 tonnes of air per tonne of concentrate.

The overall heat to be removed from the circuit is 11.8 MW. This figure includes the heats of reaction, heat generation from the agitators and air blowers, and takes into account heat losses through evaporation and adsorption by the incoming slurry. Thus the net heat generation is about 2.4 MWh per tonne of concentrate or 10.9 MWh per tonne of S^{2-}-sulfur oxidized. Waste heat is removed by a cooling water circuit with cold water circulating through cooling coils in the reactors; shell and tube heat exchangers cool the blower air. Warm water is circulated to a cooling tower and returned to the circuit. The cooling water flow is 936 m^3/h (design 1060 m^3/h), which is equivalent to 0.87 m^3/h per kg of S^{2-}-sulfur oxidized. Note, however, that the size of the cooling water circuit is very dependant on the cold water temperature achieved. At Wiluna warm water (35.6° C) is cooled to just 26° C. If this temperature could be decreased to say 23° C, the water flow could be reduced to 713 m^3/h or 0.66 m^3/h per kg S^{2-}-sulfur oxidized.

Power consumption is high in biooxidation circuits. The power is mainly used for air generation in the blowers and air dispersion in the reactors. At Wiluna, the agitator mechanisms were installed with 185 kW drives on the primary tanks and 90 kW drives on the secondary tanks. Each tank is 470 m^3 in volume. The difference in agitator power between the primary and secondary tanks is due entirely to the difference in oxygen requirements in the different stages of oxidation. Three operating blowers each with a 400 kW drive provide the air supply.

Youanmi Deeps, Western Australia. The Youanmi Deeps project is similar in size to Wiluna Sulphides, but utilized BacTech technology (Miller 1997) rather than BIOX®. Youanmi was commissioned in 1994, but ceased operation in 1998 due to high underground mining costs. The plant was designed to treat 200,000 tpa of ore, but since the sulfide-sulfur content of the ore was over twice as high as that at Wiluna, the concentrate throughput of 120 tpd was similar to Wiluna. The ore content is 15g/t Au, 7.5% S and 1.0% As. The concentrate content was 60 g/t Au, 28.0% S and 4.3% As. The Youanmi flotation circuit was comprised of a roughing and scavenging stage only; no cleaning of concentrates was required for elimination of carbonates or other acid consumers.

The plant provided a five-day residence time at a slurry density of 15-20% solids; this required reactor sizes of 500 m^3, just slightly larger than those at Wiluna. The size of these reactors is set by the throughput, slurry density, and residence time, and is unaffected by the quantity of sulfide-sulfur to be oxidized. The reactors were fabricated from SAF 2205, a stainless steel resistant to the higher levels of chloride in the waters on site. The oxidation circuit operated at 45^0 to 50°C and pH 1.2.

Youanmi required only 32% sulfide oxidation for high gold recovery efficiencies, because the gold was associated with the arsenopyrite, which represented only a third of the total sulfide-sulfur in the feed. This 32% sulfide oxidation was only about a third of that required at Wiluna, therefore, air requirements and waste heat generation at Youanmi were also a third of those generated at Wiluna. Air requirements at Youanmi were 11,400 Nm³/h; Wiluna was 31,300 Nm³/h. Because of the lower air demand, the installed power of the blower was 540 kW in two units. Wiluna's installed power was 1200 kW in three units. Similarly, the design waste heat removal from the Youanmi circuit was 4.5 MW, whereas, Wiluna's waste heat was 11.8MW. Owing to the lower waste heat removal requirements at Youanmi, the cooling circuit could be simplified, and cooling of the reactors was accomplished by a water curtain on the outside of the tanks rather than cooling coils.

The agitators at Youanmi operated in similar sized tanks to Wiluna with similar slurry densities, but because they received about a third of the airflow, the installed power on the drives at Youanmi was only 75 kW in the primary reactors and 55 kW in the subsequent reactors. This is compared with 185/90 kW (primary/subsequent reactors) at Wiluna.

The Youanmi and Wiluna circuits both required limestone addition for neutralization of the oxidation products. Wiluna also required limestone addition for pH control in the oxidation circuit. Youanmi required acid for pH control, because the sulfide that was oxidized was predominantly arsenopyrite, whose oxidation is acid consuming. A limestone milling circuit at Youanmi was included in the design with limestone addition to the oxidation products by ringmain.

Figure 2 Youanmi Deeps biooxidation project, Western Australia (photo courtesy of BacTech Environment Corporation)

Other stirred-tank biooxidation projects. Table 6 compares eight stirred-tank biooxidation plants based on their feed and percent sulfide-sulfur oxidized. Provided in this table are the design parameters (oxygen demand, reactor design and volume, and feed residence time) used and the heat generated, total power installed, specific power consumed and sulfide oxidation rate for each plant. The installed power noted in this table includes biooxidation, cooling water and air circuits only; standby units are not included.

Sao Bento operates in series with pressure oxidation. The feed rate is determined by overall process economics and is variable. Table 6 illustrates the original design, which had one reactor. Of interest in Table 6 is the relatively constant specific energy demand in terms of kWh/kg of sulfide-sulfur oxidized. This is a very useful number for a rapid economic appraisal of project economics. The figure for Youanmi is high because of the over-sizing of the second stage reactor agitators. See the comparison with Wiluna.

Table 6 Design comparison of aerated, stirred-tank biooxidation plants

Item	Unit	Fairview	Sao Bento	Harbour Lights	Wiluna	Sansu	Youanmi	Tamboraque	Kasese
Concentrate feed	t/day	35	150	40	158	960	120	60	240
Sulfide grade	%	20	18.7	18.6	22	11.4	28	30	40.9
Sulfide feed	kg/h	292	1,169	310	1,448	4,560	1,400	750	4,172
Sulfide oxidized	%	89		87	95	94	32	86	92.3
	kg/h	260	348	270	1,376	4,286	448	645	3,851
Specific heat of reaction	MJ/kgS^{2-}	26.7	29.2	29.2	29.8	33.8	35.3	70.2	24.2
Specific oxygen demand	kgO$_2$/kgS^{2-}	2.05	2.17	2.22	2.27	2.49	2.40	2.57	1.88
Total reactor volume (aerated)	m^3	913	1 374	880	3,391	18,278	3,000	1,336	5,520
Primary reactor volume	m^3	1 x 343 4 x 97	2 x 550 1 x 427	3 x 163	4 x 471	12 x 896	3 x 500	3 x 262	3 x 1 380
Number of reactors (operating)		9	3	6	8	24	6	6	4
Residence time	Days	6.1	1.0	5.1	5.0	4.4	5.0	5.3	4.0
Heat of reaction	MW	2.0	2.8	2.2	11.1	41.1	4.0	6.1	25.8
Oxygen demand	kg/h	540	755	612	3,024	10,872	987	1,656	7,220
	kg/h/m^3	0.59	1.30	0.70	0.89	0.59	0.33	1.24	1.31
Total installed power	kW	798	758	591	1,797	7,323	950		5,727
Average sulfide oxidation rate	kg/m^3/day	6.8	14.4	7.4	9.7	5.6	3.3	11.6	16.7
Specific power consumption	kWh/kg S^{2-}	1.9	1.8	1.9	1.5	1.9	2.3		1.5

Comparative operating and capital costs for the biooxidation leach circuit, counter-current decantation for washing the biooxidized residue, and the neutralization circuit of selected plants were report earlier (Brierley and Briggs 1997). These costs were related to total sulfide-sulfur in the feed and the sulfide-sulfur oxidized to achieve >90% gold recovery. The cooling circuit adds considerable capital cost, because of the need for stainless steel cooling coils, water supply pipework and cooling tower.

Stirred-Tank Bioleaching of Chalcopyrite Concentrates
BHP-Billiton (Craven and Morales 2000), BacTech/Mintek (Miller et al. 1999) and BRGM (d'Hughes et al. 2001) are all developing proprietary technologies for stirred-tank bioleaching of chalcopyrite concentrates. Chalcopyrite does not leach well using the mesophilic bacteria, as surface coatings develop on the chalcopyrite mineral phase slowing the leach rate and thwarting copper recovery. In the late 1990s reports surfaced in the technical literature that bioleaching of chalcopyrite was enhanced with mesophilic bacteria, if ferrous iron was added (Hiroyoshi et al. 1997; Hiroyoshi et al 2000; Third et al. 2000). This revelation led to studies and speculation as to why chalcopyrite leaching is enhanced by ferrous iron addition and controlling redox potential in the narrow Eh range of 615 to 645 mV (Standard Hydrogen Electrode) (Timmins and Hackl 1998; Breed et al. 2000). Increasing the temperature and controlling the redox potential in this narrow window further enhances chalcopyrite leaching. A patent (Pinches et al. 2001) was issued and assigned to Mintek, which reveals a process for controlling the redox potential to leach chalcopyrite in both tanks and heaps; the patent includes the concept of using microorganisms to control surface potential. The thermophilic microorganisms have been shown to be more effective in leaching chalcopyrite (Dew et al. 1997a; Gericke and Pinches 1999). Although the rationale for their effectiveness is not fully evident, it is likely that the thermophiles, particularly the *Archaea*, play some role in controlling redox potential at the chalcopyrite/organism interface.

These technical developments in thermophilic chalcopyrite leaching have been transformed into commercial processes. BHP-Billiton disclosed their BioCOP® process for bioleaching of chalcopyrite concentrates with the extremely thermophilic *Archaea* (Craven and Morales 2000) and BacTech/Mintek announced their chalcopyrite concentrate leaching technology using moderate thermophiles (van Staden et al. 2000).

The engineering principles of aerated, stirred-tank bioleaching of chalcopyrite concentrates at elevated temperatures (60° to 90°C) are fundamentally the same as described for pre-treating sulfidic-refractory gold concentrates at 40° to 50°C. Operating the plant at high temperatures is advantageous because of faster leach kinetics, however, there are several important considerations that must be taken into account – the effect of high temperature on oxygen mass transfer and materials of construction (Batty and Post 1999; Harvey et al. 1999) and the effect of enhanced evaporation on the water balance. BHP-Billiton pilot tested a new elevated temperature reactor design for chalcopyrite concentrate bioleaching at CODELCO's Chuquicamata Mine in Chile and have since commissioned a commercial-scale (300 m^3 reactors) plant at the Pering Mine in South Africa. Several patent applications have been filed on the technology (Tunley 1999) and engineering aspects of this technology (Dew et al. 2001; Norton et al. 2001; Basson et al. 2001; Dew et al 2001a; Dew and Miller 2001). Few details are available on the reactor design at this time, although in contrast to existing bioleach plants, the elevated temperature plant is using O_2 plant gas as opposed to air to aerate the slurry. Alliance Copper, the BHP-Billiton/CODELCO joint venture, is marketing this process.

BacTech/Mintek in conjunction with Industrias Penoles of Mexico are operating a fully integrated copper bioleach, solvent extraction and electrowinning demonstration plant in Monterrey, Mexico using moderately thermophilic microorganisms. This technology, which is the subject of several patent applications (Rhodes and Miller 2000 & 2000a; Winby et al. 2000) is being commercialized through a BacTech-Mintek-Penoles joint venture, Procesos Biometalurgicos SA de CV (PBM). No details are yet available on the engineering design of the PBM reactor.

BIOOXIDATION OF ORE AND CONCENTRATES IN HEAPS

Heap leaching of oxide copper ores with recovery of the copper by solvent extraction and electrowinning is conventional technology. The fundamentals of copper oxide heap leaching have been integrated with the principles of bioleaching to heap leach secondary copper ores. Copper heap leaching is the subject of another chapter in this book, "Copper Heap Leach Design and Practice" by R.E. Scheffel, and the reader is referred to that chapter for details. The emphasis of this section will be on the heap leaching of sulfidic-refractory gold ores with discussion on the emerging technologies of heap leaching chalcopyrite ores and sulfide concentrates. The premise of bioheap leaching is that the heap is the reactor. This means that conditions within the heap must be optimized for full participation by a suite of microorganisms that catalyze the oxidation of the sulfide minerals.

Heap Leaching Sulfidic-Refractory Precious Metal Ores

The heap leaching of sulfidic-refractory gold ores is similar in many ways to that of secondary copper ores. However, there are some notable differences and this section will only focus on those differences. For more details about heap leaching of sulfides, the reader is referred to R.E. Scheffel's chapter in this book and an earlier paper by Brierley and Brierley (1999). Sulfidic-refractory precious metal whole ore heap technology was perfected (Brierley, 1997; Brierley, 2000) and patented (Brierley and Hill, 1993, 1994 & 1998) by Newmont Mining Corporation, and is now commercially applied (Bhakta and Arthur, 2001) at the company's Nevada (USA) operations.

Microbiology and chemistry of the process. The fundamentals of the technology are same as for sulfidic-refractory gold concentrates. Microorganisms catalyze the oxidation of pyrite (FeS_2) and arsenopyrite (FeAsS), exposing gold that is locked, or occluded, within these sulfide minerals. The degradation of the sulfide minerals significantly improves the gold recovery over that of cyanide treatment alone. The principal chemical and microbial reactions are the oxidation of pyrite and arsenopyrite with microbiologically generated (Reaction 1) ferric iron

$$FeS_2 + 14\ Fe^{3+} + 8\ H_2O \rightarrow 15\ Fe^{2+} + 2\ SO_4^{2-} + 16\ H^+ \quad \text{(Reaction 16)}$$

$$FeAsS + Fe^{3+} + 2\ H_2O + 3\ O_2 \rightarrow 4\ H^+ + AsO_4^{3-} + 2\ Fe^{2+} + SO_4^{2-} \quad \text{(Reaction 17)}$$

and the microbial re-oxidation of ferrous iron (Reaction 1) to perpetuate the leaching process. Any reduced sulfur species, including elemental sulfur, that accumulates is microbially oxidized (Reaction 2).

Crushing and stacking the ore. Laboratory column studies and on-site crib and pilot test heaps are used to confirm the crush size that will provide optimum precious metal recoveries. To ensure good solution and air permeability in the heap, consideration must be given to fines generation and particle size. Newmont's process entails inoculating the crushed ore with mesophilic, moderately thermophilic or extremely thermophilic microorganisms (Brierley and Hill 1993, 1994 & 1998; Brierley 2001) employing any agglomerating method before the ore is stacked. The microbial inoculum can be prepared in tanks, ponds or can be the effluent from the heap. Applying the microbe-containing solution inoculates and acidifies the crushed ore and binds fine material to coarse ore particles. Newmont contends that agglomerating the ore with microorganisms reduces the overall residence time of the feed on the pad, because the microorganisms are distributed throughout the heap greatly minimizing the time for the organisms to reach their maximum numbers and performance.

Irrigation and aeration. The "on/off" or dynamic pad design, discussed in the R.E. Scheffel's chapter on copper heap leaching, is used for sulfidic-refractory gold heap leaching. Pad liners are constructed of a properly compacted low-permeability natural barrier covered by a HDPE liner. A 200 to 1000 mm thick gravel layer is employed to protect the liner and to allow installation of drain and air pipes below the heap.

Drainaflex pipes are installed at 2-m centers with air injection pipes immediately on top of the drainaflex pipes outside the phreatic zone to prevent flooding. Solution collection ditches are employed on both sides of the cells to direct solution by gravity to the correct pond. The agglomerated ore is stacked, aerated and irrigated.

Air is injected into the heap using a set of low-pressure high volume fans or blowers (Salomon-de-Friedberg 1998). The design normally allows for a multitude of "portable" fans to be used, so that new cells on the heap with fast oxidation rates can be provided with more air than older areas where the oxidation is almost complete. A small mobile crane moves fans. Holes (3 mm) are drilled in the bottom of 50 mm diameter air distribution pipes. The density of the holes is dependent on the amount of sulfide-sulfur to be oxidized and the oxidation rate. The greater the amount of sulfide-sulfur and the faster the oxidation rate, the greater is the density of holes. Air distribution networks typically include 500 mm diameter headers and 50 mm diameter laterals at 2-m spacing.

Solution irrigation rates vary between 2.5 and 10 l/h/m2. Rest periods are used to control the heap and solution temperatures. Solution management is a key aspect in successful heap leaching (Schlitt 1984). At a low irrigation rate, liquid percolates downward as a thin film on the rock surfaces while air moves up through the voids in a countercurrent fashion. This promotes good oxygen transfer at the film-air interface. As the irrigation rate increases, flooding occurs at pinch points between voids in the rock. This changes the airflow pattern significantly, and airflow will short-circuit through the leach material via a few large channels. The heap does not have to be completely flooded to be poorly aerated (Schlitt 2002). Modeling of the hydrodynamics of the heap leach process is an emerging discipline (Bouffard and Dixon 2001).

Heat balance. Heat gains in heaps include sulfide-sulfur oxidation and daytime solar radiation. Heat losses include:

- Evaporation – air addition leaves heap saturated in moisture
- Evaporation of irrigation solutions
- Convection from heap sides and surface
- Radiation at night
- Heat up of irrigation solutions

Temperature is controlled in the heap by:

- Heap depth – the higher the heap, the greater the heating
- Irrigation – cools the heap
- Rest periods – warms the heap
- Excess air addition may also cool the heap through evaporation

Heat builds up in the heaps due to rapid sulfide-sulfur oxidation. The temperature exceeds the uppermost limit of activity for the mesophilic, *Acidithiobacillus* and *Leptospirillum* bacteria. At about 40°C the moderately, thermophilic bacteria, such as *Acidithiobacillus caldus* and *Sulfobacillus* species, increase in numbers and perform the oxidation of iron and sulfur. At about 65°C the extremely thermophilic *Archaea* microorganisms, such as *Sulfolobus* and *Acidianus* species predominate. Sulfidic-refractory gold ore heaps heat to 70°C or higher (Bhakta and Arthur, 2001). The time required for oxidizing the sulfides in the heap varies from a minimum of about 90 days up to 250 days. The oxidation time is ore dependent (Bennett and Ritchie 2002).

Iron chemistry. As solutions are recycled, iron concentrations increase in the solution until the solubility of various iron compounds is exceeded. Iron precipitation is abundant in sulfidic-refractory precious metals. These precipitates are various jarosite compounds, including silver-jarosite, and ferric arsenate when arsenopyrite is leached. Because of iron precipitating in the heap, solution iron concentrations are not a reliable way to assess the degree of oxidation that has occurred. Sampling of the oxidized solids in the heap followed by bottle-roll cyanidation and analyses of sulfide-sulfur and total sulfur are needed for confirmation.

Nutrient addition in heaps. In heap leach operations nutrients are usually not required as sufficient amounts for the microbial population are available from the ore and ammonium nitrate blasting agents. Occasional analysis of the leach solutions should be practiced to ensure that ammonium ion, in particular, is present at a level of about 2 to 5 ppm. If needed, ammonium ion is added as $(NH_4)_2SO_4$.

Downstream processing. When sufficient sulfide mineral has been oxidized, the heaped ore may be rinsed with fresh water to remove excess acid and ferric iron, which consume cyanide, and the oxidized residue is removed from the pad. The leached residue, which still contains the gold and silver, can be agglomerated with lime and re-stacked for leaching with cyanide or limed and placed in a milling circuit with cyanide to extract gold. The latter process, called "bio-milling", is currently used by Newmont (Bhakta and Arthur, 2001). Pregnant solution from the cyanide heap leach or the bio-milling circuit is treated in a carbon adsorption circuit. Carbon loaded with gold (and silver) is desorbed in an elution circuit, regenerated in a kiln and returned to adsorption. Gold is recovered from strong eluates by electrowinning and smelting.

Some 2.4 million tonnes of ore crushed to 80% passing 1.27 cm and grading approximately 2.7 g/tonne gold were processed the first year of Newmont's Carlin, Nevada (USA) operation; recoveries ranged from 55-60% for the first three months of operation. Process enhancements are expected to improve future recoveries to about 65%.

Heap Leaching Chalcopyrite Ores: An Emerging Technology

Interest in hydrometallurgical processing of chalcopyrite is increasing, because (1) constructing and operating smelters is less appealing due to capital cost and problems in complying with air pollution standards, (2) many chalcopyrite deposits also host penalty elements, particularly arsenic and bismuth, that smelters are reluctant to accept, and (3) worldwide there are vast resources of chalcopyrite ore that are too low-grade to concentrate and process by conventional routes. Bioheap leaching is particularly attractive for treating low-grade chalcopyrite ores, because of the relatively low cost, simplicity and ability to handle arsenic and bismuth containing minerals. The basics of heap leaching chalcopyrite ore are not well understood, however, redox control (615 – 645 mV SHE) and elevated temperature ($\geq 70°C$), as employed in stirred-tank leaching of chalcopyrite concentrates, seem to be key to the process. An aerated, run-of-mine heap leach trial with 960,000 tonnes of chalcopyrite ore grading 0.27% copper at Kennecott's Bingham Canyon operation in Utah (USA) was considered a success. With just 18 months of operation, 28% of the contained copper was extracted from the test heap based on drill assay data. Temperatures in the heap exceeded 60°C with greater copper extraction noted in areas of the highest temperature (Ream and Schlitt 1997 and 1997a).

Chalcopyrite heap leaching is the subject of several patents and applications (Pinches et al. 2001) (Miller 2000). With successful demonstration trials, chalcopyrite heap leaching is very likely to be employed commercially within the next few years.

Heap Leaching Sulfide Concentrates

The GEOCOAT™ process, developed by Geobiotics, Inc., agglomerates base- or precious metal concentrates, finely ground ore, or re-ground tailings onto coarse ore particles or inert aggregate using concentrated sulfuric acid or an acidic, iron-containing leach solution. The coated coarse ore or aggregate is then bioleached in a heap configuration like that described for secondary copper ores and sulfidic-refractory gold ores (Whitlock, 1997; Johansson et al. 1999). Heaps can be inoculated with mesophilic, moderately thermophilic or hyper-thermophilic microorganisms depending on the amount of heat generated from the sulfide oxidation. The GEOCOAT™ process has been pilot tested at several locations, however, there are no commercial applications of the technology at this time. This process is protected by some 15 U.S. patents and numerous foreign patents and patent applications.

SUMMARY

Bioleaching of sulfidic-refractory precious metal and base metal concentrates and ores in tanks and heaps is commercially practiced around the world. Since commercial bioleaching practices began in the 1980s, much has been learned about the fundamentals of the technology, criteria for selection of the process, the design of tank and heap bioleach plants and the operation and performance of the plants. This chapter has explored the fundamental principles of bioleaching including the types of microorganisms used, their requirements and factors that affect their performance. Tanks and heap reactor designs were examined and the parameters that must be controlled for the microorganisms were considered.

Different suites of microorganisms are used to match conditions anticipated in tank and heap reactors. This development permits greater flexibility in plant design. Microbial leaching processes can now be operated at temperatures approaching 95°C, and this development has led to new tank reactor designs that take into consideration the issues of mass transfer of gases, evaporative losses and materials of construction at elevated temperatures. Technologists are unraveling the mechanisms that microorganisms use to catalyze reactions, and these findings coupled with engineering developments in reactor design are leading to new and important commercial applications such as the tank and heap leaching of chalcopyrite concentrates and ore and the heap leaching of concentrates.

The mining industry needs flexible, robust, cost effective, and environmentally acceptable process technologies that exhibit good metals extraction performance. Hydrometallurgical technologies top the list of preferred technologies to meet this need. Bioleaching technology and the engineering design innovations that have been coupled with this technology have come a long way toward providing an acceptable processing alternative for the mining industry.

REFERENCES

Batty, J.D. and T.A. Post. 1999. Bioleach reactor development and design. *ALTA 1999 Nickel/Cobalt Pressure Leaching & Hydrometallury Forum*. Melbourne: ALTA Metallurgical Services.

BacTech. 2002. www.bactech.com.

Basson, P., et al., International Patent Application WO 01/18266 (15 March 2001).

Bell, N and L. Quan. 1997. The application of BacTech (Australia) Ltd technology for processing refractory gold ores at Youanmi Gold Mine. *Proc. of the International Biohydrometallurgy Symposium IBS97 BIOMINE 97*. Chapter M2.3. Adelaide: Australian Mineral Foundation.

Bennett, J.W. and A.I.M. Ritchie. 2002. A proposed technique for measuring in situ the oxidation rate in biooxidation and bioleach heaps. Submitted for publication in *Hydrometallurgy*.

Bhakta, P. and B. Arthur. 2001. Heap biooxidation and gold recovery at Newmont Mining Corporation. Presented at the 2001 Annual Meeting of the Society of Mining Engineers, Denver, Colorado.

Billiton. 2002. BHP-Billiton Annual Report for 2001. www.bhpbilliton.com.

Bouffard, S.C. and D.G. Dixon. 2001. Investigative study into the hydrodynamics of heap leaching processes. *Metallurgical and Materials Transactions B* 32B:763.

Breed, A.W., C.J.N. Dempers, G.E. Searby, M.A. Jaffer and G.S. Hansford. 2000. The bioleaching of sulfide minerals: developments in understanding the mechanism and kinetics of bioleaching pyrite, arsenopyrite and chalcopyrite. *Proceedings of the SME Annual Meeting, Salt Lake City*, Preprint 00-120.

Brierley, C.L. and J.A. Brierley. 1999. Bioheap processes – operation requirements and techniques. *Copper Leaching, Solvent Extraction and Electrowinning Technologies*, ed. G.W. Jergensen, 17-27. Littleton, Colorado: Society of Mining Engineers.

Brierley, C.L. and A.P. Briggs. 1997. Minerals biooxidation/bioleaching: a guide to developing a technically and economically viable process. *After the Discovery: Proceedings of a Short Course*. Prospectors and Developers Association, Toronto, Canada.

Brierley, J.A. 1997. Heap leaching of gold bearing deposits, theory and operational description. *Biomining: Theory, Microbes and Industrial Processes*, ed. D.E. Rawlings, Chapter 5. Berlin: Springer-Verlag.

Brierley, J.A. 2000. Expanding role of microbiology in metallurgical processes. *Mining Engineering* 52(11):49.

Brierley, J.A. 2001. Response of microbial systems to thermal stress in biooxidation-heap pretreatment of refractory gold ores. *Biohydrometallurgy: Fundamentals, Technology and Sustainable Development*, Part A, ed. V.S.T. Ciminelli and O. Garcia Jr., 23-31. Amsterdam: Elsevier.

Brierley, J.A. and D.L. Hill, U.S. Patent No. 5,246,486 (21 September 1993).

Brierley, J. A. and D. L. Hill, U.S. Patent No. 5,332,559 (26 July 1994).

Brierley, J. A. and D. L. Hill, U.S. Patent No. 5,834,292 (10 November 1998).

Briggs, A.P. and M. Millard. 1997. Cobalt recovery using bacterial leaching at the Kasese Project, Uganda. *Proceedings of the International Biohydrometallurgy Symposium IBS 97 BIOMINE 97*. Chapter M2.4. Adelaide: Australian Mineral Foundation.

Brown, A., W. Irvine and P. Odd. 1994. Bioleaching – Wiluna operating experience. *BIOMINE '94*, Chapter 16. Adelaide: Australian Mineral Foundation.

Craven, P. and P. Morales. 2000. Alliance Copper: the Billiton-CODELCO strategy for commercializing copper bioleaching. *Randol Copper Hydromet Roundtable 2000*, 119-126. Golden, Colorado: Randol International Ltd.

Dew, D.W., D.M. Miller and P.C. van Aswegen. 1993. GENMIN's commercialization of the bacterial oxidation process for the treatment of refractory gold concentrates. *Randol Gold Forum Beaver Creek '93*, 229-237. Golden, Colorado: Randol International Ltd.

Dew, D., H. Marais, P. van Aswegen, and C. Loayza. 1996. Bio-oxidation of gold and copper concentrates from Peru. *Peru: Second International Gold Symposium*, 243-249. Lima: Comite Aurifero de la Sociedad Nacional de Mineria y Petroleo.

Dew, D.W., E.N. Lawson and J.L. Broadhurst. 1997. The BIOX® process for biooxidation of gold-bearing ores or concentrates. *Biomining: Theory, Microbes and Industrial Processes*, ed. D.E. Rawlings, Chapter 3. Berlin: Springer-Verlag.

Dew, D.W., C. van Buuren, K. McEwan and C. Bowker. 1997a. Bioleaching of base metal sulphide concentrates: a comparison of mesophile and thermophile bacterial cultures. *Biohydrometallurgy and the Environment Toward the Mining of the 21st Century*, eds. R. Amils and A. Ballester, 229-238, Amsterdam: Elsevier.

Dew, D.W. et al., International Patent Application WO 01/18269 (15 March 2001).

Dew, D.W., et al., International Patent Application WO 01/18268 (15 March 2001a).

Dew, D.W. and D.M. Miller, U.S. Patent No. 6245,125 (12 June 2001).

d'Hughes, P., P. Cezac, T. Cabral, F. Battalglia, E.M. Truong-Meyer and D. Morin. 1997. Bioleaching of a cobaltiferous pyrite: a continuous laboratory-scale study at high solids concentration. *Minerals Engineering* 10:507.

d'Hughes, P., P. Cezac, F. Battaglia and D. Morin. 1999. Bioleaching of a cobaltiferrous pyrite at 20% Solids: a continuous laboratory-scale study. *Biohydrometallurgy and the Environment Toward the Mining of the 21st Century*, eds. R. Amils and A. Ballester, 167-176, Amsterdam: Elsevier.

d'Hughes, P. D. Morin and S. Foucher. 2001. HIOX® project: a bioleaching process for the treatment of chalcopyrite concentrates using extreme thermophiles. *Biohydrometallurgy: Fundamentals, Technology and Sustainable Development*, Part A, ed. V.S.T. Ciminelli and O. Garcia Jr., 75-83. Amsterdam: Elsevier.

do Carmo, O.A., M.V. Lima and R.M.S. Guimaraes. 2001. BIOX® process – the Sao Bento experience. *Biohydrometallurgy: Fundamentals, Technology and Sustainable Development*, Part A, eds. V.S.T. Ciminelli and O. Garcia, Jr., 509-524. Amsterdam: Elsevier.

Gericke, M. and A. Pinches. 1999. Bioleaching of copper sulphide concentrate using extreme thermophilic bacteria. *Minerals Engineering* 12:893.

Harvey, P.I., J.D. Batty, D.W. Dew, W. Slabbert and C. van Buuren. 1999. Engineering considerations in bioleach reactor design. *BIOMINE '99*, 88-97, Adelaide: The Australian Mineral Foundation.

Hiroyoshi, N., M. Hirota, T. Hirajima, and M. Tsunekawa. 1997. A case of ferrous sulfate addition enhancing chalcopyrite leaching. *Hydrometallurgy* 47:37.

Hiroyoshi, N., H. Miki, T. Hirajima and M. Tsunekawa. 2000. A model for ferrous-promoted chalcopyrite leaching. *Hydrometallurgy* 57:31.

Huerta, G., B. Escobar, J. Rubio and R. Badilla-Ohlbaum. 1995. Short communication: adverse effect of surface-active reagents on the bioleaching of pyrite and chalcopyrite by *Thiobacillus ferrooxidans*. *World Journal of Microbiol. & Biotechnol.* 11:599.

Johansson, C., V. Shrader, J. Suissa, K. Adutwum and W. Kohr. 1999. Use of the GEOCOAT™ process for the recovery of copper from chalcopyrite. *Biohydrometallurgy and the Environment toward the Mining of the 21st Century*, eds. R. Amils and A. Ballester, 569-576. Amsterdam: Elsevier.

Johnson, D.B. and F.F. Roberto. 1997. Heterotrophic acidophiles and their roles in the bioleaching of sulfide minerals. *Biomining: Theory, Microbes and Industrial Processes*, ed. D.E. Rawlings, Chapter 13. Berlin: Springer-Verlag.

Lacey, D.T. and F. Lawson. 1970. Kinetics of the liquid-phase oxidation of acid ferrous sulfate by the bacterium *Thiobacillus ferrooxidans*. *Biotechnol. Bioeng.* 12:29.

Lawson, E.N., C.J. Nicholas and H. Pellat. 1995. The toxic effects of chloride ions on *Thiobacillus ferrooxidans*. *Biohydrometallurgical Processing*, ed. T. Vargas, C.A. Jerez, J.V. Wiertz and H. Toledo, 165-174. Santiago: University of Chile.

Miller, P.C. 1997. The design and operating practice of bacterial oxidation plant using moderate thermophiles (the BacTech Process). *Biomining: Theory, Microbes and Industrial Processes*, ed. D.E. Rawlings, Chapter 4. Berlin: Springer-Verlag.

Miller, P. 2000. International Patent Application WO 00/71763 A1 (30 November 2000).

Miller, P.C., M.K. Rhodes, R. Winby, A. Pinches and P.J. van Staden. 1999. Commercialization of bioleaching for base-metal extraction. *Minerals & Metallurgical Processing* 16:42.

Nicholson, H.M., G.R. Smith, R.J. Stewart, F.W. Kock and H.J. Marais. 1994. Design and commissioning of Ashanti's Sansu BIOX® plant. *BIOMINE '94*, Chapter 2. Adelaide: The Australian Mineral Foundation.

Norris, P.R. 1983. Iron and mineral oxidation with *Leptospirillum*-like bacteria. *Recent Progress in Biohydrometallurgy*, eds. G. Rossi and A.E. Torma, 83-96. Iglasias, Italy: Associazione Mineraria Sarda.

Norris, P.R. 1997. Thermophiles and bioleaching. *Biomining: Theory, Microbes and Industrial Processes*, ed. D.E. Rawlings, Chapter 12. Berlin: Springer-Verlag.

Norris, P.R., N.P Burton, N.A.M. Foulis. 2000. Acidophiles in bioreactor mineral processing. *Extremophiles* 4:71.

Norton, A., International Application WO 01/18267 (15 March 2001).

Perry, R.H. and C.H. Chilton (eds.). 1979. *Chemical Engineering Handbook*, 5th ed., 6-16 (equation 6-22). New York: McGraw-Hill Book Company.

Pinches, A., R. Huberts, J.W. Neale and P. Dempsey. 1994. The MINBAC™ bacterial-oxidation process. *IVth CMMI Congress*, vol. 2, 377-392. Johannesburg: SAIMM.

Pinches, T., J.W. Neale, V. Deeplaul, P. Miller, M. Rhodes and B. Hancock. 2000. The Beaconsfield bacterial oxidation gold plant. *Randol Gold & Silver Forum 2000*, 169-175. Golden, Colorado: Randol International Ltd.

Pinches, A., et al., U.S. Patent No. 6,277,341 B1 (21 August 2001).

Rawlings, D.E. 1997. Mesophilic, autotrophic bioleaching bacteria: description, physiology and role. *Biomining: Theory, Microbes and Industrial Processes*, ed. D.E. Rawlings, Chapter 11. Berlin: Springer-Verlag.

Rawlings, D.E., H. Tributsch, and G.S. Hansford. 1999. Reasons why '*Leptospirillum*'-like species rather than *Thiobacillus ferrooxidans* are the dominant iron-oxidizing bacteria in many commercial processes for the biooxidation of pyrite and related ores. *Microbiology* 145:5.

Ream, B.P. and W.J. Schlitt. 1997. Kennecott's Bingham Canyon heap leach program – part 1: the test heap and SX-EW pilot plant. *Proceedings of the ALTA 1997 Copper Hydrometallurgy Forum*, 20-21. Melbourne: ALTA Metallurgical Services.

Ream, B.P. and W.J. Schlitt. 1997a. Kennecott's Bingham Canyon heap leach program – part 2 the column leach testwork. *Proceedings of the ALTA 1997 Copper Hydrometallurgy Forum*, 46 pages. Melbourne: ALTA Metallurgical Services.

Rhodes, M. and P.C. Miller. International Patent Application No. WO 00/28099 (18 May 2000).

Rhodes, M. and P.C. Miller. International Patent Application No. WO 00/29629 (15 May 2000a).

Salomon-de-Friedberg, H. 1998. Design aspects of aeration in heap leaching. *Copper Hydromet Roundtable '98*, 243-247. Golden, Colorado: Randol International Ltd.

Sand, W., T. Gehrke, R. Hallmann, and A. Schippers. 1995. Sulfur chemistry, biofilm, and the (in)direct attack mechanisms – a critical evaluation of bacterial leaching. *Appl. Microbiol. Biotechnol.* 43:961.

Sand, W., T. Gehrke, P.-G. Jozsa, and A. Schippers. 1999. Direct versus indirect bioleaching. *Biohydrometallurgy and the Environment Toward the Mining of the 21st Century*, eds. R. Amils and A. Ballester, 27-49, Amsterdam: Elsevier.

Schippers, A., P.G. Jozsa and W. Sand. 1996. Sulfur chemistry in bacterial leaching of pyrite. *Appl. Environ. Microbiol.* 62:3424.

Schippers, A. and W. Sand. 1999. Bacterial leaching of metal sulfides proceeds by two indirect mechanisms via thiosulfate or via polysulfides and sulfur. *Appl. Environ. Microbiol.* 65: 319.

Schlitt, W.J. 1984. The role of solution management in heap and dump leaching. *Au and Ag Heap and Dump Leaching Practice*, ed. J.B. Hiskey, 69-83. Littleton, Colorado: SME-AIME.

Schlitt, W.J. 2002. Personal communication.

Slabbert, W., D. Dew, M. Godfrey, D. Miller and P. Van Aswegen. 1992. Commissioning of a BIOX® module at Sao Bento Mineracao. *Randol Gold Forum Vancouver '92*, 447-452. Golden, Colorado: Randol International Ltd.

Third, K.A., R. Cord-Ruwisch and H.R. Watling. 2000. The role of iron-oxidizing bacteria in stimulation or inhibition of chalcopyrite bioleaching. *Hydrometallurgy* 57:225.

Timmins, M. and R.P. Hackl. 1998. Bacterial leaching of chalcopyrite concentrates – prospects for a commercial process. *New Dimensions in Hydrometallurgy Seminar*, October 2, 1998, The University of British Columbia, Vancouver.

Titan Resources. 2002. www.titanresources.com.au.

Tunley, T.H., U.S. Patent No. 5,919,674 (6 July 1999).

van Aswegen, P.C. 1993. Bio-oxidation of refractory gold ores – the GENMIN experience. *BIOMINE '93*, Chapter 15. Adelaide: The Australian Mineral Foundation.

van Aswegen, P.C. and H.J. Marais. 1999. Advances in the application of the BIOX® process for refractory gold ores. *Minerals & Metallurgical Processing* 16:61.

van Staden, P.J., M. Rhodes, A. Pinches and T.E. Martinez. 2000. Process engineering of base metal concentrate bioleaching. *Randol Copper Hydromet Roundtable 2000*, 127-129. Golden, Colorado: Randol International Ltd.

Whitlock, J.L. 1997. Biooxidation of refractory gold ores (the Geobiotics Process). *Biomining: Theory, Microbes and Industrial Processes*, ed. D.E. Rawlings, Chapter 6. Berlin: Springer-Verlag.

Winby, R. et al. International Patent Application No. WO 00/23629 (27 April 2000).

12

Leaching and Adsorption Circuits

Section Co-Editors:
Dr. Chris Fleming and Michael R. Schaffner

Copper Heap Leach Design and Practice
R.E. Scheffel **1571**

Precious Metal Heap Leach Design and Practice
D.W. Kappes **1606**

Agitated Tank Leaching Selection and Design
K.A. Altman, M. Schaffner, S. McTavish **1631**

CIP/CIL/CIC Adsorption Circuit Process Selection
C.A. Fleming **1644**

CIP/CIL/CIC Adsorption Circuit Equipment Selection and Design
K.A. Altman, S. McTavish **1652**

Copper Heap Leach Design and Practice

Randolph E. Scheffel[1]

ABSTRACT

Copper heap leaching has expanded continuously the last thirty years due to the commercial development of solvent extraction. To ensure both technical and financial success, a heap leach project must follow certain disciplines. These disciplines include: (1) a proper evaluation of the resource; (2) a comprehensive metallurgical test program; and (3) an engineered design and operating plan, all of which result in achieving expected production. The key to developing a successful leaching prospect is determining the actual "leachable" mineral content and then designing, conducting, and interpreting the metallurgical test program. The principal engineering requirement is the selection of the appropriate "scaled-up" leach curve from which to design the leaching area. Additionally, the actual leach design is often dictated by site-specific constraints, and each design requires a different "operating" copper inventory, both in solution and solids. Designing flexibility for increased leach area and volume of solution is critical.

INTRODUCTION

The objective of this section is to summarize the copper heap leaching development and experience of the last thirty years. Hopefully, guidelines taken from this experience are sufficient to ensure future operators continue the recent level of improving performance. After thirty years of continued improvement, the current status of heap leach design can be characterized more as "educated guess work" than "art" – however, there is still much to learn.

One of the misconceptions of heap leaching, particularly with respect to copper, is that it is simple, straightforward, flexible and forgiving. Nothing is further from fact. Design engineers and owners anticipating such a development are well advised to seek all the counsel possible with people who have actual experience. Preferably this includes critical reviews of numerous and uniquely different operations, including other commodities. Further, consulting with investment bankers, and their third party engineers, often proves highly informative as to the cause for many of these projects under-performing initial financial projections.

HISTORY

Interest in new technological development in the minerals business generally follows the price cycle of the commodity. Such is the case with heap leaching. While the first reported copper heap or dump leaching may have been at Rio Tinto, Spain, circa 1752, the first commercial modern-day style of heap leaching was probably introduced to the uranium industry in the 1950's (Merritt 1971). Leaching of uncrushed, low-grade copper ore may have first been practiced in the US at Bisbee, AZ as early as 1923, and heap leaching was discussed by Irving in 1922 (Irving 1922). Vat leaching of finely crushed, oxide ore, combined with direct electrowinning from the impure solutions, was developed as early as 1916 at Chuquicamata in northern Chile (Eichrodt 1930, Rose 1916). This was followed by the New Cornelia Copper Company operation at Ajo, Arizona in 1917 (Tobelmann and Potter 1917). In 1925, Inspiration Consolidated Copper Company, Miami, AZ (Aldrich and Scott 1933) initiated vat leaching of mixed oxide, supergene ore. However, it was not until 1961 that the modern-day practice of heap leaching copper began its development in a significant way. It is believed the Stovall Copper Company (Miller 1967) was the first to use this process as the primary means of development at the Bluebird Mine near Miami, Arizona. At that time, copper cementation on iron was used to recover the copper, which subsequently required smelting.

---

1 Consulting Metallurgical Engineer

Gold heap leaching, having benefited from this early work in copper, was initially proven with the Cortez and Bootstrap operations in 1973 and 1974, respectively (McQuiston and Shoemaker 1981). Economic forces helped to continue its growth through the 1980s, contributing significantly to the pool of experience and operating history for later application to copper heap leaching. However, cross-pollination of these two industries was really quite limited during this period, with each separately developing their expertise. In the late 1980s, spurred by improving prices, copper again enjoyed a major expansion in the utilization of heap leaching oxide and secondary copper minerals. However, at the end of the 20th century, the hydrometallurgical treatment of primary copper, e.g. chalcopyrite, was still relegated to dreams of future commercial development.

There is no question one of the most significant developments in the copper industry in the last thirty years was the development of the ketoxime, and eventually the aldoxime, chelating extraction reagents. The ketoximes were initially developed by General Mills (House 1981 and Kordosky 1994, 1999) and commercially proven at the Bluebird Mine near Miami, AZ by Ranchers Exploration and Development Corporation (Power 1970) in 1968. The Bluebird was closely followed by Bagdad Copper Corporation's operation at Bagdad, AZ in 1970, the ammonical, scrap copper leach operation of the Capital Wire and Cable Corporation operation near Casa Grande, AZ in 1970 and ZCCM Chingola in Zambia in 1974.

The most rapid expansion of heap leach, SX/EW copper production occurred in Chile during the decade of the 1990s. Both oxide and supergene deposits were developed using fine crushing, generally less than 12-16 mm, acid agglomeration and subsequent acid or biological leaching in heaps primarily constructed via conveyor stacking 6-8 m in depth. The precursor to this development was the "thin layer leaching" concept first patented (Johnson 1975) by Holmes and Narver Inc. (H&N) for application to uranium and copper (E&MJ 1978 and Mining Magazine 1978). This process was first commercialized by Sociedad Minera Pudahuel (SMP) near Santiago, Chile in late 1980 (Domic 1981 and 1983). The Chilean rights to the H&N patent were subsequently assigned to SMP. Since this was a mixed oxide/sulfide ore, the natural progression of activities over the years resulted in SMP further developing its expertise in biological leaching. As a consequence, Cominco and Rio Algom elected to solicit SMP's assistance in the development of Quebrada Blanca and Cerro Colorado using a similar process, but stacking the ore to 6-8 m in depth. SX/EW, heap leach copper production in Chile expanded from 20,000 tpa in the 1980s to over 1,000,000 tpa by the end of the 1990s.

Absent the ability to purify copper from dilute, impure sulfate leach solutions, it would have been impossible to supply the market with the quality copper required by the downstream suppliers of refined copper products, without further expensive processing. Therefore, the development of the extraction reagents and their continued improvement were the primary reason for the growth in copper heap leaching. Today, SX/EW accounts for nearly 20% of worldwide refined copper sales. A further significant improvement in the SX/EW process was the development of copper plating on stainless steel cathode blanks. This was first practiced by Capital Wire and Cable Corporation in the early 1970s and was fully developed into an automated plating and stripping process in 1976 by Mount ISA Mines, in Queensland, Australia.

So successful have been the reagents and subsequent electrowinning practice, that it is a rare occurrence when an SX/EW facility does not routinely produce LME Grade A copper, or better.

DISCIPLINES FOR THE APPLICATION OF COPPER HEAP LEACHING

It is paramount an engineer, developing a heap leach project, understands this unit process requires an inter-disciplinary approach between geologists, mining engineers, and metallurgists. These projects can not be developed successfully with each working without regard for the others. This begins with the initial drilling and the subsequent geologic and mineralogical interpretation. The metallurgist must understand the geologic and lithologic constraints imposed by the resource. The geologist must understand the inefficiencies and implication of variable ore characteristics on the metallurgical performance of the heaps. The mining engineer must understand that moving tonnes at the least possible cost is not always the most profitable approach when total leaching performance is addressed.

The metallurgist often considers blending of different rock types to deal with ores displaying poor characteristics, such as highly clay-altered ore. However, open-pit mining generally occurs in distinct levels, and there is only limited ability, without significant stockpiling and re-handling, to conduct such blending. The metallurgist must appreciate these limitations.

The author's opinions and experience in these respects have been presented at the Randol Copper Hydromet Round Tables (Scheffel 1999 and 2000). Dicinoski, Schlitt and Ambalavaner (1998) also discuss the critical elements of developing a copper heap leach, SX/EW project. Some of the key required disciplines are:

- Resource Evaluation
- Test Work Program
- Engineering Design
- Postmortem Analysis

Each of these is discussed in detail below.

Resource Evaluation

The most critical, and often controversial, aspect of developing a successful heap leach resides with one's understanding of the resource. One perception of heap leaching is that it is a low-capital approach that can be developed in the shortest possible time in relation to other options. The other options are actually quite limited, e.g., agitation leaching of copper oxides, and most likely, flotation followed by smelting for copper sulfides. There are very distinct differences in copper mineralogy which make process selection specific to either acid leaching of oxide minerals, ferric leaching of secondary sulfide (supergene) minerals, and until a hydrometallurgical process is proven commercially, flotation and smelting of primary copper sulfides. Most likely, leaching becomes the primary option for mixed oxide/supergene deposits and for medium grade supergene deposits.

Therefore, when deciding how best to drill and evaluate a deposit, it is imperative that an early assessment be made of the probability heap leaching will be applicable. This is best accomplished by routinely conducting diagnostic assays, or extensive mineralogy, on the initial drill cuttings or core. This provides an early indication of the potential copper solubility and the type of mineralogy being drilled. If it appears the ore is potentially heap leachable, it will be necessary to consider substantial amounts of diamond core drilling, as opposed to less expensive reverse circulation (RC) drilling. It is critical the physical character, e.g., fracturing and alteration, of the ore be fully appreciated and modeled in the geologic rock model. This can only be accomplished with extensive core logging. Further, the diamond drill coring, in size, depth, and area, should take into consideration obtaining the broadest possible sampling of the resource for metallurgical testing, in addition to resource definition.

Under-capitalized companies that develop such projects with less stringent methods often find themselves in trouble. Conversely, well-capitalized companies can spend significant funds for the wrong reasons. For example, spending considerable funds in driving adits to collect large bulk samples for large-scale pilot heap or column testing can be misleading, as these are only "grab" samples, which are most often non-representative. Instead, it is often more cost effective and beneficial to spend exploration dollars on completing significant, large-diameter core sampling of the deposit, e.g., >50 mm diameter. This ensures all the different lithologies and alteration types are sampled and fully logged. Initial testing can then be conducted on individual rock types versus depth.

It is the author's observation that small columns, with diameters 4 to 6 times the largest particle size, will most often perform like 1 to 2 m square cribs or 1 m diameter column tests. Unfortunately, these will all have relationships to a commercial application that will be quite different, for reasons discussed below.

Besides the physical character of the ore, there are two additional items critical to resource evaluation -- "soluble" mineralogy and acid consumption. "Soluble" in this context means the copper minerals that can be expected to dissolve in the acid and ferric ion environment actually achievable in the heap leach system. This leads one to do considerable time-consuming and costly mineralogy on individual drill hole intervals. Another option is to develop "diagnostic" analytical techniques that allow one to estimate the "soluble" copper content quickly and economically. Recent advances in scanning electron microscopy, combined with computer software, may eventually result in economic and timely quantitative mineralogical analysis (Gottlieb et al. 2000).

Once the "soluble" copper estimate is determined, it can later be compared to column tests at various crush sizes to determine the general level of recovery of that "soluble" content. It is critical this parameter be followed for prospective development of both oxide and secondary sulfide leach projects. Oxide deposits can contain significant quantities of non-acid soluble copper oxides and other refractory compounds, which can reduce actual leachable content to <50%. Some oxide deposits can actually be poorer performers than secondary sulfide deposits for this reason.

Secondary copper deposits can be highly irregular with depth, with some sections being quite thin. Hypogene minerals such as chalcopyrite or enargite can sometimes intrude into the secondary mineralization. Also, secondary deposits can contain a complex mix of oxide and secondary transitional type minerals, including native copper. If enargite is present, this must be identified early in the resource evaluation as its low solubility in acid-ferric medium can be problematic.

A diagnostic method, which has shown reasonably broad application in this regard, is called a "sequential" assay. This was used by Inspiration in the early 1980s at its ferric-cure leach operation near Globe, AZ. This technique was further refined and reported by Parkison and Bhappu (1995). This method uses a 20°C (or heated) 5% sulfuric acid digestion for one hour followed by a 30 minute cyanide digestion at 20°C (or heated) on the acid leached and washed residue. This cyanide leach residue is then digested with aqua regia or a standard four-acid digestion for the remaining insoluble copper. The sum of the three digestions represents a "calculated" copper assay to compare directly against a standard "total" copper assay. The latter is necessary as it represents the only assay that can be reproduced with appropriate accuracy to be used for third party quality control programs required by financiers. In the "sequential" assay, the acid is believed to solubilize the non-sulfide minerals except for native copper and about half of any Cu_2O. The cyanide is assumed to solubilize the native copper and the secondary sulfide minerals, as well as bornite. The remaining residue is assumed to be the refractory oxide minerals or chalcopyrite.

It is important to understand that diagnostic methods, as well as mineralogy, are only semi-quantitative and strict conformance to procedure between laboratories is paramount, or reproducibility will be impossible. Other diagnostic methods are discussed by Hiskey (1997). It must further be emphasized that diagnostic assays can not be used without extensive initial mineralogy to substantiate the actual mineral assemblage of the specific deposit. Some copper minerals are soluble in cyanide and not in acid-ferric media and vice versa, e.g., enargite is nearly totally soluble in cyanide and only marginally soluble with ferric iron.

A second analytical parameter important to the initial resource evaluation is to develop an early estimate of the gangue acid consumption. This requires empirical methods and is subject to interpretation. However, it is critical this parameter not be overlooked in early resource evaluation.

A method that has some acceptance in this regard is a 24-hour bottle roll test at constant pH of 1.5 to 1.8 and 33% solids. This should be conducted on the assay pulp or the 10 mesh crushed ore rejects from the analytical sample preparation. Bottle roll tests on coarse material, approaching the expected crush size, are too variable due to the degradation that occurs during the test. Some have developed static testing methods to deal with this problem (Fountain 2002). The acid consumption indicated from the fine material will be much higher than will occur in actual practice. Therefore, experience is necessary in relating the results to commercially expected consumption. The actual consumption at coarser particle size, under field conditions, can be only 20-35% of the results obtained on pulp. However, potentially excessive acid consumption will clearly be indicated by this test. Conversely, some ore types can actually show very little acid consumption, which can represent a different problem. If the gangue acid consumption does not exceed the fresh acid addition, and that produced by bacterial activity in the case of sulfide leaching, acid can build in the heaps to the point it interferes with copper recovery in SX. In this situation, the fresh acid addition must be reduced or neutralization of the excess acid may be necessary.

The commercial economic limit for acid consumption is highly variable, depending on other factors such as ore grade, stripping ratio, etc. However, this consumption generally falls within 5-50 kg/t ore. A more reliable means of expressing the acid consumption of a particular project is the "specific" net acid consumption expressed as kg acid per kg of copper recovered. The generally experienced commercial acid consumption expressed in this fashion ranges from 1-7 kg acid/kg copper recovered.

Test Work Program

Ideally, a properly designed test work program should be developed only after the resource is reasonably well understood. Unfortunately, timing is generally such that the metallurgist must make certain decisions and proceed with a concurrent program. Therefore, initial test work should test samples of individual lithologic character. Once the final details of the rock model are complete, optimized column testing can be conducted on specific composites. These composites should include the major rock types within the deposit with regard to their respective distribution, both in cross-sectional area and with depth.

Once the resource evaluation suggests the ore can potentially be heap leached, the principal objectives of the test program should be to determine:

- the general level of extraction versus lighologic unit, depth, and ore grade;
- the recovery versus crush size;
- the potential production of fines with crushing;
- the degradation character of the gangue upon leaching;
- the acid consumption, with particle size affect;
- the key solution equilibrium chemistry; and
- the practical commercial heap height.

Unfortunately, no single column test, or series of column tests, can give all the answers. Therefore, a significant amount of judgment is necessary when interpreting exactly what the collective package of data provides. Scheffel and Reid (1997) and Kaczmarek et al. (1999) describe the process used in two instances to develop and interpret such test work.

Specific to developing the "Test Work Program", the following items are considered important:

- Sample Collection
- Types of Minerals
- Sample Preparation
- Types of Testing
- Heap Height
- Interpretation and Scale-up

Sample Collection. The collection of a proper suite of "representative" samples to be tested is paramount to developing a successful heap leach. To the extent this is not done, the only mitigating option is to be extremely flexible with assumptions on design pregnant solution (PLS) grade, time, and leach pad area.

Modern drilling and sampling techniques, including RC drilling, generally give sufficient statistically valid drill results to assess the "total" and "soluble" copper content and allow adequate ore grade modeling. However, only core drilling can provide the information necessary to develop the rock-type model required for developing a low-risk heap leach design. To the extent a company compromises the resource evaluation phase in this regard, the greater the risk of under-performance, or even failure, of the subsequent heap leach design and operation.

As mentioned above, the driving of adits into multi-million tonne deposits to acquire large bulk samples can be a near-fatal flaw in many large projects if these are the primary samples tested. These samples are only grab samples, which typically do not represent the actual distribution within the resource.

Types of Minerals. A proper understanding of mineralogy is critical to designing a test program. The metallurgical response, both in extraction time and acid consumption, can be drastically different for oxide minerals as opposed to secondary, transitional, and primary sulfide minerals.

Oxides. Oxide minerals, or copper minerals formed in an oxidizing environment, are far-ranging and exhibit a wide range of metallurgical performance (Baum 1996 and 1999). Some oxide minerals, such as malachite ($Cu_2CO_3(OH)_2$) and azurite ($Cu_3(CO_3)_2(OH)_2$), can leach extremely fast. However, in practice their rate of dissolution is controlled not by kinetics but by the rate acid can actually be made available from the processing system. Chrysocolla is the next most rapidly dissolved, but can also have a limitation caused by diffusion of hydrogen ion in, or copper ion out, through a non-protective silica layer (Pohlman and Olson 1976).

However, for most oxide minerals, the commercial leaching time frame in actual practice can be a factor of 2 or 3 times faster than for secondary sulfides.

Some oxide minerals, such as copper in iron compounds and "wad" (neotocite), a manganiferrous copper compound, can have dissolution rates approaching that of secondary minerals. Therefore, these minerals may not reach their ultimate economic extraction if the heap leach design considers only the fast-leaching copper oxide minerals. Manganiferous copper compounds actually benefit from reducing conditions, which are not necessarily consistent with the natural leach solution.

One of the distinguishing features between gold and copper heap leaching can be the time frame. While many gold heap leaches of 6-15 meter depth achieve their target recovery in 60-150 days, many copper oxide leaches typically require 150-300 days to attain a similar level of extraction of the "soluble" component. Supergene ore can require 300-900 days depending on ore grade. This is generally due to the impact of higher natural fines (clay) content, degradation of the ore, and higher levels of reagent required in dissolving and displacing more mass.

Oxide deposits, by the very nature of their geologic origin, can be more highly altered than supergene deposits. This leads to clay, or fines, which can significantly impact the heterogeneity of the heaps and make scale-up from laboratory column testing even more difficult. (This is not to imply that supergene deposits do not also exhibit significant alteration, as many do.)

As a result of the above, oxide deposits can be more problematic than secondary sulfide deposits -- a condition not always fully appreciated.

Sulfides. Sulfide copper mineralogy, especially the secondary copper minerals that lend themselves to possible heap leaching, is very complex. A full range of oxidation products from primary copper minerals such as chalcopyrite and enargite, or naturally occurring secondary minerals, create a complex mineral assemblage. Each mineral has a unique level of potential solubility, with its dissolution kinetics being limited by a multitude of reaction products (Bradley, Sohn and McCarter 1992).

Unlike oxide mineral leaching, the dissolution of sulfide minerals is electrochemical in nature, requiring the presence of ferric iron, generally in a sulfate medium. This chemistry can be quite complex. It requires the catalytic oxidizing behavior of aerobic *Acidithiobacillus ferrooxidans* bacteria at temperatures below 45°C and archaea, or other high temperature or extreme temperature thermophiles, when the temperature reaches 40°-70°C (Rawlings 1997 and Norris 1997). While these reactions are electrochemical in nature, they still benefit from increased temperature, but to varying degree, depending on the mineral.

Sulfide minerals also exhibit surface rest potential and can corrode by galvanic action like dissimilar metals (Hiskey and Wadsworth 1981).

In commercial secondary copper leaching operations, chalcocite (Cu_2S) is typically the most abundant mineral. It is generally accepted this mineral leaches in two stages. The first copper is nearly totally removed, leaving behind "synthetic" covellite (CuS) called "blaubleiblender" I or II (Hiskey and Wadsworth 1981). The first stage leach exhibits rapid leach kinetics in the presence of sufficient ferric iron. It also has a lesser dependence on temperature, provided there is sufficient temperature to satisfy the bacterial activity necessary to oxidize ferrous iron to ferric under natural biological leaching conditions. Recent commercial experience suggests a threshold temperature near 18°C. Ferric leaching kinetics of the "synthetic" covellite, however, are much slower and exhibit greater dependence on temperature. Naturally occurring covellite and hypogene covellite exhibit even slower leaching kinetics than the "synthetic" form. Therefore, the leaching of chalcocite is characterized by fairly rapid dissolution of 50%-60% of the copper under conditions of good bacterial activity (i.e., rapid ferrous to ferric oxidation), followed by a longer leach time to dissolve the resulting "synthetic" covellite. Therefore, in order to achieve a high recovery from chalcocite, it is necessary to provide the time required to leach the covellite. For any given pad height, as the chalcocite ore grade increases, the amount of covellite requiring dissolution increases, which generally requires a progressively longer leach cycle, provided one wishes to achieve the same level of extraction.

It must be appreciated, for the iron chemistry to work as efficiently as possible, both acid and oxygen must be present in addition to maintaining the appropriate temperature. The benefit and need for forced aeration for sulfide leaching is summarized by Schlitt (2000). The author's experience suggests this is most important for ores containing >0.5% copper as "leachable" sulfides. One must be careful in conducting column tests that are open to atmosphere and concluding that air may not be required. A column can act as a natural chimney, with thermal drafting providing sufficient oxygen to leach +2% supergene copper at a significant rate, which is not possible in actual practice without forced aeration.

Commercial experience with forced aeration has shown, in most cases, about a 3%-8% increase in extraction over a 300-500 day leach cycle for chalcocite ore grades <1.2% total copper (CuT) when compared to no forced aeration. In cases of poorer permeability, it might even be of greater benefit. Forced aeration can have even a greater impact on the level of recovery reached in the first 100-150 days during the highly ferric dependent leach stage. Girilambone Copper Company documented a significant increase in extraction on high-grade chalcocite ore (e.g., >1.5% CuT), due, not only to the use of forced aeration, but other factors favoring solution application and fines control (Walsh et al. 1997; Dudley et al. 2000). Dixon addresses some of the heat conservation considerations of sulfide leaching utilizing forced aeration (Dixon 2000).

The leaching kinetics of secondary minerals are generally such that leach cycles of 300-900 days are required on a single-lift basis to achieve the economic limit of recovery, depending on ore grade. This has a profound impact on topography and leach pad design issues. Therefore, a comprehensive test program should compare column tests of low and high "soluble" copper content to demonstrate the need for, and impact of, oxygen and time relative to ore grade.

Sample Preparation. The preparation of individual rock-type composites, which should initially be tested separately, can be critical to the interpretation and overall results of a copper leaching test program. Testing for potential recovery difference of each rock type with depth in the deposit can also be important. A comprehensive column test program generally requires parallel or duplicate column tests on identical samples to compare changes in certain process parameters. Some of these parameters include: different acid pre-treatment methods; varied solution application rates; crush size; rock type; etc. Duplicate column testing of this type requires the samples to be as close in copper grade and particle size distribution as possible. Often, the recovery difference in "duplicate" columns, where composite samples were collected from a larger master composite by conventional cone and quartering, or riffle splitting, can be >5%. In order to make valid judgment as to whether recovery differences in two comparative columns are statistically significant, the metallurgical balances for individual tests should be within 2%-3%. This is very difficult, if not impossible, to accomplish without a strict regimen of sample preparation (Keane 1998). The preferred procedure is to first screen the "master" composite into five or more sub-fractions. Then one can split each size fraction separately, collecting a weight for each fraction, so the "master" composite's natural size distribution is properly replicated. These separate weight fractions are then recombined into the test composite. A separate sample is prepared in the same manner for head/screen assay analysis for the column(s) charge.

It is during this strict sample preparation procedure when opinions and judgment can be made regarding the distribution of copper by size fraction, which can lead to indications of the potential crush size dependency on recovery. This depends to a great extent, however, on whether the copper minerals are on fractures or disseminated in the rock mass. A further clue to potential leaching success will be provided with the amount of natural fines occurring at finer crush sizes.

Types of Testing. The organization of a typical test work program is addressed elsewhere in greater detail (Scheffel and Reid 1997; Iasillo and Schlitt 1999). However, a typical recommended approach may be as follows:

1) First conduct mineralogy and diagnostic tests to determine the "soluble" copper content and acid consumption tests on assay pulp to characterize the ore by lithology and with depth and cross-sectional area (a result of the resource evaluation);
2) Conduct bottle roll tests for preliminary characterization of readily soluble species and acid consumption at coarser size (possibly static tests), in preparation for;
3) Mini-column (1.5-2 m) tests in open-cycle to test acid pretreatment options and general solubility at a couple of crush sizes;
4) Then conduct larger diameter, commercial depth column tests in "closed-cycle" with solvent extraction to better identify the crush size effect, ore grade versus recovery relationships, the acid pre-treatment scheme, pH control, iron chemistry, impurity buildup, and better overall estimate of acid consumption; and finally
5) Large-scale pilot testing, if deemed necessary.

The acid consumption testing and pretreatment schemes rely on "empirical" tests developed by individual laboratories over time. The column test procedures have significant potential for error in actual practice. Therefore, it is critical the laboratory chosen to do this work has a long history of conducting such tests with copper. If a less experienced laboratory is chosen, the owner/engineer is well advised to have a consultant experienced with this type of testing advise the laboratory and oversee the overall program.

The author, over several projects and test programs, has developed the following opinions concerning the testing program and the items that are critical to the design of a successful copper heap leach facility.

"Soluble" Copper and Iron Content and "Net" Gangue Acid Consumption. The initial analytical data package requested on drill hole intervals is critical in defining not only the "total" copper resource, but also the resource's potential to be heap leached. Obtaining an early understanding of the "soluble" mineral content, the gangue "net" acid consuming character and the acid soluble iron content are parameters of paramount importance to a viable heap leachable resource. These parameters are also required for a workable rock-type model, assuming heap leaching is the preferred approach.

Bottle Roll Tests. In the author's opinion, bottle roll tests at coarse crush size, e.g., >12 mm, have little utility in the overall picture. They are only indicative of the readily acid and, in the case of sulfide deposits, ferric soluble content. In addition, they are generally only good for finely crushed ore, as the term of the bottle roll test is generally much shorter than that required under heap leach conditions. The acid consumption results are still only preliminary and probably not much better than those obtained from the assay regime. Those experienced with both these "empirical" acid consumption tests have developed methods and judgment as to what the commercial acid consumption might be. (The pH 1.5 or 1.8 acid consumption tests on assay pulp will be 2 to 4 times higher than actually experienced in some heaps.)

Another problem with bottle roll tests at coarse particle size is the degradation that can take place, due to attrition, compromises the results.

Mini-column Tests (<1-2 m in depth). These tests are generally used primarily for scoping acid pretreatment techniques (e.g., testing acid cure versus no acid cure, and then only by one experienced in interpretation of the results) and developing a feel for crush size effects. They have the advantage of giving this information more quickly and at less expense than tall column tests.

Full Commercial Depth Tests. All testing to develop a successful heap leach should incorporate closed-cycle column testing approaching full commercial depth. Unlike gold, the solution chemistry in copper leaching changes dramatically with ore depth. The relationship of pH, free acid, iron chemistry, and Eh can only be fully appreciated by observing the full depth results with equilibrium concentrations of total dissolved solids (TDS). Generally, the rate controlling reagent concentration limitations are not realized, and therefore not appreciated, if testing stops with the top 1-2 m of a planned 6-8 m commercial depth. If one wishes to observe the actual chemistry with depth, individual column segments can be leached in series. This allows for solution collection and assay at chosen depths, or solution can simply be removed by selective sampling. This can be particularly interesting if one anticipates run-of-mine (ROM) leaching, where acidification in agglomeration is not practical and placed ore depth is generally greater.

Also, there is better history in scaling up from column tests to field heaps when the heights of each are equal. Such scale-up is the most critical objective of the metallurgical development program.

There is sufficient experience within certain analytical laboratories and consulting groups to use empirical acid consumption and agglomeration tests to adequately hit a proper target of acid addition and go straight to tall columns after minimal preliminary testing. It is preferable to test more commercial depth columns across a broad range of rock types, depths, and reagent doses than to spend this money on preliminary testing, which may have limited utility to the final result.

With respect to acid consumption and leach kinetics, it is best to conduct column tests in closed-cycle with solvent extraction of the copper and recycle of the raffinate. It is also important to simulate the expected equilibrium composition of impurities which occurs in actual practice. This can be accomplished by simple digestion tests and chemical analysis of the constituents that go into solution or by using the leach solution from a similar mine. One of the more difficult impurity levels to replicate is the iron chemistry, both total iron content and the ratio of ferric to ferrous iron. Generally, column tests ultimately reach higher levels of EMF (ferric to ferrous ratio) than is experienced in practice, at least with finely crushed, high-grade ore. This may be due to lower gaseous porosity in field constructed heaps. It must also be appreciated in batch-wise column testing the reductant, e.g., chalcocite, is not replenished as it is being dissolved. In operations, there is always fresh ore being placed under leach.

Larger Scale Pilot Heap Tests. There may be reasons, or circumstances, that warrant constructing large-scale pilot heaps of 5,000-50,000 tonnes. However, one should understand the limitations of such tests. They do not always provide any better information than a properly planned column test program, due to problems such as:

- sample representativeness;
- inability to construct the heap in the same manner as a commercial heap;
- better attention to solution application;
- better oxygenation; and
- greater attention to details.

Some advantages of large-scale pilot testing may include:

- operator training;
- testing under actual climatic conditions;
- ROM testing requires large tests;
- testing under actual conditions of ore placement method(s); and
- potentially, better "bankability" for large projects.

For on-going operations, large-scale test heaps can be much more timely and cost effective for testing alternative operating conditions, taking into consideration all the compromises discussed above.

However, given the significant expenditure required for such large-scale tests, if no commercial operation is available, this money is likely better spent in providing more detail to the column test program. This detail should include more extensive resource sampling, allowing a larger number of variables to be tested.

Heap Height. What is the proper heap height? There is no easy answer to this question. For a single lift of ore, it is best answered by assessing the implications for a given material with respect to its mineralogy, available reagent, and hydraulic character. It is believed, for example, that supergene material, which is force-aerated and relies primarily on ferric generation by bacterial activity, may be leached at greater depths than oxide material, where acid can not be generated in situ at depth.

Commercial supergene operations have successfully leached single lifts of ore at depths of 10-12 m. Multiple-lift leach designs generally leach in 5-10 m lifts stacked to greater than 45 m in some instances. However, there are compensating factors to be appreciated in all cases, as discussed below.

The uniformity of particle size distribution, which crushing and acid agglomeration provides, can extend the workable depth by improving the percolation character. If the reagent can be provided in situ, e.g., bacterial oxidation of pyrite and ferrous iron with the use of forced aeration, combined with good hydraulic behavior, a greater depth can be tolerated. High-grade oxide (2%-4%) copper, especially for malachite and azurite, may need to be leached in shallower lifts to control the pH to maintain the copper in solution. These minerals consume acid at a significant rate, nearly depleting all free acid. Excessive fines content can be another factor limiting the depth that can actually be leached, regardless of the reagent availability, due to poor solution flow characteristics.

The commercially viable depth for crushed ore heaps appears to be between 2-12 m depending on the trade-offs mentioned. Both application rate and ore depth can have profound effects on the resulting PLS grade, especially if the system is not reagent limited.

One key to understanding the benefits, or disadvantages, of different heap heights is to appreciate the actual field extraction rate and level of extraction when considering the longer residence time allowed by placing ore at a greater depth. Inefficiencies in wetting and reagent availability, which may come with greater depth, may not be offset by the longer leach time.

Interpretation and Scale-up. As mentioned above, it is nearly impossible to combine and incorporate into any single column test the equilibrium conditions that will be achieved in the field. Additionally, changes in ambient conditions, such as temperature and altitude, can not be attained unless the test work is conducted on site. And even then, the ambient conditions within a column are not what are experienced under actual leach conditions.

Many have attempted to mathematically "model" the heap leach process with little success. The knowledge of the chemistry included in these models is generally excellent, but the models fail primarily because they do not properly address the various heterogeneity factors, which are specific to each application. However, empirical models based on actual experience have generally proven to be a reasonable planning tool.

It is critical one understand the typical inefficiencies that exist between actual field heaps and columns. It is the author's experience that small diameter columns perform very close to larger diameter column or "crib" tests – both of which perform better than actual field experience. Figures 1 through 4 show typical laboratory column test results compared to actual field experience for gold, uranium, oxide copper and sulfide copper.

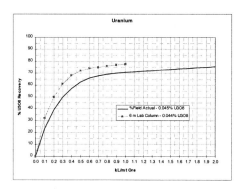

Fig. 1 Typical Uranium Recovery

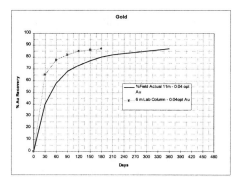

Fig. 2 Typical Gold Recovery

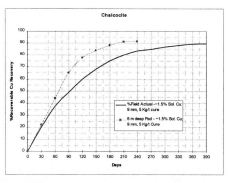

Fig. 3 Typical Chalcocite Recovery

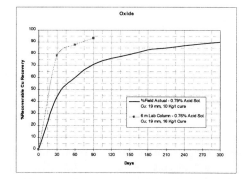

Fig. 4 Typical Cu Oxide Recovery

These inefficiencies are typical for single-lift leaching. Additional inefficiency, or inventory implications, must be added to this when leaching through multiple ore lifts, as discussed below.

Most of the inefficiencies are related to limitations of reagent in contact with minerals. This can be related to the reagent strength in contact with the ore, simply non-wetting, or long diffusion paths due to the non-homogeneity of solution flow through unsaturated medium (O'Kane 1999 and Lupo 2000).

The ability to properly design a heap leach scheme with sufficient leach area and flow rate is solely dependent on estimating a proper field extraction rate from the column testing results. In practice, a heap leach is always working against negative factors, all of which are cumulative in their impact on overall results. Therefore, a successful design must anticipate a full range of inefficiencies and uniformity factors and provide sufficient time to deal with this variability.

As illustrated in Figures 1 through 4, operating experience for four different mineral types show similar inefficiencies. This suggests there are limitations with respect to the physical conditions and uniformity of actual heap parameters, regardless of the mineral. Some of the parameters where striving for more uniformity can be beneficial are:

- ore grade;
- acid consuming character;
- solution application, especially with differing ore depths;
- near surface distribution and wetting;
- uniform permeability;
- uniform acid addition, via agglomeration, if crush size allows; and
- particle size.

Figures 5a and 5b indicate the author's recent experience with the range of recovery that can be expected for commercial leaching of 6-8 m heaps for oxide and supergene ore of about 1% "soluble" copper content.

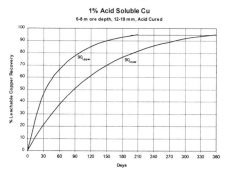
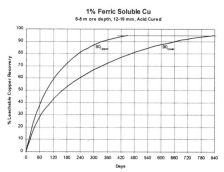

Fig. 5a Typ. 1% Ac. Sol. Cu Rec. Rate   Fig. 5b Typ. 1% Fe^{+3} Sol. Cu Rec. Rate

The upper and lower curves bound where the author believes 90% of commercial operations will fall, depending on the fines content or other parameters that limit wetting or reagent availability, e.g. acid availability for oxides and oxygen restriction due to poor air permeability in the case of supergene leaching. The upper curve is the most optimistic one should expect if the ore is free draining and the fines content is <10 wt% -150 micron (100 M) and <5% -74 micron (200 M). The lower curve is what the metallurgist hopes to avoid if at all possible. However, if fines or degradation can not be prevented, or removed, the lower limit must be analyzed relative to its impact on potential capital and operating costs and recovery.

Recent commercial experience, employing conventional heap stacking methods, suggests that for each 2 wt% increase in –150 micron sized particles, between 8-20 wt%, the recovery time to achieve a similar terminal extraction can be extended by 20%-30%. Experience at Chuquicamata on the ripios (vat tailing) re-leach project shows the importance of understanding the relationship of particle size and application rate to ensure non-saturated flow (Chahbandour et al. 2000 and Guzman et al. 2000). It is with this area of hydraulic understanding where the greatest improvements in heap leaching may eventually be developed.

Ramp-up to Full Production. A critical aspect of the final field extraction curve is the amount of ore that must be on the pads under leach to reach full production. This is not always fully appreciated, but it is very simple. If little ore is on the leach pad at the start of the SX/EW commissioning, full production will not occur on a daily basis under equilibrium operating conditions until the full ore inventory is placed under leach. For example, if the recovery curve to 90% of the "soluble" copper content is the target, and the scale-up curve suggests this will take 500 days, then there must be 500 days of ore under leach. Until this happens, production will be less than initially projected.

If one wishes to start-up at full production, once the SX/EW plant is commissioned, and then maintain that rate of production, there must be 30%-40% of the 500 day inventory of ore on the pad before any leaching begins.

This issue has not been fully appreciated in the past, and therefore, full production took longer to achieve than thought. If the actual field leach curve is not what was initially projected, then this problem is further exacerbated.

Engineering Design

With the support of a comprehensive test program, one can begin to consider the primary engineering design aspects of the heap leach. Some of the principal areas of engineering discipline requiring attention are:

- Conceptual Leach Design
- Pad and Liner Design
- Heap Construction Method
- Solution Application Method
- SX/EW Design
- Engineering and Project Implementation
- Closure Issues

While all of the above disciplines must be considered in conjunction with the other, they are discussed separately below.

Conceptual Leach Design. Once one appreciates the implications of a proper scale-up from a comprehensive test program, the mass balance becomes clear and the options for the various means of stacking and leaching ore can be ascertained.

It is important to emphasize that most heap leach projects are only as successful as the initial design is correct. It is very difficult to correct inadequate recovery response if increased leach pad area can not be readily constructed due to constraints imposed by topography or poor pad geometry. Also, one must make sure there is a long-term mine plan available, which supports the selected plant size and this plant size is compatible with allowing flexibility in operations.

There is no clear or definitive method with which to guide one in planning the conceptual leach pad design. The topography, the size of the ore reserve, operating philosophy, and environment all have to be considered in arriving at the proper choice.

In general, for higher-grade ores (i.e., generally ore grade >1%), where the leach residue can be affordably removed, the on/off, or "dynamic", leach pad is becoming the preferred choice. When the required horsepower in lifting both solution and ore, as well as the added copper inventory, are properly addressed, the on/off pad is generally favored over the permanent multi-lift leach pad. This is regardless of whether ore is leached in a single lift with intermediate liner, or leached through all lifts. The individual advantages and disadvantages of each are discussed from a conceptual design perspective below.

On/Off ("dynamic") – Single Ore Lift. The primary advantage of single-lift leaching is a higher "operating" recovery, which can approach that of the column tests for similar crush size, as long as the data has been properly scaled.

One of the initially considered disadvantages of the on/off pad is the cost of re-handling the leach residue plus the associated cost of that disposal. If a lined area for the leach residue is required for reasons of environmental isolation, then the cost may favor stacking ore on a permanent pad.

A more important hidden cost of this approach is the loss of copper production if the initial scale-up is incorrect. Historically, this has been the case with the majority of these operations. Many have resorted to leaching the residue, but this is not always practical, or possible, due to the residue being placed on un-lined areas or placed in a manner, and at a depth, where the permeability is marginal.

The choice of the type of heap leach method can be dictated solely by topography. For example, an area of initially limited flat surface containing deeper valleys or canyons can begin as a "valley-fill" operation (discussed below) and then change to an on/off operation once the proper flat area is developed.

Permanent Pad. Until recently, the most often practiced form of heap leaching was to stack ore in multiple lifts on top of one another, all over a single "permanent" liner. As mentioned above, one of the primary disadvantages of this concept is considered to be the added cost of operation with time, due to the power required in elevating both ore and water. However, an often overlooked feature of this design is the fact the base footprint of the pad must be sufficient to contain the entire resource and the liner system must be substantial enough to contain a high loading factor. In addition, the topmost lift of ore must provide sufficient "functional" leach area for the full leach cycle, plus the "dead" area required of the stacking equipment. If the base footprint is not sufficient, production in the later years can be severely hampered. An additional complicating factor is the total resource is not always fully known, requiring additional area on which to expand the pad.

Another distinguishing feature of a permanent pad, multiple-lift leach design, where solutions are allowed to flow through the underlying lifts, is the copper inventory tied up in the solution. This is discussed further below. An offsetting advantage of this method is the ore is under leach for a much longer time, until the heap reaches its full height, or an intermediate liner is installed. At this time, the copper in solution inventory can be recovered.

Once a permanent pad is decided upon as the primary design concept, there are two modes of operation – either through single lifts of ore, or through multiple lifts.

Single Lift Leaching on a Permanent Pad. Choosing to leach through a single lift of ore on a permanent pad generally means the ore is marginal in its permeability character. This may be caused by an excessive fines content, either initially or after significant degradation. Another reason to leach through only a single lift can be concerns about re-precipitation or excess solution inventory. The primary disadvantage is similar to the on/off pad, where under-estimation of the necessary leach cycle will result in having to stack over ore when recovery is below expectation.

The operating costs of this design are also higher than might initially be expected, as one has to replace all drainage piping (and forced aeration piping, if required) with each lift of ore. An added cost is also necessary to provide the intermediate liner, either in compaction of the natural leach residue or laying of synthetic geomembrane. Further, with each additional lift, the pumping and stacking costs typically increase due to the added lifting of both ore and water. (Engineered studies of the trade-off in capital and operating costs for the permanent pad versus the on/off pad generally show that the increased operating costs of the permanent pad can actually exceed the life-of-mine total costs of re-handling leach residue for the on/off concept.) Again, other factors, such as topography, can sway the decision.

Multiple-Lift Leaching on a Permanent Pad. Leaching through multiple lifts was popularized in large copper and gold ROM leach operations, especially those of low relative ore grade. This method has the advantage of being able to keep ore wetted for very long periods of time. However, it has significant disadvantages with the increase in copper values tied up in the heap moisture, and potentially, lowered permeability, due to compaction with increased load.

The author refers to this type of heap leaching as "upside-down" in that fresh ore is placed over leached ore. This reverses the chemical potential for displacement of the soluble values. This can lead to a soluble copper inventory representing from <1% of the expected "soluble" copper (that actually dissolved) to as much as 30%. The actual amount of copper tied up in the inventory depends on the combination of "soluble" ore grade, designed PLS grade and the ore's natural water retention character. This combination of PLS grade and moisture has been the primary factor in the under-performance of expected production, in many instances, where this style of operation was used. As mentioned above, this copper is recovered once no more fresh ore is stacked on top of old ore and the full leach column can be washed with raffinate or water.

Unlike gold, copper solvent extraction plants are generally more capital intensive and nearly directly proportional to the flow rate. Therefore, the tendency is to maximize the PLS grade to keep capital cost low. This can be exactly the wrong thing to do with low-grade ore utilizing a multiple-lift leach design.

One of the errors with many multiple-lift leach designs is planning for too little recovery from the top lift, expecting to get this in an underlying lift, and then losing operating control in the lower lift. This is a tricky decision.

There are many interesting questions concerning the in situ conditions in lower lifts of ore, which must be anticipated and a judgment made regarding the lower ore lifts. For oxide ore, there is the question of increased incremental acid consumption that may be uneconomic relative to the marginal copper dissolved at depth. For sulfides, the chemistry is much more complex. There can be advantages with increased depth regarding temperature development, but there can be acid balance or precipitation issues that are very complex. A flexible design is probably one that achieves an economic level of extraction in the top lift and leaves the "marginal" recovery, from say larger particle sizes, to the lower lifts. Intermediate liners at some point can be used if adverse solution chemistry or permeability problems are encountered. However, one problem with the latter option is that at the time of the placement of an intermediate liner, copper production will decline by the percentage of copper that was being dissolved in the lower lift(s), unless a system of irrigation can be provided for the lower lifts. The latter option is most likely quite problematic, and it is not known if this has been commercially proven.

The initial mining plan for cash flow estimates must recognize the added ore that must be mined on a current basis to address the inventory required of the "reversing" of the chemical potential. This increases the copper in solution inventory with each additional ore lift by a nearly constant amount. This copper can be considered "delayed" production. It can be displaced once the total ore depth is reached, generally at the end of the mine life, if the pad is monolithic in design. Historically, this additional inventory has not been provided for in advance, which causes shortfalls in production. To deal with this increased inventory and subsequent dilution effects, there can be a requirement to recycle as much as 30% of the total leach solution flow. This can also require a proportionally larger surface area – something not always fully appreciated initially.

<u>Valley-fill Leaching.</u> A hybrid of the above two leaching designs is the valley-fill design as practiced at Compania Minera Cerro Verde, Arequipa, Peru. If adequate topography is available, where the slopes of the hills or sides of broad canyons can be adequately lined, leaching can progress while ore is being placed in a fashion that ultimately fills the canyon.

One of the advantages of this concept, besides probably lowest cost in liner per tonne of ore stacked, is the leach area can increase with time. This more naturally matches the needs of flexibility in area for a maturing multiple-lift leach design. This is exactly opposite of the permanent pad concept, where the base footprint is a fixed size and leach area is lost with each successive lift. However, one must anticipate the ever-expanding drainage volumes that the initial pipe works may have to handle. Flooding of the ore must be prevented, as it can cause a significant increase in copper values tied up in solution and may result in ore instability.

An additional advantage of a valley-fill design, in special circumstances, is that if a dam is placed downstream of the toe of the heap, the natural valley can allow for storage of solution within the heap. This has advantage in colder climates and has been used mostly in gold and silver heap leaching. However, the added hydraulic head on the liner system in the area storing the solution requires a more robust liner design than typical.

Pad and Liner Design. The following discussion concerns the basic geometric design aspects and options for evaluating and designing a functional leach pad system. It does not address the actual environmental standards and options of the pad liner itself, which is a critical and necessary component of a properly designed pad. The ability to construct a "permit approved" leach pad is of paramount importance. In fact, it can be the deciding factor in the final design when the local topography is taken into consideration. The actual engineered construction of the leach pad liner system is a subject unto itself. This paper addresses primarily the mechanics and systems unique to the chemistry and physical parameters to meet the production goals. Others address the subject of proper pad liner design and stability (Breitenbach 1997, 1999, 2001).

A proper pad design must account for a multitude of considerations – all of which are different from site to site. Some of these are:

- Liner System
- Topography
- Required Pad Capacity
- Solution Management Alternatives
- Flexibility
- Climatic Factors

Liner System. The liner system is a key economic and environmental consideration. The final selection generally requires a systematic evaluation of geography, topography, local materials and overall stacking design. This requires the input of experienced geotechnical engineers, taking into consideration all environmental permitting requirements. Regardless of the environmental considerations to limit solution loss, the economic impact of lost PLS must be considered in any design.

The primary liner designs generally fall into the following categories:

- Compacted local clays or silty sediments (seldom used anymore);
- Synthetic geomembrane liners; or
- Composite liners of geomembrane laid over local clayey material.

The hydraulic head that the liner system will experience influences the liner design. Ponds are generally exposed to much higher hydraulic head, which requires the use of a double, or triple-lined system with an intermediate zone of high transmissivity for collection of solution leakage through the topmost liner. This ensures a minimal head on the lower liner(s).

<u>Compacted Local Materials.</u> Local materials containing 30%-35% fines with some plasticity can often be excavated and reapplied in multiple 3"-4" lifts, after moisture conditioning and with compaction, to attain permeability of 10^{-6} to 10^{-8} cm/sec, or better. Thicknesses of 12"-18" can produce environmentally acceptable liners. However, this application, today, is less common than the use of synthetic geomembrane and composite lined systems. The primary disadvantage is quality control and prevention of desiccant cracking, which can compromise the integrity of the initial installation.

The primary advantage can be lower cost and a more pliable liner that can better accommodate settlement. An additional advantage is that the natural permeability can allow leakage equal to the future drainage requirements, after reclamation, while allowing the soils to naturally attenuate the seepage.

Natural liners today are primarily used today for emergency overflow and low-head impoundment's to contain temporary spills.

<u>Synthetic Materials.</u> Synthetic geomembrane is, today, by far the most favored lining material. It is used individually or in combination with other synthetics or natural materials. The principal types of geomembrane lining materials are:

- PVC and CPVC;
- HDPE; and
- LDPE

PVC and CPVC. Polyvinyl chloride (PVC) was one of the primary lining materials for heap leaching prior to the advent of HDPE, but it had the disadvantage of low ultraviolet resistance. This was improved with the introduction of chlorinated polyvinyl chloride liners (CPVC).

The PVCs have excellent multi-directional elongation properties and are often used where settlement or concerns for puncturing are significant. Sheets are generally seamed with glue, and as a consequence, are considered easier for less-experienced operators to install.

The general range of thickness used is from 0.5-1.0 mm.

HDPE. High-density polyethylene has become a widely used geomembrane, due primarily to its ease of installation, with extrusion and fusion welding techniques, combined with a high ultraviolet resistance. It has a further advantage in being manufactured with varying surface roughness. This can improve heap stability with respect to interface friction properties between liner, sub-base and ore.

This material is excellent for pond and ditch installations due to its UV resistance. Also, it may provide better puncture resistance, although this is probably due more to the greater thickness generally applied, e.g., 1.5-2.0 mm.

A primary disadvantage of the HDPE is that it exhibits poor multi-directional elongation character. It is generally good in only one direction and will tear easily in the opposite direction when under stress. Large areas of exposed HDPE liner can split by stress cracking, by a combination of temperature fluctuations and cold working of the liner, or excess or poor distribution of carbon black in the manufactured sheet.

Ponds are nearly always double-lined with HDPE on top and HDPE or PVC beneath, with a leak detection and collection medium placed in between.

LDPE. Low-density polyethylene was developed primarily to address the most significant disadvantage of HDPE, e.g., the unidirectional elongation property. This material is often selected when liners are placed over compacted fill or soils that will settle significantly simply due to the load of ore being placed. These materials have similar properties to PVC in this regard. Common variations on typical LDPE sheet are very low density (VLDPE), very flexible low density (VFDPE) and linear low density (LLDPE).

A variation on LDPE is reinforced or unreinforced polypropylene (PPE), which has certain advantages in cold climates such as Alaska and Canada. This material has the best properties of HDPE and PVC, but it is also more expensive.

The primary thickness of this material is similar to HDPE at 1.5 – 2.0 mm.

Composite Liner Systems. The liner system that is most favored for use underneath heaps is a composite system where natural materials are either scarified in place, or applied in multiple lifts, moisture conditioned and compacted to a permeability of about 10^{-6} cm/sec. Then a geomembrane is placed directly on top of this natural material. The combination of the two liners provides a robust liner system. The underlying natural soil acts to plug holes that may develop in the geomembrane, minimizing leakage that might occur had there been no resistance to flow.

Whatever system is used underneath ore stacked in multiple lifts, it must be assured this system does not fail. Any failure of the liner system leads to taking a significant portion of the heap out of production, which can be fatal. The composite liner system has proven to provide long-term environmental service as long as the design of the liner system does not fail due to massive failure from ore movement. It is the latter that must be properly engineered. This is an area of design reserved for only the most experienced geotechnical engineers.

All the liner options require the placement of an over-liner material to protect the liner from puncture during placement of drain pipe and ore, to act as a drainage layer and to lower the hydraulic head on the primary liner. A network of drain lines placed on top of the geomembrane or the natural clay liner, and subsequently covered with a coarse, free-draining medium, generally provides this. It is the placement of this drainage layer of gravel, or coarsely crushed ore, that provides the greatest potential for damage to a synthetic liner. This area of pad installation is often overlooked in design, but plays a critical role in the overall performance of the total liner design and operation.

This is even more critical if the planned leach design is for a permanent pad that will leach through multiple lifts. A material must be selected of sufficient particle size and acid refractoriness so it can perform throughout the life of the project. Even if one is fortunate to find such a material locally, it is always cheap insurance to provide additional drainage pipe on top of the over-liner in case this becomes plugged.

Drain piping is generally laid on top of a protective layer of fine material placed directly on the geomembrane. The fine material is necessary to prevent puncture of the geomembrane when placing the coarse drainage blanket. The coarse drain blanket provides for a truly vertical drainage from the entire surface cross section of the ore. Both the drainage layer and the pipe works provide a high transmissive flow path from under the ore. Care should be used in the selection and application of the drain piping (Lupo 2001).

If single lift leaching is to be practiced by stacking in multiple lifts, with intermediate liners installed, e.g., Monto Verde and Escondida in northern Chile, then drainpipe must be placed in the ore and subsequent drainage blankets may not be necessary. In this case, pipe spacing must be matched to the natural permeability of the material to prevent excessive flooding of the ore. This is especially important if the ore is supergene and oxygen is an issue.

Considerable geotechnical engineering has been developed that can appropriately address the liner design options. There is plenty of experience in the industry to show what happens if these issues are not properly addressed (Breitenbach 1999).

Another key element of the drainage blanket is that it minimizes the hydraulic head on the liner, therefore reducing the driving force for possible liner leakage. Some operators of on/off pads that are operating on top of leached residue have the added advantage of providing auxiliary drainage ditches underneath the ore, in addition to the normal drain coil. This further ensures a low phreatic head on the liner and aids with aeration of sulfide ores.

Topography. The available topography can be the primary factor in the final design. Equipped with a properly scaled-up field recovery curve, it is possible to fit one of the pad construction concepts to most any topography, provided the environmental lining issues are solvable and economic. It is critical one understand the compromises on recovery that each option presents and design to mitigate that in an appropriate manner.

Leach pad area and siting should always take priority over the placement of the solvent extraction and electrowinning plants, and supporting infrastructure, which can be placed on structural fill, if necessary.

Another element of topography is the geographical location and associated ambient conditions. Ambient temperatures and precipitation may be better addressed with a specific pad design.

Required Pad Capacity. The Engineer responsible for pad design must fully analyze and provide for a functional leaching scheme through the entire life-of-mine reserve. This is a frequently observed oversight.

A square leach pad, for multiple-lift leaching (or leach residue disposal), has the greatest volumetric capacity per unit area of liner. Rectangular pads are the next best option. Many multiple-lift leach pads that conform to the local terrain and thereby become triangular in shape have been constructed. This type of pad will lose area at an accelerating rate with additional lifts, potentially making it non-functional much sooner than expected.

The principal design requirement to properly size a leach pad requires knowledge of the scaled-up field leach curve, combined with a proper understanding of the total volume of the resource to be stacked. The final pad design should then be analyzed for the latter years of operation to ensure the pad will remain functional in terms of maintaining production.

Solution Management Alternatives. Again, equipped with the proper scaled-up recovery curve, it is possible to assess the different solution management options and the implication of each option on PLS grade. Mass balances for various solution management schemes and application (or irrigation) methods are critical to developing this understanding. Different solution management schemes can impact both pad and liner design.

Once one appreciates the effect that inefficiencies from "actual" field performance has on the resulting PLS grade, it becomes clear that one must deal with a lower grade PLS than implied by a strict interpretation of column test results. If one wishes to recover a level of copper near to the terminal extraction achieved in similar depth columns, experience shows it must come from a near doubling of the volume of solution applied in the lab columns or crib tests.

Historically, most heap leach operations were of single-stage solution application design. Raffinate was applied to all the ore and all underflow solution was combined in one collection system. Some recycling of solution could also be practiced with this design. During the 1990s, staged-leaching (both counter-current and concurrent) became more accepted. This requires the ability to segregate underflow solution from different sections of ore on the pad. Some operators prefer to call this "solution stacking" or "up-grading". This is technically the result, but the primary reason for doing this is it allows for the near doubling of flow necessary to obtain the desired recovery, without doubling the SX plant capacity. This is generally the most economic alternative to the problem of heap inefficiency.

To determine if it is economic to strive for the level of extraction obtained in the column test program, it is necessary to assess the added pumping and reagent costs associated with the increased leach time and volume of solution required on an incremental basis.

Unlike gold leaching, there can be a reagent consumption reason to consider a "backward" staged-leach concept. This was first utilized on a significant commercial scale at Girilambone Copper Company in New South Wales, Australia (Lancaster and Walsh 1997). If acid consumption is a significant cost, one can apply the raffinate, which is highest in acid concentration, to the freshest ore, and then apply the resulting PLS, which is high in copper but low in acid, to the oldest ore. This minimizes the driving force to consume acid on old ore, which does not necessarily need it. The offsetting cost is the added copper inventory, which can be less expensive than the acid consumed in leaching through just one lift.

There are numerous design options that can provide the increased volume of solution necessary to achieve the targeted column recovery in actual field practice:

- one can simply add more area, continuing to apply the same solution application rate, but treat increasing volumes of solution at lower PLS grade through SX;
- one can provide for the capability to segregate solutions reporting from different ore zones that are at various times in the leach cycle or have different ore grade. Then low grade PLS from one zone can be advanced to another zone in a concurrent fashion (or counter-current depending on the solution management philosophy). This results in an up-grading of the PLS grade and significantly reduces the SX flow that would be required absent this ability;
- one can recycle a significant volume of solution, up-grading not only the exiting PLS grade, but the entire solution inventory retained by the heap; or finally,
- assuming sufficient area, and especially if leaching secondary sulfides, a rest/rinse approach for the later stages in the leach cycle can be used to maintain a higher PLS grade in less total solution flow.

The selection of one or more of the above requires a trade-off of capital cost, soluble copper inventory in heaps and operating complexity.

<u>Commercial recovery enhancement techniques</u> – A number of recovery enhancement techniques, relating to the subject of solution management, are worth mentioning. In some commercial cases, "after-the-fact" changes to operations have been necessary to achieve the expected recovery, or to improve on that.

The leaching of copper in a heterogeneous unsaturated pile can not be expected to perform better than the washing of values from an agitated pulp where all the values are already in solution. For vacuum drum filtration it takes 1.0-1.2 tonnes wash water per tonne ore to remove about 98% of the soluble values. CCD washing would require nearly 4-5 tonnes wash water per tonne ore in 3-5 stages to achieve a similar level of soluble recovery. Therefore, how can one expect both the dissolution and displacement of copper values from 1% acid-soluble ore, designed to produce 3-5 gpL copper in PLS, to be achieved in <1 tonne water per tonne ore? Yet, several large facilities were constructed during the decade of the 1990's to such a standard. Most supergene heap leach operations, with ore grade of 1.5%-3%, were designed to <4 tonnes water per tonne ore. Experience suggests the requirement is nearly double this, depending on reagent availability and ore percolation character. (The use of rest/rinse leaching on secondary ore holds promise to reduce this volume significantly, but not necessarily the time or surface area.) Ores with poor percolation character can take two to four times this amount of solution, if one wants to obtain the same level of extraction as experienced in laboratory columns of similar depth.

Nearly every large copper heap leach constructed during the decade of the 1990s initially under-performed its stated goal for recovery by 10%-20%. Nearly all these facilities adequately addressed this issue and improved "operating" recovery by essentially doubling the flow and adding the ability to stage-leach, or recycle a significant volume of solution. Unfortunately, this took most operations 2-4 years to accomplish – at a significant loss of copper and cash flow. What is important to understand is that, in every case, the wetted surface area of ore nearly doubled.

To improve the solution wetting character and to ensure better utilization of acid as a reagent, acid agglomeration provides a significant benefit. Concentrated acid agglomeration has become a standard practice for most oxide and supergene copper heap leaching projects where the ore is crushed. A primary advantage of agglomeration, in addition to uniform fines distribution, is that most of the reagent to satisfy the gangue acid requirement is intimately mixed with the ore – resulting in a more favorable pH profile occurring throughout the depth of ore. This helps sustain the copper in solution and prevents pH excursion in situ, which can cause precipitation of copper or iron.

There have been post-leaching attempts at moving the ore to establish new flow paths. Some operations have attempted to blast leached ore to improve recovery. To the author's knowledge, this is of limited value and quite site specific.

Drilling of heaps, followed by injection of solution, has also been attempted as a means of improving extraction in cases of poorly wetted ore (Figueroa, Ruiz and Ayala 1999).

Another technique, first pioneered by Girilambone Copper Company on a significant commercial basis, is to re-mine partially leached heaps in situ (James and Lancaster 1998; Weston and Dudley 2000). This may be limited to ore lifts of 6-10 meter depth, but it has proven beneficial in certain situations. A hydraulic backhoe is used to cut an initial trench in a well-drained heap. Material from the top of the next cut is placed in the open trench by gently placing the material and allowing it to roll into the trench. New aeration pipe can be installed if necessary. By turning the material in this fashion, then re-piping and re-leaching, recovery has been enhanced from 2%-10% with one or more re-mining efforts. This practice has recently been used in Chile by several operations.

Flexibility. There are two primary options available to deal with a potential under-estimation of the "field" leach rate:

- add leach surface area; or
- increase ore throughput rate.

Leach Area. This is really the only option available to obtain the projected recovery if it is not being met, assuming all conditions affecting leach chemistry and percolation uniformity have been addressed. However, the ability to increase leach area also means there must be flexibility in pumping capacity and potentially ready expansion of SX capacity, unless a rest/rinse scheme on the expanded area appropriately deals with the PLS grade. SX flow capacity can sometimes be addressed by converting an SX plant from a two-stage, series-extraction configuration to either of two parallel, single stages of extraction or a series/parallel configuration, with the addition of one additional mixer-settler unit. A flexible SX design that allows for this is generally cheap insurance.

Increased Throughput Rate. Increasing ore throughput rate is the primary way many gold mines operated during the decade of the 1980s, operated to maximize production and to lower the per ounce unit cost. It is done at the expense of "operating" recovery in a multiple-lift operation, unless more leach pad is added. It also lowers overall recovery and shortens the economic life of the operation if the leach is a single-lift design. For this option to be available, if the ore must be crushed, there must be significant flexibility in crushing, conveying and ore placement capacity -- either initially, or as a readily expandable option. This is not always practical, which is the reason why it is important to have a properly scaled-up "field" leach curve from which to design the facility.

Climatic Factors. Heap leaching is less problematic in warm, dry climates. As precipitation increases, the water balance becomes much more important to overall performance. In areas of high seasonal rainfall, PLS can be diluted and excess solutions must be retained in storage ponds to be used during high evaporation periods. Spray irrigation may have to be used to enhance evaporation, and this can be counter-productive, or incompatible with heap permeability. Site drainage of storm run-off is a critical component of any pad design. The operating plan should limit intrusion of storm water to that which actually falls on the pad area. In geographical regions where the "net" precipitation is positive, treatment and discharge of a portion of the solution inventory may be necessary.

In dry, warm climates, the type of solution application used can significantly impact evaporative water losses. Spray systems can evaporate as much as 15%-20% of the pumped volume, while certain drip emitters can reduce this to less than 1%-2% by injecting solution below the surface.

Cold climates have low net evaporation and potential net positive water accumulation, which can present special water-balance and treatment issues. An additional concern is the temperature of process solutions, which can severely restrict the extraction rate. This is often overlooked in column testing, which is typically done under warmer ambient conditions at the laboratory. This can be very significant in the overall leach pad area required.

Heap Construction Method. Heap construction practice plays an important role in the overall performance of a heap leach. It is a physical activity that deserves the same level of scrutiny and thought as the selection of the pad design itself. There are two principal options:

- Truck dumping; and
- Conveyor stacking.

Truck Dumping. In the early days, transport of ore and dumping on heaps with trucks was the primary method of heap construction. This method has the greatest flexibility, but leach results can be severely hampered by the damage truck traffic does to the magnitude and uniformity of percolation. Ores that lend themselves to ROM leach particle size have been very successfully leached using this method. Once crushing becomes the desired approach, especially with addition of moisture and acid to the ore, compaction from truck placement becomes problematic.

When truck dumping is used, the entire pad surface area requires ripping, at least twice in perpendicular directions, to relieve near-surface compaction. Studies (Urhie and Koons 2000) have shown that areas of truck traffic must be ripped to 2.4 meters to properly relieve the effects of compaction.

Conveyor Stacking. The expanded use of conveyors to stack ore was one of the more significant contributions of the gold industry to heap leaching practice. This occurred primarily during the decade of the 1980s. However, one of the first applications of conveyor stacking may have been the waste disposal system designed by R. H. Hensler for the Twin Buttes Anamax copper mine south of Tucson, Arizona in the mid-1970s. This inspired Ranchers Exploration and Development Corporation to install one of the first spreader conveyors at its uranium heap leach project in 1978 (Scheffel 1981).

There are primarily two types of conveyor stacking methods:

- mobile conveyor units, referred to as grasshoppers, used in conjunction with a radial stacker; and
- spreader conveyors, which span the entire width of the leach pad.

Mobile Conveyors and Radial Stacking. One of the primary advantages of the mobile system is that it provides some of the advantages of truck dumping with respect to flexibility. However, it is still generally limited to constructing multiple-lifts in a single monolithic fashion. (Trucks can stack in separate areas at different overall lift heights and depth.)

These systems can easily be under-powered and insufficient in structural detail, especially if multiple-lift stacking is anticipated. Therefore, an engineer is well advised to consult with suppliers that have considerable experience with this type of stacking and conveying.

Figure A illustrates a multiple-lift, permanent pad design using mobile grasshoppers and a radial stacking conveyor.

Besides the normal segregation in the vertical plane inherent with any stacking method, radial stacking has additional, subtle features of ore segregation if the radial stacker does not tram continuously. Discontinuities in ore placement, or erratic blends of ore, can cause "pipes" of coarse material to be retained against the discontinuous surface. This can result in short-circuits. For some ores of high clay content, this can actually be used to ensure flow through the ore, even though it is known to cause short-circuiting. In this case, one must rely on diffusion to leach the ore and short-circuit paths to provide the displacement stream without washing out the heap. (This method greatly increases the leach cycle, however.)

Spreader Conveyors. Spreader conveyors can have an advantage at higher tonnage rates, e.g., >1500 tph or when greater than 9-10 m ore depth is a goal. These systems are best for on/off pad designs as they become limited in ability to stack in multiple lifts. Some systems have stacked ore to about 30 m in several lifts, but elevating the ore, while feeding a continuously moving spreader, becomes more problematic with increased height.

A spreader conveyor may also have an advantage in certain topographic situations where earthworks can be minimized by using a spreader to stack over undulating surfaces, while creating a level top surface after the first or second ore lift, e.g., Compania Minera Zaldivar in northern Chile.

The spreader conveyor concept can also be used for unloading "dynamic" pads using a bucket wheel excavator, e.g., El Abra, Radomiro Tomic and El Tesoro in northern Chile. A front-end loader and mobile hopper arrangement can also be used.

Spreader conveyors do not have to be as structurally robust as than grasshoppers or radial stackers, as they generally do not have to deal with the same level of stress.

Figure B illustrates a "dynamic" leach pad utilizing a spreader conveyor to load the pad and a reclaim bucket-wheel and conveyor to unload the pad.

Solution Application Methods. Heap leaching is really an irrigation problem, and the more uniform the solution distribution -- the better the overall results. The agricultural irrigation industry has provided great service to the mining industry in the last twenty years. The multitude of application equipment and piping today is almost overwhelming.

Operators often find it necessary to test several different types and manufacturers of irrigation pipe works to arrive at the preferred method for a given site. The three principal methods are:

- flood irrigation;
- spray irrigation; and
- drip irrigation.

All have advantages and disadvantages, and the use of one or the other can become an operator-based preference, dependent on such things as maintenance, cost and ease of operation. In some cases there may be no choice, owing to ambient temperature, evaporation, percolation or degradation issues.

Flood Irrigation. Flood irrigation is rarely practiced today. It can have application in cold climates if the permeability and stability of the ore allow it. Solution can be applied under ice in small ponds to ensure flow in extremely cold climates. This technique was used at Ranchers' Naturita uranium leach operation where winter temperatures reached -30°F.

Kennecott Copper Corporation used this method for many years at Bingham Canyon prior to the discovery in the late 1960s that this method was restricting convective airflow through the dumps.

Spray Irrigation. Rainbird^{TM} impact-style sprinklers were one of the first systems of solution application that found acceptance. However, due to wind and evaporative loss, this style of solution application lost favor to the Senniger^{TM} style "wobbler". The evaporation from this style of solution application can range from 5%->20% depending on wind and climatic conditions.

Drip Irrigation. In an effort to reduce water losses and prevent particle degradation to the surface of marginally permeable ore types, various methods of drip irrigation were developed very early in the modern history of copper heap leaching.

One of the first attempts at drip irrigation was most likely the "patented" needle valve technique used by Ranchers Exploration at the Bluebird mine. Others used small diameter surgical tubing, where the pressure caused the tube to flop around and spread the solution over a limited area.

However, the primary drip irrigation system used today is that developed by the agricultural industry. Many styles and types exist, but all rely on a labyrinth installed directly in the pipe or just external to the pipe. The typical labyrinth drip designs can range from 1.5-8 l/h, or more. This allows for control of very low application rates if desired.

One of the disadvantages of the drip irrigators can be the need to filter the solution, if possible, or continuous maintenance of the drip piping is necessary to ensure uniform operation. To date, filtration of the raffinate from SX operations has not been practiced on a wide scale due primarily to high capital costs associated with materials of construction, combined with a significant operating cost. Plugging of the drip emitters, with the accompanying non-uniform distribution of leach solution, is a continuous maintenance concern for most operations. Current practice is to flush each drip line on a regular basis to remove organic and solids that accumulate with time.

Some operators appreciate the fact that drip irrigation allows for a very low application rate and very slow initial heap pre-soaking to prevent forming channels.

One of the primary advantages of drip irrigation in copper heap leaching is the conservation of heat. This is critical to supergene leaching and still of consequence to acid leaching of oxides. This has forced most operators of supergene ores to use drip irrigation and deal with the plugging issues. An additional advantage with drip irrigation is that it allows for covering of the heap with "thermofilm" plastic, which has shown significant solar heating capabilities at several Chilean operations.

Regardless whether the choice is sprinklers or drip irrigation, one of the often overlooked design features is the changing head along the length or width of a heap and the need to provide the necessary pressure regulation to ensure uniform solution distribution. Rarely does the first installation of the Engineer become the final system used, due to this oversight or changing conditions.

Another important design aspect of the irrigation and collection system is the need to minimize the suspended matter that can report to the PLS pond. This is currently dealt with by placing open drainage ditches, such that an intermediate solution (ILS) collection ditch is placed closest to the heap and the PLS ditch to the outside in staged-leach facilities. This prevents the solids washed off the heaps from reporting directly to the PLS pond. The next mitigating measure is to provide a sediment basin ahead of, or in, the PLS pond. This method, however, will not deal with the colloidal suspended fines – which may require a flocculant.

Very low drip irrigation rates may help by not mobilizing as much of the initial fine material generally washed from fresh heaps. Agglomeration helps in this regard as well.

Solvent Extraction / Electrowinning. As was mentioned earlier, the copper heap leach industry would not have grown in popularity, and to the level of significance it enjoys today, without the development of the solvent extraction reagents. In addition, the continued improvement and understanding of the electrowinning chemistry and physical equipment, especially the practice of plating on stainless steel cathodes, have contributed significantly to copper SX/EW acceptance.

While attaining the required leach recovery remains the primary factor in achieving financial success with a copper heap leach, designing and constructing flexible and workable solvent extraction and electrowinning plants is still a key consideration. Once constructed properly, the operating costs in this area have become predictable and reliable, and potential improvements in costs will pale in comparison to a 1%-2% increase in leach recovery.

Solvent extraction has become so reliable, with no reported or known actual failure of the reagent, that piloting of the solvent extraction and electrowinning system in conjunction with a column leaching system is not really necessary. It has generally been accepted with the larger capitalized plants that investment bankers require this piloting. However, this money may be better spent ensuring the representativeness of the samples tested and conducting a comprehensive column test program to ensure that recovery will be met.

The closest anyone has come to the solvent extraction reagent failing are the recent experiences at Lomas Bayas, Radomiro Tomic, and El Tesoro in northern Chile. The potential affects of nitrate on reagent degradation were recognized by reagent suppliers, but the test programs apparently did not properly identify the occurrence and magnitude of nitrate concentration in these cases. This problem was adequately addressed with the use of a reagent with greater resistance to the degradation caused by the nitrate. One of these operations conducted extensive piloting of the SX and EW and still did not appreciate this problem. This may have been due to non-representativeness of the sample or under-achievement of the eventual steady-state nitrate level during testing.

Generally, the primary concern for SX reagent suppliers is making sure there is a proper understanding of the solution chemistry, especially with respect to certain impurity levels. The principal impurities of concern are silica, chloride, fluoride and nitrate. Determining if any of these will cause problems can be established by simple analysis of the PLS produced during a properly designed test program utilizing representative samples, with columns run in closed-cycle, such that steady-state levels of impurities can be estimated.

Engineering and Project Implementation. Even if one pays attention to all the above details, a successful heap leach can still be compromised if the entire process of engineering and project implementation is not properly addressed. There are many approaches and options in this regard.

Rarely do engineering firms have the time or resources to do postmortems on previous projects. If an Engineer is contracted to do such a project, he is well advised to review the projects the firm has completed in the past. Locating the primary project engineers and then following up with the owner, if possible, can be very beneficial to the potential success of the new project. The review should determine what was done correctly, and what could benefit from additional attention to detail.

Closure Issues. No heap leach today can be permitted today without proper consideration of, not only operating environmental issues, but also those that continue beyond the life of the project. Management consideration of these issues must begin early in the resource evaluation phase by developing an appreciation for the acid generating or neutralization character of the mine waste, as well as the leach residue. Again, topography and climate are primary factors in the potential impact that closure costs may have on the viability of the project.

A reclamation plan is generally a standard requirement of permitting today. The long-term, post-operational stability of the heaps, pits, and waste dumps must be ensured. Erosion and water run-off schemes must be designed and implemented to protect the local surface and ground water from possible contamination. The case study for the Kidston Gold Mine in North Queensland, Australia (Currey and Ritchie 2001) provides an excellent summary of the issues involved with understanding the implications of mine closure very early in any heap leach project evaluation. A supplemental paper (Williams et al. 2001) addresses the engineering aspects of such a mine closure.

Postmortem Analysis

An often overlooked discipline that can be helpful to both the engineer and the owner, is a critical assessment of past operations. This is especially important when dealing with a system that many in the practice still call an "art". The use of that term should be the first alert as to the importance of comparing and benchmarking against known successful operations and to study and understand the exact reasons for failure when these occurred. Too often, engineers do not want to spend time on the negative, when in fact, it can be the best education and advice available.

Another seldom-used corporate exercise is to perform a financial, as well as a technical, "postmortem" on one's own activities. Companies are reluctant to do this as it tends to accentuate the negative, or at least there is that connotation. However, this is healthy for any company or industry.

Heap leaching, as a unit process, suffered greatly with the investment community during the decade of the 1990s when certain of these experiences from gold heap leaching were misapplied to the leaching of copper on a very large scale, in $400-$1000 million dollar projects. Some of the subtleties of heap leaching were apparently not as well understood as many companies and investment bankers thought.

Within a span of four years from 1993-1996, as many as five large-scale heap leach operations were developed in Chile alone. If one reviews these operations with the investment bankers, it is clear that they under-performed their financial projections in the eyes of the financiers. This placed many in positions of having to re-finance their loan portfolios at significant cost to the company, while placing severe constraints on operating management for the capital expenditures required to address this issue. It is to the credit of the financial community, as well as the owners and operators, that they successfully addressed the recovery issues and improved their overall performance.

Figure 6 illustrates the historic copper price trend after adjustment for inflation and identifies the start-up date of some of the major copper heap leach projects that were constructed during the 1990s. The question remains, "What would have happened to these companies or operations had their timing been nearer to the time the copper price broke toward historic lows from 1997 to the present time?"

Figure 6 Historical Copper Price Trend, 1958-2000

One economically marginal operation, unfortunate enough to be commissioned near to the peak in the cyclical copper price, has none-the-less managed to survive this severe downturn while meeting its financial obligations. This is due to the fact it achieved the performance initially projected from testing and was able to produce copper at a direct cost below that predicted in feasibility. This operation was fortunate to have met the scaled-up recovery rate, with a reasonably conservative design, combined with recognized positive leaching characteristics of the ore.

SUMMARY

To the layperson, many copper heap leach, SX/EW operations look very similar, leading one to conclude they are. As indicated herein, nothing could be further from fact, as there is a multitude of site-specific details and decisions that require individual treatment prior to arriving at the final design.

The copper industry has come a long way in the last thirty years in improving the "art" of heap leaching. However, at the beginning of the 21st century, one must still rely on experience, intuition, and well-placed judgment to a great extent, in arriving at a workable and economic design for a successful heap leach operation.

Many areas for improvement remain. Some of these are: With respect to leaching: a better understanding of the hydraulic parameters affecting unsaturated flow in non-homogeneous material; production of less fines in crushing, better control of fines in the total process, and the leaching of chalcopyrite: With respect to SX/EW; one could expect improvements in solvent extraction contactors and settlers and improvement in, or elimination of, lead anodes in electrowinning.

It is hoped that future heap leach projects continue to learn from the past and enjoy improving success as time passes – allowing the new generation of engineers to truly bring the "art" of heap leaching into the realm of "science".

ACKNOWLEDGEMENTS

The author wishes to thank and acknowledge his colleagues whom have many years of experience in their respective disciplines and who graciously assisted with editing and comment to various sections of this paper. Specifically, the author wishes to thank Mike Bernard and Ron Kelly, Terra Nova Technologies for their conveyor stacking expertise; Alan Breitenbach, Independent Consultant, for his geotechnical expertise; Joe Schlitt, Kvaerner Metals, for his peer review from an engineering perspective; and David Readett, Straits Resources Ltd., Gerald Fountain, Cambior USA, Inc, and Marcelo Jo, Compania Minera Cerro Colorado, for their world-wide peer reviews from an operating perspective.

REFERENCES

Aldrich, H.W., and Scott, W.G., "The Inspiration Leaching Plant", Rocky Mountain Edition, A.I.M.E., Vol. 106, 1933, pp. 650-677.

Baum, W., "The use of a mineralogical data base for production forecasting and trouble shooting in copper leach operations", Proceedings of Copper 99-Cobre 99 International Conference, Vol. IV – Hydrometallurgy of Copper, ed. S.K. Young, D.B. Dreisinger, R.P. Hackl, and D.G. Dixson, The Minerals, Metals and Materials Society, 1999, pp. 393-408.

Baum, W., "Optimizing Copper Leaching/SW-EW Operations with Mineralogical Data", SME Annual Meeting, Phoenix, AZ, March 11-14, 1996, pre-print No. 96-84.

Bradley, C.P., Sohn, H.Y. and McCarter, M.K., "Model for Ferric Sulfate Leaching of Copper Ores Containing a Variety of Sulfide Minerals: Part I. Modeling Uniform Size Ore Fragments", Metallurgical Transactions B, Vol. 23B, October 1992, pp. 537-548.

Breitenbach, A. J., "Overview Study of Several Geomembrane Liner Failures Under High Loads", Geosynthetics 1997 Conference Proceedings, Industrial Fabrics Association International, Long Beach, California, March, 1997, Volume 2, pp. 1045 to 1062.

Breitenbach, A. J., "Design and Construction of Heap Leach Pads". The Mining Record, Littleton, Colorado, March, 1999,Volume 110, No. 9, pp. 44 to 45.

Breitenbach, A.J., "The Good, the Bad, and the Ugly Lessons Learned in the Design and Construction of Heap Leach Pads", Copper Leaching, Solvent Extraction, and Electrowinning Technology, ed. G.V. Jergensen II, SME 1999 pp. 139-147.

Breitenbach, A. J., "Slope Stability Performance of High Heap Fill Structures for Closure", Tailings & Mine Waste Symposium 2001, Proceedings of the Eighth International Conference, Colorado State University, Fort Collins, Colorado, January 2001, pp. 137 to 144.

Chahbandour, J., Guzman, A., Filippone, C. and Carrera, I., "Hydrodynamic Characterization and Optimization of the Chuquicamata Ripios Mina Sur Aglomerados – Part 1", Randol Copper Hydromet Roundtable, September 5-8, 2000, Tucson, AZ, pp. 79-86.

Currey, N.A., and Ritchie, P.J., "Planning, People and Politics; Aiming for a Successful Mine Closure", Proceedings, "Back to the Future" 26th Annual Minerals Council of Australia Environmental Workshop, Hotel Adelaide International, Adelaide, Australia, October 14-17, 2001, 6 pp.

Dicinoski, W., Schlitt, J., and Ambalavaner, V., "A Global Engineer's Perspective of Copper Leaching, Solvent Extraction and Electrowinning", Proceedings, ALTA 1998 – Copper Hydrometallurgy Forum, Brisbane, QLD, Australia, 20-21 October, 1998, 37 pp.

Dixon, D.G., "Analysis of heat conservation during copper sulfide heap leaching", Hydrometallurgy, Vol. 58, (2000), pp. 27-41.

Domic, E.M., "Lo Aguirre's Challenge for Scaling-up New Technology in Copper Hydrometallurgy: From Concept to Industrial Application", 110th AIME Annual Meeting; Chicago, IL, February 22-27, 1981, TMS A81-46.

Domic, E.M., "TL Leaching Practice: Cost of Operations & Process Requirements", The Metallurgical Society of AIME, TMS paper A83-2, 112th Annual Meeting of AIME, 1983, 14 pp.

Dudley, K.A., Readett, D.J. and Bos, J.L., "Hydrometallurgical Processing of Copper Ores – Heap Leach, Solvent Extraction, and Electrowinning at Girilambone Copper Company", Paper presented at The AusIMM, MINPREX 2000 Conference, September 11-13, 2000, Melbourne, Victoria, Australia.

Eichrodt, C.W., "The Leaching Process at Chuquicamata, Chile", Trans. AIME 1930 Yearbook, p. 186.

Engineering and Mining Journal, "New TL leaching process yields economies for copper and uranium ores", October, 1978, pp. 100-102.

Figueroa, J.H., Ruiz, J.E., and Ayala, R., "Enhanced leaching of copper sulfide leach dumps: application at Cananea, Mexico", Proceedings of Copper 99-Cobre 99 International Conference, Vol. IV, Hydrometallurgy of Copper, ed. S.K. Young, D.B. Dreisinger, R.P. Hackl and D.G. Dixon, The Minerals, Metals $ Materials Society, 1999, pp.13-26.

Fountain, G. F., Private communication, January 9, 2002.

Gottlieb, P., Wilkie, G., Sutherland, E., Ho-Tun, E., Suthers, S., Perera, B., Jenkins, S., Spencer, S., Butcher, A. and Rayner, J., "Using Quantitative Electron Microscopy for Process Mineralogy Applications", JOM, The Minerals, Metals and Metallurgical Society (TMS), Vol. 52, No. 4, April, 2000.

Guzman, A., Filippone, C., Chahbandour, J., Srivastava, R. and Carrera, I., "Hydrodynamic Optimization of Chuquicamata's Ripios Mina Sur Aglomerados Heap Leach Project – Part 2", Randol Copper Hydromet Roundtable, September 5-8, 2000, Tucson, AZ, pp. 87-91.

Hiskey, J.B., "Diagnostic Leaching of Copper Bearing Materials", SME Annual Meeting, Preprint No. 97-83, 1997, Denver, CO.

Hiskey, J.B. and Wadsworth,, M.E., "Electrochemical Processes in the Leaching of Metal Sulfides and Oxides", Process and Fundamental Considerations of Selected Hydrometallurgical Systems, M.C. Kuhn, ed., SME 1981, pp. 304-324.

House, J.E., "The Development of the LIX Reagents", 1981 Gaudin Lecture, AIME Annual Meeting, Chicago, IL, February 25, 1981, 18 pp.

Iasillo, E. and Schlitt W.J., "Practical Aspects Associated with Evaluation of a Copper Heap Leach Project", |Copper Leaching, Solvent Extraction, and Electrowinning Technology, ed. G.V. Jergensen II, SME 1999 pp. 123-138.

Inspiration Consolidated Copper Company presentation to Arizona Section of AIME, "Hydrometallurgy at Inspiration", June 5, 1982.

Irving, J., "Heap Leaching of Low-grade Copper Ores", Engineering and Mining Journal, Vol. 113, No. 17, April 29, 1922.

James, B., and Lancaster, T., "Physical Parameters of Heap Leaching, Proceedings ALTA 1998 Copper Hydrometallurgy Forum, ALTA Metallurgical Services: Melbourne, Victoria, Australia.

Johnson, P.H., U.S. Patent # 4,017,309, "Thin-Layer Leaching Method", 1975.

Kaczmarek, A.F., Campbell, J., Schlitt, W.J. and Keane, J.M., "Designing the leach system for Cerro Negro Ore", Proceedings of Copper 99-Cobre 99 International Conference, Vol. IV – Hydrometallurgy of Copper, ed. S.K. Young, D.B. Dreisinger, R.P. Hackl, and D.G. Dixson, The Minerals, Metals and Materials Society, 1999, pp. 437-452.

Keane, J.M., "Copper Heap Leach Short Course – Commercial Ore Testing", SME Annual Meeting, Orlando, Florida, March 6-8, 1998.

Kordosky, G.A., "Solvent Extraction in the Copper Industry, A Success Story", Chemical Metallurgy – A volume in memory of Alexander Sutulov, Proceedings of The IV Meeting of the Southern Hemisphere on Mineral Technologies; and III Latin-American Congress on Froth Flotation, Volume II, Concepcion, Chile, November 20-23, 1994, Ed. I Wilkomirsky, M. Sanchez and C. Hecker, Dept. of Metallurgical Eng., University of Conception.

Kordosky, G.A., Sudderth, R.B, and Virnig, M.J., "Evolutionary Development of Solvent Extraction Reagents: Real-life Experiences", Copper Leaching, Solvent Extraction, and Electrowinning Technology, ed. G.V. Jergensen II, SME 1999, pp. 259-271.

Lancaster, T. and Walsh, D., "The Development of the Aeration of Copper Sulphide Ore at Girilambone", Proceedings International Biohydrometallurgy Symposium Biomine '97', pp. M5.4.1-M5.4.10 (Australia Mineral Foundation: Glenside, South Australia).

Lupo, J.F., "Hydraulic Considerations for Leaching Operations", SME Annual Meeting, Salt Lake City, UT, February 28-March 1, 2000, Pre-print No. 00-49.

Lupo, J.F., "Stability of HDPE Pipes Under High Heap Loads", SME Annual Meeting, Denver, CO, February 26-28, 2001, Pre-print No. 01-102.

McQuiston, F.W., and Shoemaker, R.S., Gold and Silber Cyanidation Plant Practice- Volume II, Society of Mining Engineers of American Institute of Mining, Metallurgical and Petroleum Engineers, Inc. (AIME), Port City Press, Baltimore, MD, 1981.

Merritt, R.C., The Extractive Metallurgy of Uranium, Colorado School of Mines Research Institute for the United States Atomic Energy Commission, 1971, p. 112-119.

Miller, A., "Heap leaching copper ore at Ranchers Bluebird Mine, Miami, Arizona", AIME Pre-print 67-B-339, September 6-8, 1967, presented at the SME Fall Meeting, Rocky Mountain Minerals Conference, Las Vegas, NV.

Mining Magazine, "Mining and Extraction of Copper at Lo Aguirre, Chile", July 1978, pp. 33-37.

Norris, P.R., "Thermophiles and bioleaching", Bio-mining – Theory, Microbes and Industrial Processes, Ed. D. E. Rawlings, Springer-Verlang, Berlin, 1997, pp. 247-258.

O'Kane, M., Barbour, S.L., and Haug, M.D., "A Framework for Improving the Ability to Understand and Predict the Performance of Heap Leach Piles", Proceedings of Copper 99-Cobre 99 International Conference, Vol. IV – Hydrometallurgy of Copper, ed. S.K. Young, D.B. Dreisinger, R.P. Hackl, and D.G. Dixson, The Minerals, Metals and Materials Society, 1999, pp. 409-419.

Parkison, G.A. and Bhappu, R.B., "The Sequential Copper Analysis Method – Geological, Mineralogical, and Metallurgical Implications", SME Annual Meeting, Denver, CO, March 6-9, 1995, pre-print No. 95-90.

Pohlman, S.L. and Olson, F.A., 1976, "Characteristics of Chrysocolla Pertinent to Acid Leaching", in Yannoppoulos, J.C. and Agaraval, J.C. eds., Extractive Metallurgy of Copper, New York, AIME, p. 943-959.

Power, K.L. "Operation of the First Commercial Copper Liquid Ion Exchange and Electrowinning Plant", Copper Metallurgy, R.P. Ehrlich, ed., AIME, New York (1970), pp. 1-27.

Rawlings, D.E., "Mesophilic, autotrophic bioleaching bacteria: description, physiology an role", Bio-mining – Theory, Microbes and Industrial Processes, Ed. D. E. Rawlings, Springer-Verlang, Berlin, 1997, pp. 229-245.

Rose, C. A., "Metallurgical Operations at the Chile Exploration Co.", E&MJ, Vol. 101, No. 7, 1916, p. 321.

Scheffel, R.E., "Retreatment and Stabilization of the Naturita Tailing Pile Using Heap Leaching Techniques ", Fall Meeting of the SME-AIME, September, 1981, Salt Lake City, UT.

Scheffel, R.E., "Often Repeated Errors in the Development and Design of Heap Leach Projects", Proceedings, 5th Annual, Randol-Copper Hydromet Round Table-1999, October 10, 1999, Phoenix, AZ, pp. 13-14.

Scheffel, R.E., "Disciplines Helpful for the Successful Financing and Development of Heap Leach Projects", Proceedings, 6th Annual Randol-Copper Hydromet Roundtable-2000, September 5-8, 2000, Tucson, AZ, pp. 31-34.

Scheffel, R.E. and J.B. Reid, "Development of the Andacollo Copper Project", ALTA 1997 Copper Hydrometallurgy Forum, October 20-21, 1997, Brisbane, Australia, 52 pp.

Schlitt, W.J., "The Case for Forced Aeration in Sulfide Leaching", Proceedings, ALTA 2000 - Copper 6, 2-3 October 2000, Glenelg, So. Australia, 16 pp.

Tobelmann, H.A. and Potter, J.A., "First Year of Leaching by the New Cornelia Copper Co.", Trans. AIME, Vol. 60, 1919, p.22

Uhrie, J.L. and Koons, G.J., "The Evaluation of Deeply Ripping Truck-dumped Copper Leach Stockpiles", SME Annual Meeting, Salt Lake City, UT, February 28 – March 1, 2000, Preprint No. 00-109.

Walsh, D., Lancaster, T., James, B., Braaksma, M. and Readett, D., "Developments in Heap Leaching of Copper Sulphide Ore", 6th Annual Randol at Vancouver '97' – Global Mining Opportunities - Copper Hydromet Roundtable '97', November 4-7, 1997, pp.177-180.

Weston, J.L., and Dudley, K.A., "Heap Re-mining Practices at Girilambone Copper Company", Proceedings ALTA 2000 Copper Hydrometallurgy Forum, ALTA Metallurgical Services: Adelaide, South Australia, Australia.

Williams, D.J., Wilson, G.W., Currey, N.A., and Ritchie, P.J., "Engineering Aspects of the Rehabilitation of an Open Pit Gold Operation in a Semi-arid Climate", Proceedings, "Back to the Future" 26th Annual Minerals Council of Australia Environmental Workshop, Hotel Adelaide International, Adelaide, Australia, October 14-17, 2001, 19 pp.

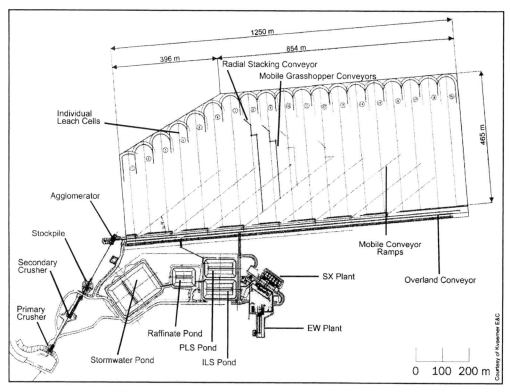

Figure A - Multiple-Lift, Permanent Pad - Conceptual Layout

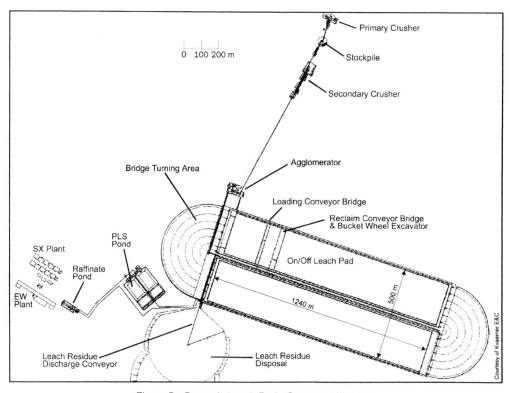

Figure B - Dynamic Leach Pad - Conceptual Layout

Precious Metal Heap Leach Design and Practice

Daniel W. Kappes[1]

ABSTRACT

Heap leaching of gold and silver ores is conducted at approximately 120 mines worldwide. Heap leaching is one of several alternative process methods for treating precious metal ores, and is selected primarily to take advantage of its low capital cost relative to other methods. Thirty-seven different heap leach operations with a total production of 198 tonnes of gold per year (6,150,000 ounces/yr.) were surveyed to determine operating practice. These operations together produce 7.4% of the world's gold. When mines not surveyed are taken into account, it is likely that heap leaching produces 12% of the world's gold. Heap leaching for silver is conducted using the same principles and operating practices as for gold, but heap leach operations produce only a small fraction of world silver production.

INTRODUCTION

Heap leaching had become a fairly sophisticated practice at least 500 years ago. Georgius Agricola, in his book De Re Metallica (publ. 1557) illustrates a heap leach with a 40-day leach cycle (Figure 1), which could pass in many ways for a modern heap leach. The Agricola heap leach recovered aluminum (actually alum) for use in the cloth dying industry. Copper heap and dump leaches in southern Spain were common by about 1700. Gold and silver heap leaching began with the first Cortez heap leach in 1969. While many projects have come and gone, Cortez is still going - their new 63,000 tonne/day South Area leach is scheduled to start up in 2002.

The largest U.S. precious metal heap leach is the Round Mountain, Nevada, operation with over 150,000 tonnes/day of ore going to crushed or run-of-mine heaps, at an average grade of 0.55 grams gold/tonne [This chapter follows the North American convention of "ton" for short ton and "tonne" for metric ton]. Worldwide, Newmont's Yanacocha, Peru, operation holds the record, with a 2002 target of nearly 370,000 tonnes/day, at an average total reserve grade of 0.87 grams gold per tonne. On the other end of the scale, some very high grade ores - up to 15 grams per tonne (0.5 oz/ton) - are being successfully processed at rates of several hundred tonnes/day (Sterling, Nevada; Hassai, Sudan; Ity, Ivory Coast). A cursory worldwide summary in late 2001 was able to identify 78 active precious metal heap leaches worldwide, of which 34 were in the U.S. (22 in Nevada). The survey no doubt missed many operations, so the worldwide total is certainly over 100. To provide a basis for this chapter, technical and/or cost data were gathered from 37 of these operations. Because many operations impose restrictions on the release of detailed data, composite results are presented.

Nevada was the "birthplace" of modern gold heap leaching in the late 1960's, and is only now giving up its dominance of this technology. Other very large gold districts - notably the pre-cambrian shield areas of Canada, Australia and South Africa - show relatively few heap leaches. There are several reasons for this geographic concentration, but the primary reason is that Nevada gold deposits tend to have been created by low-energy geologic processes - near surface hot

---

1 Kappes, Cassiday & Associates, Reno, Nevada

springs and moderate depth, hydro-thermal systems that deposited low grade gold in permeable rocks. Besides aiding gold deposition, the permeable nature of the rocks allowed uniform and deep oxidation that liberated the gold from its sulfide and carbonaceous host minerals. Shield deposits have generally had a more complicated history, which has resulted in coarse gold contained in poorly-permeable rocks. Often these ores can be successfully heap leached only after weathering has completely destroyed the rock matrix.

Figure 1. "The rocks are . . piled in . . heaps fifty feet long, eight feet wide and four feet high, which are sprinkled for forty days with water. The rocks begin to fall to pieces like slaked lime, and there originates a . . new material". Drawing and text from De Re Metallica, Herbert Hoover translation, published by Dover Publications, Inc.

WHAT IS HEAP LEACHING?

To those of us in the gold industry, the question "What is Heap Leaching?" seems to have an obvious answer. In the simplistic sense, heap leaching involves stacking of metal-bearing ore into a "heap" on an impermeable pad, irrigating the ore for an extended period of time (weeks, months or years) with a chemical solution to dissolve the sought-after metals, and collecting the leachant ("pregnant solution") as it percolates out from the base of the heap. Figure 2 is an aerial photograph showing the typical elements of a precious metals heap leach operation - open pit mine, a heap of crushed ore stacked on a plastic pad, ponds, a solution process facility for recovering gold and silver from the pregnant solution, and an office facility. For a small operation such as the one illustrated here, very limited infrastructure is required.

In a more complex sense, heap leaching should be considered as a form of milling. It requires a non-trivial expenditure of capital, and a selection of operating methods that trade off cost versus marginal recovery. Success is measured by the degree to which target levels and rates of recovery

Figure 2 Heap leach installation at Mineral Ridge, Nevada. The open pit mine is shown on the left. On the right is a two million ton heap of crushed, conveyor-stacked ore placed on a plastic-lined leach pad. Pregnant and barren solution storage ponds are located downslope from the heap. Buildings include process plant, laboratory, maintenance shop and administration offices. Photo courtesy of Tom Nimsic, American Au/Ag Associates.

are achieved. This distinguishes heap leaching from dump leaching. In dump leaching, ores are stacked and leached in the most economical way possible, and success is achieved with any level of net positive cash flow.

The bibliography of precious metals heap leaching is quite extensive, and because of time limitations a very limited bibliography has been compiled for this chapter. However, the following publications are good places to start a literature search:

- "Global Exploitation of Heap Leachable Gold Deposits", by Hausen, Petruk and Hagni, February, 1997
- "The Chemistry of Gold Extraction" by Marsden and House, 1992
- "World Gold '91", Second AusIMM-SME Joint Conference, Cairns, Australia, 1991
- "Introduction to Evaluation, Design and Operation of Precious Metal Heap Leaching Operations", by Van Zyl, Hutchison and Kiel, 1988.

Special recognition and thanks should be given to Hans von Michaelis of Randol International, Denver. Between 1981 and April 2000 Randol organized four major symposia followed by four published studies of the gold industry, and several minor meetings with their own proceedings. The combined Randol literature occupies nearly 40 volumes covering six feet of shelf space. Most modern heap leach operations are referenced.

WHY SELECT HEAP LEACHING AS THE PROCESSING METHOD?

Gold and silver can be recovered from their ores by a variety of methods, including gravity concentration, flotation, and agitated tank leaching. Methods similar to heap leaching can be employed: dump leaching and vat leaching (vat leaching is the treatment of sand or crushed ore in bedded vats with rapid solution percolation).

Typically, heap leaching is chosen for basic financial reasons - for a given situation, it represents the best return on investment. For small operations or operations in politically unstable areas, it may be chosen because it represents a more manageable level of capital investment. Some interesting examples that illustrate this issue of choice are presented below.

Capital Risk

Several years ago, the author's company was advising on a project in which the ore reserve was a few million tonnes at a grade of 7 grams of gold per tonne (0.22 oz/ton). Heap leach recovery was about 80%, well below the 92% that could be achieved in an agitated leach plant. Financial considerations strongly favored milling, and the owner was financially strong. However, the operation was located in an undeveloped country with unstable politics and socialist leanings. The owner concluded that he might lose control of a high-capital investment, whereas he could maintain control of a heap leach with an implied promise of a future larger capital investment. The operation has been running successfully and very profitably.

Lack of Sufficient Reserves

The Sterling Mine in Beatty, Nevada (Cathedral Gold Corporation) began life as an underground mine, with a reserve of 100,000 tons of ore at a grade of 11 grams gold/tonne (0.35 oz/ton). Over a fifteen year period, it mined and processed nearly one million tons, but never had enough ore reserves to justify a conventional mill. Fortunately, the Sterling ore achieves excellent heap leach recovery - the original heaps reached 90% from ore crushed to 100 mm.

Equal or Better Percent Recovery

Comsur's Comco silver heap leach at Potosi, Bolivia, showed the same recovery in both a heap and an agitated leach plant. However, the silver ore leached very slowly and residence time of up to 4 days was needed in an agitated leach plant. Although the heap leach took several months to achieve the same recovery, the economics favored the heap.

At the joint AIME/AusIMM Symposium "World Gold '91", T. Peter Philip of Newmont presented a paper "To Mill or to Leach?" in which he evaluated the decision of Newmont to build the Carlin No. 3 mill. He concluded that the mill recovery was over-estimated and the heap leach recovery underestimated, and that the decision to go with milling may have been incorrect.

Differential Recovery Is Not Sufficient To Justify Added Investment

A recent review (Kappes, 1998) concluded that for a "typical" Nevada-type ore body with ore grade of 3.0 grams gold/tonne (0.088 oz/ton), the mill recovery would have to be 21% higher than the heap leach recovery to achieve the same return on investment - and this is very seldom the case.

Of the 37 operations surveyed for this chapter, four have a head grade below 0.65 grams gold per tonne and half are between 0.65 and 1.50 grams per tonne. At these gold grades, it is usually impossible to justify the investment in a conventional agitated leaching plant.

Capital Is Very Difficult Or Expensive To Raise

Heap leaching has often provided the route for a small company to grow into a large company. A good example is Glamis Gold Corporation, which has gone from total assets of $12 million in 1984 to $112 million in 2001, based largely on its low grade heap leach projects at Picacho and Randsburg, California.

At the time this is being written in early 2002, the precious metals mining industry is experiencing a severe capital shortage and a consolidation of producers into a few large corporations. This will open up an opportunity for the creation of a new generation of junior mining companies to exploit smaller deposits, and heap leaching will play a key role in this process.

TYPE OF ORE

Heap leach recovery is very dependent on the type of ore being processed. Some typical examples are discussed below.

Carlin-Type Sedimentary Ores

These ores consist of shales and "dirty" limestones, containing very fine (submicroscopic) gold. Oxidized ores leach very well, with low reagent consumption and production recovery of 80% or better. Ores are typically coarse-crushed (75mm) but may show recovery of 70% or better at run-of-mine sizes. The largest of the northern Nevada heap leaches (Carlin, Goldstrike, Twin Creeks) treat this type of ore. Unoxidized ore contains gold locked in sulfides, and also contains organic (carbonaceous) components, which absorb the gold from solution. This ore shows heap leach recovery of only 10 to 15% and is not suitable for heap leaching. Because of the different ore types, the northern Nevada operations (for instance, Barrick's Goldstrike Mine) may employ roasters, autoclaves, agitated leach plants and heap leaches at the same minesite. Crushing is usually done in conventional systems (jaw and cone crushers) and ores are truck stacked.

Low Sulfide Acid Volcanics Or Intrusives

Typical operations treating this type of ore are Round Mountain, Nevada, and Wharf Mine, South Dakota. Original sulfide content is typically 2 to 3% pyrite, and the gold is often enclosed in the pyrite. Oxidized ores yield 65 to 85% recovery but may have to be crushed to below 12 mm (1/2 inch). Usually the tradeoff between crush size and percent recovery is a significant factor in process design. Unoxidized ores yield 45 to 55% gold recovery and nearly always need crushing. At Round Mountain, Nevada, approximately 150,000 tons per day of low grade oxide ore is treated in truck-stacked run-of-mine heaps, 30,000 tons per day of high grade oxide ore is treated in crushed (12mm), conveyor-stacked heaps, and 12,000 tons per day of unoxidized ore is treated in a processing plant (gravity separation followed by leaching in stirred tanks). Crushing is done using jaw and cone crushers; fine crushed ore contains enough fines that conveyor stacking is preferred over truck stacking.

Oxidized Massive Sulfides

The oxide zone of massive sulfide ore deposits may contain gold and silver in iron oxides. Typically these are very soft and permeable, so crushing below 75mm often does not increase heap leach recovery. The Filon Sur orebody at Tharsis, Spain (Lion Mining Company) and the Hassai Mine, Sudan (Ariab Mining Company) are successful examples of heap leaches on this type of ore. Because the ore is fine and soft, the ore is agglomerated using cement (Hassai uses 8 kg cement/tonne), and stacking of the heaps is done using conveyor transport systems.

Saprolites / Laterites

Volcanic- and intrusive-hosted orebodies in tropical climates typically have undergone intense weathering. The surface "cap" is usually a thin layer of laterite (hard iron oxide nodules). For several meters below the laterite, the ore is converted to saprolite, a very soft water-saturated clay sometimes containing gold in quartz veinlets. Silver is usually absent. These ores show the highest and most predictable recovery of all ore types, typically 92 to 95% gold recovery in lab tests, 85% or greater in field production heaps. Ores are processed at run-of-mine size (which is often 50% minus 10 mesh) or with light crushing. Ores must be agglomerated, and may require up to 40 kg of cement per tonne to make stable permeable agglomerates. Many of the West African and

Central American heap leaches process this type of ore. Good examples are Ity in the Ivory Coast, and Cerro Mojon (La Libertad) in Nicaragua. When crushing is required, one or two stages of toothed roll crushers (Stammler-type feeder-breaker or MMD Mineral Sizer) are usually employed. Conveyor systems are almost always justified; ore can be stacked with trucks if operations are controlled very carefully.

Clay-Rich Deposits
In some Carlin-type deposits, as well as in some volcanic-hosted deposits, clay deposition or clay alteration occurred along with gold deposition. The Buckhorn Mine, Nevada (Cominco, now closed) and the Barney's Canyon Mine, Utah (Kennecott) are good examples. These ores are processed using the same techniques as for saprolites, except that crushing is often necessary. Because of the mixture of soft wet clay and hard rock, a typical crushing circuit design for this type of ore is a single-stage impact crusher. Truck stacking almost always results in some loss of recovery. Agglomeration with cement may not be necessary, but conveyor stacking is usually employed.

Barney's Canyon employs belt agglomeration (mixing and consolidation of fines as it drops from conveyor belts) followed by conveyor stacking. The new La Quinua operation at Yanacocha employs belt agglomeration followed by truck stacking.

Silver-Rich Deposits
Nevada deposits contain varying amounts of silver, and the resulting bullion may assay anywhere from 95% gold, 5% silver to 99% silver, 1% gold. Silver leaches and behaves chemically the same as gold, although usually the percent silver recovery is significantly less than that of gold. Examples of nearly pure silver heap leaches are Coeur Rochester and Candelaria in Nevada, and Comco in Bolivia.

CLIMATE EXTREMES
The ideal heap leach location is a temperate semi-arid desert location such as the western U.S. However heap leaching has been successfully applied in a variety of climates:

- Illinois Creek, Alaska, and Brewery Creek, Yukon are located near the Arctic Circle and experience temperatures of minus 30°C for several months per year.
- Several heap leaches are located in the high Andes of South America (Comco at Potosi, Bolivia; Yanacocha and Pierina, Peru; Refugio, Chile) at altitudes above 4000 meters (13,000 ft). Although oxygen availability at these altitudes is only 60% of that at sea level, gold heap leaching proceeds at rates similar to that at sea level (oxygen is required for the process, but is not usually rate-limiting in a heap leach operation).
- At another extreme, Hassai, Sudan, is in the dry eastern Sahara fringes. This operation experiences normal daytime temperatures that routinely exceed 50°C in the summer, with annual rainfall of less than 20mm. One of the advantages of heap leaching over conventional cyanide leach plants and gravity recovery plants, is that heap leaching consumes very little water. With good water management practices, water consumption can be less than 0.3 tonnes water per tonne of ore.
- In tropical wet climates, rainfall of 2.5 meters per year can add over 5 tonnes of water to the leach system for each tonne of ore stacked. As discussed in a later section, this amount of water can also be handled with good water management practices.

CHEMISTRY OF GOLD/SILVER HEAP LEACHING
The chemistry of leaching gold and silver from their ores is essentially the same for both metals. A dilute alkaline solution of sodium cyanide dissolves these metals without dissolving many other ore components (copper, zinc, mercury and iron are the most common soluble impurities).

Solution is maintained at an alkaline pH of 9.5 to 11. Below a pH of 9.5, cyanide consumption is high. Above a pH of 11, metal recovery decreases.

Many heap leachable ores contain both gold and silver. Of the 28 mines that reported bullion assays, five produce a doré (impure gold-silver bullion) bar that is greater than 70% silver. Another five produce a bullion greater than 30% silver. Only five produce a bullion with less than 5% silver. Deposits in western Africa and Australia tend to be very low in silver, while those in Nevada are highly variable, ranging from pure gold to pure silver.

Silver is usually not as reactive with cyanide as gold. This is because gold almost always occurs as the metal, whereas silver may be present in the ore in many different chemical forms some of which are not cyanide-soluble. Reported heap leach recoveries (32 operations) averaged 71% gold, and ranged from 49% to 90%. Reporting run-of-mine heap leaches averaged 63%. Typical recovery for silver is 45-60%, although when silver is a minor constituent, its recovery may be only 15-25%.

The level of cyanide in the heap onflow solution ranges from 100 to 600 ppm NaCN, and averages 240 ppm for the 28 operations reporting. Forty-five percent of the operations reported cyanide strength below 200 ppm, 25% were above 300 ppm. Heap discharge solution (pregnant solution) averages 110 ppm.

Cyanide consumption, via complexation, volatilization, natural oxidation or oxidation by ore components, typically ranges from 0.1 to 1.0 kg per tonne of ore. Price of sodium cyanide is currently at a historical low of $1.00 per kg. Cement and/or lime consumption ranges from 0.5 to 40 kg per tonne of ore. Several operations use cement for alkalinity control (instead of lime) as well as for agglomeration. The price of cement or lime is $60 to $100 per tonne, $160 delivered to remote African locations.

Other leaching agents - thiosulfate, thiourea, hypochlorite, bromine - have been experimented with as an alternative to cyanide, but cyanide is by far the most effective and the most environmentally friendly leaching agent. A good discussions of the process chemistry can be found in "The Chemistry of Gold Extraction" by Marsden and House, Ellis Horwood Publishers, 1992.

LABORATORY TESTING & CONTROL

As with any processing method, it is very important to base the design on the results of a comprehensive program of laboratory testing. During the production operation, laboratory tests including column leach tests should be continued on a regular basis, since the initial ore samples are seldom representative of the entire orebody. For a heap leach, the key parameters that are defined in the laboratory are crush size, heap stability, permeability versus heap height, cyanide strength and consumption, the need for agglomeration and the amount of agglomerating agent (usually Portland cement) required, leach time, and percent recovery. Derivative parameters such as the height of individual lifts and the method of stacking are also determined.

Heap leaching has inherent risks that can be largely eliminated if the operating practices follow the results of initial and on-going laboratory testing. The risks result from the nature of the operation. The results of the process are usually not known for several weeks or months after the ore is stacked, and at this point it is not economical to reprocess the ore. Mistakes made in the initial plant design or operating practices, for instance by not crushing finely enough, or by not agglomerating or stacking properly, can result in cash flow problems that might persist for up to a year after the problem is solved.

An on-site laboratory is an important part of the infrastructure at a heap leach operation. It should include an analytical section (for ore control) and a metallurgical testing section that regularly runs column leach tests on production samples. For a small operation processing less than 5,000 tonnes of ore per day, staffing is 2-3 technicians for sample preparation and assaying and one metallurgist to conduct process tests. Large operations may have a laboratory staff of ten to fifteen people.

HEAP PERMEABILITY & FLOW EFFICIENCY

The key element in a successful heap leach project is a heap with high, and uniform permeability. In any heap there are three zones of different flow regimes:

- coarse channels, which allow direct short-circuiting of solution from top to bottom
- highly permeable zones, in which solution is efficient at contacting the rock and washing the gold downward in "plug flow"
- zones of low to zero permeability where high grade solution or unleached ore may be trapped.

Efficiency Of Solution Displacement

If the heap were "ideal" - moving in true plug flow - then when one displacement volume of solution was placed on top of the heap, it would fully replace the solution in the heap. This would be 100% wash efficiency. In practice, the "best" heap leaches exhibit a wash efficiency of about 70%. At 70% per displacement, three displacement washes are required to achieve a recovery of 95% of the dissolved metals. A fourth "displacement" is required initially, to saturate the ore. Since a typical heap contains 20% moisture, 95% recovery (of the dissolved gold/silver content) requires that 0.8 tonnes of solution must be applied to each tonne of ore. Typical practice is to apply 1.3 tonnes of solution per tonne of ore during a 70-day primary leach cycle. This suggests two things: (a) most heap leach operations are able to maintain reasonably good permeability characteristics, yielding at least 50% wash efficiency; and (b) a high percentage of the gold is solubilized early in the 70-day leach cycle.

Drainage Base

A drainage base of crushed rock and embedded perforated pipes is installed above the plastic leach pad and below the ore heap. The importance of this drainage base cannot be overemphasized. Solution should percolate vertically downward through the entire ore column, and then enter a solution removal system with zero hydraulic head. If the drainage base cannot take the entire flow then solution builds up in a stagnant zone within the heap, and leaching within this stagnant zone can be very slow.

To put this in context, a "typical" heap might run 500 meters in an upslope direction. All of the onflow solution in a strip one meter wide by 500 meters must flow out at the downslope edge of the heap through the drainage base, which is typically 0.65 meters thick. The design horizontal percolation rate through the drainage base is therefore nearly 800 times the design rate of the heap itself. This is not a difficult engineering accomplishment since flow is carried in pipes within the base.

At one Australian copper heap leach operation, three adjacent leach panels were built. The two flanking panels had a good installed drainage base but the center panel did not. Recovery in the center panel was depressed 20%. A similar effect has been seen but not quantified at gold heap leach operations.

Recovery Delay In Multiple Lift Heaps

As subsequent lifts are stacked, the lower lifts are compressed and the percentage of low permeability zones increases. The first solution exiting an upper lift may have a gold concentration of up to three times that of the ore. If impermeable zones have developed in a lower lift, high grade solution may be trapped, causing a severe reduction in recovery rate and possibly in overall recovery percentage. The highest heap leaches currently in operation are 120 meters high, with about ten lifts of ore. Hard ore, crushed or run-of-mine, can withstand the resulting pressure without significant permeability loss. Many softer ores can be agglomerated with enough cement so that they can perform under a load of 30 meters; some agglomerated ores perform satisfactorily to 100 meters. These properties can be properly evaluated in advance, in laboratory column tests, which are run under design loads.

The delay in recovery as lifts are added to the heap is partly a function of the impermeability of the lower lifts, and partly a function of the wash efficiency discussed earlier. The net effect is that average recovery is delayed as the heaps get higher, and overall pregnant solution grade decreases (requiring more solution processing capacity). Of 22 operations reporting, 18 indicated that they see a delay in time of average recovery, which ranges from 3 to 30 days per lift. The most common figure was 7 days/lift. This cash flow delay must be allowed for. Also, this number implies that for each extra lift, the capacity of the process plant should be increased in response to the decrease in gold content of pregnant solution.

Intermediate Liners

If impermeability of lower lifts becomes a serious problem, it is possible to install intermediate liners. Four of the surveyed operations reported that they install plastic intermediate liners on a regular basis. One operation reported that it regularly installs a clay intermediate liner. There are two problems with installing an intermediate liner: (a) the heap below the liner is compressed as the upper lift is placed, resulting in differential settlement and tearing of the liner; and (b) the ore below the liner cannot be washed with water, which is sometimes required as a part of final heap closure.

SOLUTION APPLICATION RATE AND LEACH TIME

With regard to sprinkling rate, the timing of gold recovery is a function of five factors:

- the rate at which the gold dissolves. Coarse gold particles dissolve very slowly, and may not fully dissolve for several months in a heap leach environment.
- the percentage of gold that exists as free or exposed particles
- the rate of diffusion of the cyanide solution into rock fractures, and of gold cyanide back out of the rock fractures. Where the gold occurs on tight fractures or in unfractured rock, the rock must be crushed into fine particles to achieve target rates and levels of recovery.
- the effect of chemical reactions within the heap, or within rock particles, which destroy cyanide and alkalinity or which consume oxygen
- the rate of washing of gold off of the rock surfaces and out of the lift of ore under leach. This is a complex issue, which depends on the overall permeability of the lift and the local permeability variations due to segregation and compaction as the lift is being constructed.

The above factors cause wide theoretical differences in the response of various ores to leaching. However, in practice most heap leach operations apply solution to crushed-ore heaps within a fairly narrow range of flows: Of 19 operations reporting, application rates for crushed-ore heaps ranged from 7 to 20 l/hr/sq. m (0.003 to 0.009 gpm/sq. ft) with an average of 11 l/hr/sq. m (0.048 gpm/sq. ft). Only three applied solution above 10 l/hr/sq. m (0.0044 gpm/sq. ft), and only four were below 8 l/hr/sq. m (0.035 gpm/sq. ft). For 17 run-of-mine heaps, the average application rate was 8.3 l/hr/sq. m (0.037 gpm/sq. ft), with only two operations above 10 l/hr/sq. m (0.044 gpm/sq. ft).

Laboratory columns always respond much faster than field heaps, for two reasons: the ore is placed in the lab column much more uniformly so that percolation is more effective; and the solution-to-ore ratio (tonnes of solution per tonne of ore in a given time frame) is generally higher in lab columns than in field heaps. For some field heaps, notably where the ore is fine crushed and the ore leaches quickly, the solution:ore ratio is a more important factor than overall leach time. However, for the majority of heap leaches, time seems as important as specific application rate.

For the 32 operations reporting, average single lift height was 8.9 meters (yielding 14.2 tonnes of ore per sq. m of top surface) and average irrigation time was 70 days during the primary leach cycle. This yields a specific solution application rate of 1.30 tonnes solution per tonne of ore. Operations with low cycle times tended to have higher application rates, suggesting that the ratio of 1 to 1.5 tonnes of solution per tonne of ore is a universal target.

For ores with very slow leaching characteristics, an intermediate pond and a recycle stream may be added to the circuit, so that each tonne of ore sees two tonnes of leach solution during an extended leach period. The process plant treats only the final pregnant stream - one tonne of solution per tonne of ore. Of 36 operations reporting, 16 had only one leach cycle, 16 had two cycles, and four had three cycles.

The use of multiple cycles is good operating practice for single-lift heaps of high grade ore. However, for multi-lift heaps this is not the case. Heap modeling indicates that once the heap attains a height of three lifts, the intermediate solution contains almost as much gold as the pregnant solution. Recycling results in a significant build-up of dissolved gold within the heap, causing a slight overall recovery loss and a cash flow delay. For multi-lift heaps, it is often possible to justify an increase in the size of the recovery plant so that only fully barren solution returns to the heap.

Some successful single-lift heaps achieve a high percentage recovery in the first leach period, but continue the leach for much longer. Sterling, with very high grade ore, leached the same ore for 18 years. The initial Cortez heaps were leached intermittently for ten years. Cost to intermittently leach old heaps may be as low as $0.10 per tonne per month.

It is extremely important to design a heap leach system so that the ore can be leached for a very long time. Unlike an agitated leach plant where the ore can be ground to a fine powder and intensively agitated, heap leaching is not a very energy-intensive process. Once a heap is built, one of the most significant variables, which the operator can employ to solve design or production problems, is leach time. Successful projects employing on-off (reusable) leach pads have been undertaken, but this is a risky practice. Some operations (such as Round Mountain, Nevada) utilized on-off pads to achieve rapid first-stage recovery, then transfer the ore to long term heaps to complete the process.

DESIGN FOR AMBIENT WEATHER CONDITIONS

Design For High Ambient Temperature
In regard to ambient temperature, high temperature is not a direct problem. In very hot desert areas where drip irrigation is used, sunlight will significantly heat the solution. Even then, because of the effect of cool night time temperatures, it is unusual to see heap effluent solution temperatures above 15°C. Mesquite at one point covered the ponds to prevent evaporation, and the resulting recirculating solution was above 30°C. Hot leach solutions dissolve less oxygen than cold solutions and this could affect the rate of gold recovery in oxygen-starved heaps. However, usually there is sufficient oxygen present, and the higher overall activity due to the higher temperature more than offsets the oxygen effect. No operating heap leaches have reported a direct problem due to high temperature of the rock or the leach solution.

Design For Low Ambient Temperature
Low temperature can be a problem. Many Nevada heap leaches report a significant recovery decrease in winter, which is offset the following summer. When a cold weather project is anticipated, column tests should be run under cold conditions. There are several reasons for a reduction in recovery rate with lower temperatures:

- Rate of reaction is slowed as solution tempeature approaches freezing. Comparative laboratory column tests show that recovery rate drops significantly when the heap temperature drops below 5°C. Solution viscosity increases significantly as temperature drops. This affects both the heap and the process plant.

- Solution flowing slowly through the normally unsaturated heaps flows via the meniscus on the surface of particles, and the thickness of this meniscus is a direct function of viscosity. Thus, cold heaps tie up more process solution (and more gold inventory) than warm heaps. In carbon columns, the rate of fluidization can be significantly affected. PICA USA Inc. (one of several activated carbon suppliers) has generated a graph of solution temperature versus percent fluidization, which is shown as Figure 3. As the graph shows, bed expansion of carbon can increase from 15% to 33% as the solution temperature decreases from 20°C to 5°C. This same viscosity effect will alter the ability of the solution to flow through the heaps.

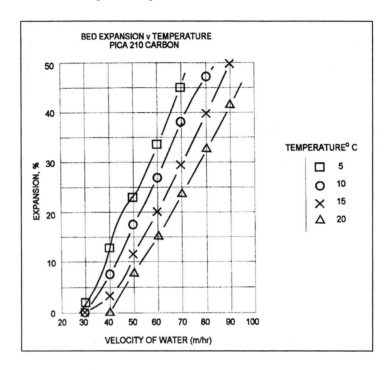

Figure 3 Effect of Temperature on Bed Expansion of Activated Carbon

- Solution surface tension drops, although not as fast as viscosity. Surface tension can affect the flow of solution through the heaps, and also it affects the ability of the solution to penetrate tight fractures within the rock. Table 1 shows the effect of temperature on the viscosity and surface tension of water.
- In very cold climates, it is possible to create a frozen wedge of solution within the heap. This occurred at Summitville, Colorado. For this reason, Brewery Creek (Yukon) stacks ore only in the summer, although they leach all year.

Temperature	Surface Tension, dynes/cm	Viscosity, centipoises
0°C	76	1.79
5°C	75	1.52
10°C	74	1.31
20°C	73	1.00
40°C	70	0.66

Table 1 Surface Tension and Viscosity as a Function of Temperature

Water Balance

Since many heap leach operations occur in desert areas where water is scarce, and others occur in environmentally sensitive areas where water discharge is not acceptable, the balance between water collection and evaporation is important. Fortunately, by adjusting the method and scheduling of solution application, it is usually possible to meet the local requirements.

Evaporation of water, regardless of its mechanism, requires a heat input of 580 calories per gram (8300 BTU/U.S. gallon) of water evaporated. A heap leach gets this heat input from three sources: direct solar heating on heap and water surfaces; latent heat in the shroud of air within the "sprinkler envelope"; and latent heat in the air that is pulled through the heap by convection.

Average 24-hr incident solar radiation on a flat horizontal surface ranges from 12,000 BTU per sq. m per day (central U.S.) to about 30,000 BTU (equatorial desert), which could theoretically evaporate 5 to 12 liters of solution per day. With a typical heap application rate of 10 l/sq. m /hr, incident solar radiation could account for an evaporation rate of 2 to 5% of applied solution when using sprinklers. Evaporation would be somewhat less when using drip irrigation (1 to 4%) because some of the solar energy is re-radiated from dry areas on top of the heap. This same heat input would result in pond evaporation of 5 to 13 mm per day.

Use of sprinklers rather than drips may result in the loss of up to 30% of solution pumped. This is because the sprinkler droplets trace an arc through a shroud of air, which is very seldom at 100% humidity. A gentle breeze of 3 km (two miles) per hour will replace the "saturated shroud" on a typical 500 m long heap with unsaturated air every 10 minutes, and the pumping action of the sprinkler droplets will cause additional rapid air replacement. A good discussion of evaporative sprinkler losses is presented in Univ. of Florida Cooperative Extension Service Bulletin 290. Typical sprinkler evaporation at operations using coarse-drop sprinklers in Nevada-type climates (arid, temperate) is up to 15% of solution pumped on summer days and 2-4% on summer nights, averaging about 7% annually.

Overall evaporative losses include the sprinkler losses, convective loss from air flowing through the heap, and losses due to heating/evaporation from ponds and from other areas not sprinkled. These have been determined at several Nevada operations to be up to 20% of total solution pumped in summer months and 10% annually. Thus, direct sprinkler loss accounts for about 60% of the total. Use of drip irrigation can reduce but not eliminate evaporative loss.

In tropical climates, noticeable losses occur even during the rainy season. KCA's in-house experience on several tropical heap leach projects where rainfall is seasonal and up to 2.5 meters per year, is that overall annual evaporative loss from all sources, when using wobbler-type sprinklers operated 24 hours/day, is about 7% of solution pumped. Typical heap application rate is 10 l/hr/sq. m, or 88 meters per year (9 inches/day). Thus, evaporative loss of 7% is equal to 6.2 meters per year on the areas actually being sprinkled. If the heap and pond systems are properly designed, the active leaching area can be up to 40% of the total area collecting rainfall; it is therefore possible to operate in water balance when rainfall is 2.5 meters/year. For these operations, very large solution surge ponds are required: at Sansu, Ashanti, Ghana (rainfall 2.5 meters/yr.), for a 3000 tonne/day heap leach, total pond volume was 60,000 cu meters.

Where rainfall is high and evaporation rate is low, some operations (such as at Yanacocha, Peru, altitude 3500 meters) cover sideslopes with plastic to minimize rain collection. Others (Rio Chiquito, Costa Rica - Mallon Minerals Corp) have tried to cover the entire heap during the rainy season, but this has not worked very well because of the mechanical difficulties of moving the cover.

In West Africa and Central America it is acceptable practice to treat and discharge excess solution during the rainy season. Typically, excess process solution is routed through a series of ponds where cyanide is destroyed using calcium hypochlorite or hydrogen peroxide, followed by adjustment of pH to near neutral. The INCO SO_2 system, using copper-catalyzed hyposulfite to destroy cyanide, is also employed for this purpose. Cyanide-free solution is further treated in controlled wetlands (swamps) to remove heavy metals prior to discharge.

The worst water balance situation occurs in cool, damp climates such as high altitude operations (for instance the Landusky-Zortman operation in Montana, now closed). In such climates, rainfall and snowfall may be significant and evaporation is minimal. Generally such heaps can stay in water balance with an aggressive program of summer sprinkling. Arctic heap leaches (Brewery Creek, Yukon; Illinois Creek, Alaska) have been able to stay in water balance because precipitation is lower than the total water requirement needed to saturate the ore.

SOLUTION APPLICATION EQUIPMENT

A variety of solution application methods have been employed, but for mainstream heap leaches the following equipment has become standard:

- Drip Emitters. Drip emitters, which issue drops of water from holes every 0.5 to 1.5 meters across the heap surface, are very common. Drip emitters are easy to maintain and minimize evaporation. The main drawback to drip emitters is that they do not provide continuous drip coverage. Thus the top one meter of the heap may not be leached very well until it is covered with the next lift. Other problems are that emitters require an intense (and expensive) use of anti-scalant, and they require the use of in-line filters.
- Wobbler Sprinklers. Wobbler Sprinklers are used at a large number of operations. Their main advantages are that they issue coarse droplets, which control but do not eliminate evaporation, and that they deliver a uniform solution distribution pattern, which ensures uniform leaching of the heap surface. The coarse droplet size has another advantage - cyanide is readily oxidized by air and sunlight, and the wobbler-type sprinkler minimizes this loss (but not as well as drip systems). Wobblers are typically placed on a 6 x 6 meter pattern across the heap surface. A disadvantage of all sprinklers is that they require continual servicing, and personnel spend extended periods working in a "rainstorm". Occasional skin contact with cyanide solution does not pose a health problem, but an environment that encourages repeated skin/solution contact is not recommended. Sprinkler maintenance personnel wear full rain gear to eliminate any exposure problem, but the working environment (especially in cold weather) is not as nice as with drip emitters.
- Reciprocating Sprinklers. Reciprocating sprinklers shoot a stream typically 5 to 8 meters long of mixed coarse and fine droplets. They are not considered ideal for heaps, but often find application for sprinkling sideslopes since they can be mounted on the top edge to cover the entire slope. If emitters and wobblers are used on sideslopes, they must be installed on the slope, which is a difficult and sometimes dangerous place for personnel.
- High Rate Evaporative Sprinklers. High rate evaporative sprinklers typically operate at high pressures with an orifice designed to produce fine droplets and shoot them in a high trajectory. Evaporative blowers using compressed air to atomize and launch the droplets can also be used. This type of equipment is not normally used at heap leach operations, but it will become more common as more heaps enter the closure mode where rapid evaporation is needed.

For the 37 operations responding, solution application methods are summarized in the list below.

- 13 use only drip emitters
- 5 use only wobbler sprinklers
- 19 use both drip emitters and wobbler sprinklers
- 10 bury the drip emitters
- 5 use buried drip emitters between lifts
- all five "tropical" leaches - rainfall above 1500 mm per year - use wobblers.

Regardless of the systems used for solution application and management, capital and operating costs for solution handling are usually small. On the heap, header pipes up to 400 mm diameter are located every 30 to 60 meters across the heap. Material of choice for these pipes is usually HDPE (High Density Polyethylene), but sometimes it is mild steel. Distribution pipes of PVC or UV-stabilized PVC, usually 75 mm to 150 mm diameter, take off from the header pipes and are placed on similar (30 to 60 meter) spacing. From these, individual drip emitter lines up to 60 meters long cross the heap on 1.0 meter centers, or sprinkler manifold pipes (25 to 50mm PVC) up to 60 meters long cross the heap on 6 to 8 meter centers.

Total piping cost including header pipes (installed) is about $0.60 per square meter, or $0.05 per tonne of ore leached.

POWER COST FOR PUMPING

For the average two-cycle leach, two tonnes of solution are pumped to the heap and one tonne to the recovery plant, for each tonne of ore leached. Typically on-heap pressure for pumping barren solution is 100 psi at the pumps, and in-plant pressure for pumping pregnant solution is 30 psi. Thus, for two tonnes of solution per tonne of ore, power for pumping is 1.8 kWhr/tonne of ore and cost is $0.14/tonne. Where heaps are very high or where evaporation is required, power consumption can approach 4.0 kWhr/tonne.

LEACH PADS AND PONDS

The leach pad below the heap is a significant element of a heap leach design. The ideal location for the heap is a nearly flat (1% slope), featureless ground surface. Usually some earthwork is required to modify contours, but it is not necessary to eliminate all undulations. It is only necessary that all solution will flow across the surface towards collection ditches on the base or sides of the heap. Where the slope exceeds 3%, the front edge of the heap (30 to 50 meters) should be graded flat to provide a buttress to prevent heap failure.

Heaps can be placed in fairly steep-walled valleys with side slopes up to 20% (12 degrees). For long slopes above 10%, careful choice of pad material is necessary. LLDPE (Linear Low Density Polyethylene) offers a good choice because it has the ability to stretch but also has a high tensile strength, and it can be heat-welded to HDPE in flatter areas.

Valley Fill Heap Leach

A "Valley Fill Heap Leach" is a heap leach that has been built upslope from an earth dam. The containment area of the dam is filled with the stacked ore. The voids in the ore provide solution containment, and this volume serves as the pregnant solution storage pond. The ore stacked in the containment area behind the dam is usually a small part of the heap, which continues upslope and above the containment area. Thus, the dam might be ten meters high and the heap 50 meters high. A good example of a Valley Fill heap leach is Rochester, Nevada, shown in Figure 4.

Valley Fill heap leaches are used where terrain is steep and the ore must be placed in a narrow valley. They are also employed in arctic or high altitude environments as a method of keeping the process solution from freezing.

In normal leach pad construction, best design practice is to spread the solution out across the liner and to minimize solution hydraulic head to a few inches in any area. With a Valley Fill design, solution flow is concentrated and hydraulic head is high. The leach pad immediately upslope from the dam (in the solution storage area) must be built very carefully, usually with extensive sub-base preparation, double liners, and extra leak detection.

PAD CONSTRUCTION COST

A typical pad consists of several layers of material, listed from bottom to top. The ideal padsite begins as a uniformly sloping area with a grade of 0.5% to 2.0% in the direction of the process ponds. However, orebodies often occur in mountainous areas. It has been general practice to place the heaps within one or two kilometers of the orebody even if this requires extensive pad

Figure 4 Valley Fill Heap Leach - Silver Heap Leach at Rochester, Nevada. Much of the process solution is stored within the heap, behind the dam which can be seen at the downslope (left) edge of the heap.

area earthworks. The Tarkwa operation in Ghana employs a three kilometer long overland conveyor to move crushed ore to the padsite. Overland conveyors of ten kilometers or longer are common in other segments of the mining industry, and can be profitably employed to move ore to a heap leach site. It is not necessary to grade the padsite to a uniform grade of one or two percent. Internal hills and valleys within the padsite can be accommodated, as well as internal slopes up to 20 percent, provided that internal drain pipes can intercept the solution and direct it downhill to the process ponds. The cost shown in Table 2 are typical installed cost per square meter of pad surface for a padsite requiring minimum preparation. If complicated earthworks are required, these may add up to $5.00 per sq m to the costs shown in Table 2.

Ponds are installed downslope from the heap to provide storage of process solution. Usually there is a pregnant solution pond, a barren solution pond, and an overflow/storm water pond. There may be one or more intermediate solution ponds (sometimes solution is recycled from older to newer heaps to build up the gold content before processing).

Ponds are sized to permit storage of sufficient process solution so that the operators do not have to closely watch the pond levels. In addition to this "operating capacity", ponds are sized to contain solution during a several-day power outage and a major rainstorm event. Pond construction is similar to leach pad construction, except there is usually a second plastic liner with leak detection between the liners.

MINING, ORE PREPARATION & STACKING

Mining of ore for heap leaching employs the same techniques and equipment as mining of ore to feed any other process method. Where uncrushed ore (run-of-mine ore) is placed on the leach pad, ore may be blasted very heavily in order to reduce rock size and improve gold recovery. In high-rainfall environments when processing clay-rich material, it is very important to practice a mining routine that minimizes the amount of rainfall absorbed by the ore.

CONSTRUCTION ELEMENT	COST, $/M²
• Preliminary earthworks - removal of topsoil, building of edge berms and collection ditches. Cost assumes minimal alterations to topography. Sometimes it is necessary to do extensive site preparation, at a cost of several dollars per square meter.	$1.00
• 150 to 300 mm of compacted clay-rich soil, engineered to a permeability of 10^{-6} cm/sec.	$1.00 - $3.00
• Limited leak detection, usually embedded small-diameter perforated pipes placed near the lower edge of the heap and in areas of solution concentration. These "daylight" to collection sumps at the front of the heap. Leakage is usually permitted up to a certain small limit provided the area is not extremely environmentally sensitive.	$0.50
• Plastic liner, usually 0.75-1.00 mm (30-40 mil) thick PVC, or 1.50 to 2.00 mm thick HDPE or LLDPE. The liner is delivered in rolls up to 2000 sq. meters each, and field-welded to form the total liner. The initial installation for a "small" heap leach may cover 100,000 square meters; large installations may install 500,000 sq. meters each year. An HDPE liner of 2.00 mm thickness has sufficient strength and puncture resistance to support a heap up to 150 meters high.	$3.00 - $5.50
• Geotextile Cover may be placed above the plastic to prevent damage of the plastic by rocks in the drainage layer. The use of the geotextile is an economic tradeoff with the crush size of the gravel.	$1.50
• Drain pipes, usually 75-100 mm perforated flexible tubing, are placed on 6 meter centers above the plastic. Where solution does not drain directly out the front of the heap, large collector pipes may also be embedded in the drainage layer.	$0.50
• Gravel cover, up to 1000mm thick, is placed next to protect the pipes and the liner, and to provide a permeable base below the heap. Cost may be low if the gravel can be produced from the ore.	$0.50 - $5.00
• TOTAL INSTALLED PAD COST	$8.00 - $17.00/M²

Table 2 Leach pad component costs

Ore preparation varies widely. Run-of-Mine (ROM) ore may be hauled from the mine and dumped directly onto the heap, as shown in Figure 5. Nineteen of 36 operations surveyed had ROM heap leaches. Of these, twelve had only ROM leaches and seven had both ROM and crushed ore leaches.

At the other extreme from ROM leaches, Comsur's Comco silver heap leach at Potosi, Bolivia, crushes and then dry grinds all ore prior to agglomeration, with a grind size of 50% passing 150 microns (100 mesh). Three operations (Ruby Hill, Barney's Canyon and Castle Mountain), grind high grade ore and reblend it with the ore stream going to the heap leach (at Ruby Hill and Castle Mountain, the high grade is leached in agitated tanks to partially recover the gold).

Ores high in clay (such as saprolites) are typically processed by two stages of crushing using toothed roll crushers, then agglomerated in drums and stacked using a conveyor stacking system. Such a system is shown in Figure 6. Many ores are crushed and then either truck-stacked or conveyor-stacked without agglomeration. For these harder ores, crushing is usually done using a jaw crusher followed by one or two stages of cone crushing.

Figure 5 Truck dumping an upper 10 meter lift of run-of-mine ore on top of a lower lift that has already been leached.

Figure 6 Agglomerating drum and conveyor stacking system with 6 meter high heap at Ity, Ivory Coast.

Agglomeration

The term "agglomeration" means different things to different operators.

- At the simplest level, the ore is hard but contains a large percentage of fines. Agglomeration means simply wetting the ore with water so the fines stick to the coarse particles, and do not segregate as the heap is built.
- At the next level, the ore contains amounts of clay or fines that begin to plug a heap of untreated ore. Belt Agglomeration may be employed. In this technique, cement and water are mixed with the ore at a series of conveyor drop points, and the mixture tends to coat the larger rock particles. The primary goal is stabilization by mixing and contact. A typical conveyor stacking system involves 10 or more drop points, so Belt Agglomeration may occur as a normal part of the process. Operations that intentionally employ drop points or slide chutes are Barney's Canyon and the La Quinua operation of Yanacocha.
- Where ores are nearly pure clays, such as the laterite/saprolite ores in tropical climates, Drum Agglomeration is usually employed. The ore is first crushed finely enough (typically 25 to 75 mm) to form particles that can be a stable nucleus for round pellets. Cement and water are then added and the ore is sent through a rolling drum. The fines and the cement form a high-cement shell around the larger particles, and the rolling action of the drum compacts and strengthens the shell. Drum size and throughput are a function of several factors, but typically a 3.7 meter diameter, 10 meter long drum can process 750 tonnes of ore per hour. A 2.5 meter diameter drum can process 250 tonnes/hr. At the Tarkwa mine in Ghana, two 3.7 meter drums are installed to process up to 20,000 tonnes ore per day.

Of the 24 crushed-ore operations responding, 11 use drum agglomeration, 5 use belt agglomeration, and 8 do not agglomerate. Fifteen use conveyor stacking systems, the remainder stack with trucks. All the operations that use drum agglomeration stack with conveyor stackers.

Truck Stacking

Where rock is hard and contains very little clay, it is possible to maintain high permeability even when ore is crushed and dumped with trucks. Truck dumping causes segregation of the ore - the fines remain on the top surface, and the coarse material rolls to the base of the lift creating a highly permeable zone at the base. To control the degree of this segregation the ore may be partially agglomerated (wetted to cause the fines to stick to the coarse material) prior to placing in the trucks. Short lifts also result in less segregation. At Sterling, Nevada the problem was avoided by stacking the ore in 1.5 meter (5 ft) lifts but leaching in 6 meter (20-ft) lifts.

Truck dumping can also result in compaction of roadways on top of the heap. Several studies have indicated large trucks noticeably compact ore to a depth of two meters. To mitigate this problem, most operations rip the ore after stacking (but prior to leaching). Number of ripper passes is important; usually it is four passes in a criss-cross pattern. Some operations (Candelaria, Nevada) practised building an elevated truck roadway that was then bulldozed away. However this requires substantial bulldozer traffic on the heap surface, which can lead to permeability problems for some ores.

Stacking the ore with trucks can result in the tie-up of a large tonnage of ore below the truck roadways. This is a bigger problem on small operations than on large ones, because the roadway width is nearly the same regardless of the daily production rate. For a heap leach of 5000 tonnes/day, the roadways on the heap can tie up one month's ore production, with a value of $1.8 million. A conveyor system that stacks ore from the base of the lift can reduce unleached inventory to a few days' production. Because of this inventory reduction, at smaller operations where the ore is crushed, it is usually less capital-intensive to install a conveyor stacking system. For operations of 100,000 tonnes/day, truck stacking is more flexible and may be less capital intensive than a conveyor system.

Conveyor Stacking

Conveyor stacking systems commonly include the following equipment:

- One or more long (overland) conveyors that transport the ore from the preparation plant to the heap. Typically these consist of conveyors up to 150 meters long. At Tarkwa, Ghana, a 3 km overland conveyor is used.
- A series of eight to fifteen "grasshopper" conveyors to transport the ore across the active heap area. Grasshoppers are inclined conveyors 20 to 30 meters long, with a tail skid and a set of wheels located near the balance point.
- A transverse conveyor to feed the stacker-follower conveyor
- A stacker-follower conveyor, typically a horizontal mobile conveyor that retracts behind the stacker
- A radial stacker 25 to 50 meters long, with a retractable 10 meter conveyor at its tip. Wheels, discharge angle, and stinger position are all motorized and are moved continuously by the operator as the heap is built.

Figure 7 shows a typical stacker system in operation. Stackers are usually operated from the base of the lift (as shown in the figure) but may be located on top of the lift, dumping over the edge. Inclined conveyors can be installed up the sides of the lower lifts, and the stacking system can be used to build multiple-lift heaps. Stackers for this purpose should have very low ground pressure tires and powerful wheel drive motors to cope with soft spots in the heap surface.

Stacking systems like the one shown in Figure 7 can be used for heaps processing up to 50,000 tonnes of ore per day, but beyond that the size of the stacker (and the bearing pressure that is exerted by the wheels) becomes prohibitive. Typical cost of a complete stacker system with a 900mm (36-inch) wide belt for a 10,000 tonne/day heap leach operation, including the stacker and

Figure 7 Stacking system for capacity of 10,000 tonnes ore per day. Elements include stacker with extendable stinger; follower conveyor; cross conveyor; and several grasshopper conveyors.

follower conveyors, and ten grasshopper conveyors, is $1.5 million (delivered and installed at a typical developing-country heap leach site). Three hundred meters of overland conveyor connecting the stacking system to the crusher/agglomeration system cost an additional $500,000.

Figure 8 Rahco stacker building a 12 meter lift by tripping the ore over the advancing edge. The stacker can climb ramps and turn sharply to fit project requirements.

For operations stacking very high tonnages, large stackers can be mounted on caterpillar tracks to reduce ground pressure. Rahco International, Inc. (Spokane, Washington) makes a unique stacker, which is ideally suited to building large heaps at high tonnage rates. The stacker, shown in Figure 8, has individual drive adjustments so that it can climb up ramps to the next level and make sharp radius turns.

RECOVERY OF GOLD AND SILVER FROM HEAP LEACH SOLUTIONS
Other chapters in this book cover the details of recovery plant operations, so this section will be limited to a brief summary of heap leach plant operating results. Basically, gold and silver can be recovered from solution by contacting the solution with granular activated carbon in columns (CIC), followed by stripping of the carbon using a hot caustic solution. This caustic solution is processed in electrolytic cells or a zinc dust precipitation vat to recover the metal, which is then melted to produce a doré (impure bullion) bar. A CIC plant is shown if Figure 9. Where the ore is high in silver, typically with a recoverable silver content of more than 10 grams per tonne (0.3 oz/ton) of ore, Merrill-Crowe zinc precipitation is used instead of carbon adsorption. In this process the solution is clarified and de-aerated, then contacted with zinc dust to precipitate metallic gold and silver. This precipitate is then melted to produce bullion.

Of 34 operations reporting, 28 use carbon in columns (CIC) for adsorption of gold and silver from leach solutions, and six use Merrill-Crowe zinc precipitation plants. Three of the six using zinc precipitation reported at least 9:1 silver:gold in the bullion. Another, with 2.6 silver to 1.0 gold, produces leach solutions that are very high grade in both gold and silver content, thus justifying the choice of zinc instead of carbon. The other two process low grade gold solutions more typical of CIC plants.

Average pregnant solution gold content at the 28 CIC operations is 0.70 grams gold per tonne of solution. For these operations, loaded carbon averages 3900 grams gold per tonne of carbon, with six above 5000 grams gold per tonne of carbon. Six of these regenerate 100% of the carbon after each strip cycle, eight regenerate only 50% of the carbon per cycle, three do not regenerate. Three "high grade" CIC operations, all in Africa, reported pregnant solution grades of 3.5, 3.0 and 11.0 grams gold per tonne. These operations reported carbon loading of 8000, 6000 and 28,000 grams gold per tonne respectively. Stripped carbon from all operations averages 90 grams gold per tonne with 50% reporting in the range of 50 to 150 grams gold per tonne.

Figure 9 Five-stage carbon adsorption column plant (CIC plant) at Glamis Gold's San Martin, Honduras project. There are two parallel column trains (one is behind the other in this view). The plant can process up to 900 cu m of solution per hour, and is sized for an operation that processes up to 20,000 tonnes of ore per day.

DESIGN CONSIDERATIONS FOR RECLAMATION AND CLOSURE

Once the heap leaching operation is completed, the facility must be closed in accordance with local environmental requirements. Closure activities are highly variable depending on the environmental sensitivity of the site, and on the regulatory regime. In general, heaps are washed for a short period of time (commonly three years), during which time one tonne of wash water or recycled treated process solution is applied. Heaps are then capped, and ponds are filled and capped.

The easiest heaps to reclaim are single-lift heaps because the older heaps are abandoned early in the life of the operation and can be washed while production operations continue. In "Valley Fill" heap leaches, nearly all the ore ever placed on the pad is situated directly under active leach areas. Thus, washing of the entire heap must wait until operations are completed. Larger operations may have two or more "Valley Fill" leach areas, and can appropriately schedule closure activities.

Environmental regulations usually applied in the United States call for reasonably complete washing of the heap to reduce pH, to remove cyanide, and to partially remove heavy metals. Cyanide is fairly easy to remove since it oxidizes naturally, but pH and heavy metals are more difficult to control. Regulators are recognizing that a better approach is to conduct a "limited" washing program and then to cap the heap with a clay cover and/or an "evapotranspiration" cover of breathable soil with an active growth of biomass. These covers are designed to prevent infiltration of water into the heap. After several years of active closure activities, the flowrate of the heap effluent decreases to a manageable level (or to zero in arid environments). Once the flowrate is an acceptably low level, heap closure is accomplished by installing a facility for recycling collected effluent back to the heap. A relatively small "cash perpetuity bond" is maintained such that the interest on the bond covers the cost of maintaining and operating the intermittent pumping facility as long as is necessary.

A two million tonne heap of ore covering 90,000 sq. meters (average thickness 14 meters), located at Goldfield, Nevada, was recently closed with a clay/soil cap. Heap effluent gradually and steadily declined to 2.0 liters/minute after 36 months. Periods of intense above-average rainfall did not affect effluent rate. While this is a small and not very high heap, scaleup of this data should be applicable for preliminary design purposes.

Worldwide practice ranges from simple washing and abandonment, to the more complex procedure described above. Environmental design is an industry unto itself, and the simplistic concepts discussed here may not be applicable in other situations. Heap closure needs to be addressed in the feasibility stage of the project.

Typical cost of closure, including three years of heap washing, is $0.50 per tonne of ore stacked. This can be accumulated as a deferred operating cost. However, for U.S. heap leaches, regulators may require a closure bond to be put up at the beginning of the project. The amount of the closure bond is calculated using "government-defined" guidelines that typically result in a bond of $1.00 per tonne of total ore to be placed. This adds a generally prohibitive line item to capital cost, which is one of the reasons why new project activity has declined in the U.S. in recent years. (This item has not been included in the capital cost summary presented in Table 3).

CAPITAL COST

Capital cost for a small "basic" heap leach (3000 tonnes/day) with minimal infrastructure at a developing-country leach site is typically $3500 to $5000 per daily tonne of ore treated, with the higher cost attributed to high logistics expenses at remote sites such as central Asia. Larger operations (15,000 - 30,000 tonnes per day) cost $2000 to $4000 per daily tonne, but may commonly reach $6000 where "corporate culture" calls for process redundancy and infrastructure. Use of a mining contractor and/or a crushing contractor is common, and may eliminate the capital costs for these line items. Capital costs for some recent installations are shown in Table 3.

- Glamis Gold's San Martin heap leach (built 1999) had a published capital cost of $27 million (Glamis 1999 Annual Report) and began operations at 13,000 tonnes of ore per day (equal to $2,100 per daily tonne). Ore was crushed, agglomerated and conveyor stacked. Mining equipment was transferred from another operation at nominal cost; the operation was designed with excess capacity to allow for rapid expansion to 20,000 tonnes per day.
- Canyon Resources Briggs Mine in Southern California, built in 1996, cost $29.9 million for 9,500 tonnes/day (Marcus, 1997). This cost included $4.2 million for permitting, and a flowsheet that included 3-stage crushing. Mining equipment was leased. Adjusted for inflation to year 2002, Briggs' capital cost equaled $3,600 per daily tonne.
- Anglo American's Yatela Project in Mali started up at an annual rate of 7,000 tonnes per day, and cost about $8,000 per daily tonne (actual published capital cost was higher, but included extraordinary items).

Capital cost breakdown is shown in Table 3 for "typical" developing-country, remote sites with minimal infrastructure and minimal redundancy. Each operation, of course, will have a unique mix of capital cost line items.

OPERATING COST

Table 4 shows the breakdown of direct cash operating costs for the 27 operations that reported results for this chapter. (Direct Cash Operating Cost as used here includes all site costs including site and local office support costs, property taxes, import duties and fees. It excludes income and severance taxes, finance costs, royalties, product marketing costs, and depreciation/depletion). These operations had an average production rate of 15,800 tonnes/day. Average mining cost per tonne of material moved was $1.16, and average waste:ore ratio was 1.68:1. Labor rates varied widely, with seven operations reporting costs below $2.00 per hour, and thirteen above $15.00 per hour. Heap leaching is not a labor-intensive process, and where labor costs are low, logistics costs are usually high. Therefore there is not an obvious correlation between labor cost per hour and total operating cost per tonne.

HEAP LEACH CAPITAL COSTS		
	3000 tonnes/day	15,000 tonnes/day
Feasibility / design studies / permitting	US$ 400,000	US$ 1,000,000
Mine equipment	2,200,000	9,900,000
Mine development	600,000	1,200,000
Crushing plant (2 stage)	1,200,000	3,500,000
Leach pads/ponds	1,000,000	4,600,000
Agglomeration/stacking system	1,000,000	3,500,000
Process pumps, plant, solution distribution piping	1,100,000	3,500,000
Laboratory	300,000	500,000
Infrastructure (power, water, access roads, site office and service facilities)	1,700,000	7,500,000
Owner's preproduction cost	700,000	2,800,000
EPCM (engineering, procurement, construction management)	900,000	2,000,000
Import duties / IVA	800,000	7,000,000
Equipment / materials transport	600,000	2,100,000
Initial operating supplies	300,000	1,500,000
Working Capital	1,200,000	3,000,000
TOTAL	14,000,000	53,600,000
CAPITAL COST PER DAILY TONNE	US$ 4,700	US$ 3,600

Table 3 Heap Leach Capital Costs

REPORTED DIRECT OPERATING COSTS			
	Mining, $/tonne	Other, $/tonne	Total $/tonne
Total 27, average	3.11	4.00	7.71
Seven lowest, avg	2.50	0.88	3.38
Six highest, avg	5.90	8.17	14.07

Table 4 Direct cash operating costs for the 27 operations which reported costs for this chapter.

Operating cost is not very sensitive to the size of operation. Published direct cash operating cost for Barrick's Pierina mine (85,000 tonnes/day) is $3.93 per tonne, including $0.87 for mining. A recent study of an on-going operation in Africa concluded that increasing production from 4,300 tonnes/day to 13,000 tonnes/day would decrease costs (excluding mining) from $5.80/tonne to $5.10/tonne.

TYPICAL HEAP LEACH OPERATING COSTS, US$/tonne			
	3,000 tonnes/day	15,000 tonnes/day	Typical Nevada 30,000 tonnes/day
Mining (Strip ratio 2.5:1, cost/tonne of ore)	3.00	2.00	1.70
Crushing, Primary	0.40	0.20	0.20
Crushing, second plus third stage	0.50	0.40	0.20
Crushing (fourth stage, to 1.7mm (10 mesh)	0.80	0.80	-.00
Agglomeration/stacking	0.20	0.10	0.10
Leach operations (incl sprinkler supplies)	0.50	0.30	0.20
Recovery plant operations	1.50	1.30	1.40
General site maintenance	0.60	0.30	0.30
Cement for agglomeration (10 kg/tonne)	1.00	1.00	-.00
Cyanide, lime, other reagents	0.30	0.30	0.30
Environmental Reclamation/ Closure	0.50	0.50	0.50
General & administrative, support expenses	1.50	0.50	0.30
TOTAL SITE CASH OPERATING COST	11.50	8.30	5.20

Table 5 Typical Heap Leach Operating Costs

Average operating costs for "typical" heap leaches can be broken down as shown in Table 5. Costs are shown for ores that need crushing, agglomerating and conveyor stacking. Not all items in the list are appropriate for all operations; the right-hand column shows costs, that are more typical of a 30,000 tonne/day, coarse-crushed, unagglomerated Nevada heap leach.

TRADEOFF BETWEEN LEACHING IN HEAPS AND IN AGITATED TANKS

The alternative to leaching of ore in heaps is to grind the ore to a fine pulp, and to leach it as a water slurry in agitated tanks. Where a large amount of cement is required for agglomeration or where the ore needs to be fine-crushed, the operating costs of agitation leaching are not necessarily higher than for heap leaching. Heap leaching normally has significant capital cost advantages, so it is favored over agitation leaching where operating factors are identical.

Combined flowsheets are also utilized. Of 37 operations reporting, four use some form of grinding or grinding/agitation leaching for part of the ore stream going to the heaps. Homestake's (now Barrick's) Ruby Hill, Nevada, operation partially leaches high grade ore in agitated leach tanks, filters the tailings and combines them with crushed ore going to agglomeration and heap leaching. Castle Mountain (Viceroy Gold) uses a similar flowsheet. Barney's Canyon (Kennecott) wet-grinds part of the ore stream, but does not leach it before adding it to the agglomerator feed. Good discussions of these combined flowsheets can be found in Lehoux (1997) and Jones (2000). As presented in the paper by Lehoux on Ruby Hill, direct operating cost of the grind/leach portion of the operation was $4.98/tonne. Analysis of the capital and operating costs presented in the paper indicate that the heap-leach-only option may have been more economic.

Comsur's Comco silver heap leach dry grinds the entire ore stream to minus 105 microns (150 mesh) prior to agglomeration. In its third year of operation, Comco switched to wet grinding but it could not control water balance in the agglomerating drum, so it switched back to dry grinding.

Six of the heap leach operations reporting also have agitated leach plants for oxide ore that run as separate "stand alone" facilities. Ore is diverted from one to the other depending on grade (or in one case, depending on sulfide content). Average nominal "cutoff grade" to the mill for these operations is 2.30 grams gold/tonne (0.067 oz/ton), and cutoff grade to the heap is 0.41 grams gold/tonne (0.012 oz/ton). In practice, the cutoff grade to the mill is a function of the ore available on a daily basis - the agitated leach plant is fed to its capacity provided the ore is of reasonable grade.

CONCLUSION

Although the concepts of precious metals heap leaching are simple, the practices have substantially evolved over the past 35 years. Early choices for pad materials, sprinkling systems, and stacker designs have been discarded under the pressure of operating experience and cost-reduction factors. Overall operating costs have continually declined as "superfluous" activities and controls have been eliminated.

In spite of the apparent simplicity of the heap leach process - or perhaps because of it - there were many failures in the early years. There is now a large resource of successful operations from which to draw the experience needed to optimize the process. Heap leaching is expected to maintain its place as one of the principal tools for extracting gold and silver from their ores for both large and small deposits. The challenge for the future will be to remember and apply the experiences of the past.

REFERENCES

Agricola. Georgius. 1556. *De Re Metallica.* Translated by Herbert C. Hoover and Lou Henry Hoover. 1912. New York: Dover Publications, Inc. 1950 edition.

Hausen, D. M., Petruk, W., and Hagni, R.D. 1997. *Global Exploitation of Heap Leachable Gold Deposits.* Warrendale, PA: The Minerals, Metals & Materials Society (T MS).

Jones, A., 2000. *Pulp Agglomeration at Homestake Mining Company's Ruby Hill Mine.* Randol Gold & Silver Forum 2000. Golden, Colorado: Randol International.

Kappes, D. 1998. *Heap Leach or Mill? Economic Considerations in a Period of Stable Gold Prices.* Randol Gold & Silver Forum '98. Golden, Colorado: Randol International.

Lehoux, P., 1997. *Agglomeration Practice at Kennecott Barney's Canyon Mining Co.* Global Exploitation of Heap Leachable Gold Deposits. Warrendale, PA: The Minerals, Metals & Materials Society (TMS).

Marcus, Jerry. 1997. *The Briggs Mine: A New Heap Leach Mine in an Environmentally Sensitive Area.* Engineering & Mining Journal. Sep. 1997.

Marsden, J., and House, I. 1992. *The Chemistry of Gold Extraction.* Chichester, England: Ellis Horwood Ltd. (div of Simon & Schuster Intl. Group).

Philip, T.P. 1991. *To Mill or To Leach?* World Gold '91. Second AusIMM-SME Joint Conference. Victoria, Australia: The Australasian Institute of Mining and Metallurgy.

PICA, USA, personal communication, Ken Thomas. website: picausa.com.

Randol International. Various Symposia Proceedings. Golden, Colorado: Randol International, Inc.

vanZyl, D., Hutchison, I., and Kiel, J. 1988. *Introduction to Evaluation, Design and Operation of Precious Metal Heap Leaching Projects.* Littleton, Colorado: Society of Mining Engineers, Inc. (SME).

World Gold '91. Gold Forum on Technology & Practice. 1991. Second AusIMM-SME Joint Conference. Victoria, Australia: The Australasian Institute of Mining and Metallurgy.

Agitated Tank Leaching Selection and Design

Kathleen A. Altman[1], Mike Schaffner[2], Stuart McTavish[3]

ABSTRACT
Design of the agitated tank leach circuit begins with metallurgical testwork to determine optimum leaching conditions for specific ore types or expected ore blends from the orebody. Metallurgical test results are used along with production requirements and site specific information to develop a circuit layout that includes tank sizes and arrangement. Equipment selection (including agitator design, tank design, tank configuration, air and power requirements) is generally based on past experience with similar ore types and application. With new ore types or unusual applications, scale-up testwork must be conducted to determine equipment specifications.

INTRODUCTION
In metallurgical applications, leaching is the process of dissolving a soluble mineral or metal from an ore (Lapedes, 1974). All metals can be solubilized, or leached, in some manner or another. However, the leaching process requires a variety of lixiviants and operating conditions, which are dependent on the mineralogy of the ore to be processed. For example, oxide gold and silver ores can be easily leached at ambient conditions in an alkaline cyanide solution. However, refractory gold ores generally require pretreatment in an acidic pressure leach or biological leach circuit. Copper leaching normally takes place in an acidic environment. Uranium leaching is done in either acidic or alkaline leach solutions depending on the specific mineralogy of the ore to be treated, such as the carbonate content. Other ores may require the use of ammonia or chloride solutions in order to solubilize the metals of interest.

For the purpose of this paper, agitated tank leaching is defined as leaching of an ore under ambient operating conditions using a recovery method that does not incorporate extraction of the metal in the same unit operation. Using this definition excludes carbon-in-pulp (CIP) and carbon-in-leach (CIL) processing. Although CIP and CIL are oftentimes considered specialized cases of agitated tank leaching, they are included in other chapters of this section that are dedicated specifically to them, which is the reason for the exclusion here.

The number of new mining projects that utilize agitated tank leaching, as defined here, has declined over the last few decades. This decline is due to the reduced demand for uranium, the advent of CIP and CIL processes for gold recovery and the limited number of high-grade oxide copper deposits that have been developed. Currently, the most prevalent use of agitated tank leaching is the recovery of precious metals in conjunction with counter current decantation (CCD) thickeners and the Merrill-Crowe zinc precipitation recovery process. This recovery technology is usually preferred when large amounts of silver are present either alone or along with gold. However, CCD/Merrill-Crowe is an established recovery process that has been used successfully for the treatment of ores that contain predominantly gold, too. For some companies, agitated tank leaching/CCD/Merrill-Crowe is still the selected gold recovery process, particularly at remote, third world mine sites.

Even though the most prevalent current application is the recovery of precious metals from alkaline cyanide leach solutions, agitated tank leaching is a recovery method that is important in a number of applications. This section deals with a generic discussion of agitated tank leaching. The discussion includes a description of the process; an outline of how specific processes are selected for new projects; and details regarding the design of agitated leach circuits.

---
1 Independent Consultant, Denver, Colorado
2 Coeur Rochester, Inc., Lovelock, Nevada
3 SNC-Lavalin Engineers & Constructors, Inc., Toronto, Ontario

The process of selecting and designing processing circuits should be a collaborative effort between the operating company, metallurgical testing facility, engineering company, and equipment suppliers. Process selection and engineering design is but part of the iterative process that takes place during the development of new projects.

During the preliminary stages of the development of a new project, cost data that were generated for similar projects in the past is modified, as required, and incorporated into a financial model in order to select the option that is most attractive from a financial perspective. It is important to understand that selection of processing alternatives is seldom done on a "stand alone" basis. All of the costs associated with the project are developed and compared in order to make a final selection. For example, the pre- and post-treatment options may be quite different depending on the recovery process that is selected. A heap leaching process may or may not require a crushing circuit, whereas agitated tank leaching circuits generally require both crushing and grinding circuits. The cost differences plus the impact on both the total metal recovery and the time of recovery must be incorporated into the financial model in order to make decisions based on accurate information.

Project decisions are generally based on successively more accurate data as a project proceeds through the development stages. Initial selections may be based on order of magnitude estimates with an accuracy level as low as plus or minus 50 percent, or less. As more detailed information is acquired, decisions are generally based on data that has been developed to successively higher levels of accuracy. The final decision to proceed with a project is normally based on a feasibility study that has accumulated sufficiently detailed data to complete an economic evaluation to a level of accuracy in the range of plus or minus 10 to 15 percent. At this stage, a process flow diagram has been selected and frozen, equipment has been selected and firm, fixed quotations have been solicited from equipment manufacturers.

PROJECT DEVELOPMENT

Ideally, the process of choosing an extraction method begins in conjunction with the process of delineating an ore body. Numerous stages of assessment are conducted along the way to operating a mining project successfully (Altman, 1999). The specific process of developing a new project, including the selection of an appropriate metal recovery process, is highly dependent upon the corporate philosophy of the company completing the development and their internal procedures and evaluation criteria. If a company relies on outside financing or listings on a public stock exchange, the development process typically becomes more rigorous and it is subjected to external reviews and legal requirements in addition to corporate requirements.

Multi-disciplinary project teams are required to develop mining projects. Key disciplines include geology, mining and metallurgy. While an ore resource may be large and interesting from a geological perspective, it is not an ore body until it is proven that the material can also be mined and processed economically.

Development of mines is an iterative process. Each assessment stage requires the acquisition and evaluation of data culminating in a specific decision to proceed with the project; delay the project development; or abandon the project. Each decision point is based primarily on an economic evaluation that is based on the available data. Early assessments may be based only upon analytical information and historical cost data for similar projects that have previously been evaluated. Typically, each time a decision is made to proceed with a new project, additional funding is allocated so that more detailed information can be generated and evaluated.

It is important to recognize that process development does not occur in isolation. As metallurgical data is generated, it is entered into an economic model along with other data specific to the project. Routinely this is a discounted cash flow (DCF) model and a decision is based on the internal rate of return (IRR) or the net present value (NPV) or both.

HISTORY

In the early 1890's, C.F. Brown developed an air-agitated tank used to leach gold ores in New Zealand (Herz 1985). This tank later became known as a Pachuca tank because of its popularity in Mexico. In the early 1900's, Pachuca tanks were the standard leaching vessels in the mining industry. By the 1960's, mechanical agitation had proven to be more economic than air agitation and existing Pachuca tanks were gradually retrofitted with agitators while new plants were designed with mechanically agitated tanks.

Over the last several decades, the use of agitated tank leaching has become much more sophisticated. The knowledge of agitator function has increased to the point where specific agitators are designed to improve the efficiency of the various agitator functions including air shear, slurry pumping, mixing, and suspension of solids. These improvements have made choosing the proper agitator for a specific ore body a more onerous task, which often requires expertise from consultants, engineering firms, and agitator manufacturers. Experience has shown that a properly designed leach circuit will operate for years with little maintenance or attention whereas a poorly designed circuit will have many operational and mechanical problems that are costly to resolve.

AGITATED TANK LEACHING – PROCESS DESCRIPTION

The concept of agitated tank leaching is fairly simple: An ore is prepared (this typically includes crushing, grinding, and pH conditioning); placed in an agitated leach tank with a leaching agent; and the metals are allowed to leach from the ore into the solution. Once leaching is complete, the "pregnant" solution is separated from the slurry with filters or CCD thickeners and processed through a metal recovery system. Merrill Crowe is typically used to recover gold and silver whereas solvent extraction and electro-winning are used to recover copper. Although the process seems simple, the design engineer must have a solid understanding of the ore body, the ore characteristics, and the leaching process in order to design a successful circuit. Some primary operating parameters, which directly affect the performance of the agitated tank leach circuit include:

- Grind – The ore must be ground to a liberation size that exposes the desired mineral to the leaching agent, as well as a size that can be suspended by the agitator. Grinds coarser than 65-mesh tend to cause excess abrasion and wear due to the degree of agitation required for suspending the coarser particle. Ores that can be successfully leached with a grind coarser than 65-mesh are typically good candidates for heap leaching.
- Slurry density – The slurry density (percent solids) determines retention time. The settling rate and viscosity of the slurry are functions of the slurry density. The viscosity, in turn, controls the oxygen mass transfer and the leaching rate.
- Number of tanks – Agitated tank leach circuits are typically designed with no less than four tanks and preferably more to prevent short-circuiting of the slurry through the tanks.
- Dissolved oxygen – Air or oxygen is often injected below the agitator to obtain the desired dissolved oxygen level.
- Reagents – Adding and maintaining the appropriate amount of reagents throughout the agitated leach circuit is critical to a successful operation. Adding insufficient quantities of reagents reduces the metal recovery but adding excess reagents increases the operating costs without recovering enough additional metal to cover the cost of the reagents.

PROCESS SELECTION

General Considerations

In the case of agitated tank leaching, the initial screening may be based entirely on analytical data. For example, the ratio of the value of a cyanide atomic absorption (AA) assay to a fire assay or an oxide copper assay to a total copper assay indicates whether the ore type is amenable to leaching or not. It also provides an indication of the recovery that can be expected in a leaching circuit.

The main extraction methods that are considered for non-refractory, i.e. amenable to leaching, ore bodies are agitated tank leaching, heap leaching, CIP, CIL, and gravity concentration. Process selection is not only related to the metallurgical performance of the ore contained in an ore body but also the overall size of the ore body. Drill samples that are used to delineate the ore body can also be used for preliminary metallurgical testing. Following an analytical determination that the ore body is amenable to leaching, the first step in a metallurgical test program, for ore bodies that are contemplating some type of leach recovery, is usually bottle roll tests. These tests begin to provide information about the relationship between particle size and recovery and time versus recovery in addition to information about reagent consumption rates.

As a rule of thumb, small ore bodies cannot support the higher capital costs associated with milling circuits. Therefore, larger-scale testing of samples may be limited to heap leach column tests until such time as sufficient ore reserves are delineated to support the cost of a milling circuit, such as an agitated tank leaching circuit. At this time, additional tests are conducted to determine if the ore is amenable to gravity concentration and to ascertain the liberation size, leach retention time, and reagent addition rates, as well as crushing and grinding characteristics, settling rates, filtration characteristics, etc.

It is important to remember that accurate test results mean nothing unless the samples that are tested are representative of the ore that will be processed. For this reason, care must be taken to ensure that all expected ore types are sampled and tested. This is one activity that must be closely coordinated between the geologists, mining engineers and process engineers.

One of the final steps in a metallurgical test program for a project that has selected an agitated tank leaching circuit is generally the operation of a continuous pilot plant that incorporates all of the unit operations that will be used in a full-scale plant. These tests are used to confirm and refine the test results that have been generated to date. Results for the operating parameters required in the engineering design and for the economic analysis are generated during this test campaign. The culmination of this testing is the selection of the process flow diagram that will be used for a full-scale plant.

The final selection usually includes consideration of project elements that are not easily quantifiable on an economic basis, such as environmental and political risks, the corporate financial position, and the company's experiences and preferences with regard to operating circuits. Therefore, the final process selection may or may not be the option with the highest NPV or IRR.

Heap Leaching versus Agitated Tank Leaching

Process Description. Modern heap leaching was first used in the North America uranium industry in 1950. The methods were adapted to the copper industry in the 1960's and the gold industry in the 1970's (Scheffel, 2002). In heap leaching, ore is placed on an impermeable liner and a leaching agent is percolated through the ore and recovered on the liner. The ore may be placed on the pad as run-of-mine, crushed, or crushed and agglomerated material. The pregnant solution from the heap is treated through a metal recovery plant, and the barren solution that exits the recovery plant is re-circulated back to the pad.

Capital Costs. The capital cost for a heap leaching circuit is generally less than the capital cost for an agitated tank leaching circuit. This is because a heap leach operation typically requires less equipment and infrastructure. A typical heap leach operation consists of a leach pad, crusher, metal recovery plant, truck shop, office complex, and solution distribution and collection pumps. A typical agitated leach tank plant consists of a crusher, grinding mills, thickeners, leach tanks, filters or CCD thickeners, metal recovery circuit, tailings facility, truck shop, and an office complex.

Operating Costs. Heap leach operating costs are generally lower per tonne of ore processed than the operating costs associated with an agitated tank leach circuit due to the reduced equipment and manpower requirements. In addition, heap leaching is typically conducted at a coarser size fraction than agitated tank leaching, thus reducing the power requirement and reagent consumption.

Recovery. Although a heap leach operation is less expensive to build and operate, metal recoveries are typically lower than recoveries in agitated tank leaching circuits due to the coarser particle sizes and less efficient contact of the leaching agent with the ore. In addition, the time required to process the ore is greater for heap leaching, which, in turn, impacts the inventory of metal in the circuits and the rate at which the metal is recovered. A typical gold heap leach will have an average leach cycle of 90-180 days to reach the targeted recovery, and copper heap leach cycles can be anywhere from 180-360 days to yield targeted recoveries. Generally less than 72 hours are required to reach targeted recoveries in agitated tank leaching circuits. This recovery-cost relationship makes heap leaching the best alternative for low-grade ore bodies. When the head grade is higher, the increased recovery and rate of recovery in addition to a lower metal inventory generally makes agitated tank leaching the better option.

Combinations of recovery circuits may also be selected. It is common for large ore bodies with both high and low grade ore to maximize the economics by installing both heap leaching and agitated tank leaching recovery circuits. In recent years some smaller ore bodies, including Kennecott's Barney's Canyon Mine and Homestake's Ruby Hill Mine, have been developed with an integrated mill and heap leach flowsheet. In this flow sheet the leached mill tailings are agglomerated with low-grade ore and placed on a leach pad. This improves recovery, allows the economic recovery of lower grade ore, and eliminates the need for a tailings pond.

Additional Considerations. Besides the obvious economic reasons, several environmental and site specific concerns should be considered when selecting between heap leaching and agitated tank leaching. These include:

- Terrain – Is there a suitable area and location for a heap leach pad or mill tailings facility?

- Ambient temperature – Will the project experience extreme cold or heat that causes freezing or excessive evaporation?
- Annual precipitation – What is the annual precipitation and precipitation cycle? Periods of excessive precipitation can cause dilution to the solution and may make it difficult to contain the increased solution volume within the processing circuit without substantially increasing the size.
- Location – Is the ore body near a national park or forest, or metropolitan area?
- Wildlife – What wildlife exists and what measures must be taken to protect it?
- Local and federal mining regulation – Are there any differences in regulations for heap leach or agitated tank leach processing circuits?
- Local and federal permitting regulations – Are there any differences in permitting requirements for heap leach or agitated tank leach processing circuits?

CIP-CIL versus Agitated Tank Leaching

Process Description. The first industrial CIP plant was installed at Homestake's South Dakota operation in 1973 (Fleming, 1998). A CIP circuit utilizes the same flowsheet as an agitated leach circuit up to the point where the slurry discharges from the final agitated leach tank. At this point, the precious metal values are recovered directly from the slurry onto granulated activated carbon in agitated CIP tanks. The carbon is retained in the tanks by any one of several different types of screens through which the slurry discharges. CIP circuits are typically designed with at least four tanks to prevent short-circuiting of slurry and allow sufficient retention time for recovery of all the metals. Although agitated leach tanks are used before CIP tanks, CIP and CIL circuits are considered as a separate process in sections D-I-4 and D-I-5.

CIL circuits are similar to CIP circuits with the exception that leaching and extraction occur simultaneously in agitated leach tanks that also contain carbon and are equipped with carbon retention screens.

The evaluation of agitated tank leaching verses CIP and CIL circuits is not as complex as the heap leach-agitated tank leach analysis. CIP and CIL circuits generally have lower capital and operating costs for gold ore bodies than agitated tank leach circuits. Silver ore bodies show better economics with agitated tank leach-Merrill Crowe circuits. This is because the volume of carbon that would have to be processed to recover economic levels of silver would increase the capital and operating cost of a CIP or CIL circuit above that of an agitated tank leaching/CCD/Merrill-Crowe circuit.

PROCESS DESIGN

Design Procedure

Once agitated tank leaching is selected, and a decision is made to proceed with the project, a detailed, project-specific engineering design is completed.

Tank Design

Time Required (Tank Sizing). For any leaching process, including agitated tank leaching, the first and foremost process design criterion is the amount of time required to complete the leaching process. The leach time required for a new circuit is determined by leach tests conducted on representative samples of the ore. If less than the optimum time is specified, the amount of metal dissolved may be insufficient. On the other hand, if too much time is specified, the incremental recovery increase may not justify the cost of larger tanks and agitators.

Slurry Density. The optimal slurry density is defined by metallurgical testing. The density of the slurry directly affects the retention time for a circuit. The higher the percentage of solids, the smaller the volume required to achieve a specified retention time. Ideally, slurry produced in the grinding circuit for an agitated tank leach circuit will report to a thickener ahead of the agitated leach tanks. The thickener helps to ensure that the feed to the leach circuit is consistent, which, in turn, ensures optimal operation of the leaching process. The optimum slurry density is directly affected by the grind size and the viscosity of the slurry.

Plant Throughput. Obviously, the tonnage of ore to be processed in a plant has a direct correlation with the size of the tanks that are required. The size of a processing plant is normally determined in conjunction with geologists who are familiar with the size of the orebody and mining engineers who have helped to select an appropriate mining method for the orebody. This information is input into appropriate financial analyses in order to select the optimal mining and processing rate for the project.

Number of Stages of Leaching. In general, a single, large tank is less expensive to fabricate than a number of smaller tanks with a combined volume equal to the volume of the large tank (von Essen and Ricks, 1999). However, the agitation process is such that short-circuiting is likely to occur if a single tank is used. As a rule of thumb, the use of four or more smaller tanks minimizes the short-circuiting problem. However, the number of leach tanks incorporated into a process design is also dependent on the profile of the leach solutions and solids in the tanks. A final important consideration in the design process is that an "extra" leach tank may be included in the design so that any of the agitated leach tanks may be bypassed for maintenance without reducing the residence time.

Sample Calculation. The nominal required tank volume is calculated by multiplying the volumetric flow rate of slurry through the circuit by the amount of retention time required. This volume is then divided by an Effective Volume Factor, which makes an allowance for the volume associated with aeration, settling, agitation, etc. Additionally, an allowance for freeboard may be made in the tank height that is used to provide the required tank volume.

For a 1,000 t/d plant with an ore specific gravity of 2.8 operated at 50% solids, and requiring a 24-h retention time, the nominal volume required is determined with the following equations.

(tonnes / day / 24 hours / day) = tonnes / hour

(1,000 t/d) / (24 h/d) = 41.7 t/h

slurry volume / hour = (solids m^3/hour + water m^3/hour) or

(solids t / h / sg ore) + (((solids t / h / % solids) - solids t / h) / sg water)

$[(41.7\ t/h)/(2.8\ t/m^3)] + [((41.7\ t/h)/0.5)-(41.7\ t/h)/(1\ t/m^3)] = 56.5\ m^3/h$

tank volume = slurry m^3/h * retention time

(56.5 m^3/h) * (24 h) = 1,357 m^3

An appropriate Effective Volume Factor is 0.92, which brings the required volume to 1,475 cubic meters [(1,357 m^3)/(0.92)]. Assuming that a total of five leach tanks provides an acceptable leach profile, the volume per tank is 271.4 cubic meters. [(1,475 m^3)/(5)]

Tank Height to Diameter Ratio. In general, the mining industry uses cylindrical tanks with a tank height to diameter ratio of 1:1. That is, the tank height, h, equals the tank diameter, T, i.e. h = T.

tank volume = $\pi r^2 h$

$(T/2)^2 T\pi = 271.4\ m^3$

$\pi [(T^3)/4] = 271.4\ m^3$

T = 7 m

In addition to this, the tank height may be increased to 7.6 m to allow for additional freeboard.

It should be kept in mind that this calculation method determines the "average" residence time for the slurry in the tank. Oldshue (1963) reminds us that all particles sizes are not necessarily retained for the same residence time. The size of a solid particle in slurry may affect the residence time. The impact of this phenomenon will be discussed further in the section on agitation requirements and design.

Tank Baffling. Generally, agitated leach tanks must be baffled. Standard baffling includes four equally-spaced baffles that are one-twelfth of the tank diameter (Oldshue, 1963). However, baffle requirements are related to mixer torque, so they are dependent on the selected agitator (Salzman, et. al. 1983).

Tank Configuration. Agitated leach tanks are usually designed so the slurry can flow by gravity from one tank to the next. Building each tank taller than the subsequent tanks and putting a false bottom in the tank to maintain the required tank volume is a common design. If the site topography allows, the tanks may be built into a hillside to accommodate gravity flow.

Materials of Construction. The materials of construction for the equipment, including tanks, agitator shafts, and impellers should be selected for the environment of the leaching process. Unlined mild steel tanks are suitable for the alkaline, cyanide leach solutions used in gold leaching for short term operations; however a chemical resistant liner should be considered for longer term operations. The impellers may be rubber-lined to minimize the abrasion due to the particles in the slurry. A leach process that utilizes sulfuric acid may require stainless steel tanks and agitator components. The choice of the materials of construction impacts the longevity of the equipment and the maintenance requirements in the future.

Reagent Additions

The reagent addition systems are dependent on the specific reagents that are required and whether the reagents are added to the slurry prior to the time it enters the leach circuit or whether the concentrations of reagents need to be trimmed as the slurry advances through the leach circuit.

For example, in a gold cyanide leach circuit, the pH of the slurry may be initially controlled through the addition of lime on the feed belt to the grinding circuit and the grinding may be done in sodium cyanide solution. In this case, it is good engineering practice to provide for trim additions of slaked lime and cyanide in the leach circuit so that the reagent additions can be optimized. The determining factors to consider in designing reagent addition systems are the sensitivity of the leaching process to the concentration of the reagents and the amount of variability that is anticipated in the ore that will be processed.

In many leaching processes, it is necessary to add oxygen, generally in the form of air, to the leach circuit. This may be done with air blowers or with low pressure air compressors depending on the volume of air and the delivery air pressure that are required. In some cases, it may even be necessary to inject oxygen into the leaching process, which results in an even more sophisticated gas production and/or storage and injection system.

Agitation Requirements and Design

Mixing Requirements and Performance. Agitators are used as a mixing device in the agitated tank leaching process. Mixing can be divided into five different areas (Oldshue and Todd, 1981):

- Liquid-solid dispersion
- Gas-liquid dispersion
- Liquid-liquid dispersion
- Blending of miscible liquids
- Production of fluid motion.

The performance of mixers is evaluated by:

- The physical uniformity of the contents of the tank
- Mass transfer or chemical reaction.

The mixer design can be divided into three elements. First, the process design, which includes the fluid mechanics of the impellers, the fluid regime required by the process, scale-up from laboratory or pilot scale to plant scale operations, and similarity. Second, the impeller power characteristics that relate power, speed, and impeller diameter. Third, the mechanical design that includes impeller design, agitator shafts, the drive assembly and the support structures (Oldshue and Todd, 1981).

In agitated tank leaching the main requirements of an agitator are:

- Solids suspension
- Oxygen transfer by gas dispersion
- Maintaining the chemical composition and physical uniformity (Fraser, et. al. 1993).

Choice of Impellers. Impellers are divided into two basic categories. These are radial flow impellers and axial flow impellers. Radial flow impeller designs include the Curved Blade Turbine (CBT), Flat Blade Turbine (FBT), Smith turbine, and Rushton turbine (Guide to impeller selection for maximum process results). As the name implies, the fluid is discharged in a horizontal or a radial direction to the tank wall (Oldshue, 1987). The flow pattern created by radial flow impellers rotates around the impeller. If low-viscosity liquids are agitated in un-baffled tanks, a vortex is formed and the liquid swirls around the vortex (Rushton and Oldshue, 1953).

Axial flow impellers include the marine propeller and the Pitched Blade Turbine (Oldshue, 1983) in addition to a number of hydrofoil designs (Guide to impeller selection for maximum process results). As the name implies, axial flow impellers create a flow pattern that is parallel to the impeller shaft (Oldshue, 1983).

Radial flow impellers have been used traditionally to disperse gas, which meets the agitated tank leaching requirement for oxygen transfer. Axial flow impellers, then, produce higher, more efficient fluid flows than radial flow impellers (Olderstein, et. al., 1989). The flow, or pumping, capability of agitators is responsible for meeting the other process requirements of agitated tank leaching, i.e. maintaining solids suspension and the physical and chemical uniformity in the leach tanks. Shear is the impeller characteristic that is responsible for gas dispersion. Because the process of agitated tank leaching requires components of both flow and shear, it is important to understand the basics of good agitator design in order that a reasonable compromise is achieved. The design must optimize the capital cost of the agitation system with the operating costs resulting from power consumption, yet assure that the process results meet the required criteria. In order to achieve the required results, the mixer design must incorporate a balance between shear and flow (Kubera and Oldshue, 1992).

The design of agitation equipment has advanced dramatically over the past half century, or so (Kubera and Oldshue, 1992; Oldshue, 1989). With the recent advent of the internet, agitator suppliers have made technical data and inquiry sheets, technical and reference information, and design calculations instantly available from our desktops. This eases and enhances coordination, cooperation and understanding between the various individuals involved in the design process.

As stated earlier, Pachucas were initially used to meet agitation requirements. Shaw (1982) provides a good description. Air lift agitation was commonly used in Pachuca tanks. A Pachuca tank has either a 90° or a 60° conical bottom and a height to diameter ratio between 2.5 and 3. An air lift tube of approximately 10% of the tank diameter is placed in the center of the tank and air is injected into the bottom of the air lift tube. This produces a flow of slurry up the tube at high velocities.

As processing rates increased and equipment capacities became larger, Pachuca systems became expensive due to the cost of cone bottom tanks, compressors, and piping. They also use a high amount of energy for the level of agitation that is produced.

The alumina industry used Pachuca tanks for precipitation in the Bayer process. In the 1960's, the alumina industry began to investigate the design of mechanical agitation systems. Mechanical draft tube circulators are the results of these efforts. The concept spread from the alumina industry to a variety of other industries including gold leaching (Shaw, 1982).

Subsequently, the role of micro- and macro-scale mixing has continued to evolve and the development of advanced materials of construction, such as composites, has removed the limitation of forming impellers from only flat stock (Oldshue, 1989). These advances have enabled a new generation of impeller design.

Flow Relationships. Oldshue (1983) explains that the power (P) applied to a mixer produces both flow (Q) and impeller head (H). The head is proportional to the velocity head of the slurry, or fluid that is being mixed. Normally, the total flow is expressed as mass flow, e.g. t/h. The total power is proportional to the total flow and the impeller head, just as it is in a pump.

$$P \propto Q H \qquad (1)$$

However, an agitated tank is not a confined channel such as a pump and piping system. Therefore, the determination of impeller flow and head for an agitation system are not as quantitative as it is for a pumping system.

At constant power, the flow to head ratio is proportional to the impeller diameter (D) and the tank diameter (T) ratio. The equation used to express this relationship is:

$$(Q/H)_p = (D/T)^{8/3} \qquad (2)$$

The impeller head is also directly related to the square root of the fluid shear rate. Therefore, it is a measure of the flow to fluid shear rate around the impeller.

Practically speaking, these relationships mean:

- A large impeller operating at a slow speed produces high flow and a low fluid shear rate.
- A small impeller operating at high speed produces a high shear rate and low flow.

Walter (1968) provides further explanation of some additional relationships. Power (P) is proportional to the speed (N) and the diameter of the impeller (D) according to the following relationship:

$$P \propto N^3 D^5 \quad (3)$$

And, the pumping capacity (Q) is proportional to the impeller speed (N) and the impeller diameter (D) according to:

$$Q \propto ND^3 \quad (4)$$

Combining equations (3) and (4) provides the following relationship:

$$(Q_1/Q_2) = (D_1/D_2)^{4/3} \quad (5)$$

These equations show that for a constant power level, a large diameter impeller has a higher pumping capacity than a smaller diameter impeller and it will run at a slower speed. Also, the larger the impeller, the lower the power required to achieve a specified pumping capacity.

However, even though the power is reduced and the speed is decreased, the torque increases. The capital cost of an agitation system is directly related to the torque requirement. Obviously, the power requirement of the system is primarily an operating cost. Therefore, an economic evaluation is needed to determine the optimum solution to be used for a specific application.

Assume that two options are available to produce identical process results. They are a 37-kW (50-hp) agitator and a 30 kW (40 hp) agitator. The 30-kW unit has a larger impeller and it operates at a lower speed. Therefore, it needs a more expensive drive due to the higher torque that is produced.

Assuming a 24 h/d, 7 d/wk operation and an operating availability of approximately 95%, the selected unit will operate for about 8,300 h/yr. For simplicity, assuming full power draw, the larger unit will consume roughly 58,250 kWh of additional electrical energy per year. [(365 d/y)(24 h/d)(0.95)(37-30 kW)] If a power cost of $0.08/kWh is anticipated, the difference in annual power costs is $4,660. This increased operating cost then needs to be used as an input to the projects economics. For a 2 year payback, the smaller, less expensive agitation unit will be attractive as long as the difference in the initial installed capital cost is more than $9,320 per unit.

Solids Suspension. In gold recovery processes, the suspension of solids predominates the agitation application. Solids suspension is a flow-controlled application, which means that the process results are directly proportional to the pumping capacity of the impeller (Kubera and Oldshue, 1992).

Oldshue (1963) presents a comprehensive review of solid suspension in hydrometallurgy. Solid suspension processes can be separated into two areas:

- Hindered settling
- Free settling solids.

If the slurry density is high enough that the particles interfere with each other during settling, it is considered hindered settling. Hindered settling includes settling velocities of less than 0.3 m/min (1 ft/min). Generally, this occurs in slurries with greater than 50% solids by weight but it is also a function of particle size and fluid viscosity. In hindered settling applications, it is more appropriate to consider the process criteria as fluid blending or fluid motion. Although this may be the case for agitated tank leaching, it is more likely that the application will be free settling solids.

In applications governed by free settling solids, the equipment design is highly dependent on the settling velocity of the solids. Percent suspension is an important consideration in solids suspension applications. It takes into account variations in the content of the leach tanks.

$$\text{Percent Suspension} = \frac{\text{Wt \% solids at the sample point}}{\text{Wt \% solids in the tank}} \times 100\%$$

It may be necessary to consider three or four separate size fractions if the particle size distribution for the process is very large.

The required process criteria for solids suspension can be subdivided into a variety of categories, including:

- Complete uniformity
- Complete off bottom suspension
- Complete motion on the tank bottom
- Filleting permitted but progressive fillet build-up not allowed
- A specified height of the suspension.

Complete uniformity implies that percent suspension is 100% at any point in the tank. However, practically speaking, it is very difficult to achieve 100% suspension in the upper layer of liquid in a tank. If the settling velocity of the particles is greater than 1.8 m/min (6 ft/min), this difficulty is confined to the upper three percent of the tank because the horizontal flow patterns that are created by the agitator cannot keep the solids with a high settling velocity suspended.

If a sampling point is placed near the top of a tank, a curve of power (kW or hp) versus percent suspension can be plotted. On this curve, there is a breakpoint where, even though there may be a large increase in the power, there is little improvement in the percent suspension. This breakpoint on the curve can serve as a practical definition of Complete Uniformity. It should be kept in mind that this point also depends on the depth of the sample point.

The other process criteria listed are relatively self-explanatory. In some cases, these criteria may be used in conjunction with one another. For a more detailed discussion, please refer to Oldshue (1963).

In solids suspension there is usually an optimum ratio of impeller diameter to tank diameter (D/T). This ratio is dependent on the slurry density (% solids), settling velocity and the process criterion required to achieve the necessary process results. In order to determine the optimum ratio, a plot is made of power versus the ratio of the impeller diameter to the tank diameter (D/T) for the selected process criterion, e.g. complete off bottom suspension of the solid particles. This plot reaches a minimum at some ratio of impeller diameter to tank diameter. It should be kept in mind that while this ratio may be the optimum to achieve the solids suspension, it may not be sufficient to achieve other process criteria and the ratio or the installed power may have to be increased in order to meet the other criteria.

The evaluation of variables becomes even more complex when a comparison is made of different types of impellers.

An important phenomenon to be considered is that with continuous flow, as in an agitated tank leach circuit, the total composition of solids in the tank is not necessarily the same as the total composition of solids in the feed and discharge streams. However, for steady state operation, the composition must be the same at the inlet and outlet of the tank. Moving the draw-off point in the tank can change the composition distribution for the tank. It is important to realize that all particle sizes in a continuous flow system may not have the same residence time, which points out the importance of the draw-off point. The draw-off point is selected by putting a down-comer in the tank. The particular draw-off point for the tank is dependent on the length of the down-comer. One reason multiple stages are used in agitated tank leaching is to minimize the chance that certain particles may be retained too long or escape too soon.

Based on the concepts that have been discussed previously, if the draw-off points were near the top of the tank, the large particles may not be suspended uniformly. Then, these larger particles would build up in the tank and may even sand in the impeller. It must also be recognized that the slurry velocity in the draw-off tube must be sufficient to keep the solids in suspension in order to avoid sand in the line.

In cases of hindered settling, the process criterion becomes fluid movement as opposed to the suspension of the solids. As the slurry density increases and hindered settling is introduced, the power requirement decreases and the uniformity of solids is generally quite good (Oldshue, 1963).

Mass Transfer. Leaching is the process of dissolving a specific component, or metal, out of a mixture of components in the solid. Therefore, leaching is the mass transfer of that component from the solid state to the liquid state. It is critical that the particles are small enough to ensure that the liquid has access to the metal of interest.

The rate of leaching is equal to the sum of the concentration driving force (ΔC), the exposed interfacial area (A) and the mass transfer coefficient (k_s). That is:

Leaching Rate = $\Delta C \, k_s \, A$

The mass transfer coefficient is a function of the mixing system.

The percentage of the specific component leached from the solids is leaching efficiency. Stage efficiency is the amount of the component that is leached compared to the amount that it is theoretically possible to leach assuming that equilibrium is reached.

The initial particle size determines the interfacial area (A) unless the particle size changes dramatically due to a large percentage of the solids being leached. The concentrations of the reagents and the concentrations of the leach components in both the solids and the leach solution determine the concentration driving force (ΔC). The power (P), the impeller diameter to tank diameter ratio (D/T) and the impeller type all have an effect on the mass transfer coefficient (k_s).

The leaching rate is generally determined by batch laboratory experiments, at least initially. In batch leaching experiments, the residence time for all particles is the same. As discussed previously, this is not necessarily true in a continuous operation because the feed is rapidly dispersed throughout the tank volume in agitated tanks. Even though a high degree of non-conformity may be present in the solid suspension, the process result is the important outcome and good leaching results can still be achieved. Running batch experiments on individual particle size fractions can provide an estimate of the individual mass transfer rates (Oldshue, 1963).

Gas Dispersion. In leaching operations, oxygen is sometimes required to complete the chemical reaction for the process. Oxygen is often supplied by the injection of air; however, oxygen gas may be used in some instances. Other gases may also be required in some instances. This process area falls into the category of gas-liquid-solid mass transfer. Leaching applications, which rely on gas dispersion, include ammoniacal leaching of nickel, the leaching of iron out of reduced ilmenite in synthetic rutile production, and neutral leaching of zinc calcine in addition to the leaching operations that have been mentioned previously (Fraser, 1992).

Traditionally, radial flow impellers have been used for gas dispersion. Radial flow impellers typically have high shear characteristics (Optimizing and scaling up mixing systems).

Once again, Oldshue (1983) provides a comprehensive discussion of gas dispersion in liquids. Dispersion of gas in a liquid is most sensitive to the design of mixing systems. This is because the mixer design influences both the interfacial area and the mass transfer coefficient.

The relationship between the mixer power and the isothermal expansion power of the gas determines the type of dispersion that is achieved. If the mixer power is less than the gas power, the gas rises unhindered and "blurps" on the surface. In this case, the mass transfer rate may be high enough even though the gas is not dispersed throughout the tank. As the impeller power increases, the gas dispersion improves. When the two power levels are about equal, minimum dispersion occurs, but the flow pattern is still gas-controlled. To achieve mixer-controlled flow, the mixer power must be two or three times higher than the energy of the gas. When the mixer power is greater than three times higher than the gas power, intimate gas dispersion is achieved. To drive the gas bubbles to the bottom of the tank, the power must be two or three times greater than the power required for intimate dispersion. At this point, the mass transfer coefficient is controlled by both the agitator power and the gas flow.

The impeller controls the bubble size and the interfacial area, so the design of the sparge ring is not a major concern. The number of holes, the size of the holes, and the direction of the holes are not important. Even an open pipe can be used if plugging occurs or corrosion is a problem.

It is not appropriate to use upward pumping axial-flow impellers. An open radial-flow impeller without a disk allows the gas to enter the low-shear area around the hub. The gas, then, passes through the impeller without going into the high shear zone at the tips of the impeller and the re-circulation of the gas occurs at the tips. If a sparge ring is used, it should be about eight tenths the diameter of the impeller so the gas can enter the high-shear zone.

The k-factor is a measure of the ratio of the power drawn by the impeller with the gas turned on, to the power drawn with the gas turned off. An impeller operating in a liquid without gas draws more power than it does when it operates at the same speed in liquid with gas because the fluid density changes with the addition of gas to the liquid. If the power draw does not decrease when gas is added to an agitated tank, it indicates that the gas is not being effectively dispersed in the tank (Rushton and Oldshue, 1953; Oldshue, 1983). The k-factor varies with the gas rate.

In order to compensate for the significantly different power demand with and without gas sparging, the agitator should be designed with an interlock to prevent operation without gas sparging or a two-speed motor should be used. The motor rating is selected for the lower power that is required when the gas is on. Then, a switching gear changes to the lower motor speed if the gas supply stops (Oldshue, 1983). A combination of impeller types may be used to achieve the desired process results.

Agitation in liquid-gas applications, such as agitated tank leaching, also involves mass transfer as discussed previously. Commonly, the design basis would be the mass transfer rate, which has units of mol/[$(m^3)(s)$]. In some simple systems the gas dispersion requirements may be adequately expressed as a quantity of gas per unit of time. When the gas phase and the liquid phase are mixed well, the concentration driving force (ΔC) is the difference between the partial pressure of the gas leaving the tank and the concentration of the dissolved gas in the liquid leaving the tank (Oldshue, 1983).

Oldshue (1983) goes on to show the following, which impact the mass transfer:

- There is an optimum impeller diameter to tank diameter ratio (D/T) for different ratios of gas flow to mixer power.
- The optimum ratio of sparge ring to impeller diameter is approximately 0.8.
- Impeller positioning has very little impact.
- As the viscosity of the liquid increases, the bubble size must increase in order for the bubbles to flow upward at a specified rise velocity.
- Higher viscosity reduces the mass transfer coefficient.
- Tall, thin tanks require less power to sparge a specified volume of air than short, squat tanks with an equal volume.

Impeller Shear. Even though the impellers used for gas dispersion typically have high shear characteristics, shear is not a particular concern in agitated tank leaching. It is, however, a major consideration in recovery processes that utilize activated carbon. For this reason, the discussion on shear is included in Section D-I-5, CIP/CIL/CIC Adsorption Circuit Equipment Selection and Sizing.

SUMMARY AND CONCLUSIONS

Agitation requirements for tank leaching require characteristics of both axial flow and radial flow impellers. Agitator manufacturers have developed mixing systems that are specifically designed to meet the process requirements for the specified application (von Essen and Ricks, 1999; von Essen, 1998; Fraser, et. al., 1993; Fraser, 1992; Kubara and Oldshue, 1992; Olderstein, et. al., 1989; Lally, 1987; Salzman, et. al., 1983; Shaw, 1982; Oldshue et. al, 1988).

Since agitated tank leaching is an established process, the design can generally be based on previous successful projects and existing data (Oldshue, 1969). However, it is important to properly define the fluid properties and to correctly specify the desired process results. A proper balance between flow and shear is required to achieve the correct design (Fraser, 1992). As discussed previously, important considerations in the design of an agitated leach circuit include the effects of particle size on the actual residence time achieved in the circuit and the relationship between the mass transfer rate and the quantity of gas that must be dispersed in a given time.

The current understanding of agitation technology allows the design engineers to select the optimum design based on an economic comparison. The best choice is the correct balance between higher power agitators that have a lower initial installed cost and lower power units that have a high initial installed cost.

REFERENCES

Altman, K.A. 1999. *Model for Developing a Metallurgically Successful Project*. Ph.D. diss, University of Nevada, Reno, Nevada.

Fleming, C.A. 1998. Thirty Years of Turbulent Change in the Gold Industry. *CIM Technical Paper* November/ December, 58.

Fraser, G.M. 1992. Gas dispersion and mixing for mineral oxidation reactors. *Conference on Extractive Metallurgy of Gold and Base Metals*. The Australian Institute of Mining and Metallurgy. 293-301.

Fraser, G.M., P.M. Kubera, M.D. Schutte, and R.J. Weetman. 1993. Process/mechanical design aspects for Lightnin A315 agitators in mineral oxidation. *Proceedings of Randol Gold Forum*. 247-253.

Guide to impeller selection for maximum process results. Philadelphia Mixers.

Herz, N 1985. SME Mineral Processing Handbook. Historical Developments in Milling of Gold Ores. 18-4.

Kubera, P.M., and J.Y. Oldshue. 1992. Advanced impeller technologies match mixing performance to process needs. *Proceedings of Randol Gold Forum*. 1-15.

Lally, K.S. 1987. A315 axial flow impleler for gas dispersion. Mixing Equipment Company, LIGHTNIN Technical Reprint.

Lapedes, D. Ed. 1974. *McGraw Hill Dictionary of Scientific and Technical Terms*. New York City, NY. St. Louis, MO. San Francisco, CA. McGraw-Hill Book Company, p. 831.

Olderstein, A.J., T.A. Post, T.J. Klimasewski. 1989. New impeller provides more efficient mass transfer in gas-liquid-solid systems. *Proceedings TMS Annual Meeting*. 1-13.

Oldshue, J.V., 1963, Solid-liquid mixing in hydrometallurgy. *Unit Processes in Hydrometallurgy*, p. 1-24

Oldshue, J.V. 1969. How to specify mixers. *Hydrocarbon Processing*.

Oldshue, J.V. 1983. Fluid mixing technology and practice. *Chemical Engineering*. 83-108

Oldsuhe, J.V. 1987. Fluid Mixing. *Encyclopedia of Physical Science and Technology, Vol. 5*

Oldshue, J.V. 1989. Fluid mixing in 1989. *Chemical Engineering Progress*. 33-42.

Oldshue, J.V., T.A. Post, R.J. Weetman, and C. Coyle. 1988. Comparison of mass transfer characteristics of radial & axial flow impellers. *Proceedings from the 6th European Conference on Mixing*.

Oldshue, J.V., and D.B. Todd. 1981. Mixing and blending. *Encyclopedia of Chemical Technology*. ed. Kirk-Othmer, 3rd ed., v. 15, New York: John Wiley & Sons.

Optimizing and scaling up mixing systems. Philadelphia Mixers.

Rushton, J.H., and J.V. Oldshue. 1953. Mixing present theory and practice, part II. *Chemical Engineering Progress*. 161-168.

Salzman, R.N., C. Coyle, R.J. Weetman, and J.C. Pharamond. 1983. High efficiency impeller for slurry storage. *Proceedings from the Eighth International Technical Conference on Slurry Transportation*. 305-309.

Scheffel, R. E. 2002 Copper Heap Leach Design and Practice. *Mineral Processing Plant Design, Control and Practice*, A. L. Mular, D. N. Halbe, D. J. Barratt, Eds., Society for Mining, Metallurgy and Exploration, Canadian Mineral Processing Division of the Canadian Institute of Mining.

Shaw, J.A. 1982. The design of draft tube circulators. *The Australian Institute of Mining and Metallurgy*. 47-58.

Von Essen, J.A. 1998. Gas-liquid-mixer correlation. *Chemical Engineering*.

VonEssen, J.A., and B. Ricks. 1999. Design agitated slurry storage tanks to minimize costs. *Chemical Engineering Progress*.

Walter, T.E. 1968. Take care choosing agitation equipment. *Pulp and Paper*.

CIP/CIL/CIC Adsorption Circuit Process Selection

Chris A. Fleming[1]

ABSTRACT

There are three basic carbon-adsorption processes used in the gold mining industry today, carbon-in-pulp (CIP), carbon-in-leach (CIL), and carbon-in-columns (CIC). The first two processes recover gold directly from pulps or slurries containing up to 55% solids, while the CIC process is used for treatment of solution, usually from a heap leaching operation. In choosing between CIP and CIL, a number of factors need to be taken into consideration. The capital cost of a CIL plant will generally be lower than CIP, but the process is not suited to all ore types, and operating costs may be higher in CIL, particularly for slow leaching gold ores.

INTRODUCTION

The hydrometallurgical process for the treatment of gold and silver ores remained unchanged for the first 70 years of the twentieth century, and consisted essentially of leaching in cyanide solution followed by solid-liquid separation, with the solid residue being washed as efficiently as possible, and the leach liquor being treated by zinc cementation to recover the precious metals. This became known as the Merrill Crowe process. While Merrill Crowe is generally extremely efficient and fairly cheap, it does have limitations in the treatment of low-grade ores and certain complex ore types. For example, ores with a high content of clay or other soft, fine minerals are usually difficult to filter, and losses of soluble gold or silver in the residues can be unacceptably high. In addition, ores with high concentrations of base metal oxides, particularly copper, are not well suited to the zinc cementation process because of high consumptions of cyanide and zinc.

The ability of activated carbon to adsorb gold from cyanide solution has been known for over 100 years (Davis, 1880). However, it wasn't until the 1950's that carbon-based processes began to attract attention in the gold mining industry, when it was demonstrated by the US Bureau of Mines that gold and silver could be recovered directly from pulp onto granules of activated carbon, thereby bypassing the costly and inefficient solid-liquid separation steps, and when methods of eluting the gold from the carbon were developed (Zadra, 1950, Zadra et al., 1952).

The first full scale CIP plant was installed at the Homestake Mine in South Dakota, USA, in the early 1970's [Hall, 1974], but the pace of change was fairly sluggish until the 1980's. By the end of that decade, however, almost the entire gold mining industry worldwide had become activated carbon disciples, to the extent that practically all new gold properties that were developed during that decade incorporated carbon adsorption in the flowsheet. Total gold production from the primary resource gold mining industry was about 1,300 tons in 1970, of which 90% (~1,200 tons), was recovered from cyanide leach solutions by cementation on zinc dust, (the remaining ~10% was recovered directly from products such as gravity concentrates and copper concentrates). By 1990, total world gold production, (from the primary sector), had increased to 2,200t, but by this time only ~30% (600t), was recovered by zinc cementation, while ~15% was smelted directly from primary gravity and copper concentrates, and a massive 55% (1,200t), was recovered by the new processes of adsorption onto activated carbon.

---

[1] *Lakefield Research Limited, Lakefield, Ontario, Canada*

Two decades is a remarkably short period of time for a new process to assume a position of such dominance, especially in an industry as historically conservative as the gold mining industry. The reasons for this were, first, the fact that the early prediction of superior economics compared to zinc cementation generally proved to be correct and, second, the carbon-based processes proved to be extremely versatile and robust, both mechanically and chemically. Carbon-based processes were applied to almost any feed, achieving higher recoveries of soluble gold, at lower capital cost and generally lower operating cost than zinc cementation. The first half dozen CIP plants, five of which were built in South Africa, treated everything from slimes fractions to whole ores, flotation concentrates, flotation tailings, roasted calcines, biologically oxidized float concentrates, re-pulped filter-plant residues and old tailings. CIP and CIL plants in North America, Australia and the Pacific Rim treated feeds ranging from low-grade, open-pit ore bodies to high grade flotation concentrates that had been pre-oxidized by pressure leaching, while adsorption on activated carbon became the process of choice for gold recovery from heap leach solutions.

Not only were carbon-based processes shown to accommodate a wide variety of feed materials, the process was also shown to be robust through temporary plant upsets, such as changes in head grade, surges in feed rate or losses in carbon activity, suffering only minor losses in gold extraction efficiency. This was an important requirement for a new process being introduced to an industry that had been lulled by 90 years' experience of the metallurgically simple and chemically forgiving Merrill Crowe process. Most gold plant operators before CIP were mechanically skilled but chemically unsophisticated, and this proved to be an adequate skill-set for the CIP process in the early days.

The new-technology pendulum probably swung too far, in that some projects that went for carbon in the 1980's might have been better-suited to the zinc cementation process. There is now a clearer understanding of the relative strengths and weaknesses of zinc and activated carbon, and there is a more open-minded approach to flowsheet evaluation in the development of new gold and silver ore bodies. For example, it is well known that the affinity of activated carbon for gold is much greater than for silver, and that ores that contain either high silver concentrations or a high silver to gold ratio, are probably better suited to zinc cementation than carbon adsorption. In addition, because of the relatively high cost of carbon elution and reactivation, as well as limitations in the extent to which gold can be loaded onto activated carbon (technical and economical), it is also preferable to treat very rich gold cyanide solutions by zinc cementation, rather than with activated carbon.

CARBON ADSORPTION PROCESSES

Carbon adsorption processes have become universally accepted as the standard method for gold and silver recovery from solutions or pulps. The only exceptions, as pointed out above, are solutions or pulps that contain:

- high Ag to Au ratios
- very high Au concentrations
- species that interfere with carbon adsorption (typically organics).

There are three basic carbon-adsorption processes used in the gold mining industry today:

- carbon-in-pulp (CIP),
- carbon-in-leach (CIL)
- carbon-in-columns (CIC).

The first two processes recover gold directly from pulps or slurries containing up to 55% solids, while the CIC process is used for treatment of solutions (typically from heap leach operations). The main differences between pulp and solution treatment are physical, and relate to the separation and recovery of activated carbon granules from pulp or solution, after gold adsorption has occurred. In CIP and CIL, this is done by screening in a wide variety of mechanical or static (air-cleaned) screening devices, which have a screen mesh size that allows the gold-depleted pulp to pass through, while retaining the carbon granules. In CIC adsorption, on the other hand, the gold and silver-bearing solution generally flows upward through a series of columns that are packed with activated carbon, at a flowrate that is sufficient to lightly fluidize the bed of carbon, but not so fast that the carbon flows from the top of the column. Therefore, physical separation of the carbon from the solution is not an issue. In all three processes (CIP, CIL, CIC), solution or pulp flows by gravity from one contactor to the next, in a cascade of between 4 and 8 adsorption tanks, while the carbon in the tanks is periodically pumped in a countercurrent direction to the flow of solution or pulp.

CIP and CIL versus Merrill Crowe

The fundamental difference between CIP or CIL and the zinc-cementation process lies in the fact that, in order to recover gold by reduction on zinc dust, it is first necessary to produce a clarified filtrate, whereas in CIP and CIL it is possible to extract gold cyanide directly from the slurry. In this way, the costly and inefficient solid-liquid separation processes of a Merrill Crowe gold plant are replaced by the relatively simple and inexpensive screening procedures that are used in CIP and CIL to recover the carbon granules from the leach slurry.

Early estimates of the capital and operating costs associated with the CIP process (Potter et al., 1974, Bhappu et al., 1974, Bailey, 1987), indicated that savings of 20 to 50 percent could be expected, compared to the conventional process, and these estimates have been vindicated, for the most part, on the great many CIP and CIL plants operating around the world today. This has meant that lower-grade gold ores can be economically treated than would previously have been the case, and this was a major stimulus to new gold mine development around the world in the latter part of the 20th century, leading to an annual increase in gold production of about 3% from 1970 (~1,300 tons total), to 2000 (~2,800 tons). Another important benefit of CIP and CIL processes is the generally improved efficiency of gold extraction that can be achieved. This stems from two factors. Firstly, gold losses from a filtration or CCD plant usually amount to about 1 percent of the dissolved gold in the pregnant solution (i.e., 0.03 to 0.05 mg/L), in the case of easily filterable solids, while even higher soluble losses are suffered in the case of poorly filtering or settling solids. By contrast, soluble losses of less than 0.01 ppm can be achieved on a well-managed CIP or CIL plant. Secondly, the additional 5 or 6 hours leaching time in the CIP adsorption contactors usually results in extra gold dissolution. This extra dissolution can be considerable if there are *preg-robbing* constituents in the ore. Evidence for extra dissolution is obtained by an analysis of the washed solids in the feed and the discharge from a CIP plant. In most instances, gold concentrations in the residues are lower than in the feed solids, despite the presence of fine abraided carbon from the CIP tanks in the residue.

The final advantage of the CIP and CIL processes is that they are far less vulnerable than zinc cementation to impurities such as sulphide, arsenate and antimonate in the leach liquor, while parameters such as cyanide concentration, pH and oxygen concentration do not have to be as carefully controlled, because their influence on the loading of gold on carbon is fairly minor.

CIC versus Merrill Crowe

While the major cost advantage of processes using activated carbon over zinc cementation lies in the ability of granular carbon to extract gold and silver directly from pulps, there are also compelling arguments for using carbon rather than zinc for the recovery of gold and silver from clarified or unclarified solutions.

The choice between carbon columns and zinc cementation is based on analyses of capital and operating costs and consideration of the metallurgical efficiency. As a broad generalization, treatment in carbon columns is more economical for large volumes of low-grade solutions containing mainly gold, whereas zinc cementation is preferred for relatively small volumes of high-grade solution, particularly those rich in silver.

Obvious examples of large-volume, low-grade solutions are tailings ponds of existing gold plants where low concentrations of gold and silver are present in solution, either because of inefficient solid/liquid separation and recovery in the main plant, or because of additional leaching that has occurred in the tailings dam. There are a number of plants around the world recovering gold from sources such as this, in packed carbon columns, and the payback period for this sort of scavenging operation can be as short as a few weeks! Another example of a situation in which carbon is usually preferred to zinc is in the treatment of solutions from heap-leaching operations. Here too, the main advantage of carbon over zinc cementation is that the pregnant solution need not be clarified or deaerated prior to feeding to a carbon column. Another important advantage of carbon over zinc for heap leaching is the fact that the carbon process is "clean" and does not introduce new ions into the recirculating leach liquor. In the cementation process zinc cyanide, which is a by-product of the zinc cementation reaction, builds up in solution with continuous recycling of barren solutions to the heap. In extreme cases, the build up of zinc must be controlled by removing a bleed stream of leach liquor and treating it to destroy cyanide and precipitate zinc. At best, the build up of zinc in heap leach liquors adds significantly to the time and cost of washing and detoxifying a spent heap. In fact, even when Merrill Crowe has been selected as the method of recovering gold and silver during a heap leach operation, carbon columns are often installed at the end of the mine's life, to recover precious metals from the wash solutions. This is because of the need to have free cyanide in solution for effective Merrill Crowe operation (which would defeat the objective of washing cyanide from the exhausted heap). Activated carbon does not suffer from this problem, because the loading of gold and silver on carbon actually improves as the cyanide in solution decreases (Fleming, 1984).

For these and other reasons, CIC has become the standard process for the recovery of gold from solutions generated by heap leaching.

CIP versus CIL

The difference between CIP and CIL lies in the extent to which the gold and silver are leached from the ore prior to carbon adsorption. In CIP, most of the leachable gold and silver (typically >98%) have been leached before the first carbon adsorption stage, and gold adsorption is carried out in specially designed tanks that are situated after the leaching tanks in the overall process flow. The adsorption tanks are typically a quarter to a tenth the size of the leaching tanks. In CIL, carbon is added to the leaching tanks, and adsorption occurs simultaneously with leaching. In some CIL plants, there are one or two stages of leaching before the first CIL adsorption tank, and the extent of gold leaching before carbon adsorption can be as high as 95% - in other words, not much different from CIP. In other CIL plants, carbon is present in all the leaching tanks, and both the leaching and adsorption processes occur simultaneously throughout.

In choosing between CIP and CIL for a particular application, a number of factors must be weighed up.

Advantages of CIL

- The major advantage of CIL over CIP is the lower capital cost of the plant, which results from the removal of five or six CIP adsorption tanks and ancillary equipment from the flowsheet. The additional operating costs associated with the CIP section would also be eliminated.

- Because CIL leaching tanks are much bigger than CIP adsorption tanks, the pulp surface area available for interstage screening is much greater in CIL, and screening is less of a bottleneck in the process.

Disadvantages of CIL

- On the basic assumption that the total inventory of carbon in CIL and CIP plants would be the same (for the same overall metallurgical performance), the carbon concentration in the pulp would be up to 5 times lower in CIL than in CIP. Hence, more pulp is transferred upstream in CIL, and back-mixed during countercurrent transfer of the carbon, which leads to metallurgical inefficiencies and places an increased load on the carbon advance pumps and the interstage screens. In a CIL plant, up to 25% of the pulp flow could be back-mixed with countercurrent carbon advancement, whereas in CIP back-mixing will generally be less than 5% of the total pulp flow.
- The kinetics of gold adsorption on carbon are slower in CIL than in CIP, because leaching is incomplete when the pulp encounters carbon in the adsorption stages of CIL, and the concentration of gold in solution is lower. To compensate for a decrease in the kinetic driving force, more carbon is needed to match the metallurgical performance achieved in CIP. Even for an ore from which the gold leaches at a relatively fast rate, the carbon inventory in CIL should be 10 to 15 per cent higher than in CIP, for the same metallurgical performance.
- CIP has a subtle advantage over CIL in that extra carbon adsorption tanks can be retrofitted into an existing CIP plant at relatively low cost, by simply converting the last few leach tanks to adsorption tanks. Many of the early CIP plants were designed on the basis of data from laboratory tests and small pilot plants and, for a number of reasons, the full scale plants tended to perform below design specification on startup. In such cases, it was possible to bring the plant performance back in line with design specification by increasing the inventory of carbon in the plant or increasing the number of adsorption stages. A CIP plant obviously has far more flexibility than a CIL plant in this regard.

General

The mechanisms by which gold and silver cyanide load onto carbon, as well as their equilibrium loading capacities, are believed to be the same for solutions and pulps, although the rate of loading is significantly faster from solution than from pulp. This is because the average size of the carbon granules is usually smaller in CIC – owing to the different screening requirements, which leads to faster mass transfer kinetics. This typically allows CIC plants to operate at 50 to 100 percent higher gold (and silver) loadings than CIP or CIL plants, for the same concentrations of gold and silver in the leach solution, and the same contact time between solution and carbon. The processing methods for eluting gold and silver from carbon and regenerating the carbon for recycling, are also the same for CIP, CIL and CIC.

OPTIMIZATION OF ADSORPTION

It is now well established that the gold cyanide ion ($Au[CN]_2^-$) loads on to activated carbon by a reversible adsorption mechanism, without undergoing any chemical change, probably as an ion pair [Adams, et al, 1989, Fleming and Nicol, 1984]. Verification of the reversibility of adsorption was achieved in an experiment in which loaded carbon from the first stage of a CIP plant was isolated in a basket in the last stage of the same CIP plant. [Fleming and Nicol, 1984] Gold on the carbon in the basket slowly desorbed (with $t_{1/2}$ ~48 hours), in contact with the low gold tenor solutions of the pulp in the last CIP contractors (Figure 1).

It is expected that gold on very fine carbon granules would desorb far more rapidly than depicted in Figure 1, and an important practical consequence of this is that fine carbon generated by abrasion in a CIP circuit will fortuitously desorb most of its gold values before discharging with the barren pulp from the circuit.

On the other hand, coarse carbon granules will desorb slower ($t_{1/2}$ ~48 hrs), than the average residence time of pulp in a CIP or CIL contacter (1 to 6 hours), so loss of coarse carbon (through screen breakage or other mechanical failures), could result in significant gold loss.

The kinetic characteristics of aurocyanide adsorption and desorption on activated carbon are quite unique in fact, and stem from the unusual physical properties of activated carbon. For example, activated carbon is a highly porous substrate, with an enormous total surface area of over 1000 m^2 per gram of carbon. Evidence suggests that as much as 90% of this surface area is in very small micropores (i.e. with a cross sectional diameter of the pores of less than 100 Å), and as such are either inaccessible to the fairly large aurocyanide molecule, or becomes accessible only after tortuously slow diffusion. This property helps to explain the observed kinetic phenomena in gold cyanide processing.

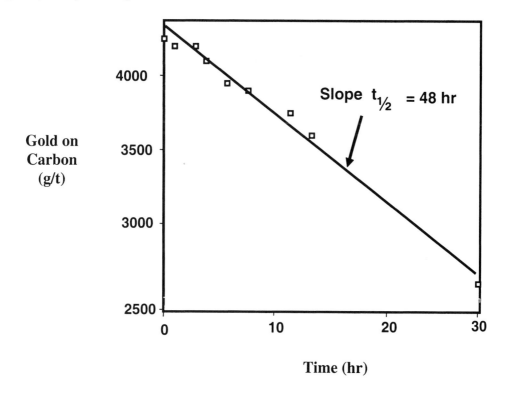

**Figure 1: Themodynamic aspects of adsorption
Reversibility of adsorption**

Plant experience and numerous laboratory investigations have shown that the initial rate of aurocyanide adsorption onto carbon is fairly rapid, and that it is controlled by the hydrodynamics in the adsorption contactor. This initial, film-diffusion controlled reaction, which presumably involves adsorption of gold cyanide molecules in the carbon macropores (which can be as large as

00,000 Å in diameter), continues for 24 per 48 hours.[2] Subsequently, gold cyanide continues to be adsorbed slowly onto the carbon almost indefinitely and, in practice, the establishment of a true equilibrium is difficult.

During the latter period of pore-diffusion-controlled adsorption, gold cyanide presumably diffuses slowly into the carbon micropores, and, as the cross-sectional area of the micro-pores approaches that of the aurocyanide ion, the resistance to mass transfer becomes infinite.

From a practical point of view, therefore, the aurocyanide-carbon interaction can be considered to possess two thermodynamic regimes: the macropore-mesopore pseudo equilibrium and the total equilibrium, and the value of the former could be 10 times lower than the value of the latter.

This aspect of the kinetics of gold cyanide loading on activated carbon is illustrated in Figure 2, which depicts the result of an experiment conducted at the Western Areas CIP plant in the early 1980's, (Nicol et al., 1984).

In this experiment, fresh activated carbon was placed in an agitated basket inside the pulp of the first CIP adsorption stage of the plant. The basket was made of 35 mesh stainless steel screening material, which allowed slurry to flow through, but isolated the carbon in the basket from that in the bulk of the adsorption tank. The experiment continued for 3 weeks, during which time samples of carbon were frequently taken from the CIP tank and from inside the basket, and analyzed for gold. The data in Figure 2 depicts the analytical data.

The loading of gold on the carbon in the CIP tank cycled upwards and downwards from ~2,000 to ~4,000 g/t (with an average of ~3,000 g/t), as carbon was periodically advanced in and out of the tank during normal CIP operations. The gold on the carbon in the basket, however, continued to increase over the 3 week period, reaching ~11,000 g/t by the end of the test.

The gold grade in solution in the feed to the Western Areas CIP plant at that time was ~2.5 mg/L, so the experiment showed that an upgrading ratio of 5,000:1, can be achieved.

In practice, the operating conditions in most CIP and CIL plants are set to achieve an upgrading ratio in the range 750:1 to 1,500:1, which is similar to that shown for the plant operating conditions in Figure 2. This range of up-grading ratio will likely be the correct operating strategy in most cases. Not only is the rate of extraction relatively fast and responsive to good mixing efficiency in the adsorption contractors, but the rate of elution and ultimate elution efficiency under a given set of conditions are also maximized. In addition, lock up of gold on carbon in the adsorption plant is minimized, and the risk of gold losses through screen breakages or carbon theft is also minimized.

However, the data in Figure 2 shows that these operating conditions are quite conservative, and much higher gold loading could be achieved, which would benefit process economics by reducing the amount of carbon reporting to the expensive carbon elution and regeneration processes. If higher gold loadings on carbon were targetted for an operating CIP or CIL plant, it would be necessary to increase the inventory and concentration of carbon in the adsorption tanks to compensate for the slower kinetics of loading at higher concentrations of gold on carbon. This could become a limiting factor in CIP adsorption tanks, but not in CIL.

Finally, the unusual kinetic loading properties depicted in Figure 2 explain the rugged efficiency of CIP, CIL and CIC processes, and their sluggish response to plant upsets. The fact that the gold on the carbon is way below the equilibrium loading for each of the adsorption stages, allows carbon adsorption plants to be operated very efficiently (most of the time), with low levels of process control and plant surveillance.

---

[2] *This initial rate of loading responds to factors such as pulp density, pulp viscosity, mixing efficiency, carbon particle size, and the presence, in the pulp, of species that adsorb onto and poison the carbon.*

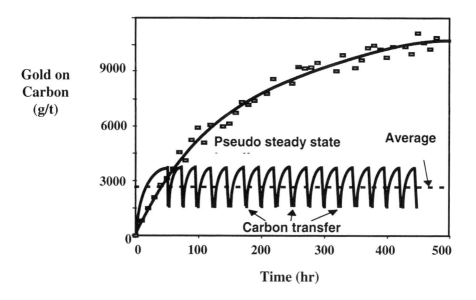

Figure 2 Kinetic aspects of adsorption

REFERENCES

Adams, M.D & Fleming, C.A. 1989. The mechanism of adsorption of aurocyanide onto activated carbon. Metallurgical Transactions B, Vol. 20, pp. 315-325.

Bailey, P.R. 1987. Application of activated carbon to gold recovery. In: G.G. Stanley (Editor), The Extractive Metallurgy of Gold in South Africa. S. Afr. Inst. Min. Metall., Monogr. Ser., M7, 379-614.

Bhappu. R.B., Lewis, M.F. & McAllister, J.A. 1974. Leaching of low-grade ores, economic evaluation of available processes. AIME Annu. Meet. (Dallas, Texas).

Davis, W.N. 1880. Use of carbon for the precipitation of gold from solution and subsequent burning. US Pat. 227,963.

Fleming, C.A. & Nicol, M.J. 1984. The adsorption of gold cyanide onto activated carbon. III. Factors influencing the rate of loading and equilibrium capacity. JS Afr Inst Min Metall, Vol. 84, No. 4, pp.85-93.

Hall, K.B. 1974. Homestake uses carbon-in-pulp to recover gold from slimes. World Mining, 27(12), 44.

Nicol, M.J., Fleming, C.A. & Cromberge, G. 1984. The adsorption of gold cyanide onto activated carbon. I. The kinetics of adsorption
from pulps. J. South African Inst. Mining Metall., 84(2), 50-4.

Potter, G.M. & Salisbury, H.B. 1974. Innovations in gold metallurgy. Min. Congr. J., 54.

Zadra, J.B. 1950. A process for the recovery of gold from activated carbon by leaching and electrolysis. USBM RI 4672.

Zadra, J.B., Engel, A.L. & Heinen, H.J. 1952. Process for recovering gold and silver from activated carbon by leaching and electrolysis. USBM RI 4843.

CIP/CIL/CIC Adsorption Circuit Equipment Selection and Design

Kathleen A. Altman[1], Stuart McTavish[2]

ABSTRACT
Design of a CIP/CIL/CIC circuit begins with metallurgical testwork and data acquisition. The metallurgist who conducts this testwork must work closely with the process engineer to develop an optimum design. This design must also include secondary factors such as plant capacity, site-specific geographic constraints and environmental considerations. Equipment selection is based on standard plant practice unless there is reason to believe the feed will be substantially different from the norm. In this case, scale-up tests are conducted in order to obtain more rigorous design data, and the design is modified based on this information.

INTRODUCTION
The design of carbon-in-pulp (CIP) and carbon-in-leach (CIL) circuits is closely related to the design of agitated leach circuits with a few additional considerations. The primary consideration is the use of granulated activated carbon (GAC) to recover gold from the leach circuit. This means that carbon attrition is a major concern in the selection of the equipment and additional equipment is required to handle the carbon that is utilized in the circuit.

Carbon losses due to attrition, or deterioration of the GAC, are important not only because of the cost of the carbon itself, but also because the fine carbon that results from the attrition continues to adsorb precious metals, and then, cannot be contained in the recovery circuit. The associated loss of fine carbon that is precious metals-laden results in reduced recovery of the precious metal, which detrimentally affects the profitability of the operation.

In CIP/CIL adsorption circuits, the slurry that is being leached and the carbon being loaded with precious metals flow counter-current to one another. This includes a "forward flow" of the pulp, or slurry, and a "reverse flow" of the carbon (Rogans, et. al. 1998). The slurry generally flows by gravity from the first tank to each successive tank just as it does in an agitated leach circuit. The carbon, then, is advanced from the last tank towards the first in the counter-current fashion. The slurry flow is continuous and the carbon is typically advanced on an intermittent schedule using pumps. The selection of an appropriate recovery process is based on an economic evaluation of the proposed project. In order to quantify the capital and operating costs to use in these evaluations, a certain amount of engineering is required, including the selection and basic design of appropriate equipment for each option under consideration.

As a general rule, CIL circuits provide better project economics and are, therefore, the preferred treatment option, particularly when preg-robbing ores are leached. This is due primarily to an increase in gold recovery. In theory, at least, the GAC has a higher affinity for the gold in solution than the carbonaceous material that is contained in the ore itself. Thus, the gold losses in the solid residues, or tails, are reduced and the overall recovery is increased, which results in an economic advantage. CIL circuits generally have larger carbon inventories and larger elution circuits as a consequence.

Carbon-in-column (CIC) circuits are typically used to recover precious metal from solutions, as opposed to slurries. This recovery method is most commonly used in precious metal heap leaching applications, although it is sometimes used to recover gold (and/or silver) from thickener overflow solutions.

---
1 Independent Consultant, Denver, Colorado
2 SNC-Lavalin Engineers and Constructors, Inc., Toronto, Ontario

As with most recovery processes, there have been dramatic improvements in the design of equipment over the course of time. For example, agitators have been specifically designed for CIL and CIP applications, and carbon handling concepts and designs have evolved from EPAC and vibrating screens to Kambalda, NKM and MPS interstage screens and, most recently, to pump cell technology.

PROCESS DESIGN

Design Procedure

The process of designing CIP/CIL/CIC adsorption circuits is similar to the procedure used for agitated leach circuits. It requires a collaborative effort between the operating company, the metallurgical testing facility, the engineering company, and the equipment suppliers just as it does for the agitated leach circuits. Please refer to section D-I-3, Agitated Tank Leaching Selection and Design, for a more detailed discussion.

The number of stages and carbon movement are determined based on the solution feed grade, desired barren solution grade, design loaded carbon grade and precious metal – carbon adsorption characteristics of slurry (solution). The capacity of the elution circuit is then selected based on the calculated carbon movement.

CIP/CIL Circuits

Tank Design. The tank design is nearly the same as it is for agitated leaching. This includes:

- Tank sizing
- Slurry density
- Plant throughput
- Number of stages of leaching/adsorption
- Tank height to diameter ratio
- Tank baffling
- Tank configuration
- Materials of construction.

The slurry density in a carbon adsorption circuit (CIL and CIP) is usually lower than it is in an agitated leach circuit. It must be controlled carefully in order to keep the GAC properly distributed throughout the tanks. A range of 35 to 50 percent solids is typical, although it is dependent on the slurry characteristics of the ore being processed. If the slurry density is too high, the carbon floats near the surface of the tank. If it is too low, the carbon sinks to the bottom of the tank. In both cases, mass transfer of gold to the carbon particles is impaired, and the interstage transfer of carbon is constrained.

Tank sizing for the leach section of a CIP plant and for the leach/adsorption vessels of a CIL plant are based on the slurry flow rate, the retention time required to leach the gold from the ore and load it onto carbon, and maintaining the required number of stages. The retention time in leach and CIL circuits is typically 24 to 48 hours, and in CIP circuits is typically 0.75 h to 1.0 h per stage, usually with 5 to 7 stages.

Reagent Additions. In CIP/CIL cyanide leach circuits, the pH of the slurry may be initially controlled through the addition of lime on the feed belt to the grinding circuit and, particularly in CIP circuits, the grinding may be done in cyanide solution. In this case, it is good engineering practice to provide for trim additions of slaked lime and/or caustic soda and cyanide in the leach circuit so that the reagent consumptions can be optimized, i.e. reagent costs minimized and metal recoveries maximized. The determining factors to consider in designing reagent addition systems are the sensitivity of the leaching process to the concentration of the reagents and the amount of variability that is anticipated in the ore that will be processed. When ores contain high concentrations of copper, high free cyanide levels are maintained to minimize adsorption of copper onto the carbon.

In CIP and CIL carbon adsorption processes, it is necessary to add oxygen, generally in the form of air, to the circuit. This may be done with air blowers or with low pressure air compressors, depending on the volume of air and the delivery air pressure that are required. In some cases, either liquid or gaseous oxygen may be introduced into the leaching process, which results in even more sophisticated gas production and/or storage and injection systems. By definition, CIP is for adsorption only. No air is normally added to CIP tanks for leaching, although some leaching does occur in CIP circuits.

The amount of carbon added to the circuit is dependent on the plant throughput and the quantity of recoverable precious metal contained in the ore. The amount of carbon moved through the adsorption circuit is also coordinated with the capacity of the carbon elution and carbon regeneration circuits.

For a 1,000 t/d plant operated at 45% solids, the solution flow rate is 1,222 t/d [(1,000 t/d/0.45) – 1,000 t/d)]. If the recoverable precious metal grade in the ore is 5 g/t, the total amount of metal recovered daily is 5,000 grams (1000 t/d * 5 g/t) and the solution grade would be 4.1 g/t [(1,000 t/d * 5 g/t)/1,222 t/d]. Using the calculation method presented in the agitated tank leaching section, each tank would be approximately 343.3 m^3 based on the criteria presented in the table below. These are design criteria for CIL.

Table 1: Sample CIL Design Criteria

Plant throughput	1,000 tpd
Ore specific gravity	2.8
Slurry density	45% solids
Retention time	24 h
Effective volume factor	0.92
Number of stages	5
Tank volume	343.3 m^3

The concentration of carbon required to achieve a certain target gold extraction efficiency can be determined by metallurgical testwork and mathematical modeling. If the carbon concentration is 10 g/L, the total amount of GAC contained in each tank is 3.4 t for a total inventory of 17 tonnes [(343.3 m^3*1,000 L/m3)*(10 g/L)/(1000g/kg)/(1,000 kg/t)].

Assuming that a carbon processing rate of 1 t/d has been selected, the anticipated carbon loading would be about 5,000 g/t [(5,000 g/d)/(1 t/d)]. This is well below the maximum amount of precious metal that can normally be loaded on GAC. However, it is important that the carbon loading be checked against the theoretical precious metal loading at the design operating conditions. The selected carbon gold loading has to be consistent with a reasonable retention time in the CIP/CIL circuit to minimize carbon poisoning from prolonged loading.

Agitation Requirements and Design. The agitation requirements and design for CIP/CIL circuits are the same as they are for agitated leach circuits with the additional emphasis on minimizing carbon attrition. The main criteria include the following, which are explained in detail in section D-I-3:

- Mixing requirements and performance
- Choice of impellers
- Flow relationships
- Solids suspension
- Mass transfer
- Gas dispersion.

Impellers used for gas dispersion typically have high shear characteristics. Shear is a major consideration in CIP and CIL recovery circuits because it is an underlying cause of attrition of GAC. However, some shear is required for leaching to take place.

Shaw and McDonough (1982) provide a comprehensive discussion of mechanical agitation in the CIP process. The impeller type, impeller diameter and speed, as well as the geometrical relationship between the impeller and the tank will determine the split between flow energy and shear energy. Large impellers running slowly convert more energy into flow energy than shear energy. Or, another way of looking at it is that large impellers can produce the same flow as smaller impellers while using less energy. However, larger impellers also have a higher capital cost. It is possible, with careful design, to operate at higher speeds and, hence, reduce the capital cost without suffering from a large shear penalty.

Shear rate is a change in velocity with distance. It has the units of "m/sec." Several different factors are important in agitation systems:

- Maximum shear rate in the impeller zone
- Average shear rate in the impeller zone
- Maximum shear rate in the tank
- Average shear rate in the tank
- Frequency of circulation through the high shear zone.

These factors are related to impeller diameter, rotational speed and impeller type. Although impeller tip speed is sometimes used as an indication of shear rate, different impeller types can have dramatically different shear rates even though they have the same tip speed. Maximum shear rate results from velocity gradients generated in the liquid by the impeller. The maximum shear rate often occurs some distance from the impeller itself.

Shear stress is dependent on both the shear rate and the viscosity of the liquid. It is shear stress that causes particle damage. Although it is difficult to quantify carbon attrition, Shaw and McDonough (1982) provide some guidelines.

- Draft tube agitation systems provoke less carbon attrition than open impeller systems.
- Aerofoil impellers provoke less carbon attrition than flat plate impellers in both open impeller and draft tube agitation designs.

Both of these reductions appear to be due to the velocity gradients generated by each of the respective designs.

In general, draft tube circulators have higher capital costs but consume significantly less power than open impeller agitator designs. Since draft tube circulators are preferable in some cases, a description of key design factors is warranted.

Shaw (1982) provides details about the design of draft tube circulators. Historically, draft tube circulators were first designed for the alumina industry, which had previously used Pachuca systems, just as the gold industry had.

A draft tube circulator includes a flat-bottomed tank and a draft tube. The diameter of the draft tube is on the order of 20- to 40-percent of the tank diameter. An axial flow impeller is positioned near the top of the draft tube. The impeller pumps down the draft tube and this flow, in turn, produces an upward flow in the annulus.

Two distinct velocity criteria are considered in the design of draft tube circulators:

- The velocity down the draft tube is determined by the bottom-sweeping design, which is a function of the particle size distribution of the slurry, the pulp rheology and the shape of the tank bottom.
- The average upward velocity in the annulus must be greater than a minimum, which is dependent on the free-settling velocity of the largest particle to be circulated.

Power requirements are the lowest at the ratio of draft tube diameter to tank diameter where both of these velocity criteria are met.

The operation of a draft tube circulator is similar to the operation of an axial flow pump. Energy requirements depend on the hydraulic efficiency and the power-flow relationships. Also, the stability of the draft tube circulator is dependent on a head versus flow curve. Draft tube circulator designs may be very simple, almost primitive, or quite sophisticated. Although more sophisticated systems are generally more expensive, they may be justified due to lower energy consumption.

Trash and Safety Screens. CIP and CIL circuits rely on screens to control the movement of carbon. In order to prevent the interstage screens from being impacted by wood chips or plastic, the feed to a CIL or CIP circuit is first passed through a trash screen. This screen can be a vibrating screen or a linear screen that is the same mesh size as the interstage screens. On the discharge of the circuit, the slurry is generally passed over a safety screen. This screen can also be either a vibrating screen or linear screen and is usually one mesh smaller than the interstage screen. The safety screen recovers any carbon that has been attritioned to a small enough size to pass through the interstage screen and recovers any full size carbon should the last interstage screen develop a hole.

Carbon Control and Handling. Screens and pumps are the main components of the carbon handling systems associated with CIP and CIL operations. Screens are used to contain the carbon in the slurry tanks while the slurry is allowed to gravitate through the screens on a continuous basis. Then, some sort of pumping is required to transfer carbon from one stage of the circuit to the next stage upstream. Initially airlifts were used for inter – tank slurry transfer. However, this method was quite inefficient, and in some cases was found to result in excessive carbon abrasion. In the mid – 1980's recessed impeller pumps replaced airlifts for slurry transfer. Vertical recessed impeller pumps are used in CIP/CIL tanks while horizontal recessed impeller pumps are used to transfer carbon within elution/regeneration circuits.

Interstage screening has been one of the biggest problems to overcome in carbon adsorption circuits (Rogans, et. al., 1998). One reason for the difficulty is that the particle size distribution is small for both the solids in the slurry and the carbon. Plus, the size of the carbon particles is continually reduced due to the attrition mechanisms. Because interstage screening has been a major concern, it has gone through a continual development process since the inception of CIP and CIL processing. Early circuit designs utilized equalized-pressure air cleaned (EPAC) screens or external vibrating screens (Rogans and Cartner, 1996a). EPAC screens had high maintenance costs that resulted in high operating costs. The screen clothes were the subject of frequent failures. One of the first developments in interstage screen technology was the replacement of air swept screens with mechanically swept screens. Derrick in – tank vibrating screens replaced the EPAC screens in North America and were the standard in the late 1980's to the mid 1990's in large CIL/CIP plants. In South African gold plants, Kambalda screens replaced many of the EPAC screens. However, Kambalda screens still had major difficulties. In order to perform maintenance on the screen, the pulp level had to be lowered, which required bypassing of the tanks. Kambalda screens also had high power requirements because they used horizontal mechanical pulse blades. This led to the development of vertical, mechanically swept interstage screens, which were developed for the Daggafontein CIL plant in South Africa. The idea came from a similar screen in Australia so they were named North Kalgoorlie Mines (NKM) screens.

NKM screens are formed in the shape of a basket. They have rotating pulse blades around each screen surface. These blades dislodge the carbon that is retained in the screens. The NKM design provided a higher pulp flow through the screens than previous designs. Rogans and Cartner (1996a) provide a comparison of throughput rates for the various screen designs.

Table 2: Interstage Screen Throughput Rates

Screen Type	Pulp Flow Rate $m^3/m^2/h$
Vibratory	54
EPAC	20 to 40
Kambalda	40
NKM	70

Even though NKM screens dramatically improved interstage screening in CIP and CIL plants, they, too, were not without problems. The internal conical wedge was subject to plugging, which results in loss of open area and a subsequent increased flow rate, as well as associated increased wear rate that reduces screen life. Also, phase separation occurs when the flow is reduced or stopped because the dampening effect of the screen on the pulse blades is high enough that there is no agitation in the annular area. Then, even after normal flow is resumed, the NKM screen still operates in a clogged or partially clogged condition. Again, this results in increased wear rates caused by higher flow rates in the open areas of the screen. This resulted in the development of the Mineral Process Separating (MPS) and the Mineral Process Separating (Pumping) [MPS(P)] screens. The MPS screen is used in conventional, gravity flow CIP/CIL circuits. The MPS(P) screen incorporates an up-pumping mechanism that lifts the pulp inside the screen and places it in a launder above the pulp level in the tank.

Rogans and Cartner (1996a) and A.A.C. Mineral Process Separating Screens, describe details of the design of MPS and MPS(P) screens. The MPS screens eliminate the problems encountered with NKM interstage screens. Addition of the pumping mechanism in MPS(P) screens allows the leach tanks to be placed at the same elevation, which reduces the capital cost of the leach tanks due to reduced civil and construction costs. Benefits of using MPS and MPS(P) screens include the following, as outlined by Rogans and Cartner (1996a):

- Increased throughput rates, e.g. 70 to 100 $m^3/m^2/h$
- Reduced power requirements, i.e. 5.5 kW per 500 m^3/h for MPS screens versus 21 kW for the same throughput through Kambalda screens
- Good tolerance for surging flows
- Cleaning can be accomplished in a matter of minutes without reducing the slurry level in the leach tank
- MPS(P) screens can also transfer highly viscous pulp through the interstage screens, even though this material cannot normally be processed in conventional CIP/CIL plants.

One final advantage of the MPS(P) screen is that it has allowed the development of the Anglo American Corporation (AAC) Pump-cell Adsorption circuit, which relies on rotating the pulp feed and tailings discharge positions through a series of contactors that are all placed at the same elevation. This design results in the countercurrent movement of carbon without physically moving the carbon from one contactor to the next (The A.A.C. Pump-cell Adsorption Circuit).

Rogans, et. al. (1998) use the results of simulations to quantify the advantages of pump-cell carousels as compared to more traditional carbon adsorption circuits. Rogans and Cartner (1996b) and Schoeman, et. al. (1996) compare pump-cell technology with CIL and CIP plants and describe installations at two South African gold mines. Operating data from five pump-cell installations and two CIP plants is presented by Macintosh, et. al. (2000).

CIC CIRCUITS

The design and equipment selection for CIC circuits is based on the required solution flow rate and the solution grade, which directly affects the carbon loading concentration that can be achieved. Calculating the quantity of gold to be recovered on a daily basis and dividing it by the grade of the loaded carbon determine the carbon capacity required in a CIC circuit. The quantity of carbon to be advanced daily is nominally the capacity of each cell in a carbon column. This is also generally matched to the capacity of the carbon elution circuit. Cascading columns are the most prevalent design for CIC circuits, although alternative designs such as stacked vertical columns have been used successfully. The columns are operated in an up-flow mode, and the design is such that the solution velocity provides fluidization of the carbon bed in the cell. Obviously, the sizing of the carbon cells must allow for this bed expansion. Published curves are available, which show the relationship between fluidization and velocity for carbon of various size ranges.

Stationary screens are provided to ensure that the carbon does not escape from the cells of the carbon columns in the event of surges in the solution flow rate.

Carbon is advanced from cell to cell in CIC circuits, counter current to the solution flow rate, similar to the process used in CIP and CIL circuits. The movement can be accomplished using eductors. However, carbon attrition is an important consideration in the selection of carbon handling equipment, and recessed impeller pumps have been shown to reduce the rate of carbon attrition in CIC circuits. Therefore, the carbon movement design is similar to that used in CIP/CIL circuits.

SUMMARY AND CONCLUSIONS

CIP/CIL/CIC circuits are common gold recovery circuits. CIP circuits include a separate agitated leach circuit and a carbon adsorption circuit, while a CIL circuit provides simultaneous leaching and carbon adsorption. The design process for CIP and CIL circuits is similar to that used for agitated leach circuits, with the added considerations required when granulated activated carbon is used in the recovery circuit.

Tank designs are dependent on the residence time required for leaching and adsorption, and the number of stages required to achieve an appropriate leach profile. Successful agitator designs use either open impeller or draft tube circulator designs.

Interstage screen designs have, perhaps, changed the most since the initial development of carbon-in-pulp and carbon-in-leach processing for the recovery of gold. Current design technology minimizes operating difficulties and reduces operating costs. The newest technology for interstage screens has the added advantage of allowing the development of pump-cell carbon adsorption circuits.

CIC circuits are similar to CIP and CIL circuits except that they recover solubilized gold from solutions instead of slurries. This allows a simpler design that depends primarily on the solution flow rate and the gold tenor in the solution.

CIP/CIL and CIC are all proven technologies that have now been successfully used for a number of years. Therefore, there are a number of operating plants that can serve as a design basis for new plants.

REFERENCES

A.A.C. Mineral Process Separating Screens. Kemix.

The A.A.C. Pump-cell Adsorption Circuit. Kemix.

Macintosh, A., D. McArthur, and R.M. Whyte. 2000. Process Choices for Carbon Technology, *Proceedings Randol Gold & Silver Forum*. Vancouver, B.C. Canada.

Rogans, E.J., and W.N. Cartner. 1996a, The Development of Mineral Process Separating Interstage Screen (MPS) from the NKM Interstage Screen, *Proceedings Randol Gold Forum*. Squaw Creek, CA, USA

Rogans, E.J., and W.N. Cartner. 1996b. The Pump-cell Adsorption Circuit for In Pulp Applications, *Proceedings Randol Gold Forum*, Squaw Creek, CA, USA

Rogans, E.J., A. J. Macintosh, and N. Morrison. 1998. Carbon-In-Pulp and Carbon-In-Leach Adsorption Circuits – Optimisation of Design Using the Carousel System, Kemix Technical Paper, 23 pp.

Schoeman, N., E. J. Rogans, and A..J., MacIntosh. 1996. AAC Pump-cells: A Cost-effective Means of Gold Recovery from Slurries, *Proceedings SAIMM Hidden Wealth Conference,*.Johannesburg, S.A. p. 173-179.

Shaw, J.A., 1982, The Design of Draft Tube Circulators, *Proceedings of the Australasian Institute of Mining and Metallurgy*, No. 283, p. 47-58

Shaw, J.A., and R.J. McDonough. 1982. Mechanical Agitation in the Carbon-in-Pulp Process. Technical Reprint. LIGHTNIN.

13
Extraction

Section Co-Editors:
Paul G. Semple and Ron Bradburn

Zinc Cementation—The Merrill Crowe Process
A.P. Hampton **1663**

Selection and Design of Carbon Reactivation Circuits
J. von Beckmann, P.G. Semple **1680**

Selection and Sizing of Elution and Electrowinning Circuits
P. Hosford, J. Wells **1694**

Selection and Sizing of Copper Solvent Extraction and Electrowinning Equipment and Circuits
C.G. Anderson, M.A. Giralico, T.A. Post, T.G. Robinson, O.S. Tinkler **1709**

Zinc Cementation – The Merrill Crowe Process

A. Paul Hampton, P.Eng.

ABSTRACT

The classical method for the recovery of precious metals from cyanide leach solutions is cementation using zinc powder, the Merrill Crowe process, which was developed in the 1890's and used almost exclusively until the introduction of carbon adsorption processes in the 1970's and 1980's. The unit operations of the Merrill Crowe process include solid liquid separation, clarification, vacuum deaeration in packed towers, zinc addition and filtration of precipitated gold and silver using pressure filters. This paper discusses Merrill Crowe process application and circuit design including solution chemistry, preliminary process design parameters, equipment selection and sizing, process performance and relative costs.

INTRODUCTION

Zinc cementation, which later became known as the Merrill Crowe Process, was the original method chosen for the recovery of precious metals from solutions generated by the cyanidation process, during its developmental stages. The zinc cementation process was first applied in the 1890's and was used, with subsequent improvements, until the introduction of carbon adsorption processes in the 1970's and 1980's. This paper discusses the application and design of the Merrill Crowe process beginning with a brief history of the process, the basic chemistry and the development of basic process design criteria for the development of a Merrill Crowe process flow sheet.

The Merrill Crowe process consists of the following main steps, which follow cyanide leaching:

- Solid liquid separation, using CCD thickeners, or vacuum filtration;
- Clarification of pregnant solution to approximately 1 ppm suspended solids;
- De-aeration of the clarified solution using packed towers (Crowe) under vacuum;
- Addition of powdered zinc using a mixing cone;
- Precipitation in the pipeline between the press feed pumps and the filter presses;
- Filtration of the precipitated metals in a filter press;
- Smelting of the precious metal precipitate in the refinery to form dore;

Figure 1 is a simplified flow sheet of the Merrill Crowe process for low temperature cyanide leach solutions, requiring deaeration. A Merrill Crowe flow sheet designed to treat high temperature carbon elution streams, would be similar, but would not require the deaeration step (Marsden, 1990).

History

Cyanidation was first applied in the extraction of gold and silver from low grade ores by J. S. McArthur and Doctors Robert and William Forrest of Scotland in the mid 1880's. A British patent was issued to the group in 1887 and U.S. patents were issued in 1889 for the process, which included agitation of pulp in the presence of air followed by the precipitation of gold with zinc in a separate solution. The first mining application of cyanidation for gold recovery was at the Crown Mine in New Zealand in 1889. The process was subsequently applied in South Africa, the United States and mining districts around the world. The zinc cementation process was introduced in 1890 and became an integral part of the cyanidation process (Dorey, van Zyl and Keil, 1988; Fleming, 1998). The first application of the zinc cementation process in the United States was by C.W. Merrill at the Homestake Mine in Lead South Dakota in 1897 (Chi, 1992).

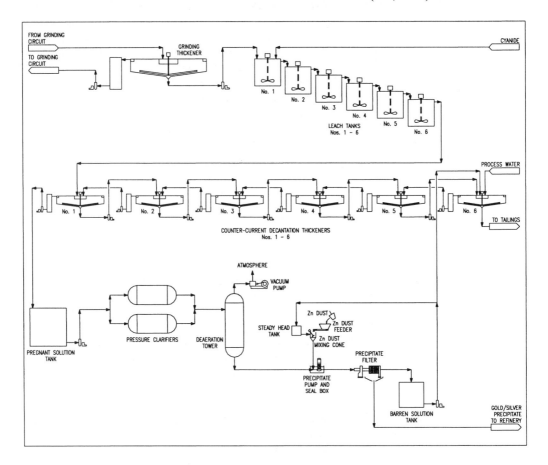

Figure 1. Simplified flow sheet for a typical leaching, CCD and Merrill Crowe circuit.

Zinc cementation was initially performed using long sloping boxes filled with bundles of coarse zinc shavings. Gold bearing solutions were passed through sand filters to remove suspended solids and then through the zinc boxes for metal precipitation. Vertical plates were installed in the boxes forming chambers to direct the flow of solutions through the beds of zinc shavings to improve the contact of the solutions with the zinc. The method proved to be effective but inefficient due to coating of the coarse zinc surfaces with deposited metals or insoluble zinc hydroxide. (Atwood, 1985; Wood, 1996)

Lead salts were introduced in 1894 to address the passivation problem. The bundles of zinc shavings were dipped in solutions of lead acetate before placement in the zinc boxes (Wood, 1996). The lead deposits on the zinc surfaces formed cathodic areas for preferential precipitation of precious metals, leaving the adjacent anodic zinc surfaces exposed for dissolution (Chi, 1992).

Clarification was found to be a very important stage, which affects both metal recovery and precipitate grade. When suspended solids were present, the rate of precipitation was found to decrease, which lowered recoveries. The zinc boxes acted as sand filters trapping fine materials that would potentially blind the beds of zinc and dilute the grade of the precious metal precipitate. (Atwood, 1985)

Through operating experience, it was also identified that some sands trapped in the zinc boxes, such as pyrite and marcasite, tended to improve precipitation and to lower zinc consumption. The improvement was attributed to the reaction of these sulfide minerals with oxygen and cyanide resulting in the consumption of dissolved oxygen. The reduced oxygen was found to improve the efficiency of the process by both reducing the passivation of the zinc surfaces and also by reducing the direct leaching of zinc, which requires oxygen (Atwood, 1985).

The next significant step in the development of the zinc precipitation process was substitution of zinc powder for zinc shavings in 1907-1908 by C.W. Merrill, while studying the effect of surface area on the rate of precipitation. Merrill added zinc dust (fume) to leach solution and then pumped the slurry through a filter press. The precipitate and zinc remained in the filter press and the barren solution was recycled to the process (Wood, 1996). The result was an increase in rate and recovery and improved zinc utilization. The increased surface area and reactivity also resulted in higher zinc consumptions as the rates of the passivation and zinc dissolution reactions increased, making it all the more important to remove oxygen from the system. In 1916, the vacuum deaeration tank was introduced by T.B. Crowe and incorporated into the Merrill Crowe process. (Atwood, 1985; Chi, 1992; Wood, 1996).

The Merrill Crowe process has generally remained in tact since 1916. The main improvements in the process since that time have been in the design and efficiency of the equipment and automation systems; clarifiers, vacuum towers with modern packing and filter presses. There is also a better understanding of the process chemistry, aiding in optimization of reagent usage and consumption resulting in reduced operating costs.

PROCESS SELECTION – CHOICE OF MERRILL CROWE VS CARBON ADSORPTION

The two main processes currently in use for the recovery of precious metals from cyanide leach solutions are zinc precipitation and carbon adsorption. Fleming reports that the carbon in leach process has, in most cases, proved to be more efficient and to have 20 to 50 percent lower capital and operating costs than Merrill Crowe. Currently over 70 percent of world's gold production is recovered using carbon adsorption processes (Fleming, 1998).

The Merrill Crowe process has an advantage over the carbon adsorption process in cases where the metal concentrations are high, such as ores containing significant amounts of silver; high silver to gold ratios. Zinc cementation can also be applied to the recovery of precious metals from carbon strip solutions as an alternative to direct electrowinning.

Carbon in leach has the advantage over Merrill Crowe when ores contain significant levels of organic carbon, high base metal concentrations and when the ore contains high clays, which are difficult to filter.

Examples of operations, which utilize zinc cementation include:

- Newmont Mining Company - Minera Yanacocha mine in Peru
- Barrick Gold's Pierina Mine in the same district in Peru

- Placer Dome's La Coipa Mine in Chile
- Goldfields Operating Company, Chimney Creek- Carbon Strip Circuit
- FMC Paradise Peak – Carbon Strip Circuit
- Equity Silver Mine, British Columbia – Carbon Strip Circuit

PROCESS CHEMISTRY

Cyanide Leaching

Leaching of precious metal ores in alkaline cyanide solutions yields metal bearing solution containing relatively low concentrations of gold, silver, copper, zinc, and other metal cyanide complexes depending on the composition of the ore and the type method of leaching employed.

The dissolution of gold and other metals in alkaline cyanide solutions is generally accepted to be an electrochemical reaction composed of two half-cell reactions. The oxidation of gold from Au^0 to Au^+ represents the most common anodic reaction. The cathodic half cell reaction consists of the reduction of oxygen and water. The reactions are corrosion type and occur in adjacent areas on the surfaces of gold particles. The gold (I) or aurocyanide complex is very stable and was reported by Finkelstein to be the predominant species formed in cyanide leach solutions. The stability of aurocyanide, $Au(CN)_2^-$, is such that it remains stable in the absence of free cyanide and at very low pH. (Hedley, 1958; Finkelstein, 1972)

Gold is typically present in gold ores in elemental form or as an alloy with silver; electrum. The overall reaction for the dissolution of gold is represented by Eisner's equation (Hedley, 1958):

$$4Au + 8CN^- + O_2 + 2H_2O = 4Au(CN)_2^- + 4(OH)^-.$$

The anodic half reaction, the oxidation of gold, is represented by:

$$4Au + 8CN^- = 4Au(CN)_2^- + 4e^-$$

The cathodic half reaction, the reduction of oxygen and water, is represented by:

$$O_2 + 2H_2O + 4e^- = 4(OH)^-$$

Cyanide soluble silver minerals, typically found in gold-silver ores, include electrum, an alloy of gold and silver, and argentite, a single sulfide mineral. More complex ores contain minerals such as tetrahedrite, which contain varying amounts of antimony and arsenic, can also be present. The overall reaction representing the dissolution of Argentite, silver sulfide, is represented by the following (Hedley, 1958):

$$Ag_2S + 3CN^- + O_2 + 2H_2O = 2Ag(CN)_2^- + CNS^- + 4(OH)^-.$$

The base metal reactions are of similar form. The dissolution reaction for chalcocite is as follows:

$$Cu_2S + 7CN^- + O_2 + 2H_2O = 2Cu(CN)_3^{2-} + CNS^- + 4(OH)^-.$$

Some of the important metal complexes produced during the cyanide leaching of gold and silver ores include:

- $Au(CN)_2^-$

- $Ag(CN)_2^-$
- $Hg(CN)_4^{2-}$
- $Cu(CN)_3^{2-}$
- $Zn(CN)_4^{2-}$
- $Fe(CN)_6^{4-}$

Antimony and arsenic form soluble oxide compounds in alkaline solutions and are significant consumers of oxygen.

Cementation Chemistry

The zinc cementation reaction in alkaline cyanide solutions is an electrochemical displacement reaction, which involves the reduction of gold and silver, which occur as cyanide complexes, $Au(CN)_2^-$ and $Ag(CN)_2^-$, the reduction of oxygen and water and the oxidation and dissolution of the zinc metal which forms a cyanide complex, $Zn(CN)_4^{2-}$. The gold and silver metal forms coatings on the surface of the zinc particles and the zinc in turn corrodes and dissolves into solution. The reactions are driven by the differences in electrochemical potential between the more noble precious metals and zinc. Metals, which are more electropositive than zinc, will reduced to their metallic states, while zinc will be dissolved. Table 1 contains a selection of reduction potentials for metals typically found in cyanide leach solutions.

Table 1   Electrochemical Series – Standard Reduction Potentials (Vanysek, 1984, Finkelstein, 1972, Fang, 1992).

Reaction	E^0, V
$2H^+ + 2e^- = H_2$	0.00
$2H_2O + 2e^- = 2OH^- + H_2$	-0.828
$Au^+ + e^- = Au^0$	1.692
$Au(CN)_2^- + e^- = Au^0 + 2CN^-$	-0.473
$Au^{3+} + 3e^- = Au^0$	1.498
$Ag^+ + e^- = Ag^0$	0.8
$Ag(CN)_2^- + e^- = Ag^0 + 2CN^-$	-0.269
$Cu^+ + e^- = Cu^0$	0.521
$Cu^{2+} + 2e^- = Cu^0$	0.342
$Cu^{2+} + 2CN^- + e^- = Cu(CN)_2^-$	1.103
$Zn^{2+} + 2e^- = Zn^0$	-0.762
$Zn(CN)_4^{2-} + 2e^- = Zn^0 + 4CN^-$	-1.260
$O_2 + 2H_2O + 4e^- = 4OH^-$	0.401
$Pb^{2+} + 2e^- = Pb^0$	-0.126

The relationship between the change in Gibbs free energy, ΔG^0, of an electrochemical reaction and the standard reduction potential, E^0, is represented by:

$$\Delta G^0 = -zFE^0,$$

where, z is the number of electrons transferred and F is the Faraday constant. To combine half cell reactions, the free energy for each reaction is determined and then summed. The free energy of the combined reactions can then be used to determine the new standard potential,

$$E^0 = -[\Delta G^0/zF].$$

The reaction will proceed spontaneously if the free energy is negative (Moore, 1983).

The cell potential for a system, which is not at unit activity, is determined using the Nernst equation,

$$E = E^0 + 2.303\, RT/zF \log a_{ox}/a_{red}$$

where E is the cell potential, R is the gas constant, T is the absolute temperature, z is the number of electrons transferred, F is the Faraday constant and a_{ox} and a_{red} are the activities of the of oxidized and reduced species. The resulting cell potential is then used to calculate the change in Gibbs free energy for the reaction,

$$\Delta G = -zFE.$$

The reduction of gold by zinc metal is represented by the following overall reaction (Finkelstein, 1972),

$$2Au(CN)_2^- + Zn = 2Au + Zn(CN)_4^{2-}$$

Assuming that free cyanide is available and a direct transfer of cyanide ions is not necessary (Marsden, 1990),

$$Au(CN)_2^- + Zn + 4CN^- = Au + Zn(CN)_4^{2-} + 2CN^-$$

The half-cell reactions representing the reduction of gold and the oxidation of zinc are:

$$Au(CN)_2^- + e^- = Au + 2CN^-$$

$$Zn + 4CN^- = Zn(CN)_4^{2-} + 2e^-$$

Water and dissolved oxygen are also reduced by zinc:

$$2H_2O + 2e^- = 2OH^- + H_2$$

$$O_2 + 2H_2O + 4e^- = 4OH^-$$

The efficiency of precious metal cementation is dependent on the effective dissolution of zinc, however if sufficient oxygen and free cyanide are available, zinc will dissolve independently, resulting only in increased zinc consumption. If the concentration of free cyanide becomes low, due to insufficient cyanide addition or excessive zinc addition, zinc will react to form zinc hydroxide, which may passivate zinc surfaces and plug filters.

Competing reactions include (Finkelstein, 1972; Fang, 1992):

$$2Zn + 8CN^- + O_2 + 2H_2O = 2Zn(CN)_4^{2-} + 4(OH)^-$$

$$Zn + 4CN^- + 2H_2O = Zn(CN)_4^{2-} + 2(OH)^- + H_2$$

$$2Au(CN)_2^- + Zn + 3(OH)^- = 2Au + HZnO_2^- + 4CN^- + H_2O$$

$$Zn^{2+} + 2OH^- = Zn(OH)_2 + 2e^-$$

Base metals including copper are precipitated along with the precious metals, though in practice it has been shown that copper precipitation can be minimized by maintaining and excess of free cyanide.

Kinetics

The rate of the cementation reaction is first order with respect to the concentration of aurocyanide complex, $Au(CN)_2^-$ and directly proportional to the surface area available for reaction as represented by the following equation (Finkelstein, 1972):

$$\log (C_0/C_i) = k\,A\,t$$

where: C_0 = initial concentration of aurocyanide
  C_i = concentration of aurocyanide at time t
  k = reaction rate constant
  A = zinc surface area available
  t = time

The rate is controlled by the diffusion of aurocyanide to the surface of the zinc particle under all conditions. The extent of reaction is limited by the amount of available cathodic surface. The anodic reaction is the limiting step only when the zinc surface is blocked by insoluble precipitate (Finkelstein, 1972).

Effects of Operating Parameters on Zinc Precipitation

The zinc precipitation reactions are affected by the following parameters:

- pH
- CN^- concentration
- Suspended solids concentration
- Dissolved oxygen
- Temperature
- Metal concentrations
- Lead addition
- Scale forming compounds such as calcium sulfate and sodium silicate

The effects of each of these parameters with respect to operating efficiency follow:

pH

- The pH must remain high to prevent the formation of volatile HCN gas.
- An increase in pH, in the absence of precious metal cyanide complexes, will tend promote the formation of insoluble zinc hydroxide. (Finkelstein, 1972; Fang, 1992)

CN^- Concentration

- The concentration of cyanide must be optimized along with the addition of zinc powder,
- Excess free cyanide will reduce the tendency for zinc hydroxide formation, but will lead to an increase in zinc consumption through independent zinc dissolution.
- Very low CN^- concentration results in the formation of $Zn(OH)_2$ on zinc surfaces inhibiting cementation.

Oxygen Concentration

- The rate of precipitation decreases with an increase in oxygen concentration;
- The rate of independent zinc dissolution reactions increases with an increase in oxygen concentration;
- The potential for redissolution of gold increases as the oxygen concentration increases and the zinc concentration decreases;

Temperature

- An increase in solution temperature results in an increase in the rates of reaction for both precious metals precipitation and the dissolution of zinc, however the reactions are very fast at ambient temperatures;
- As the temperature increases, the solubility of oxygen decreases, which reduces the need for dearation;
- Deaeration is not required when applying zinc precipitation to hot carbon strip solutions (Marsden, 1990);

Zinc Concentration

- The amount of zinc required, to precipitate the precious metals, increases as the precious metal concentration decreases. The rate controlling step is the diffusion of metal cyanide complexes to the zinc surfaces (Finkelstein, 1972);
- Excess zinc addition must be optimized, as it will consume cyanide, which will allow formation of zinc hydroxide.

Metal Concentrations

- The efficiency of cementation increases with an increase in metal concentration (Finkelstein, 1972);
- Base metals inhibit the precipitation of precious metals by forming coatings on the zinc particles;
- Base metals are large zinc consumers;
- Base metals precipitated in Merrill Crowe will have to be removed in the refining stage of the process.

PbNO3 Addition

- Lead nitrate additions of 10 to 15 parts per million increase the activity of zinc by forming a lead zinc couple. The lead precipitates on portions of the zinc surface forming cathodic areas and adjacent anodic zinc areas. Precious metals are preferentially precipitated on the cathodic lead surfaces and zinc dissolution occurs at the exposed zinc surfaces. Excessive lead addition will result in complete coating of the zinc surfaces, inhibiting cementation (Finkelstein, 1972; Chi, 1992; Fang, 1992);

- Lead increases the activity of zinc as described previously and reduces the tendency to form passivating layers of zinc hydroxide as cyanide concentrations decrease. If the cyanide concentrations fall too low, the lead will not prevent zinc hydroxide blinding (Fang, 1992);

- SEM work by Fang shows a change in the crystal structure of silver deposits on the zinc particles when lead is added. Without lead, a smooth coating of silver effectively covers the surface of the zinc particles. With lead, the silver deposits form dendrites or clusters of crystals allowing contact of reagents with the zinc surface. Gold was found to form smooth deposits both with and without lead addition. (Fang, 1992);

Zinc Particle Size

- The rate of the cementation reactions increases with a decrease in zinc particle size.
- The rate of zinc dissolution increases as the zinc particle size decreases,
- Filtration becomes more difficult as the size of the zinc particle decreases.

Scale Formation

- Zinc dust will act as a seed particle for the precipitation of scale formers such as calcium sulfate and calcium carbonate, which will encapsulate the zinc particle.

CHARACTERISTICS OF COMMON CYANIDE LEACH SOLUTIONS

Cyanidation is applied to a variety of ore types, which require different leaching methods and produce pregnant solution chemistries, which range from simple to very complex. The Merrill Crowe process can be applied to all of these systems, however there are those in which carbon adsorption, the chief competitor to Merrill Crowe, would prove to be more economic.

The most commonly used cyanide leaching systems include heap and vat leaching of coarse (greater than 8 mm) crushed ore, agitated tank leaching of finely ground (80 percent passing 74 microns) ore and flotation tailings, and agitated tank leaching of sulfide concentrates, which may be very finely ground (80 percent passing 45 microns). Zinc precipitation can also be applied to the recovery of metals from high temperature eluant from carbon stripping circuits (Marsden, 1990).

Solution characteristics for each system will differ due to the differences in grade, mineralogy, preparation and type of solid liquid separation applied. Solution quality will differ in suspended solids load, types of suspended solids, concentrations of precious and cyanide soluble base metals, solution temperature, free cyanide concentration, the presence of flocculants and scale forming compounds. Carbon, strip solutions are characterized by high temperature and high precious metal concentrations. Vacuum deaeration is not required (Marsden, 1990).

PROCESS DESCRIPTION AND DESIGN PARAMETERS

This section provides a more detailed description of the Merrill Crowe process. The flow sheet assumed for the purpose of this paper is given in Figure 1, and example ranges of process design criteria are presented in Table 2. Development of actual process design criteria during engineering will require substantiation though metallurgical testing. Final equipment selection and sizing should be determined using the metallurgical test data and information from equipment manufacturers.

Solid Liquid Separation

Pregnant leach solution reporting from the solid liquid separation step will flow to a pregnant solution storage tank. The pregnant solution storage should provide sufficient volume to allow operation of the plant steadily and independently from the rest of the mill. The main issue will be scheduled and unscheduled mill down time. The pregnant solution storage tank also provides additional time for the settling of coarse solids which may carry over from the CCD or filtration sections. If flow is lost to the precipitate filters, the cake will be dropped and should be removed before restarting the plant to prevent blinding.

Table 2 – Merrill Crowe Process – Order of Magnitude Process Design Criteria

DESCRIPTION	UNIT	VALUE/SPECIFICATION	REFERENCE
Solid Liquid Separation		CCD/Filtration	
Pregnant solution storage tank	hrs	4	Maintenance
Clarification			
Filter type		Horizontal pressure, US	
		Vertical pressure, Funda	
		Vertical tubular, Stellar	
Specific solution flow rate	$m^3/h/m^2$	1.5 – 2.0	Atwood, 1985
Number of filters		2	Minimum
Operating/Standby		1/1	
Filter cloth type		Polypropylene	
Filter aid types		Diatomaceous earth or Perlite	
Suspended solids in feed	mg/L	100 – 300	
Suspended solids discharge	mg/L	1	
Deaeration			
Vessel types		Packed tower	
Packing types		Rings or tellerettes	
Tower specific flow rate	$m^3/h/m^2$	50 - 85	Atwood, 1985
	$m^3/h/m^2$	70	Design
Tower aspect ratio	ht to dia.	2:1 – 3:1	
Vacuum required for deaeration	mm Hg	500	
	Pa	67,500	
O_2 conc. in pregnant solution	mg/m^3	6	Varies w/ T and P
O_2 conc. in barren solution	mg/m^3	1	Design Target
Precip filter feed pump type		Vertical centrifugal	
Zinc Addition			
Zinc feeder type		Variable speed auger	
Zinc addition rate	Stoichiometric		
Gold	g Zn/g Au	0.33	Calculated
Silver	g Zn/g Ag	0.61	Calculated
Mercury	g Zn/g Hg	0.33	Calculated
Copper	g Zn/g Cu	1.03	Calculated
Excess zinc addition rate	Typical	150 – 300%	Design
Metal Concentration vs. Excess			
100 ppm		10%	Atwood, 1985
5 ppm		200%	Atwood, 1985
1 ppm		1,500%	Atwood, 1985
Zinc induction		Agitated mixing cone	
Very high grade solution	200–300ppm	Dry zinc to reaction tank	Mansanti, 1989
Lead nitrate addition	ppm	10 – 15	Fang, 1990
Precipitate Filter			
Type		Plate and Frame	
		Horizontal leaf (Funda)	
		Tubular (Stellar)	
Specific Flow Rate	$m^3/h/m^2$	4.5	Typical
Operating cycle time	days	7	Typical
Precipitate Composition			Ore Dependent
Gold + Silver	%	30 – 80	
Zinc	%	5 – 30	
Lead	%	0.2 – 2	
Copper	%	0.1 – 2	
Mercury	%	0 – 2	
Insol	%	5 – 15	
Barren Set Point	g Au/t	1.7	Mansanti, 1989

Clarification

The pregnant solution, typically thickener overflow solution will be pumped from the pregnant solution storage tank through pressure filters, to reduce the solids concentration from a typical 100 to 300 part per million range to less than 1 part per million. Clarification is achieved by the use of pressure filters containing either horizontal or vertical leaves, which can be automatically washed. At least two pressure filters should be available which can be operated in parallel with one operating and one on standby to allow continued operation through wash cycles.

The pressure filters consist of horizontal tanks, which contain a series of filter disks mounted on the solution discharge pipe, which extends through the center of the tank. The filter elements or disks are covered with a polypropylene cloth, which is must be precoated with a layer of fine silica to create a bed of filter media to trap very fine particles. The fine silica consists of either diatomaceous earth or perlite and can be purchased with varying particle size distributions. The blend of precoat materials used for a given operation is selected to provide the required solution clarity, the highest flow rate, lowest pressure drop and longest filtration life. The optimization of these variables will be unique to each operation and procedures will be developed based upon each specific ore type. In addition to precoating the filters, a continuous addition of the diatomaceous earth may be added directly to the pregnant solution as body feed, when the suspended solids are very fine or clay rich.

Pregnant solution is pumped through the pressure filters until the pressure drop increases to a preset limit based on the pressure rating of the filter and the flow rate through the filter, which will decrease as the pressure increases.

At the end of the filtration cycle, a second parallel filter is placed on line to maintain the desired process flow rate, and the filter to be cleaned is taken off-line. The filter tank is drained along with filtered solids that will slump by gravity from the filter clothes when the process solution flow is discontinued. The filter cloths are then washed with built in high pressure sprays which remove filter cake that may stick to the cloths. The filter discs rotate past the sprays. Upon completion of the wash cycle, the vessels are closed and filled with barren solution and the precoating process is initiated.

The precoat system consists of a precoat mix tank and circulation pumps. The appropriate amount of precoat is added to the precoat mix tank and the precoat slurry is then circulated in a closed loop through the filter vessel until the returning solution becomes clear. The filter is then put back into operation by introducing pregnant solution while closing the precoat circuit, being careful to maintain differential pressure on the filters so that the layer of precoat will remain in place. If differential pressure is lost across the filter the precoat will fall to the bottom of the vessel and the process must be repeated.

The sizing of the filters is based on the flow rate of pregnant solution and the amount and type of suspended solids contained in the solution. Typically the filters are designed for a specific flow rate in the range of 1.5 cubic meters per hour per square meter of filter area.

Blinding of the filter cloth is the leading problem, which affects the performance of the clarification filters and may determine the ultimate cycle time. Some of the sources of blinding include:

- Flocculent from the CCD circuit;
- Very fine suspended solids; clays;
- Calcium carbonate precipitation or scaling can cause premature blinding of filters. Antiscalants are used to prevent the scaling problem.

- Precipitation of sodium silicate gel on the filters can be a problem if significant amounts of sodium hydroxide are used in the leaching circuit. Potential sources of sodium hydroxide in the leaching circuit include cyanide addition, especially when using low concentration, recovered cyanide, and final pH control.
- Precipitation of iron hydroxides.

Deaeration – Crowe Tower

Deaeration of the clarified pregnant solution is typically accomplished using a packed tower under vacuum. Clarified pregnant solution discharging the filters flows to the top of the deaeration tower, where the solution is distributed over a bed of packing, which provides surface area for thin film formation and release of dissolved oxygen. A vacuum pump is used to reduce the pressure within the vessel to approximately 500 mm Hg or 67,500 Pa. Evolved gases including oxygen and ammonia are exhausted through the vacuum pump. Deaerated solution discharges from the tower, by gravity, through a vertical discharge pipe directly into the suctions of the product filter press feed pumps. The height of the discharge pipe, which remains full of solution (the barometric leg), is determined by the amount of head required to overcome the vacuum developed by the vacuum pump and to provide the required net positive suction head for the filter press feed pumps.

Various types of towers have been used for deaeration, from the original Crowe tower, which contained triangular section, wooden boards, to modern packed towers that use efficient polypropylene rings, saddles or tellerettes. In addition to packed towers, bubble cap tray or baffle towers could be used. Baffle towers consist of plate arrangements within the towers. The solution will cascade over the edge of one plate to the next forming a thin curtain of droplets providing surface area for transfer. Baffle tower arrangements include the disk and donut type and opposing inclined plate. The advantage of baffle towers is that they are less susceptible to blockage by carbonate or gypsum scale than the more efficient dumped packing (Fair, J.R., 1984).

In sizing a packed tower, the diameter is determined by the liquid flow rate, the capacity of the packing and the gas flow rate, which moves countercurrent to the solution. These rates along with the characteristics of the liquid and gas streams can then be used to estimate the pressure drop and predicted flooding points for a packed tower using the pressure drop correlation by Eckert (Fair, J.R., 1984). The pressure drop correlation helps to define the acceptable operating range of a tower and provides a means to optimize the tower diameter.

The height of the packing within a tower can be determined from the overall volumetric liquid phase mass transfer coefficient, $K_L a$, for the process, which is found experimentally (Edwards, W.M., 1984). The relationship between the mass transfer coefficient and the packing height is given by the equation:

$$K_L a = n_A / (h_T S \Delta x^*_{lm}),$$

where n_A is the overall rate of transfer of solute A, h_T is the total packed depth in the tower, S is the tower cross sectional area, and Δx^*_{lm} is the log mean concentration difference given by:

$$\Delta x^*_{lm} = (x^* - x)_2 - (x^* - x)_1 / \ln[(x^* - x)_2 / (x^* - x)_1]$$

where subscripts 1 and 2 designate the bottom and top of the tower respectively. x^* is the mole fraction of gas in the solution which is in equilibrium with the bulk gas concentration, and x is the mole fraction of gas in solution (Edwards, W.M., 1984).

The equilibrium gas concentration in the solution at the specified pressure can be determined using Henry's Law,

$$p_A = Hx_A,$$

where H is the Henry's law constant, p_A is the partial pressure of A in the gas phase and x_A is the mole fraction of A in solution. If A is assumed to be air and pA is the total pressure, the concentration of oxygen can be estimated from the chemical composition of air (Edwards, W.M., 1984).

The desorption of oxygen, hydrogen and carbon dioxide from water was studied by Sherwood and Holloway and correlations were developed for the mass transfer coefficient and the height of a transfer unit for the system. The correlations can be used along with the generalized equation by Cornell to determine the packing height for various types of packing (Fair, J.R., 1984).

The best source for mass transfer data including the overall mass transfer coefficient would be from existing commercial operations. Performance data on various packing types and materials can be obtained from the packing and tower internals manufacturers.

The specific flow rate for preliminary design of packed, vacuum, deaeration towers is approximately 70 to 85 $m^3/h/m^2$. The height to diameter ratio for reported towers ranges from 1:1 to 3:1. The 3:1 ratio is more common and is thought to be more efficient as long as the tower is not allowed to operate in a flooded condition. The vacuum pump is sized for the maximum amount of gas to be transferred and the target absolute pressure of 67.5 kPa.

Cementation

Zinc powder is metered into the deaerated pregnant solution using various types of feeders. Some commonly used feeders include, variable speed auger type feeders and rotating disk. In most cases it is necessary to use a vibrating feed hopper to prevent bridging.

Zinc powder is typically fed into a small agitated mixing cone containing barren solution, which is positioned above the suction of the filter press feed pumps. The zinc slurry will flow by gravity, through a control valve, into the pump suction. The zinc cone should be installed on a platform adjacent to the deaeration tower and above the liquid level in the tower to allow gravity flow into the pump and prevent overflowing when the filter press, feed pumps are shut down. A steady head tank will be provided to maintain a constant level in the mixing cone.

The zinc addition rate will be the stoichiometric amount of zinc required to precipitate the precious and base metals in solution plus an excess amount, which will be dependent on the metal concentrations of the solution. The excess zinc required with respect to metal concentration in solution has been reported to be ten percent for solutions containing 100 parts per million metal, 200 percent for solutions containing 5 parts per million metal and 1500 percent for solutions containing 1 part per million metal (Atwood, 1985). (Marsden, 1991) reported that five to 10 times the stoichiometric requirement would be needed for carbon eluants.

The cementation reaction occurs very rapidly and sufficient retention time is available for the reaction to take place in the pipeline between the filter press feed pumps and the filter presses. The key issue in the design of the filter press feed pumps is the prevention of air ingress. Leakage of air through the pump shaft seal will allow oxygen to enter the system. The standard method of coping with this problem is to use inline vertical centrifugal pumps, which are submerged in barren solution above the shaft seal.

Operating problems reported in adding zinc slurry to the system include wetting of the zinc, control of the zinc solution feed rate, plating of metals on the wetted parts of pumps and build-up of zinc and precipitate in the pumps and pipelines. Extreme cases are reported in the treatment of very high grade carbon strip solutions (Mansanti, 1989). In the Chimney Creek case, the plating and build-up problem lead to the addition of a separate pumping system to inject the zinc slurry into the press feed pump discharge line. One recommendation to minimize the plating problem or at least contain it was to add dry zinc to an agitated reaction tank to permit the precipitation reaction to occur before entering the piping system (Mansanti, 1989).

Precipitate Filtration

Filtration of the precipitate has been accomplished in various types of pressure filters. A common type is the plate and frame filter press. Others include enclosed, automated, horizontal leaf filters and candle or tubular type filters. Plate and frame filter presses can be fitted with canvas or polypropylene filter clothes. In some cases a combination of canvas filter clothes covered by filter paper has been used. The filter presses are typically precoated with diatomaceous earth or perlite at the beginning of the filtration cycle to prevent blinding. Zinc can be added along with the precoat to insure that the gold is completely precipitated during the start-up of the precipitation/filtration cycle. During steady state operation, the presses will contain excess zinc.

The filters will be operated for specified periods of time determined by the maintenance schedule, the accounting schedule, the time it takes for the pressure to reach the maximum recommended operating level or the flows decrease to an unacceptable level. At the end of the selected operating cycle, the filters are taken off line and drained. Compressed air is blown through the filters to force as much of the liquid out of the filter cake as possible. The presses are then opened and the filter cake is dropped into carts, which can be transported to the refinery. This operation is sensitive to security and is typically performed under the supervision of the refinery personal and plant security staff.

The filter clothes are scraped and washed, the presses are reassembled and the system is put back in operation. The precipitate is then transported in the same cart to the refinery area for smelting.

Clean-up and Smelting

There are different approaches to the treatment of Merrill Crowe precipitates in the refinery depending on the composition of the precipitate and the types of equipment available. The three main types of furnaces referred to in the literature include gas or diesel fired reverberatory, submerged arc and induction type furnaces. Induction furnaces are typically not the best choice for the smelting of zinc precipitate unless the precipitate is acid washed prior to smelting. Induction furnaces only heat the metal and so the slag is heated indirectly. The high quantities of fluxes result in reduced crucible life. The reverberatory furnace has been used successfully in operations requiring high flux to charge ratios. The reverberatory furnace is either gas or diesel fired and provides heating directly to the top of the melt allowing the use of higher temperature slag mixtures.

The precipitate clean-up and smelting processes at the Nerco DeLamar Mine and at Coeur Rochester were similar. The precipitate at DeLamar in the late 1980s and early 1990's was filtered using plate and frame filter presses. The filter cloths were covered by filter papers, which were removed along with the precipitate during cleanup. The wet precipitate was mixed with fluxes and loaded into the 2000 lb copper reverberatory furnace. The smelting process was performed in a single step and dore bars were poured. It should be noted that the precipitate at DeLamar was relatively free of base metals.

Treatment at the Rand refinery was much more involved. The precipitate was washed from the Stellar filters into vats in which sulfuric acid was introduced to dissolve the zinc. After acid washing, the remaining gold leach residue was filtered using plate and frame, drum or belt filters and the filter cake was packed into calcining trays. The residue was calcined at a temperature ranging from 550 to 700 degrees C for 16 hours. The resulting calcine was then smelted in submerged arc furnaces.

The treatment process at Paradise Peak included an initial sulfuric acid leach step followed by filtration and retorting of the filter cake in order to remove and capture the mercury. The retorting process was performed over 24 hour period at a temperature of 730 degrees C (1350 F) (Mansanti, 1989).

Equity Silver acid leached the precipitate using hydrochloric acid. The residue was filtered, drying and fired in a 225kg induction furnace. The buttons from the first pour were then remelted in a 45 kg induction furnace. Initially, sulfuric acid was used to leach the precipitate, however, sulfates and sulfides remaining in the leached residue resulted in the formation of significant sulfide matte layers during melting (Semple, 1987).

OPERATING COST VARIABLES

Consumables

The operating costs associated with the Merrill Crowe process are associated with operating and maintenance labor and consumable items. The preliminary design criteria presented in Table 2, include estimated unit consumptions for some of the consumable items. The following is a summary of the key operating costs

Manpower
 Operating Labor
 Maintenance Labor

Consumables
 Power
 Cyanide
 Filter aid
 Filter cloth
 Propane
 Zinc
 Lead nitrate
 Antiscalant
 Filter paper (if used)
 Freight

Actual operating costs for a given project should be estimated based on the results of metallurgical test work, local labor costs, local power and quotes for all consumables including freight. In very remote regions, the availability of operating and maintenance supplies may dictate the type of plant constructed.

CONCLUSIONS

The zinc precipitation or Merrill Crowe process has proved to be a flexible, time tested method for the recovery of precious metals from cyanide leach solutions and can be applied to the majority of ore types except those containing organic carbon. The key parameters for the successful operation of zinc precipitation include:

- Low concentrations of suspended solids
- Low concentrations of dissolved oxygen
- Low base metal concentrations
- Optimized free cyanide concentration during precipitation
- Optimized zinc addition rate
- 10 to 15 ppm lead nitrate addition

The carbon adsorption processes have steadily improved over the last twenty years and are now reported to be more efficient and less costly than zinc precipitation for many applications. Cases in which zinc precipitation remains more cost effective than carbon adsorption include the treatment of ores containing high silver to gold ratios and ores containing significant mercury concentrations. Carbon adsorption is required for processing ores containing organic carbon and carbon adsorption has an advantage over zinc precipitation when high base metal concentrations are present and where high clay is present preventing effective solid liquid separation.

To obtain the benefits of both processes when treating high silver ores, operators have applied a combination of Merrill Crowe and carbon in leach. Initial pregnant solution or No. 1 thickener overflow solution is treated in a Merrill Crowe circuit to remove the majority of the silver and the thickener underflow and remainder of the slurries and solutions are treated using either carbon in pulp or carbon in leach processes.

REFERENCES

1. Atwood R.L. and R.H. Atwood. 1985. Design Considerations for Merrill-Crowe Plants. Society of Mining Engineers of AIME. Preprint Number 85-353.

2. Chi, G. 1992. *Study of Merrill Crowe Processing: Solubility of Zinc in Alkaline Cyanide Solution*. Master of Science Thesis, University of Nevada-Reno, Nevada.

3. Dorey, R., D. van Zyl and J. Kiel. 1988. Overview of Heap Leaching Technology, In *Introduction to Evaluation, Design and Operation of Precious Metal Heap Leaching Projects*, Chapter 1, ed. van Zyl, D.J.A., I.P.G. Hutchison, and J.E.Kiel, Chapter 1. Society of Mining Engineers Inc.

4. Edwards, W.M., 1984. Mass Transfer and Gas Absorption, In *Perry's Chemical Engineers' Handbook*, ed. Perry, H.P., D.W. Green, and J.O. Maloney, Sixth Edition, Section 14. New York: McGraw-Hill Book Company.

5. Fair, J.R., D.E Steinmeyer, W.R. Penney, and B.B. Crocker. 1984. Liquid-Gas Systems, In *Perry's Chemical Engineers' Handbook*, ed. Perry, H.P., D.W. Green, and J.O. Maloney, Sixth Edition, Section 18. New York: McGraw-Hill Book Company.

6. Fang, M. 1992. *Solubility of Zinc in the Merrill Crowe Process. Master of Science Thesis*. University of Nevada-Reno, Nevada.

7. Finkelstein, N.P. 1972. The Chemistry of the Extraction of Gold from its Ores, In *Gold Metallurgy in South Africa*, ed. Adamson, R.J., Chamber of Mines of South Africa, Johannesburg. 284-347.

8. Fleming, C.A. 1998. Thirty Years of Turbulent Change in the Gold Industry. *Proceedings 30th Annual Operator's Conference of the Canadian Mineral Processors*. Paper No. 3.

9. Haggerty, S.1991. An Overview of Corona's Nickel Plate Mine. *Proceedings 21th Annual Meeting of the Canadian Mineral Processors*. Paper No. 16.

10. Hedley, N. and Kentro, D.M. 1945. Copper Cyanogen Complexes in Cyanidation. *The Canadian Institute of Mining and Metallurgy Transactions*, Volume XLVIII, 237-251.

11. Hedley N. and H. Tabachnick. 1958. Chemistry of Cyanidation. In *Mineral Dressing Notes Number 23*. New York. American Cyanimid Company.

12. Lindeman, D. and R. M. Nendick, 1992. A Comparison of South African and North American Practices in the Extractive Metallurgy of Gold. *24th Canadian Mineral Processors Conference*, Ottawa, Ontario.

13. Mansanti, J.G. and M.F.Gleason.1989. Funda Filters for Zinc Precipitation, Start Up and Operation. *2nd Annual Intermountain Mining & Processing Operators Symposium*, Elko, Nevada.

14. Marsden, J.O. 1990. Practical Aspects of the Chemistry of Zinc Precipitation from High Temperature Carbon Eluates, *Randol, Phase IV, Innovations in Gold and Silver Recovery*, Volume 11, Chapter 33, pages 6357- 6361.

15. Milligan, D. A., O.A. Muhtadi, and R.B. Thorndycraft. 1988. Metal Production. In *Introduction to Evaluation, Design and Operation of Precious Metal Heap Leaching Projects*, Chapter 9, ed. van Zyl, D.J.A., I.P.G. Hutchison, and J.E.Kiel, Chapters 1. Society of Mining Engineers Inc.

16. Moore, W.J. 1983. Electrochemical Cells. In *Basic Physical Chemistry*, Chapter 17. New Jersey: Prentice Hall, Inc.

17. Muhtadi, O. A., 1988. Metal Extraction (Recovery Systems), In *Introduction to Evaluation, Design and Operation of Precious Metal Heap Leaching Projects*, Chapter 8, ed. van Zyl, D.J.A., I.P.G. Hutchison, and J.E.Kiel. Society of Mining Engineers Inc.

18. Nendick, R.M. 1983. An Economic Comparison of the Carbon-in-Pulp and Merrill-Crowe Processes for Precious Metal Recovery. *Process Economics International*, Vol. III, No. 4.

19. Peter, E.L., R.C. Reid, and E. Buck. Physical and Chemical Data. In *Perry's Chemical Engineers' Handbook*, ed. Perry, H.P., D.W. Green, and J.O. Maloney, Sixth Edition, Section 3.New York: McGraw-Hill Book Company.

20. Shantz, R. 1976. Leaching Chalcocite with Cyanide. *Engineering and Mining Journal*, October.

21. Semple, P.G. 1986. Equity Silver Mines Scavenger Circuit. *Proceedings 19th Annual Meeting of the Canadian Mineral Processors*. Paper No. 7.

22. Vanysek, P. 1983-1984. General Chemistry, Electrochemical Series. In CRC Handbook of Chemistry and Physics, ed. Weast, R.C., M.J. Astle and W.H. Beyer, 64th Edition, Section D, Florida: CRC Press Inc.

Selection and Design of Carbon Reactivation Circuits

Dr. Joerg von Beckmann[1] and Paul G. Semple[2]

ABSTRACT

This paper deals with the selection and sizing of various components in carbon reactivation circuits.

Carbon reactivation includes the acid washing and the thermal reactivation of activated carbon. The acid wash circuit design addresses the flow sheet design, process control and materials of construction aspects of the acid wash circuit.

The thermal reactivation circuit includes design consideration for feed bins, quench tanks, screens, carbon feeders, and carbon regeneration kilns. Comparative operating costs of different energy sources and kiln efficiencies are discussed. The need for off-gas particulate scrubbing and mercury vapour handling is addressed.

In addition to the design of the carbon reactivation circuits, some procedures for monitoring the performance of the circuit are discussed.

INTRODUCTION

The carbon reactivation circuit is series of unit processes designed to restore the activated carbon's ability to recover precious metals from cyanidation circuit solutions. Since each circuit will be treating a unique solution, which will result in unique carbon fouling problems, the reactivation circuit design must consider several variables, which includes the preference of the plant operator.

The main unit operations within the reactivation circuit are acid washing, elution and thermal reactivation. This paper will focus on the design considerations in the acid wash and thermal reactivation circuits, although it should be noted that all three operations must be operated efficiently to ensure proper carbon reactivation and the resultant low soluble gold losses from the cyanidation circuit.

A typical flow-sheet for a carbon reactivation circuit is shown in Figure 1.

ACID WASH CIRCUIT

The first stage in the carbon reactivation circuit is the acid wash circuit. Early carbon circuit designs often considered acid wash after carbon elution but recent designs have typically been based on acid washing prior to elution. Acid washing prior to elution has become prominent due to the following:

- Hydrochloric acid has become the predominant acid used and precious metal loss, especially silver, when nitric acid was considered has been minimized,
- The acid wash tank can be used to measure loaded carbon batch sizes,
- Most operators believe that removing scale build-up prior to elution improves the overall elution efficiency.

Several operators, especially in Australia, utilize a common acid-wash elution tank. In this instance the loaded carbon is recovered in a holding tank and then transported to the elution tank for further treatment. This practise is less common in North America, and the design considerations for this alternative are discussed later.

---

[1] Lochhead Haggerty Engineering. & Mfg. Co. Ltd., Delta, B.C., Canada
[2] Penguin Automated Systems Inc., Oakville, Ont, Canada

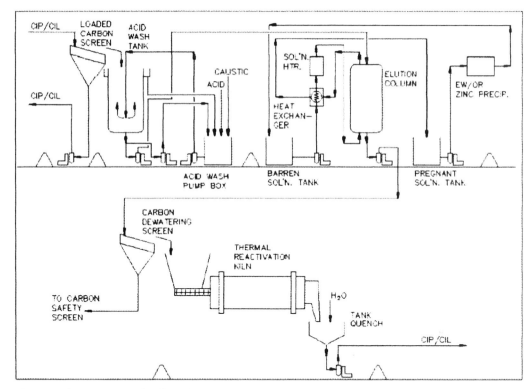

Figure 1 Carbon Reactivation Flowsheet

The objective in the acid wash circuit is to remove scale build-up on the carbon, thereby opening up the activated carbon micro-pores. The step is important to maintain the carbon's ability to recover additional gold as well as maximize the surface exposure that will improve the overall gold elution efficiency in the downstream processing.

Early test-work in acid washing also considered the removal of base metal contaminants during acid wash. The acid washing circuit will remove some base metals but the overall efficiency is low since activated carbon has a high affinity to base metals at a low pH. Once these mechanisms were understood, it became apparent that base metals, especially copper, could be more effectively removed in the elution circuit.

Process Flowsheet

As previously mentioned there are two predominant options for acid washing. The first option utilizes a stand alone acid wash tank while the second option utilizes the elution column as the acid wash vessel. The selection of the option depends on the operator preference, and the second option is predominantly used in Australia. In either case the circuit will include an acid wash pump, pump box and acid metering pump.

The loaded carbon will be recovered from the loaded carbon screen and transported to the acid wash tank. The acid, typically hydrochloric acid, will be metered into the acid wash pump box and the acidic solution pumped through the acid wash tank. The overflow solution from the acid wash tank will return to the acid wash pump box where additional acid will added and the solution is recycled through the acid wash tank.

Once the acid wash cycle is complete the acidic solution is drained back to the acid wash pump box where it is either neutralized and discarded to tailings or retained as the solution for the next acid wash cycle.

Typically hydrochloric acid is utilized in the acid wash circuit although nitric acid has been used at several operations. Although the use of nitric acid did simplify the materials of construction, there were claims made by some carbon manufacturers that the use of nitric acid did attack the cell structure of the carbon and lead to excessive carbon fines losses after numerous cycles through the acid wash circuit. These claims have caused most operators to utilize hydrochloric acid in the circuit.

Process Design Criteria

Acid Wash Tank
Carbon Bulk Density	440 gm/liter
Carbon Batch Size	1-10 tonnes, defined by carbon loading and movement rate
Cycle Time	4 hours per day
Acid Wash Cycle	One cycle per day
Tank H:D ratio	Typically >4:1
Carbon Bed Expansion	50-100%
Freeboard	0.5-1.0 meter
Tank Bottom	Conical

Acid Wash Pump Box
Geometry	Flat bottomed tank
H:D Ratio	Typically 1:1
Volume	Equal to acid wash tank
Pumping Rate	2 Bed Volumes per hour

Acid Addition
Acid Used	Hydrochloric
Acid Consumption	25-50 kg/tonne carbon
Acid Pump	Metering or barrel pump
Acid Storage	Barrels
Neutralizing Agent	Caustic

Materials of Construction

The use of hydrochloric acid limits the use of stainless steel alloys, and most acid wash circuits utilize either plastic or FRP tanks. The Australian design where the elution column is used for acid washing incorporates a butyl rubber insert liner into the elution column. This design was also initially used in South African gold plants but has been largely replaced with stand-alone acid wash tanks. The main issue around the liner in the elution column was maintaining the liner when exposed to high acid concentrations and subsequent high temperature conditions. The Australian operators tend to accept this maintenance aspect and replace the column after the liner fails and the structural integrity of the column is compromised. Most North American operators will not accept this inconvenience and acid wash circuits are designed as stand alone unit operations.

Process valves within the circuit must be selected based upon their location and the duty to which they are exposed. Several options of introducing the solution into the acid wash circuit have been utilized. Each option has different requirements for the material of construction of valves and screens if required. Since screening material under acid conditions typically requires expensive acid resistant materials most recent circuit designs attempt to avoid screens when ever possible.

Process Control

Once the acid wash circuit is ready to be operated the process control is typically limited to a metered addition of acid into the acid wash tank and circulation of acid through the column for a fixed period of time, typically 4 hours. Some operations control the acid flow into the system based upon a pH reading from the acid wash tank and control the solution pH in this tank at a set-point of 1.0. Other operations monitor both the pH in the acid wash tank and the pH of the tank overflow solution. In this instance the circuit is operated until the two pH readings converge indicating that no additional acid is being consumed and the acid wash benefit is diminishing.

Containment

The acid wash circuit is located within a facility that contains numerous solutions containing cyanide and care must be taken to avoid accidental mixing of these solutions. The acid wash circuit will be designed with its own separate containment area that is typically lined with a polymer coating. This coating is required to prevent accidental spills from attacking the concrete and degrading the floors and curbs. The volume of the containment must be sufficient to hold the contents of all tanks contained within the circuit.

Since the loaded carbon screen may be located over the acid wash circuit the screen undersize must be directed to a pump outside the acid wash area and within its own containment.

ELUTION CIRCUIT

The design of the elution circuit is beyond the scope of this paper, but it is important to note that proper elution plays a significant role in carbon reactivation. In addition to removing the precious metals, thereby re-establishing the carbon's ability to recover additional precious metals, the elution circuit can be utilized to remove base metals, especially copper. Base metals such as copper can be removed during a pre-soak with a high cyanide solution. This approach has proven to be extremely effective and is much more efficient than attempts to remove copper in the acid wash circuit.

The elution process must be monitored and operated in a manner where the maximum precious metals and base metals are removed from the carbon. There have been operations that were not effectively removing copper from the carbon during elution and this copper was contributing to poor carbon reactivation and ultimately high soluble loses. This can be especially important when silver is present since inefficient copper and silver removal can lead to alloys of copper-silver building up on the carbon during thermal reactivation. These alloys, once formed, will not be removed in the next carbon elution cycle and will lead to escalating stripped carbon assays and poor carbon activation.

In summary, elution does play an important part in carbon activation, and during operations it is important to survey the carbon assays to make sure the elution circuit is not only effectively removing precious metals but also ensure that the maximum base metal removal is also being achieved.

THERMAL REACTIVATION

A number of factors dictate the selection and sizing of equipment for the thermal carbon reactivation circuit, including the frequency of operation, the most economical energy source and the presence of mercury in the circuit. The selection of type of kiln, whether horizontal, vertical or other, often depends on a combination of past experience, capital cost, life expectancy of the equipment, and personal preference. Horizontal kilns have proven to be the most reliable and trouble free and are the most prevalent in the industry today. The main focus of discussion will be with reference to horizontal thermal carbon reactivation kilns.

There are several papers (Avraaides, Miovski, and Van Hooft 1989, Urbanic, Jula and Faulkner 1985) discussing the optimum environment for the reactivation process. The generally accepted conditions for reactivation are carbon temperatures in the range of 650 to 750°C for a period of 15 to 30 minutes, in a non-oxidizing environment. Steam is the most economical inert atmosphere to blanket the carbon and prevent oxidation of the carbon at the higher temperatures. Whether the steam is produced in the kiln by the introduction of wet carbon, or whether it is produced separately and injected into a kiln with dry carbon, is from an energy consideration, academic. However, there is evidence suggesting that the expulsion of steam from within the carbon particle is beneficial in the reactivation process, hence it is better to produce the steam from wet carbon than to inject steam into the kiln.

Feed Bins, Quench Tanks and Screens

Carbon is transported throughout the circuit using water, either by pumping or educting. The carbon concentration in the slurry is low, and the excess water must be removed by a dewatering screen before being introduced into the kiln.

The optimum moisture content of the carbon as it enters the kiln depends on the type of kiln, the condition of the seals which prevent air from entering the kiln (or steam leaving). Successful reactivation can be achieved with moisture contents as low as 20% by weight with good seals and without the injection of supplementary steam. Most mechanical methods of dewatering can achieve moisture contents in the range of 50% to 35% without the use of preheating the carbon.

It is important to keep in mind that the mechanical dewatering of carbon is much less costly than thermal dewatering. Any excess water which is not required for reactivation or to produce the inert atmosphere, should be removed before it enters the kiln. It takes 0.63 kWh to convert 1 kg of water to steam and a carbon circuit reactivating 100 kg carbon per hour at 50% moisture content (100 kg water + 100 kg carbon) would save the equivalent of 21 kW by reducing the moisture content from 50% to 40%. At $0.05/kWh, for example, this would equate to $9,200/year savings and it would be more economical to spend these initial savings on a high quality dewatering screen to guarantee 40% moisture.

The feed bin should be sized equivalent to the carbon elution batch or larger. Excess water that can be drained through lower screens will provide additional savings to the cost of evaporating. If the dewatering screen is capable of reducing the moisture content to 40% or less, there will be minor savings, while at 50% moisture the carbon will continue to drain in the feed bin.

Quench tanks are often undersized, when they are batch operated. The quench tank needs to have carbon storage capacity equal to the production rate multiplied by the time between emptying cycles. Whether the quench tank is operated on a batch or continuous process, the rate of water removal from the tank must be equal to the water entering the tank. The level in the tank must not drop below the depth of the discharge chute, else the air seal is lost, and the carbon drops into an empty tank at over 600°C. When water is added to the tank, quantities of steam dangerous to personnel will be produced for a short period of time.

Another consideration in the quench tank is that carbon adds 0.24 kWh/kg carbon to the quench tank water. This heat must be removed through the introduction of cool water, or the circulating water volume must be sufficient to absorb this energy without overheating.

In many installations, building height is limiting in the design of the feed bin and quench tank. Often the feed bin is sized for a full batch, but the quench tank size will be reduced since insufficient height is available.

Feed bins having included angles of 60 degrees or greater, and quench tanks having included angles of 45 degrees or greater will empty completely.

Carbon Feeders
Feeding wet carbon into a kiln requires that the feeder act as a seal, or air lock, between the feed bin and the kiln inert atmosphere. At the same time, carbon attrition is a consideration for all mechanical handling systems. An inclined feed chute will not work by itself, since there is no way to control the feed rate to the kiln. A knife gate is not suitable, since it tends to bridge and not provide uniform flow.

An inclined feed screw or auger having variable pitch and incline backwards in the direction of flow provides an excellent seal, and with additional dewatering screens below the screw, the screw will provide supplementary water removal. The screw should not be operated at high speed. Feed screw diameters vary from 50 mm to 200 mm.

Energy Sources Available and Cost
The most common sources of energy for operating kilns are electric power and fossil fuels. Included in the common fossil fuels are natural gas, diesel fuel oil, and propane.

In arriving at a decision as to which energy source is most economical, the unit price of the energy source is not the only consideration. An electric furnace and a fossil fuel furnace do not have the same thermal efficiencies, and this must be factored into the decision.

The major difference between electric and fossil fuel furnaces, is that the electric furnace does not require the combustion of air in order for heat to be released or transferred. An electric furnace is refractory lined with resistance type electric elements which surrounds the kiln shell, and heat is transferred to the shell though radiation and free convection. There is no air flow required, and there is no air required entering or leaving the furnace for the heat transfer process to be completed.

The fossil fuel kiln, although having the same configuration as the electric furnace, generally has a series of tangentially aligned burners firing to the underside of the kiln shell. Heat is released through combustion, and combustion requires a continuous supply of air being introduced into the furnace to oxidize the fuel. The hot gases produced lose much of their heat as these products of combustion come in contact with the cooler kiln shell. Here the heat transfer process is partly through radiation and mostly through forced convection.

Combustion of air, without enhanced oxygen, will under stoichiometric concentrations of air and gas (or oil) produce flame temperatures in the order of 1900°C. The temperature of the combustion products as they leave the furnace in conjunction with the flame (or hot gas furnace) temperature determines the efficiency of the furnace. A furnace having a stoichiometric flame temperature and an exiting flue gas temperature of 15°C would have an efficiency of 100%. That is, all the heat that was released by the combustion of fuel has been removed, and the exiting products of combustion have the same temperature as the entering combustion air. Boilers almost achieve this efficiency, since they are never allowed to run out of water (the heat sink) and have flue discharge temperatures which approach the inlet combustion air temperatures. One key to this efficiency is the constant water load. A boiler which runs out of water and continues to operate, rapidly turns to molten metal. A kiln operating under stoichiometric conditions risks a similar fate.

Kiln furnaces cannot be operated at boiler efficiencies since the carbon load cannot be guaranteed at all times and the maximum flame temperature cannot, from a practical consideration, exceed the melting point of the metal shell.

Since the heating of the kiln shell is indirect, the furnace temperature must be equal to or exceed the process temperature. In this case, the flue gas temperature cannot be much less than 700°C if the product temperature is expected to be 650°C. With a hot mix temperature of 1200°C, the efficiency of this furnace, neglecting heat losses, is approximately 40% (Figure 2). In other words, 60% of the heat is which is generated through combustion cannot be recovered in the process. An electric kiln, by comparison would have a 100% efficiency. It is assumed in this comparison that the radiant and convective losses through the furnace walls for both the electric and fossil fuel furnace can be considered equal, since each operates at the same temperature differentials and insulating properties.

Using these figures, the electric furnace is two and a half times more efficient in converting the source energy to heat. For cost comparison, therefore, the unit energy cost of electricity would need to be two and a half times more expensive than the fossil fuel before one would select fossil fuel as the preferred energy source. For example, if electric power is available at $0.05/kWhr, diesel fuel must be available at $0.21/L.

Heat recovery techniques are available that allow the preheating of combustion air or preheating of the carbon, but few of these alternatives will achieve the energy efficiency of the electric kiln.

Both the electric kiln (Figure 3) and the fossil fuel fired kiln (Figure 4) have a process stack that discharges the steam and organics (flotation reagents, cyanide, oils, etc.) liberated by the reactivation process. For an electric kiln, this is the only stack. The fuel-fired kiln has in addition to the process stack a furnace products of combustion stack.

This extra stack may require a separate emissions permit, and the initial permitting and maintenance of this permit may influence the decision. The most common sources of energy for operating kilns are electric power and fossil fuels.

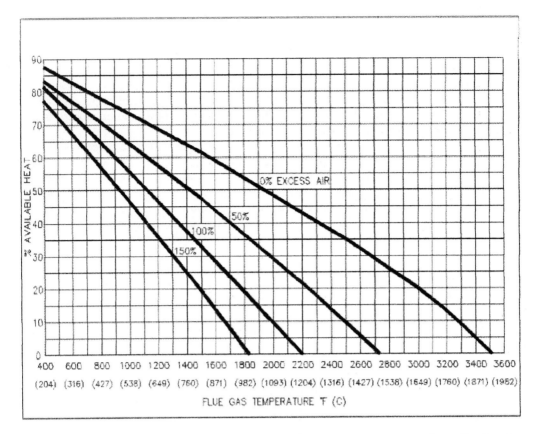

Figure 2 Available heat from fuel fired furnace (taken from North American Combustion Handbook, Volume 1, 3rd Ed., 1986).

Example: If the maximum flame temperature is 1200°C, corresponding to 100% excess air curve, follow the curve up to flue gas temperature, where the curve intersects with 700°C. Read left to 40% available heat.

Regeneration Kiln Selection

There are a number of different designs of carbon reactivation kilns available, and these may be divided into three main groups – vertical, horizontal rotary and horizontal rotary kiln/dryer combinations. In all cases, the basic heat transfer mechanism is through conduction, radiation, or both.

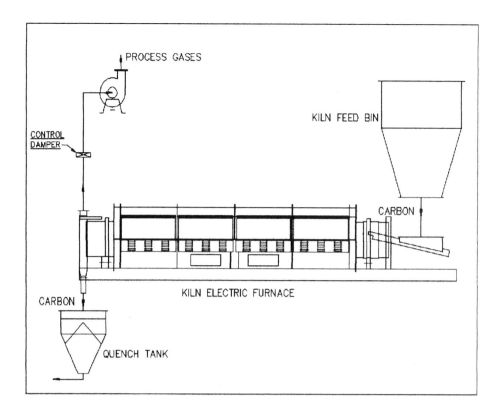

Figure 3 Typical Horizontal Electric Kiln

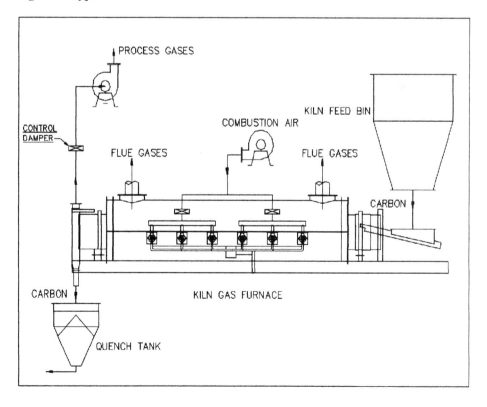

Figure 4 Typical Standard Horizontal Fossil Fuel Fired Kiln

Vertical Kilns

The vertical kiln (Figure 5) has been used successfully in some applications, and not so successfully in others (Semple 1987). The kiln consists of a vertical shaft, annulus or series of tubes through which the carbon flows by gravity. The heat is applied to the outer surface of the carbon transport mechanism, usually supplied by a separate furnace attached to the kiln assembly. Carbon can also be directly heated using electrical current passing though the carbon bed, using electrodes. The carbon is fed into the top of the kiln, usually after some form of predrying from furnace off gases. The transport of the carbon is a plug flow. Some designs have used spoiler devices within the tubes or annulus to cause mixing of the carbon, to allow more contact with the heated surfaces.

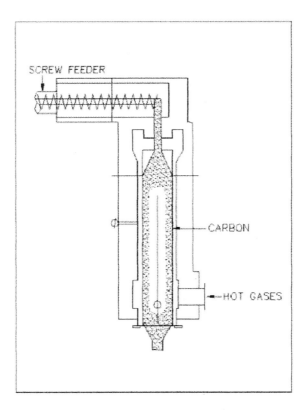

Figure 5 Typical Vertical Kiln

In the vertical kiln, the heated gases flow countercurrent to the carbon, and the carbon is tightly packed in the tubes or annulus. The flow of steam and organics released in the reactivation process is also countercurrent to the flow of carbon, with the hot steam and organics coming in contact with the coldest carbon. The net effect is a column type distillation where compounds evaporate only to be recondensed further up the column, resulting in fouling of the carbon, or a buildup of compounds in a section of the kiln. Removal of the steam and organics is not immediate, since the gases must flow through the packed carbon bed.

Horizontal Rotary

Horizontal rotary kilns require that the carbon be transported along the inside of a cylindrical tube by rotating the tube. This rotary shell is normally positioned at a slight incline, or slope, to facilitate the movement of the carbon from the feed to the discharge end. Heat is supplied to the outside of the rotary shell, though a furnace surrounding the major length of the shell. The shell extends beyond the furnace at both the feed and discharge ends, and is supported on machined supporting rings fixed to the shell. The rings, or tires, are supported on two smaller trunnion wheels located under each tire. The kiln rotation is achieved usually through a chain drive, since kiln rotation speeds are quite slow, normally in the range of 1 - 6 rpm.

At each end of the rotating drum a rotating seal is required to seal the process from the ambient air, and prevent steam and organics from discharging into the plant. The rotary shells must be manufactured from high temperature alloys, containing nickel, which have large linear thermal expansion coefficients. Between ambient and operating temperature of a 10 m long kiln, the change in shell length can be 100 mm. The seals must be able to accommodate this change in length. Normally, the shell expansion is taken up at the discharge end of the kiln, with the drive being located at the feed end, where the feed end tire position is fixed between two thrust rolls. The shell is allowed to expand only towards the discharge end.

The furnace has two (sometimes three) heating zones, each controlled by a separate control loop. Thermocouples or infrared sensors located in each furnace section sense the air or shell temperature, and provide a constant temperature in each furnace section. Carbon bed temperature can be monitored using a thermocouple projecting into shell from the discharge end of the kiln, but due to thermal lag between the furnace and the carbon bed this does not provide a good opportunity for process control.

The feed end of the kiln is the "cold" or drying zone and the discharge end the "hot" or reactivation zone. Furnace setpoint temperatures are higher in the reactivation zone and vary between 100 and 200°C, with the feed end being set at say 550°C and the discharge end set at 700°C. The first half of the kiln acts as a dryer and preheater of the carbon where the steam and low boiling point organics are removed from the carbon. The major heat load is at the feed end, where the water evaporates. The lower setpoint temperature at the feed end allows the process to be more uniformly distributed along the length of the kiln. If the temperature at the feed end is set too high, the feed end may never achieve setpoint, and will be operating at 100% output constantly, while the reactivation zone will be operating at around 15% output.

The kiln diameter and length are designed to provide optimum heat transfer to the carbon, and provide the necessary residence time at reactivation temperatures. Carbon volumetric shell fillage the kiln is in the range of 5-10%, and typical length to diameter ratios are in the order of 7:1 to 10:1.

The steam being produced and organics being liberated in the horizontal kiln are readily removed from the carbon bed, since the dryer end of the kiln has lifters attached to the shell to both increase heat transfer surface area and mix the carbon. The large free volume in the kiln shell, 85-90% which is not filled with carbon, allows the steam and organics to be quickly removed without recombining with the carbon. There is some debate as to whether the steam should be removed at the feed end or the discharge end. Although it would seem advantageous to remove steam and organics as soon as possible, at the feed end, a steam blanket is required over the entire carbon bed, and if steam is removed at the feed end, additional steam injection may be required. Furthermore, the problem of low boiling point organics coming in contact with colder parts of the kiln or the cold carbon results in the same problems found in the vertical kilns. Since the temperatures increase along the shell length from the feed to discharge end, the steam becomes superheated and the organic vapors remain at high enough temperatures not to condense in the process, if the gases are allowed to travel the length of the kiln and are removed at the discharge end.

Once the carbon leaves the furnace section, it passes though a short cooling zone. The cooling will reduce the carbon to below a visible color temperature, after which the carbon drops directly into the quench tank water.

The steam and organic vapors are drawn from the kiln by an induced draft fan which maintains a slight negative pressure (0.2 – 0.5"WC) in the kiln. Although negative pressures risk ingress of air into the kiln, due to poor seal maintenance, positive pressures risk steam, organics or mercury vapor discharging into the work area, which is even less desirable.

Horizontal Rotary Kiln/Dryer Combination

Direct heat rotary dryers have been used for decades to dry many products from sand to pharmaceuticals. The rotary dryer uses hot gases (air) in a cocurrent or a countercurrent configuration to dry materials by making direct contact between the particles and the hot gases. The rotary dryer showers the product across the cross section of the cylindrical drum to produce a curtain of material through which the hot gases pass, resulting in a high thermal drying efficiency. This process is much more efficient than the indirect heating drying process in the carbon reactivation kiln.

A recent introduction of a patented piggy back combination of a rotary predryer and a fossil fuel fired horizontal carbon reactivation kiln (Figure 6) uses the furnace gases from the kiln to dry the carbon in a cocurrent directly heated dryer. The 650 – 700°C furnace gases make contact directly with the cold 40-50% moisture carbon. The carbon discharges from the dryer at 15-20% moisture, sufficient to produce the inert atmosphere in the kiln, and is preheated to about 90 °C. Most of the steam and lower boiling point organics are removed in the dryer, and do not find their way into the reactivation kiln.

The predryed carbon is fed by gravity to the carbon reactivation kiln below the dryer, and the kiln and predryer atmospheres are separated by a rotary gate valve. The valve allows the carbon to flow through without damaging the carbon, since the gate valve fillage is less than 5%. The dryer is operated at a negative pressure, similar to a kiln, however the setpoint is slightly more negative (-0.75"WC) causing any leakage between the two atmospheres to favor movement of gases from the kiln to the dryer.

The kiln section is no different from the standard kiln, except that fillage can be as high as 15%, since the heat transfer is reduced. Typically, a standard rotary kiln designed for 100 kg/hr would process 200 kg/hr carbon if fitted with a rotary predryer. Table 1 compares the thermal efficiencies of the electric kiln, the fossil fuel fired kiln and the kiln/dryer combination.

The capital cost of the kiln/dryer combination is higher than a standard kiln, and the capital cost must be weighed against the operating cost of the conventional kiln. Capital cost recovery for the higher efficiency is typically 6 months on a 125 kg/hr kiln operating 24 hrs/day.

Process Emissions

Common to all emissions will be those products released by the reactivation process, which include steam, organic vapors, carbon fines, and perhaps mercury vapor if found in the ore.

In addition, fossil fired kilns will have products of combustion, including CO_2, NO_x, CO, SO_2 and water vapor. In a standard kiln the process gases and the products of combustion have separate stacks. Although permitting may be required for both stacks, treatment of the products of combustion will not generally be necessary. Treatment may be required for the process gas.

An electric kiln will have no products of combustion, and therefore the only emissions will be those produced during reactivation.

The kiln/dryer combination may require all of the gas to be treated, since it comes directly in contact with the carbon in the dryer.

Particulate scrubbing may be required, depending on the hardness of the carbon. Softer carbon, or as previously mentioned, nitric acid washed carbon, can produce more fines. Particulate scrubbing involves the removal of carbon fines, which can be achieved with a number of common scrubbers. Consideration must be given to the high temperature of the steam, 300–350°C, and the fact that the majority of gas flow from the standard kiln is superheated steam and not air. Wet scrubbing will remove many of the organics.

Residual mercury on the carbon will be liberated in the carbon reactivation kiln since reactivation temperatures are much higher than the vaporization temperature of the mercury (357°C at sea level). The mercury vapor will condense in the ducting, exhaust fan and stack if not immediately removed once it leaves the kiln. Mercury can be removed by condensing the vapor in a wet scrubber followed by treatment in sulfur or iodine treated carbon columns to bring concentrations to below acceptable levels. Other methods of condensing mercury can be used; however, these alternatives typically require pre-treatment to remove particulates.

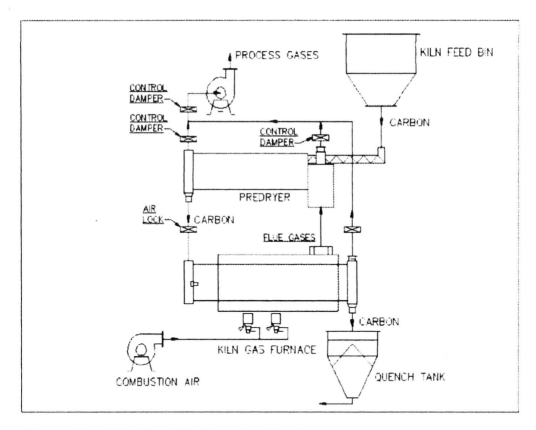

Figure 6 Fossil Fuel Fired Kiln/Dryer Combination

Table 1. Net heat input required for different kilns, in kW/kg dry carbon and relative thermal efficiencies. Heat input based on 50% carbon feed moisture. Fuel units kW. Net heat is the actual operating requirement, and does not include reserves.

	Electric Horizontal	Std. Fossil Fuel Horizontal	Fossil Fuel Kiln/Dryer Combo
Net Process Heat Required	1.46	1.46	0.72
Net Heat Input Supplied	1.46	3.65	1.80
Heat Lost in Flue Gases		2.19	1.08
Heat Recovered in Predryer			1.08
Efficiency relative to Electric	100%	40%	81%
Efficiency relative to Std. Fossil Fuel	250%	100%	203%
Efficiency relative to Kiln/Dryer Combo	123%	49%	100%

Carbon column design is controlled by two variables; free stream velocity and contact time. Free stream velocity determines the cross section of the columns and contact time determines the column height. The resulting volume determines the weight of carbon needed to fill that space. If carbon volume is selected on the basis of adsorption capacity of the treated carbon only, the columns will be undersized. Typical free stream velocities and contact times range from 100 – 200 mm/sec and 8 to 20 seconds, respectively (Calgon Carbon Corp.). Pressure drops across the carbon columns must be taken into account in the selection of the exhaust fan.

QUALITY CONTROL

For each batch of carbon sent through the carbon reactivation circuit, grab samples should be obtained at each stage. The samples should include loaded carbon, acid washed carbon, stripped carbon and thermally reactivated carbon. These samples will be used for both precious and base metal analysis to determine an approximate material balance as well as a quality control for the reactivation circuit.

The standard procedures developed for testing the adsorption characteristics of activated carbon are based upon highly controlled laboratory conditions and the modeling of the results to provide a loading profile that approximates a Freundlich isotherm relationship. In general these procedures are too time consuming to be practically applied to the routine testing of carbon quality and should be utilized if significant problems arise with carbon activity.

Several companies have developed simple carbon testing procedures that allow the relative carbon activity to be measured and monitored for each batch of carbon. These tests are typically used to measure the carbon activity as a percentage of fresh carbon activity, allowing the relative efficiency of each stage to be gauged.

The tests developed are usually based upon contacting activated carbon samples with a stock cyanide solution obtained from the leach circuit. A fixed amount of carbon, typically 10 grams, is added to a one-liter leach solution sample and the carbon slurry is contacted either mechanically or on a bottle roll for one hour. The solution is assayed for precious metals and the gold recovery is expressed as a percentage of the activity of the results from a control test using fresh carbon.

This simplified procedure will provide information on the relative carbon activity for each stage of the circuit. It is also important to note that carbon specific gravity will change significantly during the various stages of the carbon circuit. Although a fixed carbon sample size is desired, it is important to utilize a fixed volume of carbon, thereby eliminating the specific gravity effect. Typically a graphical representation of carbon volume versus fresh carbon weight is developed and the carbon sample size is based upon the initial graphical relationship.

In addition to the carbon activity throughout the circuit, this procedure can be used to monitor the quality of all new carbon added to the circuit. The quality of activated carbon is dependent on numerous factors and in certain instances it may be prudent to routinely test the quality of fresh carbon added to the circuit to ensure it performs to the level indicated during the initial evaluation and selection.

REFERENCES

1. Aguayo S., S., Diaz G., H. "Fundamentals of activated carbon", First English Ed. 2000, Universidad de Sonora, Hermossilo, Sonora, Mexico.

2. Avraaides, J., Miovski, P. and Van Hooft, P. "Thermal reactivation of carbon used in the recovery of gold from cyanide pulps and solutions", Research and Development in Extractive Metallurgy – 1987; The Aus. I.M.M; Adelaide Branch, May 1987.

3. Calgon Carbon Corporation, Product Bulletin, HGR-LH impregnated activated carbon.

4. Semple, P., Equity Silver Mines Scavenging Circuit, paper No. 7, Session B, Proceedings - 19th Annual Meeting of the Canadian Mineral Processors Convention, 1987.

5. Urbanic, J. E., Jula, R.J., and Faulkner, W.D. "Regeneration of activated carbon used for recovery of gold", Min. and Metal Process., 2(4) 193-198, Nov. 1985.

Selection and Sizing of Elution and Electrowinning Circuits

Paul Hosford[1] and John Wells[2]

ABSTRACT
This paper will discuss the different aspects of gold elution and electrowinning. In terms of elution, it will review the two main approaches to elution, namely "AARL" and "Zadra", and discuss the main operating variables, such as temperature, time, pressure and column dimensions. Electrowinning will be discussed, with particular emphasis on circuit design and operating parameters.

INTRODUCTION
The commercial use of carbon in gold recovery plants (CIP and CIL) became widespread from the late 1970s, early 1980s, and within the space of two or three years had become the technology of choice for most of the many new gold projects built during that period. The Merrill Crowe process, based on the precipitation onto zinc of gold and silver from clarified solutions, thereafter became generally restricted to projects with high silver values, or for high tonnage, high dore production heap leach projects. The first generation of carbon plants were usually small tonnage operations, and the quantity of loaded carbon produced was typically 1-2 tonnes per day. Thus the first generation of carbon elution and electrowinning plants were based upon relatively small equipment sizes. However, these initial plants had to overcome many developmental problems in all aspects of the process. As these problems were resolved, and with the increasing acceptance and confidence in the carbon based processes, the plants became exponentially larger, requiring elution systems that could treat 20-30 tpd of carbon, or more, and electrowinning plants that could treat increasingly larger volumes of pregnant solutions.

ELUTION
There have generally been two approaches to the elution of gold from carbon, that have stood the test of time from both the commercial and technical points of view. These are generally known as:

- The Anglo American Research Elution Process
- The Zadra Elution Process

THE ANGLO AMERICAN RESEARCH (AARL) ELUTION PROCESS
Development of this elution process commenced in the late 1970s in response to the growing awareness of the capital and operating cost advantages of the carbon process compared to the conventional Merrill Crowe process. Anglo American Corp. investigated a process that would "decouple" elution and electrowinning, as compared to the Zadra process, where the two unit operations run in series, in a continuous manner.

The AARL system is based on batch elution, over typically a 6-10 hour period. Initially, the AARL elution cycle incorporated acid washing and elution in a single elution column. However, due to the necessity for special materials of construction, the general practice nowadays is to use separate acid wash and elution vessels. The loaded carbon from CIP or CIL is typically transferred to a loaded carbon pulp dewatering screen, where the carbon (-6 + 16 mesh or similar) is separated from the slurry (or solution in

---
[1] Hatch Associates Ltd., Vancouver, B.C. Canada V6G 1A5
[2] Hatch Associates Ltd., Vancouver, B.C. Canada V6G 1A5

the case of CIC). These screens are usually fitted with molded polyurethane panels, with apertures of 14-18 mesh. The carbon is washed with water sprays and then falls by gravity into the column feed bin, or into the acid wash tank, if separate acid wash and elution vessels are used.. The column feed bin is usually up to 1-2 times the volumetric capacity of the column, allowing surge capacity between absorption and elution. The mild steel bins have typically a circular, vertical segment with a cone, (cone included angle at 90°). It is a simple operation to flush this carbon from the bin into the elution column, which is normally placed directly beneath the bin. A simplified diagram of the AARL Elution System is shown in Figure 1 (*Stanley*)

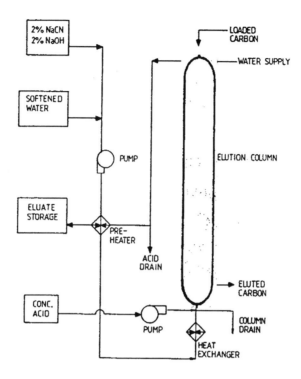

Figure 1: Simplified flowsheet of the AARL elution process

In the early days of development of the process, contaminants in the loaded carbon such as slimes, wood fiber and plastic material (particularly prevalent from hard rock underground mines such as those in South Africa) caused severe problems in elution, particularly blocking of the internal screens and nozzles. Elutriation was attempted, with some degree of success, to remove this trash material. With the advent of better trash removal ahead of CIP/CIL using trash screens such as horizontal belt filters and more efficient screening and washing of loaded carbon, this problem has been more or less eliminated.

After loading the column, the carbon is usually first acid washed with one bed volume of dilute (3%v/v) hydrochloric acid, followed by 1-2 bed volumes (BV) of water. The temperature in the acid wash process varies from ambient to an elevated temperature of typically 50-90°C, (the latter is technically more efficient and is probably cost effective). The temperature of the water rinse raises the system temperature to 110° - 120°C. The spent acid is usually discharged to the absorption circuit. In the original AARL plants, the acid wash and gold elution were done in the same column. This required special column liners and construction materials to handle the range of operating conditions (both acid and alkaline) and temperatures. The circuit development quite rapidly moved to a two-column approach, one for acid washing and one for elution. The authors are aware of operations that have not applied the acid wash step on a regular basis, preferring to acid wash every third batch for example. This appears to allow acceptable elution, but is probably only effective for particularly clean ores, and is generally not recommended. When using a separate vessel for acid washing, FRP is usually a satisfactory material of construction although rubber lined mild steel is also appropriate. The acid washed carbon is then pre-heated with ½ - 1 BV of a strong caustic-cyanide solution (typically 2% NaOH, 3% NaCN) at 110°C for thirty minutes. Some recent experience suggests that the solution can have a much lower cyanide content. After this soak the gold and silver is then eluted with 5-6 BV of high quality softened water (less than about 300 g/t sodium), at 120°C and at a flow rate of about 2 bed volumes/hour. Thus, each batch elution can be completed in an 8-hour operating shift, using an automatic control sequence of reagent pumps and valves. The gold is eluted in a high pH solution and is typically recovered by electrowinning as described later.

OPERATING VARIABLES IN THE AARL ELUTION PROCESS
Acid Washing
Initially it was considered that the acid wash (generally with dilute, 3% hydrochloric acid, but nitric acid is sometimes used) is more effective at an elevated temperature. Cold acid removes calcium and zinc from carbon, while hot acid at about 60°C - 90°C effectively removes calcium, zinc, nickel, and iron. However, common practice today is not to acid wash at elevated temperature. For highly contaminated carbon, the benefits of a hot acid wash in terms of subsequent elution efficiency and carbon activity could be justified.
Water Quality for Elution
The initial development of the AARL process focused upon elution with water with low ionic strength. This required the installation of water softening/demineralization equipment. Subsequent work and plant results suggest that following efficient acid washing the elution water quality is not as critical, and good quality regional water may be suitable.
Column Geometry and Design
The first generation of carbon plants was designed for daily carbon production of typically 1-2 tpd, or less. At a carbon density of 0.5 t/m^3, this represented a carbon volume of 2-4m^3/day. Handling this small quantity of material was relatively easy, and a high column height to diameter ratio could be maintained without the column height becoming impractical. Early development work had illustrated the higher elution efficiency that could be achieved with a H:D ratio of between 6:1 and 10:1 and a ratio of about 8:1 became the desired value. Thus, in these early plants, column dimensions of about 1 metre diameter and heights of 6-8 metres were selected. As the plant sizes increased, and the daily carbon movement increased accordingly, it became more difficult to maintain these column H:D ratios. With a lower H:D ration, the importance of good solution distribution became more critical.

The flow in a column can be either:

- Upflow, or
- Downflow

Upflow is generally preferred as it dilutes (expands) the bed, rather than compacting the bed as would occur in a downflow mode. Solution distribution across the bed is achieved by a series of nozzle caps. These have generally been replaced by the use of tubular wedgewire screens that can be removed, cleaned and replaced without opening up the column. Four (or more) of these screens are placed on a manifold to distribute the incoming eluting solution. The use of the manifold system provides a major advantage in that eluted carbon can be removed from the bottom of the column, as compared to side exit above a nozzle distributor plate.

The initial AARL process contemplated a single column for both acid wash and alkaline cyanide-caustic elution. The advantage of this was a simpler layout, and less carbon movement. The disadvantage was the need for more exotic (and high cost) materials of construction, such as Hastelloy, or liners of butyl or ebonite type rubber. The industry quite rapidly adopted a two column approach. Acid washing is generally carried out in a rubber lined/mild steel or FRP vessel, and elution vessels are typically fabricated from mild steel, rubber lined mild steel or stainless steel.

In the design of the original plants in South Africa, various methods were considered for moving carbon through the process, from absorption to regeneration. Eductors were the early method of choice. These required fairly precise sizing and were prone to choking. These have normally been replaced by more simple, and easy to operate systems based upon recessed impellor pumps, compressed air or pressurized water.

Temperature
Initial development concluded that elevated temperature in the 115-125°C range would allow elution to be completed (95% or more precious metal recovery) with as little as 4-6 BV's of eluate. At temperatures below this and particularly below 100°C the number of BV's increased rapidly (10-12 BV or more). Most AARL operations have elected to run at elevated temperatures. The columns use pressure rupture discs under these conditions. A few operations used or considered low temperature/ambient pressure elution, particularly in remote areas with a need to use simple technology. However, the high pressure/temperature elution system is essentially simple and robust, and can be controlled using low cost PLC type systems.

Temperature is usually achieved and controlled by means of heat exchangers (see Figures 1 and 2). The early generation of plants used low cost plate and frame heat exchangers, which were adequate for the relatively low volumetric flowrates. However, they were susceptible to scaling and choking and more recent, larger plants have preferred shell and tube heat exchangers. Heat to the process can be provided by electrical power or burning gas. Thermopacs (*Davidson and Schoeman 1991*) can be used to heat a thermic oil ring, which feeds the shell and tube heat-exchangers.

ELUATE SOLUTION
The initial plants used a soak solution of 1-2% NaCN and 2-3% NaOH. The eluate (EW feed) was therefore a dilute solution of NaCN/NaOH. Recent plant experience has reported lower reagent concentrations yielding acceptable elution results, and in some cases the NaCN has been eliminated.

ZADRA ELUTION
As stated previously, Zadra elution is different to the AARL system in that the elution and the electrowinning operate simultaneously by continuous circulation of the eluate through the column and the electrowinning cell(s) in series (as shown in Figure 2).

The eluate solution is typically 0.2-0.5% of NaCN and 1-2% NaOH. All of the operating variables that apply to AARL are also valid for Zadra elution. A question often asked by gold metallurgists is which elution system is best, AARL or Zadra? (along with other common carbon technology questions such as CIP vs. CIL vs. and Carbon vs. Merrill Crowe). Several technical papers have attempted to answer these questions. In the case of Zadra vs. AARL, the general consensus is there is not a great deal to choose between the two and it will frequently come down to operator experience. Zadra has tended to predominate in North America where it was initially developed, whilst AARL quite clearly dominates (but not exclusively) in the Southern Hemisphere.

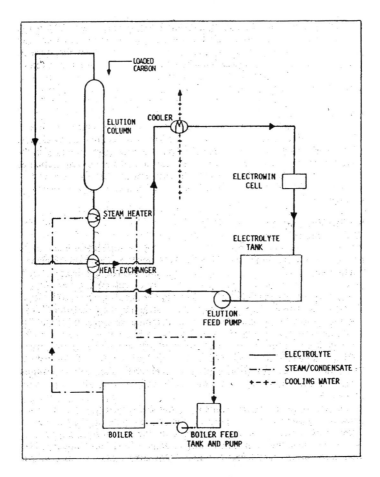

Figure 2. Simplified flow diagram of the Zadra elution system.

Advantages of AARL are its relative speed, and the decoupling from electrowinning may have benefits (although these are somewhat difficult to quantify). It also requires less heat input. Advantages of Zadra are a (perceived) more simple flowsheet, and less need for high quality water. One paper (*Costello et al 1988*) claimed that while Zadra had overall economic benefits, the AARL system could treat 2-3 batches of carbon per day in a single column, (as long as sufficient tanks and EW cells are provided). Whichever system is used, it is important to elute efficiently, and to obtain an eluted carbon value of 50 g/t Au or less. This carbon is returned to the final absorption stage, and to ensure low soluble losses of gold, the equilibrium conditions in this stage require a carbon with less than 100 g/t Au throughout the tank, (*Davidson and Schoeman 1991*). Eight operations were reported (*Richards and Wells 1987*) where loaded carbon was eluted from 5500 g/t Au (average) to 150 g/t Au (average), representing an elution gold recovery (efficiency) of 97.3%.

The authors of this paper would suggest that there is little to choose between the two processes. Both are relatively simple, rugged and fairly forgiving of operational problems. Automatic valve sequencing using a PLC can be equally effective for both systems, requiring only minimal operator interface once the column is loaded. Elution represents a small part of the overall gold plant costs, certainly less than 5%. Hence the impact on overall total project capital (and operating) costs of selecting Zadra or AARL is negligible.

ELECTROWINNING

Principles and Chemistry
Like any other electrowinning process, oxidation reactions taking place at the anode generate electrons, which are consumed at the cathode to deposit the metal. The following electrode reactions take place during electrolysis of an alkaline gold cyanide solution:

- Cathode: $Au(CN)_2^- + e^- \rightarrow Au + 2\ CN^-$
- Anode: $2\ H_2O \rightarrow 4\ H^+ + O_2 + 4e^-$

In cyanide solutions, gold is present as a stable auro-cyanide complex anion with a comparatively high cathodic potential (E_o). This cathodic shift demands higher cell voltage and consequently, other cathodic reactions like the evolution of H_2 by discharge of H^+ and the reduction of O_2 can also take place. These additional reactions consume current and reduce the current efficiency of the gold electrowinning process.

The equilibrium potentials for various metal cyanides are listed in Table 1 below. The cyanide complexes of Hg, Pb, and Ag are nobler than that of gold, so that these metals will deposit preferentially to gold. The concentration of metal in eluates has an influence on electrowinning performance although the cyanide complex of copper is less noble than gold, it must be appreciated that 10^{-4} mol/l of dissolved copper is only 6,3 g/t. Every 10x increase in concentration causes the equilibrium potential to shift positively by 0,06/e V. The equilibrium potential for 630 g/t dissolved copper will therefore be –0,63V, which is the same value for that of gold at 20 g/t. As a result, copper will also deposit with gold if the concentration is sufficiently high. The remaining ions in Table 1 (i.e. Fe, Ni, and Zn) will not normally co-deposit with gold. However, if their concentrations are extremely high in relation to that of gold, a small amount of iron and possibly nickel may be deposited.

Table 1. Equilibrium potentials for the reduction of various metal cyanide ions (metal ion concentration = 10^{-4} mol/l; 0,2% NaCN)

Reaction			e	E_{eq}, V
$Hg(CN)_4^{2-}$	$\rightarrow$	Hg	-2	-0,33
$Pb(CN)_4^{2-}$	$\rightarrow$	Pb	-2	-0,38
$Ag(CN)_2^-$	$\rightarrow$	Ag	-1	-0,45
$Au(CN)_2^-$	$\rightarrow$	Au	-1	-0,63
$Cu(CN)_3^{2-}$	$\rightarrow$	Cu	-2	-0,75
$Fe(CN)_6^{4-}$	$\rightarrow$	Fe	-4	-0,99
$Ni(CN)_4^{2-}$	$\rightarrow$	Ni	-2	-1,07
$Zn(CN)_4^{2-}$	$\rightarrow$	Zn	-2	-1,22
O_2	$\rightarrow$	$2\ OH^-$	pH = 13	0,45
$2H_2O$	$\rightarrow$	$H_2 + 2\ OH^-$	pH = 3	-0,78

Also listed in Table 1 are the equilibrium potentials at a pH value of 13 for the reduction of dissolved oxygen (0,45 V), and water to hydrogen (-0,78 V). The reduction of dissolved oxygen is the most favourable reaction, and consumes more than 50 percent of the total cathode current under normal conditions. As the reduction of water to hydrogen is not limited by mass transport, the production of hydrogen consumes a considerable proportion of the remaining current. Current efficiencies for the production of gold are therefore very low, and values of 0.5 to 20% are typical. It is important to also note that the evolution of hydrogen and reduction of oxygen at the cathode results in a localized increase of the pH at the cathode surface, whereas the oxidation of water at the anode results in a localized fall in the eluate pH.

Types and Features of Electrowinning Cells

Because the reactions occurring in any electrowinning process are heterogeneous, involving the exchange of electrons between a solid electrode and ions or molecules dissolved in solution, the rate of any reaction will depend upon the electrode potential, the electrode area, and the rate of mass transport of electroactive species to the surface of the electrode. If the electrode potential is sufficiently negative that all electroactive species undergo reaction as soon as they reach the electrode surface, the overall rate of reaction is determined only by the available electrode area and the hydrodynamic conditions in the electrolyte. (*Paul, Filmer, and Nichols 1982*).

In order to increase the rate of an electrowinning process it is necessary to increase the electrode area, or the mass transport characteristics of the electrolyte, or both. The standard rectangular electrowinning cells, which are well proven in Industry, exploit the very high surface area of steel wool or stainless steel mesh as the active cathode material. Other cell designs based on steel wool cathodes include the cylindrical Zadra and membrane AARL cells, which have since become obsolete in the Industry.

The rotating tubular bed reactor, the impact rod reactor, and the EMEW cell have been developed for the recovery of metals from the rinse waters of electroplating operations, and are in use commercially. It has been suggested that the EMEW cell could also be used for recovery of gold from carbon strip eluates (*ElectroMetals 2001*).

Developments in cell design include features to facilitate operational convenience. The Kemix sludge reactor allows both electrowinning and cathode wash cycles to be performed in an enclosed vessel. Currently, five units are in commercial operation at two mines in South Africa (*Proudfoot 2002*). The Summit Valley sludging cell allows cathode washing to occur with the cathodes in place. Precious metal sludge is collected in the cell bottom and is then pumped to a filter. (*Weldon 2002*).

STANDARD ELECTROWINNING CELL

Features

The typical standard cell consists of a rectangular, stainless steel tank containing typically between 14-33 cathodes and 15-35 anodes, for cells ranging in nominal capacity from $1m^3$ to $3.5m^3$. Cathodes and anodes are mounted alternately along the length of the cell, so that the cathodes are "sandwiched" between the anodes. The design concepts for this type of cell dates from work carried out by Mintek in South Africa and the USBM in the mid-1970's to late-1980's, and later modified and commercialized by a number vendors in the USA, South Africa and Australia. A schematic of a typical standard electrowinning sludging cell is shown in Figure 3.

The cells are fitted with stainless steel lids, which are usually connected to a gas extraction fan to remove hydrogen, ammonia, and oxygen gases evolved during the electrowinning process.

Operation

Originally, most cells were designed and operated to electroplate the precious metals onto the cathodes. The cathodes were periodically removed and either digested in hydrochloric acid to remove most of the steel wool or simply calcined and smelted. The gold deposited onto the cathodes is very fine grained and usually appears almost black in colour. These cells are equipped with stainless steel mesh anodes and cathodes consisting of a polypropylene frame, or basket, packed with steel wool. Cathode loadings of 4 to 5:1 metal to steel wool mass are typical but ratios of 18:1 can be attained. The cathodes closest to the feed inlet experience the highest solution grades, and load the most rapidly. Typical operating procedure is to remove these highly loaded cathodes for harvesting, advance the following cathodes and replace the fresh cathodes at the discharge end of the cell. The harvesting and maintenance of these cells can take 4 to 24 hours of operation time per week, depending on the gold production rate.

Presently, most new operations favour operating cells so as to promote the deposition of gold as sludge rather than plating. The electrodeposited gold forms as fine grains on the cathode surface and are readily dislodged by the co generated hydrogen gas bubbles and the velocity of the eluate solution and accumulates on the bottom of the tank below the cathodes as a black sludge. The bottom of a sludging cell is sloped to a drain point. Periodically, the cell is taken off line and drained to expose the cathodes. The cathodes are washed in place using a pressure washer to dislodge the deposited gold. The gold sludge is typically filtered through a filter press or a sock filter, dried and smelted. These cells generally use punched plate stainless steel anodes and stainless steel mesh cathodes, so called basketless cathodes. The harvesting and maintenance of these cells typically takes 30 minutes to 1 hour of operation time per week, depending on the gold production rate and the extent of metal "bonding" to the cathodes.

The main reasons for this change to sludging operation are significantly decreased operator time required to service and maintain the cathodes and reduced operator handling of the cathodes. This minimizes operator exposure to toxic metals if these are present in the strip solution (i.e., Hg, Cd, As) and reduces the potential security risk. (*Weldon 2002*)

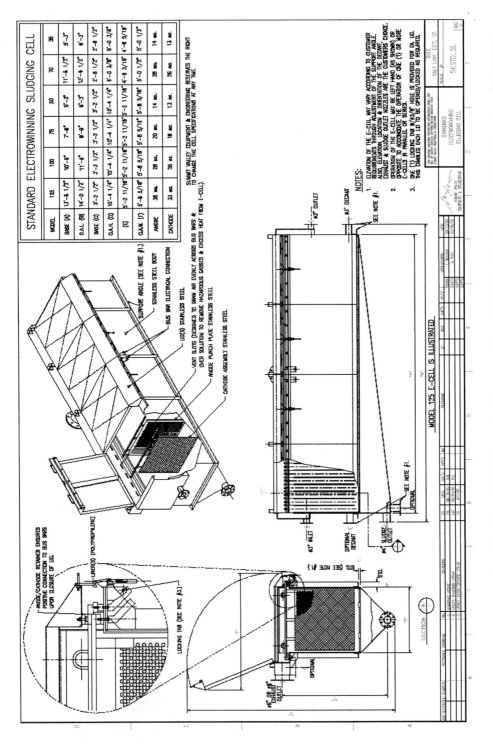

Figure 3: Standard Electrowinning Sludging Cell Schematic, Courtesy Summit Valley Equipment and Engineering, SLC, USA

The main variations in operating parameters between these two modes of operation are current density and fluid superficial velocity. Operation in sludging mode generally requires higher levels of both, as

- The higher current density promotes a random growth of electrodeposits which loosely adhere to the cathodes and are readily dislodged by the greater evolution of hydrogen gas.
- The higher fluid superficial velocities in the cell reduce the metal ion depleted zone in the immediate vicinity of the cathodes, and reduces the thickness of the electrical double layer of the cathodes (*Weldon 2002*). If the superficial velocity is too low, the reduction reaction is limited by the diffusion rate of the metal through the electrical double layer.

Typical curves for extraction efficiency (percent) vs. effective cell retention time (R_f E/W, minutes) are shown in Figure 4 (Courtesy of SVEE), and demonstrates the general improvement in extraction efficiency attained as the cell design and operation of the cell has been improved.

Current design and innovation includes a new cell that is completely enclosed and can be automatically washed down (*Weldon 2002*), which should provide both operating and security benefits.

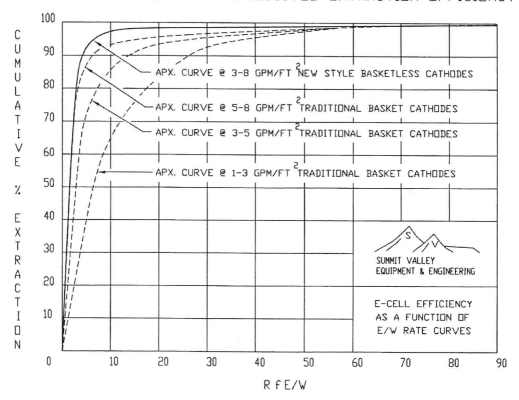

(*Note: 1 gpm/ft^3 = 2.4 m/h*)

Figure 4: Courtesy of Summit Valley Equipment & Engineering, SLC, USA

ELECTROWINNING CIRCUIT DESIGN AND OPERATION

The process variations to be considered in electrowinning circuit design and operation relate to flowsheet configuration, the design of the cells and the chemistry of the eluate solution.

Configuration of Strip/Electrowinning Circuit

Two main stripping procedures are commonly used as discussed in the previous section of the paper, the Zadra and AARL processes. Each places different constraints upon the electrowinning circuit.

In the Zadra process the electrowinning process is integral with the stripping process. The electrowinning cells must be able to cope with an electrolyte which is very caustic (pH above 13), which contains relatively low concentrations of gold (less than 50 g/t) for most of the elution period, and which is at a temperature of 80 to 90°C. It is desirable that the cell should have a single pass extraction as high as possible in order to ensure rapid elution of gold from the loaded carbon. Strip and electrowinning feed flowrates are typically in the order of 2 bed volumes per hour.

Stripping using the AARL method requires a separate electrowinning circuit, independent of the strip process. The eluate is circulated continuously through the electrowinning cell(s) until the cell effluent is acceptably low.

The concentration C_t of an electroactive species at any time, t, after the start of a multi-pass electrowinning operation can be calculated from the equation –

$$C_t = C_o \exp \frac{(-U.E.t)}{V} \qquad (1)$$

Where, Co is the initial concentration of species in the reservoir at the start of the operation, ppm
U is the eluate feed flowrate to the electrowinning cell, m^3/h
E is the single-pass extraction, percent
t is the electrowinning cycle operating time, hrs
V is the eluate volume to be processed, m^3.

In contrast to the Zadra elution procedure, which requires a high value of E to ensure rapid elution, the electrowinning cells employed for treatment of AARL eluate do not necessarily require a high single pass extraction provided that they can be operated at a high circulating flow rate; i.e. it is the product UE in equation (1) which will determine the overall rate of gold recovery. However, in practice, by balancing eluate flowrate and operating time, single pass cell extraction efficiencies in the range 90 to 97% range are readily achievable with standard cells.

Sizing of the electrowinning circuit requires consideration of a number of factors, which are discussed below. It is prudent for the design metallurgist to discuss in detail with the selected cell vendor, the duty requirements for each particular unit.

1. Cell hydraulic loading. Typical standard hydraulic loadings for optimal cell extraction efficiency are in the range of 0.1 to 0.35 m^3/min/m^2 (m/min), with a maximum of 0.4 m/min. The cells shown in Figure 4 have an approximate submerged cross sectional area of 0.7 m^2 and hence, based on the hydraulic loading discussed above, are suitable for flowrates of up to approximately 14 m^3/h per cell. If electrowinning at higher flowrates is required, it is necessary to utilize a number of cells operated in parallel.
2. Cell extraction efficiency. The size of cell required for a particular duty is estimated from the cathode retention time for the type of cell considered. For example, Fig 2 shows that for a basketless sludging standard cell with a retention time of 10 minutes, where the cell hydraulic loading is between 7 – 19 m/h (3-8 gpm/ft2), an extraction efficiency of about 97% can be attained. However, efficiency is influenced by site specific strip solution chemistry, such as concentration of competing metals and pH.
3. Metal electro-deposition rate. Rectifiers are initially sized on the theoretical current requirements for the mass of metals to be electrowon in a particular time period, and a current efficiency estimated from experience of similar or typical strip solution chemistry. Current densities ranging from 20 – 70 A/m^2 are typical.

4. Cell production capacity. Operating experience with SVEE standard cells has indicated that metal loadings in the range of 9000 to 18000 oz/yr/m^2 of cathode area can generally be attained. The number of cathodes per cell can be estimated initially from the required metal production capacity of the circuit and the current requirement for the system.

Operational Variables

Other important operating variables that significantly affect the performance of the electrowinning cells are:

- Eluate temperature – significantly higher extraction efficiencies are achieved at eluate temperatures in excess of 70°C, likely due to a combination of lower dissolved oxygen content, reduced solution viscosity and increased ionic mobility at the higher temperatures. Previous versions of cells operated at lower temperatures due to problems with warping of the polypropylene cathode baskets, whereas the present all metal cathode cells can operate at temperatures up to 90°C.
- Solution chemistry - generally, at the elevated pH ranges typical of AARL and Zadra elution processes, solution conductivity is not a problem. It is important to maintain eluate pH in the range of 12-13 to achieve adequate ionic mobility and electrolyte conductivity. However, some operations have experienced problems with silicate dissolution during elution at high pH's, which has lead to subsequent severely reduced extraction efficiency in the electrowinning circuit. The silicates originated as sand entrained in the loaded carbon fed to the strip column. This serves to emphasize the need to consider the elution and electrowinning circuits as an interrelated system.

A survey of the system and operating parameters for a number of gold mines worldwide are presented in Figure 5, to illustrate the range of operating practices in the Industry.

Practical Considerations

The electrowinning plant designer has also a number of specific practical issues to consider:

- Provision of adequate ventilation to minimize the fire and explosion hazard resulting from the generation of hydrogen gas in operating cells. Typically, a metal extraction hood encloses the top of the cell and is ducted to an extraction fan.
- Materials of construction. Initially, cells and hoods were fabricated from polypropylene or polyurethane, but the practice has ceased resulting from a number of accidents caused by fire and explosion in these cells. Presently, most standard cells are fabricated from stainless steel.
- Plant layout issues. Consideration has be given to:
    - Height above the cell in order to remove cathodes
    - A hoist above the cell in order to remove loaded cathodes
    - Sludge collection and handling equipment
    - The location of the rectifier, which should be as close as possible to minimize the size of the cables, yet far enough away from the cell for accessibility and safety due to accidental wetting.
    - Personnel safety from accidental exposure to hot eluate pipelines.

Survey of Current Gold Industry Electrowinning Practice

Strip Process		Zadra	AARL	AARL	Zadra	Zadra	Zadra	Zadra	Zadra
Location		Australia	Australia	S. America	N. America	N. America	N. America	N. America	N. America
Electrowinning									
Operating hours per day		12	20	24	Dep on strips	23	23	12	12
Number of operating cells		3	4	6	1	2	2	3	2
Operating cells in series (i.e., stages)		1	2	3	1	1	1	1	1
Cell size	m³	1.90	3.85	2.83	3.54	3.54	2.12	3.50	3.54
Design solution flowrate/cell	m³/hr	18.0	17.0	22.8	6.5	5.0	1.7	7.0	5.7
Actual solution flowrate/cell	m³/hr	10.8	17.0	23.0	5.5	5.7	2.3	4.5	5.7
Pregnant solution grade: Au	g/t	80.0	1000.0	80.0	250.0	85.7	85.7	720.0	154.3
Barren solution grade: Au	g/t	15.0	5.0	4.0	8.0	5.1	5.1	1.6	30.9
Percentage Au recovery	%	99.0	98.0	97.3	95.0	94.0	94.0	99.8	80.0
Cell solution temperature	°C	90	80	85	80	88	88	70	74
Type of cathode		Woven Steel Wool	Stainless Steel 16ply	Stainless Steel	Stainless Steel Mesh	316 ss Wool	316 ss Wool	Stainless Steel Wool	Stainless Steel
Type of anode		316 ss Perf Plate	Stainless Steel Mesh	Stainless Steel	Stainless Steel	316 ss Punch Plate	316 ss Punch Plate	Stainless Steel Mesh	Stainless Steel
Cathode sludge removal method		Manual drain once/wk	Water spray 3000 psi	Wash cathodes	High pressure Washer	High pressure Washer	High pressure Washer	Water Wash	Pump to press
Sludge gold grade	%	44		18	80	49	49	91	56
Operating voltage	V	3.0	7.5	12.0	2.8	4.0	4.0	3.2	3.1
Operating amperage draw	A	1350	3400	2000	1000	600	300	725	700
Power consumption	kWh/t	0.38	1.50	1.04	0.50	0.42	0.53	0.52	0.38
Residence time per cell	min	10.56	13.59	7.39	38.61	37.39	56.09	46.67	37.39
Total Residence for circuit	min	10.56	27.18	22.17	38.61	37.39	56.09	46.67	37.39
Filter Data									
Filter type (plate and frame, candle filter, etc.)			P&F		Filter Bags	No filter press	No filter press		
Filter press feed pump type			Diaphram						
Operating hours per day			5						
Installed filter area	m²		3.1						
Filter cake moisture	%		20.0						

Figure 5: Survey of Current Gold Industry Electrowinning Practice (*Ritson 1998*)

Survey of Current Gold Industry Electrowinning Practice									
Strip Process		Zadra	Zadra	Zadra	Zadra	Zadra	Zadra	Zadra	Zadra
Location		S. America	S. Africa	S. Africa	S. Africa	N. America	N. America	N. America	N. America
Electrowinning									
Operating hours per day		16	24	20	24	20	24		
Number of operating cells		8	3	24	18	4	3		2
Operating cells in series (i.e., stages)		1	3	1	1	2	1		1
Cell size	m³	0.40		0.40	0.40	3.54	3.54		2.12
Design solution flowrate/cell	m³/hr	10.0		10.0	10.0	40.9	15.9		6.8
Actual solution flowrate/cell	m³/hr	7.5	1.0	5.0	7.5	31.8	13.6		4.5
Pregnant solution grade: Au	g/t	250.0	60.0	250.0	250.0	102.2	171.4		308.6
Barren solution grade: Au	g/t	<5	1.1	1.0	<5	9.4	27.4		1.7
Percentage Au recovery	%	98.0	98.2	97.0	98.0	90.8	95.0		98.0
Cell solution temperature	°C	85	80	90	90	91	82		82
Type of cathode		Polyprop basket with mild steel	Polyprop basket with mild steel	Polyprop basket with mild steel	Polyprop basket with mild steel	Stainless Steel Wool	Punch Plate SS wool		Punch Plate SS wool
Type of anode		Stainless Steel	Stainless Steel	Stainless Steel	Stainless Steel	Stainless Steel Punch Plate	Punch plate & screen		Punch plate & screen
Cathode sludge removal method		High press wash	Manual	High press wash	High press wash	Pressure wash	High press wash		High press wash
Sludge gold grade	%		12	28		85	70-80		70-80
Operating voltage	V	5.5		13.0	13.5	3.1	4.0		4.0
Operating amperage draw	A	550	600	600	600	1150	1100		1100
Power consumption	kWh/t	0.40		1.56	1.08	0.11	0.32		0.97
Residence time per cell	min	3.20		4.80	3.20	6.68	15.58		28.05
Total Residence for circuit	min	3.20		4.80	3.20	13.36	15.58		28.05
Filter Data									
Filter type (plate and frame, candle filter, etc.)		P&F		P&F	P&F	P&F	P&F		P&F
Filter press feed pump type		Centr.		Centr.	Centr.	Diaphram			
Operating hours per day		3		3	8	1-2			
Installed filter area	m²	3.0		3.0		32.0	50.0		50.0
Length of air dry cycle (if applicable)	hrs	0.50		0.50		0.5			
Filter cake moisture	%	20.0		20.0	20.0	10.0	10-40		10-40

Figure 5 (Continued): Survey of Current Gold Industry Electrowinning Practice *(Ritson 1998)*

REFERENCES

1. Davidson, R.J. and N. Schoeman. June 1991 The Management of Carbon in a High-tonnage CIP Operation. *Journal SAIMM*,

2. ElectroMetals Technologies Ltd. November 2001 Private communication., Australia.

3. M. Proudfoot. March 2002 Private communication. Kemix Ltd., South Africa.

4. M.C. Costello et al. 1988 Carbon Absorption, Elution, and Electrowinning of Gold Ores with up to 4:1 Silver to Gold Ratios. *Perth Gold*.

5. Paul R.R., A.O. Filmer, and M.J. Nichols. 1982 The Recovery of Gold from Concentrated Aurocyanide Electrolytes. Hydometallurgy – Research, Development and Plant Practice. *Proc. 3rd International Symposium Hydromet,* Metal Voc. AIME

6. Richards, R.H. and J.A. Wells. September 1987 Contemporary Practices and Innovative Features of Gold Recovery Installations in Canada, *American Mining Congress*.

7. Ritson, G. May 1998 World Gold Survey, *Technical Report*. Hatch Associates. Canada.

8. Stanley, G.G (Edited by). *The Extractive Metallurgy of Gold in South Africa, Vol. I.*

9. Weldon, T. March 2002 Private communication. Summit Valley Equipment and Engineering Inc. (SVEE), USA.

SELECTION AND SIZING OF COPPER SOLVENT EXTRACTION AND ELECTROWINNING EQUIPMENT AND CIRCUITS

Corby G. Anderson[1], Mike A. Giralico[2], Thomas A. Post[3], Tim G. Robinson[4] and Owen S. Tinkler[5]

ABSTRACT

Since the late 1960's copper solvent extraction (SX) coupled with electrowinning (EW) has been a growing technical application for production of copper metal. The industrial scale copper solvent extraction process is a counter-current multi-stage contacting operation. This usually consists of an extraction and a stripping section involving the use of kerosene based solvents and sulfuric acid to increase the copper concentration and purity of a copper rich solution. Typically, copper recovery from the purified copper solutions is done by electrowinning. This paper will outline this technology and the fundamentals involved in the selection and sizing of copper solvent extraction and electrowinning circuits.

INTRODUCTION

Feed solutions to SX circuits typically arise from leaching processes such as heap leaching. The complete solvent extraction chain typically involves two extraction steps (E1 & E2). Copper is extracted from the pregnant leach stream (PLS) into the organic stream. Then one or two stripping steps (S1 & S2) follow where copper is stripped from the organic phase into a depleted copper solution bearing sulfuric acid from the electrowinning tankhouse. Copper is recovered by electrowinning, which is precipitation of a metal by electrolytic reduction on a cathode while using an inert insoluble anode.

THE MCCABE-THIELE DIAGRAM

The solvent extraction process, as applied to metals extraction, can be defined as: "The selective transfer and concentration of metal ions from an impure aqueous phase via an organic phase to a second pure aqueous phase from which the desired metal can be recovered in a usable form."

Figure 1 illustrates a typical solvent extraction and electrowinning circuit. Oxime based reagents typically form the basis of copper extraction and they are illustrated in Figure 2. The fundamental reversible chemical reaction for solvent extraction is noted as equation 1.

$$2RH + Cu^{++} + SO_4^{-2} <---> R_2Cu + 2H^+ + SO_4^{-2} \qquad (1)$$

To design a solvent extraction circuit and to predict its performance we require the use of McCabe-Thiele stage construction methodology. This allows us to make predictions on extraction and stripping performance under imperfect conditions.

In order to construct a McCabe-Thiele diagram, distribution curves or equilibrium isotherms must first be generated in the laboratory. For the extract isotherm (Fig 3), aliquots of feed liquor are mixed with a measured amount of organic reagent, such that a range of say 6-9 samples is

---

[1] Center for Advanced Mineral and Metallurgical Processing, Montana Tech of The University of Montana, Butte, Montana
[2] LIGHTNIN, Rochester, New York
[3] ConsultDrPost, Rochester, New York
[4] CTIANCOR, Scottsdale, Arizona
[5] Avecia, Phoenix, Arizona

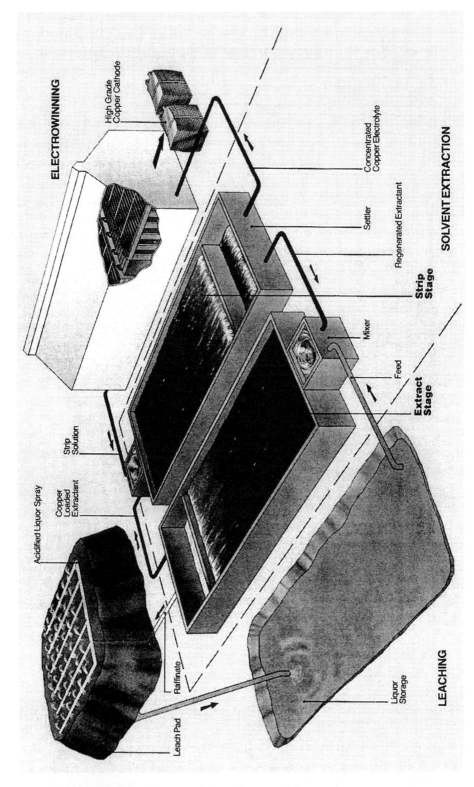

Figure 1. Typical industrial copper solvent extraction and electrowinning circuit.

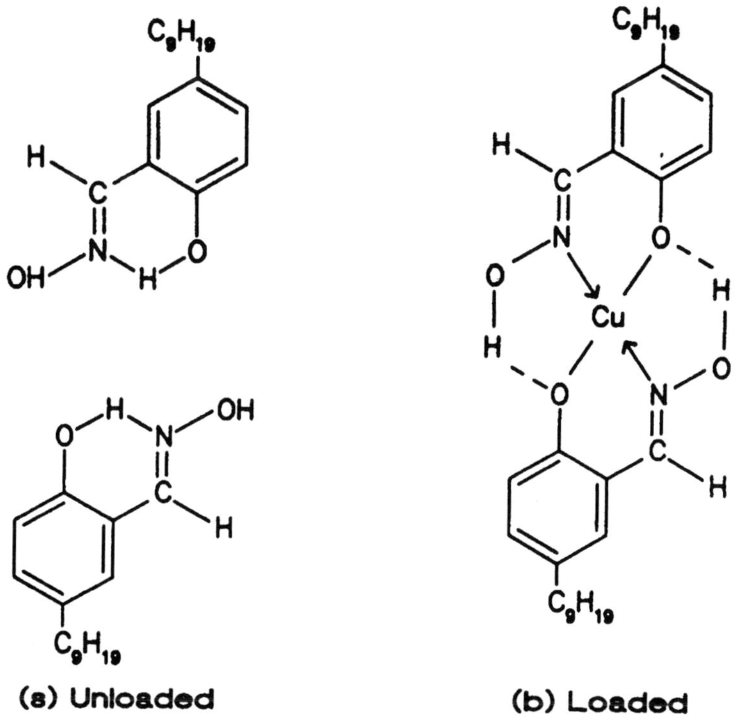

Figure 2. Nonyl salicylaldoxime molecules and copper complex. Two salicylaldoxime molecules complex with a Cu^{++} cation and release two H^+ ions (Biswas and Davenport, 1992).

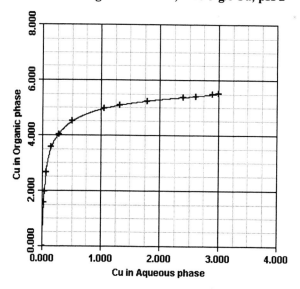

Figure 3. Typical extraction isotherm.

available representing various O/A ratios. The ratios typically cover a range from around 2:1 to 1:10. Mixing should be vigorous and for about15 minutes, to ensure that equilibrium is reached in each case before the phases are allowed to separate for analysis.

The exact reagent concentration required to match the concentration of copper in the feed solution will depend on factors such as solution pH, circuit configuration (i.e. the number of extraction and stripping stages) and electrolyte composition. For the purpose of a preliminary evaluation, however, the approximate reagent concentration required (expressed as percentage by volume) can be calculated simply by multiplying the concentration of copper in the feed (expressed in g/l) by three. Current commercial reagents load from 0.52 - 0.58 g/l copper per volume percent (v/o) reagent and are able to transfer around 60% of this available copper during the stripping process.
A strip isotherm (Fig. 4) is generated in a similar fashion by contacting loaded organic phase with various proportions of the strip liquor (electrolyte). In this case 4-5 points are adequate, typically in the range 5:1 to 1:1. The isotherms, of the types shown in Figures 3 and 4, form the basis of preliminary design for any solvent extraction plant. They are used in conjunction with each other to produce a McCabe Thiele construction, the metallurgical profile of an SX circuit. An example is shown in Figure 5.

10 vol% Acorga M5640 TM, Lean Electrolyte 35 g/l Cu 180 g/l Sulfuric Acid

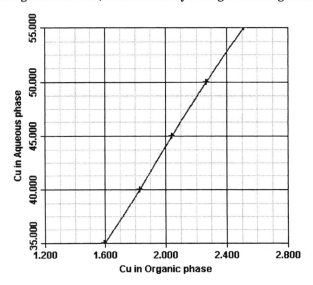

Figure 4. Typical strip isotherm.

The first principle to consider is that of the operating line. The slope of the operating line defines the organic to aqueous flow ratio required to achieve a desired recovery e.g. if the volumes of organic and aqueous were to be equal, then an increase in organic tenor of 1 gpl copper would be matched by a decrease in the aqueous tenor also of 1 gpl; if say we had twice the volume of organic then it would only need an increase of 0.5 gpl in the organic to effect a decrease of 1 gpl in the aqueous tenor. In the first case, the operating line has a slope of 1 and in the other it has a slope of 0.5. Therefore, it is obvious that by choosing the O/A ratio, the concentration in the organic layer will be varied for any aqueous feed tenor. The origin of the operating line is important to locate and this is defined by the stripped organic level and the raffinate level, which latter is obtained by trial and error (if done by hand), or iteration (simulation software will be discussed later).

In the example of Figure 5, an O/A ratio of 1:1 is illustrated so that the slope of the operating line is defined accordingly, passing through the stripped organic value of 2.34 gpl copper and a raffinate level of that required. In order to predict stage performance, a vertical line is drawn from the aqueous feed tenor to meet the operating line, and then horizontally to meet the equilibrium isotherm. This point on the isotherm represents the gpl copper in the aqueous and organic phases leaving the first stage of extraction.

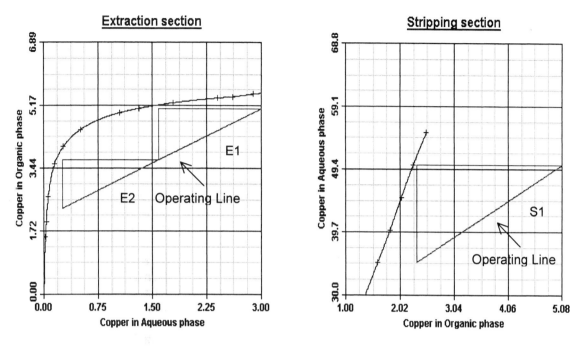

Figure 5. McCabe-Thiele diagrams for extraction and stripping.

In practice, as stated, the reaction will not reach equilibrium exactly and say 90-98 percent efficiency can be assumed for both extract and strip. In Figure 5, a stage efficiency of 95% was used in the extract section and 95% in the strip section. Graphically this is represented by the diagonal of the rectangle completed about the first stage, relative to the extended diagonal touching the isotherm. When done manually, this naturally is also a matter of trial and error. The example shows that the first (E1) stage of extraction would result in a loaded organic of 5.08 gpl copper and an intermediate aqueous raffinate of 1.59 gpl copper. Continuing the stepping-off process to a second extract stage (E2), again allowing for a stage efficiency correction, it can be seen that the aqueous copper tenor drops from 3 gpl in to 0.26 gpl, while the loaded organic rises from 2.34 gpl to 5.08 gpl. Naturally, if this is done by hand, it will be unlikely that the chosen raffinate level will be achieved in an integer number of stages and therefore a trial and error method must be used, changing the raffinate and hence the origin of the operating line. Alternatively, a different O/A ratio could be chosen to change the slope of the operating line. Figure 6 illustrates a typical industrial solvent extraction 2x1 circuit derived from the McCabe Thiele methodology.

Clearly this stepping-off procedure, allowance for stage efficiency, and iteration regarding raffinate levels and O/A ratios, can be very tedious to other than a skilled technologist and with this in mind, computer modeling programs have been developed by engineers and suppliers such as Avecia. The benefits and descriptions of a computer model will be described in the following section.

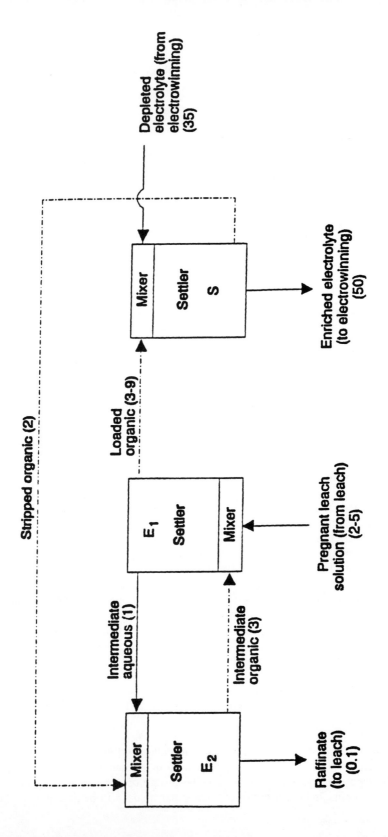

Figure 6. Industrial solvent extraction 2x1 circuit with typical copper tenors noted in brackets (Biswas and Davenport, 1992).

Avecia Computer Modeling Software

The software can operate in design mode or performance mode. In either case the extraction and strip isotherms are needed as input and then the computer will either decide the O/A ratio, number of stages, etc., to achieve a chosen raffinate, or alternatively will estimate the best achievable raffinate for a given set of plant parameters. The End User Module (MEUMTM) enables users to manipulate data as rapidly as binary file data exchange, via the Internet. The software package includes expert system technology and database and plant control capability. Features include:

1. Isotherm construction from laboratory-generated equilibrium data using the curve-fitting facility integral to the software package. Isotherm prediction.
2. Prediction of copper recovery and generation of corresponding mass balance data for a range of plant configurations.
3. The ability to vary the number of extraction and strip stages employed and arrange them in conventional or "parallel/series" mode as required by the design engineer.
4. Automatic recalculation of isotherm data as the reagent concentration is varied.
5. Automatic analysis of plant performance data and calculation of operating stage efficiencies and O/A ratios from metallurgical data supplied by the user.
6. Display of output data in the form of mass balances, circuit mimic diagrams or constructed McCabe-Thiele plots.
7. Local and "Master" database storage and retrieval of isotherms i.e.: flowsheets.
8. Spreadsheet links to allow plant design data to be calculated directly from flowsheet results.

SX Testing

The metallurgical engineer who has been commissioned to develop a metal recovery process flowsheet using SX technology will typically complete a laboratory study to assess the feasibility of the envisaged treatment route as the first step in the process. This study will identify possible problem areas. The scale on which pilot plant testing should be carried out is dependent on many factors. Some of these are:

1. The quantity of aqueous feed solution available.
2. The data required including whether this data will be required for SX plant design purposes.
3. Funding available.
4. Waste disposal facilities available.

It should be noted for a pilot plant trial in which the SX operation is required to be integrated with the leach and EW operations that the scale of these latter operations will determine the size requirements of the SX pilot plant.

For laboratory scale pilot plant testing, suitable equipment is readily available from a variety of sources. One common design is the so-called "Bell rig" which consists of mixer-settler equipment fabricated in clear UPVC or Polycarbonate to allow inspection of dispersion bands and solids activity at the O/W interface. "Bell" units typically have small square mixer boxes of 125 - 150 ml capacity and a settler design which largely resembles that of commercial plants. However, the dispersion distribution system tends to be fairly basic compared to that on a commercial plant. It has been our experience that this scale of equipment is best used along with good quality metering pumps to better control flow rates and prevent "phase flipping" in the mixers. The purpose of carrying out tests on this scale is to increase confidence generally in SX process technology and more specifically in the effectiveness of the flowsheet under development following the original bench tests. The power input to the mixers on this type of equipment is far in excess of that required. For this reason, testwork on this scale will not provide reliable information for design purposes on either the flow rate per unit of settler area (Specific Flow), the entrainment characteristics or the stage efficiency of the system under test. Useful data, however, may be obtained on performance factors such as metal recovery, selectivity or basic problems e.g.: phase separation under dynamic operating conditions

For the production of more accurate data for preliminary plant design purposes a pilot plant capable of handling 10-40 liters per minute of combined aqueous and organic flow is considered advisable. Industry experience has shown that pilot plant scale test work can be used effectively to evaluate reagent performance, confirm the suitability of the envisaged flowsheet and define more closely the physical operating characteristics to be expected.

The features of this size of pilot plant are:

1. It allows collection of operating information on a scale which can be used for plant design.
2. The settler walls can be fabricated from transparent plastic to allow operators to assess visually what is happening in the dispersion band and how solids are being controlled. The structure of the plant is sufficiently strong to withstand all but the very worst weather conditions.
3. Standard pumps, valves and fittings are readily available in sizes suitable for this type of operation.
4. It can be integrated easily into a commercial plant circuit to carry out investigational work on problems arising in existing operations.

Where long term effects are being studied, it is preferable that this type of pilot plant operates continuously, although most of the useful data is likely to be gathered during daylight hours.

An important part of the information to be gathered from this scale of operation is entrainment data. Although not strictly representative of the values likely to be obtained on the full-scale plant entrainment values determined will indicate whether the system under evaluation is likely to have major problems in this area. Certainly any improvements observed on this type of pilot plant should translate to the full scale plant.

Samples of exiting phases for entrainment measurement should be collected in equipment specially modified for this purpose. These are designed to fit in a laboratory centrifuge and the amount of entrained phase measured directly on a marked scale calibrated to give results expressed in ppm. The quality of the information which can be obtained on this scale with regard to % metal recovery, selectivity, band depths, entrainment and solids distribution against variables such as staging, phase continuity, phase recycle, temperature, specific settler flow (settler sizing), mixer retention time (mixer sizing) and impeller speed (power input) has been proved to be very reliable. Care should be taken, however, in scaling up from some of this data as solution linear velocities, for instance, on this type of pilot plant tend to be higher than on commercial plants due to differences in settler dimensions.

SOLVENT EXTRACTION MIXERS AND SETTLERS

While the concept of solvent extraction has existed for centuries, it is the engineering design and application which brought it to prominence in the copper industry. The backbone of this development has been the successful full scale practice of phase mixing and phase disengagement. As illustrated in Figure 7 this is generally accomplished in so called mixer-settlers. The design and operational concepts behind these unit operations is discussed in the following sections of the paper.

SOLVENT EXTRACTION MIXING CONCEPTS

Since the late 1960s, the Holmes&Narver pumper design has been the standard, serving the SX-market faithfully. Then the Davy BB, with its characteristic draft tube, made some inroads in the late 1970s. For nearly 3 decades nothing changed until the requirement came for reducing organic entrainment and increasing throughputs. In 1986 LIGHTNIN introduced the A310 hydrofoil as an efficient mixer in the auxiliary mix boxes. Then they came up with a new curved bladed pumper called the R320 in 1995. Since then numerous modifications have been made to reduce entrainment, decrease power and increase pumping capacities, while maintaining the head requirement of the system. These new designs greatly

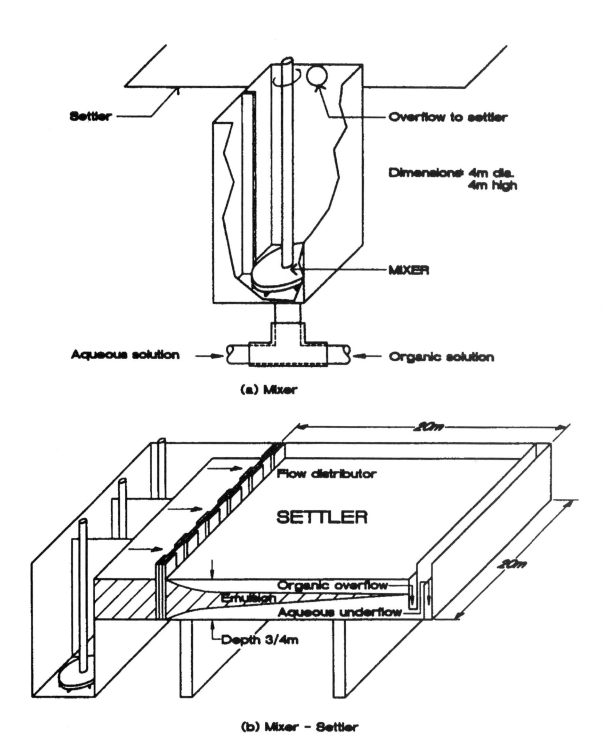

Figure 7. Conventional solvent extraction mixer and mixer – settler (Biswas and Davenport, 1992).

improve the hydraulic efficiency of these open pumpers. Most of the mix boxes now consist of both pumper and hydrofoil combinations whereas the hydrofoils in the auxiliaries are now up-pumping. The mix boxes aren't even straight sided anymore. The best designs now call for cylindrical tanks. Figure 8 illustrates a modern mixing system design.

These optimized impeller designs reduce aqueous and organic entrainment losses significantly, reduce or eliminate air entrainment and crud formation, enhance mass transfer and maximize the overall copper yield. When compared to the earlier pumper designs, these new designs require less installed power and better process results. The auxiliary hydrofoil impellers further assist the curved bladed pumpers by maintaining a good dispersion and mass transfer, without increasing the head requirement. The designer should select impellers that require minimal installed power while maintaining good dispersion and mass transfer to minimize entrainment and maximize copper production. Both the pumper and auxiliary impellers can be fabricated in any material including Derakane 470 Vinyl Ester FRP Composite. Currently, these new impeller designs are operating worldwide in over 60 installations with flow rates from 50 to 22,000 GPM and developed heads up to 50 inches.

Process Requirements of SX Pumpers and Auxiliary Mixers
The pumpers are required to provide the desired head and flow through the plant, while creating the initial liquid-liquid dispersion. The auxiliary mixers are intended to maintain this dispersion and extend the stage residence time to improve mass transfer stage efficiency. Once plant design conditions are met, fine-tuning can make a conventional plant run much more efficiently, in terms of operating cost and lower entrainment losses.

Pump-Mixer Requirements
SX-pumpers must develop a head high enough to achieve the desired flow rate through the SX plant. If the pumpers fail to provide this key requirement, all other features of the SX-plant are meaningless. This flow rate must also be variable, due to the changing concentrations of dissolved copper in the PLS. The best way to vary the flow rate is with variable speed frequency drive controllers, not by adjusting pipe valves, which increase turbulence. It is also desirable that provisions be made during the design phase that would allow future increases in the flow rates, since most mines will investigate increased production rates.

The pumpers must also create a liquid-liquid dispersion. Ideally, the droplet distribution is uniform; creating droplets small enough to promote mass transfer but large enough to inhibit entrainment losses. Not only must the pumpers create the dispersion, they must maintain a stable dispersion in the primary stage. Too much shear and turbulence creates a wide distribution of droplet sizes. A large percentage are fines; so small that they end up in the other phase and as entrainment. Too little shear creates unstable large droplets that tend to separate quickly into phases. Phase separation in the mixing stages causes less than optimum copper recovery and limits production capacity.

The droplets are created by the bursting mechanism. This means that large droplets or agglomerates of one phase burst into many smaller spherical droplets in localized areas behind the blades where the pressure drop is the greatest. Elongated (stretched) droplets do not exist in copper SX. Excess power dissipation, as described by power per unit volume or P/V, is responsible for this bursting mechanism. Depending on the local P/V, this bursting of droplets can cascade downward to create very fine and sometimes unrecoverable droplets as entrainment losses. A minimum P/V is needed to maintain a stable dispersion. The local P/V in the shedding vortices is typically an order of magnitude higher than the overall P/V and this inhomogeneity creates the droplet distribution. Inefficient pumpers create very high local P/V and high overall P/V, which causes a wide distribution in droplet sizes, fines, haze, and entrainment losses.

Ideally, the pumper should not operate so violently as to cause air entrainment through vortices and rough air-liquid surface motion. Air that reaches the pumper blades will also disperse into fine bubbles. They attach themselves to aqueous droplets and ruin the function of the settlers and increase entrainment

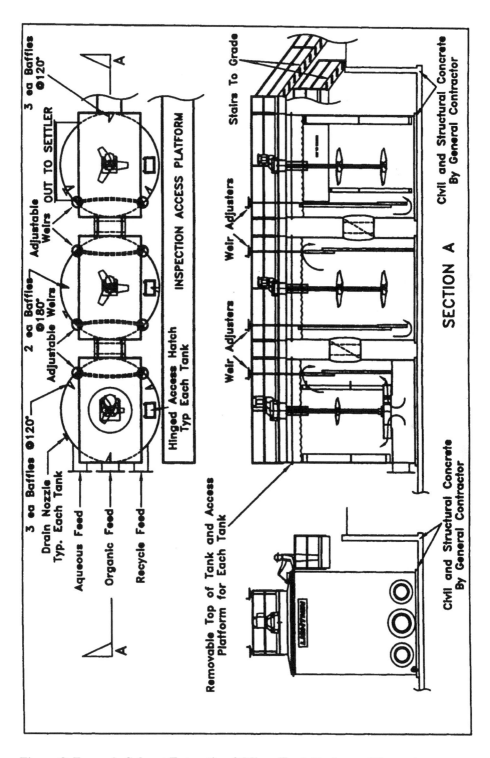

Figure 8. Example Solvent Extraction 3 Mixer Tank Design and Layout.

losses. If silicates are present, air-liquid-solid mixtures create crud, which is another leading cause of entrainment loss. Foam has also created problems with organic carryover to the auxiliary stages. Air entrainment should be avoided at all costs. Corners in mix boxes are notorious for initiating vortexes. This is one reason to move away from mix boxes and toward stages made of cylindrical tanks. Also, baffles protruding through the liquid surface form vortices behind the baffles. Careful design of the baffles is also important.

Auxiliary Mixer Requirements
Auxiliary mixers need to maintain the dispersion created in the primary stage. The auxiliary box provides more residence time and increases the stage efficiency based on the isotherm of the extractant. Proper placement and operation of the auxiliary impeller can reduce the head requirements for the pumper. These mixers create a mixing flow superimposed on the flow through the system created by the pumper. This additional flow creates a better contacting pattern for the two phases, so that maximum concentration gradients can be achieved. This enhances mass transfer and stage efficiency. To achieve a well-mixed dispersion, a CSTR (continuous stirred tank reactor) model can be achieved if the flow generated by the auxiliary mixer is at least three times the flow created by the pumper. To achieve the best results, the auxiliary mixer should also be equipped with a variable speed device. As the flow rate is varied in the pumper, the developed flow of the auxiliary should be changed, too.

A minimum overall P/V must also be obtained in the auxiliary mix stage to avoid phase separation. Radial turbines require similar power levels as the radial pumpers, because they are designed to transform the power into turbulence and shear and not into flow. To get sufficient flow in all areas of an auxiliary mix box, the P/V of radial turbines must be relatively high. This causes further decreases in the droplet size population and creates emulsions or phase inversions. Axial turbines concentrate their power into developing flow and not turbulence and shear. Thus, axial turbines require less power to activate all areas of the mix box. Properly designed hydrofoil impellers are similar to airplane wings, which reduce the drag (turbulence) and create more lift (flow). Installed in an up-pumping mode in an SX-plant, they can reduce the minimum overall P/V to an order of magnitude less than the pumpers and still achieve 3-13 times the flow rate. Variable speed devices should be installed so the P/V can be adjusted, to allow the droplets to coalesce into larger stable sizes. This is ideal, because it preconditions the dispersion before entering the settler.

Another reason to install variable speed devices on the auxiliaries is to account for changes in temperature and the so-called viscosity effect of these dispersions at cold temperatures. As the temperature gets colder, the viscosity-effect increases the separation times. Thus, the likelihood of phase separation in the mix stages decreases and less P/V is required to maintain that dispersion. The same is valid when discussing the difference between organic continuous and aqueous continuous operations. Typically, a conventional plant is designed for the worst case, i.e., organic continuous operation, when determining settler dimensions and aqueous continuous operation, which is overkill on power demands for the mixers most of the time. With variable speed, the entire unit can be fine-tuned. Like the pumpers, excessive power should be avoided in order to reduce air induction through vortices or surface splashing. The reasons are the same as for the pumpers. Surface splashing can be minimized by careful design of the tanks internals, tank liquid depth to diameter ratio, and proper mixer to tank diameter ratios.

Advantages of Conventional SX Mixer-Settler Designs
Conventional SX mixer-settler designs have been around since the late 1960s. As commercial trends push towards larger plants and higher production rates, these designs have shown definite weaknesses in the generation of excessive shear and air entrainment at the higher power levels required to meet production targets. This has left some to believe that the only way to improve performance is to design completely new technologies. Now that the requirements for the mixers have been clearly stated, reassessing the value of the conventional design is important.

The greatest advantage of the conventional design is in its simplicity. Most of the piping is easily accessible as well as the mixing stages and settlers. If a pumper needs maintenance, a complete pumper-shaft-gearbox unit can be exchanged with a replacement in less than 20 minutes. The down time is reduced. Splash walls, covers and tank lids do not need to be removed first. An auxiliary mixer can be replaced without any down time at all, if required.

A change in performance can be easily detected. Where sight glasses in the newer technologies give a very limited view (the dispersion is essentially black) the mixing stages and settlers of the conventional units are open and extend the view over the entire surface. Changes can be detected visually. Detecting surface organic pockets is very easy, indicating the onset of phase disengagement. The effect of changing the mixer speeds of either the pumpers or auxiliaries on the dispersion and air entrainment can be easily monitored. Thus, if a cooler weather period approaches, the impeller speed of the auxiliary mixers can be lowered to deliver the minimum power required to maintain the dispersion. If the pumper speed is lowered to decrease the overall flow rate, the speed of the auxiliary mixers can be increased, since the increased residence times require more power to maintain the dispersion. The open architecture allows immediate visual response of the changes made.

Similarly, the open architecture allows easy access to the fluids being mixed. Samples can be easily taken to detect such things as O:A ratios, fluid continuum, and separation or split times. The open architecture is a great asset for mixers designed to avoid air induction. As flow rates have increased the basic designs of pumpers and auxiliary mixers have required higher tip speeds and excessive power requirements. The additional P/V will cause the fluids to look violent. As a result, the surface begins to splash, and vortices appear behind baffles. Placement of lids on the surface diminishes air induction, but the mixers are still imparting excessive power to the volume. The lids have only solved part of the problem, and they have made access to the dispersion and the mixers more difficult. The high P/V is still creating a wide variation in droplet sizes, including those very small droplets (haze or fines) responsible for entrainment losses.

Finally, simple designs are more economic by definition. The plant design in well known, trusted, and has a much lower capital cost due to the simplicity. Using the high efficiency impellers in a standard design produces the same result as if an alternate design is chosen. As such, the conventional design is an overall good investment.

Optimizing the Design of Pumpers and Auxiliary Mixers

Conventional pumpers are still in use today. They are called the General Mills or the Holmes & Narver pumper. Attached to the underside of a disk, are six straight blades that extend from the edge of the disk to the center. The inner edge is tapered toward the disk. The height of the blade is always 1/8 of the impeller diameter. The corresponding orifice is without exception between 0.33 and 0.37 times the impeller diameter.

In the late 1970s Davy introduced the Davy BB. This impeller is still in use today and has found wide acceptance. It is placed on top of a draft tube, so that it can be placed in the natural media of which is the desired continuum. It has an upper and lower disk with eight curved blades between them. The lower disk is required to achieve head. The height of the blades is always 1/7 of the impeller diameter. The draft tube opening is always ½ of the impeller diameter.

Since the relative dimensions of these pumpers are fixed, increasing the speed (N), increases flow (Q), tip speed (TS=πND), power (P) and power/volume (P/V) proportionately. A scale-up criterion was used for these pumpers that was N^3D^2. It can be shown that, for a given impeller type, N^3D^2 is proportional to P/V. The critical value of N^3D^2 is dependent on pumper type, power number and tank dimensions. Tip speed limits were imposed on pumpers based on operating experience without understanding the underlying mechanism of droplet breakup. Today, we know that tip speed is only part of the whole design criterion and P/V plays a much bigger role in pump-mixer selection than previously thought.

Characterizing pumper design

The primary requirement of a pumper is to develop enough head to achieve the desired flow rate of an SX plant. Each pumper has a set of characteristic head-flow and power-flow curves (Giralico Post, and Preston, 1995). Since a 50:50 mixture of PLS and Organic has a mixed SG of approximately 1.0, these curves have been developed using only water. Using Bernoulli's law, the hydraulic power (P_{Hyd}) required to pump a fluid of average density (ρ) at a given flow rate (Q) over a given head (H) is:

$$P_{Hyd} = \rho g H Q \qquad (2)$$

This is the minimum power needed to generate just the head and flow, without any turbulence or shear. Using the power-flow curve, the power (P) needed by the pumper can be calculated. On existing plants, it can be measured. The hydraulic efficiency (ε) of a pumper system can be determined by:

$$\varepsilon = \frac{P_{Hyd}}{P} \qquad (3)$$

The excess power (P_e) is defined as the power above P_{hyd}.

$$P_e = P - P_{Hyd} \qquad (4)$$

P_e is the power responsible for the droplet breakup mechanism, along with the shear gradients emitting from the blade tips. This power results in turbulence. Most of this power is emitted in close proximity of the pumper's blades, so that once the droplets are formed they usually do not decrease any further in size unless they are recirculated back to within this high shearing zone.

Obviously, pumpers with high values of hydraulic efficiencies are directing their power more toward developing head and flow instead of creating very small droplets. LIGHTNIN's research has shown hydraulic efficiencies achievable up to 67% with a normal operating range of 35-45% with the new R320 family of impellers. Among the conventional pumpers, the Holmes&Narver and other straight bladed turbines are typically in the 15-20% range, while the Davy BB lies between 20-30%. Generally, entrainment losses are indirectly proportional to hydraulic efficiency. Common among all types of pumpers is that hydraulic efficiency decreases as scale decreases. In pilot plants, the range of hydraulic efficiencies is very small, although they compare in the same relative order. Straight bladed pumpers may have efficiencies as low as 5% and the best practical curved bladed pumpers may obtain a maximum of 20%. Much of this is due to poor box geometry design and higher residence times than on full scale. Another reason is that pilot plants do not need to create as much head as their full-scale counterparts. In large enough pilot plants, this can be defeated with novel mixer designs as discussed above.

On the other hand, as scale increases, the range of hydraulic efficiencies increases dramatically as stated above. A mere 6% increase in hydraulic efficiency decreased overall organic-in-aqueous and aqueous-in-organic entrainment by one half over a three-month trial period[1]. Comparison was made at a full-scale copper SX-plant operating two different impeller designs in parallel.

P/V is not the answer to everything. Shear gradients are also important. Determination of shear gradients ($\gamma = \partial v/\partial r$) requires the knowledge of local velocity (v) profiles in the radial (r) direction. A laser operating in a back scattering mode (Giralico, Post, and Preston, 1995). conveniently determines this. Two types of impeller shear gradients are important for scale-up (Weetman and Oldshue 1988) the average shear gradient γ_{avg} and the maximum shear gradient γ_{max}. γ_{max} is responsible for the smallest droplets. If a mixer is designed with a uniform exit velocity profile, the droplet distribution will also be uniform. For scale-up, the following rules are typical for radial turbines (Oldshue 1984).

$$\gamma_{max} = \xi_1 \cdot N \cdot D \quad and \quad \gamma_{avg} = \xi_2 \cdot N \tag{5}$$

ξ_1 and ξ_2 are constants, which are impeller dependent. Equation 5 shows that γ_{max} is proportional to tip speed. ξ_1 has a lower value for long curved bladed impellers than straight bladed impellers[1]. Upon scale-up, the average and maximum shear gradients obviously diverge. Whereas tip speed is often constant, impeller speed always decreases with scale-up. Thus, the drop size distribution is always wider for larger scale equipment unless both gradients are similar, or if P/V is lower in the larger scale.

These concepts show that scale-up is not simple, and that entrainment losses determined in pilot plants do not necessarily predict large-scale entrainment losses. Taking ε, P/V, γ and residence times into account is important. Finding higher entrainment losses upon scale-up is not unusual. A conventional system that develops the desired head and flow with curved bladed pumpers having the highest possible hydraulic efficiency and still maintaining phase stability (minimum P/V) without air induction (maximum P/V) at tip speeds below 1000 ft/min results in the best possible SX-system. Fulfilling all these requirements is possible, provided the mix stage geometries are optimized with the pumper selection.

Optimized Pumper Designs
No single pumper design is optimized for all operating conditions. Some generalities can be made, though, to help select the best pumper for each SX plant. To increase the flow rates of conventional pumpers, only two variables can be changed, either speed or pumper diameter. This is because the geometrical settings of the pumpers are fixed.

$$Q = Nq \cdot N \cdot D^3 \tag{6}$$

Upon scale-up N decreases. Therefore higher flow rates are only possible by increasing D (since Nq (the dimensionless flow number) is fairly constant for geometrically similar impellers). When the impeller to tank diameter ratio, D/T, > 0.7 the radial flow becomes throttled resulting in higher than expected power draws and lower hydraulic efficiencies. At the same time, the power increases dramatically, by the fifth power of D.

$$P = Np \cdot \rho \cdot N^3 \cdot D^5 \tag{7}$$

Since Np, the dimensionless power number, is also relatively constant for a geometrically similar pumper, increasing D to increase flow rates also increases the power.

Other methods are available to increase flow. For example, within certain limits, increasing the blade width can increase the flow rate. This increases both Nq and Np, but because the diameter can be decreased, tip speed and P (and P/V) can be decreased. This invariably increases hydraulic efficiency. Curving the blades does not necessarily increase the developed head over a straight bladed pumper, but Np can be halved. The net effect is higher hydraulic efficiencies at the desired head and flow. The blade length of curved bladed impellers affects the developed head and flow, but has relatively little effect on power. Obviously the curvature of the blades is crucial. Constant radius pumpers are easier to build and can operate over a wider range of flow rates. Pumpers with variable radii are more complicated to build but can achieve the very high hydraulic efficiencies in a narrow range of flow rates.

A big factor in improving the process can be achieved by modifying the other very important part of the pump, the orifice. The conventional straight bladed pumper design is typically accompanied by an orifice to pumper diameter ratio (D0/D) of 0.35. By increasing this ratio, more flow and head can be generated without increasing the power draw (provided the pumper is not throttled by design). This happens up to a ratio of approx. 0.5. Further flow increases require larger D0/D's and higher power but, by increasing the orifice diameter, the systems head-flow performance increases and hydraulic efficiency stays in the 36-45% range.

Dimensionless Numbers

Dimensionless numbers are essential for the purpose of scale-up. The two more common dimensionless numbers have already been described. Rearranging equations 6 and 7 give:

$$Nq = \frac{Q}{N \cdot D^3} \quad \text{and} \quad Np = \frac{P}{\rho \cdot N^3 \cdot D^5} \qquad (6')\,(7')$$

During small-scale experiments, where D is constant, the major influence on fluid dynamics is the impeller speed, N. By changing N and measuring the flow rate Q, generated by the system, N_q is determined. If the small-scale unit is attached to a torque cell, the measured torque and the speed yield the impeller power. From this, N_P is determined. Plotting $N_P = f(N_q)$ is convenient (Giralico, Post, and Preston, 1995). This results in a curve characteristic to the pumper design. If the pumper and its corresponding orifice are built geometrically similar, the small-scale curve can be used to design the full-scale system with equations 6' and 7'.

Equation 2 shows how the hydraulic power is related to the head, but does not show how to determine it. LIGHTNIN developed a dimensionless head number for this purpose. It is basically the hydraulic power divided by the tip speed squared.

$$Nh = \frac{2 \cdot g \cdot H}{TS^2} = \frac{2 \cdot g \cdot H}{\pi^2 \cdot N^2 \cdot D^2} \qquad (8)$$

During the small scale testing, the liquid will rise directly with the impeller speed. The difference in liquid level from N=0 to N is the head, H. Plotting $Nh = f(N_q)$ is also characteristic of the pumper design.

Problems with Over-Efficient Pumpers

If a pumper is designed merely based on optimizing hydraulic efficiency, the SX-plant may run into problems. This is particularly the case when retrofitting an existing unit, because all pump stage dimensions are given, and very difficult to change. Therefore, at a given flow rate, the residence time in a pump stage is also given. The minimum power required to maintain a stable dispersion is dependent on pumper type, location, organic solution properties, tank dimensions and mostly on residence time. Generally, as residence time increases, more power is required to hinder coalescence. Once this minimum power level is determined, the hydraulic efficiency cannot go above a maximum level. Otherwise, the dispersion will already begin to separate in the primary stage, leading to reduced mass transfer. As mentioned earlier, efficiencies of $\varepsilon = 67\%$ are achievable. However, this is often too efficient and can result in phase separation. To maximize hydraulic efficiency, the residence times in the pump stages must be lower than one minute, sometimes as low as 20 seconds. The rest of the residence time must be carried out in the auxiliary mix stages. These design considerations are being carried out on new installations where the SX extractant kinetics is taken full advantage. For existing units, the hydraulic efficiency may need to be artificially reduced so that enough power is available for phase stability. This can be done in several ways; i.e., higher off-bottom clearances, narrower blades, or the addition of radial vanes to the top disk of the impeller.

Pumper Design for less than Minimum P/V Operations

If the existing unit has a high liquid depth to tank diameter ratio (Z/T>0.7) the best retrofit can actually exceed the maximum hydraulic efficiency required to maintain a stable dispersion. Tall tanks can be equipped with one or more A510 auxiliary up-pumping mixers in the upper portion of the primary stage. This patented feature (Post, Howk and Preston 1996) has been installed successfully in several full-scale installations, which had suffered from acute phase separation problems. It can maintain the dispersion because the A510 auxiliary mixer requires about 1/10 of the power of a typical pumper. It can keep the surface active, without affecting the head and flow characteristics of the pumper. Thus, the pumper is designed just to deliver the desired head and flow at the lowest possible power consumption. The upper, built-in, auxiliary impeller maintains the dispersion in the otherwise quiet zones away from the pumper's active zone. Retrofits have been so successful, that current design practice is to make the primary stage taller.

Table 1: Some installations with the dual impeller concept.

Customer	Country	Location	Application	Pumper
Sociedad Minera Cerro Verde	PERU	Arequipa	Copper	R321/A310
Mexicana del Cobre	MEXICO	ESDE II	Copper	R320/A310
Soquimich	CHILE	Antofagasta	Iodine	R320/A310
Quiborax	CHILE	Santiago	Borax	R320/A310
IMC	USA	California	Boric Acid	R320/A6000
Phelps Dodge	USA	Morensi	Copper	R320/A310
CS Metals	USA	Louisanna	Vanadium	R320/A310
Cal Energy	USA	Salton Sea	Zinc	R320D/A6000

Optimizing Auxiliary Mixer Design

The optimized auxiliary mixer must achieve a high internal flow rate with the least amount of power. 100% efficient impellers would be ideal, but don't exist. Excess P_e is not required, because the droplet distribution has already been formed in the pump stage. Hydrofoiled impellers are the most efficient flow generating impellers available today. Composite impellers like the A6000 are the most efficient hydrofoils since they have true hydrofoil profiles including variable blade cross-section thickness and optimum twist and camber. They can generate the same flow rate as its metal counterpart, the A510, at about half the power. The composites are also ideal for solutions high in chlorides.

All other types of auxiliary mixers require too much power to achieve high internal flow rates. In fact some radial devices can cause an otherwise aqueous continuous operation to flip over to organic continuous, even at O/A ratios less than one. Too much power will also further decrease droplet size and increase the settling time. This is not necessary, since more than 90% of the equilibrium was achieved in the primary stage.

Because of the low power consumption of hydrofoils, tip vortices are greatly reduced, and instead of further decreasing the droplet size, they can actually achieve some coalescence of the smallest droplets. Pointed in the upward direction, the internal flow generated by the A510 or A6000 goes upward near the shaft and down at the walls. This flow pattern supports the flow coming from the pump stage through the underflow weir or downcomer, reducing the head requirements of the pumper.

The hydrofoil cannot be too small, because the power of the thrust is then focused at the center of the surface. The resulting flow pattern would look like an upward surging spring, which could induce unwanted air. The concentrated local power means that other regions may be underpowered with resulting organic puddles. The hydrofoil cannot be too large either. If the impeller diameter is too large, there will be no room for the downward flow near the walls. The spreading of the upward flow collides with the downward recirculation resulting in a thrashing and violent surface motion with unwanted air induction. This collision of the flows causes an increase in power demand and thus unnecessary turbulence. This chaotic flow pattern is also the result of any sized pitched bladed turbine, because the discharge angle of the flow is not axial like a hydrofoil, but off to an angle. It is not possible to get a top-to-bottom flow pattern with a pitched bladed impeller.

Auxiliary mixer units should have variable speed drives. The added expense is well worth it. Since conditions vary from summer to winter and high grade to low grade PLS, a variable speed mixer is also intuitive. In addition, the simplicity of the conventional mixer settler design makes this addition really worthwhile. Once installed, the optimized operating philosophy should be to adjust the speed on the auxiliary mixers to provide just enough power so that no organic pockets are visible on the surface. This is usually the case when the upward flow in the center of the auxiliary stage is visible but not overpowering. This adjustability allows for the best control of the SX plant and will contribute significantly in reducing not only entrainment losses, but may make it possible to decrease the size of the settlers further. It also eliminates air induction and reduces crud formation.

Entrainment
With proper consideration to an efficient pumper and auxiliary mixer design, numerous large-scale operators in the copper industry have achieved entrainment levels in the 6 to 20 ppm level. Examples are: Codelco Salvador, Cerro Colorado, Mante Verde, Mantos Blancos Santa Barbara, Aberfoyle/Gunpowder, Girilambone Copper Co, Pasminco Metals/BHAS, El Abra, Compania Minera Zaldivar, Mexicana del Cobre/Nacozari, Southern Peru Copper/Toquepala, BHP Copper/San Manuel, Phelps Dodge/Tyrone, Asarco Silver Bell, El Tesoro, Andacollo, and CS Metals.

Current Trends in Copper SX-Mixing
Three obvious paths are followed to optimize SX-EW performance with conventional systems (Giralico Post and Preston, 1996). One is retrofitting existing plants to the most optimized mixing solutions outlined here. Usually, opening the orifice will be necessary to achieve optimum performance. In any case, auxiliary mixers should be outfitted with hydrofoils like the A510 or A6000 impellers and variable speed drives. The other two paths are reserved for new installations. One is to reduce the size of the primary stage to achieve lower residence times, lower absolute power and higher hydraulic efficiencies. The remainder of the residence time would be mixed in the auxiliary stage or stages. Since the P/V in the auxiliaries is an order of magnitude less than the pumpers, the overall power dissipation to the dispersion is less. This results in less turbulence and less organic entrainment. The other path is to achieve the entire mass transfer in a single tank of larger size, which satisfies the desired residence time. The approach is to size the most hydraulic efficient pumper just to meet the desired head and flow requirements, and install up-pumping hydrofoils on the same shaft to maintain phase stability in the upper regions of the tank. This method reduces the number of mixers required for each mixer-settler while the overall efficiency of the system could well allow either higher specific settler loading (more throughput) or decrease the settler area (capital cost) for a given flow rate.

The conventional mixer-settler design has not yet outlived its usefulness. The proper design of the pumpers and auxiliary mixers, coupled with new settler technology available from consulting engineers, enable conventional plants to meet the requirements of the process without the need for costly and complicated new technologies. The open architecture of the conventional design is ideal for process monitoring, optimization, and control. The documented (Giralico Post and Preston, 1995) lower organic entrainment losses from optimized curved bladed pumpers and hydrofoil auxiliaries with their lower overall power consumption, makes the conventional mixer-settler design the most economical solution for copper solvent extraction.

SOLVENT EXTRACTION SETTLER CONCEPTS
Settlers serve the function of providing quiescent conditions for effective phase separation after mixing. In application this occurs on a continuous basis and occurs in a series of two stages. First there is the so called 'breaking' of the organic phase from the aqueous phase. This is the natural separation of phases due to density differences. Then, the second stage is direction of the separated phases via an adjustable full width weir overflow and underflow system. A conventional settler is illustrated in Figure 7 (b), Figure 9 and Figure 10. The key elements of

design are not complicated. The idea is to provide even quiescent flow with enough residence time to allow complete phase separation. Settlers are built on a low profile using stainless steel, concrete lined with HDPE, or reinforced fiberglass. Settlers are almost always square or nearly square. When needed, they are provide with covers. The key component to understand is that the settler width determines the solution or phase velocity. Thus the key concepts are settling rate, or settling specific area and average solution, or space, velocity. These design criteria are normally

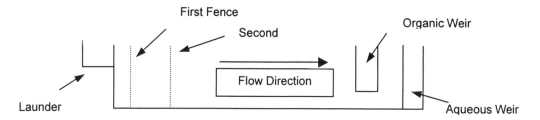

Figure 9. Conventional settler plan view

established for settler designs at about 4 to 5 $M^3/hr/ M^2$ and 3 to 6 cm/sec respectively. Table 2 (Liguori, A. J. et. al 1997) illustrates a typical design calculation sheet and the typical design criteria for designing and specifying a settler system. It is important to note that the 'total flow' includes both the aqueous and organic settler flows combined together. This calculation produces the general settler size where a maximum settler width is set at about 30 meters. The trade off in any settler design is the higher the settler phase velocity, the narrower the settler will be. This lowers the capital cost and constant flow distribution through the settler is also easier to maintain. But, there is an increased risk of entrainment. As well, it is beneficial to design for a separated organic depth of about 180-300 mm while having the aqueous phase depth at 450 mm to 600 mm. An additional aspect of the design is the inclusion of a settler distribution fence system also known as 'picket fences'. This encourages quiescent and even flow conditions within the settler. Normally two sets of picket fences in series are utilized within the settler.

The organic solution is the key ingredient in the SX plant. It is also a very expensive ingredient. Thus important detail should be taken to avoid an extensively large inventory, include protection against overflows and spillage, include reclaim, clean-up, and spillage measures in the design, and minimize the potential of contamination Settlers should be sized realistically with reasonable safety factors to base the initial inventory. Include a loaded organic surge tank. Settlers will have adjustable aqueous discharge weirs to enable operators to optimize the organic layer thickness. Install level alarms and high level cutoffs to protect inventory. Lastly, every operation must protect against contamination. Designs must minimize contaminants sources in PLS, water make-up, acid make-up, anti-corrosion chemicals, oils, dust, change in ore types, and insects. When calculating initial organic inventory, be sure to include an allowance for the crest on settlers, 50% of each mix box, and piping , as well as more obvious loaded organic tank and settlers inventory.

The operational concept of the settler requires phase separation of the immiscible liquid-liquid dispersion of the organic (kerosene) and aqueous (sulfuric acid) components. Common configurations feed the dispersion through a launder at the center of the settler, followed by an angled picket fence to evenly distribute the inlet stream and followed by a second picket fence located at approximately one third of the total settler length to assure even flow distribution further downstream. The exit section of the settler consists of an overflow weir for the lighter organic phase and an underflow-overflow weir for the heavier aqueous phase. Typical volume ratios of organic to aqueous in the settler are 1:2 to 1:1. Conventional design methodology for rectangular settlers is based on hydraulic theory and empirical results from pilot plants and full scale installations.

Table 2. Typical settler sizing and design calculation sheet (Liguori, 1997).

SETTLER SIZE DESIGN

1.	Circuit Configuration	2 x 1
2.	PLS Flow, Total	1,000
3.	No. of SX Trains	1
4.	O/A Ratio	1:1
5.	Total Flow Per Settler, (M^3/hr.)	2,000
6.	Settling Rate, (M^3/hr/ M^2)	4.4
7.	Organic Depth, (mm)	300
8.	Average Aqueous Depth, (mm)	450
9.	Settler Area (total flow/settler/settling rate), (M^2)	454.5
10.	Average Solution Velocity, (cm/sec)	3
11.	Settler Width (flow/settler/tot. solution depth/solution velocity), (M)	24.7 (say 25.0)
12.	Initial Settler Length (settler area/settler width), (M)	18.2
13.	Adjust Length for Distribution Fence, (M)	1.0
14.	Adjust Length for Weirs, (M)	2.5
15.	Total Settler length, (M)	21.7 (say 22.0)

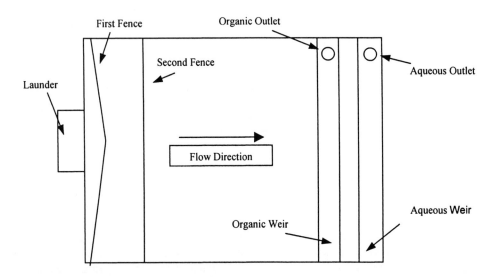

Figure 10. Conventional settler elevation view

Generally speaking, the basic settler dimensions are derived from open channel flow theory to develop linear velocity targets, specific flow rate (total flowrate per settler footprint area) and picket fence design. Empirical results have refined the linear velocity, specific flowrate, and total settler depth guidelines derived from theory. The commercial use of solvent extraction technology in copper mining applications supports the claim that the simple rectangular arrangement commonly in use is successful both technically and economically. Nonetheless several issues remain that offer the opportunity for improvement in settler performance. Organic entrainment in the aqueous phase is a cost consideration. Organic lost into the aqueous phase during stripping and wash is sent to the tank house and lost during electro-winning. While there are several ancillary technologies available to recover the entrained organic after the settler, any improvement in the settler itself would be beneficial.

A second concern is the generation of crud or scum. This is a three phase semi-stable emulsion consisting of solid particulate, organic liquid and entrained air. Depending on the continuous phase of the dispersion this emulsion may collect at the organic weir between the organic and aqueous phases and block transport between the two layers. It can thus lead to a shortened effective length of the settler, it also provides a source of organic material that erodes from the crud layer into the aqueous phase. As the degree of crud generation varies between mine sites there are multiple approaches to dealing with this problem ranging from doing nothing to regular removal using specialized suction devices.

A third and less obvious issue is the flow pattern in the settler itself. The flow in a settler is not completely laminar and channeling occurs. Specifically, the flow velocity in the center of the settler can be different than at the edges and reverse flow can often be observed, particularly behind the second picket fence. It is difficult to assess the entrainment impact of this behavior without major modifications to either the shape of the settler or its internals. At minimum however, reverse flow negates the assumption of plug flow used in coalescence review and indicates that for a given total design flow rate existing rectangular settlers exceed the minimum foot print requirement for process.

Settling Phenomenon
After the organic and aqueous phases have been mixed as a dispersion for a sufficient time to achieve the desired mass transfer, they are pumped into the settler and begin to separate. The dispersion band shrinks until it nearly disappears and there may be left just a thin film between the organic and aqueous layers. This separation is called the primary break time and is a measure of the time, which is required to achieve this primary break, determines the size of the settler. There are different basic settler designs, tank internals, and additives that will reduce the time it takes for the aqueous and organic to separate. It is necessary to understand some of the hydraulics, which determine the depth of the organic layer floating on the top of the aqueous. Any surge in the system can result in one cell being starved of solvent while another cell is overflowing. The level of the aqueous outlet weir governs the depth of the solvent layer, for each settler. To insure smooth operation of the settler, the level of the aqueous outlet weirs should be changed only a little at a time. This is because a change of one inch in the level of the water crest flowing over the weir while a settler is operating will change the depth of the solvent layer in that settler by nearly six inches flooding a great deal of solvent from the system. Conversely, lowering an outlet weir by a large amount will cause solvent to pile up to a much greater depth, limiting the flow of solvent from that stage for a period of time. The best word of advise here is to be patient and make small changes. After each change, let the system reach equilibrium before making further changes.

Settler Internals
Designs for inlet end presently consist of a launder or a pipe to kill velocities into the settler. Companies like LIGHTNIN are presently evaluating a distribution outlet design that will point the flow to the corners of the settler resulting in smoother plug flow pattern, eliminating reversing flows and channeling. Picket fences usually are supplied in either a straight design or a Chevron

type design. Chevron designs develop higher head loss. A coalescence rail further slows down the flow, but more importantly controls the emulsion band between aqueous and organic layer. The band-controlling interface can be adjusted up and down the coalescence rail to locate the optimum setting allowing clean filtering. Some plants even incorporate additional picket fences or coalescence media above the standard two. These should not be needed if the inlet/first picket fence is well designed.

Designs for the discharge end are affected by the piping and layout considerations. The organic weir sets the static liquid depth. The adjustable aqueous weir sets organic thickness. Fixed advance aqueous weir could split aqueous recycle and advance aqueous flows. Other designs used to achieve the same results. Piping connections located at the bottom fitted with vortex breakers. This will eliminate excess air entrainment. The height of the advance weir usually is set slightly below the normal setting of the adjustable weir to minimize splash and air entrainment.

Computational Fluid Dynamics (CFD) Use as a Design Tool
CFD can be used as an invaluable tool in understanding what actually happens inside a settling tank. This tool is utilized to understand what flow patterns actually take place as it is almost impossible to visually determine what is happening as a result of physical characteristics of the dispersion, then breaking phases. Some examples follow. Figure 11 shows what can happen if redirection vanes are installed to the inlet of the settler. Without the vanes, there are considerable areas in the corners that see almost no flow at all. If this space can't be utilized, it is just additional holdup volume of chemicals that is expensive and could be eliminated. Adding the redirection vanes let us utilize the maximum volume of the settler by getting better flows in the corners.

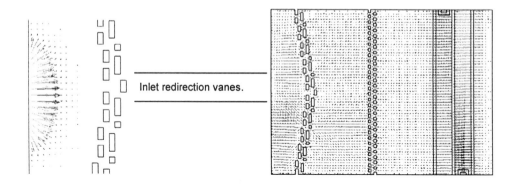

Figure 11. CFD study of redirection vanes to the inlet of the settler.

Figure 12 shows the affect of angle on the picket fence. One being an industry standard typical angle, the second being an increased angle, and the third showing why a concave picket fence may be used. The concave picket fence results are interesting, as what happens is the flow is initially split more behind the picket fence, but as it flows through the fence the flow is redirected and focused. This results in an uneven flow distribution past the fence and is not recommended.

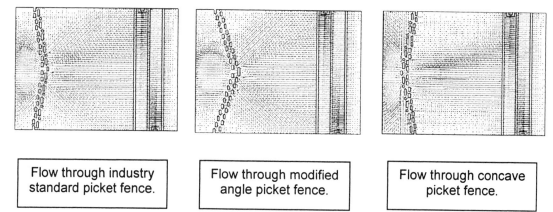

Flow through industry standard picket fence.

Flow through modified angle picket fence.

Flow through concave picket fence.

Figure 12. CFD study of affect of angle on picket fences.

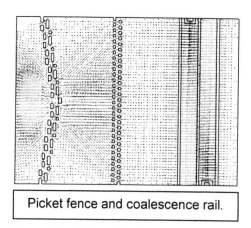

Picket fence and coalescence rail.

Figure 13. CFD study of coalescence rail.

Figure 13 shows the need for a coalescence barrier between the picket fence and the overflow weirs. They straighten the flow past the barrier feeding the weirs. Also the emulsion interface between the aqueous and organic phases can be controlled, benefiting from cleaner split solutions.

Individual Feature Design
The current inability to model the entire flow and process features of a settler requires a reassessment of the usefulness of CFD for application in this process. A number of observations in both small scale experimental modeling and full scale settler analysis indicate that CFD may offer the opportunity to optimize individual features of a settler. Observation of the flow pattern in the inlet region of the settler shows strong channeling toward the center and inlet launder and first picket fence modeling from a fluid dynamic perspective appears useful. Recirculation in the flow field in the section after the second picket fence puts the overall settler shape and second picket fence design into question. Observation of the flow in front of the organic weir points to potential entrainment sources stemming from the weir shape. Finally, CFD allows relatively quick analysis of the flow pattern changes that can be expected from changes in operating flow rate.

Inlet Launder
Figure 14 models a 9.3 m wide by 0.9 m deep settler from the 2 m wide inlet launder to the front end of first picket fence located 6.9 m into the settler. Inlet flow velocity is 1 m/s corresponding to common full scale application. The model is run 3-D, laminar steady state. Note that only a single phase is modeled. This represents a relatively accurate picture of the flow before the first and

second fence as phase separation often has not occurred until after the second fence. Since the inlet launder is assumed to be located in the center of the settler only one half is modeled, the mirror image half is inferred along the centerline symmetry axis.

The positive x-direction velocity field distribution of this arrangement is shown in Figure 14. Note that clear areas indicate reverse flow in the settler. Here, a plane located 0.35 m off bottom is highlighted showing strong recirculation of the flow within the front section of the settler not only in the horizontal plane but also vertically (plot not shown for brevity).

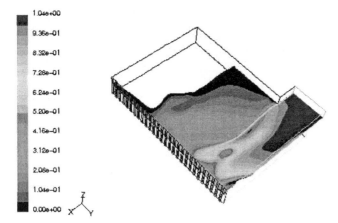

Figure 14: Horizontal slice, velocity magnitude – positive X

Plug flow does not hold in the front end of the settler.

The axial distribution of the velocity field can be avoided either through the use of an inlet launder spanning the entire height of the settler or by angling the launder from the inlet to the settler floor. The recirculation in the horizontal plane can be avoided through the use of inlet launder directional vanes or through modified picket fence design, or a combination of both.

Overall Settler Shape
Figures 15 and 16 show two settler designs with identical overall settler length, launder width, weir width, and liquid depth. The sole difference lies in the trapezoidal shape of the settler in Figure 16 where significant portions of the foot print has been removed. The inlet velocity is set at 1 m/s, overall dimensions of the rectangular settler are appropriate for operating settler design. Figures 15 and 16 show x-direction velocity field distribution in the horizontal plane located 0.35 m off bottom. Note that clear areas reflect flow in the negative x-direction, or flow reversal.

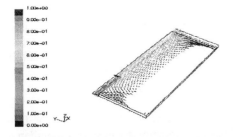 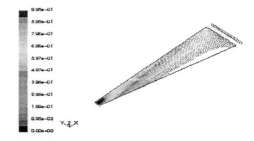

Figure 15: Positive X-direction Figure 16: Positive X-direction
Flow field in rectangular settler Flow field in trapezoidal settler

Both designs show recirculation in the outside portion of settler thus necessitating picket fences for flow distribution. The trapezoidal settler shows a lesser degree of recirculation coupled with smaller foot print. While the average residence time of the trapezoidal settler will be shorter due to the reduced volume, the distribution will be closer to plug flow than the rectangular settler of equal weir width.

Fence Spacing and Design
In order to facilitate the evaluation of picket fence designs in a settler this section of the work is done in 2D. Due to the relatively small size of the pickets and inter-picket spacing in relation to the overall settler dimensions a 3-D analysis is computationally not reasonable for the scope of this paper.

A 2-dimensional plot of a rectangular settler without picket fence is equivalent to Figure 11. Figure 17 adds a single picket fence located roughly 1/3 into the settler. The model pickets are 0.1 m wide with 0.025 m spacing. Overall settler dimensions are 25 m wide by 20 m long, inlet velocity is set at 1 m/s. The resultant flow pattern shows the distribution across the width of the settler. This particular picket arrangement is not optimized as the flow preferentially channels to the outside of the settler. Figure 18 represents the flow field generated in a settler with symmetric double picket fence. For Figure 18 all pickets are centered and completely symmetric. The resultant flow field is uniform and raises the question if single picket fence arrangements are feasible full scale installations.

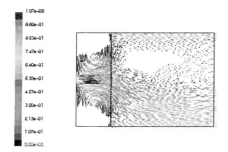 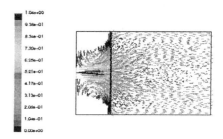

Figure 17: Rectangular settler – single picket fence   **Figure 18: Rectangular settler double symmetric picket fence**

Weir Design
As we progress through the settler the final major feature in the design is the weir section. While the basic concept of an overflow weir is simple care must be taken to avoid cross phase entrainment as well as air entrainment. The modeling of air entrainment in the overflow section of the weir requires multiphase modeling that is not addressed here but other features of the weir can be evaluated using single phase modeling. Figure 19 shows a plain rectangular organic weir cross section. By nature of the design the organic-aqueous interface lies on the vertical wall thus forming a stagnation point some distance upstream of the weir itself. Transport of fluid from front to back in the settler locates any accumulated crud in the interfacial region right at this stagnation point. Note the downward flow velocity of the liquid from the interfacial region toward the organic weir underflow. This phenomenon has been validated experimentally and can lead to crud layer erosion and entrainment losses of organic in aqueous. Specific design modifications to alleviate this problem are not addressed here but their evaluation via CFD is a logical starting place.

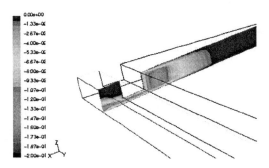

Figure 19: Organic weir cross section

To date attempts to develop a single comprehensive computational design tool for solvent extraction settlers have met with very limited success. A single phase, multiple species model has shown good agreement with experimental results in small-scale open models. Nonetheless, CFD can be used to develop optimized solutions for individual settler components such as inlet launders, picket fence design and location, overall settler shape and overflow weir design. Generalized results are that a trapezoidal shape of the settler offers reductions in overall footprint coupled with improved plug flow performance and reduced organic inventory.

SIZING AND SELECTION OF COPPER ELECTROWINNING PLANTS

Whereas the combination of leach, solvent extraction (SX) with electrowinning of copper has been used commercially for the past 30 years, direct electrowinning of copper has been implemented for over 85 years. (Davenport, Jenkins, Kennedy, and Robinson, 1999, Robinson, Davenport, and Jenkins, 1997)

Chemically the process can be described as;

Cathode reaction $\quad Cu^{++} + 2e^- \rightarrow Cu^0 \quad$ (9)

Anode reaction $\quad H_2O \rightarrow \frac{1}{2} O_{2(g)} + 2H^+ + 2e^- \quad$ (10)

The sizing of copper electrowinning plants in this section will take into consideration the following:

- Faraday's law
- electrode active area
- current density
- current efficiency
- time efficiency
- number of cathodes per cell
- rectifier size
- electrode handling machine size

A typical electrowinning cell and its components are shown in Figure 20.

Plant Sizing
Essentially sizing of the copper electrowinning plant can be easily calculated by using Faraday's Law. The equation is below:

$$\frac{dm_{Cu}}{dt} = \frac{C_D * A * TE * CE * n}{F} \quad (11)$$

where:

$\dfrac{dm_{Cu}}{dt}$ = rate of copper produced (capacity), tonnes / hour

C_D = current density, A/m^2

A = cathode area, m^2

TE = time efficiency, %

CE = current efficiency, %

n = number of cathodes in cellhouse

F = Faraday's constant equal 96, 485 coulombs per gram equivalent mass.

Faraday's Law states that 96, 485 coulombs will theoretically plate one gram equivalent mass of copper (or any metal). For example, copper metal is 8.4 x 10^5 Amp hr per tonne of Cu.

An alternative is to look at electrical equivalent which for Cu^{2+} is 1.18576 gms per amphour

M_{Cu} = F * A * CE * TE * t   (12)

where   t = time (hrs)

F = 1.18576 gms per amp hours

A, CE, TE are as before.

Electrode Active Area (A, m^2)
Typical cathode area for copper electrowinning is two times one meter square (on both sides). Due to current 'edge effects', the anode is smaller. Overlaps of 30-50 mm on the cathode edge are typical for a smooth plating finish at the edges.

Current Density (C_D, A/m^2)
Traditional starter sheet operation design current densities were around 200-240 A/m^2. With the development of permanent cathode systems with straighter electrodes, design current density can range from 280-300 A/m^2 and plants can operate at well over 300 A/m^2.

Current Efficiency (CE, %)
Current efficiency can be assumed to be 90% for design purposes even if operations can run at 95%.

Loss of current efficiency is due to short circuits between anode and cathode and ferric iron concentration in the electrolyte, which competes with the copper electrowinning reaction. Iron can come from the leach solution.

Time Efficiency (TE, %)
Time efficiency in the electrowinning cellhouse is close to 100% as cathodes are loaded and unloaded 'live' with current still running through the cell. Capacity loses can occur during the clean out of anode sludge from cells .Two cells per 100 in service are normally cleaned per day. Every third cathode is harvested from the cell.

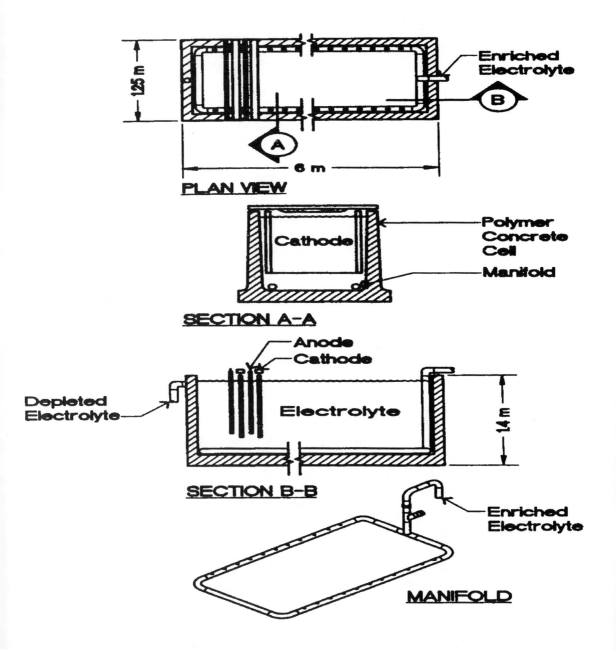

Figure 20. Typical electrowinning cell components and design (Biswas and Davenport, 1992).

Number of Cathodes Per Cell

Number of cathodes per cell or length of cell is dependent on the amount of copper produced in the plant.

For capacities of

5-15 ktpa	30-36 cathodes/cell
20-30 ktpa	45 cathodes/cell
40 and above	60-66 cathodes/cell

In colder climates where building space is at a premium low capacity plants can have cathodes of 60 to 80 each in the cell. Standard cathode to cathode spacing is 100 mm.

The number of cells in cell house times the number of cathodes in the cell equals the total number of cathodes.

Rectifier Size
Rectifier size is based on total current and total voltage.
Total current is calculated

$$I_T = C_D * A * n \quad (13)$$

I_T = current, kA
C_D = current density, A/m^2
A = plating area, m^2
n = number of cathodes per cell

Total voltage is calculated

$$V_T = V_C + V_L \quad (14)$$

V_T = Total voltage, V
V_C = Cell section voltage, V
V_L = busbar and losses voltage, V

Cell section voltage is calculated by the total number of cells in the rectifier section by the individual cell voltage (approximately 2.0 V per cell). Voltage losses and busbar voltages can also add 10-15% to design.

Rectifiers may be initially sized for proposed future expansions.

Electrode Handling Machine Size
Size of permanent cathode machine can be simply calculated by the following

$$Mc = n/(D * S * E) \quad (15)$$

Mc = machine capacity, cathodes/hr.
n = total number of cathodes in cellhouse
D = number of production stripping days/week
S = number shifts per day
E = effective working hours/shift

The target is to strip all of the cellhouse cathodes in one week. Typically in copper electro refineries weekdays from Monday to Friday are the stripping days but in the mine location of a leach SX-EW plant stripping can occur on a virtually continuous basis as suits the labor availability. Total number of cathodes is calculated by multiplying total cells by cathodes per cell.

Copper Electrowinning Process Selection

Selection of technology includes

- cellhouse technology
- cells
- ventilation
- contacts

Cellhouse Technology

Before the late 1970's copper cellhouses used starter sheets technology these separate starter sheets were plated (1-2 days old), harvested, stripped, trimmed, straightened and loops attached for commercial production. Disadvantages of this process were high labor demand required to prepare the starter sheets, high machinery maintenance cost and lower productivity.

Permanent cathode systems such as the ISA Process (Robinson, O'Kane, and Armstrong, 1995) in the late 70's and Kidd process in late 80's were developed for productivity and cathode quality reasons. The introduction of the ISA Process removed the need to grow starter sheets and so eliminated that whole section from the cell house. Central to the technology were the 316 stainless steel cathode plates and specialized machinery used to strip the cathode. Cathodes would be grown from 6 to 14 days depending on design and operating regime in the commercial cells and harvested from the cells to be stripped in the specially designed machines.

Features of each technology are as follows:

Isa Process

- copper plated 304 stainless steel hanger bar
- 3.25 mm thick 316L stainless steel cathode plate.
- "U" shaped rectilinear cathode washing and stripping machine design
- separate cathode sheets from each cathode plate

Kidd Process

- solid copper bar hanger bar welded to cathode plate
- 3.0 to 3.125mm thick 316L stainless steel plate
- circular / rotating cathode washing and stripping machine.
- Single enveloped cathode from each cathode plate

ISA Process technology previously required the use of wax to mask the edges of the cathode but now both technologies offer a waxless technology.

Recently other technology derivatives such as an electrometallurgical process developed by Outokumpu are being marketed.

Anodes

The first copper electrowinning plants used cast antimonial lead anodes. These anodes would warp, cause short circuits and could contaminate the copper cathode. Cold rolled and alloyed lead anodes have now become the standard anode technology in copper electrowinning.Cold rolling makes the anode stronger, stiffer and straighter than a cast anode. Alloying with calcium and tin also assist with strength and produce a more consistent corrosion product. Life of lead anode now can last six and in some cases ten years. Typical anode blade thickness is 6 mm . The trend to higher current densities has seen the introduction of thicker anode blades.The blade dimensions of an anode are smaller than the cathode due to 'plating overlaps' with overlaps of 30mm on the sides and 30 to 50 mm on the bottom being ideal.

Electrowinning Cell Technology
Traditional cell construction material in the past has been monolithic re-enforced concrete and liner systems. Liners were made of lead, FRP and PP, HDPE plastics. Over time liners would wear, break and get damaged and acid would attack the concrete leading to costly maintenance.

Polymer concrete cells were first tested in Zambia in the 1970's but it was in the 1990's where the polymer concrete unicell concept became accepted as the universal standard for copper electrowinning design. (Robinson, Karcas and Weatherseed, 1999)

Unicell's acceptance is due to the following reasons:
- monolithic single casting results in acid resistant homogenous structure
- integrated inlet and overflow piping
- acceptance of the cell to stand as a structural unit where walkways, piping, valves can be attached to inserts embedded in Unicell structure.

Benefits include reduced maintenance and overall cellhouse capital cost.

Ventilation Mist Suppression
Methods of mist suppression available in copper electrowinning today include:
- mist suppression foams
- mist suppression balls and beads
- open tankhouse designs
- cell top hoods
- positive flow ventilation
- combinations of the above

Busbar Contacts/Cell Top Insulators
Busbars that transfer current from one cell to another can vary in design. These include:
- triangular
- 'dogbone'
- double contact

Triangular is the most common and is simple, low cost, easy to replace and requires simple cell top furniture. However it uses an asymmetric anode hanger bar and has low current carrying capacity.

The 'dogbone' contact is a cross section of copper busbar that connects two triangular (most common) contacts and allows for symmetrical anode design and can carry more current. It is a more complicated design and so more expensive and is more difficult to clean.

Recent busbar design developments have lead to the double contact design, which include equalizing bars for anodes or cathodes or both. The equalizing bars equalize current between the electrodes, raising current efficiency and potential current density.

TYPICAL COPPER SX AND EW PLANT DESIGN CRITERIA LISTING
As a service to the reader the following pages list the pertinent design criteria typical for a proposed 13,500 tonne per annum copper cathode production facility utilizing a 2x1 SX circuit. (Taylor, 1995).

SUMMARY
Since the late 1960's copper solvent extraction coupled with electrowinning has been a growing technical application for production of copper metal. This paper outlined this technology and the fundamentals involved in the selection and sizing of copper solvent extraction and electrowinning circuits.

ACKNOWLEGMENTS
The authors gratefully acknowledge the expert assistance of Ms. Tami J. Cashell as well as thank Dr. Courtney A. Young, Dr. Hsin H. Huang and Mr. G. Andrew Hadden for their careful review of the manuscript.

Typical Copper SX EW Design Criteria

PLANT DESIGN CRITERIA

GENERAL

PLANT CAPACITY t/a CATHODE		13,500

PRODUCTION

OPERATING SCHEDULE

	h/d	24
	d/y	365
OVERALL UTILISATION/AVAILABILITY	%	95

ASSOCIATED PLANT (EXCLUDED)

ORE TYPE		OXIDISED COPPER
LEACH SYSTEM		HEAPS
COPPER GRADE	%	1
NOMINAL CRUSH SIZE	mm	25

SOLVENT EXTRACTION

EXTRACTION SECTION

FEED SOLUTION

FLOWRATE	m^3/h	500
COPPER	g/l	3.5
TOTAL IRON	g/l	5.0
FERRIC IRON	g/l	1.0
pH		2.0
CHLORIDE (MAXIMUM)	mg/l	75
SOLIDS (MAXIMUM)	mg/l	20
COPPER RECOVERY	%	93
TEMPERATURE - DESIGN	°C	20
- MAXIMUM	°C	30
RAFFINATE COPPER	g/l	0.245
RAFFINATE FLOWRATE (EXCL. IRON BLEED)	m^3/h	500
NO. OF STAGES		2
MIX BOXES PER STAGE		2
MIXER RETENTION PER STAGE	MIN	
MIXER O/A RATIO		1:1
SETTLER SPECIFIC AREA	m^3/h/m^2	5.0
ORGANIC FLOWRATE	m^3/h	500
STAGE EFFICIENCY	%	95

STRIPPING SECTION

ELECTROLYTE FLOWRATE	m^3/h	108.3
SPENT ELECTROLYTE		
- COPPER	g/l	33
- ACID, MAX.	g/l	170
- IRON, MAX.	g/l	2
STRONG ELECTROLYTE		
- COPPER	g/l	48
NO. OF STAGES		1
MIX BOXES PER STAGE		2
MIXER RETENTION PER STAGE	MIN	2
MIXER O/A RATIO		1:1
SETTLER SPECIFIC AREA	m^3/h/m^2	5.0
ORGANIC FLOWRATE	m^3/h	500
TEMPERATURE - DESIGN	°C	20
- MAXIMUM	°C	30
STAGE EFFICIENCY	%	95
ORGANIC PHASE		
REAGENT TYPE		ACORGA M 5640, LIX 984,OR EQUIVALENT ALLIED CHEMICALS MOC 100
CONCENTRATION (CONFIRM WITH SUPPLIERS)	%	13
DILUENT TYPE		HIGH FLASHPOINT KEROSENE SHELLSOL 2046, ESCAID 100 OR EQUIV.
FLASHPOINT (ASTM D-93)	°C	80
AROMATICS, MAX	%	20
COPPER: IRON TRANSFER RATIO		1000 (EXPECTED)
		500 (DESIGN)
ORGANIC ENTRAINMENT IN RAFFINATE AND STRONG ELECTROLYTE	ppm	30-75
AQUEOUS ENTRAINMENT IN LOADED ORGANIC PPM		250-500
AQUEOUS ENTRAINMENT IN STRIPPED ORGANIC	ppm	300-600

Typical Copper SX EW Design Criteria - Continued

MIXER – SETTLERS			
NO. OF STREAMS			
- EXTRACTION			1
- STRIPPING			1
TYPE			CONVENTIONAL
PRIMARY MIX IMPELLER			PUMP-MIX
SECONDARY MIX IMPELLER			TURBINE
MIXING BOX PHASE CONTINUITY			
- E1			AQUEOUS
- E2			ORGANIC
- S1			ORGANIC
RECYCLES			AQUEOUS ON ALL STAGES
ELECTROLYTE PREPARATION			
ELECTROLYTE CLEAN-UP			
TYPE			FLOAT COLUMNS
NO. OF STAGES			2
DESIGNS AVAILABLE			JAMESON
			MAGMA
			MINFLOAT
			COMINCO
			PYRAMID
TARGET FINAL ORGANIC CONTENT		ppm	5 OR LESS
ELECTROLYTE HEAT EXCHANGER			
TYPE			PLATE AND FRAME
STRONG ELECT. TEMP IN/OUT		°C	20/40
SPENT ELECT. TEMP IN/OUT		°C	45/25
ELECTROWINNING			
CIRCUIT DESIGN			
CURRENT DENSITY			
- NORMAL		A/m²	280
- MAX.		A/m²	300
CURRENT EFFICIENCY		%	90
COPPER QUALITY			LME GRADE A
DESIGN CAPACITY T/D			39
OPERATING CYCLE		DAYS	7
METHOD			1/3 CELL LIVE LOAD
PRODUCT BUNDLES		t	2-3
CELL DESIGN			
NUMBER OF CELLS: SCAVENGERS			11
TOTAL			53
CATHODES PER CELL			64
ANODES PER CELL			39
CATHODES - TYPE			316 STAINLESS STEEL
- ACTIVE AREA			1 m x 1.1m
CATHODE SPACING		mm	95
ANODES - MATERIAL			Pb/Ca/Sn ALLOY
- THICKNESS		mm	SOLID ROLLED SHEET 6
ELECTRICAL			
NO. OF RECTIFIER CIRCUITS			1
CELL VOLTAGE:			
- NORMAL			2.0
- MAX.			2.2
CELL CURRENT:			
- NORMAL		A	24,000
- MAX.		A	26,000
BUSBAR - MATERIAL			COPPER
SOLUTION CIRCULATION			
FLOWRATE/CELL:			
- SCAVENGERS		m³/h	10.3
- COMMERCIAL		m³/h	8.5
COPPER DROP PER CELL:			
- SCAVENGERS		g/L	2.5
- COMMERCIAL		g/L	3.0
COBALT DOSING		ppm	100
IRON CONTENT, MAX		g/L	2

Typical Copper SX EW Design Criteria - Continued

CHLORIDE CONCENTRATION, MAX	ppm	30
MIST SUPPRESSION METHOD		POLYPROPYLENE BEADS
CELL OPERATING TEMP.	°C	35-45

TANK AND POND SIZING CRITERIA

RETENTION TIMES:

SX FEED POND	HR	24
RAFFINATE POND	HR	24
LOADED ORGANIC TANK	MIN	20
STRONG ELECTROLYTE TANK	MIN	30
SPENT ELECTROLYTE TANK	MIN	30
SOLUTION HOLDING TANK		CONTENTS OF 1 COMPLETE SX SETTLER
DILUENT STORAGE*		1 ½ TANKER LOADS
EXTRACTANT STORAGE		3 MONTHS SUPPLY
COBALTOUS SULPHATE STORAGE		3 MONTHS SUPPLY

(* MAY NEED 2-3 MONTHS FOR REMOTE SITE)

UTILITIES

EW MAKE-UP WATER

CHLORIDE CONTENT	ppm	<30
SOLIDS CONTENT	ppm	<10

EW MAKE-UP ACID

QUALITY		WHITE ACID

HOT WATER

DUTIES		CATHODE WASHING AND WASH DOWN
TEMP.MAX	°C	80

REFERENCES

Biswas, A.K., and W.G. Davenport. 1992, <u>Extractive Metallurgy of Copper</u>, 3rd Edition, Elsevier Science Inc., New York.

Davenport, W. G., J. Jenkins, B. Kennedy, T. Robinson. 1999, "Leach, Solvent Extraction, Electrowinning of Copper - 1999 World Operating Data", Copper 99, Phoenix, Arizona, USA. October.

Davy International, undated, <u>Qualifications for Copper Leaching Solvent Extraction Elecrowinning Facilities</u>, Section 4:Technology – SX-EW.

Gigas, B., and M.A. Giralico. 2000, "SX: CFD Settler Optimization", Randol Copper Hydromet Roundtable 2000, September, Tucson, Arizona.

Gigas, B., and M.A. Giralico. 2002, "Advanced Methods for Designing Today's Optimum Solvent Extraction Mixer Settler Unit", ISEC 2002, Johannesburg, South Africa, March.

Giralico, M.A., T.A. Post, M.J. Preston. 1995, " Improve the performance of your copper solvent extraction process by optimizing the design and operation of your pumper and auxiliary impellers.", SME Annual Meeting, Preprint Number 95-189,Denver, Colorado, March 6-9,

Giralico, M.A., T.A. Post, M.J. Preston. 1996, "Optimizing the performance of new and existing SX plants through the use of high efficiency impellers for pumper and auxiliary mixers." SME Annual Meeting, Phoenix, Arizona, March 11-14.

Jorgensen, G.V. Editor, 1999, <u>Copper Leaching, Solvent Extraction, and Electrowinning Technology</u>, pp 239 ff, Society for Mining, Metallurgy, and Exploration, Inc.; Littleton, Co, USA

Liguori, A. J. 1997, <u>Copper Heap Leach, SME Short Course Notes</u>, Denver, Colorado.

Lo, T.C., M.H.I. Baird, and C. Hanson. Editors: 1991, Handbook of Solvent Extraction; Section 9, Krieger Publishing Company, Malabar, Florida.

Miller, G. 2001, "Design Tools To Control Transients In Solvent Extraction Plants", SME Annual Meeting 2001, Denver, March.

Oldshue, J.Y. 1984, <u>Fluid Mixing Technology</u>, Chapter 2: Impeller fluid shear rates and pumping capacity, McGraw-Hill, New York.

Post, T.A., R.A. Howk, and M.J. Preston. 1996, "Impeller System and Method for Enhanced-Flow Pumping of Liquids", United States Patent Nr. 5,511,881, Apr. 30.

Robinson, T., J. O'Kane, and R.W. Armstrong. 1995, "Copper Electrowinning and the ISA PROCESS", Copper 95, Santiago, Chile.

Robinson, T., W.G. Davenport, and J. Jenkins. 1997, "Electrolytic Copper Electrowinning and Solvent Extraction World Operating Data", Randol Copper Hydromet Roundtable Vancouver, November.

Robinson, T., G. Karcas, and M. Weatherseed. 1999, "Polymer Concrete Cell Applications in the Electrometallurgical Industry", ALTA 1999 Copper Hydrometallurgy Forum, Queensland, September.

Taylor, A. 1995, <u>Copper Leach/SX/EW Project Development</u>, ALTA METALLURGICAL SERVICES, Brisbane Australia.

Taylor, A. 1995, <u>Copper SX/EW Seminar: Basic Principles, Alternatives, Plant Design</u>, ALTA Metallurgical Services, Melbourne, Australia, Section 13

Weetman, R.J. and J.Y. Oldshue. 1988, "Power, flow, and shear characteristics of mixing impellers.", 6th European Conference on Mixing, Pavia, Italy, 24-26 May.

14
Bullion Production and Refining

Section Editor:
Dr. Corby Anderson

Bullion Production and Refining
C.O. Gale, T.A. Weldon **1747**

Platinum Group Metal Bullion Production and Refining
C.G. Anderson, L.C. Newman, G.K. Roset **1760**

Fundamentals of the Analysis of Gold, Silver, and Platinum Group Metals
C.G. Anderson **1778**

BULLION PRODUCTION AND REFINING

Charles O. Gale[1] and Todd A. Weldon[2]

ABSTRACT

Bullion production and refining for the present day precious metal mine is generally limited to the melt refining of gravity separated concentrates, electrowinning sludges, and the product from zinc precipitation processing. Guidelines, given as basic rules-of-thumb, are presented for equipment selection and sizing. Some industrial compositions of electrowinning sludges and zinc precipitates are given, as are typical flux compositions used for melt refining these precious metal concentrates. The art of ultra-refining dore' is not included in this paper. Retorting mercury, cadmium, and selenium is discussed.

EXTRACTION OVERVIEW

Gold and silver extraction and concentration from mining ores and/or process solutions has been traditionally accomplished through mercury amalgamation, gravity concentration and separation, and sodium cyanide leaching coupled with carbon adsorption or zinc precipitation, independently or in combination. The precious metals recovered from these initial concentrating processes are upgraded by refining. For this paper, the term refining refers to both hydrometallurgical and pyrometallurgical processing. For a brief review of the extraction and concentrating methods:

(1) Mercury amalgamation is one of the oldest, yet least used today, methods to recover fine particle size free gold and silver by alloying it with mercury.
(2) Gravity concentration and separation is also one the oldest methods of free gold and silver recovery. Gravity concentration recovers fine to nugget size particles.
(3) Sodium cyanide leaching of gold and silver is relatively new as an extraction medium, and leaching has allowed microscopic particles of gold and silver to be recovered.
(4) Sodium cyanide leaching is often coupled with carbon adsorption and electrowinning. The carbon is contacted with precious metal laden leach solutions where the metals load onto the carbon. The carbon is then stripped with a hot, sodium cyanide solution, which becomes highly concentrated with the precious metals. The precious metals are then plated onto the cathodes, either loosely or intimately, of electrolytic cells and recovered through downstream processing.
(5) Sodium cyanide leaching is less often coupled with carbon adsorption and zinc precipitation. The strip solution as in (4) above is contacted with fine particle size zinc dust wherein the precious metals are precipitated and collected in a filter.
(6) Sodium cyanide leaching may be coupled with direct electrowinning if the precious metal tenor is sufficiently high and deleterious impurities are not present.
(7) Sodium cyanide leaching is often coupled directly with zinc precipitation when the solution tenor is sufficiently high. Zinc precipitation as a concentration method dominates in the silver mining industry.

---

[1] Summit Valley Equipment & Engineering, Inc., West Bountiful, Utah, USA.
[2] Summit Valley Equipment & Engineering, Inc., West Bountiful, Utah, USA.

REFINING

In addition to electrowinning sludge and zinc precipitation concentrates, precious metal dusts, bag house and scrubber residues, floor sweepings, melt crucible residues, and furnace liners are also upgraded by refining. Refining is usually performed in two stages (Marsden and House 1993):

(1) Treatment at the point of production to produce crude bullion that is easily handled and accurately sampled.
(2) Refining of crude bullion to produce high purity gold and silver for market.

Since the metallurgical practice of high purity refining of gold and silver is treated in detail in many fine publications, this paper primarily focuses on refining the precious metal concentrates to the crude bullion state. The authors' guidance for equipment selection and sizing to produce crude bullion is given.

REFINERY

The gold mine "Refinery" frequently contains ore or solution treating equipment for the production of the first precious metal concentrate, equipment to intermediately process the concentrate, and equipment for processing the concentrate into crude bullion. Thus, a refinery can house electrolytic cells, zinc precipitate filters, gravity concentration and separation tables, drying ovens, retorts, flux mixing equipment, melting furnaces and molds.

The refinery layout varies greatly with the type of concentrating equipment selected. For electrowinning, corresponding valves and manifolds to the cell are located in the refinery area. For zinc precipitation, corresponding valves and manifolds are commonly located outside the refinery area to facilitate the changing of filter presses without entering the secure refinery. When a precipitation filter press is full, the flow stream is redirected to an empty filter press from outside the refinery area. Typical floor plans for a silver/gold mine with a zinc precipitation circuit and a gold mine with a carbon/electrowinning circuit are illustrated on the following page.

ELECTROLYTIC RECOVERY

In today's gold extraction plants, electrowon metals are often produced as a loose cathode deposit or sludge. The sludge is washed from the cathode or may simply fall from the cathode to the bottom of the electrowinning cell. This sludge is flushed from the cell and pump into suitable filters where it is collected for further processing.

The sludge is comparatively low in volume and can be relatively pure, 70 to 90 percent precious metals. Impurities often encountered are the base metals, mercury, cadmium, selenium, Al_2O_3, SiO_2, MgO, and CaO. Once filtered, a typical bulk density of the sludge is 1.5 grams/cm^3 (dry basis) (Hall, 2001).

By judiciously controlling the anode to cathode voltage potential at the cell, solution impurities such as copper can be minimized in the sludge product. The electrowinning cell can be considered as the first refining step.

For electrowinning sludge filter sizing, the authors suggest using a conservative filtered sludge density estimate of 1,100 kg/m^3 (dry). Note that downstream processing equipment is volumetrically controlled by the filtered sludge volume.

ZINC PRECIPITATION

Large silver producing mines, and gold mines with a silver to gold ratio of +5:1 or greater, utilize the zinc precipitation process. By current standards, a clarified and deaerated pregnant solution should have less than 5 ppm total suspended solids and a dissolved oxygen concentration of less than 0.5 ppm. Zinc powder is added in relation to precious metal content and oxygen and impurity concentrations. At the zinc addition point, diatomaceous earth is often added to the process stream and is used as a filtering aid. The filter aid provides body to the precipitate and decreases the rate of filter press blinding for some process solutions.

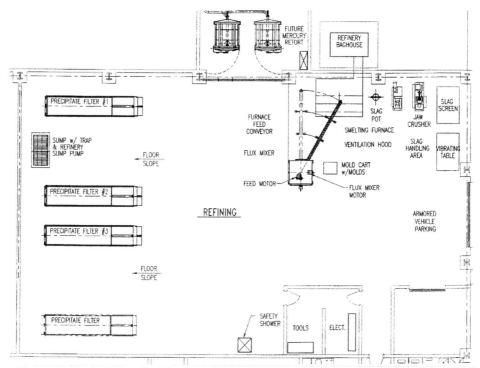

Figure 1 Typical silver/gold mine zinc precipitation circuit refinery

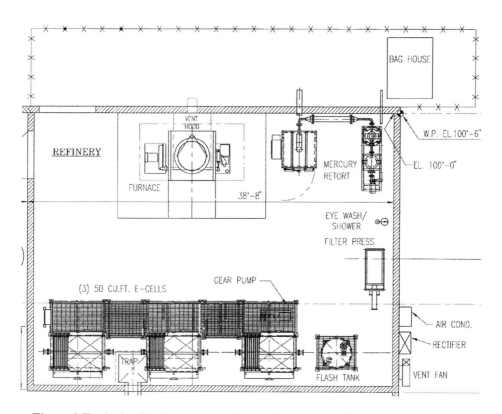

Figure 2 Typical gold mine carbon/electrowinning circuit refinery

Zinc precipitate quality can vary from 30 to near 85 percent precious metals. Start-ups and new operations often have low precious metal concentration precipitates due to process upsets and unrefined process control. Suspended solids not removed in the clarification filters, excess zinc (often added due to high oxygen concentrations), precipitating impurities, and body feed all lower the precipitate grade.

By economy, the precious metal production from silver operations must be greater than gold operations. The silver metal concentrate is by production more voluminous as compared to gold. Silver refining equipment is necessarily larger in volumetric capacity versus gold plants.

Precipitate handling equipment must be sized conservatively. The authors use a precipitate bulk density of 1,100 kg/m^3 (dry). This density is often considered too conservative, but has actually been measured at more than one location. Precipitate moisture after filtering will be 30 to 40 percent. When sizing drying ovens, the wet weight of the precipitate must be considered since most oven racks are weight limited.

For silver mining zinc precipitates, retorts are larger and are generally designed to be loaded with fork trucks or mechanized pan-handling lifts. A large mercury retorting operation in a silver refinery can handle more than seven cubic meters of zinc precipitate per day varying in mercury concentration from one to twenty percent (dry basis).

MERCURY, CADMIUM AND SELENIUM REMOVAL

Retorting is universally accepted as the refining step for the removal of mercury, and may be used for cadmium and selenium. Vapor pressure curves for Hg, Zn, ZnO, Cd, CdO, Se, SeO, Te, and Sb are shown on the following pages. The vapor pressure unit of measurement for the curves is Torr, or mm Hg absolute pressure. Distillation or sublimation occurs when the retort atmosphere is of lower absolute pressure than the specie vapor pressure at a given temperature.

Mercury

The fundamental physics of mercury retorting have been known for a long time, but due to ever more stringent plant hygiene requirements, the equipment has been driven from the simple to the robust. To retort mercury, electrowinning sludge or precipitate is heated to 425°C/550°C in a vacuum environment ranging from 50 mm vacuum (50 mm below atmospheric pressure of 760 mm at sea level) to the low absolute pressures of 150 torr (150 mm pressure absolute). The mercury is removed from the sludge and/or precipitate as a vapor, transported from the heating chamber, and is condensed in water cooled heat exchangers.

Mercury boils very easily (70.65 Kcal/Kg) (127.4Btu/lb), but also condenses very easily. Mercury will often condense in some very inconvenient places within the retort equipment. Care must be taken to maintain retort system surface temperatures in contact with mercury vapor above the mercury dew point.

Mercury vapor is very heavy and can be difficult to transport from the retort chamber to the condenser. A controlled sweep gas is often used to assist in transporting the mercury to the condenser. The sweep gas reduces the partial pressure of mercury vapor, and thus, aids the distillation process. With a sweep gas, the condenser efficiency is reduced and must be sized accordingly.

Equipment designed and operated at low absolute pressures (150 to 300 torr) is generally more successful in meeting the process requirements in a hygienically acceptable manner. To operate at the low absolute pressures, the equipment must be maintained nearly leak free. In these systems, the mercury vapor has less opportunity to condense outside of the condenser. In low pressure systems, fewer non-condensables will enter the condenser and the theoretical condenser efficiency will be higher. Mercury and water are the major condensable vapors. Air is the major non-condensable.

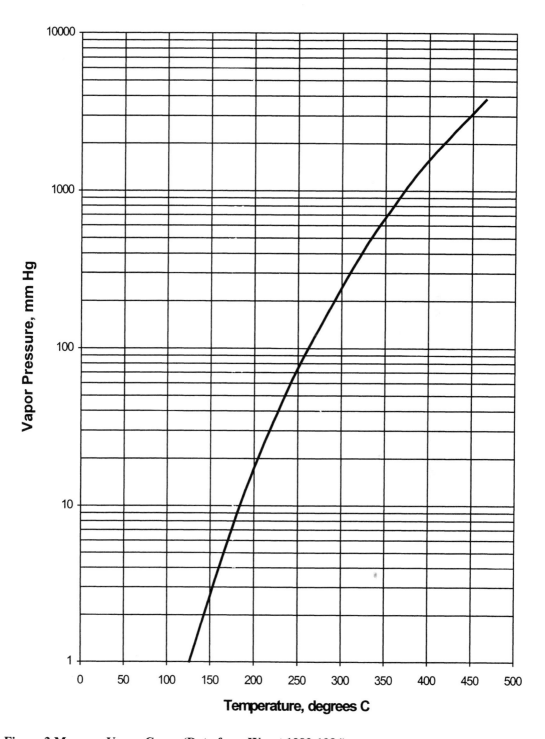

Figure 3 Mercury Vapor Curve (Data from Weast 1983-1984)

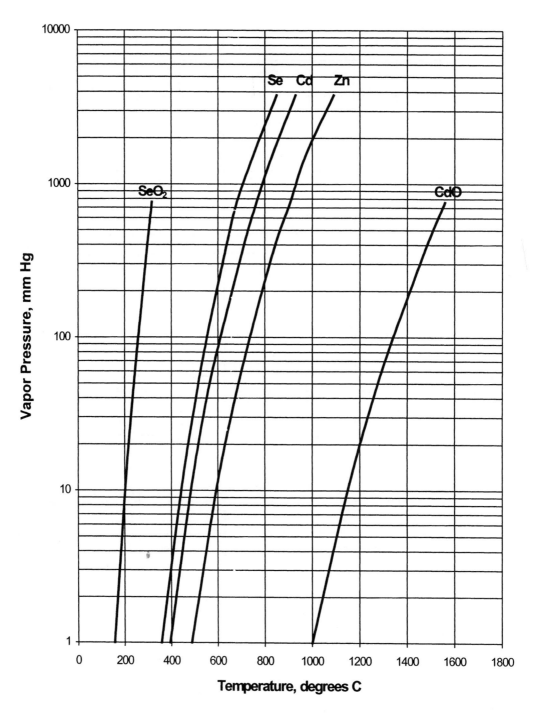

Figure 4 Vapor Pressure Curves (Data from Weast 1983-1984)

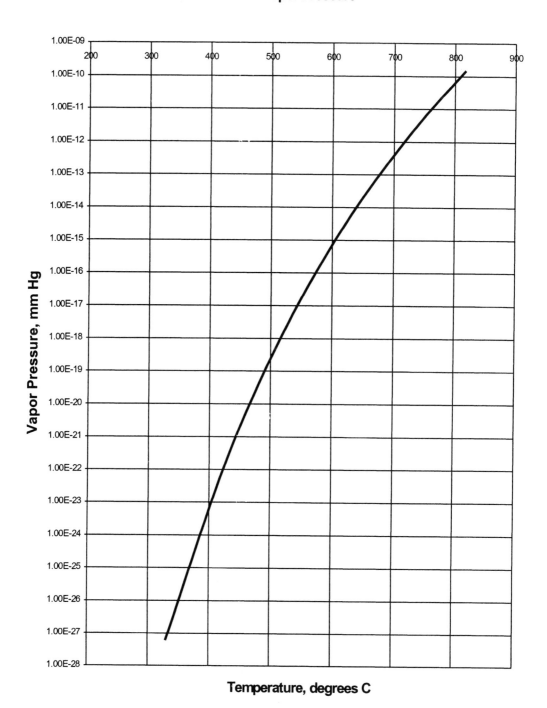

Figure 5 Zinc Oxide Vapor Pressure (Data from Weast 1983-1984)

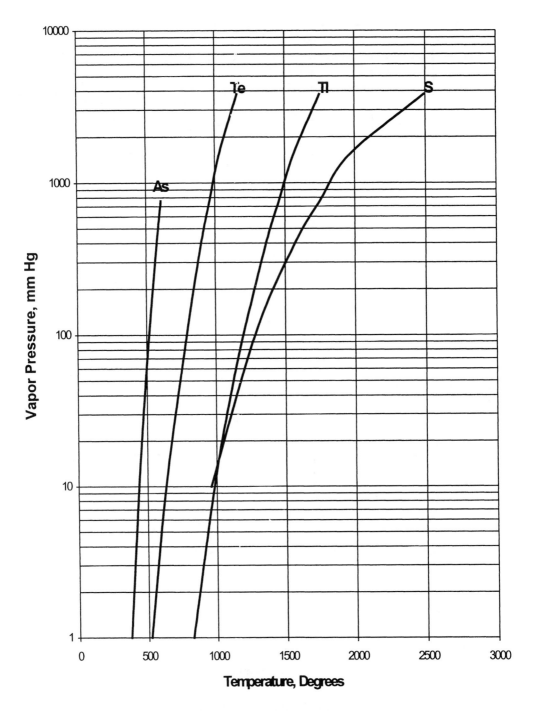

Figure 6 Vapor Pressure Curves (Data from Weast 1983-1984)

Mercury containing sludges or precipitates should not be heated rapidly in a vacuum. Rapid and/or violent boiling of water contained in the sludge or precipitate will cause precious metal dusts to transport out of the retort towards the condensing system. A drying step, slow and controlled heat-up, is generally incorporated in mercury retorting systems to prevent excessive dusting. The sludges or precipitates must be placed in pans of 100 mm to 150 mm maximum depth for retorting to be successful.

Cadmium

Cadmium metal removal via distillation is thermodynamically possible, as is cadmium oxide distillation. Equipment constraints restrict the distillation of cadmium to the metal specie only. The cadmium oxide vapor pressure is too low to complete an economically viable distillation. During the distillation of cadmium metal, cadmium oxidation must be prevented.

Cadmium distillation equipment has been placed into operation, but no operating data is available to the authors. The equipment is designed to heat the product to 700°C under a vacuum of 5 mm in a nitrogen atmosphere. At 535°C, cadmium will begin to distill. The cadmium is collected as a sublimate, which is manually removed from the system.

Selenium

Selenium can be successfully removed via distillation as documented by FMC Paradise Peak (Armburst 1989). Selenium is more commonly associated with zinc precipitated high silver products. Selenium must first be oxidized to selenium dioxide in order to distill. The process kinetics are slow. Lengthy holding periods at temperature are needed to allow oxygen to diffuse and react, and the resulting oxide gas to migrate to the condensers (sublimer).

MELT REFINING

The impurities commonly removed during melt refining are iron, zinc, calcium, magnesium, sodium, alumina, and silica by reaction with the flux and/or atmosphere. Mercury, selenium, zinc, and cadmium may volatilize during melting and escape into the furnace exhaust system. The exhaust system and scrubbing equipment must account for the collection of these metal fumes. Silver and gold also may volatize during melting, but if the melt temperature is maintained below 1250°C, the volatilization rate is generally low for gold. Silver has a higher vapor pressure than gold and its loss can be greater. Precious metal volatilization losses increase as the volatile impurity concentrations increase, notably tellurium, antimony, and mercury (Marsden 1993). Table 1 contains selected refinery feedstock from electrowinning and zinc precipitation processing facilities.

Melt Refining Electrowinning Sludge

If mercury, cadmium, and/or selenium are not present in the electrowinning sludge, the retorting step is eliminated. A process drying cycle of the sludge may not be required if the sludge cake has a low percent moisture. At one facility, filtered electrowinning gold sludge at 18 to 20 percent moisture is sometimes charged to the melting furnace without drying (Hall 2001). Flux and gold sludge are mixed (frequently by hand) and charged to an empty melting furnace.

Fluxes for processing electrowinning cell sludge are generally kept to a minimum. Flux composition can be as simple as 100% borax ($Na_2B_4O_7$), which supplies a cover during melting, to the more common mixtures of borax, niter ($NaNO_3$), silica (SiO_2), fluorspar (CaF_2), and sometimes soda ash (Na_2CO_3). Generally, niter is used if base metals are to be oxidized. If alkaline and alkaline earth metals are the only impurities, niter is often eliminated.

As a rule of thumb, a designer should allow for a flux volume equal to the electrolytic sludge volume being melted. If soda ash is used, a boiling slag often results during melting, thus sufficient crucible volume must be available to contain this boil. Table 2 contains some selected flux compositions from industry for electrowinning sludges.

Table 1 Selected refinery feedstock from electrowinning and zinc precipitation processing (Kahl 2002; McMillen 2001; Hall 2002; Haldane 2002; Frentress 2002, Anonymous)

Constituent	Electrowinning Cell Sludge	Electrowinning Cell Sludge	Zinc Precipitate	Zinc Precipitate	Electrowinning Cell Sludge
Au	74.3%	82%	21%	80-85%	18%
Ag	14.8%		61%	with Au	12%
Mg+Ca	0.95%				
Fe	0.01%	Yes			Yes
Co+Ni	0.08	Yes			
Cu	0.06%	Yes	0.6%	Yes	Yes
Zn	0.21%		1.2%	5-10%	
Mo	0.11%				
Cd	Trace				
Sb	Trace				
Te	Trace				
Pb	Trace				
Al	0.11				
Se	Trace				
Si			9%	Yes	Yes
Hg				2-5%	

Table 2 Selected flux compositions from industry (Kahl 2002; McMillen 2001; Hall 2002; Haldane 2002; Frentress 2002, Anonymous)

| Constituent | Zinc Precipitate | Electrowinning | | | |
		Sludge 1	Sludge 2	Sludge 3	Sludge 4
Niter	0.2x Base Metal Wt	3 %	20%	21%	
Borax	2x Base Metal Wt	30%		48%	85%
Soda Ash	0.24x Base Metal Wt				5%
Silica		7%		17%	10%
Fluorspar				14%	
Basis for Total	Wt of base metals	Wet sludge weight	Wet sludge weight	Substantial base metals, Sludge/slag is 2:1	40% of sludge

Furnace Refining Zinc Precipitate

After drying and/or retorting, the precipitate is mixed with fluxes and melted. The fluxes generally contain substantial borax for fluidity, enough niter to oxidize all of the zinc and any other base metals, fluorspar (CaF_2) for fluidity, and possibly soda ash to flux any excess silica and to lower the slag melting point. Table 2 contains some selected flux compositions from the industry.

Flux mixing in large volume silver plants is generally mechanized. The mixing equipment varies from .75 m^3 mortar mixers to 4 m^3 cement mixers to specialized "paddle mixers" of 3 to 6 m^3 capacity. The mixers are sized to match the batch-melting requirement. As a rule of thumb, when sizing zinc precipitate mixing and melting equipment, the flux charge volume should range from 2 to 3 times the precipitate volume.

Furnaces may be charged, melted and poured in one batch. In some operations the furnaces are alternately charged and melted, charged and melted, until sufficiently full of refined metal and slag to pour. To trap slag, cascade-casting arrangements are often used. Simple carousels with bar molds are also used, as are cone shaped slag pots.

Acid Digestion

Acid digestion of zinc precipitates can be advantageous to remove excess zinc and limey precipitates prior to melting. The amount of slag will be reduced which reduces the cost of flux, gold/silver losses in slag, and the energy required to melt.

Sulfuric acid is generally used for precipitate digestion. The digestion will produce small amounts of hazardous hydrogen cyanide (HCN), and voluminous amounts of flammable hydrogen (H_2). If arsenic or antimony are present, arsine (AsH_3) and stibine (SbH_3) can be produced and are highly toxic. At times, nitric acid is added to reduce the production of arsine gas (Marsden 1993). To assure effective zinc removal, the acid digestion is finished with about one percent excess free acid concentration. The ventilation system must be efficiently designed to capture and dilute all the evolved gases.

The acid digested precipitate must be filtered and washed, then dried and calcined or retorted prior to melting. Residual sulphates from acid washing promote equipment corrosion during retorting, and the residual sulphates can produce a "matte" during melting.

SLAG TREATMENT

In many operations, slag is ground and returned to the leach circuit. Some operations recycle small amounts of slag into the next melt, some sell the slag to a slag processor, and very few seriously process the slag in the refinery to recover precious metal fines. On-site slag processing can consist of crushing, grinding and screening (the oversize is re-melted), followed by gravity separation.

FURNACE SELECTION

Melt furnaces can vary extensively, as noted in the paper "Precious Metal Refinery Process Selection – An Overview" by G. Warren (Warren 2001). The choice between gas-fired, oil-fired or induction heated furnaces is made on the basis of fuel source economics, productivity, environmental considerations and refinery hygiene.

In general, when highly concentrated gold sludge is being processed, crucible style melt furnaces are preferred. Fuel fired crucible furnaces are considerably less expensive than induction heated furnaces and the power consumption of induction furnaces can be quite high.

The refinery will be cooler and require less ventilation with induction furnaces and the off-gas system can be much smaller. Induction furnaces can melt faster and crucible life should be longer than fuel fired furnaces. Induction stirring also assists metal to slag contact and reactions.

In large silver refineries, the fuel-fired, refractory lined reverberatory furnace appears to be the most common type of melting furnace. A few plants have large fuel fired crucible furnaces. The melting area ventilation systems are sized to handle the combustion exhaust at maximum firing rates.

VENTILATION

Industrial hygiene is very important in precious metal refineries. Heavy metals fumes and toxic gases can be produced during the refining processes. Mercury fumes are likely the most common problem, but other heavy metal fumes and dusts such as zinc, cadmium, silver, lead and selenium can be present.

Extraction products such as zinc precipitates and electrowinning cell sludges that have any measurable mercury should be stored in covered containers and processed in encapsulated or well ventilated equipment. Slag mixing equipment should be ventilated. Melt furnaces may be partially enclosed and hooded. Retort systems should be essentially leak free with no vapor losses to the environment. Acid digestion equipment must be hooded and/or encapsulated.

Electrowinning Cells

Electrowinning Cells must be ventilated to prevent hazardous vapors and byproduct gases from entering the refinery. Hydrogen is produced at the cathode and ammonia is produced by the oxidation of cyanide near anode. Both gases can react explosively under the right conditions and should never be allowed to collect in unventilated equipment or work areas. Mercury vapor is found in some electrolytic cell vent gases.

To prevent hazardous gas and vapor buildup, a cell ventilation system should sweep fresh air uniformly across the entire electrolytic compartment. Preferably fresh air should ingress from the operating service-side and exhaust from the opposite side. The cell exhaust should be routed to a suitable scrubbing system. Ventilation flow rates as high as 1.7 m^3/hr (1 cfm) per designed buss ampere are used. A 2,000 ampere cell can have a ventilation flow of 3,400 m^3/hr (2,000 cfm).

Retorts

Mercury retorts should have exhaust hoods located near the charge doors. The hoods must be designed to capture heavy mercury vapors that fall as they exit the retort door, i.e. the hood is more effective at the bottom of the door on a mercury retort. If retort doors leak, mercury droplets will collect on the door near the leak as a tell tale sign. The hoods should be designed to capture and contain mercury droplets.

Refinery Ventilation Systems

Refinery ventilation systems should be designed to affect a floor ventilation sweep. Ideally, supply air should sweep down the walls on one side of the refinery, travel across the room at floor level to the other side of the refinery, and exit via a duct or curtain wall on or near the floor. Low lying areas, such a recesses or pits, should be adequately exhausted by adding ventilation system pickup ducts is these areas. Metal fumes are very dense, and transport with difficulty.

SUMMARY

In the mining industry, the precious metal concentrating processes used throughout the world are gravity separation, electrowinning, and zinc precipitation. Products from these unit processes are often upgraded via retorting and acid digestion, albeit careful control of the metal concentrating processes can limit contamination. Oxidizing base metal contaminants in the concentrates via calcining is not commonly practiced and was not discussed. Melt refining is used as the final dore' bullion production unit process at most mines. Ultra-refining of bullion is usually completed by others after the bullion has been sold. Concentrate processing equipment should be sized conservatively to allow for new project startup process complications. To safeguard worker hygiene, careful attention must be paid to the design and operation of ventilation systems.

REFERENCES

Anonymous. Private Communication, Coeur Rochester.

Armburst, W., D. Jensen, C.H. Sheerin, and R.A. Smith. 1989. *Removal of Selenium From the Dore' at FMC Gold Company's Paradise Peak Mine*. Presented at the 2nd Annual Elko Operators Convention.

Frentress, R. 2002. Private Communication. Round Mountain Gold.

Haldane, T. 2002. Private Communication. Glamis Gold Ltd.

Hall, D. 2001. Private Communication. Placer Dome: Musselwhite Mine.

Kahl, T. 2002, Private Communication. Barrick Goldstrike Mines.

Marsden, J., I. House. 1993. *The Chemistry of Gold Extraction*. New York: Ellis Horwood Limited.

McMillen, G. 2001. Private Communication. Florida Canyon Mining Co.

Warren, G., 2001. Precious Metal Refinery Process Selection – An Overview. *Proceedings-33rd Annual Meeting of the Canadian Mineral Processors*. 69-80.

Weast, R.C., Editor-in-Chief. 1983-1984. *The Handbook of Chemistry and Physics*. Florida: CRC Press, Inc.

PLATINUM GROUP METAL BULLION PRODUCTION AND REFINING

Corby G. Anderson [1], Lance C. Newman [2], and Greg K. Roset [3]

ABSTRACT
In recent years, the production and refining of platinum group metals (i.e. PGM's) has become increasingly important. The variety of processes used in the recovery and refining of platinum group metals are classified into two major process categories. These process categories are based on the differences in the raw material sources: one, on the recovery and refining of precious metals from the concentrates obtained from mined platinum bearing copper and nickel sulfide ores, and the other from secondary raw material sources such as recycled industrial products. Examples of the latter include spent catalysts, electronic scrap, spent electrolytes, and jewelry scrap. This paper focuses on the primary production and refining of platinum group metals. However, the basic principles utilized are analogous for the recovery and refining of secondary and recycled PGM's.

INTRODUCTION
There are two primary sources for the platinum group metals: those found in deposits in such as South Africa's Bushveld complex, at North American Palladium in Canada or at Stillwater Mining in the USA; and those derived as a by-product of primary copper and/or nickel electrorefining as from anode slimes such as at INCO in Canada or in the former Soviet Union near Noril'sk and other sites (Flett, 1984, Hunt and Lever, 1969, Kirk-Othmer, 1968, Dayton and Burger, 1981, Dayton and Burger 1982, Loebenstein, 1983, Minty, 1999). Alluvial deposits of native platinum or alloys such as osmiridium are so scarce that they are not included. The composition and grade of raw materials differ, and depend on producing mine locations from which they are obtained, as do the methods of pretreatment and of preprocessing. Platinum, Pt, and palladium, Pd, are considered the major, primary platinum group metals. Rhodium, Rh, iridium, Ir, ruthenium, Ru, and osmium, Os, are considered the minor, secondary platinum group metals. The following annotated material excerpts (Benner, 1991) summarize primary PGM recovery and refining.

PGM'S FOUND WITH NICKEL AND COPPER SULFIDE CONCENTRATES
Platinum group metals can concentrate in sulfide matte that is produced during the smelting of nickel and copper ore concentrates. Historically, in the INCO Orford process, sodium sulfide is added to the copper-nickel sulfide matte, thereby forming two separate copper and nickel matte phases. The respective metals are recovered from those mattes by a converting process that yields a crude metal product suitable for electrorefining. The anode slimes obtained by electrorefining crude nickel contains the major portion of the platinum group metals. The slime concentrates are air roasted. Copper and nickel are leached out with sulfuric acid, to yield PGM-enriched residues from which the platinum group metals are further concentrated by melting them with charcoal and litharge to form a lead button. The resulting lead alloy is nitric acid leached. The acid insoluble residue becomes a raw material for further refining (Hougan and Zacharisen, 1975). The

---

[1] Center for Advanced Mineral and Metallurgical Processing, Montana Tech of The University of Montana, Butte, Montana
[2] Stillwater Mining Company, Columbus, Montana
[3] Stillwater Mining Company, Columbus, Montana

pressurized carbonyl process practiced by INCO Canada strips nickel as gaseous nickel carbonyl, and leaves residues that contain platinum group metals (Head, et al, 1976)

PGM's from the J-M Reef at Stillwater Mining in the United States of America

The platinum, palladium and other platinum group metals are associated with nickel-copper-iron sulfides in the J-M Reef within the Stillwater complex. Stillwater is the only primary source of PGM's in the United States (Sharratt, 1994). The ore is mined and milled to produce a bulk nickel-copper sulfide concentrate. The concentrate is transported about 65 km to the Stillwater smelter, located in the town of Columbus, Montana, where it is smelted to produce a nickel-copper matte (Hodges, Roset, Matousek and Marcantonio, 1991). At the start-up of the smelter in 1990, the nickel-copper matte, containing approximately 42% Ni, 27% Cu, 22.5% S and 2.1% PGM, was shipped to a custom refiner in Europe. Later, Stillwater, in conjunction with Sherritt, developed process flowsheet for refining of the Stillwater matte to produce a high-grade PGM bullion concentrate, Sherritt's two-stage acid pressure leaching technology (Weir, Kerfoot and Kofluk, 1986). As this is a fundamental example of primary PGM production, pertinent process details follow and are illustrated in Figures 1-4.

At this time, Stillwater operates two PGM ore milling facilities. One is at the East Boulder mine while the other is at the mine at Nye. The operation of each is similar and the older Nye facility will be focused on as an operating example. The Stillwater Nye concentrator has been expanded from 450 to 2700 tonnes per day of ore capacity since it's startup in 1987 (Turk, 2000). The feed material comes from a jaw crusher system designed to crush the mined ore to minus 150 mm. Further comminution is performed via a closed circuit SAG mill, pebble mill and ball milling circuit. This produces a cyclone sized ore product of p80 145 microns that reports to rougher flotation and flash flotation based on size. The Nye concentrator uses four major flotation reagents. These are potassium amyl xanthate promoter, Cytec 3477 promoter, CMC for talc suppression and sulfuric acid for pH adjustment. In essence the operating philosophy is focused on maximum recovery of all sulfides that bear the PGM's. The rougher concentrate is subjected to two stages of cleaning followed by final column cell flotation. This produces a bulk copper and nickel sulfide concentrate containing the PGM's. The tails from the rougher circuit are reground and subjected to middling and scavenger flotation. The final mill tails report to the mine paste backfill plant or the tailings pond. The Nye concentrator flowsheet is shown in Figure 1.

Concentrate is transferred in bins from the East Boulder and Nye mills to the smelter as a filter cake. The concentrate is approximately 10% moisture and contains 0.2% of platinum and palladium per tonne in an approximate ratio of 3 to 1 Pd to Pt. The bins of filter cake are sampled, dried, and pneumatically conveyed to the concentrate bin for smelting in the electric furnace (Bushman and Roset, 2002, Hodges, Roset, Matousek and Marcantonio, 1991). Recycled PGM bearing automotive catalysts are also fed through this pneumatic system to the electric furnace. The electric furnace feed consists of dry concentrate, limestone, and dust and is fed through an airslide. Converter slag and crushed reverts are fed through two bins located above the furnace. The electric furnace is rectangular with 3 in-line prebaked electrodes. The furnace lining consists of magnesia-chrome refractory bricks, with copper coolers installed in the slag zone. The furnace produces two products, slag and matte. Furnace slag contains oxide materials (mostly SiO_2 and FeO) and is removed several times a day into a pit for air-cooling. The slag exits the furnace at 1500-1550°C (2700-2800°F) and when cool is transferred back to the ore milling facilities for recovery of any metals left in the slag. Furnace matte contains about 0.8% per ton of platinum and palladium as well as copper, nickel, sulfur and iron. It is tapped from the furnace approximately every 8 hours into ladles and granulated in preparation for converting. Granulated furnace matte is converted in Top Blown Rotary Converters (TBRC's) using oxygen. Converter slag is poured into ladles, granulated, dried and returned to the electric furnace. Converter matte is poured into ladles, granulated, dried and transported to the Base Metals Refinery (BMR). The converter matte now

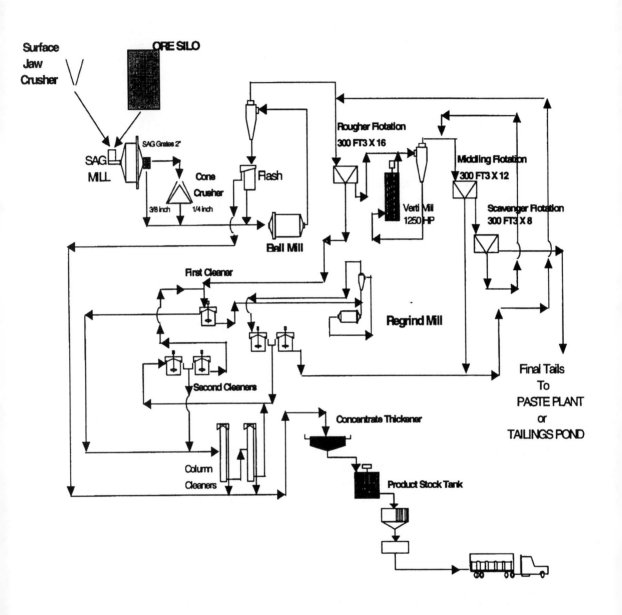

Figure 1. The Stillwater Mining Company Nye Operation PGM Concentrator flowsheet.

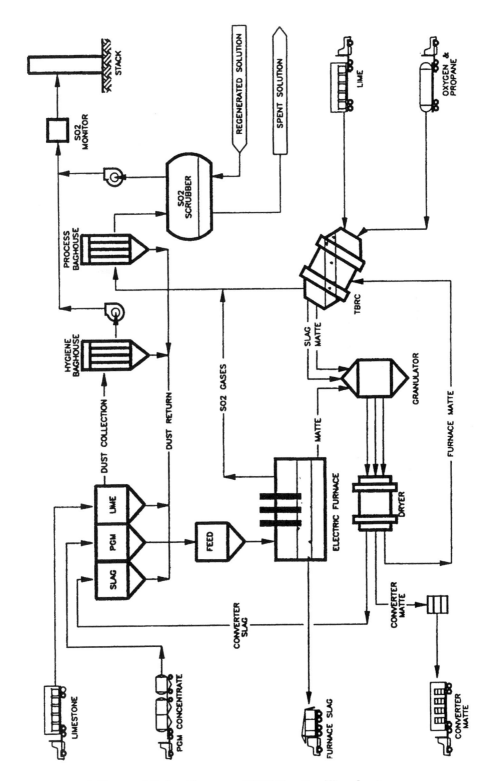

Figure 2. The Stillwater Mining Company PGM Smelter Flowsheet.

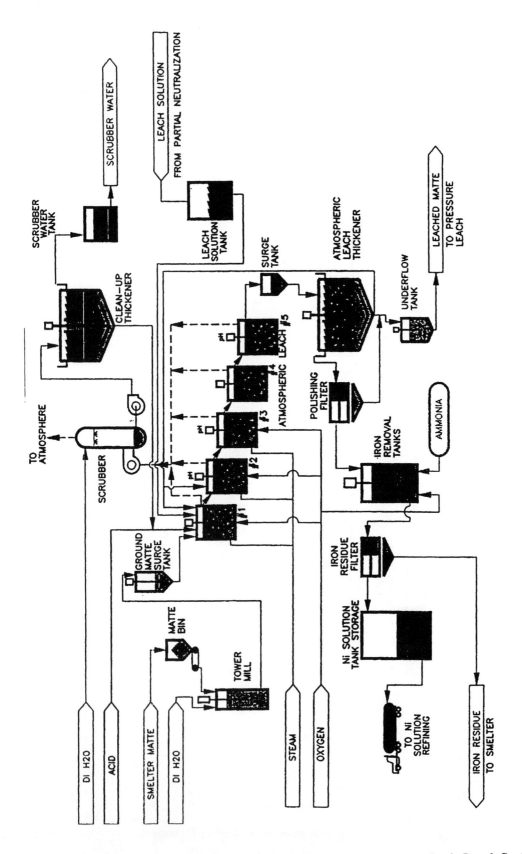

Figure 3. The Stillwater Mining Company Base Metals Refinery Atmospheric Leach Section Flowsheet.

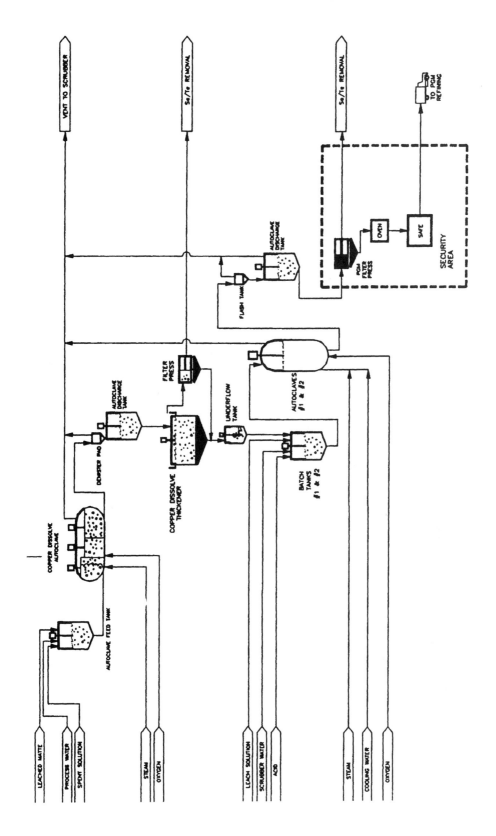

Figure 4. The Stillwater Mining Company Base Metals Refinery Pressure Leach Section Flowsheet.

contains approximately 2.0 % per tonne of platinum and palladium. All off-gases first pass through baghouses for particulate capture before being cleaned of sulfur dioxide (SO_2) in scrubbers. Over 99.7% of the sulfur dioxide is captured from the off-gases. A sodium regeneration circuit provides caustic for the scrubbing system. Gypsum generated in the regeneration process is pressure filtered into low moisture cake and is trucked away for resale to the cement or agricultural industries. The Stillwater smelter flowsheet is illustrated in Figure 2.

The Stillwater Base Metals Refinery (i.e. BMR) process consists of matte grinding, atmospheric leaching, pressure leaching, PGM concentrate separation and iron precipitation. (Newman and Makwana, 1997). Figure 3 illustrates the atmospheric leach process flowsheet. First, the converter matte from the smelter is ground batch-wise in a tower mill to yield a 70% solids slurry in water, with 80% of the solids passing 74 pm. The ground matte is leached with the recycled acidic pressure leach solution and oxygen in a series of five cascading agitated tanks. Some of the nickel and iron are extracted from the matte, while some of the copper present in the pressure leach solution is precipitated according to the following typical reactions:

$$Ni^o + H_2SO_4 + 0.5\ O_2 \rightarrow NiSO_4 + H_2O \qquad (1)$$

$$Ni_3S_2 + H_2SO_4 + 0.5\ O_2 \rightarrow NiSO_4 + 2\ NiS + H_2O \qquad (2)$$

$$Ni_3S_2 + 2\ CuSO_4 \rightarrow Cu_2S + 2\ NiSO_4 + NiS \qquad (3)$$

$$Fe^o + H_2SO_4 + 0.5\ O_2 \rightarrow FeSO_4 + H2O \qquad (4)$$

The dissolved iron remains essentially in the ferrous state under the prevailing low pH (2.0 to 2.2) condition of the atmospheric leach process. Any PGM present in the feed solution are also co-precipitated with the copper. The unleached residue from the atmospheric leach process is separated by thickening and then leached further under elevated temperature and oxygen pressure. Figure 4 illustrates the pressure leach process flowsheet. The atmospheric leach residue consists essentially of millerite (NiS), digenite ($Cu_{1.8}S$), djurleite ($Cu_{1.96}S$) and iron in the form of magnetite and hydrated ferric oxide. The principal reactions in the pressure leach process are of the type shown below.

$$NiS + 2\ O_2 \rightarrow NiSO_4 \qquad (5)$$

$$Cu_2S + H_2SO_4 + 2.5\ O_2 \rightarrow 2\ CuSO_4 + H_2O \qquad (6)$$

$$Fe_2O_3 + 3\ H_2SO_4 \rightarrow Fe_2(SO_4)_3 + 3\ H_2O \qquad (7)$$

Magnetite, if present, does not dissolve at the relatively low acid concentrations (20 to 25 g/L) and temperature (130 to 135°C) prevailing in the pressure leach process.

The atmospheric leach thickener overflow solution is polish filtered and then treated to precipitate most of the iron as ammonium jarosite, which is filtered and returned to the smelter. The iron precipitation process is necessary to meet the requirements of the nickel-copper sulphate solution customer. The iron precipitation process also acts as a backstop to precipitate or catch any soluble and insoluble PGM that may be present in the atmospheric leach solution,. The iron is precipitated as ammonium jarosite according to the following reactions.

$$2\ FeSO_4 + 0.5\ O_2 + H_2SO_4 \rightarrow Fe_2(SO_4)_3 + 3\ H_2O \qquad (8)$$

$$3\ Fe_2(SO_4)_3 + 2\ NH_3 + 12\ H_2O \rightarrow 2\ NH_4Fe_3(OH)_6(SO_4)_2 + 5\ H_2SO_4 \qquad (9)$$

The matte grinding, pressure leaching, PGM concentrate separation and iron precipitation circuits are designed to operate batch-wise, while the atmospheric leach circuit is designed to operate on a continuous basis during each shift. The plant produces a final readily refined bullion concentrate of about 60%PGM content. As well, by-product nickel sulfate and copper metal are now produced from this zero discharge hydrometallurgical facility.

PGM's from the Merensky Reef in South Africa

The platinum-bearing iron-chromite reef layers within the basic pyroxene deposit of the Bushveld igneous rock complex are accompanied by small particles of the sulfides of iron, copper and nickel. The platinum, partly in the form of native metal, is invariably found as a ferro-platinum alloy and/or as the sulfide, and arsenide of iron, copper or nickel. The platinum group metal content of the crude ore that ranges from 4 to 6 grams per ton, consists of 50 to 60 percent platinum and 20 to 25 percent of palladium. The crude ores are crushed and pulverized. Sulfide concentrates containing these metals are separated and recovered by flotation processing (Flett, 1984, Hunt and Lever, 1969, Kirk-Othmer, 1968, Dayton and Burger, 1981, Dayton and Burger 1982). Native platinum and their ferro-alloys are separated by gravity separation techniques. The sulfide concentrates are smelted in an electric furnace and produce a 25 percent copper-nickel matte. This matte is further desulfurized by oxidizing in a converting process, to produce high grade 75 to 80 percent copper-nickel matte which contains as much as 2 kg of platinum group metals per ton of matte. This high-grade matte is water granulated and finely crushed, then the base metals copper and nickel are leached from it by acid solution to leave a residue that contains most of the platinum group metals. These then become the raw materials for further recovery and refining of, the platinum group metals. This process is analogous to the previous in depth descriptions of the processes in use at Stillwater and further details will not be elucidated in this paper.

INCO CANADA CONVENTIONAL PGM BULLION CONCENTRATE REFINING METHOD

A process flow chart employed at INCO Canada (Shin Kinzoku, 1977, Kirk-Othmer, 1968), shown in Figure 5, is illustrative of the conventional methods based on dissolution and precipitation processing. Other conventional processes differ only in detail from the one shown in this figure. The chemistry differences are related to raw material characteristics. By following the process flow chart, the respective unit chemical process steps are outlined. The raw material is attacked with aqua regia to dissolve gold, platinum and palladium. The aqua regia solution is treated with ferrous sulfate. Gold is precipitated and separated for its recovery and further purification. Saturated ammonium chloride solution is added to the gold free solution to precipitate ammonium hexachloroplatinate. This precipitate is separated and calcined to form platinum metal sponge of 98 percent purity. This sponge is further refined by repeated dissolution and reprecipitation. The filtrate from the ammonium hexachloroplatinate is treated with ammonium hydroxide solution to which is later added hydrochloric acid. Diammine palladium dichloride is precipitated, separated, and calcined in a hydrogen atmosphere. The crude palladium metal obtained is further purified by repeated dissolution and reprecipitation. The original insoluble residue from the aqua regia attack is fused with fluxes composed of litharge, soda ash, borax and carbon. The lead alloy formed collects all the precious metals including the secondary platinum group metals. The lead alloy is leached in hot nitric acid solution to dissolve silver and lead. The lead in the leach liquor is precipitated as lead sulfate, and the silver, as its chloride. The nitric acid insoluble residue is fused with sodium bisulfate, forming a water-soluble rhodium sulfate salt. A water solution of this salt is treated with ammonium nitrite solution which precipitates ammonium rhodium nitrite, $(NH_4)_3Rh(NO_2)_6$. The nitrite salt is later treated with hydrochloric acid to form ammonium rhodium chloride, $(NH_4)_3RhCl_6$, which is further refined before conversion to rhodium metal. The water-insoluble residue from the sodium bisulfate fusion is next fused with a mixture of sodium hydroxide and potassium nitrate. Osmium and ruthenium form their respective water-soluble

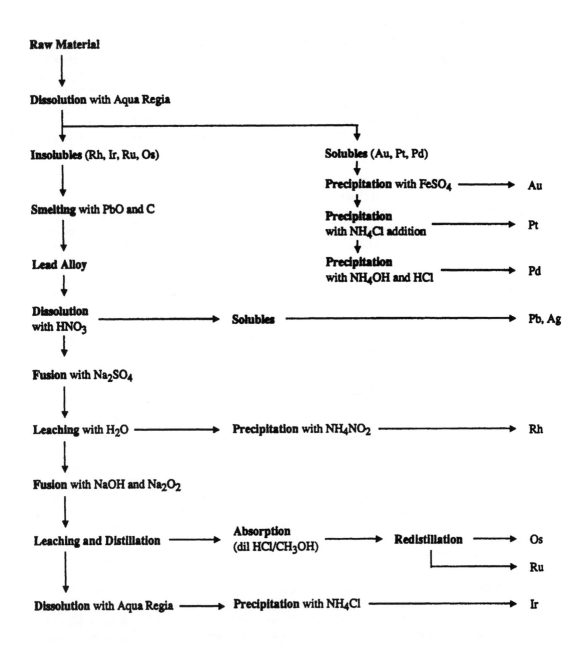

Figure 5. INCO Canada Conventional PGM Bullion Concentrate Refinery Flowsheet.

potassium salts, (potassium osmate and potassium ruthenate). Iridium forms an oxide that is leachable with aqua regia solution. After the potassium salts of osmium and ruthenium are dissolved away in water, the solution is treated with hydrochloric acid and chlorine gas. Volatile tetroxides, OsO_4 and RuO_4, are evolved and they are absorbed into a mixture of dilute hydrochloric acid and methanol. When the hydrochloric acid/methanol solution is heated, only osmium tetroxide is evolved. This is absorbed into a sodium hydroxide/methanol solution, from which ammonium osmium chloride is precipitated by the addition of ammonium chloride. This chloride salt is then calcined in a hydrogen atmosphere to produce crude osmium. This is further refined by repeated distillation and precipitation separation. Ruthenium remains as ruthenium oxychloride in the residue of the previous evaporation step. This is reduced by heating in a hydrogen atmosphere to obtain crude ruthenium. This is further treated for purification. The iridium oxide that is formed during the previous fusion treatment is dissolved in aqua regia and precipitated with ammonium chloride solution as ammonium iridium chloride. Crude iridium is obtained by calcining the ammonium iridium chloride compound in a hydrogen atmosphere. Further purification of the ammonium iridium chloride requires repeated dissolution and reprecipitation.

This summary of a typical conventional process flow chart for platinum group metals describes the separation and production of the respective crude metals. A complete separation of the respective metals cannot be made in a simple dissolution-precipitation step. Co-precipitation occurs and other metal elements remain as impurities. To obtain pure metals of the respective elements, the crude metal requires further refining. Generally, refining processes involve many repeated steps of dissolution, conditioning and precipitation operations.

PGM BULLION CONCENTRATE SOLVENT EXTRACTION REFINING TECHNOLOGY

The degree of separation attained with the conventional and traditional separations used in refining platinum group metals cannot be considered to be efficient in terms of the yields, the complexity of the operation and the labor expended. Since the early 1970's, considerable research and development efforts have been directed toward replacing traditional processes with new solvent extraction technology (Warshawsky, 1983, West, 1984, Edwards, 1984) . Flett 1983, Flett 1982, Dhara, 1984). Commercial refineries have incorporated this technology into their operations, employing various solvent extraction schemes. Major advantages of these solvent extraction processes are. Lower inventory due to the reduced overall processing time; higher separation efficiency; higher product purity; improved yields; flexibility; versatility and capability of continuous operation with process control.

The separation and refining processes that employ solvent extraction technology are based on physico-chemical characteristics of precious metal solution chemistry such as the nature of the complex ionic species and their redox potentials. Extraction schemes vary with the kind of extractant used, extraction rates, distribution ratios, separation efficiency, etc. For the commercial and industrial operation of solvent extraction processes to be successful, various operational conditions and considerations have to be established with respect to efficiency and rate of back extraction (stripping), degradation and loss of extractants, flexibility and versatility of process, etc. Depending on differences in the forms of raw material, contents and grade of precious metals in them, and pretreatments, the respective commercial refineries have developed and applied different solvent extraction technologies and processes. Currently, there are three large refineries using solvent extraction technology, and future large-scale refineries will probably employ solvent extraction technology.

The aqueous chemistry of precious metals as their chlorides in hydrochloric acid is complex. Thermodynamic and kinetic behaviors influence the solvent extraction scheme selected. The following are brief summaries of the respective element chlorocomplex behavior that have been studied and understood for application to solvent extraction. These include oxidation and coordination numbers, number of d-electrons and rates of ligand exchange reactions by amine extractants. The detailed information is contained in the cited literature.

Gold, platinum and palladium dissolve relatively easily in hydrochloric acid/chlorine mixtures. Osmium, ruthenium, rhodium and iridium and their oxides have slow or near zero dissolution rates in such mixtures. Alkaline fusion is often required before these elements will dissolve in hydrochloric acid/chlorine solutions. Various basic amine extractants are capable of forming ion-pair compounds in the organic phase. The degree of ion-pair formation depends on the size of anionic chlorocomplex ions and the ionic charge. These increase from $MCl4^{1-}$, $MCl6^{2-}$, to $MCl4^{2-}$ to $MCl6^{3-}$. The basicity of amine extractants increases from primary, secondary, tertiary to quarternary. The more basic the amine extractant, the stronger is the ion-pair formation. This will result in good extraction capability but it will be difficult to strip. Where neutral organic solvation extractants involve the extraction of chlorocomplexes, the extractability increases from TPP, TBP to TOPO, following the basicity-increase of the esters. (TPP = Triphenyl phosphite, TBP = tributyl phosphite, TOPO = trioctylphosphine oxide.) Ligand d^4 exchange rates depend on d-electron confirguations of the octahedral complexes. The order of the exchange rate is $d^5 > d^4 > d^3 > d^6$. With d^4 square planar complexes, the order is palladium(II)> gold(III) > platinum(II). Redox reaction rates also depend on d-electron configurations. It is well known that the reduction of iridium(IV) to iridium(III) is much faster than that of platinum(IV) to platinum(II).

Chlorocomplexes of gold, platinum and palladium formed in hydrochloric acid/chlorine solution are generally those of higher oxidation states. Those of iridium, rhodium, osmium and ruthenium are complex. The latter two elements particularly present many kinds of mixed aquochlorocomplexes. The forms of chlorocomplexes or aquocomplexes depend on solution potentials, chloride ion concentration and acid concentrations.

This summary is just a short abstract of the aqueous chemistry in solvent extraction technology. The process flow charts of three commercial refineries are described in the next section.

INCO ACTON BULLION CONCENTRATE REFINERY
Concurrent with the implementation of its new pressurized carbonyl process, employed to replace conventional nickel production from crude nickel metal, INCO examined its traditional lengthy and less efficient precious metals separation, recovery and refining process in light of solvent extraction. INCO developed and now operates a solvent extraction process for precious metals recovery refining (Barnes and Edwards, 1982, Rimmer, 1974). The process flow chart is outlined Figure 6.

The raw feed material is copper electrorefining anode slimes from primary copper production. These, in turn, are derived from nickel carbonyl gasification residues that were reprocessed in a copper converter. The process is as follows: The raw material is attacked at 90-98°C with hydrochloric acid/chlorine gas. The undissolved residues are attacked with nitric acid for silver and lead removal, and then fused with sodium hydroxide at 500 to 600°C. The alkaline fusion salt is then dissolved in hydrochloric/chlorine solution for the following treatments. After removal of excess chlorine from the solution, it is neutralized with sodium hydroxide. Sodium bromate, $NaBrO_3$, solution is added to it and the vapors of tetroxides of ruthenium and osmium are distilled away and absorbed in weak hydrochloric acid solution. Hydrolysis with sodium hydroxide follows and base metals such as copper are filtered out as their solid hydroxides. The solution is adjusted to contain a hydrochloric acid concentration of 3 to 4 mole/liter. Gold is extracted from the solution into dibutyl carbitol (DBC). The organic phase is scrubbed with dilute, 1-2M hydrochloric acid to remove base metals. Gold is recovered from the scrubbed organic phase by direct reduction with aqueous oxalic acid. Palladium is extracted into an organic phase with

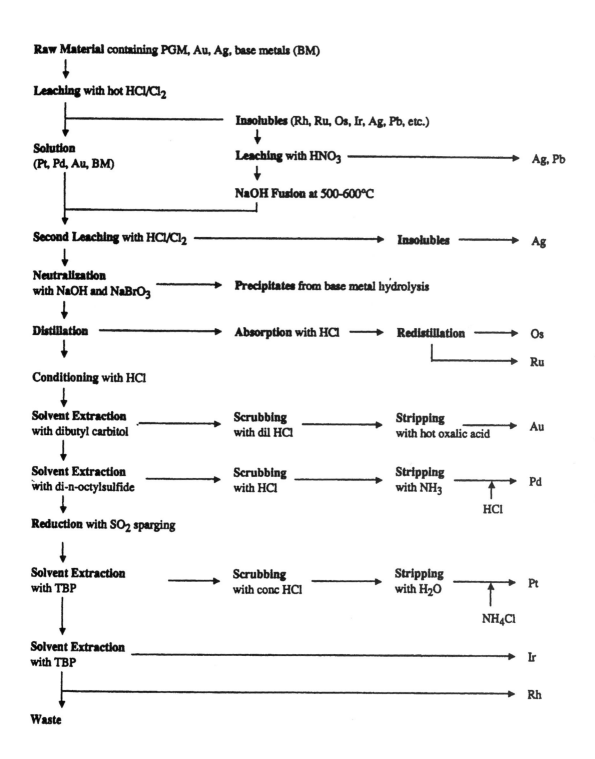

Figure 6. INCO Acton PGM Bullion Concentrate Refinery Flowsheet.

di-n-octyl sulfide, (DOS, R_2S), which leaves a palladium free aqueous raffinate phase. This process utilizes the lability of the $PdCl_4^{2-}$ ion with respect to its ligand exchange rate. The extraction exchange is described as follows:

$$[PdCl_4^{2-}]_{aq} + 2[R_2S]_{org} \rightarrow [PdCl_2(R_2S)2]_{org} + 2[Cl^-]_{aq} \quad (10)$$

The organic phase is scrubbed with dilute hydrochloric acid, then palladium is stripped from the organic phase with an aqueous ammonium hydroxide solution to form $[Pd(NH_3)_4^{2+}]_{aq}$. This ammoniacal solution is neutralized with hydrochloric acid which precipitates $Pd(NH_4)_2Cl_2$. The stripping reaction is as follows:

$$[PdCl_2(R_2S)_2]_{org} + 4[NH_3]_{aq} \rightarrow 2[R_2S]_{org} + [Pd(NH_3)_4^{2+}]_{aq} + 2[Cl^-]_{aq} \quad (11)$$

Platinum is recovered from the palladium free aqueous raffinate phase. The hydrochloric acid concentration in it is adjusted to 5 to 6 mole per liter. Iridium (IV) is reduced to iridium (III) by sparging sulfur dioxide gas into the solution. Platinum is then extracted into tri-n-butyl phosphate (TBP) by the following reaction:

$$2[H^+]_{aq}[PtCl_6(H_2O)_2^{2-}]_{aq} + 2[TBP]_{org} \rightarrow [H_2PtCl_6(TBP)_2]_{org} + 2[H_2O]_{aq} \quad (12)$$

After scrubbing the TBP organic phase with hydrochloric acid solution, it is stripped with water. Platinum is recovered as $(NH_4)_2PtCl_6$ with ammonium chloride.

Iridium in the platinum free aqueous raffinate phase is then oxidized from the (III) state back to its (IV) state. It is extracted from this aqueous phase with TBP into an organic TBP phase. Iridium is then recovered from this TBP phase in a process similar to that described above for platinum. The separation of rhodium, is still under development according to the cited literature.

MATTHEY RUSTENBURG PGM BULLION CONCENTRATE REFINERY
The Matthey Rustenburg Refinery (Cleare, Charlesworth and Bryson, 1979, Reavell and Charlesworth, 1980 and Charlesworth, 1981) having similar reasons to those of INCO, to overcome the demerits of traditional separation and refining process, have started a solvent extraction process that uses oxime/amine extractants. This technology is based on the "straight chain process". The process flow chart is shown in Figure 7. The precious metal raw materials are dissolved in hydrochloric acid/chlorine solution. Silver forms insoluble AgCl which is separated and recovered here. Gold, in solution in the form of $AuCl_4^-$, is extracted by either of the solution extraction schemes with TBP or with methyl isobutyl ketone (MIBK), into the organic phase. Its extraction reaction may be described as:

$$[H^+(AuCl_4^-)]_{aq} + [TBP]_{org} \leftrightarrow [HAuCl_4][TBP]_{org} \quad (13)$$

Other impurities, Fe, Te, etc., are extracted as well so that the organic phase is next scrubbed with hydrochloric acid for impurity removal. Gold is reduced and recovered from the organic phase with iron powder. Palladium is extracted by a beta-hydroxyoxime, making use of the ligand exchange reaction as follows:

$$[PdCl_4^{2-}]_{aq} + 2[RH]_{org} \rightarrow [PdR_2]_{org} + 2[H^+]_{aq} + 4[Cl^-]_{aq} \quad (14)$$

Since this extraction rate, i.e., the ligand exchange rate, is small, an organic amine reagent is added to accelerate the extraction. After the organic phase has been scrubbed with weak hydrochloric acid to remove base metal impurities, palladium stripping is done with aqueous ammonium

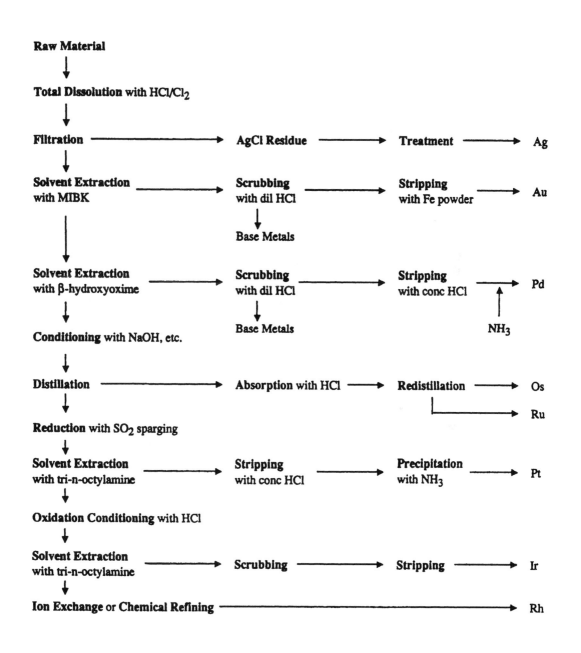

Figure 7. Matthey Rustenberg PGM Bullion Concentrate Refinery Flowsheet.

hydroxide solution. Palladium is precipitated from it as $(NH_4)_2PdCl_6$ with hydrochloric acid. Osmium and ruthenium are separated as their volatile tetroxides after the solution has been neutralized with alkaline hydroxide. The tetroxides are absorbed into dilute hydrochloric acid solution and this is later redistilled for separation of osmium from ruthenium. After reducing the iridium (IV) to its (III) state, platinum is extracted by the tertiary amine, tri-n-octyl amine into the organic phase, as:

$$[PtCl_6^{2-}]_{aq} + 2[RH^+]_{org} \rightarrow [(RH)_2PtCl_6]_{org} \qquad (15)$$

Platinum is stripped from the organic phase with 10-12M hydrochloric acid. The platinum is precipitated from it as $(NH_4)_2PtCl_6$ with ammonium chloride. After oxidizing iridium (III) back to its (IV) state, and adjusting the acid concentration to about 4 mole per liter, iridium is then extracted into the organic phase of the tri-n-octyl amine phase in the same way as the platinum. The organic phase is scrubbed with dilute acid. The iridium is stripped into the aqueous phase after reducing it to its (III) valency state and recovered. Finally, rhodium is separated and recovered from the solution containing other element impurities. The detailed information on the operation and the chemistry rhodium recovery is not available at present.

LONRHO PGM BULLION CONCENTRATE REFINERY

The Lonrho Refinery solvent extraction process flowsheet is show in Figure 8 (Berry, 1979, Edwards, 1979). This process is a modified solvent extraction technology based on a process scheme developed at the National Institute of Metallurgy, South Africa. The Impala Platinum Refinery is said to have installed a similar process at Springs, South Africa. The raw material at Lonrho has a relatively high content of secondary platinum group metals, i.e. metals other than platinum and palladium, when compared with raw material feeds at other refineries. The raw material is first treated for base metal removal by leaching with acid solution. Then it is reduced with carbon and aluminum to produce an aluminum precious metal alloy. This alloy, upon dissolution in hydrochloric acid/chlorine solution solubilizes rhodium, iridium, ruthenium and osmium in a shorter period of time. The alloy is dissolved in hydrochloric acid. The remaining insolubles are dissolved in hydrochloric acid/chlorine solution. Silver is precipitated from the solution by dilution with water. Gold is reduced with a sulfur dioxide sparge. The solution is adjusted to a 0.5 - 1.0 M acid concentration. The secondary platinum group metal valence states are adjusted to three. Platinum and palladium are extracted from the conditioned solution by the acetic acid derivative of a secondary amine, i.e. R_2NCH_2COOH. Both metals are stripped from the organic phase with hydrochloric acid. The principle of the separation of platinum from palladium is based on the fact that $PdCl_4^{2-}$ is much more labile in its ligand exchange reactions than is $PtCl_6^{2-}$. Only palladium is extracted from the raffinate into an organic sulfide phase, a ligand exchange reaction to form $PdCl_2(RSR)_2$. This is stripped from its organic phase with aqueous ammonia as follows:

$$PdCl_2(RSR)_{2\,org} + 4[NH_3]_{aq} \rightarrow 2[RSR]_{org} + [Pd(NH_3)_4^{2+}]_{aq} + 2[Cl^-]_{aq} \qquad (16)$$

Palladium in the form of $Pd(NH_3)_4Cl_2$ is precipitated by the addition of hydrochloric acid to the aqueous ammonia strip. The acidified palladium free raffinate is made alkaline and osmium tetroxide vapor is distilled from the solution and absorbed in dilute hydrochloric acid. Ruthenium is separated for recovery as an ionic ruthenium nitrosyl complex, formed by the addition of nitric acid to the filtrate. The nitrosyl complex is transferred to an organic phase by a tertiary amine extractant. The ruthenium is stripped with 10 percent sodium hydroxide and precipitated as hydroxide. This is recovered and amine extractant is washed with hydrochloric acid and returned to the extraction step. Iridium complex ions in the solution are separated by the adsorption on to strongly basic resins. After desorption by a solution saturated with sulfur dioxide, followed by oxidation of the iridium to valence (IV),

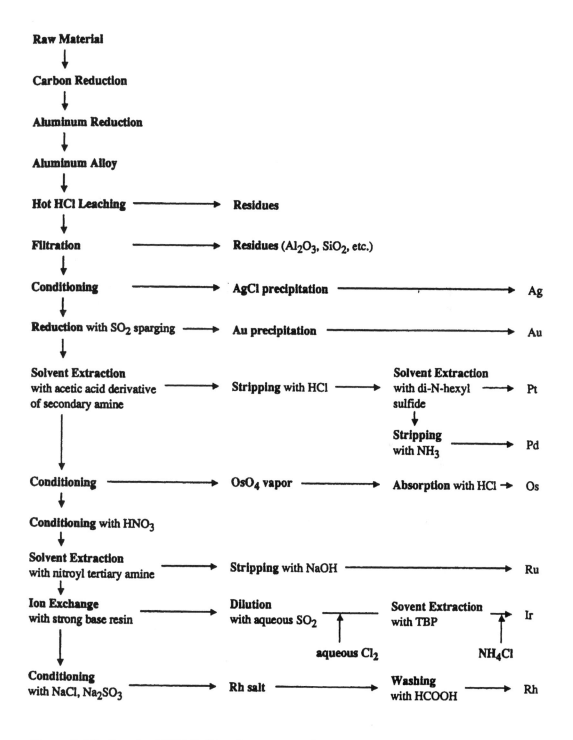

Figure 8. The Lonrho PGM Bullion Concentrate Refinery Flowsheet.

its chlorocomplex ions are extracted by TBP. The TBP phase is stripped with water and iridium in the water phase is precipitated as $(NH_4)_2IrCl_6$ by addition of ammonium chloride solution. Rhodium is precipitated out as its sodium salt by addition of sodium chloride and sodium sulfite solution and separated. This precipitate is then dissolved in acid solution and recovered in the form of $(NH_4)_3RhCl_6$ by the addition of ammonium chloride solution.

SUMMARY
In recent years, the production and refining of platinum group metals has become increasingly important. This paper focused on the primary production and refining of PGM's. However, the basic principles utilized are analogous for the recovery and refining of secondary and recycled PGM's.

ACKNOWLEDGMENTS
The authors gratefully acknowledge the expert assistance of Ms. Tami J. Cashell as well as thank Mr. G. Andrew Hadden for his literature search and careful review of the manuscript.

REFERENCES
Barnes, J.E., and J.D. Edwards. 1982, "Solvent Extraction at INCO's Acton Precious Metal Refinery", Chemistry and Industry, March 6,151.

Benner, L.S., T. Suzuki, K. Meguro, and S. Tanaka. 1991, Precious Metals Science and Technology, International Precious Metals Institute, Allentown Pennsylvania.

Berry, R.I. 1979, "Refining Precious Metals", Chemical Engineering June 18, 90.

Bushman, B., and G. Roset. 2002, Stillwater Mining Company, Personal communication, March.

Charlesworth, P. 1981, "Separating the Platinum Group Metals by Liquid-Liquid Extraction", Platinum Metals Review, 25(3), 106.

Cleare, M.J., P. Charlesworth, and D.J. Bryson. 1979, "Solvent Extraction in Platinum Metal Processing", Journal Chem. Tech. Biotech., 29, 210.

Dayton, S.H. and J.R. Burger. 1981, "Resouces, Rare Formations in Abundance", Engineering & Mining Journal, 182(11), 67.

Dayton, S.H., and J.R. Burger. 1982,"Platinum Mining Profiles", Engineering & Mining Journal, 4, 182(11), 64.

Dhara, S.C. 1984, Solvent Extraction of Precious Metals with Organic Amines, "Precious Metals", p. 199, The Metallurgical Society, AIME.

Edwards, R.I. 1976, "Refining of the Platinum Group Metals", Journal of Metals, 28(8), 4.

Edwards, R.I. 1979, "Selective Solvent Extraction for the Refining of Platinum Metals", Proceedings of International Solvent Extraction Conference, Canada.

Flett D.S. 1984, Visit Minute to Western Platinum Mines Ltd., Private Communication, Akio Fuwa.

Flett, D.S. 1982, Solvent Extraction in Hydrometallurgy, Proceedings of International Seminar, EMU, London., 9, 39.

Flett, D.S. 1982, Solvent Extraction in Precious Metals Refining, Proceedings of International Seminar, EMU, London.

Head, M.P. 1976, "Nickel Refining by the TBRC Smelting and Pressure Carbonyl Route", Paper Presented at the AIME Meeting, NV.

Hodges, G.J.,. G.K. Roset, J.W. Matousek, and P.J. Marcantonio. 1991, "Stillwater Mining Co.'s Precious Metals Smelter: From Pilot to Production," Mining Engineering, Vol. 43, No. 7, 724-727.

Hougan, L.R. and. H. Zacharisen. 1975, "Recovery of Nickel, Copper and Precious Metal Concentrate from High Grade Precious Metal Mattes", Journal of Metals, 27(5), 6 .

Hunt L.B. and F.M. Lever. 1969, "Availability of the Platinum Metals", Platinum Metals Review, 19(4), 126.

Kirk-Othmer.1968, "Encyclopedia of Chemical Technology", 2nd Ed., Vol. 15, p. 832, John-Wiley and Sons Inc.

Loebenstein, J.R. 1983, "Platinum Group Metals in the USSR", Platinum Group Metals 1983, IPMI, Williamsburg, Virginia.

Minty, K.C. 1999, "North American Palladium Ltd., Lac Lles Mines Ltd., -An Update"; PRECIOUS METALS 1999, Proceedings of the 23rd Annual Conference, IPMI, Acapulco, Mexico.

Newman, L. and M. Makwana. 1997, "Commissioning of the Stillwater Mining Company Base Metals Refinery", <u>Hydrometallurgy and Refining of Nickel and Cobalt</u>, Proceedings of the 27th Annual Hydrometallurgical Meeting, The Metallurgical Society of CIM, Sudbury, Ontario, August

Reavell, L.R.P., and P.C. Charlesworth. 1980, "The Application of Solvent Extraction to Platinum Group Metals Refining", Proceedings of International Solvent Extraction Conference, Belgium.

Rimmer, B.F. 1974, "Refining of Gold from Precious Metal Concentrates by Liquid-Liquid Extraction", bid, January 19, 63.

Sharratt, J.M., 1994, "Stillwater Mining Company", International Precious Metals Institute. Vol. 18, No. 1, Jan-Feb 3 - 6.

Shin Kinzoku Handbook, 1977, "New Metals Data Handbook", p. 282, Agne Publishing Company, Japan.

Turk, D.J., 2000 Stillwater Mining Company Nye Concentrator Operation, SME Annual Meeting, Preprint 00-10, Salt Lake City, Utah.

Warshawsky, A. 1983, Integrated Ion Exchange and Liquid-Liquid Extraction Process for the Separation of Platinum Group Metals, " Hydrometallurgy Research, Development, Plant Practice", p. 517, The Metallurgical Society, AIME.

West, R.C. Ed. (1984),"CRC Handbook of Chemistry and Physics", p. D-156, CRC Press Inc.

Weir, D.R., D.G. Kerfoot, and R.P. Kofluk R.P. 1986, "Recovery of Platinum Group Metals from NickelCopper-Iron Matte", United States Patent. No. 4,571,262, February 18.

FUNDAMENTALS OF THE ANALYSIS OF GOLD, SILVER AND PLATINUM GROUP METALS

Corby G. Anderson[1]

ABSTRACT

The analysis of silver and gold content in ores, concentrates and other materials is generally determined by classical lead based fire assay. As well, platinum group metals (PGM's) such as platinum, palladium, rhodium, iridium, ruthenium and osmium are often separated or concentrated by fire assay. This paper outlines the basic concepts involved in this proven methodology. Included are general procedures for selecting appropriate fire assay charges based on approximate chemical compositions of the samples. Also included are descriptions of modern instrumental methods for the determination of the noble metals with a focus on the analysis of PGM's.

INTRODUCTION

The determination of gold, silver, platinum, palladium, rhodium, ruthenium, osmium and iridium metals is often an important aspect of the discovery, development, design and operation of a precious metals mine, as well as the associated concentrator or metal production plant. As well, an abundance of financial issues in the mineral processing industry rely on the accurate and precise determination of the precious metals. The basic procedures and technologies involved in this discipline are discussed in this paper.

FIRE ASSAY FUNDAMENTALS

Fire assaying is an established form of quantitative chemical analysis by which metals are separated and determined in ores and metallurgical products with the aid of heat and dry reagents. Traditionally, the object is to form a melt of at least two phases - a complex liquid borosilicate slag and a liquid lead precious metal collection phase of a controlled size. For PGM's, collectors other than lead such as copper or nickel sulfide are often utilized. The high degree of solubility of the noble metals such as Ag and Au in molten metallic lead plus the great difference in specific gravity between the lead and slag permit the separation of the noble metals from the slag as lead alloys. The subsequent removal of the lead as lead oxide in a porous vessel known as a cupel by a carefully controlled oxidizing fusion separates the lead from the noble metals. The remaining metallic bead is then analyzed for the noble metals. By analogy, fire assay may be looked upon as 'pyrometallurgical solvent extraction' whereby a precious metal is collected and concentrated in a selective 'solvent'. Then the solvent is stripped from the precious metal by cupeling.

The mixture of chemicals and ore used for the pot fusion provides a complex system. By analogy with simple systems such as a metal oxide with borax or silica, one may make reasonable guesses concerning some of the reactions, but a complete explanation of the reaction for even one ore composition must await an extensive examination of these multicomponent systems. Thus, the technique of a fire assay collection of the noble metals is largely an empirical process assisted to a degree by some fundamental principles. Initially the ore to be assayed must be in an exceedingly fine state of division and thoroughly mixed with the flux constituents. These conditions are necessary to ensure the intimate contact of each ore particle with particles of the melting flux. Ideally this contact must be maintained during the early stage of the fusing process. This is necessary in order to ensure in-situ a sufficiently complete reaction between ore and flux and the simultaneous production of the fine

---

[1] Center for Advanced Mineral and Metallurgical Processing, Montana Tech, Butte, MT, USA.

globules of lead by the reduction of litharge (PbO). To bring about this condition, the composition of the flux, the temperature, and its rate of increase must be carefully arranged. The optimum viscosity is somewhat dependent upon the proportion of borax and the character of the borates and silicates that are formed at or near incipient fusion.

In the absence of this juxtaposition of lead and noble metal, one must depend upon the ability of the high-density noble metal to settle to the bottom of the pot during the subsequent fusion process. This settling process is facilitated by the increase in fluidity of the mixture at the elevated temperatures, and for platinum, palladium, and gold at least, it is also facilitated by the alloying tendency with the lead finally collected in the bottom of the crucible. Thus, for certain of the noble metals, a reasonably acceptable assay may be achieved even under unfavorable conditions. This opinion is supported by the fact that the classical lead fire assay does not accomplish the direct recovery of iridium through the formation of a homogeneous lead alloy. It is not unlikely that practically all of the iridium, and iridosmine, when the latter is present, are recovered by the fall of these high density metals through the low viscosity liquid in the later stages of the fusion. However, there is ample evidence that in general, all six of the platinum metals and silver and gold can be quantitatively recovered by careful use and manipulation of lead collection.

The beginnings of fire assaying can be traced to the finds in Troy II (about 2600 B.C.) and in the Cappadocian Tablets (2250 -1950 B.C.). These finds prove that very pure silver was made in the twenty-fifth century B.C. From this evidence we must conclude that the cupellation process, and therefore fire assaying, was invented in Asia Minor in the first half of the third millennium B.C. shortly after the discovery of the manufacture of lead from galena (Forbes, 1950). The first convincing evidence of the production of silver from lead ores is the cupel buttons found at Mahmatlar in the late third millenium B.C. and that are now in the Hittite Museum in Ankara, Turkey. Now, most methods combine fire assay precious metals collection with subsequent instrumental determination. The advantage of fire assay techniques for the determination of noble metals is the ability to use a relatively large ore sample from which to concentrate these metals, in addition to eliminating virtually all the associated gangue minerals.

The chemical reactions which take place in an assay fusion are very complex and beyond the scope of this paper. Detailed treatment of the theory involved can be found in textbooks on fire assaying (Bugbee, 1940; Shepard and Dietrich, 1940). However, one must initially determine whether the ore is neutral, reducing, or oxidizing in character. Neutral ores have no reducing or oxidizing power and usually are the siliceous, oxide, and carbonate ores. Reducing ores decompose litharge to form metallic lead and usually consist of sulfides and carbonaceous matter. Oxidizing ores contain ferric oxide and manganese dioxide which, when fused with fluxes, oxidize lead or reducing agents. Ores with considerable oxidizing power are comparatively rare. The fundamental overall chemical equation for the process of litharge reduction can be expressed as:

$$2 PbO + C \rightarrow Pb + CO_2 \qquad (1)$$

The fundamental overall chemical equation for the process of lead oxidation is expressed as:

$$2 Pb + O_2 \rightarrow 2PbO \qquad (2)$$

Fire assay is better facilitated if the chemical and mineralogical composition of the ore is known or can be obtained. For this reason it is advisable to make a semi-quantitative analysis via a suitable instrumental method along with a coarse mineralogical examination of the sample either visually or with x-ray diffraction. From this information one can usually determine whether an ore is neutral, reducing, or oxidizing. All this information is necessary to prepare an optimum charge so that maximum recovery of the noble metals may be realized. If the description and an analysis of a sample or ore indicates the presence of sulfides, a preliminary fusion is recommended to establish the "reducing power" of the ore. Figure 1. illustrates this (Ammen, 1984).

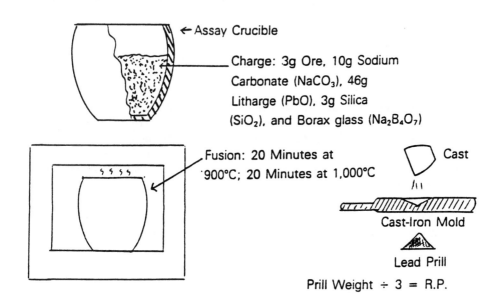

Figure 1 Establishing the reducing power of an ore

The term "reducing power," as used in fire assaying, is defined as the amount of lead that 1 g of the ore will produce when fused with an excess of litharge. Conversely, oxidizing power is the amount of lead 1 g of ore will oxidize to lead oxide. When the need arises to establish the reducing power of an ore, a preliminary fusion is employed typically using the following charge constituted in a 10-g fireclay crucible: 3 g ore, 10 g Na_2CO_3, 46 g PbO, 3 g SiO_2, and 1 g $Na_2B_4O_7$. The fusion is performed at a temperature of 900°C to 1000°C for 40 minutes. The lead button produced from this is weighed and the reducing power is then calculated. For example, if 3.00 g of ore is used and a lead button is obtained weighing 15.00 g, the reducing power of the ore is 15.00/3.00 = 5.00. In some cases, a small amount of flour is added in the preliminary fusion to ensure that, in the case of neutral or oxidizing ores, a lead button is produced.

The approximate reducing power (R.P.) of some common minerals and reagents are listed in Table 1.

Table 1. Approximate reducing power (R.P.) of some common minerals and reagents

Mineral or reagent	R.P.
Flour	10-11
FeS_2 Pyrite	11
PbS, Galena	3-4
Cu_2S, Chalcocite	5
FeAsS, Arsenopyrite	7
Sb_2S_3 Stibnite	7
$CuFeS_2$ Chalcopyrite	8
ZnS, Sphalerite	8
FeS, Pyrrhotite	9
Fe, Metallic iron	4-6
C, Carbon	18-25

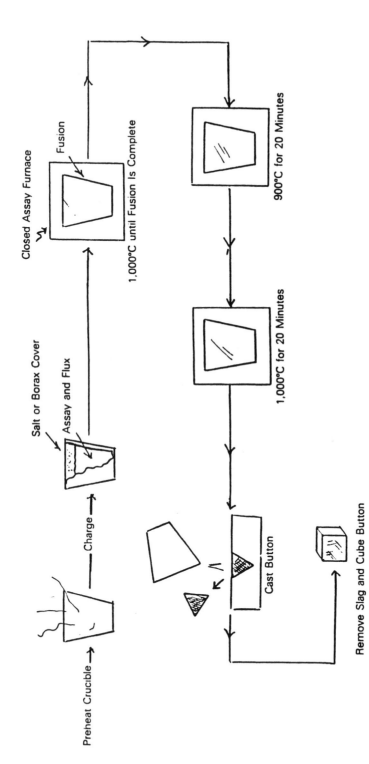

Figure 2 Flow chart for crucible fusion

An oxidizing ore is one that will not reduce PbO. Flour or another reductant must be used to produce a lead button. Oxidizing ores seldom need more than a slight increase in the amount of flour used. Table 2. lists some oxidizing ores and reagents and their oxidizing power (O.P.):

Table 2. Approximate oxidizing power (O.P.) of some common minerals and reagents

Mineral or reagent	O.P.
Fe_2O_3, Hematite	1.3
MnO_2, Pyrolusite	2.4
Fe_3O_4, Magnetite	0.9
KNO_3, Niter	4.2
Magnetite-ilmenite	0.4-0.6

Having established the reducing power of the ore, the following calculations are used to determine the amount of oxidant in the form of nitre, required to obtain a desirable-sized button (28 -30 g) when 15 g of sample is used:

Lead Total reducing effect of ore 15 x 5.00 =	75.0 g
Lead button desired =	30.0 g
Difference, ore equivalent that must be oxidized by niter =	45.0 g
1 g of niter oxidizes (from Table 2)	4.2 g
Niter required = 45/4.2 =	10.7 g

Dry reagents or flux components are added to the pulverized ore in a fireclay crucible to effect a fusion at an easily attained temperature. Each reagent serves a specific purpose in the fusion process, as follows:

Sodium carbonate (soda). - Na_2CO_3, is a powerful basic flux and readily forms alkali sulfides during the crucible fusion. Some sulfates are formed in the presence of air, and for this reason Na_2CO_3 can be considered a desulfurizing and oxidizing agent. It melts at 852° C; when heated to 950°C it undergoes a slight dissociation with the evolution of a small amount of CO_2 and the liberation of about 0.4 percent of free alkali. Both the free alkali and sodium carbonate react to form silicates and aluminates

Litharge.- PbO, is also a basic flux and acts as an oxidizing and desulfurizing agent. It melts at 883°C and on being reduced it provides the lead necessary for the collection of the noble metals. Litharge has such a strong affinity for silica that, if the crucible charge does not contain enough silica, the PbO will attack the crucible walls and, if left long enough, will eat a hole through the crucible.

Silica. - SiO_2, is a strong acidic reagent which combines with the metal oxides to form silicates, the foundation of almost all slags. It is added to the charge when the ore is deficient in silica to give a more fluid melt and to protect the crucibles from the corrosive action of litharge.

Borax glass. - $Na_2B_4O_7$, is extremely viscous when melted, but at a red heat it becomes fluid and a strong acid, dissolving and fluxing practically all the metallic oxides both acidic and basic. In addition, the fact that it fuses at a low temperature facilitates slagging of the ore and lowers the fusing point of all slags. For these reasons it is used in almost every crucible fusion.

Calcium fluoride. - CaF_2, is used especially when the aluminum content of the sample is 1 percent or more. It increases the fluidity of almost any charge.

Flour. - Flour is a reducing agent because of the carbon that it contains and is commonly used in the crucible charge.

Potassium nitrate. - KNO_3, commonly known as niter, is a strong oxidizing agent. It melts at 339°C, but at a higher temperature it decomposes giving off oxygen which oxidizes sulfur and many of the metals. Potassium nitrate is used chiefly to oxidize sulfide-bearing ores. It is advisable to establish the reducing power of the sample because excessive amounts of nitre may cause boiling-over of the charge.

The most important factor for a successful crucible fusion is the proper selection and amounts of the flux components. A good flux will produce a slag with the following characteristics:

1. It must have a formation temperature within the temperature range of the assay furnace.

2. It must remain sufficiently thick at or near its formation temperature to allow any precious metals present to be released from their chemical or mechanical bonds with the gangue before the flux allows the collector particles of lead to drop down and alloy with the precious metal values.

3. It should become sufficiently thin when heated above its formation temperature to allow the reduced lead globules to settle through it easily.

4. It should completely decompose the gangue to a fluid slag, and it should have a very low affinity for gold and silver.

5. Its chemical composition should be such that it does not excessively attack or flux away the crucible.

6. Its specific gravity should be low enough to give a good separation between the lead and the slag.

7. When the slag is cold, it should be homogeneous and easily removed from the button.

8. It should not retain higher oxides (oxides containing more than 2 0 atoms per molecule) of metal, and yet, at the same time, should contain all the impurities present in the gangue.

9. It should be free of sulfides.

When the desired flux has been worked up based on the preliminary fusion to determine the reducing power of the ore and on the examination of the ore the fusion can proceed. Because the selection of flux components is based primarily upon a knowledge of the chemical or mineralogical composition of the sample, it is advisable to perform an analysis of the sample prior to constituting the flux. By comparing the results for elements (of 1 percent or more) obtained from the semi quantitative analysis with the approximate chemical composition of the various types of samples listed in Table 3 (Haffty, Riley and Goss, 1977), the proper charge can generally be constituted. In depth flux calculations are beyond the scope of this paper and are covered in other sources (Shepard and Dietrich, 1940)

Experience has shown that for basic and ultra basic rocks, including mineralized basic rocks, a flux high in borax should be used. This flux is also used for silicates where the Fe and Mg are each 5 -10 percent or more. A convenient flux to use for these types of samples consists of 30 g Na_2CO_3, 35 g PbO, 4 g SiO_2, 35 g $Na_2B_4O_7$, 1 g CaF_2, and 3.2 g of flour for a 15-g sample. One component, SiO_2, may be varied from 4 to 8 g depending upon the silica content of the sample.

Table 3. Charges for various types of samples

[G, Greater than 10 percent to maximum for the compound, or greater than value shown. R.P., Reducing power of the sample. Leaders (....) indicate charge component was not used]

Sample type and auxiliary applications	Typical analyses, in percent		Charge, in grams							
			Sample weight	Na$_2$CO$_3$	PbO	SiO$_2$	Na$_2$B$_4$O$_7$	CaF$_2$	Flour	KNO$_3$
Aluminum[1] oxide (reagent)	Al	G	15	30	70	15	10	8	3.0	0
Arsenopyrite R.P. = 2.82	Fe Ca As Co	1 3 G G	15	25	90	8	10	1	0	3.5
Basalt	Si Al Fe Mg Ca	7 — G 5 — 10 5 — 10 5 — 10 5 — 10	15	30	35	4–8	35	1	3.2	0
Calcite	Ca	G	15	20	50	14	12	0	3.0	0
Ceramic[1] (Al$_2$O$_3$)	Si Al Mg	5 G 1 — 2	15	30	70	15	10	8	3.0	0
Chromite concentrate (refractory) 5 times for 15 g	Si Al Fe Mg Cr	5 10 G G G	3	30	35	10	30	1	3.5	0
Auxiliary flux 5 times for 15 g			....	10	....	5	25	....	....	....

Table 3. Charges for various types of samples - continued

Sample type and auxiliary applications	Typical analyses, in percent		Sample weight	Charge, in grams						
				Na$_2$CO$_3$	PbO	SiO$_2$	Na$_2$B$_4$O$_7$	CaF$_2$	Flour	KNO$_3$
Chromitite (chromium-rich ores)	Si Al Fe Mg Cr	1 — G 1 — G 7 — G 7 — G — G	15	30	35	10	30	1–2	3.2	0
Auxiliary flux			...	10	...	5	25	...	...	...
Copper oxides and carbonates	Cu	G(26.3)	15	20	100	8	5	1	3.0	0
Copper sulfides	Cu	G(20.4)	15	20	100	10	5	1	0	7.5
Diabase	Si Al Fe Mg Ca	7 — G 7 — 10 7 — 10 2 — 7 1 — 5	15	30	35	4	35	1	3.2	0
Dolomite	Mg Ca	7 — 10 — G	15	30	34	6	35	0	3.4	0
Dolomite (impure)	Mg Ca	3 — 7 7 — 10	15	25	50	10	8	0	3.0	0
Dunite	Si Fe Mg	7 — G 3 — 5 — G	15	30	35	4–8	35	0	3.2	0
Fluorite rock	Si Al Fe Ca	<1 — 7 1 — 2 1 — 2 — G	15	25	40	15	8	0	3.0	0
Galena R.P. = 3.41	Pb	G	15	20	50	5	3	0	0	5.3

Table 3. Charges for various types of samples - continued

Sample	Components									
Galena and sphalerite, composite R.P. = 3.16	Si 10 Fe 2 Pb G(15.2) Zn G(15.6)		15	25	55	13	5	1	0	4.0
Greenstone	Si 7 Al 5 Fe 1 Ca G5 K	G	15	25	45	3	10	1	3.0	0
Hematite (iron ore)	Si 2 Al 1 Fe — Ca 1	G	15	25	60	12–15	7	1	3.8	0
Jasperoid	Si	G	15	20	50	1	3	0	2.8	0
Kaolin rock[1]	Si 7 Al — Fe <1 Ca —	G G 3 G	15	20	50	12	10	5	2.8	0
Magnetite ore	Fe	G	15	25	50	15	8	4	4.0	0
Manganese-rich ores	Si <1 Al <1 Fe 7 Mg <1 Ca 1 Na <1 K <1 Mn G2	G 7 10 3 7 3 7	15	30	35	6–12	35	1	3.2	0
Norite	Si 3 Al 7 Fe 3 Mg 3 Ca 3	G G G G G	15	30	35	4–8	35	1–2	3.2	0

Table 3. Charges for various types of samples - continued

Sample type and auxiliary applications	Typical analyses, in percent		Sample weight	Charge, in grams						
				Na_2CO_3	PbO	SiO_2	$Na_2B_4O_7$	CaF_2	Flour	KNO_3
Peridotite	Si 7 – Al <1 – Fe 3 Mg Cr <1 –	G 7 G G1	15	30	35	4–8	35	0–1	3.2	0
Pyrite R.P. = 10.56	Fe	G	15	30	60	12	10	0	0	30
Pyroxenite	Si 1 – Al 3 – Fe 7 – Mg 5 Ca 10	G G G	15	30	35	4	35	1–2	3.2	0
Pyrrhotite R.P. = 5.89	Si 7 Al 1 Fe Mg 1 Ni 1	G	15	35	70	10	10	0	0	15
Quartz	Si	G	15	20	40	0	3	0	3.0	0
Quartzite	Si	G	15	20	50	1	3	0	2.8	0
Rhyolite	Si Al 3 –	G 5	15	20	50	1	3	1	2.8	0
Black sand	Si 5 Al 2 Fe G2 Ti Cr 1	G	15	30	40	20	8	5	2.5–3.8	0

Table 3. Charges for various types of samples - continued

Sample	Composition									
Serpentinite (serpentine rock)	Si 2 — Fe 5 — G 7 Mg G	15	20	30	6	30	2	4.0	0	
Siliceous material	Si G Al 10 Fe <3 Mg <3 Ca <3 Na <4 K <4	15	20	50	0–3	3–5	1	2.8–3.0	0	
Sphalerite R.P. = 7.00	Si 1 Fe 3 Pb 1 G(59) Zn	15	30	90	12	10	0	0	18	
Sulfides (massive) twice for 15 g R.P. = 8.07	Fe G Cu 5 Ni 5	7.5	35	70	12	10	0	0	7.0	
Sulfides (massive) in gabbro R.P. = 4.28	Si 7 Fe 10 Mg 1 — Ca 5 As 1 Cu G2 2 Ni 2 Sb 3	15	25	100	14	5	0	0	8.5	
Sulfide vein twice for 15 g R.P. = 4.99	Si 3 Al 1 — 2 Fe — G Mg 1 — 2 Ca 1 — 2 Cu 7 Ni 10	7.5	35	70	10	10	0–1	0	2.2	

Table 3. Charges for various types of samples - continued

Sample type and auxiliary applications	Typical analyses, in percent	Sample weight	Na$_2$CO$_3$	PbO	SiO$_2$	Na$_2$B$_4$O$_7$	CaF$_2$	Flour	KNO$_3$
Tuff	Si 7 G Al 3 Na 3 K 3	15	20	50	1	3	1	2.8–3.0	0
Auxiliary applications:									
Preliminary fusion to determine R.P. of sample		3	10	46	3	1	0	0	0
"Wash" for shot in slag		...	25	35	10	3	0	2.9	0
Litharge for Ag or Au		...	20	100	15	10	0	3.0	0
Determination of Ag and Au in inquarts (run 2–4 singly)		1 inquart	20	40	15	3	1	3.0	0
Determination of R.P. for flour		...	5	60	5	0	0	2.5	0

[1] Also add 10 g of K$_2$CO$_3$ to the charge.

CaF_2 is another flux component which may vary widely (0 -15 g) depending upon the sample being fused. As mentioned previously, it increases the fluidity of most charges. When the elemental aluminum concentration of the ore or sample is less than 1 percent, CaF_2 is not added; if the aluminum concentration is 1-10 percent, 1 g of CaF_2 is added to the charge; if it is 10-20 percent, 2 g is added. For high-grade aluminum bearing samples as much as 8 g may be used. Other samples not necessarily containing high aluminum but requiring an excessive amount (4 g or more) of CaF_2 are black sands, magnetite, and calcium phosphate (bone ash).

For difficult samples like chromite a high borax charge in addition to an auxiliary flux is used. The main charge contains 30 g Na_2CO_3, 35 g PbO, 10 g SiO_2, 30 g $Na_2B_4O_7$, 1 g CaF_2, and 3.2 g flour plus 15 g of sample and an added noble metal (Ag or Au) that serves to collect other noble metals during cupellation. These components are mixed well in a "30-g" fireclay crucible. The auxiliary flux consists of 10 g Na_2CO_3, 5 g SiO_2, and 25 g $Na_2B_4O_7$, added to and mixed well in a "20-g" fireclay crucible. After fusion of the contents of both crucibles, the melt in the "20-g" crucible is added to the main charge.

When copper, nickel, or manganese is present in sulfides in appreciable amounts (2 -5 percent), all components of the flux are increased. This is to increase the volume of the melt so that the constituents of the sample will dissolve more readily. For example, in the analysis of a sample described as a sulfide-bearing fissure vein, the percentages of the following elements were determined: Fe > 10; Zn > 10; Mn >.2; Pb, 3; Cu, 1; Si, 2; and Al, 0.2. By a preliminary fusion the "reducing power" of this sample was determined to be 5.37. The first flux, used for a sphalerite, was made to contain 20 g Na_2CO_3, 60 g PbO, 8 g SiO_2, 6 g $Na_2B_4O_7$, and 12 g KNO_3 for a 15-g sample. After fusion, this flux produced an undesirable stony slag. A repeat fusion, where all components of the flux were increased with the exception of KNO_3, contained 30 g Na_2CO_3, 90 g PbO, 12 g SiO_2, 10 g $Na_2B_4O_7$, and 11 g KNO_3. This flux produced an acceptable slag and lead button. The preceding example shows that a flux used for a mineral such as sphalerite (ZnS) would not necessarily produce an acceptable fusion when other elements or compounds are present in sufficient amounts.

Siliceous samples or ores containing 60 percent silica or more and low in the ferromagnesian silicates usually give good fusions using the following flux: 20 g Na_2CO_3, 50 g PbO, 0-3 g SiO_2, 3-5 g $Na_2B_4O_7$, 1g CaF_2, and 2.8 g flour for 15 g of sample. Samples high in aluminum content are difficult to fuse and for this reason K_2CO_3, in addition to the other flux components, is added to the charge. The mixture of K_2CO_3 and Na_2CO_3, lowers the fusion temperature more than would either compound alone and, therefore, allows more time for reaction to take place. From the data given in Table 3 one can also adjust the charge if a mixture of minerals is contained in one sample. For example, Table 4 lists the charges for calcite and quartz, and an adjusted charge if the sample were to contain half calcite and half quartz; all quantities are in grams.

Table 4. Fire Assay Fusion Flux Charge for Various Minerals and Mineral Mixtures.

	Calcite	Quartz	Half Calcite, half quartz
Ore,	15g	15g	15g
Na_2CO_3	20g	20g	20g
PbO	50g	40g	45g
SiO_2	14g	0g	7g
$Na_2B_4O_7$	12g	3g	8g
CaF_2.	0g	0g	0g
Flour	3g	3g	3g

A seen, the adjusted charge is obtained by calculating the difference in the amount of each flux component and dividing by two. The result is then added to the smaller figure.

Fire Assay Sample Preparation

In preparing the sample for analysis one must constantly be on guard to prevent contamination as well as provide a representative sample. Many samples vary in form as well as composition. However, the basic procedure for pulverizing most samples is essentially the same. For example, if the samples are received in huge chunks, they are reduced to pieces of about 5 cm with a sledge hammer and steel plate. They are then passed through a large jaw crusher which reduces the size to about 1.9 cm. Following this, the samples are passed through a small jaw crusher which reduces them to pea size about 0.5 cm or smaller. When the amount of crushed sample is large, it is next passed over a Jones splitter one or more times to obtain enough representative sample to fill a 120-ml capacity container. The 120-ml sample is passed through a vertical ceramic pulverizer that grinds the sample to about 100% passing 100 mesh. In the fire assay, 15 g of sample is usually used. The concentration of each noble metal in the sample is reported in parts per million. In countries where the metric system is used, values are reported in grams per metric ton, which is the equivalent of parts per million. The latter can be easily converted (using appropriate conversion factor) to various units for expressing concentration whether it is North American, English, or other systems. In North America, it is customary to use a factor weight of sample such that each milligram of a noble metal in the sample is equivalent to one troy ounce in one avoirdupois ton (2000 lb) of ore. As one ton contains 29,167 troy ounces, the assay ton containing 29.167 g is the factor weight normally used. Milligrams per assay ton are reported as troy ounces per short ton of ore. In England and Australia the long ton of 2240 lb is used, and the factor weight becomes 32.667 g. These factor weights are termed assay tons.

Theory of The Fire Assay Crucible Fusion

Most ores are by themselves infusible, but if finely pulverized and mixed in proper proportion with flux components in a fireclay crucible the mixture will fuse at an easily attained temperature. The ore and flux components are so intimately mixed that each particle of ore is in contact with particles of the flux components. As the temperature of the mass is gradually raised, part of the litharge (PbO) is reduced to lead commencing at about 550-600° C by the carbon in the flour, as previously illustrated in equation 1, or by the sulfides innate to sample. The mist of lead droplets produced collects or alloys with the noble metals released from the surrounding particles of decomposed sample. Part of the PbO forms slag-miscible compounds, such as the lead silicates, which are absorbed by the slag. The conditions should be such that the slag remains viscous until the ore particles are thoroughly decomposed and every particle of the noble metals has been taken up by the adjacent suspended droplets of lead. After this point has been reached, the temperature is raised to ensure that the slag is thoroughly fluid. This condition allows the lead droplets to accrete and fall like fine raindrops through the slag to form the lead button in which the noble metals are concentrated.

Fire Assay Fusion Procedure

The furnace is brought to a temperature of 1000°C. Using heat protective equipment, the crucibles with their contents are now placed in the furnace in rows of four using a charging fork and starting at the rear of the furnace. A space of about 0.5 cm between crucibles is maintained in the event of boil over. After the furnace door is closed, the temperature is turned down to 900°C for about 20 minutes. In the meantime, the iron molds in which the fusions are to be poured are placed on the steel table with their aluminum covers. After 20 minutes the temperature is raised to 1000°C for an additional 20 minutes. At the end of this time, the furnace door is opened, and the fusion crucibles are removed singly using the crucible-scorifier tongs. After each removal, the door of the furnace is closed by tripping the switch with one hand, being careful to clench the fusion crucible firmly with the crucible-scorifier tongs in the other hand. This is illustrated in Figure 2 (Ammen, 1984).

The bottom of the fusion crucible is tapped lightly on the steel table and the melt is swirled to ensure that it is liquid and of the right consistency. The melt is then poured into the iron mold, after which the crucible is rapidly rotated vertically so that the liquid does not run down the sides of the crucible. After every two fusions are poured, the cover of the mold is advanced to prevent the solidifying slag from ejecting. After all fusions have been poured, the surface of the furnace is raked and smoothed with hot powdered bone ash stored at the rear of the furnace.

After the crucibles and melts have cooled, the crucibles are examined for shot, and the slag of the cooled melt is broken with a steel rod and hammer. The small amount of slag remaining on the lead button is removed by tapping it with the steel rod. The button is then shaped into a cube with a hammer and anvil. The corners are rounded slightly for convenience of handling. If the button appears somewhat brittle, it should not be hammered more than necessary or it may flake and crack. The buttons are marked in pencil with the last three digits of the sample number and placed on a button tray, ready for cupellation.

Fire Assay Refusions
Repeat fusions of a sample are conducted (usually with a different flux) when an undesirable fusion was obtained and sufficient sample is available. When the sample is insufficient, the slag from the first fusion can be reprocessed. Indications of an unsuccessful fusion are (a) the slag contains shot or globules of lead; (b) the melt is viscous and gives a sloppy pour; (c) the charge is too siliceous as is indicated by glassy streamers, or too basic which gives a muddy pour and a stony appearance. If conditions are too basic slag is high in litharge and an insufficient amount of lead was reduced. This is recognized by its higher-than average specific gravity and even a crystalline character. There are also unsuccessful fusions where the sample has not even decomposed.

Unsuccessful fusions usually fall into one or more of the categories just listed based on their pour and the nature of their slag. Knowing which flux components to alter and the proper amounts to use to obtain a successful fusion may require considerable thought and experimentation. In some instances generalities may be stated, but in others the problem is unique.

No specific set of conditions are known for the cause of shot in the slag. Consequently, no generalities can be stated to avoid the production of shot. However, some hypotheses or observations may be or have been suggested. For example, if the melt is viscous it may cause the globules of lead to remain in suspension and prevent them from combining to form the lead button. Or, the globules of lead may become coated and thus prevented from combining. Making the melt more fluid in the former instance may aid in the combining and collection of the lead globules. In the latter instance, the causes for coating of the lead globules require more investigation. However, the addition of a considerable amount of $Na_2B_4O_7$ (30 -35 g), as in fusing chromite ores, seems to prevent coating of the globules in many cases.

When a viscous pour is evident, steps must be taken to increase its fluidity by increasing one or more flux components, frequently Na_2CO_3 or PbO or $Na_2B_4O_7$, but sometimes SiO_2 or CaF_2. If a highly siliceous pour is indicated, the usual treatment is to increase the Na_2CO_3 by 5-10 g, and when the pour is too basic, to increase the SiO_2 and $Na_2B_4O_7$. Both pours need additional PbO (about 20 g).

If, during the pour and after cooling, material similar to the sample appears on the side of the fireclay crucible and in the slag, the sample obviously has not decomposed. This condition suggests that the proper flux may not have been selected. Further study and experimentation to select the proper flux will usually solve this problem. However, in a few instances, such as when fusing chromite concentrates, it may be necessary to decrease the size of the sample from the routine 15 g to 3 g to obtain an acceptable fusion.

Fire Assay Scorification
Scorification may sometimes be used as a substitute for a fusion assay. It is used extensively for certain types of silver and gold assays such as bullion or recycled high-grade precious metals. In these instances it involves mixing the ore sample with lead, and covering with borax. The reactions involved are similar to the pot fusion, but there is also a series of reactions between air and the constituents of the charge. The oxidized lead

forms part of the fusion mixture, and the residual lead acts as a collector. The scorification assay is not, in general, recommended for accurate noble metal determination in minerals. The process is essential, however, for the reduction of button size or for cleaning a badly contaminated button. In this instance the button is transferred to a scorifier, and covered with a few grams of borax and silica. At the required temperature of 1050-1100°C and in the presence of air the melted lead is oxidized, and together with the borax and silica forms a slag with the oxidized base metal contaminants. The slag moves progressively to the periphery of the scorifier, and the molten lead forms the center or *eye* of the fusion. The melt may then be poured as in the crucible fusion. This is illustrated in Figure 3 (Ammen, 1984).

Fire Assay Cupellation
The lead button obtained from the crucible fusion or scorification is treated by a process called cupellation to separate the noble metals from the lead. This consists of an oxidizing fusion in a porous vessel called a cupel. The lead oxidizes rapidly to molten PbO, 98.5 percent of which is absorbed by the cupel and 1.5 percent of which is volatilized. The cupel is a shallow cup typically made of compressed bone ash (calcium phosphate) with or without a binder added. A high-grade bone ash cupel will absorb its weight in litharge. When this process has been carried to completion, the noble metals are left on the cupel in the form of a bead. The cupel itself may be regarded as a membrane permeable to molten litharge and impermeable to lead and the noble metals. First, the size of the cupel is selected based on the weight of the lead button. For buttons weighing 32 g or less, 3.8-cm-diameter cupels are used, and for buttons weighing 32-48 g, 4.4-cm-diameter cupels are used. Due to the intense heat of the furnace and for convenience of handling, no more than 18-24 buttons should be cupelled at a time. The cupel should weigh approximately one-third more than the weight of the lead button being cupeled. Cupels can be made or purchased. There are several substitutes for bone ash-such as cement and magnesia. The end product of cupellation is a small dore bullion bead composed of the less-readily oxidized metals-generally, silver and gold, but also any metals of the platinum family that were present in the collection button.

The cupellation procedure is simple, and if followed step by step, good results should result with few problems. However, without care, in excess of 5% of the silver values can be lost during cupellation. Thus, cupellation must be done methodically as follows:

Step 1: Heat the cupellation furnace to 850°C.

Step 2: Set out some cupels that are sound, dry, and free from dust. Purchased cupels are usually the proper density and are sound. The cupel has to be sufficiently permeable to allow the litharge to be easily absorbed, but at the same time not be so porous as to allow the bead or lead to sink in. The surface tension of the lead and of the resulting dore bead is such that neither will penetrate into the cupel. Place the cupels in the furnace with an extra row of empty cupels between them and the door. This row is to prevent the cupels in use from the thermal variances and shocks to which they would be subjected when the door of the furnace is opened. Heat the cupels for 15 to 20 minutes with the door closed and the temperature at 850°C.

Step 3: Hammer the lead fusion buttons into rough cubes or "prills". Open the door, quickly place the prills in the hot cupels, and reclose the door. Use a peep hole in the furnace door occasionally to observe what is occurring inside. Each lead cube quickly melts and spreads out. The top of the lead pool in the cupel is covered with a black layer of lead oxide, slag, dirt, and other solids that have floated to the top. In a minute or two, this covering layer flows to the edge of the cupel and starts to soak into the cupel. The metallic lead, which had been covered, is then exposed. This is called "opening up" or "uncovering." At the opening-up stage, the remaining solids are swept aside and the exposed bare surface of the lead begins to burn. The lead oxide continues to drain off and soak into the cupel, although a small percentage of the lead vaporizes.

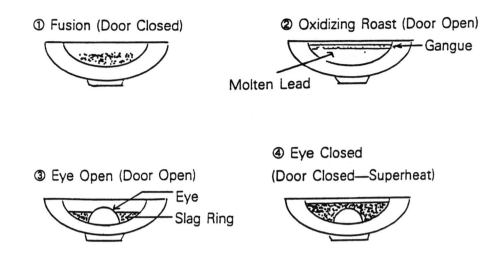

Figure 3 Scorification

(These lead fumes should be avoided at all costs as lead can be an extremely dangerous substance. Fire assayers should be tested regularly for lead contamination). A cupel that fails to "open" is called a "frozen cupel," and a frozen cupel usually results from a cupel's having too thick a layer of material over the lead to get rid of it all, or from a cupel's being too cool. Thus during the entire operation the door is kept closed until all the cupels have opened.

Step 4: When the cupels have all opened and the lead is being oxidized and absorbed by the bone ash, the oxidizing reaction of the lead is exothermic, and the cupel acts as an insulator. This significantly raises the temperature of the lead being oxidized, and, unless the temperature of the furnace is lowered from 850°C, the loss of silver will be promoted. In this case, the lead becomes considerably brighter than the furnace walls. This is called "driving the cupel" and is a poor practice. Once the cupels have opened, the furnace temperature must be lowered to a point at which the forming litharge is maintained slightly above its melting temperature. At this point, the burning lead is red, instead of yellow or dazzling yellow-white. As long as the lead and litharge are kept at a temperature at which the lead continues to burn and at which the forming litharge remains molten long enough to soak into the bone ash cupel, satisfactory results will be obtained. If some of the litharge that vaporizes condenses as dendritic crystals on the cupel above the burning lead, it is a very good sign that any silver loss will be extremely minor. These dendritic crystals are called "feathers."

Step 5: Near the end point, when the lead is nearly gone and only a thin coating is left on the dore bead, raise the temperature of the furnace to keep the molten alloy of silver, gold, and other values in a liquid state and to drive off the last bit of lead. When the end point is close, there is a play of colors in the cupel, and the bead appears to spin about the cupel wildly. At the very end, the bead may become very bright for a split second; this action is called the "blick" or "wink." The cause of this is that the liquid metal super cools at the instant the last of the lead leaves it, and, when it goes from a liquid to a solid, the latent heat raises the temperature to produce the "wink" or "blick." If the dore bead is quite large and contains much silver, the cupel should be moved toward the front of the furnace and covered with a hot cupel so that the top surface will remain liquid (not crust over) as the bead solidifies. Silver, when molten, will release any dissolved oxygen released at the moment of solidification-the familiar "spit." In cupellation, if a solid crust forms around a liquid core, as the core then solidifies, the "spit" of released oxygen remains trapped by the crust, and builds up pressure until a miniature explosion occurs. The explosion fragments the crust (an occurrence known as "sprouting"), resulting in physical loss of some of the value.

Step 6: Carefully clean and weigh the bead recovered from cupellation.

Figure 4 illustrates cupellation (Ammen, 1984).

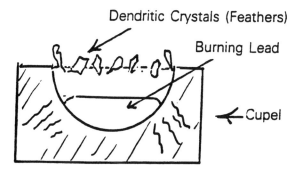

Figure 4 Cupellation

With the platinum metals the final stage of cupellation is not usually clearly defined. The platinum metals, if present, will display a variety of effects on the surface of the cold silver bead. It may be emphasized here that the silver bead collection and its subsequent wet treatment, when properly applied, serves as an excellent method for the determination of gold, palladium, and platinum together with traces of rhodium, iridium, and ruthenium. It is quite unacceptable for osmium. For larger proportions of the more insoluble platinum metals, and where the amount of osmium is required, the direct wet treatment of the lead button is preferable. Osmium is largely volatilized during the cupellation step. Ruthenium, rhodium, and iridium have only a limited solubility in the silver and may be partially mechanically lost unless special precautions are taken.

The Fire Assay Silver and Gold Parting Process
Parting is the process of separating silver from gold by dissolution in nitric acid. In the traditional fire assay process for gold and silver, when gold is alloyed with less than three times its weight in silver, it is difficult to accomplish a good clean parting because the gold protects the silver from attack by the nitric acid. When faced with such a bead, enough 99.9% pure silver must be added to make the silver in the sample weigh over three times as much as the gold. Usually this is done in a process known as inquartation by adding the pure silver at the initial stage of crucible fusion or first cupellation. The word inquartation means one quarter. Of course deduct the weight of inquarted or added silver must be deducted from the final totals.

For the classical parting process, one part nitric acid mixed with seven parts distilled water is used. This is placed in a small beaker or porcelain dish, and brought to almost boiling. The gold/silver dore bead is then dropped into the acid solution. The acid will immediately start to dissolve or 'part' the silver from the dore bead. If the bead fails to start parting, full-strength nitric acid, drop by drop, is added until the bead starts to react with the acid solution. The nitric acid used for parting must be CP reagent grade. If the nitric acid contained any chlorine, it would in reality be a mild aqua regia, putting some of the gold from the dore bead into solution and coating the bead with a precipitate of insoluble silver chloride. If the gold sponge is allowed to disintegrate into small particles it becomes a problem. Therefore, the acid solution should be hot during the entire parting operation. If the bead to be parted weighs more than a few milligrams, it should be flattened on a polished anvil with a polished hammer and rolled or flattened to about ten-thousandths of an inch thick. A very large bead might have to be annealed several times to avoid being split or cracked during flattening. This is accomplished by heating the bead (while holding it in platinum tweezers) to a red heat and dropping it into cold water between flattenings on the anvil or rolls. The metal is then rolled into a spiral or cornet for parting. The resultant gold sponge should be washed with distilled water. The sponge is then annealed by heating to a red heat, at which point it will assume its gold color.

Alternative Fire Assay Collection Media
Besides the use of classical lead collection, other fire assay collection media have been suggested for collecting small amounts of precious metals. Some involve the use of iron-nickel-copper alloys and tin (Faye and Inman, 1961, Whitney 1982, De Neve 1986). Because of the lower fusion temperature required, copper alone has also been suggested as a collector of the precious metals. (De Neve, 1986, Banbury and Beamish, 1965, Agrawal and Beamish,1965, Banbury and Beamish, 1966, Diamantatos, 1987). The collection of the precious metals into nickel sulfide is also being used to an increasing extent. (Robert, van Wyk, and Palmer, 1971, Kruger and van Wyk, 1972, Kallmann and Maul, 1983). This approach has many advantages over the conventional lead-based fire assay system and will be described later in more detail.

A new technique involving the collection of the precious metals into copper sulfide has recently been described and has several advantages over the nickel sulfide technique (Kallmann, 1986). First, copper sulfide collects gold quantitatively. Another advantage is the possibility of dissolving the copper sulfide in hydrobromic acid, thus allowing the separation of the Precious metals from large amounts of silver and lead. In both the nickel sulfide and copper sulfide collection schemes, the base metals are removed by acid treatments and the precious metals contained in the acid-insoluble residue are eventually determined instrumentally, or where justified, gravimetrically. The fire assay procedures previously mentioned provide separations from most matrix elements. Dissolution of the lead button in nitric or perchloric acids, dissolution of nickel sulfide in hydrochloric acid and of cuprous sulfide in hydrobromic acid allow the rapid removal of the collecting media.

If a lead button has been subjected to cupellation, the silver bead containing the precious metals can be treated with dilute nitric acid to dissolve the silver and most of the palladium. Platinum will also dissolve if the sample contains at least an equivalent amount of gold. (The silver bead naturally must be large enough to hold any rhodium, ruthenium, and iridium mechanically; minimum ratio of Ag to PGM is 1000:1). These three elements, together with gold, remain insoluble in dilute nitric acid and can be determined instrumentally after fusion of the insoluble residue with sodium peroxide and acidification of the leached melt with hydrochloric acid. In another recently introduced fire-assay technique, cupellation is interrupted when the weight of the lead button has been reduced to less than 1 gram. This lead bead is then analyzed for the precious metals by standard optical emission spectrometry techniques (ASTM, 1982). Platinum and palladium can be determined instrumentally after removal of the silver by precipitation as the chloride. If present in sufficient quantities, platinum and palladium can be determined gravimetrically after precipitation with ammonium chloride (for Pt) or dimethylglyoxime (for Pd). Also of interest is the selective extraction of the palladium complex with chloroform (Fraser, Beamish and McBride, 1954).

For PGM's a flux consisting, of a 2-to-1 mixture of sodium carbonate and sodium tetraborate with nickel sulfide as the collector can be utilized. The presence of at least 10 g of silica is desirable for a good fusion. The silica is usually contributed by the sample, but, where it is not, it can be added as powdered silica. With a flux-to-sample ratio of 1-to-1, 20 g of nickel oxide and 10 g of sulfur, a satisfactory button is obtained at 1000°C. For chromites a suitable fusion required a 3-to-1 flux-to-sample mixture. For samples of concentrate an increase in the flux-to-sample ratio had little or no effect on the recovery of noble metals. The quantity of flux necessary for a consistently high recovery of all the noble metals lies between 90 to 105 g; below 90 the recovery is low and an excess above 105 the recoveries are erratic. The minimum weight of button is about 25 g. Buttons much larger do not increase the recovery of noble metals. Excess of sulfur produced a grayish-yellow button which sometimes disintegrated on standing. The authors of the nickel sulfide collection investigated the efficiency of the method relative to the classical lead collection and a wet separation involving selective extraction and recoveries with tellurium. The results show that the collection of noble metals with nickel sulfide is, for all the noble metals except gold, superior or equal to collection by the lead method. However, the higher results obtained for platinum by the acid-extraction procedures indicate that its recovery is incomplete by both fusion procedures. Ruthenium, osmium, and iridium are not determined by the acid-extraction procedures. A further disadvantage of the integrated nickel sulfide method is that it is time consuming compared to the lead assay. However claims for its superiority for some sample types justify its inclusion here.

By comparison, using nickel sulfide as the collector has several advantages over the lead-collection method. For instance, a smaller flux to sample ratio and a lower fusion temperature (1000°C as against 1200°C) are used. Furthermore, the method is applicable to all six platinum-group metals and can be applied to samples high in nickel and sulfur without the pretreatment that is required in the lead method. No change in flux composition is required for different types of samples, except for chromite ores, where the quantity of flux used must be higher. The advantages of the lead collection procedure are that it requires less time for analysis, particularly in determinations of total platinum-group metals, and that the gold recovery is some 10 to 20% higher than with the nickel sulfide method. Apart from the possible incomplete collection of platinum and gold, the nickel sulfide procedure offers a precise and accurate method for the concentration and isolation of the noble metals in samples of ores, concentrates, and mattes. Its applicability to the different samples thus far encountered, together with the simplicity of the technique renders it an extremely useful procedure for the analysis of the noble metals.

The method is applicable to the determination of all six of the platinum group metals in ores, concentrates, and mattes. The recovery of gold appears to be incomplete, and the lead collection method is therefore preferable when gold is to be determined. Samples containing sulfur need not be roasted before the fusion, although the sulfur content must be taken into account when the required amount of flux is calculated. If the sulfur content is unknown and the determination of osmium is not required, the sample can be roasted.

If difficulty is encountered in the obtaining of a satisfactory fusion and button (as has been experienced with samples containing unusual amounts of zinc), the sample can be leached in concentrated hydrochloric acid before the fusion. (This procedure results in a satisfactory fusion and button from Merensky Reef samples.)

For samples containing more than 10 ppm of total noble metals, platinum, palladium, rhodium, ruthenium, and iridium can be determined in one button. If the total concentration is less than 10 ppm, a separate button must be prepared for the determination of iridium. For all samples, a separate button is prepared for the determination of osmium.

DETERMINATION OF THE PRECIOUS METALS BY INSTRUMENTAL METHODS
Instrumental methods introduced during the last 20 years have undoubtedly revolutionized the repertoire of the precious metals analytical chemist. Optical emission and x-ray fluorescence methods were already used in the precious metal industry to a limited extent. About 20 years ago AAS methods made their triumphant entry, followed in rapid succession by many other techniques

designed to determine the composition and the properties of the precious metals, particularly those used to an increasing extent in the electronics industry. A brief, annotated, excerpted description of these follows (Benner, Suziki, Meguro, Tanaka, 1991, Furuya, and Kallmann, 1991). As well, the determination of PGM's with instrumental analysis will be illustrated.

Optical Emission Spectroscopy (OES)

This technique is particularly well suited for the determination of impurities in pure precious metals. In addition, when combined with preconcentration techniques based on chemical or fire-assay preconcentration techniques, it allows the determination of platinum and palladium concentrations in complex matrixes down to 0.03 ug or 0.001 oz/ton. In this particular procedure, 20 mg of gold is used as a collector and the Pt and Pd content of the gold is compared with that of gold standards containing known quantities of the two PGM's. With minor modifications, the method can be applied to the determination of trace amounts of rhodium and iridium. One attractive feature of OES with photographic recording is its capability of providing simultaneous qualitative and/or quantitative information on many elements (typically, 20-40 or more). The direct current spectral source. Although the high voltage ac and interrupted dc are useful for some applications, they have largely been replaced by newer techniques, such as AAS and PES.

Spark-Source Mass-Spectroscopy (SSMS)

Spark-source mass-spectroscopy (SSMS) is a semi quantitative technique with ultrahigh sensitivity, which has detection limits in the low ug/g and ng/g ranges. When used instead of OES for final measurements, it allows determination of PGM content of complex matrices at the 1 ng/g level. In this technique, the sample is sparked in vacuum by a high-energy radio frequency spark to produce positive ions of the sample elements. A double-focusing spectrometer separates the ions according to their "mass-to charge" first in an electrostatic, than in a strong magnetic field. The ions thus separated are recorded photographically on an ion-sensitive photographic plate or are measured by means of photo multipliers. Accelerator mass spectrometry has recently been used to determine the osmium in terrestrial samples.

X-Ray Fluorescence (XRF)

XRF is based on measurements of the secondary x-rays emitted by the constituents of a sample excited by primary x-rays. Two different types of XRF instruments are available, energy-dispersive and wavelength dispersive. Considerable progress in the instrumentation, particularly, the precision of the detection and measuring devices, allows rapid determination of the Precious metals with excellent precision. Mufti-channel spectrometers facilitate the simultaneous determination of all the precious metals and many base metals. The main advantage of XRF over the various atomic emission or absorption techniques (OES, DCP, ICP, AAS) described separately, is that it is non-destructive, allowing recovery of the original sample after the determination. Its main disadvantage is that for quantitative work standards with the same chemical composition and physical characteristics as the sample must be available or prepared. If solid samples are to be analyzed, facilities are required to prepare a set or sets of PM-bearing alloys with highly polished surfaces. Thus, one laboratory routinely analyses platinum alloys containing 5-10% of palladium and/or rhodium, another determines the PGM's in a tin button obtained fire assay. A solution technique used by another organization simplifies the preparation of standards, but at the expense of sensitivity. Microgram amounts of all precious metals have been determined by absorbent-pad-and cellulose pellet techniques. Computer programs have been designed to correct for the positive or negative effects of other Precious metals or those of base metals on the result of the PM being determined. As far as sensitivity is concerned, with wavelength-dispersive instruments and using a tungsten tube as the primary source of x-rays, the radiation of the K-lines of Ru, Rh, Pd, and Ag is 2-3 times more intense than that of the L-lines of Os, Ir, Pt and Au.

Flame Atomic Absorption Spectrometry (FAAS)

FAAS has largely replaced spectrophotometry as the work-horse in the precious-metals analytical laboratory. There are several reasons for this. (a) FAAS is virtually element specific. Thus a PM which cannot be determined spectrophotometrically at all in the presence of certain other Precious metals or base metals, can often be determined with comparative ease by FAAS. (b) In many instances, there is no interference by moderate concentrations of base metals, and even where this is, it can be dealt with by the standard-addition technique

The limitations to the use of FAAS in precious-metals analysis mainly arise from the sensitivity which is poor for some PGM's, particularly iridium. The relative concentrations of PGM required to match the AAS response of a unit concentration of silver are: Pd 4, Rh 5, Ru 9, Os 100, Ir 150, Pt 38, and Au 5. A relative value of 40 or more would indicate that the element should not be handled by FAAS (Os, Ir) unless present in substantial quantities or isolated from the matrix. It must also be remembered that FAAS is a solution technique and therefore requires that the element(s) can be dissolved with relative ease and without introducing too much extraneous matter. This, of course, is difficult, if not impossible, in the case of many samples containing the Precious metals in a complex matrix.

Electrothermal Atomic Absorption Spectrometry (ETAAS)

This technique supplements FAAS, in as much as it offers greatly enhanced sensitivity. Unfortunately, the precision is much poorer because of the small volume of sample solution (5-50 itl) used and the difficulties encountered in reproducible sampling and control of the atomization conditions. In addition, since ETAAS also generally relies on the preparation of solutions of samples and standards, it is subject to the same dissolution limitations as FAAS. Attempts have been made to use solid samples, but with limited success. An interesting application of ETAAS for geochemical exploration work was recently described in which diantipyrylmethane was used for isolating the PGM's by extraction into chloroform.

Plasma Emission Spectrometry (PES)

Plasma emission spectrometry is a variant of atomic emission spectrometry (AES), based on the use of a plasma for excitation. Plasmas are increasingly used as a spectral excitation source for determining the precious metals. Generally, the sample is introduced in the form of a solution that is atomized by the carrier gas in various fashions. Two types of plasma are used: the direct current plasma (DCP) and the inductively-coupled plasma (ICP). Lasers are also used as an excitation source. Moderately priced instruments are readily available, based on the sequential principle (measurement of one precious metal at a time). There are also more complex instruments based on the use of multi-channel detectors (ICP) or cassettes (DCP) allowing the simultaneous determination of all the Precious metals and many base metals. There is even a fast sequential DCP instrument.(99-102) PES is decidedly superior to FAAS in regard to sensitivity and dynamic range while AAS has certain advantages over the arc and spark emission methods for PM analysis. PES methods tolerate the presence of moderate amounts of alkali metal salts. This will often permit fusion of a sample or a residue with alkaline fluxes such as sodium peroxide and sodium carbonate. The linear dynamic ranges (ltg/ml) of DCP for the eight precious metals at interference-free wavelength are: Ru 0.5-50, Rh 0.1-30, Pd 0.1-30, Ag 0.4-60, Os 0.5-100, Ir 0.5-5, Pt 0.3-75, Au 0.3-100. The corresponding ranges of ICP are similar, but depend to some extent on the instrumentation used and the availability of interference-free wavelengths.

Inductively Coupled Plasma-Mass Spectrometry (ICP-MS)

This technique has recently been introduced, but not yet fully examined as to its suitability for precious metal determinations. In this technique ICP is used instead of the spark-source, and ions can be extracted from the plasma (which is at atmospheric pressure) and introduced (at greatly reduced pressure) into a spectrometer for mass resolution and detection. ICP-MS may be of value to laboratories having no access to SSMS. Laser Mass Spectrometers have also been built but are still in the experimental stage.

Inductively Coupled Plasma-Atomic Fluorescence Spectrometry (ICP-AFS)
This technique has recently been advocated as an efficient tool for precious metal analysis. In atomic fluorescence, the plasma does not function as an excitation source but solely as an atomization cell to produce ground state or low energy excited state atoms. Excitation is mainly by resonance absorption of light from an external light source, and the fluorescence emitted by return to a lower energy state is viewed at an angle to the excitation beam. The sensitivity of the method is said to be comparable to that of FES techniques. The cost of the equipment compares favorably with that of AAS.

Activation Analysis (NAA)
Gamma rays, charged particles and particularly neutrons react with isotopes of the precious metals to produce radioactive nuclides. The characteristic radiation emitted by the nuclides produced can be used for the detection and determination of the Precious metals. In some instances, neutron activation is more sensitive than any other technique. Though instrumental NAA (INAA) can be applied effectively to gold and silver determinations, in the case of the PGM's some separations or concentration may be necessary, either before or after irradiation. Recently, particle induced x-ray emission spectrometry (PIXE) has been developed. It has been used in conjunction with preconcentration of the Precious metals by the nickel sulfide fine-assay technique.

Controlled-Potential Coulometry (CPC)
In controlled-potential coulometry, the substance is electrolyzed at a working electrode with the potential controlled or kept constant during the electrolysis by means of potentiostat. The current is integrated with an electronic integrator or coulometer. A reference electrode and a two-electrode electrolysis cell are employed. This technique has been shown to be very effective for the determination of all the Precious metals. Unfortunately, it has not yet received the attention from precious metal analysts that it richly deserves.

Other instrumental techniques that may be of value to the precious metals analyst for specific applications include polarography (now rarely used), differential pulse polarography, anodic stripping voltametry, ion-selective electrode potentiometry and most recently, ion chromatography. The applicability of the latter technique to PGM analysis was recently discussed by Heberlin.

The compositional change on the surface of a cold-worked silver-palladium alloy has been observed by x-ray photoelectron spectrometry. Many studies on the surface composition of homogeneous alloys of gold-silver and of silver-palladium have been reported by scanning electron microscopy-energy dispersive x-ray detection (SEM-EDX), Auger electron spectrometry (AES), SIMS and XPS. Studies on the surface absorption of PGM's on the surface of catalysts have been reported by a number of different techniques, such as UPS XPS, LEEDS and AES. Since such techniques go far beyond the scope of this short review, the appropriate literature should be consulted for more details.

The Instrumental Analysis of Minor PGM Levels
Instrumental techniques are the mainstay of analysis for low (ppm) levels of PGM's. These methods include inductively coupled plasma (ICP) spectroscopy, direct current plasma (DCP) spectroscopy and atomic absorption spectroscopy (De Neve and Hofmans, 1989, Homeier and Smith, 1989, Shore, 1989, Skrabak and Demers, 1988, Blumberg, 1997, Hofmans and Adriaenssens, 1993). Recently combination techniques such as ICP–mass spectrometry have been used to reached sub-ppm levels of quantification. The "plasma" techniques are performed on

solutions derived from either direct acid dissolution of a sample or from the parting solution resulting from fire assay of a sample. Atomic absorption spectroscopy, best suited for aqueous solutions, is used for the analysis of parting solutions where silver is precipitated. Many elements, including Pt and Pd, are not precipitated with the silver. Often buffers are added to increase measurement sensitivity. Direct acid dissolution requires matrix matching for the major elements present in solutions. Analysis of parting solutions requires precipitating Ag as the chloride or taking suitable precautions if the solution is analyzed directly. The accuracy of these comparative techniques is based on the use of solution standards traceable to NIST primary standards. The ability of modern instruments to scan wavelength regions and correct for background, matrix and spectral interferences increases accuracy. The use of internal standards improves the precision of the analysis. Error sources include incomplete sample dissolution, dilution problems, cross contamination of glassware, improper selection of concentration ranges and unmatched enhancement effects of sample or reagent preparation matrices.

The Instrumental Analysis of Major PGM Levels
Ideally suited for low-level analysis, instrumental techniques lose their advantage when concentrations reach the point where dilution and instrumental errors become large. PGM concentration ranges above 6% necessitate a different methodology to obtain the required accuracy and precision. Gravimetric wet chemical methods have traditionally been used. These methods encompass the time-honored techniques of separation, pH adjustment, precipitation, filtration, evaporation, reduction, drying and other laboratory manipulative skills. However, X-ray fluorescence spectroscopy (XRF) has been used successfully for the analysis of many PGM alloys (Wissmann and Nordheim, 1993). The method is quick and sample preparation is minimal requiring only surface milling. The main drawback is the need for exacting, well-characterized standards. This can be prohibitive when dealing with a multitude of unknown samples. However XRF using standardless techniques based on "fundamental parameters" can serve the important function of providing fast and accurate preliminary assays necessary for the optimization of the final analysis technique (Savolainen, 1999). With standardless techniques, XRF instruments can check for all elements from Na to U.

Recently, an instrumental method with the accuracy and precision of a fire assay has been developed by combining some of the best features of instrumental and gravimetric methods. The procedure incorporates mass dilutions with the internal standardized, drift corrected, multi-repetitive ICP technique developed at NIST. (Salit, 2000, Salit, 2001). Precision is improved by: 1) making all dilutions by mass; 2) measuring analyte and internal standard simultaneously; 3) analyzing four separately prepared sample weighings and standards ten times with each analysis consisting of five integrations; and 4) drift correcting all results using six-figure fitted polynomials. (Salit, and Turk, 1998). Internal standards and their approximate mass fractions have been determined for all the PGM elements. High accuracy requires the use of NIST certified solutions and matrix-matched standards. The only error to be cognizant of is incomplete dissolution of the sample and the analyte. The method is robust in that there are numerous checkpoints to determine if a problem occurred. For example, if precision levels from repetitive analyses are out of range, this indicates the presence of unacceptable instrumental error.

The Instrumental Analysis of High Purity PGM's
High purity PGM's, 99% and above, are analyzed by determining the concentration levels of impurity elements. There are several instrumental techniques available including ICP or DCP, spark emission and mass spectroscopy. (Arniaud and Liabeuf, 1992). For the plasma methods, samples must be dissolved with an acid that requires proper cleaning techniques and acid blank matching. Standards are easily made from certified solution standards. Spark emission techniques require initial calibration for each of the elements and this, in turn, may require certified standards containing 35 to 40 elements to cover all likely impurities. Once an instrument is calibrated, only

a check standard and recalibration sample is necessary. The analysis then becomes very fast since sample preparation consists simply of surface milling. Mass spectroscopic analysis of solid samples, although very sensitive (ppb levels) and not requiring specific standard development, is not suited for daily routine analyses because of the time involved in instrument setup.

SUMMARY

The analysis of silver and gold content in ores, concentrates and other materials is generally determined by classical lead based fire assay. As well, platinum group metals (PGM's) such as platinum, palladium, rhodium, iridium, ruthenium and osmium are often separated or concentrated by fire assay. This paper outlined the basic concepts involved in this proven methodology. Included were general procedures for selecting appropriate fire assay charges based on approximate chemical compositions of the samples. Also included were descriptions of modern instrumental methods for the determination of the noble metals with a focus on the analysis of PGM's.

ACKNOWLEDGEMENTS

The author is grateful for the assistance and expertise provided by Ms. Tami Cashell in the preparation of this paper. As well, Dr. Dave Kinneberg and Mr. Arnold Savolainen are acknowledged for their technical input and review.

REFERENCES

Agrawal, K.C, and F.E. Beamish. 1965, Z. Anal. Chem., 211, 265.

Agricola, Georgius. 1556, De re metallica, translated by H. C. Hoover and L. H. Hoover, 1950 [reprinted from The Mining Magazine, London, 1912]: New York, Dover Publications, 638 p.

Arniaud, D. and N. Liabeuf. 1992, "Analysis of Impurities in Palladium and Platinum Matrices by Inductively Coupled Plasma Optical Emission Spectroscopy (ICP-OES), Rhodium/Sampling and Analysis.

R.C. Kaltenbach and L. Manziek, eds., International Precious Metals Institute, Allentown, PA pp. 221- 234.

Ammen, C. W. 1984, Recovery and Refining of Precious Metals, Van Nostrand Reinhold, New York.

Annegarn, H.J., C.C. Erasmus, J.P.F. Sellschop, and M. Tredous. 1983, "Sensitivity Amplification by Sample Preconcentration in Ion Beam Analysis", Nucl. Instr. Methods; Phys. Res., 218, 33.

ASTM, 1982, Annual Book of ASTM Standards, Part 42- E 400, 531.

Banbury L.M., and F.E. Beamish. 1965, Z. Anal. Chem., 211,178

Banbury, L.M., and F.E. Beamish. 1966, Z. Anal. Chem , 218,263

Barnett, P. R., W.P. Huleatt, L.F. Rader, and A.T. Myers, A. T. 1955, "Spectrographic determination of contamination of rock samples after grinding with alumina ceramic.", Am. Jour. Sci., v. 253, no. 2, p. 121-124.

Beamish, F.E., and J.C. van Loon. 1977,Analysis of Noble Metals, Academic Press, NY

Beamish, F.E., 1966, The Analytical Chemistry of Noble Metals, Pergamon Press, Oxford.

Beamish, F.E., and J.C. van Loon. 1972, Recent Advances in the Analytical Chemistry of the Noble Metals, Pergamon Press, Oxford.

Benner, L.S., Suzuki, T., Meguro, K., and Tanaka, S., 1991, Precious Metals Science and Technology, International Precious Metals Institute, Allentown Pennsylvania.

Blumberg, P. 1992, "Quantitative Analysis of Rhodium in Material Containing Iridium," Rhodium/Sampling and Analysis, R.C. Kaltenbach and L. Manziek, eds., International Precious Metals Institute, Allentown, PA, pp. 91-96.

Blumberg, P. et al., . 1997, "ICP Evaluation of Precious Metals," Precious Metal Recovery/Refining Seminar, j. Vogt, ed., International Precious Metals Institute, Allentown, PA, pp. 101-112.

Bright, J. translator, 1965, Jeremiah, v. 21 of The Anchor Bible: Garden City, N. Y., Doubleday & Co.

Bugbee, E. E., 1940, A Textbook of Fire Assaying [3rd ed.]: New York, John Wiley & Sons, 314 p.

Clarke, F. W., and W.F. Hillebrand. 1897, Analyses of rocks, with a chapter on analytical methods, laboratory of the United States Geological Survey, 1880 to 1896: U.S. Geol. Survey Bull. 148, p. 9.

Covell, D. F., 1959, "Determination of gamma-ray abundance directly from the total absorption peak.", Anal. Chemistry, v. 31, p. 1785-1790.

Dahood, Mitchell, S. J., translator, 1966, Psalms I: 1-50, v. 16 of The Anchor Bible: Garden City, N.Y., Doubleday & Co.

De Neve, R. and H. Hofmans. 1989, "Determination of Precious Metals by Emission or Absorption Spectrometric Techniques at MHO," Precious Metals Sampling and Analysis, S. Kallmann, ed., International Precious Metals Institute, Allentown, PA, pp. 41-52.

De Neve, R., 1986, "Analysis of Precious Metals Combining Classical Collection Procedures with Modern Instrumentation.", paper presented at 10th Annual IPMI Meeting, Lake Tahoe, NV.

Diamantatos, A., 1986, Analyst, 111, 213.

Diamantatos, A., 1987, Talanta, 8, 34.

Egan, A. 1986, Atomic Energy of Canada Limited, private communication.

Emmons, S. F., 1886, Geology and mining industry of Leadville, Colorado: U.S. Geol. Survey Mon. 12, 770 p. in neutron activation analytical determinations by uranium fission: Jour. Radioanalytical Chemistry, v. 10, p. 137-138.

Faye and Inman, 1961, Anal. Chem., 33,278

Forbes, R. J., 1950, "Metallurgy in antiquity.", Leiden, Netherlands,

Brill, E.J., 1964, "Studies in ancient technology.", v. VIII, 2d revised ed.: Leiden, Netherlands

Furuya, K. and S. Kallmann. 1991, Chapter 5 – Determination of the Precious Metals, Precious Metals Science and Technology, L.S. Benner, et al., ed., International Precious Metals Institute, Allentown, PA, pp. 223-246.

Fraser, J.G., F.E. Beamish, and W.A.E. McBride. 1954, Anal. Chem.

Greenland, L. P., J.J. Rowe, and J.I. Dinnin. 1971, Application of triple coincidence counting and of fire-assay separation to the neutron-activation determination of iridium: U.S. Geol. Survey Prof. Paper 750-B, p. 13175-13179.

Grimaldi, F. S., and M.M. Schnepfe. 1970, Determination of iridium in mafic rocks by atomic absorption: Talanta, v. 17, p. 617 -621.

Haffty, Joseph, and L.B. Riley. 1968. Determination of palladium, platinum, and rhodium in geologic materials by fire assay and emission spectrography: Talanta, v. 15, p. 111-117.

Haffty, Joseph, and L.B. Riley. 1971, Suggested method for spectrochemical analysis of geologic materials by the fire-assay preconcentration-intermittent d-c arc technique, in Methods for emission spectrochemical analysis, 6th ed.: Am. Soc. Testing and Materials, p. 1027-1031.

Haffty, J., L.B. Riley and W.D. Goss. 1977, A Manual on Fire Assaying and Determination of the Noble Metals in Geological Materials, US Geological Survey Bulletin 1445, Washington D.C..

Harrar, J.E. and M.C. Waggoner. 1981, Pap. Surf, 68,41.

Herberlin. S. ,1987, Paper presented at IPMI Western Regional Analytical Symposium, San Jose, CA.

Hillebrand, W. F., and E.T. Allen. 1905, Comparison of a wet and crucible-fire methods for the assay of gold telluride ores: U.S. Geol. Survey Bull. 253, 30 p.

Hofmans, D. and E. Adriaenssens. 1993, "Determination of Precious metals by ICP at UM: Advantagges and Quality Assurance," Precious Metals 1993, R.K. Mishra, ed., International Precious Metals Institute, Allentown, Pa., pp.235-246.

Homeier, E.H. and D.W. Smith. 1989, "Determination of Platinum in Catalyst Residues by Inductively Coupled Plasma Atomic Emission Spectroscopy," Precious Metals Sampling and Analysis, S. Kallmann, ed., International Precious Metals Institute, Allentown, PA, pp. 67-75.

Huffman, Claude, Jr., J.D. Mensik, and L.F. Rader. 1966, "Determination of silver in mineralized rocks by atomic-absorption spectrophotometry", U.S. Geol. Survey Prof. Paper 550-B, p. 13189-13191.

Huffman, Claude, Jr., J.D. Mensik, and L.B. Riley. 1967, "Determination of gold in geologic materials by solvent extraction and atomic-absorption spectrometry", U.S. Geol. Survey Circ. 544, 6 p.Koda, Y., 1970, Determination of radioruthenium using a polyethylene film: Jour. Radioanalytical Chemistry, v. 6, p. 345-357.

Kallmann, S. 1986, "Sampling and Analysis of Spent Automotive Catalyst," Platinum Group Metals Seminar 1985, E.D. Zysk, ed., International Precious Metals Institute, Allentown, PA, pp. 233-240.

Kallmann, S. 1983," Interdependence of Instrumental and Classical Chemical Methods of Analysis of Precious Metals", paper presented at Eastern Analytical Symposium, New York, NY.

Kallmann, S. and C. Maul. 1983, Talanta, 20,2.

Kallmann, S., 1986, Talanta, 22,75.

Kruger, M.M. and E. van Wyk. 1972, Nat. Inst. Met. Rep. S. Afr. Rept., 1432.

Kudo, M., Y. Nihei, T. Machiyama, K. Furuya, and H. Kamada. 1977, "Some Problems on Quantitative Surface Analysis of Copper-Nickel and Palladium-Silver Alloys by Means of X-Ray Photoelectron Spectroscopy (XPS-ESCA)", Bunseki Kagaku, 26, 173.

Lancione, R. 1983, "Precious Metal Analysis by Inductively Coupled Plasma-Atomic Fluorescence Spectrometry, paper presented at Eastern Analytical Symposium, NY.

Lenahan, W.C. and R. de L. Murry-Smith, 2000, Chapter XVIII – The Determination of the Platinum Group Metals, Assay and Analytical Practice in the South African Mining Industry, Chamber of Mines of South Africa, Johannesburg, pp. 431-505.

McLaughlin, R. E., 1992, "Modifications of Tradituional Methods in the Analysis of Precious Metals," Rhodium/Sampling and Analysis, R.C. Kaltenbach and L. Manziek, eds., International Precious Metals Institute, Allentown, PA, pp. 235-244.

Millard, H. T., Jr., and A.J. Bartel. 1971, "A neutron activation analysis procedure for the determination of the noble metals in geological samples.", in A. O. Brunfelt and E. 0. Steinnes, eds., Activation analysis in geochemistry and cosmochemistry, NATO Advanced Study Inst., Il jeller, Norway, Sept. 7-12, 1970, Proc.: OsloBergen-Tromso, Norway, Universitetsforlaget, p. 353-,358.

Moreland, John, and A.T. Myers. 1973, Notes on use and maintenance of vertical pulverizers for geologic materials: U.S. Geol. Survey open-file report, 6 p.

Myers, A. T., and R.G. Havens. 1970, "Spectrochemistry applied to geology and geochemistry by the U.S. Geological Survey in the Rocky Mountain region." in Proceedings of the second seminar on geochemical prospecting methods and techniques, Ceylon, 1970: U.N. ECAFE Mineral Resources Devel. Ser. 38, p. 286-291

Robert, R.V.D., E. van Wyk, and R. Palmer. 1971, Nat. Inst. Met. Rep. S. Afr. Rept., 1371.

Rowe, J. J. 1973, "Determination of gold in phosphates by activation analysis using epithermal neutrons", U.S. Geol. Survey, Jour. Research, v. 1, no. 1, p. 79-80.

Rowe, J. J., and F.O. Simon. 1968, "The determination of gold in geologic materials by neutron-activation analysis using fire assay for the radiochemical separations.", U.S. Geol. Survey Circ. 599, 4 p.

Salit, M.L., et al. 2001, "Single-Element Solution Comparisons with a High-Performance Inductively Coupled Plasma Optical Emission Spectrometric Method," Analytical Chemistry, Vol. 73, No. 20, pp. 4821-4829.

Salit, M.L., et al. 2000, "An ICP-OES Method with 0.2% Expanded Uncertainties for the Characterization of LiAlO2," Analytical Chemistry, Vol. 72, No. 15, pp. 3504-3511.

Salit, M.L. and G.C. Turk. 1998, "A drift Correction Procedure," Analytical Chemistry, Vol. 70, No. 15, pp.3184-3190.

Savolainen, A.M., et al. 1999, "Preliminary Analysis and Method Optimization Based on X-ray Fluorescence – Application to Precious Metals," Analytical Technologies in the Mineral Industries, Cabri, L.J., et al., eds., The Minerals, Metals and Materials Society, pp 55-71.

Shepard and Dietrich. 1940, Fire Assaying, McGraw-Hill Book Company, Inc. New York.

Shore, L. 1992, "The ICP Determination of Impurities in Rhodium," Rhodium/Sampling and Analysis, R.C. Kaltenbach and L. Manziek, eds., International Precious Metals Institute, Allentown, PA, pp. 97-120.

Shore, L. 1989, "ICP vs. DCP: More Than Just a Coin Toss," Precious Metals Sampling and Analysis, S. Kallmann, ed., International Precious Metals Institute, Allentown, PA, pp. 115-143

Skrabak, J.W. and D.R. Demers. 1988 , "Applications of ICP-Atomic Fluorescence Spectrometry to the Analysis of Precious Metals," Precious Metals1988, R.M. Nadkarni, ed., International Precious Metals Institute, Allentown, PA, pp.559-570.

Smith, E.A. 1987, The Sampling and Assay of the Precious Metals, 2nd ed. (reprint), Met-Chem Research Inc., Boulder, CO, pp. 408-424.

Wissmann, F. and U. Nordheim. 1993, "Precise Quantification of Precious Metals by ICP – A comparison of Sophisticated ICP Methods Versus Classical Wet Chemical Analysis with regard to Accuracy, Ease of Use and Cost-Effectiveness.", Precious Metals 1993, R.K. Mishra, ed., International Precious Metals Institute, Allentown, PA, pp.117-130.

Whitney, J. 1982, "X-Ray Fluorescence Analysis of Precious Metal Concentrates Melted with Tin", paper presented at Eastern Analytical Symposium, NY.

15
Tailings Disposal, Wastewater Disposal, and the Environment

Section Co-Editors:
James R. Arnold and Dr. George W. Poling

Management of Tailings Disposal on Land
B.S. Brown 1809

Design of Tailings Dams and Impoundments
P.C. Lighthall, M.P. Davies, S. Rice, T.E. Martin 1828

Hazardous Constituent Removal from Waste and Process Water
L. Twidwell, J. McCloskey, M. Gale-Lee 1847

Treatment of Solutions and Slurries for Cyanide Removal
M.M. Botz, T.I. Mudder 1866

Strategies for Minimization and Management of Acid Rock Drainage and Other Mining-Influenced Waters
R.L. Schmiermund 1886

Environmental and Social Considerations in Facility Siting
B.A. Filas, R.W. Reisinger, C.C. Parnow 1902

Management of Tailings Disposal on Land

Bruce S. Brown. PhD., P.E.
Principal, Knight Piesold Consulting

INTRODUCTION

The management of the disposal of mill tailings is of critical importance to the success of any mining project. Failures of tailings facilities have resulted in loss of life, devastating environmental damage, the closure of mining operations, dramatic declines in share value and, in some countries, the personal liability of the mine management. The issues surrounding the development of tailings facilities are therefore often the main focus of regulatory scrutiny during the permitting process for a mine development.

Successful management of tailings disposal requires a good understanding of the complete life cycle of a tailings facility. Only in this context can decisions be made that will result in overall minimization of risk for all stakeholders. This paper is designed to give people who are responsible for the management of the development, operation and/or closure of tailings facilities an understanding of what is required for each step in the lifecycle of a tailings facility.

Life Cycle of a Tailings Facility

The typical life cycle of a tailings facility starts with the recognition of the need for a tailings facility for a new or existing mining operation. The steps in the cycle will include most or all of the following:

- Identification of all regulatory requirements and laws governing the design, operation and closure of a tailings facility.
- Definition of the quantity and physical and chemical characteristics of the tailings to be stored.
- Siting study.
- Environmental baseline studies.
- Scoping level design for prefeasibility.
- Preliminary design for feasibility.
- Environmental impact assessment and permitting.
- Detailed design, construction drawings and specifications.
- Regulatory review.
- Construction of Stage 1 of the facility.
- Startup and commissioning.
- Operation and monitoring.
- Ongoing staged construction.
- Safety reviews and risk management.
- Closure and decommissioning.
- Reclamation.
- Post closure monitoring.

Each step in the life cycle is closely related to the others and decisions made in the early stages of the life cycle will have a profound effect on the options available in the later stages. The term "designing for closure" relates to this whereby the options for closure are fully considered in the initial stages of development.

Regulatory Framework

The development, operation and closure of tailings facilities are, to some extent, regulated in most areas of the world. In the USA, regulation is generally carried out at the state level and in some instances at the county level. Regulation is typically prescriptive whereby specific design features are mandatory for given waste types. In Canada, the regulatory process has both provincial and and federal components. In countries that have little or no enforceable regulations covering the design and construction of tailings facilities, World Bank standards are commonly used as a basis for design.

The first step in the development of tailings facility is to determine all regulatory requirements and laws that are applicable. In some countries, there are no such regulations or the regulations that exist are not enforced. In these cases it is often the guidelines and standards of the lending agencies that finance the project that are applied.

Notwithstanding the existence or lack of regulations in a given jurisdiction, responsible mining companies apply appropriate measures to international standards. Some developed countries have laws that require companies to apply as a minimum, the standards and regulations of the home country when mining in other countries.

DEVELOPMENT

Site Selection

The selection of the most appropriate site for a tailings facility requires the evaluation of many and often conflicting criteria. In addition to determining the "best' site, there is often a regulatory requirement that it be demonstrated that all potential sites have been fully considered. A formal site selection process is often the best way to do this. A desk study of regional topographic mapping is the first step where areas that can reasonably contain the required volume of tailings are identified. An outline for a facility is laid out at each site to the level that the major dimensions and bulk quantities of major works can be determined. The outline should also delineate the impacted areas, the hydrological catchment areas, and show the relationship of the site to the mining project as a whole.

The selection of a preferred site or combination of sites for one or more mine facilities is often accomplished using various decision-making tools and presentation techniques. After the initial identification of alternative sites, a fatal flaw or exclusion criteria assessment is often conducted to eliminate those alternatives with clear feasibility problems, such as a major fault located through the facility footprint, a shortage of available water, or an alternative that cannot be permitted (e.g. submarine tailing disposal). The remaining feasible alternatives may then be compared in physical terms, including, for instance, impacts on the surrounding environment, required construction materials, and pumping and hauling distances and elevations.

At this stage, appropriate decision criteria may be selected and categorized in major accounts such as 'environmental', 'engineering' or 'economic' and subsequently weighted to allow the numerical ranking of the feasible alternatives in a decision matrix. From the selection of preferred alternatives, combinations of preferred mine components are identified, resulting in a limited number of feasible, overall mine development plans. Overall mine development plans, comprising, for example, preferred combinations of open pits, waste rock storage areas and tailings storage facilities may then be compared using a larger scoped multiple accounts evaluation method. In addition to the physical considerations discussed above, a multiple accounts evaluation will consider non-physical accounts, sub-accounts and indicators, such as socio-economic impacts, aboriginal issues, financial or economic analyses and the comparison of project risks to arrive at a defensible and balanced project development plan for permitting and construction.

Site Characterization

Climate and Hydrology. The climate and hydrology of a region are strongly linked, so it is important to understand both topics in order to fully appreciate the water management challenges facing any particular mine development. In this context, climate generally refers to patterns of temperature and atmospheric moisture, while hydrology refers to patterns of surface runoff. The total amount and timing of the precipitation runoff at a site dictates many aspects of a mine's tailings disposal system, ranging from environmental concerns such as what level of impact the tailings disposal might have on background water quality, to engineering concerns such as how much live storage is required in a tailings pond or whether or not diversions or a spillway are needed, and if so, how large they should it be.

Typically, little or no site-specific climatic and hydrologic data is available for most mine development sites, particularly at the time of initial design. In most instances, a data collection program is initiated during project feasibility and environmental impact studies, but to a large degree, these data sets are very short term and of limited use at the time of project design. As a result, the hydrometeorologic characterization of a project area is generally conducted on the basis of regional data sets, which are extrapolated to the project site according to known or suspected weather patterns, similarities of watershed characteristics, and an understanding of the fundamentals of hydrometeorologic systems, including lapse rates, orographic effects and rainfall/runoff mechanisms.

The hydrometeorologic characterization of a site typically involves estimates of temperature, evaporation, precipitation (rain and snow) and runoff, on both average monthly and annual bases, including some measure of variation. Also required are values of extreme precipitation and flow, which are usually presented in terms of likelihood of occurrence. In addition, in colder climates, patterns of snow accumulation and melt are needed.

Regional Geology. The regional geology identifies the bedrock geology and structural setting in the vicinity and under the tailings facility site. The findings from this element of site selection will classify large scale features. Specific site investigations will be carried out based on the regional features to accurately map the features on a more concentrated scale. The location of these large features is a very important component of site characterization. The type of bedrock or presence of regional faulting greatly affects the location and type of tailings facility to be built.

Regional geology maps can usually be obtained from various government agencies. Generally, large maps in the order of one to a million scale are easily obtained. The availability of smaller scale maps, such as one to two hundred and fifty thousand or one to one hundred thousand, is much more variable due to the economical, political and geological characteristics of the area in question. Many industrialized countries have government geology departments at the federal and provincial or state level, while many developing countries only have limited geological data at the federal level. In addition, the size and population of the country, state or province may control the map scale. For example, Canada is a large, industrialized country; however, the scale and accuracy of regional mapping is large and incomplete in many areas because these areas are vast and the populations are small. Another source that is becoming useful for gathering regional information is the internet. Many countries have maps in digital form for viewing and printing.

Once regional maps have been obtained, it is recommended that a qualified person carry out field reconnaissance to verify the mapping. Smaller scale mapping should be carried out if these maps do not exist or there is a complicated geological or structural setting at the proposed site. It is very important to identify weak bedrock geology units or active regional faults at this stage for site characterization. These identifications will aid designers and form the framework for detailed site investigations.

Terrain Analysis. A terrain analysis is carried out to assist in characterizing a tailings facility site. This analysis is designed to identify terrain units and natural hazards around the site that could impact the tailings facility or the tailings deposit and to identify potential materials for construction. The terrain analysis will not necessarily accurately determine the characteristics in the footprint of the facility but is a useful tool for defining where detailed site investigations are required.

There are two levels of terrain analysis. The first level comprises air photo interpretation and the second level involves ground reconnaissance to verify the observations made by air photo interpretation. The detail of each of these two levels of study is based on the following 14 main terrain attributes: bedrock geology, quaternary geology, geomorphology, weathering, erosion and deposition, climate, vegetation and pedology, hydrogeology, geotechnics, volcanic activity, neotectonics and seismicity, natural dams, human activity and land use. This list is exhaustive and many of these attributes will not be present at a given tailings facility site. However, it is prudent to check for all these attributes initially to determine potential natural hazards. In areas of steep, wet terrain, it is recommended that 1:15,000 scale air photographs be used for the first level of reconnaissance. Steep terrain is generally characterized with significant portions of the area having slopes greater than 50% or 27°. If drier, flatter terrain is present in the vicinity of the site, 1:40,000 scale air photographs are adequate. If reconnaissance flights and air photo production can be acquired, this exercise can be completed in conjunction with topographic surveys.

The second level comprised ground proofing of the air photo interpretation. The requirements for ground reconnaissance verification of the air photo interpretation is based on the terrain attributes identified in the first stage. Generally, areas where the slopes are greater than 50% require foot traverses, while areas of flatter terrain is completed by ground checks supported by helicopter or vehicle. Unstable or potentially unstable areas identified during the air photo interpretation must be checked in detail by foot traverses and 1:5000 scale mapping. The person who carried out the air photo interpretation should complete the field checking.

The results of the terrain assessment should be made into a terrain hazard map, outlining areas of potential natural hazards such as landslides, debris flows and snow avalanches. This information can be used for risk/fatal flaw analyses associated with the tailings facility and to quantify safety concerns during construction and operation. In addition, the information from the analysis can be used to delineate possible local borrow materials for embankment construction.

Hazard Classification. The hazard classification of a tailings facility is required to characterize potential hazards and the consequences of failure of the tailings facility. This enables appropriate design parameters to be selected, including design earthquake events and design flood events.

For new facilities, a hazard classification should be carried out during feasibility design studies and confirmed for final design. In the United States there is not just one set of guidelines in use. The hazard classification guidelines adopted for a project will depend on the project location and maybe also the owner. Typically, each state has an agency that regulates most of the dams (water-retaining and tailings dams) within its jurisdiction, but various other agencies oversee a significant number of projects. These include the US Army Corps of Engineers and the US Bureau of Reclamation. In Canada, the hazard classification for a tailings dam is typically carried out using the Dam Safety Guidelines (1999) of the Canadian Dam Association. Internationally, the International Commission on Large Dams provide similar guidelines (ICOLD 1989).

A hazard classification is carried out by assessing the hazard potential and consequences of failure of a dam. Several factors are considered, and typically include the size (storage capacity) of the tailings facility, height of dam, the potential economic loss, environmental impact and potential loss of life that would result from failure. Once a hazard rating has been assigned to a tailings dam, appropriate design earthquakes and flood events are selected. A high hazard classification requires more stringent design criteria.

For seismic design of a tailings facility, two levels of design earthquake are generally considered: the Operating Basis Earthquake (OBE) and the Maximum Design Earthquake (MDE). The OBE is typically determined using probabilistic seismic hazard methods, and represents the level of earthquake shaking at the dam site for which only minor damage is acceptable. A tailings dam is expected to function normally after an OBE. The MDE represents the maximum ground shaking for which the dam is designed. For large tailings dams classified with a high hazard, the MDE is often characterized as an event corresponding to the Maximum Credible Earthquake. Damage to a tailings dam is acceptable from the MDE, provided the integrity and stability of the dam is maintained and the release of impounded tailings is prevented (ICOLD, 1995).

A similar approach is used to determine the design flood for a tailings facility. The resulting design flood will range from 1 in 50 year recurrence period to the Probable Maximum Flood depending on hazard rating and size of dam.

Seismicity. The seismic stability of a tailings dam is an important component in the design and operation of a tailings facility. In regions of high or even moderate seismicity, it is often the seismic loading that controls dam stability. Consequently, a seismicity review of the region where the project is located should be carried out during the initial stages of project development. Initial studies at the conceptual or prefeasibility level may only include a review of existing information regarding the regional seismicity, and preliminary seismic design parameters may be obtained from seismic hazard maps, if available, for the region. However, for the feasibility and final design stages of a tailings facility, more sophisticated methods of analysis are required. These typically include both deterministic and probabilistic methods of seismic risk analysis.

A probabilistic analysis is carried out to define a unique probability of occurrence for each possible level of ground acceleration experienced at a site. The methodology used for the probabilistic analysis is based on that presented by Cornell (1968). The likelihood of occurrence of earthquakes within defined seismic source zones is determined by examining seismicity data. Using historical earthquake records for the region, magnitude-frequency recurrence relationships are established for each potential earthquake source or fault zone. The seismic design parameters selected for the design of a tailings facility are dependent on the level of seismicity in the region and the geologic and tectonic conditions at and in the vicinity of the site.

Unlike the probabilistic analysis, the deterministic method does not account for the likelihood of occurrence of a predicted ground acceleration. Seismic source zones or fault systems in the region are defined and maximum earthquake magnitudes assigned to each source. The resulting deterministic acceleration at the study site for each source is considered to be the maximum acceleration that can occur, on the basis of geologic and tectonic information. The maximum acceleration produced by this procedure is referred to as the maximum credible acceleration and the corresponding earthquake as the Maximum Credible Earthquake (MCE). The MCE is defined as "the largest reasonably conceivable earthquake that appears possible along a recognized fault or within a geographically defined tectonic province, under the presently known or presumed tectonic framework" (ICOLD, 1989).

Site Investigations.

Geotechnical site investigations for a tailings facility need to be carefully planned with an initial design concept in mind, and should be conducted to an increasing level of detail as the facility moves from scoping level to preliminary and on to detailed design. Further, site investigations should be planned to investigate the geologic interpretation of the site developed during the site characterization phase. The key objectives of site investigations for tailings facilities are:

- Confirm any potential natural hazards identified during the terrain analysis phase.
- Characterize the foundation materials through sampling and laboratory index tests such as particle size distribution and plasticity.
- Characterize the existing groundwater conditions through drilling investigations including in-situ hydro-geologic properties of the foundation soils such as permeability.
- Determine the geotechnical properties of the foundation soils, such as shear strength and compressibility.
- Confirm the availability and characteristics of the earth or rock fill materials required to construct the facility to the proposed design concept, including mine waste materials from open pit development.

Typically test pits, drill holes and seismic refraction surveys are sufficient tools for site investigations for detailed design of a facility. Test pits are used to investigate potential construction material borrow sources and the shallow foundation conditions for the tailings dam and the basin. Hydraulic excavators with minimum 5 m (16.5 ft) depth of excavation are typically used to excavate test pits. Test pits alone may be sufficient for scoping level designs.

Drilling investigations of the foundation are required for design of the tailings embankment. A properly planned and executed site investigation of the foundation soil or rock includes careful selection of the drilling method(s) to ensure the necessary data is collected. A geotechnical engineer needs to supervise all drilling activities, to ensure sampling and testing are completed at appropriate locations. Electronic piezo-cone penetration tests (CPTU) are very efficient at estimating in-situ soil properties in sands, silts and clays, but cannot be pushed into gravelly material. CPTU's can extend the data from more traditional drilling, sampling and Standard Penetration Testing (SPT). For gravelly materials, the Becker Drill is a common method that provides samples and penetration data correlated to the SPT.

Seismic refraction surveys include seismic lines that provide vertical profiles of depth to harder foundation soils or bedrock. Seismic refraction with both compression and shear wave velocity measurements are particularly valuable for design in seismically active areas.

Samples collected from the investigations are tested in a soils laboratory for index, strength and compressibility characteristics, and undisturbed samples of foundation materials are recommended for the latter two types of tests. For the materials from potential borrow areas, index tests must include compaction testing to confirm their suitability for placement in the tailings dam, basin liner or drainage systems. Strength and compressibility tests are also required for the fill materials, and should be performed on remolded samples compacted to the density criteria that will be required by the construction specifications.

Waste Characterization.

Characterization of both the solid and solution portion of the mill tailings is an essential step in the development of the tailings management plan. The acid generating potential and metal leaching characteristics of the solid tailings mass and the chemistry of the liquid effluent will effect the design of the tailings containment facility and be important considerations in solution management and reclamation planning. Acid rock drainage (ARD), metal leaching, and contaminant/metal release from mine tailing facilities are recognized environmental concerns. To ensure that natural aquatic systems are not significantly degraded, it is important to fully understand the tailings material by conducting a thorough waste characterization program.

A phased approach to waste characterization is often a prudent way to proceed. The relatively inexpensive static tests can be performed on a large number of tailings solids samples, with the results of the initial testing used to plan for additional, more detailed testing requirements. This stage of the characterization program will include testing to determine the trace element content of the waste, acid base accounting (ABA) or equivalent to determine the relative balance of potentially acid generating and potentially acid consuming minerals, and leach extraction testing to measure the soluble components of the samples. Whole rock analysis and mineralogical descriptions may also be conducted.

The trace element testing will usually consist of a full suite of metal analyses (ICP-MS/ICP-ES). The trace element concentrations will indicate which metals are naturally high in the waste and may be a concern for future leaching.

Acid base accounting, or the determination of the relative amounts of acid generating and acid neutralizing minerals in a sample, can be accomplished through a number of test procedures, with the Sobek Acid Base Account Test and the Modified Acid Base Account Test being the most common. These tests measure the acid potential (AP), also called the maximum potential acidity (MPA), of the sample based on its sulfur or sulfide content. The samples neutralizing potential (NP) is determined by titrating a pulverized sample of the material with an acid, with the resulting NP representing the acid neutralizing capacity of the sample. The samples net neutralizing potential (NNP = NP – AP), ratio of NP to AP, and paste pH are also determined. The results of the ABA testing are compared to general guidelines to assess whether the samples are likely to generate acid. Guidelines may vary for different areas but are generally based upon the net neutralization potential (NNP) and the ratio between neutralization potential and acid potential (NP/AP).

Leach extraction testing measures the soluble component of the sample. Some commonly used tests include the B.C. Special Waste Extraction Procedure (SWEP), the U.S. EPA 1312 procedure, and the U.S. EPA TCLP leach test. Leach extraction test results will determine the leaching potential of the waste and highlight those metals that are likely to leach from the tailings under neutral pH conditions.

Static testing provides valuable insight into the acid generating and metal leaching potential of the waste. Static testing does not, however, provide information about the rate of acid generation or neutralization and should not be used to predict drainage water quality in the field.

The next step in the characterization process for the tailings solids is kinetic testing. If static testing results show high variability between samples a relatively high number of samples should be considered for kinetic testing, while if samples were largely consistent, fewer samples are needed for the kinetic testing phase. Kinetic testing is used to confirm the samples' acid generating or neutralizing characteristics while determining the rates of reaction for acid generation and neutralization. Kinetic tests are most often conducted in a laboratory, where chemical weathering is simulated over time in cells or columns. The composition of the drainage collected from the test containers can be used to predict drainage water quality in the field if concentrations are corrected for the effects of temperature, flushing, and particle size. The test container drainage is analyzed for total and dissolved metals (ICP-MS), conductivity, total sulfur, sulfate, sulfide, and pH. Kinetic testing procedures include humidity cell tests, humidity column tests, column leach tests, Soxhlet extraction test, and field plot tests.

It is equally important to characterize the tailings solution or the aqueous portion of the tailings slurry. A sample of the tailings solution from the pilot test work should be analyzed for a range of parameters including total and dissolved metals (ICP-MS), nutrients, and reagents and reagent by-products used in the process. For example, if cyanide is used in the process, the effluent should be tested for the full range of cyanide species (total cyanide, free cyanide, WAD cyanide, strong acid dissociable cyanide, cyanate, thiocyanate). The speciation of any metals of concern should be determined since the species of metal present has a significant effect on the metal's availability and toxicity. Turbidity and suspended solids are also measured. Effluent concentrations are compared to effluent guidelines such as Canada's Metal Mining Liquid Effluent Regulations (MMLER) or the U.S. EPA Effluent Guidelines. A toxicity test of the effluent is also often required to ensure that the effluent is not acutely toxic to aquatic life. The toxicity tests are commonly performed on *Daphnia* or rainbow trout. Effluent concentrations, coupled with the site hydrology, can also be used to perform water quality modeling exercises if discharge to nearby waterways is considered. Predicted water quality concentrations are then compared to regional water quality guidelines or criteria for the protection of aquatic life.

Design Options

Conventional Tailings. Tailings produced by most conventional milling processes comprise a slurry of solids and solution with a solids content ranging from ~30-50% depending on the process. The conventional process of depositing the tailings is to transport the slurry by pipeline to the tailings facility and to discharge it into the tailings facility . The slurry is deposited on beaches or into a pond where the solids settle leaving supernatant process water that collects in a supernatant pond. The supernatant water, along with rainfall runoff that collects in the pond is usually recycled back to the mill for reuse in the process.

Thickened Tailings. The discharge of thickened tailings is a concept that has been in use for over 20 years in the mining industry. Thickened tailings disposal requires dewatering of the tailings slurry to about 50 to 60 percent solids, at which the tailings behave more like a highly viscous fluid, compared to a liquid slurry. The main advantage of this disposal method is that impoundment embankments may be significantly reduced in size or even eliminated, due to the reduced requirements to handle supernatant water.

Paste Tailings. The term Paste Tailings is generally used to describe a tailings slurry that does not segregate during mixing, transportation or placement and has a working consistency similar to wet concrete. Paste tailings can be transported through a pipeline but, unlike a slurry, the paste has no critical flow velocity below which solids settle out in the pipe. In recent years there have been significant advances in dewatering technology and a number of processes are now available to produce paste tailings. These include conventional thickener/filter methods and the Paste Production Storage Mechanism (PPSM).

The amount of water required to maintain a paste consistency depends on the type of tailings. Generally, as the fines content of the solids fraction increases, the amount of water retained by the paste also increases. Paste tailings typically consist of approximately 60 percent solids for a fine tailings, up to about 80 percent for a coarse tailings material.

The properties of a tailings paste can be modified as required, by adding materials such as Portland cement and fly ash to alter chemistry and strength characteristics. For above ground disposal this may be useful for stabilization of a paste tailings berm, erosion protection or reclamation of the tailings surface by providing improved trafficability.

Dewatered tailings have been used as backfill in underground mining for well over a century. The main uses of dewatered tailings in underground mining have been and still are for stope support, to provide a working floor and for waste disposal.

The design of paste backfill using dewatered tailings requires an understanding of the physical-chemical as well as the mechanical properties of the tailings materials. Backfill design requires the determination of the fill composition and water requirements for fill preparation to produce an acceptable, cost-effective mix. Compressive strength and permeability are recognized as the most important mechanical properties. Backfill strength is commonly enhanced for high early strength requirements by additives such as Portland cement, ground slag or fly ash. Chemical additives to improve permeability include flocculants, accelerators and retarders. The most commonly used forms of backfill are high density slurries and paste fills which can be tailored for high early strength with reduced cement consumption.

The design of a successful backfill operation, based on dewatered tailings, requires the assessment of backfill preparation, storage and transport, bulkheads, quality control, fill material preparation, placement methods, maintenance, labour and void preparation. Backfill pilot plant testing is commonly initiated for trials prior to the implementation of full-scale operations. Full production systems can often be designed utilizing most of the components of an ordinary run-of-mill tailings disposal system.

Paste tailings have been used as underground mine backfill for a number of years. The surface disposal of paste tailings is, however, a relatively new concept which only recently has been considered as a viable alternative to conventional tailings disposal.

Dewatering of tailings reduces or eliminates the initial consolidation phase of a slurry. In certain cases, dewatering prior to surface disposal may offer advantages in terms of material handling, placement strategy and environmental impact when compared to conventional tailings disposal. These advantages could include the following:

- Decreased time required for the tailings mass to achieve its ultimate density and volume due to the higher initial solids content.
- Reduced storage requirements due to more rapid consolidation and reduced volumes of supernatant water release.
- Reduced seepage water from ongoing tailings consolidation.
- Reduced dam construction requirements (smaller embankments and reduced fill volumes).
- Reduced geotechnical hazards associated with containment structures due to improved material stability and strength and reduced excess pore pressures.
- Reduced short and long-term environmental liability due to reduced seepage.
- Increased operational and design flexibility for tailings facilities due to improved handling and storage characteristics.

Dry tailings. For "dry" tailings the moisture content of the tailings is typically reduced by filtration methods to about 20 to 30 percent of the dry solids weight. At these low moisture contents, the tailings no longer exhibit the flow characteristics of a slurry or paste and form a "dry cake" material. This can usually be handled using conventional earth moving equipment and transported by truck or conveyor system. New technology is being developed that may allow the transportation of the "dry" tailings by froth transport in pipelines. Current filtering technology allows large tonnages of tailings to be dewatered more efficiently and economically.

Co-disposal of tailings with other mine wastes. In some cases other mine waste products such as waste rock can be combined with dewatered tailings for co-disposal within the same tailings facility. For paste and dry tailings materials a composite material can be created with the waste rock for transport to the facility and placement. Potential advantages of the co-disposal of dewatered tailings and waste rock include:

- stability of the composite material when compared to the tailings alone
- Decreased storage volume with the tailings filling the voids in the waste rock.
- Decreased ARD potential when waste rock is encapsulated in paste tailings.
- Reduced disturbance area for waste storage.

Materials for the construction of containment structures. There are many types of structures built to contain the tailings solids and solutions and to protect the surrounding environment from contamination. The structures or embankments are classified by the materials used to construct them and by the shape of the embankment cross section.

The use of waste products from the mining process including waste rock from the mine and the tailings from the milling process in the construction of the embankment can be economically and environmentally beneficial. For example, waste rock can be directly hauled from the mine to the embankment for the cost of the overhaul. This can often be done at a relatively low cost. Tailings can be processed using hydro cyclones or controlled spigotting to classify the tailings into a course and fine fraction. The course fraction can be used for embankment construction, again at a relatively low cost. The use of waste products for embankment construction can reduce the need for excavating natural construction materials and thereby reduce the overall area of land disturbance. It is necessary to fully characterize the waste materials before using them in embankment construction. Mineralized waste rock may be subject to oxidation and release of metals that could cause environmental impacts. Oxidation can also lead to mechanical breakdown of rock particles that can result in a loss of strength. The course fraction of tailings can also be subject to oxidation and acid drainage and can be subject to water and wind erosion.

The alternative to using waste materials is the use of borrowed natural materials. The availability of different types of materials such as low permeability soils, sand and gravel, durable rock, etc., dictates the design of the containment structures. Excavation of these materials from within the storage basin of the facility can have the added advantage of increasing the storage capacity of the facility.

Embankment types. There are four basic types of embankments shown on Figure 1. These are as follows:
- Downstream
- Upstream
- Centerline
- Modified Centerline

The **downstream embankment** is constructed in stages so that the centerline of the embankment crest moves downstream with each stage. This embankment type uses the most construction materials and is therefore the most expensive option for embankment design. The downstream embankment is completely independent of the physical properties of the tailings deposit. There are some jurisdictions such as California where this is the only option available for embankment design due to dam safety regulations.

The **upstream embankment** is constructed in stages where each stage is constructed on the tailings beach immediately upstream of the previous stage. The centerline of the embankment crest therefore moves upstream with the construction of each stage. This embankment type uses the least amount of construction materials and is therefore the least cost option for embankment design. In the past, a large proportion of tailings facilities have been constructed using the upstream method combined with spigotting of tailings to produce a course fraction for embankment construction. There have been some major failures of embankments constructed using these methods. The failures have in the most part been due to a lack of drainage in the embankment section resulting in high water levels. Earthquakes have caused liquefaction of the saturated tailings and resulting loss of strength has caused flow slides or large scale deformations and loss of freeboard. These failures have led to the complete ban of upstream construction methods in some jurisdictions, most notably, Chilé.

Figure 1 – Types of Tailings Embankments

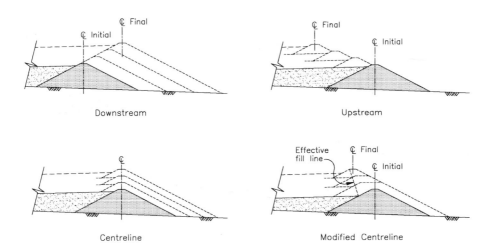

It is interesting to note that no upstream dam has failed that has been rigorously designed using modern engineering principles to ensure that the embankment is adequately drained and the phreatic surface is controlled.

The **centerline** embankment is constructed in stages so that the location of the centerline of the embankment crest does not change with each stage. The upstream toe of each embankment stage is constructed over the tailings beach but the majority of each new stage is founded on the previous embankment stage. This method relies on some strength and structural support for the upstream slope but does not rely on the tailings characteristics for overall stability. Liquefaction of the tailings as a result of earthquake loading could result in, at worst, some localized instability of the upstream slope of the embankment stage. This would not result in significant damage. The centerline method is a compromise between the higher cost downstream embankment and the higher risk upstream embankment.

A variation of the centerline embankment is the **modified centerline embankment**. This method allows the embankment crest centerline to move slightly upstream optimizing the quantity of construction materials required in the downstream shell zone of the embankment. Modern analytical techniques are used to determine how far the embankment crest centerline can be moved upstream with staged construction and still be independent of the physical characteristics of the tailings materials for overall stability. These techniques have resulted in significant cost savings without compromising embankment stability. Modified centerline embankments have been designed, permitted and constructed in USA, Canada and South America.

Geotechnical Design

The goal of the geotechnical design of a tailings facility is to provide a sound, stable facility at a minimum construction cost. As the design moves from scoping level to detailed design, an increasing level of design detail accompanies an increasing level of confidence in the construction cost estimates. Levels of contingency to the estimated cost of a tailings facility can be decreased as the design level increases.

With due consideration of local and national regulatory requirements, the geotechnical designer must address the following key issues:

- Structural stability of the tailings dam and appurtenant structures under the key loading conditions: construction, static long-term, seismic loading (if applicable) and under loads resulting from rapid draw-down of the water level if that is possible.
- Stability against seepage flow through core zones, basin liners, drains and filters that prevent the migration of tailings solids or fill materials through the dam or into the basin drains.
- Control of seepage rates below allowable limits through the life of the facility. Seepage flow rates to the surface pond and to the drainage systems resulting from tailings consolidation need to be included in addressing this issue.
- Stability of the facility against surface erosion by water or wind.

For scoping level designs, these stability issues may be addressed using experience and relatively simple calculations. For preliminary design, computer analyses for structural and seepage stability are typically employed. Limit equilibrium stability analyses and finite element seepage analyses are easy to perform these days and are normally provided in a preliminary design. For detailed design, additional analyses such as numerical modeling of structural stability are often employed, particularly where the assessed risk for the facility is moderate or high.

For seismic structural stability of tailings dams, it is most efficient to stage the stability analyses such that complexity of the analyses increases at each design level. The ICOLD bulletin on tailings dams and seismicity (1995) states:

"Seismic stability assessment of tailings dams.... is often carried out in a staged approach, which involves starting with a simpler analysis and progressing to more complex analyses as required by the specific case. In ascending order of cost and complexity, these are:

Static Limit-equilibrium Stability Analysis using Steady-State Strength;
Simplified Seismic Stability Analysis; and
Finite Element Seismic Stability Analysis"

The geotechnical characteristics of the tailings can only truly be determined after deposition takes place for a period of 1 or 2 years and in-situ testing techniques are used to measure design parameters. Hence, design of Stage 1 of a tailings facility should incorporate reasonably conservative assumptions on the tailings characteristics that will be confirmed by site investigations after construction of the first stage or two of the facility. Also, finite element analysis is usually not justified until after the in-situ investigations of the tailings

Basin Liner Systems. The requirements for lining the tailings facility basin depend on the waste characterisation of the tailings to be stored, the hydrogeologic characteristics of the basin and regulatory requirements. Regulatory requirements can be either performance or prescriptive in nature. Performance regulations usually set limits to leakage rates or levels of impact on receiving waters. Prescriptive regulations mandate a specific liner design on the basis of the waste characterization. In either case, the waste characterization of the potential seepage water from the tailings facility is the driving criteria in liner design. It is possible to find sites that are naturally impermeable and would therefore not require constructed liners. The difficulty is in conclusively demonstrating during the site investigation phase of the design that no discrete leakage pathways exist.

Constructed liners range in cost and complexity from simple single layer plastic membranes to multi-layer compound liners with drainage layers, friction layers and leak detection layers. A range of liner types is shown schematically on Figure 2.

Low permeability **soil liners** have some distinct advantages. Firstly there is a significant time for the travel of the initial seepage front through the liner. Secondly, there may be a reduction in the concentration of waste constituents in the seepage water as a result of dispersion, diffusion and adsorption by the soil. Thirdly, the liner will consolidate under loading by the tailings deposit and this will result in reductions in permeability. A disadvantage of soil liners is that seepage takes place over the whole area of the liner. For large facilities, even if the permeability of the soil liner is low, the total seepage rate can still be significant. Other disadvantages are that soil liners are erodable, are subject to desiccation cracking, frost heave, and osmotic consolidation.

Synthetic liners comprise thin plastic membranes. Typical plastics used for these liners include HDPE, LDPE, PVC, etc. The liners are placed in sheets and the seams are welded in the field to create a continuous membrane. Although the plastic materials are effectively impermeable, the liners invariably have a finite permeability due to leakage through pin holes and seaming defects. Typical leakage rates for a well installed membrane liners are 30 – 300 gallons per day per acre of liner. (0.3 to 3 cubic meters per day per hectare).

The leakage through a soil liner comprises a small unit leakage over a large area while leakage through a membrane liner comprises high unit leakage over very small areas (pin holes).

Figure 2 – Liner Systems

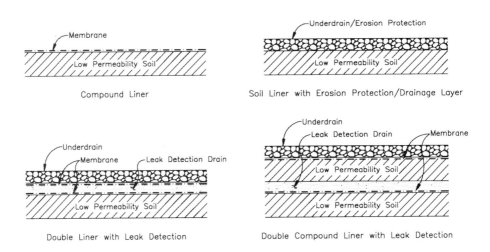

A **compound liner** comprises a membrane liner placed in direct contact with a soil liner. This combines the best of both liner types (a small unit leakage rate over a very small area) and results in leakage rates substantially lower than either individual liners.

The main factor controlling the leakage rate through a liner is the hydraulic head on the liner. For a single liner the head can be significant. The head on the primary liner can be controlled by the installation of a secondary or inner liner separated from the primary liner by a drainage layer. The drainage layer is designed to have sufficient capacity to remove any leakage through the inner liner without significant build up of hydraulic head. This restricts the head acting on the primary or outer liner and thereby limits leakage out of the facility.

The drainage layer between liners is often used as a leak detection system to measure leakage through the inner liner. Many regulatory agencies have limitations on the leakage flow from this drain. In order to control this leakage rate through the inner liner, a compound liner is sometimes required. A more meaningful performance specification is to limit the head buildup in the drain as this directly controls the leakage rate through the outer liner and from the facility.

An important consideration in the design of liner systems is the interface shear strength between adjacent liner components. Interface strength between the membrane liner and soil liner, membrane liner and drainage media, etc, can be very low. This can result in instabilities even on relatively gentle slopes. It important to identify the most critical interface shear strength in the proposed liner system for inclusion into stability analyses. Specialized laboratory testing of interface shear strengths is generally required.

With a liner in place, the tailings deposit will need an underdrainage system if the deposit is to become fully consolidated. An under consolidated deposit has a lower density and therefore requires a larger storage capacity for a given tonnage. The under consolidated deposit also has a low strength that can cause problems during construction of centerline or upstream embankment raises and when placing cover systems for reclamation. An underdrainage system provides a pathway for the tailings deposit to drain and reduces the hydraulic head at the base of the deposit. The effect of this is to halve the drainage path for consolidation. This reduces the time required for the deposit to consolidate by a factor of four. It reduces the hydraulic head in the liner system ant therefore will reduce the leakage from the facility.

Construction Considerations

Detailed designs for the construction of tailings facilities comprise technical specifications for the work and detailed drawings issued for construction. These documents can be included in a contract for the construction of the facility by a third party contractor or be used by the mine to carry out the construction using their own equipment and staff. The design of tailings facilities, like that of most large civil structures are based on the available topographic mapping and information generated by the site investigation. It is inevitable that there will be conditions encountered in the field during construction that differ to some degree from those assumed during the design. It is essential that a representative of the design engineer be onsite to provide technical direction for construction. This will ensure that significant variations in site conditions are recognized and that the design is modified as required to meet the design intent. This service is often integrated with the quality assurance/quality control (QA/QC) role. QA/QC is required during the construction of a tailings facility so that conformance of the work with the technical specifications and drawings is measured and recorded. This continuity from the design through construction is also important from a liability perspective. A designer that has no involvement with the construction can take the position that any shortcoming is a result of the construction or due to changed conditions that were only evident during construction. The constructor conversely can take the position that there was shortcomings in the design.

Continuity of design engineer over the life of a facility can also be important as, by it's nature, the mining industry generally has a high staff turnover at a given mine site. Often the design engineer provides the only continuous presence over the mine life.

An important and often overlooked aspect of construction of a tailings facility is the preparation of detailed as built drawings of the completed facility. The preparation of as built drawings takes place at the end of the project when budgets are exhausted and enthusiasm is low. These drawings however are invaluable as the basis for ongoing staged construction and if there are any problems with the performance of the facility.

OPERATIONS

Tailings Deposition

The method of placement of tailings into the facility depends on the form of the tailings and the desired tailings and water pond configuration. All tailings except for dry or dewatered tailings are generally conveyed to the tailings facility by pipeline. Paste tailings are included in this group even though the paste viscosity is very high, and the friction losses are significant, paste tailings can be conveyed in a pipeline using high head, positive displacement pumping systems.

Conventional and thickened tailings are usually spigotted onto a tailings beach. The solids settle onto the beach and supernatant water is released. The supernatant water flows down the beach to a surface pond. The spigotting of tailings generally result in segregation of the tailings solids. The courser fraction of the tailings will settle out first resulting in a steeper beach. The finer fraction is conveyed further from the discharge point with the finest fraction settling out beneath the surface pond. The degree of segregation can be controlled by the energy of the tailings discharge stream and the solids content of the slurry. A single discharge point can result in a turbulent, high energy tailings stream that can carry the course fraction of the slurry further from the discharge point. This reduces segregation and flattens the tailings beach. Spreading the discharge over multiple spigots or the use of spray bars results in a low energy, lamina flow of tailings onto the beach. The settlement of the courser fraction is faster and the beach slope is increased. The solids content affects segregation as slurries with higher pulp densities are more viscous and the settling rate of the larger particles is impeded. Thickened tailings, as described above, minimizes segregation but the increased viscosity results in a steeper beach with slopes up to 2%.

Paste tailings require special deposition techniques due to the very high viscosity of the tailings. The paste is non segregating and does not produce significant supernatant water. Tailings deposit slopes of up to 10% or greater are possible and therefore the discharge point for the tailings must be moved often to place the tailings into the facility. Two methods of placement of paste tailings are the top-down and the bottom-up methods. The top-down method is shown schematically on Figure 3.

Figure 3 – The Top-Down Method of Paste Placement

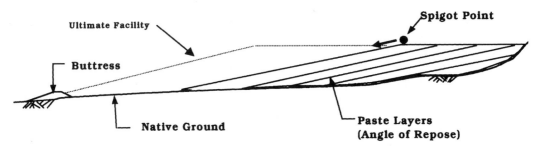

This method deposits the paste tailings downslope from a point upslope from the final toe of the deposit and the discharge is progressively moved towards the final toe.

Figure 4 – The Bottom-Up Method of Paste Placement

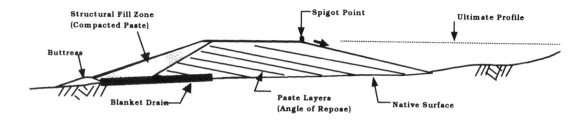

The bottom up method is shown on Figure 4. This method deposits the paste tailings upslope from the final toe of the deposit. A structural fill zone is required to contain the deposit and provide a platform for spigotting. The main advantage of the top-down method is in water management. Runoff from the deposit reports to the downstream toe where it can be collected and managed. In the bottom-up method, the low point where runoff accumulates is always moving upslope making collection and management more problematic. The main advantage of the bottom-up method is that the downstream extent of the paste deposit is well defined. Any erosion or sloughing of the paste deposit is upstream and will always be confined. With the top-down method the toe buttress is the only containment to prevent the migration of tailings by erosion or sloughing outside the limits of the tailings facility. If this migration is excessive then it may be necessary to raise the buttress.

When hydocyclones are used for segregating course tailings for embankment construction, the underflow has many of the characteristics of paste tailings. The course underflow can be directly placed in the embankment section, conveyed to the embankment by pipeline or launder or stockpiled and placed using earthmoving machinery. The cyclone overflow comprises the fine tailings fraction and the majority of the water from the feed slurry. The overflow can be deposited by spigotting onto beaches in a conventional manner.

The deposition of dewatered tailings or dry stacking requires earthmoving methods and can be achieved by truck haul dump and spreading or by conveyor stacking. The main issue with dry stacking is trafficability of the surface of the deposit. This can be problematic in areas of high rainfall and snowfall. In these cases, temporary storage facilities may be required to store tailings until such time as trafficability can be established after a rainfall event or until snow can be cleared from the area of tailings placement.

Water Management

A key part of the operation of a tailings facility is the management of water in the facility. Water enters the tailings facility as process water in the tailings slurry, direct precipitation, and runoff from surrounding undiverted catchments. Water is stored permanently in the tailings deposit as pore water. Water is also lost to evaporation and seepage. In almost all cases, water is recycled to the mill for use in the process. Generally, if the total quantity of the water lost to permanent storage, evaporation and seepage is greater than the water into the facility from precipitation and runoff, the facility is in water deficit and makeup water is required to sustain the milling operation. If this is not the case then the facility is in water surplus and the excess water will need to be treated and released or adequate storage will be required to contain the water.

Precipitation and evaporation usually vary seasonally. This results in some months producing water surpluses and some months requiring makeup water. Water can be stored in the wetter months and used as makeup water in the dryer months. A detailed water balance is required for a tailings facility that accounts for the water inputs and outputs on a month by month basis. The water balances should consider climatic variation to ensure that there is sufficient makeup capacity for dry periods and sufficient treatment or storage capacity for wet periods. The water balance is used to predict the variation in size of the supernatant pond on a seasonal basis as the facility is filled.

A further aspect of water management is to ensure that the tailings facility has adequate capacity at all times to store, route, or otherwise deal with the runoff from extreme precipitation events. Overtopping as a result of storm runoff is one of the most common causes of tailings facility failure. It is imperative that adequate storage is provided to contain the supernatant pond resulting from the design wet season plus the runoff from the design storm plus required freeboard for wave run up. It is often prudent to include an emergency spillway as a last line of defense. This will enable water to spill from the facility if the design storms are exceeded and prevent overtopping and potential failure of the facility.

Staged Construction

In almost all cases, the construction of tailings facility is staged over its service life. The facility is staged so that each stage provides the minimum volume for storage of tailings, the supernatant pond and extreme storm runoff for the period up to the construction of the next stage of the facility. This defers capital expenditure for tailings facility construction thereby minimizing initial capital investment.

The other advantage of staging the construction of a tailings facility is that an observational approach can be taken to the ongoing design of the facility. When the initial design is done, many assumptions are made regarding the physical characteristics of the tailings including the expected deposit density, strength characteristics, liquefaction potential etc. These assumptions are necessarily conservative. After several years of operations, the performance of the tailings deposit can be investigated producing hard data on these characteristics. Staged construction allows for the design to be modified and optimized to incorporate these data often resulting in reduction in construction and operating costs for tailings management.

Monitoring and Surveillance

Designs for tailings facilities are based on basic assumptions as to the behaviour of the embankments, the tailings deposit, the hydrogeology and the water management. It is essential that adequate instrumentation is installed and that a comprehensive monitoring and surveillance program is implemented. The monitoring program should include a formal inspection carried out at least annually by a senior engineer with full knowledge of the design and operating criteria for the facility. The results from the program should be reviewed on a regular basis, preferably by the design engineer to confirm that the design requirements are being met. A regular review of monitoring data can provide early warning of developing problems. This allows remedial action to be taken before the problem develops into a major problem or worse, a failure.

Many regulatory jurisdictions require that a regular report be prepared that fully describes the operation of the tailings facility over the reporting period and records the volumes and types of wastes that have been deposited, the findings of the formal inspections and the resulting recommendations, the data from instrumentation and interpretations thereof.

Safety Review and Risk Management

Detailed safety reviews, should be carried out for tailings facilities at the initial design stage, at a stage during operations and at closure. These reviews and risk assessments are best carried out by third party experts unrelated to the owner or the design engineer. The review should cover the adequacy of the design, an audit of the QA/QC program, and a review of the construction documentation.

Similarly, a formal risk analysis should be carried out at least once in the early stages of the operation of a tailings facility. In an engineered risk assessment, the risk associated with each failure mode and for each considered structure is considered to arise as a result of two sufficient and necessary components: likelihood and consequences. Mathematically, risk is simply the product of likelihood and consequences expressed in units of probability and dollars respectively. A life, resource or operation may only be considered to be at risk if both the potential (likelihood) for a failure and an impact (consequence) of that failure exist. Risk may be increased due either to an increased likelihood of occurrence and/or increased consequences of occurrence.

A quantitative risk assessment is typically considered to be more robust and thorough due to the finite expected values that are determined for each failure mode associated with the structure being considered. In other words, if it is determined that there is a 10 percent chance (likelihood) of a hydrometeorological event causing one million dollars of erosion damage (consequence) over the life of the mine, then the expected value (risk) of that failure mode or hazard is $100,000. The expected value of this hazard may then be compared directly to other identified failure modes in order to assist with decision making during operations or design stages of a project. The obvious challenge is the estimation of probabilities or likelihoods and the quantification of consequences in terms of dollars. Is the likelihood of an event 1×10^{-6} or 1×10^{-7} ? What value is placed on a pristine lake?

In contrast, a qualitative risk assessment relies on a subjective rating of perceived likelihoods and consequences that are combined according to descriptive codes and compared with one another. Although apparently simplistic, a qualitative risk assessment may provide valuable insight to the vulnerability of various project components where insufficient or potentially misleading quantitative data is available.

Regardless of the method used, the results provide information on which risk management decisions can be made. The overall goal of risk management is to ensure that efforts are focused to minimize the over all risk of operating the facility and that priorities are set accordingly..

CLOSURE

Planning for Closure

The main requirements for closure of a tailings facility is to provide long term secure and stable storage for the tailings materials. The closure works must ensure that the tailings solids are not transported from the facility by wind or water erosion and that the contaminants in the tailings deposit are not dispersed into the environment by seepage, vegetation uptake etc.

The design for closure of a tailings facility commences at the earliest stages of conceptual design during initial development. A conceptual remediation and closure plan is developed at this stage and the objective of this plan is to provide a systematic approach to decommissioning the facility and to return the disturbed lands to a habitat capability similar to pre-mining conditions, or to an acceptable alternative. To achieve this, a closure plan must be flexible enough to allow for future changes in the mine plan and to take advantage of information obtained from ongoing reclamation research. An alternatives analysis is often carried out to select the best option to suit both current conditions and potential changes. As well, measures to progressively reclaim the site during operations are often included to reduce the environmental and safety risks at the site. The closure plan is often the basis for estimating the cost of closure and setting the level of bonding required by regulatory agencies from mining companies to ensure that the reclamation is funded. Ongoing development and refinement of the closure plan along with progressive reclamation during the operation of the tailings facility can be the basis for reduction of the reclamation bond.

The land use prior to development provides a guide to the criteria for post closure land use for the site. Tailings facilities located in a desert location will have different criteria for closure than facilities sited on productive farm or range lands. The first step in designing a tailings facility is therefore to determine how the facility is going to look after closure and reclamation. This includes how the final site drainage will be established, how the facility will be covered, revegetated, stabilized to prevent erosion etc. Once these concepts are finalized, the design of the facility including the tailings deposition, location of the supernatant ponds liners and drainage systems can be carried out to complement closure concepts.

When planning closure, another consideration is if the mine closes earlier than planned due to economic or other circumstances. This can leave the tailings surface a long way below the planned final spillway location and can result in costly works to establish a workable closure configuration.

To meet the requirements for closure, the tailings surface generally requires stabilization. Stabilization can range from direct revegetation of the tailings surface, if possible to multiple layered cover systems. Direct revegetation is an option where the available nutrients and climate make establishment of sustainable vegetation possible. This will prevent wind erosion but will generally do little to prevent infiltration of precipitation that will eventually seep from the deposit. If there are any contaminants such as heavy metals, the plants may take these up and make them accessible to grazing animals. Simple revegetation is therefore only applicable where the tailings are benign.

Cover systems for tailings deposits are designed to carry out a number of functions. These can include providing a capillary break to prevent the migration of contaminants out of the deposit, a barrier to limit the infiltration of precipitation into the tailings deposit, an erosion resistant layer to prevent wind or water erosion and growth media for the establishment of vegetation. Cover systems can cover very large areas and can require significant resources for construction. It is important to identify the source of materials at the earliest stages and to ensure that the environmental impact of accessing these materials is not excessive. It is very important to conserve potential materials such as topsoil during the construction of the facility so that they are available for reclamation works.

Flooded tailings surfaces have been designed and constructed for tailings where acid drainage is an issue. Flooding of the tailings surface maintains the tailings in a state of saturation and prevents the oxidation of sulfide minerals. This method of closure is very effective in preventing oxidation but has significant associated issues. The flooded deposit is a constant source of seepage and can result in ongoing transport of contaminants. The flooded deposit is a perpetual liability that will require ongoing inspection and maintenance.

Mine tailings that are unconsolidated or only partially consolidated at closure will continue to consolidate after closure, resulting in on-going settlement of the tailings surface and the generation of consolidation seepage. Estimating the degree of consolidation of a tailings deposit at closure is therefore a key consideration for reclamation, particularly when surface covers are to be constructed or estimates of post closure consolidation seepage rates are required. Predictions of the consolidation rate, the magnitude of surface settlement and seepage flows after closure can be determined using data from in-situ field testing, laboratory testing of tailings samples and computer modeling of the tailings consolidation process. If the ongoing consolidation of the tailings and the time frame for this are problematic, then measures can be designed to significantly reduce the time for consolidation. Some under consolidated tailings deposits will take decades or even centuries to fully consolidate and this may postpone final reclamation indefinitely. The installation of vertical drains into the tailings deposit can reduce the time for consolidation of the tailings deposit to a few years.

Closure planning requires a water management plan. Once tailings deposition and recycle of process water stops, precipitation can cause water to accumulate in the facility. In addition, underdrainage from the tailings deposit, seepage recovery, etc. continues after the mill has stopped. Methods of dealing with the water depends on the level of contamination. Treatment can be as simple as sediment control and release or require complex biological treatment to remove a variety of contaminants. It is sometimes possible to dispose of excess water using controlled land application. Water treatment plants are usually costly to build and require significant ongoing operating costs. It is very important for closure planning to predict the level of water treatment and the length of time that the treatment will required.

Post Closure Maintenance and monitoring

All tailings facilities require a degree of maintenance and monitoring after reclamation is complete. The level of effort required depends on the complexity of the reclaimed structure and the requirement for ongoing treatment of effluents from seepage and surface runoff. As previously stated, flooded closures require water retaining structures that require frequent inspections and maintenance to ensure that the facility continues to be safe. Diversions, drainage ditches, spillways, cover systems etc. are all subject to deterioration. Root systems from vegetation can penetrate liners, animals can burrow through embankments and covers, beavers can dam diversion ditches and spillways. A post closure maintenance program should anticipate the requirements to deal with these issues and provide the funds necessary to carry these out.

CONCLUSIONS

This paper provides a glimpse of the scope and complexity of work required to successfully manage a tailings facility through it's complete life cycle. Tailings facilities are non revenue generating and money spent on development, operations and closure comes directly from the bottom line. The successful operation of a tailings facility however is of fundamental importance as potential failure can represent the greatest risk to the environment and a mining company's reputation, viability and share value.

ACKNOWLEDGEMENTS

This paper was prepared with input from a number of engineering and environmental specialists in the Knight Piesold Consulting group whose contributions are gratefully acknowledged.

REFERENCES

Canadian Dam Association (CDA), (1999), "Dam Safety Guidelines".

Cornell, C.A. (1968) "Engineering Seismic Risk Analysis" Bulletin of the Seismological Society of America, Vol. 58, p.1583-1606.

ICOLD (1989) "Selecting Seismic Parameters for Large Dams", International Commission on Large Dams, Bulletin 72.

ICOLD (1995) "Tailings Dams and Seismicity", International Commission on Large Dams, Bulletin 98.

Design of Tailings Dams and Impoundments

Peter C. Lighthall, Michael P. Davies, Steve Rice and Todd E. Martin

ABSTRACT

The state of practice for tailings dam and impoundment design is summarized. The design process, which embraces construction, operational and closure issues together with requisite technical aspects, has evolved over the past several decades though the engineering principles have remained the same. The design process has evolved to meet the demands of a regulatory environment that has become increasingly stringent together with more challenging mining economic conditions, as well as alternative tailings management technologies have been developed; including filtered dewatered, stacked deposits; thickened dewatered systems; frozen tailings deposits; and paste disposal. In response to a number of highly publicized tailings impoundment failures, tailings management systems, dam safety programs, and risk assessment techniques have been created and are now part of standard practice and appropriate dam/impoundment stewardship. The concept of environmental sustainability is now an integral component of the tailings dam/tailings impoundment design process and appropriate project conceptualization followed by stewardship of the design are essential to this sustainability.

INTRODUCTION

Environmental factors have increasingly driven tailings dam designs in recent years, but not necessarily to the benefit of engineering safety of tailings dam structures. For while design tools have improved and designs may have become more rigorous with technological improvements, the safety record of tailings dams has not markedly improved. Highly publicised failures continue to occur, resulting in a negative image for the mining industry, and particularly for Canadian mining companies, which have been involved in several of the more prominent failures. Unfortunately, the failure statistics do not tell the real story, which is that design, construction, operation and management of tailings facilities has advanced tremendously over the past thirty years. The security of tailings facilities is now a recognized priority at a corporate level in most large mining companies and the concept of sustainable mining, which clearly involves appropriate mine waste stewardship, is an accepted part of the modern industry.

Tailings, and waste rock, are the waste products of the mining industry. Their disposal adds to the cost of production, and consequently, it is desirable to accomplish their disposal as economically as possible. This requirement for low cost led to development the upstream method of tailings dam construction, which was the standard for tailings disposal up to the mid-1900's, irrespective of site conditions. With the advent of sound engineering practice, it became recognised that there are significant weaknesses and risks in the upstream method of construction under many site conditions. To augment the upstream method, embankment designs were developed using downstream and centreline construction methods. Sound civil engineering designs for embankment slopes, transition zones and filters were applied to tailings dams on an industry wide basis for the first time, beginning in the 1960's. Once designed and constructed based on an empirical and experience-based approach, tailings dam design has since evolved into a formal specialist engineering discipline.

Over the past 30 years, the most significant change in tailings disposal technology has been the recognition of the long-term geochemical risks, particularly the potential for acid drainage and metal leaching. The tailings dam designer of 30 years ago was not tasked with understanding and addressing geochemical issues. Today, geochemical characterisation is one of the most important aspects of tailings disposal planning, and designs for operation and closure are focussed on geochemical issues. These issues often govern not only the type of tailings dam, but also may govern tailings dam site selection. Contaminant loading analysis is now required on practically all tailings dam design projects. Closure strategies to prevent long-term geochemical impact to receiving environments are often the driving factors in design of tailings impoundments.

The other significant trend over the last 30 years has been the considerable number of highly publicised failures of tailings dams that have continued to occur with alarming frequency over the past three decades. In response to these large numbers of failures, mining corporations, financial institutions, environmental groups, government regulators and even the general public have instituted much more rigorous scrutiny of tailings management systems. This paper attempts to describe how the state of practice of tailing impoundment design, construction and operation has changed, and to outline some of the technologies that have been developed in recent years.

MINE TAILINGS

Tailings are the finely ground barren minerals left following ore extraction processes. They are typically a product of milling, although in several industries, e.g. the oil sands mining activity in Northern Alberta, large volumes of tailings can be produced without mechanical crushing. In milling, the process begins with crushing mine-run ore to particle sizes generally in the range of millimetres to centimetres. Crushed ore is then further reduced by grinding mills to sizes less than 1 mm in ball mills, rod mills, and semi-autogenous (SAG) mills. Water is added to the ground ore, and the material remains in slurry form throughout the remainder of the extraction process.

The grain size distribution of tailings depends upon the characteristics of the ore and the mill processes used to concentrate and extract the metal values. A wide range of tailings gradation curves exist for various mining operations and consequently, tailings may range from sand to clay-sized particles. For most base metal mines 40% to 70% of the tailings will pass a No. 200 sieve (74 μm). However, some milling processes such as gold extraction may grind the ore so that 90% or more of the tailings pass the No. 200 sieve. Tailings may also include metal precipitates from neutralisation sludges or residues from pressure leaching processes. Such materials may exhibit long-term chemical stability concerns and need to be disposed in secure, lined facilities.

BACKGROUND AND HISTORY OF TAILINGS DAMS

Mining has been carried out in some form for at least 5 000 years. In forms more similar to modern mining, crude millstone crushing and grinding of ore were initially practised in the New World in the 1500's, and continued through the mid-1800's. The largest change over those centuries was the introduction of steam power, which greatly increased the capacity of grinding mills and hence, the amount of barren by-product (tailings) produced.

Minerals of economic interest were initially separated from crushed rock according to differences in specific gravity. The remaining tailings were traditionally routed to some convenient location. The location of greatest convenience was often the nearest stream or river where the tailings were then removed from the deposition area by flow and storage concerns were largely eliminated. Later in the 1800s, two significant developments, which changed mining dramatically, were the development of froth flotation and the introduction of cyanide for gold extraction.

Flotation and cyanidation greatly increased the world's ability to mine low-grade ore bodies, and resulted in the production of still larger quantities of tailings with even finer gradation (i.e., more minus 74 μm material). However, tailings disposal practices remained largely unchanged and, as a result, more tailings were being placed and transported over greater distances into receiving streams, lakes and oceans.

Around 1900, remote-mining districts began to develop, and attract supporting industries and community development. Conflicts developed over land and water use, particularly with agricultural interests. Accumulated tailings regularly plugged irrigation ditches and "contaminated" downstream growing areas. Farmers began to notice lesser crop yields from tailings-impacted lands. Issues with land and water use that led to the initial conflicts then led to litigation in both North America and Europe. Legal precedents gradually brought an end to uncontrolled disposal of tailings in most of the western world, with a complete cessation of such practices occurring by about 1930.

To retain the ability to mine, industry fostered construction of some of the first dams to retain tailings. Early dams were often built across a stream channel with only limited provisions for passing statistically infrequent floods. Consequently, as larger rainfalls or freshet periods occurred, few of these early in-stream dams survived. Very little, if any, engineering or regulatory input was involved in the construction or operation of early dams.

Mechanized earth-moving equipment was not available to the early dam builders. As a result, a hand-labour construction procedure (the initial upstream method) was developed. A low, dyked impoundment was initially filled with hydraulically-deposited tailings, and then incrementally raised by constructing low berms above and behind the dyke of the previous level. This construction procedure, now almost always mechanized, remains in use at many mines today.

The first departure from traditional upstream dam construction likely followed the failure of the Barahona tailings dam in Chile. During a large earthquake in 1928, the Barahona upstream-constructed dam failed, killing more than 50 people in the ensuing, catastrophic flowslide. The Barahona dam was replaced by a more stable downstream dam, which used cyclones to procure coarser-sized material for dam construction from the overall tailings stream. By the 1940's, the availability of high-capacity earthmoving equipment, especially at open-pit mines, made it possible to construct tailings dams of compacted earthfill in a manner similar to conventional water dam construction practice (and with a corresponding higher degree of safety).

The development of tailings dam technology proceeded on an empirical basis, geared largely to the construction practices and equipment available at the time. This development was largely without the benefit of engineering design in the contemporary sense. Nonetheless, by the 1950's many fundamental dam engineering principles were understood and applied to tailings dams at a number of mines in North America. It was not until the 1960's, however, that geotechnical engineering and related disciplines adopted, refined, and widely applied these empirical design rules. The 1965 earthquake-induced failures of several tailings dams in Chile received considerable attention and proved to be a key factor in early research into the phenomenon of liquefaction. Earthquake-induced liquefaction remains a key design consideration in tailings dam design.

Issues related to the environmental impacts from tailings dams were first seriously introduced in the 1970's in relation to uranium tailings. However, environmental issues related to mining had received attention for centuries. Public concerns about the effects of acid rock drainage (ARD) have existed for roughly 1,000 years in Norway. Public concerns were similarly expressed hundreds of years ago in Spain and in Greece.

In the early 1970's, most of the tailings dam structural technical issues (e.g. static and earthquake induced liquefaction of tailings, seepage phenomena and foundation stability) were fairly well understood and handled in designs. Probably the only significant geotechnical issue not recognised by most designers was the static load-induced liquefaction (e.g. the reason for many previously "unexplained" sudden failures). However, issues related to geochemical stability were not as well recognised, and tailings impoundments were rarely designed with reclamation and closure in mind.

Over the past 30 years, environmental issues have grown in importance, as attention has largely turned from mine economics and physical stability of tailings dam to their potential chemical effects and contaminant transport mechanisms. Physical stability have remained at the forefront, as recent tailings dam failures have drawn unfortunate publicity to the mining industry, with severe financial implications in many cases. In response, a great many mining companies, at a corporate level, have identified safe tailings management as a priority, and have made resources available to address that priority. A significant tailings impoundment failure will almost certainly have a direct cost in the tens of millions of dollars and indirect costs, including devaluation of share equity, often many times the direct costs. In all of the tailings dam failure cases, a few examples of which are noted later in this paper, relatively simple, well-understood structural failure mechanisms were found to be at fault in causing the incidents.

THE ROLE OF GOVERNMENT REGULATIONS AND WORLDWIDE STANDARDS

As a new century unfolds, regulators and non-government organisations worldwide are becoming increasingly educated about tailings dam design and stewardship requirements. Lending agencies have also dramatically increased their technical requirements prior to funding new or expanded projects. This trend in education is a welcome development as candid discussions on risk levels for any given technical issue can be carried out with good understanding from all stakeholders. However, an unwelcome development has been the significant amount of non-technical and often misguided opposition and it is this latter situation that causes the most grief for the mining industry. No longer can a mine development proponent simply agree to meet the criteria of the senior governing authority and provide evidence of credible design. Often, several levels of government and non-government organisations must be satisfied with the proposed mining development. The tailings impoundment is often the most critical component of a mine development in the eyes of regulators and third party interest groups.

Many developing countries, where the international mining industry is focussing considerable attention, have only recently enacted regulations pertaining to tailings disposal. These regulations typically are based on those in place in more developed jurisdictions. Regulators in developing countries, however, all too often lack the resources and the expertise to fully implement these regulations. As a result, regulations in developing countries are often technically prescriptive. Such technically prescriptive regulations provide a false sense of security, since failures are so often the result of a combination of design flaws and improper stewardship. Many failures have occurred at facilities that conformed with all regulations, except for the most important of all (the dam failed).

The problem of limited resources for regulators is by no means unique to developing jurisdictions. Government budget cutbacks in jurisdictions such as Canada and North America mean that the mining industry is striving to a condition of "co-regulation", in partnership with regulators. This is a welcome trend, placing the initiative for continuous improvement and safety of tailings disposal facilities squarely with the mining industry.

Although visible exceptions continue to arise, as noted later in this paper, the modern tailings facility is typified by a well-designed and constructed facility that has met several levels of regulatory and non-regulatory scrutiny and has received corporate attention to the highest level. Most of the world's mining companies have multi-national operations, and to continue operation, and to attract share capital, must be seen to have exemplary environmental and safety records. Regulators, mining companies and international environmental organisations have developed numerous programs to ensure a high level of security of tailings disposal systems. Examples of some of these programs include:

- The Mining Association of Canada (MAC), has recently published a document entitled "A Guide to the Management of Tailings Facilities" (MAC, 1998);
- The Canadian Dam Association (CDA) recently updated its dam safety guidelines (CDA, 1999). The update focussed in large part on incorporating elements specific to the safety of tailings dams;
- The International Committee on Large Dams (ICOLD), and related organizations, have published numerous materials with regards to tailings dams;
- The United Nations Environment Programme, Industry and Environment (UNEP), and the International Council on Metals and the Environment (ICME) have been active in recent years in sponsorship of seminars, and publication of case studies (UNEP-ICME, 1997 & 1998), related to tailings management;
- Several major Canadian-based mining companies have established corporate policies and procedures to ensure that all personnel involved in stewardship of tailings facilities, from the corporate level to the operators, clearly understand their roles and responsibilities (e.g., Siwik, 1997);
- Numerous mining companies, including Syncrude, Kennecott Utah Copper, and Inco, retain a board of eminent geotechnical consultants to provide independent review and

advice in terms of the design, operation, and management of their tailings facilities. These programs are described in McKenna (1998), Dunne (1997) and (McCann, 1998); and
- Many mining companies have regular third party risk assessment programs for their tailings facilities, in which experienced consultants, usually teamed with the owner's personnel, carry out audits of tailings facilities.

Of all of the above measures, the authors consider the last two to be of most importance.

SITE CHARACTERIZATION

A major factor in managing tailings impoundment failure risk is to carry out adequate site characterisation for siting and designing tailings impoundments. Experience in tailings impoundment design and construction has served to emphasise the critical importance of a proper understanding of the geology of impoundment sites, and of an appreciation of how geology will affect the design, construction, and performance of the tailings facility. Many of the catastrophic structural failures have occurred as a direct result of inadequate site characterization.

Site characterization has always relied on carrying out thoughtful site investigations to develop a thorough understanding of site geology. The judgement of experienced engineering geologists should play a major role in site characterization. Traditional tools of geological mapping, air photo interpretation, test pitting, geotechnical drilling continue to form the basics of site investigation. However, significant changes and technological improvements have been made, so that tailings designers have a much wider range of tools from which to choose. As well, the needs of site characterization have added new demands. Some significant advances/changers in recent years include:

- Generally improved technology in site investigation techniques;
- Increased emphasis on water management and environmental characterization of the site, particularly with regards to hydrogeology;
- Greater importance being stressed on recognising geochemical issues, such as acid rock drainage (ARD), and in terms of attenuation of groundwater contaminant transport;
- Greater emphasis on closure and reclamation considerations and the site investigation required to support that design; and
- Development of site investigation methodologies for characterization of liquefaction potential that have evolved significantly.

Technological Advances

The technology available for use in execution of site investigation programs has advanced greatly in recent decades. There are more and better-equipped geotechnical drilling contractors, with more powerful and efficient drilling equipment. Examination of aerial photographs represents an essential method of site assessment, and the quality of aerial photography has improved considerably. Remote sensing, and satellite imagery techniques are now available and prove invaluable on many projects. Geographic Information Systems (GIS) have been developed and represent a major advance in the collation and ultimate usefulness of site data. The world earthquake database, and the means for using it in probabilistic characterisation of site seismicity, a significant consideration for tailings impoundment design, has also advanced. Geophysical methods have also become more useful as techniques have become more reliable and analysing the data within complex solutions more readily available with the dramatic increase in computer processing power.

Perhaps the most significant advance has been the development and widespread application of electronic piezocone technology for geotechnical and environmental characterization of clay, silt, and sand soils. The piezocone provides end bearing, friction, and porewater pressure data on a near-continuous basis as the probe is advanced. From these data, such geotechnical information as soil type, shear strength, in situ state, sensitivity, relative density/consistency, and liquefaction susceptibility can be determined using semi-quantitative relationships. Resistivity measurements are also possible, enabling contaminant plumes in groundwater to be delineated. The piezocone can be used effectively as a piezometer to characterize porewater pressure gradients, and for in situ estimation of hydraulic conductivity and consolidation parameters. Piezocone technology is particularly well suited to geotechnical and environmental profiling of mine tailings deposits.

Emphasis on Environmental Site Characterisation
The increasing emphasis on environmental protection in siting, design, construction, operation, and closure of tailings impoundments has placed increased emphasis on environmental aspects of site investigations.

In terms of siting and design of a tailings impoundment, groundwater quality protection is perhaps the most significant environmental protection aspect requiring investigation. To evaluate potential groundwater quality impacts, the following are essential:

- Establish baseline (pre-development) groundwater quality conditions, by collecting surface water and groundwater samples (monitoring wells).
- Identify principal hydrogeologic units (overburden and bedrock), and develop a hydrogeologic model of the site. In most cases, a larger, regional hydrogeologic model is also required.
- Model tailings impoundment development and estimate contaminant loadings. This may require characterization of attenuation capacity in the hydrogeologic units. Based on that model, and on parametric (sensitivity) analyses, determine compliance versus non-compliance at appropriate locations.

Seepage modelling, and contaminant transport modelling, can now be carried out using very powerful yet simple to use finite element and finite difference computer models. These models include both saturated and unsaturated flow regimes, which allow better prediction of behaviour at the important interface between the tailings and the atmosphere, where much of the geochemical activity is occurring. The development of these tools has to some degree driven the need to obtain the data that allows their effective use. The foundation of these models is a reasonable hydrogeologic model, and the foundation of a hydrogeologic model is an understanding of local and regional geology.

Geochemical Characterization
The role of geochemical issues in tailings impoundment design, and particularly closure, is equally as important as geotechnical issues driven primarily by the critical issue of acid rock drainage (ARD). Tailings that are potentially acid generating require closure strategies, and therefore impoundment designs, that will prevent/control acid generation. This issue is a major component of initial mine studies involve addressing the acid generation potential of tailings and/or waste rock in a comprehensive manner.

Susceptibility of dam fill and foundation materials to structural change due to the effects of ARD generation in the tailings deposit must also be considered. For example, if an impervious core dam is being considered, then the mineralogy of the core material should be checked, as dissolution of carbonates within the material could greatly increase the permeability of the core. Similarly, geochemical effects on materials being considered as a clay liner for the impoundment must also be considered. As another example of the importance of this issue, there have been many documented case histories of ARD resulting in clogging of internal drainage zones within tailings dams, requiring in many cases extensive remedial measures.

NEW DEVELOPMENTS IN TAILINGS DISPOSAL
A number of improvements have been made in tailings disposal technology and tailings dam design, both to improve on the weaknesses of previous practices and also to take advantage of tailings processing technologies. Geochemical aspects now largely drive the siting, of a tailings

impoundment, the design of retention structures, and tailings disposal technology. There have been technologies put forward as panaceas for tailings disposal problems, which have turned out to be flawed in practice. These improvements can be categorized as changes in basic management practices and changes in tailings characteristics through pre-discharge dewatering.

Designing for Geochemical Issues

Geochemical issues have become highly prominent as severe acid generation problems became apparent at a number of mature mines around the world. Some of these mines, which had been operated by smaller mining companies, became orphan sites, leaving significant legacies for future generations. The majority of the acid drainage mine sites have become very expensive legacies for the major mining companies that owned them. It has been necessary to develop and operate acid drainage collection and treatment systems for continued operation and closure of numerous mines. Capital costs for ARD collection and treatment systems have been in the several tens of millions of dollars, with ongoing operating costs up to several millions of dollars annually. As a result, companies developing new mines have focussed on methods to predict and prevent or reduce acid generation from tailings.

Considerable research, for example CANMET's Mine Effluent Neutral Drainage (MEND) program, was carried out in the 1980's and 1990's, to assess viable methods of acid drainage control. The most significant conclusion of the past 20 years is that it is far easier (economic) to prevent ARD in the first place than to control it. From a number of existing sites where tailings had been placed in lakes in northern Canada, it was concluded that long-term submergence of acidic wastes was probably the most effective means of ARD control. Considerable work has also been done on placement of impervious closure covers over tailings to prevent ingress of air and water. Sophisticated designs of multiple-layer covers, incorporating impervious zones, pervious capillary barriers and topsoil for vegetation growth, have been developed. Covers have been found to present the risk of long term cracking or erosion, and to be ineffective in excluding air, so are less favoured solutions than submergence from the geochemical standpoint. Some of the main technologies for reduction of ARD potential from sulphide bearing tailings are the following:

1. **Design for submergence by flooding the tailings at closure.** This is a solution, which is being increasingly encouraged and accepted by regulators. However, the authors are concerned that flooded impoundments may create a risky legacy. The more traditional closure configuration for tailings impoundments has been to draw down water ponds as completely as possible, to reduce the potential for dam failure by overtopping or erosion. To raise water levels in impoundments formed by high dams could present considerable long-term risk. One of the reasons that closed tailings impoundments have traditionally proven to be generally more safe, from the physical stability perspective, than operating impoundments is the relatively more "drained" condition of closed impoundments that do not include a large water pond. The flooded closure scenario represents an "undrained" condition that does not allow this improvement in physical stability to develop, so the risk does not decrease with time.

2. **Treatment of tailings to create non-acid generating covers.** To avoid the necessity of flooding impoundments, non-reactive covers of tailings can be placed on the top of the impoundment on the last few years of operation. It has been shown in several mining operations, for example at the Inco Ontario Division central milling operation in Copper Cliff, Ontario, that by the relatively inexpensive installation of some additional flotation capacity, pyrite can be removed to the level that the tailings can be made non-acid generating. The upper non-acid generating tailings placed on top can be left as a wide beach for dam safety, while the underlying mass of potentially acid generating tailings remains saturated below the long-term water table in the impoundment. Normally, the small amount of pyrite removed by flotation can be disposed as a separate tailings stream, placed in the deepest part of the impoundment where it can be left flooded.

3. **Lake or ocean subaqueous disposal.** The surest, safest and most cost-effective solution to prevent ARD is sub-aqueous disposal in a lake or the ocean. Tailings will remain permanently submerged and have shown to be non-reactive under water and to have few permanent environmental impacts. The challenge for this solution is that regulators have become reluctant to permit lake or ocean disposal, and there are not always appropriate sites available. In addition, the public often reacts emotionally and negatively to the concept of such disposal, despite the considerable benefits of these approaches. The authors are aware of at least two examples where public pressure incited regulators to demand that existing operations switch from ocean and lake disposal to on-land impoundments, with the result that environmental problems actually increased. The authors do note a slight trend to re-acceptance of subaqueous disposal, particularly in the marine environment, as the true environmental impact of the technique can be demonstrated to be almost negligible in certain instances. Moreover, the corporate risks and environmental liabilities associated with surface tailings storage on many projects grows to the point where project viability is threatened without looking to environmentally acceptable alternatives including subaqueous disposal.

Improved Basic Design Concepts

Improved Upstream Construction. Considerable attention has been given to improving traditional upstream dam construction to make the technique not only economical but also stable under both static and dynamic conditions. Numerous failures of upstream constructed dams have occurred. The failures have been the results of earthquakes, high saturation levels, steep slopes, poor water control in the pond, poor construction techniques incorporating fines in the dam shell, static liquefaction, and failures of embedded decant structures. Most failures have involved some combination of the above weaknesses.

Based on the above experiences, and through the use of improved analytical tools (computer programs for stability, seepage, and deformation under both static and seismic conditions), safe, optimised designs have been developed. Some of the key design features that have been added include:

- Underdrainage, either as finger drains or blanket drains, to lower the phreatic level in the dam shell;
- Beaches compacted to some minimum width to provide a stable dam shell. Beaches are compacted by tracking with bulldozers, which are also used for pushing up berms for support of spigot lines;
- Slopes designed to a lower angle than was used for many failed tailings dams. Slopes are generally set at 3 horizontal to 1 vertical or flatter, depending on the other measures incorporated into the designs. Steeper slopes, without an adequate drained and/or compacted beach, create the potential for spontaneous static liquefaction - a phenomenon not widely recognized in 1972 but one responsible for a number of major tailings dam failures.

Figure 1 below shows a typical section of an improved upstream design.

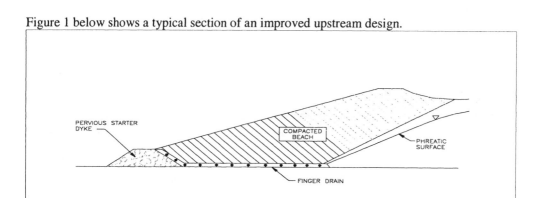

Figure 1 Typical section of improved upstream tailings dam design

Lined Tailings Impoundments. With the advent of larger gold mining operations, and the almost universal use of sodium cyanide as an essential part of gold extraction, the need came about to develop impervious impoundments to contain cyanide solutions. Although cyanide is in most forms an unstable compound that naturally breaks down on exposure to air, it can be very persistent and migrate long distances in groundwater. As well as cyanided gold tailings, other types of tailings may also be considered potentially contaminating. For protection of aquifers, where tailings impoundments are not sited over impervious soils or bedrock and embankment cut-offs are not sufficient to reduce seepage, it is often necessary to design and construct a liner over the base of a tailings impoundment. Great progress has been made in liner design and construction practise.

Liners may be as simple as selective placement of impervious soil to cover outcrops of pervious bedrock or granular soils, or may need to be a composite liner system constructed over the entire impoundment. Where geomembrane liners are used, it is normal practise to incorporate a drainage layer above the geomembrane, to reduce the pressure head on the liner and minimise leakage through imperfections in the liner. Another benefit of such under-drainage is that a low pore pressure condition is achieved in the tailings, giving them a higher strength than would exist without such under-drainage. The drainage layer typically consists of at least 300 mm of granular material, with perforated pipes at intervals within the drainage layer. The pipes are laid to drain water extracted from the base of the tailings deposit and to discharge to a seepage recovery pond. Figure 2 below shows two typical configurations of lined impoundments. Figure 2a shows a liner extending up the face of the embankment, requiring special detailing of drainage pipe penetrations through the liner. In Figure 2b, the liner extends beneath the embankment. In the latter case, care must be taken to design for lower foundation shear strength for the downstream slope of the embankment, as the liner may form a plane of weakness.

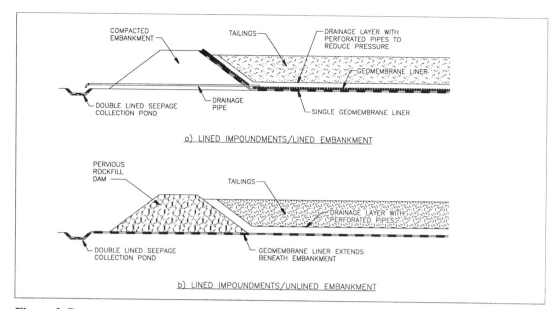

Figure 2 Conceptual sections of lined impoundments with underdrains

Dewatering Technologies. As shown on Figure 3 below, the basic segregating slurry is part of a continuum of water contents available to the tailings designer in 2000. Although tailings dewatering was previously practised for other purposes in the mining process, until recently the only form of tailings for most tailings facilities was a segregating, pumpable slurry with geotechnical water contents of well over 100%.

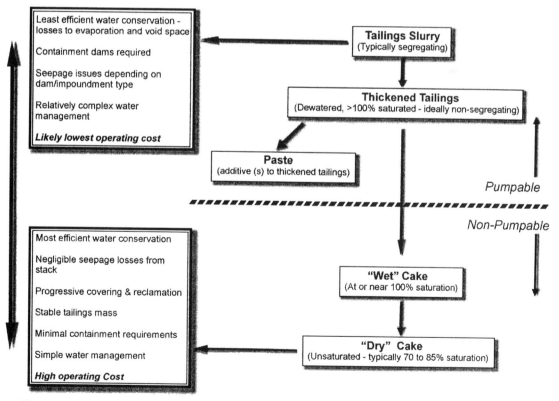

Figure 3 Classification of Tailings by Degree of Dewatering (after Davies and Rice, 2001)

There are several candidate scenarios where dewatered tailings systems would be of advantage to the mining operation. However, dewatered tailings systems have less application for larger operations for which tailings ponds must serve dual roles as water storage reservoirs, particularly where water balances must be managed to store annual snowmelt runoff to provide water for year round operation.

"Dry" Cake filtered tailings disposal. Development of large capacity, vacuum and pressure belt filter technology has presented the opportunity for disposing tailings in a dewatered state, rather than as a conventional slurry. Tailings can be dewatered to less than 20% moisture content (using soil mechanics convention, in which moisture content is defined as weight of water divided by the dry weight of solids). At these moisture contents, the material can be transported by conveyor or truck, and placed, spread and compacted to form an unsaturated, dense and stable tailings stack (often termed a "dry stack") requiring no dam for retention. While the technology is currently considerably more expensive per tonne of tailings stored than conventional slurry systems, and would be prohibitively expensive for very large tonnage applications, it has particular advantages in the following applications:

- **In very arid regions, where water conservation is an important issue.** The prime example of such system is at the La Coipa silver/gold operation in the Atacama region of Chile. A daily tailings production of 18,000 t is dewatered by belt filters, conveyed to the tailings site and stacked with a radial, mobile conveyor system. The vacuum filter system was selected for this site because of the need to recover dissolved gold from solution, but is also advantageous for water conservation and also for stability of the tailings deposit in this high seismicity location; and
- **In very cold regions, where water handling is very difficult in winter.** A dewatered tailings system, using truck transport, is in operation at Falconbridge's Raglan nickel operation in the arctic region of northern Quebec. The system is also intended to provide a solution for potential acid generation, as the tailings stack will become permanently frozen. A dry stack tailings system is also being planned for a new gold project in central Alaska.
- **Relatively low tonnage operations.** A separate tailings impoundment can be avoided all together by having a tailings/waste rock co-disposal facility.
- **Regions where a "dry landscape" upon closure is required.** The tailings area can be developed and managed more like a waste dump and therefore avoids many of the operation and closure challenges of a conventional impoundment.

Moreover, filtered tailings stacks have regulatory attraction, require a smaller footprint for tailings storage (much lower bulking factor), are easier to reclaim and close, and have much lower long-term liability in terms of structural integrity and potential environmental impact. Figure 4 below shows a photograph of a large dry-stack tailings system. Davies and Rice (2001) present a state-of-practice overview of dry stack filtered tailings facilities.

Figure 4 Example Dry-Stack Tailings Facility

Thickened/paste technologies. It is critical before basing mining operations on new technologies to carry out adequate engineering studies to demonstrate feasibility. Several tailings disposal technologies have been introduced to the mining industry that, over the past 30 years, have not proven out to be as effective as may have been hoped. While all have contained good ideas, they have often been wholly or partially unsuccessful, or have not found extensive application to date. However, two developments will likely see renewed emphasis over the coming years.

1. **Paste disposal.** The development of improved thickener technology has led to tremendous advances in paste tailings for underground backfill operations. Paste tailings are essentially the whole tailings stream, thickened to a dense slurry (previously only the coarse fraction of tailings was separated from tailings for use as backfill). Cement is added to the paste and the material is pumped underground to use as ground support in mined out stopes. Advocates of paste technology have promoted its use for surface tailings disposal, claiming that it can be placed in stable configurations with the cement providing adequate strength. However, the technology has not been shown to be economically pragmatic at a large scale for many surface applications.
2. **Thickened disposal.** Thickened tailings are paste without the additives. Thickened disposal is a technique that has been proposed for over 25 years and has been implemented in a few operations. The main premise of thickened disposal is that tailings may be thickened to a degree that they may be discharged from one or several discharge points to form a non-segregating tailings mass with little or no water pond. In the most classical connotation, thickened tailings is assumed to form a conical mass with the tailings surface sloping downwards from the centre of the cone. A thickened tailings system, if successful, should require lower retaining dykes, as storage is gained by raising the centre of the impoundment. It had been proposed that, at the start of operations, the tailings could be thickened to a lesser degree, when flatter slopes on the impoundment would suffice, then later thickened to a higher degree as it became necessary to raise the central point of the cone. In most instances where thickened tailings was implemented, thickening technology was not capable of producing a consistent non-segregating

material, so fines would form a very flat slope and require additional dyking at the toe. As well, flatter than projected slopes were experienced, and it was not possible to steepen these slopes to avoid extensive land use impact. From the above experiences, the thickened disposal did not become accepted. It has, however, been very successful in very arid regions, such as the gold mining districts of Australia. In recent years, high density thickening technology has been developed which make it useful to re-examine thickened disposal. The authors are aware of several major mining projects considering thickened tailings as an alternative management practice.

INSTRUMENTATION AND MONITORING

General
The reasons for instrumentation and monitoring of tailings dams are as set out by Klohn (1972). Reworded, these reasons are as follows:

1. To check that the environmental performance of the facility is meeting design intent, with no downstream out-of-compliance impacts on surface water quality and groundwater quality.
2. To confirm that the dam and impoundment are safe from the physical standpoint, from construction through operation and closure.
3. To provide the data required for confirmation and/or optimisation of design and construction through successive stages of impoundment construction and development.

The key consideration to be accounted for in instrumentation and monitoring of tailings impoundments is that they are dynamic structures, changing nearly continuously, constructed over periods of several to many years, often under the stewardship of changing personnel. An important consideration that has come into focus in recent years is the realisation that monitoring must continue into the closure phase, and that only in rare and favourable circumstances can a truly "walk-away" closure scenario, requiring no ongoing monitoring or maintenance, be achieved.

Technology and Communication Improvements
Technology improvements in instrumentation and monitoring, and in the ability to communicate the results, have been remarkable over the last 30 years. Instrumentation has become more advanced, robust, accurate, and reliable. It can be read more quickly and efficiently due to improved data logging equipment. It is now possible to monitor instrumentation remotely, using automatic data loggers and telemetry to transmit data to the office. Personal computers make plotting of the data in graphical form simple and rapid. Results can be emailed to the design engineer's office for quick review. Threshold (alert) levels can be included on these plots to visually define where the data plots in relation to safe versus unsafe conditions. Photographs can likewise be emailed for quick review by other mine personnel and/or the designer. Video surveillance cameras may even have some application at very remote sites, particularly for monitoring during the closure phase.

Despite these favourable trends, there is a significant caveat: technology advances are not yet at the point where they can replace visual inspection of the structure by a qualified, experienced engineer. Nor can technology serve as a substitute for application of imagination and judgement to the interpretation to monitoring data.

Training of Operations Personnel and Documentation of Monitoring Program
The international mining industry, and the mining industry in Canada in particular, has become increasingly focused on tailings dam safety issues. Instrumentation and monitoring represent important tools in tailings dam safety programs. It is becoming increasingly common for mining companies to require that their employees responsible for tailings dam operation receive training in dam monitoring.

The instrumentation and monitoring program for a tailings dam is normally included in a comprehensive Operations Manual (a document required by legislation in an increasing number of jurisdictions). Having an Operations Manual in place is now considered to be state of practice for tailings facilities management, and provides the following benefits:

1. It provides a concise, practical document that can be used by site operating personnel for guidelines on operation and monitoring of the tailings facilities.
2. It serves as a useful training document for new personnel involved in tailings management and operations.
3. Its existence provides reassurance to senior level management, and to regulators, that formalised practices are in place for the safe operation of the facility.
4. It demonstrates due diligence on the part of the owner.

Environmental Performance

Requirements for environmental performance have become more stringent, paralleling environmental regulations. Environmental performance monitoring can include the following:

- Surface water quality, downstream and upstream of the tailings impoundment.
- Groundwater quality, downstream and upstream of the tailings impoundment.
- Tailings pond supernatant water quality.
- Acid Base Accounting (ABA) testing of waste rock and tailings to determine susceptibility to acid rock drainage generation.
- Air quality (dust).
- Fisheries resources, and maintenance of minimum flows to fish bearing watercourses.
- Progress of revegetation where test plots and/or progressive reclamation are underway.

Most mining operations have on staff an Environmental Superintendent or Coordinator, who typically takes a keen interest in the operation and monitoring of the tailings impoundment.

Dam Safety Monitoring

The increasing focus on tailings dam safety brings with it an increasing awareness of the importance of a good monitoring and instrumentation program to confirm that the tailings dam is in a safe condition. Guidelines in terms of monitoring of tailings dams are provided by ICOLD (1994) and the Canadian Dam Association (1999), among others.

Monitoring to confirm tailings dam safety involves the following components:

- Periodic, detailed visual inspections of the tailings dam and its associated appurtenant structures (spillway, decants, diversion ditches). These inspections are carried out and documented by mine personnel who have received training in potential modes of dam failure and their warning signs. Any unusual conditions or concerns must be immediately reported.
- Definition of "green light" (safe), versus "yellow light" (caution) and "red light" (unsafe) conditions, and a pre-determined course of action (who to contact, what to do, increased frequency of monitoring, etc.) if yellow or red light conditions are noted.
- Reading of instrumentation (piezometers, survey monuments, inclinometers, seepage weirs, etc.) according to a set schedule, presentation of the results in graphical form, and interpretation and reporting of the results to the appropriate personnel.

The scope and frequency of dam safety monitoring will change through the life of a tailings impoundment, depending on the phase (construction versus operation versus closure) and on the consistency of monitoring results.

Confirmation and Optimisation of Design and Construction

Tailings dam engineering practice places considerable reliance on monitoring of the structure performance to confirm satisfactory performance, and to confirm design assumptions. Tailings dam construction, because it happens on a near-continuous basis, provides the opportunity to optimise design and construction over the life of the facility. This is the basic tenet of the observational method (Peck, 1969), a risk management method accepted and used in geotechnical engineering to avoid initial designs that may be overly conservative and overly expensive. As tailings disposal represents a cost rather than a profit centre to mining operations, the advantages of this method to optimise design and construction are obvious.

The elements of the observational method are illustrated schematically on Figure 5. Monitoring and instrumentation represent an integral component of the method. The observational method has a number of limitations, as follows:

- The method is not suitable for failure modes that can develop very quickly, with little or no warning, examples being static liquefaction of loose tailings, or a brittle, over-consolidated clay deposit that undergoes significant loss in strength with minimal straining. Such failure modes can only be properly addressed by good design.
- Once unfavourable conditions are noted, there must be sufficient time, and resources available, to react, putting in place measures that are pre-determined, an essential requirement of the method.
- The method cannot compensate for an inadequate site investigation program.
- The monitoring program in support of the observational method must be properly designed, not just to confirm anticipated conditions, but, more importantly, to detect unanticipated, unfavourable conditions.

Failure to recognise these limitations represents an abuse of the process, which then goes from being the observational approach to the "hope for the best" approach.

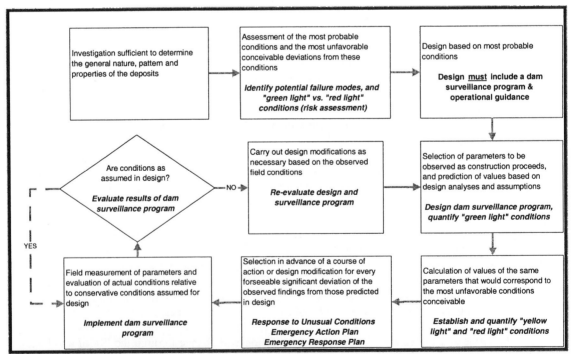

Figure 5 Elements of the Observational Method (adapted from Peck, 1969)

Operational Monitoring

Tailings and water management plan must be formulated for any tailings impoundment, and these must be monitored regularly. Elements of such plans include:

- Tailings deposition schedule;
- Storage versus elevation relationship for the impoundment;
- Operation of diversion structures;
- A mass balance model; and
- Pond filling and dam raising schedule.

Tailings and water management plans are projections, and require updating and calibration against actual conditions on a regular basis, usually no less frequently than annually. Operational monitoring data required for this purpose are as follows:

1. Measured precipitation, evaporation, runoff and snowpack data (mass balance models typically assume average annual conditions, broken down to a monthly basis). Runoff data is particularly useful in confirmation and adjustment of assumed runoff coefficients.
2. Regular tailings beach surveys and soundings to determine above and below water tailings slopes, and to allow the elevation versus storage volume curve for the impoundment to be updated.
3. Recording of tailings discharge points, elevations, and tonnages of tailings discharged from each point.
4. Pond level measurements, no less frequently than monthly.
5. Operation of reclaim barge, decants, spillways, etc.
6. Other water inflows to the impoundment (e.g. mine water) and outflows (e.g. water discharged directly, or discharged following treatment).

The Operations Manual should describe the data required, the frequency with which it is to be collected, and the manner in which it is to be collated and reported.

CONSTRUCTION AND OPERATION PROBLEMS

Davies et al. (2000) note that if one becomes a student of tailings impoundment case histories, an interesting conclusion arises. Tailings dam failures, each and every one, are entirely explainable in hindsight. These failures cannot be described as unpredictable accidents. There are no unknown loading causes, no mysterious soil mechanics, no "substantially different material behaviour" and definitely no acceptable failures. There is lack of design ability, poor stewardship (construction, operating or closure) or a combination of the two, in each and every case history. Tailings impoundment operational "upsets" or more catastrophic failures are a result of design and/or construction/operational management flaws - not "acts of god".

Should an severe upset or breach failure occur, several ramifications can be expected including, but not limited to:

- Extended production interruption;
- Possible injury and, in extreme cases, loss of life (there are more than 1100 documented fatalities attributable to tailings impoundment failures, Davies and Martin, 2000);
- Environmental damage;
- Damage to company and industry image;
- Financial impact to mine, corporate body and shareholders; and
- Legal responsibility for company officers

For perspective, a few recent failures are presented with their summary characteristics and impacts.

Stava, Italy, 1985
Static liquefaction collapse of two dykes, with release of 190,000 m^3 mud wave, travelling 10 km down a valley and killing 268. Legal ramifications for mine owners.

Bafokeng, 1974 and Merriespruit, 1994, South Africa
Tailings paddocks overtopped as a result of high rainfalls on ponds with excess water storage. Release of 3 million m^3 and 600,000 m^3 respectively, with 12 and 17 deaths and immense property damage.

Omai Gold Mine, 1995, Guyana
Dam failure due to inadequate internal filter design/construction led to release of cyanide solution to river. Although minimal short-term environmental impact was indicated, extended production loss, loss of share value and worldwide outrage resulted.

Los Frailes, 1998, Spain
Shallow foundation failure led to translation of dam shell and subsequent breach of one "cell" of tailings impoundment resulting in the release of 4 to 5 million m^3 of water and tailings slurry. Slurry inundated some significant areas of agricultural land.

DESIGN FOR CLOSURE AND RECLAMATION OF TAILINGS IMPOUNDMENTS

General
Many old mining operations, and the environmental problems associated with them, are now in the care of governments. Regulators, the public, and lending agencies are no longer willing to accept mining operations resulting in long term environmental legacies for governments. Therefore, closure and reclamation considerations have become perhaps the most important driving factors in siting, design, and permitting of mining projects. Successful closure and reclamation of tailings impoundments to a secure condition are an obvious necessity. Keeping closure in mind from day one is very much in the best interest of a mine operator, as it facilitates the most effective closure and reclamation of the site, and also avoids high costs at the end of the mine life. Regulators increasingly demand that mining companies post reclamation bonds for existing or proposed new mining developments. Impounded tailings so often represent the most significant closure risk, and so it follows that these issues are best dealt with when a project is in its conceptual stages.

Design Criteria for Closure
The design criteria (flood, earthquake, static stability) to which tailings dams must be designed depend on the time of exposure to the hazard. Because the exposure period of closed impoundments is perpetuity, the tailings dam must in most cases be designed to endure the most extreme events, such as the Maximum Credible Earthquake (MCE) and the Probable Maximum Flood (PMF). Modern tailings dam design must account for these requirements at the outset, so that expensive retrofit measures are not necessary at the end of the mine life. Design must also account for the cumulative effects of major floods (e.g. damage to riprap and/or spillways) and multiple earthquakes in high seismic zones.

Other design aspects that must be considered for closure are, for example, the erosion resistance of the dam, and the durabililty (weathering resistance) of the materials of which the dam is constructed.

"Dry" Closure
Where closure does not involve submergence of tailings below a water cover, modern practice requires either direct revegetation of the tailings surface, and/or covering of the tailings surface with a material that is more erosion resistant than tailings. Mining companies now typically carry

out extensive reclamation studies (appropriate species, topsoil availability and requirements, etc.) during the feasibility study phase of projects.

The "dry" closure scenario can be used for closure of tailings with potential for generation of ARD, in concert with collection and treatment of seepage and (if necessary) surface runoff from the impoundment. Eventually, the sulphides would be depleted, and/or the contaminant generation rate would become sufficiently low as to eliminate the need for continuing collection and treatment. The "short term" disadvantage of having to collect and treat must be weighed against the longer term advantage of improved dam safety and reduced failure consequences associated with now "inert" tailings.

Submerged Closure of ARD Tailings Impoundments
Permanent submergence of tailings behind a water retaining dam to prevent ARD resolves the key geochemical issue, but places more stringent requirements on the safety of the tailings dam. The safety of a dam will not increase over time because the "drained" closure condition is not achieved. The closure spillway(s) now represents a particularly critical component of the closed impoundment, and, like the dam, must be inspected and maintained on a regular basis. The dam safety program implemented through the closure phase should be no less comprehensive than one that would be implemented for a conventional water retaining dam with similar failure consequences. In fact, the failure consequence of a tailings dam impounding sulphide tailings would be higher than that for a conventional water retaining dam. Failure of the tailings dam would, besides uncontrolled release of water, also represent an environmental failure in the release of sulphide tailings to the environment. Submerged closure of ARD tailings impoundments, therefore, is not a "walk-away" closure scenario.

Other ARD Control Alternatives
Other ARD control alternatives include:
- Permanent neutralization of the ARD source by adding/enhancing neutralization potential within the tailings mass, such as could be achieved by mixing in finely crushed limestone with the tailings; or
- Sulphide removal and separate storage towards the latter stages of the mine life, such that the portion of the tailings containing sulphides remains saturated and/or under very low oxygen flux.

As both understanding of the ARD problem and the technologies available for mitigation of the problem continue to advance (e.g. improving water treatment technologies), it is likely that closure options less adverse to dam safety than permanent submergence will continue to be developed and become increasingly cost effective.

CONCLUSIONS
A review of the state-of-practice in tailings dam/impoundment design and construction shows that great technical progress has been made. Better investigation and design tools are available. New technologies in thickening and filtration of tailings have provided the opportunities for alternatives to disposing of tailings as conventional slurries. New concepts include "dry-stack" tailings, thickened tailings and paste tailings. Geomembrane liners are commonly used where tailings may present a risk of groundwater contamination, and design and construction methods for lined impoundments have been developed. Improvements have been made to the traditional upstream construction method to reduce stability risks.

Environmental considerations have become increasingly more important in tailings dam design and permitting. Closure planning has become an integral part of initial design and permitting. Designs must address ARD and include measures for long term control and/or prevention of ARD.

In spite of the improvements in tailings disposal practices, a number of highly publicised failures have overshadowed the advances. These failures have led to increased scrutiny of mining projects by regulators, environmental groups, and financial institutions. Numerous guidelines and risk assessment programs have been developed by the industry to reduce tailings dam failure risks.

REFERENCES

Canadian Dam Association (CDA) 1999. *"Dam Safety Guidelines"*.

Davies, M.P. and S. Rice (2001). An alternative to conventional tailings management – "dry stack" filtered tailings. . *"Tailings and Mine Waste 2001,"* Fort Collins, Colorado, pp. 411-420.

Davies, M.P., T.E. Martin and P.C. Lighthall (2000). Tailings dam stability: essential ingredients for success. *Chapter 40, Slope Stability in Surface Mining,* SME, pp. 365-377.

Dunne, B. 1997. Managing design and construction of tailings dams. *"Proceedings of the International Workshop on Managing the Risks of Tailings Disposal"*. ICME-UNEP, Stockholm, pp. 77-88.

International Committee on Large Dams (ICOLD) 1994. *"Tailings Dams – Design of Drainage – Review and Recommendations"*, Bulletin 97.

International Council on Metals and the Environment (ICME) and United Nations Environment Programme (UNEP) 1997. *"Proceedings of the International Workshop on Managing the Risks of Tailings Disposal"*, Stockholm.

International Council on Metals and the Environment (ICME) and United Nations Environment Programme (UNEP) 1998. *"Case Studies on Tailings Management"*.

Martin, T.E. and M.P. Davies (1999). Trends in the stewardship of tailings dams. *"Tailings and Mine Waste 2000,"* Fort Collins, Colorado.

McCann, M. 1998. Sustaining the corporate memory at Inco's Copper Cliff operations. *"Case Studies on Tailings Management"*, UNEP-ICME, pp. 55-56.

McKenna, G. 1998. "Celebrating 25 years: Syncrude's geotechnical review board". Geotechnical News, Vol. 16(3), September, pp. 34-41.

Mining Association of Canada 1998. *"A Guide to the Management of Tailings Facilities"*.

Peck, R.B. 1980. "Where has all the judgement gone? The 5th Laurits Bjerrum Memorial Lecture", Canadian Geotechnical Journal, *17(4)*, pp. 584-590.

Siwik, 1997. Tailings management: roles and responsibilities. *"Proceedings of the International Workshop on Managing the Risks of Tailings Disposal"*. ICME-UNEP, Stockholm, pp. 143-158.

Hazardous Constituent Removal from Waste and Process Water

Larry Twidwell[1], Jay McCloskey and Michelle Gale-Lee[2]

ABSTRACT
A review of technologies appropriate for removing contaminant constituents from wastewater and plant effluents is presented. Emphasis of this presentation is placed on the removal of oxyanions of arsenic and selenium with additional consideration given to the removal of heavy metals such as copper, lead, zinc, and thallium. Current technologies and potentially new technologies appropriate for achieving present day environmental requirements are summarized and discussed.

INTRODUCTION
It is indeed a formidable undertaking to summarize appropriate technologies capable of removing oxyanions and heavy metals from waste and process waters to the µg/liter concentration level in a relatively brief presentation. Therefore, the emphasis of this paper has been placed on the removal of oxyanions of arsenic and selenium. Literature review publications will be noted for the removal of heavy metal contaminants.

Environmental Standards
Maximum concentration level (MCL), secondary drinking water standards and aquatic life standards are presented in Table 1.

Table 1. Environmental considerations for toxic element constituents

Species	Standards, µg/L		
	MCL or Secondary	Aquatic Life[1]	
Oxyanions		Acute	Chronic
As	$10^1, 25^{2,3}$	360	190
Sb	6^1		
Se	$50^1, 10^{2,3}$	20	5
Heavy Metals			
Cd	$5^{1,2}$	3.9	1.1
Cu	$1300^1, <1000^2$	18	12
Pb	$15^1, 5^2$	82	3.2
Zn	$5000^1, <5000^2$	120	110
Tl	2^1		

[1]US (arsenic now 50 µg/l; becomes 10 in 2006); [2]CAN; [3]WHO

SUMMARY OF TECHNOLOGIES
Extensive literature reviews have been published that identify potential technologies for removing arsenic, selenium, and heavy metals from waste waters, e.g., arsenic (Nriagu 1994, MWTP 1994,

---

[1] Montana Tech of the University of Montana, Butte MT ltwidwell@mtech.edu
[2] MSE-Technology Applications, Butte MT jmcclosk@mse-ta.com

Twidwell et al. 1999, Harris 2000, Welham et al. 2000, Riveros et al. 2001), selenium (MWTP 1999, Twidwell et al. 1999a, 2000), heavy metals, especially copper, lead, zinc (Young 2000, SenGupta 2002), and thallium (Grzetic and Zemann 1993, Nriagu 1998, MWTP 1999a, Williams-Beam and Twidwell 2001). The identified treatments include technologies now being utilized and technologies that show potential for being implemented in the near future, examples include: precipitation/adsorption, treated activated carbon, tailored ion exchange, and reduction processes. Detailed discussions for each of these technologies are presented elsewhere (as referenced above). An abbreviated annotation of the literature is presented in the following sections.

Arsenic

Precipitation. The classic work of Chukhlantsev (1956) on arsenate solubility as a function of pH is often referenced in the literature. Plots of his data as log solubility versus pH have been extrapolated to predict extremely low solubilities for individual metal arsenates. Some of these extrapolations have been shown by to be greatly in error, e.g., calcium, iron, lead and barium (Robins 1981, 83, 88; Nishimura and Tozawa 1978, 85; and Essington 1988). The extrapolations are in error at higher pH values because of the relative thermodynamic stability of hydroxides and carbonates (compared to arsenates) and the extrapolations predict arsenic solubilities that are much too low in outdoor storage environments. Great care should be taken when using metal arsenate extrapolated solubility values.

Calcium Arsenate-There have been a number of published investigations (Laguitton 1976, Nishimura et al. 1978, 85, 88: Plessas 1993; Robins 1984; Bothe and Brown 1999; and many others) that have demonstrated that arsenic can be effectively removed by lime precipitation, i.e., near drinking water standards are achievable if the Ca/As mole ratio >3 at pH's >9 (Nishimura and Itoh 1985). Lime precipitation of arsenic with subsequent outdoor storage was the accepted industry disposal method until the middle 1980's. Robins (1981, 84) demonstrated in the early 1980's that calcium arsenate is not stable in an outdoor storage environment in the presence of air at pH levels above 8.3. Carbon dioxide in air converts the calcium arsenate to calcium carbonate with the subsequent release of arsenic back into the aqueous phase. Therefore, lime precipitation of arsenic with subsequent outdoor storage is not presently viewed as an appropriate disposal technology. However, Valenzuela et al. (2000) and Castro et al. (2000) report that lime precipitation (with subsequent calcination to crystallize the calcium arsenate compounds) is practiced at several Chilean copper smelters.

Ferric Arsenate-The use of autoclave conditions allows for the formation of ferric arsenate (scorodite). The solubility of crystalline scorodite at ambient temperatures has been reported by Krause and Ettel (1989) to be <50 µg/L (at pH ~4). The long term stability of pond stored products have been demonstrated by INCO at their CRED facility in Sudbury, Ontario, e.g., ferric arsenate is precipitated under autoclave conditions, producing a product that is 0.5-2% As. The final arsenic bearing filter cake is transported to a tailings storage area. Krause (1992) reports that the outfall from the storage area has shown arsenic concentrations, in general, to be <20 µg/L. The Campbell Mine at Red Lake Ontario also uses autoclave formation of ferric arsenate salts. This product is stored (since 1991) along with flotation tailings. The final outfall arsenic concentration is approximately 100 µg/L (Riveros et al. 2001). Several recent studies on the Campbell tailings have shown that anaerobic reduction of ferric arsenate by sulfidic materials and bacterial activity results in increases in the pore water concentration of arsenic (Riveros et al. 2001). Detailed reviews of the stability of ferric arsenate are presented by Riveros et al. (2001) and Welham et al. (2000).

Mineral-Like Arsenates-Several studies have been conducted investigating the removal of arsenic from solutions by formation of mineral-like arsenate compounds (Porubaev et al. 1972, Nikolaev et al. 1974, 76, Comba 1988, Bothe and Brown 1999, 99a). Porubaev et al. and Nikolaev et al. demonstrated effective removal of arsenate by lime addition in the presence of phosphate. Comba (1988) investigated the formation of mimetite, ($Pb_5(AsO_4)_3Cl$). The arsenic solubility was minimal at pH 5-6, e.g., 4 µg/L. The compound was shown to be stable in the presence of carbon dioxide by thermodynamic calculation and by experimental verification. The author suggests that mimetite may be a method of removing arsenic from process solutions because of the effectiveness of the precipitation and the ease of filtration (because of the morphology of the precipitates). However, mimetite will not pass the EPA Toxicity Characteristic Leach Procedure (TCLP) for lead because of the complexing of lead by acetate. Therefore, it cannot be considered an outdoor storable waste form. Bothe and Brown (1999, 99a) demonstrated that calcium hydroxyarsenate compounds form at ambient temperature. Arsenic solubilities at near neutral conditions were 10 μg/L and 500 μg/L for $Ca_4(OH)_2(AsO_4)_2 \cdot 4H_2O$ and $Ca_5(AsO_4)_3OH$, respectively. Khoe et al. (1991) have reported the formation of ferrous arsenate. This compound shows a minimum solubility (at pH 7.5) of ~8 μg/L. The authors suggest that ferrous arsenate would be environmentally stable if stored in sub-surface anoxic conditions (where ferrous iron would not be oxidized).

Adsorption. A large number of investigations have focused on adsorption of arsenic on ferric oxyhydroxide (referred to here as ferrihydrite) and on alumina/aluminum hydroxide surfaces. A relatively small portion of the literature describing the adsorption of arsenic onto these surfaces is annotated below. A detailed literature review is published for arsenic by the EPA Mine Waste Treatment Program (MWTP 1994).

Ferrihydrite-An important detailed review of conditions for formation and the stability of ferrihydrite is presented by Jambor and Dutrizac (1998). Some of the characteristic features of amorphous ferrihydrite formation and its conversion to more crystalline forms (goethite and hematite) include the following. The approximate formula for ferrihydrite is generally considered to be $5Fe_2O_3 \cdot 9H_2O$ (Schwertmann et al. 1982, 83, 91; and Eggleton 1987, 88). The surface area of freshly precipitated ferrihydrite is 180-300 m^2/g (Schwertmann et al. 1983, 91). The rate of crystallization of ferrihydrite to hematite or goethite at 25°C is a function of time and pH, e.g., conversion was half complete in 280 days at pH 4. Conversion of ferrihydrite to goethite results in a relatively large change in surface area, i.e., freshly prepared ferrihydrite showed a surface area of about 150 m^2/g; when converted to goethite at 25°C the area was reduced to 92 m^2/g; when converted to goethite at 90°C the area was reduced to 9 m^2/g (Schwertmann et al. 1983).

The fact that conversion of ferrihydrite occurs reasonably rapidly and that the conversion results in a significant decrease in surface area may hold important negative consequences for long term outdoor storage stability for surface adsorbed arsenic. However, the ferrihydrite conversion rate may be mitigated (changed from days to years) by the presence of other species and solution conditions during precipitation. Factors that decrease the rate of conversion of ferrihydrite to more crystalline forms include: lower pH (Schwertmann and Thalmann 1976, Schwertmann and Murad 1983), lower temperatures (Cornell and Schwertmann 1996), presence of silicate (Schwertmann and Thalmann 1976, Schwertmann and Fechter 1982, Cornell et al. 1987a), aluminum (Schwertmann 1984, Schulze and Schwertmann 1984), manganese (Davies 1984, Cornell and Giovanoli 1987, Scott 1991) and organics (Schwertmann 1966, Cornell and Schwertmann 1979, Cornell 1985).

Ferrihydrite/Arsenic-Studies investigating the adsorption of anions on ferrihydrite surfaces are not new. Municipal water treatment systems have utilized ferric, ferrous and aluminum hydroxide precipitation for many years to cleanse heavy metals and phosphates from solution as a final polishing stage. In fact, doping manuals are available: ferric chloride (AWWA 1988), aluminum sulfate (AWWA 1988a). The US EPA has chosen ferric precipitation as the Best Demonstrated Available Technology (BDAT) for arsenic bearing solutions. EPA notes, however, that arsenic bearing ferrihydrite still leaches and may require further treatment, such as vitrification. EPA still has concerns about the stability of arsenic bearing ferrihydrite precipitates under long term storage conditions (Rosengrant and Fargo 1990). Several recent and detailed reviews of arsenic adsorption and stability considerations are presented by Twidwell et al. (1999), Harris (2000), Welham et al. (2000), Riveros et al. (2001), and SepGuta (2002a).

Pierce and Moore (1980) have reported that ferrihydrite has a tremendous adsorption capacity for aqueous As(V), i.e., greater than 1.0 mole As(V)/mole Fe (maximum at pH 4). Puls and Powell (1991) also reported significant As(V) adsorption on ferrihydrite, e.g., up to one percent arsenic. Fuller, et al (1993) have also confirmed the significant capacity of ferrihydrite for arsenate adsorption, e.g., 0.7 mole As(V)/mole Fe was achieved when iron was hydrolyzed and precipitated in the presence of As(V) in the pH range 7.5-9.0. Their work also showed that adsorption of As(V) on previously precipitated ferrihydrite was very effective but much less effective than when the iron was precipitated in-situ with the As(V), e.g., 0.25 mole As/mole Fe was achieved by solid slurry adsorption. The authors also reported that the rate of As(V) removal was much faster for in-situ iron hydrolysis and precipitation (<5 minutes) than for the slurry adsorption (which showed a rapid pick-up, then a continued slow adsorption for days). One must be aware that the test work demonstrated that there was a release of a portion of the adsorbed arsenic with time, e.g., desorption at pH 8 resulted in an increase in the Fe/As mole ratio in the solid from 5.3 to 8.3 over a 750 hour test period. This represents a significant decrease in the solids arsenic content. Also, Belzile and Tessier (1990) add to the concern that adsorbed As(V) on ferrihydrite may not be a long term stable storable waste form. They have compared the data (over the pH range 4-8) of Pierce and Moore (1980) for adsorption on ferrihydrite with the data of Hingston (1968) for adsorption on crystalline goethite. The adsorption equilibrium constants for the two materials are orders of magnitude apart, i.e., adsorption is much more favorable on ferrihydrite than on goethite. Their conclusion was that "goethite cannot efficiently compete with amorphous Fe oxyhydroxides for adsorbing arsenate in natural waters."

An important unknown at this time is whether the product from ferrihydrite adsorption of arsenic will be stable if storage conditions become anaerobic. Brannon et al. (1987) have demonstrated that anaerobic lake sediments convert As(V) to As(III) (pH 5-8.0). However, when the anaerobic conditions were shifted by aerobic leaching the previously reduced As(III) was reconverted to more immobile As(V) which was associated with aluminum and iron oxyhydroxides. Masschelegn (1991) found that arsenic concentration in moderately reducing soils (0-100 mV) was controlled by the dissolution of oxyhydroxides and that the arsenic concentration in solution increased by a factor of 25 times when the redox potential was reducing, e.g., -200 mv (pH 5). Also unknown is the effect of bacteria in buried storage systems.

Removal of arsenic from solution by ferric precipitation has been and is presently practiced at numerous facilities, e.g., the Noranda Horne Smelter, the Giant Mine, the Con Mine, and the Teck-Corona mine (Riveros et al. 2001); the Kennecott Utah Smelter, Placer Dome Lonetree and Getchell mines (on a periodic basis), and Barrick's gold mining operations in Nevada (McCloskey 1999). An example at a mine water site (pilot scale demonstrations) is the Susie Mine, Rimini, MT (MWTP 1997).

A case study is presented by Banerjee (2002) to illustrate the optimization and utilization of the ferrihydrite adsorption technology for the removal of arsenic from a groundwater at a northeastern US industrial facility. The water contained approximately 30 mg/L arsenic as As(III), As(V), and organic arsenic. A full scale treatment plant is presently in operation and has demonstrated consistent arsenic removal from 18-48 mg/L to <5 µg/L. The plant treats 550,000-750,000 liters/day groundwater using a Fe/As weight ratio of 10, hydrogen peroxide/Fe weight ratio of 0.5, and pH of 7.5-8.0. Other examples for the treatment of wastewater and/or effluent plant water by ferrihydrite precipitation and adsorption are presented by the US EPA (EPA 2000, 01).

Sengupta and Greenleaf (2002) present a case study based on the work of Chwirka et al. (2000) for arsenic removal from Albuquerque, NM drinking water. Chwirka et al. investigated the use of Ion Exchange (IX), Activated Alumina (AA), Nanofiltration (NF), and Ferric

Coagulation/Microfiltration (C/MF) for lowering the As(V) concentration from 100 to 10μg/L. They concluded that C/MF was the least expensive option for the treatment of 8,700 m^3/d and that option has been selected for implementation in Albuquerque. The cost estimate for each process was: capital cost (1999 US$); IX $5.2 million; AA $4.6 million; NF $3.9 million; C/MF $4.1 million and annualized capital plus O&M yearly cost; IX $447,000; AA $444,000; NF $390,000; C/MF $273,000. Chwirka and Narasimhan (2002) present a spreadsheet that is available to the reader to compare the capital and operating cost of the four technologies presented above. The details for using the spreadsheet (which can be downloaded) are presented in the reference. Other example publications illustrating the application of ferrihydrite adsorption in drinking water treatment plants are presented by EPA (2000), Hering et al. (1996), McNeill and Edwards (1995), and Cadena and Kirk (1995).

*Ferrihydrite/Arsenic Summary-*Two ferric precipitation arsenic removal technologies are presently practiced by industry, i.e. ambient temperature ferrihydrite precipitation and elevated temperature autoclave precipitation of ferric arsenate. The adsorption technology is relatively simple, can be performed at ambient temperatures, and the presence of commonly associated metals (copper, lead, zinc) and gypsum have a stabilizing effect on the long term stability of the product. The disadvantages of the adsorption technology are the formation of voluminous waste material that is difficult to filter, the requirement that the arsenic be present in the fully oxidized state (arsenate), and the question as to long term stability of the product in the presence of reducing substances. The autoclave technology produces less waste product, the product is easily filtered into a dense, compact filter cake, and less iron is required to sequester the arsenic. The disadvantages of the ferric arsenate precipitation are that the treatment process is more capital intensive, the compound may dissolve incongruently if the pH is >4, and it may not be stable under reducing or anaerobic bacterial conditions (Riveros et al. 2001, Rochette et al. 1998).

Riveros et al. (2001) concluded from their extensive review of the literature that "for practical purposes, arsenical ferrihydrite can be considered stable provided the Fe/As molar ratio is greater than 3, the pH is slightly acidic and that it does not come in contact with reducing substances such as reactive sulphides or reducing conditions such as deep water, bacteria or algae." Swash, et al (2000) have concluded that from the point of view of safe disposal of arsenic, that there is no clear experimental evidence yet favoring the low temperature precipitates over the high temperature precipitates or vise versa (Riveros et al. 2001). A detailed review of the stability of scododite, ferrihydrite and ferrihydrite/arsenate adsorption is presented by Welham et al. (2000). Their conclusions include "there are significant problems with the use of jarosite and scorodite as phases for the disposal of iron and/or arsenic from metallurgical systems. Neither phase is stable under typical atmospheric weathering conditions with transformation to goethite predicted to occur. The currently permitted discharge level of arsenic is only achieved due to the slow kinetics of the transformation releasing arsenic over time. Crystalline scorodite is two orders of magnitude less soluble than amorphous iron (III) arsenate precipitate often formed in low temperature systems."

*Alumina/Aluminum Hydroxide-*Alumina and aluminum hydroxide have been investigated for arsenic removal from solution by adsorption, example publications include: Hingston et al. (1970, 72), Gulledge and O'Connor (1973), Anderson (1974, 76), Leckie et al. (1980), Diamadopoulos and Benedek (1984), Ghosh and Teoh (1985), Goldberg (1986), and Lake (1990).

Activated alumina is widely utilized as an adsorbent in treatment of drinking waters. Example publications covering this subject include: Ghosh and Teoh (1985), Rozelle (1986), Fox and Sorg (1987), Fox (1989), and Lake (1990). Lake (1990) concluded that activated alumina is competitive in effectiveness for As(V) adsorption with other drinking water purification adsorbents and is the most cost effective adsorbent for drinking water treatment. Stewart and Kessler (1991) conducted a pilot study for treatment of six million liters of water by activated alumina. Adsorption reduced the As(V) concentration to analytical detection limits; filters lasted for over a hundred days and operating costs were estimated to be $0.20/1000 liters.

The use of activated alumina has not been extensively studied for application to dilute solution effluent wastewater or contaminated groundwater. Some studies have been conducted that demonstrate it may be applicable to these waters. Sisk et al. (1990) evaluated the use of activated alumina on a pilot scale to treat As(V) bearing groundwater. Alumina and aluminum hydroxide were shown to be ineffective for As(III) species except at pH> 9. The authors found that activated alumina was capable of extracting As(V) (at pH 5) to <50µg/L. The adsorption process is greatly affected by the presence of competing anionic species, especially sulfate. As(V) is much more effectively adsorbed by amorphous aluminum hydroxide than by alumina but ferrihydrite adsorption is much more effective than either of the two.

Activated Carbon-Activated carbon adsorption (ACA) has been used extensively for treating drinking waters but only when the waters contained concentrations near the drinking water standard and when the arsenic is present as As(V) (Fox and Sorg 1987). ACA treatment of arsenic bearing multicomponent wastewater has been shown to be relatively ineffective (Fu 1983, Huang and Fu 1984, Sisk et al. 1990, Wolff and Rudasill 1990, Rajakovic 1992, and many others).

Rajakovic (1992) reported on the use of metal impregnated AC. He found that copper was the most effective additive for enhancing (pH 4-9) the adsorption of arsenic. The arsenic adsorption was increased from essentially no adsorption to 48 mg As/g C. Huang et al. (1984, 89) have also investigated carbon adsorption and metal ion-doping as a means of enhancing arsenic extraction, e.g., these investigators doped activated carbons with various metals; barium, copper, ferrous and ferric iron. Ferrous iron was found to be greatly superior to all the other metals tested. Enhancement in As(V) adsorption was ten-fold (for pH range 4-5) over acid washed AC. Sulfuric acid was used to strip the adsorbed arsenic and the carbon was then regenerated by soaking in a ferrous salt solution. Essentially complete arsenic extraction was achieved from a 15 mg/L solution by the stripped and regenerated carbon.

Ion Exchange (IX). Ion exchange has been evaluated by a number of investigators for polishing drinking waters and dilute solution groundwater (Ghosh 1985, Rozelle 1986, 87, Fox 1989, Lankford 1990, Rajakovic 1992, SenGupta et al. 2002, SenGupta and Greenleaf 2002, Zappi 1990, and many others). The evaluations have often involved comparative removals with other drinking water clean up technologies, e.g., RO, AA, IX, NF, Ferric C/MF, and ACA. In general ion exchange has been shown to be competitive in effectiveness for removing arsenic from drinking waters. However, other treatment processes have been shown to be more economical, especially ferric C/MF (Chwirka and Narasimhan 2002).

There have only been a few investigations concerned with IX treatment of As(V) bearing multicomponent waste water. Chanda et al. (1988) investigated the use of ferric form resin exchangers (Dow XFS-4195, Chelex 100) and found that the resins were several fold more effective than ferrihydrite for As(V) adsorption. Arsenate was adsorbed to below drinking water standards, was effectively stripped from the resin, and the resin could be reactivated and repeatedly reused. The saturation loading capacity of the resins varied between 45-70 mg/g of wet resin. This study was conducted on synthetic arsenic bearing solutions and did not contain competing anions (such as sulfate, phosphate, humic). Hauck et al. (1990) also investigated ferric form ion exchangers and found complete arsenic removal from wastewater. Clifford et al. (1990) investigated strong-base chloride anion-exchange resins. They demonstrated essentially complete extraction of As(V) in the presence of As(III). Sisk et al. (1990) investigated the removal of arsenic from groundwater (on a pilot scale basis) by two ion exchange resins (Amberlite IRA 402 and Ionac A-641). Both ion exchangers reduced the arsenic content to below the drinking water standard.

Reverse Osmosis (RO)/Membrane Separation. RO is the most universally used treatment technology for point-of-use/point-of-entry treatment of drinking waters (Huxstep and Sorg 1981,

88, Rozelle 1987, Fox 1989, Clifford 1990, Gandilhon et al. 1992, and many others). Rogers (1989) studied point-of-use RO using polyamide membranes at 78 sites. Arsenic (0.059 mg/L), iron, manganese, chloride and total dissolved solids were lowered to below the drinking water standard over a twenty month test period. Fox and Sorg (1987) evaluated five different membranes (in a small continuously operating pilot scale test system) on natural Florida groundwater containing 15 elements. They found all the membranes removed >95% of all dissolved species. Fox and Sorg (1987) report that five techniques are in use for point-of-use treatment: reverse osmosis, activated alumina, ion exchange, granular activated carbon, and distillation. They state that EPA has approved the first three of these for removal of inorganic contaminants as Best Available Technologies (BAT). Fox (1989) found that low pressure RO was not completely effective at high arsenic concentrations (i.e., at 1.08 mg As/liter). However, high pressure units handled all concentrations from 0.1-1.0 mg As/liter). There are numerous other references illustrating the application of RO to drinking waters that are not quoted here.

Reduction Processes. Removal of arsenic by reduction with iron has been reported by several investigators (Santana 1996, McCloskey 1999, Dahlgren 2000, Nikolaidis 2002, Cockhill 2002, and Hadden 2002). Arsenic has been stripped to less than analytical detection limits from laboratory samples utilizing iron reduction technology (Santana 1996, Dahlgren 2000, Hadden 2002). MSE has utilized a proprietary catalyzed cementation process to strip arsenic from a large variety of process and mine waters to less than analytical detection limits. The process has been demonstrated on a pilot scale (1-5 gallons/min) at an industrial site (McCloskey 1999).There are a number of studies investigating the use of iron bearing reactive permeable barrier walls to treat groundwater plumes. Most of the reported work has dealt with the destruction of organic compounds, reduction of Cr(VI) to Cr(III), or radionuclides (O'Hannesin and Gillham1992, Kaplan et al. 1994, Cantrell et al. 1995, Powell et al. 1995, Appleton 1996). The use of zero valence iron appears to hold promise for future applications for arsenic, selenium and heavy metal removal. The advantages of the technology include: the process is independent of the arsenic species valence state; it is not influenced by the present of sulfate and other anions; and heavy metals more noble than iron are coextracted.

Selenium
Precipitation. Precipitation of selenate and selenite compounds as a water treatment technology to lower selenium to <10 µg/L is ineffective because of the relatively high solubility of the compounds as a function of pH (Twidwell et al. 1999a).

Adsorption. A large number of investigations have focused on adsorption of selenium on ferric ferrihydrite and alumina/aluminum hydroxide surfaces. A relatively small portion of the literature describing the adsorption of selenium onto these surfaces is annotated below. Detailed literature reviews are published for selenium by the EPA Mine Waste Treatment Program (MWTP 1999).

Ferrihydrite-Ferrihydrite adsorption has been selected as the BDAT for selenium removal from wastewater by the USEPA (Rosengrant and Fargo 1990). Many investigators have studied the adsorption of selenium on ferrihydrite (Howard 1972, 77, EPRI 1980, Benjamin et al. 1981, 82, Merrill et al 1986, Balistrieri and Chao 1987, 90, Hayes et al. 1987 Pengchu and Sparks 1990, Smith and Jenne 1991, WSPA 1995, Parida et al 1997, and many others).The commonality between these studies is that: Se(IV) is adsorbed much more effectively than Se(VI); ferrihydrite provides the best surface for Se(IV) adsorption, i.e., amorphous ferrihydrite is better than crystalline geothite; the higher the initial concentration of Se(IV) the more effective the adsorption. The best pH for effective Se(IV) adsorption is 4-6 (85-90% removal), adsorption decreases slowly until about pH 7 (80-85% removal), then decreases drastically at higher pH (20-40% removal). Se(VI) adsorption is poor (<10% removal) at all pH levels and the adsorption is strongly affected by the presence of sulfate and aqueous silica species.

Parida et al. (1997) investigated the adsorption of Se(IV) onto various surfaces. They found that ferrihydrite adsorbed significantly more Se(IV) than the other phases. The order of adsorption was ferrihydrite (225 m2/g) > δFeOOH (117 m2/g) > γFeOOH (61.1 m2/g) > βFeOOH (77.8 m2/g) > αFeOOH (70.8 m2/g), e.g., at 20 mg Se(IV)/L, ferrihydrite was loaded to 48 mg/g while αFeOOH was only loaded to 10 mg/g. Approximately ninety percent removal of Se(IV) was achieved from a 20 mg Se(IV)/L solution by using 1.2 g ferrihydrite/L at a pH of 3. Whereas, αFeOOH removed only about 25% of the Se(IV) under the same conditions. Adsorption was a relatively strong function of solution pH, i.e., adsorption fell rapidly as the pH was increased from 3.5 to 9.5. There was essentially no adsorption at or above pH 9.5. Note that even at pH 3 removal to the µg/L was not achieved. MSE piloted the ferrihydrite adsorption process for the EPA MWTP on a smelter wastewater that contained two mg/L Se(VI) at a pH of 5-8. Removal to <50 µg/L required extremely high additions of ferric ions (MWTP 2001).

Balistrieri and Chao (1990) investigated the role of the presence of other anions on the adsorption of selenium by ferrihydrite. The order of adsorption at pH 7 was phosphate > silicate = arsenate > bicarbonate/carbonate > citrate = selenite > molybdate > oxalate > fluoride = selenate > sulfate. Therefore, the adsorption of selenium is an anion that is in competition with other anions.

Hayes et al. (1987) studied the adsorption of selenium oxyanions on αFeOOH surfaces. Their conclusion was that Se(IV) adsorbs in a bidentate manner (inner-sphere adsorption which is much stronger than outer-sphere adsorption). Se(VI) adsorbs as an outer-sphere hydrated complex, which is why it can be easily replaced by other solution anions such as sulfate.

Merrill et al. (1986, 87) and EPRI (1980, 85) found that the optimum pH for Se(IV) removal was 6.5 and optimum iron dosage was 14 mg/L for a water initially containing 40-60 µg/L Se(IV). The effluent selenium concentration resulting from the treatment of 115 liters/minute in a continuous pilot facility was < 10 µg/L. WSPA (1995) reported similar results for the treatment of a petroleum biotreater effluent (the solutions contained cyanide as well as selenium and sulfate), e.g., the optimal pH was 5-7; optimal ferric chloride dosage was 14-28 mg iron/L.

Selenium adsorption occurs on precipitated ferrihydrite. However, the use of ferrihydrite adsorption is not utilized industrially because: even though Se(IV) is effectively removed at pH<8 rarely can µg/L concentrations be achieved; Se(VI) is poorly adsorbed at any pH (which mean that reduction of the Se(VI) prior to adsorption is required); the presence of other aqueous species in the solution influences the removal of both Se(IV) and Se(VI); and long term stability of this product in outdoor storage is questionable.

Alumina/Aluminum Hydroxide -Activated alumina has been studied for selenium adsorption by several investigators: Trussell et al (1980, 91), Yuan et al. (1983), Hornung et al. (1983), Altman and Hegerle (1993). Trussell et al. (1980) investigated the adsorption of Se(IV) and Se(VI) on activated alumina. Se(IV) was effectively adsorbed to analytical detection limits over the pH range 3-7 (for 100-200 µg Se/L, one hour exposure). The loading capacity was 90 mg Se/L of alumina. Se(VI) adsorption was much less effective (loading capacity was 7 mg Se/L of alumina). The order of selectivity for anions over the pH range 3-7 was hydroxide > arsenate > selenite > sulfate > selenate > arsenite. Sulfate and bicarbonate had little effect on Se(IV) but greatly affected Se(VI) adsorption. Trussell et al. (1991) demonstrated that selenium adsorbed on alumina could be effectively stripped using 0.5% NaOH. A cost estimate was presented for treating 3.8 million liters/day containing 100 µg Se/L. The cost estimate was 6¢/1000 liters for Se(IV) and 21¢/1000 liters for Se(VI).

The above studies were conducted on groundwater not on mine waters. Investigators Altman and Hegerle (1993) and Batista and Young (1994, 97) have applied alumina adsorption to an actual mine effluent (FMC Gold Paradise Peak Mine in Nevada). They demonstrated that mine waters that have appreciable silica present cannot be effectively treated by alumina adsorption. The alumina adsorption capacities for selenium and silica were determined to be 0.2 mg Se/L alumina, 3.6 mg Si/L alumina, respectively. In the absence of silica the alumina adsorption capacity for selenium was 0.9 mg/L alumina.

Activated Carbon-Activated carbon adsorption of either Se(IV) or Se(VI) has been shown to be ineffective, e.g., Se(IV) or Se(VI) at concentrations from 30-100 µg/L showed < 4% removal using dosages of activated carbon up to 100 mg/L (Sorg 1978).

Ion Exchange. IX is used widely for treating drinking waters, dilute metal bearing solutions, wastewater, and groundwater. However, its application for removing selenium has some successes and some failures (Maneval et al. 1985, Boegel and Clifford 1990); see the MWTP (1999) for a more detailed discussion.

The Western States Petroleum Association (WSPA 1995) investigated the use of IX (strong base anion resins) for treating refinery wastewater (sourwater and biotreater effluent, containing 11-4,870 µg Se/L). Their conclusions included: IX is very inefficient for selenium because other anionic species may load in preference to the selenium, especially sulfate. Se(IV) loaded well but after 200-600 bed volumes sulfate displaced the selenium. IX is not able to routinely produce effluents with <50 µg/L total selenium. Efforts to identify selective resins were unsuccessful.

Tailored chelating polymer resins have been investigated (Ramana and Sengupta 1992) for extracting selenium in the presence of high concentrations of sulfate. The resins investigated included Dow 2N (loaded with copper, 1.7 meq/g) and IRA-900 (no copper loading). The resins were evaluated for extraction in a solution containing 250 mg sulfate/L. The selectivity sequence at pH 9.5 was selenate > sulfate > selenite > nitrate > chloride. Resin loading capacities and cost evaluations were not presented by the authors.

Virnig and Weerts (1993) considered the use of liquid ion exchange as a way to treat spent gold heap leach effluent. The ion exchange reagent was CyanoMet R, a proprietary reagent. The reagent extracted anion complexes including gold, silver, nickel, copper, zinc, and selenium. The extracting reagent contained 30 weight percent CyanoMet R. There were three stages of extraction followed by two stages of strip (four percent NaOH). The selenium removal from the influent feed (11 mg Se/L) was very good, e.g., the final raffinate solution contained 70 µg Se/L. Pilot studies were conducted at a Nevada gold mine. The feed rate was 1 gallon/minute. The selenium removal from the influent feed (1.7 mg Se/L) was very good, e.g., the final raffinate solution contained 36 µg Se/L. This technology was not implemented at the mine site.

Membrane Separations.
Reverse Osmosis (RO)-RO is listed by EPA as one of the Best Available Technologies (BAT) for selenium removal (Pontius 1995, Kapoor et al. 1995), i.e., the removal effectiveness is quoted as being >80% irregardless of valence state. Reverse osmosis/membrane ultrafiltration processes require that the solutions to be treated contain very dilute concentration of solids. Therefore, pretreatments are normally required in order not to foul the separating membrane. Whereas, RO is readily applicable to drinking waters it is doubtful that this separation technique is applicable to acid mine waters except as a final polishing step.

Nanofiltration (NF)-NF appears to be a technology on the horizon for treating some low metal containing selenium bearing mine waters (Kharaka et al. 2000). NF has been used commercially for sulfate removal from seawater prior to injection into oil field reservoirs (Bilstad 1992); for sulfate removal from concentrated brines (Kharaka et al. 2000); for organic compound removal from paper plant effluents (Afonso et al. 1992); and for organic compound removal from groundwater (Fu et al. 1994). Publications demonstrating the application of NF to mine waters were not found but the technology has been applied to agricultural waters (high in Se(VI), sulfate and total dissolved solids) taken from San Joaquin Valley drainage (Kharaka et al. 2000).

Nanofiltration is based the use of membranes constructed of a porous inert layer of polysulfone and a negatively charged hydrophobic rejection layer. These membranes reject multivalence anions, including sulfate. The technology is similar to RO but the NF system is operated at pressures that are about one-third of that required for RO. Kharaka (2000) demonstrated >95% removal of selenium for waters (pH 6.3-8.5) containing 24-308 µg Se(VI)/L; 2,080-26,100 µg sulfate/L; and 780-38,800 µg TDS/L. Similar recoveries were demonstrated for Se(VI) concentrations up to ~1000 µg/L.

Reduction Processes. There are a number of proposed reduction processes in the literature. However, only metallic and biological reduction appear to be capable of lowering the selenium to <10 µg/L in a cost effective manner. Neither of the reduction processes are sensitive to interference by sulfate, carbonate, or nitrate (as are the adsorption processes). Metallic reduction has been applied on an industrial scale by McGrew, et al (1996) and recent detailed laboratory investigations have been completed by Montana Tech graduate students (Dahlgren 2000; Cockhill 2002; Hadden 2002). Selenium removal to <2 µg/L has been achieved.

Iron Reductant-McGrew et al. (1996) have developed a selenium recovery process using iron powder as the reductant. McGrew notes that mine waters are usually extremely high in sulfate compared to selenium (and compared to drinking waters and groundwater), e.g., the ratios are often of the order 108 sulfate/selenium. Therefore, a process is required that is not influenced by sulfate. The success of this process compared to other proposed iron reduction processes is that the rate of Se(VI) reduction by iron to selenium is catalyzed by the presence of copper.

McGrew demonstrated their process on a gold mine heap leach effluent (Brewer Pit water, 257 µg Se/L). The water was tested (pH 3-3.5) using 1 to 10 g Fe/L, and copper levels between 10 and 81 mg/L. The optimum conditions resulted in selenium concentrations less than detection limit (<5 µg/L). Residence times of 10-20 minutes were required for a copper content of 10 mg/L at an iron dose of 10 g/L. The consumption of iron was ~3 pounds/1000 gallons. The product could be further treated for selenium recovery. Simple conventional hydrometallurgical equipment would be required to implement this technology, i.e., mix chambers, dewatering settlers, and clarifiers. This process has been applied at the Brewer mine since 1994.

Biological Reduction-There are several patents and studies in the literature that propose potential biological processes for removing selenium (Kauffman et al. 1986, Altringer et al. 1989, 91, Ergas et al. 1990, Oremland 1991,93, Koren et al, 1992, Adams et al. 2000).

Biological reduction has been extensively studied at the lab and pilot scale by Adams (2000). The bacterium that appears to offer great promise is *P. Stutzeri*, which can reduce both Se(IV) and Se(VI) species. Adams has conducted test work on mine waters, e.g., a mine water containing 620 µg Se(VI)/L was treated in a single-stage aerobic bioreactor. The live biofilms produced effluent waters containing <10 µg Se/L for about nine months without breakthrough. Further test work was conducted to investigate the possibility of simultaneously removing cyanide and selenium from mine waters. The results showed that simultaneous destruction of cyanide (102 mg/L) and reduction of selenium (from 31.1 mg/L to <10 µg/L) could be successfully accomplished. Koren et al. (1992) also investigated the use of *P. stutzeri* for Se specie reduction to elemental selenium. The authors demonstrated that bacteria could survive and operate effectively in Se(VI) solutions containing over 3500 mg Se/L. The rate of reduction was ~32 mg Se/L/hr. The bacteria could also survive and operate effectively in Se(IV) solutions to approximately 1500 mg/L. Maximum reduction rates occurred in the pH range 7-9.5; optimal temperature was 25-35°C. The presence of sulfate, nitrate, and nitrite ions had no harmful effect on either Se(VI) or Se(IV) reduction.

Heavy Metals

Precipitation. Banerjee (2002) has reviewed the literature on heavy metal removal by compound precipitation. His conclusion is that hydroxide precipitation of heavy metals (copper, lead, zinc), in a properly designed system, can achieve metal removal to about 500 µg/L and if sulfide precipitation is utilized removals to about 100 µg/L can be achieved. However, for achieving the removal of heavy metal concentrations to <few µg/L requires the utilization of a precipitation/adsorption technology, i.e., precipitation of ferrihydride and adsorption of heavy metals on the newly precipitated surfaces.

Adsorption. *Ferrihydrite*-Iron precipitation/adsorption is an often used technology for removing heavy metals to <10 µg/L (Banerjee 2002). Definitive studies for trace heavy metals (and oxyanions) removal by this technology has been published by Leckie et al. (1980, 85), Rai et al. (1984), Merrill et al. (1985), Runnells et al. (1990), and Dzombak and Morel (1990). The conclusion of these publications is that metals are readily adsorbed on ferrihydrite surfaces. The effectiveness of the adsorption is dependent on solution metal concentration, presence of associated ions, iron/metals ratio, and pH. The pH of the adsorption edges has been shown to decrease in the order lead < copper < zinc < cadmium and selenate < chromate < selenite < arsenate.

A case study is presented by Banerjee (2002) to illustrate the optimization and utilization of this technology for the removal of copper, lead and zinc. A full scale steel making wastewater treatment plant is presently in operation (since 1997) and has demonstrated consistent heavy metal removal from Cu (400-600 µg/L), Zn (100-200 µg/L), and Pb (50 µg/L) to <5 µg/L for each metal. The system presently treats 230 liters/minute, at pH 8-8.5, using 10 mg Fe/L, and a 10 minute adsorption time. The author emphasizes that each water system must be optimized for pH and additive iron/metal weight ratio (usually 20-30) based on the desired final metal concentration, mix of metals present, and solution pH.

Activated Carbon-Reed (2002) has reviewed the literature and has concluded that activated carbon adsorption has not been widely commercially accepted. Reed has also concluded that the use of activated carbon will see increased future use for the removal of heavy metals from waste solutions. Corapcioglu and Huang (1987) have investigated metal adsorption on fourteen different activated carbons. Their research resulted in demonstrating that activated carbon may be applicable to metal bearing wastewater by careful selection of the carbon substrate and by adjustment of the wastewater pH to an appropriate level.

Other Technologies. Because coverage of the voluminous literature describing technologies for removing heavy metals from waste and process waters is beyond the scope of this presentation the reader is referred to the following summary publications: SenGupta (2002), SenGupta and SenGupta (2002), Franzreb and Watson (2002), Banerjee (2002), Reed (2002), Sanjay et al. (2002), Williams-Beam and Twidwell (2001), Watson (1999), Nriagu (1998), Cornell and Schwertmann (1996), Grzetic and Zemann (1993), and Leckie et al. (1980, 85).

SUMMARY

Various technologies have been evaluated for removing arsenic from aqueous solutions. Several of the technologies appear appropriate for treating drinking water where the initial arsenic concentration is already relatively low (<100 µg/L) and where there are few competing anions present, e.g., ferrihydrite precipitation/coagulation; alumina adsorption, membrane, and ion exchange technologies are all capable of achieving the new US EPA MCL of <10 µg/L. However, only a few technologies appear to be appropriate for treating multicomponent arsenic bearing waste and mine waters. The US EPA BDAT and the most commonly utilized technology is ferric arsenate precipitation and/or ferrihydrite adsorption. Both of these technologies are capable of achieving arsenic concentrations that are <10 µg/L. It appears that the use of these technologies will continue to dominate into the near future. However, because long term stability of the ferric products in outdoor storage environments is still a concern, other technologies must continue to be investigated.

The other nonadsorption technologies discussed above must also be scrutinized with respect to final arsenic bearing product disposal or storage. The membrane and ion exchange processes produce a concentrated brine solution that still must be further treated. The metallic reduction process appears to be capable of lowering arsenic concentrations to analytical detection limit levels, however, the final product will be an arsenic bearing metallic product that must also be further treated or stored in an environmentally acceptable manner.

The US EPA BDAT for removing selenium from wastewater is ferrihydrite adsorption. The technology is inappropriate for achieving selenium concentrations of <10 µg/L unless the selenium exists exclusively as Se(IV). Those technologies that appear to be potentially appropriate for achieving selenium concentrations of < 10µg/L include: membrane technologies, such as nanofiltration; tailored ion exchange; and reduction processes such as metallic iron reduction and biological reduction. The membrane and ion exchange processes produce a concentrated selenium bearing solution that must be further treated. The reduction processes produce either a selenium bearing metallic sludge or a selenium bearing biological sludge material. Both these product contain elemental selenium at relatively high concentrations. The disposal options for handling these products must be further considered but one of the options includes the recovery of elemental selenium as a potentially marketable feedstock.

The final solution to the problem of how to safely dispose of arsenic and/or selenium bearing waste forms has yet to be determined.

ACKNOWLEDGMENTS
Literature reviews for arsenic and selenium removal technologies were conducted for the US EPA, Office of Research and Development, Mine Waste Technology Program. The work was performed under an Interagency Agreement between the EPA National Risk Management Research Laboratory and the US Department of Energy National Energy Technology Laboratory (NETL). Montana Tech of the University of Montana is a subcontractor to MSE Technology Applications, Inc. which manages the MWTP through the NETL at the Western Environmental Technology Office under contract No. DE-AC22-96EW96405.

REFERENCES
Adams, J., T. Pickett, J. Hogge. 2000. Biotreatability for Se removal process, In: *Minor Elements' 2000.* (Ed) C. Young, SME, Littleton, CO. 53-66.

Afonso, M. et al. 1992. Nanofiltration removal of chlorinated organic compounds from alkaline bleaching effluent in a pulp and paper plant. Wat. Res. 26, 1639-43.

Altringer, P., D. Larsen, K. Gardner. 1989. Bench-scale process development of selenium removal from wastewaste using facultative bacteria. In: *Biohydromet. 1989*, CANMET, 643-57

Altringer, P., R. Lien, K. Gardner. 1991. Biological and chemical selenium removal from precious metals solutions, In: *Symp. Env. Management for the 1990's.* SME-AIME, Littleton, CO. 135-42.

Altman, K., K. Hegerle. 1993. Se removal processes considered for the heap leach detoxification project paradise peak mine, FMC Gold. In: *Gearing for a New Era in Gold*, Beaver Creek, CO. Randol Gold, Denver, CO. 349-53.

Anderson, M. 1974. Arsenate adsorption on amorphous Al hydroxide. *MS Thesis* (74). *PhD Thesis* (76) John Hopkins Univ.; and, 1976. Anderson, M., J. Ferguson, J. Gavis. Arsenate adsorption on amorphous Al hydroxide. *J. Colloid and Interface Sci.* 54, (3), 391-99.

Appleton, E. 1996. A nickel-iron wall against contaminated groundwater, E.S.&T. 30 (12) 537-39.

AWWA . 1988. AWWA standard for Fe chloride--liquid, ground, or lump. *AWWA*, VIII, B407-88, 15p; and, 1988a. AWWA standard for Al sulfate--liquid, ground, or lump. *AWWA*, IX, B403-88, 17 p.

Balistrieri, L., T. Chao. 1987. Selenium adsorption by geothite. *Soil Science Soc. Am. J.* 51, (5); and, 1990. Adsorption of Se by amorphous iron oxyhydroxides and manganese dioxide. *Geochim. Cosmochim. Acta*, 54, 739-51.

Banerjee, K. 2002. Case studies for immobilizing toxic metals with iron coprecipitation and adsorption. In *Env. Separation of Heavy Metals*. (Ed) SenGupta, Lewis Publ., NY, NY. 181-204.

Batista, J., J. Young. 1994. The influence of aqueous silica on the adsorption of Se by activated alumina. *AAWA Water Research*, 167-81; and, 1997. Removal of Se from gold heap leachate by activated alumina adsorption. *TMS*, Warrendale, PA. 29-36.

Belzile, N., A. Tessier. 1990. Interactions between As and iron oxyhydroxides in lacustrine sediments, *Geochim. et Cosmochim. Acta*. 54, 103-10.

Benjamin, M., N. Bloom. 1981. Interactions of strongly binding cations and anions on amorphous iron oxyhydroxide. In: *Adsorption from Aqueous Solutions*, (Ed) P. Tewari, N.Y., N.Y. Plenum.

Benjamin, M., K. Hayes, J. Leckie. 1982. Removal of toxic metals from power-generation waste streams by adsorption and coprecipitation. *JWPCF*. 54, (11), 1472-81.

Bilstad, T. 1992. Sulfate separation from seawater by nanofiltration, In: *Produced Water*. (Eds) J. Ray, F. Engel, Plenum Press. NY, NY. 503-09.

Boegel, J., D. Clifford. 1986. Se oxidation and removal by ion exchange, *EPA/600/2-86/031*, Washington, DC.

Bothe, J., P.W. Brown. 1999. As immobilization by calcium arsenate formation, *Env. Sci.Technology*. 33(21). 3806-11; and, 1999a. The stabilities of calcium arsenates at $23\pm1°c$, *J. Haz. Materials*. 69(2). 197-207.

Brannon, J., W. Patrick. 1987. Fixation, transformation and mobilization of arsenic in sediments, *Env. Sci. Tech.* 21, 450-59.

Cadena, F., T. Kirk. 1995. Arsenate precipitation using ferric iron in acidic conditions. WRRI Rep. 293. *New Mexico Water Resources Research Institute*. Las Cruces, NM.

Cantrell, K., D. Kaplan, T.Wietsma. 1995. Zero valence iron for the in-situ remediation of selected metals in groundwater. *J. Haz. Materials*. 42, 201-12.

Castro, S., P. Munoz. 2000. Removal of As(III) in solution by co-precipitation of calcium arsenite and calcium hydroxide. *Fifth Int. Conf. on Clean Technologies for the Mining Industries*. (Eds) M. Sanchez, et al. Univ. Concepcion, Chile. 121-29.

Chandra, M., et al. 1988. Ligand exchange sorption of arsenate and arsenite by chelating resins in ferric ion form: iminodiacetic chelating resin Chelex 100. *React. Polym., Ion Exch., Sorbents*. 8, (1), 85-95; and, 1988a Ligand exchange sorption of arsenate and arsenite by chelating resins in ferric ion form: weak-base chelating resin XFS-4195. 7, (2-3), 251-61.

Chukhlantsev, V. 1956. The solubility product of metal arsenates, *Zhur. Neorg. Khim.*, I, 1975-82, II, 529-35; *Zhur. Anal. Khim.*, 11, 529-35; *J. Anal. Chem.*, 11, 565-71.

Chwirka, J., B. Thomson. 2000. Removing As from groundwater. JAWWA. 92 (3), 79-88.

Chwirka, J., R. Narasimhan. 2002. AWWA arsenic treatment cost estimating tool. *Web Site*: www.awwa.org/govtaff/AWWAArsenicCostsRev.xls.

Clifford, D., 1990. Ion Exchange and Inorganic Adsorption, Water Quality and Treatment, McGraw-Hill Inc. NY, NY. 561-640.

Cockhill, L. 2002. Se removal by column cementation, *MS Thesis*. Montana Tech, Butte, MT.

Comba, P., L. Twidwell. 1988. Removal of As from process and wastewater. *Proc. As Metallurgy Fundamentals and Applications*, (Eds) R. Reddy, et al. TMS, Warrendale, PA. 305-19.

Corapcioglu, O., C. Huang. 1987. The adsorption of heavy metals onto hydrous activated carbon. *Water Research*, 21, 1031-44.

Cornell, R. 1985. Effect of simple sugars on the alkaline transformation of ferrihydrite into goethite and hematite. *Clays Clay Minerals*, 33, 219-27; and, Cornell, R., R. Giovanoli. 1987. Effect of Mn on the transformation of ferrihydrite into goethite and jacobsite in alkaline media. 35, (1), 11-20; and, Cornell, R., R. Giovanoli, P. Schnidler. 1987a. Effect of silicate species on the transformation of ferrihydrite into goethite and hematite in alkaline media. 35, (1), 21-8.

Cornell, R., U. Schwertmann. 1979. Influences of organic anions on the crystallization of ferrihydrite. *Clays Clay Minerals*. 27, 402-10.

Cornell, R., U. Schwertmann. 1996. The Iron Oxides. VCH. NY, NY. 574 p.

Davies, S. 1984. Mn(II) oxidation in the presence of metal oxides. *PhD Thesis*, Calif. Inst. Tech., Pasadena, CA, 170 p.

Dahlgren, E. 2000. Sonic enhancement of Se removal by cementation from wastewater, *MS Thesis*. Montana Tech, Butte, MT, 169.

Diamadopoulos, E., A. Benedek. 1984. Al hydrolysis effects on P removal from wastewater. *JWPCF*, 56, (11), 1165-72.

Downing, et al. 1988. Removing Se from water. US Pat. 4,725,357.

Dzomabak, D., F. Morel. 1990. Surface complexation modeling: hydrous ferric oxide. John Wiley and Sons, NY, NY.

Eggleton, R. 1987. Noncrystalline Fe-Si-Al-oxyhydroxides. *Clay, Clay Minerals*. 35 (1), 29-37; and, Eggleton, R., W. Fitzpatrick. 1988. New data and a revised structural model for ferrihydrite, 36 (2), 113-18.

EPA. 2000. Technologies and costs for removal of As from drinking water. EPA-R-00-028. EPA Office of Solid Waste and Emergency Response, Washington, DC.

EPA. 2001. Treatment technologies for site cleanup: annual status report. EPA-542-R-01-004. EPA Office of Solid Waste and Emergency Response, Washington, DC.

EPRI. 1980. Adsorption/Coprecipitation of trace elements from water with iron oxyhydroxide. *Electric Power Res. Inst.* EPRI CF-1513, Palo Alto, CA.

Ergas, S. et al. 1990. Microbial process for removal of Se from agricultural drainage water, Univ. Calif. At Davis, Report to USBOR, Contract 9-FC-20-07720.

Essington, M. 1988. Solubility of barium arsenate. *Soil Sci. Soc. Am.*. 52. 1566-70.

Fox, K. 1989. Field experience point-of-use systems for As removal. *JAWWA*. 81, (2), 94-101; and, Fox, K., T. Sorg. 1987. Controlling As, F, and U by point-of-use treatment. 79, (10), 81-8.

Franzreb, M., J. Watson. 2002. Elimination of heavy metals from wastewater by magnetic separations. In: *Env. Separation of Heavy Metals: Engineering Processes*. Lewis Publishers, NY, NY. 97-135.

Fu, P. 1983. Treatment of As(V) containing water by activated carbon process. *MS Thesis*, Univ. Del., Newark, Del.

Fu, P. et al. 1994. Selecting membranes for removing precursors, *JAWWA*, 86, 55-72.

Fuller, C. et al. 1993. Surface chemistry of ferrihydrite: ii. kinetics of arsenate adsorption and coprecipitation. *Geochim. et Cosmochim. Acta.* 57, 2271-82.

Gandilhon, A., N. Tambo, 1992. Experimental study on a membrane ultrafiltration process for drinking-water production. *Aqua*. 41 (4), 203-08.

Ghosh, M., R. Teoh. 1985. Adsorption of As on hydrous aluminum oxide. In: *Toxic and Haz. Wastes: Seventh Mid-Atlantic Ind. Waste Conf.* Technomic Publ. Co. Lancaster, PA, 139-55.

Goldberg, S. 1986. Chemical modeling of arsenate adsorption on Al and Fe oxide minerals. *Soil Sci. Soc. Am. J.* 50, 1154-60.

Grzetic, I., J. Zemann. (Eds). 1993. Neues jahrbuch fur mineralogie, thallium chemistry, In: *Proc. Int. Symp. On Tl Chemistry, Geochemistry, Mineralogy, Ores and Environmental*. 166. (1), E. Schweizerbart'sche Verlagsbuchhandlung, Stuttgart, Germany. 43-52.

Gulledge, J., J. O'Connor. 1973. Removal of As(V) from water by adsorption on Al and Fe hydroxides. *JAWWA*, 548-52.

Hadden, A. 2002. Galvanically enhanced removal of Se from wastewater. *MS Thesis*. Montana Tech, Butte, MT.

Harris, B. 2000. The removal and stabilization of As from aqueous process solutions: past, present, future. In: *Minor Elements 2000, Processing and Env. Aspects of As, Sb, Se, Te, and Bi.* (Ed) C. Young, SME. Littleton, CO. 3-21.

Hayes, K., et al. 1987. In Situ x-ray absorption study of surface complexes: Se oxyanions on alpha FeOOH. *Science*, 238, 783-86.

Hering, J., et al. 1996. Arsenic removal by ferric chloride. JAWWAI, 88 (4), 155-67.

Hingston, F. 1970. Specific adsorption of anions on goethite and gibbsite. *PhD Thesis*, Univ. W. Australia; and, 1972. Anion adsorption by goethite and gibbsite. *J. Soil Science*. 23, 2.

Hingston, F., et al. 1968. Adsorption of selenite by goethite. In: *Adsorption from Aqueous Solution*. ACS, Advanced Chemistry Series. NY, NY. 79, 82-90.

Hornung, S., J. Yuan, M. Ghosh. 1983. Se removal in fixed bed activated alumina adsorbers. In: *Create a New Excellence*. AWWA Annual Conference. 299-318.

Howard, J. 1972. Control of geochemical behavior of Se in natural waters by adsorption on hydrous ferric oxides. *Trace Substances in Env. Health, V*, 485-95; and, 1977. Geochemistry of Se: formation of ferroselite and Se behavior in the vicinity of oxidizing sulfide and uranium deposits. *Geochim. Cosmochim. Acta*, 41, 1665-78.

Huang, C., P. Fu. 1984. Treatment of As(V) containing water by the activated carbon. *JWPCF*, 56, (3), 80-88.

Huang, C., L. Vane. 1989. Enhancing As(V) removal by a Fe(II)-treated activated carbon. *JWPCF*, 61, (9), 1596-1603.

Huxstep, M. 1981. Project summary: inorganic contaminant removal from drinking water by reverse osmosis, *NTIS PB 81-224 420, Government Printing Office*, Washington, DC. 2; and, Huxstep, M., T. Sorg. 1988. Reverse osmosis treatment to remove inorganic contaminants from drinking water, EPA/600/2-87/109, *NTIS PB88-147780*. 50.

Jambor, J., J. Dutrizac. 1998. Occurrence and constitution of natural and synthetic ferrihydrite, a widespread iron oxyhydroxide. *Chemical Review*, 98, 2549-85.

Kaplan, D., K. Cnatrell, T. Wietsma. 1994. Formation of a barrier to groundwater by injecting metallic-iron colloids: effect of influent colloid concentration. In: In Situ Remediation: Scientific Basis for Current and Future Technologies, *Battelle Press*, Columbus, OH.

Kapoor, A., T. Tanjore. 1995. Removal of Se from water and wastewater, *Env. Sci.Tech*. 49, (2), 137-47

Karim, Z. 1984. Characteristics of ferrihydrite formed by oxidation of $FeCl_2$ solutions containing different amounts of silica. *Clays Clay Minerals*. 32, 181-4.

Kauffman, J., W. Laughlin, R. Baldwin. 1986. Microbiological treatment of uranium mine waters, *Env. Sci. Tech*. 20 (3), 243-48.

Kharaka, Y., J. Thorsden, R. Schroeder. 2000. Nanofiltration membranes used to remove Se and other minor elements from wastewater, *Proc. Minor Elements'2000*. (Ed) C. Young, SME, Littleton, CO. 371-80.

Khoe, G., J.C.-Y Huang, R. Robins. 1991. Precipitation chemistry of the aqueous ferrous-arsenate system. EPD Congress. (Ed) D.R. Gaskell, TMS, Warrendale, PA. 103-16.

Koren, D., W. Gould, L.Lortie. 1992. Se removal from waste water, In: *Proc. Waste Processing and Recycling in Min. and Met. Industries*, CIM, Edmonton, Alberta, Can. 171-82.

Krause, E. 1992. Arsenic removal at INCO's Cred plant and disposal practice, *Proc. Arsenic Workshop*, EPA Office of Solid Waste and Emergency Response, Washington, DC.

Krause, E., V. Ettel. 1989. Solubilities and stabilities of ferric arsenate compounds, Hydromet., 22, (3), 311-37.

Laguitton, D. 1976. As removal from gold-mine waste waters: basic chemistry of lime addition method. *CIM Bull.*, 69, 105-9.

Lake, L. 1990. Alumina for POU/POE removal of F and As water treatment. *Point-of-Use/Entry Treatment of Drinking Water*, Noyes Publ., Park Ridge, New Jersey, 88-9.

Leckie, J., D. Merrill, W. Chow. 1985. Trace element removal from power plant waste streams by adsorption/coprecipitation with amorphous iron hydroxide. *Separation of Heavy Metals and Other Trace Contaminants*. (Eds) R. Peters, B. Mo Kim, *AICHE*, 28-42.

Leckie, M., et al. 1980. Adsorption/coprecipitation of trace elements from water with iron hydroxide. EPRI Document CS-1513, *Electric Power Res. Inst.*, Palo Alto, CA.; and, M. Benjamin, K. Hayes, J. Leckie. 1982. Removal of toxic metals from power generation waste streams by adsorption. *JWPCF*, 54, (11). 1472-81.

Maneval, et al. 1985, Se removal from drinking water by ion exchange, *EPA/600/2-85/074*. US Government Printing Office, Washington, DC.

Masscheleyn, P., R. Delaune, W. Patrick. 1991. As and Se chemistry as affected by sediment redox potential and pH. J. Env. Qual. 20, 522-27; and, Effect of redox potential and ph on arsenic speciation and solubility in a contaminated soil. *Env. Sci. Tech*. 25, 1414-19.

McCloskey, 1999. Private communications for one of the authors, MSE-TA, Butte, MT.

McGrew, K., J. Murphy, D. Williams. 1996. Se reduction via conventional water treatment, *Randol Gold Forum '96*, Randol Gold, Denver, CO. 129-41.

McNeil, L., M. Edwards. 1995. Soluble As removal at water treatment plants. JAWWA, 105-13.

Merrill, D., et al. 1986. Field Evaluation of As and Se removal by iron coprecipitation. *JWPCF*, 58, (1); and, 1987. *Environ. Progress*, 6, (2), 82-8.

Merrill, D., P. Maroney, D. Parker. 1985. Trace element removal by coprecipitation with amorphous iron oxyhydroxide: engineering evaluation. *Electric Power Research Institute Coal Combustion Systems Division*, Report EPRI, CS 4087. Palo Alto, CA. 264.

MWTP. 1994. Issues identification and technology prioritization report: As. *EPA Mine Waste Treatment Program*. Activity I, Vol 5. MWTP-41. MSE-TA, Butte, MT., 169.

MWTP. 1997. Arsenic oxidation demonstration project-final Report. *EPA Mine Waste Treatment Program*. Activity III. Project 7. MWTP-84. MSE-TA, Butte MT.

MWTP. 1999. Issues identification and technology prioritization report: Se. *EPA Mine Waste Technology Program*, Activity I, Vol.VII, MWTP-106. MSE-TA, Butte MT., 118.

MWTP. 1999a. Issues identification and technology prioritization report: Tl. *EPA Mine Waste Treatment Program*. Activity I, Vol VIII. MWTP-143. MSE-TA, Butte MT., 76.

MWTP. 2001. Final report—Se treatment/removal alternatives demonstration project. *EPA Mine Waste Treatment Program*. Activity I. MWTP-143. MSE-TA, Butte MT., 69.

Nikolaev, A., et al. 1974, 76. As(III) oxidation in aqueous solutions by air oxygen in the presence of a catalyst. Izv. Sib. Otd- Akad. Nauk SSSR, *Ser. Khim, Nauk.*. 5. 36-40; and, 1976. Oxidation of As(III) in aqueous solutions in the presence of native pyrolusite. 3, 48-51; and 6, 36-40.

Nikolaidis, N., J. Lackovic, G. Dobbs. 2002. As remediation technlogy-AsRT. *US Pat. Appl.* 60/050,250. 9.

Nishimura, T., I Itoh. 1985. The calcium-arsenic-water-air system. In: *Proc. Impurity Control Disposal*, 15th CIM Hydromet. Meeting. Vancouver, BC. 2.1-2.19.

Nishimura, T., K. Tozawa. 1978. On the solubility product of ferric, calcium and magnesium arsenates. *Bull. Inst. Min. Dress. and Met.* 34, 202-6.

Nishimura, T., K. Tozawa. 1985. Removal of As from wastewater by addition of calcium hydroxide and stabilization of As precipitates by calcination. In: *Proc. Impurity Control and Disposal*, 15th Hydromet. Sym. CIM, Vancouver, CAN. 3.1-3.18.

Nishimura, T., K. Tozawa, R.G. Robins. 1983. The calcium-arsenic-water system. *MMIJ/Aus IMM Joint Symp.*, Met. Session JD-2-1, Sendai, Japan.

Nishimura, T., K. Tozawa, R. Robins. 1988. Stabilities and solubilities of metal arsenites and arsenates in water and effect of sulfate and carbonate ions on their solubilities. In: *Proc. As Met. Fundamentals and Applications*, (Eds) R. Reddy, et al. TMS. Warrendale, PA., 77-98.

Nriagu, J.O. (Ed.). 1994. <u>Arsenic in the environment: Part 1 cycling and characterization</u>, *Advances in Env. Sci. and Tech.*, Wiley and Sons, NY, NY. 448.

Nriagu, J.O. (Ed.). 1998. <u>Thallium in the environment</u>, vol. 29, *Advances in Env. Sci. and Tech.*, Wiley and Sons, NY, NY. 284.

O'Hannesin, S., R. Gillham 1992. A permeable reaction wall for insitu dehalogenating organic compounds", Proc. 5th Can.Geotechnical Soc. Conf., Toronto, Canada.

Oremland, R. 1991, 93. Selenate removal from waste water. US Patent 5,009,786; 5,271,831.

Parida K., et al. 1997. Studies on ferric oxide hydroxides. iii. adsorption of selenite on different forms of iron oxyhydrides. *J. Colloid Interface Science*. 185, (2), 355-62.

Pengchu, A., D. Sparks. 1990. Kinetics of selenate and selenite adsorption/desorption at the goethite/water interface. *Env. Sci. Tech.* 24, (12), 1848-56.

Pierce, M., C. Moore. 1980. Adsorption of arsenite on amorphous iron hydroxide from dilute aqueous solution. *Env. Sci. Tech.* 14, (2), 214-16.

Plessas, K. 1992. Removal of As from wastewater. *MSc Thesis*, Montana Tech, Butte, MT. 118.

Pontius, F. 1990. An update of the federal drinking water regs. JAWWA. 87 (2), 48-58.

Porubaev, V., et al. 1972. Use of phosphates for the purification of waste waters to remove nonferrous metal ions and arsenic, *Rud Tsvet. Metal.* 10, 110-18.

Powell, R., R. Puls, C. Paul. 1994. Innovative solutions for contaminated sit management, *Water Env. Fed.* Miami, FL. 485-96.

Powell, R., R. Puls, S. Hightower. 1995. Coupled iron corrosion and chromate reduction: mechanisms for substance remediation. *ES&T*, 29 (8), 1913-22.

Puls, R., R. Powell. 1991. Colloidal ferric oxide transport studies in laboratory model systems using shallow aquifer material. EPA/600/D-91/098. *US Printing Office*, Washington, DC. 5 p.

Rai, D. et al. 1984. Chemical attentuation rates, coefficients and constants in leachate migration, vol. 1, a critical review, *EPRI* Rep. EA-3356. Ele. Power Res. Inst., Palo Alto, CA.

Rajakovic, L. 1992. Sorption of As onto activated carbon impregnated with metallic silver and copper. *Separation Sci. Tech.* 27, (11), 1423-33; and, As removal from water by chemisorption filters. *Env. Pollution.* 75, (3), 279-87.

Ramana, A., A. Sengupta. 1992. Removing Se(IV) and As(V) Oxyanions with Tailored Chelating Polymers, *J. Env. Engn.* ASCE, 118 (5), 755-75.

Reed, B. 2002. Removal of heavy metals by activated carbon. In *Env. Separation of Heavy Metals*. (Ed) A. SenGupta, Lewis Publishers, NY, 181-204.

Riveros, P., J. Dutrizac, P. Spencer. 2001. As disposal practices in the metallurgical industry. *Can. Met. Quart.* 40, (4), 395-420.

Robins, R.G. 1981. The solubility of metal arsenates. *Met. Trans. 12B.* TMS. Warrendale, PA. 103-9.

Robins, R.G. 1983. The stabilities of As(V) and As(III) compounds in aqueous metal extraction systems. In: *Proc. Hydrometallurgy Research, Development and Plant Practice*, (Ed) K. Osseo-Asare, J.D. Miller. TMS. Warrendale, PA. 291-310.

Robins, R.G. 1984. The stability of As in gold mine processing wastes. *Precious Metals*, (Eds) V. Kydryk, D. Corrigan, W. Liang. TMS. Warrendale, PA, 241-49.

Robins, R.G. 1988. Arsenic hydrometallurgy. *Proc. As Metallurgy Fundamentals and Applications.* (Ed) R. Reddy, et al. TMS. Warrendale, PA. 215-48.

Rogers, K. 1989. Point-of-Use treatment of drinking water in San Ysidro, NM. EPA/600/2-89/050, NTIS PB90-108838, *US Printing Office*, Washington, DC. 65 p.

Rochette, E., G. Li, S. Fendorf. 1998. Stability of arsenate minerals in soils under biotically generated reducing conditions. *Soil Sci. Soc. Am. J.* 62, 1530-37.

Rosengrant, L., L. Fargo. 1990. Final best demonstrated available technology (BDAT) background document for K031, K084, K101, K102, As wastes (D004), Se wastes (D010), and P and U wastes containing As and Se listing constituents. EPA/530/SW-90/059A. EPA, Wash., DC. 124.

Rozelle, L. 1987. Point-of-Use water treatment for removal of chemical contaminants from drinking water. In: *AWWA Sem. on Water Quality Concerns in the Distribution System.* AWWA, Denver, CO. 121-38.

Runnells, R., R. Skoda. 1990. Redox modeling of As in the presence of Fe: applications to equilibrium computer modeling. In: *Proc. Env. Res. Conf. on Groundwater Quality and Water Disposal*, EPRI, Palo Alto, CA. 22-1-11.

Sanjay, H., A. Fataftah, D. Walia. 2002. Humasorb: a coal-derived humic acid-based heavy metal sorbent.In: *Env. Separation of Heavy Metals: Engr. Processes.* Lewis Publishers, NY, 347-74.

Santina, P. 1996. Sulfur-Modified iron (SMI) process for As removal. *US Pat. 5,575,919*.

Schulze, D., U. Schwertmann. 1984. The influence of Al on iron oxides: x. properties of Al-substituted goethites. *Clay Minerals*, 19, 521-39.

Schwertmann, U. 1966. Inhibitory effect of soil organic matter on the crystallization of amorphous ferric hydroxide. *Nature*, 212, 645-46.

Schwertmann, U. 1984. The influence of Al on iron oxides: ix. dissolution of Al-goethite in 6 M HCl. *Clay Minerals*, 19, 9-19.

Schwertmann, U., R. Cornell. 1991. <u>Iron oxides in the laboratory: preparation and characterization</u>. *VCH Verlagsgesellschaft mbH*, Weinheim, Germany, 137 p.

Schwertmann, U., H. Fechter. 1982. The point of zero charge of natural and synthetic ferrihydrite and its relation to silicate. *Clay Minerals*, 17, 471-6.

Schwertmann, U., E. Murad. 1983, 85. Effect of pH on the formation of goethite and hematite from ferrihydrite. *Clays, Clay Minerals.* 31 (4), 277-84; and, 1985. Properties of goethites of varying crystallinity. 33, (5), 369-78.

Schwertmann, U., H. Thalmann 1976. The influence of Fe(II), Si, and pH on the formation of lepidocrocite and ferrihydrite during oxidation of aqueous $FeCl_2$ solutions. *Clay Minerals.* 11, 189-200.

Scott, M. 1991. Kinetics of adsorption and redox processes on Fe and Mn oxides: reactions of As(IIII) and Se(IV) at goethite and birnessite surfaces. *Ph.D. Thesis*, CIT, 266 p.

SenGupta, A. 2002. Principles of heavy metal separation: an introduction. In: *Env. Separation of Heavy Metals: Engineering Processes.* Lewis Publishers, NY, 1-14.

SenGupta, A. (Ed). 2002a. Environmental separation of heavy metals: engineering processes. Lewis Publishers, N.Y. 380 p.

SenGupta, A., J. Greenleaf. 2002. Arsenic in subsurface water: its chemistry and removal by engineered processes. In: *Env. Separation of Heavy Metals: Engr. Processes.* Lewis Publishers, NY. 265-306.

Sengupta, S., A. SenGupta. 2002. Trace heavy metal separation by chelating ion exchangers. In: *Env. Separation of Heavy Metals: Engineering Processes.* Lewis Publishers, NY, NY. 45-91.

Sisk W., et al. 1990. As contaminated groundwater treatment pilot study. In: *Superfund '90.* 11th Natl. Conf. Washington, DC. 901-6.

Smith, R., E. Jenne. 1991. Recalculation, evaluation, and prediction of surface complexation constants for metal adsorption on iron and manganese oxides. *Env. Sci. Tech.* 25, (3), 525-30.

Sorg, T., G. Logsdon. 1978. Treatment technology to meet interim primary drinking water regulations for inorganics: part 2. *JAWWA.* 379-93.

Sparkman, J., et al. 1990. Adsorption of oxyanions by spent western oil shale: selenite. *Env. Geol. Water Science*, 15, (2), 93-9.

Stewart, H., K. Kessler. 1991. Evaluation of As removal by activated alumina filtration at a small community public water supply. *J. New England Water Works Assoc.* 105, (3), 179-99.

Stiksma J., et al. 1996. Iron addition for impurity control at Sherritt's nickel refinery. In: *Proc. Iron control and disposal.* (Eds) J. Dutrizac, G. Harris. CIM. Montreal, Quebec, 287-98.

Swash, P., A. Monhemius, J. Schaekers. 2000. Solubilities of process residues from biological oxidation pretreatments of refractory gold ores. *Minor Elements 2000.* (Ed) C. Young. SME, Littleton, CO. 115-22.

Trussell, R., A. Trussell, P. Kraft. 1980, 91. Se removal from groundwater using activated alumina. EPA-600/12-80-153. Washington, DC; and, 1991. Se removal with activated alumina. *AWWA Res. Found. Water Quality Res. News.* 19, 4-5.

Twidwell, L., et al. 1993. Removal of As from wastewater and stabilization of As bearing waste solids: summary of experimental results. *J. Haz. Mat.* 36, 69-80.

Twidwell, L., et al.1999. Technologies and potential technologies for removing As from process and wastewater, In: *Proc. REWAS'99, Global Symp. on Recycling, Waste Treat. and Clean Tech.* (Eds) I. Gaballah, J. Hager, R. Solozaral, San Sabastian, Spain, TMS Warrendale, PA. 1715-26.

Twidwell, L. et al. 1999a. Technologies and potential technologies for removing Se from process and wastewater. In: *Proc. REWAS'99, Global Symp. on Recycling, Waste Treat. and Clean Tech.* (Eds) I. Gaballah, J. Hager, R. Solozaral, San Sabastian, Spain, TMS Warrendale, PA, 1645-56; and, 2000. Technologies and potential technologies for removing Se from process and waste water: update. In: *Proc. Minor Elements 2000*, (Ed) C. Young, SME, Littleton, CO, 53-66.

Valenzuela, A., K. Fytas, M. Sanchez. 2000. As management in pyrometallurgical processes. part II. recovery and disposal. *Int. Conf. on Clean Technologies for the Min. Industries.* (Eds) M. Sanchez, F. Vergara and S. Castro. Univ. Concepcion, Chile. 107-21.

Virnig, M., K. Weerts. 1993. CyanoMet R-A process for the extraction and concentration of cyanide species from alkaline liquors, *Randol Gold Forum '93*, Beaver Creek, CO. 333-36.

Watson, J. 1999. Separation methods for waste and environmental applications. NY. Marcel Dekker. 600 p.

Weir, D., I. Masters. 1980, 82. Removing As from aqueous solutions. *Can. Pat.358,966*, (1980); *U.S. Pat. 4,366,128*, (1982); *Fr. Pat. 2,488,869*, (1982).

Welham, N., K. Malati, S. Vukcevic. 2000. The stability of iron phases presently used for disposal from metallurgical systems-a review. *Minerals Engineering*, 13 (8-9). 911-31.

Williams-Beam, C., L. Twidwell. 2001. Potential technologies for removing Tl from mine and process wastewater: an abbreviated annotation of the literature. *JEMEP*, March, 2002

Wolff, C., C. Rudasill. 1990. Baird and McGuire Superfund Site: investigation of As and Pb removal from groundwater. *Superfund '90,*. 11th Natl. Conf. EPA. Washington, DC. 371-75.

WSPA. 1995. Se removal technology study - final report. Western States Petroleum Association, Concord, CA.

Young, C. (Ed) 2000. Minor Elements 2000, Processing and Environmental Aspects of As, Sb, Se, Te, and Bi. In: *Proceedings SME Annual Meeting*, Salt Lake City, UT. SME, Littleton, CO. 408.

Yuan, J. et al. 1983. Adsorption of As and Se on activated alumina. In: *Proc. ASCE, Env. Engr. Division Specialty Conf.* (Eds) A. Medine, M. Anderson, Boulder, CO, July 6-8. 433-41.

Treatment Of Solutions And Slurries For Cyanide Removal

Michael M. Botz, P.E. [1] and Terry I. Mudder, Ph.D. [2]

ABSTRACT
A variety of proven and reliable chemical, physical and biological treatment processes have been developed for the removal and recovery of cyanide from mill tailings and process solution. The purpose of this chapter is to discuss various cyanide treatment processes, their common areas of application and treatment performances that can be reliably achieved at full-scale. Emphasis is placed upon treatment processes with proven field success, as well as those processes exhibiting significant potential for specific application at mine sites.

WATER AND CYANIDE MANAGEMENT
An integral and key component of water management systems at precious metals mining sites is the approach adopted to manage cyanide-containing solutions and slurries. Excluding the bulk storage of cyanide reagents such as sodium cyanide, most cyanide present at mining sites will be present in water solutions. Therefore, to a great extent the management of water and the management of cyanide can be considered as one and the same and should be simultaneously considered when developing a water management plan.

All mining sites that utilize cyanide for metals recovery should have a comprehensive and well-maintained cyanide management plan. A good cyanide management plan will include descriptions of how cyanide-containing solutions and slurries are to be handled, stored, contained and monitored, and in many cases the plan will also include a description of treatment plants used to remove cyanide from solutions or slurries. At sites where natural cyanide attenuation is important, the cyanide management plan should address the specifics of predicting and monitoring the effectiveness of the attenuation processes.

Despite the critical importance of having a formal written cyanide management plan, many mining operations have not developed such a plan. The lack of a cyanide management plan, in some cases, has contributed to adverse environmental incidents involving cyanide (Mudder and Botz 2001a). Attempts to remedy this situation have been recently made by the United Nations Environment Programme (UNEP), which is developing an international code for the management of cyanide (www.cyantists.com/cyanide.htm). Implementation and adherence to this code, augmented by experienced scientific and engineering judgment, will help reduce both the number and severity of environmental incidents involving cyanide.

---

1 Elbow Creek Engineering, Inc., Joliet, Montana.
2 TIMES Ltd., Sheridan, Wyoming.

BACKGROUND

In the mining industry, cyanide is primarily used for extracting silver and gold from ores, but cyanide is also used in low concentrations as a flotation reagent for the recovery of base metals such as copper, lead and zinc. At these operations, cyanide treatment systems may be required to address potential toxicity issues in regard to wildlife, waterfowl and/or aquatic life. This may include the removal of cyanide from one or more of the following:

- Slurry tailings from milling operations
- Bleed solution from Merrill-Crowe operations
- Excess solution from heap or vat leaching operations
- Supernatant solution from tailings impoundments
- Seepage collected from ponds or tailings impoundments

Cyanide treatment is classified as either a destruction-based process or a recovery-based process. In a destruction process, either chemical or biological reactions are utilized to convert cyanide into another less toxic compound, usually cyanate. Recovery processes are a recycling approach in which cyanide is removed from the solution or slurry and then re-used in a metallurgical circuit.

Selection of an appropriate cyanide treatment process involves the consideration of many factors, but generally the number of candidate processes for a particular application can be narrowed following an inspection of the untreated solution/slurry chemistry and the desired level of treatment. The common applications for cyanide treatment in the mining industry are the following:

1. <u>Tailings slurry treatment</u> is employed when the cyanide level must be lowered prior to being discharged into a tailings storage facility. In this application, the initial tailings slurry WAD cyanide level typically ranges from about 100 to 500 mg/L and treatment to less than 50 mg/L WAD cyanide is commonly established as the goal for wildlife and waterfowl protection.

2. <u>Solution treatment</u> is employed when the cyanide level in a decant or process solution must be lowered prior to being discharged into the environment. Treatment of WAD cyanide to below 1.0 mg/l is occasionally required to ensure the protection of aquatic ecosystems. Treatment technologies for decant solution commonly employ chemical oxidation and polishing processes, which are applicable to relatively low concentrations of cyanide and generate a high quality effluent.

CYANIDE ANALYSIS

The term "cyanide" generally refers to one of three classifications of cyanide, and it is critical to define the class of cyanide that is to be removed in a treatment plant. The three classes of cyanide are: (1) total cyanide; (2) weak acid dissociable (WAD) cyanide; and (3) free cyanide as shown in Figure 1. Each of these forms of cyanide has a specific analytical methodology for its measurement, and it is important that the relationship between these forms be understood when analyzing cyanide-containing solutions. As indicated in Figure 1, for a given solution the total cyanide level is always greater than or equal to the WAD cyanide level, and likewise, the WAD cyanide level is always greater than or equal to the free cyanide concentration.

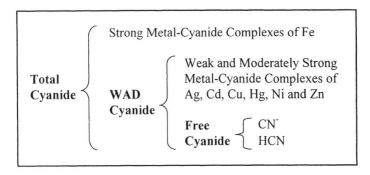

Figure 1 Classification of Cyanide Compounds

The appropriate approach to assessing the quality of untreated and untreated samples in most situations is to analyze for WAD cyanide since this includes the toxicologically or environmentally important forms of cyanide, including free cyanide and moderately and weakly complexed metal-cyanides. Total cyanide includes free cyanide, WAD cyanide plus the relatively non-toxic iron-cyanide complexes. Complete characterization of a cyanide solution generally includes analyses for pH, total cyanide, WAD cyanide, thiocyanate, cyanate, ammonia, nitrate and base metals such as copper, iron, nickel and zinc.

CYANIDE DESTRUCTION

Most cyanide destruction processes operate on the principle of converting cyanide into a less toxic compound through an oxidation reaction. There are several destruction processes that are well proven to produce treated solutions or slurries with low levels of cyanide as well as metals. In the following sections, several cyanide destruction processes are discussed along with their typical areas of application. With all of these processes, laboratory and/or pilot testing is required to confirm the level of treatment achievable and to evaluate the associated reagent consumptions.

The INCO Sulfur Dioxide and Air Process

The sulfur dioxide (SO_2) and air process was developed by INCO Limited in the 1980's and is currently in operation at over thirty mine sites worldwide. The process utilizes SO_2 and air at a slightly alkaline pH in the presence of a soluble copper catalyst to oxidize cyanide to the less toxic compound cyanate (OCN^-).

$$SO_2 + O_2 + H_2O + CN^- \xrightarrow{Cu^{+2}\ Catalyst} OCN^- + SO_4^{-2} + 2H^+$$

The theoretical usage of SO_2 in the process is 2.46 grams of SO_2 per gram of CN^- oxidized, but in practice the actual usage ranges from about 3.5 to 5.0 grams SO_2 per gram of CN^- oxidized. The SO_2 required in the reaction can be supplied either as liquid sulfur dioxide or a sulfur salt such as sodium metabisulfite ($Na_2S_2O_5$) or sodium sulfite (Na_2SO_3).

Oxygen (O_2) is also required in the reaction and is generally supplied by sparging atmospheric air into the reaction vessels. The reaction is typically carried out at a pH of about 9.0 in one or more agitated tanks, and lime is added to neutralize the acid (H^+) formed in the reaction to maintain the pH in this range. Lime usage is generally on the order of about 3.0 to 5.0 grams per gram of CN^- oxidized. As indicated, copper (Cu^{+2}) is required as a catalyst, which is usually added as a solution of copper sulfate ($CuSO_4$-$5H_2O$) to provide a copper concentration in the range of about 10 to 50 mg/L, depending upon the corresponding cyanide concentration. In solutions where sufficient copper is already present, supplemental addition of copper may not be required. A flowsheet for the sulfur dioxide and air process is shown in Figure 2.

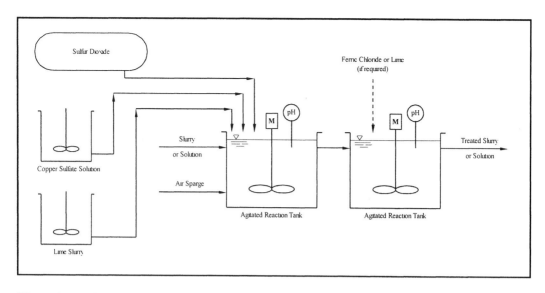

Figure 2 The Sulfur Dioxide and Air Cyanide Destruction Process

Upon completion of the cyanide oxidation reaction, metals previously complexed with cyanide, such as copper, nickel and zinc, are precipitated as metal-hydroxide compounds. Iron cyanide removal is affected through precipitation according to the following generalized reaction where 'M' represents copper, nickel or zinc:

$$2M^{+2} + Fe(CN)_6^{-4} \rightarrow M_2Fe(CN)_6 \text{ (solid)}$$

The primary advantage of the sulfur dioxide and air process is with slurry tailings, but it is also effective for the treatment of solutions for the oxidation of free and WAD cyanides. Representative results for treatment of solution and slurry with the sulfur dioxide and air process are shown in Table 1 (Ingles and Scott 1987).

Table 1 Treatment Results Using the INCO Sulfur Dioxide and Air Process

Parameter	Solution		Tailings Slurry	
	Untreated (mg/L)	Treated (mg/L)	Untreated (mg/L)	Treated (mg/L)
Total Cyanide	450	0.1 to 2.0	115	0.1 to 1.0
Copper	35	1 to 10	17	0.2 to 2.0
Iron	1.5	<0.5	0.7	0.02 to 0.3
Zinc	66	0.5 to 2.0	18	<0.01

As indicated in Table 1, the sulfur dioxide and air process is capable of achieving low and environmentally acceptable levels of both cyanide and metals. Generally, the best application of this process is with tailings slurries containing low to moderately high initial levels of cyanide and when treated cyanide levels of less than about 5 mg/L are required. In some cases, solutions treated with this process may be of suitable quality to permit their discharge. The process does not remove thiocyanate quantitatively, although a few percent of this cyanide related compound are typically removed during treatment.

The Copper Catalyzed Hydrogen Peroxide Process

The hydrogen peroxide treatment process chemistry is similar to that described for the INCO process, but hydrogen peroxide is utilized rather than sulfur dioxide and air. With this process, soluble copper is also required as a catalyst and the end product of the reaction is cyanate.

$$H_2O_2 + CN^- \xrightarrow{Cu^{+2} \; Catalyst} OCN^- + H_2O$$

The primary application of the hydrogen peroxide process is with solutions rather than slurries due to the high consumption of hydrogen peroxide that occurs in slurry applications. The process is typically applied to treat relatively low levels of cyanide to achieve effluent quality that may be suitable for discharge. The hydrogen peroxide process is effective for the treatment of solutions for the oxidation of free and WAD cyanides, and iron cyanides are removed through precipitation of insoluble copper-iron-cyanide complexes. As indicated in the above reaction, hydrogen peroxide reacts with cyanide to form cyanate and water, a process which limits the build-up of dissolved solids in the solution being treated.

The theoretical usage of H_2O_2 in the process is 1.31 grams H_2O_2 per gram of CN^- oxidized, but in practice the actual usage ranges from about 2.0 to 8.0 grams H_2O_2 per gram of CN^- oxidized. The H_2O_2 used in the process is typically provided as a concentrated liquid in 50% or 70% strength.

Although the reaction can be carried out over a wide pH range, it is usually conducted at a pH of about 9.0 to 9.5 for optimal removal of residual metals such as copper, nickel and zinc initially complexed to cyanide. If iron cyanide must also be removed to low levels, then a lower pH is needed to maximize the precipitation of copper-iron-cyanides at the expense of lowering the removal efficiencies of copper, nickel and zinc. It is common in these instances to use a two-stage system with intermediate removal of the precipitated iron cyanide. As indicated, copper (Cu^{+2}) is required as a soluble catalyst, which is usually added as a solution of copper sulfate ($CuSO_4 \text{-} 5H_2O$) to provide a copper concentration in the range of about 10 to 50 mg/L, depending upon the initial cyanide and copper concentrations. Upon completion of the indicated reaction, metals previously complexed with cyanide, such as copper, nickel and zinc, are precipitated as metal-hydroxide compounds.

Representative results for treatment of solution with the hydrogen peroxide process are shown in Table 2 (Mudder et al. 2001b).

Table 2 Treatment Results Using the Hydrogen Peroxide Process

Parameter	Solution	
	Untreated (mg/L)	Treated (mg/L)
Total Cyanide	19	0.7
WAD Cyanide	19	0.7
Copper	20	0.4
Iron	<0.1	<0.1

As indicated in Table 2, the hydrogen peroxide process is capable of achieving low levels of both cyanide and metals. Generally, the best application of this process is with solutions containing relatively low initial levels of cyanide and when treated cyanide levels of less than about 1 mg/L are required. Oftentimes, solutions treated with this process may be of suitable quality to permit their discharge. As with the INCO process, this process does not remove thiocyanate quantitatively, but does remove a few percent of this cyanide related compound during treatment.

The Caro's Acid Process

Peroxymonosulfuric acid (H_2SO_5), also known as Caro's acid, is a reagent used in a recently developed cyanide treatment process that has found application at a few sites. Caro's acid is produced from concentrated hydrogen peroxide and sulfuric acid in an exothermic reaction (Norcross 1996):

$$H_2O_2 + H_2SO_4 \rightarrow H_2SO_5 + H_2O$$

Due to its instability, Caro's acid is produced on-site in-situ and used immediately for cyanide detoxification with only minimal intermediate storage. At room temperature, Caro's acid is stable for several hours, however at elevated temperature it is stable only for several minutes, decomposing to liberate oxygen, water and sulfur trioxide (SO_3). Production of Caro's acid is typically conducted with 1.5 to 3.0 moles of H_2SO_4 per mole of H_2O_2 to yield a product of up to 80% purity. Normally, 70% hydrogen peroxide solution and 93% sulfuric acid solution are used to generate Caro's acid.

The overall oxidation reaction of Caro's acid with cyanide is shown below.

$$H_2SO_5 + CN^- \rightarrow OCN^- + SO_4^{-2} + 2H^+$$

The theoretical usage of H_2SO_5 in the process is 4.39 grams H_2SO_5 per gram of cyanide oxidized, but in practice 5.0 to 15.0 grams H_2SO_5 per gram of cyanide oxidized is required. Acid produced in the reaction (H^+) is typically neutralized with lime, if necessary, and the reaction is normally carried out at a pH in the range of about 7.0 to 10.0.

Caro's acid is used in slurry treatment applications where the addition of a copper catalyst is not desirable, which is typically only in situations where the sulfur dioxide and air process is not suited. In solution applications, other destruction processes, such as the hydrogen peroxide process, are preferred to the Caro's acid process. A flowsheet for the Caro's acid process is shown in Figure 3.

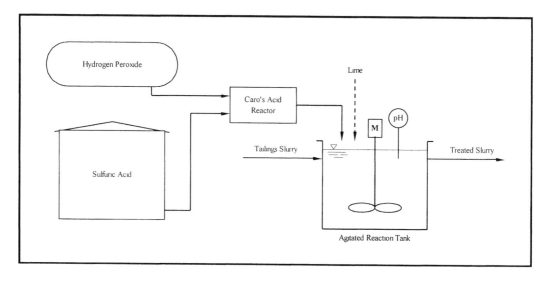

Figure 3 The Caro's Acid Cyanide Destruction Process

Representative results for treatment of slurry with Caro's acid are shown in Table 3 (Mudder et al. 2001b).

Table 3 Treatment Results Using Caro's Acid

Test Number	Slurry WAD Cyanide Concentration	
	Untreated (mg/L)	Treated (mg/L)
1	44.5	8.5
2	37.5	4.2
3	46.0	14.0
4	39.8	4.0
5	115.0	27.1
6	113.1	16.3
7	101.5	18.7

As indicated in Table 3, the Caro's acid process is capable of achieving levels of WAD cyanide below 50 mg/L, which are generally suitable for discharge into tailings impoundments. Generally, the best application of this process is with tailings slurries containing low to moderate initial levels of cyanide and when treated cyanide levels of less than about 10 to 50 mg/L are required.

The Alkaline Chlorination Process

Alkaline chlorination at one time was the most widely applied of the cyanide treatment processes, but it has gradually been replaced by other processes and is now only used occasionally. Alkaline chlorination is effective at treating cyanide to low levels, but the process can be relatively expensive to operate due to high reagent usages. The cyanide destruction reaction is two-step, the first step in which cyanide is converted to cyanogen chloride (CNCl) and the second step in which cyanogen chloride hydrolyzes to yield cyanate.

$$Cl_2 + CN^- \rightarrow CNCl + Cl^-$$

$$CNCl + H_2O \rightarrow OCN^- + Cl^- + 2H^+$$

In the presence of a slight excess of chlorine, cyanate is further hydrolyzed to yield ammonia in a catalytic reaction.

$$OCN^- + 3H_2O \xrightarrow{Cl_2\ Catalyst} NH_4^+ + HCO_3^- + OH^-$$

If sufficient excess chlorine is available, the reaction continues through "breakpoint chlorination" in which ammonia is fully oxidized to nitrogen gas (N_2).

$$3Cl_2 + 2NH_4^+ \rightarrow N_2 + 6Cl^- + 8H^+$$

In addition to reacting with cyanide, cyanate and ammonia, the alkaline chlorination process will preferentially oxidize thiocyanate, which in some cases can lead to excessively high consumptions of chlorine. It is the removal of thiocyanate that makes this cyanide treatment process unique when compared to other chemical oxidation processes.

$$4Cl_2 + SCN^- + 5H_2O \rightarrow SO_4^{-2} + OCN^- + 8Cl^- + 10H^+$$

The primary application of the alkaline chlorination process is with solutions rather than slurries due to the high consumption of chlorine that occurs in slurry applications. The process is typically applied to treat low solutions flows initially containing low to high levels of cyanide to achieve cyanide levels that may be suitable for discharge. The process is effective for the treatment of solutions for the oxidation of free and WAD cyanides, but a lesser amount of iron cyanides are removed depending on the levels of other base metals in the solution being treated. As can be seen in the above reactions, a significant increase in the treated water dissolved solids concentration may result, particularly with chloride.

The theoretical usage of Cl_2 to oxidize cyanide to cyanate is 2.73 grams Cl_2 per gram of CN^- oxidized, but in practice the actual usage ranges from about 3.0 to 8.0 grams Cl_2 per gram of CN^- oxidized. The Cl_2 used in the process can be provided as a liquid Cl_2, as a 12.5% solution of sodium hypochlorite (NaOCl) or as a solution of calcium hypochlorite ($Ca(OCl)_2$). Chlorine consumptions for the oxidation of ammonia and thiocyanate can be calculated from the above reactions. In addition, the above reactions generate varying amounts of acid (H^+) which is typically neutralized by adding lime or sodium hydroxide to the reaction vessels.

The reaction is carried out at a pH of greater than 10.5 to ensure potentially irritating cyanogen chloride is rapidly hydrolyzed to cyanate. An advantage of the process is that copper is not required as a catalyst as with the sulfur dioxide/air and hydrogen peroxide processes. Upon completion of the cyanide oxidation reaction, metals previously complexed with cyanide, such as copper, nickel and zinc, are precipitated as metal-hydroxide compounds.

Representative results for treatment of solution via alkaline chlorination are shown in Table 4 (Ingles and Scott 1987).

Table 4 Treatment Results Using the Alkaline Chlorination Process

Parameter	Solution	
	Untreated (mg/L)	Treated (mg/L)
Total Cyanide	2,000	8.3
WAD Cyanide	1,900	0.7
Copper	290	5.0
Iron	2.4	2.8
Zinc	740	3.9

As indicated in Table 4, the alkaline chlorination process is capable of achieving low levels of both cyanide and metals. Generally, the best application of this process is with low flows of solutions containing high to low initial levels of cyanide when treated cyanide levels of less than about 1 mg/L are required. Oftentimes, solutions treated with this process may be of suitable quality to permit their discharge.

Iron-Cyanide Precipitation

Free, WAD and total cyanides will all react with ferrous iron to yield a variety of soluble and insoluble compounds, primarily hexacyanoferrate (III) ($Fe(CN)_6^{-3}$), Prussian blue ($Fe_4[Fe(CN)_6]_3$) and other insoluble metal-iron-cyanide ($M_XFe_Y(CN)_6$) compounds such as those of copper or zinc (Adams 1992).

$$Fe^{+2} + 6CN^- + \tfrac{1}{4}O_2 + H^+ \rightarrow Fe(CN)_6^{-3} + \tfrac{1}{2}H_2O$$

$$4Fe^{+2} + 3Fe(CN)_6^{-3} + \tfrac{1}{4}O_2 + H^+ \rightarrow Fe_4[Fe(CN)_6]_3 + \tfrac{1}{2}H_2O$$

These reactions act to lower the free and WAD cyanide concentrations by converting them to stable iron cyanide compounds (soluble and insoluble), while the iron-cyanide concentration is lowered as a result of precipitation reactions.

The iron-cyanide precipitation process is limited in its suitability to situations where the precipitation reactions can be controlled and the precipitated solids can be separated and properly disposed. Proper handling and disposal of the cyanide precipitates generated in the process is important and represents the major disadvantage of this non-destructive process. In the past, this process was widely used to convert free and WAD cyanides to less toxic iron-cyanide compounds, but its present utility is primarily as a polishing process to reduce total cyanide concentrations to less than about 1 to 5 mg/L. There are a number of environmental drawbacks to this process, including the generation of cyanide precipitates and the formation of stable and soluble iron-cyanide compounds that will persist for many years and may require further treatment.

The process is optimally carried out at a pH of about 5.0 to 6.0 and iron is added as ferrous sulfate ($FeSO_4 \cdot 7H_2O$). Ferrous sulfate usage ranges from about 0.5 to 5.0 moles Fe per mole of CN^- depending on the desired level of treatment (Adams 1992; Dzombak et al. 1996). As indicated in Table 5, the iron-cyanide precipitation process is capable of achieving relatively low levels of total cyanide at pH 7.0 and an Fe:CN molar ratio of about 4:1 (Dzombak et al. 1996).

Table 5 Treatment Results Using the Iron-Cyanide Precipitation Process

Parameter	Solution	
	Untreated (mg/L)	Treated (mg/L)
Total Cyanide	8.8	0.89

Effluent Polishing with Activated Carbon
Activated carbon has a high affinity for many metal-cyanide compounds, including the soluble cyanide compounds of copper, iron, nickel and zinc. Activated carbon is suitable for use as a polishing treatment process to remove cyanide to low levels, when the initial cyanide concentrations are already below about 2.0 mg/L. This is a simple and effective process, convenient for installation at sites where activated carbon is used in metallurgical processes for precious metals recovery. At these sites, newly purchased carbon can be used for water treatment, and then when the carbon breaks through and is no longer suitable for water treatment, it can be transferred to the metallurgical circuit for continued use. This has been done at a number of sites to produce high quality effluents without impacting gold recovery operations. At inactive sites, regeneration and/or disposal of loaded carbon must be considered.

Representative results for treatment of solution using activated carbon adsorption are shown in Table 6 (Botz and Mudder 1997).

Table 6 Treatment Results Using the Activated Carbon Adsorption Process

Parameter	Solution	
	Untreated (mg/L)	Treated (mg/L)
Total Cyanide	0.98	0.20
Copper	0.02	<0.02
Iron	0.22	0.02
Nickel	0.15	0.15
Zinc	0.02	<0.02

Biological Treatment
Biological treatment processes have become more widespread in the mining industry due to the success of the plant installed at Homestake Lead, USA in the 1980's. In this plant, an aerobic attached growth biological treatment is used to remove cyanide, thiocyanate, cyanate, ammonia and metals from tailings impoundment decant solution prior to discharge into a trout fishery. The plant has been operating successfully for over fifteen years, producing high-quality effluent.

A multiple stage suspended growth biological treatment plant was installed by Homestake Nickel Plate, Canada in the mid-1990's to treat tailings impoundment seepage. This plant is a suspended sludge system with both aerobic and anaerobic treatment sections to remove cyanide, thiocyanate, cyanate, ammonia, nitrate and metals.

Another biological treatment process was developed for the Homestake Santa Fe, USA mine to treat draindown from the decommissioned heap leach operation. This process, known as the passive Biopass process, is suitable for solution flows of less than about 10 m^3/hour for the removal of cyanide, thiocyanate, cyanate, ammonia, nitrate and metals.

The applicability of biological processes for the treatment of cyanide solutions in the mining industry has been somewhat limited, but is growing again with several applications being developed throughout the world. Their applicability is primarily with continuous solution flows with temperatures above about 10°C. The key advantage to biological treatment is the ability to simultaneously remove several compounds in a single process, often at a much lower cost than would be encountered with other treatment processes. In situations where cyanide and one or more of its related compounds of cyanate, thiocyanate, ammonia and nitrate must be removed, biological treatment should be considered. Representative results from the above three biological treatment plants are presented in Table 7 (Given et al 2001; Mudder et al. 2001a; Mudder et al. 2001c).

Table 7 Treatment Results Using Biological Processes

Parameter	Homestake Lead		Homestake Nickel Plate		Homestake Santa Fe	
	Untreated (mg/L)	Treated (mg/L)	Untreated (mg/L)	Treated (mg/L)	Untreated (mg/L)	Treated (mg/L)
Total Cyanide	3.39	0.37	1.04	0.44	--	--
WAD Cyanide	2.34	0.03	0.33	0.04	14	<0.2
Thiocyanate	--	--	379	0.08	--	--
Ammonia	5.31	0.27	25.3	0.15	--	--
Nitrate	--	21.9	2.8	0.13	55.6	0.8
Copper	0.49	0.04	0.02	0.005	10.3	<0.5
Iron	0.1 to 5.0	0.27	0.06	0.02	--	--
Nickel	0.01 to 0.04	0.03	--	--	--	--
Zinc	0.01 to 0.1	0.01	--	--	--	--

Other Cyanide Treatment Processes

There are a number of other treatment processes that have been applied at full scale to treat cyanide, but implementation of these processes has been limited for a number of reasons. Ion exchange and reverse osmosis are frequently considered for treatment of decant solution, but with both of these processes waste brine is generated as a by-product. Disposal or further treatment of this brine is difficult and expensive, and in some cases the brine may be hazardous and require special handling. The amount of waste brine generated in by ion exchange and reverse osmosis typically ranges from about 10% to 30% of the volume of water treated. Ion exchange and reverse osmosis processes are also relatively expensive and complex to construct, operate and maintain. Due to these drawbacks, ion exchange and reverse osmosis are limited to situations where waste brine can be easily disposed or treated, or where very high quality effluent is required. An advantage of these processes, particularly reverse osmosis, it that in some cases simultaneous removal of cyanide, cyanate, thiocyanate, ammonia, and nitrate can be affected. Ion exchange is occasionally used to target ammonia or nitrate removal from decant solution, and additional information in this regard is provided later in this chapter.

Ozone is a strong oxidant and capable of oxidizing free and WAD cyanides to cyanate, ammonia and nitrate. However, ozone initially oxidizes thiocyanate to cyanate or cyanide depending on the solution pH (Botz et al. 2001). The reaction rate is rapid and generally only limited by the rate at which ozone can be absorbed into the solution. Low effluent cyanide concentrations can be achieved with ozone, but treatment may result in the formation of cyanate, ammonia and nitrate. Ozone is relatively expensive to produce and this has limited its use for cyanide destruction, particularly for large water flows, but may find application in small-volume polishing applications. Iron cyanides are also oxidized by ozone, but the reaction rate is too slow at ambient temperature for practical application. At elevated temperature and in the presence of ultraviolet radiation, iron cyanides are converted to cyanate by ozone. Ammonia can be oxidized to nitrate by ozone, but an alkaline pH is required.

CYANIDE RECOVERY

A number of cyanide recovery processes have been investigated over the previous one hundred years, but only two have found widespread application, as described in the following sections. The need for development and implementation of processes for recovery of cyanide will be important to the success of current and future mining operations. The requirement stems from concerns over low metal prices and the realization that more stringent environmental regulations will be developed, thereby restricting the concentrations of cyanide that can be discharged into tailings impoundments. With increasing concern over groundwater and wildlife issues, there will be increasing pressure to regulate cyanide entering tailings impoundments more closely. The advantages of cyanide recovery include the lowering of cyanide and metals levels entering a tailings impoundment, the economic benefit of recycling cyanide and the potential reduction in downstream treatment requirements for cyanide and its related compounds of cyanate, thiocyanate, ammonia and nitrate.

Stripping and Absorption

The stripping and absorption approach to recovering cyanide, also known as the acidification-volatilization-reneutralization (AVR) and Cyanisorb processes, remove cyanide from solution as hydrogen cyanide gas. At a pH of less than about 8.0, free cyanide and some WAD cyanide compounds are converted to hydrogen cyanide gas, which can then be air-stripped from solution. Once removed from solution as hydrogen cyanide gas, the hydrogen cyanide is easily absorbed into an alkaline solution of sodium hydroxide or lime.

The three main reactions involved with the cyanide recovery process are as follows.

$2CN^- + H_2SO_4 \rightarrow 2HCN_{(aq)} + 2SO_4^{-2}$ (acidification)

$HCN_{(aq)} \rightarrow HCN_{(g)}$ (stripping)

$HCN_{(g)} + NaOH \rightarrow NaCN + H_2O$ (absorption)

This basic process has been used at about ten sites worldwide to affect WAD cyanide recoveries ranging from about 70% to over 95% with both slurries and solutions. Its application is primarily with solutions or slurries with moderate to high concentrations of cyanide under both high and low-flow conditions. The presence of dissolved copper in untreated solution can cause difficulties with conventional cyanide recovery processes, and in some cases pre-treatment of the solution with sodium sulfide (Na_2S) may be required to precipitate copper sulfide (Cu_2S) prior to cyanide recovery.

The flowsheet for the Golden Cross cyanide recovery plant that operated in New Zealand during the life of the mine is shown in Figure 4. This plant processed tailings slurry and reduced the overall cyanide consumption at the site by about one-half. A key benefit of operating this cyanide recovery plant was after closure of the mine when tailings impoundment seepage was directly discharged to a trout fishery. This was possible because of the low concentrations of cyanide and related compounds present in the impoundment during mine operation.

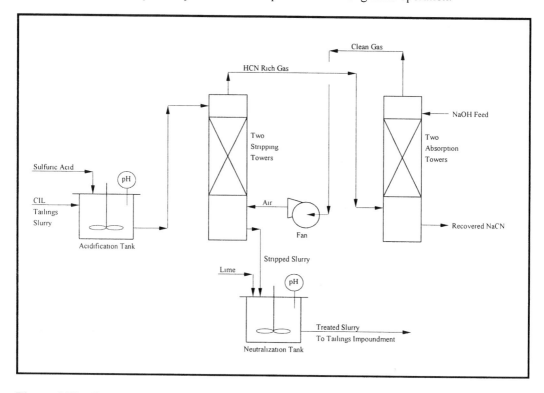

Figure 4 The Golden Cross Cyanide Recovery Plant Flowsheet

Table 8 lists representative treatment results from several cyanide recovery processes (Moura 2001; Goldstone and Mudder 2001; Omofoma and Hampton 1992).

Table 8 Treatment Results Using the Stripping and Absorption Cyanide Recovery Process

Plant	WAD Cyanide		
	Untreated (mg/L)	Treated (mg/L)	Percent Recovery
Golden Cross, New Zealand (Slurry)	~250	<30	85% to 90%
DeLamar, USA (Solution)	~300	<25	>90%
Cerro Vanguardia, Argentina (Solution)	~300	<30	>90%

Cyanide Recovery through Water Recycle and Tailings Washing

Cyanide can effectively be recovered and re-used by recycling cyanide-containing solutions within a metallurgical circuit. This is commonly conducted using tailings thickeners or tailings filters to separate solution from tailings solids, with the solution being recycled in the grinding and/or leaching circuits. This approach to recovering cyanide should be evaluated for all operations utilizing cyanide, and its performance can be determined by a simple mass balance calculation.

Cyanide recovery affected in this manner is purely a physical process, with the recovery of cyanide accompanying the recovery of water from slurry tailings. In some cases, thickeners or filters can be used in conjunction with a chemical-based cyanide recovery process (Botz and Mudder 2001b) to increase the overall cyanide recovery percentage. Cyanide recovery in water recycle and wash circuits are capable of achieving 90% cyanide recovery, but their implementation must be preceded by a careful examination of the site water balance.

NATURAL CYANIDE ATTENUATION

It is well known that cyanide solutions placed in ponds or tailings impoundments undergo natural attenuation reactions which result in the lowering of the cyanide concentration. These attenuation reactions are dominated by natural volatilization of hydrogen cyanide, but other reactions such as biological degradation, oxidation, hydrolysis, photolysis and precipitation also occur. Natural cyanide attenuation occurs with all cyanide solutions exposed to the atmosphere, whether intended or not. At several sites, ponds or tailings impoundments are intentionally designed to maximize the rate of cyanide attenuation, and in some cases resultant solutions are suitable for discharge. Advantages of natural attenuation include lower capital and operating costs when compared to chemical oxidation processes.

Two approaches have been developed to predict the rate of cyanide attenuation in ponds and tailings impoundments. The first method is empirical in nature and uses experimentally derived rate coefficients to estimate the rate of attenuation using a first order decay equation (Simovic et al. 1985). This approach is relatively simple to apply, but its applicability at a given site must be verified by conducting field testwork and the results may not be accurate under changing weather, pond/impoundment geometry or chemistry conditions.

The second approach to modeling natural cyanide attenuation was developed by Botz and Mudder (2000). This approach utilizes detailed solution chemical equilibria and kinetic calculations to predict the rate of cyanide losses from ponds and impoundments through a variety of reactions. The reactions of cyanate, thiocyanate, ammonia and nitrate can also be modeled with this approach. This approach can be time-intensive to apply at a given site, but the results are accurate under a wide variety of weather, pond/impoundment geometry and chemistry conditions.

Examples of natural cyanide attenuation in tailings impoundments are presented in Table 9 as observed at several mines in Australia (Minerals Council of Australia 1996). These data correspond to WAD cyanide reductions of ranging from about 55% to 99%, reflective of the varying tailings chemistries, climatic conditions and tailings impoundment geometries at these sites. An additional example of natural attenuation of both cyanide and its related compounds is given in Figure 5 (Schmidt et al. 1981).

Table 9 Examples of Natural Cyanide Attenuation

WAD Cyanide in Tailings Discharge (mg/L)	WAD Cyanide in Tailings Impoundment Decant Solution (mg/L)	Percent WAD Cyanide Reduction
210	94	55%
48	10	79%
57	0.5	99%
150	20	87%
125	22	82%
186	20	89%
82	12	85%
99	9	91%

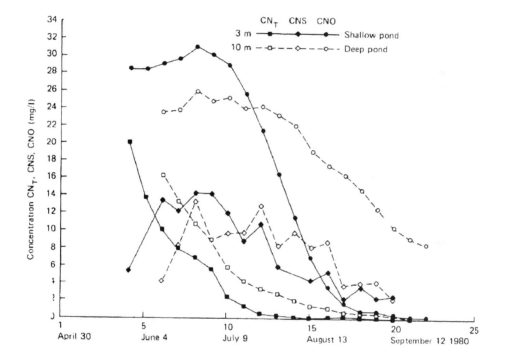

Figure 5 Example of the Natural Attenuation of Cyanide and Related Compounds

TREATMENT OF CYANIDE RELATED COMPOUNDS

The primary constituents of concern in cyanidation solutions include not only the various forms of cyanide, but the cyanide related compounds thiocyanate, cyanate, ammonia and nitrate. In many cases, these compounds are important from both a water quality and toxicity standpoint and low levels must be achieved in treated waters. The following sections provide general background information on the removal cyanide related compounds.

Thiocyanate Treatment

Thiocyanate is formed through the interaction of cyanide with sulfur-containing compounds, particularly sulfide minerals such as pyrite, chalcopyrite or arsenopyrite. As with cyanate, thiocyanate is not a cyanide compound but related to cyanide and usually only found in solutions that also contain cyanide. Thiocyanate is far less toxic than cyanide and exhibits unique chemical, analytical and treatment characteristics. Thiocyanate removal from decant solution is not routinely practiced in the mining industry, but there are full-scale treatment plants that remove thiocyanate. In the case of biological treatment of cyanide, thiocyanate is also removed as part of the process, and may be required if land application of the solution is contemplated. There are no documented cases of thiocyanate resulting in adverse environmental impacts to aquatic life in the mining industry.

Removal of thiocyanate from decant solution can be accomplished with one of several available destruction methods (Mudder and Botz 2001b). It is possible to chemically regenerate cyanide from thiocyanate, however this process has not been implemented on a full-scale basis (Botz et al. 2001). The chemical destruction methods utilize an oxidant such as chlorine to convert thiocyanate to cyanide at an alkaline pH, and then cyanide oxidation rapidly continues to yield cyanate, ammonia and nitrate. Several oxidants are capable of oxidizing thiocyanate, but only chlorine and ozone yield suitably rapid reaction kinetics.

Alkaline chlorination and ozone efficiently destroy thiocyanate and are capable of lowering thiocyanate concentrations to a few mg/L. If the residual chloride or chlorine content in treated solution is of concern, then the ozone process may be preferable since ozone dissociates to oxygen or water. Both chlorine and ozone can be used to simultaneously oxidize cyanide, cyanate, and thiocyanate, and often the choice of the appropriate oxidant is based on considerations of cost, by-product generation and process efficiency.

A lower cost alternative in many cases for thiocyanate destruction is biological treatment. Microorganisms in an aerobic environment readily oxidize thiocyanate and the reactions are rapid at temperatures above about $10°C$ to $15°C$. As with the chemical destruction methods, biological thiocyanate treatment processes can readily be configured to simultaneously remove cyanide, cyanate, and ammonia along with thiocyanate. At a North American site, a biological treatment plant is used to lower cyanide, cyanate, thiocyanate and ammonia from initial concentrations of about 0.4 mg/L, 300 mg/L, 500 mg/L and 40 mg/L, respectively, to final concentrations of about 0.08 mg/L, <5 mg/L, 0.3 mg/L and 0.5 mg/L, respectively (Given et al. 2001). The advantages of biological processes over chemical processes for thiocyanate removal are that capital and operating costs are relatively low and the concentration of reaction by-products is low. However, biological processes are kinetically slow at colder temperatures and do not respond well to rapid fluctuations in solution flow or chemistry.

Cyanate Treatment

The compound cyanate is related to cyanide and is often found in waters that contain cyanide. Cyanate originates from the oxidation of cyanide but exhibits different chemical, analytical, treatment and toxicity characteristics. Treatment of solutions for cyanate removal is uncommon because it is far less toxic than cyanide, is generally present in metallurgical solutions at low concentrations and does not persist in the environment for long periods of time. In some cases the cyanate concentration in decant solution may be sufficiently high as to warrant implementation of a cyanate removal process, and typically this is the result of cyanate produced in a cyanide destruction process. This would be limited to situations where decant solution were to be discharged to the environment and the concentration of cyanate in untreated solution would be toxic to aquatic organisms.

The authors are not aware of any full-scale water treatment facilities operating in the mining industry that specifically target the removal of cyanate or contain a limit on the level of cyanate that can be discharged. However, there are full-scale water treatment plants that incidentally remove cyanate along with their intended purpose of removing cyanide or other related compounds. Most notable are several biological treatment plants where processes to remove cyanide, thiocyanate, and ammonia also result in the removal of cyanate. An example is a mine in North America where decant solution is biologically treated for cyanide, thiocyanate, and ammonia removal (Given et al. 2001). These species are removed to low levels in the aerobic treatment system, although the cyanate level is also reduced from about 300 mg/L to less than about 5 mg/L.

The biological mechanism of cyanate removal is first the oxidation of cyanate to ammonia, and then ammonia removal proceeds through a biological process termed nitrification. Subsequent nitrate removal through a biological denitrification process may also be warranted depending upon the resultant nitrate concentration in solution. Information regarding the removal of ammonia and nitrate through chemical, biological, and physical means is presented later in this chapter.

Cyanate may also be removed from solution using chemical oxidation or hydrolysis processes. Chemical oxidation with chlorine at a slightly alkaline pH will convert cyanate to ammonia, though the chlorination process can easily be configured to complete the oxidation of ammonia into nitrogen gas through a process termed "breakpoint chlorination". Ozone at an alkaline pH is capable of converting cyanate directly to nitrate, thereby avoiding the intermediate formation of ammonia. The advantage of the ozone process is that the concentration of dissolved species in treated solution is not increased significantly since ozone dissociates into oxygen and water. Cyanate also can be hydrolyzed to ammonia at an acidic pH, though the reaction is relatively slow at low temperatures, which in some cases may require solution heating. With the oxidation or hydrolysis reactions, cyanate is converted either into ammonia or nitrate and subsequent removal of these compounds may be required depending upon their resultant concentrations.

Ammonia Treatment

Ammonia is toxic to aquatic organisms, particularly fish, but is usually present in metallurgical streams at low concentrations. The two sources that are responsible for the majority of ammonia that may be present in decant solutions are the following:

1. A mixture of ammonium nitrate and fuel oil (ANFO) is often used as a blasting agent at mining operations. A small percentage of ANFO used in blasting will remain unreacted and report as ammonia and nitrate is slurry tailings discharged into a tailings storage facility. The concentration of ammonia originating from this source is generally low, though in some circumstances ammonia removal from decant solution is required as a direct result of ANFO usage.

2. Ammonia is one of the breakdown products of thiocyanate and cyanide, and forms through the hydrolysis of cyanate in a tailings storage facility. If cyanate is present in decant solution at an elevated concentration, then often there will be a correspondingly elevated concentration of ammonia.

Through a combination of these two sources, ammonia removal from decant solution is occasionally required at mining operations, particularly if treated solution must be discharged into the environment. Primarily the concern is with toxicity to aquatic organisms since ammonia is generally not present in decant solution at concentrations that would be toxic to wildlife or waterfowl. If natural attenuation in a tailings storage facility is not sufficient to limit the concentration of ammonia, then implementation of a decant solution treatment system may be required for water to be discharged to the environment. Treatment options considered typically include biological, chlorination and ion exchange processes, though in some cases air stripping may be considered.

Ammonia is readily oxidized through biological nitrification by an aerobic biological treatment process. This process is practiced at many municipal wastewater treatment plants throughout the world. The reaction product from this process is first nitrite and then nitrate, which is less toxic than ammonia but may also require removal depending upon its concentration. It is not uncommon for biological treatment plants to reduce ammonia to concentrations below 1 mg/L, however applications are generally limited to situations where the solution flow and chemistry do not fluctuate rapidly. Biological conversion of ammonia to nitrate can be conducted at relatively low temperatures, but reactions rates are higher at temperatures above about $10°C$.

Ammonia removal through the breakpoint chlorination process converts ammonia directly into nitrogen gas, thereby avoiding nitrate formation as would be encountered with other ammonia destruction processes. The chlorination process is efficient at removing ammonia to low levels and can also be configured to affect the simultaneous removal of cyanide, cyanate and thiocyanate as well as ammonia. A disadvantage with chlorination however, is that as ammonia concentrations increase, larger quantities of chlorine may be required depending upon the solution flow rate. In addition, chlorine added to solution will ultimately convert to chloride and increase the dissolved solids concentration in treated solution. If the concentration of either chloride or total dissolved solids is of concern in the treated water, then alternatives to chlorination should be considered. Chlorine is also highly toxic to aquatic organisms and a dechlorination process must follow any chlorination process used to treat water for discharge to the environment.

In some cases, ion exchange is used to reduce ammonia concentrations without causing a significant increase in the dissolved solids concentration. Ion exchange is also not as prone to process upsets as a result of flow and chemistry fluctuations in comparison to biological treatment processes. Consideration of ion exchange is appropriate when concentrations of interfering species such as sodium, calcium and magnesium are relatively low and when the solution pH is less than about 9.0. Ion exchange resins are only suitable for solutions and not slurries.

Under these conditions, ion exchange resins can be selective towards ammonia removal and the process may be economical for full-scale implementation. A significant disadvantage of ion exchange is that resins must be periodically regenerated using concentrated solutions of sodium chloride or sulfuric acid. These solutions along with all ammonia removed from solution will be present in the waste regenerant solution, and disposal of this waste solution is often difficult and expensive. In addition, ion exchange resins can become fouled due to the presence of certain dissolved metals in solution, and resin fouling can lead to high costs for purchasing new resin and for disposing fouled resin. Because of these disadvantages, ion exchange has only been practiced to a limited extent at mining operations.

Air stripping of ammonia from solution at a pH above about 11.0 is effective at reducing its concentration and in select applications this process may be economical. For example, if the solution to be treated has a pH near 11.0, then stripping of ammonia can be conducted with little or no initial pH adjustment. The disadvantage arises when the initial pH is below 11.0 and must be adjusted to the alkaline region using lime or sodium hydroxide. This adds to the cost of the process and increases the concentration of dissolved solids in the treated solution. In addition, if the pH of stripped solution must be lowered to less than 9.0 before being discharged into the environment, then sulfuric acid addition may be required. This also adds both to the cost of the process and to the concentration of dissolved solids in treated solution. Scale formation in process equipment may also be problematic due to carbon dioxide absorption from atmospheric air that will occur at elevated solution pH values. Because of these drawbacks, ammonia stripping is limited to applications when other treatment approaches would not be feasible or economical.

Nitrate Treatment
Nitrate is a relatively non-toxic compound at the concentrations typically observed in decant solutions and usually is not of concern relative to wildlife, waterfowl or aquatic organism toxicity. The primary concern with nitrate generally is related to drinking waters where elevated nitrite and nitrate concentrations can be harmful to humans, particularly young children and infants. In addition, nitrate is a biological nutrient and in some cases can lead to accelerated algae growth in waters, thereby consuming dissolved oxygen and impairing the ability of fish to survive. Nitrate is a relatively stable compound in surface waters and because of this, its removal from waters discharged to the environment is often required. There are relatively few treatment technologies that can be implemented on a full-scale basis to reliably lower nitrate levels, though the few that are available are effective and economic in many cases.

The most widely applied nitrate treatment technology is biological denitrification which proceeds under anoxic conditions. In this process, nitrate is converted to nitrogen gas, which is then vented to the atmosphere. Like most biological process, denitrification is best suited for situations where the solution flow and chemistry do not fluctuate rapidly and where the solution temperature is above about 10°C to 15°C. The process does not significantly increase the concentration of dissolved solids, but does require the addition of a supplemental food source such as methanol or molasses. A key advantage of biological denitrification is that it can be coupled with an aerobic biological process to affect the removal of cyanide, cyanate, thiocyanate, ammonia and nitrate. Under many circumstances, biological treatment systems are inexpensive to construct, operate and maintain and will provide high-quality effluent.

Ion exchange can also be used to remove nitrate to low levels, but as described for ammonia removal, the disadvantages of waste brine disposal and resin fouling have limited its application in the mining industry. Under conditions where the concentration of interfering compounds such as chloride and sulfate are relatively low, ion exchange may be an economical approach.

SUMMARY
As indicated in the previous discussion, there are several treatment processes that have been successfully used worldwide for cyanide removal at mining operations. The key to successful implementation of these processes involves consideration of the following:

- Site water and cyanide balances under both average and extreme climate conditions
- The range of cyanide treatment processes available and their ability to be used individually or in combination to achieve treatment objectives
- Proper testing, design, construction, maintenance and monitoring of both water management and cyanide management facilities

By carefully considering these aspects of water and cyanide management before, during and after mine operation, operators can reduce the potential for environmental impacts associated with the use of cyanide.

Another aspect of cyanide treatment to be considered is the potential environmental impact of the cyanide related compounds cyanate, thiocyanate, ammonia and nitrate. These compounds may be present in mining solutions to varying extents and may require treatment if water is to be discharged. Each of these cyanide related compounds is affected differently in the treatment processes discussed and this should be considered when evaluating cyanide treatment alternatives for a given site.

Table 10 provides a simplified summary of the general applications of various treatment technologies for the removal of iron cyanide and WAD cyanide. This table represents a very simplified summary, but can be used as a conceptual screening tool when evaluation cyanide treatment processes.

Table 10 Preliminary Guide to Selecting Cyanide Treatment Processes

Treatment Process	Iron Cyanide Removal	WAD Cyanide Removal	Slurry Application	Solution Application
SO$_2$/Air	✓	✓	✓	✓
Hydrogen Peroxide	✓	✓		✓
Caro's Acid		✓	✓	
Alkaline Chlorination	✓	✓		✓
Iron Precipitation	✓	✓	✓	✓
Activated Carbon	✓	✓		✓
Biological	✓	✓		✓
Cyanide Recovery		✓	✓	✓
Natural Attenuation	✓	✓	✓	✓

REFERENCES

Adams, M.D. 1992. The Removal of Cyanide from Aqueous Solution by the Use of Ferrous Sulphate. *Journal of the South African Institute Mining & Metallurgy*. 92:1. 17-25.

Botz, M.M. and T.I. Mudder. 1997. Mine Water Treatment with Activated Carbon. *Proceedings Randol Gold Forum*. 207-210.

Botz, M.M. and T.I. Mudder. 2000. Modeling of Natural Cyanide Attenuation in Tailings Impoundments. *Minerals and Metallurgical Processing*. 17:4. 228-233.

Botz, M.M. and T.I. Mudder. 2001a. An Overview of Cyanide Treatment and Recovery Methods. In *The Cyanide Monograph*, Ed. T.I. Mudder and M.M. Botz. London: Mining Journal Books Limited.

Botz, M.M. and T.I. Mudder. 2001b. Cyanide Recovery Applications for CCD Circuits. In *The Cyanide Monograph*, Ed. T.I. Mudder and M.M. Botz. London: Mining Journal Books Limited.

Botz, M.M., W. Phillips, T. Polglase and R. Jenny. 2001. Processes for the Regeneration of Cyanide from Thiocyanate. *Minerals and Metallurgical Processing*. 18:3. 126-132.

Dzombak, D.A., C.L. Dobbs, C.J. Culleiton, J.R. Smith and D. Krause. 1996. Removal of Cyanide from Spent Potlining Leachate by Iron Cyanide Precipitation. *Proceedings WEFTEC 69th Annual Conference & Exposition*.

Given, B., B. Dixon, G. Douglas, R. Mihoc and T. Mudder. 2001. Combined Aerobic and Anaerobic Biological Treatment of Tailings Solution at the Nickel Plate Mine. In *The Cyanide Monograph*, Ed. T.I. Mudder and M.M. Botz. London: Mining Journal Books Limited.

Goldstone, A. and T.I. Mudder. 2001. Cyanisorb Cyanide Recovery Process Design, Commissioning and Early Performance. In *The Cyanide Monograph*, Ed. T.I. Mudder and M.M. Botz. London: Mining Journal Books Limited.

Ingles, J. and J.S. Scott. 1987. *State-of-the-Art of Processes for the Treatment of Gold Mill Effluents*. Environment Canada.

Minerals Council of Australia. 1996. *Tailings Storage Facilities at Australian Gold Mines*.

Moura, W. 2001. Private Communication. AngloGold South America.

Mudder, T.I. and M.M. Botz. 2001a. A Global Perspective of Cyanide. In *The Cyanide Monograph*, Ed. T.I. Mudder and M.M. Botz. London: Mining Journal Books Limited.

Mudder, T.I. and M.M. Botz. 2001b. An Overview of Water Treatment Methods for Thiocyanate Removal. In *The Cyanide Monograph*, Ed. T.I. Mudder and M.M. Botz. London: Mining Journal Books Limited.

Mudder, T.I., F. Fox, J. Whitlock, T. Fero, G. Smith, R. Waterland and J. Vietl. 2001a. Biological Treatment of Cyanidation Wastewaters: Design, Startup, and Operation of a Full Scale Facility. In *The Cyanide Monograph*, Ed. T.I. Mudder and M.M. Botz. London: Mining Journal Books Limited.

Mudder, T.I., M.M. Botz and A. Smith. 2001b. *Chemistry and Treatment of Cyanidation Wastes*, 2nd Edition, London: Mining Journal Books Limited.

Mudder, T.I., S. Miller, A. Cox, D. McWharter and L. Russell. 2001c. The Biopass System: Phase I Laboratory Evaluation. In *The Cyanide Monograph*, Ed. T.I. Mudder and M.M. Botz. London: Mining Journal Books Limited.

Norcross, R. 1996. New Developments in Caro's Acid Technology for Cyanide Destruction. *Proceedings of Randol Gold Forum*. 175-177.

Omofoma, M.A. and A.P. Hampton. 1992. Cyanide Recovery in a CCD Merrill-Crowe Circuit: Pilot Testwork of a Cyanisorb Process at the NERCO DeLamar Silver Mine. *Proceedings Randol Gold Forum*. 359-365.

Schmidt, J.W., L. Simovic and E.E. Shannon. 1981. Development Studies for Suitable Technologies for the Removal of Cyanide and Heavy Metals from Gold Milling Effluent. *Proceedings 36th Industrial Waste Conference, Purdue University*. 831-849.

Simovic, L., W. Snodgrass, K. Murphy and J. Schmidt. 1985. Development of a Model to Describe the Natural Attenuation of Cyanide in Gold Mill Effluents. In *Cyanide and the Environment, Vol. II*. Ed. D. Van Zyl. 413-432.

Strategies for Minimization and Management of Acid Rock Drainage and Other Mining-Influenced Waters

Ron L. Schmiermund[1]

ABSTRACT

Long-term treatment of non-process, mining-influenced waters (MIWs) is a significant economic, legal, and public relations liability. Problematic waters may not resemble classic ARD. Onerous treatment requirements can often be traced to shortcomings in characterization and/or handling of waste rock. A holistic and non-prescriptive approach to waste management and MIW is recommended, including consideration of genetic and weathering aspects of subject ore deposits, flexible mine plans, testing programs adapted to the deposit and its environment, and recognition that future water treatment costs must be justified by present-day mining practices. This chapter identifies aspects of the approach and provides a framework for decision-making.

INTRODUCTION AND SCOPE

The adverse effect of mining on natural waters has been documented for nearly 2000 years and was almost certainly widely recognized long before that. The same causes and, unfortunately, many of the same consequences remain today to greater or lesser extents, depending on the history of a given mining district and its location in the world. In fact, water-related problems are frequently cited as the most obvious environmental issues associated with mining (Evans, 1996). These problematic waters are collectively referred to here as MIWs (Schmiermund and Drozd, 1997).

A discussion of this topic may seem out of place in this volume, but it can be argued that contamination of natural water by mining is rapidly becoming the most significant technical challenge to mining and milling. Without an environmentally acceptable mining and processing plan, the most sophisticated mechanical and chemical techniques for recovering resources are wasted because the mine will not be permitted in a growing percentage of the world's countries. Without an economically and environmentally viable method for closing a mine, all operational profits may well be consumed in very long-term post-closure maintenance.

Many readers may automatically equate acid-rock drainage (ARD), typically with high sulfate concentrations and often with elevated metals, with MIWs. Although ARD may be the most common and most visible MIW, the topic is considerably broader than ARD. MIWs include non-acidic, even hyper-alkaline waters, with or without high sulfate or common toxic heavy metals. Nitrate, cyanide, chloride, fluoride, or suspended sediments may constitute the offending constituents. The complexity of ARD alone, let alone the full spectrum of MIWs, is extreme and has been extensively explored by numerous authors. It is beyond the scope of this chapter to address the technical aspects of ARD or other MIWs except where they have direct bearing on the discussion, but the reader will be directed to key publications.

It is the intent of this chapter to provide a framework for anticipating water-related problems as early as during exploration, throughout development, and during mine planning. Various mine components and their potential for contaminating water are discussed along with strategies for timely intervention in the processes that affect water. Finally, geologic, laboratory and field

---

[1] Knight Piésold and Co., Denver, Colorado.

methods for predicting impacts of mining-related materials on natural waters are discussed. Throughout, prevention is the preferred management approach.

A similar scope of subject matter has been addressed in much more detail, but still in an abbreviated form, by Morin and Hutt (1997), and the reader is directed there for practical theory and case studies. Indeed, a significant problem with these subjects is their breadth and variety and the need to translate knowledge at the microscopic scale to practice at the huge scale of modern commercial mining.

Summary of Relevant Geochemical Processes

A myriad of chemical reactions impact MIWs. The following list attempts to include the generally most influential ones.

- Inorganic oxidation of simple iron sulfide minerals (pyrite, marcasite, pyrrhotite) by oxygen. Reaction products include protons, sulfate, and ferrous iron.
- Bacterially-catalyzed oxidation of simple iron sulfide minerals by ferric iron. Reaction products as above.
- Further oxidation of pyrite oxidation products, specifically ferrous iron to ferric oxyhydroxide solids (e.g., goethite, limonite, "yellow-boy").
- Inorganic or organically-catalyzed oxidation of other metal or semi-metal sulfide minerals. Reaction products include sulfate, ± oxidized metal ions, ± metal or semi-metal oxyanions (e.g., molybdate, arsenate, selenate), and ± protons.
- Reaction of sulfide oxidation products, especially protons, with carbonate minerals and select silicates to neutralize acidity.
- Concentration of sulfide-mineral oxidation products by evaporation with subsequent precipitation of secondary sulfate minerals.
- Dissolution of secondary sulfate minerals derived from sulfide-mineral oxidation.
- Dissolution of primary evaporite minerals or other relatively high-solubility minerals associated with an ore deposit (e.g., gypsum, sylvite, trona).
- Attenuation processes that limit mobility and/or toxicity of dissolved constituents, especially metals, including sorption, precipitation, co-precipitation, plant uptake, complexation with volatilization, photooxidation, and bacterial metabolism.
- Mechanisms that enhance mobility, including complexation, sorption to suspended sediments and colloidal-facilitated transport.
- Incorporation of components of industrial process into natural waters by dissolution or mixing, including cyanides, nitrates, organic flotation reagents, lixiviants, etc.

For further information about the mechanisms listed above, the reader is directed to several chapters appearing in Plumlee and Logsdon (1998) and Filipek and Plumlee (1998) but especially Nordstrom and Alpers (1999) and Alpers and Nordstrom (1999). In addition, Perkins, et al. (1995) and other reports published by Mine Environment Neutral Drainage, or MEND, provide extensive discussions and references. See Tremblay and Hogan (2001) for an overview and summary of the MEND program. U.S. Geological Survey websites and InfoMine (www.infomine.com) provide additional information and portals to numerous other information sources.

MINING SITUATIONS THAT MAY ADVERSELY AFFECT WATER AND POTENTIAL INTERVENTION STRATEGIES

The following situations or operational components may be found in modern metal mines and with a few limitations can be applied to coal mining and other non-metal mines. With the exception of direct discharges or accidental leakage from process facilities such as ponds, tanks, or pipelines, the sources of most natural water contamination can be traced to these situations or components.

In Situ Exposed Rock in Surface and Underground Mines

Most ore minerals and many associated gangue minerals are inherently chemically unstable at the earth's surface. Consequently, when deposits are exposed during mining either in surface or underground mines, the contained minerals begin to chemically re-equilibrate in response to the new environment. This is especially relevant to many hypogene metal sulfide deposits, and sulfidic over and underburden associated with coal, evaporates, and chemical precipitate deposits. The products of such re-equilibration may become contaminants of surface waters, including pit lakes, and groundwaters and areas of exposure in mines are sources of those contaminants. Chief among them are the decomposition products of ore and gangue sulfide minerals (acidity, sulfate, and metals). In addition, artifacts of mining activities may be present in waters originating from areas of exposed rock, including nitrogen compounds from blasting agents, salts used for dust suppression and deicing, and general suspended sediments.

Important variations to the foregoing are found in deposits that have either already re-equilibrated, or partially re-equilibrated, with the surface environment (i.e., they have already undergone weathering) or consist of minerals that react very slowly under surface conditions. These include oxidized portions of originally sulfidic rock masses, supergene-enrichment zones of ore deposits (typically originally sulfidic), primary non-sulfidic ore deposits (e.g., iron, titanium, chromium, aluminum, and tin oxides), silicate mineral deposits, and most placers. The single most important commonality among deposits in this group is the relative paucity of sulfide minerals and thus the decreased capacity to produce ARD and associated phenomena. However, deposits in this group should not automatically be considered inert, and care should be taken in evaluating data from characterization tests designed for fresh rock (see Prediction Techniques).

Open-pit high walls are obvious locations of exposure and weathering of ore and associated mineralized waste rock. Underground openings of all descriptions, but especially rubbleized or block-caved areas, are potential sites for intense weathering activity due to the focusing of air and water flows and possibly higher temperatures and relative humidities. At various times, all portions of a an ore body being actively mined will be exposed to weathering, but mining usually proceeds rapidly enough so that the main concern lies with material that will be exposed in the final pit wall or underground openings and thus be subject to weathering for long post-mining periods. An especially problematic and symbiotic situation results when underground and surface mines come in contact via shaft and drift penetrations of pit high walls.

Block models employed in ore reserve estimation are convenient methods for characterizing the final pit walls, but provisions must be made early for generating and including the necessary data in the model. Unfortunately, the kind of data required for environmental predictions is not commonly collected during mine planning because it usually relates to non-ore minerals (e.g., pyrite, pyrrhotite, arsenopyrite, etc.) or non-ore elements. It is not uncommon that the database for a very well characterized deposit is virtually devoid of any information useful for estimation future environmental impacts.

Examples of information that would be useful for environmental purposes and might be collected from exploration/developmental drilling programs are given below. An early evaluation of the deposit from an environmental standpoint can better define the important variables to be collected.

- Distribution and abundance of gangue sulfide minerals, especially pyrite, marcasite, or pyrrhotite.
- Pyrite morphology (e.g., disseminated or in veinlets, coarsely crystalline or sooty).
- Total sulfur assays and a correlation to sulfide sulfur and visual estimates of pyrite in cores.
- Assays of important potential contaminants (e.g., arsenic, selenium, thallium, non-commercial copper, zinc, or lead).
- Distribution and abundance of reactive non-sulfide gangue minerals, especially carbonate minerals with distinctions made between calcite, dolomite, and siderite.
- Distribution and identity of secondary sulfate and oxides, aided by use of infrared spectrophotometer, e.g., alunite, jarosite, and scorodite).
- Distribution of rock types that may be relevant to chemical and physical weathering (especially carbonates and reactive silicates).
- Distribution of hydrothermal and supergene alteration zones with clear documentation of the mineralogic characteristic minerals of each zone.

- Information on primary and secondary porosity, fracture density, and fracture characteristics.
- Any information on water depths encountered during drilling, artesian flow rates from drill holes.
- Chemical analyses of water flowing from drill holes.
- Retain complete cores or cuttings (including overburden) from representative holes, especially near peripheral areas of the pit (i.e., near the ultimate pit limit).
- Retain assay pulps for later environmental-related analyses.

Efforts to quantify the processes associated with weathering of pit high walls have resulted in complex conceptual and numerical models that consider reaction rates, surface areas, fracture densities, diffusion of gases and liquids, and moisture movement. Alpers and Nordstrom (1999) review the complex geochemical aspects and existing models that pertain to reaction in this environment. Physical aspects of the wall-rock environment are complicated by the fact that pit walls are extremely irregular, and the presence of broken rock and the effects of blasting are difficult to quantify (Morin and Hutt, 1997). Tunnels and shafts penetrating the pit wall may dominate the recharge of groundwater to the pit. Because of the inherent complexity, any model of pit wall or underground face weathering behavior may require considerable calibration prior to accepting the results.

One such model is MINEWALL 2.0 developed under the MEND program to address a wide range of situations. It is available in extensive documentation and demonstrated applications (MEND, 1995a; MEND, 1995b; MEND, 1995c). Implementation of MINEWALL requires empirical inputs that characterize in situ sulfide oxidation rates and information on the production of solutes from exposed rock over time. To supplement in-field measurements that can estimate these variables or to substitute for them if provisions were not made for their collection, laboratory data (especially kinetic test data [see paragraph entitled "'Kinetic' Laboratory Methods for Anticipating Weathering Behavior"]) have been used in various ways. Extrapolation of small-scale, accelerated laboratory experiments will always carry risks of yielding misleading information. Morin and Hutt (1997 and 1999) argue the value of using observational data as the source of rate and yield data for models and provide examples of data generated from field exposure experiments (see paragraph entitled "Site-specific Field Methods"). However, considerable foresight and time are required to obtain such data. Commissioning of field experiments should occur as soon as bulk rock becomes available, but valuable information can also be gained from undisturbed natural outcrops.

More theoretical approaches to obtaining oxidation rates and solute production rates have been developed based on the Davis-Ritchie (D-R) equations (Davis and Ritchie, 1986a and 1986b), which assume that oxygen diffusion is the rate-limiting step in sulfide oxidation. One of the models/codes incorporating the D-R conceptual equations is PYROX (Wunderly, et al., 1996). Other proprietary codes incorporating the equations exist, but it should be recalled that the D-R equations are mainly about oxygen diffusion. They were originally applied to heap leach piles and subsequently to dumps and tailings where oxygen diffusion through pore spaces can be shown to be rate-limiting. Care must be taken to correctly address inherent differences in diffusion characteristics at exposed rock faces. In addition, the D-R equations do not always consider water films on reactive grains that may be significant in some dumps but perhaps not on exposed rock faces in arid climates. Fennemore, et al. (1998) argues that the D-R approach overestimates oxidation progress in arid climates.

Intervention Strategies
For surface mines:

- Collect data on distributions of potentially problematic minerals and elements in much the same way that ore grades data are collected.
- Evaluate the costs of removing as much reactive rock (including non-ore) as possible; compare handling costs to incremental increases in water treatment costs.

- Identify problematic rock masses and control blasting in those areas; consider installation of long-hole drains and grouting during mining while equipment and access are available.
- Consider the real costs of long-term water treatment relative to benefits of extending a mining operation into and exposing problem material.
- Remove rubble from benches and design for rapid draining and reduced infiltration, especially in areas of more reactive rock.

For underground mines:

- Consider mine designs that allow for post-mining flooding and oxygen exclusion.
- Consider mine designs that dewater reactive rock passively.
- Modify surface recharge areas where possible to limit flows into mine.
- Position shafts to maximize post-mining flooding.
- Install critical permanent bulkheads while access and equipment are available.
- Evaluate backstoping with beneficial material capable of reversing undesirable chemistry.

Waste Rock Dumps and Ore Stockpiles

Waste rock dumps in conjunction with ore stockpiles present the same issues for contamination of waters as do pit walls or tunnels because they can contain the full variety of rock types present in a given ore deposit. However, the reactivity of the rock dumps can be higher than the corresponding in situ lithologies because of increased surface area and the potential for enhanced exposure to air and water. In addition, waste rock dumps may be relegated to environmentally unfortunate locations, including natural drainages, for logistical reasons. Dumps generally do not receive the level of engineering study necessary to control their internal structure, and thus water and air permeability, because dumps generate no revenue. However, failure to properly design and operate waste rock dumps can consume the revenues generated by other mining units in the form of long-term treatment of contaminated seepage.

Waste rock dumps should be thought of as very large and exceedingly complex (bio)chemical reactors beyond the scope of this discussion. The interiors are difficult to investigate and are typically poorly known both physically (Smith, et al., 1995) and chemically (Ritchie, 1994). In addition to the waste rock material itself, reactants include infiltrating water from precipitation, groundwater seepage or runoff, and atmospheric oxygen. Although the boundary conditions for these external reactants can be easily quantified (e.g., precipitation amounts and atmospheric partial pressure of oxygen), their fluxes into the dump and distributions within the dump are known to be complex functions of many variables. Ongoing products of the dump reactors (i.e., seepage) represent their risk to natural waters and are easily quantified but difficult to predict over the life of the dump. Although problematic effluents from dumps are most commonly acidic, sulfate- and metal-enriched (i.e., ARD), they can be representative of every type of MIW. MEND (1996) describes individual waste rock dumps and discusses methods of predicting their environmental behavior.

From the operator's perspective, it is most useful to understand the aspects of waste rock dumps that can be controlled to minimize production of contaminated seepage. Fortunately, preventative measures are reasonably well understood and technically feasible. Probably the greatest benefit can be realized by properly selecting and preparing the dump site, applying construction techniques that segregate and encapsulate the most reactive wastes in less reactive material, and creating an internal structure that limits circulation of reactants. The greatest obstacle appears to be recognition of the necessity for investing capital and operating funds before and during construction to minimize closure costs and to prevent perpetual water treatment costs.

Operators should be keenly aware that prevention of MIW drainage problems, especially ARD generation, is the only demonstrated viable long-term option. Once initiated and established, the reaction sequences that create ARD are virtually unstoppable. They may be retarded by oxygen deprivation or application of biocides, but reversal is highly unlikely. Elimination of infiltration would eventually stop seepage but is essentially impossible to achieve and/or maintain. Peak contaminant (un)loading from waste rock dumps may not occur until after mine closure, so early favorable indications should not be assumed to preclude later adverse behavior.

Intervention Strategies

- Characterize footprint of proposed waste rock dumps with respect to existing surface and subsurface water flows and potential post-dump flow paths (e.g., faults, karst features).
- Evaluate interception/diversion ditches and sub-dump culverts or drains to isolate waste rock from water as much as possible.
- Design outer surfaces to minimize infiltration. Different climates and precipitation characteristics will dictate different designs.
- Coordinate dump design with mine plan based on block model containing environmentally-relevant information. Objective should be to isolate acid-generating (or other problematic) waste from external reactants and to associate it with acid-consuming (or other beneficial) material. Careful planning is required to minimize excess handling. Block-model data can be enhanced through rapid analysis of blast hole cuttings for critical or indicative environmental parameters (e.g., total sulfur, leachable sulfate, whole-rock arsenic, etc.). Real-time GPS-based systems for directing haul trucks based on ore grade can be similarly used to direct waste rock dumping.
- Identify waste rock or other waste materials with best contaminant-attenuating or reaction-inhibiting characteristics for preferential placement in the dump. Waste rock with high ANP can be placed below or mixed with rock with high APP to neutralize acid that might be produced. Materials with capacity to sorb contaminant metals can be placed downgradient of metal-rich waste rock to act as a primary barrier to migration. Organic material such as wood chips can be mixed with sulfidic material to act as a sacrificial reductant or oxygen "getter."
- Possible application of biocides or sulfide mineral passivators to especially reactive sulfide material to inhibit reactions.
- Evaluate techniques for placing and otherwise modifying dump material to limit infiltration and vertical percolation of water and circulation of air. An effective internal permeability structure may be engineered through creation of fine-grained layers and/or capillary breaks. Selective dumping, truck traffic pattern, and ripping and dozing may create the desired effects.
- Limit air infiltration by eliminating or covering coarse-grained toe deposits.
- Create accurate as-built records of dump's internal construction in the event that drilling for sampling or injection of biocide is required.
- Consider constructing sampling access ports to demonstrate performance and/or provide early warning of problems.

Tailings and Tailings Storage Facilities (TSFs)

Modern mines manage tailings through storage in engineered facilities or controlled placement in natural structures with favorable characteristics. Many of mining's worst and sustained environmental legacies are sourced in uncontrolled tailings disposal into rivers and coastal waters. Engineered TSFs are most commonly designed to drain to allow compaction and physical stabilization of the tailings and thus will be sources of waste water that must be managed long-term. Other facilities simply leak. Waters exiting TSFs evolve over time from pure process water with a wide range of characteristics, depending of the beneficiation process, to rain water that has equilibrated with the tailings. Stability of the impoundment structure itself has also proven to be of great concern in both metal- and coal-mining operations. Severe high-profile failures have resulted in very significant impacts to natural waters and extremely high costs for remediation.

Tailings are obviously derived from the same materials that require consideration in the cases of in situ exposures and waste rock dumps, but they differ in many important ways that can influence their chemical stability and thus the risk that they pose to natural waters. Tailings represent concentrations (relative to the overall deposit) of gangue minerals, in particular non-ore sulfide minerals that may have potential for ARD production and may be hosts of toxic metals. Tailings have increased and frequently high surface areas resulting from size separation and/or comminution and as a result may experience increased reaction rates.

Tailings are typically placed in TSFs as a slurry and retain high moisture contents for long periods, thus potentially facilitating the initiation of water-rock reactions and a cycle of water contamination. In addition, the liquid fraction of the tailings slurry may contain artifacts of the beneficiation process such as cyanide, acids, caustics, dissolved metals, and organic compounds. Depending on the placement of the TSF relative to the surrounding topography and water table, tailings may be a source of recharge to groundwater or a point of discharge. Robertson (1994) discusses this and other aspects of water movement associated with TSFs and notes that specifics are highly site-dependent.

Predicting the potential impacts of TSFs on surrounding surface and groundwaters must consider the processes that render solutes mobile within the tailings mass as well as how those solutes might be transferred to the environment. Solutes may exist in the tailings slurry liquid fraction when it arrives at the TSF or arise from further in situ reactions of the tailings solids with the process fluids or new reactions with the atmosphere and/or with the pore fluids as they evolve over time. As is the case of waste rock dumps, the most important of these reactions is likely to be pyrite oxidation. Elberling, et al. (1994) discusses the fundamentals of a model for pyrite oxidation in tailings that are incorporated into computer codes such as RATAP (Scharer, et al., 1994) and WATAIL (Scharer, et al., 1993) and integrated approaches as described by Nicholson, et al. (2000). Wunderly, et al. (1996) discusses the PYROX model that was incorporated with PLUME2D into MINTOX (Molson, et al., 1997), another model for TSF evolution (Alpers and Nordstrom, 1999) for a more complete history of the evolution of MINTOX.

From the foregoing discussion, it is obvious that tailings and TSFs require careful design and maintenance to avoid realization of the various potentials for contamination of natural waters. However, there are several characteristics of tailings masses that can be taken advantage of and perhaps optimized to limit risks. The fine-grained natures of tailings and the associated retention of moisture serve to impede the diffusion of oxygen from the atmosphere, thus dramatically limiting the rate sulfide mineral oxidation. Further, when the sulfide minerals are abundant throughout the tailings mass, oxygen is rapidly consumed very near the surface. In dry climates, oxidation of sulfides near the surface of the tailings mass may result in precipitation of secondary sulfate minerals that further limit oxygen infiltration, tend to stabilize the tailings by forming a crust, and may be self-repairing after rainfall events.

Intervention strategies include:

- Optimize siting of TSFs to contain seepage and constrain/focus its exit from the facility.
- Design tailings placement to maximize efficiency of drainage.
- Selectively place problem materials
- Separate facility for waste streams containing problem materials (as opposed to co-mingling prior to disposal).
- Physically stabilize tailings surface to prevent erosion and exposure of fresh surfaces.
- Choose material for final layer that optimizes weathering retardation.

Heap, Valley-Fill, and Dump Leach Facilities

Both heap and dump leaching can pose risks to ground and surface water for the following reasons:

1. Excursions of pregnant solutions occur even in cases of well-constructed pads with adequately designed liners. Dump leaching, especially when initiated as an afterthought, may pose higher risks because liners are not present and formal pregnant solution collection systems were never installed.
2. Handling of large volumes of pregnant and barren solutions, especially where large retention ponds are involved, incurs risk of leakage.
3. Rinsing, as may be required and appropriate for closure, generates volumes of solution that must be disposed of and/or detoxified. Treatment of rinsates may result in sludges requiring safe disposal to protect natural waters.
4. Post-closure infiltration of precipitation can result in seepage from spent leach piles.

5. Leaching of sulfidic material is chemically equivalent to creating ARD and, as such, can be a very difficult process to stop.

Intervention Strategies

Engineered pads offer distinct environmental advantages; consequently, intervention scenarios can be more proactive. Because a pad that functions as designed is likely to have the lowest operating and the best environmental performance, many of the intervention scenarios below will have multiple benefits.

- Fully characterize the material's leaching characteristics, including rinse-down, prior to initiating leaching operations. Use sufficiently large apparatus to assure meaningful results.
- Anticipate disposal of rinse-down solutions.
- Engineer leak-detection systems that provide the earliest advice of a leak and information on the location of the leak.
- Design segmented pads that would allow for partial shutdown in the event of a localized leak.
- Require and preserve accurate as-built drawings of leach pads and pre-dump topography.
- Consider chemical reactions that might occur within the pad and block drainage systems, resulting in unacceptable head pressures on the liner and leakage. This may be especially important where a separate oxidant phase, in the form of air, is injected at the base of the pad and moves counter-current to solution flow paths.
- Install sampling access to during pad loading. This might consist of a vertical array of lysimeters to allow sampling of pore waters and gases during and operation and rinsing.
- Design sampling access into the leachate collection systems upstream of mixing boxes and manifolds to allow sampling of individual cells

In Situ Leaching/Solution Mining

Examples of the application of in situ mining methods include:

- Surface application of acidic solutions to disseminated sulfide and oxide deposits with solution recovery via pumping from abandoned underground workings
- Subsurface injection of acidic solutions in copper sulfide deposits with solution recovery via pumping wells
- Injection of alkaline solutions into sandstone-hosted uranium oxide deposits with recovery via pumping wells
- In situ solubilization of trona and other evaporites left in pillars by injection of undersaturated process solutions into old underground workings and pumping from resultant mine pools

The introduction into an aquifer of any solution that differs chemically from native groundwater risks contaminating the receiving groundwater. However, the risks may be mitigated by the nature of the aquifer, unusable ambient groundwater quality, or lack of reasonable access to or demand for the water. Groundwater contamination may occur in aquifers not specifically targeted for mining via leakage from injection or recovery wells or migration of solutions along structures. Solution mining may also pose risks to surface water because of the need to manage pregnant solutions and raffinates on the surface. Solutions unfit for reconditioning and reinjection must be disposed of, and land-application and infiltration galleries are often among the preferred options. Careful consideration must be given to the fate of all dissolved constituents present in solutions intended for land application. Depending on their mobility and attenuation, they may be taken up by plants, emerge as constituents in seep and spring waters, or become incorporated into shallow groundwaters.

General Land Disturbance and Watershed Impacts
Disturbance of the land surface is inherent in mining, and the associated water quality and water supply issues can be extremely important to stakeholders over large areas. Generally, the water quality issues arising from simple road building and other construction are limited to increased suspended sediments and turbidity. More serious impacts may be realized by diversion of runoff from one watershed to another or by interrupting the path of canals and ditches that supply water to downgradient areas.

Intervention Strategies
- Establish a complete inventory of existing water supplies, sinks, and conveyances in the area of a planned disturbance.
- Evaluate impacts to downstream users and apply appropriate industry-standard methods for sediment control.

PREDICTION TECHNIQUES
Recognition that water contamination issues associated with mining, especially ARD, required management and should be prevented if possible became commonplace about the middle of the 20th century. It was probably inevitable that laboratory methods would be sought to quantify the future behavior of geologic materials in response to weathering and to predict the effects on natural waters. Industry and regulators have come to rely upon those methods that are collectively considered the tools of "waste characterization" studies. Unfortunately, problems abound, and there are distinct conceptual and operational biases toward one aspect of MIWs – prediction of the generation of low-pH waters resulting from the oxidation of pyrite. It is important to realize those biases and thus the limitations inherent in "standard" waste characterization. This chapter only addresses the major categories of laboratory tests. The references provide a much broader treatment.

White III, et al. (1999) traces several methods in common use today to 1974 and the eastern U.S. bituminous coal region. The fact that those methods were developed for coal and are now routinely used to evaluate all mining waste materials, often with little or no recognition of the underlying assumptions, is indicative of shortcomings in current industry-standard practice. This seems to be the case despite extensive and credible scientific work exposing those shortcomings, and wide distribution of the findings.

Detailed descriptions of testing methods are outside the scope of this chapter, but references are provided. The material addressed here is sufficiently complex that the advice of an experienced practitioner is recommended before committing to a program of waste characterization or interpreting the results of a previous one. The following paragraph entitled "Using Economic Geology to Anticipate Water-related Problems" discusses a non-laboratory approach that may well provide the most reliable guidance but, unfortunately, would be considered unconventional and lacks the appearance of quantitative credibility offered by laboratory tests.

Using Economic Geology to Anticipate Water-related Problems
A concept possibly introduced by Kwong (1993), expanded upon by Plumlee, et al. (1994) and Plumlee, (1999), and applied to a wide variety of mineral deposits by du Bray (1995) is remarkable in its logic and usefulness. This method relies on the extensive economic geology and environmental knowledge of existing mineral deposits around the world and typical mining and beneficiation methods associated with those deposits. That knowledge is then used to predict the behavior of other analogous deposits and mines. Each class of mineral deposit has been associated with a "geoenvironmental model." The basic models invite customization that considers site-specific climates, geologic variations, and specifics of the beneficiation process. The data and observations that form the basis for these geoenvironmental models have clear advantages over laboratory data because they inherently include all variables, are full-scale, and accurately reflect the influence of time.

Geoenvironmental models of mineral deposits may be applied at all stages of development, ranging from regional conceptual exploration to expansion or closure planning for an active mine.

However, early application would seen to offer the greatest benefit if it applies to decisionmaking schemes as outlined by Lord, et al. (2001). Application of geoenvironmental models of ore deposits is essentially free and should prevent most environmental surprises.

Static Laboratory Methods for Predicting Acid Production and Neutralization

Considered the most fundamental of the waste characterization laboratory tests is a group of procedures known as the "static tests" or "acid-base accounting" (ABA) tests. In general, the procedures can be divided into two types – those that estimate acid generation or production potential (AGP or APP) and those that estimate acid neutralization potential (ANP). Results are typically expressed in terms of tonnes of $CaCO_3$ per 1,000 tonnes of rock and often assessed by calculating the difference or net acid neutralizing potential (ANP – APP = NANP or NNP) or the ratio (ANP/APP).

Paste pH

Although simple and very inexpensive paste pH can provide much useful, albeit qualitative, information, COASTECH Research Inc. (1991) provides details for the simple procedure. The resultant pH in the high solids/liquid reflects the influence of only the most rapidly reacting, typically most soluble, solids. Generally, hydrolysis associated with selected sulfate, carbonate, hydroxide, chloride, and nitrate salts will dominate such reactions. Redox reactions, especially heterogeneous ones, are expected to have no effect because they are slow relative to the duration of the test. Accordingly, fresh sulfide minerals have no direct effect, but reaction products from prior oxidation of sulfides may be responsible for low paste pHs and generally indicative of ARD potentials. Minerals in this group include secondary sulfate minerals as discussed by Nordstrom (1982). Where evaporative conditions are influential and result in abundant secondary sulfate minerals after sulfide mineral oxidation, the paste pH will predict the runoff pH from the next rainfall event. Paste pH values near and above neutrality clearly reflect the absence of acid sulfate minerals but may not accurately reflect the ultimate equilibrium pH because of relatively slower reaction rates for hydrolysis of carbonate minerals.

Estimating APP. Estimating APP appears simply but is really quite complex. It relies on measures of total sulfur in a material and the sulfur contained in one or more chemical extractions designed to liberate specific fractions of the total sulfur. These are referred to as sulfur forms or species, the most relevant of which is sulfide-sulfur. The objective is to quantify the sulfur contained in those sulfide minerals that will liberate protons when oxidized. According to standard procedures, sulfide-sulfur concentration is assumed to be entirely contained in pyrite and is then mathematically converted to APP according to the following simple formula:

$$APP\ (T\ CaCO_3/1000\ T\ rock) = wt\ \%\ S_{sulfide} \times 31.25$$

The multiplicative factor 31.25 is referred to below as "the factor." The following points pertain to determining and interpreting APP values:

- Measures of sulfur species are only estimates defined by the extraction procedure. Independent confirmation should be sought.
- Multiple and potentially significantly different "standard" methods for estimating sulfur species exist, including Sobek, et al. (1978) and ASTM (1988). Unfortunately, results are not assured to be comparable, and many laboratories have modified the "standard" methods making inter-laboratory comparisons questionable; not all laboratories are candid about such modifications.
- Both standard methods are specifically designed to discriminate between sulfur found in the common sulfide minerals in coal (pyrite or marcasite), organically-bound sulfur (probably only significant in coal), and those readily-soluble sulfate minerals found in coal. No provision is made for non-pyrite/marcasite sulfide minerals, and it is left to the consumer to request tests that might elucidate this issue.
- Not all sulfide minerals are subject to dissolution by the standard methods, so not all true sulfide-sulfur may report to sulfide-sulfur as defined by the methods.

- A wide variety of oxidation stoichiometries apply to sulfide minerals found in base and precious metal deposits, but the 31.25 factor described above is strictly based on:

$$FeS_2 + \frac{7}{2}O_2 + H_2O \rightarrow Fe^{2+} + 2SO_4^{2-} + 2H^+$$

$$CaCO_3 + H^+ \rightarrow Ca^{2+} + HCO_3^-$$

In fact, it can be shown that the factor ranges from negative values in the case of sulfide minerals that actually consume acid upon oxidation (e.g., chalcocite and pentlandite), zero for chalcopyrite, to nearly 100 for realgar. The prevailing stoichiometry may also vary if oxidants other that oxygen are introduced into the system. If ferric iron is introduced in quantities greater than what might be achieved by the oxidation of locally available pyrite, the applicable factor could exceed several hundred.
- The basis of the factor in stoichiometry implies the assumption that the reactions go to completion, thus the calculated APP represents a maximum capacity that may not be realized or may be realized only very slowly.

Estimating ANP. Considerably more attention has been given to ANP determinations relative to APP determinations and, in fact, ANP is even more subject to the details of the testing procedure. Unfortunately, numerous methods exist with subtle operational differences that can produce not-so-subtle differences in the results. Three fundamental approaches to ANP estimations exist:

- The most commonly used methods involve mixing a known mass of rock with a known quantity of acid, allowing them to react with or without boiling, followed by a back-titration with a standardized base to determine the amount of acid consumed. Results from this seemingly simple method are subject to variations sourced in the grain size of the specimen, the molarity of the acid, temperature and duration of the reaction period, redox conditions of the experiment, and the endpoint pH used for the back-titration. An especially sensitive variable involves the use of "the fizz test," a highly subjective method of choosing the appropriate molarity and volume of acid.

    White III, et al. (1999) describes five methods that follow this approach. COASTECH Research Inc. (1991) provides detailed methods for three. Lawrence and Wang (1997) discuss the sometimes dramatic influence of method details on the outcome of ANP tests. Lawrence and Wang (1996) explore in great detail issues related to the various ANP techniques.

    Methods in this group are variously known as Standard Sobek, Modified Sobek, the Lawrence Method, B.C. Research Method, Standard and Modified NP (pH 6). It is incumbent upon the consumer to obtain a complete description of the method as practiced by an individual laboratory and to not rely on the advertised name of the method. No one method is universal, and no method currently available is without merit in some application. In general, it is not possible to choose the best ANP method a priori. Rather, it is strongly recommended that any significant waste characterization program initially include a comparison of ANP methods and reconciliation of the results with the deposit mineralogy (see below).

- A second method tends to reinforce a popular oversimplification of ANP that presumes that simple carbonate minerals are the sole, or only important, basis for ANP. It employs measurements of total carbon and organic carbon to estimate total inorganic carbon by difference. Calculations are then made based on the assumption that all inorganic carbon is present as simple calcium and magnesium carbonate minerals that are available for acid neutralization. Although, this approach may be applicable and highly efficient in specific cases where simple carbonates are predominant, it is not recommended for general use. This method has been submitted for inclusion by ASTM. See White III, et al. (1999) for further discussion and references.

- A third approach involves direct, quantitative determination of mineral abundances using microscopy, X-ray diffraction in conjunction with X-ray fluorescence, mineral separations, or other applicable techniques. A theoretical acid consumption could then be assigned to each mineral. This method is probably only rarely used alone but should be performed on representatives of any unfamiliar group of materials for validation of assumptions about the identity of minerals responsible for acid neutralization.

"Kinetic" Laboratory Methods for Anticipating Weathering Behavior

The so-called "kinetic" procedures are intended to extend the information accumulated from the static tests. While static tests purport to measure the ultimate potential of a rock for acid generation, kinetic tests attempt to provide information about whether that potential will be realized. The format of most kinetic tests is to induce accelerated weathering by enhancing the conditions thought to increase weathering rates, specifically pyrite oxidation rates. No consideration is given to acid-neutralizing rates. Testing typically consists of alternately exposing crushed rock to flows of air with higher and lower relative humidity during a one-week period and flushing the material at the end of the week to remove soluble reaction products. The one-week cycle is then reinitiated. With few exceptions, the tests do not record and/or control air flow rates, relative humidity, or temperature. In addition, grain size and chamber configurations are variable between laboratories.

Test methods following the procedures outlined above are considered "conventional" and are conducted in various configurations of chambers but were originally conducted in shoebox-sized boxes (humidity cells). ASTM (1996) describes an analogous apparatus and method using a column and a far more rigorous procedure for controlling testing conditions. Other, less commonly used procedures have been devised but are in many ways more abstract relative to field conditions. These include a single-pass column extraction, the B.C Research Confirmation Test that forces bacterial (*Thiobacillus ferrooxidans*) sulfide mineral oxidation, Shake-Flask where all reactions take place in solution under agitation with or without bacteria, and Soxhlet Extractions that re-circulates extraction fluids through a small rock sample at elevated temperature. COASTECH Research Inc. (1991) provides detailed procedures for humidity cells and the latter four types of kinetic tests.

A conventional kinetic test can provide indications of the relative rates of acid-producing and acid-neutralizing reactions. However, it is a common misconception that kinetic tests provide information on absolute rates of reaction under field conditions. In general, they cannot. This can only be accomplished when extensive empirical correlations exist between true field-based measurements of rates and laboratory tests; in that case, why perform kinetic tests? Kinetic testing, as currently practiced, is far from being sufficiently sophisticated to predict absolute rates.

One benefit of kinetic testing is that liberation of constituents (other than protons, iron, and sulfur) resulting from sulfide mineral destruction and acidic attack of other minerals can be evaluated. All the procedures described above generate solutions that may be analyzed for their dissolved constituents and essentially represent a leaching experiment under conditions affected by sulfide mineral oxidation.

Kinetic tests, as originally proposed by Sobek, et al. (1978) took the form of "humidity cells." Their use was recommended as a follow-up to acid-base accounting when the results of those tests were "inconclusive," i.e., could not be interpreted as clearly predicting net acid-producing or acid-neutralizing behavior. This recommendation for using kinetic testing as a "tie-breaker" seems to have given rise to another misapplication of the procedures by implying that the tests would demonstrate whether or not the material would "go acid." Kinetic testing used in that mode will always be either indefensible or nearly worthless.

Results are indefensible if testing reveals no net acid production because it can always be argued that the testing time was insufficient. Confirmation of this problem is offered by the fact that the recommended duration of the original tests (for coal-related materials) was ten weeks. The accepted time has steadily increased through 20 and 40 weeks, and examples of rocks finally "going-acid" after several years of continuous testing have been published.

Results are nearly worthless if net acid production is finally achieved because static testing would have likely already predicted that the total acid-neutralizing capacity was less than the total acid-producing capacity. Further, as discussed above, the time required to achieve net acid production is of extremely limited use for predicting field behavior. Even comparing the apparent relative rates of acid production and acid neutralization can be misleading because the experiment was designed to increase sulfide mineral oxidation only.

Routine use of any of the above "kinetic" tests in a waste characterization program should be carefully considered because of their relatively high costs, potentially long times required, and the ambiguous results. Only a very focused program with narrowly defined objectives and supporting field data is likely to make significant and definitive use of kinetic testing.

Laboratory Methods for Predicting Metal Leachability
In addition to an evaluation of a material's potential to produce ARD or to liberate metals and other constituents in conjunction with ARD production, there is concern for constituent leachability under non-ARD conditions. Laboratory tests addressing this situation subject crushed rock to agitation with mildly acidic solution designed to simulate natural precipitation or solutions containing dilute organic acids from degradation of organic matter.

Tests simulating leaching by natural precipitation include the Synthetic Precipitation Leaching Procedure (EPA, 1994a) and the very similar, old version of the Nevada Meteoric Water Mobility Procedure. Both use a bottle-roll apparatus to mix a relatively small mass of crushed rock and artificial precipitation for 18 to 24 hours. Leachates are decanted, filtered, and analyzed. A revised Nevada Meteoric Water Mobility Procedure (Standardized Column Percolation Test Procedure) re-circulates a mass of artificial precipitation through an equal and relatively large mass of coarsely crushed rock (NDEP, 1996). Another procedure commonly but potentially inappropriately used is Toxicity Characteristic Leaching Procedure, or TCLP (EPA, 1994b).

Although leachability tests are straightforward in their execution and the results useful for comparison of materials, a quantitative interpretation of the data for field applications is very difficult. The concentration of a given metal in the leachate from any one of the above procedures is a function of the concentration of the metal in the solid, its extractability, and the volume of the water relative to the mass of the solid. What is an appropriate solid/liquid ratio for simulating field conditions? Furthermore, it is unlikely that a human receptor would attempt to drink the equivalent of a direct leachate from a mining waste material. How much dilution should be incorporated into the calculation of concentrations at the point where a receptor would drink the water? The only standards for evaluating a leachate directly are assigned to just eight "RCRA" metals and are associated with the TCLP test only. The standards in that case are inexplicably based on 100 X drinking water standards and may be irrelevant to a given mine situation.

To be realistic and meaningful, leachability studies should address actual field conditions. Data should be obtained from on-site exposure experiments that determine actual ratios of rainfall to runoff for a given mass of rock at the actual size anticipated for a dump or other situation. Runoff concentrations should then be evaluated in the context of a site-wide water balance model that includes meteorological conditions, location, flow and mass of receiving water bodies, and locations of receptors.

Site-specific Field Methods
In addition to laboratory tests, field weathering or kinetic tests may be conducted at the mine site in advance of mining and continued throughout operations. Among the obvious advantages of on-site experiments over laboratory tests are that environmental conditions are far more representative of expected post-mining conditions, time is not compressed in an unquantifiable way, and physical scales (especially particle sizes and rock masses) can be more appropriate to actual conditions. Further advantages are the low cost and ability to use existing personnel to monitor the experiments. Disadvantages include the time required for acquisition of useful data and the requirement for greater amounts of sample material. Non-ideality remains and quantification and upscaling can sill be challenging, but laboratory data must be reconciled to such empirical data.

Morin and Hutt (1997) discuss two basic types of on-site weathering experiments that can be operated under ambient conditions to provide non-laboratory information on ARD production and characteristics. These include piles, cribs, or bins of rock exposed to ambient weather and equipped with a drainage collection system to intercept all runoff. A second type is "minewall stations" that collect runoff from known areas of rock faces exposed in mine settings or natural outcrops. Depending on the location and setting of these rock faces, the runoff may be derived from atmospheric precipitation or groundwater sources from either the saturated or vadose zones.

Extrapolation of information gained from either type of experiment can be problematic. Rock piles may differ from actual conditions of concern in terms of exposed surface areas and thus the efficiency of wetting by a given precipitation event. Not only is runoff volume a non-linear function of precipitation volumes, solute concentrations in runoff are related to runoff volumes, and concentrations in a given runoff event are influenced by the nature of the previous event. In the case of rock face stations, similar issues exist, and additional efforts must be made to separate and quantify solution contributions from various sources.

Despite the shortcomings and interpretive complexities inherent in any experiment with a large number of variables, there can be no excuse or rationale for not establishing a robust on-site experimental program early in the operation and maintaining it throughout the mine life and possibly beyond. The program should include an annual review of the data with consideration of what impacts the results might have on future mining practices at the site. In addition, careful consideration should be given to the applicability of the experiment design – will it provide the required information for closure design, runoff prediction, long-term tailings behavior, or pit lake chemistry prediction? Considerable financial advantages may be derived from a well designed and operated program.

REFERENCES

Alpers, C.N and D.K. Nordstrom. 1999. "Geochemical modeling of water-rock interactions in mining environments." *The Environmental Geochemistry of Mineral Deposits Part A: Processes, Techniques and Health Issues, Reviews in Econ. Geology, Vol. 6B*, edited by G.S. Plumlee and M.J. Logsdon, Littleton, CO: Soc. Econ. Geologists, 289-323.

ASTM. 1988. "Method 4239-85." *American Society for Testing and Materials Annual Book of ASTM Standards, Vol. Sec. 5., Vol. 05.05: Standard test methods for sulfur in the analysis of coal and coke using high temperature tube furnace combustion method*, 385-90.

ASTM. 1996. *Standard Test Method for Accelerated Weathering of Solid Materials Using a Modified Humidity Cell*. Tech. Report No. D-5744-96. Am. Soc. Testing Materials.

COASTECH Research Inc. 1991. *Acid rock drainage Prediction Manual – A manual of chemical evaluation procedures for the determination of acid generation from mine wastes*. Prepared for CANMET – MSL Div. and Dept. of Energy, Mines and Resources, Canada No. MEND Project 1.16.1b. Ottawa, Ontario, Canada: Mine Environment Neutral Drainage (MEND).

Davis, G.B. and A.I.M. Ritchie. 1986a and 1986b. "A model of oxidation in pyritic mine wastes: Part 1: Equations and approximate solutions." *Appl. Math. Modeling*, 10 (October): 314-22.

du Bray, E.A. (ed.). 1995. "Preliminary compilation of descriptive geoenvironmental mineral deposit models." *U.S. Geol. Survey*.

Elberling, B., R.V. Nicholson, E.J. Reardon, and P. Tibble. 1994. "Evaluation of sulfide oxidation rates: A study comparing oxygen fluxes and rates of oxidation product release." *Can. Geotech. J.*, 31: 375-83.

EPA. 1994a. "Toxicity Characteristic Leaching Procedure." *Test Methods for Evaluating Solid Wastes, Physical/Chemical Methods Laboratory Manual, SW-846*. U.S. Environmental Protection Agency.

EPA. 1994b. "Synthetic Precipitation Leaching Procedure." *Test Methods for Evaluating Solid Wastes, Physical/Chemical Methods Laboratory Manual, SW-846*. U.S. Environmental Protection Agency.

Evans, B. 1996. An environmental perspective on the national and global mining industry. In *Maintaining Compatibility of Mining and the Environment*, edited by G.H. Brimhall and L.B. Gustsfson, 17-21. Littleton,CO: Soc. Econ. Geologists.

Fennemore, G.G., W.C. Neller, and A. Davis. 1998. "Modeling pyrite oxidation in arid environments." *Environ. Sci. Technol*, 32, No. 18: 2680-87.

Filipek, L.H., and G.S. Plumlee, Editors. 1998. "The Environmental Geochemistry of Mineral Deposits." *Reviews in Economic Geology, Vol. 6B*. Littleton, CO: Soc. Econ. Geologists.

Kwong, Y.T.J. 1993. *Prediction and prevention of acid rock drainage from a geological and mineralogical perspective*. Tech. Report No. 1.32.1. Ottawa, Ontario, Canada: Mine Environment Neutral Drainage (MEND).

Lawrence, R.W. and Y. Wang. 1996. *Determination of neutralization potential for acid rock drainage prediction*. Tech. Report No. 1.16.3. Ottawa, Ontario, Canada: Mine Environment Neutral Drainage (MEND).

Lawrence, R.W. and Y. Wang. 1997. "Determination of neutralization potential in the prediction of acid rock drainage." *Prediction*. Fourth International Conference on Acid Rock Drainage, Proceedings. Vancouver, B.C., Canada: 449-464.

Lord, D., M. Ethridge, M. Wilson, G. Hall, and P. Uttley. 2001. "Measuring exploration success: An alternative to the discovery-cost-per-ounce method of quantifying exploration effectiveness." *SEG Newsletter*, 45 (April).

MEND. 1995a. *MINEWALL 2.0: User's Manual*. Tech. Rept. No. MEND Project 1.15.2a. Ottawa, Ontario, Canada: Mine Environment Neutral Drainage. 55 pp. plus diskette.

MEND. 1995b. *MINEWALL 2.0: Literature review and conceptual models*. Tech. Rept. No. MEND Project 1.15.2b. Ottawa, Ontario, Canada: Mine Environment Neutral Drainage. 97 pp.

MEND. 1995c. *Application of MINEWALL 2.0 to three minesites*. Tech. Rept. No. MEND Project 1.15.2c. Ottawa, Ontario, Canada: Mine Environment Neutral Drainage. 193 pp.

Molson, J.W., D.W. Blowes, E.O. Frind, J.G. Bain, and M.D. Wunderly. 1997. *Metal transport and immobilization at mine tailings impoundments*, Waterloo Center for Groundwater Research, Univ. Waterloo no. MEND Associate Proj. PA-2. Ottawa, Ontario, Canada: Mine Environment Neutral Drainage.

Morin, K.A., and N.M. Hutt. 1997. *Environmental Geochemistry of Minesite Drainage: Practical Theory and Case Studies*. Vancouver: Minesite Drainage Assessment Group (MDAG Publishing).

Morin, K.A., and N.M. Hutt. 1999. "Prediction of drainage chemistry in post-mining landscapes using operational monitoring data." Paper presented at the Ecology of Post Mining Landscapes, Cottbus, Germany.

NDEP. 1996. *Meteoric Water Mobility Procedure (MWMP), Standardized Column Percolation Rest Procedure*. Nevada Division of Env. Protection.

Nicholson, A. D., M.J. Rinker, , P. Tibble, G Williams, and M. Wiseman. 2000. "An integrated approach to assess acid generation and metal release from sulfide tailings using oxygen consumption measurements, porewater chemistry, and geochemical modeling." Paper presented at the Fifth International Conf. on Acid Rock Drainage (ICARD 2000). Denver, Colorado: Soc. Mining, Metal and Explor. (SME).

Nordstrom, D.K. 1982. "Aqueous pyrite oxidation and the consequent formation of secondary iron minerals." *Acid Sulfate Weathering, SSSA Spec. Pub. No. 10*, edited by D.M. Kral, 37-56. Madison, WI: Soil Science Soc. Am.

Nordstrom, D.K., and C.N. Alpers. 1999. "Geochemistry of acid mine water." *The Environmental Geochemistry of Mineral Deposits Part A: Processes, Techniques and Health Issues, Reviews in Econ. Geology, Vol. 6B*, edited by G.S. Plumlee and M.J. Logsdon, 133-60. Littleton, CO: Soc. Econ. Geologists.

Perkins, E.H., H.W. Nesbitt, W.D. Gunter, L.C. St-Arnaud, and J.R. Mycroft. 1995. *Critical review of geochemical processes and geochemical models adaptable for prediction of acidic drainage from waste rock*, Noranda Technology Center, MEND Report 1.42.1. Ottawa, Ontario, Canada: Mine Environment Neutral Drainage (MEND).

Plumlee, G.S. 1999. "The environmental geology of mineral deposits." *The Environmental Geochemistry of Mineral Deposits Part A: Processes, Techniques and Health Issues, Reviews in Econ. Geology, Vol. 6B*, edited by G.S. Plumlee and M.J. Logsdon, 71-116. Littleton, CO: Soc. Econ. Geologists.

Plumlee, G.S., and M.J. Logsdon, Editors. 1998. *The Environmental Geochemistry of Mineral Deposits. Reviews in Economic Geology, Vol. 6A*. Littleton, Colorado: Soc. Econ. Geologists.

Plumlee, G.S., K.S. Smith, and W.H. Ficklin. 1994. "Geoenvironmental models of mineral deposits, and geology-based mineral-environmental assessments of public lands." *U.S. Geol. Survey Open File Report*, 94-203.

Robertson, W.D. 1994. "The physical hydrology of mill-tailings impoundments." *The Environmental Geochemistry of Sulfide Mine-wastes.Short Course Handbook, Vol. 22, Waterloo, Ontario, May, 1994*, edited by J.L. Jambor and D.W. Blowes, 1-17. Nepean, Ontario: Mineralogical Assoc. Canada.

Scharer, J.M., W.K. Annable, and R.V. Nicholson. 1993. "WATAIL – A tailings basin model to evaluate transient water quality of acid mine drainage." Report to Falconbridge, Ltd. and MEND-Ontario. Waterloo, Ontario: Institute for Groundwater Research, University of Waterloo.

Scharer, J.M., R.V. Nicholson, B. Halbert, and W.J. Snodgrass. 1994. "A computer program to assess acid generation in pyritic tailings." *Environmental Geochemistry of Sulfide Oxidation, ACS Symp. Ser. 550*, edited by C.N. Alpers and D.W. Blowes, 132-52. Washington, D.C.: Am. Chemical Soc.

Schmiermund, R.L., and M.A. Drozd. 1997. "Acid mine drainage and other mining-influenced waters (MIW)." *Mining Environmental Handbook*, edited by J.J. Marcus, 599-617. London: Imperial College Press.

Smith, L., D. Lopez, R. Beckie, K.A. Morin, R. Dawson, and W. Price. 1995. *Hydrogeology of waste rock dumps*. Tech. Rept. No. MEND Associate Proj. PA-1. Ottawa, Ontario, Canada: Mine Environment Neutral Drainage.

Sobek, A.A., W.A. Schuller, J.R. Freeman, and R.M Smith. 1978. *Field and laboratory methods applicable to overburden and minesoils*. U.S.E.P.A.

Tremblay, G.A., and C.M. Hogan, Editors. 2001. *MEND Manual*. Ottawa, Ontario, Canada: Mine Environment Neutral Drainage (MEND).

White, III, W.W., Lapakko.K.A., and R.L. Cox. 1999. "Static-test methods most commonly used to predict acid-mine drainage: Practical guidelines for use and interpretation." *The Environmental Geochemistry of Mineral Deposits, Part A: Processes, Techniques and Health Issues, Reviews in Econ. Geology, Vol. 6B*, edited by G.S. Plumlee and M.J. Logsdon, 352-38. Littleton, CO: Soc. Econ. Geologists.

Wunderly, M.D., D.W. Blowes, E.O. Frind, and C.J. Ptacek. 1996. "Sulfide mineral oxidation and subsequent reactive transport of oxidation products in mine tailings impoundments: A numerical model." *Water Resources Res.*, 32, No. 10: 3173-78.

Environmental and Social Considerations in Facility Siting

Barbara A. Filas, Robert W. Reisinger, and Cynthia C. Parnow[1]

ABSTRACT

The environmental and social setting of a project plays an increasingly significant role in the siting of contemporary mining and processing facilities. While the mine location and development options are relatively fixed, there is far more flexibility in the siting of the plant site, waste repositories, and ancillary facilities. This paper explores the various environmental and social concerns to be taken into account in the siting of various project components. It discusses the analytical and evaluation processes by which impacts are quantified and priorities are set. These elements are integrated to arrive at a final project layout that is not only technically and economically feasible but also sensitive to important environmental and social variables.

INTRODUCTION

Selecting the best sites for the major support facilities at a mine is critical to the long-term success of the project. Moving material from one location to another for processing, shipping, and waste disposal is expensive, and the further the distance that material will need to be moved, the more costly the operating expenses will typically be. However, proximity of support facilities to the mine itself is not the only siting consideration that will affect the economics of a project. Environmental and social considerations can also have a profound effect on the overall costs and profitability of a project if their effects are not balanced with other siting considerations.

While the nature and extent of the mineral resource dictates where the mine will be located, there are usually many options available for siting support facilities. Support facilities may include primary crushing facilities, concentrating and refining facilities, coarse and fine waste disposal sites, fresh water supply reservoirs, and a variety of support buildings including offices, truck shops, and warehouses. None of these support facilities is fixed by the location of the mineralization, and, as such, site selection must consider the environmental effects, social ramifications, and project economics – commonly referred to as the "triple bottom line" – as an integrated and balanced decision.

Every project is unique. There is no single set of rules to define the perfect layout to accommodate every design instance. It is important to prioritize the elements that contribute to a siting decision, fully understand the cost ramifications of each element, and then balance those factors into the final siting decisions. It is not enough to simply look at the engineering aspects any more as the environmental and social costs of a project are as much a part of the contemporary mining project feasibility as the act of mining itself.

The remainder of this paper discusses how strategically locating mining and processing facilities can enhance the environmental and social aspects of a project. Environmental factors are presented first, followed by social factors.

ENVIRONMENTAL FACTORS

The location of major project components can have a significant effect on air, water, and land resources. Because every site and every project is unique, each of the environmental disciplines

---

[1] Knight Piésold and Co., Denver, Colorado.

must be considered on their own merit and relative to the regulatory purview under which the project will operate. For example, the most important consideration for a cement-producing operation may be the ability to meet air quality requirements at major point sources such as the cement kiln. For a gold mine, it may be the control of cyanide solutions from a heap leach pad or tailings storage facility. For a mountaintop removal coal mine, the ability to reclaim the site to an acceptable landform may be the most important consideration. In each of these examples, other environmental factors may contribute to the decision balance, but the critical issues will most likely drive the decisions in terms of public and/or regulatory acceptance of a particular project design.

Most regulatory jurisdictions in the world now require some sort of environmental impact evaluation for mining projects. Even the major international lending institutions have promulgated policies and guidelines for conducting environmental impact assessments of the projects they finance. In the past, it was common for the engineering planning and facility siting to occur exclusive of environmental considerations and the final layout then supplied to environmental analysts for impact evaluation. Recently, as various project alternatives have been evaluated in the impact assessment process and the public has become more involved in the regulatory or lending institution decisions, changes to site selection and design decisions due to environmental factors are becoming increasingly common. As a result, factoring environmental issues into the upfront planning and design process has become more the norm than the exception.

The environmental impact assessment process is relatively straightforward and well understood. It is a rigorous process of establishing a detailed profile of the existing site conditions in the project area prior to project development. Baselines are typically established that profile physical resources such as climate, air quality, landform, geology, soils, surface water, and groundwater. Ecological disciplines typically profile the terrestrial and aquatic flora and fauna in the project area and how the habitats factor into local and regional ecosystems as a whole. These conditions establish the background against which project development plans and designs are overlain and evaluated. The evaluation considers, resource by resource, the impacts that project development will have on each receiving media. As significant impacts are identified, mitigation measures are considered as needed to reduce the level of impact to each receiving media to acceptable levels.

Environmental scoping is a process by which the key environmental considerations are identified in advance of the detailed baseline studies and environmental impact assessment. While scoping does not circumvent the need for a detailed environmental impact analysis for a project, it is often used to determine the level of significance that the project impacts may have on the receiving media in the early planning stages. It can also be an important tool in determining whether any environmental, permitting, or regulatory fatal flaws exist before significant project expenditures are made.

Scoping typically consists of a site reconnaissance by a qualified environmental professional and a preliminary assessment of the level of significance that project development will have on the various environmental receiving media. This preliminary assessment of impacts and their significance can then be factored into the planning and design process to limit the number of surprises that may be identified in the detailed environmental assessment process.

Elements of the scoping process are typically evaluated by resource much as they are for the impact assessment. In performing the environmental reconnaissance of a project area, the environmental professional must ascertain relevant site information about each resource, the regulatory purview, and the project plans, for they will all fit together. The following presents ideas that may be significant to the environmental scoping for siting project facilities. It is not intended to be a comprehensive listing but to give the reader a sense of the types of information that should be drawn out in the scoping process in order to meaningfully participate in the facility siting and design effort.

Climate/Meteorology

Significant climatic events such as hurricanes or extreme cold can affect the types of facilities that can be safely operated at a particular site. These events should be considered not only in terms of how they will affect the engineering aspects of a project but also in terms of how access could be restricted, thereby increasing the probability of risk situations during periods of inclement weather.

Major projects may disturb significant areas of land and/or burn significant volumes of carbon fuels. For these projects, a more global perspective to climate impacts may need to be considered in light of greenhouse gas effects.

Air Quality

The presence of other industrial emission sources proximate to the project area can affect the airborne pollutant load. Some regulatory jurisdictions, such as the United States, allow air pollutant loads up to a particular level, after which no additional pollutant sources are allowed. Understanding whether the project is located in a designated or specially regulated air quality attainment area is important to determining the volume of pollutant that will be allowed in any point source discharge.

Air quality control system performance is generally measured at the property boundary or at the closest point of public access. Wind patterns and dispersion characteristics may influence a project's ability to meet air quality expectations. Siting facilities with due consideration of the compliance points relative to wind patterns and dispersion characteristics can save on control equipment and compliance issues.

Landform

Landforms that constitute a mining project area may include topographic features such as mountains, valleys, lakes, creeks, rivers, and wetlands. The nature of the landform can affect where facilities can safety be sited during mine operations and closure. For example, an undulating landform can be used to advantageously shield facilities from major view sheds by locating them in lower areas. In locating facilities in such depressions, however, it is important to consider the potential for temperature inversions that may entrap air pollutants.

Geology and Seismology

Geology and seismology can play significant roles in facility siting. Unstable geologic formations may lead to the shifting, cracking, or collapse of structures. For example, karst terrain may lead to increased sinkhole formation that in turn may lead to structural failures. If possible, facilities should not be located in areas of unstable geologic features and seismic activity. For facilities that must be located in these areas, design and construction practices that are more stringent must be taken.

The geochemical characteristics of the ore, waste rock, and mine exposures (for example, pit walls and underground workings) also can impact facility siting. The basic question to be addressed concerning geochemistry is whether the geochemical properties of a project-generated material will potentially result in acid generation and leaching of metals. If the material is potentially acid generating, emphasis should be placed on isolating the facility to the extent possible to avoid or reduce the potential for environmental impacts. This is generally done by situating the facility away from surface water drainages and by best practices that may include, for example, covering/capping the facility.

Soils

The availability of soils needed for operations and closure is often limited. As such, the conservation and efficient use of site soils is required. The siting of facilities should consider the need for and the availability of soils to serve as plant growth media, low-permeability materials, and other soil materials that may be needed throughout the life of the project. A soil material balance will provide a plan for meeting the soil requirements for the project.

Locating facilities in strategic areas can help optimize the project soil material balance. For example, an area that includes non-beneficial soils may offer a preferred location for a facility. As another example, it may be feasible in some instances to backfill mine wastes into pits or underground workings, thereby reducing surface and soil disturbance as well as the amount of soil needed for reclamation covers.

Certain soil profiles offer increased protection of groundwater resources and should be considered in siting mine waste facilities. Low-permeability soils, soils with a high attenuating capacity for metals, and an increased depth to the water table all are advantageous to the protection of groundwater resources.

Hydrology
The potential disturbance of surface water and groundwater resources is a significant environmental issue associated with mining projects. The site hydrologic characterization serves as a tool to locate facilities in areas that would avoid or minimize potential environmental impacts associated with water resources.

The primary means for protecting a water resource is to prevent the resource from coming into contact with potential contaminants. Only after performing a comprehensive evaluation of the potential options for isolating the water resource should treatment options be evaluated as they tend to be more costly than prevention.

Logically, facilities should be located in areas that are not susceptible to flooding. The site hydrologic characterization should include locating the extent of the design flood event as it may apply to the site. In addition, mine waste facilities should not be located in areas where surface water drainages exist in order to limit the potential for geochemical reactions. In cases where development of a project in mountainous topography may require the placement of waste materials in valleys containing intermittent or perennial streams, run-on diversions and underdrains should be considered to prevent water from contacting the facilities.

Depth to groundwater is an important environmental consideration for locating waste facilities when a variation in the depth to the water table in the project area exists. Locating waste facilities in an area where the water table is deeper is preferable to an area where the water table is shallower in that the overlying soil column may provide for the attenuation of potentially adverse contaminants, thereby reducing the environmental risk of the project.

An adequate source of water for the various aspects of the mine project may be developed from surface water and groundwater resources. The location of facilities that require substantial water inputs – for example, a concentrating facility – should consider the location of water supplies.

Ecosystems
Facilities should be located to minimize their impacts on sensitive ecosystems as identified during the site reconnaissance. For example, mine access roads may be located to minimize the impact associated with wildlife migration routes. Where impacts to ecosystems are unavoidable, the creation of conservation areas may provide a viable solution to mitigating degradation of habitat and the potential loss of biodiversity.

SOCIAL FACTORS
Public acceptance of a mining project is fundamental to its success. A project that is technically sound may fail without public acceptance. The best engineering and environmental evaluations cannot overpower public opinion, which is usually driven by public concerns and how they are factored into project siting and design decisions.

Social factors affecting project component siting decisions can take many forms. Social impact assessments have long been included as part of the environmental impact evaluation process; however, the concept of sustainable development has only in recent years become a major focal point of many mining initiatives. Where development banks like the World Bank are involved, social issues and sustainable development are crucial as they represent the institution's fundamental reason for project investment.

The social impact assessment has evolved from being a component of the environmental assessment into a stand-alone evaluation. Baselines typically consist of profiling the local population and demographics, social and political structures, land use and natural resources management, livelihood systems and employment opportunities, social services and infrastructure, public health care, vulnerable groups and indigenous peoples, and cultural and historic resources. Like the environmental baseline for the project, the social baseline establishes the background against which project development plans and designs are evaluated. The social impact assessment considers the impacts that project development will have on individuals, communities, customs, and historical features of the area. For any identified significant social impact, mitigation measures are considered to reduce the level of impact to acceptable levels.

Public consultation and disclosure is a fundamental part of effective project planning. Informing the local community and interested stakeholders of the tentative project plans early and often in the planning and design process is an important step in establishing effective communications between the project proponent and stakeholders. The initial project scoping will typically identify key social issues that might affect project component siting. Preliminary siting and design concepts for major project components should be presented to the public through information meetings in the earliest phases of the project in order to solicit public input regarding performance concerns or site preferences. This public input must be factored into siting and design decisions along with the economic and environmental considerations to assure project success.

The following presents several social considerations as they relate to siting project facilities. As with the preceding discussion on environmental considerations, the list is not intended to be exhaustive, but rather should to give the reader a sense of the types of information that should be collected during the project scoping phase.

Land Use
The baseline evaluation should identify traditional land uses in the project area. Many new mining projects occur in third world countries in areas that may have historically been used for agricultural or pasture lands. For this reason, it is necessary to site facilities taking into consideration what may not appear to be an important agricultural area but may indeed have a significant impact on local income sources.

Resettlement
Among the more important social impact issues that a mining project may need to undertake is the relocation of households or entire communities to accommodate project development plans. Resettlement issues are extremely sensitive and can cause significant project opposition if not handled fairly, expeditiously, efficiently, and transparently. While resettlement should be avoided to the extent possible, when it is necessary, its goal should be to provide each affected household with an equivalent or better living situation. In the event of resettlement, a community consultation program should be implemented in the early stages of the project, possibly during the exploration phase, so that the needs and concerns of affected populations in the project area can be properly assessed. Early input from the community often results in a better public opinion of the project and eases project acceptance along ensuing phases.

In-Migration
Another major social impact that may result from the mining project is in-migration. In-migration occurs for a variety of reasons, not the least of which is an increase in employment opportunities. These employment opportunities are the result of direct project employment as well as the increased need for goods and services to accommodate project employees and their families.

Local communities may not have the capacity to accommodate the personnel requirements of a new project. In this case, there may be an influx of job seekers as the project gets underway. The social assessment must include an evaluation of the ability of local communities to provide the required work force and to accommodate newcomers. Communities that are targeted to receive an influx of newcomers must be identified and a set of actions developed to ensure that the newcomers are preferentially directed to locations where they can be adequately accommodated. This is typically accomplished by a combination of "push" and "pull" factors. "Push" factors will discourage settlement in certain areas and may include security provisions. "Pull" factors will encourage job seekers to settle in targeted locations by providing attractive features such as low-cost housing opportunities or improved goods and services.

Public Health and Safety
Mining projects often supply infrastructure in the form of clinics and hospitals that are made accessible to the general population. Many mining projects will provide these services if they are unavailable in near proximity to the project site. Providing this infrastructure in locations that provide easy public access can result in favorable public opinion of the project.

Sustainability
Often mining projects are viewed as a short-term use of non-renewable natural resources. This has resulted in a poor public opinion of mining in areas that have historically been non-industrial and whose inhabitants are often in lower income categories. Promoting a long-term, responsible use of natural resources will help ensure that there are resources available for sustained industrial growth far into the future. Mining projects can provide opportunities for long-term sustainable development in areas where there traditionally has been no infrastructure or employment related to industry. Training programs included in the community development program as well as on-the-job employee training provide skills in areas where training might otherwise not be available.

An important aspect of the initial project scoping process is to assess the needs of the community and the sustainable development programs that would be most beneficial. For instance, offering agricultural classes that instruct communities on modern agricultural practices as part of the community development program can lead to higher regional productivity long after the mining operation has ceased.

Cultural and Historic Resources
Prior to preliminary siting and design concepts for major project components, a baseline evaluation of the cultural and historic resources in the project area must be performed. Identifying cultural and historic resources as part of the baseline evaluation will reduce the possibility of accidentally disturbing these resources during project activities, which could result in adverse public reaction. As many archeological or cultural resources cannot be quantitatively evaluated for significance due to either lack of supporting evidence or abundance of sites, alternative methods of protection of these resources are becoming an important mitigation method. If the baseline evaluation results in cultural or historical discoveries, in situ protection is generally preferred; however, data recovery or advancement of cultural interest programs can often serve as adequate mitigation.

CONCLUSIONS
The siting of facilities with due consideration for the environmental and social factors can greatly enhance the overall environmental and social performance of the project. While facility siting offers an opportunity to mitigate environmental and social impacts, it is recognized that best practices are integral to the overall mitigation of environmental and social impacts. Some examples of best practices include mitigating surface water impacts through the strategic placement of run-on diversion structures, mitigating air impacts by watering haul roads, and reducing surface disturbance by salvaging soils where facilities will be located.

A properly designed geographic information system (GIS) offers an excellent tool for visually integrating important environmental and social factors into the siting of project facilities. Ultimately, an evaluation of the environmental and social risks associated with each facility siting alternative should be performed to optimize the siting of mining and processing facilities.

There is rarely a downside to factoring the environmental and social issues into facility siting decisions. Environmental and social consideration will usually support site selection in close proximity to the mine. This is because the more spread out the project components are, the greater the potential for environmental impacts and risk.

By integrating project economics, environmental considerations, and social issues in a balanced siting decision, the layout options that emerge will likely prove to be the most economic in the long term.

16

Construction Materials for Equipment and Plants

Section Co-Editors:
Dr. Rod McElroy and Wesley Young

Selection of Metallic Materials for the Mining/Metallurgical Industry
G. Coates 1911

Elastomers in the Mineral Processing Industry
P. Schnarr, L.E. Schaeffer, H.J. Weinand 1932

Plastics for Process Plants and Equipment
G.W. McCuaig 1953

Commercial Acceptance and Applications of Masonry and Membrane Systems for the Process Industries
R.E. Aliasso Jr., T.E. Crandall, D.M. Malone, R.J. Storms 1962

Selection of Metallic Materials for the Mining/Metallurgical Industry

Gary Coates[1]

ABSTRACT

The paper is broad-based and describes properties, applications and limitations of various metals (irons and steels, alloy steels, stainless steels, nickel, aluminum, and copper alloys, etc.) used in the mining and metallurgical industries. Mechanical properties, corrosion resistance, wear resistance, and high temperature properties will be discussed. Areas of applications will be discussed. Some guidelines for materials selection and application, and resources are included.

INTRODUCTION

Metallic materials are used in the mining/metallurgical industry often with very little thought. We are generally familiar with them, knowing their advantages and disadvantages. We may use them for their strength, corrosion or wear resistance, for their ability to withstand high temperatures, as a barrier for solids, liquids or gases, etc. What is normally used are alloys of metals. The use of high purity iron, nickel, aluminum etc. is very limited. Alloys, mixtures of at least one metal and one other element, are the commonly available metals. Steels and irons are iron-based, but usually with intentional additions of carbon, manganese, silicon, etc. "Stainless" steels contain a minimum of about 11% chromium, usually with other elements such as nickel, molybdenum, etc. to give them special properties. Some guidelines will be given to aid in materials selection. Selecting the right material also involves issues such as price and availability considerations.

The higher alloyed materials generally require more sophisticated fabrication and especially welding techniques. This means that some attention should be paid to the selection of the fabricator, the inspection of the equipment and the supporting documentation. Some examples will be covered.

In this paper, most of the alloys will be classified by their UNS (Unified Numbering System) number, an attempt to classify the chemical composition of all metals and alloys by a unique identifier composed one letter and 5 numbers. These numbers are used today by ASTM (American Society of Testing and Materials) for identifying alloys in their specifications. However in many cases the older designations are better known, so that for example, AISI (American Iron and Steel Institute) numbers will be also used.

It is not possible in a paper such as this to cover all metals and types of applications. More exotic metals such as zirconium, tantalum, platinum, etc. which do find occasional use are not covered. Coatings (metallic, plastics, paint, rubber) will not be covered in any great detail – each is a major topic in itself.

As metallurgical processes often involve aggressive and perhaps toxic or explosive chemicals, materials selection needs to concern itself also with health, safety and environmental issues. A pipe leaking toxic chemicals can have severe legal consequences on the operation of a plant, and for the management of the plant too.

The paper is divided into 3 parts. The first section called Alloys deals with a broad description of the classes of alloys with general properties and applications. The second part gives a few key principles that are important when selecting and working with alloys. In the final section, the use of available resources to aid in selecting materials is discussed.

---

[1] Nickel Development Institute, Toronto, Canada

METALS AND ALLOYS
STEELS AND IRONS

A number of the common names given to steels and irons are more traditional than accurately reflecting metallurgical classifications. In the UNS classification, carbon and alloys steels begin with either G, H or K, carbon or alloy steel castings with J, and cast irons begin with F. Steels are available in a wide range of strength levels, both inherent and from heat treatment. In many applications, this group of materials are the first that should be considered, because of their high availability, low price, ease of fabrication, familiarity, etc. Despite their comparatively low corrosion resistance, it is sometimes possible to build a piece of equipment using thick material to compensate for a high corrosion rate. However, other factors have to be taken into account, especially when safety and environmental issues are at stake. Difficulty in inspecting a critical component to know when it is nearing its end of life is a significant factor that can lead to a decision to use a more corrosion resistant but more expensive material.

The corrosion resistance of these materials can be improved through various methods. Various coatings can be applied, both galvanic types such as zinc, or barrier types such as paint, plastic, rubbers, etc. (In many cases, it is more appropriate to view the coatings, especially rubber and plastics, as the prime material supported by a substrate of steel.) Steels underground or in water can be cathodically protected by application of a electrical current either directly from a power source or via special anodes of magnesium, aluminum, etc. These subjects are all outside the scope of this paper.

Carbon steels. These are the workhorse alloys of constructional materials, piping, plate and sheet. Carbon steels are iron with small amounts of carbon, manganese, and a few other steel-making elements such as silicon. A high carbon steel such as AISI 1045 (UNS G10450) may be easily hardenable with heat treatment. If the carbon content is relatively low, the steel will not easily harden, so for example, welding can be done without any special precautions. These steels do not become brittle at cold temperatures, so that a killed steel might be suitable down to $-30°C$ ($-20°F$). Carbon steels do offer a certain degree of high temperature and fire resistance. However the weathering steels (below) are more commonly used at temperatures above about $300°C$ ($570°F$) because of their improved oxidation resistance and higher strength. The main alloying elements of a few common carbon steels are included in Table 1. With iron-based alloys, the balance of the composition not stated is iron.

Table 1 Nominal composition (weight %) of some common carbon and alloy steels

NAME	Type	UNS number	C	Mn	Cr	Ni	Mo
AISI 1010	Low carbon steel	G10100	0.10	0.45			
AISI 1045	High carbon steel	G10450	0.45	0.75			
ASTM A 36	Carbon structural steel	K02598	0.20	1.0			
2¼ Cr / ½Mo	Alloy steel for high T	K21930	0.10	0.45	2.2		1.0
3½% nickel (LF3)	Alloy steel for low T	K32025	0.15	0.7		3.5	
AISI 4340	Alloy steel hardenable	G43400	0.40	0.7	0.8	1.8	0.2
B-1 cast	Austenitic Manganese	J91119	1.0	13			

Alloy steels. This class of steel falls into different categories. There is alloying to improve the strength of the material, both at ambient temperature and at high temperature, and especially the heat treated strength and hardness. The chromium-molybdenum grades, e.g. 2¼ Cr / ½Mo (UNS K21930) have greatly improved high temperature strength. The addition of nickel to a steel improves low temperature ductility, so that an alloy such as K32025 with 3½% nickel is good for many cryogenic services. Abrasion resistant steels, many of which are based on the AISI 4340 (UNS G43400) composition, are quenched and tempered to give a very high hardness. They tend to also be quite brittle, and can fracture under impact loading. Such steels are often used as abrasion resistant liner plates. Where water or other corrosives are involved, it should be remembered that these materials are not particularly corrosion resistant, with the net result that high wear rates can occur. In water, the steel forms a thin oxide layer, which is easily removed by abrasive particles, and the process is repeated again and again resulting in high rates of metal loss. In that sense, corrosion and abrasive wear are synergistic. An improvement in the corrosion resistance of the metal may improve the total wear resistance considerably.

Weathering steels. These are low-alloy high strength structural steels that are often separated out from the alloy steel category because they have markedly increased corrosion resistance. They are alloyed with copper and usually a few other elements, primarily chromium, nickel, and vanadium, usually less than 1.5% total, and have good corrosion resistance in normal atmospheric exposure conditions. They are not hardenable by heat treatment. They form a semi-protective layer of oxide or "rust". This layer is not protective under abrasive conditions, nor totally protective in marine or heavily polluted industrial atmospheres. A typical weathering steel (4 grades are covered in A 588) will have a 345 MPa (50 ksi) minimum yield strength and a 485 MPa (70 ksi) minimum tensile strength. They have been used for building exteriors, structural elements on bridges, and other structural applications where a rusty appearance is tolerated. Mining/ metallurgical applications include structural applications, and also certain equipment like fans, blowers, and ductwork, especially where sulfurous gases are present. Weathering steels are suitable for hot gas ducting with appropriate temperature limitations and other precautions (*Andrew 1997*).

Austenitic manganese steels. These are a series of cast or wrought steels with 10% or more manganese. They are soft, unlike most abrasion-resistant alloy steels, but cold work rapidly to resist abrasion. Thereby the surface is wear-resistant, while the interior remains ductile.

Cast irons. This is a family of materials broken out into unalloyed cast irons and alloyed cast irons. The carbon content varies from as low as 0.7% to as high as 4.5%. They are mostly used as valves, pumps, pipes, fittings, and occasionally as bar. Table 2 lists some representative cast irons.

Table 2 Nominal composition (weight %) of some representative cast irons

Type	UNS number	C	Mn	Si	Cr	Ni	Mo	Cu
Gray cast iron	F10004	3.5	0.7	2.5				
White cast iron	F45001	2.7	1.0	0.6	2.5	4.0	0.5	
Malleable cast iron	F20000	2.5	0.8	1.5				
Silicon cast iron	F47003	0.9	1.2	14.5	0.3		0.3	0.3
Nickel Cast Iron D2B	F43001	2.6	1.0	2.3	3.5	20		

Unalloyed cast irons include gray cast iron, ductile cast iron, malleable iron, and white cast iron. Gray cast iron is inexpensive, easily cast, and machinable. Ductile cast iron, made with a small nickel-magnesium addition, is often preferred for improved resistance to shock. They can also be modified to have excellent low temperature ductility. White cast iron is a controlled chemical composition gray iron that is rapidly cooled to make it very hard and very brittle. It is virtually unmachinable, and extremely difficult to weld. It has excellent abrasion resistance. Chilled iron is so produced to have hard white cast iron on the surface, and a more ductile gray cast iron interior. An extended heat treatment of white cast iron results in a type called malleable iron. Malleable iron has improved ductility and strength to gray cast iron, and is commonly used in tools, machinery, automotive parts, etc. There are also a number of proprietary chemical addition treatments that are made to improve the properties of the unalloyed cast irons.

Alloyed cast irons have significant alloying additions of silicon, nickel, molybdenum, chromium or copper. A silicon cast iron which contains about 14.5% silicon performs well in hot concentrated sulfuric acid. Variations of this alloy with chromium and molybdenum give high corrosion resistance in chloride-containing media. A number of proprietary versions exist. Nickel can also be added in amounts of about 0.5% to 6% to improve common versions of gray cast iron.

At about 4.5%, a martensitic alloy is produced with outstanding resistance to wear. Austenitic gray cast irons contain 14% to 38% nickel. Some of these alloys, such as F43001 have very good corrosion resistance and are suitable for moderately high temperatures.

STAINLESS STEELS

Stainless steels are a family of steels which all show passive behavior, resulting from a minimum chromium content of about 11%. Other elements such as molybdenum, nickel, copper and nitrogen are added to increase corrosion resistance to different environments. The passive oxide layer on the exposed surface is chromium rich, but also contains the other alloying elements. The stainless steel family is composed of several different sub-families based on the crystal structure of the grains. Each of the families is discussed separately - austenitic, duplex, , ferritic, martensitic and precipitation hardenable. Stainless alloys span the range of corrosion resistance from low to very high, and also strength levels from relatively low to very high.

Austenitic stainless steel. The majority of the worldwide production of stainless steels is of the austenitic type, also known as the "300 series" stainless steels. These grades contain nickel and other alloying elements that promote an austenitic structure. These alloys offer many different advantages, such as ease of fabrication and weldability, excellent high temperature strength, and excellent low temperature ductility. Those alloys with a fully austenitic structure are essentially non-magnetic, an important property in some applications. This family of alloys offers the widest range of corrosion resistance, from the base grade 304 up to high performance alloys such as S32654 and N08031, which are bordering on the nickel-base family. Table 3 lists a number of more common austenitic stainless steels, but there are numerous more alloys produced, some for very restricted applications.

Table 3 Nominal composition (weight %) of some representative austenitic stainless steels

Family / Alloy	C, H, or both*	UNS number	C_{max}	Cr	Ni	Mo	N	Other
AUSTENITIC								
304	C&H	S30400	0.08	19	9	-	0.06	
304L	C	S30403	0.030	19	9.5	-	0.06	
304H	H	S30409	0.04 - 0.10	19	9	-	0.05	
316L	C	S31603	0.030	17	12	2.2	0.06	
317L	C	S31703	0.030	19	13	3.3	0.06	
317LMN	C	S31726	0.030	18.5	15	4.3	0.15	
321	H	S32100	0.08	18	10	-	0.04	Ti
347	H	S34700	0.08	18	10.5	-	0.04	Nb
309S	H	S30908	0.08	23	13	-	0.04	
310S	H	S31008	0.08	25	20	-	0.03	
Alloy 253	H	S30815	0.05 - 0.10	21	11	-	0.17	Ce, Si
330	H	N08330	0.08	18.5	36	-	0.03	Si
High Performance Austenitic								
Alloy 20	C	N08020	0.07	20	35	2.5	0.04	Nb+Ta, Cu
Alloy 904L	C	N08904	0.02	20	25	4.3	0.04	Cu
Alloy 825	C	N08825	0.05	21	40	3.0	0.04	Cu, Ti, Al
6% Mo (18%Ni)	C	S31254	0.020	20	18	6.2	0.20	Cu
6% Mo (25% Ni)	C	N08367	0.030	21	24	6.3	0.22	Cu
Alloy 654	C	S32654	0.020	25	22	7.5	0.5	Cu, Mn
Alloy 31	C	N08031	0.015	27	31	6.5	0.2	Cu

* Primary end use: C = corrosion resistant, H = high temperature resistant, C&H = both

Nomenclature. The UNS system uses S for stainless steels and N for nickel alloys. That is based on a definition that if an alloy has 50% or more iron, and meets the minimum chromium content requirement, then it is a stainless alloy, and is given an S number. If the alloy has less than 50% iron, and there is more nickel than any other element other than iron, it will then be given an N number. If less than 50% iron and more cobalt than any other element, it would be classified with the cobalt alloys in the R series. Another definition, used today by ASTM and in this paper, is based on the highest element in the alloy. If the minimum chromium content restriction is met, and there is more iron that any other element, then it is classified as a stainless steel. If there is more nickel than iron, than it is a nickel alloy.

In the old AISI classification system, the suffix "L" (as in 304L) stands for low carbon, most often 0.030% carbon maximum. Low carbon alloys should always be used whenever welding is involved for corrosive service. The UNS number for these old standard grades should end in "03", indicating the maximum carbon level. This system is not followed for the newer or proprietary alloys, for which almost all exist only with low carbon. An alternative to using low carbon is to stabilize with elements such as titanium, niobium, or tantalum. AISI 321 (S32100) or AISI 321 is essentially 304 with a titanium addition to compensate for the higher carbon content to achieve proper corrosion resistance after welding. These stabilizing elements have other drawbacks both in steelmaking and in welding, so the low carbon stainless steels have predominated for many years, at least in North America. Not so long ago, the stabilized versions predominated in many countries of Europe, but more recently they have also adopted the low carbon stainless steels for corrosive services. However, when on an old drawing a stabilized stainless grade is being specified for a corrosive service, rather than replace it with the same, it is best to ask whether a low carbon stainless could be used. Alloy 20(N08020) is an example though of an older corrosion-resisting alloy produced only in a stabilized version. It is very important to use the low carbon or stabilized stainless steels in corrosive oxidizing conditions, such as weak sulfuric acid solutions with oxidizing ions such as cupric or ferric.

The "H" (as in 304H) specifies both a minimum and maximum carbon content, normally 0.04 to 0.10%. The last 2 digits of the UNS number are normally "09". H grades are meant to be used at elevated temperatures, where carbon gives the alloy increased high temperature strength. The ASME Boiler and Pressure Vessel Code gives stainless steels with 0.04% minimum carbon significantly higher allowable stresses at virtually all elevated temperatures. At one time, it was very expensive to lower the carbon content in stainless steels to the low carbon levels, and consequently, if one did not specify a low carbon stainless steel, one was virtually assured of getting one with a higher carbon level. Today it is easily and inexpensively achieved and even advantageous for most modern steelmakers, so one is most likely to receive a low carbon stainless if one does not otherwise specify. The prudent way however is to always specify the exact grade that one needs, depending on the application.

"Dual-grade" stainless steels are ones that meet the specifications of two alloys. It is possible to produce dual grade 304/304L (UNS S30400/S30403). It is certifiable to both grades, and acceptable to the pressure vessel authorities. That allows one to design a pressure vessel using the allowable stresses of 304, but ensuring low carbon for corrosion resistance after welding.

In the older AISI designations, the suffix "S" was a way of designating a low carbon version of some of the high carbon heat resistant alloys. For example, 310S (UNS S31008) contains 0.08% maximum carbon, a limit set to have improved weldability. Since these alloys are only intended for high temperature usage, steelmakers do not usually produce them with low carbon. However, it is possible to specify 310H (S31010) with 0.04% to 0.10% carbon.

Corrosion resistance. Unlike many other materials, it is not common for stainless steels to fail by general or uniform corrosion. It is more likely that they fail by localized corrosion, e.g. pitting, crevice corrosion, stress corrosion cracking, or intergranular corrosion.

As mentioned earlier, the corrosion resistance of austenitic stainless steels spans a wide range. This includes strong acids to strong bases and most chemicals in-between. It should not be assumed that the higher the alloying content, the more corrosion resistant it is. In nitric acid and in 98% sulfuric, 304L is superior to 316L and most high molybdenum stainless steels. That is because the chromium content is the most important factor in these highly oxidizing conditions and molybdenum can be detrimental. It is important to check a particular alloy's corrosion resistance to the specific environment. In sulfuric acid at lower concentrations which tends to be reducing in nature, molybdenum alloys are preferred. 316L is normally considered the base grade, i.e. one would never use 304L in dilute sulfuric acid. Even higher alloyed materials are often required. However, there is no guarantee that a 6% molybdenum stainless is better than a 4% molybdenum stainless steel. Alloys like 904L and Alloy 20 are often used, and in high chloride containing acid solutions, the 6% Mo grades give good performance. Much good corrosion data exists for stainless steels; references will be discussed in a later section. A small copper alloying addition has been found especially useful to improve corrosion resistance in intermediate concentration ranges of sulfuric acid.

Austenitic stainless steels are also useful in chloride media, with molybdenum and nitrogen being two powerful alloying elements. Many users are aware that austenitic stainless steels are susceptible to chloride stress corrosion cracking (SCC). This is especially true of the 304L and 316L, whereas the high performance grades with significant nickel and molybdenum additions are highly resistant, although not immune (*Arnvig et al 1998*). Immunity normally is assured only when the nickel content exceeds 40%. The necessary preconditions for SCC are chlorides, tensile stresses, and a certain minimum threshold temperature. In most cases the minimum temperature is around 60°C (140°F), although under special conditions, SCC is possible even at room temperature.

It is fairly easy to roughly compare the pitting resistance of various austenitic stainless steels. The formula is based on the effectiveness of the important alloying elements on resisting the initiation to pitting, and gives a number called the pitting resistance equivalent, or PRE.

$$PRE = \%Cr + 3.3 \times \%Mo + 16 \times \%N$$

There are a couple of variations in this formula, so it is important when comparing numbers to know which formula was used. The formula doesn't take into account many factors such as surface finish, impurities in the steel, welds, and other critical factors, and it does not allow one to determine whether pitting will or will not occur in a particular alloy. Nonetheless it is useful for comparing materials. It is clear from the formula that the high performance austenitic stainless steels with high molybdenum and nitrogen contents have far greater pitting resistance. For example, the 6% Mo grades are far better than Alloy 904L, which in turn is better than Alloy 20. This ranking holds up in real industrial environments where pitting is a factor. Do not try to use it for other environments.

There are some environments where one generally avoids using stainless steels. Chemicals such as wet chlorine, hydrochloric acid, sodium hypochlorite, and hydrofluoric acid are ones where stainless steels have only a very limited usability.

The thin passive oxide layer on stainless steels has an amazing tolerance to velocity. In waters and other less corrosive chemicals, velocities up to 30 metres per second (100 feet per minute) present no problem. In chloride-containing waters, stainless steels are most likely to be attacked in stagnant conditions. A high flow velocity prevents pitting from occurring, therefore a typical minimum velocity is specified in heat exchanger tubes using brackish or sea water of 1.5 metres per second (5 feet per second).

One problem often encountered is a reduced corrosion resistance of the machinable grades of stainless steel. For example, 303 (S30300) is a version of 304 with high sulfur (0.15% minimum), and machines very easily. It may not perform as well as 304 in some environments.

Mechanical properties. Austenitic stainless steels are for most applications supplied in the annealed condition, and as such have reasonable ambient temperature strength levels and excellent ductility, as shown in Table 4.

Table 4 Minimum mechanical properties for some austenitic stainless steels according to ASTM A 240

Alloy	UNS Number	Yield Strength MPa (ksi)	Tensile Strength MPa (ksi)	Min. Elongation %
304L	S30403	170 (25)	485 (70)	40
304	S30400	205 (30)	515 (75)	40
304H	S30409	205 (30)	515 (75)	40
316L	S31603	170 (25)	485 (70)	40
317L	S31703	205 (30)	515 (75)	40
Alloy 904L	N08904	215 (31)	490 (71)	35
Alloy 20	N08020	241 (35)	551 (80)	30
6% Mo (18% Ni)	S31254	310 (45)	655 (95)	35
Alloy 654	S32654	430 (62)	750 (109)	40

Low carbon stainless steels have slightly lower strengths than their higher carbon counterparts, although today this can be improved by a intentional, small (less than 0.10%) nitrogen addition. That is what in effect has happen to obtain dual grade 304/304L. Higher nitrogen levels in some of the high performance alloys improves their strength levels, although the nitrogen is primarily present for other reasons, primarily corrosion resistance and structural stability.

Although austenitic stainless steels cannot be heat treated to improve their strength, they can be easily cold worked to a very high strength and hardness. For example, 301 stainless in fully hard condition has a minimum yield strength of 965 MPa (140 ksi or 140,000 psi) and tensile strength of 1270 MPa (185 ksi). In such a condition, the material is a powerful spring. There are some special applications for strip steel in this condition, for example a thin sole plate in safety shoes.

Austenitic stainless steels maintain their good ductility down to very low temperatures. The ASME code permits the use of some grades down to as low as −253°C (-423°F) without even requiring impact tests to check their ductility, others to only -196°C (-320°F). All these grades also get slightly stronger at cryogenic temperatures. Welds and cast material may not have quite the same ductility, and it is necessary for example to impact test welds for service below -101°C (-150°F).

High temperature properties. The austenitic stainless steels in general offer excellent oxidation resistance and high temperature strength. Other grades such as 330 (N08330) are excellent in carburization environments. In the standard high temperature austenitic stainless steels, 304 has useful oxidation resistance up to about 900°C (1650°F), whereas 309S and 310S are suitable up to about 1030°C (1900°F) and 1100°C (2000°F) respectively. Special grades such as alloy 253 (S30815) are formulated to not only have excellent oxidation resistance up to 1100°C (2000°F), but combine that with improved high temperature strength. One disadvantage of the austenitic grades is a relatively high thermal expansion coefficient. In designing equipment for high temperature use, allowance must be made for movement of the material. At temperatures over about 550°C (1000°F), 304, 321, or 347 type alloys are often preferred over other ferritic materials because they are substantially stronger in creep strength. The standard yield strength measurements do not describe the slow yielding that goes on above this temperature.

Fabrication. Most shops are familiar with fabricating the standard austenitic stainless steels. Welding at first is a bit more complicated than carbon steel, machining calls for different tools and feeds and speeds, and there is more springback in forming operations, but the learning curve is not very difficult. There are several good handbooks available with good information on this subject. When it comes to the high performance stainless steels, there can be significant differences related to welding and fabrication. These alloys are relatively expensive and are being used for very corrosive environments, so it is imperative that the shop be familiar with any necessary special practices. Most steel mills and some suppliers have booklets on fabricating their special alloys. It would be wise to choose a fabricator that has had experience with that particular or a similar alloy. If not, then it would be prudent to make sure the fabricator has the information and that the information has been disseminated to the shop floor. It is also important that during installation of the equipment, the installers should be aware of what they can and cannot do. And of course, a mill's own maintenance department should be aware of how to properly weld and repair such alloys. Each alloy basically has its own special filler metal, although in some cases, a more universal filler metal may be possible.

When a piece of equipment in austenitic stainless steel is delivered to site, or sits out in the open for awhile, there often appears rust marks on the surface. These marks were usually not visible when it left the fabricators shop. The cause is almost always surface contamination with iron, from tools, lifting chains, clamps, etc., sometimes during fabrication, sometimes during transportation and off-loading. In the cases where the free iron on the surface has originated in fabrication, it does not become visible until it comes in contact with water. That can be rain, night time condensation, etc. Although often only a cosmetic issue, in certain corrosion applications, this rusting can initiate other types of corrosion in service, especially pitting. It is always best to try to remove the contamination. Brushing and grinding are not particularly effective ways, as the free iron tends to smear on the stainless surface. It is better to chemically remove the contamination, using either phosphoric acid, or a special pickling mixture of nitric/hydrofluoric acid. The latter is available as a paste, and while quite nasty, is very effective. Heat tint from welding is also best removed this way.

Duplex stainless steels. A duplex stainless steel is one that has a structure made up of two different phases, in this case austenite and ferrite. This is usually accomplished by reducing the content of nickel, a strong austenite former, to intermediate levels. Although the first duplex stainless steel, AISI 329, was developed in the 1930's, duplex stainless steels fell out of favor in the 1950's because of some shortcomings. These have been mostly overcome today, and there has been a resurgence in use as industry has found them to be very useful engineering materials, economically solving problems with some of the austenitic stainless steels. There are many different duplex stainless steels grades available today, all containing nitrogen which solves many of the earlier shortcomings. They range from low alloyed to high alloyed, with many of them being proprietary. This can present a problem with availability. The most common duplex stainless steel today is called 2205. It originally had a UNS number of S31803, but it has been upgraded slightly, and only the new version, UNS S32205, should be used. Table 4 gives the chemical composition of 4 representative duplex grades.

Table 5 Nominal composition (weight %) of some representative duplex stainless steels

Alloy	C, H, o both*	UNS number	C_{max}	Cr	Ni	Mo	N	Other
2304 (low alloy)	C	S32304	0.030	23	4	0.3	0.12	
2205 (intermediate)	C	S32205	0.030	22.5	5.5	3.2	0.17	
Alloy 255 (superduplex)	C	S32550	0.040	25	5.5	3.3	0.18	Cu
Alloy Z100 (superduplex)	C	S32760	0.030	25	7	3.5	0.25	Cu, W

* Primary end use: C = corrosion resistant, H = high temperature resistant, C&H = both

All of these grades are meant for corrosion-resisting applications. In fact, welded constructions in duplex stainless steels should only be used up to about 240°C (460°F) and unwelded up to 270°C (520°F). Above those respective temperatures, welds and parent material can suffer from impaired ductility with time. The guidelines given about fabrication of high performance austenitic stainless steels are equally if not more relevant with duplex stainless steels. They are generally cost effective materials, but their corrosion resistance and ductility can be easily impaired. Usually it is only after being in-service that these "mistakes" are discovered.

ASTM A 923 is a specification for ensuring wrought mill products in a couple of the duplex grades are delivered with the high level of quality that is expected. It has been also applied to cast products and fabrications, although since the standard wasn't written with those products in mind, care must be taken to properly interpret the tests and results.

Duplex stainless steels have a lower thermal expansion coefficient than the austenitic grades, closer to carbon steel. This property can be used, for example, if replacing carbon heat exchanger tubes. Duplex tubes have been often used this way, welding them to the carbon steel tubesheet.

Corrosion resistance. One of the major uses of duplex stainless steels has been to replace 304L, 316L, or 317L that has suffered chloride stress corrosion cracking (SCC). All of the duplex alloys have better chloride SCC resistance than the above three 300 series alloys. Many of the high performance austenitic stainless steels have similar stress corrosion cracking resistance to the duplex - the differences are very complicated.

In terms of pitting and crevice corrosion resistance, the 2304 grade is often compared to 316L, the 2205 is close to 904L, and the superduplex stainless steels e.g. 255, Z100 and others, compare to the 6% Mo stainless steels. The PRE formula discussed for austenitic stainless steels is also valid for duplex stainless steels.

There are certain acids where duplex stainless steels outperform their austenitic counterparts, e.g. organic acids such as acetic and formic. In dilute sulphuric acid solutions mixed with oxidizing metal ions such as cupric and ferric, there have been some mixed reports. In some cases the duplex alloys perform very well, but in other cases, the ferrite phase is selectively attacked. It is strongly advised that in-plant corrosion testing be performed prior to any large scale installation. The superduplex alloys do seem to work reasonably well in certain autoclave applications, although the actual corrosion rates vary from very low to very high. Again, care is advised and in-plant testing is virtually mandatory. It has been said that when the duplex alloys are good, they are very good, and when they are bad, they are very bad.

Mechanical properties. The strength levels at ambient temperature of the duplex grades are generally higher than the austenitic stainless steels, but the ductility suffers slightly as shown in Table 6. These high strengths are quite important when designing pressure vessels, pressure piping systems, etc. as much of the wall thickness is only there for pressure-retaining purposes, and not for the corrosion resistance.

Table 6 Minimum mechanical properties for some duplex stainless steels according to ASTM A 240

Alloy	UNS Number	Yield Strength MPa (ksi)	Tensile Strength MPa (ksi)	Min. Elongation %
2304	S32304	400 (58)	600 (87)	25
2205	S32205	450 (65)	620 (90)	25
Alloy 255	S32550	550 (80)	760 (110)	15
Alloy Z100	S32760	550 (80)	750 (108)	25

Duplex stainless steels will become brittle at low temperature. With proper precautions, duplex stainless steels can be used down to -40°C (-40°F) or slightly lower, however only with impact testing.

Fabrication. Welding has often been the critical issue, however special filler metals combined with proper quality wrought products have solved most of the potential problems. It is still possible through ignorance for a fabricator to cause serious impairment to duplex stainless steel.

Ferritic stainless steels. These alloys form part of the AISI 400 series stainless steels, to which martensitic stainless steels also belong. The ferrite phase is achieved to removing all or most of the nickel content. In theory, ferritic stainless steels are not hardenable by heat treatment, or at least not intended to be hardened. Table 7 lists some of the ferritic stainless steels. Like the other families, there is a wide range of ferritic stainless steels with corrosion resistance varying from very low to very high. The 409 grade, used in very large quantities in automobile exhaust systems, is barely a stainless steel with only 11.2% chromium. The 3CR12 alloy, originally developed in South Africa, has just slightly higher corrosion resistance. There exist different variations of this grade from different producers. 430 is just slightly less corrosion resistant than 304, and is commonly used in less expensive housewares. The 444 alloy compares roughly corrosion-wise with 316L. The 29-4-2 alloy compares roughly with the 6% Mo grades. The ferritic stainless steels all are virtually immune to chloride stress corrosion cracking.

Table 7 Nominal composition (weight %) of some representative ferritic stainless steels

Alloy	C, H, or both*	UNS number	C_{max}	Cr	Ni	Mo	N	Other
409	C&H	S40900	0.08	11.2	0.1	-	-	Ti
3CR12	C	S41003	0.030	11.5	0.5	-	-	Ti
430	C	S43000	0.12	17	-	-	-	-
444	C	S44400	0.025	18.5	0.5	2.1	-	Nb+Ti
Alloy 29-4-2	C	S44800	0.010	29	2.2	3.8	-	-
446	H	S44600	0.20	25	-	-	-	-

* Primary end use: C = corrosion resistant, H = high temperature resistant, C&H = both

One of the difficulties with ferritic stainless steels relates to grain size and poor ductility. As the chromium content increases, the ductility for the same thickness decreases. In addition, significant weldability issues arise where welds can have very poor ductility in the heat affected zone. In practice the higher the alloy, the less the maximum thickness available. Some of the highest alloy ferritic alloys are available in a maximum thickness of only 0.6 mm (0.024 inch). This still makes these alloys suitable for applications such as heat exchanger tubes and sheets. Ferritic stainless steels have better thermal conductivity than austenitic stainless steels.

The lower chromium ferritic stainless steels find application where an improvement is required over carbon steel, but perhaps cannot justify the cost of 304, or where 304 is not suitable, e.g. for reasons of chloride stress corrosion cracking. Ferritic stainless steels can be embrittled by hydrogen however. They have been used for structural applications as well as some wear applications, e.g. for ore chutes, buckets, rail cars, conveyor decking plates, structural applications in and around water, and in fume and dust extraction equipment. 409 is also used in hot exhaust ducts - the very low chromium content means that it does not embrittle in the 370-510°C (700-900°F) range like the other ferritic stainless steels.

446 is a high chromium ferritic stainless steel sometimes used in smelters for high temperature high sulfur-containing gases, or whether there is some splashing contact with liquid copper. Not only does 446 embrittle in the 370-510°C (700-900°F) temperature range, it also embrittles due to sigma phase formation in the 540-870°C (1000-1600°F) temperature range, where the material can become as brittle as glass fairly quickly.

The mechanical properties of the ferritic stainless steels are quite reasonable compared to regular carbon steels, although ductility is not as good as for the austenitic stainless steels, as shown in Table 8.

Table 8 Minimum mechanical properties for some ferritic stainless steels according to ASTM A 240 or A 176

Alloy	UNS Number	Yield Strength MPa (ksi)	Tensile Strength MPa (ksi)	Min. Elongation %
409	S40900	205 (30)	380 (55)	22
430	S43000	205 (30)	450 (65)	22
444	S44400	275 (40)	415 (60)	20
29-4-2	S44800	415 (60)	550 (80)	20

Martensitic stainless steels. The main characteristic of this family of alloys is that they are hardenable by heat treatment, typically a quench and temper type. Some grades obtain a very high hardness, with 440C achieving typically 56-60 on the Rockwell C scale. Of course the ductility of the material suffers at such high hardness. Properties can be modified by a change in tempering temperature, a compromise between hardness and ductility. The best of the martensitic stainless steels has a corrosion resistance far inferior to 304 stainless. These alloys are stocked in the annealed condition for easy machining. Not only are these alloys soft in this condition, but will "rust" with exposure to water and air. In the hardened condition, they have slightly better corrosion resistance and will not rust. This author has visited several end users who had specified a martensitic stainless steel for a particular application, but forgot to specify that they wanted it in the hardened condition! Table 9 lists two martensitic stainless steels as representing the range of alloys. There are numerous others, many of which are proprietary.

Table 9 Nominal composition (weight %) of two representative martensitic stainless steels

Alloy	C, H, or both*	UNS number	C_{max}	Cr	Ni	Mo	N	Other
410	C	S41000	0.15	12.5	0.2	-	-	
440C	C	S44004	1.1	17	-	-	-	

* Primary end use: C = corrosion resistant, H = high temperature resistant, C&H = both

These alloys are primarily used for their hardness, or especially 410S, for their strength. They offer some abrasion resistance, but like the quenched and tempered alloy steels, they are not so successful in corrosive environments. They are produced mainly as round bar for shafting, but some plate and sheet are also available.

Table 10 lists some representative heat treatments and resulting properties.

Table 10 Nominal mechanical properties after hardening and tempering for two martensitic stainless steels, 25 mm (1 inch) diameter round bar

Alloy	UNS Number	Heat treatment T = tempered	Yield Strength MPa (ksi)	Tensile Strength MPa (ksi)	Elong. %	Hardness Rc
410	S41000	As Hardened				43
		T 200°C (400°F)	1000 (145)	1310 (190)	15	41
		T 315°C (600°F)	965 (140)	1240 (180)	15	39
		T 650°C (1200°F)	590 (85)	760 (110)	23	B97
440C	S44004	As Hardened				C60
		T 315°C (600°F)	1970 (285)	1900 (275)	2	C57

Precipitation hardenable stainless steels. The outstanding characteristic of this group of alloys is that they have the best corrosion resistance of any hardenable stainless steel. In this group, there are actually different families (martensitic, austenitic, semi-austenitic) as well as grades. The hardening mechanism is different from the martensitic stainless steels – these alloys form precipitates. Instead of a quench and temper heat treatment, these alloys are heated to a suitable temperature for a certain period of time. The hardenability is not dependent on cooling or quenching rate, but on time at temperature. Since there is no quenching, there is little risk for distortion. Parts do change size slightly (shrink) at a calculable rate. Table 11 gives the composition of two more common grades. Alloy 17-4 is by far the most commonly available and most used within the mining/metallurgical industry. Alloy 450 is a proprietary grade. In another oddity, these grades are hard in the annealed (as-delivered) condition, and in the hardening heat treatment, they can be made either harder or softer, according to the desire. These grades can be used in the annealed condition, although they will have better ductility in the precipitation hardened condition. Mechanical properties for both grades in some heat treated conditions are included in Table 12.

Some other precipitation hardenable grades are soft in the as-delivered condition.

Table 11 Nominal composition (weight %) of two precipitation hardenable stainless steels

Alloy	C, H, or both*	UNS number	C_{max}	Cr	Ni	Mo	N	Other
Alloy 17-4	C	S17400	0.07	16	4	-	-	Cu, Nb
Alloy 450	C	S45000	0.05	15	6	-	-	Cu, Nb

*Primary end use: C = corrosion resistant, H = high temperature resistant, C&H = both

Table 12 Minimum mechanical properties after different heat treatments for two precipitation hardenable stainless steels, according to ASTM A 564

Alloy	UNS Number	Condition	Yield Strength MPa (ksi)	Tensile Strength MPa (ksi)	Elong. % L/T	Hardness Rc
17-4	S17400	H900	1170 (170)	1310 (190)	10	40
		H1025	1000 (145)	1070 (155)	12	38
		H1150	720 (105)	930 (135)	16	28
450	S45000	Annealed	660 (95)	900 (130)	10	-
		H900	1170 (170)	1240 (180)	6/10	39
		H1150	520 (75)	860 (125)	12/18	26

Applications include shafting, bolts, pins, and other articles made from round bar. Alloy 17-4 is also available as plate and sheet, and is weldable.

Although a grade like 17-4 might cost more initially than a martensitic stainless steel, the cost of heat treatment is less, and likelihood of distortion is minimal. The improved corrosion resistance over the martensitic alloys is often a major advantage.

Cast Stainless Steels. The above discussion was about wrought stainless steels, but for the most part, comparable cast alloys are produced. They are given different designations, by both the American Casting Institute (ACI) and by UNS. The latter uses different numbers because the chemical composition does differ. See the section "Cast versus Wrought" under Selection of Materials – A Few Key Principles.

In Table 13, the names and family groupings of some more common cast stainless steels with similar wrought alloys are given. Some cast alloys have no ACI number, but have a UNS number, and vice versa.

In both the duplex stainless steels and the high temperature austenitic stainless steels, there are far more cast alloys that don't have a wrought equivalent. It should also be noticed that only one ferritic stainless steel is represented in the table below. Due to their low ductility in heavy sections, totally ferritic castings are not common.

Table 13 Identification of selected similar wrought and cast stainless steels

Wrought Alloy	Wrought UNS Number	Similar Cast Alloy	Cast UNS Number
Austenitic			
304	S30400	CF8	J92600
304L	S30403	CF3	J92500
304H	S30409	(CF8)	J92590
316	S31600	CF8M	J92900
316L	S31603	CF3M	J92800
317L	S31703	CG3M	J29999
321	S32100	---	J92630
347	S34700	CF8C	J92660
309	S30900	CH20	J93402
310	S31000	CK20	J94202
High Performance Austenitic			
Alloy 20	N08020	CN7M	N08007
6% Mo (18%Ni)	S31254	CK3MCuN	J93254
6% Mo (25% Ni)	N08367	CN3MN	J94651
Duplex			
2205	S32205	CD3MN	J92205
Alloy Z100	S32760	CD3MWCuN	J93380
Ferritic			
446	S44600	CC50	J92615
Martensitic			
410	S41000	CA15	J91150
420	S42000	CA40	J91153
Prec. Hardenable			
17-4	S17400	CB7Cu1	J92180

NICKEL ALLOYS.

Nickel alloys for the mining/metallurgical industry can be broken down into 4 main categories - commercially pure nickel, chromium-containing alloys (which will be broken further into 2 subgroups), nickel-copper alloys, and nickel-molybdenum alloys. The chemical compositions of some of the more significant alloys are included in Table 14.

Commercially pure nickel. Alloy 200 and its variations are used commercially used for high temperature caustic, but find little use in the mining industry. Nickel plating, whether by the electroless or electrochemical methods, is outside of the scope of this paper, but does find use in this industry.

Chromium containing alloys. This group has the largest number of alloys and the broadest usage in the nickel alloy family. It can be subdivided into 2 groups, the nickel-chromium alloys and the nickel-chromium-molybdenum alloys.

Nickel-chromium alloys. The main alloy in this group is Alloy 600, with Alloy 601 being an important variation strictly for high temperature applications. Alloy 600 is used for corrosive services, mostly in caustic, but also for high temperature halogen service. With 76% nickel, Alloy 600 is immune to chloride stress corrosion cracking, but will pit.

Nickel-chromium-molybdenum alloys. This group of alloys is the most useful to the mining/metallurgical industry because of their high resistance to acids, both oxidizing and reducing. They perform well in highly acidic chloride containing environments, and are used extensively in wet flue gas scrubbers. Alloy 625 is useful in a broad range of applications, but is being replaced in many applications by the "C" family of alloys. C-276 is still the most common, but various new improvements have been commercialized with C22, Alloy 59, and now C2000. There is some data to suggest that these alloys suffer from transpassive corrosion in some oxidizing autoclave environments, and more work needs to be done with some of the higher chromium alloys such as G-30, which may perform better.

Table 14 Nominal composition (weight %) of some selected nickel alloys

Alloy	C, H, or both*	UNS number	Ni	Cr	Mo	Fe	Cu	Other
Alloy 200	C	N02200	99.0 min					
Alloy 600	C&H	N06600	76	15.5		7		
Alloy 601	H	N06601	61	23		13		Al
Alloy 625	C&H	N06625	61	21	9	4		Nb, Ti, Al
Alloy C-276	C	N10276	57	16	16	5		W, V
Alloy C-22	C	N06022	56	22	14	4		W, V,
Alloy G-30	C	N06030	43	30	5	15	1.5	Nb, W
Alloy 400	C	N04400	66				31	
Alloy K-500	C	N05500	66				29	Al, Ti
Alloy B-2	C	N10665	69		28			

* Primary end use: C = corrosion resistant, H = high temperature resistant, C&H = both

Nickel-copper alloys. The two main alloys in this group are Alloy 400 and a precipitation hardenable variation K-500. The latter is useful for shafting, and for parts requiring increased hardness or resistance to galling. These alloys are used extensively in brine systems, and certain applications in fresh water systems. Like most copper-containing alloys, they are subject to corrosion by ammonia. They have poor resistance in oxidizing media.

Nickel-molybdenum alloys. This group consists of Alloy B-2 and other variations. These alloys show excellent performance in pure hydrochloric acid up to the boiling point, but very small quantities of metallic oxidizing ions, even a few parts per million, or even oxygenated acid, can result in very high corrosion rates. This author has heard of a few occasions where these alloys have been used where they shouldn't have, and where this most expensive alloy had lasted only weeks. Where it is good, it is usually very good. In-plant testing is strongly advised before specifying such alloys.

Cast nickel alloys. There exist roughly similar cast versions of the wrought alloys, but often there are sufficient differences. Cast nickel alloys are found in ASTM A 494.

ALUMINUM

Aluminum falls into the "passive" group of materials, meaning that they depend on a passive oxide layer for their corrosion resistance. They can exhibit good corrosion resistance in the middle pH ranges, but not in strong acids or bases. This makes them a useful material in many air and water environments. Although pure aluminum is soft, most commercial alloys are either strain hardened (indicated with an H suffix) or heat treatable (indicated with a T suffix). This gives them a high strength to mass ratio, an advantage used in many structural applications. Aluminum alloys have good low temperature ductility, making them useful for certain cryogenic applications. Aluminum alloys are fairly easily fabricated.

Table 15 lists the UNS classification of the alloy types.

Table 15 Wrought aluminum alloy UNS number structure

UNS Number	Major Alloying Element
A91xxx	None, 99.0% Al minimum
A92xxx	Copper
A93xxx	Manganese
A94xxx	Silicon
A95xxx	Magnesium
A96xxx	Magnesium and silicon
A97xxx	Zinc
A98xxx	Miscellaneous

Note: Cast alloys are numbered similarly, but with the first number as 0, e.g. A02000

Corrosion resistance. Most aluminum alloys have similar corrosion resistance, although the copper containing alloys are the least corrosion resistant. A rough pH guide for use is 4.5 to 9.5. This qualifies for most natural waters and external atmospheric conditions. Mine waters will often fall outside this range, and the presence of metallic ions such as iron, copper, lead, and mercury will cause accelerated corrosion. Aluminum is used for specific chemicals, e.g. hydrogen peroxide, highly concentrated nitric acid (>90%), and many organic chemicals. However, certain chloro-organic compounds such as ethylene dichloride or certain alcohols can rapidly corrode aluminum generating hydrogen, creating a fire hazard.

Applications. Aluminum is used for many structural applications, such as roofing, building panels, hand railings, ladders, fencing, and shaft man cages. Sometimes pre-painted aluminum is used for buildings not only to improve lifetime, but to lower reflectivity. Other applications include underground ventilation ducting, heat exchangers for mine refrigeration plants, ore hoppers, and air and conduit piping (*Andrews 1997*). Use of aluminum as well as magnesium and titanium underground in certain mines, especially coal mines, may be forbidden or restricted, due to the possibility of fires or explosions related to a thermite reaction (*Forrester and Bonnell, 2001*). When these light metals come in frictional contact with iron oxide e.g. rust on steel, a violent reaction occurs that can ignite a flammable atmosphere.

COPPER AND COPPER ALLOYS

Some use of copper and copper alloys is made in the mining/metallurgical industry. The copper alloys can be divided roughly into categories of brasses (copper-zinc alloys), copper-nickel alloys, and bronzes (copper primarily alloyed with any elements other than the previous two, e.g., tin, silicon, aluminum). Table 16 lists a number of these grades. Nickel-copper alloys were covered under nickel alloys.

Table 16 Nominal composition (weight %) of some common copper and copper alloys

Name	UNS number	Cu	Zn	Sn	Al	Ni	Other
Wrought Alloys							
Copper (ETP)	C11000	99.9					
Red Brass	C23000	85	15				
Yellow Brass	C27000	69	31				
Admiralty Brass	C44300	72	27	1			As 0.1
Phosphor Bronze	C52400	90		10			P 0.3
Aluminum Bronze D	C61400	90	0.2		7		Fe 3
High Silicon Bronze	C65500	95	1.5			0.6	Si 3
90-10 Cu-Ni	C70600	86	1			10	Fe 1.5
70-30 Cu-Ni	C71500	68	1			30	Fe 1
Castings							
Leaded Red Brass	C83600	85	5	5			Pb 5
Manganese Bronze	C86500	57	40	1		1	Mn 1
Tin or G Bronze	C90500	87	2	10		1	

Corrosion resistance. Copper is resistant to most natural waters, except for soft acidic waters, and within certain velocity limitations. It is especially useful in stagnant potable waters. The brasses which contain tin and usually another element such as arsenic (admiralty brasses) are also used in more aggressive waters. The copper nickels are useful also in more aggressive waters, and can tolerate a higher flow rate than other copper alloys. Some of the bronzes can give adequate corrosion resistance in weak sulphuric acid solutions. A good source of information about copper and its alloys is the Copper Development Association (www.copper.org)

Applications. In addition to use in various types of water for piping, heat exchangers, etc., there are some copper alloys used in special applications where certain properties are required. Beryllium bronzes are hard and used for their non-sparking properties in tools. Other copper alloys have excellent anti-galling properties.

TITANIUM

Titanium and its alloys were rarely used in the mining/hydrometallurgical industry previously, except where very aggressive chemicals such as hydrochloric acid were used. It was also used to some extent for seawater applications. Titanium belongs to the reactive class of metals, to which zirconium and tantalum also belong. It has become very useful in the newer hydrometallurgical processes involving high temperature and high pressure. Titanium is often alloyed with small amounts of other elements (primarily palladium and ruthenium) to improve its corrosion resistance, especially in reducing acids. Table 17 lists some of the more corrosion resistant titanium alloys. The very expensive Titanium Grade 7 with 0.15% Pd used to be the only step up from Titanium Grade 2 (unalloyed titanium), but now a large number of alloys exist which are lower in cost than Grade 7.

Table 17 Specific alloying additions (weight %) of some titanium alloys

Name	UNS number	Pd	Ru	Mo	Ni
Grade 2 (unalloyed)	R50400				
Grade 7	R52400	0.15			
Grade 12	R53400			0.3	0.7
Grade 16	R52402	0.06			
Grade 17	R52252	0.06			
Grade 26	R52404		0.11		
Grade 27	R52254		0.11		

Titanium alloys can offer excellent corrosion resistance in a wide variety of oxidizing and reducing environments, which cannot be covered in this paper. There are many excellent sources of information, such as the Titanium Development Institute (www.titanium.org).

A few words of caution about titanium. Titanium can burn in strongly oxidizing conditions. Titanium is successfully used in wet chlorine, but can burn in dry chlorine. It can burn in highly oxygenated conditions, which has occurred in some autoclave circuits. Titanium should also not be used in liquid oxygen nor red fuming nitric acid. Titanium becomes brittle when exposed to hydrogen ions, which can even originate from corrosion, causing hydriding. Other seeming innocuous chemicals such as methanol (which may be used as a cleaning solution) can result in degradation of the metal. Even small traces of fluorides may cause hydriding and accelerated corrosion rates.

OTHER METALS AND ALLOYS

Other metals such as lead, cobalt, zinc, tin, magnesium, tantalum, zirconium, silver, gold, and platinum may be used for specific purposes, but are not covered in this paper.

SELECTION OF ALLOYS - A FEW KEY PRINCIPLES
Operating Conditions

In order to select the proper material, it is necessary to know the exact conditions. In the author's experience, most of the time there is not sufficient data given by engineering staff or plant staff to make anything other than general statements. Following is a list of things to consider in order to help to determine what the metal will actual experience and will hopefully be able to handle.

1. Impurities. Plant and process personnel often know what chemicals are in the process that are important to the process, but are not always aware of the total makeup of the solutions. Knowing levels of impurities such as chlorides and fluorides are critical. Even parts per million of certain oxidizing ions can either destroy some alloys or help keep passive other ones. A full analysis is a very good starting point. It is important to know if a solution is aerated or de-aerated.

2. Heat flow. Often the temperature in a tank will be set at one level, but there are areas that may have a much higher temperature.

3. Upset conditions. Although it is not possible to consider every possibility, nor to guard against them, it is a good exercise to look at the various scenarios. For example, if a valve is jammed open or shut, or if a tank is only partially filled, or if one part of the process has to be bypassed, what is the affect on the temperature, concentrations, etc. This is especially important if hazardous chemicals are involved.

4. Process changes. Again, one cannot look at every possibility, but often process engineers know that they are likely to modify the process, often to produce more product. Will higher temperatures or higher velocities be used. What about different ores – will they result in a change in contaminants and will that have a positive or negative affect on corrosion rates.

Design

Each material has its strengths and weaknesses. Often one is tempted to replace materials using exactly the same sizes. Although sometimes this works, sometimes it doesn't and often money is spent needlessly. Carbon steel components are often made thick to allow for a measurable corrosion rate. If replacing with stainless steel, it may be that there will be no measurable corrosion and the material thickness can be reduced significantly. Carbon steel and copper often have velocity limitations, so heat exchangers and equipment are designed for low velocity flow. Stainless steels often suffer localized attack in stagnant conditions, so a high velocity is preferred to keep the surface clean and avoid pitting. Detailed discussion of design considerations for stainless steels is given in NiDI Publication No. 9014.

Cast versus Wrought product form

There are distinct differences between nominally the same alloy grade in cast and wrought forms, although sometimes the difference doesn't mean very much, and one is not necessarily better than another. However it is important to be aware that there are differences, and when they may become a factor.

Differences based on production methods. Typically wrought stainless steel alloys are melted and refined in modern equipment in batches of 20 to 150 tonnes. The refining is well controlled, impurity levels generally very low, and the difference between aim and actual chemical composition is very small. One should remember that this aim composition is the aim of the mill, not necessarily what is best for e.g. the corrosion resistance of the steel. As an example, the molybdenum content of 316L according to UNS S31603 is 2.0 - 3.0%. Steel makers today can control the molybdenum content quite tightly to between 1.95 and 2.05% (and remember that 1.95% meets the requirement of 2.0% minimum!) Wrought products are produced in series on highly automated lines, meaning less chance for human error and a high degree of consistency. In contrast, stainless steel foundries produce castings virtually individually by the piece. The furnace size can vary from tens of kilos to 50 tonnes. There is often no refining stage for the steel at a foundry, the impurity levels are determined by the choice of raw materials, and the quality can vary considerably from foundry to foundry, and to some extent, from piece to piece. With castings, there can be considerable weld repair at the manufacturing stage to correct problems from the pouring stage. Some castings specifications call for heat treatment after the weld repairs, e.g. ASTM A 744, others e.g. ASTM A 743 do not, although often there is an option for that as a supplementary requirement. In critical applications, heat treatment of the casting after all welding may be necessary. In castings, some degree of porosity is normal, whereas in wrought products, there should be no porosity. Various non-destructive examination methods are available for castings, but it can be difficult to know which ones to specify. A new ASTM specification, ASTM A 990, "Specially Controlled for Pressure Retaining Parts for Corrosive Service" is intended for the most severe services.

Differences in alloys. Because castings are often melted in smaller quantities, there is the potential to customize the chemical composition for a particular application. Where it might be virtually impossible to interest a wrought mill in producing 10 tonnes of a special chemistry sheet, there would be many foundries eager and willing to quote on the same quantity of castings. There are many unique casting alloys that are virtually impossible to make via the wrought route. This includes alloys that have poor hot or cold ductility, and alloys that cannot be easily welded into components, but can be cast to their final shape.

With wrought alloys, with the exception of a number of specialty alloys, the names of the grades are standardized and easily identifiable. With cast alloys, there are a host of trade names that can make identifying alloys and finding substitutes a chore.

There can be some significant differences in performance between nominally identical wrought and cast alloys, for example between wrought and cast "Alloy 20". The cast version contains less nickel and is niobium-free, which means that it should not be welded. Other differences relate to metallographic structure in the austenitic stainless alloys. Wrought 304 stainless steel is normally close to 100% austenitic in structure, whereas the cast version, CF8, usually contains between 10 and 20% ferrite. These differences in structure are intentional, controlled via both chemistry and heat treatment. The ferrite component of most austenitic castings is intended to reduce defects during casting, especially hot tears. A magnet is strongly attracted to the ferrite in these alloys, whereas a magnet has only faint or no attraction to wrought austenitic alloys. However, for the same reasons as castings, welds have some ferrite and a magnet will be fairly strongly attracted. However, certain chemicals can preferentially corrode this delta-ferrite, and especially in the higher molybdenum alloys such as CG3M, a 317L (3.0-4.0% Mo) type alloy. Some of the even higher molybdenum cast stainless alloys are fully austenitic, e.g. CK3CuMN and CN3MN, both with over 6% Mo. The fully austenitic structure creates some casting challenges, and as a word of warning, discussions should be had with the selected foundry regarding heat treatment procedures for these alloys. The ASTM specifications for these grades specify too low a minimum heat treatment temperature.

The larger grain size of castings can be an advantage in high temperature applications with respect to creep strength. Castings grades typically have a higher carbon content, also an advantage for high temperature strength, although a disadvantage in corrosive applications. For example, 310S wrought product has a carbon maximum of 0.08%, whereas the rough cast equivalent, CK20, has 0.20% carbon maximum. There is also a higher carbon cast version, HK40, which is more common, and that has 0.45% carbon max. The higher carbon does decrease room temperature ductility, but castings are usually cast to shape, so this is not usually important.

Cost and Availability

One engineer the author knows is constantly reminding people that availability of a material is a material property just like strength or modulus of elasticity. Certainly price and availability do go hand in hand. 304L and 316L are the most commonly available stainless steels, produced by dozens of producers around the world. There is a wide variety of product forms and much competition which keeps the price down. Conversely, proprietary grades available from only one supplier can not only be difficult to obtain, they can be very expensive. Try to specify alternative alloys where possible. For example, there are 4 or 5 different variations of the 6% Mo stainless steels, all with roughly (but not exactly) the same corrosion resistance and mechanical properties. One supplier might be good on plate, another on sheet, and another on bar. By giving options, it is possible for fabricators to put together a reasonably priced package by having the suppliers compete against each other. On the other hand, if one supplier is actively working with you in supplying information and test coupons etc., it is only fair to try to work with them to resolve the commercial issues that may arise. Put a value on the technical support that they give you.

RESOURCES
Prior experience

If the process is not new, but exists at another plant, much can be learned there about materials selection. One should be aware that there can be significant differences based on compositional differences in the ore body, differences in climate, water quality, availability of skilled labor, and even differences in the operating philosophy of the company. Therefore it is impossible to say that if Alloy Q worked at Plant A, then Alloy Q will work at Plant B. It is not sufficient to just look at the current and past history of materials' performance; it is important to thoroughly evaluate the solution chemistries, abrasive qualities, and other factors such as changes to the process, process upsets that may have occurred, and maintenance practices. Materials selection data for existing plants needs to be used with caution. It is valuable to know that a pipe has lasted a certain number of years, but it is far more valuable to know if it has its original wall thickness left, or if it is ready to be replaced.

Much valuable information can be gathered from test coupons exposed to the actual solutions. It takes time and effort to put together a test program and then collect and evaluate the data afterwards. It is almost becoming a lost art. Test coupons should be exposed for a minimum of 3 months, and preferably a full year. When a corrosion failure occurs, most engineers wish they had installed test coupons a year earlier.

The comparison process requires dedicated teamwork from process specialists, design specialists, maintenance specialists, materials specialists, fabrication specialists, etc.

Selection of materials for a new plant using a new process is an extremely difficult task. So many factors have to be taken into account, with little experience to rely upon. However, important clues can be gained from similar if not identical plants if they exist, pilot plants, and perhaps even chemical plants that might have similarities.

Corrosion Handbooks

There are many good Corrosion Tables and Handbooks available. In the hands of an experienced materials specialist, they provide clues to what materials might work. In the hands of the inexperienced, they have often led to very bad materials selection errors. It works both ways – sometimes an expensive material is chosen when a more cost-effective one would have worked as well, whereas most of the time, a material is chosen that fails quickly, and sometimes catastrophically. Corrosion tables have many limitations. They are most often based on single, reagent grade chemicals, not what one has in a metallurgical plant. Most tables report only a general corrosion rate, valid if uniform corrosion is the corrosion mechanism. With stainless steels and nickel alloys, localized corrosion is more likely to occur and lead to failure. Corrosion tables rarely take into account velocity effects, abrasive effects, effect of oxygen concentration in the corrosive, vapor phase affects, sensitivity to contaminants, etc. However, they are valuable and can give us clues to which type of alloys to consider.

Internet

Much useful information is available directly from the Internet from reliable sources. Reliable sources are sometimes the same organizations and companies that you would go to anyhow for information, but there are new only net-based suppliers. A good search engine such as Google is useful in identifying unknown grades and trade names.

SUMMARY

This paper attempts to discuss in general terms the properties of various metals used in the mining/metallurgical industry, with both advantages and disadvantages of the various types. A few tips are given about selection of materials, as well as some resources for further information.

REFERENCES

R.H.C. Andrew. 1997. Practical Guidelines for Corrosion Protection in the Mining and Metallurgical Industry. NACE International, Houston TX.

P.E. Arnvig, H. Andersen, et al. 1998. SCC of Stainless steel Under Evaporative Conditions. Corrosion 98, Paper 251. NACE International, Houston, TX.

D.J. Forrester, G.W. Bonnell. 2001. The use of light metals and their alloys in underground coal mines. CIM Bulletin, Vol. 94, No. 1054, p76ff

STANDARDS AND SPECIFICATIONS

ASTM Standards. Books of standards are published annually by ASTM International, West Conshohocken, PA. Individual standards are updated on a regular basis, some as many as several times a year. Individual standards are available on a download bases from the ASTM website, www.astm.org

UNS. Metals and Alloys in the Unified Numbering System. 9th Edition. 2001. Society of Automotive Engineers, Warrendale, PA

SELECT RECOMMEND HANDBOOKS

Printed paper editions are given, but note that many of these are available in electronic version, either on CD or downloadable from web sites.

CASTI Handbook of Stainless Steels & Nickel Alloys. S. Lamb, editor. CASTI Publishing, Edmonton, AB 1999

Steel Products Manual: Stainless Steels. H. Cobb, editor. The Iron and Steel Society, Warrendale, PA 1999

Metals Handbook, Desk Edition. 2nd Edition. J.R. Davis, editor. ASM International, Metals Park, OH. 1998

The Metals Black Book, Ferrous Metals. 4th Edition. J.E. Bringas & M.L. Wayman, editors. CASTI Publishing, Edmonton, AB. 2000

The Metals Red Book, Volume 2, Nonferrous Metals. 3rd Edition. J.E. Bringas & M.L. Wayman, editors. CASTI Publishing, Edmonton, AB. 2000

Avesta Sheffield Corrosion Handbook for Stainless Steels. 8th Edition. Then Avesta Sheffield, now AvestaPolarit, Avesta Sweden 1999

Corrosion Data Survey, Metals Section. 6th Edition. D.L. Graver, editor. NACE International, Houston, TX. 1986

Woldman's Engineering Alloys. 9th Edition. J.P. Frick, editor. ASM International, Metals Park, OH. 2001

Stahlschlüssel (Key to Steel). 2001 edition. Verlag Stahlschlüssel Wegst GmbH, Marbach, Germany. 2001

SOME USEFUL BROCHURES AVAILABLE FROM THE NICKEL DEVELOPMENT INSTITUTE

Available either as pdf downloads from www.nidi.org or in print. Many others available.

No. 12 014 Copper-Nickel Fabrication
No. 12 007 Copper-Nickel Alloys - Properties and Applications
No. 11 022 Castings - Stainless Steel and Nickel-Base
No. 11 021 High-Performance Stainless Steels
No. 11 018 Properties and Applications of Ni-Resist and Ductile Ni-Resist Alloys
No. 11 017 Ni-Hard Material Data and Applications
No. 11 012 Guidelines for the Welded Fabrication of Nickel Alloys for Corrosion-Resistant Service
No. 11 011 Machining Nickel Alloys
No. 11 007 Guidelines for the Welded Fabrication of Nickel-Containing Stainless Steels for Corrosion-Resistant Service
No. 11 006 Nickel in Powder Metallurgy Steels
No. 10 088 Nickel Plating and Electroforming - Essential Industries for today and the Future
No. 10 086 Corrosion and Heat-Resistant Nickel Alloys , Guidelines for Selection and Application
No. 10 081 Properties and Applications of Electroless Nickel
No. 10 075 Selection and Use of Stainless Steels and Nickel Bearing Alloys in Nitric Acid
No. 10 074 The Corrosion Resistance of Nickel-containing Alloys in Hydrofluoric Acid, Hydrogen Fluoride, and Fluorine
No. 10 068 Specifying Stainless Steel Surface Treatments
No. 10 064 Clad Engineering
No. 10 063 Selection and Use of Stainless Steels and Nickel Bearing Alloys in Organic Acids
No. 10 057 Selection and Performance of Stainless Steels and Other Nickel Bearing Alloys in Sulphuric Acid
No. 10 021 Procurement of Quality Stainless Steel Castings
No. 9014 Design Guidelines for the Selection and Use of Stainless Steels

Elastomers in the Mineral Processing Industry

Paul Schnarr[1], Leon E. Schaeffer[2], and Hans J. Weinand[3]

ABSTRACT

Elastomers have been used extensively in the mineral processing industry since the 19th century. They are used, cost effectively, mostly because of their resistance to wear, impact, flexing, and corrosion, but also for their ability to control vibration and noise. Elastomers are used in process operations from mining, comminution and separation through to final product handling. The most appropriate elastomer for a given application is selected for its particular properties.

INTRODUCTION

The definition of an elastomer is "a polymeric material, such as a synthetic rubber or plastic, which at room temperature can be stretched under low stress to at least twice its original length and, upon immediate release of the stress, will return with force to its approximate original length."[1] This property allows elastomers to be used for many products which are subject to deformation or compression and must not be destroyed by such forces.

Abrasion resistance is often the main feature in choosing an elastomer based product over alternate products. In the mineral processing industry, abrasion often results from a slurry's suspended particles coming in contact with the elastomer in a combination of impinging and sliding action. The erosive wear of a metal alloy depends on the alloy's microstructure and hardness; an elastomer's resistance to abrasion, especially of the impinging type, depends on the resilience of the elastomer. When a particle contacts the wear surface, an elastomer deflects, then returns to its original position, with little or no wear. This unique property of elastomers generally gives them a substantial increase in abrasion resistance over metals, provided the particle size and speed are not too great.[2]

Another main reason for using an elastomer product is the chemical resistance of the elastomer. In general, the rule "like dissolves like" applies. This means that nonpolar elastomers such as natural rubber will handle most chemicals used in the mineral processing industry, which are typically water based polar solutions. When nonpolar solvents or oils are encountered, elastomers such as nitrile-butadiene rubber, must be used. Field experience, collected over many years, provides the best basis for successful material selection. Chemical resistance charts are also useful. In difficult cases, especially those involving mixed and/or unknown chemicals at varying temperatures, immersion testing in the customer's actual plant is the best basis for a choice of materials.

Most elastomers used in mineral processing are actually used in composites. They can be bonded to textile fibers such as nylon, polyester, polyaramid and others to increase strength and stiffness, with a loss in elongation. They can be bonded to metals to combine the strength and rigidity of the metal with the elasticity of the elastomer.[3] They can be bonded to various plastics, often to get a surface with a very low coefficient of friction.

An elastomer compound development is usually a compromise, which can be shown as an equilateral triangle.

---

[1] Technical Director-Rubber Engineering, Salt Lake City, UT
[2] Rubber Engineering, Salt Lake City, UT
[3] Rubber Engineering, Germany

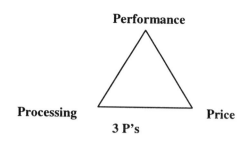

3 P's

The rubber compounder tries to achieve the best compromise by choosing the elastomer, or blends of elastomers, and then adding various fillers and chemicals, of which there are an infinite number of combinations. The end customer sees only the Performance and Price part of the triangle but the compounder has to deal with processing through various mixing, forming and vulcanization steps.

HISTORY

Ancient mesoamerican people were processing rubber by 1600 B.C. The latex was combined with a species of morning glory vine to give products with enhanced elastic behavior, such as rubber bands to haft stone ax heads to wooden handles.[4] It is easy to speculate that such an ax could have been used in a mining operation. Christopher Columbus was probably the first European to see natural rubber. The mention of rubber trees is to be found in the eighth (decade) of De Orbe Novo by Pietro Martire d Angliera, published in Latin in 1516.

Very little use was made of rubber until 1823 when Charles MacIntosh patented the use of coal tar naptha to dissolve natural rubber and use it to produce waterproof garments or "macintoshes."[5] These useful garments were undoubtedly used in wet mining conditions. About 1839 Charles Goodyear discovered that rubber chains could be linked together, i.e. vulcanized, by heating with sulfur and white lead. In this process, sulfur linkages form bridges between the rubber chains.[6] Thus the first true elastomer was produced. Goodyear made little commercial progress. In 1843 Thomas Hancock of London took out a patent similar to Goodyear's and was more commercially successful. One of his first products was thin sheets of rubber and these were probably used in some form of mineral processing. By 1850 a whole range of rubber articles was available[5] and though it is doubtful many of these were specifically designed for mineral processing, it is likely many did find use.

Advances in elastomer technology have been essentially continuous since the early discoveries. Most were made with the tire industry in mind, as it is by far the largest user of elastomers. Since many of these advances involve basic properties such as abrasion, tear, and flex resistance they can be readily adapted by compounders, who are designing elastomer products for the mineral processing industry.

POLYMER TYPES-ASTM D2000

ASTM D2000 is the standard classification system for rubber products in automotive applications. ASTM D2000 3.1: Purpose: states " The purpose of this classification is to provide guidance to the engineer in the selection of practical, commercially available rubber materials, and further to provide a method for specifying these materials by the use of a simple "line call-out" designation." It also serves the needs of other industries in arranging rubber products into characteristic material designations. These designations are determined by types, based on resistance to heat aging, and classes, based on resistance to swelling in oil. Table I categorizes some of the more common types used in mineral processing. The abbreviations are per ASTM D1418-01.

Heat and Oil Resistances of Elastomers

ASTM D-2000 Designations

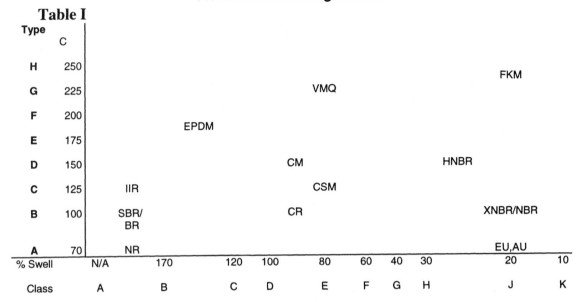

Table I

There are also grade numbers, suffix letters, and numbers to describe necessary qualities beyond the basic requirements. There are also standards groups in various other countries.

COMMONLY USED ELASTOMERS

Natural Rubber (NR)

This is probably the biggest single polymer used in mineral processing. The latex from the trees of Heveas basiliensis is the only important commercial source. The hydrocarbon portion of NR is cis-1,4 polyisoprene, which comprises more than 10% of a mixture with naturally occurring resins, proteins, sugars, etc.[5]

As to the Price part of the 3 P's, in early 2002, (all future price comparison in this paper will be based on early 2002 prices) NR is the lowest priced polymer. It is even lower priced than carbon black, the most important filler, on a volume basis, so that high polymer content "pure gum" compounds are relatively inexpensive. Because of the supply and demand, prices can fluctuate quite widely year by year. Several years ago the prices were approximately three times current prices. It was also more expensive than most common synthetic "tire grade" polymers.

As to the Processing part of the 3 P's, NR is very good. It mixes, calendars and extrudes well. It has good green (uncured) strength and very good building "tack," which means it adheres to itself and other rubber compounds very well and tends not to stick to the metal surfaces of the processing equipment. This building tack and green strength makes it very suitable for composites such as tires, hose, belts, etc where different components have to be adhered to each other before vulcanization. It also makes for good seam quality in tank lining. NR is often used as a tie gum in halobutyl lining for that purpose. It also makes it relatively easy to line the ID of pipes, especially small diameter ones, where a tube is formed and expanded and must adhere to the pipe before and after curing.

As to the Performance part of the 3 P's NR is often the preferred polymer. "Pure gum" compounds (often defined as those with a specific gravity of less than 1.0) with a hardness of approx. 40 Shore A, are often used in sheeting, pump liners, pipe lining, etc. because of their very high resilience, combined with good strength. Strength can also be characterized as tear resistance or as cut growth resistance. This high strength of soft natural rubber is certainly due to its ability to undergo strain-induced crystallization.

The physical properties of NR compounds (and synthetic rubber compounds) are affected by the type and amount of fillers used. Carbon black is the most commonly used filler. Increasing amounts of carbon black increases the hardness and modulus of the vulcanizates. Resilience and resistance to impinging type abrasion decreases along with elongation. Tensile and tear strength and resistance to sliding type abrasion increase, with increased loading, up to certain point, and then decrease. The black loading is adjusted to maximize the desired properties. Fine particle sized silica can also be used with much the same response as carbon black. The relatively new highly dispersible forms, in combination with organo-silane coupling agents are the basis of the "green tire" tread, which combines good abrasion resistance along with low rolling resistance.[8] These advantages also translate to products used in mineral processing. In NR compounds, this technology can be used to produce compounds with the abrasion resistance of a "pure gum" compound but with the cut and tear resistance of a harder black loaded compound. In applications, such as mill circuit discharge pumps handling slurries with up to 12mm diameter particles, the service life is increased considerably for the pump liners.[9]

The favorable elastic properties of NR manifest themselves in very low damping (low hysteresis) and low heat build up in dynamic deformations.[3] This makes NR especially useful in dynamic applications such as vibration elements used in centrifuges and most importantly in very large off-the-road tires, which also use mostly NR because of its low heat build up, dynamic fatigue resistance and excellent resistance to cutting and chipping.

NR vulcanizates are not as heat resistant as most synthetics. 70 C is normally considered the maximum service temperature in dry applications. Special compounding techniques using very low sulfur or no sulfur cure systems, along with protective antioxidants can extend this limit to over 100°C, especially in wet service. Non-protected NR vulcanizates are very prone to ozone cracking, especially in non-black compounds. If they are stored improperly, under stress, they may even crack before being put into service. Black compounds which contain proper amounts of wax and modern straining antiozonants are far less prone to cracking.

Since NR is nonpolar, its vulcanizates can swell to several times their original volume in nonpolar solvents such as mineral oil, toluene, gasoline, diesel, etc. NR has good resistance to polar fluids such as many mild acids and bases typically found in the mineral processing industry.

The compression set resistance of NR can be very good, with proper compounding, at relatively low temperatures (<70°C), so it is widely used in load bearing applications.

The low temperature properties of NR vulcanizate are excellent without any special compounding and exceeded only by polybutadiene or silicone.

NR can be blended with various synthetic rubbers with a diene component. Synthetics can be added to NR to improve heat resistance, sliding abrasion resistance, and give moderate oil resistance. Natural rubber can be added to synthetics to improve building tack and green strength and to lower cost.

Polyisoprene (IR)

Initial attempts to make synthetic rubber date back to the mid 1800's. Two samples labeled "Isoprene" and "Artificial Rubber" were recently discovered at the University in England, where Sir William Telden conducted his experiments in 1892.[10] Commercialization of synthetic polyisoprene was started in 1960.

The performance of synthetic polyisoprene (IR) is somewhat similar to its natural counterpart.

Processing is better in some ways in that it does not require mastication and generally mixes, calendars and extrudes faster and with more consistency. Processing is poorer in that it lacks the green strength and tack of NR.

The main reason it is now not used more is that it is currently priced much higher than NR. A number of plants processing it have been shut down and a lot of material is coming out of Russia, which had a large capacity for strategic reasons.[11]

Styrene Butadiene Rubbers (SBR)

Large-scale production of SBR started in WWII because of the shortage of NR. It was known in the U.S., as GR-S (Government Rubber-Styrene).

It is a major polymer in passenger and light truck tires but is little used in large off-road tires because it's compounds have higher heat build up than NR and less resistance to cutting and tearing. It does have good resistance to sliding abrasion and impact so it can be used in products such as belts, sheet rubber, truck box liners, and mill liners. The heat resistance is somewhat better than NR and special compounding can take the maximum temperature rating up to approximately 110°C.

Processing is generally good, but SBR lacks the tack and green strength of NR. It does have lower nerve and flows better than NR and can be blended with NR to improve flow in large transfer molded products.

The price of SBR is currently a little higher than NR, but historically has averaged a little less, and does not fluctuate as widely.

Butadiene Rubber (BR)

Most of the butadiene rubber used is solution polymerized with a very high cis-1, 4 configuration.

BR is seldom used as the sole polymer in any mineral processing application as the strength is very low and processing very difficult. BR is most commonly blended with NR or SBR, with most of the technology coming from the tire industry.

When blended with NR, or SBR in compounds containing filler to give hardness, usually in the 50-80 Shore A range, BR improves abrasion resistance, heat aging, resilience, reversion resistance, fatigue resistance and low temperature flexability.[12]

Sidewalls of passenger tires are commonly based on blends of natural rubber and Cis-polybutadiene. They have outstanding resistance to catastrophic crack growth under severe conditions, such as when a tire runs into a curb or hits a deep chuckhole; but also a sidewall must resist crack growth over long times at smaller strains experienced during "normal" rolling.[13] This resistance to different types of crack growth is very important for many products used in mineral processing other than tires, such as belts, especially those with flanges, hose and mill liners, which often use blends of NR and BR.

The processing of BR blends is generally good, NR/BR blends have good tack and green strength so they can be used for composites.

The price of BR is currently lower than that of NR and the same as SBR, but historically it has commanded a premium over both.

Butyl Rubber (IIR)

IIR is a copolymer of mainly isobutylene and a small portion of isoprene. This largely saturated chain determines its main properties; good resistance to oxidative and ozone degradation, to chemicals, and a low gas permeability[3] It is very nonpolar and therefore resistant to polar chemicals, especially acids and bases at higher temperature and concentration than NR compounds. Mechanical properties are fair. Resilience at room temperature is very low and this leads to low resistance to impinging abrasion. Its main use in mineral processing is the lining of tanks, pipes, and pumps.

Processing is fair in general but it is very incompatible with diene rubbers such as NR, SBR, and BR so small amounts of an IIR compound can contaminate other compounds or be contaminated by other compounds, so the manufacturing facility must be careful to avoid such contamination.

IIR polymers and compounds are considerably more expensive than NR, BR or SBR.

Chlorobutyl (CIIR) and Bromobutyl (BIIR)

CIIR and BIIR are prepared by the halogenation of IIR. Either halogen gives increased cure reactivity. As a result improvements occur in vulcanization rates, the state of cure, and reversion resistance and covulcanization with other diene rubbers is also possible.

CIIR and even more so BIIR vulcanizates have lower gas permeability, better weather and ozone resistance, higher hysteresis, better resistance to chemicals, better heat resistance, better adhesion to other rubbers than those of IIR.[3]

CIIR and BIIR vulcanizates are used in much the same places as IIR and have replaced IIR in many applications such as belts, hose, and tank lining.

There is very little difference in processing BIIR and CIIR or IIR except BIIR or CIIR do not contaminate or are contaminated, by other diene rubbers. The price of BIIR and CIIR is only slightly more than IIR.

Ethylene – Propylene Rubber (EPM and EPDM)

The first polymers commercialized were merely the copolymers of ethylene and propylene (EPM). Because the polymers are totally saturated, the compounds could only be cured with peroxides. The limitations of peroxide cures did not satisfy the needs of the rubber industry. If a third monomer, a diene, is added during polymerization the resulting rubber (EPDM) can be vulcanized with sulfur, giving more flexibility in curing, without a significant loss in the stability of the original copolymers.[6] The rest of this paper will deal with only EPDM, as it is much more commonly used in mineral processing.

As EPDM's have fully saturated backbones, (the unsaturation of the diene is in a side chain) the resistance to ozone and oxygen is excellent.

As nonpolar hydrocarbon elastomers, with an amorphous nature, EPDM polymers have good electrical properties.

The nonpolar nature gives resistance to polar materials such as phosphate esters, many ketones and alcohols; many acids and bases, water and steam. Resistance to chlorinated solvents is fair. Resistance to nonpolar oils, gasoline etc is very poor, although compounds with high loading of black and oil have lower volume swell compared to other hydrocarbon elastomers.

Resilience is only fair so resistance to impinging abrasion is much less than natural rubber vulcanizates but much better than butyl vulcanizates and therefore EPDM has replaced butyl in many applications requiring such abrasion resistance. Resistance to sliding abrasion is good, as well as tear resistance.

Compression set resistance, particularly at high temperatures is good, if properly compounded. Heat resistance of up to 200°C can be achieved for special applications.

Processing is generally good except for a lack of building tack. This severely limits EPDM's use in hand lay applications such as tank lining.

The price of EPDM compounds varies widely. High filler and oil loading are more widely used than in most polymers to achieve relatively low cost compounds, but these compounds tend to have poor strength and abrasion resistance. High quality stocks especially those with peroxide cures, are considerably more expensive than NR, BR, or SBR stocks.

EPDM is used in hot material belts, and hose because of its excellent heat resistance. Excellent electrical properties often make it the preferred polymer for cable insulation and electrical connectors. Its chemical resistance often makes it the preferred polymer for specialty belts, hose and pump liners.

Polychloroprene (CR)

Neoprene is the generic name for polychloroprene. It has been produced commercially since 1931 and had rapid acceptance because it is much superior to natural rubber for heat resistance and oil resistance.

Heat resistance is better than NR, BR, or SBR but cannot approach that of EPDM. When heated in the absence of air neoprene withstands degradation better than many elastomers normally considered more heat resistant and retains its properties 15 times longer than the presence of air.

Compression set at higher temperatures is better than natural rubber and 100°C is typically the test temperature rather than 70°C.

Abrasion resistance is not as good as natural rubber but generally better than most heat and oil resistance polymers. This is also true for tear strength and flex resistance. The resilience of gum neoprene vulcanizates is a little lower than natural rubber but it decreases less with increased filler loading,[14] so the resilience of most practical neoprene compounds is higher than that of natural rubber with comparable volume loading.

Because of the chlorine in the polymer, products made from neoprene resist combustion to a greater degree than products made from non-halogen bearing polymers.[14] This means neoprene can be compounded to meet the flammability requirements of the Mine Safety and Health Administration (MSHA), Conveyor Belt Flammability Program, and various other flammability tests, without massive loadings of soft mineral fillers, and special flame-resistant plasticizers. Such loading in non-halogen bearing polymers leads to much poorer properties, such as abrasion resistance and tear strength.

While the oil resistance is not as good as some highly polar elastomers such as nitrile butadiene, neoprene compounds are generally considered moderately oil resistant. Compounded with red lead, neoprene compounds are very water resistant, as well as moderately oil resistant. This leads to use in belts, hose, pump liners, and other products which are handling oily water, especially at temperatures approaching boiling, and also require good abrasion resistance. These same types of compounds are also resistant to many acids at higher temperature than natural rubber can handle. Neoprene shouldn't be used in parts, which are bonded to metal for hydrochloric acid service because acid migration can cause bond failures.

Processing is generally fairly good but compounds can be relatively fast curing at lower temperature, (scorchy) so care must be made in mixing, calendering and extrusion to prevent premature vulcanization. Tack is generally quite good so building composites is very practical in belts, hose and tank lining. Good tack and green strength also make neoprene very common in contact cements and in 2-part room temperature vulcanizing cements, which can be used to bond various substances such as rubber/rubber or rubber/metal.

The cost of neoprene is quite high compared to NR, BR, or SBR, and even other heat and ozone resistant polymers such as EPDM, so where EPDM can be used, it is usually preferred because of price considerations.

Nitrile Elastomers
Simple nitrile elastomers are copolymers of acrylonitrile (ACN) and butadiene in monomer ratios ranging from 18/82 to 50/50. (NBR)

The basis for selection of an elastomer having a particular monomer ratio is usually the oil/solvent resistance, as well as the low temperature performance required in the final vulcanized article. Higher ACN gives better oil resistance but poorer low temperature properties. Using the same base formula a 45% ACN polymer will give a compound with 0% swell in ASTM #3 after immersion for 70hrs at 149°C, while a 18% ACN polymer leads to 30% swell. The 45% CAN compound has a brittle point of -8°C while the 18% ACN compound is -58°C. As the ACN content increases, tensile strength and hardness increase while the resilience and compression set resistance decrease.[6]

Oil resistance is usually the reason for use of nitrile compounds in belts, hose (especially tubes), o-rings, seals, gaskets and various other products. Heat resistance is generally good and can be enhanced with special compounding techniques, and by using polymers with built in antioxidants. For ultimate heat resistance a hydrogenated nitrile (HNBR) is used. Compounds made from HNBR have been used successfully in flapper seals in a smelter operation at 200°C, handling very abrasive and corrosive fly ash.[15]

Resistant to impinging abrasion is only fair, as resilience is generally low with normal ACN levels. Resistance to sliding abrasion is also fair but can be dramatically improved by using a carboxylated nitrile (XNBR); HNBR is also excellent.

Flex resistance for NBR products is fair but can be improved with special antioxidants. HNBR elastomers have outstanding flex resistance.

Resistance to polar fluids decreases with increasing ACN levels and there are usually better choices for resistance to acid, bases and water than NBR. HNBR has much better resistance to oxidizing fluids and can be used in hot acids and bases. HNBR has been used in water service, under high pressure, at up to 175°C.

Ozone resistance, except for HNBR, is generally poor but can be dramatically improved by blending with polyvinl chloride (PVC). This leads to use in such products as hose covers, cable covers and conveyer belts where oil resistance and fuel resistances are required. NBR/PVC can also be blended with NR or SBR to give moderate oil resistance and abrasion resistance. The tack of the NR portion allows such blends to be used to line tanks, pipes and other hand lay products.

Processing characteristics are good for all the nitriles, except for low tack. This limits their use in land lay applications.

The cost of NBR is between NR and neoprene. XNBR is a little more. HNBR compounds are roughly 10 times the cost of NBR, which limits their use to products which absolutely require their unique properties.

Chlorosufonated Polyethylene (CSM)

HYPALON® chlorosulfonated polyethylene was first introduced by Du Pont in 1952.[6] CSM compounds have good heat, oxygen and ozone resistance, moderate oil resistance and excellent electrical properties but their main feature for use in mineral processing is their resistance to strong oxidizing acids.

Tanks, pipes, and pumps can be lined with CSM and hose tubes can handle corrosive chemicals including 66° Baumé sulfuric acid.[6] Good colorability, along with the other good properties, make CSM compounds a good choice for colored cable jackets.

CSM compounds can be used for geomembranes for lining reservoirs. They are installed in an uncured state for simple seaming and repairs, if necessary, and then slowly cure during subsequent aging, giving an increase in toughness and durability.[6]

Processing is a little different than most polymers, in that CSM is more thermoplastic and softens more with heat. Compounds tend to be "scorchy." Tack is not very good but hand lay items can be made with care. When making items such as hose, neoprene is often used as a cord coat, because its better tack makes building easier, and adhesion to the cord and the CSM tube and/or cover is good.

The cost of CSM compounds is generally a little more than neoprene compounds.

Silicone Elastomers (Q)

Silicone rubber has both excellent low (-65°C) temperature properties and can stand exposure to 315°C.

Poor properties such as tear strength and abrasion resistance limit silicone rubber in most mineral processing application. Liquid RTV (room temperature vulcanizing) compounds are useful for small repairs and sealing applications and have been used for poured-in-place gaskets.

Polyurethanes (AU or EU)

Most polyurethanes are different from other elastomers in that they are cast. Two components are mixed together. One of the components is a prepolymer, which consists of two major chemical structures. One is an isocyanate, usually MDI (methylenebisdiphenyl diisocyanate) or TDI (tolulene diisocyanate). The other is a polyol, either a polyether (EU) or a polyester (AU). The other compound is a curative, which contains hydroxyl or amine groups.

Urethanes are used in general because they have excellent physical properties including high tensile and tear strength, high resilience, abrasion resistance, and excellent resistant to nonpolar oils and fuel and ozone. Compared to other elastomers polyurethanes tend to have their optimum properties at much higher hardness and modulus. This leads to higher load bearing capabilities and higher tip speeds in pump impellers, than softer elastomers.

As a general guideline, esters are better for tensile strength, tear strength, sliding abrasion oil resistance and heat resistance. MDI-ethers are better for rebound, low temperature properties, impingement abrasion and hydrolysis resistance.[16]

The limitation of polyurethanes are chiefly three.[16] Owing to a certain thermoplasticity their upper temperature limit is around 110°C. Polyurethanes are subject to hydrolysis in the presence of moisture and high temperature. At low temperatures most polyurethanes can withstand continuous contact with water for years. None can withstand steam for prolonged periods. In between these extremes polyurethanes may or may not be suitable for use. The MDI-ethers are much preferred for hydrolysis resistance. Sometimes a polyurethane part used in the wrong conditions will appear to be performing better than the same part made from a more water-resistant elastomer, such as natural rubber, and then it will fail rapidly. Certain chemical environments (strong acids and bases and polar solvents such as ketones or esters) are also unsuitable for polyurethanes because of their polar nature.

Processing is different than other elastomers in that most polyurethanes are cast into a mold, as previously discussed. The mixing of the prepolymer and curative can be done with specially designed machines in large shops, or by hand in small shops, or for very small runs. The polyurethane is often bonded to metal in the casting operation. Fabrication errors may induce failure even in properly designed parts. Errors include incorrect proportions of prepolymers and curative, overheating the prepolymers during storage and improper application of adhesive.[17]

The cost of polyurethanes varies with the type but is generally considerably more than general purpose elastomers such as natural rubber. This can be made up for, especially for small runs, by the rather low cost of the tooling and equipment used.

Polyurethane is also available in millable gum form, which can be handled on conventional rubber processing equipment and cured in similar molds. It comes in various grades much like the castable types and offers similar properties at a little higher price.

Lack of knowledge, poor technical advice, and consequently the wrong choice of grade or inadequate processing are the main causes of comparatively low use of millable gum polyurethane in the world wide rubber industry.[18]

Fluoroelastomers

Practical fluoroelastomers, introduced in the 1950's, provide extraordinary levels of resistance to chemicals, oils and heat.

VITON®, made by DuPont, is probably the name most familiar to the mineral processing industry, although other companies in the U.S.A., Europe and Asia produce a wide variety of different fluoroestomers.

Few fluoroestomers product are specifically designed for mineral processing but some are used, in hose and pump liners, to handle aromatic solvents such as toluene and very strong acids, and o-rings, gaskets and seals to handle aggressive chemicals and solvents, and extreme heat.

Processing tends to be difficult. Tack is very poor, contamination can be a problem and high temperature post cures are often required.

The cost of even the lowest price fluoroestomers polymer is roughly sixty times that of natural rubber. The most expensive parts are costed by the gram.

Other Elastomers

There are many other elastomers available but products from them are seldom designed for the mineral processing industry, even though some are used. A sampling of them are polyacrylates, (ACM, ANM, AEM,) chloro-polyethylene, (CM) ethylene/vinyl acetate, (EVM) Epichlorohydrin (CO or ECO), epoxidized natural rubber (ENR), acrylonitrile- isoprene (NIR) and stryene-isoprene rubber (SIR).

There is also a whole class of thermoplastic elastomers (TPE). These behave more or less like other elastomers at room temperature but like plastics above their melt temperatures. They are processed on equipment typically used in the plastic industry. Many or even most rubber processors don't tend to think of TPE's as true elastomers, but more like flexible plastics, although they do have some use in mineral processing, in items such as hose and cables.

USES

As previously stated elastomer products are used extensively in mineral processing, mostly because of resistance to wear, flexing and corrosion. Some of the specific uses are:

Truck Box Liners

These are often the first elastomer product to see ore in an open pit mining operation. They are molded in large, thick sections and fastened by T-bolts or welded studs. They often outwear metal liners considerably.[19] Their elasticity during loading reduces chassis shock. This can lead to reduced wear of the tires and lower maintenance of drive train components of the of the truck. Driver acceptance is also good because of this reduced shock. Disadvantages, compared to metal liners, include reduced volume, but elastomers have the advantage if the total weight being carried is the limiting factor. In cold climates where the engine exhaust is run through the tank bed the heat is too great for elastomer liners. The most common elastomers used are SBR, NR, and BR compounded for maximum resistance to impact, cutting and abrasion.

Tires

Most of the tires specifically designed for the mineral processing industry are off-the-road (OTR) types. These tires are exposed to very rough, sharp surfaces with the result being chipping/ chunking of treads.[20] In haulage equipment, where large loads are carried at relatively high speeds, heat generation is a major factor in designing the compounds used in the tires.

The needs of the OTR tires are usually satisfied by natural rubber compounds, or blends of NR/IR, NR/SBR or NR/BR.

Belts

The needs of elastomer compounds used in standard flat belting are similar to many other applications. Cover compounds must be abrasion resistant, resistant to cutting and chipping, and resistant to cracking caused by repeated flexing and exposure to ozone and low temperatures. The textile or steel reinforcement must be coated with a compound, which must bond to the textile or steel and also to the cover compound and not delaminate after continuous flexing, much like a tire. For use in normal conditions these needs are met by NR, SBR, or blends of NR, SBR, and BR.

There are also many special situations where resistance to oils or solvents of various kinds may be required and appropriate compounds must be chosen. This can be complicated in extremely cold climates, as found in the oilsand mining industry, where special low ACN NBR compounds are often chosen as giving the best compromise of low temperature flexibility and oil resistance.

Flame resistance is often required by various regulators, such as MHSA and special compounds must be used to meet their requirement. Oil resistance and flame resistance are often needed, along with normal belt requirements. Neoprene or NBR / PVC special compounds are often used.

High temperatures are sometimes encountered and specially compounded EPDM's are often used.

Some belts are used to convey material up steep grades and these belts will often have molded in cleats or nubs to provide the lift, and molded in flanges to contain the material. These flanges are subject to more strain, than the main cover of the belt, and must be designed to prevent flex cracking. Special requirements of other conveyor belts are also encountered and must be compounded for. EPDM has been used successfully at 150°C.[21] These and other belts may be required to be electrically conductive to prevent static build up and sparking. This is usually accomplished by using special carbon blacks.

Belts are also used for filtering applications. Horizontal filter belts are endless belts with grooves leading to holes in the center of the belt. The belt supports a filter cloth and a vacuum is used to draw fluid through the filter cloth and the belt. The fluids are often fairly strong acids or bases, at elevated temperatures, so chemical resistance is often the main criteria for choice of an elastomer. Belts are usually designed for a particular application and immersion testing of various elastomers is often performed at the users site before an elastomer compound is chosen.[23]

Cables

Low voltage mining cables generally contain two elastomeric components, namely, an insulation and a jacket. The insulation must be specially compounded for electrical properties, usually using mineral fillers, such as calcined clay treated with silane coupling agents. The jacket must protect the rest of the cable components. Compounds tend to be similar to oil and flame resistant belt compounds, except that they also must have surface resistivity (non-conducting). High voltage mining cables also utilize two electrically conducting elastomeric compounds that are extruded over the conductor and insulation and are part of the electrical shielding system.[22]

Pumps

There are many uses for elastomer lined pumps in mineral processing.

Vertical sump pumps use hand lay compounds extensively on the submerged portion of the pumps, such as the pedestal and the piping. Chemical resistance is usually the main basis for the selection of elastomers. Inside the pumps, the liners and impeller are often molded elastomers, chosen for abrasion resistance as well as chemical resistance.

The mining industry's continual request for larger processing equipment has placed new design demands on suppliers of heavy-duty slurry pumps. Such pumps (mostly horizontal) are often selected for some as the toughest jobs in a hard rock concentrator, such as mill discharge, cyclone feed or tailing transport.[24]

Soft natural rubber is usually selected in purely abrasive application. Larger particles are handled even better with compounds which combine the tear strength of a harder compound with the resilience of a pure gum.[9]

Slurry pumps usually have thicker liners than other elastomer lined pumps as experience has shown that when lining thickness is increased by a factor, wear life is increased by several times that factor.

Elastomer compounds can be bonded to materials such as fiber glass reinforced plastic (FRP) or thermosetting phenolic / nylon cloth to stiffen liners to prevent collapsing during disruptions such as cavitation or surges. They can be bonded to ceramics to take advantage of the best of both materials.

Materials other than natural rubber may be used, usually for fluid resistance, much the same as previously described for other applications.

Positive displacement pumps are used in feeding autoclaves in the pressure oxidation of gold and other ores. The valves are a critical item in these pumps and subject to abrasive and corrosive slurries, and flexing at very high temperatures (up to 210°C). Work is being done (by this author and others) to develop proprietary compounds for this extremely demanding application.

Wear Liners

Various items that can loosely be called wear liners are extensively used. These include liners for chutes, mine cars, skips, crushers, feeders, launders, and others. Abrasion resistance is usually the most important consideration. This is affected by factors previously discussed and also by the impact angle. Most abrasion damage is dome by the shearing force whose vector is parallel to the rubbers impact surface – particularly at impact angles under 30°.[19]

Screens

Rubber screens are used for scalping and dewatering. Scalping screens are usually molded with steel reinforcement. Other screens may have a textile reinforcement. They can be molded in large sheets and the holes can be cut using a high pressure water jet to prevent "hourglass" shaped holes, which can occur from die cutting. Rubber's major advantage over steel screens is greater life span, while other advantages include reduced "binding" or clogging, a noise level reduction of up to 15 dB and increased life of the screening unit.[19]

Hose

Much hose is specifically designed for mineral processing, especially those for handling slurries. The tubes tend to be similar to pump liners and other abrasion resistant products. Covers tend to be very similar to those used as cable covers.

Froth Flotation

Elastomer compounds are used extensively in froth flotation units which allow mining of low grade and complex ore bodies, which would have otherwise been regarded as uneconomic. In earlier practice the tailings of gravity plants were of a higher grade than ore treated in modern flotation plants.[25]

Launders and tanks are rubber lined, as well as star rotors, dispersers and hoods. Natural rubber is usually used and the life of the parts is usually measured in years because abrasion in the relativity slow moving rotors is not too severe. Other elastomers are chosen, usually for oil resistance. These include neoprene, NBR, and urethane although urethane has had some problems with hydolysis in long term service. The units tend to get bigger for greater efficiency. Star rotors are now being transfer molded over 1.1m high,[23] using over 200kg of rubber compound.

Handlay Linings

Important applications where rubber is hand laid, in uncured sheets, and bonded to metal during vulcanization, include pipe, fittings, filter drums, and tanks. Numerous other products can also be hand laid.

For pipes and fittings, pure gum natural rubber is usually the choice, especially for tailings, where it typically gives service for many years. It is very easy to install. Thickness can vary and up to 25 mm is common. Double layers or more can be applied as almost every handlay job is a custom design and is very easy to change, because no expensive molds are involved. Moderate oil resistance can be achieved in natural rubber compounds by blending with oil resistant polymers. Other polymers such as neoprene, CSM or CIIR are used for chemical resistance.

Tanks and filter drums are often lined with chemical resistance in mind. Soft natural rubber is still preferred if conditions warrant, as it is the most economical. Hard rubbers (ebonite) or combinations of soft and hard rubbers are sometimes used for heat resistance and chemical resistance, sometimes with graphite fillers, specifically for chlorine resistance. Since ebonite is hard, and not a true elastomer anymore, it can crack due to impact or drastic temperature changes.

Various other elastomers are used for chemical (often acids) resistance, with the most common being neoprene, CSM and CIIR. CIIR is often used with natural rubber tiegum to make the lay up easier and the bond better.

Modern adhesive systems have been specially designed for hand lining.[26] They usually consist of a metal primer, an intermediate coat, and a tacky coat to hold things together before and during cure. Rubber tearing bonds are usually obtainable.

If possible, an autoclave cure using steam pressure to give temperatures of approximately 140-150°C, is preferred for maximum adhesion, both rubber to metal and rubber to rubber. Demand for pipe over 18m long has led to the construction of very long autoclaves.

If the parts are too large for an autoclave an exhaust steam cure is usually the next choice. The cure time is long and there is more tendency to get blisters, and therefore to have minor repairs to the lining. For on site lining where steam is not available, a chemical cure (which involves toxic and flammable chemicals) or a very fast curing compound, which cures at ambient temperatures, can be used. Those systems tend to be much more difficult and therefore expensive to apply.

Grinding Circuit Applications

The following sections, based on a recent paper by Leon E. Schaeffer and Hans Weinand,[27] will discuss the detailed applications for rubber in a typical grinding mill circuit with emphasis on the use of rubber in the mill itself.

In the 50's and 60's rubber was installed in grinding mills on a trial and error method. This has changed in the last three decades of the 1900's with use of computers.

In the future rubber will replace more traditional metal liners. There are industries other than mining which use grinding mill circuits, such as the power industry. Computers will assist in assuring that rubber is successful in the expanded use of rubber in these circuits in the year 2000 and beyond.

Grinding Mill Circuit (closed circuit)
A grinding mill circuit normally starts with a coarse crushed feed material and finishes with a predetermined particle size, which is referred to as the liberation size in a mineral processing plant. The complete circuit uses rubber as an acceptable lining material.

The coarse material that is referred to as 'feed' is directed to the grinding mill via a rubber lined feed pipe or feed chute. Water is added at this point or prior to this point to make a slurry. The feed pipe enters the center of the feed end of the grinding mill and requires a sealing arrangement to prevent slurry from leaking from the mill. There are several designs available utilizing rubber.

From this point the feed material goes through a rubber-lined trunnion. There are also several designs to consider here. Some are better for rubber. After this the feed material enters the rubber lined grinding chamber of the mill. After some retention time the ground feed material, which is now called 'product', discharges through a rubber lined discharge trunnion liner. From here the product passes through a discharge trommel screen with nominal 10mm slots. The product passes through the opening in the trommel screen and only a small amount of tramp or oversized material is discharge from the circuit. The minus 10mm product then passes by gravity into a rubber-lined sump and is pumped to a rubber-lined cyclone via a rubber-lined pump. The fine material or product passed on for use or further processing while the coarse fractions pass back to the feed material of the circuit for further reduction. The connecting piping is also rubber lined.

In this typical circuit, as outlined above and shown in figure #1, all components are rubber lined but other lining materials could be used.

Feed chute or pipe
Feed material is fed to the grinding mill via a feed pipe or chute. In a fine or medium grinding application a feed pipe is used and is lined with 40 to 50 durometer natural gum rubber. The rubber lining has a nominal thickness of 12mm with a double thickness on the outer radius of the pipe to take impact and sliding abrasion. In some cases a molded urethane liner is used especially if it is a secondary or tertiary grinding circuit.

In primary grinding circuits, such as autogenous or semi-autogenous grinding, a heavy-duty feed chute is used. Here molded rubber bars in the 60 to 70 durometer range are used. Partially worn mill lining lifter bars can be used for this application.

A rubber hose can also be used for this application.

Mill Feed pipe or chute seal
During the operation of a grinding mill, there is some slurry, which splashes at the feed entry area. Some type of sealing device is required here to prevent the slurry from being discharge from this feed area of the mill. There are several designs that have been used.

One design uses a small bucket wheel to lift the slurry up and dump it into a top opening of the feed pipe where it is fed back into the mill. A drip ring is attached to the pipe so slurry in this area drops into the bucket wheel area. A 6mm flat rubber gasket is attached to the outside area preventing any splashing from getting out of the mill. (See Fig. #2)

Another method is to attach a rubber or urethane molded wedge shape to the feed pipe. This would match with another wedge surface and create a seal. (See Fig. #3)

Still another method is to attach a flat 12mm thick gum rubber gasket with a metal back-up ring to the outside of the mill. The inside diameter of flat rubber would be 12 mm smaller than the outside diameter of the feed pipe. When the feed pipe is pushed through the smaller opening, a seal is created. Since the mill is running at a relatively slow speed, the wear on the outside of the pipe is minimal.

However, some mill maintenance people have elected to put rubber on the outside of the pipe as well for extended life. (See Fig. #4)

Mill feed trunnion area

The feed trunnion area is used to transport the slurry from the feed pipe or chute to the inside of the grinding mill chamber. For years cast metal liners in a horizontal or conical shape have been used with advancing spirals in this area. (See Fig. #5 & #6) A rubber lined trunnion liner of the same design in some cases has lasted 4 to 5 times longer than metal. Rubber in the 40 to 50 durometer range would be used in light and medium duty mills. However, in the heavy-duty autogenous and semi-autogenous mill a 60 to 70 durometer rubber would be used.

In some cases because of the mill and piping configurations, a horizontal trunnion liner with advancing spiral has to be used. The spirals create turbulence, which increase the wear. If the arrangement permits, a smooth conical trunnion would be preferred. A rubber lined smooth conical feed trunnion liner would have reduced wear and provide a longer life. (See Fig. #7)

Grinding mill liner

Rubber is an excellent lining material for grinding mills. It can be used in most grinding mills.

The heads of the grinding mills do not contribute to the actual grinding, except if the heads have a large angle. So the rubber head linings are used to protect the actual mill head from wear. Various designs are being used to get the longest possible life. A bar and plate design is common where the bar gets most of the wear. But the bar does create turbulence and it is therefore desirable to have smooth linings around the trunnion areas.

The shell or cylinder lining design is very important since it transmits the power into the charge. It is partially responsible for the type of grinding action in the mill. Here again a bar and plate design is commonly used where the shape of each can be modified to get the optimum performance of the grinding mill. (See Fig. #8) The trial and error method has been replaced with computer generated simulations to show the effects of a liner design.

Towards the end of the 1900's research groups and universities got more involved with theoretical and graphical explanations of what actually happens inside a grinding mill. The use of DEM, Discrete Element Method, has helped to determine the forces and trajectories of individual components of the mill charge. This results into a fairly accurate simulation of the actions inside a grinding mill.

By using the various programs available, a mill lining can be designed with some assurance of a successful operation, thereby eliminating the previous trial and error method. Currently these programs work in one plane of the mill or 2D, two-dimensional. However, 3D simulations are rapidly being developed and will be used in the years ahead. (See Fig. # 9)

The rubber used for these lining is between 55 and 70 durometer. The lower durometer rubber is used for lighter duties while the higher durometer for heavy-duty applications.

Discharge trunnion liner

The discharge trunnion liner is normally horizontal with a reverse spiral to return grinding media to the inside of the mill. A rubber covered trunnion liner can last 3 to 4 times longer than metal. (See Fig. #10)

Discharge trommel screen assembly

The discharge trommel screen assembly consists of a steel frame, some screening material and an advancing retention spiral.

In order to protect the trommel frame for corrosion and erosion, it is rubber covered with a 40-durometer rubber in a nominal 6mm thickness or less.

The discharge trommel screen is used to protect the mill discharge pump from tramp or oversized material. In some cases it is used to separate the oversized material which is returned for further reduction. The screen material can be rubber, however urethane has also been used successfully in many applications.

The advancing spiral can be a rubber covered steel or it can be made from urethane in segmented components. (See Fig. #11)

Product discharge sump
The sump receives all of the product material that is discharged through the trommel screen. The sump is sized to have a certain amount of retention time or surge capacity depending on the system design.

A sump will typically have a 12 mm thick 40-durometer rubber lining. A trommel screen normally discharges most of the material at the beginning of the screening area. It is desirable to double the rubber thickness in the sump area where this large volume of material hits the sump surface. (See Fig. #12)

Mill discharge pump
The mill discharge pump pumps the mill discharge product to hydro-cyclones.

The pump can have many different lining materials. The pump casing can be solid metal or have metal, rubber or urethane linings. The pump impeller can be metal, rubber or urethane. There may be a combination of material used inside the pump. For instance, the pump could have a rubber lining in the casing and a rubber, metal or urethane impeller. Each application has to be analyzed for the appropriate materials to get the best economy. (See Fig. #13)

Hydrocyclone
The cyclone is used to continuously separate the product according to the desired cut in the particle size distribution. The overflow or fine material goes to flotation or processing outside of the grinding circuit. The underflow or coarse particles are recirculated back though the mill with new feed material.

The lining materials used in cyclones are rubber, urethane and ceramic. These materials may also be used in combinations. The inlet head could use rubber, the body and cone could use urethane, and the apex could be made from ceramics. The choice of materials is based upon the best wear life and economy. (See Fig. #14)

Piping
The piping system connects the pumps, cyclones, and return together. Because the slurry is abrasive, the pipe should be rubber lined. This is also an excellent application for rubber slurry hose because of its flexibility. The rubber lining material for both the pipe and hose is normally 40 to 45 durometer natural gum rubber. Some circuits may have chemicals or oils which would require a different type of rubber.

Normal Applications for grinding mill circuits
The normal application for a grinding mill circuit is the mineral or ore processing plants, which handle iron, lead, zinc gold, copper, silver, phosphate etc. There are many processing plants throughout the world. Metal is still the dominant mill lining material used throughout the world. A lot of these are good applications for rubber. Even thought the growth of the grinding mill circuits is minimal because of the low metal prices, there are many existing metal lined mills, which could use rubber. The big grinding mills which go up to 40 foot diameter are using metal but there are still some portions of these which can use rubber.

Applications other than mining
Normally, when someone mentions a grinding mill circuit it is thought to be in mineral processing or mining operation. However, with the current and future trends to clean up the environment, power plants are using a grinding mill circuit to reduce the particle size of limestone and mix it with water to produce slurry. The slurry is pumped to a scrubber, which reduce the sulfuric acid in

the flue gas emissions in coal burning plants. Even though the limestone grinding circuit is not in a mining operation, it is a processing operation for reducing limestone and is an excellent application for rubber. These systems are referred to as FGD (Flue Gas Desulfurization) systems.

The United States has installed many of these scrubbing systems at coal burning plants over the last three decades. Europe and the rest of the world have installed some systems over the last decade. However, as local air pollution laws become more stringent, more of these systems will be installed in the future and there will be more applications for rubber.

Conclusion

With the proper design and rubber selection, rubber can be economical used in many grinding mill circuits. With the use of computers with such programs, as mill simulations, the designs will be economical from the beginning. In cases where the mill grinding circuit is existing, it is advisable to examine the areas where the major wear takes place and redesign the area to reduce wear. This redesigning and the proper use of rubber can make an installation very successful.

Mining is the major use of the grinding mill circuit but there are other industries that use this to reduce particle size of materials. The power plant with the FGD system is one of these. These applications will provide some future growth for rubber in the years to come.

ACKNOWLEDGMENTS

The author would like to thank the various people in the references for their helpful discussions; Joseph Wagner for help in obtaining literature, Leon Schaeffer and Hans Weinand for permission to use their paper and Tony J. Valdez for putting this all together.

REFERENCES

1. McGraw Hill Dictionary of Scientific and Technical Terms, Forth Edition, 1989.

2. B. Betts and P. Schnarr, 1997, Metals and Elastomers for Pump Application, AICHE, Lakeland FL.

3. Werner Hofmann, 1980, Rubber Technology Handbook, Oxford University Press.

4. Science 06/19/99 Vol. 284 Issue 5422, p1988: Prehistoric Polymers: Rubber Processing in Ancient Mesoamerica.

5. C.M. Blow, 1971, Rubber Technology and Manufacture, Butterworth and Co. (Publishers) Ltd.

6. Robert F Ohm, 1990, The Vanderbilt Rubber Handbook, RT Vanderbilt Co. Inc.

7. Maurice Morton, 1987 Rubber Technology, Van Nostran Reinhold Company Inc.

8. Product Information, Ultrasil® 7000GR, 3/1999, Degussa-Huls Corporation.

9. Conversations with Ron Bourgeois and Alex Roudnev, Weir Slurry Group.

10. Joan C. Long, July – August 2001, Rubber Chemistry and Technology, Rubber Division, American Chemical Society Inc. The History of Rubber. A Survey of Sources about the History of Rubber.

11. Conversation with Douglas N.Hartley, Interex World Resources, Ltd.

12. H. Fries, B. Stollfub, Oct 18-21, Rubber Division, American Chemical Society, "Structure and Properties of Butadiene Rubber."

13. G.R. Hamed, HS Kim, September- October 2000, Rubber Chemistry and Technology, Rubber Division, American Chemical Society Inc. On the Reason That Passenger Tire Sidewalls are based on Blends of Natural Rubber and Cis-Polybutadiene.

14. R.M. Murray, D.C. Thomson, 1963, The Neoprenes, E. I. Dupont de Nemours and Company

15. Conversations with Keith J. Hart, Rubber Engineering.

16. Dr. Ronald R. Fuest, What Polyurethane? Where? Selecting the Right Polyurethane for Various Applications, Uniroyal Chemical Co.

17. Kenneth R. Oster, March 27, 1995, Rubber and Plastic Nerves, Polyurethanes: Achieving Top Performance, Air Products and Chemicals Inc.

18. Jim Ahnemiller, Nov 1999, Rubber World, An Introduction to the Chemistry of Polyurethane Rubbers, TSE Industries.

19. Richard A. Thomas, Diverse Use in an Expanding Market for Rubber in Mining and Processing, reprinted from Engineering and Mining Journal.

20. John B. Habec, Floyd A. Walker, Oct 23-26, 1984, Rubber Division, American Chemical Society, Improved Durability in OTR Mining Tires.

21. Conversations with Mark Schmidt, Cambelt

22. Daniel H. Jessop Jr., Steve Bunish, Robert M. Wade, Nov. 1985, Rubber World, Elastomers for Power Cables in Mining.

23. Conversations with John Alexander, Rubber Engineering; Jerry Hunt, Eimco Process Equipment.

24. Russel A. Cartes, March 1999, Engineering and Mining Journal, Vol. 200

25. B.A. Wills, 1992, Mineral Processing Technology, Pergam Press.

26. Product information, Chemlok®, Adhesive Guide for the Rubber Lining Industry, Chemical Products Group / Lord Corporation

27. Leon E.Schaeffer and hans J. Weinand, Rubber Applications in the Mining Industry, IRC 2000 Rubber Conference, Helsinki, Finalnd, June 2000

Grinding mill circuit

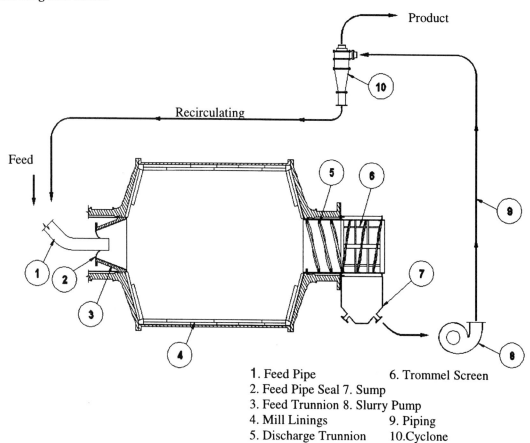

1. Feed Pipe
2. Feed Pipe Seal
3. Feed Trunnion
4. Mill Linings
5. Discharge Trunnion
6. Trommel Screen
7. Sump
8. Slurry Pump
9. Piping
10. Cyclone

Figure 1

Feed Pipe Seal

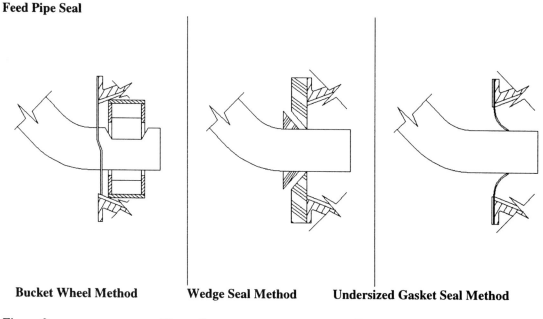

Bucket Wheel Method **Wedge Seal Method** **Undersized Gasket Seal Method**

Figure 2　　　　　　　Figure 3　　　　　　　Figure 4

Feed Trunnion Liners

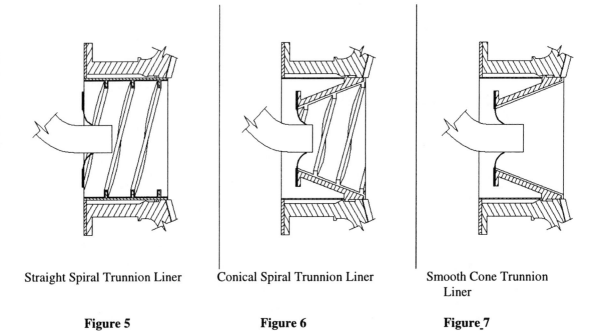

Straight Spiral Trunnion Liner

Figure 5

Conical Spiral Trunnion Liner

Figure 6

Smooth Cone Trunnion Liner

Figure 7

Typical Mill Liner Section

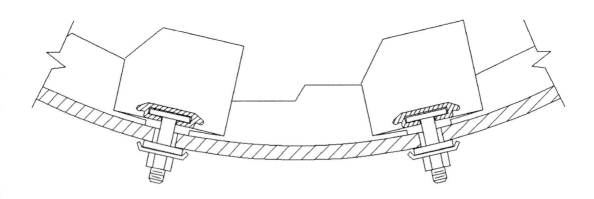

Figure 8

MILL SIMULATION

Figure 9

Typical Discharge Trunnion Liner

Typical Discharge Trommel Screen

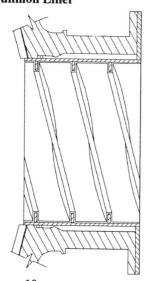

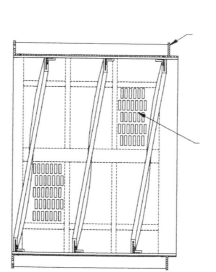

Rubber-covered Frame

Urethane or Rubber Screen Panels

Figure 10

Figure 11

Rubber Lined Sump

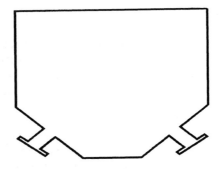

Figure 12

Slurry Pump

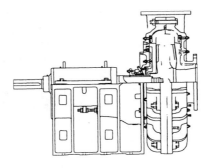

Figure 13

Slurry cyclone

Rubber Lined, Also Urethane, Metal or Ceramic

FIGURE 14

Plastics For Process Plants & Equipment

Guyle W.McCuaig[1]

ABSTRACT

Thermoplastic Thermoset and Dual Laminate piping, ducting, and vessels have been used world-wide, primarily in chemical and electrochemical processes for mineral processing, for the past 40 years. Use of thermoplastics and dual laminates as lightweight and corrosion resistant alternatives to metals particularly in electrochemical and by-product gas cleaning operations, is now standard practice for many liquid, slurry, and gas-liquid mass transfer operations in mineral processing.

INTRODUCTION

The design and fabrication of Thermoplastic, Thermoset, and Dual Laminate equipment in hydrometallurgy is a major growth industry worldwide *(Bertelmann 1992)*. New Thermoplastic welding standards, such as the North America AWS.G.1.10 and the new Dual Laminate Vessel Standard, American Society of Mechanical Engineering (ASME)-Reinforced Thermoset Plastics (RTP-1), Dual Laminate Appendix M-14, combined with existing European standards British Standards Organization (BS) 4994 and Deutsche International Norm (DIN) 16965 Part Two, allows today's designers and engineers to be confident with their choice of plastic materials for hydrometallurgy processes (See Table 1).

Table 1 Existing International Standards for Dual Laminate equipment

1	DIN 16965 Part 2: Type B pipes dimensions
2	DIN 16964: General quality requirements and testing
3	DIN 53769 Part 1: Determination of adhesive shear strength of Type B pipeline components
4	DIN 16966 Part 2, DIN 16966 Part 4, DIM 16966 Part 5
5	DIN 16966 Part 8: Laminated joints dimensions for Type B piping
6	BS 6464: 1984 – British Standard specification for reinforced plastics pipes, fittings and joints for process plants
7	PRN 88: Swedish pressure piping code for plastics
8	DIN 16962 Part 4-12: Polypropylene pipe and fittings, dimensions, fusion and general quality requirements
9	DIN 16963 Part 4-10: Polyethylene pipe and fittings, dimensions, and fusion
10	DVS 2207 Part 15: Heating element butt welding of PVDF and ECTFE pipe and fittings
11	CGSB 41-GP-22: Canadian Standard for: process equipment; reinforced polyester; chemical resistant, custom contact moulded pipe and fittings (NBS PS-15-69 American equivalent)
12	ASME B31.3-1996: Process piping, ASME code for pressure piping B31.3
13	ASTM C1147-95: Standard practice for determining the short term tensile weld strength of chemical resistant thermoplastics
14	ANSI/AWS G1.10: 2000 –Evaluation of hot gas and heated tool thermoplastics welds
15	CEN (Europe): In process standard for approval testing plastics welding personnel in Europe

---

1 Prolite Plastics Ltd., Vancouver, Canada

GLOSSARY

Thermoplastic Liner - Thermoplastic sheet, pipe or rod used as corrosion resistant liner in Dual Laminate constructions. Injection moulded, extruded or press-laminated from Thermoplastic resins.

Thermoset - Thermosetting resin of polyester or epoxy vinylester used to manufacture tanks or piping; mainly used as the structural shell.

Dual Laminate - Appliance manufactured with a Thermoplastic liner fully bonded to Thermoset structural layer.

ASME: American Society of Mechanical Engineers

AWS: American Welding Society for Plastics G.I.A Committee.

BS: British Standards Organization

DIN: Deutsche International Norm

ECTFE: Ethylene Chlorotrifluoroethylene

FRP: Fiberglass Reinforced Plastic

Furan: Thermosetting resin of furfuryl alcohol

HDPE: High Density Polyethylene

MFA: Methylfluoroalkoxy Fluoropolymer

PE: Polyethylene

PFA: Perfluoralkoxy Fluoropolymer

PP: Polypropylene

PP-H: Polypropylene Homopolymer

PTFE: Polytetrafluorethylene (teflon)

PVDF: Polyvinylidene Fluoride

RTP: Reinforced Thermoset Plastic

PLASTICS FOR PROCESS PLANTS AND EQUIPMENT

Thermoplastics, thermosets, and dual laminates *(Wegener 1991)* are widely used in 3 major areas of solution metallurgy, including a) Hydrometallurgical acid leaching, b) Electrochemical metallurgy, and c) By-product gas cleaning (see Table 2).

Table 2 Process Plant Applications of Plastic Materials

Metal	Process	Conditions	Materials
Zinc:	Calcine Leach	90° C, H_2SO_4	PP-H/FRP
	Electrowinning	70°C, H_2SO_4	PP/FRP
Copper:	Electrowinning	70°C, H_2SO_4	PVC/FRP
	Heap Leaching	ambient, dilute H_2SO_4	HDPE
Magnesium:	Leaching	110°C, HCl	PVDF/FRP
Silver	Electrolysis	80° C HNO_2	PVDF/FRP
	Selenium reduction	SO_2/H_2SO_4	ECTFE/FRP
Columbium & Zirconium:	Solvent Extraction	HCl/Benzene	Furan; MFA/FRP
Fertilizer	Phosphoric Acid	H_2SO_4/H_3PO_4	FRP; PP/FRP
Gold Ore Roasting	Flue-Gas Scrubbing	SO_2/H_2SO_4	PP/FRP; PVC/FRP
Fume Hoods/Ducts		$HClO_4/ H_2SO_4$	PVC-FRP

Electrochemical processes lend themselves to dual laminate construction due to the ability of the thermoplastic liner to resist attack by the high concentrations of acids used, the ability to resist the permeation of acids *(Van Amerongen 1964)*, and the flexibility to resist elevated temperatures up to 95° C in acid solutions.

Dual laminate piping (see Figure 1) can use the strength of the Fiberglass Reinforced Plastic (FRP), which is at the same order of magnitude of some metals to withstand long pipe span distances without extra support, large surge pressures (based on the 10:1 safety factor of the FRP casing), and a minimum expansion/contraction in a restricted environment such as is found in anchored pipe racks (see Figure 2).

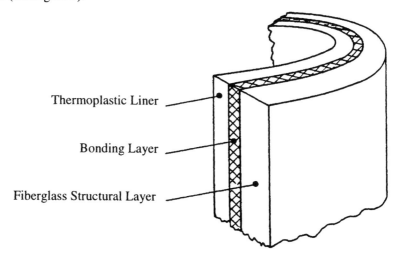

Figure 1- Cut-out of dual laminate pipe section to include thermoplastic liner, bonding mechanism (chemical bond or fabric embedded mechanical bond) and FRP structural

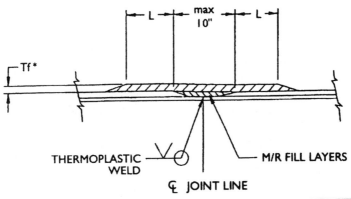

SIZE (Inch)	Nominal Length (in.) L (min)	HLU Standard 150 psi Tl
1	4	3/16
2	4	3/16
3	4	1/4
4	4	1/4
6	4	3/8
8	4	3/8
10	6	1/2
12	6	5/8
14	6	3/4
16	6	7/8
20	6	1
24	6	1-1/4

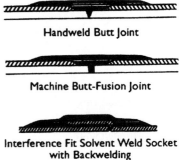

Handweld Butt Joint

Machine Butt-Fusion Joint

Interference Fit Solvent Weld Socket with Backwelding

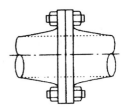

Flat Face Drilled Flange
150 lb. ANSI Standard Drilling

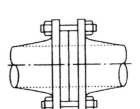

Stub End with Steel Backing Flange
150 lb. ANSI Standard Drilling

NOTES:
1. Material to be PVC/CPVC, PP/PE, PVDF, ECTFE, MFA
   1.1. Liner to be Thermoplastic.
   1.2. Structural laminates:
      1.2.1. Resin: Premium grade vinylester resin
      1.2.2. Cure system: MEKP-CONAP-DMA
      1.2.3. Reinforcement: Type E-Glass ECR-Glass 1-1/2 oz/sq.ft. chopped strand mat (M)
         Type E-Glass 24 oz/sq.yd. woven roving (R)
         Type E-Glass continuous roving for filament winding
      1.2.4. Surface coat resin: c/w C-Veil and UV resistance agent
2. Structural FRP Lamination:
   2.1 Filament wound pipe: FW, CV
      2.1.1. Filament winding continuous roving wound at +/- 55° to thicknesses required
   2.2 Hand Laid-up Pipe, Fittings and Joints:
      2.2.1. Structural hand laid-up layer to be composed of layers of mat and roving to approximate thicknesses required

Figure 2: Dual Laminate Pipe Joining Methods

Dual laminate vessels (see Figure 3) are lightweight when compared to brick or ceramic linings, and can be built in one piece to withstand the corrosion of hydrometallurgical processes at roughly half the capital cost.

Recent advances in world standards have brought credibility and safety to many higher temperature and load bearing applications that were once the domain of metals, alloys, and acid brick lined or ceramic components..

Critical services for corrosive media cover an extreme range of chemical compositions in inorganics.

Figure 3: PVC/FRP Dual Laminate Tower

One common use of dual laminates in hydrometallurgy is electrolytic zinc manufacture using Polypropylene/FRP (see Fig. 4), which gives excellent corrosion resistance and mechanical properties up to 95° C and is resistant to chlorides (often a problem with stainless steel). Polypropylene/Fiberglass Reinforced Plastic (PP/FRP), is unaffected by any electrochemical currents in the cells and will handle the abrasion/corrosion problems of slurries and solids due to the nature of polypropylene liners (which are similar to polyethylene – *Conde 2000*). The liner thickness can be varied to provide appropriate allowance for erosion/abrasion.

Figure 4: PP/FRP Zinc Cell Feed Tank

Silver electrolysis (and gold parting) in nitric acid again lends itself to Polyvinylidene Fluoride/Fiberglass Reinforced Plastic (PVDF/FRP) construction with the corrosion of the concentrated nitric acid handled by the PVDF liner and the structural FRP able to withstand the elevated temperature.

Strong acid chloride solutions at temperatures up to 110°C are handled by various fluoropolymer/FRP combinations as noted in Table 3. At lower temperatures, HDPE and PP are satisfactory.

Acid leaching and solvent extraction technologies use a variety of inorganic acids and aggressive organic solvents which are best handled by thermoplastics at lower temperatures or dual laminates at elevated temperatures (see Table 3).

Table 3 Suggested Thermoplastic & Dual Laminate pipe temperature and service ranges

Liner	Max. Temp Plastic only	Min. Temp.	Material Spec.	Max. Temp Service Dual Lam.	Min. Temp. Service Dual Lam.	DIN 16964 Bond Strength N/MM^2
PVC	60°C	-15°C	ASTM 1784	80°C	-15°C	7
CPVC	90°C	-15°C	ASTM 1784	100°C	-15°C	7
PP-H	90°C	-15°C	ASTM D4101	100°C	-30°C	3.5
PVDF	140°C	-40°C	ASTM D3222	121°C	-40°C	5
ECTFE	180°C	-76°C	ASTM D3275	121°C	-50°C	5
MFA	260°C	-190°C	ASTM D6314	121°C	-50°C	5
HDPE	80°C	-50°C	D1248	85°C	-50°C	3.5
PFA	260°C	-190°C	D3307	121°C	-50°C	5

The most widely practiced application of low concentration sulphuric acid distribution leaching is for copper leaching pads, where sulphuric acid solution is distributed over a wide surface area using high density polypropylene pipe and collection of the liquid with a flexible liner of PVC or PE underneath the mass of copper ore. Plastic piping is cost effective in this service (see Figure 5).

Plastics find application under reducing acid conditions, such as in SO_2 reduction of selenium in copper and silver refining (Table 2) where alternatives (ceramics, exotic alloys) are significantly more expensive.

Another major application for PP/FRP vessels is leaching of uranium ore that is generally extracted in sulphuric acid. Historically, wood stave tanks were used in this service. Old plants have been retrofitted with HDPE or PP liners, bolted to the walls. One-piece vessels in PP/FRP construction can now handle this process with a liner thickness designed for the abrasion requirements of the vessel.

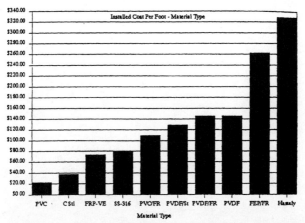

Figure 5: Process Piping Cost Comparison for Thermoplastic, FRP, Dual Laminate, Steel, Lined Steel, and Alloys (1st Quarter 2002 in U.S. Dollars)

In the processing of ores to produce zirconium and columbium a leaching and extraction process using mixed acids and organics such as benzene and toluene is used. This is a very difficult corrosion system for any thermoplastic or thermoset, but the most commonly used material in this case is furan, which is the only thermoset with suitable resistance to aromatic organic solvents as well as inorganic acids.

Recent trials have shown fluoropolymers such as MFA, teflon or PFA in dual laminate construction to effectively resist permeation in many lining techniques (see Figure 6).

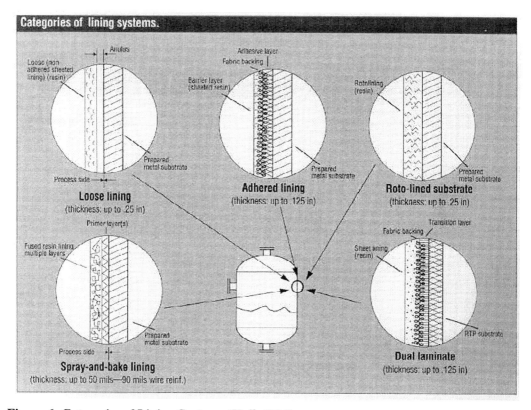

Figure 6: Categories of Lining Systems *(Hall, 1994)*

Off-gas handling of roasting processes comprises the third major use of plastics in process plants. The most common applications would be zinc or gold processes where SO_2 is collected and converted to H_2SO_4 in acid plants. Recently constructed acid plants designed in Europe have used polypropylene/FRP gas ducting and scrubbing towers. Some North American technologies use thermosets of straight FRP construction, with limited lifespan due mainly to SO_2 attack. These materials have replaced acid brick and ceramic lined steel towers in most cases.

In Noranda's new magnesium production process, magnesium is leached from asbestos waste using azeotropic HCl near 110^0C; PVDF/FRP ducting is a good material selection with care to be taken to avoid excess permeation.

In design of high temperature acid or 2 phase (e.g. $H_2SO_4 + SO_2$) systems care must be taken to review not only the fluoropolymer which would be most corrosion resistant but also the permeation effects on the bonding system. If hydrochloric acid or a mixture of severe gases permeates through the fluoropolymer sheet, care must be taken to choose the correct backing or mechanical bonding cloth. For high temperatures the best choice may be a glass backing rather than a synthetic backing such as polyacrylic nitrile or polyester. Recent research has found this to be an important issue in the long-term service of fluoropolymers in laminate construction as in lining of steel reactor vessels for pressure or vacuum service (see Figure 7).

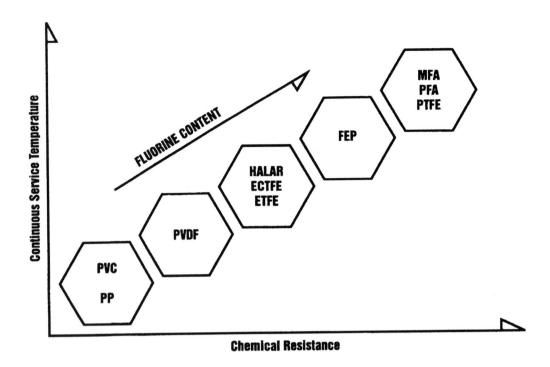

Figure 7: Thermoplastic Products- Service Temperature vs. Chemical Resistance

CONCLUSIONS

Thermoplastic, thermoset, and dual laminate use for equipment and piping has been growing in the world-wide mineral processing industry. Successful applications range from liquid slurry lines to gas transfer lines to high pressure piping and vessels.

Plastic materials have non-metallic/non-electromagnetic properties, which is important in relation to electrochemical corrosion, grounding, and safety, as well as for eliminating the need for costly instrumentation.

In leaching processes, plastics have lightweight properties so that piping can be moved easily in outdoor operations, particularly in difficult terrain. Plastics can resist temperatures as low as -50^{0}C, and will withstand corrosion and ultra-violet attack for long periods of time.

In gas cleaning operations, high temperature fluoropolymers are becoming competitive with lead lined, acid brick lined, or alloy materials.

New applications will develop and in response to both developments in material technology and wider recognition of engineering properties of plastic based materials.

Selection of plastics for mineral processing applications can now be based on a combination of international standards and industrial practice.

REFERENCES

Bertelmann, L. 1992: Pipework in Plant Construction Kunstoffe 82 No.6, pp.510-514.

Dow Chemical 1994: 7th University of Witwatersrand Composites Conference, Johannesburg, Rep. of South Africa.

Conde, Maria 2000: Swedish Corrosion Institute - Hydrochloric Acid and Water Permeability in Fluoropolymer Tubes, NACE Conference.

Glein, Gary 1996: Metals, Fiberglass, and Thermoplastic Tanks and piping, a review of costs and other decision factors, NACE Conference.

Hall, Nelson L.:Dupont Engineering, Using Fluoropolymers to Resist Permeation of Corrosives.

Lueghamer, Albert 1977: Polypropylene Comparison of Types PP-H, PP-B, PP-R - Private Communication, AGRU.

McCuaig, G. 2000: Prolite 2000 Prolite Plastics Ltd., Prolam Piping Handbook.

Pankow, Virginia R. 1986: Dredging applications of high density Polyethylene Pipe - Hydraulics Laboratory, U.S. Army Engineer Waterways Experiment Station, Vicksburg, MS 39180-0631.

Schommer, R. 1987: Thermoplastic Liner in Tank Construction - Troplast AG, Troisdorf, Germany.

Van Amerongen, G.J. 1964: Diffusion in elastomers, Rubber Chemistry and Technology 37 pp. 1064-1152.

Wegener, M. 1991: Strategies for the Correct Design or Components in Fiber Composite Materials, Thesis, Aschen.

Weib, Dr. Johann 1999: PVC-C Materials Purity and Stress Relieving - Private Communication, Troplast.

Commercial Acceptance and Applications of Masonry and Membrane Systems for the Process Industries

Robert E. Aliasso, Jr.[1] Thomas E. Crandall,[2] David M. Malone[3] and Robert J. Storms[4]

ABSTRACT
The use of ceramic and membrane materials is a well-established engineering approach to corrosion control, with over 100 years of commercial history. The concepts of a composite, corrosion proof, ceramic lining system are reviewed to allow the mechanics of a proper design to be understood. The primary function of each of the composite parts is reviewed in relation to properties of the composite vessel or lining system. A review of design factors to be considered when choosing applications for ceramic and membrane lined process vessels is presented. Past and recent commercial applications of composite ceramic lining systems are presented to illustrate life cycle reliability. The economics of ceramic and membrane lining systems are briefly discussed.

INTRODUCTION
The use of masonry and membrane lining systems for process vessels is a well-established engineering approach to corrosion control. This approach was first applied in the late 1800s to the digesters of pulp manufacturers in the developing paper industry.

The initial and current applications are based on the excellent engineering properties of masonry and membrane materials, including:

- Chemical resistance
- Thermal shock resistance
- Abrasion resistance
- Good thermal and electrical insulating properties
- Strength in compression
- Extreme mechanical durability
- Low life cycle costs
- Versatility in physical shape and application.

In many cases, a masonry and membrane lining system is used when no other materials of construction can be found to resist the corrosive/erosive environment. The properties of masonry and membrane lining systems have contributed to the commercial viability and growth of many new applications throughout various industries.

COMPOSITE CERAMIC LINING SYSTEMS
Before technical and economic discussions can begin, a composite vessel or ceramic lining system or structure must first be defined. The corrosion or abrasion resistant vessel or ceramic lining system consists of three main components:

- Structural body – the containment vessel; usually metal (always for elevated pressure applications) or concrete for atmospheric operation

---

[1] The Stebbins Engineering and Manufacturing Company and Stebbins Africa (Pty) Ltd
[2] The Stebbins Engineering and Manufacturing Company
[3] The Stebbins Engineering and Manufacturing Company and Stebbins Australia Pty Ltd
[4] The Stebbins Engineering and Manufacturing Company

- Primary corrosion barrier - an organic (e.g. FRP/FRF or rubber) or inorganic (e.g. lead or alloy) membrane; and
- Masonry barrier - the masonry lining (e.g. carbon, caustic resistant, or acid brick).

Each layer in a composite vessel ceramic lining system has a primary function that is essential to the success of the corrosion protection. To more fully understand the concept of the composite system, we need to understand the functional purpose of each part.

A properly designed ceramic lining system and structure accounts for each component to be designed into a composite system. A complex analysis is completed to confirm that each component of the system operates safely within its mechanical properties. The containment vessel is designed to withstand the hydraulic and physical loads imposed by the process conditions while the ceramic lining system is designed to withstand the chemical, thermal and abrasion conditions of the process. It is important to design the ceramic lining system to account for the interactive forces between the containment vessel and the ceramic lining.

Component #1 - Structural Body

The structural body is designed to carry the physical load of the process and of any ancillary equipment. Most commercially practiced design codes for the structural body ignore the structural requirements of a masonry and membrane lining system. Since the ceramic lining system is interactive, it is essential to involve the masonry and membrane designer in the design of the structural body. Details of design and fabrication for the structural body must be altered to accommodate the requirements of the masonry and membrane lining system. These include:

- Elimination of flat surfaces and ledges;
- Restrictions in yield/strain to comply with masonry and membrane mechanical properties; and
- Elimination of surface defects and geometry that is not compatible with developing intimate and total contact of the membrane and masonry lining with the substrate.

Component #2 - Primary Corrosion Barrier

Corrosion resistance is predicated on the theory of a tight composite lining system. This theory is dependent on the membrane being in intimate contact with the structural body and the masonry barrier being in intimate contact with the membrane. Any voids between the membrane and structural body that occur provide cells that will result in a pool of corrosive liquid against the structural body, due to osmotic pressure.

The primary corrosion barrier is selected to prevent corrosion of the structural body. The membrane can prevent corrosive reactions from occurring by:

- Acting as an impermeable layer that keeps the corrosive chemicals from the substrate;
- Acting as an electrical insulator to prevent the classical Galvanic Cell from forming;
- Acting as a sacrificial barrier which neutralizes any corrosive chemicals before it reaches the substrate; and
- Acting as a catalyst that sets up a passivating atmosphere adjacent to the substrate.

The membrane choice must be optimized for thickness, cost and durability. In addition, the membrane must be physically tough enough for the masonry lining to be installed without worry of damage. Occasionally, membranes are chosen that require such carefully controlled installation conditions for proper application that the choice becomes impractical. In addition, it can disrupt the work of others nearby due to safety or environmental reasons. It is typical that the nonmetallic membranes are spark tested before ceramic lining application.

In addition to known metallic membrane materials such as lead, many nonmetallic materials can function as primary corrosion barriers. Materials such as fibre reinforced plastic (FRP), fibre reinforced furan (FRF), rubber and even Portland cement can have outstanding corrosion resistance when used behind a masonry barrier.

Component #3 – Masonry Barriers

The masonry barrier can consist of many materials ranging from very expensive silicon carbide, high alumina, or carbon to cost effective fireclay or red shale. The selection of the masonry material and the thickness depends primarily upon the thermal, chemical and abrasion conditions that exist within the vessel.

The masonry barrier is used to limit/alter the exposure of the primary corrosion barrier (membrane) from temperature, process chemistry, and abrasion. The masonry barrier performs many functions to help the primary corrosion barrier perform its designed function"

- Masonry barriers should be designed to provide thermal insulation. This allows the primary corrosion barrier to operate at temperature levels required to achieve optimum corrosion resistance.
- Most masonry barriers have low porosity and can be designed to decrease the corrosive chemical exposure of the primary corrosion barrier. The masonry barrier prevents direct and free exchange of the corrosive chemicals with the primary barrier.
- Most masonry barriers develop controlled forces that push against the primary corrosion barrier (membrane), holding it in intimate contact with the substrate.

Properties of most masonry materials exhibit non-reversible physical growth due to chemical swell. This chemical growth is completely independent of thermal expansion. The chemical growth, in a properly designed composite ceramic lining system, aids in lining stability, when coupled with proper configuration. Chemical growth also eliminates bond dependency of the lining to the substrate, since it provides a radial thrust moving the lining into more intimate contact with the structural body.

HISTORY/APPLICATIONS

First used in North America in the 1880s, masonry and membrane lining systems have a long and successful history. Pulp manufacture for papermaking utilized sulfite digesters which operated at high temperature, elevated pressure and contained very corrosive chemicals. In the 1880s, limited corrosion control choices existed to protect mild steel process vessels such as pulp digesters. Lead sheets were the typical choice to protect process vessels from corrosion. However, fatigue often caused failure of the lead. Also, the exposed lead sheets commonly sagged, causing failure. Henry Stebbins, a pioneer in the field of masonry and membrane systems for corrosion control first applied masonry over the lead for support. It was later discovered that this composite lining system offered greater reliability than exposed lead.

From these first applications, masonry and membrane lining systems played a role in the growing process industries.

In the 1930s, a monolithic chemical resistant masonry and concrete design for construction of process tanks was first applied, by Henry Stebbins' company. The first monolithic reinforced concrete and masonry tanks were rectangular and had concrete covers. The walls and covers of these tanks became part of the plants' walls and operating floors. This construction technique is unique in that the concrete shell and the corrosion-resistant lining are built simultaneously without forms. This unique formless construction is suitable for a wide range of atmospheric chemical processes.

During the first half of the 1900s process chemistry increasingly became more aggressive as more corrosion control methods became available. Masonry and membrane linings already had more than 50 years of success at this time. Masonry and membrane lining systems allowed many processes to become commercially viable during this period by offering a reliable means of corrosion protection of critical process equipment where no other materials would offer proven reliable protection.

The (non-aluminum) extractive metallurgy industry first applied pressure leach autoclaves in 1947 for tin reactors at Wah Chang in Texas City, TX. Again, Henry Stebbins' company was called upon to perform a design for these "new-age" reactors. Innovative material selection and design techniques were employed based on the 60 years of pressure vessel experience from their ongoing involvement with batch pulp digesters.

A ten-year lapse preceded the next large step in metal refining pressure leach application. In the 1950s, a large complex for nickel laterite processing was planned in Moa Bay, Cuba and Braithwaite, LA by Freeport Nickel. These plants brought new challenges for masonry membrane lining design. All critical process equipment and nearly all process vessels were equipped with masonry and membrane lining systems for corrosion control. Autoclaves operating at temperatures in excess of 400° F and pressures over 500 psi were designed. The original masonry linings were installed in 1959. Many of the vessels have their original base course linings and operate today! Needless to say, the design and material selection was a great success.

Over the last 20 years, many pressure leach processes have been employed for gold, nickel, copper, cobalt, and other metals. In many of these applications, masonry and membrane lining systems have been chosen, and have shown very high reliability when properly designed.

During the same period, atmospheric metal leaching has also grown. Masonry and membrane linings or monolithic construction continue to play an increasing role in corrosion control techniques for these atmospheric applications. There are existing applications where 20+ years of maintenance free service have been demonstrated.

PRESSURE VESSEL COMPOSITE DESIGN

Tight vs. loose design concepts
Now that we have established all of the components of the composite, it is important to note the different design concepts and the reliability comparisons for different applications.

For brick lined vessels in general, two design methods exist - **tight** and **loose** linings. The tight lining design is the most reliable corrosion control design for hydrometallurgical systems. A loose lining is common to dry, high temperature refractory, where voids are made part of the lining design to allow for thermal expansion. These voids are detrimental to long term lining success for pressure vessels containing corrosive liquids.

Unlike high temperature refractory materials, corrosion resistant masonry exhibits chemical swelling, in addition to reversible thermal expansion. Chemical swelling is irreversible and is a function of the raw material and manufacturing process of the finished masonry product. The most common masonry used for corrosion protection in hydrometallurgical process vessels is a fireclay acid brick, although graphite, carbon and specialty masonry are also specified and used.

Fireclay acid brick can be classified as standard duty acid (SDA) brick or pressure vessel grade acid (PVGA) brick. SDA bricks are usually tight-bodied material that exhibit very good acid resistance and high chemical swell rates. These materials are very resistant to wear but commonly spall when used in pressure vessels due to the high chemical swelling rates. PVGA bricks offer very good acid resistance but have low chemical swelling rates. When used in pressure vessels, PVGA brick materials have shown excellent spall resistance when compared to SDA brick.

PVGA bricks are used only in a tight-bonded lining system in which controlled swelling allows the entire masonry to be supported under low stress and remain stable. Membranes are typically noncompressible and do not flow under pressure. Brick spalling is typically insignificant. Membranes, since they are better protected, offer longer service.

Process inputs to pressure vessel design
Successful applications of pressure hydrometallurgy require effective collaboration between the process design team and the vessel/lining design team. Process inputs to design of a pressure leach vessel, often in the general form of a performance specification, will typically include:

- Slurry flowrate and retention time - to determine process dimensions of vessel.
- Operating pressure and temperature range – to determine mechanical design of vessel.
- Chemical conditions – to determine lining components and nozzles design.
- Number and relative size of compartments
- General and specific gas-liquid mass transfer requirements (e.g. tonnes/h of oxygen, agitator power, dimensions, weight)
- Temperature control strategy (e.g. flash/recycle vs. dilution water cooling for exothermic feeds).

Once established by the process design team, this data provides the fundamental design criteria that the vessel/lining design team must integrate into the design of the vessel, lining, partitions, and nozzles.

Pressure vessel design requirements

When lining a mild steel pressure vessel with masonry and membrane systems, owners and consulting engineers must pay close attention to coordinating the design requirements of the masonry and membrane lining with those of the steel vessel. Depending on the lining contractor's material selection and process conditions, various special requirements may exist.

It is very important that the steel is not purchased ahead of lining contractor's completion of the masonry, membrane and nozzle designs. The designs should be completed by a lining contractor that specializes in corrosion consulting which is focused on the design, production and installation of masonry and membrane systems. This approach has a proven high success rate in overall lining reliability, performance and economic optimization. Until the lining thickness can be determined and the nozzle designs are completed by the lining designer, the steel dimensions will not be known or optimized.

DEVELOPMENTS

Continuing areas of general development in pressure hydrometallurgy are identified and briefly discussed as follows:

Masonry

The most significant general developments have been in the production of consistent quality PVGA. This material was initially developed over 70 years ago for batch pressurized processes in pulp manufacturing for papermaking. There is an extremely high success rate when utilizing PVGA for pressure vessel applications in mineral processing.

Nozzle Designs

A variety of changes have taken place in this critical area. New seal arrangements have been developed for high pressure sealing. New materials have been tested and used to allow better corrosion resistance and thermal protection of the membrane under newer, high process temperatures.

Organic Membranes

Lead has been a common choice as the membrane in pressure vessels, with some alloy used in nozzle areas. However, new high temperature, high chloride or other high severity processes may make lead an inappropriate selection. New masonry and membrane lining designs which utilize organic membranes with excellent resistance to high temperature and chlorides are being utilized in pressure vessels. The organic membranes should be carefully selected to conform to a tight lining design so as to remain hard, well bonded and not flow under hot conditions to form voids.

Vapor Zone Mortars

In pressure oxidation autoclaves, silicate mortars have a short life in the vapor zone before repointing is required. The use of inorganic lead based mortars has solved the repointing problem, but poses new challenges for safe handling during installation and especially removal when dust is unavoidably generated.

Owners and consulting engineers must look carefully at a lining contractor's safety record and performance history when handling lead. If handled properly and installed correctly, low maintenance can be expected through the life of lead based mortar applications

With newer higher temperature processes, especially for nickel laterite production, more temperature resistance is required for mortars. Commercial applications exist where new vapor zone materials are used in acidic conditions in excess of 550° F.

COMMERCIAL APPLICATIONS
When an owner or consulting engineer performs a feasibility study, various potential applications exist for masonry or membrane systems. Applications requiring consideration of masonry options include:

Pressure Leach Applications	High (>230°C) pressure acid leach (i.e. laterites)
	Sulfide pressure oxidation (high, intermediate or low severity)
	Steam recovery ("flash and splash") and final (flash) pulp discharge
	Precipitation reactors
Atmospheric Operations	Mixed acid or otherwise highly aggressive chemical conditions
	Insulation for endothermic reactions

Example – Pressure Oxidation of Refractory Gold Ores/Concentrates
Pressure oxidation of refractory gold ores and concentrates has been a major source of new gold production since commissioning of Homestake's McLaughlin mine autoclave circuit in 1988; over 20 autoclaves have been installed for similar service worldwide. All of these vessels (and many of their auxiliary flash and heat recovery units) employ masonry/membrane/steel shell construction

For competitive reasons, the authors are not prepared to present current design information. Readers requiring further descriptive information are referred to the open literature (Thomas, 1994 and cited references) with the caution that all areas of masonry/membrane technology continue to evolve rapidly.

Example - TiO$_2$ Ore Autoclave
In the beneficiation of ore for one manufacturer's chloride process for the production of pigment grade TiO$_2$, the ore is upgraded from 60 to 95% TiO$_2$ by leaching with hot azeotropic hydrochloric acid. The process unit for this reaction is a rotating, nonmetallic membrane and ceramic lined mild steel autoclave.

The reduced ore is fed to the rotating autoclave followed by a charge of hot HCl (18 %). The contents are steamed up to 148.9°C (300°F) at 206.9-275.9 kPag (30-40 psig).

The ceramic lining has lifters (brick ledges) that cascade the ore as the autoclave rotates. This action causes a slow mixing to promote the dissolution of soluble metallic compounds that are removed in solution to upgrade the ore. The movement of the ore creates a highly abrasive and corrosive condition. Few metals can survive in this process. This process has been successfully contained in a spherical ball autoclave, lined with an acid resistant nonmetallic membrane and a two course brick lining. In this application we have all of the basic components:

- The carbon steel spherical pressure vessel (structural body);
- The membrane (primary corrosion barrier); and
- The ceramic acting as a thermal, chemical, and abrasive shield for the membrane.

None of these materials could limit corrosion by itself. In combination, they function as an economical solution to contain a highly corrosive and erosive pulp, which allows this process to be commercially viable.

General Examples
Over the last thirty years new processes involving that composite masonry and membrane lining systems have shown great success. Notable processes include flue gas desulphurization (FGD) systems for fossil fired power plants, atmospheric leaching reactors for metal refining and pressure leach autoclave circuits for metal extraction.

The majority of pressure leach applications throughout the world have used masonry and membrane lining systems with excellent success. Due to this success, many of the same companies are now utilizing masonry and membrane lining systems in their atmospheric leach vessels.

COMMERCIAL BENEFITS
Several significant factors make ceramic lining systems commercially beneficial to various process industries.

Process Flexibility
During the pre-feasibility and feasibility stage of a project, process characteristics and equipment costs are initially reviewed. It is typically at this stage of the project where various types of corrosion resistant liners for the critical vessels are considered.

Of the various lining options available, composite ceramic lining systems offer a high degree of process flexibility. This is mainly due to the ceramics ability to be tolerant of significant changes in process conditions.

Fundamentally, ceramics have excellent engineering properties that allow them to perform in many different process applications. Ceramics are: abrasion resistant, chemically resistant, thermal shock resistant, very strong in compression, extremely durable, economical compared to other materials and provide good thermal and electrical insulation.

The unique physical characteristics of many types of ceramics such as acid brick, caustic resistant brick, carbon brick, silica brick, silicon carbide brick, zirconia brick, borosilicate brick, high alumina brick, porcelain brick, and alumina silicate brick make ceramic linings extremely versatile.

Due to the wide range of masonry and membrane material choices, there are very few situations for which a composite masonry and membrane combination cannot be engineered to allow the operator a high degree of process flexibility. Also due to the engineering properties, a masonry and membrane combination can be reliably applied to contain very severe corrosive/erosive conditions.

Vessel Reliability
Ceramics tolerance of variable process conditions is a significant factor in providing reliability. Most ceramics, even if they are not completely resistant to a specific set of chemical conditions, function as a sacrificial lining with a useful life while preventing catastrophic damage to the substrate. This is due to the large mass of chemical resistant material present and the multi-component nature of the lining.

Economic Impact
Similar to process requirements, which are established during the pre-feasibility and feasibility stage of a project, the economics are also investigated and defined. During this stage the cost of the major equipment and vessels is estimated. The feasibility of a project depends greatly on favorable economics for capital costs, and expected operating and maintenance costs.

In reviewing cost information, ceramic lining systems can range from under US$20/ft^2 to over US$250/ft^2 depending upon the vessel and its operating conditions. Overall economics have been strongly in favor of ceramic lining systems for many applications.

CURRENT OPPORTUNITIES
Many opportunities exist for plant designers and the operators to save capital and maintenance costs by using composite ceramic linings systems more extensively in their process vessels.

A comprehensive technical and economic review of alloy, ceramic, and other corrosion resistant technologies available for each project is a sound business approach. Considerable savings in life cycle costs of capital equipment may be identified during the early stages of a project by scrutiny of the materials of construction.

CONCLUSION
It should be considered basic philosophy that all materials including ceramics, alloys, rubber and coatings are designed and selected based on their ability to resist the chemical and physical conditions of a particular process. We hope we have provided basic information to understand the engineering design and material selection parameters for composite ceramic linings and structures.

The selection, and structural analysis of composite ceramic lining systems for atmospheric and pressure vessels is a well-established and proven science. Design criteria have been proven by the test of time and the actual in-service performance has been demonstrated. Because of many years of development, there is now in service a wide range of composite masonry and membrane lining system applications. However, today's engineering graduates have not been exposed to the above types of materials and construction, due to the lack of standard textbooks that deal with these subjects.

We must continually strive to increase our knowledge of all the available choices to develop the most economical solutions to our corrosion problems. The use of composite ceramic lining systems provides the engineer with a reliable method for providing an economical means to contain the most aggressive chemical processes. Ceramic linings and structures are very versatile, and usually competitive with alloys, rubber linings and high grade coatings, when evaluated on a life cycle basis.

REFERENCES

E.F.Tucker. January 1950. Digester Linings for Soluble-Base Sulphite Pulping. *TAPPI* Vol. 33 No.1

Beaumont Thomas. March 1950. Non-Metallic Lining Materials for Process Vessels in the Pulp Paper Industry. *Corrosion*

R. Hancock, D. Malone, and G. Charlebois. November 1990. Design and Quality Control of Composite Ceramic Lining System for Corrosion Control. *Proceedings NACE, Canadian Region, Eastern Conference*

Gary W. Charlebois. January 1991. Chemical Resistant Ceramics for the Process Industries. *Materials Performance* Vol. 30, No. 1, p. 71-75.

David J. Malone, Robert J. Storms, and Thomas E. Crandall. June 1995. Commercial benefits of ceramic lining systems used in atmospheric and pressure leach vessels. *Proceedings ALTA Conference*. Hydrometallurgy 39, p. 163-167.

The Stebbins Engineering and Manufacturing Company, *et al*, unpublished internal research and memorandum. The Stebbins Engineering and Manufacturing Company, Watertown, NY USA

R. Flynn. 1981. Construction, Inspection and Maintenance of Tile Tanks and Linings. *Proceedings TAPPI Engineering Conference*. Book II

Gary W. Charlebois. January 1991. Chemical Resistant Ceramics for the Process Industries. *Materials Performance* Vol. 30, No. 1, p. 71-75.

Robert E. Aliasso. September 1996. The Use of Ceramics as a Highly Reliable Means of Protecting FGD Equipment. *Proceedings NACE Northeast Region, 34th Annual Corrosion Conference*

K. G. Thomas, Research, Engineering Design and Operation of a Pressure Hydrometallurgy Facility for Gold Extraction; CIP Gegevens Koninkluke Bibliotheck, Den Haag; ISBN 0-9698067-0-1, 1994.

17

Power, Water, and Support Facilities

Section Co-Editors:
M.N. Brodie and Charles R. Edwards

The Development of an Electric Power Distribution System *M.N. Brodie*	1973
Selection of Motors and Drive Systems for Comminution Circuits *P.F. Thomas*	1983
Selection of Metallurgical Laboratory and Assay Equipment: Laboratory Designs and Layouts *P.F. Wells*	2011
On-Line Composition Analysis of Mineral Slurries *T.F. Braden, M. Kongas, K. Saloheimo*	2020

The Development of an Electric Power Distribution System

Malcolm N. Brodie. P. Eng.[1]

ABSTRACT
The planning of an industrial electric power distribution system involves several stages of study. The first stage is establishing the peak power demand of the projected facility. The second stage is establishing an acceptable source for the power. The third stage is the development of the economic trade-offs for first costs and losses, first costs and the value of any loss of production following a first electrical failure, and first cost and the value of any loss of production that may be attributable to unsatisfactory power quality. The final stage is the development of a composite that incorporates all of the requirements and choices that have been established in the initial stages.

INTRODUCTION
The successful exploitation of an ore body depends on a number of factors including the development of a suitable electrical source and distribution system. This in turn requires knowledge and consideration of the size of the intended operation, the planned operating life, the cost of power, the cost of down time, and the availability and cost of capital. These different items should be evaluated first in isolation then in concert.

DEVELOPMENT OF THE ELECTRIC POWER DISTRIBUTION SYSTEM
The electric power distribution system must satisfy a number of criteria. Those to be examined include:

- the capacity to allow all process equipment to operate in concert as well as independently
- conformance with the requirements of the inspection authority having jurisdiction (ANSI, CSA, etc.)
- minimization of the energy losses
- ease with which the system can be maintained
- minimization of the down time following a first failure

The concept of a single source and a multitude of diverse loads leads to the development of a radial distribution system. This is the normal system starting point and it is modified to suit individual loads, the physical location of the loads, and to enhance the security of selected parts of the system.

The balancing of the incremental cost of increased reliability, and the probability of an outage with the attendant loss of production, is commonly done using statistical data and weighting factors based on experience and judgement. These can be assigned for each system component to achieve a numerical comparison.

As part of such a study, the agreement on values to be used for short interruptions, scheduled interruptions, and deferred production is important because production lost during or because of

---
[1] Sr. Staff Consultant, Electrical, Fluor Daniel Wright Ltd.

any interruption is effectively deferred until the end of mine life. In addition, where mine life is expected to be short in terms of the normal life of electrical equipment, the requirement for preventative maintenance is very limited. As an example, a high voltage power circuit breaker has a mechanical operation expectation that could easily exceed the expected life of the mine. The same circuit breaker has a fault operating limit that restricts it to relatively few operations at rated fault current. In this application, a circuit breaker might not see a rated fault current operation during the life of the mine.

The incorporation of stand-by generators for critical process areas and Uninterruptible Power Supply (UPS) equipment for critical control, protection, and communication functions facilitates re-starting following an interruption. The next stage of development of the system must look at the effect of a first fault anywhere in the system, and the acceptance of the consequences or the selection of suitable measures for mitigation. As an example, a transformer failure could, of itself, cause an outage of a couple of months in the area it supplies but the availability of a spare would reduce this to a day. Where a day is considered unacceptable, the next option is a parallel unit that can carry the combined load at its ONAF 65°C rating. This would bring an outage down to a few hours with a small increase in losses. If this is still considered to be unacceptable, the next stage of development requires automatic switching which should reduce the interruption to momentary.

The assignment of the costs of each type and duration of outage must recognize both the loss of income and the avoidable and unavoidable costs. This information is usually provided by the mine in question.

The statistical availability of the various items of electrical equipment in the envisaged configuration can be taken from historical data and adjusted to suit the particular application.

Where guidance from the mine is limited, the selection of a system configuration usually starts with a simple radial system then, as required by judgement, sufficient extra equipment is added to limit any loss from a first fault to an amount considered appropriate for the service. To ensure that the basis and the expected result are understood and agreed, it is most desirable to have the pertinent factors tabled.

There are two system frequencies in common use in different parts of the world. The north American practice of using 60 Hertz (Hz) results in motors and transformers that are smaller, lighter, and less costly than the corresponding 50 Hz equipment that is used in Europe. Other parts of the world have variously followed one practice or the other. In many cases, consideration should be given to the use of 60 Hz equipment even in an area where 50 Hz is present, except when political influence mandates otherwise.

There are standardized voltage levels in both systems that are related to both the size of an individual load and to the aggregate load in a small area. In general, both systems describe V<1000 volts as low, 1000<V<15 kV as intermediate, and V>15 kV as high. In both systems loads will be served at the lowest voltage that does not result in more than 500 A. Similarly, when a load would exceed 1000 A it would normally be transferred to the next higher voltage. In between it could be either depending on the type and size of the adjacent loads.

The most common 50 Hz low voltage is 400 V, although some installations are now using 690 V. This does not affect either motors or control but allows a significant saving in conductors. The most common 60 Hz low voltages are 480 V in the United States and 600 V in Canada. Again the 600 V level results in a significant saving in conductors. For very small loads that can be served by a single phase supply, the common 50 Hz voltage is 220 V and the 60 Hz voltage is 120/240 V.

The most common 50 Hz intermediate voltages are 3.3 kV, 6.6 kV, and 11 kV. The corresponding 60 Hz voltages are 4.16Y/2.4 kV, 7.2Y/4.2 kV, and 13.8Y/8 kV. In both systems, there are existing systems that operate at voltages different from these, but the contemporary approach for minimizing current is to use a voltage near the upper limit for each class of equipment.

The sub-transmission and transmission voltages range through 25 kV, 34 kV, 46 kV, 69 kV, 138 kV and 230 kV. The selection of voltages in these ranges is sensitive to the maximum load and the distance over which it must be carried. As an example, it would be possible to carry 100 MVA for 1 mile at 69 kV, but to carry it for 100 miles consideration would have to be given to 230 kV.

SIZE (CAPACITY OF THE ELECTRICAL SYSTEM)

The size is directly related to the desired rates for mining and mineral processing. It is affected by the process selected, the mining plan; e.g.,-open pit or underground, the availability of water, the selected location for storage of tailings, and the need for residential accommodation.

A common starting point is an allowance of 1 kVA per ton processed per day for a conventional crushing, grinding and flotation plant. Other plant types; e.g., crushing, leaching, SX/EW, can have significantly different values. The process designer will refine the plant requirements to recognize harder or softer ore, simple or complex processes, single or multiple separations and the associated materials handling; e.g., feed, concentrate, and tailings. The requirements for power to provide water, tailings disposal, and residential accommodation will all depend on the site. The full range from least required to most required is of the order of -10%, to +15%. The need to handle material over a significant difference in elevation can make a further difference.

The utilization of variable frequency drives and/or large rectifiers has caused serious power quality concerns because of the generation of harmonics. The effects can be significantly reduced by phase shifting to simulate a high pulse number. If the basic unit is a full wave three phase switching assembly, the current will have the classic 6 pulse form. Selected phase shifting causes the associated harmonics to add vectorially instead of arithmetically. The effect of this shifting with similar loads is the capping of the sum near the value of a single unit for any combination from one unit to all units. The options are:

Table 1 Options for phase shifting

No. of units	Pulses	Shift, °
1	6	0
2	12	30
3	18	20
4	24	15
5	30	12
6	36	10
7	42	8.6
8	48	7.5
9	54	6.7
10	60	6

As an example, a system based on 48 pulse ultimate development could start with two units at $0 \pm 7.5°$ and have 0° as a common spare. It could add 30° and $30 \pm 7.5°$ and finish with $0 \pm 15°$ which would give 7 units with a common 0° spare. If 7 units or less are installed, the spare can replace any transformer with little effect on the system. If 8 units are installed, the spare will match one other unit and, unless that is the unit being replaced, will cause an increase in the harmonic level. This will necessitate replacement by the original unit to bring the harmonic level back to the original level.

OPERATING LIFE

This is the basic design life in years. It can range from as little as five years to more than fifty years. It is normally selected by the Owner to optimize his return, having in mind his knowledge of the ore body and his assessment of the potential for future political interference with his operation.

COST OF POWER

The cost considered here is the cost of power available at the intended plant. It may be either purchased from an existing source or produced on site. If purchased, the cost will be the purchased cost plus the cost to amortize the cost of connection, plus the cost of the losses that occur in the connection. If produced on site, it will include the cost of fuel at the site plus the cost to amortize the generating equipment cost and the cost to operate it. In either case, consideration must be given to the effect of inflation over the life of the plant.

COST OF DOWN TIME

This cost is the cost that results from a first failure. It is not the cost of routine scheduled down time that is a factor in plant annual capacity. The cost to be used is usually taken as a combination of foregone income, plus the unavoidable continuing costs, plus the costs associated with the restoration of the damaged component.

COST AND AVAILABILITY OF CAPITAL

The availability of capital is normally the Owner's concern. The cost of capital is normally defined as the rate of return desired by the Owner.

CONSIDERATION OF FIRST FAILURE

Electrical equipment when properly selected, installed, and maintained, routinely exhibits acceptably long life. In spite of this, failures do occur and the consequences must be evaluated so that appropriate insurance spares can be kept on hand. The times to restore service that follow are offered only as a guide: the particular circumstances of each installation, including site conditions, site labour skills, and the presence of spares must be part of the evaluation if the financial consequence of any first failure is to be minimized.

The worst failure is the loss of the source. Where power is generated on site, the source should include sufficient engines to carry the load plus two additional units. This allows for a unit as spare while a unit is being overhauled. Where power is purchased from a remote source, the receiving transformation should be able to carry the full plant with one unit out of service. As a minimum, with two transformers each should be able to carry the full load at 65°C rise with one stage of forced cooling.

The working transformer in Figure 1A will be fully loaded and working at rated temperature. A spare must be rated for the full capacity so the combined capacity is 2.0 p.u. Failure of the unit in service will cause an outage of a few days.

The transformers in Figure 1B will each be loaded to 0.75 p.u. of its self-cooled rating and will be working at about 0.6 p.u. of its rated temperature rise. This will statistically increase the life of each by four times. No additional spare is required so the combined capacity is 1.5 p.u. Each unit would be rated at 1.0/1.5 p.u. kVA, ONAN/ONAF, 55/65°C rise. Failure of either unit will cause an outage of about one day.

The transformers in Figure 1C will be operating as in Figure 1B but the addition of the secondary switchgear would reduce an outage to about 4 hours.

These approaches do not repair the fault but limit the downtime to that required for switching. Repair of a unique transformer could take 6 or more months depending on the damage and access to plant, materials and labour.

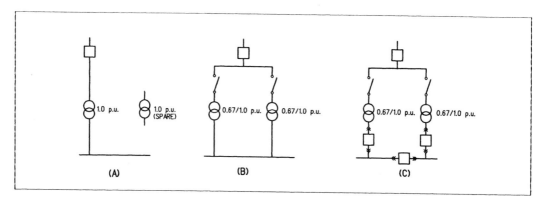

Figure 1 Transformer - high voltage

Overhead lines can be spliced in hours, a pole can be replaced in a day, or a transmission tower replaced in weeks. In most cases a single tower can be replaced temporarily by a wooden structure in days.

A failure of a power cable can usually be located and spliced in a couple of days depending upon whether it is direct buried or in ducts.

Provision for interconnection can frequently be used to advantage. Secondary substations may be double-ended, switchable between two feeders, or interconnected by secondary cables, depending upon the desired level of protection.

The configuration in Figure 2A is a typical radial arrangement. The availability of a suitable spare unit would limit an outage to a couple of days.

The configuration in Figure 2B is a typical double-ended arrangement. A transformer failure should be isolated in a few hours.

The configuration in Figure 2C is similar to that in Figure 2B but is dependent on the interconnecting cable. A transformer failure should be isolated in a few hours.

Both Figures 2B and 2C use extra transformer capacity to reduce the duration of an outage that would result from the failure of a transformer.

The arrangement in Figure 2C might allow beneficial location of the two units if the loads form two separate groups.

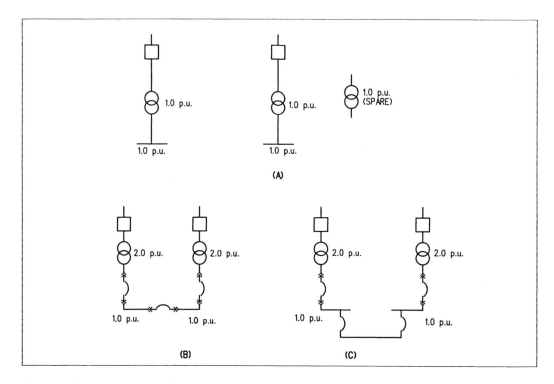

Figure 2 Transformer - medium voltage

Analysis has shown that cables are the most failure-prone component of a system but also the most easily repaired. Transformers are not readily repaired so they must be backed up by the capacity of another, the availability of a spare, or by interconnection. The consequences of each potential failure must be weighed, the appropriate response selected, and the necessary insurance spares be made available.

The most likely failure in a motor is a bearing. If an impending failure is recognized before the bearing collapses, the time to replace the bearing should be a matter of hours. If a bearing is left until it has collapsed, there is a real risk that rotor and stator will meet and the laminations may be damaged beyond repair. The next likely failure is electrical breakdown of the winding insulation. In some cases, a knowledgeable technician can isolate the failure by cutting out the damaged coil and returning the motor to service. This does not repair the damage but allows the motor to continue in service while arrangements are made to either rewind or replace it. It has to be recognized that the modern motor with vacuum impregnated windings is hard to repair with a partial rewind: the coils are so firmly secured in the slots that they are almost impossible to remove without damage.

Most control is an assembly of small components. Except in case of a power component failure that causes extensive arc damage, the smaller control components can be replaced in hours.

It must be noted that, if there are no appropriate insurance spares on hand at the time of failure, the down time will increase to that required to effect a repair or to arrange for a replacement. For transformers and motors this is considered to be far too long to even be considered. Any first failure should be covered by either excess capacity in another unit or backed up by a spare of equal or greater capacity. The more unique the item is, the more unlikely it is that a reasonable replacement will be available on short notice.

EVALUATION OF LOSSES

All energized electrical equipment suffers from losses. These losses cannot be eliminated but can be adjusted for least owning cost for any load factor and period of time. The most direct approach to evaluating the losses is the establishment of the present value of a 1 kW loss under agreed conditions.

The factors are:

- the incremental cost of electricity (e) ($/kWh)
- the anticipated inflation in the cost of electricity (f) (%/annum)
- the selected interest rate(i) (%/annum)
- the period under consideration (i.e., project life) (N) (years)
- the anticipated annual load factor (ELF) (p.u. rated)
- the anticipated annual load factor increment (g) (p.u. rated)
- any associated load related losses (e.g., forced cooling) (kW).

The formulae used to give effect to these factors are shown in Appendix 1. An example of the result is shown in Table No. 2 which gives typical default values for the factors and the corresponding present value of a kW for three different power costs.

Table 2 – Present value of losses under selected conditions

Condition	Default Value
Inflation rate	5%/annum (0.05 p.u.)
Interest rate	10%/annum (0.10 p.u.)
Project life	12 years
Annual load factor	0.8
Annual load increment	2%/annum (0.02 p.u.)
Cost of Electricity $ /kWh	**Present Value of 1 kW**
$0.05	$3,926
$0.07	$5,496
$0.09	$7,067

SELECTION OF CONDUCTOR SIZE

There are a number of criteria to consider when selecting conductors including the requirements for insulation and handling.

Insulation is rated by the maximum temperature to which it can be continuously exposed, the environment in which it can be used (dry only or wet and dry), and its ability to withstand the activities associated with installation and subsequent use. Minimum first cost tends to favour minimal conductor size which, in turn, favours insulation with a high temperature rating. The insulations most commonly used at this time are cross-linked polyethylene (X-link) and ethylene-propylene- rubber (EPR), both of which may be used in wet or dry locations at temperatures up to 90°C.

The prime requirement of the normal installation codes (ANSI NFPA 70, CSA C22.1) is that the continuous service temperature of a conductor must not exceed that for which the insulation is rated. This gives the minimum conductor size that can be considered. There are three other factors that can have a significant effect on the selection of the size of any particular conductor.

The first factor is the length of the run. Voltage drop is proportional to the length of the run and is usually limited to 3% of the nominal voltage for the circuit. As the length increases it becomes necessary to use a larger conductor.

The second factor is the value of the losses that will occur in the cable. This is proportional to the length of the run also, but is evaluated separately from the voltage drop because it is affected by the cost of power as well as by the length. It is commonly found that this alone will justify at least one conductor size larger than would be required to satisfy the temperature rating.

The third factor is related to an assessment of the probability of having to increase motor size on any particular drive. An owner may require each motor starter to be equipped with load conductors that correspond to the maximum rating of the starter rather than the initial motor where it is smaller so that each starter can be used with any motor up to its capacity without other change except overload setting.

Installations made using armoured cables; e.g., Teck, usually make these comparisons by considering only the incremental cost of the cables because the associated incremental costs for supports and labour are insignificant. Installations made using conduit and wire must consider the incremental cost of the conduit each time a size change is necessary.

GROUNDING AND BONDING

Grounding refers to a permanent continuous conductive path to the earth. It must be achieved in such a manner that the path can carry any current that can be imposed on it while limiting the consequent voltage rise. Where the current is intentionally limited, the permitted current must be able to actuate the associated protective devices. It provides a reference for all system voltages when established at the source.

Bonding refers to the extension of the ground connection to all non-current-carrying metal parts under the same general requirements.

Grounding and bonding conductors are both to carry current only in case of a failure. This is quite separate from a neutral, which normally carries residual (unbalance) current, although the neutral will be connected to ground at its source.

The other condition to be recognized is the potential requirement for isolated grounding of some electronic equipment. This can apply to Programmable Logic Controllers (PLCs) and to associated computer installations. It can be effected by using isolating UPS equipment and/or marked outlets, and carrying an insulated grounding conductor from the outlets all the way to the system ground electrode without any other interconnection with the power system grounding or bonding conductors. In effect, this is similar to the required separation of lightning arrester grounding conductors and power system grounding conductors except at the ground electrode.

EMERGENCY POWER

Emergency power may be required to be available in any case of loss of the normal supply to provide lighting for safety, to prevent equipment damage, or to facilitate restarting of the process. UPS equipment may be used to provide for the continuation of control functions but not the operation of process equipment (see Figure 3(A)). The supply for process equipment is normally provided by engine driven generators that are started automatically on sensing loss of the normal supply. The engines can be expected to pick up load within 10 seconds of start initiation. To do this dependably, each engine must be exercised regularly; e.g., for one hour once a week, with a load of not less than 1/3 of its rating. For small engines the exercise load is usually independent of the normal power system (see Figure 3(B)). Large engines can parallel with the normal system so that the fuel is not wasted. These engines can be started by loss of the normal source, by a peak shaving controller, or by an exercise timer. In the first case, load must be shed to suit the engine(s). In the other cases, it will operate in parallel with the normal source at a preset load level (see Figure 3(C)).

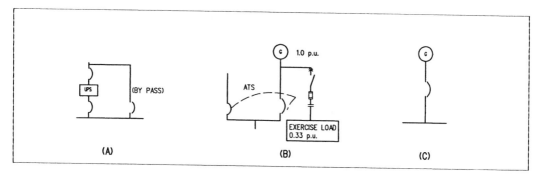

Figure 3 Emergency power

INSTALLATION

The various codes covering electrical installations and the interpretations by the authorities having jurisdiction allow some optional approaches to the installation of equipment.

Motor enclosures range from open to totally enclosed. It is common practice to have all squirrel cage induction motors totally enclosed without regard for the environment. The other motors (wound rotor induction and synchronous) tend to be the larger units and will generally have site specific enclosures.

Motor control, both LV and MV, is most commonly arranged into assemblies of NEMA 1 enclosures and placed in electrical rooms that are kept clean and cool. It has to be expected that maintenance will be compromised when control units are mounted outside of an electrical room even when the enclosures are changed to NEMA 12, NEMA 3R, or NEMA 4 as may be appropriate.

Switchgear is affected in a similar manner. It has to be expected that maintenance will be best in an electrical room, next best when assembled with a walk-in enclosure, and most difficult with a weatherproof enclosure.

Transformers that are large will usually be oil-filled and installed outdoors. These should have valved, removeable radiators and be braced for vacuum filling. They should have a control compartment with a thermostatically controlled heater, a light, and a convenience outlet. They should be installed with provision to contain and cool all of the insulating oil in case of a rupture.

Transformers that are small may be oil-filled or dry types. This choice commonly depends upon whether they can be mounted outdoors, the site elevation, and the environment.

High current interconnections as between a transformer secondary and the associated switchgear may use insulated bus, cable duct, or cables. All other connections may use armoured cables supported by tray or individual conductors in conduit. The armoured cable known as Teck was pioneered by a mine but has spread to all the other industries including pulp and paper, and petro-chemical, and to commercial work. There are still installations being made with wire and conduit but they are generally limited to either additions or some special circumstances.

Appendix 1 Evaluation of Losses

The incremental cost of electricity is a combination of standby costs and energy costs whether the power is purchased or generated on site. If the power is purchased, the incremental cost is a combination of the applicable rate schedule demand charge, the applicable rate schedule energy charge, and the associated load factor. The relationship is:

Incremental cost (e) = (Demand charge) $((\text{Load Factor})(720))^{-1}$ + energy charge \$/kWh

where Load Factor = $\dfrac{\text{hours in service per month}}{720}$

An example might be: $\quad e = \dfrac{4.101\$/\text{kW}}{0.9\,(720)} + 0.0437\$/\text{kWh} = \$0.0500/\text{kWh}$

The inflation rate (f) is assumed to be constant and should be rounded to an agreed figure (e.g., 5%/annum).

The selected interest rate (i) may be either the current cost of money or a designated rate of return (e.g., 10%/annum).

The project life (N) will be the period selected for the study (e.g., 12 years).

The anticipated annual load factor (ELF) affects the combination of no-load and load losses. It is also affected by the design margin for growth. A typical value for an industrial plant might be 0.8.

The anticipated annual load increment (g) is assumed to be small but a constant positive value. It might be 2 %.

If the device being evaluated is continuously energized but intermittently loaded it becomes necessary to allow for the no-load losses and for the incremental load losses. This is typical of a transformer where the No-Load losses and the incremental Load Losses are identified.

To simplify the calculations it is advantageous to combine the interest rate and the inflation factor into the net discount rate (r) where:

$$r = (1+i)(1+f)^{-1} - 1 \qquad \text{where i and f are per unit}$$

The value of No-Load losses for a continuously energized unit then becomes:

$$8760\,(e)\,(r)^{-1}(1 - (1+r)^{-N})(NL)$$

The value of Load related Losses becomes:

$$8760\,(e)\,(ELF)^2((1+g)^{2N}(1+r)^{-N} - 1)(1 - (1+r)(1+g)^{-2})^{-1}(LL)$$

As an example, the present value of a 1 kW loss in a power cable using the typical values suggested above would be calculated as follows:

$$r = (1 + 0.10)(1 + 0.05)^{-1} - 1 = 0.048$$

$$\text{PV of 1 kW} = 8760\,(0.05)\,(0.048)^{-1}(1 - (1 + 0.048)^{-12})\,(1) = \$3926$$

Selection of Motors and Drive Systems for Comminution Circuits

Peter F. Thomas[1], P. Eng.

ABSTRACT

The selection of a mill drive has been complicated by three distinct trends. All mills (Autogenous Grinding (AG), Semi-Autogenous Grinding (SAG), Ball, and Pebble) are getting larger. All process control is becoming more complex. The desired electrical system capacity is frequently not available.

AG and SAG mills up to 40 ft (12.2 m) in diameter are in service and designs are available up to 44 ft in diameter (13.4 m). Process requirements usually require adjustable speed drives for these mills.

Ball mills up to 26 ft (7.9 m) in diameter are in service. The majority of the smaller sizes operate at a fixed speed but adjustable speed is becoming more common for the larger mills.

There are many options available for both fixed and variable speed drives. These options are defined, the features explained, the limitations discussed, and the costs compared to assist with the selection for any particular case.

Initial cost, cost of installation, cost of commissioning, and cost of operation and maintenance have to be taken into account for each drive option considered. The impact of each drive system on the design, cost and operation of the overall plant electrical system is discussed. Also, consideration must be given to mill drive availability, and the potential cost resulting from delays in plant production.

INTRODUCTION

Over the past 30 years a number of papers have appeared discussing mill drives (see references). The primary emphasis has been on AG and SAG mills, particularly the variable speed options for larger mills. Recently, an interest in variable speed for large diameter ball mills has developed. The intent of this review is to present the latest technology considered for AG and SAG mills, and ball mill drives, plus provide some guidelines on when to select fixed and variable speed drives. For the variable speed options, gear and gearless drives are included.

Single-pinion fixed-speed gear drives offer the simplest design and the lowest capital and operating costs. In the past these drives were restricted to smaller mills, however recent technological changes in gear manufacturing have extended the range of mill diameters. Similarly, dual-pinion fixed-speed drives offer reliable and simple designs with proven load sharing capabilities. These fixed-speed drive systems can be used on the largest ball mills considered to date.

With variable speed, particularly for AG and SAG mills, the drive system became more complex and expensive. In the not too distant past the variable-speed choices were few (namely dc motors), but improvements in solid state technology and controls has resulted in a number of higher-efficiency variable-frequency alternatives that have pushed dc motors out of the picture. Likewise, modern slip-energy recovery systems for wound rotor motors have overcome high-energy losses generally associated with liquid rheostats.

---

[1] Marketing Manager, Mineral Processing, GE Industrial Systems, Peterborough, ON, Canada

A guide to contemporary application practice based on the common North American diameter/length ratios of 2:1 for SAG mills and 2:3 for ball mills is shown in Figure 1.

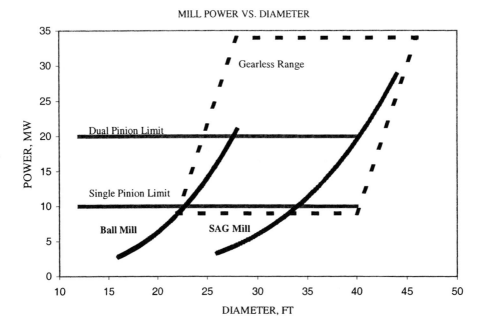

Figure 1 Mill power vs mill diameter

Taken as a whole, the engineer is faced with many choices for a drive system. The selection process involves quantitative (economic) and qualitative (subjective) factors. Choosing the correct factors is difficult and, for most projects, represents one of the most important choices in selecting the proper system. Fortunately, the engineer has a number of analytical tools to resolve the economic issues. Given the correct (and sometimes incorrect) assumptions, some subjective factors can be transformed into economic factors. However, in the end it may be the unresolved subjective factors that will determine the selection, even after exhaustive analysis of the capital and operating costs.

PROCESS CONSIDERATIONS

Why Variable Speed?

The operator (or control system) can rapidly react to changes in ore characteristics, be it ore hardness or feed size distribution. Soft and/or fine ore can result in a low total charge volume leading to liner damage and accelerated ball and liner wear. This condition can be corrected by increasing the circuit feed rate only when downstream conditions permit such changes. Otherwise, it must be corrected by reducing the speed of the mill to force the balls to impinge on the charge and not the liners. When grinding out a SAG mill, variable speed is valuable for the same reasons. Variable speed drives also provide the advantages of slow starts (stops) of the mill and, for some systems, inching of the mill and protection against damage by movement of a cemented mill charge.

However, for most overflow ball mill circuits, variable speed is only valuable when circuit feed rate control is required downstream. Without downstream constraints, ball mills are typically operated with a maximum ball charge and a fixed speed, thus negating any need for variable speed. SAG and ball mill circuits are designed such that the circuit capacity is ball mill limited over most of the range of ore hardness expected from the ore body. Most variations are compensated for by the variable speed drive on the SAG mill. In rare cases, the ore may be so variable that the speed (or power) range for SAG mill cannot compensate for an extremely hard component, resulting in a need to reduce the power (or speed) of the ball mill to avoid over grinding and affecting downstream processes. In a few circuits, the operator can balance the SAG mill and ball mill by directing a portion or all of the crushed SAG-mill pebbles to the ball mill. Other plants allow partial cyclone underflow recycle to the SAG mill to further improve the balance of power.

It should be noted that mine planning, ore blending and blasting practices play an important role in the design of milling circuits. Modern plant designs recognize that feed characteristic control is not perfect and that variable-speed SAG mills are necessary for any successful operation. In nearly all cases, regardless of the SAG mill diameter, variable speed is not an issue for evaluation and is accepted as the standard design. On the other hand, variable-speed ball mills require careful evaluation and in most cases cannot be justified based on process considerations. It is only the cases where gearless drives are considered that trade-off evaluations are required between variable-speed and fixed-speed drives for ball mills and between gearless and gear-driven variable-speed drives for SAG mills.

ELECTRICAL CONSIDERATIONS

Table 1 gives 16 different configurations for the range of mills covered in this paper.

Table 1 Basic arrangement of mill power transmissions

Drive Type	Driver Primary		No. of Pinions		Motor Speed	Motor Type	Power Supply
1	Fixed	Ring Gear	Single	Reducer	900–1200	W.R.	
2				Reducer	900–1200	Ind.	
3				N/A	180–200	Syn.	
4			Dual	Reducer	900–1200	W.R.	
5.				N/A	180–200	Syn.	Quadra.
6	Vari.	Ring Gear	Single	Reducer	900–1200	W.R.	SER
7				Reducer	900–1200	Ind.	PWM
8				N/A	180–200	Syn.	LCI
9				N/A	180–200	Syn.	PWM
10				N/A	180–200	Syn.	CCV
11			Dual	Reducer	900–1200	W.R.	SER
12				Reducer	900–1200	Ind.	PWM
13				N/A	180–200	Syn.	LCI
14				N/A	180–200	Syn.	PWM
15				N/A	180–200	Syn.	CCV
16		Gearless	N/A	N/A	9–15	Syn.	CCV

The induction motors, both squirrel cage and wound rotor type, will be high speed design with 6 or 8 poles, depending on the choice of gears.

The synchronous motors will be low speed with 30 to 40 poles depending on frequency and gear ratio. Synchronous motors used with cycloconverters on geared systems will have 8 – 10 poles.

In selecting the type of drive to be used, several factors have to be considered before coming to a final decision, and these are discussed below.

Enclosure and Ventilation

The shape of a motor is subject to a number of constraints. The normal starting point for a design of an induction motor makes the axial length of the active material equal to the diameter of the rotor. This will also apply to high speed synchronous motors.(i.e., motors with 4 – 14 poles), Typical length to diameter ratios range between 0.6:1 to 1.5:1.

With induction motors the rotor diameter is constrained by the centrifugal force that can be allowed without exerting undue stress on the cage or wound rotor winding, and/or without lowering the interference fit between the shaft or spider, and the laminations to an unstable level.

With high speed Synchronous motors, the rotor diameter is again limited by the centrifugal force acting on the poles, the field winding and the fit between the shaft and the spider.

Another constraint on rotor diameter may occur when WK^2 must be limited to achieve fast response to speed regulation.

Low speed synchronous motors have to provide sufficient space to accommodate the multiplicity of poles so the physical dimensions generally end up with the rotor diameter being much greater than the axial length of active material.

Design criteria for Synchronous motors is based on the ratio of rotor axial length of active material to the distance between the centres of ajoining pole tips (pole pitch). With low speed machines the optimum value is 1.6 to 1, and can be as high as 4 to 1.

High speed machines will be between 1 to 1 and 2 to 1.

The low speed synchronous motor will have a length to diameter ratio of 0.15:1 to 0.35:1.

Manufacturing practice will use a series of diameters, typically increasing in 10% intervals, with a range of axial lengths for each diameter.

These proportions gives rise to two distinctly different requirements for enclosure and ventilation. The high speed machine will have relatively high pressure drop through its ventilation passages such that dirt particles entering the machine will tend to precipitate within the machine, thereby restricting the airflow, and causing increase in temperature in the machine. This condition is further aggravated by the relatively small area available for air inlet, thereby creating a high air velocity and drawing heavier dirt particles into the airstream.

By contrast the low speed motor with 30 – 40 poles, enjoys a low pressure drop through its ventilation passages, and the large area available for air inlet, gives rise to a low inlet air velocity. This results in only the finer dirt particles being drawn into the motor, and because of the low pressure drop, most of these are blown through the motor with very little left within the machine.

Fixed Speed Operation

For the low speed motor it is common practice to enclose the air inlets down to the motor base, and allow the ventilating air to be drawn from the pit below the motor through openings in the pit wall, further reducing the air velocity entering the motor enclosure, and allowing it to discharge into the mill bay at the top of the motor frame. (See Fig. 2)

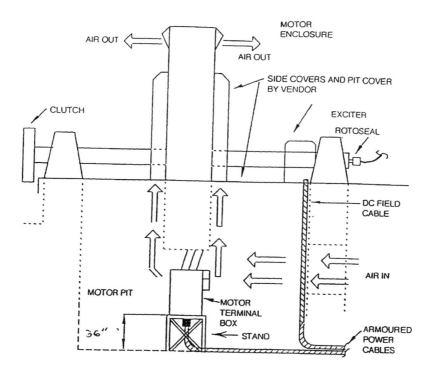

Figure 2 Updraft enclosure and ventilation system

However the high speed motor must have added protection from the entry of dirt, and the choice will lie between an open motor with filtered air inlet, a weather protected type II enclosure, a totally enclosed motor with an air to air heat exchanger, or a totally enclosed motor with an air to water heat exchanger. The WP II enclosure is usually specified to also have inlet air filters. Whichever of these options is selected, it will increase the initial cost and add increased maintenance expense to the installation.

Adjustable Speed Operation
Grinding mills are constant torque devices, however rotor driven fans on electric motors follow a cube law with the result that at reduced speed the motor is unable to ventilate satisfactorily if the speed reduction is greater than 10%–15%, and/or if it is required to use the motor for inching the mill. Because of this, it becomes necessary to add forced ventilation to the motors. If the motors are of open construction for ventilation from the surrounding air, then a filtered air supply should be ducted to the motor air inlets, with sufficient static pressure to overcome the system resistance of the ductwork, and the motor pressure drop requirement. It should be noted that forced ventilation cannot be applied to a WPII enclosure. Where machine geometry permits, the ventilation system can be equipped with gravity dampers, to allow self-ventilation when the motor is running at or close to full speed. If the motors are totally enclosed with some form of heat exchanger, then separately motor-driven internal air circulating fans should be employed. For motors with air-to-air heat exchangers, the external fans should be separately driven. Gearless

motors are usually totally enclosed and equipped with a number of heat exchangers and internal air circulating fans.

When calculating losses (to determine overall efficiency of a drive system) the power to drive the fans should be added to the system losses.

The gearless motor will also be equipped with a small filtered air pressurizing system to maintain positive pressure in the motor enclosure to assist in keeping dust, and liquid contaminants from the motor enclosure, along with carefully designed running seals.

Motor Construction

High speed induction and synchronous motors are usually supplied with end bracket mounted bearings. Low speed synchronous motors, however, are usually supplied with pedestal-supported bearings and a fabricated steel base having sufficient axial length to permit axial shifting of the stator for cleaning and maintenance purposes.

In order to provide adjustment of the motor for alignment purposes, it has become practical to install soleplates between the motor base and the foundation, thereby allowing adjustment of the motor position without disturbing its own alignment to the base.

With wound rotor induction motors, the sliprings should be located outside of the motor enclosure to prevent brush dust from entering the motor windings, alternatively a separate internal ventilated enclosure for the sliprings may be supplied.

The gearless motor stator will be split into three, four or six pieces depending upon shipping clearances, and lift capacity. Heat exchangers are either distributed around the stator frame, or are located in banks below the stator, each accompanied by an internal air circulating fan. The rotor of the gearless motor is comprised of field poles which are attached to the mill, usually at the shell/head interface onto a flanged extension of the mill head, this being the stiffest part of the mill structure. In some instances, with shell-type bearing construction, the rotor components may be mounted on a torque tube extension of the mill shell, or onto a flange ring on the shell located close to the shell bearing.

Bearings

Both high and low speed motors, induction and synchronous, will have sleeve type bearings. For low speed motors the bearings will usually be self-oil lubricated. High speed bearings will be flood lubricated, and flood lube units with heat exchangers will be required to cool and circulate the oil. The power consumption of the flood lube unit should be added to the motor losses. The gearless motor is, of course, supplied without bearings, and relies on the mill bearings for support of the rotor.

To assist in installation and commissioning, high-pressure lift pumps are supplied with self-lubricated bearings. These are also required for inching if the motor is used for this purpose.

Soleplates

For ease of installation and alignment soleplates are supplied with both bracket bearing and pedestal bearing design motors. These are particularly helpful when installing twin pinion drives.

Clutches

The air clutch is used extensively with single and twin pinion low speed drives when using synchronous motors. This device permits the motor to be started uncoupled, thereby enabling the use of low inrush current design motors, and also avoiding any torque amplifications during the acceleration of the motor.

With weak power systems, reduced voltage starting of the motors is possible, and on twin pinion drives each motor can be started separately, thereby reducing the impact on the power system.

A major benefit of the air clutch is that it will act as a shearpin in the event of a power fault at the motor. A sudden short circuit on the terminals of any motor can produce very high instantaneous torques, and depending on the type and design of the motor this torque can be up to

10–12 times rated torque. This can have a devastating effect on the coupling, and/or gearing connected to the motor shaft, but once the air clutch has accelerated the load, and locked up, its air pressure can be slightly reduced to limit its torque capability to a value less than 200%. If a fault should occur the clutch will slip, thereby protecting the connected equipment from damaging torques.

With adjustable speed drives using low speed motors, the air clutch again comes into use a) as a shear pin against excessive short circuit torques, b) in the case of a twin LCI or PWM drive, both single pinion or twin pinion, it is required to provide bypass operation of the mill at fixed speed, by starting the single pinion motor, or one of the twin pinion motors across the line, and c) in the case of LCI, in order to avoid torque amplifications in the zero to 10% speed range, where the inverter bridge is forced commutated, the clutch is left open and only closed when the inverter is load commutated and torque pulsations have reduced to non threatening values.

It should be noted that the air clutch is not used to accelerate the load with high speed motors due to the excessive wear which takes place during acceleration. However the air clutch can be installed as a shear pin on the high speed motor, or between the low speed output shaft of a gearbox and the pinion driving the ring gear to allow an unloaded motor start.

Installation and Alignment

Correct alignment of drive train components is critical to good performance of an installation. Also it is important to check alignment regularly since foundations can settle differently, and an initially sound motor to pinion, and pinion to gear alignment can move significantly during the first few months of operation. Misalignment between motor and pinion is a major contributor to wear in a clutch. Once a drive train has been aligned statically, it must be checked in the dynamic state of driving a loaded mill. This has often proved difficult to achieve because of pressure to keep up production once the mill has been put into service, with resultant excessive clutch and gear wear. Continuous infra red monitoring of pinion tooth mesh is recommended since it would warn against development of a meshing error. Motor to pinion alignment can also be checked with the mill running and records kept which will show any change over time.

One major cause for concern in alignment occurs with a mill designed for operation in both directions of rotation. With a single pinion drive, the location of the pinion is such that the pinion tooth is lifting the girth gear and the pinion therefore presses down onto its bearings. If the rotation is reversed, the pinion tooth is now pushing down on the girth gear and the pinion is pushing upwards on its bearings. As a result, the pinion will lift upwards by the amount of clearance designed into the pinion bearings. The minimum clearance is typically 0.005" (0.127 mm), but can be as large as 0.012" (0.304 mm), which is enough to increase wear in the clutch. Also by reversing rotation, the axial centre line of the mill will shift as the centre of gravity of the charge moves from one side of centre to the other. This problem also occurs with twin pinion drives where one pinion moves up whichever rotation is selected. The solution is to set the alignment at the mid point of the bearing clearance, which reduces the error by half and brings the alignment error to a more acceptable level.

With gearless mills, which are invariably reversible, the axial centre line of the mill will move towards the centre of gravity of the charge, thereby creating an error in the airgap between the stator and the rotor poles. The worst-case condition of this occurs when the mill is starting to turn, and the charge is being lifted prior to cascading. As the airgap changes, the imbalance in the gap is reflected in the magnitude of the magnetic pull between the stator and the rotor. Consequently, the stiffness of the stator frame, mill bearings and the foundation must be designed to achieve stable operation with this magnetic pull acting on the components. The motor designer can reduce the maximum pull and the required structural strength and also reduce the total weight of the machine if he can use multiple paths or circuits in the design of the stator winding. (e.g., with a four circuit winding the pull will be reduced to about 40% of that exerted in a machine with a single circuit).

Drive Configurations

The use of squirrel cage induction motors is limited to small single pinion drives. This is due primarily to the high inrush current of the motor. The squirrel cage motor has not been considered for the twin drive option primarily due to the high inrush current problem, and the lack of adequate controls to maintain reasonable levels of load sharing between the drives.

Fixed Speed Single Pinion

Single pinion drives utilizing one pinion are currently available up to 10.0 MW. A high speed motor will drive the pinion through a main gearbox, and will be either a 6 or 8 pole machine.

A low speed, leading power factor synchronous motor, driving the pinion directly through an air clutch is the preferred drive configuration since it offers high efficiency, reactive power capability, overload capacity, rugged construction, low inrush current, minimum maintenance, and competetive evaluated cost.

The single input to a power-splitting pinion stand drive using a high speed wound rotor motor can be used up to approximately 13.5 MW. The motor will utilize a liquid rheostat for starting purposes, and will drive through the gear reducer to two pinions driving a single ring gear.

A synchronous motor can be applied to this drive configuration using a hydroviscous clutch to allow an unloaded start for the motor, thereby avoiding torque oscillations during acceleration, and enabling the motor inrush current to be controlled within the constraints of the power system. Because of the small installed base for this drive system, performance data regarding availability and maintainability is not available.

Fixed Speed Twin Pinion

Twin high speed wound rotor induction motors can be used to drive through gearboxes to twin pinions. The motors will share load on average within about 5% and can share a common starting resistor. Gear runout can give rise to load swings between the two motors during a revolution, which can lead to accelerated gear wear. Also motor characteristics may not match perfectly, which can cause an offset in load share between the two motors. To overcome these conditions, a permanent slip resistor may be installed between the two rotors. This will improve the load share capability, but it will be at the expense of drive efficiency, which will be reduced by 1.2%–1.5%.

A twin low speed synchronous motor drive has been successfully developed utilizing air clutches to allow the motors to start individually. Additionally, this system has a special winding built into the quadrature axis of the motor rotors (see Figure 3) to allow precise load sharing at all times; i.e., on average, and during each revolution of the mill, compensating for gear runout. Air clutches are installed between each motot and its associated pinion, which allows the motors to be started one at a time thereby lowering the impact on the power system. Once the motors are synchronized, the mill is accelerated using the air clutches. After the clutches lock up, the load share between the two motors is monitored, and any error is adjusted by applying current to the quadrature axis winding in the motor rotors. If this error exceeds a pre determined amount, an automatic clutch pulsing system comes into play and brings the two rotors closer to exact load sharing. The regulator controlling the quadrature axis current then brings the two motors into exact load sharing, not only on average, but during each revolution of the mill compensating for run out of the gears. This system is known as Quadramatic^{tm}. A typical layout of the motor pit is shown in Figure 4 where all cable connections to the motor are made in the motor pit, leaving clear access at the mill floor operating level. It should be noted that this drive imposes less stress on the pinions and ring gear than any other geared drive, regardless whether they are single or twin pinion. This twin pinion approach enjoys all the benefits and features of the low speed single pinion synchronous motor described above.

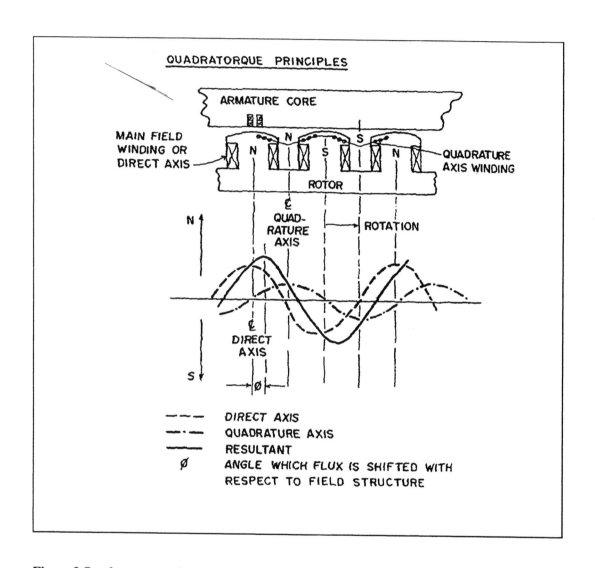

Figure 3 Quadratorque principles

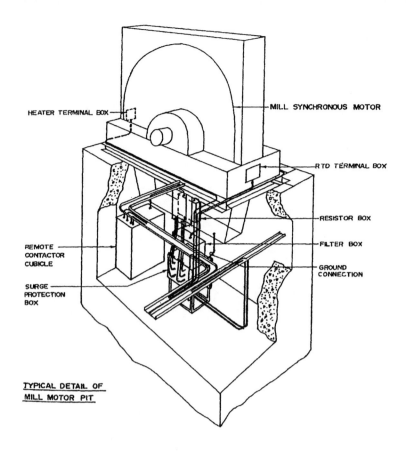

Figure 4 Typical detail of mill motor pit

VARIABLE SPEED DRIVE OPTIONS

From Table I, it will be seen that eleven adjustable speed options have been identified. These cover single pinion geared, twin pinion geared, and gearless.

Wound Rotor Motor(s) with Slip Energy Recovery (SER)

This drive system can be applied to both single and twin pinion geared drives. Since wound rotor induction motors are not cost competitive at low speed, the scheme generally employs high speed motors driving through reduction gears to the pinion(s).

The drive is started with the use of liquid rheostats providing adequate starting and accelerating torque to bring the mill up to speed, with a low inrush current of approximately 200% to 250%.

When the motor reaches full speed it will be driving the mill at its maximum speed, typically 80% of critical. In order to run at base speed (76% critical) and further to 60% critical, the motor slip must be increased accordingly. This can be achieved by inserting resistance in the rotor circuit, and dissipating this energy into the liquid rheostat. This is very inefficient, so rather than using the rheostat, the slip energy is converted to direct current, inverted to the frequency of the power system feeding the motor, and then fed back into the power system through a step up transformer.

Most mill configurations will have natural frequencies in the region of 2 Hz–4 Hz, 8 Hz–10 Hz, 20 Hz, and 40 Hz. In reducing the speed of the wound rotor motor, the slip energy recovery equipment will generate forcing frequencies at multiples of 6 times the slip frequency, depending on the number of pulses built into the equipment. A higher number of pulses equates to increase in cost. For example in reducing the motor speed by 5%, the SER will excite all natural frequencies between zero and 18 Hz (6 pulse) or 36 Hz (12 pulse), etc. Because of this there is a high probability that certain speeds in the operating range of the mill will need to be deadbanded (unable to operate within a particular speed range).

Wound rotor motors have low electromagnetic damping, and will tend to accelerate gear wear.

Advantages of SER drive:

- Low first cost (unless 18 or 24 pulse SER is used)
- Motors will operate (at 80% C.S.) if SER is out of service
- Low speed inching available with liquid rheostat
- Smaller filter sized for slip energy only
- Can load share twin drive.

Disadvantages of SER drive:

- Deadbanding required to avoid resonance
- Accelerated gear wear
- Poor overall efficiency
- Need for special attention to motor enclosure and ventilation
- Need to maintain precise alignment of motor/gear/pinion.
- No shear protection against transient torques caused by short circuit faults.
- High maintenance of sliprings, brushes and liquid rheostats
- No automatic protection against a cemented charge.

A variation of this drive involves connecting the rotor windings to a low frequency source. This variation allows the motor speed to be reduced or increased. If PWM power is used for this purpose, then smooth speed variation is possible. This system is limited to relatively small speed variations.

Pulse Width Modulated (PWM) Drive

This drive system is a relatively new system which can be used for variable speed mill drives.

It can be applied to both Induction and Synchronous motors.

The PWM was originally introduced for small low voltage drives, but has now been developed in various configurations for medium voltage applications.

There are now several configurations and combinations of semiconductors available to produce the power required for mill drives up to at least 20 MW.

These include:

- IGBT (Integrated Gate Bipolar Transistor)
- IGCT (Integrated Gate Commutated Thyristor)
- IEGT (Injection Enhanced Gate Transistor).

The device draws its power from a dc bus and inverts it to alternating current by chopping blocks of dc at high frequency.

The IGBT is the original transistor power supply and multiple modules are connected in series to increase voltage, and in parallel to increase current. Switching frequency is typically 1600 Hz

The IGCT is a higher powered thyristor which can handle much higher current than the IGBT, thereby reducing the number of devices required for a given power output. Switching frequency is typically 500 Hz.

The IEGT is a higher powered transistor than the IGBT, with a very low gating voltage and with a high rate of change in gating voltage (dV/dt) requiring minimal snubber circuitry. It is switched at 500 Hz. This device enjoys the power level of the IGCT with far lower complexity and power requirements in the gating circuit.

Because of the high dc content in the PWM output, a step up or down transformer should not be placed between the PWM and the motor. Unless the Utility voltage matches the PWM output voltage, the motor cannot be transferred to the bus without a transformer that matches the Utility voltage to that of the motor.

The PWM can be configured for two or four quadrant operation. Since grinding mills do not regenerate, two quadrant operation will suffice. This will allow the converter side to be a straight diode bridge.

The input transformer will be designed for a minimum of 12 pulse, or alternatively 18 or 24 pulse, thereby reducing the harmonic content to the Utility to below that required by IEEE 519.

This configuration will operate at a power factor of 0.95–0.96 lag.

If a PWM converter bridge is substituted for the diode bridge, then the power factor can be improved further to unity or leading.

The PWM requires an input transformer with a large number of windings to feed the relatively large number of circuits.

The PWM is a voltage source inverter and can be used on induction motors, as well as on synchronous motors, so where initial cost is of prime importance, then the combination of PWM, squirrel cage induction motor, and gearbox can provide the lowest initial cost. To protect gears from the high transient torques that can occur in the unlikely event of a cable or terminal fault at the motor, a disengaging device should be installed between the motor and the gearbox.

The PWM power supply is usually rated at its maximum capability, so in order to provide sufficient torque to accelerate the mill through its cascade point, the PWM, as with all electronic power supplies, must be capable of producing sufficient power to accelerate the mill through the cascade point. If the mill is accelerated slowly, then the maximum amount of torque required should not exceed 130%. Similarly, this or greater torque can be required when inching the mill to check for a cemented charge. For inching, the maximum torque could be required for approximately 30 seconds. This capability should be confirmed by the supplier for a range between 150% and 160% of the required steady state mill load for a minimum of 30 seconds. This capability should be confirmed by the supplier.

With its high inrush current, the squirrel cage induction motor may have difficulty starting across the line should the PWM be out of service.

Larger drives, in conjunction with low speed synchronous motors for both single and twin pinion, have been considered, and like all other drive types there are benefits and disadvantages. They compare quite closely to the LCI drive and, where the power requirement matches the output capability of the drive, they are competitive with the LCI. At the time of writing, there is insufficient installed base to comment on reliability and availability, but they do offer another choice.

The PWM can also be considered for the power supply for gearless drives where it has the distinct advantage of having a high front-end power factor, harmonic content to meet IEEE 519, and very low risk of creating damaging short circuits.

The PWM can also be used on a twin pinion drive.

Advantages of the PWM system:

- Can work with either induction or synchronous motor
- Synchronous motor operates at 1.0 power factor

- High power factor to the Utility
- Eliminates need for a harmonic filter
- Does not generate significant torque pulsations
- Can inch at very low speed
- Cemented charge protection available
- Reversible
- Line side power interruptions do not cause faults on load side.

Disadvantages of the PWM system:

- Relatively complex electronics
- Cannot transform output voltage
- Motor may need transformer in order to run on the Utility.

Synchronous Motor(s) with Load Commutated Inverter (LCI) Drive

The LCI can be configured in any one of three ways:

- 12/6 pulse
- 12/12 pulse with two winding motor
- 12/12 pulse with summing transformer and single winding motor.

Most LCI systems will utilize a three winding transformer to connect to the Utility, with a 30° phase shift (Wye - Delta) between the secondary windings and with each secondary feeding into a 6 pulse converter bridge. This configuration will cancel out most of the 5th and 7th harmonics, thereby presenting a 12 pulse load to the Utility, and reducing the size of the required filter to meet IEEE 519. The LCI is a current source inverter, and it uses the back EMF generated within a synchronous motor to commutate the inverter bridge so that it will invert at the frequency at which the motor is running. Between zero and 10% of rated speed, there is insufficient back EMF to commutate the inverter, so the bridge is force-commutated over this range.

The curves in Figure 5 show the ripple torque effect for both 6 pulse and 12 pulse inverters, during forced commutation and load commutation.

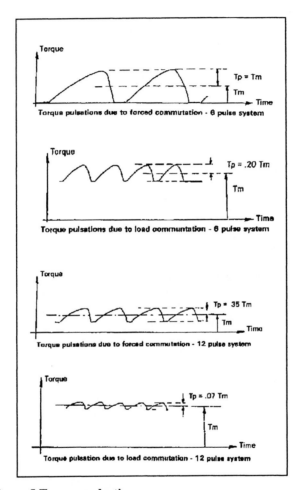

Figure 5 Torque pulsation curve

The 12/6 Pulse LCI. This configuration (see Figure 6) will cancel most of the 5th and 7th harmonics fed into the system as stated previously. However, the inverter will feed 5th and 7th harmonics (as well as higher order 11th and 13th; etc.) into the motor requiring it to be made larger to accommodate the heating effect of these harmonics; and for satisfactory commutation, the motor subtransient reactance must be kept below 18%–20%. Other than with relatively small mills, below 2200 kW, the 6 pulse system is not recommended.

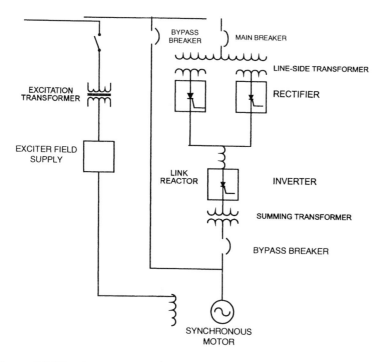

Figure 6 12/6 pulse LCI with bypass capability

The 12/12 Pulse LCI. As well as cancelling the 5th & 7th harmonics on the line side, this configuration (see Figure 7) will also cancel out most of the 5th and 7th harmonics on the inverter side. With high speed motors, it may be convenient to feed the two 6 pulse outputs from the inverter into two phase shifted windings in the motor stator. If this is done, it would not be possible to operate the motor on the Utility at fixed speed in the event of an outage of the LCI. With low speed motors, a two winding configuration would not be practical due to the large number of stator slots that would be required to accommodate the winding. To overcome this, and in the case of the high speed motor, the outputs of the two 6 pulse inverter bridges may be summed together through a three winding transformer to provide 3 phase 12 pulse power to the motor. The output voltage of this transformer can be selected to match the distribution voltage so that the motor could bypass the LCI and run at fixed speed if required. By cancelling out the 5th and 7th harmonics, the motor will run much cooler and the ripple torque produced by the motor will reduce to a much lower value than that produced with a 6 pulse supply. The motor can be designed with much higher reactances, which results in a low inrush current if the motor is required to start across the line.

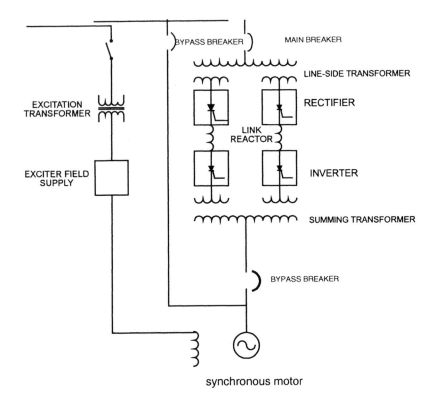

Figure 7 12/12 pulse LCI with bypass capability

This configuration would allow one design of motor to drive both fixed and variable speed mills of like power and base speed, thereby saving on spares and cost of foundation design.

A disconnect switch between the output of the summing transformer and the motor, supplemented by a bypass motor starter, would enable the motor to run at either adjustable speed or fixed speed.

In order to protect the gears from the damaging effect of transient torques which can be generated in electric motors under cable or terminal fault conditions, air clutches are provided between the motor(s) and the pinion(s). The air clutches are disengaged while the motors accelerate through the forced commutation speed range (0%–10%), thereby avoiding any risk of torque amplification at natural frequencies. Further, if the drive is to operate in bypass (see Figure 8), the air clutch allows both motors to be brought up to synchronous speed uncoupled.

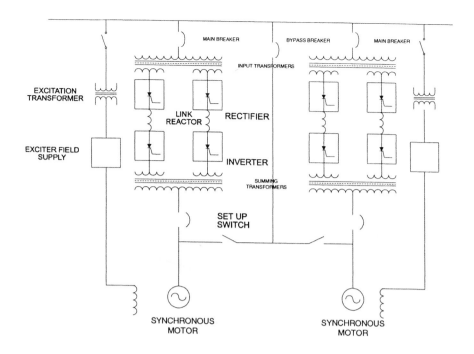

Figure 8 Twin 12/12 pulse LCI with bypass capability

The LCI operates at a lagging power factor of 0.85 at full load, but this will be improved to approximately 0.95 when a filter is added. The filter will increase the system losses by approx. 0.5%.

Advantages of the LCI system:

- Current source inverter. Inverter fault currents are controlled by the dc reactor in the LCI
- With summing transformer, motor can match Utility voltage, allowing interchangeability with fixed speed motors on Ball Mills
- Motors can run at fixed speed on Utility in the event of drive outage
- Low cost, low loss filter can reduce harmonics
- Inching available at 10% of rated speed
- Inching by position available
- Cemented charge protection available
- Initial cost competitive with other systems above 3000 hp
- Proven system
- Drive is reversible and regenerative
- Very high availability.

Disadvantages of the LCI system:

- Increase in footprint to accommodate summing transformers, and bypass set-up switches
- Unsuitable for continuous operation below 10% speed
- 12/12 with summing transformer premium priced on smaller mills.

Cycloconverter (CCV) Fed Systems

The CCV drive is a low frequency drive supplying power at frequencies up to 30 % of line frequency.

It has been the drive of choice for gearless drives where the motors operate at frequencies between zero and 7 Hz. However, with the introduction of the PWM power supply, the two drives will compete with each other for the engineers favour.

The CCV can also be applied to geared drives both single and twin pinion but, these configurations, since the output frequency is less than 30% of line frequency, the motor will have a fewer number of poles in order to run at pinion speed.

The CCV is a voltage source inverter, and since the frequency conversion occurs without going through a conversion to dc and inversion to ac, a fault occurring on the line side can result in a sudden short circuit on the motor side. Special techniques can be provided to ensure an orderly shutdown of the CCV bridge prior to the drive breaker opening, however opening an upstream breaker can cause such a fault (see Figure 9). To cater to this condition, it is necessary to provide extra bracing to the motor windings, and/or add high speed interrupters between the CCV and the motor.

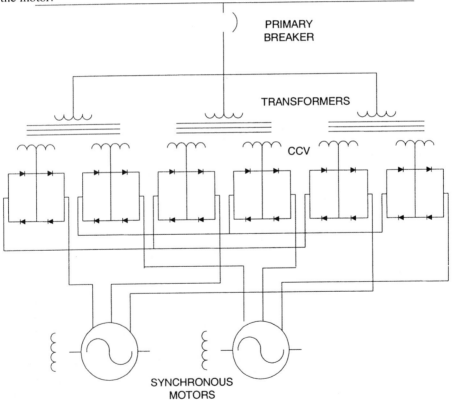

Figure 9 Twin CCV geared drive

To protect gears from the high transient torques that can occur with the CCV, disengaging devices such as shear pins or air clutches should be inserted between the motors and the pinions. Because the motor(s) must run at low frequency, it is not possible to operate a CCV drive with bypass capability.

The CCV normally requires an encoder for operation at very low speed, with vector control above about 5% of rated speed.

The CCV is fed from three, three winding transformers, with their secondaries connected wye and delta to provide some harmonic cancellation. This works adequately when the output of the CCV is fed into a single motor (as in the case of the gearless drive). However in the case of a twin pinion drive, the cancellation will only be partial because the motor rotors will depart from true alignment to each other during a revolution of the mill due to gear run out.

Since the harmonics generated by the CCV are multiples of the operating output frequency of the drive, it follows that the total amount of harmonics generated will be greater than with a LCI, but these harmonics are of smaller magnitude and are spread over a larger range of frequencies. The need for a filter, and the size of the filter are both dependent upon the short circuit capacity of the power system relative to the drive capacity. Generally speaking, if the system short circuit MVA is less than 20 times the converter MVA, then it is probable that a broad band filter will be required to control harmonics. The design of the filter is more costly than the tuned filter that is used with a LCI, and its losses are about three times those of a tuned filter for the same size drive.

The CCV provides a clean quasi sinusoidal current waveform to the motor(s), and can be operated at frequencies down to zero without torque pulsations.

The output voltage of the CCV for a twin drive is in the order of 1500 volts.

Advantages of the CCV drive:

- Inching available down to zero rpm
- Cemented charge protection available
- Inching by position available
- No torque pulsations
- Marginally better motor efficiency than other drives
- Reversible and regenerative
- Charge let down capability.

Disadvantages of the CCV drive:

- Motor(s) dedicated to the CCV, no bypass capability
- Susceptible to power supply transients
- Broad band filter required, with higher cost and higher losses
- Higher first cost
- Low voltage output increases cable costs
- Need "shear protection ".

CCV and PWM Driven Gearless Drives

With the gearless drive, the rotor of a synchronous motor is mounted directly on to the mill, usually on a flanged extension of the mill head. The stator is made in segments, generally three for smaller diameter mills and four for larger diameter mills. Where lifting capability and /or shipping clearances demand, six segments may be considered. The stator winding may comprise one or two windings, using single- or multi-turn coils that are deployed in one or more circuits, depending on the design philosophy of the manufacturer (see Figure 10). Special attention should be given to this design since it will have significant impact on the reliability, availability, and repairability of the machine.

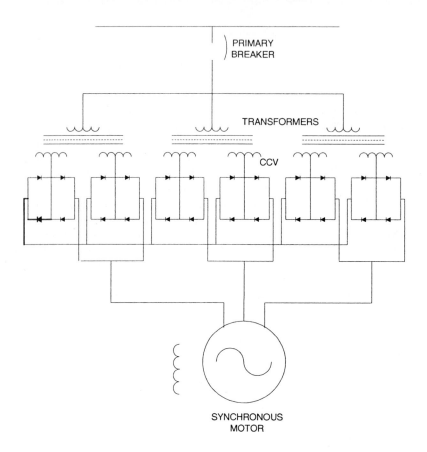

Figure 10 Gearless drive with single 3 phase CCV supply

Motor voltage can vary from less than 1500 volts to close to 5000 volts, depending on the design of the stator winding.

As with the twin CCV drive, the gearless motor is subject to sudden short circuits arising from unscheduled interruption in the power supply due to upstream breaker trips. These occurrences have caused significant damage to installed equipment and its foundations. These forces must be taken into consideration in the design of the motor and its foundation.

This kind of event will not occur with a PWM fed mill (see Figure 11).

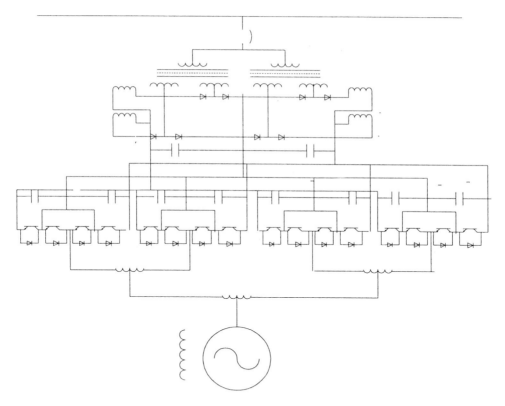

Figure 11 Gearless drive with single 3 phase PWM IEGT supply

Because of the low frequency at which these motors operate, typically 0 Hz–7 Hz, the core losses only comprise approximately 10% of the total motor losses, leaving the stator and rotor copper losses as the prime source of machine losses. By adding copper to the machine, the winding resistance can be lowered, thereby achieving remarkably high motor efficiency. However it must be understood that nothing is free, and the cost increment in achieving an efficiency improvement of 1% may well exceed the loss evaluation in dollars for the kW saved.

Once the motor is properly aligned the airgap is continuously monitored by sensors, which will alarm and shutdown the machine if the airgap error exceeds the allowed tolerance.

Special running seals are provided between stator and rotor to protect against the ingress of water, mud, and dust.

Advantages of the gearless drive:

- Elimination of gears reduces quantity of components
- Gear losses are eliminated
- Can achieve high efficiency
- Inching available down to zero rpm
- Cemented charge protection available
- Inching by position available
- No torque pulsations
- Reversible and regenerative
- Charge let down capability
- No filter required with PWM.

Disadvantages of the gearless drive:

- Dedicated to CCV or PWM power supply
- Susceptible to power supply disturbances CCV only
- Broad band filter required with CCV, with higher cost and higher losses
- Higher first cost on smaller units
- High installation cost
- Long outage if motor damaged

OTHER CONSIDERATIONS

When a mill comes to rest there is a tendency for the charge in the mill to settle and cement into a solid or semi-solid mass. If an attempt is made to start the mill in this condition, there is a great risk of carrying the charge through close to 180º at which point the charge will break away from the shell and crash across the mill with potentially destructive results.

To avoid this kind of occurrence, it is good practice to inch the mill to ensure that the charge is loose and able to cascade.

Mechanical inching drives are normally portable so that one machine can serve several mills.

A disadvantage of the mechanical incher is the time taken to install and remove it before running the mill.

Electric inching has been available for some time.

Wound rotor motors can be turned slowly using their liquid rheostats. This will enable the mill to be positioned for maintenance purposes as well as check for a cemented charge prior to starting the mill.

To inch synchronous motors with sliprings, the field is excited, and a dc source is applied through a set of commutating contactors to the stator winding. These contactors synthesize a low frequency ac voltage. A newer version of this system replaces the contactors with static switches to achieve the same result.

If this system is to be considered with synchronous motors having brushless exciters, then the motors must be specified to have an ac fed exciter to allow full excitation at zero speed. This system also allows mill positioning for maintenance as well as checking for a cemented charge prior to starting the mill. If the motor has a conventional dc fed exciter, then this kind of inching cannot be used.

Methods have been devised to check for a cemented charge by:

- Reducing air pressure into the clutch to a level just above that required to cascade the mill, and closing the clutch. Either the mill will accelerate, showing that the charge was not cemented, or the mill will stall with clutch slippage showing that the charge was in fact cemented. This method has two flaws: a) there is a lot of clutch wear, and b) if the mill is only partially loaded, then there may be sufficient torque available from the clutch to carry the cemented charge up to the top of the mill and allow it to fall under gravity.
- Noting the position of a liner bolt, applying the clutch for 2–3 seconds, then tripping the clutch. If, when the mill rolls back to the rest position, the liner bolt has moved forward from its original position, then the charge is unlikely to be cemented. Conversely, if the position of the liner bolt is unchanged, then the charge is likely to be cemented. This method also adds to clutch wear.
- Using a combination of degrees of rotation of the mill, and acoustics and/or vibration, it is possible to determine whether the charge has cascaded within an allowed number of degrees of rotation. This method is independent of the amount of fill in the mill, and allows the mill to continue to accelerate if cascade is detected, thereby minimizing wear in the clutch. None of these methods should be used for positioning the mill.

If the mill has a variable speed drive on it, then the drive can be used for both inching and detection of a cemented charge.

It should be noted that if the motor is connected to the drive train during inching, then high pressure lift pumps must be installed at the motor bearings because the inching speed will be too low for the bearings to maintain an oil film.

Also if the motor is used for inching electrically, then care must be taken to ensure that the time taken for inching is within the thermal capacity of the motor, unless forced air ventilation is used.

Power Factor

Large electrical power users purchase their electrical energy according to a tariff negotiated with the Utility.

This usually takes the form of $"X" per kVA (kilo volt-ampere) of maximum demand, plus $"Y" per kWh of energy consumed.

In order to keep the demand charge as low as possible, it is desirable to keep the power factor of the site as close to unity as possible.

Traditionally mine sites have relied on their relatively large installed base of synchronous motors to supply the required power factor correction. With the introduction of adjustable speed drives, some of the inherent reactive power availability has been removed, and it has become necessary to rely on the power factor improvement obtained from the capacitors in the drive filter, supported by any synchronous motors used in the installation. In some cases, banks of capacitors have to be added to limit any power factor penalties.

Example: 1 SAG Mill 15 MW with uncorrected power factor of 0.84 lag plus two ball mills each 7.5 MW fixed speed 0.8 pf lead single pinion geared synchronous motors. SAG mill corrected to 0.95 lag using filter capacitors. Total MVAr's (Mega volt amperes reactive) contributed to the system = 6.32 MVAr. lead.

If such an installation is repeated with adjustable speed drives on the ball mill(s) then the installation will draw a total of 9.86 MVAr's from the system. Therefore, to supply the same amount of power factor correction for the plant 16 leading MVAr's are required. The cost of this must be added to the total cost of the second case.

In cases like this, it may be better to consider a 15 MVAr synchronous condenser to supplement the MVAr's obtained from the required filters.

Overload Capability

Unless it is specified, and included in the price of the equipment, no overload capability is available from any adjustable speed drive system, or from a fixed speed induction motor drive.

Fixed speed synchronous motors, however, do have inherent overload capacity since it is possible to exchange their reactive power capability for active power, so a 0.8 leading power factor synchronous motor can supply a 25% increase in its kW output without exceeding its temperature guarantees.

Power Distribution Considerations

For large installations, it is sometimes desirable to use a primary distribution voltage higher than 13.8 kV and 24 kV is commonly selected.

This voltage can be fed directly into the primary transformers of adjustable speed drives using LCI, PWM or CCV power. As previously shown, twin LCI drives can obtain bypass capability by installing a 24 kV/motor voltage transformer. This transformer can also be used as a spare for the LCI drive.

Wound rotor fixed and adjustable speed drives along with all fixed speed drives will require a lower voltage and a separate medium voltage distribution system. This can be at 13.8, 6.6, or 4.16 kV and will supply all the medium voltage loads at the site. This approach to the power distribution does not cause additional transformer loss penalties for medium voltage equipment below 24 kV.

Altitude Considerations

As altitude increases the air density decreases. The heat generated by electrical equipment becomes more and more difficult to dissipate with altitude increases. Equipment that depends directly on cooling air to remove its losses has to be overdesigned so that the cooling air can remove the heat while not exceeding the equipment's thermal capacity. Where air-to-water heat exchangers are employed for equipment cooling, the equipment will require less derating if the heat exchangers are made larger to increase the heat collection surface. Also voltage, BIL and partial discharge margins are reduced.

Maintenance Considerations

Electrical equipment today is designed and built to be as free from maintenance as possible. Electronic solid state equipment has virtually no wearing parts, and is not expected to fail or wear out in normal operation. However it is customary to carry an inventory of components for an adjustable speed power supply. For motors of a geared drive, spare bearings, brushes, brushholders, brushless exciter components, and often a spare wound stator and rotor are mandated. Where more than one motor of the same rating is installed, one set of spares will be shared between them. For a gearless drive, the cost of a spare stator would be prohibitive so this is not called for. Spare rotor poles can be carried, and with some designs stator winding components can also be carried. Because of design differences, the costs of these capital spares have not been shown.

Generally speaking, routine maintenance on geared drives can be done during normal planned mill outages. Unplanned outages due to component failure can usually be handled in a relatively short time. Back up systems can provide for continued operation until a planned outage can be arranged. With gearless motors, a stator winding fault will cause an extended outage which can range from 2–3 days to several weeks, depending on the type of construction used by the motor manufacturer.

Equipment Cost

In compiling costs, the following criteria has been used:

- An allowance for the losses is added to the first cost in an attemp to demonstrate the effect of efficiency for each drive option
- Fixed speed single and twin pinion drives do not include an electrical house since their control equipment is normally located in an electrical room supplied for this and other equipment. All adjustable speed drives include an electrical house, with the usual complement of equipment such as drive controls, motor control center for mill motor and mill auxiliaries, graphics package, etc.
- Fixed speed drives are fed from a medium voltage distribution system
- Adjustable speed drives are fed at the design distribution voltage, except wound rotor drives can only accept power at 15 kV or less. No transformer penalty (cost or loss evaluation) has been applied to the wound rotor drive since a distribution voltage of 15 kV or less will normally be available as well as higher voltage which can feed directly into CCV, PWM, or LCI drives
- Primary circuit breakers for each drive are included
- The equipment costs are reflections of the efficiencies used to compile the loss evaluations. In general, the efficiency values used are at the level where the cost of improving the losses matches the loss evaluation number of $3,000 per kW
- Motor speeds of 200 rpm for geared low speed synchronous motors and 900 rpm for wound rotor induction motors are used
- The bypass equipment (switchgear, transformer, and set ups) as shown earlier in Figure 8, increases the first cost of the electrical equipment by approximately 5%.

Figures 12, 13, 14, and 15 show the cost relationship between the various drive systems. Note: These costs are shown on a "per unit" base; i.e., they relate to the cost of one drive relative to another without using actual costs frozen in time. Figure 12 covers adjustable speed drives for SAG mills using first cost, and this is followed by Figure 13, which takes into account the drive efficiency. Since the geared LCI PWM and CCV cost and performance are very close to each other, a single curve for low speed synchronous (LSS) is used.

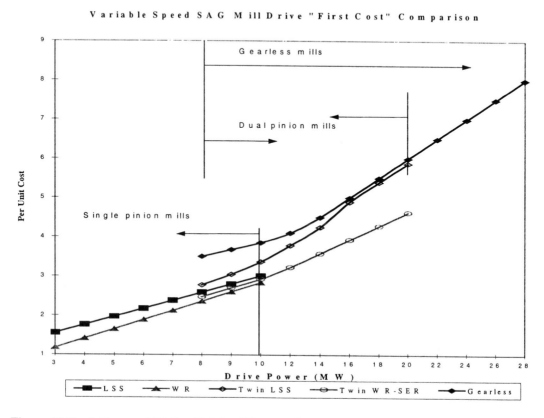

Figure 12 Variable speed SAG mill drive "first cost" comparison

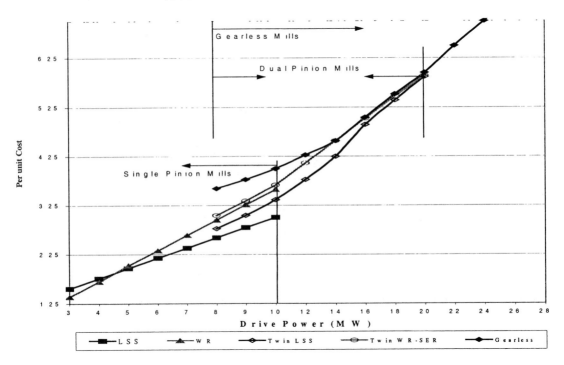

Figure 13 Evaluated variable speed SAG mill drive cost (losses evaluated at $3000/kW)

Figures 14 and 15 show drive first and evaluated costs for fixed speed ball mills.

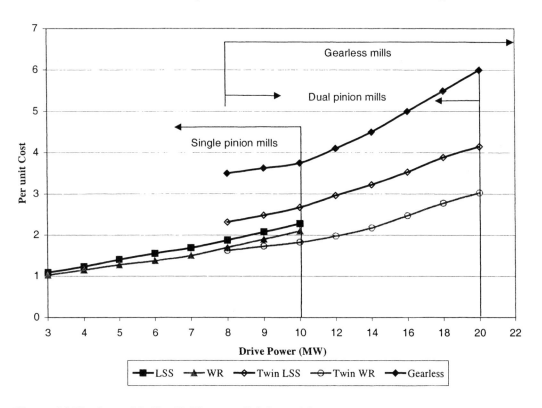

Figure 14 Fixed speed ball mill "first cost" drive pricing

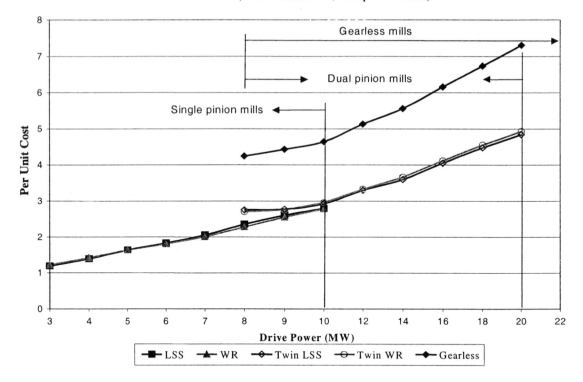

Figure 15 Evaluated ball mill drive cost comparison (loss evaluated at $3000 per kW loss)

In all cases the loss evaluation has a significant impact on the total cost.

A more detailed analysis of costs can be found in **"Selection and Evaluation of Grinding Mill Drives "** C.D. Danecki, G.A. Grandy and P.F. Thomas Paper DB-14 CIMM-SME Symposium Vancouver Oct 2002.

ACKNOWLEDGEMENTS

The author wishes to thank Stuart Walters for his help in formulating this document, Mac Brodie for his helpful and encouraging critique as the document developed, George Grandy, and Craig Danecki for providing process and mechanical supporting information.

In conclusion, I would like to thank GE Canada for their support and assistance during the preparation of this paper.

REFERENCES

Barratt D.J., Brodie M.N., and Pfeifer M. 1996. SAG Milling Design Trends, Comparative Economics, Mill Sizes and Drives. *Proceedings International Autogenous and Semi-Augotenoug Grinding Technology 1996,* eds. A.L. Mular, D.J. Barratt, and D.N. Knight, III : 1228.

Bassarear, J.H., and Thomas, P.F. 1985. Variable Speed Drives for Semiautogenous Mills. *SME Meeting.*

Danecki C.D., Grandy G.A., and Thomas P.F. 2001. Mill Drives in the Third Millenium. *SME Annual Meeting.*

Selection of Metallurgical Laboratory and Assay Equipment; Laboratory Designs and Layouts

Peter F. Wells,[1]

ABSTRACT

The design of laboratories for metallurgical and chemical studies in mineral processing operations has to consider

    1) the needs for safety and health of the staff operating the laboratory,

    2) the requirement to reduce the samples to a mass and particle size that is suitable for metallurgical test work or chemical analysis,

    3) the need to ensure that samples do not become oxidized or contaminated during processing,

    4) the requirements to be able to efficiently carry out the test work or analysis and transmit the results to the required parties.

The necessary equipment, the layout of laboratories and the data acquisition and transmittal systems are considered based on the above requirements.

INTRODUCTION

Design of metallurgical control and development laboratories has received very little attention in the metallurgical literature with only the description of the Western Australian Mineral Research Centre (Bagshaw 1996) being published in the last ten years. The Laboratory Design Handbook (Crawley Cooper 1994) provides a wealth of information on research laboratory design but, while essential reading to provide general background, has no specific information for the designer of a metallurgical laboratory. Prior to that, the paper in the previous edition of this manual – "Design and Installation of Concentration and Dewatering circuits" (McKenzie and Haig 1986) provided a comprehensive survey of the state of the art at that time. Other works, which should be consulted, are "Design of a Mine-site Laboratory Facility"(Shelton 1981), and "A Guide to Laboratory Design"(Everett and Hughes 1981). The U.S. National Research Council book "Laboratory Design" (Coleman 1951) is very comprehensive but is now mainly of historical interest.

While the essentials of work flow and efficiency of operation have been well covered in the literature, the important change that has now become apparent is the need to emphasize the safety and health of the metallurgist and his or her staff and the great need to make the metallurgical laboratory a pleasant place to work. It is increasingly difficult to attract and retain highly competent people in a field that is, in certain quarters, viewed as a sunset industry and often involves a fly in/fly out work schedule. Since concentrators are intrinsically noisy and dusty and metallurgical personnel must, of necessity, spend a substantial part of their working day in that environment, it is essential that the metallurgical laboratory, in contrast, is a pleasant place to work in to maximize the operation's ability to keep professional staff.

---

[1] Section Head, Mineral Processing ,INCO Technical Services Limited, Mississauga, Ont. L5K 1Z9, email: pfwells@inco.com

Most of the discussion in this paper is directed towards the design of metallurgical laboratories associated with operating plants. However many of the criteria are also applicable to research and development facilities and the requirements for the latter will be covered in a separate section.

LABORATORY PLANNING

Involvement of the metallurgist, who will be responsible for the operation of the metallurgical laboratory, in the design process will be the most critical parameter in ensuring that a satisfactory design is achieved provided that the individual has the breadth of experience in various operations to know what does and doesn't work. The metallurgist's first task will be to determine the expectations of management with respect to the functions and goals of the metallurgical department and then state them explicitly in the form of a mission statement. This will then lead to agreement on the number of staff and the equipment and, only when these parameters have been established, should the architect and engineering company become involved.

In most metallurgical plants, the location of the metallurgical laboratory will be dictated by the following considerations:

1) noise, dust and vibration levels in the immediate area

2) the retrieval and processing of the daily (shift) metallurgical accounting samples

3) the retrieval and processing of metallurgical development samples

4) the need for the metallurgical personnel to work closely with operations management and personnel

5) the need to monitor, calibrate and correct the operation of an on-stream analyser

In base metal plants, the on stream analyzer and gang sampler should be centrally located and having the metallurgical laboratory adjacent to this equipment helps ensure that the analyzer receives timely attention. The metallurgical laboratory should also be close to the control room thus allowing the easy interaction of the metallurgist and metallurgical technicians with operations staff. A central location may make providing outside windows to the laboratory a challenge but the latter is an important consideration when trying to maximize the aesthetic appeal of the working environment.

In addition to choosing a location not immediately adjacent to sources of dust and noise the laboratory environment can be improved by the use of adequate sound insulation and double doors to create an airlock with boot cleaning facilities by the outer door.

SAFETY EQUIPMENT

The safety station will include an emergency shower and eyewash, fire blanket, "kill" switch, portable fire extinguishers and protective glasses, monogoggles, face shields and gloves. Gas testing equipment, cyanide and other antidotes, resuscitation equipment and a spill kit should be located in the same place.

Vented cupboards (usually incorporated into dust hoods) are required for acid and organic liquid storage. All electrical equipment that could come into contact with these vapours must be rated explosion-proof. A separate cupboard for personal protective equipment adjacent to the main exit should be provided. A second exit from the laboratory is required for reason of safety as well as convenience.

Sprinklers for fire protection should be added although Fire Code requirements will likely make them mandatory in any case.

LABORATORY SERVICES

Ideally inter-floor space of at least 1.2 meters will be allowed for services. Air make-up is a large item as the laboratory must be maintained at a positive pressure with respect to the concentrator to prevent infiltration of dust. A typical 1.4 meter long dust hood will withdraw 0.57 m^3/sec (1200 cfm) of air resulting a significant heating or air conditioning load and to prevent noise in the feed duct system, oversize air feed systems and inlet ducts will be needed.

Ceilings at least 3.1 meters (10ft) high allow for use of flotation columns in the laboratory as well as giving an uncluttered feel to the workspace. Natural lighting from a full-length window on one side of the laboratory with a well designed artificial lighting plan will further enhance the working environment.

Water requirements are for hot and cold domestic water, demineralized water and addition of a carefully marked faucet to deliver process water will assist the metallurgist when carrying out flotation testing. A tempered water supply is required for the shower and eyewash.

Dried compressed air should be supplied for flotation air, for hoses in dust hoods and for pressure filters if fitted. The vacuum pump and trap for the filters must be located where it can be easily serviced but far enough from the laboratory that the noise from the pump will not be heard. A water sealed pump is normally specified.

Drains can report to the final tailings sump provided that there is a separate drain for acidic residues that reports to a holding tank for neutralization before discharge.

Power available in the laboratory should include some multiphase lines for powering crushers, mills and magnetic separators .

REQUIRED EQUIPMENT

Sample Preparation

The equipment for a metallurgical laboratory will be somewhat dependent on the particular industry but in almost all cases the requirement to efficiently process accounting and development samples will dictate the laboratory layout. A rotary sample splitter with vacuum filters and sample bucket cleaning facilities will be the first station. Samples will require drying in an oven, which must have very close temperature regulation for sulfide samples to prevent oxidation. The ability to control the temperature between 40 and 100 degrees Celsius will cover the range required for most dry operations. Hot plates may be used to dry samples provided that the technician agitates the cakes to prevent local overheating. The dried filter cakes must be weighed and then screened with the oversize reduced to –75 microns in a roller mill or similar in preparation for most assay procedures. The roller mill or pulveriser should be well insulated to reduce noise and the soundproofing system should be convenient for the technician to install and remove. Modern roller mills are extremely heavy and a pneumatically operated cantilevered arm should be used to move the mill from the grinding cabinet to the dust hood. Rolling, riffling to further reduce the samples size, bagging and addition of an identifying bar code strip on the sample bag complete the process.

Hot plate drying, if used, and final sample preparation must be carried out in dust hoods. Ideally, there will be separate hoods and pulverizing equipment for feed, concentrates and tailings to minimize contamination. Face velocities in dust hoods should exceed 0.5m/sec (100 ft/min) and the design must cause the dust-laden air to descend to a trap. Care must also be taken to ensure that the noise level created by the air movement and fan does not exceed 75 DBA to provide a satisfactory working environment for the metallurgical technicians. Flooring should be sloped slightly to a drain to allow for wash down and should be "cushion floor" tiled. If ceramic tiles or concrete floors are specified, rubber cushion mats must be provided. A storage area for "reject" samples should be included in the immediate area. Not keeping additional samples from both production and development work has caused much uncertainty when there is a failure to balance and a well organized reject sample storage area allows the metallurgist to retrieve samples for mineralogical study or re-assay with minimal expense. A similar storage room should be included for storage of equipment and supplies.

Test Samples

Whatever the process used to concentrate the ore in the plant, the metallurgist will require a laboratory equivalent to carry out plant performance monitoring and development studies. An area for air drying samples of ore and equipment for crushing and screening the ore to the size acceptable for a laboratory rod mill (normally between 3.35 –1.7 mm (6 and 10 mesh)) is essential. This should be in a separate room in order to minimize contamination and localize the noise. The room should have a roll up door or at least high double doors to the plant area to allow samples to be brought in on a fork lift truck. Riffles or preferably a rotary splitter will be needed to split the ore into aliquots for grinding and a sample for head analysis. Bar coding of all samples assists in maintaining an easily retrievable inventory. Scales for weighing large samples and the aliquots and a freezer to store samples complete the equipment requirements for the preparation room.

Comminution

Grinding of ores for flotation is usually carried out with a 2 kilogram charge in a 200mm (8 inch) diameter rod mill with provision in the mill for addition of air or oxygen in order to keep the Redox potential of the ore similar to that in the full-scale mill. It may be necessary to use a mixture of stainless and mild steel rods to adjust the potential to that observed in the plant cyclone overflow. If the plant operation includes a SAG or AG mill, use of ceramic balls as media may be necessary to obtain the correct Redox potential. Mills should be in a counter balanced framework so that no lifting is involved when charging or emptying the mill and complete soundproofing around the mill with a convenient means of removal is a must.

Ultrafine grinding such as often required in plants where leaching is the main separation technique are best carried out with an attrition grinders such as those supplied by Metso Minerals (Davey 2002). Care that the Redox potential of the ground product is similar to that obtained in the plant is even more important with such grinding devices. Attritors (sometimes called detritors) should also be used for simulating fine regrind operations.

Flotation/Leaching/Gravity Separation

It is preferable to carry out test flotation under a dust/fume hood because of the fine aerosol containing solids produced by a flotation cell. The dust hood must be of adequate height so that the metallurgist will not be leaning forward when using the flotation machine. A vacuum filter station should be located on a bench next to or opposite the dust hood to allow for rapid dewatering of products. Continuous monitoring of physical and chemical parameters such as temperature, pH, Redox and conductivity during flotation testing has become an accepted part of the procedure and facilities to hook the instruments to a datalogger or to the plant network should be provided. There are now several automated laboratory flotation cells available commercially and such an investment will pay back quickly from the improved reproducibility obtained.

For leach testing, a miniplant line is preferable to the traditional enclosed container on rolls as it is possible to measure and control the leaching environment much more closely. The miniplant system should be hooded and have large enough feed and product tanks to allow the system to duplicate the residence time of the operation.

Where plants employ gravity separation, setting up a laboratory scale unit may be feasible depending on the particle size of the feed. In case where spirals are used, a full-scale spiral, agitated stock tank and feed pump can be set up if sufficient ceiling height is allowed.

Additional Equipment

Sizing equipment comprising wet and dry screening facilities and a cyclosizer will be required. Sieve shakers should be well sound-proofed and be located next to a dust hood for sample weighing. An ultrasonic screen cleaner should be located close to the shaker.

Often missing from modern laboratories is a polished-section making station and optical microscope. While the use of such instruments as the QEM Scan and the Mineral Liberation Analyser (MLA) has injected a much needed quantitative element into mineralogy, the importance of the plant metallurgist being able to identify the average state of liberation of the ore and then being able to identify when changes in mineralogy cause a variation in metallurgical response cannot be overemphasized. In remote locations the polished sections can be prepared, photographed and the photographs transmitted electronically to a consulting mineralogist if more specialized assistance is required.

ANALYTICAL LABORATORY

Assay laboratory requirements vary widely depending on the industry. Operations with sulfide ores will likely rely on Ion Carbon Plasma (ICP) or Atomic Adsorption (AA) spectroscopy which means that the dissolution of samples in perchloric acid will be required. Hoods where perchloric acid is used must be completely washed down periodically and the wash down solution and other acidic waste from the laboratory must be disposed of in such a way that it does not come into contact with sulfide because of the danger of hydrogen sulfide or sulfur dioxide formation. The same safety equipment is required as for the metallurgical laboratory. Vibration is of major concern for analytical balances and although anti-vibration platforms can minimize the problem, it is still advisable to locate balances close to a major support column and preferably close to an exterior wall.

If high grade concentrates, 50% Zinc or 70% lead for instance, are to be analyzed, very great care has to be exercised because of the imprecision introduced by a large number of dilutions or very small sample size. In some instances, especially where an inexperienced staff is required to carry out these determinations, titrimetric or gravimetric procedures should be used.

LABORATORY INFORMATION MANAGEMENT SYSTEM

The most significant way in which metallurgical laboratories have changed since 1987 when the previous design manual was published, is in the way the data generated in the plant and laboratory is processed and transmitted to management. In most concentrators the operating data are now captured and stored on a server using software such as the PI system (OSIsoft of San Leandro, California, USA) (Bascur and Kennedy 2002) which allows downloading of any data into spreadsheets or presentation of the data on a personal computer in real time. This software has been used to process laboratory derived data such as assays which are input directly from the instrument to a 'Tag" on the server. The tonnages, also obtained directly from scales, and the assays are verified by the metallurgist and transferred into a new set of tags with the daily or shift balance being calculated from the verified data. The balance can be transmitted to management electronically as a spreadsheet so that the process requires no clerical assistance; entry and transcription errors are eliminated but there is a substantial investment in computer software.

LABORATORY LAYOUT AND LOCATION

The suggested layout for a metallurgical laboratory shown in the Figure 1 would be expected to service up to a 10,000 tonne per day operation employing flotation as the primary separation process with an uncomplicated flowsheet. The expected complement of a chief metallurgist, metallurgist and two metallurgical technicians would have sufficient space for all routine metallurgical procedures. Office space for the metallurgist and metallurgical technicians should be close to the laboratory; either between the metallurgical laboratory and the analytical laboratory or on the floor above next to the control room. Figure 2 shows a suggested laboratory location below the control room and metallurgist's offices and next to the analytical laboratory. The on-stream analyzer would be located in the center of the plant on the flotation floor adjacent to the metallurgical laboratory.

METALLURGICAL RESEARCH AND DEVELOPMENT LABORATORY

The major difference between a laboratory for research and development as compared to that in a production environment is the provision of facilities for all of the major separation techniques and also sufficient space and services to operate a continuous miniplant. The requirement to provide estimates of circulating loads in new or revised circuits, which has up until recently been attempted, with limited success, by use of locked cycle tests, is now achieved using a miniplant with feed rates in the 10 to 20 kg/h range. Such plants have equipment which is similar in size to the typical laboratory batch rod mill or flotation cell. At the low feed rates, it is possible to obtain satisfactory design data for feasibility costing of a process with perhaps 2 to 3 tonnes of ore, a quantity that can be obtained from drill core.

Environmental requirements dictate that discharges from the laboratory must be retained in a tank or lagoon. Drainage reports to a holdings tank were it is treated and assayed before release or disposal. Solids are removed by allowing them to settle in drums with the latter transported to a suitable disposal site –normally a company tailings area.

A requirement of any metallurgical laboratory is a freezer and research laboratories should have sufficient storage to keep several tonnes of samples. This necessitates the installation of a walk-in freezer which should have the capability to maintain $-30°C$.

A design for a mineral processing laboratory that has been found to work well is shown in Figure 3. The layout for this laboratory resulted from the input of all of the mineral processing staff who had dealt with the limitations of their former laboratory in some cases for close to 30 years. The company where this laboratory was installed gave permission for publication of the design but asked that their name not be revealed.

DESIGN AND CONSTRUCTION

During the design phase, the metallurgist as the customer, should be expected to monitor the engineering closely to ensure that his/her specifications are adhered to. Similarly, there should be frequent visits to site to ensure that changes initiated during construction do not impair the overall operation of the laboratory. Commissioning of the laboratory must include checking of hoods to ensure that specifications are met, faucets have the correct water, that water is not contaminated and that drains report to the correct sump or sewer.

ACKNOWLEDGEMENTS

I would like to thank Andrew Kerr of Inco, Clarabelle Mill for reviewing the document and making many helpful suggestions. I wish to thank Inco Limited for permission to publish this paper.

REFERENCES

A.N. Bagshaw 1996. The Western Australian Mineral Research Centre: A Resource for Metallurgical Testwork and R&D, *Proceedings of the AusIMM, Annual Conference* 24-28 March 1996, Perth, Western Australia, Australia 253-257

E Crawley Cooper 1994. *The Laboratory Design Handbook*, CRC Press, Inc, Boca Raton, Florida, USA

C.L. McKenzie and R.A. Haig 1986. Metallurgical Laboratory Design and Operation, *Design and Installation of Concentration and Dewatering Circuits*, Eds A.L. Mular and M.A. Anderson, Soc of Mining Engineers Inc, Littleton, Colorado, USA, 655-666

D.C. Shelton 1981. Design of A Mine-Site Laboratory Facility, *Proceedings of Asian Mining '81, Singapore, 23-25 November*, Institution of Mining and Metallurgy, London, England, 275-284

K Everett and D. Hughes 1981 *A Guide to Laboratory Design*, Butterworths, London, England

6) H.S. Coleman 1951. *Laboratory Design*, Ed. Reinhold Publishing Corporation, New York, USA

7) Graham Davey 2002, Ultrafine and fine grinding using the METSO stirred Media Detritor (SMD), *Proceedings of 34th Annual Meeting of the Canadian Mineral Processors, January 2002*, Editor Jan Nesset, Canadian Mineral Processors Division of Canadian Institute of Mining Metallurgy and Petroleum, Montreal, Canada

Osvaldo A. Bascur and J. Patrick Kennedy 2002. Web Enabled Industrial Desktop To Increase Overall Process Effectiveness In Metallurgical Plants, *SME Annual Meeting*, Feb. 25-27,. PrePrint 02-134, Soc of Mining Engineers Inc, Littleton, Colorado, USA

Key for Figure 1

A	Roll up door to ore lay down area
B	Jaw crusher
C	Gyratory crusher
D	Screen
E	Roll Crusher
F	Rotary Splitter for dry coarse solids
G	Scale
H	Freezer
J	Rotary splitter for slurry samples
K	Filter stations
L	Hooded hot plates
M	Drying Oven
N	Sample Preparation hoods
O	Pulverizer
P	Sample weighing and Screen Cleaning
Q	Dual Screen Shaker (sound-proofed)
R	Cyclosizer
S	Laboratory rod mill (sound-proofed)
T	Flotation Hoods
U	Attritor
V	Store and Reject Sample Storage
W	Instrument bench
X	Computer work station
Y	Polished section polishing and light microscope
Z	Exterior window
1	Floor drains
2	Sinks
3	Door to Analytical Laboratory or Metallurgists' offices
4	Safety supply cabinet

Figure 1, Metallurgical Laboratory layout

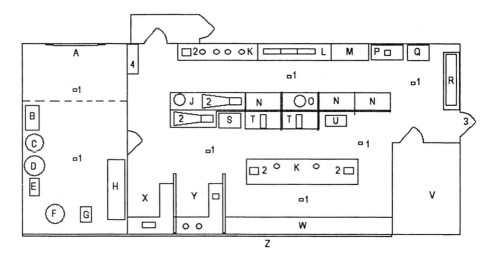

Figure 2, Suggested Location for Metallurgical Facilities in Concentrator

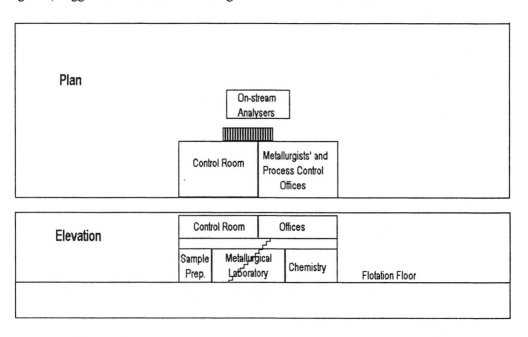

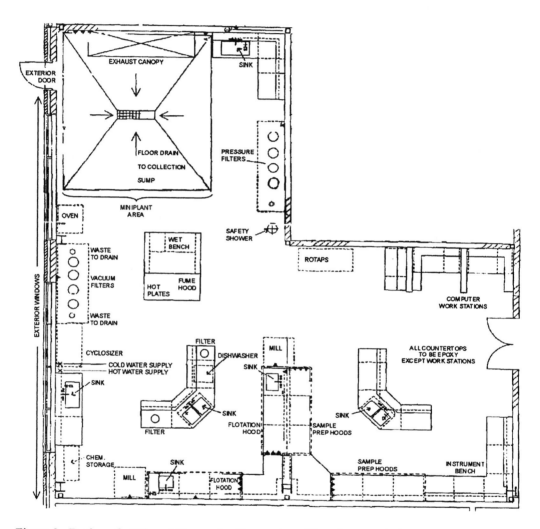

Figure 3, Design of a Mineral Processing Research and Development Laboratory

On-Line Composition Analysis of Mineral Slurries

T.F. Braden, Outokumpu Technology USA Inc.
M. Kongas, Outokumpu Mintec Oyj
K. Saloheimo, Outokumpu Mintec Oyj

ABSTRACT

In the past 25 years, on-line composition analysis (OCA) of mineral slurries has become a vital part of not only design but also the operation and control of base metal mineral processing plants. Operating plants need OCA information in order to optimize their profitability under varying ore feed and metal market conditions. A description and brief status of OCA methods such as prompt gamma neutron activation (PGNAA), nuclear magnetic resonance (NMR), laser induced breakdown spectroscopy (LIBS), froth image analysis and x-ray diffraction will be presented with a more detailed discussion developed for x-ray fluorescence (XRF). Return on investment for implementation of OCA utilizing XRF is presented.

INTRODUCTION

The prime objective when managing a mineral beneficiation process is to ensure that all sections of the available ore body are processed in a manner that maximizes plant profitability. On-line composition analysis (OCA) plays a crucial role in helping to achieve maximum profitability. It must be noted, up-front, that OCA by itself cannot accomplish the goal of maximum profitability, but requires a very close partnership with a process control/management system and other instrumentation. However, irrespective of whether a plant is manually or computer controlled the availability of adequate real-time information on the dynamic behavior of a process is of vital importance to its operating efficiency. When control decisions are based on only a limited amount of process information, it can seriously impair the ability of the production personnel to formulate and implement sound control strategies. Likewise, when reliance has to be placed on outdated assay data that fails to accurately characterize the short-term trends that exist within a process, it is safe to predict that money is being lost (Barker et al. 1987).

BACKGROUND

In years past, it was common practice as a plant level operator or foreman to have a vanning plaque and microscope handy to help illuminate changes in ore type, losses into the tails, or bad actors in the final concentrate. The use of a vanning plaque and microscope required a knowledgeable head, a steady (shaky for the purpose of vanning) hand and a calibrated eyeball. Operating decisions based on the use of a vanning plaque and microscope required an obvious, not subtle, change in some aspect of a process stream. If the eyeball and head were properly calibrated then a change in the process could be recognized and plant operation modified to account for the change. Unfortunately, ore bodies have become lower and lower grade making many changes almost imperceptible to the calibrated eyeball even though the knowledgeable head can almost make-out the culprit. Individual opinions on the plant behavior, which were not based on facts, made consistent operation impossible.

Likewise, it was common in many plants to train operating personnel to accomplish "slop assays" by cutting grab samples from the process, filtering and drying the sample and then performing a simplified mimic of the wet chemical method employed in the analytical laboratory to determine assay. At best this method gave a limited trend of the actual assay and at worst was wrong and misled the operator about a change in the process.

Since the day that mining was first begun, it was apparent to all operating and management personnel that there was a need for fast and accurate analysis of the metal composition of a process stream. If such an instrument could be found then not only would plants be able to improve their process control but also the economics of processing lower and lower ore grades would improve. Suffice it to say, as Hales and Marchant (Hales and Marchant 1980) so aptly stated, "the single most important on-line instrument is one that will measure metal concentrations." The need to measure has always existed but it was not until the last 25 years that we were able to find an instrument that could successfully measure metal concentrations.

DISCUSSION OF ON-LINE COMPOSITION ANALYSIS METHODS

OCA instrumentation come in very many types and are based on many different measurement methods. Since the last review of OCA methods, (Cooper 1984), most of these instruments have not changed nor have their methods. Few have found real application in mineral plants around the world. However, it is worthwhile mentioning some of the old and some of the new because of their potential bright future with further development and refinement.

PGNAA

Prompt gamma neutron activation analysis (PGNAA) is based on a nuclear reaction occurring between neutrons and atomic nuclei in the sample. Absorption of neutrons by the nuclei releases one or several gamma-ray photons. The gamma radiation is measured using an energy dispersive detector. The radiation intensity at the element specific characteristic energies is dependent on the concentration of that element in the sample. (Tran & Evans, 2001)

In principle PGNAA is suitable for analyzing the whole periodic table of elements, with a few exceptions. There is, however, quite large variation in the sensitivity of different elements to the activation process.

The neutrons travel long distances through matter without absorption and thus the rate of the prompt gamma reactions is low. The deep penetration of the neutrons and measured gamma rays allows analysis of bulk material on a conveyor belt or in a chute. On the other hand, the sample volume to be measured must be relatively large as compared to other methods, in order to get sufficient sensitivity. With mineral slurries, maintaining a homogeneous suspension in a large volume presents a technical challenge.

At low concentrations PGNAA requires relatively long measurement times to achieve sufficient accuracy. Minimum detection limits can reach down to 0.01% for the most detectable elements at a 10 min counting time with 0.1% being a typical value.

The most common neutron source used in commercial systems is a Californium-252 isotope source. The 2.6 years' half-life creates a need for periodic replacement of the sources to maintain the analytical performance. In recent years there has been quite intensive development on neutron tubes as an alternative source (Lebrun et al. 1998). However, the lifetime of neutron tubes is still fairly short for continuous on-line operation. The radiation shields required by operational safety make the systems quite large in size and weight.

PGNAA applications in OCA of concentrator plants is limited to cases where XRF does not provide all required analysis information. This is usually the situation when assays of light elements (silicon (Si), aluminum (Al), potassium (K), sodium (Na), magnesium (Mg), phosphorus (P) and sulfur (S)) are of crucial importance to the flotation control.

Typical PGNAA systems cost upwards of $500,000 not including the cost of installation.

NMR

Nuclear magnetic resonance (NMR) is based on absorption of radio frequency electromagnetic waves by atoms that are polarized in a strong electromagnetic field. The method is sensitive to any element isotopes having an odd number of protons and neutrons in the nucleus. The absorption of the resonant frequency, or emission after a short RF pulse, in a sample chamber is dependent on the concentration of the absorbing element. The structure of the absorption is sensitive to the molecular form of the sample material. NMR spectroscopy is widely used for studying molecular structures in the laboratory.

The only working application for NMR analysis in concentrator processes is phosphorous assaying in phosphate plants (Shoniker et al., 1998). Fluorine analysis has been tested.

LIBS

Laser Induced Breakdown Spectroscopy (LIBS) uses a focused Laser light beam pulse directed to the sample surface, causing a small short-lived plasma spark. The plasma atoms emit characteristic photons in the visible light and near-UV region (200-800 nm). The emitted light is collected and transmitted via a fiber optic cable to a spectrophotometer and detected using a CCD element. The laser pulses can be repeated at typically 10 Hz rate, and several spectra are usually averaged to reduce the random variation in the measurement results. Like all optical emission methods, LIBS can in principle detect any elements in the periodic table. (Laughlin et al, 1999; Rosenwasser et al, 2000)

The reported minimum detection limits for different elements in liquid or solid samples range from 0.1 to 1000 ppm depending on the application, sample matrix and laser source/spectrometer setup.

One of the basic characteristics of the technique is its limitation to a very small sample volume at a time. In a practical setup, the atomized sample in a single spark is of the order of micrograms. The small sample volume can be overcome by scanning the sample surface with a large number of laser sparks.

When applied directly to slurry, LIBS has practical representivity problems since the spark occurs only at a very shallow depth. Aqueous systems are difficult for the technique due to strong absorption of the laser energy by water.

LIBS has evolved in environmental applications to measure low heavy metal concentrations in soil and metallic particle pollutants in air. So far, the OCA applications have been field tests for iron ore concentrate (Barrette et al., 1999) and copper concentrate.

Froth Image Analysis

The visible color, or reflectance of light from the froth surface has been applied to OCA of concentrates in connection with some flotation froth image analysis research. The method is indirect and in practice requires an elemental on-line analysis for calibration. The applications include the prediction of molybdenum (Mo) concentration in a copper (Cu)-Mo concentrate as well as zinc (Zn) content in a zinc rougher concentrate (Ylinen et al, 2000).

X-ray Diffraction

X-ray diffraction (XRD) measures the concentrations of minerals not elements. Minerals are made up of atoms and molecules arranged in an orderly three-dimensional array – a crystalline lattice. When a x-ray beam is directed onto the mineral surface, radiation is diffracted at known angles characteristic of the crystal structure. Each mineral has its own unique crystal structure and therefore the angle of the diffracted radiation tells what mineral is present. The intensity of the diffracted radiation is directly proportional to the concentration of the mineral.

Detection limits for XRD analyzers installed online are typically 250 ppm. XRD accuracy is extremely sensitive to particle size variation since XRD is primarily a surface phenomenon. X-ray diffraction (XRD) has been used to measure mineral or indirectly calculate elemental concentrations in phosphate and potash processes (Saarhelo et al, 1990).

X-ray Fluorescence

The OCA methods previously in this section have not found general acceptance in mineral processing plants. Whether it is due to their high capital investment or their poor performance, none of these methods have been accepted as being able to successfully measure metal concentrations on a on-line analysis basis.

Not mentioned, so far, is the OCA method of X-ray fluorescence (XRF). X-ray fluorescence has found general acceptance in base metal mineral processing, especially in plants utilizing flotation for ore concentration. It has become the dominant and decidedly more successful OCA method. OCA-XRF deserves a more detailed explanation and description in order to do it justice.

WHY OCA-XRF ANALYSIS?

When reliance is placed upon traditional sampling and laboratory analysis techniques, delays of 2-24 hours are common between sample collection and the availability of an assay. The value of the laboratory assay information is further diluted when it is based upon a composite shift sample; as composite samples completely mask the effect of short-term process changes. In contrast, when an on-line x-ray fluorescence (XRF) analyzer is used for continuous real-time, multi-stream surveillance, adverse process trends are detected in minutes rather than hours. This permits the timely implementation of corrective control measures, based upon assays that accurately reflect the prevailing process conditions. Figure 1 compares the performance of the Laboratory versus an On-line XRF Analyzer.

Why On-Line Composition Analysis? (using XRF)		
	Laboratory	Analyzer
Two ways to produce assays		
Cost per assay	$50 - $100	$0.02 - $0.1
Delay from sampling	4 - 48 hours	1 - 5 min
Frequency	2 - 24 hours	10 - 20 min
Manpower	1 – 4 per shift	0.2 per shift

Figure 1.0 Comparison of Laboratory versus On-line XRF Analyzer

On-line XRF analysis of slurry samples is based on some of the the same measurement principles as the assaying of manually prepared samples in the laboratory. The major differences are:

- On-line samples are taken in the same way day and night. Laboratory samples are inconsistent because of manual sampling and preparation.
- With an on-line XRF analyzer the sample contains the full range of original particles. Thus, particle size affects the accuracy of coarse slurry measurements when compared to a homogenized sample in the laboratory. On-line XRF analyzer assays reflect the present status of the process. Laboratory assays are out of date and thus may have large errors, if used for process control.

When considering all the effects, on-line XRF analysis is more appropriate for process control purposes than laboratory assays. The vast majority of OCA installations in the last 25 years have been XRF analyzers. Based on personal observation, it can be said that there remain very few base metal operating plants that do not have an OCA-XRF analyzer

HISTORY OF OCA-XRF ANALYSIS

The history of the development of the OCA-XRF analyzer is well documented by (Cooper, H.R. 1976). In the 25 years since this last history was published, the installation of OCA in mineral plants (primarily base metal plants) has started to approach the saturation point, where almost every open and operating plant has an OCA-XRF analyzer installed. The past and recent history of OCA-XRF analysis can be summarized in the following Table 1.0

PERIOD	TECHNOLOGY	INDUSTRY
Pioneering (1955 – 1959)	Changes rapidly due to research	Speculation leads to few beta test sites
Development (1960 – 1969)	Changes less rapidly, research and development lead to new discoveries	Less speculative, no industry wide acceptance only plant by plant acceptance
Growth (1970 – 1985)	Stable, fewer new innovations, focus on packages that add value for money	Confidence increases, industry wide acceptance, companies start to standardize
Maturity (1986 – 2002)	Stable, no new innovations, quality becomes primary issue	Approaches saturation with obsolescence becoming issue
Future (2003 - ????)	New breakthrough offers restart to Pioneering or Development Period	Speculation leads to few beta test sites

Table 1.0 Lifecycle of XRF-OCA Technology

XRF TECHNOLOGY DESCRIPTION

X-ray fluorescence (XRF) technology involves the use of an excitation source (x-ray tube or radioactive isotope) to provide radiation used to excite a sample and cause the elements in the sample to fluoresce. Fluorescence (see Figure 2.0) is caused by the radiation from the excitation source, called X-ray quanta, which knock out inner shell electrons in the sample atoms ("excitation"). An electron from a higher shell fills the electron vacancy to keep the atom stable. The energy difference between shells is emitted as an X-ray fluorescence photon. The energy of the emitted radiation photon is characteristic to each individual element in the periodic table. An XRF analyzer contains a detector or series of detectors to measure and convert the emitted radiation photons into characteristic energy intensities corresponding to elements, for example, copper (Cu), nickel (Ni) or zinc (Zn). This intensity information is then converted to assays using a calibration model.

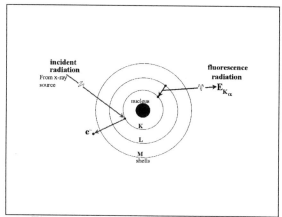

Figure 2.0 X-ray Fluorescence Principle

The x-ray fluorescence intensities can be measured using fixed crystal wavelength dispersive spectrometers (WDX) for each element and solids content. The alternative is to use energy dispersive (EDX) detectors, which measure the whole fluorescence spectrum. A comparison of WDX and EDX has previously been presented in, "On-stream X-ray Analysis" by Harrison Cooper (Cooper H.R. 1976).

The assays from the analyzer indicate the percentage by weight of the concentration of an element of the total solids. To compensate for solids content variation in the sample, the measured element specific intensities are corrected by measuring the backscatter signal from water in the sample.

Owing to the absorption of water and solids, the measured fluorescence signal comes only from a thin 1mm layer just in front of the plastic window separating the continuously flowing sample from the x-ray source and detector (See Figure 3.0). This window must stay clean and the sample in the sensitive area must be representative and free from air. Sample flow against the window prevents scaling and fills the sensitive layer continuously with fresh representative sample.

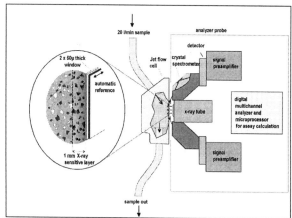

Figure 3.0 Description of XRF Analyzer Flow Cell and Analyzer Probe

Analysis of light elements in slurry using XRF (sulfur (S) for example) is not possible because the fluorescent radiation from the light element is absorbed by water, the analyzer windows and finally the air path to the detector.

On-stream versus In-stream Systems

On-stream and in-stream analysis refer to the method in which the process stream to be analyzed is presented to the x-ray system. The choice as to whether to install an on-stream or in-stream system requires special consideration. The viewpoint as to which type of system to pursue will largely depend upon what is demanded from the system and how the assay data is utilized. In a plant where the analyzer system is employed to automatically implement assay-based control, the operating staff will demand accurate assays and be committed to keeping the system properly calibrated. However, when a system will be used simply to furnish the operating staff with analytical trends, there is often a tendency to place less importance on accuracy and more emphasis on minimizing operating costs.

Three important factors should be taken into consideration when deciding which system to install: (1) representativeness of the sample that is analyzed, (2) ease and quality of calibration, (3) analytical performance of the XRF technology.

Representativeness of Sample

When the in-stream method is used, a separate analyzer probe is installed at each process stream; refer to Figure 4.0. The active area of the probe is immersed in the stream and analyzes the material flowing past it. In contrast the modern on-stream system shown in Figure 5.0 can employ one or more analyzer probes, each of which is capable of measuring 1 to 24 streams.

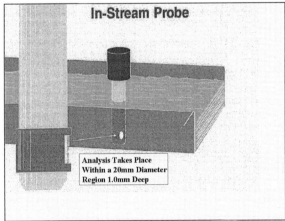

Figure 4.0 In-stream Probe

Ruggedly constructed devices manufactured out of steel, lined with rubbed or ceramic, are used to cut continuous samples of 100-300 liters per minute from process flows as large as 5,000 liters per second. These devices are know as primary samplers and are designed to extract a truly representative sample that is characteristic of the whole process flow. Each primary sampler is specifically designed to handle a particular process application and flow conditions. The sample flow is then transported via gravity or pumped to a device known as a multiplexer.

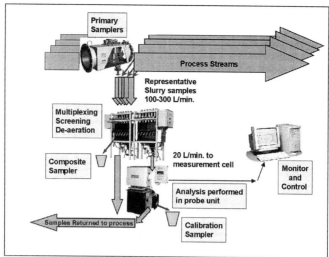

Figure 5.0 On-stream Analyzer System

The multiplexer reduces the primary flow to the analyzer's measuring cell to approximately 20 liters per minute and controls which of the 6 to 18 streams is passing through the system for measurement. The multiplexer, piping and the measuring cell are automatically flushed with water between sample measurements. The multiplexer also ensures that the material being analyzed is free of air bubbles, oversized particles and flotsam such as plastic and wood chips; any of which produce inaccuracy in the assay results. A level control device within the multiplexer ensures that the measuring cell receives a constant flow.

The flow to the cell, and its design, are of fundamental importance to the overall accuracy of the system. The sample is continuously analyzed as it passes through the measuring cell. A thin polymer (mylar) film (window) separates the x-ray tube and detectors from the sample. XRF is a surface measuring technique in that the analysis region is restricted to a volume of sample very close to the window. (More detailed discussion of this phenomenon will come later in this document.) The cell design must, therefore, ensure that all sizes of particle within the slurry pass through the narrow region at the surface of the window, otherwise, inaccuracy will result due to biased sampling. The cell geometry is designed to produce a high degree of sample mixing by creating turbulence within the sample flow.

Scale build-up on the window can create major inaccuracy. One major source of scale can be the reagent chemicals used in the process. Good cell design helps to minimize this serious problem. The turbulence within the cell not only creates good sample mixing, it also results in the window being cleaned as a result of the particles abrading the surface. Window ruptures are sensed by the system and alarmed for operators to replace. The latest technology employs a moving window film like a camera that moves continuously between the measuring cell and the analyzer probe.

The in-stream method as shown in figure 4.0 reduces the capital and operating costs associated with sampling, sample transport and pumping. However, care should be taken to minimize slurry particle segregation in the process stream. Some type of external mixing device should be introduced into the process to eliminate, or minimize, particle segregation.

As in the case of the previously described on-stream method, an in-stream probe has similar sized window and only analyzes a thin layer of sample. The window separates the process stream from an inner protective window and the radioactive source and detector.

In order to generate an accurate assay, it is essential that the material that passes the surface of the window is representative of the total process flow. In the case of a slurry, accuracy is affected by variations in particle size, sample composition, mineralization, the slurry density and variable flow conditions. When immersing a probe into a process stream, there is always a large degree of uncertainty as to the representativeness of the material passing through the thin region at the surface of the window. In the case of a wide launder, the various particle size fractions tend to separate into layers and the stream may frequently meander. Sample bias can also occur in a large vessel where segregation through settling, flotation and/or poor mixing results in non-representative material passing the probe window.

Because variations in process flow change the conditions at the window and this impacts upon the analyzer calibration, it is often necessary to minimize these effects by passing the process stream, or part of it, through a steel construction known as an analysis zone, refer to Figure 6.0. When large flows are involved, a primary sampler may also be needed.

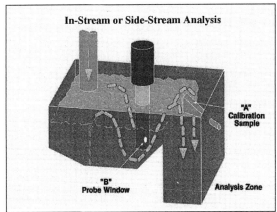

Figure 6.0 In-stream Analysis Zone

Whereas factors such as inadequate sample mixing and variations in process flow, entrained air, coarse particles, flotsam and window buildup can be easily controlled in an on-stream system; they present fundamental problems for an in-stream system requiring careful engineering to avoid or eliminate.

Ease and Quality of Calibration

Calibration will be described in more detail later in this document. For the purposes of this discussion calibration is the process of calculating the parameters which the analyzer will use to calculate assays from the x-ray intensities measured.

The initial calibration procedure requires a suite of samples to be collected from each point in the process that is assayed. The number of samples required will depend upon the number of elements being measured and the sample matrix.

In the case of an on-stream analyzer, the process flow that is being sampled for calibration is the actual sample flow passing through the measurement cell. After leaving the measurement cell, a calibration sample device diverts the sample into a bucket. The calibration sample is then analyzed in the laboratory, where it is weighed, filtered and dried to zero moisture. After sample preparation it is then chemically analyzed. The x-ray intensities measured by the analyzer can then be compared with the laboratory results. The analyzer's on-line modeling and regression programs are used to establish the most suitable calculation model for each assay; i.e. conversion of x-ray intensities into metal concentrations.

A fundamental difference exists in the procedure for collecting calibration samples for an in-stream system. Due to the longer measurement time of an EDX probe, it is necessary to measure a tailing stream for 5-10 minutes in order to obtain acceptable statistics. During this time period, a series of individual grab samples or a continuous sample must be cut from the process. If the composition or particle size distribution of the sample that is sent to the laboratory differs from the material that flowed within the thin region of the probe window, then the quality of the calibration will be substandard.

If the probe is used with an analysis zone, figure 6.0, the calibration sample would be extracted at point "A", while the analysis is measured at point "B". Hence the need to insure proper mixing at the probe window and the taking of a complete cut across the overflow at point "A".

Quality of Analysis EDX versus WDX

Most in-stream probes utilize a radioisotope source for generation of x-rays and use the measurement technique known as Energy Dispersive X-ray Fluorescence (EDX). This analysis technique has historically used cryogenically cooled solid state detectors that require liquid nitrogen or room temperature Germanium detectors. However, a new Peltier cooled silicon PIN diode detector is available that does not require liquid nitrogen.

The wavelength dispersive x-ray fluorescence (WDX) technique typically uses a x-ray tube to generate x-rays and uses a crystal spectrometer with room temperature proportional counter detector for each element to be measured. Most WDX systems require either air-cooling or water-cooling for the x-ray tube and a high voltage power supply.

High resolution is very important when the x-ray lines of the measured elements are closely spaced, as in the case of cobalt (Co) and iron (Fe). Figure 7.0 compares the resolving power of an EDX solid state detector with that of a WDX system. It can be seen from the top spectrum that an EDX detector cannot separate the Co and Fe peaks; whereas, the WDX system resolves the x-ray peaks. At 5.9 KeV the resolution of a WDX detector is 30 eV; whereas, the liquid nitrogen cooled solid state EDX detector is 170 eV. The PIN type EDX detector's resolution has been reported to be 280 eV.

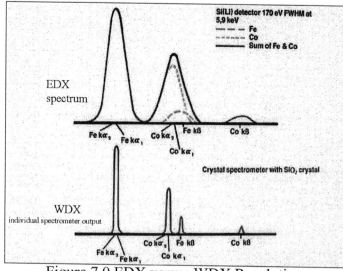

Figure 7.0 EDX versus WDX Resolution

Consideration should be given to the relative accuracy and sensitivity differences between WDX and EDX systems. For example, Figure 8.0 shows the measured spectrum for a zinc (Zn) tails stream containing .12% Zn. From the figure you can see the Zn peak is barely visible and corresponds to 2 counts per second of Zn (less background). In comparison, Figure 9.0 shows a WDX measurement for the same material where the Zn peak corresponds to 52 counts per second. For the EDX system to achieve a 10% relative accuracy (based on counting statistics only) 1,000 counts must be accumulated. This will require 500 seconds or 8.3 minutes measurement time in order to accumulate enough counts to offer a 10% relative accuracy. For the WDX system to achieve a 10% relative accuracy would require 19.2 seconds measurement. As the concentrations to be measured increase, the difference in the ability of the EDX system to offer an accurate analysis in a reasonable time becomes more achievable.

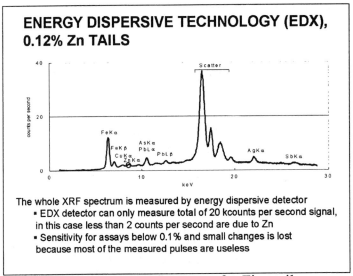

Figure 8.0 EDX Spectrum for Zinc tails

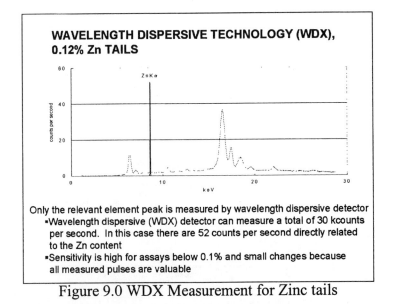

Figure 9.0 WDX Measurement for Zinc tails

Another consideration when comparing EDX and WDX systems is the sensitivity of the system as measured by the minimum detection limit (MDL). MDL is defined as the smallest concentration the analyzer can correctly discriminate from zero concentration. WDX technology can be used to analyze concentrations typically down to 10 ppm in slurries. In contrast, the relative performance of an EDX probe is a magnitude lower, i.e. 100 ppm. In the case of the PIN detector based EDX analyzer the MDL may be 200 ppm or as high as 500 ppm. The above factors need to be considered if a system is to measure very low concentrations. This is especially important if the samples contain elements whose atomic numbers are closely spaced.

For analyzing elements lower in atomic number than titanium (Ti), EDX is required. EDX devices can analyze slurries containing elements as light as chlorine (Cl) without special geometry and nitrogen purging. Both EDX and WDX probes can analyze elements up to uranium (U).

XRF ANALYZER SYSTEM COMPONENTS

A typical on-stream XRF analyzer system (refer to Figure 10.0) consists of:

Primary sampling system

Secondary sampling system

Analyzer probe

Calibration sampler

Probe control set

Analyzer management station

Connection to process control system (digital control system, DCS)

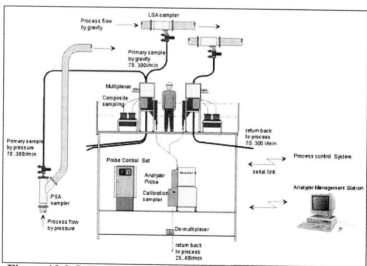

Figure 10.0 On-stream XRF Analyzer System with Components

Primary Sampling System

The primary sampling system consists of standard proven samplers installed in the process flow, and pipes and pumps for transporting the primary sample flow to the multiplexers and back to the process. The multiplexers are part of the secondary sampling system. The primary sampling system will be discussed in more detail later in this document.

Secondary Sampling System

The secondary sampling system includes both multiplexers and demultiplexers. Each multiplexer has 1-6 inlets, and there can be up to four multiplexers connected to one analyzer probe. The multiplexers also include an automatic composite shift sampler, which cuts shift samples from the sample streams. The user can define the time interval for the composite sampling. A sample filtering unit can be included as an optional device. For the sample measurement the multiplexers cut a smaller secondary sample, remove air and trash from the sample, stabilize the sample flow, and direct it to the analyzer probe and into the measurement cell. The demultiplexers direct the sample flow back from the calibration sampler to the return pipes.

Analyzer Probe

In the analyzer probe, the element intensities are measured by using the X-ray fluorescence method. The analyzer probe contains the measurement cell, measurement channels, X-ray tube, high voltage supply, pulse processing electronics, safety interlocks, a cooling device, and an Automatic Window Changer as an optional feature. From the measurement cell, the sample flows to the calibration sampler.

Calibration Sampler

The calibration sampler is a cross-cutting sampler used for collecting representative samples for laboratory analysis during the calibration intensity measurements. From the calibration sampler, the demultiplexers direct the sample flow back into the return pipes.

Probe Control Set

The Probe Control Set (PCS) is a field user interface for the analyzer system, which includes a display, indicator lights, and control switches. It also includes processors for controlling the measurement sequence, and the sampling equipment (on demand control is accomplished here), and for transferring data to external systems. The PCS offers a user-friendly approach for monitoring and controlling the system functions.

Analyzer Management Station

The Analyzer Management Station is a PC, which includes tools for calibrating the analyzer and managing the parameters of the analyzer. The AM Station displays the assays, operational state of the analyzer, and the active alarms. The AM Station also includes service and maintenance tools that allow the support personnel to view the logged diagnostic data from the analyzer.

Connection to Process Control System (Digital Control System, DCS)

The DCS connection is a point-to-point serial communication line, which gives information on the statuses, alarms, and concentrations. The user can also make a measurement request in the DCS system; i.e., you can program the analyzer to measure a certain sample.

Plant Ethernet network is used more and more to communicate assay information for monitoring and control.

SLURRY SAMPLING TECHNOLOGY

A well-engineered sampling system can result in the reliable generation of a representative sample of the process that will operate without blockages at all process conditions. In addition, a properly designed sampling system can result in low operating costs with minimal maintenance requirements. The capital investment required for a reliable and accurate sampling system can be minimized by investing in knowledgeable and experienced sample engineering prior to the installation and startup of the system.

The location of the analyzer in the plant has a decisive effect on the investment and maintenance costs of the sample transport system. Therefore, selecting a place for the analyzer is the most important aspect in the design of the sampling system. Likewise, consideration should be given to not only sample flow to the analyzer but also the sample reject flow from the analyzer back to the process.

The turbulence of the slurry flow in process pipes and launders mixes the slurry well in horizontal directions. This has made it possible to develop simple and small stationary (static) samplers, which are placed in the process where the slurry is well mixed. The errors in assays are insignificant compared to the errors caused by changes in particle size and mineralogy. In a well-engineered sampling system the use of static samplers can give good quality samples for runtime material balances. For really accurate material balance calculations shift composite samples collected by cross-stream cutter samplers give samples correctly mirroring variable process flows and solids contents.

Static samplers are chosen with the following criteria in mind:

1. Provide the correct sample flow rate for the analyzer secondary sampling system

2. Provide a representative sample of the process flow.

For good representivity, flow velocity into a stationary cutter or nozzle should be engineered to be about the same as the velocity of the bulk flow around the cutter or nozzle. This principle is called **isokinetic sampling**. If the velocity of slurry into the cutter is too high, more water is sucked from around the cutter. Only the finer, lighter particles are preferentially captured with that flow, causing a deficiency of coarse particles in the sample. If the velocity of slurry into the cutter is too slow, the finer, lighter particles follow the flow around the cutter causing an abundance of coarser, heavier particles in the sample.

Static samplers can be divided into two groups. The first group uses the process pressure at the sampling point which is typically generated via a pump. The second static sampler group, the gravity flow static samplers, have no process pressure at the sampling point. Figure 11.0 shows an example of a typical pressure pipe sampler (PSA) from the first group which uses process pressure to generate a sample. The PSA sampler, when properly engineered, can use the process pressure to transport the sample to the analyzer.

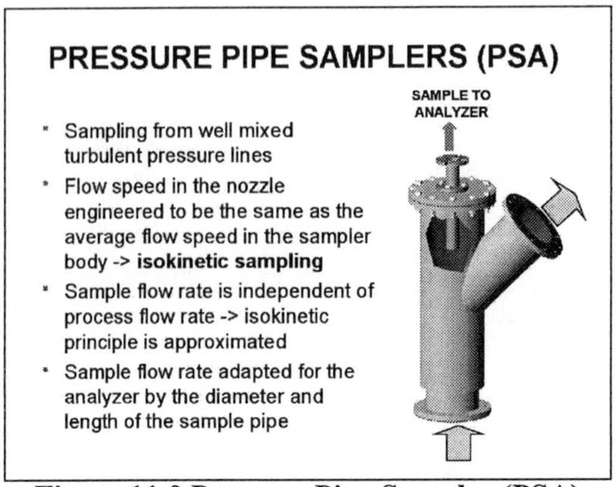

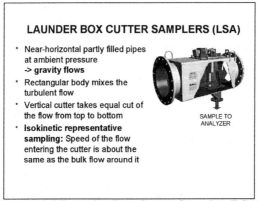

Figure 11.0 Pressure Pipe Sampler (PSA)

The second type of static sampler is called a launder box cutter sampler (LSA). Figure 12.0 explains the principles behind this type of sampler.

Figure 12.0 Launder Box Cutter Sampler (LSA)

Sample flow from the static samplers can be continuous or controlled "on demand". Controlled sampling means automatic stoppage of the sample flow and flushing of the cutter or nozzle and sample pipes, when the analyzer does not need the sample for measurement. All static samplers should be equipped with a shut off valve for the sample flow and flushing valves for upstream and downstream flushing of the cutter or nozzle and sample pipe. In the case of on demand control all valves should have pneumatic or electric actuators.

EXAMPLE OF AN XRF ANALYZER PROJECT

Purchasing, installing, and starting up an XRF analyzer is referred to as an XRF Project. This section describes the entire project in a chronological order, from collecting the required data and the feasibility study to making the Customer Support Agreement. The time schedule for this project can be found in Figure 13.0

The project consists of several predefined phases:
- Collecting the application data
- Checking the feasibility
- Defining the project scope and tender
- Making the business contract
- Planning

- Delivering the analyzer
- Installing
- Commissioning and startup
- Calibration
- Training
- Customer Support Agreement

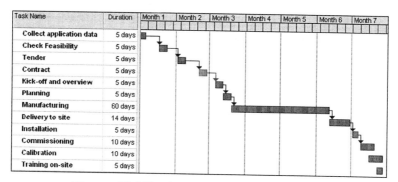

Figure 13.0 Typical XRF Analyzer Project

A prerequisite for the successful implementation of an XRF analyzer project is an effective co-operation between the supplier and the future users of the analyzer. The project should be led by an experienced project manager from the supplier, an XRF analyzer expert, who ensures that reliable support is provided throughout the XRF analyzer project.

Collecting The Application Data

Correct application data is needed for checking the feasibility and ensuring a smooth start of the project. The collected data contains information on the samples to be measured such as: element concentrations, slurry densities, particle size distributions, and environmental information such as electric voltages used on site, ambient temperatures and altitude above sea level.

Checking The Feasibility

Before making the business contract, the supplier should study how the requirements of the application can be fulfilled. On the basis of this feasibility check, the supplier and the customer determine how the analyzer should be set up to best suit the particular purpose.

Defining The Project Scope and Tender

The project scope and tender are defined on the basis of application data and the feasibility check.

Contract

The business contract is signed when both parties have agreed on the modules and options required for constructing an analyzer system suited for the purpose.

Planning

Before installing the system, the local infrastructure of the system needs to be planned carefully, including power supply, pressurized air, water supply, and sampling system.

There are two important phases in the XRF analyzer project; installation of the sampling and calibration. A well-planned installation is crucial for the proper operation of the sampling system, and a careful calibration ensures accurate and reliable measurement results.

The sampling needs to be planned with special care. It is recommended that a supplier expert takes care of the basic engineering. The supplier expert considers carefully what kind of samplers to install, and how to place them in the plant, because this has an effect on the sample quality. The sample pipes and their installation are crucial for the reliable sample flows. It should also be considered whether to use pumps or have the sample flow by gravity or process pressure.

The planning takes place on site, and includes both supplier and customer personnel. A proper primary sampling system guarantees a reliable and representative sample, and a suitable sample flow rate for your analyzer system. There are a wide range of different primary sampling components for different sampling situations and process flows, which make it easy to find the most suitable solution for the plant.

Delivery

The delivery of the system takes place about ten (10) weeks after the signing of the business contract. During this time, the analyzer system is engineered, manufactured, tested, and shipped to the customer location. The delivery time varies according to the content and extent of the delivery.

Installation

There are two installation practices used with an XRF Analyzer:

The system can be installed entirely by supplier (this includes installation of both the analyzer and the sampling system)

The customer can install the system on its own and then supplier checks the installation.

Most XRF analyzer systems have a modular structure, and therefore, can be installed in several different configurations. When installing the system, consider the location of the analyzer carefully:

The analyzer should be placed in the middle of the selected sampling points to allow easy sample transportation without using costly pumping systems.

Insure that the analyzer is easily accessible: there should be enough working space in the multiplexer area and on both sides of the analyzer probe to enable trouble-free operator access and maintenance.

Avoid placing the analyzer near the mills due to the noise and vibration, which can cause disturbances in the detectors and also makes for an unpleasant working environment.

A typical on-stream XRF analyzer can be installed with three different layouts:

Two Level Layout (Figure 14.0) In the basic, two level installation layout, the multiplexers are located on the second level, and the analyzer probe and the Probe Control Set (PCS) on the first level.

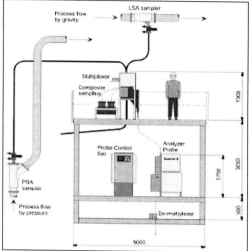

Figure 14.0 A two level installation layout of the analyzer system.

Low-Head Layout (Figure 15.0) In a low-head installation, all components of the XRF analyzer are on the same level.

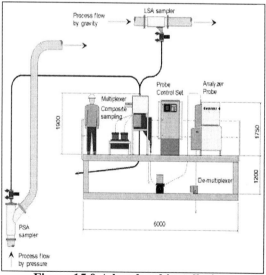

Figure 15.0 A low-head installation.

Shelter Installation (Figure 16.0) For the convenience of the operating and maintenance personnel it is often better to have the analyzer probe and the probe control set (electronics) located inside an analyzer shelter, and the multiplexers on the roof of the shelter. The Analyzer Management Station can be placed in the shelter or the control room. The shelter can be delivered with an air cooler, which stabilizes the ambient temperature in the analyzer probe area. Please note, modern OCA-XRF analyzer systems have been designed to meet NEMA 4 and NEMA 4X standards with corrosion resistant materials of construction thereby eliminating the need for a shelter except in the case of operator convenience.

Figure 16.0 Shelter installation

Commissioning and Startup

When the XRF analyzer system has been installed, the supplier's commissioning engineer should check that all parts of the system are properly installed and configured, which is a prerequisite for correct operation and accurate measurements. The commissioning engineer should also assist with the configuration.

When the analyzer installation has been approved, the commissioning engineer will start the system, and check that all parts of the analyzer function properly.

Calibration

Calibration contains two simultaneous processes; i.e., collecting the information on the intensities and assays and determining the mathematical formulas for calculating the concentrations from the intensities. Each analysis of each sample flow requires its own calibration.

In the calibration process, the measured intensities are registered, and the measured samples are analyzed in the laboratory. This is repeated approximately twenty times for each process line. The results of both the laboratory and the analyzer are compared to find a mathematical correlation formula using a calibration tool like the one shown in Figure 17.0. The samples should be taken from a varying range of circumstances, i.e., from low and high element concentrations.

In the implementation project, calibration is divided into two different phases; a preliminary and a final calibration. The preliminary calibration includes finding the first calibration model for the analyzer. The final calibration includes collecting observations from all process situations of a normal range, and calculating the equations for the calibration.

Preliminary Calibration: During the training, a preliminary calibration takes place. 10-20 calibration samples are taken and analyzed by the customer. The time period between the startup of the analyzer and preliminary calibration is approximately 1 to 2 weeks. The preliminary calibration includes finding the first calibration model for the analyzer.

From the preliminary calibration onwards, the customer staff is fully capable of taking the control of the analyzer, and the analyzer can already be used normally in the concentrator, although the measurement accuracy has not reached its peak yet.

Final Calibration: During the final calibration, the customer continues taking calibration samples. The final calibration includes collecting observations from all normal process situations (approximately 40-70 observations from each stream). The samples should cover the variation found in the process. On the basis of these observations, the equations for the final calibration are calculated. There is remote assistance available for these calculations. Final calibration can take several months to be completed, depending on the plant and the stability of the process. After the final calibration, the analyzer reaches the best measurement accuracy. The peak performance is maintained by checking the calibration for process changes.

Maintaining the Calibration: In maintaining the calibration, new observations ought to be collected approximately once a week in order to ensure accurate measurement results. In a year, adequate observation material is collected to recalculate totally new coefficients of the calibration equations. In case of a reformed process or a new ore deposit, the recalibration is performed as necessary. An important part of maintaining the calibration is the availability of a dedicated software tool for calibration of the analyzer. Figure 17.0 shows some typical outputs of such a tool. As can be seen in the figure, all the regression variables are visible and a relative error plot shows the comparison of the laboratory and the analyzer assays.

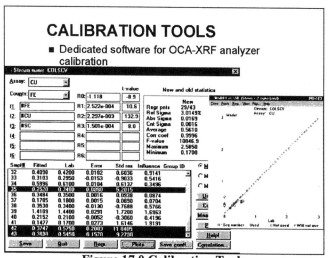

Figure 17.0 Calibration Tools

Training

The supplier should provide XRF analyzer training either on-site or on supplier's premises. The training is very practical; the analyzer is operated normally already during the training period. The training includes collecting the first samples for the calibration. The training is based on continuous interaction between users and the trainer, and so, users are able to influence the contents of the training.

Customer Support Agreement

The XRF analyzer customer support agreement insures that the analyzer system will be checked twice a year to prevent degradation of the analyzer system performance. This can be considered a preventive service to insure the analyzer performs the same after two years or two days of operation. The agreement should also include a Remote Diagnostic System (see Figure 18.0) which is provided by using a modem connection for direct communication with the Analyzer Management Station. Regular training conducted onsite should also be included in the contract. Support of calibration after completion of final calibration should be included in the customer support agreement to insure that the best possible calibration is maintained throughout the life of the analyzer system.

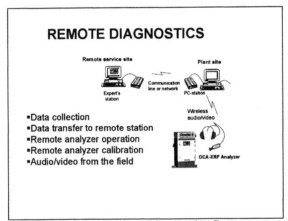

Figure 18.0 Remote Diagnostics Support

OPERATION AND MAINTENANCE OF XRF ANALYZERs

The XRF analyzer has been designed for trouble-free operation and easy maintenance. An XRF analyzer does not require a full-time operating individual for daily support, but only regular check-ups to ensure that the analyzer is functioning properly. Operation and maintenance tasks are generally shared between four different user types in the following way:

Metallurgist

> The metallurgist maintains the calibration of the analyzer. Responsibilities include reporting on the shift values, and observing as well as maintaining the calibration.

Process operator

> The process operator monitors assays and uses the measurement results for running the process. The measurement results are available in the Analyzer Management Station and the Process Control System. The alarms and status of the analyzer are also available for the Process operator in the Analyzer Management Station.

Floor operator

> The floor operator is responsible for the maintenance of sampling equipment. The tasks include cleaning the trash screens daily, maintaining primary sample flows, changing the measuring cell window and unplugging the sample lines and samplers, if necessary.

Instrumentation staff

> The instrumentation staff is responsible for the maintenance and troubleshooting of any operating problems with the analyzer. Specific training on component replacement is an important part of training received from the supplier. The responsibility of maintaining the analyzer is typically shared between the supplier and the customer instrumentation staff.

OPERATION SAFETY

A typical XRF analyzer includes a reliable security system. It has a number of internal sensors that indicate if abnormal situations occur in the analyzer. In the case of an abnormality, the measurement cell flow is automatically bypassed, the measurement functions are disabled, and the X-ray radiation is shut down. The operator safety is guaranteed if, the regulations and safety instructions are followed at all times.

Safety and licensing requirements differ between the EDX and WDX systems. When a radioactive source is used rather than a x-ray tube, special licensing and periodic inspection are required in most countries. With a x-ray tube, there is no radiation when there is no power to the tube, so no special licensing or inspection is typically required. It is important to check with local officials to insure that the correct licensing and inspection is done for the type of system used. Likewise, handling, replacement and disposal of radioactive sources requires special consideration. Check with the supplier of the source to quantify the requirements for disposal.

WHAT TO EXPECT FROM MODERN OCA—TRACE

The current state-of-the-art in OCA, as we have been discussing so far, is the on-stream XRF analyzer system. Using the characteristic of this system we can build a picture of what is now possible in OCA. The factors that can be used to build this picture are:

- Timely assay information

- Reliable system performance

- Accurate measurement of metal concentrations

- Cost Effective investment as measured by capital costs, operating and maintenance costs and balanced by payback or return on the costs applied

Timely Assay Information:

Measurement time of the XRF technology is a key component of timely generation of assay information. Typical performance runs from 1 minute to 5 minutes per stream analyzed for the simultaneous measurement of up to 6 to 10 elements. The absolute best achievable at this time is 15 seconds for high concentration streams like the final concentrate. The feed and the tails stream would require 30 seconds measurement time. Please note, that the physics of the XRF measurement and the type of equipment used tie measurement time and accuracy together. For the sake of this discussion the measurement times used are those that can achieve the instantaneous accuracies mentioned later in this document.

Another important component of this timely generation of assays is the multiplexing of more than one stream past the same analyzer. Multiplexing makes the overall system more cost effective because for a minimal increase in capital cost more potential payback can be added to the system through measurement of additional streams. Multiplexing means that after each stream is measured there must be a draining time to allow the majority of the previous stream to remove itself from the multiplexer. Then a water flush must be made to clean-out the rest of the previous material, followed by a filling time required to change over to the next stream and reach a stable environment to make the next measurement. Typically this drain, flush and fill cycle requires from 20 seconds for tails and feed streams to 40 to 60 seconds for frothy concentrate streams. A 18 stream multiplexed system is capable of measuring all 18 streams in under 15 minutes and a 12 stream system capable of 10 minutes measurement time.

All of these system times (measurement, drain, etc.) must be accomplished within the feedback needs of the overall plant control system in order to provide timely information for control purposes. If the analyzer does not complete the measurement fast enough then the control system has to either wait for the information or develop its own calculated measurement.

Reliable System Performance:

Reliable system performance is measured by the total availability of the XRF analyzer system components. Components such as primary sampling, secondary sampling and analyzer probe have their own individual availability that influences the total system availability. In practice, total system availabilities of 95% or better have been achieved, (Jensen D.L. 1999). In order to achieve a total system availability of 95% the component system availabilities must also be very high. Experience has shown that analyzer probe availability is 99% over a period of one year. Secondary sampling system availability is also 99% due to the simplicity of the design and abrasion resistant materials of construction. The primary sampling system availability is the most variable. This is most often due to poor sample system design, improper installation, and the typical wear and scaling of lines and failure of components such as pinch valves, solenoid valves and coils to drive the valves. Once design and installation problems have been eliminated then the sampling system availability can approach 95%. "On demand" control of sample line flushing offers to maximize the sample line life and built-in system alarms for flowrate changes allow for quick determination and repair of sampling system failures and can minimize downtime for replacement of a failed part. XRF analyzer system availability of >95% is possible and achievable when such automatic flushing controls and flow alarms are utilized.

Accurate Measurement of Metal Concentrations:

The accuracy of a measurement is a function of the sample parameters such as matrix composition, mineralization and particle size. For the purpose of using the measured results for process control the accuracy's being discussed are instantaneous accuracy's. That means that if you were to take a calibration sample at the exact time the measurement is being made by the analyzer, then the relative error between the calibration sample and the analyzer would define the instantaneous accuracy of the analyzer. Under normal operation conditions, the following relative standard deviations can be achieved for individual slurry sample measurements of concentration levels well above the relevant minimum detection limits of the analyzer:

Minor concentrations 3-6%
Major concentrations 1-4%

Cost Effective Investment:

The typical capital investment* for an on-line XRF analyzer system on a per stream basis is:

24 stream system = $22,000 - $40,000
6 stream system = $50,000 – $68,000
1 stream system = $82,000 - $102,000

Consumables costs* for an on-line XRF analyzer system on a per stream basis are:

24 stream system = $1,500 - $2,500
6 stream system = $1,000 - $2,000
1 stream system = $500 - $750

Operating costs* for an on-line analyzer system on a per stream, per year basis are:

24 stream system = $250 - $780
6 stream system = $900 – $1,125
1 stream system = $1,100 - $1,600

*Depending on the system type, physical plant layout, pumping requirements and sampling requirements. Consumables does not include the cost for liquid nitrogen.

BENEFITS OF XRF

The benefits of having an OCA-XRF analyzer in a plant have been documented in a few cases (Jones et al. 1991; Jensen 1999; Kongas et al. 2001; Lahteenmaki et al. 1999). Remember it's not the OCA-XRF alone that gives the benefits but a control system to automatically act on the assay information must be considered. Looking at these documented results shows, at least, in one case (Jones et al. 1991), the payback for the investment was on the order of 4.8 to 5.3 months. Benefits such as improvements in plant metallurgy (grade and recovery) are typical (see Figure 19.0) and are necessary to insure a good payback for the investment. Other benefits that become obvious after implementation are:

- better understanding of the process, i.e. coarser grind leads to lower recovery or high cleaner circuit circulating loads result in significant losses
- calculation of material balances leading to determining process bottlenecks
- ability to on-line test chemicals or process changes in a before and after scenario

- helpful training tool for new operating personnel
- lower reagent costs when combined with an assay based control strategy
- provide on-line metallurgical accounting

When speaking about the benefits of having an OCA-XRF analyzer it is amazing that only a few documented cases of actual calculated payback can be found. Either most of the systems installed did not achieve any payback or else no one has bothered to calculate the payback or publish it. There is a large positive benefit of doing a payback analysis. It serves as a confirmation of the economic justification of the purchase and serves as a future reference when the OCA-XRF system becomes obsolete.

XRF SYSTEM OBSOLESCENCE

Since the introduction of on-stream XRF analyzers over 35 years ago many operating plants have experienced the phenomenon known as analyzer obsolescence. This phenomenon relates to the fact that technology, hardware and software changes quickly and often components cannot be replaced because manufacturing has moved on to the latest technology or hardware component. This leaves a plant in the position of having increased maintenance costs and having to replace not just a component but to upgrade the whole analyzer. Currently the economical lifetime of an XRF analyzer is between ten and fifteen years. It is important to have a proactive plan to either upgrade the obsolete parts or replace the analyzer system before reaching the point of analyzer death. The benefit of being proactive is to maintain the original benefit of the analyzer system and not lose it to higher operating costs and lower system availability.

WHERE ARE WE? A PHILOSOPHICAL PERSPECTIVE

We are at a point where the distribution of plants that have OCA-XRF analyzers and are using them for plant control can be described by the distribution shown in Figure 19.0. On the lower end where a plant does not have an OCA-XRF there are only a handful of plants. On the upper end of where plants are utilizing their OCA-XRF in a plant-wide optimizing or economic based control philosophy there are only a handful.

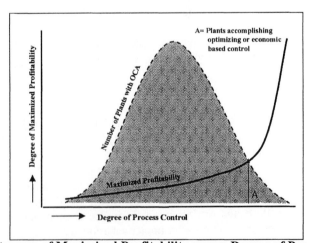

Figure 19.0 Histogram of Maximized Profitability versus Degree of Process Control

In between these extremes are 90% or more that are not utilizing their OCA-XRF to the fullest extent for one reason or another. Some of those reasons are:

1. Ore body is simple and therefore requires only simple process control, only need assay trends for emergency actions

2. Process control strategy/system too complicated or requires major investment in time and/or manpower to change

3. OCA-XRF system project failed to deliver a properly designed sampling and analyzer system, resulting in maintenance problems and low system availability

4. OCA-XRF system calibration not maintained properly

5. The overall operating philosophy of plant management does not support a high level of process control

The Future of OCA

There is still room to improve the reliability of existing sampling systems by utilizing "on demand" flushing and control. Likewise, calibration and analyzer probe maintenance can be improved via remote diagnostic support relieving the plant from having to rely on a high level of local expertise. New hybrid systems that combine the benefits of EDX and WDX are available. These systems allow each installation to optimize the cost versus benefits associated with whether WDX or EDX is used.

If we polish our crystal ball and look into the near future, it will likely show that OCA will, still, be based primarily on XRF technology. However, XRF technology will be complimented with particle size analyzers, video imaging, and specialized statistic and control packages. The combination of XRF technology and video imaging has already been demonstrated (vanOlst et al. 2001). Likewise, on-line mineral balances and net smelter return calculations have been accomplished with the right combination of technologies (Holdsworth 2002). Improving the measurement of OCA via combining OCA-XRF assays with other faster process measurements will probably be the wave of the near future.

In the slightly more distant future the direct on-line measurement of minerals and perhaps even the on-line measurement of the degree of liberation of minerals will be possible. Already some success off-line using a digital scanning electron microscope to characterize ore and mineral types has helped improve plant efficiency (Lotter et al. 2002). In another case, results from the digital scanning electron microscope have been used to calculate the phase-specific surface area (PSSA) and have correlated it to metallurgical results in laboratory scale tests (Winckers A.H. 2002). Now we're talking liberation! Wouldn't it be ironic if the next breakthrough in on-line composition analysis would be based on using the old microscope on the operating plant floor?

REFERENCES

Barker, D.R., H.J. Melama, T.F. Braden 1987. The use of on-stream analysis and computer-based process management for concentrator control applications. Paper presented at American Mining Congress convention 1987.

Barrette L., S. Turmel, J.A. Boivin, M. Sabsabi, T. Martinovic, G. Quellet, On-line iron ore slurry monitoring using laser induced plasma spectroscopy, in Control and Optimization in Minerals, Metals and Materials Processing, 38th Annual Conference of Metallurgists of CIM, Quebec City, Quebec, Canada 1999.

Cooper, H.R. 1976. On-stream x-ray analysis, Published in Flotation A.M. Gaudin Memorial Volume, ed. M.C. Fuerstenau, Chapter 30, New York:AIME.

Cooper, H.R. 1984. Recent development in on-line composition analysis of process streams. Published in Control '84 Mineral/Metallurgical Processing, ed. J.A. Herbst, Chapter 4, New York:SME.

Hales, L.B., G.R. Marchant 1979. Instrumentation. Published in Computer Methods for the 80's in the Mineral Industry, ed. A. Weiss, Chapter 5.2, New York:SME.

Holdsworth M., M. Sadler, R. Sawyer 2002. Optimizing concentrate production at the Greens Creek mine, SME preprint 02-063.

Jensen, D.L., Flotation supervisory control at Cyprus Bagdad. Published in Advances in Flotation Technology, ed. B.K. Parekh and J.D. Miller, Section 6, 433-440.

Jones, J.A., R.D. Deister II, C.W. Hill, D.R. Barker, P.B. Crummie 1991. Process control at the Doe Run Company. Proceedings Plant Operator's Symposium, SME Annual Meeting 1991.

Kongas M., K. Saloheimo 2001. When is the XRF assay good enough for process control. SME preprint 01-189.

Lahteenmaki S., J. Miettunen, K. Saloheimo 1999. 30 years of on-stream analysis at the Pyhasalmi mine. SME preprint 99-147.

Laughlin A. W., C. R. Mansfield, D. A. Cremers, M. J. Ferris, Real-time, in situ analysis of exploration samples using LIBS, Preprint 99-105, SME Annual Meeting, Denver, Colorado, 1999.

Lebrun P., P. Le Tourneur, B. Poumarède, H. Möller, P. Bach, On-line analysis of bulk materials using pulsed neutron interrogation, 15th Int. Conf. on Applications of Accelerators in Research and Industry, Denton, Texas, 1998.

Lotter N.O., P.J. Whittaker, L. Kormos, J.S. Stickling, G.J. Wilkie 2002. The development of process mineralogy at Falconbridge Limited, and application to the Raglan mill. Proceedings 34th Annual Meeting of the Canadian Mineral Processors. Session I. Paper 1.

vanOlst M., N. Brown, P. Bourke, S. Ronkainen 2001. Improving flotation plant performance at Cadia by controlling and optimising the rate of froth recovery using Outokumpu FrothMaster. Proceedings 33rd Annual Meeting of the Canadian Mineral Processors. Session I. Paper 3.

Rosenwasser S., G. Asimellis, B. Bromley, R. Hazlett, J. Martin and A. Zigler, Development of a Method for Automated Quantitative Analysis of Ores Using LIBS, LIBS 2000 1st International Conference on Laser Induced Plasma Spectroscopy and Applications, Pisa, Italy, 2000.

Saarhelo K., U. Paakkinen and P. Pennanen, On-line Analysis in Industrial Mineral Applications, 8th Industrial Minerals International Congress, 1990.

Shoniker Joe and Ronald Vedova Roger L. Vaughn, PCS Phosphate White Springs automatic control and on-stream analysis innovations have pay-off in big gains, Engineering Foundation Conference on Phosphate Beneficiation, White Springs, Florida, 1998.

Tran K. C. and M. Evans, Intelligent instruments for process control of raw materials in the basic industries, Preprint 01-34, SME Annual Meeting, Denver, Colorado, 2001.

Winckers A.H. 2002. Metallurgical mapping of the San Nicolas deposit. Proceedings 34th Annual Meeting of the Canadian Mineral Processors. Session I. Paper 3.

Ylinen R., J. Miettunen, M. Molander, E.-R. Siliämaa, Vision- and model based control of flotation, Future Trends in Automation in Mineral and Metal Processing, IFAC Workshop, Finland, 2000.

18

Process Control and Instrumentation

Section Co-Editors:
Robert Edwards and Aundra Nix

Introduction to Process Control
B. Flintoff 2051

Well Balanced Control Systems
T. Stuffco, K. Sunna 2066

The Selection of Control Hardware for Mineral Processing
R.A. Medower, R.E. Cook 2077

Basic Field Instrumentation and Control System Maintenance in Mineral Processing Circuits
J.R. Sienkiewicz 2104

Strategies for Instrumentation and Control of Crushing Circuits
S.D. Parsons, S.J. Parker, J.W. Craven, R.P. Sloan 2114

Strategies for the Instrumentation and Control of Grinding Circuits
R. Edwards, A. Vien, R. Perry 2130

Strategies for the Instrumentation and Control of Solid–Solid Separation Processes
G.H. Luttrell, M.J. Mankosa 2152

Strategies for Instrumentation and Control of Thickeners and Other Solid–Liquid Separation Circuits
F. Schoenbrunn, L. Hales, D. Bedell 2164

Strategies for Instrumentation and Control of Flotation Circuits
H. Laurila, J. Karesvuori, O. Tiili 2174

Pressure Oxidation Control Strategies
J. Cole, J. Rust 2196

Introduction to Process Control

B. Flintoff[1]

ABSTRACT

Process control for the stabilization and optimization of mineral processing circuits first emerged in the late 1960's, and over the past three decades has evolved into one of the most capital effective investments available to mill management. Consequently, the art and science of applying the technology has emerged as a core competence in the industry. This chapter provides an introductory overview of process control, including elements of both practical and theoretical interest. The intent is to provide a systems oriented framework for thinking about process control by exploring the constituent elements of: process; measurement/modulation; hardware; strategies, users; and, maintenance/development.

PREAMBLE

Process control in mineral concentration has been around for many decades. Over time, the balance between manual versus automatic regulation and/or optimization of the constituent processes has changed very significantly. While it is a little difficult to point to an event, or even a period that signaled the move by operators to accept automatic process control, the author believes that the early '70's marked the point of a profound change in thinking. The catalyst was the minicomputer, which combined I/O and HMI devices with high level programming languages (e.g. BASIC, FORTRAN), all in a relatively inexpensive package. By today's standards these machines were extraordinarily primitive, but for the first time process engineers had a means to easily develop, test, and modify process control strategies. Moreover, operators also had a tool that marshaled and presented relevant process data and information in a form that would enable them to make better, faster, and more systems-oriented decisions. An interesting byproduct of the introduction of this new technology was that it provided the common ground for operators and engineers to discuss and implement ideas aimed at process improvement – a pre-cursor to what later became know in the "Quality culture" as self-directed teams. The returns associated with these process control investments were often very impressive (see Chapter 2, Flintoff and Mular, 1992, for example).

The emergence of the PLC in the late 1970's and the DCS in the early 1980's as design standards for new plants was perhaps the concrete indication of a wide scale embrace of automation by the mineral processing community. In this period, university curricula were also modified to include courses on process control along with related technologies such as modeling and simulation. These were necessary to prepare graduates to fill an implementation void that the rapid advance of instrumentation and control hardware had created at the process interface. Paradoxically, the old succession model was inappropriate, as for the first time the young staff were required to mentor the older staff on the art and science of digital control.

The 1990's and the first few years of the new millennium have seen steady progress in process control development. From a technology perspective, the principal drivers have been: (a) the PC, which has changed the thinking and expectations around control hardware architecture; (b) advanced control , which has emerged in robust and effective implementations; (c) digital bus technology, which is changing the way we think of implementing process control; and, (d) the Internet, which while it has not yet had any direct impact on process control, has dramatically and very quickly raised the level of computer literacy and HMI expectations of operators and engineers alike.

There is little doubt that management now recognizes process control as one of the most capital effective investments available in the pursuit of shareholder value. Moreover, as automation technologies evolve in mining, one now sees growing interest in the digital integration

---

[1] *Metso Minerals – Minerals Processing Business Line, York, PA*

of the enterprise, enabling the application of business controls, the next level of "automation." Process control is a key enabling technology in this new field and it promises to be an interesting time for those challenged to bridge the plant floor with the board room.

The process control journey is destined to continue, as we explore improvements in existing systems and approaches, as well as new applications. That makes this volume an important contribution, as it benchmarks the current status of a technology that is ever in flux.

INTRODUCTION TO PROCESS CONTROL

This introduction is intended to provide a top-down conceptual overview to one of the most important technologies process engineers have at their disposal today - process control[2]. More specifically, it aims to present a framework for the reader to think about this subject. To do this, the author has chosen to begin by examining process control in the broadest context, and then to drill down to the fundamental elements that will be discussed in the companion chapters.

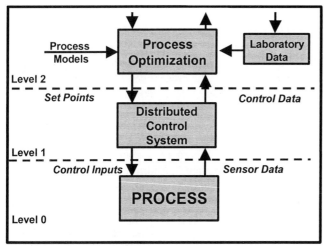

Figure 1 A Categorization of process control

Most of us working in the area are familiar with control categorizations (segmentation by level) such as illustrated in Figure 1. That is, we think of process control as the measurement and pre-processing of field data, which is then presented to operators and/or algorithms, that then decide whether manual/automatic changes are required to final control elements, or other variables outside of the control system. In other words, process control is focused on the regulation and optimization of equipment or circuits in "normal" operation, and actions are based on objectives set periodically by management. Figure 2 is based on an adapted form of the Computer Integrated Manufacturing (CIM) model by Bhatt, 1992. It illustrates process control in the context of a much more holistic information technology hierarchy. In this figure the information derived from, and actions taken by the control system are now synchronized with asset management (e.g. maintenance planning and scheduling, procurement, etc. functions) and with business systems (e.g. mine planning/dispatch, concentrate marketing & sales, shipping logistics, etc.).

---

[2] The reader more interested in process control theory is referred to the standard college text books on the subject, including Stephanopoulus, 1984 and Seborg et al., 1989.

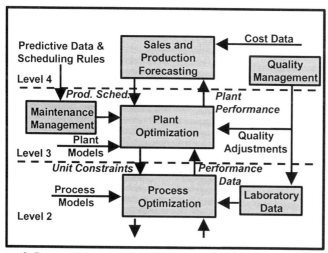

Figure 2 Connecting process control with business controls

Succinctly, Figure 1 is more concerned with Process Management, while Figure 2 is representative Production Management, and for the purposes of this chapter the working definitions for these two terms are as follows.

Process Management: That set of activities focused on the management of operating equipment and/or circuits that comprise a unique process in the production value chain. These activities are necessitated by changes in the processing characteristics of the feedstock(s) and/or changes in production targets. This is achieved by the regulation of certain material inputs, process operating conditions, and product quality through the optimal deployment of people and technology. (Process control is an integral part of process management.)

Production Management: That set of activities focused on the overall management of the constituent processes in the production value chain. These activities are necessitated by such factors as the maintenance requirements of the process(es), and by changing markets and/or corporate business goals. This is achieved by the development of enterprise wide operations and maintenance schedules as well the establishment of process specific production targets and the communication of this information to the process and asset management subsystems.

The focus of this chapter is "process control," as generally defined by Figure 1. Figure 3 provides a more detailed look at the elements that comprise an effective industrial process control system. As the phrase implies, "process control" begins with process. Process can have as significant an impact on the performance of a control system as any of the other (control) elements. The instrumentation layer includes sensors to make measurements of process variables, and final control elements used to manipulate variables in the field. The control hardware layer comprises all of the Input/Output (I/O) subsystems, computer gear and operating software, as well as the peripheral devices for data presentation and storage. Control strategies consist of a blend of more traditional regulatory and advanced controls to develop a robust and effective package. This layer often includes some *ad hoc* functionality unique to the process, and intended to enhance the robustness of the application. The user group layer comprises: the instrumentation and electrical technicians who support the instrumentation, hardware and regulatory controls; the process and control engineers who support and develop the strategies; and, the operators who manage the application and provide expert input as required. Last, but certainly by no means least, are the maintenance, training and system development programs that are essentially intended to sustain and improve the various elements of the system. This latter element is probably the Achilles' heel of a control system, and while things seem to be much better these days, the 1970's and 1980's included numerous examples of well performing systems that soon fell into disuse because of the lack site champions and formal programs of support. Finally, the structure of Figure 3 is strongly correlated with the chronological development of the discipline in mineral processing. This starts with on-line composition analysis (beginning in the 1960's), moves through the minicomputer era (beginning in the 1970's), to the control strategy development (beginning in the 1980's), and on to the user group (beginning in the 1990's).

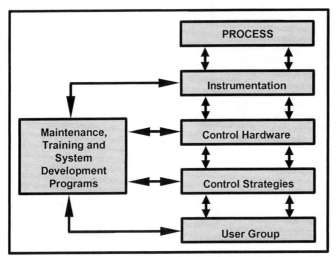

Figure 3 The elements of an effective industrial control system

Academics generally focus on the first three layers (instrumentation + hardware + strategies), which collectively have been termed the Control Triad, depicted in Figure 4. It is clear from the collection of papers in this session that these are very important points. From an operator's perspective, perhaps the most important of these is the apex dealing with strategies, as these are arguably the most unique, depending upon the process, the field instrumentation complement, the nature of plant disturbances, and the business goals of the corporation. Just to illustrate this point, even the advanced control of similar grinding circuits with similar objectives shows at least 20% customization in the application.

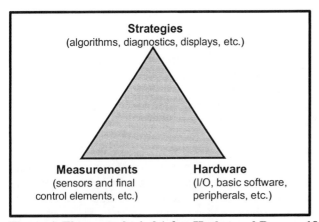

Figure 4 The control triad (after Herbst and Bascur, 1984)

Figure 5 drills down into strategies and there are several interesting points to be made. The first is that strategies are hierarchical in nature, e.g. an effective supervisory control strategy can only be built on a foundation of effective regulatory controls, and so on. What is different is that the level of infrastructure investment (in field instrumentation and computer hardware, etc.)

decreases as one moves up the hierarchy, whereas the intellectual investment increases. Secondly, there are performance related benefits realized by each level of control strategy. In the past, one of the more difficult choices was to determine when one was approaching the point of diminishing returns on the development input/output curve. That is, when does one abandon the tools at one level, and move on to the next. *De facto* standards are emerging based on the industry's collective experience.

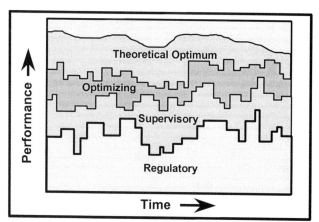

Figure 5 Levels of process control (strategies)

Finally, process control (e.g. as seen through research publications, undergraduate courses, etc.) can sometimes give one the impression that this is a highly mathematical, typically complex, often counter-intuitive science. All of that can be true, particularly as one moves up the hierarchy shown in Figure 5. However, we would all do well to remember that the general goals of process control are really quite straightforward, and as illustrated in Figure 6, they are simply to: (a) squeeze the variance – i.e. demonstrate that the process can be controlled; and then (b) shift the target – i.e., exploit the benefits of control by maintaining optimum targets.

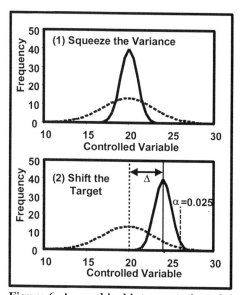

Figure 6 A graphical interpretation of the goals of process control

The balance of the paper takes a closer look at some of the issues and trends in process control, using the elements of Figure 3 as the framework. The maintenance and developments aspects will be addressed within the other layers.

PROCESS

Over the past couple of decades there has been a move to simplify the design of mineral processing flow sheets. Two indicators of this trend would be very large equipment (e.g. SAG mills of 12.2 m diameter) and a move to more open circuits in flotation. The first indicator dictates the need for very effective process control because the operation depends more and more on fewer parallel circuits, and in many cases just a single very large process line. The second indicator results from what has been termed in industry as the "Back to the Basics" movement (e.g. Stowe, 1992), and an example is presented in this section.

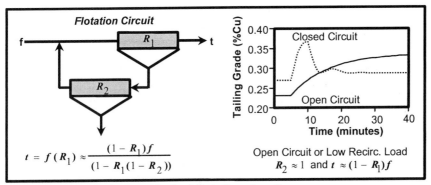

Figure 7 Control of a hypothetical flotation circuit

To introduce the problem, consider the hypothetical flotation circuit illustrated in Figure 7. In this instance we suppose that it is desirable to regulate the rougher tailing grade, perhaps by the manipulation of airflow. The existence of a recirculating stream from the cleaner tails can induce control problems. The algebra in the figure demonstrates the relationship between tailings grade, feed grade and the recoveries in the rougher and cleaner banks. If the recirculating load is very low ($R_2 \approx 1$), the response of the circuit to, say, a decrease in air flow in the rougher bank shows a "first order" like increase to the new steady state, as illustrated in the graphic. If, on the other hand, the recirculating load is very large, then a decrease in air flow will initially cause much of this material to be rejected in the tailings stream, but with increased rougher residence time the rougher tailing will eventually stabilize at a level quite close to the value prior to the change in air flow, again as shown on the graphic. Intuitively, one would expect that process control would be easier to implement and more effective in the case of a very low recirculating load. This simple contrived example illustrates a couple of points. Firstly, it helps to explain why older design procedures tended to favor flotation circuits with a fairly high degree of recirculation. They were inherently self-regulating, except in the presence of a sustained disturbance, which would ultimately necessitate a compromise on grade and/or recovery. Secondly, it explains the recent move to more open circuits which often provide superior metallurgical performance, but which demand a greater degree of process control to maintain operational efficiency.

Morari, 1983, has used the term resilience (see definitions below) to capture this notion of the interaction between process and control.

- "Resilience – Describes the ability of the plant to move fast and smoothly from one operating condition to another (including start-up and shut-down) and to deal effectively with disturbances."
- "Dynamic Resilience – the quality of the regulatory and servo behavior which can be obtained for the plant by feedback ."

Not surprisingly, the impact of the process on the quality of process control has been addressed by leading control theoreticians: e.g.– *"However, it has long been recognized by both industry and academia that modifications of the physical system itself can sometimes affect the resilience significantly more than changes in the controller."* (Morari, 1983) or *"The relation between process control and design is also important. Control systems have traditionally been introduced into a given process to simplify or improve their operation. It has, however, become clear that much can be gained by considering process control and design in one context. The availability of a control system always gives the designer an extra degree of freedom, which frequently can be used to improve performance or economy. Similarly, there are many situations where difficult control problems arise because of improper design."* (Åström and Wittenmark, 1984)

To conclude this section, and in keeping with the flotation theme, Figure 8 illustrates the circuit re-engineering effort at a Canadian copper-zinc concentrator (Stowe, 1992). As suggested above, the much more open circuit on the right was easier to operate and control, leading to significant performance improvements. It is important to note that in this case there was no new process equipment, sensors or hardware, simply a redeployment of existing assets.

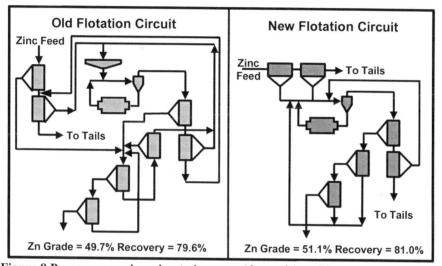

Figure 8 Process re-engineering to improve (dynamic) resilience

INSTRUMENTATION

Field instrumentation includes sensors (e.g. flow, temperature, density, composition, pressure, level, etc.) and final control elements (e.g. valve position, variable speed drives, etc.), as depicted in Figure 9. Given the preponderance of devices in the sensor category, it is not uncommon to hear sensors and instrumentation used synonymously. Accordingly, this section will focus principally on sensors.

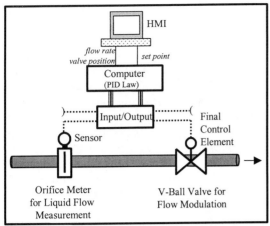

Figure 9 A flow control loop starting and ending with instrumentation

Many of the sensors required to develop a solid foundation of regulatory controls exist, as will be illustrated below. Although the slurry systems frequently encountered in mineral processing have presented some design challenges, quite a large number of the sensors (e.g. flow meters, level monitors, pressure gauges, etc.) in use are common to other process industries. Despite this broad usage, sensors are not always trouble free. The fact that there are still arguments over the general merit of slurry flow measurement underscores the point, i.e. you can still find people who claim success and others who claim failure using magnetic or ultrasonic flow meters. About the only thing one can be sure of is that this is generally not the fault of the instrument. Pareto's rule applies, as ~80% of the problems arise from improper application and/or installation. The point being that one must take care in selecting, installing, calibrating and maintaining field instrumentation.

Table 1 is intended to illustrate the range of sensor technologies that are commonly applied in grinding circuit process control. Clearly, there are many choices of technologies, and within each technology there is generally a good choice of suppliers. However, what one really needs from this rather extensive list is dictated by the circuit disturbances and the strategies employed to reject them.

Table 1 Sensor technology choices for comminution systems (source Herbst *et al.*, 2002)

Measurement	Technology Employed
• Bin Level (solids) • Tank Level (slurry/water)	• ultrasonic devices, laser devices, load cells, mechanical devices • ultrasonic devices, capacitance probes, differential pressure devices, conductivity probes, mechanical devices
• Motor Power	• current transducer (+ conversion), power transducer, torque meter
• Solids flow • Slurry Flow • Water Flow	• electronic belt scale, nuclear belt scale, impact meter • magnetic units, ultrasonic units • vortex shedding devices, turbine meters, differential pressure devices
• Moisture (dry solids) • Moisture (slurries)	• microwave units • radiation gauges, U tubes, differential pressure devices
• Pressure	• diaphragm devices
• Vibration/Sound	• accelerometers/microphones
• Temperature	• thermocouples, resistance thermal devices, IR imaging
• Particle Size (dry solids) • Particle Size (slurries)	• image analysis techniques • ultrasonic devices, mechanical (caliper) devices, soft sensors

• pH	• specialized electrodes, conductivity probes
• Tramp Metal	• magnetic field devices
• Mill Load	• power based devices, acoustics, load cells, strain gauges, soft sensors, conductivity
• Speed	• tachometer

One of the exciting developments related to the more generic instrumentation is the emergence of Foundation Fieldbus (or Profibus). Broadly speaking, this is a standard digital communications protocol that has permitted supply side firms to imbed intelligence in their sensors and final control elements. For example, one can buy a flow sensor that not only measures flow (the only signal available on the old analog [e.g. 4 – 20ma output] gear), but using embedded diagnostics can perform condition monitoring functions to indicate whether the estimated flow has been corrupted by mechanical or process problems. Moreover, this device can have control functionality (sampling and signal filtering, PID controllers, loop diagnostics, etc.) and can communicate with neighboring smart devices. Significant reductions in project costs arise from savings in detailed engineering (~20%)[3], configuration (~30%), and installation and commissioning (~40%). Operational savings arising from predictive maintenance and remote support (~70%) are also very significant.

Perhaps motivated by Galileo's (1564-1642) challenge - "We have to measure everything that can be measured, and make measurable everything that is not yet measurable." - sensor development continues to be a focal point in control research. One can differentiate between technology that is finding different applications, and applications seemingly still looking for a technology. For example, in the case of the former there has been a rebirth of interest in acoustics for monitoring AG/SAG mill performance, and in particular to avoid ball on liner collisions from a cataracting charge in SAG mills (e.g. Valderrama et al., 200, Campbell et al., 2001, Pax, 2001). Vision systems, and especially those based on image analysis have also seen relatively fast acceptance in coarse particle sizing (e.g. WipFrag – Maerz, 2001; Split – Girdner et al., 2001; and T-Vis – Herbst and Blust, 2000), and froth monitoring (e.g. JKFrothcam – Kittel et al., 2001; Frothmaster – Brown et al., 2001; Visiofroth[4]).

With respect to applications looking for a technology, mill load estimation is a good example. This is a critically important variable in throughout optimization for AG and SAG mills, hence the research activity over the past decade or so. As one can see from Table 1, there are power-based devices (e.g. Pontt et al.,1997; Koivistoninen and Miettunen, 1989), acoustic devices (e.g. Barrientos and Telias, 1997; Spencer et al., 2000), load cells units (e.g. Jones and Wright, 2001), strain gauge systems (e.g. Herbst et al., 1990; Dupont and Vien, 2001), conductivity units (e.g. Marklund and Oja, 1996; van Nierop and Moys, 1995) and soft sensors (e.g. Herbst et al., 1989), to choose from. All have strengths and weaknesses, and with the possible exception of soft sensors, none have a large enough installed base or long enough operating history to be established as the technology leader. In fact, as more is learned, it seems likely that some combination of technologies will emerge as the best means to measure mill load and other in-mill properties.

Finally, there have been some interesting statistical developments over the past few years looking at new and superior empirical modeling tools with which to develop more robust empirical calibrations (e.g. Clustering – Ginsberg and Whiten, 1992) or process models (e.g. Partial Least Squares – Hodouin et al., 1993 and Bartolacci and Boujila, 2000).

---

[3] The cost savings have been distilled from numerous vendor presentations on this subject, and should probably be considered as best case estimates.
[4] Personal communication with Thierry Monredon, Cisa, Orleans, France

HARDWARE

Control hardware is a very important part of a process control system, and often accounts for ~ 20% - 25% of the total capital cost in a relatively well-instrumented mineral processing plant. It is also an area that is quite well serviced by the major vendors, which generally helps to make selection a rather low risk step.

Despite the long heralded convergence of functionality between the two main platforms - the Programmable Logic Controller (PLC) and the Distributed Control System (DCS) - there remain some technical differences, and these often require careful consideration in the context of the overall application. Interestingly, the philosophical loyalties that evolved around these two options in the 1980's and 1990's (the electrical camp vs. the instrumentation camp) appear to remain more or less solidly entrenched. A consequence of both factors is the hybrid control system, a very common architecture in plants built in the 1980's and 1990's. Figure 10 provides an illustration of a typical architecture for a control system designed in the '90's. In this instance the PLC is generally used to manage the discrete I/O and safety interlocking, while the DCS handles the analog I/O and Human Machine Interface (HMI).

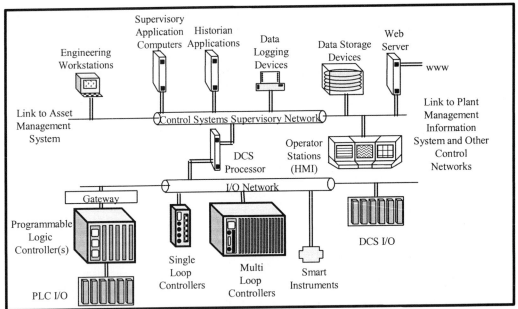

Figure 10 Components of a plant control system (adapted from Flintoff & Mular, 1992)

The evolution of smart instruments is beginning to change the role of the control hardware, as I/O, signal conditioning, single loop control, etc. migrate to the field devices. Increasingly, the emphasis of the hardware suppliers is on data management and presentation, and on advanced applications including plant management information systems and advanced controls. This is perhaps better developed in what was once the analog world, so it is the DCS suppliers who are leading the change through an increased emphasis on the functionality depicted in Figure 2. One can expect this trend to continue, as most of the more robust control and condition monitoring applications will also migrate to the plant floor. More specifically, they will be embedded in the process equipment yielding what has been termed "intelligent processes", conceptually similar to smart instrumentation.

STRATEGIES

Much will be said on this topic in the companion papers. For the purposes of this introduction we can categorize strategies into essentially two levels – regulatory and advanced control. At the regulatory level the strategies are typically implemented through single-input single-output controllers, almost always using the Proportional Integral Derivative (PID) control law. Advanced control strategies are generally multiple input – multiple output (usually set points), and although a number of more analytical techniques have been applied (e.g. Decoupling Control – Hulbert and Woodburn, 1983 and Model Predictive Control – Vien et al., 1991), the industry standard appears to be a blend of artificial intelligence and analytical techniques, known as model-based fuzzy expert control.

On the regulatory level, the strategies are generally implemented through PID control laws and the manipulated variables are flows, speeds, etc. on the plant floor. Despite the wealth of application history in this area, Bialkowski's (1992) rather unsettling findings in the pulp and paper industry were found to be applicable to mineral processing. To briefly summarize: *"If you have been keeping score, only 20% of the loops surveyed actually decrease variability in automatic over manual mode of operation, in the short term."* This prompted an interest in getting back to basics on such as issues as signal conditioning and loop tuning (e.g. Vien et al., 2000) and on performance diagnostics (e.g. Perry et al., 2000), all of which have helped to improve the quality of control at the regulatory level. In addition, effective solutions have been established for some of the more challenging regulatory loops, and these have necessarily evolved beyond pure Proportional Integral Derivative (PID) controllers.

A good example of the latter is the multiple feeder control problem for large SAG mills, which is illustrated in Figure 11. In this instance, the size of the equipment generally combines to lower dynamic resilience as a result of the significant dead time between the feeders and the weigh scale, even though this sensor is placed as close as possible to the feeders. A further complication is that effective dead time, and controller gain is a function of the combination of feeders running (in automatic, vs, manual vs off). Another layer of complexity is added if the feeder speeds require some bias to reflect ore flow issues arising from, say, stockpile segregation. Vien et al., 2000, have described a very robust regulatory algorithm that is well suited for this problem.

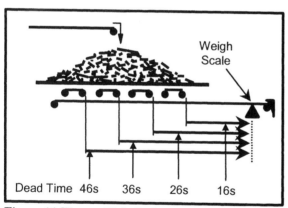

Figure 11 The multiple feeder control problem in AG/SAG mill feed regulation

One can find the first evidence of advanced control back in the 1970's, when process engineers using control minicomputers began to supplement PID logic with 'If-Then-Else' heuristics to optimize circuit performance. Efforts to use analytical (i.e. mode-based) advanced control techniques have been problematic, partly because of their dependence on the reliability of sensor inputs, partly because the models themselves are simplifications of the process, and partly because the models require continuous adaptation as things change. Sandvik, 1985, provides a particularly good description of some of these issues. The introduction of the real-time expert system (with fuzzy logic) in the late 1980's provided a platform for implementing advanced control strategies that circumvented some of the limitations of the model-based approach. However, it isn't surprising that over the past few years the two approaches have been blended to exploit the strengths of both.

Figure 12 is a schematic of a model-based fuzzy expert control structure. In this particular instance, the model can be based on process phenomenology and the adaptation achieved with an extended Kalman filter. It is also possible to employ statistical models, specifically neural networks, and frequently retrain the model routinely, or as new process conditions and/or poor predictive accuracy are encountered . Of course, within this structure both can be used. This author's opinion is that whenever possible, phenomenological models should be employed, as they are more faithful to the engineering principles, physics and chemistry underlying the process. Moreover, the adaptation involves parameters with a physical significance, which makes error checking relatively easy.

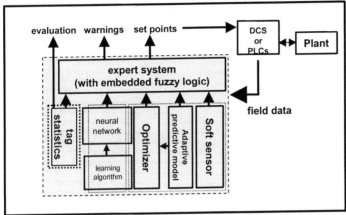

Figure 12 An advanced control structure for mineral processing

Broussaud et al., 2001, have shown that the structure in Figure 12 can be applied to plants at essentially either end of the field instrumentation spectrum. Their general philosophy is to use the phenomenological models within a soft sensor to estimate parameters such as SAG load, ore hardness, etc., which are not directly measured or measurable. Moreover, the adapted model from the soft sensor can then be used in optimization. Depending upon the optimization approach, the model can be used in its dynamic or steady state form to deduce new set points that will drive the process toward the optimum operating condition. Of course, depending upon the quality of the input data, the optimizer may make unrealistic recommendations, and one of the roles of the expert system is to filter these values prior to implementation. Not everything can be done quantitatively, and heuristics are employed to first ensure process stability and then to optimize performance. This control logic is implemented with crisp and fuzzy rules. It is of some interest to note than the major developmental focus in advanced control appears to be directed at embedded advanced measurement technologies (e.g. the image analysis systems mentioned earlier), which can enhance controller performance.

To conclude, it is of some interest to note that we can distinguish a third level of control strategies, and these have been called either watchdog control or shell (or jacket) software (referred to as *ad hoc* functionality in the Introduction). The intent of these strategies is essentially to improve the robustness of the control applications by ensuring that the more traditional controls are protected from unusual operating conditions (e.g. start-ups & shutdowns, very large disturbances, etc.), where they would be unstable and require excessive operator intervention. Flintoff and Edwards, 1992, provide an illustration of the use of watchdog control in crushing. Astrom described this rather well (Astrom et al., 1986) and noted that given the application specificity, this has been an area largely ignored by control researchers and developed exclusively by practitioners.

USERS

It is essential that each operation have the necessary complement of people to support the process control system from the field instrumentation through to the users. The numbers of technical people generally increase with the size (~I/O count) and complexity (~ "analog" I/O), and generally decrease with the level of sophistication of the system (e.g. diagnostic and asset management software) and training. In the constant pursuit of doing more with less, a major focus for the users of the process control systems is essentially one of training and education. Other drivers include a greater emphasis on improved control performance, and succession.

Much has been written on the subject of training and education for users in process control (e.g. Vien et al., 1994, Flintoff, 1995, Rybinski et al., 2001). Suffice it to say that this too is a process that requires management commitment to sustain the return on the process control investment.

Downsizing and succession issues are combining to create an interesting dilemma for operators. As more and more sites demand improved process control performance through better regulatory and advanced control, the need for deeper subject matter expertise grows. Recruiting and retaining such people has been a challenge for the mining industry, as we compete with all of the process industries. Typically, the other sectors offer very competitive salaries and usually superior work locations. The end result is that something that could be classified as a necessary core competence for every mineral processing operation, is a responsibility that has to be increasingly outsourced.

Successful outsourcing models exist in other disciplines, especially Information Technology (IT). However, given the global nature of mining, effective and timely remote communications is an essential element of such a relationship. This may be the single biggest bottleneck, but that is changing through the rapid advance of IT, and especially Internet technologies. Specialty consultants, engineering firms and equipment suppliers are all moving to establish themselves as "Application Service Providers" with 24x7 remote technical support. The evolution of these new business models means that the technical people at site must adopt a leas technical and more managerial role toward process control. However the irony is that to be effective in the latter, they must remain quite familiar with the former.

CONCLUSIONS

Process control has been an important part of mineral processing for many decades, but automatic process control has really established itself over the past 30 years. The '70's can be thought of as the embryonic period of development, where the first efforts at integration involving computers were completed. The '80's was a growth period, with some of the attendant growing pains, and, among other things, featured the development of better tools and techniques for implementing process control strategies. The 90's were a also a period of some growth, but also a period of maturation as proven approaches began to emerge in both hardware architectures and especially in software structure. Exciting development in process control continues on all fronts: from field instrumentation (new sensors); to control hardware (smart instruments); and strategies (improved modeling and optimization methods); to the users (new business models, educational programs, etc.). Equally exciting is the growing profile of process control as a cornerstone of the enterprise business systems, drawing it more into the realm of production management. This latter transformation will have some interesting trickle down effects as such activities as condition monitoring/predictive maintenance and real-time enterprise modeling become deeply entangled with control. The future is bright for those with an interest in this discipline.

On the business side, and where it has been properly applied and supported, process control has consistently delivered on the financial projections, with paybacks of from weeks to months and returns on investments of from 20% to 200%. One can only assume that with increasing regulatory interest in resource utilization, and the increased business interest in shareholder value and sustainable development, there will be increasing demand placed on all facets of operation. This is especially true for process control, since process management is effectively implemented through the elements of field instrumentation, hardware, strategies, users and maintenance programs that together comprise the process control system. The process control value chain is only strong as its weakest link, and successful operations must necessarily explore the options for continuous improvement in all elements of Figure 3.

REFERENCES

Åström, K., Wittenmark, B., 1984, Computer Controlled Systems: Theory and Design, Prentice Hall, pg. 157

Åström, K., Anton, J., Arzen, K., 1986, Expert Control, Automatica, Vol. 22, No. 3, pp. 277 - 286

Barrientos R., Telias M., 1997, Nuevos Sonsores en el Circuito SAG, Proc. SAG Workshop, Vina de Mar, Chile

Bartolacci, G., Boujila, A., 2000, Application of Multivariate Tools to Mineral Processing Data Analysis and Modeling – Flotation Case, Proc. IFAC Workshop, Automation in Mineral and Metal Processing, Finland, August, pp. 188 - 193

Bhatt, S., 1992, The Control Connection, Chemical Engineering, May, pp. 91 – 94.

Bialkowski, W., 1992, Dreams vs. Realities: A View From Both Sides of the Gap, Control Systems '92: Modern Process Control in the Pulp and Paper Industry, Whistler, B.C., pp. 283 – 295

Broussaud, A., Guyot, O., McKay, J., Hope, R., 2001, Advanced Control Of SAG And FAG Mills With Comprehensive Or Limited Instrumentation, Proc. International Autogenous and Semi-autogenous Grinding Technology, ed. Barratt, Allen and Mular, Vancouver, B.C., Canada, Sept., Vol. II, pp. 358-372

Brown, N., Bourke, P., Ronkainen, S., van Olst, M., 2001, Improving Flotation Plant Performance at Cadia by Controlling and Optimizing the Rate of Froth Recovery Using Outokumpu Frothmaster, Proc. 33rd. Can. Min. Proc., CIM, Ottawa, pp. 25 - 38

Campbell, J., *et al*., 2001, SAG Mill Monitoring Using Surface Vibrations, in Proc. Int'l AG and SAG Grinding Technology 2001, eds. Barratt, Allan and Mular, Vol. 11, pp. 373 - 385

Dupont J., Vien A., 2001, Continuous SAG Volumetric Charge Measurement, Proc. 33rd AGM Can. Min. Proc., CIM, pp. 52 – 67

Flintoff, B., Mular, A, 1992, A Practical Guide to Process Control in the Minerals Industry, Gastown, Vancouver

Flintoff, B., Edwards, R., 1992, Process Control in Crushing, in Comminution – Theory and Practice, ed. Kawatra, SME, pp. 505 – 515

Flintoff, B., 1995, Control of Mineral Processing Systems, Proceedings of the XIX Int. Min. Proc. Cong., San Francisco, Vol. 1, pp. 15 - 23

Ginsberg, D., Whiten, W., 1992, The Application of Clustering to the Calibration of On-Stream Analysis Equipment, Int. J. Min. Proc., Vol. 36, pp. 63-79

Girdner, K., Handy, J., Kemeny, J., 2001, Improvements in Fragmentation Measurement Software for SAG Mill Process Control, in Proc. Int'l AG and SAG Grinding Technology 2001, eds. Barratt, Allan and Mular, Vol. 11, pp. 250 – 269

Herbst, J., Bascur, O., 1984, Mineral Processing Control in the '80's – Realities and Dreams, in Control '84, ed. Herbst, SME, pp. 197 – 215

Herbst J., Pate W., Oblad E., 1989, Experiences in the Use of Model Based Expert Control Systems in Autogenous and Semi Autogenous Grinding Circuits, Proc. Advances in AG and SAG Grinding Technology, eds. Mular and Agar, Vancouver, Vol. 2, pp. 669- 686

Herbst J., Hales L., Gabardi T., 1990, Continuous Measurement and Control of Charge Volume in Tumbling Mills, Control '90, Ed. Rajamani and Herbst, SME, pp. 163-171

Herbst, J., Blust, S., 2000, Video Sampling for Mine-to-Mill Performance Evaluation: Model Calibration and Simulation; in Control 2000, ed. Herbst, pp. 157 - 166

Herbst, J, Lo, Y., Flintoff, B., 2002, Size Reduction and Liberation, in Principles of Mineral Processing, ed. Fuerstenau and Han, SEM, 2002

Hodouin, D., MacGregor, J, Hou, M., Franklin, M., 1993, Mulitvariate Statistical Analysis on Mineral Processing Plant Data, CIM Bull., Nov/Dec, pp. 23 - 33

Hulbert, D., Woodburn, T., 1983, Multivariable Control of a Wet-Grinding Circuit, J. AIChE, Vol. 29, No.2, pp. 186 - 191

Jones, R., Wright, A., 2001, Selecting and Configuring Load Cells for AG/SAG Mill Grinding Applications, in Proc. Int'l AG and SAG Grinding Technology 2001, eds. Barratt, Allan and Mular, Vol. 11, pp. 227 – 239

Kittel, S., Galleguillos, P., Urtubia, H., 2001, Rougher Flotation in Escondida Flotation Plant, SME Preprint 01-53

Koivistoinen, P., Miettunen, J., 1989, The Effect of Mill Lining on the Power Draw of a Grinding Mill and its Utilization in Control, Proc. Advances in AG and SAG Grinding Technology, eds. Mular and Agar, Vancouver, Vol. 2, pp. 687- 695

Maerz, N., 2001, Automated On-Line Optical Sizing Analysis, in Proc. Int'l AG and SAG Grinding Technology 2001, eds. Barratt, Allan and Mular, Vol. 11, pp. 250 – 269

Marklund U., Oja J., 1996, Grinding Control at Aitik: Optimization of Autogenous Grinding Through Mill Filling Measurement and Multivariate Statistical Analysis, Proc. Intl. AG and SAG Grinding Technology, eds. Mular, Barratt and Knight, Vancouver, Vol. 2, pp. 617- 631

Morari, M., 1983, Design of Resilient Processing Plants – III, Chemical Engineering Science, Vol. 38, No. 11, pp. 1881 – 1891

Pax, R., 2001, Non Contact Acoustic Measurement on In-Mill Variables of a SAG Mill, in Proc. Int'l AG and SAG Grinding Technology 2001, eds. Barratt, Allan and Mular, Vol. 11, pp. 386 – 393

Perry, R., Supomo, A., Mular, M., Neale, A., 2000, Monitoring Control Loop Health at P.T. Freeport, Control 2000, ed. Herbst, SME, pp. 71 - 81

Pont, J., Valderama, W., Magne, L., Pozo, R., 1997, MONSAG: Un Sistema para el Mointereo On-Line de la Carga en Molinos SAG, Proc. SAG Workshop, Vina de Mar, Chile

Rybinski, E., Zunich, R., Grondin, M., Flintoff, B., 2001,Operator Education: An Important Element of the Corporate Knowledge Management Effort, SME Preprint 01-156

Sandvik, K., 1985, Limitations to Advanced Control in Complex Sulphide Flotation Plants, in Flotation of Sulphide Minerals, ed. Forssberg, Elsevier, pp. 433 – 446

Stephanopoulos, G., 1984, Chemical Process Control Process: AN Introduction to Theory and Practice, Dynamics and Control, Prentice Hall, New Jersey

Seborg, D., Edgar, T., Mellichamp, D., 1989, Process Dynamics and Control, Wiley & Sons, New York

Stowe, K., 1992, Noranda's Approach to Complex Ores – Present and Future, AMIRA Technical Meeting

Valderrama, W., et al., 2000, The Impactmeter, A New Instrument for Monitoring and Avoiding Harmful High-Energy Impacts on the Mill liners in SAG Mills, Proc. IFAC Workshop, Automation in Mineral and Metal Processing, Finland, August, pp. 286 - 289

Van Nierop M., Moys M., 1995, Measurement of Load Behaviour in an Industrial Grinding Mill, IFAC Automation in Mining, Minerals and Metals Proc., Sun City, South Arfica, 1995

Vien, A., Fragomeni, D., Larsen, C.R., Fisher, D.G., 1991, MOCCA: A Grinding Circuit Control Application, SME Ann. Mtg., Denver, Colorado, February

Vien, A., Edwards, R., Perry, R., Flintoff, B., 2000, Making Regulatory Control a Priority, Control 2000, ed. Herbst, SME, pp. 59 – 70

Vien, A., Palomino, J., Gonzalez, P., Perry, R., 2000, Multiple Feeder Control, Proc. An Gen. Mtg. Can. Min. Proc., CIM, Ottawa, pp. 298 - 312

Vien, A., Willett, D., Flintoff, B.C., Hendriks, D.H., 1994, Mill Operator Training - Where Do We Go?, Proc. 26th Annual General Meeting, Canadian Mineral Processors, Ottawa 1994, Paper no. 16, 15 pages

Well Balanced Control Systems

Tom Stuffco[1] and Khaled Sunna[2]

ABSTRACT

Successful control systems are characterized as reliable, relatively uncomplicated, easily expandable, and capable of supporting higher layers such as enterprise systems and advanced controls. This paper will explore the ingredients necessary to build a healthy control system foundation. Perhaps technically less satisfying then the specter of advanced techniques, the essential components are basic and include standards for configuration of controls, graphics, communication gateways, documentation, and personnel support.

INTRODUCTION

A well-balanced, successful control system has many human characteristics that are naturally desirable. Just as many humans have endearing qualities and features that make them distinct and successful, so goes the control system. A successful control system will communicate openly and easily with the rest of the world, "articulate" information clearly and concisely, demonstrate a degree of independence and be highly reliable. While it is difficult to find the right balance between reliability, scalability, and integration, these characteristics are essential to the overall success and acceptance of the system.

Upon installation, the control system must successfully meet the fundamental requirement of monitoring the process while suppressing disturbances. In order to sustain this foundation, functional system maintenance practices must be well established. Establishing this maintenance function is governed by two paramount events: system design and system selection.

A team with diverse talents must deliver the design essentials for both project start-up and subsequent support. End users must seek some level of assurance that the proposed system can support their immediate project needs as well as their future needs. If properly equipped, the system has good prospects for performing well throughout its useful life. If essential ingredients are missing, high aspirations for achieving world-class controls are reduced to resentful ongoing system support.

DESIGN

The quality with which the initial control system design was/is performed has a significant effect on future achievements. It is imperative that the design team develops certain standards prior to embarking on a new installation project.

Standards are a crucial component of the maintenance efforts that will ensue once the control systems are active. A system that lacks standards, but prefers the "flavor of the day" approach is unsupportable and unsustainable. Individuals responsible for the maintenance role under these conditions spend a significant portion of their time developing and implementing standards; which

---

[1] Operations Technical Support Manager, P.T. Freeport Indonesia Company
[2] Senior Project Manager, EMA Consulting

are often sub-standard because, out of necessity, they will develop them to minimize the effort required for implementation.

With reference to the following diagram (Figure 1), standards are most critical in the fundamental layer, including the low-level regulatory and discrete controls as well as the human-machine interface (HMI) and data acquisition areas. The design team's first goal must be to develop acceptable support systems for these controls. Each object within the fundamental layer should be given due consideration to ensure that resources exist to address vital issues.

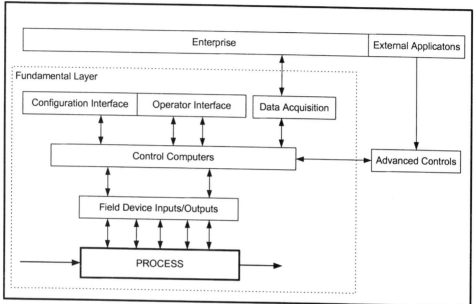

Figure 1: Control system Model

Process

A thorough understanding of the process itself needs to exist, before any of the upper levels can be effective. Expertise in the minerals processing field needs to be apparent in developing the control strategies because process controls will only achieve the level of stability allowed by the process. Understanding these process limitations and dynamics is necessary in order to define adequate performance and identify potential gains. Control engineers need to work in conjunction with operators and process engineers to ensure that control solutions can be achieved.

A firm understanding of the process is necessary for the proper development of the higher control layer.

Field I/O

The devices within the electrical and instrument (E&I) disciplines connect the process to the control layer. Obviously the system will be severely handicapped without accurate inputs or proper manipulation of outputs. Qualified E&I engineers are key to the overall project success as they provide expertise in the selection, design, configuration, and implementation of field gear.

This E&I group ensures that field devices are placed correctly to provide adequate access, allowing for uncomplicated maintenance and calibration. In addition, the group also serves to verify that sound practices are employed during the design and installation activities associated with field cabling. Conformance to high quality standards for cable routing, cabinet wiring and terminations will have a lasting effect on ownership of the system. People tend to react very differently when faced with a cabinet that has been laid out meticulously versus one where the wires are simply "flung in" and long enough to get the door closed. Internal cabinet appearance is a solid indicator of the overall system health.

During the design phase, I/O assignments should contain sufficient spare capacity for future additions. After installation, these spares can be readily managed through a system that will maintain the cabinet integrity.

Because of the added complexity of sampling, on-line analyzers (e.g., density, size, assay, and moisture) present a particular challenge. Provisions for sampling are occasionally overlooked when locating these instruments in the design stage, but they are critical for proper calibration. Despite the expense, these instruments will only perform as well as their calibration allows. A group effort consisting of participants from instrumentation, process engineering, and the laboratory is necessary to correctly design sampling procedures that will minimize errors during calibration. Systems personnel are also needed if these instruments include computer interfaces, which should be incorporated with the larger control system maintenance activities.

The importance of proper and up-to-date documentation is critical and the installation project must conclude with the hand-over of a proper set of "as-built" drawings. The lack of proper documentation will eventually lead to an increase in downtime and can become very costly. This documentation, as shown in the example in Figure 2, represents the final milestone marking the completion of a successful design.

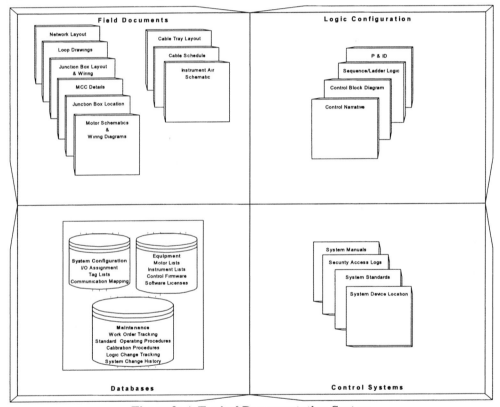

Figure 2: A Typical Documentation System

Logic

Basic standards included in the logic component are centered on documentation and control templates, using a combination of configuration application(s) and supporting databases. Control engineers must balance their desire to create "art" against maintaining consistency that ensures the fundamental layer is sustainable. The addition of any control algorithm should be justified through an improvement in process stability that warrants the added complexity. This is not to say that elegant and sophisticated control strategies aren't necessary, but rather, that their application should be tempered with the equally valuable application of the "KISS" (Keep It Simple) principle. The less time spent on understanding inconsistent logic arrangements, the more time available for optimizing performance.

Sustainability of the control logic requires proper documentation. Simple feedback controllers are easy to comprehend with loop drawings and block sheets. A control narrative that clearly explains the objectives and inner workings of the strategy, however, should accompany complex configurations like cascade loops, adaptive controls, and extensive switching arrangements.

Interface

The interface level should focus on meeting the needs of those individuals who have to continually interact with it. Operators will have habit patterns where controls are expected to move in certain ways and the thoughtful application of properly designed templates that provide consistency are key. Standards should cover symbols, object sizes, text, color schemes, animation, measuring units, etc. "Poorly conceived standards have sometimes been established casually, or without meaningful consideration of operator needs. Without attention to these needs, the operator's performance may suffer" (Considine 1993).

Data Acquisition

The *data acquisition* object in Figure 1 has been included in the fundamental control layer to cover the basic need of operational reporting. After the significant investment in a control system, the operation should not continue to rely on manual data entry for production reports. A validation method needs to be in place before automated data collection can be used, however and it is a good idea to separate "raw" from "official" data. The development of these databases is typically done with MIS support so operating statistics can be integrated with other departments.

Even after the introduction of a control system, many managers prefer to have control room operators continue with manual logs as a means of verifying that key performance indicators get checked and mentally noted with a deterministic frequency.

Communication Networks

Essential information can flow over various assorted networks. These networks must meet the highest standards of reliability and availability through redundant, fault tolerant network design. Correct system's support will result in compliant communication paths that are unobstructed and reliable.

On contemporary networks, constrained bandwidth may be a thing of the past, but most legacy networks are pushing the limits. Standards at this layer are aimed at preserving signal transmission times within acceptable limits and maintaining network availability. On the soft side, this is accomplished through configuration of proper phasing, update rates, and grouping of memory segments. On the hard side, this is achieved through sound management practices including the separation of critical and non-critical communications.

In addition to the absolute requirement for healthy control communications, data flow is at the heart of the system's capability for expandability. "The need to transfer data to other systems is simply recognition that no one package can do all things equally well. In many applications, it makes sense to pass a processing task on to an expert. The ability of a package to exchange data is the main feature of so called open systems."(Levine 1996).

Miscellaneous Design Considerations

Remote monitoring provisions, such as X-terminals, greatly enhance troubleshooting and reduce downtime by allowing technicians fast access to systems while off-site. Also, managers and supervisors that are "plugged in" have a much easier time directing resources to cover situations and keep abreast of operational conditions. There may be added security concerns with remote access, but sound security management policies and practices should overcome these. Adding the capability for remote access is worthwhile investment that can pay for itself many times over.

A tagging convention must be established during the design phase. This convention should have both the flexibility to grow beyond the current scale, and the ability to uniquely identify

process and system components. A tagging convention will typically cover the following components:
- Sub-plants and process circuits
- Equipment
- Control system clusters or nodes
- Control processors and gateways
- Instruments
- Input/Output points

Not only do tagging conventions assign a unique identifier to process components, but they also establish a common grouping system which enables support personnel to quickly pinpoint the location and type of a given point, device, or instrument.

A successful tagging system will require little or no modification when crossing system and platform boundaries. For example, the same tag name would be assigned to an I/O point in the field, the control system, the advanced control system, the historian, the data acquisition and reporting system, and finally on the engineering drawings.

A proper nomenclature that is established during the early stages of a plant's life will allow great expansion flexibility and reinforce operating standards. Without this common convention, troubleshooting and design efforts become extremely complex.

Overall Design Team Considerations
As noted earlier, the design team is multi-disciplined and consists of in house representatives from: management, operations, control and process engineering, computer and business systems, instrumentation, and electrical disciplines. External resources such as vendors, consultants and control engineering firms may also be on the team, depending on the size of the project and the availability of in house resources. Either way, it is imperative that the team has well-established standards to follow.

Often mining companies will not have the necessary resources to undertake such a project and they need to seek outside engineering assistance. If standards for each element do not exist or cannot be given to the engineering firm, then ensure that they are at least reviewed by the design team prior to the development phase. It is an awful feeling when you discover that the entire set of loop drawings have been created using a loop numbering system that you don't agree with!

If the standards aren't reviewed, you are at the mercy of the group that has been employed to design the system. This can either be very profitable or a very costly endeavor: very profitable if the standards acquired are well thought out; very costly if the standards are ill conceived. The qualifications, track record and experience of the engineering firm along with a review of previous installations and their standards used should suffice to establish whether or not their capabilities meet your needs.

SYSTEM SELECTION

Without question, one of the single largest decisions impacting the control system's future is the initial system selection. This single decision will influence (or dictate) every aspect of system development for the remainder of the mining project's life. Once the system goes on line and is functional at some regulatory level, it is next to impossible to justify ever replacing it. If the system chosen does not continuously evolve over time through the concerted efforts of the supplier, vendor, and end user, the entire enterprise to some degree will suffer. This decision should not be made lightly or without some appreciation for the ever-evolving field of information and computer technology.

When selecting a control system for the first time, you will be confronted with numerous opinions on what and whose system is best for your application. If you are on an expansion or retrofit project you are likely already locked into a supplier, which on one hand, is beneficial because the effort involved in evaluating competing supplier's product lines is daunting. On the other hand, it can be detrimental because the slate is not clean and you will inherit any past sins (if that is the case).

Define Requirements

The selection process should start with clearly identifying the mine's initial and future requirements. That is, select a system that is appropriate for the purpose sought. Requirements can be as little as a few simple regulatory loops with minimal discrete controls to as large as a multiple network system with hundreds of loops spanning many miles. It is also critical to look beyond the initial needs of the department and include any corporate enterprise goals. Corporate expectations will most certainly include some form of reporting of operational results, but can expand much further.

Options

The choices on platforms include what is collectively referred to as "industrial systems" containing any one or combinations of PCs, PLCs, and DCS architectures. Market demands and technological advancements have and continue to blur any distinction between these architectures. For small to mid-range systems the largest remaining difference lies in their historical roots and product development paths. The DCS to a large extent continues to dominate analog controls while PLCs are the standard for discrete controls. PLC manufacturers were enabled with the introduction of PC based SCADA (Supervisory Control and Data Acquisition) software, which allowed them to provide very cost effective means of supplying operator interfaces and data collection. These solutions were in turn made possible through the introduction of stable operating systems such as Windows NT and its derivatives.

PCs are simply more microprocessors that can and are being employed at different layers in the architecture. The need for durability at the field and control layers, however, often prohibits the use of PCs in the field where industrially hardened equipment is a prerequisite. These are typically only available from the PLC and DCS vendors. PC based systems combined with solid field interfaces can offer an acceptable alternative, however. The advent of field bus standards and the increasing popularity of Ethernet will continue to drive these market changes.

Open Architecture

To some extent, DCS architectures are still vertically integrated (one vendor supplies every layer from field I/O through to application solutions). This has been a necessary path of evolution because dependable, reliable methods of integrating the various elements were, in the past, non-existent. These proprietary, closed architecture systems, however, are at the end of their life cycle as users are demanding the application of open standards. This transformation to horizontal structures is a welcome change for the industry. DCS vendors are supplying scalable packages for both low and high end users, while the PLC manufacturers are improving their product lines in the analog control field. Gates (1999) describes the benefits of this evolution best: "Horizontal integration makes for high volume and low price. The independence of each layer means that competition drives each layer to evolve at maximum speed... This delayering will increase competition and customer choice".

Part of the benefits that accrue with the horizontal integration are the realization of "best-of-breed" applications from vendors who specialize in specific target areas. Data acquisition, alarm managers, loop performance managers are a few of these. Advanced Control Applications are another.

Horizontal Integration and Communication Issues

As open standards improve and become more common, the tasks associated with the maintenance field should become less onerous. This is particularly true for disparate systems that have independent control configuration data structures, such as the PLC and DCS. Communications between these systems are commonly funneled through serial links with limited capacity. Constraints here dictate the need to have ultra-clean communications because they are typically transferring control signals. These communications are most efficiently dealt with using contiguous memory (a grouped block of memory registers) areas. It is both annoying and dangerous when an operator or control sequence requests a motor to shutdown or startup, and the command is missed because of overloaded communication I/O buffers.

Integrated configuration software that can span the boundaries of these disparate systems will be welcome indeed. Currently offline relational databases that document the memory ranges for communications are the only means of supporting the signals. Consider the transmission path of a single field input in the hypothetical example in Figure 3.

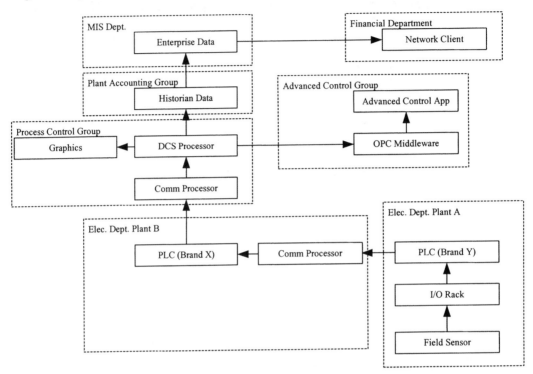

Figure 3: Hypothetical Communications Signaling

The number of "hops, skips, and jumps" associated with a single input extends across multiple systems as well as multiple departments. Each of these systems has a configuration database that needs to be updated if the source is altered. The maintenance chore associated with modifying a point is huge when system boundaries are crossed. Imagine being the electrician on a routine job of reassigning an input point from reading kWh to Amps. You'd have to phone the electrical department in plant B, the control engineer, the system engineer maintaining the historian, the MIS technician, and the accountant using the signal! In addition, the variable being tracked will be out of commission for some time while everyone realigns their readings. Hopefully as the industry becomes more horizontally integrated through open standards, other standards such as self-defining-data-formats (e.g., XML) will at the same time enable the integration of their configuration structures.

System Selection Summary
In summary, regardless of the control system's size, choose a system from a reputable supplier with a well-established track record for support and development, while insisting on some degree of openness. Computer based control systems continue to evolve rapidly and for most applications any one or combination of the major system vendors can meet your requirements. It is key to identify both your current and future needs and then find a vendor that can meet them.

MAINTENANCE

"Industrial automation is becoming increasingly interdisciplinary in nature. Indeed it is difficult to define what really is the hardcore of instrumentation and control technology." (Considine 1993). This quote best illustrates the gray area between the root disciplines responsible for support and maintenance, and clearly accentuates the need to have demarcated lines of responsibility.

Solutions for managing the control system must be designed and implemented to fit the various elements together in an efficient system. Table 1 presents a method of systematically categorizing development and support activities for the control system fundamental elements. Basic issues need to be addressed for each of the activities as they apply to each element. For instance, what resources are required, who will be responsible, where will the resources be located, and when will the activity be carried out?

Table 1: Simplified System Architecture

Control Element				Develop			Maintain	
				Design	Build	Document	Monitor	Repair
1	Information Presentation							
1.1		Graphical Display Management						
1.1.01			Operator Displays					
1.1.02			Trends					
1.2		Process Alarm & Annunciator Management						
1.3		Access Privileges & Security						
2	Control Logic							
2.1		Continuous Analog						
2.2		Sequential Discrete						
2.3		Advanced						
3	Data Acquisition							
3.1		Instruments						
3.1.01			Analog Sensors & Final Control Elements					
3.1.02			Discrete Sensors & Final Control Elements					
3.1.02			Analyzers					
3.2		Historian						
3.3		Network Communication						
3.3.01			Field					
3.3.02			Peer to Peer					
3.3.02			Enterprise					

The following lists are by no means exhaustive. They are only intended to provide examples of how to cover basic issues for an element's associated activities.

"What" hardware and software has been chosen to realize each of these elements? Each element may be viewed as having a hardware component and/or a software application that is used for the various activities of: designing, configuring/building, documenting, monitoring, and repairing/calibrating.

The size and nature of the hardware inventory depends on the location of the plant and on the availability of local vendors and suppliers. Industrial plants that operate in a remote location for example would prudently retain an adequate supply of devices and accessories that are required to maintain normal daily operation. On the other hand, the size of the inventory can be reduced to critical devices only if the plant is located nearby a reliable supplier.

It is however advisable to always retain all critical items in stock. Critical items are those that have a direct impact on daily production and the lack of which would translate into a definite loss of production.

Two maintenance tools that provide solid value and are worthwhile investments are alarm managers and loop performance managers. An alarm management system goes beyond the basic alarming packages that are provided by control system vendors. Generally replacing control room alarm printers, alarm managers historize all discrete alarm events on a separate computer and make them readily available for analysis. Several prepackaged tools then work collectively to provide the control engineer with the ability to analyze process alarms and system events. Searches and queries, frequency analysis, data manipulation, and even expert downtime-cause analysis through the study of preconfigured patterns are some of capabilities alarm management systems have. Automated reports from these systems aid daily maintenance functions by highlighting the more frequent alarms and distinguishing between "true" and "false" alarms. Alarm managers are also indispensable for alarm rationalization and analyzing discrete sequences during a shutdown. Support crews can retrieve an accurate picture of the events that took place prior to the shutdown and therefore better identify the contributing factors. With the proper configuration, alarm managers can provide some predictive insight on the state of control system devices. Unhealthy devices can be isolated and the problem corrected at an early stage thus preventing a total device failure and the associated loss of production.

Loop Performance managers, on the other hand, use data from the data acquisition system to monitor up to hundreds of control loops and detect deteriorating performance. Since controllers are naturally error tolerant, this deterioration can extend over a long time without detection. Automatically tracking the performance of controllers is a significant aide for prioritizing tuning activities. It is also extremely useful for developing and testing alternate control strategies.

"**Who**" or which department is responsible for carrying out each of the activities such as configuring and documenting access privileges? Both the software applications and hardware components used to maintain system security and access privileges require assignment of responsibility. Individuals will have to ensure that the security system is working, maintained, and upgraded as needed.

If not done during the design stage, begin demarcating areas of responsibility by element and activity. This will ensure that scheduled maintenance practices exist for all areas. Identifying who will conduct this maintenance will in turn determine the skills and training needs.

SKILL	Joe	Harry	Frank	Sue	James
Display Configuration	X			X	X
Limit Switch Repair		X	X		
Controller Performance Check			X		X
Extend DH+ Network	X			X	
Install Fiber Optic Panel		X			X
Configure historian Point			X	X	
Change Access Passwords	X				X
Tune Dahlin Algorithm				X	
Calibrate OSA		X	X		
Reboot Control Processor	X	X	X	X	

Figure 4: Sample Skills Matrix

Adequate on-site skill inventory levels will depend on staff turn-over rates, the time required to develop the skill level, and the availability of skilled resources. An up-to-date skills matrix (as shown in Figure 4) is helpful to ensure adequate resources exist and to identify training requirements for succession planning. Critical skills require some form of redundancy and can be used to establish on-call lists and vacation scheduling constraints. The system cannot solely rely on any one individual.

For highly specialized skills, cross training between departments may provide an attractive option. When deciding upon structural redundancies it is best to look at the areas most at risk. An unplanned plant outage is absolutely unacceptable and cross training, or some other form of redundant back up plan is necessary in the areas of communications, system diagnostics and hardware failures.

For system support, suppliers can play a pivotal role that becomes practically compulsory at some larger system size. Formalized vendor support contracts are a common means of providing additional plant coverage. These agreements typically cover technical and logistic areas that cannot be effectively addressed by the plant's fulltime staff. This perhaps would cover the "monitor" activity for the "control logic hardware" and "peer to peer network communication" elements identified in Table 1. A typical maintenance and support agreement offers repair services on hardware devices, dial-in services for system and process diagnostics, software version upgrades, expert advice concerning technical matters, system health checks and performance measurement.

"When" or what frequency will performance monitoring (e.g., loop performance, alarm analysis, etc.) and system checks (e.g. backups) be done? Each piece of hardware and software will require inspection or evaluation at some pre-determined interval. Establishing these frequencies is based on the component's rate of failure or rate of degradation. These basic maintenance functions could then be incorporated into a computerized maintenance planning application if one exists.

"Where" will the resources be stationed for the repair or maintenance activities such as the sensor calibration area, PLC configuration terminals, etc.?

The configuration work area should be separated from the operations' area. That is, it is a good idea to have a separate configuration room where a documentation library can be established and configuration work can be carried out without distracting operations. Ideally, the locations should be close together to enable natural, easy interactions between the two groups, but they are separate functions and they need their individual space.

During an emergency situation, reliable documentation at your fingertips is indispensable. As with an airplane analogy, 99.9% of the time is carefree travel followed by a few seconds of shear terror. During these periods it is absolutely essential that you have your diagnostic information available and ready to isolate the problem.

CONCLUSIONS

Successful control systems are marked by the application of sound standards and practices for managing the essential elements of the fundamental control layer. These standards are much easier to implement and adopt if they are enforced during the initial system design. Commitment to the standards by the entire team will provide a solid foundation upon which advanced techniques may be leveraged.

The importance of proper and up-to-date documentation cannot be emphasized enough. The lack of proper documentation will undoubtedly lead to an increase in downtime and can become very costly. There are several disciplines that share the responsibility of supporting a control system. Without an adequate record of system and process modifications, normal support functions develop into an overwhelming task.

Information technology is transforming the control industry on an almost daily basis. It is essential to select a system that will meet the current needs of the operation as well as evolve over time to meet future requirements. Control systems that are continuously cultivated and developed will provide a high degree of performance over their life.

There is undeniably a broad range of experience and skill required to fully develop and support a contemporary control system. The contents of this paper merely skim the surface of these topic areas. Hopefully though, the reader is left with a greater appreciation for the diverse nature of control systems in the minerals industry. If inclined to do so, readers are strongly encouraged to read the materials listed in the reference list.

ACKNOWLEDGEMENTS
The authors wish to thank the management of PT Freeport Indonesia for first, insisting on nothing less than the application of world-class standards during each of the several mill and control system expansions over the past decade and second, for the opportunity to write this paper. The entire control team at Freeport is also acknowledged for their effort and dedication to developing and maintaining a high level of standards.

REFERENCES
Considine, D.M. 1993. *Process/Industrial Instruments & Controls Handbook, 4th Edition*, McGraw-Hill Inc.

Flintoff, B.C., Mular, A.L., 1992. *A Practical Guide to Process Controls in the Minerals Industry*, Gastown Printers Ltd.

Gates W.M., 1999. *Business at the Speed of Thought.* Warner Books, USA, pg. 420.

Levine, 1996. *The Control Handbook*, CRC Press, pg. 435

Mark Brewer, Kristine Chin, 1999. "Keep Advanced Control Systems Online", Chemical Engineering, August.

Morin, M.A., 2002. "The Case for Open Data Format Standards", CIM Bulletin Vol. 95, January.

Neale, A.J., Veloo, C., 1997. "Process Control at P.T. Freeport Indonesia's Milling Operations", Society of Mining Engineers, Littleton CO.

The Selection of Control Hardware for Mineral Processing

Robert A. Medower[1] and Robert E. Cook[2]

ABSTRACT

The selection of a cost effective control system that will meet current and future process requirements is a challenge, especially in this era of rapidly changing technology. Our objective is to provide the reader with information that will assist with the "control hardware selection" decision making process. We also include a section on "The future of process control technology" based upon current development trends.

INTRODUCTION

Minerals' processing sites are historically long lived, 20 year life of mine is not uncommon and 75 years is not unheard of. Considering that hardware technology advancements are introduced every six months, or less, it is prudent to select process automation systems that adhere to "open" industrial standards over proprietary based systems. "Open" systems are generally regarded as having the ability to embrace any new or existing technology that is perceived to have a positive impact on profitability, with minimal cost.

The selected process automation system architecture must be able to support a wide variety of communication protocols, hard wired and wireless, to bring data into the system, convert the data to information and supply that information to business and regulatory agency information systems.

Mineral deposits require substantial amounts of energy to transport and process into useable form and present a real processing challenge in that the deposits are never homogeneous and ore body compositional characteristics are all over the map! Extracting the desired mineral(s) from the deposit at a cost that will produce sufficient profit margin, despite severe fluctuations in market price, requires process automation architecture capable of supporting an integrated enterprise operation. Such an operation encompasses the activities of individuals, process units, process areas, plant level activities, mine management and business planning and accounting systems.

In the minerals processing industry, the key to survival is driving the cost of production per unit volume to the lowest possible level and hope that the market price will provide enough profit to sustain operations. Business information technology is rapidly becoming an important tool in the quest for productivity and achieving product quality targets. Getting key dynamic business performance information at the right time to the right person or process unit is crucial to successfully achieving corporate goals. The process automation architecture selected must therefore be able to support multiple network communication protocols.

---

1 Industry Consultant, Invensys Process Systems Inc., Eden Prairie, MN
2 Industry Marketing Director, Invensys Process Systems Inc., Foxboro, MA

A BRIEF INTRODUCTION TO PROCESS CONTROL

Some of the readers of this document may not be familiar with process control, so we present the following figures (1 and 2) as a brief introduction to the subject.

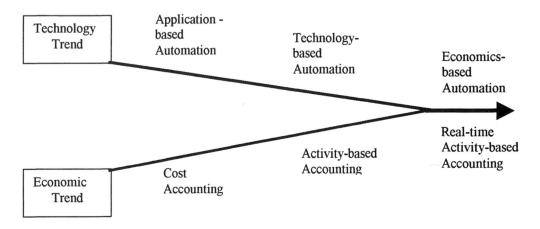

Figure 1 Automation/economic focus trends

Until the mid 1980's, minerals processing companies employed large numbers of technicians and engineers to proactively maintain and improve operations to achieve production and quality goals. The focus was on increasing production and improving product quality to meet expanding market demands. Since then, advances in technology have improved productivity, enabling companies to substantially reduce the number of people required to achieve production and quality targets. These companies no longer have sufficient personnel to proactively devise control strategies and applications to optimize the use of assets. Increasingly, minerals processing companies are relying more heavily on their vendors to provide the expertise, products and services required to remain competitive in their market place.

From the 1940's through the 1980's, purchasing decisions for process measurement devices, actuators and controllers was strongly influenced by technicians and engineers. Selection of measurement devices and controllers was heavily influenced by the "latest technology" forcing vendors to embrace and develop new technologies. The control vendors pitched leading edge technology, accuracy and reliability in their efforts to win projects.

The low cost of microprocessors has resulted in accuracy and reliability of field devices to improve dramatically. The focus has shifted to the amount of process and self diagnostic information a field device can provide over a variety of communication protocols. In the near future, self validation capability will also be an important factor.

Economics has been primarily a "transactional" effort. Quarterly results were compiled from data derived manually. These results were used to devise strategies for improving performance in the next quarter. With the introduction of Local and Wide Area Communications Networks and advances in business software, companies began to move from simple cost accounting, towards activity based costing. In the mid 1990's, technological advancements have permitted automation and economics to merge, allowing Real-time Activity-based Accounting.

Figure 2 is intended to provide a review of control product development evolution from the 1950's through 2003.

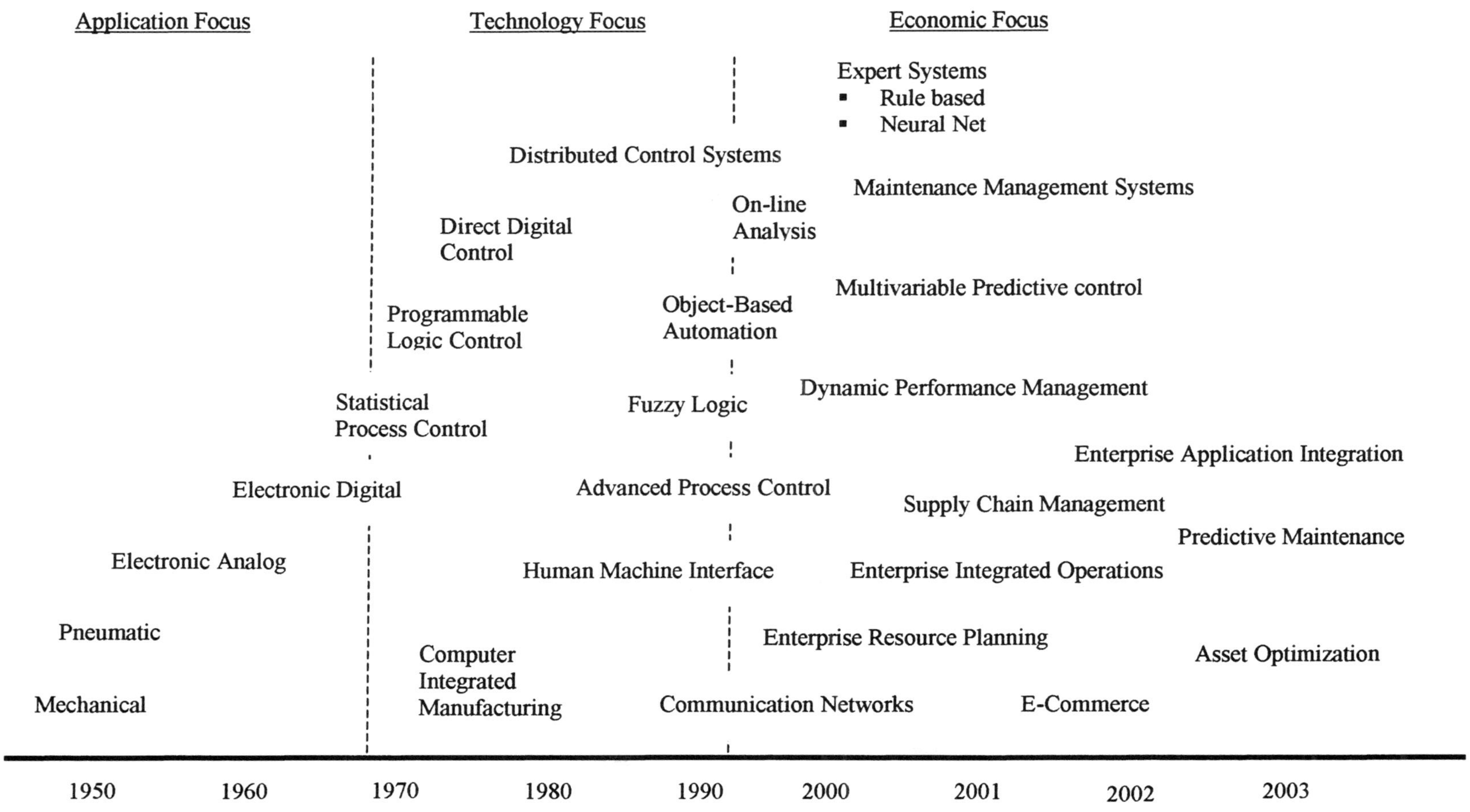

Application Focus | Technology Focus | Economic Focus

		Expert Systems
		■ Rule based
		■ Neural Net
		Maintenance Management Systems
		Multivariable Predictive control
		Dynamic Performance Management
		Enterprise Application Integration
		Predictive Maintenance
		Asset Optimization
	Distributed Control Systems	
	On-line Analysis	
	Object-Based Automation	
	Fuzzy Logic	
	Advanced Process Control	
	Human Machine Interface	
	Supply Chain Management	
	Enterprise Integrated Operations	
	Enterprise Resource Planning	
	Communication Networks	
	E-Commerce	
	Direct Digital Control	
	Programmable Logic Control	
Statistical Process Control		
Electronic Digital		
Computer Integrated Manufacturing		
Electronic Analog		
Pneumatic		
Mechanical		

1950　1960　1970　1980　1990　2000　2001　2002　2003

Figure 2 Evolution of Process Automation

The justification for investing in process automation has changed dramatically. Economics and production have always been the underlying factors, however, until computer technology became powerful enough to meet processing requirements and cost effective enough to implement in field devices, users relied on human labor to meet their objectives. Current and future control system selections will be more heavily based on economics and "computer" labor.

Selecting the right control hardware for minerals processing projects should be based on an anticipated return on investment. Typically, the reason proposed for investment in control technology is to improve productivity, quality and/or meet regulatory requirements at the lowest possible cost. Two relatively new justifications include asset management and predictive maintenance.

ASSET MANAGEMENT AND PREDICTIVE MAINTENANCE

In the year 2002 and beyond, predictive maintenance and asset management programs will have an impact on the level of integration between machine protection and advanced process control strategies. Machine manufacturers will imbed microprocessors into their products that will communicate via an information network. The available information will be accessible for use in advanced control strategies, asset management, predictive maintenance and by the manufacturers from remote locations for fault diagnosis and performance monitoring.

Asset management is increasingly important to the survivability of minerals processing companies. In the 1920's through 1970's high grade natural resources were in abundance, energy prices were controlled, environmental regulations were weak, manual labor was the primary vehicle to achieve productivity, wages were relatively low and the market could absorb all that could be produced. Since then, available ore grades are lower, product quality specifications are becoming more stringent, energy prices are being de-regulated, environmental regulations are increasingly restrictive, productivity per work hour is an important financial indicator, hourly wages in developed countries have increased substantially, and market demand is volatile.

Minerals processing companies are realizing that the path to profitability lies in controlling costs by minimizing the impact on the environment, reducing "head count," eliminating waste and increasing efficiency throughout the production and distribution process.

Process automation systems must have the ability to integrate rapidly changing technology, be a vehicle for real time bi-directional information between the process and the enterprise, run highly sophisticated process applications and become more predictive in controlling the process and maintaining equipment.

EMERGING TECHNOLOGIES

There are many emerging technologies in 2002 that will have a positive impact on process automation, asset management and the availability and accuracy of real time information. MEMS, SEVA™ and .NET are three of those technologies.

MEMS. This technology will increasingly be used to produce sensors to be embedded in all types of process equipment to provide real time status information. The automotive industry and machinery manufacturers are already employing MEMS for this purpose.

MicroElectroMechanical systems are microscopic electrical/mechanical devices with three-dimensional moving parts acting as sensors and actuators. They are relatively inexpensive, robust and easily manufactured using current integrated circuit manufacturing technology. They are widely used in a number of industries but most pertinent to minerals processing is their use in ore haul vehicles.

To see the "future" (beyond year 2000) one only needs to look at the level of sophistication applied to mine haul trucks in the late 1990's. On-board receivers collect data inputs from microprocessors embedded in tires, engines and the truck frame and send that information via wireless communications to operations management. In addition, mine dispatch systems, using information from GPS units installed on trucks, drills and trains, efficiently monitor vehicle location and permit dispatchers to control the ore characteristics presented to the primary crusher. The benefits include improved safety, increased vehicle availability, improved maintenance scheduling and lower operating costs. Technological advances will allow other process machinery such as crushers and grinding mills to become more intelligent.

SEVA™ A self validation technology that will improve process data quality by expanding the scope of valid data available from process sensors. This, in turn, will allow the recipient of the data to make better decisions regarding control of the process.

SEVA™, self validating intelligent sensors (the term "sensor" in this case, includes the transducer and transmitter as a unit) have the ability to not only digitally transmit information about the health of the "sensor," as intelligent transmitters do, but process information as well. The benefits include higher process uptime, higher quality products, lower maintenance and operating costs and improved safety.

.NET. One of the "Instant Messaging" products introduced to the market in 2002 as a "Web services" application. It provides instant, validated, information transfer between microprocessor based systems with very little human intervention.

Two "Instant messaging" applications pertinent to mineral processing are alarm handling and predictive maintenance.
- The control system host would be able to communicate alarm information to selected wireless handheld displays.
- Process units would be able to provide a "health" report to a maintenance host computer including a list of current faults as well as wear parts that have neared their service life.

THE SELECTION PROCESS

How do we begin the control hardware selection process? The first steps include generation of a bid document by; determining the scope of the project, the affect of the addition on up and down stream process units, the impact on staffing, division of labor responsibilities, return on investment and how to measure the anticipated results. Accountability and "head count" are increasingly important factors in the process automation selection process.

The next step is to determine who will choose the control system. Typically, it is the project manager or purchasing agent whose primary incentive is to be under budget. Alternatively, an engineering/design firm is hired and given authorization to make the decision. The engineering/design firm's primary incentive is to supply the minimum technology required to meet the project objectives within budget and schedule constraints. Neither of these scenarios will necessarily be in the best interest of the owner, from either a short or long term profitability point of view. The budget is the overriding factor and is usually established by the owner without the influence of a qualified control engineer who could provide input on the best level of control to implement to obtain the fastest return on investment.

It is interesting to note that the cost of an installed process control system, as a percentage of the total project, is typically the smallest number but is the first to be negatively affected by a budget shortfall. This happens, in spite of the fact that the control system has substantial influence on short and long term profitability.

The Investment Structure

Figure 3 illustrates a typical process automation investment structure. Beginning with the "Process" as the foundation, each subsequent level builds on the one below it. Without a firm foundation, advanced and Expert control applications are of very limited benefit and are typically detrimental to profitability.

The first level above the Process employs intelligent field devices to provide accurate process information and process variable manipulation. The second level, basic regulatory control, provides process stabilization. The third level utilizes intelligent process analyzers to provide quality control. The fourth level utilizes advanced regulatory, multivariable predictive and Expert control software strategies to provide process and profit optimization. The fifth level provides an integrated, bi-directional information path between the process and appropriate enterprise individuals for timely decision making.

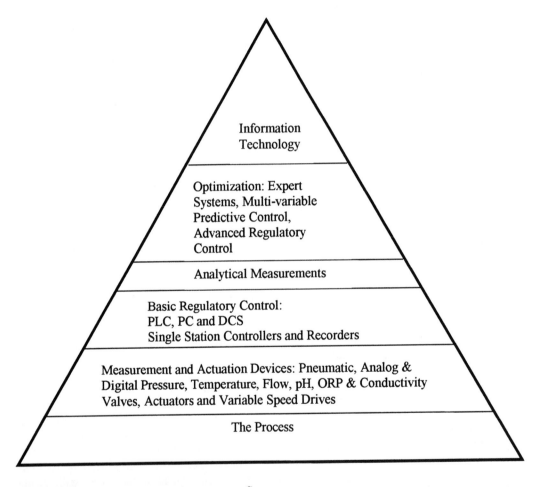

Figure 3 Process Automation Investment Structure

The Impact of Variability

Processing minerals is difficult due to the wide variability in ore body characteristics. Compositional analysis of the ore body is used to design the physical plant. To optimize mine life, the physical plant design is a compromise that requires blending the various ore grades to fit within the constraints of available process machinery. Mine life can be shortened if the easier to process high grade ore is presented to the primary crusher without blending the lower grade ores. If the mine operator presents a large volume of low grade ore to the primary crusher, mine life may increase, however, the cost per ton of product substantially increases.

The compromises stated above, plus severe environmental conditions associated with mineral processing, present a challenge to process units, automation strategies, process measurements and actuation devices. Technological advances in products utilized by the process control industry are providing solutions to those problems.

Measurement and Actuation Devices

Process measurement and actuation devices are the "eyes and hands" of the control system. In the 1920's and 1930's the "eyes and hands" were close coupled with the "controller" by necessity. These devices were based primarily on mechanical technology, i.e. filled systems, levers, bellows, etc. Control was accomplished by manual adjustment. The full range of the physical process sensor was constrained by mechanical and physical limitations. The accuracy of the process information available from these sensors was dependent upon the transducer technology (pneumatic, dc current or digital), process measurement type (pressure, temperature or flow), desired operating range, repeatability, turndown ratio and process constraints. Turndown is the range over which the process measurement will be acceptably accurate. For example, a standard d/p cell connected to an orifice plate has a flow measurement turndown ratio of 3:1. Rangeability (the range over which the instrument meets the stated linearity of uncertainty requirements) and uncertainty (the range of values within which the true value lies) become much more important when self evaluating technology is employed.

Sensor rangeability has, in the past, been substantially greater than the transducer attached to it. Transducer rangeability used to be fixed at a 5:1 ratio, i.e. 3 to 15 psi or 4 to 20 milliamps, and had to be bench calibrated to the desired operating range. Intelligent transmitters speak digitally (i.e. 32 bit floating point), can easily match the range of the sensor to which they are attached and can be remotely re-ranged to accommodate process changes without negatively affecting measurement accuracy.

The majority of field measurement and actuation devices purchased through the 1990's were of the 5:1 rangeability type. Intelligent devices have gained popularity because they can be easily re-ranged from remote terminals without negatively affecting the accuracy of the information conveyed, can communicate self diagnosed faults to process control systems utilizing communication bus technology and substantially reduce installation/startup costs. In the future, sensor based microprocessors will contain control algorithms permitting implementation of local control strategies.

The same types of sensors used to monitor and control the process are used for process unit protection. Each process unit has physical constraints such as; bearing temperatures, lubrication, vibration, speed, power, etc., which must be monitored and controlled. Minerals processing units are large pieces of machinery that are costly to purchase, operate and maintain. Typically, the machine vendor supplies or specifies the protective devices required along with instructions on how the unit is to be protected. The owner's personnel or a systems integrator programs the protective supervisory system based on manufacturer's recommendations. In the future, the machine vendor will program intelligent sensors with embedded microprocessors with a preprogrammed protection strategy for the specific process unit.

Device accuracy (the effects of non-linearity, hysteresis and non-repeatability at reference conditions) is only one of the factors in the level of uncertainty in the measurement. Installation and process variability also have an effect on uncertainty.

Communications Bus Technology

A modern control system may have four distinct communications levels:

- The first level includes communications between process measurement devices, actuators, other devices and the control system Input/Output modules (Sensorbus, Fig. 4).

- The second level includes communications between Input/Output modules, other devices and the control system controllers/integrators/gateways (Controlbus).

- The third level includes communications between the control system controllers and the host processor, operator and engineering workstations (Nodebus).

- The fourth level includes communications between the host processor/workstation processor and enterprise information networks, Local Area Networks, Wide Area Networks, the Internet and other devices (IT bus).

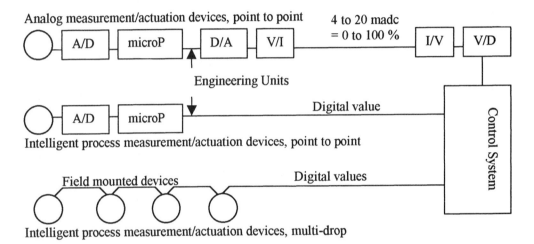

Figure 4 Typical level 1 Sensorbus Communications

Level 1 communications. The minerals processing industry primarily employs single channel, point to point communications between analog/digital process devices and the control system. Most mineral processes are spread over large physical areas with relatively low, mixed function, I/O density at any given location, are subject to frequent electrical disturbances and subject all process and control equipment to harsh environments.

Level 1 is the most difficult level to address, from a communications perspective, because it needs to accommodate; process measurements, process actuators/positioners, process analyzers, bar code readers and on-off devices like motor starters, on-off valves, switches, etc. Life was simpler when the only types of signals to deal with were 4 to 20 madc, standardized voltage values and relay contacts. The only choice was to run twisted pair, shielded copper wire from the control system to each device in the process. The same wires that carried the process signal were also used to power the process measurement/actuation devices. This wiring made control systems susceptible to disruption from electrical disturbances such as lightening, high voltage power cables, large motor operations, hand held communications devices and ground loops.

Fortunately, all vendors of intelligent point to point and multi-drop field devices designed their products to be compatible with existing wiring. In this case, the same twisted pair copper wire carries the power to the transmitter and the digital information provided by the field device(s).

Some important issues to be aware of when specifying point to point or multi-drop technology for projects include:
- How the selected technology powers field devices
- Interoperability
- The number of configuration tools required (if multiple vendors products are selected)
- The security of cables running from field devices to the control system.

Interoperability is the capability to substitute a field device from one manufacturer for that of another manufacturer without loss of functionality. The primary benefit is the freedom to choose the right device for an application, irrespective of the chosen control system.

Level 1 Communication Protocol Comparison

The most popular protocols employed through 2002 were; 4 to 20 madc, HART, FOXCOM, Foundation Fieldbus H1 and H2, and Profibus PA/DP.

4 to 20 madc. This industry standard has been in existence for many years and all major measurement, analyzer, actuation and control system vendors products adhere to this standard providing a high degree of interoperability.

Devices communicate point to point, must be bench calibrated, are individually powered from the control system and provide a continuous, single channel, analog signal.

The maximum distance between devices under ideal conditions, using 16 Ga. shielded cable, is one mile. Realistically, in minerals processing applications, the maximum distance is substantially shorter due to the large installed quantity of high voltage equipment. Signal cabling must be isolated from high voltage cabling to minimize electrical disturbance. This requirement increases the installation costs substantially.

HART. This protocol was introduced in 1989 and has gained wide acceptance in the minerals processing industry. Currently (2002) there are approximately 150 vendors supplying products that adhere to this protocol. It uses analog 4 to 20 madc for the measurement signal on which a digital frequency is superimposed to carry information and status reporting data.

Hart was developed as a transmitter protocol rather than a control system protocol, therefore, a stand alone HART M275 intelligent field device configurator is required. Digital data is transmitted at 1200 bps. Update rate is dependent upon the device and whether the signals are multiplexed through a single modem on a multiple channel I/O module or if each channel has it's own modem.

Interoperability is very good, as all vendors submit their products to HART for testing and certification but there is no guarantee that all of the features of the standard are available in each vendor's product. Every device implements Universal Commands (read output, tag number, etc.) and the Common practice Commands (rerange, change tag name, etc.) but not all of the configurable parameters in the standard are necessarily available to the chosen control system.

All diagnostics in the HART protocol are "device specific" commands (i.e. Command #48). Therefore, when you integrate a Foxboro or Rosemount HART device to a Honeywell control system, for example, the system would know that there was a problem due to the diagnostic message, but it may not be capable of interpreting exactly what it is.

Devices communicate point to point, can be remotely reranged, are individually powered from the control system and provide a continuous, single channel, analog signal. Point to point supports up to three devices (2 masters + 1 slave). Maximum distance is 1524 m. Wiring from these devices must be isolated from high voltage cables to minimize electrical disturbance. Installation costs are as high as for 4 to 20 madc, however, startup and maintenance costs savings can be substantial.

HART protocol supports multi-drop (2 masters + 15 slaves) but most major control system vendors do not support it.

FOXCOM. Foxboro introduced this protocol in 1988 for use by Foxboro manufactured intelligent field devices. Foxboro intelligent devices can be configured for full digital (4800 bps) with 10 times per second digital updates, or for analog 4 to 20 madc, with digital updates (600 bps) 2 times per second. When field devices are configured for analog mode, all information, including the measurement, is still digitally transmitted to the Foxboro control system. The digitally transmitted information is not included in the loop loading calculation, so the analog output may be wired to other devices in the control loop.

Devices communicate point to point bi-directionally, can be remotely configured, modified, maintained and operated from any system console, have real time diagnostics and are individually powered from the control system. Point to point shielded, twisted pair supports up to eight devices per I/O module in a star configuration. Maximum distance is 610 m at 4800 bps, 1829 m at 600 bps. Wiring from these devices must be isolated from high voltage cables to minimize electrical disturbance. Installation costs are as high as for 4 to 20 madc, however, startup and maintenance costs savings can be substantial.

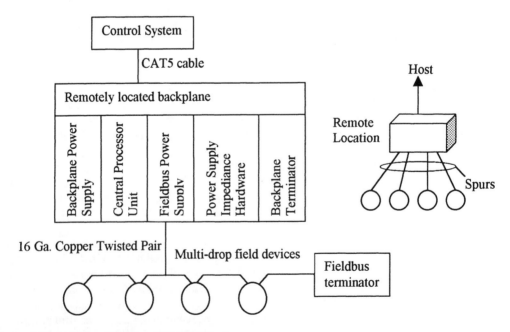

Figure 5 Foundation Fieldbus H1 Architecture

Foundation Fieldbus H1 and H2. Fieldbus H1 and H2 are all-digital protocols (including signal and data transmissions). H1 was introduced in 1995 and H2 (Hi Speed Ethernet) was introduced in 2000. FF H1 transmits data at 31.25 Kbps over shielded twisted pair or fiber optic cables, up to 2000 m with type A cable, including spurs. Figure 5 shows a typical FF H1 architecture.

Approximately eight devices share power transmitted on the twisted pair bus. The power supplies can be embedded in the I/O subsystem or external. Bus topology is multi-drop with long spurs (see figure 5).

FF H2 (HSE) operates at 100 Mbps, at this speed cabling media becomes a critical factor and maximum distances for twisted pair CAT5 cable will be short, 100 m max., and fiber optic 2000 m max. Devices must be externally powered, as fiber optic cable is incapable of carrying electrical energy.

Profibus PA/DP. Profibus PA/DP is an all digital protocol developed by the German Government and Siemens as the primary proponent. DP was introduced in 1994 and PA in 1995. Profibus PA is an extension of the remote I/O capability of Profibus DP. Over 300 vendors supply products available with profibus protocol as of 2002.

Either twisted pair or Ethernet can be used to transmit signal and data between the I/O and control system. Maximum cable length for DP is dependent upon the speed of transmission; 100m/segment at 12 Mps to 1,200 m/segment at slower transmission speeds. Maximum cable length for PA is 1,900 m/segment. The signal/data transmission cable powers devices.

Profibus DP transmits data by rotating a set of values called tokens around all of the connected devices on the network, which works well with the scan cycle concept of PLC programming. A problem arises when the amount of data increases to the point where the bus becomes overloaded.

When designing a Profibus network, keep in mind that there are data transmission speed and cable length limitations to be considered.

Level 2 communications. At the controlbus level are primarily between a variety of multiple vendors' control and analytical products connected to a control system. Most mineral processes employ hybrid systems that are a mix of PLC's, Analyzers and other devices that connect to the installed process automation system through a variety of protocols. Popular protocols include Modbus, Data Highway, Profibus DP, and Ethernet.

Modbus (introduced in 1978) and Modbus+ are proprietary protocols designed for Modicon PLCs. Modicon published the simple, straight forward, easy to use protocol standard and offered it free of charge to anyone who wanted to use it. It became very popular with a wide variety of vendors because almost any device with a microprocessor and serial port could use it. The positive is that almost every vendor supports this protocol. The negative is that data transmission is slow and non-deterministic. The more deterministic a network is, the easier it is to predict its accuracy and the more repeatable its performance will be in actual operation. This speaks to a protocols ability to efficiently send event or time based data between a device and a controller.

Data Highway and Data Highway + are proprietary protocols developed for Allen Bradley (AB) PLCs. AB has worked with a number of DCS control system vendors to develop gateways, co-processing modules and integrators to create bi-directional communications between AB PLCs and multiple vendors DCS systems.

Industrial Ethernet is gaining favor as the communications network of the future. Ethernet became the prominent protocol for information technology networks and its use is expanding to include levels 1, 2 and 3 communications. It is unlikely that Ethernet will be the "sole survivor" in the area of process control networks for future minerals processing projects. A combination of wireless, Foundation Fieldbus and Ethernet will be the best choice for future (2002 to 2007) new projects or large control system replacement projects.

Levels 3 and 4 communications. Ethernet was introduced in the early 1970's as an information tool for office and business applications. It has the ability to move large blocks of data at high speed, peer to peer between workstations, host computers and PCs.

Standards for file transport, e-mail and hypertext exist for Internet applications but Ethernet application standards for automation are not as readily available. These issues are being addressed through development of standards by a number of organizations and as of 2002 there are three products that are potential survivors; EtherNet/IP, Fieldbus Foundation Hi Speed Ethernet (HSE) and Interface for Distributed Automation (IDA). All three are based on transmission control protocol/internet protocol (TCP/IP). Other Ethernet based products vying for market share include Interbus, Profinet and Modbus TCP/IP.

A typical Ethernet network architecture for process automation (see figure 6) integrates control stations and workstations with a variety of Ethernet compatible control products. As with any technology, there are limitations and as the volume and speed of transmission increase, collision avoidance becomes increasingly important.

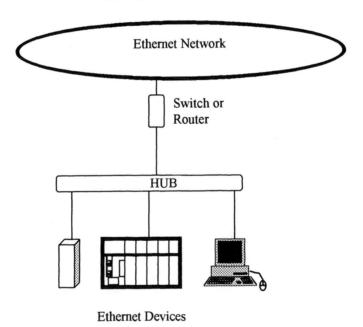

Figure 6 Typical Ethernet Architecture

When planning an Ethernet network, first determine the number of stations to be interconnected and their physical locations. This information is required for step two, selecting the proper indoor and outdoor cabling for the network. CAT5 shielded twisted pair (10/100 Base-T) cabling for a maximum distance of 100 m or Multimode fiber optic cable for a maximum distance of 2 km. The distance between the farthest stations on the network cannot exceed 4.2 km. The maximum number of attached devices is dependent on the amount of data throughput required.

Mineral processing site communication and control system networks require a large number of connections and must process a large amount of data. Hubs are devices used to provide network layout flexibility and switches prevent network traffic on one segment from impacting performance on another segment.

A key feature of an Ethernet Network is that all the devices in the network should be able to communicate with one another. To enable this feature, the "Three Hop Rule" must be adhered to. A path between any two devices in the network should have no more than three switches en route. This rule is best achieved by employing a trunk/star topology, the recommended topology for mineral processing plant bus networks.

Cable Selection

Selection of fiber optic cable, wherever possible, is highly recommended for all levels of communication in a mineral processing plant. Fiber optic technology is capable of sending and receiving information at high speed over great distances (up to 150 km without using a repeater) using light as the data carrier. The signal is not disrupted by outside sources like electricity, rain, humidity or radio transmissions. They are ideal for transmitting information because they are highly secure (do not induce or emit any external energy), transmit data at high speed and signal loss can be detected almost immediately if monitored.

There are two types of optical fiber cables, singlemode and multimode. Singlemode is used for data transmissions between 8 and 150 km. Lasers are used as the light source because they produce focused, parallel light to limit losses. Multimode (recommended for industrial applications) is used for data transmissions between 8 to 10 km. Light emitting diodes are used as the light source for these applications.

The following fiber optic cable characteristics are recommended for harsh environments: multimode, graded-index fibers with 62.5 micron core/125 micron cladding and maximum signal losses of 1dB/km at a wavelength of 1,300 nm and 3.5 dB/km at a wavelength of 850 NM. Two fibers are required for each communications path, one to transmit and the other to receive.

Fiber optic systems are more cost effective than copper wire, they are lighter, less costly to maintain and do not require repeaters for distances up to 150 km. Fiber optic cables require more expertise to install, however, the benefits substantially outweigh the added one time cost.

For the next few years (beyond 2002) all new mineral processing control system installations will utilize copper wire and emerging wireless technology in combination with fiber optic cables. Copper wire, along with emerging wireless technology, will be used to connect remotely mounted I/O modules to locally mounted field devices and fiber optic technology will be the norm for longer communication runs and to interconnect automation subsystems.

An important issue that will impact the use of fiber optic technology at level 1 is powering the field devices. In most minerals processing plants, power is readily available in all process areas so cost to power the field devices from sources in close proximity should be minimal, however, most modern transmitters are two wire devices designed to accept twisted pair copper only. Twisted pair copper carries the signal as well as supplying power to the device. If the field device were designed with separate signal and power connections, an alternative would be to combine twisted pair copper wire and fiber optics in a single multiconductor cable to address the loop power issue. In the future, wireless technology may eliminate the need for level 1 twisted pair wire in some applications.

Figure 7 is an example of one possible, cost effective, process automation installation architecture for most minerals processing plants. In this case, a relatively short level 1 shielded copper twisted pair (sensorbus) is used to connect intelligent field devices to a remotely mounted I/O rack. A redundant level 2 multi-conductor fiber optic cable (controlbus) is used to connect the I/O rack to the control system.

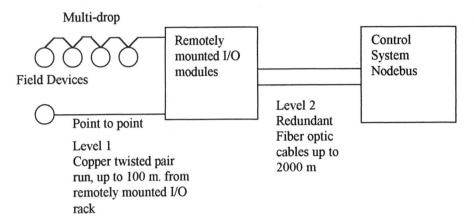

Figure 7 Level 1/Level 2 Communications Bus Architecture Example

Basic Regulatory Automation System Selection

The type of automation system selected for a given project is dependent upon many factors and there is no single vendor that manufactures all of the devices required to properly control a mineral process. In addition, mineral processes are considered to be large, from a control point of view, due to the large number of physical I/O points installed with a high percentage of complex analog loops as well as the large processing area footprint. These criteria in 2002 still point toward selecting a DCS (distributed control system) as the primary control system.

The process of extraction and concentration can present complex problems. The desired mineral is embedded in the earth's crust and must be extracted and concentrated before it can be brought to the market place. Sometimes there are multiple minerals that need to be extracted and separated which adds to the complexity. Some portions of the liberation process require advanced control and/or Expert systems. This also, in 2002, points towards selecting a DCS system as the primary process automation system.

Control SystemTypes

Overview. There are three types of control systems competing for mineral processing projects, DCS, PLC and Hybrid [Personal Computer (PC) based control systems]. PLC's were introduced in 1969 and DCS systems in the 1980's and Hybrid systems began to gain popularity in 2000 in limited applications.

Figure 8 illustrates the market sectors, as of 2002, in which the three types of control systems are positioned. Small to midsize hybrid systems are gaining market share, as all industries gravitate from hardware to software based control applications.

Both DCS and PLC systems have incorporated PC's for specific functions to increase functionality and reduced costs. DCS systems utilize PC's with Windows operating systems as Operator's workstations and to connect to information networks. PLC's use PC's as application workstation's, Human Machine Interfaces (HMI's) and for application configuration/maintenance stations.

Hybrid and DCS vendors have developed reusable, prepackaged, industry specific applications to minimize engineering costs. Vendors of both system types have developed extensive libraries of process control applications that can be quickly integrated into large projects or provided with specific processing machinery to permit "plug and play." In either case, the engineering time gap between programming PLC's vs. Hybrid and DCS systems is growing larger.

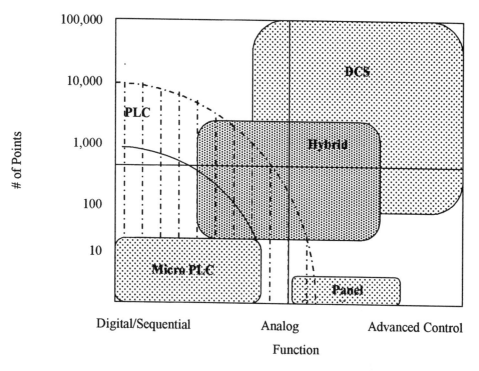

Figure 8 Market Sectors by Control System Type as of 2002

PLC's. PLC's, in this comparison, are stand alone PLC's that do not incorporate a personal computer as a host. They were designed to replace relays, timers, switches and hard wired control panels. They rapidly gained favor with the automotive industry. The automotive industry is now (year 2000) moving away from hardware based systems to software based systems.

Architecturally (see figure 9), all PLC's have the same basic components; Input/Output modules, a Central Processing Unit (CPU), solid state memory and a power supply. A programming device with a hard drive that is loaded with proprietary configuration software provided by the manufacturer of the PLC or by a third party vendor. PLC's are programmed using ladder logic (see figure 10), structured text (Boolean logic), ladder logic with advanced function blocks or sequential function chart (SFC).

Note that each input and output is tagged with an identifier, i.e. I0001and Q0001 for digital inputs and outputs, AI000X and AQ00X for analog inputs and outputs, etc. Other programs that require bit status or data must reference the tag identifier to obtain the information. Process variable scaling, i.e. analog tank level, requires tag identifier register manipulation.

Programming efforts can become lengthy if the applications are complex or require a large number of ladder rungs. The number of rungs required to implement an application strategy also affects control response time, troubleshooting time and startup time. PLC application programming efforts typically require 30% or more engineering time than programming identical applications on Hybrid or DCS control systems.

PLC's process inputs, logic and outputs sequentially in a "loop" or scan cycle. It runs through the loop as fast as it can and response time is determined by the worst case loop time. External events, such as interrupt driven schedulers, cannot interrupt the scan.

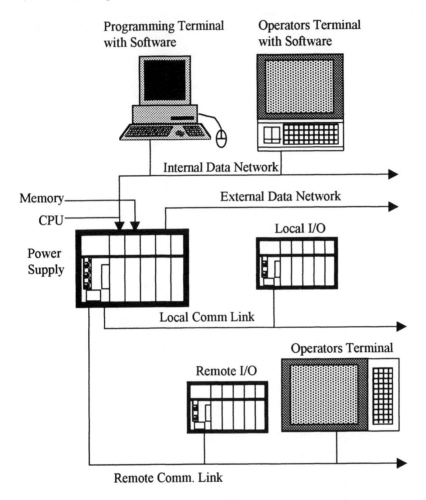

Figure 9 Typical PLC Architecture

The I/O modules provide a physical connection to process equipment. Inputs include analog and digital signals from pushbuttons, switches, relays, etc. Outputs include digital signals to motor starter relays, solenoid valves, etc. and analog signals to valves, variable speed drives, etc.

The CPU consists of a variety of microprocessors that are programmed to perform logic and memory functions described by the application program. Program and data files reside in the CPU memory. Program files store the control application, subroutine and error files. Data files store data received from the I/O modules, status bits, counter and timer presets, accumulated values and other stored constants or variables for use by the program files. Memory size is specified in kilobytes (1 KB = 1,024 words) of storage space. PLC memory capacity ranges between less than 1 KB to 64 KB. The programming device normally is connected to the system only for programming, start-up or troubleshooting purposes.

The power supply converts either 240 vac or 120 vac to +5, -15 and +15 vdc for use by the rack components. All I/O devices are externally powered.

Modern PLC systems can do almost anything that a DCS system can do but, intelligent devices, data management, fieldbus applications and integration with other vendor's products are still better handled by Hybrid/DCS systems. Hybrid systems are defined as personal computers connected to a variety of available I/O substructures, including PLC's.

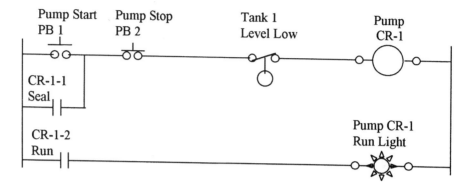

Figure 10 Ladder Logic Diagram Example

DCS. DCS systems were originally designed to handle large, complex control applications. Initially, the chemical, oil and gas industries benefited most from this technology. These systems become more price competitive as project scope increases and system component prices decrease.

Architecturally (see figure 11), DCS systems are software oriented and the hardware, including Application Workstations, Operator Workstations, Control Processors, Gateways, Device Integrators and Intelligent devices are designed to support the system's software. Operating software can be UNIX or Windows and both can connect to a common communications bus.

System resources can be expanded to whatever is required for the total project. By design, power supply capacity and mounting structure capacity increase as input and output devices are added.

The communication node is the backbone of a DCS system. Multiple nodes can be interconnected to provide a plantwide control network. For example, the primary crusher, concentrator, smelter, waste treatment, acid plant control systems and any other process area can be interconnected by fiber optic network cables.

Each node can handle a large number of workstations and network communications to corporate information systems. The number of devices connected to the node is scaled to project requirements and additional devices can be easily added, incrementally, at any time, to accommodate requirement changes.

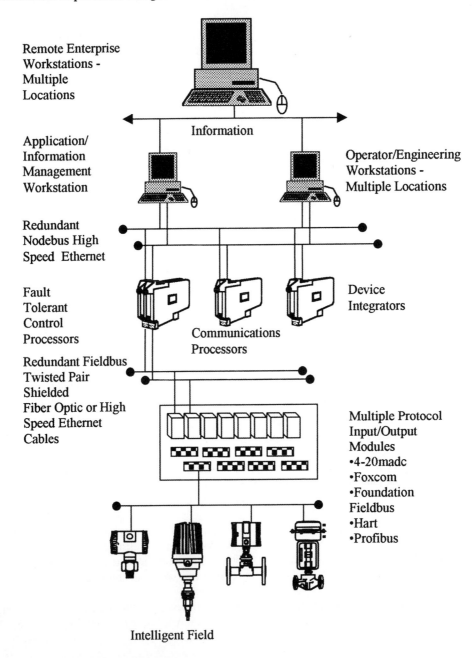

Figure 11 Typical DCS Architecture

Cost effectively scaling down to handle smaller projects, especially those that require a large number of discrete and small numbers of analog I/O points, can be a challenge. The "overhead" of software licensing costs makes it difficult for DCS systems to compete with hardware based systems. Decreasing hardware costs, prepackaged application software, interoperability with third party devices, the growing importance of information technology, 24 hour, 7 days per week support availability and the end user's desire to reduce operating costs are off setting system selection cost factors.

Software solutions are flexible and well suited to solving mineral processing asset management problems, see figure 12. DCS systems use an object oriented global database, permitting any system or networked hardware device to attach to the bus and have full access to all process data and control applications without having to preprogram the device.

The system configuration software is always available and accessible from any local or remote workstation. Access to the configuration software is controlled by password to prevent unauthorized personnel from modifying system or application programs.

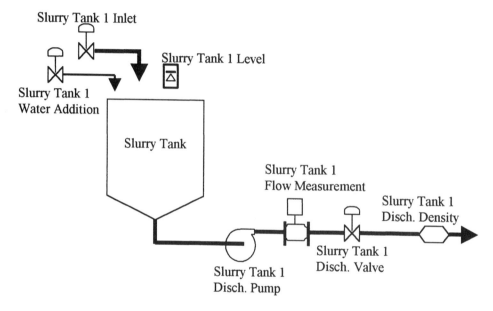

Figure 12 Typical Slurry Tank Control Application

Note that each process device in figure 12 is identified by name, as an object. Any program that requires status or data need only state the object name and parameter desired. For example, if slurry tank density data is required by operating personnel or by another software program, the human or software program requestor need only specify the object, Slurry_Tank_1: Disch_Density.Meas, and the system will locate that information and transport it to the appropriate location.

The above application is a complex control problem because the objective is to keep the slurry tank from either overflowing or running empty while maintaining the proper density of the pumped slurry. The process variables include tank level, water addition, pump speed and density. The solution to this control problem requires a control strategy that consists of multiple cascade control loops plus a calculator block, all standard control blocks within the standard DCS software. See figure 13 for "Control Block" illustration.

Control blocks were developed in the 1960's to make it easier for control engineers to use computers for process control application development. All the traditional functions, analog in, analog out, digital in, digital out, calculation, Boolean, sequencing etc. were programmed as tools called "blocks." Developing control applications was simplified because these blocks did not have to be programmed each time prior to use; they only had to be drawn from a library of preprogrammed blocks and configured to suit the application.

The control block structure developers realized that programming complex control problems on computers could be intimidating. The solution to the programming dilemma was to develop tools that allowed the programmers to "configure" complex applications by grouping a number of control blocks into an entity, giving the entity a unique name and storing it in the library for future use.

Each component in the control application is identified as a unique object that would be entered only once into the system database and reused as often as needed in multiple control strategies. This approach was named "Object Oriented Programming."

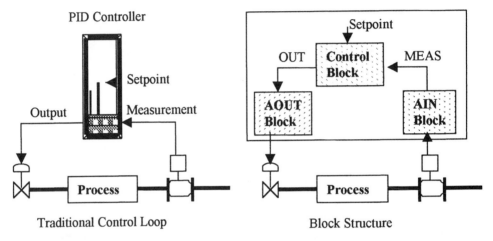

Figure 13 Control Block Structure

DCS systems are developed, sold and serviced by the manufacturer. The manufacturer may also offer project engineering, training and installation services. These systems are sometimes referred to as "Integrated Systems" because the DCS vendor is the single responsible source and is the primary provider of short and long term system and/or application support.

Hybrid. Hybrid PC based systems are relatively new competitors in the process control industry. PC's have been on the market for a long time, but inadequate software, poor reliability, interoperability issues and slow processing speeds kept them from being seriously considered for process control. All of these issues, for the most part, have been addressed but development is still a work in progress.

Hybrid systems are normally assembled and programmed by contractors or System Integrators. The architecture (figure 14) consists of a Personal Computer with either Windows or Linux operating systems, an HMI (Wonderware and Intellution are popular HMI vendors in 2002) and an I/O substructure (either generic or PLC) with power supplies. These systems are comprised of multiple vendors' products and an issue yet to be addressed, is future interoperability as each product evolves separately. In addition, responsibility for long term support of the hardware and application engineering transfer to the end user once the control system contractual objectives are met.

A primary reason that Hybrid systems are attractively priced is that all of the components, including software, are priced "ala Carte." The vendor's strategy is to provide the user with the specific functionality required for each plant area. To this end, the software licenses are offered "unbundled" and the user must decide what role each workstation will play so that the appropriate licenses are purchased. These licenses are usually defined as "Systems" or "Suites" and are based on functionality, i.e. Development, Run Time, Viewer, Information, etc.

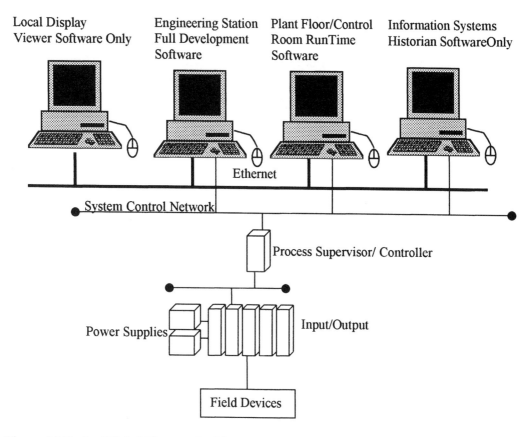

Figure 14 Typical Hybrid System Architecture

License Types

Development. Normally consists of a PC with the full complement of software installed to do system development plus a license for the number of "tags" or points anticipated to be connected to the system. The price point for each license is usually based on a specific number of points; 500, 1000, 2,500, 5,000, etc. A historian can be optionally furnished with this suite.

Run Time. Includes all of the software required to run the required applications but not development software. A Run Time server gets its data through its own database over the control network or over Ethernet from the development station. All process alarm detection and acknowledgment is done at the workstation level, and not at the function block level as is done in a DCS.

Link. The user configures the tags that each station uses for communications. "Link" software allows these stations to supply tag data to other computers running the selected HMI software, however, computers not loaded with this software will not see the tag data. In an unsecured mode, reliable data transfer rates of approximately 5000 points per second are possible. In secured mode (extensive communication handshaking plus error checking to verify the transaction took place and the correct data was transferred) the reliable data transfer rate drops to approximately 500 points per second.

Operations Viewer. Does not sit on the control system network and is used for viewing operations only. It consists of a PC with preloaded software for viewing operations in plant process areas or offices. The exchange of data is established using Ethernet to a Run Time or Development station.

Information Manager. Consists of a PC connected to the network loaded with software required to create the point or tag database. The license is priced based on the number of points, i.e. 500, 2,500, 5,000, 25,000, etc. This server is a dedicated historian connected to the control network or through Ethernet to a Run Time and/or Development station.

The above suites are also offered as software packages only, permitting the Systems Integrator to purchase PCs separately.

The contractor or Systems Integrator is responsible for the packaging, configuration and application work as well as the design and purchasing of the appropriate cabling and communication devices, such as Ethernet switches, Hubs, etc. to permit communications between stations.

DCS vendors also provide hybrid systems for entry level projects. They package these systems using unbundled elements of their DCS software and furnish low cost I/O substructures of their own, or third party manufacture to compete in this market sector. The I/O substructures employed are frequently purchased from PLC vendors.

The primary advantage of a DCS vendor's hybrid system is single source system responsibility, guaranteed interoperability between all hardware and software components, long term application support and 24 hour, 7days/week service support.

The control domain (Basic regulatory control)

From 1908 to the mid 1980s the control domain shifted from the process unit to the control room. As microprocessors migrated to field devices, the trend is driving the control domain back to the process unit. In 2000 the first field devices, supported by Fieldbus Foundation technology, were offered with embedded control blocks. This technological trend means that the need for traditional Input/Output module subsystems will decline as field device microprocessor power increases.

Although we have focused on process measurement and actuation devices, mineral processing requires a large number of motors. As motor control system manufacturers move towards embedding powerful microprocessors into their products the need for relays, timers and switches will also diminish.

As the control domain shifts to embedded systems, computers and intelligent field devices, traditional I/O subsystems will be bypassed. Hybrid and DCS systems provide more computing power, storage capacity, programming flexibility and are in better position to deal with direct connections to fieldbus and information technologies due to their software orientation.

"Plug and play" is a popular theme that can apply to process units when the control domain for a given process unit is integral with the unit. For example, a crusher or grinding mill could be furnished with the protective measurement/actuation devices and controller pre-installed and configured. In place of wiring, wireless temperature and pressure devices could communicate with a local control processor that was able to communicate with the "global" community via high speed information networks. Up and down stream process units would also communicate with the network forming a process control area.

When selecting process control systems for mineral processing it is important to consider system life cycle costs. Will the system selected have to be totally replaced in five years or less? Or is it flexible enough to accommodate technological changes needed for the enterprise to remain profitably in business for many years.

The control domain (Analytical Devices, Advanced and Expert systems)

We were once advised by a contractor's project manager that mineral processing is not "rocket science" so the simpler the control strategy the better. The statement about rocket science may be accurate, however, reality does not support the "simpler the better" portion of the statement. Every mineral processing company we've visited has either implemented, or is planning to implement, advanced regulatory control and/or Expert systems to improve operating efficiency. Grinding and flotation are two primary beneficiaries of advanced and Expert control strategies.

There are three levels of control strategy that are employed above basic regulatory control; Advanced Regulatory, Multivariable Predictive and Expert systems. Expert systems are Rule Based, use Neural Net technology or a combination of the two. These systems require robust computing power and input from process analytical devices to be optimally effective. The control system selected must have the capacity to incorporate or integrate with these tools.

Analytical devices are steadily improving in their ability to provide accurate results. The primary nemesis is the sampling system. As the analyzers move from the lab to "in-line," the sampling system problems will hopefully diminish.

Another problem with analyzers is the time lag between acquiring data and transmitting results. This issue has been addressed at one mine site by developing and implementing a Predictive Algorithm that provides a predicted value every minute vs. the normal analyzer cycle of every 15 to 20 minutes. Process performance improved measurably after implementation of the predictive algorithm.

Analytical devices have been developed for each process area. Optical analyzers have been developed for crusher and grinding mill feed analysis. Crossbelt analyzers are available for compositional analysis of conveyed ore. X-ray analyzers for elemental composition and fine particle size analyzers for mill discharge are in wide spread use. The control system selected must be able to cost effectively interoperate with these devices.

Advanced regulatory control's primary contribution to operating efficiency is reduction of deviation around desired process operating setpoints. The primary tools used at this level are; feed-forward, cascade, ratio, lead/lag, etc.

Multi-variable Predictive control is a matrix based, modeling, predictive optimizing process controller. Primary tools include; Adaptive Controller, Fuzzy Logic, Constraint Controller, Optimizer, Multi-model Director and Neural Net.

Rule based Expert systems are based on process operating knowledge. A set of rules is developed to emulate the "best" operator.

A Neural Net Expert system is software that has the capacity to learn how to react to process deviations from setpoints and provide guidance to the process controller to prevent or minimize upsets.

Evolution of the "Control Room"

Until recently, control rooms were located in each process area to minimize process automation installation costs and because of technological constraints. With the introduction of remotely mounted industrial I/O subsystems, remotely controlled video systems, inexpensive wireless technology and fiber optic cables, the industry is moving toward centrally located, process wide, operations centers.

It is interesting to note that the control industry began selling "distributed control" products in 1908. Pneumatic technology of the 1940's required that monitoring, control and actuation devices be in close proximity to the process unit over which they had influence. Advances in pneumatic technology in the 1950's permitted process controllers to be consolidated on control panels in close proximity to a group of process units. When electronic single station controllers, recorders and indicators were introduced in the 1960's control panels were consolidated into large control rooms so fewer operators could take responsibility for a larger portion of the process. The introduction of computers in the 1970's substantially reduced the need for control panels.

Until recently (year 2000) the focus has always been on controlling the process in real time based on transactional guidance from corporate business management. With the introduction of high speed communications systems and sophisticated business and asset management software, what was once "transactional" is rapidly becoming "real-time." The primary drivers for real-time process business management include; fluctuations in market demand, increased operating/production costs and market demand for higher quality, custom products.

The question is; is there still a need for "control rooms?" Many mineral processing companies have already consolidated their crusher and concentrator control rooms into one location. The traditional primary function of the "control room" is expanding to include real-time process management so perhaps the "control room" is dead! Long live the "Process Management Operations Center!"

The process management automation center is primarily a hub that links key process information to the corporate information network. This center does not necessarily need to be located at the site. In the late 1990's, while attending a National Crushed Stone Association conference in California, we witnessed a demonstration of remotely stopping and starting a stone crushing circuit located in a quarry on the East Coast. A video system scanned the process area prior to stopping/re-starting the circuit and the video signal was transmitted to California so we could observe the process. Neither a human nor the proverbial dog to keep the human from touching anything was on site. A laptop computer was used to remotely control the process equipment via the Internet. We were advised that this company operated the facility "lights out" from 4:30 PM to 7:30 AM each day. Safeguards are in place to allow the process to shut itself down if a fault occurs. Maintenance crews arrive in the morning to address any problems that occurred over night and to maintain the process equipment while operating crews replenish the stockpile.

The rapid advance of technology will have a major impact on how, and from what location(s) processes are operated. Gordon Moore, physical chemist, Co-founder of Intel and author of Moore's Law: Computer processing power will double approximately every 18 months. In 2002 a personal computer with a 2+ Gigabit Pentium processor is available on the open market for approximately $1,000. In October of 2001, Intel announced a new optical networking subsystem designed to deliver 10 Gigabit Ethernet and the worlds first complete CMOS physical medium dependent chip set for 10 Gigabit per second applications. This level of computing power will permit the operation of mineral processes from virtually any location. The primary process automation system selected must be able to rapidly handle high volumes of information, be able to be programmed using object names accessible by any workstation, easily interface with multiple vendors products and connect with a wide variety of communications networks.

THE FUTURE OF CONTROL TECHNOLOGY

Throughout this document we have strived to include some idea of the future as it pertains to each element of a process control system. Perhaps a review of the mineral process from mining to shipping, and what we view the future to be in each area, is the most efficient way to summarize.

Mining. For open pit and underground operations, video systems and robotics will be more heavily utilized in mineral extraction and material handling to enhance safety. The machines employed for extraction and material handling will have embedded intelligence to provide current health and predictive maintenance information. Wireless communications will provide web access so that mine planning, the dispatch system, down stream process control unit and machine manufacturer can obtain remote access. On board analyzers will be able to determine the composition of the ore being transported and provide that input to down stream process units.

Crushing. Video systems will monitor feed size and automatically adjust the gap to obtain the desired discharge size. Vibration analyzers, video systems, temperatures, pressure measurements, lube systems, and drive motor power monitors will wirelessly connect to an on-board microprocessor configured to optimize production and protect the crusher from damage. The microprocessor will also connect to surrounding process equipment via a fiber optic network.

Concentration. Video systems and high speed information networks, in a trunk/star configuration, will interconnect all of the process machinery in the concentrator. Each process unit will have intelligent process measurements, analyzers and actuators wirelessly connected to an embedded microprocessor designed to optimally control the unit, communicate with other process units and permit remote access. The microprocessor will also transmit current health and predictive maintenance information to the process management operations center.

Process Management Operations Center. The host controller in this location will maintain all pertinent process and maintenance data and store an image of the current control strategy for automatic down load to each unit in the event that a local unit microprocessor is replaced. With this technology, process units could be easily added or removed from service. The host system could easily be reconfigured to accommodate process flow modifications and new equipment.

SELECTION GUIDE

For those responsible for establishing the project budget, process automation systems are functionally no longer limited to process control. They are an integral part of an Enterprise wide information network that will have a major impact on short and long term profitability. Whether the project is of limited scope, i.e. addition of a process unit, or a new process plant, it is important to consider the level of investment required to achieve the desired production and quality targets in the shortest possible time.

A few pertinent questions prior to system selection and budgeting decisions.
- What is the scope of work?
- What constitutes a successful implementation and how will it be measured?
- How will up and downstream process units be affected?
- What impact will contractual division of labor have on control system selection?

- What will be the impact on staffing?
- Who will specify, select, purchase, configure and install the system?
- What are their qualifications?
- To what degree are they accountable?
- Will the selection be based on the total installed price of the entire process automation system (best approach)? Or will it based on traditional methods of evaluating field devices, control system and installation as separate entities? The latter choice precludes considering installation cost savings associated with remotely mounted I/O substructures.

We've divided table 1 into two basic function types, PLC and DCS/Hybrid because Hybrid and DCS systems are rapidly merging as microprocessors become more powerful. The actual hardware selected should be based on the most cost effective system available that will meet or exceed project expectations and anticipated return on investment. The "X" indicates the best functional system type to meet the requirements of the stated variable.

Table 1 Functional system type selection guideline (2002)

Variable	DCS/Hybrid	PLC
Application is digital (discrete) I/O intensive, >70%		X
Application is discrete logic intensive (simple sequencing, Boolean, etc.)		X
Desired response speed not faster than 2ms		X
Lowest number of engineering hours to develop applications	X	
Global database	X	
Common system wide database	X	
Interoperability with multiple vendors products	X	
Requirement for advanced and Expert control strategies	X	
Prepackaged applications availability	X	
Information networking requirements	X	
Multi-tasking/ Multi-user requirements	X	
Desired response time is 0.1 ms or less	X	
Sophisticated alarming is required	X	
The process is highly interactive, process unit to process unit	X	
Lowest life-cycle cost	X	
Lowest Installed cost	X	
Adheres to the open industrial standards (OIS) model	X	
Choice of operating systems	X	
Supports WEB browsers and wireless communications		X
Supports multiple fieldbuses and legacy I/O	X	
Accepts any Foundation Fieldbus device without proprietary resource file	X	
Simple graphical configurator	X	
Ability to configure and run PID algorithms in field devices	X	
Ease of "Change Management"	X	
Supports object oriented naming structure	X	
Computer aided design software for configuration and back documentation	X	
Integration if intelligent field devices	X	
Redundant system wide communications	X	
Full asset management, including predictive maintenance	X	
Workstation Configuration and diagnostics for system and field devices	X	
Remote system access and diagnostics from factory support centers	X	
Long term support of legacy products	X	

An additional item for consideration is migration from legacy products to current technology. The life cycle cost of the system will increase substantially if the selected process automation system can not be cost effectively upgraded to current standards.

Table 1 provides some guidance to process automation type selection on a functional basis instead of a hardware basis. Anticipated technological advancements in communications and electronics will eliminate some of the elements currently found in today's (2002) process automation systems. In the near future, hardware based PLC and I/O substructures will be functionally replaced by embedded software.

CONCLUSION

The vision of the future is, for the most part, based on technology all ready available. BMW announced the availability of their top line model with "Drive by Wire" technology. The throttle, brakes and transmission are all controlled by computers and are disconnected mechanically from the vehicle operator. Lexis offers smart cruise control that automatically adjusts speed to accommodate varying traffic conditions. Large aircraft are being controlled from remote locations with precision, using wireless high speed communications. Video systems attached to these aircraft allow the operator to view the aircraft's surroundings in real time.

The control system of the future for minerals processing is a large capacity information system connected to individual process units by high speed fiber optic networks. The optimizing control strategy is embedded in microprocessors located in the intelligent process measurement and/or actuation devices and wirelessly connected to other pertinent devices in the control loop.

Sam Walton, founder of Wal-Mart, understood technology's role in building a company. He realized that technology was not an end in and of itself, but a tool to be applied by people who knew how to use it. He invested heavily in computer technology and hired people who understood that technology was a powerful tool for boosting productivity, understanding customers' needs and allowed employees to perform their jobs better and faster.

His investments in technology and people who knew how to use it, created a company valued at $220 Billion (in 2001) that consistently grew at an annual rate of 14%.

When presented with the opportunity to select a process control system, choose wisely.

ACKNOLEDGEMENTS

We are grateful to Peter Martin, Chief Marketing officer, Invensys Process Systems, Inc. and author of "Dynamic Performance Management" and "Bottom Line Automation," for his input to this document. In addition, we thank Kevin Fitzgerald, Director of Measurement Integration, Invensys Process Systems, Inc. for is input on the direction of communications bus technology and measurement products.

BASIC FIELD INSTRUMENTATION AND CONTROL SYSTEM MAINTENANCE IN MINERAL PROCESSING CIRCUITS

Joseph R. Sienkiewicz, PE[1]

ABSTRACT

The mining and minerals processing industry presents unique challenges in the selection, application and maintenance of process measurement and control devices. Because of the typically harsh environments and difficult process conditions, special consideration must be given to the ruggedness, reliability and maintainability of these devices. This paper will explore various aspects of the device selection process, considering the degree of instrumentation, measurement types and methods, materials, and signal types. Fundamental aspects of instrument maintenance will be covered, such as complexity, accessibility and process equipment provisions. Asset management and preventative maintenance concepts will also be discussed.

ENVIRONMENT

Mining and minerals processing plants are often located in some of the most remote locations on earth. From high rugged mountain areas, to hostile and hot dry deserts, to wet, humid rainforest regions, these plants must operate round-the-clock. Resources such as water and power, as well as essential services like transportation and telecommunications, are usually in short supply and costly to obtain. In order to survive in today's economy and be profitable, these facilities must often be self sufficient and extremely efficient. Personnel must also be minimized, not only for economics, but also because of the difficulty in attracting qualified people to work at these locations. To meet these challenges, modern plants must be highly instrumented so that centralized control and advanced automation capabilities are facilitated. Having a strong, well-organized plant instrumentation maintenance department, with trained and qualified personnel, is essential.

Many mine sites are days away from supporting infrastructure, so extra planning is required to assure adequate availability of technical support and timely access to spare parts for repair. Arranging maintenance contracts with major instrument suppliers or independent representatives of multiple manufacturers is fairly common. These contracts often include guaranteed response times for on site availability of support personnel. Arrangements can also be made for spare parts availability on a consignment basis, either on-site, or in a nearby off-site warehouse, so those spares don't have to be purchased.

Environmental and meteorological conditions are often extreme, and their effects can adversely affect the operation, accuracy and reliability of instruments. A number of mining facilities are located between 10,000 and 14,000 feet altitude. High altitudes, due to the thin atmosphere, affect cooling efficiency. Supplemental cooling may be required in order to keep

---

[1] Kvaerner E&C, San Ramon, California

instruments within their operation ranges. Likewise, extra cooling may be required in hot desert environments for obviously different reasons. Conversely, instruments subject to freezing temperatures must be suitably insulated and heat traced. Instrument cabinets subject to freezing should include a space heater. One condition that is often overlooked is the effect of direct sunlight on instruments and enclosures. Sunlight can overheat exposed equipment in a surprisingly short time, especially if enclosures have dark exteriors and relatively large surface areas. Sun shades or pre-fabricated environmental instrument enclosures are typically used to mitigate this problem. If equipment installed outdoors could be subject to electrical storm (lightening) activity, then those associated externally wired circuits should be provided with lightening and surge protection devices. Instrumentation circuits should also be protected against electrical noise and damage from electrostatic discharge, radio frequency interference, switch contact bounce, and power supply disturbances caused by load switching and lightning. Protection must be provided against windblown dust and rain, periodic washdowns, and also the corrosive effects of chemical pollutants often present at minerals processing facilities. Instruments and enclosures should always be provided with a NEMA 4X enclosure rating, which provides the necessary protection. NEMA 4X may not be required if there is no corrosive atmosphere, in which case, NEMA 4 will suffice.

It should not be assumed that if stainless steel instrument tubing were used, it would withstand a corrosive atmosphere. For instance, if chloride ions are present, stainless steel instrument tubing will quickly pit and leak. Materials must therefore be selected which are suitable for the specific corrosives present. This may require an evaluation through the use of analytical instruments or corrosion coupons.

PLANT CONTROL PHILOSOPHY

The quantity, type, and features of plant instrumentation and control devices depend, to a large extent, on the operating philosophy of the plant, and the degree of automation and advanced control desired. In the past, most minerals processing facility operations were labor intensive. Instrumentation was relatively simple and provided to facilitate decentralized local operating floor oriented control. With the advent of modern computerized Distributed Control Systems (DCS) the focus changed to centralized control from a main control room. Through the computer displays, operators could monitor and control all the equipment in a facility. Additional instrumentation was required to act as the eyes and ears for the operator and to provide data for process analysis.

Fundamental process measurements such as temperature, pressure, flow, level, density, pH, etc., connected to the DCS through electronic signals, have to be both accurate and reliable. Sophisticated computer controlled analytical systems, such as particle size analyzers and multi-stream x-ray analyzer systems, connected to the DCS through data links, provide additional real-time analytical information to the operator, previously available only from laboratory analysis. The promise of advanced supervisory control, employing model based predictive control, expert systems, neural networks, and fuzzy logic techniques, in order to increase efficiency and profitability, often require even more process measurements to be effective.

Instrumentation maintenance support is of paramount importance; its value should not be underestimated. In the world of centralized control, it can easily make the difference between profit and loss if a plant is operating inefficiently due to missing or erroneous process information.

MINERAL PROCESSES

The processes encountered in the minerals industry present very demanding requirements of measurement and control devices. In ore crushing, conveying, stockpiling and storage operations, instruments are exposed to rocks, dust, dirt, water sprays and high vibration. In milling operations, instruments must deal with highly abrasive slurries, which cause severe abrasion and erosion of piping and related components. In flotation circuits, they must endure a variety of specialized chemical reagents and foaming fluids. In smelters, high temperatures, molten metals, corrosive gasses, and sulfuric acid are present. Autoclaving presents the additional challenges of high pressures, temperatures and highly corrosive slurries. In solvent extraction and electrowinning (SXEW) circuits, highly acidic solutions are routinely dealt with. Instruments must be rugged enough to handle all these challenges and continue to operate reliably for a reasonably long period without repair or replacement. They also must be carefully selected for their specific application to properly take the process fluid properties into account and to apply the appropriate scientific measurement method and apply the proper materials of construction.

PROCESS MEASUREMENT DEVICE SELECTION

There are many issues involved in the proper selection and application of process measurement instruments. Some fundamental considerations include: environmental constraints, the nature and properties of the fluid, the condition of the fluid, geometry and orientation of pipes and vessels, installation requirements and restrictions, location and access requirements, accuracy, complexity, routine maintenance requirements, and capital and operating costs. For each type of measurement, there are usually a variety of scientific principles available; the object of the selection process is to choose the one that best matches most of the selection criteria.

The majority of field mounted transmitters are two-wire 4-20 mA types. Certain transmitters, requiring more power than available from a two-wire loop (magmeters, density, sonic level, etc.), have a supplemental power source requirement (usually 120VAC) which also powers the isolated 4-20 mA output for that device. "Smart" 4-20 mA transmitters with HART (Highway Addressable Smart Transmitter) protocol have become a current de-facto standard for most applications. The HART digital data are superimposed onto the usual 4-20 mA signal. They are preferred because they are more accurate, cover a wider measurement range, are easier to calibrate, and provide a wealth of information not available with the analog signal alone. Smart Fieldbus transmitters, which communicate with the plant DCS via a digital communications network, eliminating the analog signals altogether, are becoming more common as manufacturers increasingly offer this technology in their products, and they gain wider acceptance by users.

There is usually a struggle between keeping the initial cost of an instrument within budget for a given purchase, and minimizing the cost to operate and maintain that device. For example, a valve may be selected for an application based on the lowest qualified bid, but it might have to be re-built every three months in operation, while another valve, costing twice as much, may last a year in the same service. Unfortunately, many suppliers are reluctant to offer performance guarantees because of the potential variability of process properties and conditions. Sometimes only experience, and not technical compliance, will determine the better choice.

The following sections are not intended to be a comprehensive tutorial or guideline on the proper selection and sizing of field instrumentation; there are many books and literature that cover this in great detail. Rather, this is a summary of common use and practice for this industry.

Flow

Technologies include electromagnetic, vortex shedding, coriolis mass flow, thermal mass flow, weigh scales, ultrasonic, open channel weir (with level sensing), and differential pressure (orifice plates, flow nozzles, venturi tubes, averaging pitot tubes), vane/target, turbines, positive displacement and variable area rotameters.

Flow measurements in relatively clean liquids are often made using orifice plates. Vortex shedding type of flow instruments are usually used for applications requiring measurements with turndown ratios (The ratio of the maximum measurable flow rate to the minimum measurable flow rate) of more than 10:1. Turbine meters are used where high accuracy is required, however they are usually high maintenance items due to the wear of moving parts.

Thermal mass flow transmitters are often used in column flotation cell sparging systems and other low velocity flow applications for gasses and liquids.

Ultrasonic flow meters, both Doppler and time-of-flight are seldom used in this industry due to their unreliability and inaccuracy in harsh environments. They may be considered for measurement of highly corrosive fluids, since the transducers are mounted outside the pipe.

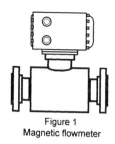

Figure 1
Magnetic flowmeter

Electromagnetic flowmeters are preferred for slurry service and fluids containing suspended particulate like plant process water. They have the advantage of being obstructionless and have a good turndown ratio. They must be installed in piping runs which assure a full pipe (a vertical upflowing section is ideal) and require a conductive fluid and sufficient fluid velocity.

Averaging Pitot tube elements with differential pressure transmitters are normally used for gas flows and offer low differential pressure drops. An air purging system is used where there are entrained solids in the gas, which can cause a plugging of the Pitot tube.

Gravimetric conveyor belt scales with four idlers are usually used to weight solids flow when high accuracy is required for inventory control or custody transfer. Four idler belt scales can provide a minimum of 0.5% accuracy. Less accurate weigh scales with fewer idlers are normally used for relative measurement, for process control of mill feed and material handling systems. The belt scale electronics are normally microprocessor based and belt speed compensated. Nuclear type belt scales are used where physical constraints preclude the use of gravimetric scales. These scales are also microprocessor based and belt speed compensated.

Level

Technologies include bubbler, hydrostatic (pressure), capacitance, ultrasonic, radar, laser, nuclear radiation, time domain reflectometry, buoyancy, float displacement, load cells, and strain gauges.

Level measurements of relatively clean liquids are usually made with pressure transmitters measuring hydrostatic head. Diaphragm seals are used if corrosive liquids or entrained solids are present.

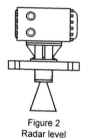

Figure 2
Radar level

Level measurements of liquids with varying densities in vented tanks, including tanks with agitators, are normally made using sonic level sensors and transmitters.

Capacitance level probes are used for level measurement in high temperature and/or pressurized tanks or vessels, or where the process fluid will tend to foam.

Sonic level sensors also are used to monitor storage bin continuous level inventories, and pulp levels (using a float target in a stillwell), and froth height.

Radar level sensors are often used in stillwell applications where high ranges are measured.

Laser level gauges can be used for both liquids and solids. Although not yet widely used by this industry, this technology shows great potential for solving some of the most difficult problems encountered in bin and silo level measurement. These devices typically employ a time-of-flight measurement of near-infrared diode laser pulses. They have the advantage of high accuracy and reliability, good interference immunity, long range, and no beam divergence, which can cause false echoes. They have the ability to measure from oblique angles and are relatively unaffected by temperature, solids coning, acoustically absorbing materials (dust) or low dielectric constants.

Nuclear radiation level transmitters are used for measurement in difficult process and vessel applications like autoclaves, flash vessels and gyratory crusher discharge surge hoppers. The source of radiation is usually Cesium 137 (30 year half-life) or Cobalt 60 (5.3 year half-life), and the detector is either a Geiger-Muller or scintillation type. A scintillation detector is preferred because it is more sensitive, and requires a smaller nuclear source. The owner must obtain a local or national regulatory agency license in order to use these devices on site.

Feed bin loading and loss-in-weight measurement systems are usually monitored by high accuracy load cells or strain gages with transmitters for measurement of varying vessel weight which is directly proportional to inventory level.

Pressure
Measurements include gauge and absolute, differential, draft, and hydrostatic.

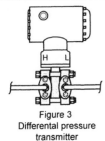
Figure 3
Differental pressure transmitter

Diaphragm seals are used in pressure measurements of slurry lines, corrosive liquids, or entrained solids. Pressure measurement of lines with thick slurries are preferably made by using a rubber-lined, in-line spool pressure sensor.

Differential pressure type transmitters are also used to indicate dirty air filters, including bag house dust collectors, as well as filters in liquid lines.

Temperature
Technologies include thermocouples, RTDs, and infrared optical pyrometers.

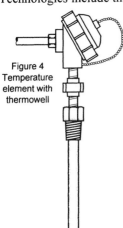
Figure 4
Temperature element with thermowell

Temperature measurements up to 800°F, requiring remote transmission of the signal, are normally made by using a platinum resistance temperature detector, (RTD) calibrated to 100 ohms at 0°C, with a scale factor of 0.00385 ohms/ohm/°C.

Thermocouples with temperature transmitters will be used for temperatures above 800°F. The following thermocouple types are typically used for the maximum operating temperature indicated:

1. Type K - 2400°F
2. Type R - 3000°F
3. Type N - 2400°F for Oxygen rich environment

All temperature sensors on pipes, tanks, and vessels should be installed in suitable thermowells.

Density
Technologies include nuclear radiation, weight, and hydrostatic head (differential pressure).

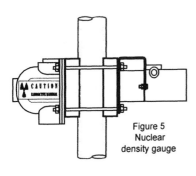
Figure 5
Nuclear density gauge

Nuclear radiation type density gauges are almost always used for density measurement. These devices are clamped to the outside of a pipe and require a full pipe for proper measurement. Like magmeters, installation in a vertical upflowing section is ideal. The source of radiation is usually Cesium 137 (30 year half-life) or Cobalt 60 (5.3 year half-life). The detector is either a Geiger-Muller or scintillation type. A scintillation detector is preferred because it is more sensitive, and requires a smaller nuclear source. The owner must obtain a local or national regulatory agency license in order to use these devices on site.

Analysis
Among the vast array of real-time analytical measurements available, the most common types used by this industry include pH/ORP, conductivity, oxygen content and combustion efficiency, flammable and/or toxic gas and vapor detectors, multi-stream x-ray florescence, pulp particle size, and flue and stack gas emissions analyzers.

Most of these complex devices require constant service, maintenance, and calibration, often in association with laboratory analysis comparisons. It is not unusual for them to require a full time attendant, whose salary must be factored into the overall operating cost for these systems.

PROCESS CONTROL DEVICE SELECTION

Final control devices are very important elements of fluid and material handling systems because they regulate the process flows, in order to maintain throughput and material balances, under constantly changing conditions. Three fundamental methods are used: throttling control valves, variable pulse modulation, and variable speed prime movers (pumps, feeders and conveyers).

Valves

Types include globe, gate, plug, ball, butterfly, angle, choke, dart, diaphragm, pinch, annular orifice, needle, regulator, and relief. Valves work on the principle of dissipating a portion of the process fluid energy. When selecting control valves, a number of factors must be considered, such as required flow capacity, body material, body pressure rating, necessary seat leakage class, body size and style, pipe connection method, trim material, (the internal parts of a valve which are in contact with the flowing process fluid) and desired flow characteristics.

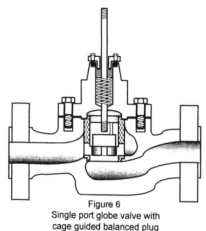

Figure 6
Single port globe valve with cage guided balanced plug

Valves are sized to handle the requirements of minimum and maximum flow rate, inlet and outlet pressure, and fluid properties. Based on the process fluid properties and conditions, a required flow capacity factor, Cv is calculated, which then serves as a basis for valve size and type selection. Control valves are normally selected to absorb 33% of the total system friction head at design flow. This places them at a control point in the range of variability, balancing between pressure drop and subsequent energy loss, and control rangeability. If a valve is too large, it will tend to operate in a narrow range near closed. In almost all cases, a properly sized control valve will be one or two line sizes smaller than the pipe, or have a reduced port.

In selecting the materials of construction, for both the valve's body and trim, special considered must be given to factors which could reduce the useful life. Among those are high fluid differential pressure and velocity, potentially causing cavitation, flashing, pitting, erosion, noise, and vibration. Entrained particles and slurries, as well as corrosive fluids also need special consideration. Expensive special alloys may be required, like titanium or super duplex stainless steel for severe corrosive applications, while harder materials, like ceramics and coatings, might be used for slurries and erosive fluids. Elastomer type valves like diaphragm or pinch valves are typically used for slurry throttling service. All valves for oxygen service exposed to fluid velocities greater than 200 feet per second should be constructed of copper base alloys, preferably Monel. 316 stainless steel can be used where velocities are below 200 feet per second. Instruments for oxygen service should be properly cleaned and prepared prior to installation.

The geometry of the valve trim determines the valve flow characteristic curve. Three primary characteristics are linear, quick opening and equal percentage. The flow through the valve will thus follow a distinctive curve in relation to stem position. The desired valve characteristic usually is determined by examining the overall process gain and response conditions. The most common valve characteristic is equal percentage, which minimizes large flow changes when the valve is near closed, and mitigates

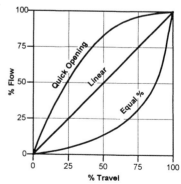

Figure 7 Valve characteristic curves

subsequent wear and controllability problems. As general rules of thumb, quick opening valves are used in on-off and pressure relief applications, linear characteristics are often used in slow processes for level, low flow, and temperature control. Equal percentage characteristics are applied in fast processes for pressure control or where highly variable pressure drops in flow applications are encountered.

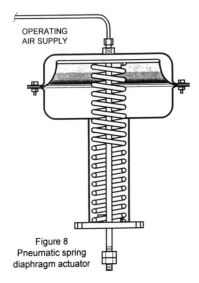

Figure 8
Pneumatic spring diaphragm actuator

Valve actuators are used to provide the mechanical force necessary to stroke an automatic valve. Typical types are pneumatic (spring diaphragm, spring cylinder, and double acting cylinder), air motor, electro-hydraulic, and motorized. Some considerations affecting selection and sizing are maximum differential pressure across the valve and its port size, stroke speed, stem stiffness, performance characteristics, seat leakage class requirements, and failure action. By far, the majority of actuators in use are pneumatic. It should be noted that large valves and / or high differential process pressures require very large actuators, and special consideration must be given for their space allowance, mechanical supports, and maintenance access. Motorized valves are often used where plant air is not readily available, but as they typically use gear reducers for torque multiplication, they usually have a comparatively slow speed of response.

Positioners are used on throttling valves to precisely position the valve stem in relation to the control signal, by using a stem position feedback to correct the error between desired and actual. Positioners are used, especially on larger valves, to overcome the friction and subsequent hysteresis in the stem mechanism, in order to maintain proper position under dynamic process conditions, to provide better accuracy for control, and to increase overall speed of response.

Each type of valve has an effective range of pressure drop they can handle, with some overlap. For example, for low pressures, butterfly valves can be used. For increased pressure, a plug or ball valve would be appropriate. Globe valves would be a good choice for high-pressure applications. Special anti-cavitation trim would be required to prevent seat damage at high fluid velocities. Cost is often a major factor in selecting a control valve type. Globe valve designs are popular for relatively small lines up to 3-4", but become very costly in larger sizes. Ball and plug valves are normally used in lines up to about 6". Butterfly valves would be the choice for lines larger than 6". For slurry throttling service, diaphragm valves are typically used in up to 3" lines, while pinch valves are common for 3" and larger sizes. Of course, these rules of thumb only apply if the selected valve satisfies the process conditions.

Variable Speed
Variable speed (DC) drives and Variable Frequency (AC) drives are commonly connected to mechanical equipment such as conveyors, rotary valves, screw feeders, etc., to effect flow modulation of prime movers for solids handling. They can also be used in fluid control on pumps and fans instead of modulating valves. The use of these on fluid flow applications is usually more energy efficient, but usually at a higher initial capital cost. A variable speed drive in slurry applications is often preferred over the use of valves because of the high abrasive wear effects on throttling valves, and subsequent high maintenance costs, especially in larger line sizes.

Pulsed
Variable pulse width and pulse duration solenoids, either direct acting, or pilot duty (a small device that controls a larger device, usually with signal conversion and force multiplication),

in combination with pneumatically actuated diaphragm valves or pumps, are often used to modulate slurry flows. Milk of lime addition for pH control is a common application of this. When the slurry fluid velocity is too low, the lime settles out of solution and cakes readily, so static piping and standard throttling valves plug quickly. A re-circulation pipe loop is used to keep the lime in suspension and deliver it to taps at the use points. A diaphragm valve/solenoid combination is then used at the tap, and pulsed at a frequency proportion to a pH control demand.

MAINTENANCE CONSIDERATIONS

Proper maintenance and calibration of instruments and control devices has a direct affect on product quality and throughput, and can have a significant impact on profitability. Reducing downtime is also a critical issue, which dramatically affects profitability.

Safety

Safety has become one of the most important issues in plant operations today. Improperly operating or failed process systems can cause personnel injury and even death. Careless or improper maintenance can be the root cause and must be avoided at all cost. In addition to the potential for injury or loss of life, negligent operating or maintenance practices can result in heavy fines and costly lawsuits.

Good Housekeeping

Good housekeeping practices are important for preventing damage to instrumentation devices, wiring, and control cabinets. This may seem obvious, but is often overlooked and contributes to failures and subsequent costly shutdowns.

Make sure all cabinet doors and instrument enclosures are securely closed. Even a small crack exposes these devices to the weather, intrusion of dust and dirt, potential corrosive environments, and even rodent damage. Locks should be used sparingly because they prevent access, especially during critical events, and keys may be difficult to find. It is far better to adhere to procedures.

Rooms containing electrical and instrumentation equipment should be equipped with automatic door closures and should be regularly checked to make sure they are not blocked open. Make sure that any new room wall penetrations are properly sealed and that any HVAC equipment or positive pressure systems are operating properly. If chemical filters are used in the ventilation system because of a corrosive environment, be sure they are checked and serviced regularly.

Instrument Installation

Accessibility of field instruments is extremely important, both for observation of process conditions and proper device operation, and to safely facilitate their maintenance and servicing.

All instruments and control valves should be located so as to be easily accessible from grade or from elevated platforms that already form part of the required access routes of the plant, with instrument indicators visible to process operators. As an example, the use of remote diaphragm seals with capillary connections to conveniently located pressure transmitters can be used.

Process connections to field mounted instruments generally use 1/2 in. National Pipe Thread (NPT) pipe taps, and should include a ball or gate isolation root valve for servicing the instrument without de-pressurizing the line. High temperature and pressure process connections must be provided with double block and bleed valves for safety reasons. Siphons or pigtails are used for pressure gauges in high temperature applications for protection from temperature damage.

Differential pressure transmitters are normally provided with a factory installed 3-valve manifold to allow isolation and calibration of the device on line. Control valves are often provided with a bypass manifold, including inlet and outlet isolation valves. This allows the valve to be safely repaired without shutting down the line. The manual bypass valve is normally sized the same as the control valve to allow manual throttling of the fluid while the control valve is out of service. Handwheels for manual operation are sometimes included on the control valve for temporary operation if the actuator mechanisms fail.

Process connections for slurries or fluids which are corrosive or contain suspended solids, even when employing a diaphragm seal, are often purged or flushed with a compatible fluid, utilizing a suitable back-pressure regulator, to prevent line plugging and diaphragm seizing.

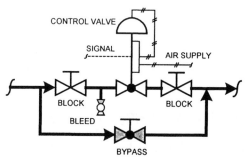

Figure 9 3-valve bypass manifold

Spare Parts and Consumables

The plant warehouse usually manages the storage, retrieval and replacement of spare parts. The quantity and types of parts kept on hand is determined, to a large extent, on the effect on plant production if they are not readily available when failures occur, and how quickly they can be obtained off-site. Good design practice dictates the standardization of instrument manufacturers and minimization of variations in measurement and control techniques. Using smart transmitters, having wider measurement ranges, will cover more applications. This then allows maximum interchangeability of parts and reduces the required inventory to cover the same risk level.

Instrument Shop

The instrument shop is a key element in any successful maintenance program. The desire to increase product quality and yields fuels the demand for improved accuracy and reliability. The overall goal is to maximize maintenance cost effectiveness.

The primary tasks for the shop are cleaning, inspecting and testing, calibration, repair, and rebuilding of instrumentation devices per manufacturer's instructions and calibration data sheets. The extent of a shop's capabilities is usually in relation to the remoteness and size of the site. If the shop cannot repair an instrument due to lack of specialized equipment, or if is a warranty issue, then they will handle the arrangements for off-site servicing. The shops must be well stocked with test equipment and machinery to effectively fulfil their duty. Process calibrators must be significantly more accurate than the instruments they calibrate and should be re-calibrated themselves on a regular basis, against standards traceable to the National Bureau of Standards. As a minimum, instruments are periodically calibrated at 0%, 50%, and 100% of span using appropriate test instruments to simulate inputs and to read outputs. Smart transmitters are calibrated either from a hand-held field communicator or the DCS, per manufacturer's instructions.

Training

Proper personnel training cannot be over-emphasized. Technical training programs, conducted either on or off-site, keep maintenance personnel up to date with the latest equipment and repair techniques. Having a person trained on equipment before confronting its failure will save valuable time in not having to take a crash course from the manual while attempting to repair that device. Conducting regular training on proper procedures and safety is a must.

Documentation

A good maintenance program is dependent on access to complete, accurate and up-to-date reference material. An organized and well-run document control department will pay for itself many times over in timesaving alone. Maintaining equipment on erroneous or out of date information is sometimes worse than no maintenance at all. If reference material is missing, it must be obtained from off-site sources (assuming it is still available). Originals should never be removed from the record storage; only copies, preferably with a date stamp, should be allowed.

Electronic record management systems are becoming an essential tool of document control departments. These systems facilitate the document identification, logging, organizing, storage, and retrieval of critical process systems and equipment reference material. These documents include vendor manuals and instructions, engineering drawings and data, operating and maintenance procedures, operating manuals and safety programs.

Personnel should always be sure that the documents they use are the latest, up-to-date version. If there is any doubt, they should be checked against the document control's master copy. When authorized changes are made to the process, equipment, or instrumentation (like adjustments to calibration ranges or control strategy modifications), the changes should be reflected in as-built information markups, on the appropriate master documents, in an organized and controlled manner. This is all too often not done, because each incremental change seems insignificant or is forgotten, until someone realizes that nothing matches the documentation anymore. Then a costly as-built remediation program must be undertaken to bring reference material up to date.

Computerized Maintenance Management Tools

Instrument management software and systems provide powerful tools and utilities to manage the thousands of instruments typically installed in a large plant. The "Don't fix it if it ain't broke" philosophy is no longer acceptable. Predictive and preventative maintenance, and regular calibration intervals can help avoid unplanned shutdowns or unsafe process conditions.

HART and Fieldbus smart devices can deliver multiple parameter capability and make those data available for automatic logging and computer analysis. Data such as tag, make, model and serial number, installation / calibration date, process parameter data, process values and alarms, as well as out of tolerance and fault and failure data can be obtained by uploading from the instruments.

Asset management and instrumentation management support software can automatically gather, store, and analyze these data in real-time to keep track of an individual instrument's history, provide alerts of faults and failures, evaluate relative performance, determine valve performance and wear, and schedule preventative maintenance intervals.

Even without smart instruments, preventative and predictive maintenance routines can be provided through the statistical analysis of process event and alarm records, equipment running hours, production totals, energy use, frequency of failures, etc.

CONCLUSION

These are exciting times for plant instrumentation and control technology. New digital capabilities afford unprecedented opportunities for maximizing plant profitability, not only from the aspect of more accurate and reliable instruments, but in plant availability through the application of advanced computerized maintenance management tools. The potential for increased maintenance and repair efficiency and reduced plant downtime is great. Embracing this technology does not come without cost. Personnel must be trained or recruited to work with and be comfortable with this advanced technology. Management must also recognize the importance of it and be supportive, not only organizationally but also financially. The potential benefits will surely outweigh the investment.

Strategies For Instrumentation and Control of Crushing Circuits

Stephen D. Parsons[1], Susan J. Parker[2], John W. Craven[3] and Robert P. Sloan[4]

ABSTRACT

In recent years instrumentation and control advancements have imparted valuable on-line information for crusher control. These advancements have facilitated enhanced decision making from both a production and a maintenance perspective. This paper provides an overview of the instrumentation and process control strategies used within the mineral industry for primary crushing and multi-stage crushing plants. Crusher instrumentation as well as regulatory and advanced control strategies will be examined focusing specifically on strategies to improve product quality and plant throughput.

INTRODUCTION

The continuous demand for high quality crushed product and subsequent high downstream costs associated with below specification product highlights the need for effective crusher control and system efficiency. As such, new crushing plants as well as older plants have embraced control and instrumentation initiatives at both the regulatory control and advanced control levels.

Irrespective of whether we are referring to primary crushing or multi-stage crushing, the main objective of a crushing plant is to maintain operating conditions that result in an optimum throughput-product size relationship; this may be either maximum throughput at a constant product size or the finest size possible for a given throughput. Additionally, constraints such as continuity of flow between circuits must be maintained, and disturbances such as ore feed rate and hardness must be compensated for (Herbst and Oblad, 1985). The specific control strategies used to accomplish this depends on the downstream requirements and the circuit configuration.

In terms of the mechanics of crushing, maximum throughput can be realised by ensuring that instrumentation and control strategies are implemented with the goal of maximising crusher energy utilisation. Operationally, maximum throughput can be achieved through the optimisation of plant availability. Irregular feed rates to primary crushing plants render optimisation, specifically the maximisation of crusher energy, problematic. As such, instrumentation and control strategies for primary crushing tend to focus more on fault detection and to a lesser extent on optimising the crusher mechanics. With the inclusion of recirculating loads, surge bins and subsequent feedrate control, multi-stage crushing plants employ a broader process control functionality than that of primary crushing plants. Consequently, this paper has a stronger focus on the control of multi-stage crushing plants.

In general, the focus of secondary and tertiary crushing circuits is to maximise plant throughput while crushing to a specified crushed product size. Where the crusher produces a saleable product (e.g. road-stone quarries), the control objective is usually to maximise the production of certain size fractions from each tonne of feed (Wills 1997).

---

[1] MinnovEX Technologies, Toronto, Ontario, Canada
[2] MinnovEX Technologies, Toronto, Ontario, Canada
[3] MinnovEX Technologies, Toronto, Ontario, Canada
[4] MinnovEX Technologies, Toronto, Ontario, Canada

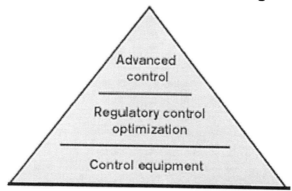

Figure 1 : Crusher process control infrastructure

The process control infrastructure consists of crusher control equipment (i.e. plant instrumentation), regulatory control and advanced control. Refer to Figure 1.

Regulatory control provides four main functions:

1. Allows the process to operate at a chosen target;
2. Minimizes the effects of disturbances;
3. Reduces the effect of ore variability; and
4. Provides for safe and efficient start-up, operation, and shutdown of the process.

Hence, the regulatory control system's function is to facilitate the consistent execution of control actions under dynamic conditions. Regulatory control can be provided by either independent (local) controllers or a centralised control system with the latter being adopted as the standard for new crushing plants. The control system uses either Programmable Logic Controllers (PLC's) or a Distributed Control System (DCS). Advanced Process Control Systems can be incorporated to provide added optimisation. To date, expert systems have been the most common means of deploying advanced control solutions for crushing plants.

This chapter is organised into two main sections. The first focuses on instrumentation requirements and applications for crushing plants. The second reviews process control strategies, including fault detection, stability control and optimisation control. Instrumentation and control strategies are documented for both primary and multi-stage crushing plants.

INSTRUMENTATION

Accurate and robust instrumentation is critical to the long-term sustainability and success of a control strategy, particularly given the harsh operating environment of a crushing plant. The following conditions complicate the continuous control of crushing circuits:

- Variability in feed (i.e. size distribution, hardness, feedrate);
- Abrasive nature of feed material;
- Noisy crusher power signals; and
- Unmeasured disturbances (i.e. tramp metal, chute plugging etc.).

Table 1: Process control instrumentation – standard equipment for crusher design

Manufacturer's Instrumentation – generally included with crusher packages	
1. Crusher unit	
Vibration Transmitters	– positioned on the adjustment ring to detect excessive forces in the crushing chamber
RTD's	– monitors crusher bearing temperatures
Pressure transmitter	– provides guidance on the crusher clamping pressure and/or the air dust seal pressure
Proximity Switches	– positioned close to the adjustment drive pinions on crushers to help gauge the crusher closed side setting
Flow switch	– positioned on water lines where water is used as the medium for the crusher dust seal
2. Crusher Drive unit	
Power Transducers	– measures the motor power draw
Amp transmitter	– measures the current draw on the crusher motor
RTD's	– monitors crusher motor and bearing temperatures
3. Crusher Lube Package	
Flow switch	– positioned on the crusher lube lines to detect oil flow
RTD	– monitors lube oil temperature (reservoir, return line)
Differential Pressure transmitter	– detects plugging across lube filters filter
Level switch	– provides guidance on the oil tank level
Standard Plant Instrumentation – generally not included in crusher package	
Level Transmitters	– measures the level of material in the crushing cavity and/or in surge bins
Belt Weightometers	– measures the mass of material on conveyor belts
Tilt Switches	– positioned in feed chutes to detect plugging events
Amp Transmitter	– measures the current draw on auxiliary equipment motors including crushers, conveyors, screens and feeders
Metal Detectors	– positioned before the crusher to detect the presence of tramp steel on the crusher feed conveyor
Video Monitoring Equipment	– cameras positioned above conveyors and crusher cavities to provide visual feedback for operating personnel
Power Transducers	– measures the motor power draw for feeders and vibrating screens
Flow Switch	– detects restriction or valve failure in water or oil lines
RTD's	– continuously monitors motor, bearing and oil temperatures

Table 2: Process control instrumentation – non-standard instrumentation and technological advancements

On-line Image Analysis	– positioned in primary and multi-stage crushing plants for particle size monitoring
Internal equipment sensors	– equipment manufacturers are placing smart sensors (with programmable microprocessors) within equipment to impart more data for fault detection purposes

Table 1 provides a list of commonly used crusher instruments, or what is considered the standard instrumentation for control purposes. The standard instrumentation table is broken down further into what is typically provided with the manufacturer's crusher package and what is typically added to the design by either the operation or engineering company. Table 1 presents only the main instrumentation that should be considered for direct crusher control and does not extend to peripheral process functions. Non-standard instrumentation (application specific equipment) is listed in Table 2. This presents the higher-end or technologically advanced equipment that is not considered standard but has proven to add control value for specific applications. The instrumentation included in both the standard and non-standard tables was selected from the perspective of throughput maximization, product quality control and crusher protection.

Level Sensing

Level sensors (usually ultrasonic) are used throughout crushing plants in surge bins, crusher dump pockets, crusher cavities and slurry pump boxes. Level sensors impart valuable on-line data for throughput control via mass balancing and for ensuring that adequate operating levels are maintained to mitigate equipment damage from excessive impact.

Ultrasonic transmitters can be problematic in the presence of high dust concentrations. Dust clouds will reduce the transmittance of sound signals, and dust fouling or build-up on the transmitters can severely impact the reliability and/or accuracy of the instrument. As a result, radar based level sensors are gaining popularity. Proximity and nuclear density switches are often used for high level detection in dusty bins.

Level sensors provide an on-line measurement of the ore level in the crushing cavity, thereby providing an indication of whether the crusher is being 'choke fed' (a crusher is considered 'choked' when the crusher cavity is full). It is generally accepted that choke feeding secondary and tertiary crushers produces an increase in fines production and a higher overall throughput. Figure 2 illustrates the typical positioning of a level sensor above the cavity for crushers that are operated under choke fed conditions.

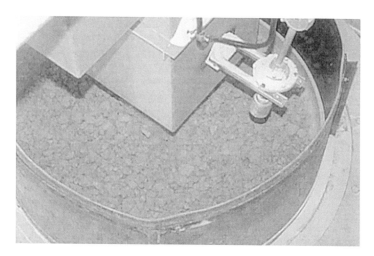

Figure 2 : Positioning of an ultrasonic level sensor (courtesy of Nordberg)

In consideration of water flush crushers, two level sensors are typically used, one to monitor the ore level and the other to monitor the water level in the crusher cavity. Maintaining an adequate water level ensures that sufficient water is present to flush the fines from the crusher. The water level is typically measured in a protected stillwell.

With respect to ore bed levels in crusher dump pockets and surge bins, the bin level and often the rate of change of the bin level, is used to regulate upstream feedrates via mass balancing, with the step change to the feeder commensurate with the severity of the condition. This approach is more specific to advanced process control algorithms. The step changes could either cascade back sequentially to the primary plant feeder or create a simultaneous cut to all upstream feedrates. The objective in both cases is to balance the plant feedrate and maximise throughput, subsequently adding stability to all downstream processes. The control strategy employed is a function of the plant configuration and circuit intricacies.

Bin level sensors coupled with plant interlocks provide an equipment protection function. Level sensors provide an indication of when the bed level is dangerously low in the bin, to the point where equipment is exposed and susceptible to excessive impact from the feed material. Plant interlocks are configured within the regulatory control system to trip the feeder well before the bin empties, thus ensuring the equipment is always under a protective layer of ore. Low-level interlocks are especially important on crusher dump pockets where unreduced ore drops directly onto an apron feeder or vibrating feeder.

Conveyor weightometer

Conveyor weightometers are positioned strategically across the plant to weigh and totalise tonnage at "key" locations, such as:

1. the crusher product conveyor belt, i.e. screen undersize;
2. the plant feed belt; and
3. the crusher discharge (jaw discharge, cone crusher discharge).

The weightometers should be positioned such that the crusher feedrate, crusher discharge rate and plant circulating loads can be directly measured or easily calculated.

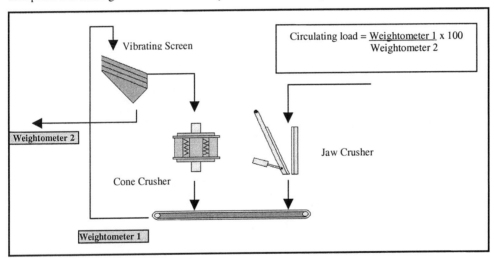

Figure 3 : Positioning of weightometers in a 2-stage crushing circuit

Figure 3 illustrates the typical placement for conveyor weightometers in a two-stage crushing circuit. Weightometer-2 weighs the tonnage of the crushed product and Weightometer-1 weighs the sum of the discharge from the jaw crusher and cone crusher. The circulating load is then determined using the two values (as per Figure 3).

Conveyor Cameras

Cameras are used as an operations tool to monitor problematic transfer points and/or plant equipment in an effort to detect problems prior to equipment damage and subsequent plant or equipment downtime. Common camera locations are as follows:

- Conveyor head pulleys;
- Belt magnets;
- Above jaw crusher cavities;
- Transfer chutes; and
- Screening equipment such as double deck screens and grizzlies.

Proximity Switches (Crusher Setting)

The crusher setting can be inferred through the implementation of proximity switches positioned on the crusher pinion teeth (as shown in Figure 4). The proximity switches count the number of pinion teeth during a bowl rotation – adjustment and the direction of rotation. Since each incremental change translates to a known "gap" increase, the crusher setting can be monitored automatically. The calculated crusher setting is re-calibrated when the bowl is tightened to the point where the mantle and bowl are touching.

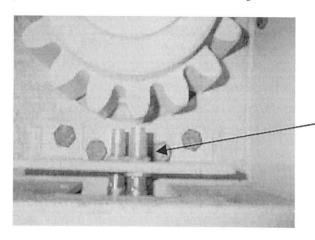

Two non-contacting proximity switches counting pinion teeth during adjustment

Figure 4 : Crusher gap automation – proximity switches (courtesy of Nordberg)

Metal Detectors

Some of the tramp metal entering a crusher may be manganese steel, which is non-magnetic and does not get picked up by the conveyor magnet. Consequently, metal detectors are positioned after the conveyor magnets to flag the non-magnetic tramp metal before it enters the crusher cavity.

Power Measurement

Diligently monitoring the crusher power draw via a power transducer is the most common approach used to ensure that the crusher is operating as close to and within a realistic tolerance of the maximum operating power. It is generally accepted that crushers should operate at approximately 85% of the full load amps.

Vibration Sensors

Vibration sensors can be mounted on the crusher adjustment ring (see Figure 5) to continuously measure adjustment ring movement and provide an alarm signal when the crushing force design limit has been exceeded due to the presence of uncrushables (such as tramp steel or wood chips) or because of changes in the feed material. Via the signal trending either the operator or the advanced control system can infer the nature of the problem and subsequently execute the appropriate control action. For example, instances where the vibration is reoccurring at equal intervals and at a relatively high frequency suggests the presence of a recirculating load of uncrushables in the circuit.

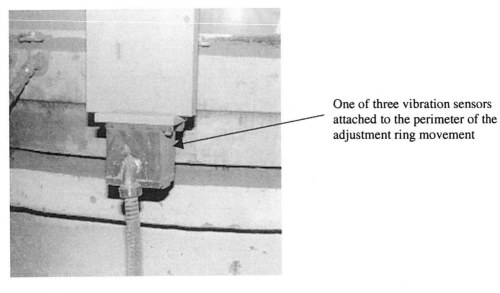

Figure 5 : Crusher vibration sensors (courtesy of Nordberg)

Vibration sensors can also be used for fault diagnosis by measuring the amplitude and frequency of the bowl vibration. Figure 6 is an example of a vibration reading trend; the peaks indicate overload conditions. Historical data, including signal magnitude and trending, can aid the detection and/or diagnosis of the problem, whether this be an overload condition or a crusher maintenance issue.

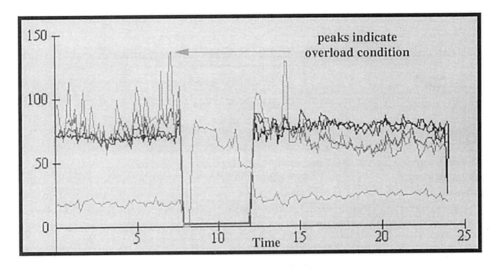

Figure 6 : Vibration sensing readout (courtesy of Nordberg)

Currently, vibration sensors are only sensitive enough to differentiate movement in the adjustment ring from normal operating vibration.

Image Analysis
Digital image analysis is continuing to gain acceptance within the industry as a means to calculate the size distribution of material on conveyor belts. There are various packages on the market, and most systems perform the following functions: acquire the digital image, perform pre-processing

on the image, delineate the individual fragments in the image using digital image processing techniques and then apply statistical algorithms to determine the particle size distribution. Figure 7 provides an example of the delineated image in comparison with a direct photograph.

Figure 7 : photograph of a primary crusher product (top) and associated delineated image with fines identification (courtesy of Split Engineering)

The majority of applications to date have been used to calculate the particle size distribution of the primary crusher product and feed to autogenous grinding and semi-autogenous grinding circuits. In recent years image analysis has also extended to secondary crushing plants.

Primary crusher applications:
- Located on the feed and discharge conveyors for in-pit crushers, the feed and discharge size distributions are used together to assess the relative hardness of the fragmented rock and to provide data for the monitoring of the crusher performance and crusher wear. It is also possible to integrate the Dispatch data to identify the location of the ore on each truck as they feed to the primary crusher and correlate the run of mine (ROM) fragmentation information to the blasting parameters. These techniques are currently being used at the Phelps Dodge Sierrita in-pit crushers (Keremy, J. et al 2001).

Secondary crusher applications:
- Measurement of the crusher feed ore size distribution.
- Located on the crusher feed belt, to provide additional information for inferencing the dynamics of the crushing plant. The measured size distribution can be used together with the feed tonnage and recirculating load measurements to classify the ore hardness (as discussed in the conveyor weightometer section).

Internal equipment sensors

Equipment manufacturers are placing internal equipment sensors in key strategic locations. For example, new crushers from Metso Minerals are being fitted with internal RTD's to continuously monitor the countershaft bearing temperature. It is expected that the countershaft bearing temperature sensor will provide critical data to help flag problematic conditions before a catastrophic failure occurs.

CONTROL STRATEGIES

Fault detection, stability control and optimisation control are three key elements of crusher control strategies. An effective plant-wide control strategy should be an appropriate blend of the three. Typically, the regulatory control system performs the fault detection, stability control and lower level optimisation control, with the advanced control system performing the higher-level optimisation control and fault diagnosis. The control strategies are broken down into two sections, those conducted by regulatory control systems and those conducted by advanced control systems.

A. Regulatory Control

Fault Detection. Fault detection, such as plant interlocks, is used as a means to mitigate damage to plant equipment. Crushing plants are susceptible to downtime since crushing equipment, by nature, is very expensive and standby units or capital spares are frequently not included in the plant design. Therefore, process control, specifically fault detection, is critical for the maximisation of plant availability and the protection of both plant equipment and plant personnel. Table 3 provides a list of conditions requiring a plant interlock.

Table 3 : Examples of crusher fault detection for a secondary crushing plant

Disturbance Event	Control Action
High crusher oil temperature	Crusher trip
Tramp metal trip	Crusher feed conveyor trip
Dust seal water – no flow	Crusher trip
Crusher amp overload	Crusher trip

Stability Logic. Stability process control strategies, designed to add stability to the process with respect to throughput and a consistent product size, are first introduced as part of regulatory control, with further optimisation conducted via either higher level regulatory control or advanced control. Stability logic must be designed to effectively manage disturbance events, such as crusher peak overloads and chute or crusher plugging. Crusher disturbance logic generally employs large step changes, with the objective to quickly stabilise the process. Table 4 provides a list of operational events that promote circuit instability and necessitate disturbance logic.

Table 4 : Crusher disturbance events

Disturbance Event	Control Action
Change in ore characteristics (size distribution, ore hardness, moisture content)	Assess the criticality of the situation. Adjust the tuning sets by moving to a regime with either more aggressive or more conservative step changes to the feed rate. If the ore is harder, for example, the tuning sets should shift automatically to a regime with smaller, more conservative, step changes
Irregular feedrates to the crushing plant	Employ feedrate control logic utilising bin levels and the rate of change of bin levels to regulate feedrate throughout the plant.

Feedrate Control. Changes in ore characteristics, including variations in the crushing work index, size distribution and feed moisture, cause fluctuations in the crusher power draw. The variations in ore characteristics are often sufficient to necessitate the manipulation of either the crusher feedrate or crusher setting, generally the only two manipulated variables available in a traditional crushing plant.

While the crusher feedrate is often controlled in consideration the bin level trending, specifically imbalances between crushing stages, the determination of the maximum/optimal feedrate is typically a function of the crusher power draw, assuming that the crusher is already being choke fed.

The ability to maintain the crusher power draw within a tight band of the power setting is a function of the event response time and the selection of an appropriate control action. Surge bins with variable rate feeders or conveyors should be positioned in close proximity to the crusher to diminish lag times in the system, and appropriate control logic must be formulated in consideration of the following:

- the nature of the power excursions - cycling, spikes;
- the level of ore in the crusher cavity;
- an ore change inferred via process modelling or using on-line imaging; and
- the crusher setting.

Figure 8 provides an illustration of actual real-time data for a tertiary crusher with poor feedrate control. The amp draw fluctuations are a result of feed rate variations. Note that while the average amp draw is far below full-load, the peak levels still exceed the crusher full load power rating.

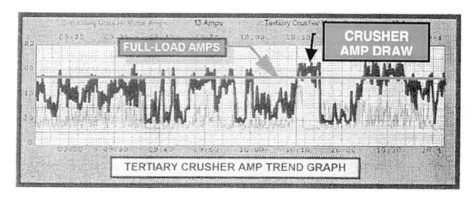

Figure 8 : Crusher trend chart, where heavy line is crusher amps, illustrates poor feedrate control (courtesy of Nordberg)

Figure 9 illustrates a tertiary crusher with continuously high power draw attributed to good feedrate control. In this case a surge bin is located directly upstream of the crusher and the belt feeder is directly controlled to maintain the crusher power draw setpoint. This configuration and the control strategy utilised facilitate choke feeding and the elimination of overload peaks.

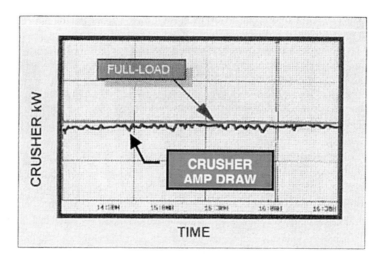

Figure 9 : Trend Chart – good feedrate control (courtesy of Nordberg)

Most vendor-provided crusher control packages are equipped with 3 types of control modes:

1. Auto power setpoint – crusher feedrate is manipulated to run to a power setpoint;
2. Auto level setpoint – crusher feedrate is manipulated to run to a cavity level setpoint; and
3. Manual – crusher feedrate is operated to a percent output.

Water flush crushers utilise an auto water setpoint as well, where the water flowrate is a volumetric addition based on the ore feedrate to the crusher.

Typical crusher control strategies use the auto power mode to maintain a power setpoint up to the point where the level in the crusher cavity exceeds a predefined limit. The crusher is subsequently operated in auto level mode until there is an adequate level in the crusher cavity, which triggers the switch back to auto power mode. In both the auto power and the auto level modes the manipulated variable is the crusher feedrate. This approach should maximise the feedrate to the crusher, but it is the plant-wide stability and optimisation logic which regulates the feedrates throughout the plant while not exceeding the circuit constraints, such as transfer chute limitations and conveyor limits.

Despite adequate choke feeding, it may still be difficult to track the power setpoint over the duration of a shift if the ore is very abrasive and contributes to high liner wear rates. It is not uncommon for plants to regap their crushers every 12 hours to maintain the product quality and crushing efficiencies.

Optimisation control. Optimisation logic can be executed via the regulatory control system, but it is generally considered "lower level" control relative to that provided by an advanced control system. The following section provides examples of optimisation control strategies implemented via a regulatory control system. "Higher level" optimisation logic implemented via advanced process control systems is discussed in the next section. This level forms a "grey area" between regulatory control (DCS/PLC) and the common understanding of Advanced Process Control (APC).

In the following examples the optimisation logic considered the impact of each phase on the downstream process. The impact is quantified and considered in the overall control strategy. Attempts are being made to optimise the whole process by controlling each step, not only the restraints of the single process but each successive process. For example, the Phelps Dodge Sierrita Mine is using imaging systems on the primary crusher feed and product to monitor the crusher performance and crusher wear. The information is also being used to gauge the secondary crusher feed size and estimate the work index. In addition, the truck-by-truck basis of the feed imaging allows the possibility to trace the size information back to the bench position for fragmentation models and hole-by-hole work indices.

The impact of crusher control on downstream processes has been noted at both Mount Isa Mines (Anon. 1973) and the Majdanpek Copper Mine (Grujić 1996).

At Mount Isa Mines improved crusher control strategies have led to increased grinding throughput. The crusher motor current measurements have been filtered to remove spiking caused by variable ore size and hardness, to allow the installed crusher control system to stabilize power. Reducing the magnitude of fluctuations allows the crusher to operate at higher current setpoints without increasing the probability of an overload trip. The control system on this multi-stage process balances the primary and secondary crushing stages and ensures that at least one is operating at maximum capacity. The secondary bin level is used to determine whether the primary or secondary circuit is the constraint.

By ensuring balanced, near-choke-fed crusher operation, 10 to 20 percent of the total ore previously reporting above 9.5 mm now reports below 9.5 mm (but predominantly above 4.7 mm). Reduction in the 9.5 mm material is of significance in maximizing grinding throughput without loss of flotation feed sizing.

The goal of crusher control at Majdanpek Copper Mine was to minimize particle size and optimise ore flow using motor stress regulation. The data processing is integrated, i.e. automatically controls and stabilises at setpoints within individual loops. The crusher feed control is based on the level in the crusher feed bin and the closed-side setting is based on motor power. Crusher throughput was increased while the average crushed product size decreased after implementing the optimisation control (refer to Table 5). In addition, liner wear decreased. The effect on the downstream grinding is indicated in Table 6.

Table 5 : Effect of Automatic Process Control on the Crushing Circuit at Majdanpek

Crushing Process	Q(t/h)	P80(mm)	W(kWh/t)	Steel Linings(g/t)	No. of Months
Without auto. Process control	2580	9.0	2.76	9.12	11
With auto. Process Control	2740	6.8	7.30	7.30	10

Table 6 : Effect of Automatic Process Control on Grinding at Majdanpek

Grinding and Float Process	W(kWh/t)	+208µm(%)	-74µm(%)
Without auto. Process Control	20.8	21.2	54.0
With auto. Process Control	19.3	14.1	58.1

These studies indicate the positive impact of optimisation crusher control on downstream processes.

B. Advanced Process Control

The group of solutions collectively referred to as Advanced Process Control (APC) constitute a wide range of applications from the more traditional Internal Model Control (IMC) and Statistical Process Control (SPC) to new age algorithms including expert systems and non-linear adaptive multi-variable controllers. Whatever the form, APC algorithms share a common trait – they are strategic as opposed to tactical in nature. That is, as opposed to making sub-second manipulations and changes at the controller level, such as maintaining a setpoint, these applications sit on a higher plane, determining and then sending setpoints to the regulatory system with the goal being the "optimisation" of the process. Each of the algorithms operate in a slightly different manner but share the similar concept of developing a "global" model, whether that is fundamental, phenomenological, empirical, stochastic or heuristic. This model can then be drawn upon to provide insight into the process and, with that knowledge, facilitate decision-making.

While there are many secondary benefits derived from advanced process control, it is generally accepted that there are three demonstrable results: stabilisation, optimisation and education. Through a combination of increased vigilance, consistency of application and (with some of the algorithms) the ability to consider information and relationships not readily available to an operator, an advanced control system can both stabilise and push the process to the operational limits. Additionally, APC algorithms result in an intensive and comprehensive educational process. As the ability to control the plant at its operational limits is demonstrated, the people within that operation gain a better appreciation of where those limits exist and why. In many cases the process of developing the APC strategies provides fundamental insight into the operation.

The benefits of such solutions appear obvious, but there were and are a number of factors that work against successful implementation. This includes the difficulty in motivating projects as crushing is seen as a "basic" unit operation with little room for improvement (and therefore limited economic benefit).

There are generally three main components to an advanced control system. They are as follows:

1. Sensors/data :   i. measured (automatic input);   ii. measured (manual input);   i. inferred
2. The execution platform :   i. hardware;   ii. software
3. Integration: i. design;   ii. execution of the application.

Sensors/data is the most critical component of any solution. The key to this aspect, other than the obvious GIGO (garbage in - garbage out) is that much of the critical data is actually inferred as opposed to measured. As a result, data is even more critical to the outcome and success of the solution due to propagation of error. Without the ability to accurately, or at a minimum precisely, measure the variables having the most profound impact on crushing performance (i.e. ore hardness), the ability to successfully execute an APC solution had been compromised. Suffice it to say that the more empirically based the model, the higher the "domain" accuracy and the less extensible it is beyond its calibrated or understood range of application. For this reason non-heuristic-based systems have not been widely embraced to date.

Heuristic expert systems, which strive to emulate operator action, have proven successful in numerous applications. They are more reliant on trends and the precision of a measurement than on the accuracy of a reading and a complete understanding of the process. In the future, as more of the key parameters within the crushing plant become available on-line, the authors can envision a model-based solution enhancing the heuristic foundation now being established.

In expert systems control logic has been applied to manipulate:

1. The rate of feed to the crusher;
2. The crusher closed-side discharge setting;
3. The water addition rate for water flush crushers; and
4. Future blast pattern designs.

An appropriate example of APC are control strategies that leverage the identification of operating regimes thereby facilitating a multi state (or variable tuned) solution that responds in a manner that is consistent with the characteristics of that operational point. This is particularly valid within the crushing plant where the difference in ore hardness has a direct impact on the controllability of the circuit, with harder ore requiring a more conservative approach and the recognition of softer ore permitting the exploitation of the circuits potential.

Control Options

Within a crushing circuit, ore storage coupled with an effective control scheme greatly facilitates material flow between stages. In a 3-stage crushing circuit, for example, imbalances in capacity show up in the secondary and tertiary bin levels. A sustained high level in a tertiary bin suggests excess capacity in the secondary crusher whereas a sustained low level suggests a secondary crusher capacity limitation. In an advanced control system the imbalance would be quantified and then considered in the determination of the appropriate control strategy, one where the overall process economics is maximised.

Bin levels and power draw may be used to balance the sections while ensuring that at least one of the stages is operating at maximum capacity. Table 7 provides an overview of typical conditions amenable to advanced process control with the control action for each.

Table 7 : Typical advanced control applications for secondary crusher

Condition	Possible Action
Power draw increase	Reduce the feed rate or open the closed side setting
Unequal motor loads in multi-stage process	Adjust closed side settings for maximum size reduction and throughput
Power increases as crushed product size increases and throughput decreases	Adjust feed size via blending or changing blast fragmentation
Crusher bowl level high (lower probe)	Reduce the feed rate
Crusher bowl level extreme (upper probe)	Stop feeder
Excessive vibration	Open closed side setting

Figure 11 illustrates the feedrate control logic implemented as part of the Placer Dome Zaldivar Mine secondary crusher expert system (Craven 2000). Figure 10 provides an overview of the Zaldivar production process.

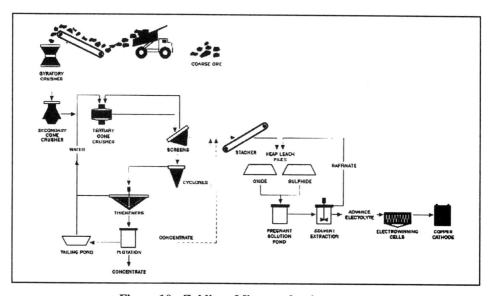

Figure 10 : Zaldivar Mine production process

The expert system logic presented in Figure 11 represents only one of many logic sets utilised within the framework of the expert system. The rectangular blocks represent the status of the condition while the rounded blocks represent control actions. The logic proceeds sequentially starting with an evaluation of the status of the secondary bin. If the bin is not empty, or contains material, then the logic evaluates the status of the crusher power. If the power is extremely high, or high and increasing, the logic will unequivocally execute step change decreases to the crusher feedrate commensurate with the severity of the condition. The feedrate control relies preferentially on the crusher power and then on the level in the crusher cavity. Only when the crusher power is not increasing does the logic include an evaluation of the level in the crusher cavity. Lower levels in the cavity utilise larger step changes in an effort to produce "choke fed" conditions.

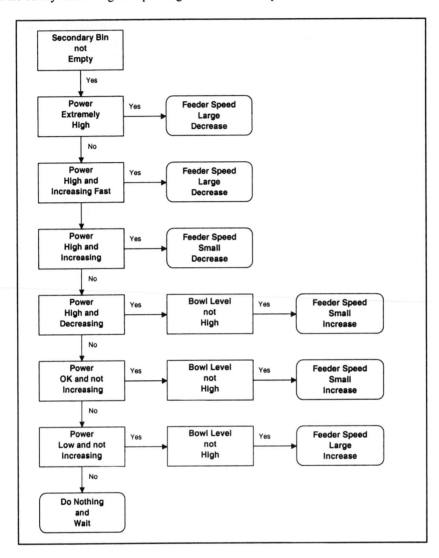

Figure 11 : Secondary Crushing Plant Advanced Control Logic (courtesy of Placer Dome - Zaldivar Mine)

In this case fuzzy logic was used to quantify the severity of the condition and to determine the magnitude of the step change. On site tuning is required to determine the actual magnitude of the step change and the required timing.

The logic utilised is always site specific. For example the following advanced control logic differs slightly from the Zaldivar logic, although the ultimate objectives are the same – to maximize crusher power and consistently choke feed the crusher. The control logic for an advanced control system at Brenda Mines (Flintoff and Edwards, 1992) was based on the tertiary bin level and the crusher power, as follows:

1. Low tertiary bin level

    a. increase secondary crusher power setpoints;
    b. temporarily decrease tertiary throughput as long as the condition holds;
    c. if the secondary crusher is operating at maximum power or condition is prolonged, then request an increase in the secondary closed-side setting; and
    d. if the condition is prolonged further request that the tertiary crusher be stopped.
2. High tertiary bin level
    a. decrease the power setpoints on the secondary crushers;
    b. if the condition prolonged and any of the tertiary crushers are down, then request that a tertiary crusher be started; and
    c. if the condition is prolonged further request a decrease to the secondary crusher closed-side setting.

In summary, advanced crusher control systems can be used to:

1. Automate crusher closed-side setting;
2. Implement adaptive control modes for a) constant setting and b) maximum power;
3. Maintain a balance between crushing sections;
4. Diagnose faults online;
5. Provide online operation manual; and
6. Trend and analyse performance indicators.

CONCLUSION

This paper has provided the main instrumentation and control strategies for crusher circuit design. Although all circuit configurations and peripheral instrumentation were not considered, the information should provide insight into the main instrumentation and control strategies needed to satisfy the crushing objective - to promote and maintain operating conditions that result in the optimal throughput-product size relationship.

The ability to detect faults and provide feedrate control has been the primary determinant for instrumentation and control strategy selection. It is expected that the continued development of new and smarter sensors coupled with the shift towards more of a mine/mill operating philosophy will continue to expand the control possibilities for crushing plants.

ACKNOWLEDGEMENTS

The authors would like to thank Tom Bobo, Split Engineering; Jennifer Abols, Metso Minerals; and Placer Dome's Zaldivar Mine, in particular Jim Whittaker, for their contribution to this chapter. In addition, we thank various personnel within MinnovEX Technologies Inc. for both their input and valuable discussion, in particular Glenn Dobby and Michael Schaffer.

REFERENCES

Anon. 1973. Crushing Control Systems at Mount Isa Mines Limited, Case Study 7
Craven, J.W. 2000. MET (MinnovEX Expert Technology) Application Manual for the Zaldivar Crushing Plant Expert System
Flintoff, B.C. and Edwards, R.P. 1992. SME Proc. Phoenix, Az.
Grujić, M.M. 1996. Technology improvements of crushing process in Majdanpek Copper Mine, Int.J.Miner.Process, 1996, p. 44
Herbst, J.A.. and Oblad, A.E. 1985. Modern Control Theory Applied to Crushing Part 1, IFAC Automation, p. 301
Kemeny, J., Mofya, E., Kaunda, R., Perry, G., Morin, B. 2001. Improvements in Blast fragmentation models using digital image processing, Proc.38th Rock Mechanics Symposium, Washington, D.C.
Wills, B.A. 1997. Mineral Processing Technology, Sixth Edition, Butterworth-Heinemann, Oxford

STRATEGIES FOR THE INSTRUMENTATION AND CONTROL OF GRINDING CIRCUITS

Robert Edwards, André Vien and Rob Perry[1]

ABSTRACT

Process control is critical to the optimized operation of all grinding circuits. Being one of the major cost centres in a mineral processing plant, and often one of the production limiting stages, it is essential that the grinding circuit run not only smoothly, but also as close to its theoretical optimum as possible. Advances in instrumentation, the continued practical application of regulatory controls, and the maturing of the advanced control tools, such as model-based expert systems, fuzzy logic, and neural networks, have provided the right conditions for optimum grinding circuit operation.

Effective and practical advanced controls, well-tuned and robust regulatory controls, and a solid layer of instrumentation are the keys to optimum circuit control. This paper discusses the techniques being employed in controlling today's mineral processing grinding circuits.

INTRODUCTION

Grinding circuits are designed to reduce material to a size suitable for treatment in subsequent separation and recovery processes. Their performance dictates the efficiency with which breakage energy is imparted, thereby determining the overall efficiency of this comminution stage. Judiciously implemented process controls are an important part of ensuring cost effective operations.

The grinding process is the most energy intensive, and thereby usually the most expensive, operation in a typical mineral processing concentrator. Circuit measurements and the knowledge of their relationship to the processes occurring indicate how this energy is being applied (Lynch 1977, Napier-Munn et al. 1996, Stanley 1987, and Wills 1988). The application of the energy can be considered as a function of two types of variables:

- Equipment variables that are essentially fixed (mill size, ball charge, cyclone geometry)
- Manipulated variables that can be changed continuously (feed rate, dilution water, mill speed, pump speed)

Grinding circuit design is the selection of the equipment variables while grinding circuit *control* is selection of values for the manipulated variables. Grinding circuit control in its many forms basically comprises the following four steps repeated continuously:

- A grinding circuit production objective is defined
- Measurements from the circuit are used to deduce if the circuit is meeting its objective

---

[1] Metso Minerals, Business Line Mineral Processing – Process Technology, Kelowna, B.C., Canada

- Values of measurements required to meet objectives are defined
- Variables are manipulated to drive measurements towards the desired values

In order to best understand grinding circuit control, and certainly prior to assembling an integrated control strategy, it is important to understand the process and the control tools available. In this paper, and in an attempt to better relate the processes and relationships to the measurements and manipulated variables, we will first look at each component of a typical mineral processing circuit. Important details on the structure of control loops are given before examining the typical basic circuit controls. Pertinent comments on the application of instrumentation to grinding control are given, but detailed instrument discussions can be found elsewhere in this volume. We conclude the paper by examining the interactions of the various circuit components and the application of control to find the optimum process operating range.

GRINDING CIRCUIT CONTROL OBJECTIVES

A clear objective is an essential component of a grinding circuit control strategy. Without clear direction, even the best controls cannot help a grinding circuit achieve optimal performance and may even negatively impact production. That being said, control objectives are dynamic targets based on overall plant economics (e.g. net smelter returns) and downstream process requirements (e.g. grind versus recovery), and require continuous re-evaluation in the face of changing downstream equipment performance (e.g. flotation circuit capacity) and metal prices.

The control objectives need to provide unique, realistic and attainable targets. There must be at least one degree of freedom to accommodate ore type variations; such as changes in feed size, hardness, etc. For example, it is unrealistic to expect a grinding circuit to achieve maximum fineness AND maximum throughput – either throughput OR fineness is achieved at the expense of the other. Examples of grinding circuit control objectives are:

- Maximum throughput whilst keeping product density above 42% solids and cyclone overflow size below 80% passing 75 microns
- Finest 80% passing size possible, at a minimum of 30% passing 20 μm, while maintaining a fixed 2000 tph feed rate.

The objectives define boundaries within which the circuit can safely operate both physically and economically. Accordingly, these boundaries become the limits within which control strategies such as "constraint based optimization" can work.

BASIC GRINDING CIRCUIT CONTROL ELEMENTS

Grinding circuits are applied in a wide variety of applications and take on numerous configurations to best suit the overall downstream process needs. Focusing on the most widely used components in the metallic mineral processing industry, this discussion is restricted to common single and multi-stage milling circuits, both in open or closed circuit with product size classifiers. The control strategies will vary depending on the circuit configuration, but are always comprised of a combination of common *process* and *control* elements.

Process Elements

The scope of a grinding circuit control strategy typically includes all equipment between the fine ore bins (or stockpiles) and the flotation feed (or other downstream process). The circuits are comprised of ore feeders, conveyors, grinding mill(s), classifiers, and classifier feed pump(s) and depending on ore and breakage characteristics may or may not incorporate (pebble) crushing.

Common circuit configurations include primary mills in closed circuit with classifiers (Figure 1), or primary mills followed by secondary mills closed by classifiers (Figure 2). In either case the process elements are similar – grinding mill, classification, and slurry transport (e.g. pumping). An understanding of the behaviour and interaction of these components is key to effectively implementing grinding circuit controls. [Earlier works by Carriere 1982, Chang 1982 and Brown 1982, though dated in terms of control hardware, contain good discussions on primary ball milling and AG/SAG grinding circuit control.] Following is a summary of grinding circuit components and their control requirements.

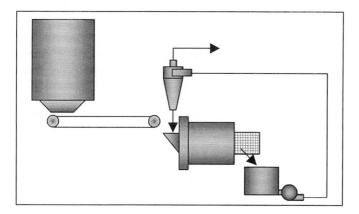

Figure 1 – Single Stage Closed Circuit Milling

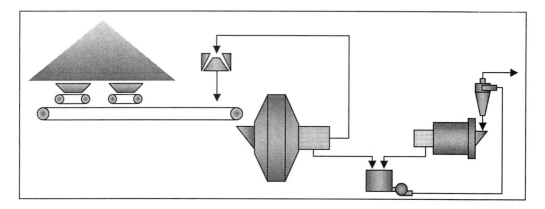

Figure 2 – Two Stage Closed Circuit Milling

Primary Mills. The primary grinding mill is usually the biggest single consumer of power in the concentrator and plays an important role in conditioning the concentrator feed for downstream processing. Primary mills are typically autogenous (AG), semi-autogenous (SAG), ball or rod-mills with the latter being relegated to lower throughput, or multiple circuit plants. Rod and ball mill circuits are typically preceded by secondary and tertiary crushing circuits and considered relatively stable to operate. On the other hand, AG and SAG mills are often only preceded by a primary crusher and, being inherently sensitive to feed size and hardness, are much less stable. The efficiency of primary grinding mills in general is a function of mill filling and charge motion and in AG and SAG milling, unlike primary rod and ball mills, these parameters vary considerably and need much closer attention.

Intermediate (Pebble) Crushers. Crushers are used ahead of or in closed circuit with primary mills to accelerate the breakage of intermediate or 'critical' sized material. Crushing is more energy efficient than grinding but the capital and operating costs per ton for installing a crushing circuit are generally higher. Judicious use of crushing can increase circuit capacity up to 15% or more. Standard or small gyratory crushers have been used to pre-crush primary mill feed whereas shorthead crushers are often used to crush the oversize of primary mill discharges (Needham and Folland 1994).

Secondary Mill. It is common to require a second stage of grinding to achieve the final product specification for a downstream process. Rod mills and primary ball mills have typically been followed by a second (and sometimes third) stage of grinding. And whereas some AG mills have been able to achieve the final product size in a single stage, most AG/SAG circuits have two or more stages to increase efficiency. Secondary mills are almost always in closed circuit with a classifier, so that the control of a secondary mill requires the optimization of the mill and classifier *circuit*. Secondary mill and classifier interactions play a critical part in the overall optimization of the grinding circuit.

Classification. Classifiers are use to separate the fine material already at or below the product size specification from that requiring further breakage. Classifier fines typically report to downstream processes, while classifier coarse typically reports back to the mill for further comminution. Particles can be selected for classification based on size and/or density. For example, size classification in a hydrocyclone can be complicated by the presence of secondary high specific gravity sulphide particles. Typical classifiers are screens, hydrocyclones and spiral classifiers, though the latter are becoming less common.

Auxiliary Equipment. The significance of auxiliary equipment such as pumps, pump boxes, piping, conveyors and chutes is often overlooked in grinding circuit control strategies. Pumps and pump boxes play an integral role in delivering slurry between mills and classifiers. Unstable performance at this intermediate stage of processing can lead to flow surges and spills (with their associated uncontrolled circuit water addition) that have significant negative impacts on overall circuit performance. Chutes, pipes or conveyors that are undersized, damaged or blocked can constrict material flow, exacerbate spillage problems and further negatively impact circuit performance. The cost and effort to maintain these simple, but crucial process components are small compared to production losses they can cause. It is vital that throughput or performance limits are not due to the auxiliary equipment.

Control Elements

There are three basic elements to a control loop; the measurement (sensor based process measurement), the controller (computer applied algorithm), and the manipulated variable (final control element). In a simple flow loop (Figure 3) this would be a flow meter, a PID controller and a control valve. In a more complex mill control loop these elements might include a dynamic SAG mill model, an expert system "controller", and a tonnage feed rate controller.

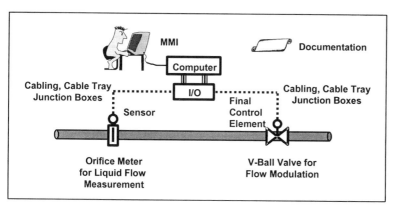

Figure 3 – Example Instrument - Control Loop

A detailed quality analysis of the measurements should be the first step in any grinding circuit work, whether it be commissioning, development, or analysis. Following is a brief discussion on signal conditioning followed by guidelines on the application and use of process measurements.

Measurements. Reliable, accurate and precise measurements are the foundation of good control. In all cases the quality of the measurement must be known to ensure it is used appropriately within the control strategy; i.e. imprecise, unreliable measurements have to treated cautiously. Additionally, the conditioning of the input signal must be appropriate for the application. For example, in a simple flow loop the digital filtering of the flow meter reading must ensure the controller does not respond to high frequency noise and only the underlying "real" process measurement. In a more complex model-based loop the relative reliability of the inputs to the mill model will be given more or less weight to preferentially bias the result towards that of the most reliable measurements. Signal conditioning is therefore an integral step in achieving reliable process measurements.

Signal Conditioning. Process sampling time, digital filtering and anti-aliasing are three important techniques for signal conditioning (Vien, Edwards and Flintoff 1998). Selection of the proper process sampling time is key to effectively determining accurate dynamic process performance information.

With feed rates, and thereby volumetric flow rates, varying considerably in AG/SAG circuits it important to select a digital sampling time suitable for all operating conditions. For example, as feed rates increase the process time constants in pump boxes (e.g. retention time) will decrease potentially requiring a higher sampling frequency for adequate level control. A sample time that is too short will place unnecessary demands on the data acquisition subsystem, and when dealing with noisy signals may cause the controller to react to high frequency noise instead of the underlying process signal (as previously mentioned). Sampling at a rate too slow for the process will inaccurately represent the process dynamics resulting in poor controller performance. It is recommended that a sufficiently high frequency be selected to handle all foreseeable circumstances.

Filtering is a technique used to reduce the effects of process and measurement noise while retaining the important information contained in the signal. Filtering can be found at the instrument level where analog filters are typically used to remove high frequency noise, and at the controller level where digital filters are used to remove mid-frequency noise (though the move to "Smart" instruments is moving the digital filter into the field instrumentation level as well). Common forms are: the "exponential" filter, which places more weight on the most recent measurement; and, the "moving average" filter, which treats all past values with the same weight. Both techniques effectively remove high frequency noise, but can over-damp the process when too much filtering is applied (i.e. "over-damping" makes the process more sluggish by increasing the apparent process time constant).

Signal "aliasing" is a problem that arises from the discrete sampling nature of digital controllers. A signal can be sampled too slowly for the frequency of the process variation and can make the process appear to have a much slower variation than the true underlying process. This is the basis of aliasing and if not detected will cause the improper tuning of digital filters and PID controllers resulting in poor controller performance. Moreover, the PID controller will react to compensate for the aliased signal instead of the true signal. Aliasing is a potential problem in using moving average filters with incorrect sampling times.

Texts and papers addressing controller basics provide implementation guidelines for signal conditioning.

Field Measurements. Following is a discussion of field measurements focusing not on the specific instrument, but on the issues influencing the control signal. Detailed discussions of suitable instrumentation may be found elsewhere in this text and in previous publications (e.g. Flintoff and Mular 1992, Hathaway 1982.)

Weightometers - Tonnage measurements from belt scales always require some form of filtering. Highly variable raw measurements may contain tonnage spikes exceeding the scale measurement limit (called *saturation*) leading to under-reading of the feed rate. Feeders generally have a response time in the range 4 to 10 seconds so variations with frequencies more than 0.25 cycles per second should be removed with a suitable filter. Regular calibration checks are key to reliable belt scale measurement.

Flowmeters - Magnetic flowmeters are widely used for measuring flow rates of water and slurry additions. The instruments are usually reliable and require little maintenance if installed in the correct location. Flow control valves are usually fast acting so flow measurements do not usually require much filtering. Small changes in flow can be immediately corrected by small changes in valve position. However, on large lines eliminating these many small position changes will extend the life of the larger control valves without affecting the relatively robust grinding process. The same is true for flow signals used in mass flow calculations where rapid variations in valve position can and should be avoided.

Density - Nuclear density gauges are the most common devices used to measure slurry density in grinding circuits though differential pressure cells are sometimes used in sumps, spiral classifiers and flotation columns. The slurry density is converted to percent solids by mass using an assumed average solids density. Cyclone feed density is commonly measured, and though a useful signal, measuring cyclone overflow density can be difficult. Successful cyclone overflow measurements have been achieved however, by taking a thief stream through an on-line device such as a particle size monitor, etc. Calculated steady-state cyclone overflow density is also used in some control schemes.

Power – Accurately monitoring (and controlling) mill power is important for virtually all mill control schemes, and critical in most AG/SAG applications. Electric or electro-mechanical meters are typically used to measure mill and crusher power. The mill power signals are often somewhat noisy and require some degree of filtering, while crusher power signals are typically very noisy and must be filtered to be usable.

Bearing Pressure - Mill bearing pressure (i.e. hydrostatic lube system back pressure) is a critical reading for AG and SAG applications and requires special attention, as early detection of changes in an AG/SAG mill load is vital. Bearing pressure readings are affected by both feed and mechanical condition of the mill and can show variations due to the mill rotation (Perry and Anderson 1996). If the feed and mechanical variations are filtered correctly any true change in bearing pressure is revealed. If too much filtering is applied the changes in bearing pressure are dampened resulting in an unacceptable and unnecessary apparent time delay. The main noise frequency on a bearing pressure signal will likely be twice the rotational speed of the mill. If this signal is sampled too slowly there will be aliasing and subsequent control problems. Successful filtering techniques are frequency (FFT) and noise cancellation based. As bearing pressure is sensitive to temperature and oil quality (viscosity effects) logic involving bearing pressure should compensate for drift.

Mill weight – Mill weight is a key measurement in classical AG/SAG control strategies (Mular and Burkert 1989). Mill weight is now commonly measured with load cells strategically placed under the mill bearings and is proving a reliable replacement for the bearing pressure signal, though often both signals are used. The advantage of load cells over bearing pressure is their measurement of absolute weight and the reduction in signal noise, independent of lube viscosity, temperature and other effects (Evans 2001). They are most common on newer mills as the retrofit cost and effort on older mills can be high. The disadvantage is their relative cost compared to the readily available pressure signal, and replacement expense should a sensor fail. The installation must be done carefully as incorrect installation of items such as the load cell cables will cause significant measurement error (e.g. keep cables away from high tension power cables and use consistent wire lengths for equal signal attenuation) (Jones and Wright 2001).

In large SAG mills load cells are installed under the non-drive end of the mill (Marshall 2000). This "end weight" is used to calibrate the mill load. Prior to startup the mill is filled with water and the cells calibrated to the known mass. During this initial calibration the proportion of weight on the cells (i.e. at the non-drive end) is assumed constant throughout the mill life and used to determine the mill contents under the various loading conditions encountered during operation.

The data from load cells also contains information about the condition and performance of lifters. The extraction, analysis and use of this information for maintenance and control is an interesting area of development (Marshall 2000).

Sound - The wide spectrum of sound emanating from an operating mill contains a wealth of information about the internal processes and is a promising area for future mill control developments. In the past the sound intensity over a certain frequency band was used to control mill feed or mill speed, but not without some problems (e.g. the sound signal could be similar when a mill is running empty, as compared to running well, and is a function of the frequency monitored, ore type, etc.).

Recent research and commercial products analyse the frequency spectrum more intensely (e.g. Pax 2001). Modern computing equipment allows for continuous digitizing and decomposition of the sound signals and allows the determination of both the position and nature of particle impacts on the liners. In one case, the characteristic spectrum of each type of event in a SAG mill (rock on rock, ball on liner, etc.) is used to determine the relative amount of each event. In another, product intensity peaks in certain frequency bands are counted to estimate the number of ball on liner impacts per second. Suitable microphones and conditioning of the signals are therefore required to isolate the important frequencies.

Image - Vision systems are being implemented in AG/SAG and crushing circuits to qualitatively characterize the feed materials (Norbert 2001, Girdner, Handy and Kemeny 2001, Edwards, Flintoff and Perry 1997). The size distribution of material on a conveyor belt can be measured with one of several commercially available image based systems. The systems vary in the techniques used to manipulate the two-dimensional video image to produce a three-dimensional size distribution. The conversion must account for differing orientation of particles, stratification on the conveyor belt and the problems associated with overlapping particles. In all cases the 2-D image is first processed to delineate the particle boundaries, where; in one system an equivalent ellipse is fit to each particle outline and an empirical correction used to convert this to the size distribution, and in another the chord lengths across particles are used to determine their size and distribution. The discussion continues on the merits of each, and their usefulness for control, but in both cases several images need to be analyzed to produce a statistically significant number of measurements.

Models (phenomenological) - Process models comprise another form of process "measurement". Mathematical models, so called "soft sensors", can be used to provide estimates of process values either difficult to measure, such as ore grindability or mill filling, or key physical measurements requiring modeled redundancy (Broussaud, Guyot, and McKay 2001). The former is based on phenomenological models, which are developed from deep process understanding. In the case of AG/SAG grinding they can describe the relationship between power, pressure, mill speed and grinding rates. This technique also provides a model structure that can be used to describe the entire circuit. Such models are currently in use (e.g. Samsog et al. 1996) and have been used by optimization algorithms to determine the input values needed to best meet a predefined performance criteria. The models can also be used to predict near-future circuit behavior.

The Kalman filtering technique provides a mechanism to adjust the model parameters according the latest actual process measurement (Herbst 1989). The model predicted values and measured process values are weighted according to their relative reliability and used to update the model coefficients. The newly calibrated model is used until another new process reading is available, the cycle then repeats. The updating process is continuous and is particularly useful if the time between process readings is long (e.g. a multiplexed particle size analyzer might provide a particle size reading every 30 minutes with estimated values calculated every minute between readings). The Kalman filter, in combination with a phenomenological model, also provides "measurement" redundancy should field instruments fail. Again, using the particle size analyzer example, should the analyzer feed lines plug causing a loss of signal then the model estimates can be used in the interim until the analyzer is repaired.

Models (neural networks) - A neural network (NN) is a mathematical model that tries to match the functionality of the brain in a very simplified manner. It consists of processing elements analogous to biological *neurons*. The neurons each have a number of internal parameters called weights. Altering the weights will change the network response to a stimulus (e.g. inputs, measurements, etc.). The processing elements are usually organized into groups called layers with a typical network consisting of one or more layers. The network interacts with the outside world through interconnections of some of the layers with input and output buffers. The input buffer holds the data presented to the network and the output buffer holds the corresponding response of the network. The goal is thus to choose the weights for all the neurons in the network to achieve the desired input/output relationship. This process is done automatically by the NN and is known as training or learning.

For control applications two networks may be required. The first network is trained to model the plant. It is typically used to infer the state of the process. A second network may be trained to control by using the first network to back-calculate the required control action. This control action then becomes the desired output for the back-propagation training of the "controller" network (Nguyen and Widrow 1990).

Flament et al. (1990) in their conclusion, summarize well the problems and expectations of neural networks:

Due to their ability to approximate any relationship between variables, artificial neural networks (ANN) carry on the expectations of many people. In mineral processing, ANN could be applied in modeling, simulation and automatic control of processing plants. However, due to their non-linear features on one hand and to the difficulties encountered in designing a network on the other hand, this new technology should be preferred only when traditional techniques fail. ANN are at their best when non-linearities or complexity are the main features of the problem to be solved. Intelligent sensors, when real devices are unavailable, sensor recordings filtering, when traditional filters do not work, and process modeling, when simple transfer functions are not good enough, are some examples of such applications.

Rates of change – A calculated *rate of change,* though strictly speaking not a field measurement, can be a very useful control signal. For example, ball mill viscous overloads are characterized by rapid drops in mill power that need to be differentiated from acceptable long term trends and inherent process noise (Austin and Flintoff 1987). Monitoring the rate of change of the mill power draw provides a key indicator that can be used to trigger corrective action before the physical manifestations of the overload are seen (e.g. excessive chip and steel rejection from the mill discharge). Similarly, monitoring the rate of change of AG/SAG power and load is the key to implementing the classical power/load mill feed controller. Rates of change may be calculated using the simple difference between values over time, the difference between current value and a heavily filtered value, a linear regression through a set of points versus time, and the cumulative sum of differences with respect to a reference.

Actuators

There are a limited number of manipulated variables in a grinding circuit. Whilst there are many possible measurements and inferred values, control schemes can only manipulate one of the following values:

Tonnage - The feed to a grinding circuit is one of the primary control mechanisms. Depending on the grinding circuit objectives, fixed tonnage or a maximum tonnage strategy may be required. In AG/SAG mills tonnage may be the only effective control mechanism. Typically, production considerations dictate that the supervisory control strategies increase tonnage as one of the first remedial steps, but decrease tonnage as one of the last. Usually, feed tonnage is manipulated by adjusting the speed of reclaim feeders under fine ore bins or coarse ore stockpiles. Feeder dynamics and conveyor belt length may dictate the use of more sophisticated control algorithms should the time lag be significant and/or if multiple feeders are present (Anderson, Perry and Neale 1996).

Water – Water additions to the grinding circuit are used to maintain primary mill feed density (e.g. water ratio to feed rate), maintain secondary mill rheology, manipulate ball mill recirculating load (e.g. to hold the cyclone feed density setpoint), and/or control cyclone overflow density. Water is often controlled to a flow setpoint ratioed or cascaded from a primary controller, e.g. feed water flow rate from the feed solids tonnage controller, or cyclone feed water flow from a cyclone feed density controller.

Pump speed - Pump speed is often used to control pump box level in the face of changing feed flow rates and/or recirculating loads (e.g. cyclone feed pumps and pump boxes). They are particularly useful in AG/SAG circuits where feed rates will necessarily change to accommodate mill load conditions.

Mill speed - Variable speed mills are viewed as a necessity by some, but an indulgence by others. The scale is often tipped by the capital cost of the variable speed motor weighed against its initially perceived benefits. Variable speed is a tremendous advantage where ore hardness varies, as it will affect both the mill's grinding rate and discharge rate. It may be the only alternative when significant changes in ore hardness or feed size cannot be accommodated by changes in tonnage and/or water; i.e. to maintain either mill throughput or a safe mill charge.

The most flexible and expensive variable speed drive is the wrap-around motor which allows speed to be changing continuously from 0 rpm upwards. The ring motor is favored for large mills where mechanical constraints preclude the use of gear drives. A ring motor can also serve as an inching drive. More limited but less expensive drives, such as the load-commutated inverter (LCI), allow speed changes over a much narrower range such as 65% to 80% critical.

Large *ball mills* have also been fitted with ring motors. In situations where the friability and hardness of the ore is expected to vary considerably over the mine life, variable speed drives have been used to accommodate the differing balance in grinding load between SAG mill and ball mill circuits.

Controllers

The controller executes algorithms to manipulate the actuator based on process measurements, and attempts to hold the process at the prescribed setpoint (ratioed, cascaded, remotely set by a supervisory routine, or operator entered). The most common type is the Proportional-Integral-Derivative (PID) controller, though other types are seeing more use; e.g. the dead-time compensating routines.

PID Control Theory. The PID controller is the standard control algorithm used in virtually all regulatory control systems. It is implemented in many, slightly differing forms which can give rise to some confusion. The PID is so ubiquitous that a good understanding of its use and limitations is essential for optimum control. The reader should refer to a practical process control textbook for a detailed analysis (Stephanopoulos 1984), but the important principles are described below.

Given a setpoint and a measurement, the PID calculates the value of its output as a function of any combination of the following three values:
- **Proportional**: size of the difference between setpoint and measurement, called the error.
- **Integral**: the integral of error over time.
- **Derivative**: the rate of change of either the error or measurement.

It is instructive to look at one form of the PID controller, which is given by the equation:

$$u_t = u_{t-1} + K_c \left[\Delta e_t + \frac{T_s}{\tau_i} e_t + \frac{\tau_d}{T_s} \left(\Delta e_t - \Delta e_{t-1} \right) \right]$$

where: u_t = PID controller output at time t
K_c = Controller gain
e_t = Error (setpoint – measurement)

T_s = Sampling time
τ_I = Controller integral time
τ_d = Controller derivative time
Δ = difference operator e.g. $\Delta e_t = e_t - e_{t-1}$

The derivative action is sensitive to small variations in the measurement so it is infrequently used in practice unless correct filtering is employed.

PID Sampling Interval. All measurements will be contaminated by variations of one kind or another and it is the frequency of these variations that needs to be considered. It is important to understand whether the controller and/or process can remove variations of a particular frequency range (e.g. 20 seconds per cycle or slower), or whether the variation is too fast to remove and thus can be ignored. Selecting the appropriate measurement filtering and tuning constants is therefore critical and is dictated by three factors (Vien et al. 1998 and 2000):

- The actual dynamics of the process to changes in manipulated variable.
- The desired dynamics of the process to changes in manipulated variable.
- The amount of fast and slow variations in the raw measurement compared to the dynamics of the process.

One factor that is commonly overlooked is the execution frequency of the PID loop or sampling interval of the measurement (T_s). Most control systems allow this time to be set for each loop, but it is most often left at the default value or set according to processor loading rather than process control considerations. Setting T_s too short makes the controller react mainly to high frequency noise rather than low frequency process changes. Setting T_s too long and the controller may miss important dynamic information. Selection of T_s remains subjective but general guidelines are:

- Based on physical variables of flow, $T_s = 1$ second or less; and level, $T_s = 5$ seconds;
- Based on open loop process response then:
  - $0.1 < T_s/\tau_{max} < 0.2$ where τ_{max} = dominant process time constant
  - $0.2 < T_s/\tau_d < 1.0$ where τ_d = process deadtime
- Based on controller settings:
  - $T_s > 0.01\tau_i$ where τ_i = controller integral time

GRINDING CIRCUIT PROCESS CONTROL

The level of process control; i.e. the degree of sophistication and complexity, can be loosely tied to process performance.

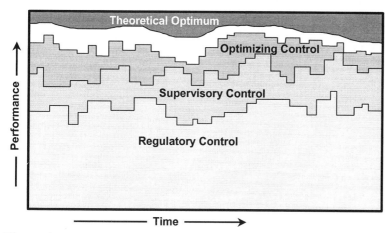

Figure 4 – Increasing Performance with Level of Control (after Rogers 1985)

As suggested by Rogers (1985) and shown in Figure 4, increasing sophistication generally means an increase in process performance. Regulatory control provides a large proportion of the benefits to be derived from improved control. Successively higher levels of control contribute, but perhaps to a lesser extent, and may be required to attain the highest circuit performance. The ultimate *theoretical optimum* can understandably never be achieved (e.g. 100% energy efficiency).

BASIC REGULATORY CONTROL LOOPS

The minimum acceptable level of process control is the regulatory control level. At this level the control system simply maintains setpoints to compensate for process disturbances and setpoint changes. Typical regulatory control loops in a grinding circuit are:

- Mill feed tonnage control (to reclaim feeder speed control)
- Mill feed water control (ratioed from mill feedrate control)
- SAG Mill sound control
- Crusher cavity level control
- Pump box level control (often with a variable speed pump)
- Classifier feed water flow rate control (to maintain a cyclone feed density)
- Cyclone feed pressure control (discretely opening and closing cyclones)
- Ball mill feed water flow rate control (to maintain internal mill rheology)
- Reagent flow control

Mill Feed Tonnage Control

Mill feed tonnage control is usually a simple PI control loop, though because of time lags and noise can be difficult to tune. Where there is a recycle stream many operators prefer to control fresh feed rather than total mill feed, allowing a supervisory loop to make changes based on recycle tonnage. The deadtime between the feeders and the weightometer is a common problem. A PID controller must be *de-tuned* (e.g. tuned to react slowly) if there is significant deadtime. In theory, a different type of control should be used if the deadtime exceeds the process time constant. In practice, if the deadtime is more than 5 times the response time of the feeders it is better to use a deadtime compensation controller such as a Dahlin Algorithm or Smith Predictor (e.g. if it takes 10 seconds for a feeder to complete a speed change then a deadtime of more than 50 seconds should be handled by the Dahlin algorithm).

The deadtime between the feeders and the mill may appear quite long, sometimes up to 2 minutes. However, the significance of this time must be compared with the response time of the mill (i.e. to changes in feed tonnage) before deciding if any special control is justified. Variable speed SAG mill feed conveyors have been installed on several large mills in recent years with the hope of using belt speed and feeder speed changes together to accelerate changes in actual mill feed. These do not necessarily improve control because for control stability the scheme must be detuned such that the response of the feeders *and* conveyor is similar to that achievable with a single control loop on a fixed speed belt. For example, with both feeder and conveyor being speed controlled there would be a controller with a process time constant of 100 seconds and 0 seconds of deadtime; whereas with a fixed speed conveyor there would be a controller with a ~15 second time constant and 75 seconds of deadtime – it is a tradeoff between the process time constant and deadtime, but deadtime can be handled effectively with special algorithms previously mentioned, a long process time constant requires a slower control response.

Mill Feed Water Control

Mill feed water is used to control the grinding conditions in the mill. Adding water can be used to increase the slurry discharge rate from the mill. The sensitivity of an AG or SAG mill to changes in water addition is ore dependent, with softer ores typically being less responsive. Water is also usually added at the mill discharge to give the correct conditions for downstream processes such as screening, classification or further grinding.

Mill feed water is added to the primary mill feed chute in ratio to the feed tonnage. To ensure consistent mill density it is important to note that in this case the "feed tonnage" is the actual mill feed; e.g. fresh feed plus recycle (as pebbles or classifier underflow, etc.). Additionally, the weightometer may be a significant distance from the mill so the tonnage measurement may need to be delayed in the calculation. The feed water control can be configured to act as either a simple ratio or as a calculated mill discharge density.

SAG Mill Sound/Speed

A sound controller can be used to regulate SAG mill speed (Perry and Anderson 1996). The sound control loop cascades the speed setpoint to the drive controller. In some installations the sound controller cascades a setpoint to a speed controller in the mill control system, which, in turn, sends a setpoint to a controller in the mill drive. In this latter triple cascade configuration however, the "extra" loop unnecessarily serves as an additional filtering stage and slows overall control performance. It should be excluded to simplify tuning of the sound controller.

Sound to speed control can be used as mill shell protection and as a mill load adjustment (e.g. adjust speed before resorting to reducing feed tonnage). Increasing mill speed will typically increase mill sound and vise versa for decreasing mill speed.

Crusher Level Control

Shorthead crushers are often used as recycle crushers in AG/SAG grinding circuits. Crushing efficiency is highest when the crusher feed cavity level can be maintained for choke feed. Being able to regulate recycle crusher federate, though more expensive, is always justified. Power monitoring and control of crusher feed is used as a safety override. Mechanical devices to bypass the crusher feed are needed to divert tramp metal (that has made it passed the belt magnet and other safe guards) and to provide a release should the crusher cavity begin to overfill. A surge bin with feeder controls provides a buffer and temporary relief from high recycle rates, but cannot maintain a lower feed rate when the bin fills. A feed bypass should always be available. Manual or automatic adjustment of the crusher gap provides a further adjustment if choke feeding cannot be maintained, or the cavity level gets too full.

Pump Box Level Control

Pump box level control is required to maintain balanced grinding circuit operation. It is not necessarily important to control the level at a precise setpoint, but rather it is important not to let the pump box overflow or run dry. An overflowing pump box creates spillage that needs to be cleaned up and clean-up introduces unmeasured, uncontrolled water into the circuit while at the same time increasing the possibility of chemical contamination (e.g. oil into flotation circuits). A very low pump box level allows the pump to cavitate, disrupting the slurry flow and negatively effecting cyclone performance. It is typical to allow the level to "drift" anywhere in the upper portion of the pump box (e.g. 70% level setpoint plus or minus ~25%). The level is often detected using an ultrasonic device and controlled with a variable speed pump. A lower limit must be placed on the pump speed to stop it from slowing too much and allowing the slurry to settle in the lines.

In cases where a fixed speed pump is used, water addition is manipulated to maintain level. This option is satisfactory only if downstream units can readily compensate for changes in density. If the pump is in closed circuit with a hydrocyclone then manipulation of water for level control complicates circuit operations because of its impact on cyclone feed density and the circulating load (i.e. classifier efficiency changes with changes in feed density). It is preferable to control level with pump speed in this case as it provides more direct "measurement" to "actuator" response.

Classifier Feed Water Flow Rate Control

Water is added to the classifier feed to manipulate/maintain the classifier feed density (especially with hydrocyclones) and the circulating load. It is often added to an operator entered setpoint, but can be cascaded from a (cyclone) feed density or mass flow controller. As a cautionary note, a great deal of water may be required to maintain the density setpoint if the circuit is unstable or the density setpoint is even slightly too high or too low. Control logic should be included to protect the water flow rate setpoint from ramping up too high or too low.

In some instances, the density control is tied directly to the control valve (e.g. a direct acting loop and not a cascade loop). This practice can cause a breakdown in control if the density gauge becomes unreliable or the circuit is too unstable. Should the density controller need to be disabled for any reason, the water flow control reverts back to manual manipulation of the valve position with no feedback for the flow rate. Clearly, a flow rate setpoint is preferred over a valve position setting, especially if water header pressures are know to vary.

Where classifier product (e.g. cyclone overflow) percent solids must be maintained, the classifier feed water setpoint is typically adjusted based on an online circuit solids mass balance. The classifier product is not often easily instrumented and it is usually not possible to obtain a reliable slurry percent solids reading due to piping and instrumentation considerations for direct density control.

Cyclone Feed Pressure Control
Cyclone feed pressure control is necessary if wide swings in circulating load are expected (as can be anticipated with AG/SAG circuits). Cyclones operate most effectively with a steady feed flow rate and steady operating pressure. When the pressure gets too high an additional cyclone needs to be opened and when the pressure gets too low a cyclone needs to be closed. This process can be automated through the use of air-actuated knife gates tied to a cyclone header pressure controller. To affect this control, the cyclone header must have enough available cyclones so that the addition or removal of one cyclone does not create too much disturbance. Typically five or six operating cyclones are the minimum. A controller can be configured to open a cyclone above a pressure setpoint plus a deadband and close a cyclone below the setpoint minus a deadband. This deadband is important in order to ensure that the cyclones do not open and close too often. Cyclone pressure control may also be linked to cyclone feed pump speed to provide a constant flow per cyclone.

Ball Mill Feed Water Flow Rate Control
Sometimes secondary ball mill feed water is necessary to maintain a suitable mill rheology; e.g. if direct classifier underflow contains too little water causing viscous overloads. This can happen when treating ores with significant clay content. Ball mill water is typically added to an operator entered setpoint at a level just enough to stop the onset of overload. Too much water can decrease the mill density to a point that grinding efficiency is lost. Water addition to the classifier feed includes any mill feed water and a total water feed controller is used to control water addition in the classifier feed sump.

Reagent Flow Control
In many installations reagents are added to the grinding circuit. A typical example is lime addition to maintain a certain flotation feed pH. Reagents can be added dry on the mill feed conveyor or in liquid form at the mill feed water addition point. Reagent can be ratioed to a feed tonnage or to a feed metal content; e.g. frother to feed tonnage and collector to feed metal. A pH probe can used to provide a feedback signal to the pH controller. The dynamics of pH response and each lime addition point must be measured. In many operations lime is added in ratio to SAG mill feed and then to ball mill feed by the pH controller. The dynamic response of pH to lime from SAG feed to cyclone overflow is very similar to that for ball mill feed to cyclone overflow. In this case there are advantages to controlling lime valves in parallel rather than in series.

SUPERVISORY CONTROL

Multiple feeder control

The feed to a primary mill is controlled by changing the speed of one or more feeders. The tonnage control loop must provide both disturbance rejection (regulatory control) and response to setpoint changes (servo control). In larger operations ore is typically fed by two or more feeders at a time. The ratio of speeds between these feeders is manipulated to achieve a desired ore blend or stockpile profile. It is important to ensure that changes in the feeder ratio do not disturb the mill feed tonnage control. This includes starting and stopping feeders, which is equivalent to changing a feeder's ratio to and from zero. Multiple feeder control (Vien et al. 2000) is used to make the tuning of the tonnage control independent of the number of feeders running and their ratio.

As an example of multiple feeder control consider a system of three feeders where a 1% change of speed of a single feeder was found to give an 11.2 tph change in tonnage. If each feeder's speed were simply a tonnage controller (WIC) output multiplied by its ratio (e.g. the proportion of feed to be supplied by the given feeder), the gain of the process would appear differently to the controller depending on the feeder ratios used, as shown in Table 1. This would require changing the tuning of the tonnage controller depending on the feeder ratios, a clearly unacceptable proposition that would result in all feeders being conservatively tuned and likely unnecessarily slow.

Table 1: Example of feeder ratio control without compensation

Change in WIC output	Ratio of Feeder #			Change of speed of Feeder #			Change in tonnage
	1	2	3	1	2	3	
+1%	2	3	4	+2%	+3%	+4%	78.4 tph
+1%	0.5	0.8	1.2	+0.5%	+0.8%	+1.2%	28 tph

In order to make WIC tuning independent of feeder configuration, the speed of a feeder is calculated using the following equation:

$$Speed_i = (WIC.CO \times N) \times \frac{Ratio_i}{\sum_{i=1}^{N}(Ratio_i \times Run_i)}$$

where: $Speed_i$ = speed output to feeder 'i'
WIC.CO = Output of tonnage controller (%)
N = Number of feeders installed
$Ratio_i$ = Ratio of feeder 'i'
Run_i = Running status of feeder 'i' (1=running, 0=stopped)

Using this equation, the example in Table 1 can be recalculated as shown in Table 2 so the tonnage controller tuning does not need to change. Note that this scheme relies on a constant feeder gain (tonnage per % speed). The feeder gain may vary with speed so linearization can be added to make the feeder gain appear constant to the controller. In addition, a delay and filter can be added so that the response of the 2nd and 3rd feeders appears to be the same as for the 1st feeder. For example, consider the installation shown in Figure 5 where the time taken for a section of belt to travel between feeders (t_{1-2} and t_{2-3}) is significant compared to the time taken for a feeder to change speeds.

Table 2: Example of feeder ratio control with compensation

Change in WIC output	Ratio of Feeder #			Change of speed of Feeder #			Change in tonnage
	1	2	3	1	2	3	
+1%	2	3	4	+0.7%	+1%	+1.3%	33.6 tph
+1%	0.5	0.8	1.2	+0.6%	+1%	+1.4%	33.6 tph

If the ratio of feeder 1 is decreased, the speed of feeders 2 and 3 must increase to deliver the same tonnage, which is shown as the height of material on the belt. Feeders 2 and 3 should wait until the section of belt that was under feeder 1, when it was slowed down, reaches them before changing their speeds.

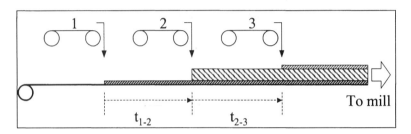

Figure 5 - Multiple feeder discharge onto belt

SAG Mill Power/Pressure Tonnage Control

SAG mill performance is a function of the mill charge. The volume and mass of the charge are seldom measured in industry, but are typically inferred from the power draw and weight of the mill either directly, by using a soft sensor, or indirectly in the control logic. The amount of charge in the mill is controlled by changing the feed tonnage (material in) or the discharge rate (material out). The discharge rate can be manipulated by changing feed water or mill speed. Characterization of any given mill is required to determine which control structure is suitable.

The most common method of controlling mill charge is by using a weight (or pressure) and/or power controller to manipulate mill feed tonnage. The choice between power and weight/pressure depends on the process and mill conditions. Various schemes have been tried with combinations of power and weight/pressure. Mular and Burkert (1989) discuss triple cascade loops of power/load/tonnage or power controls with rate of change of power and pressure monitoring. All of which have been successful to a certain degree, but are restrained in their effectiveness by the limitations inherent in the PID control structure employed.

Direct PID control of power is typically problematic due to the classical AG/SAG overload relationship (i.e. a decrease in AG/SAG power when in an overload condition) and to the fact that maximum power draw does not always correspond to maximum throughput, especially for softer ores. In this case it may be preferable to control load (at the optimum level). On variable speed mills power control is used as protection. For either type of control, the response is usually slow so an appropriate control type and execution frequency must be used. PID control is often not suitable.

SAG Overload

Detection and prevention of SAG mill overload is one of the most common supervisory control routines. A classic SAG mill overload is indicated by falling power draw with a simultaneous increase in mill load (e.g. increasing weight or bearing pressure). Under overload conditions so much material has accumulated in the mill that the motion in the charge is inhibited causing a drop in grinding efficiency and further accumulation of material. As the charge volume increases, its center of gravity moves towards the mill centerline, producing the drop in power draw. The larger charge volume, however, gives a rise in bearing pressure (and mill mass) at the same time.

In large SAG mills treating softer/finer ores the drop in power may be not be evident as it occurs only at very high loads (possibly beyond the mill operating range) and the only indication of overload is the net gain in mill weight. If an overload condition persists the charge volume may grow until it spills out of or pushes back into the feed chute, resulting in severe mechanical damage. The exact conditions that identify an overload for a given mill must be determined by observation. The actions required to correct an overload condition will also be specific for each mill.

Typically control logic examines the rate of change of power and pressure (or weight) to detect an impending overload and automatically triggers corrective action; i.e. cutting feed tonnage by half. The method of calculation of rate-of-change is important.

SAG Tonnage and Speed Control
The load in a SAG mill will change with varying ore conditions and may develop into a serious underload or overload condition if corrections are not made. The relative effect of tonnage, mill speed and feed water on load for any mill must be determined from mill observations prior to designing a suitable load control strategy. A common control strategy in SAG mills, where the objective is maximum tonnage, is to try and add tonnage as the first reaction to decreasing load. Similarly, cutting tonnage is the last reaction to increasing load.

Crusher Gap
Crushers are used in grinding circuits to accelerate size reduction of fractions that are ground inefficiently in a mill (e.g. the proverbial "critical size"). The crusher throughput is controlled by the crusher gap setting. A supervisory loop evaluates the accumulation of material in the grinding circuit to determine if problems can be rectified by changing crusher gap. Modern crushers have the capacity to automatically adjust the closed side setting. These crushers usually have logic from the manufacturer to open the gap based on excessive vibration or power draw and close the gap based on power. Optimizing crusher use can have a significant effect on throughput. As an example, for a semiautogenous-ball mill-crusher (SABC) circuit the control strategy will determine if the gap should be adjusted to accommodate all recycle tonnage or if the crusher is held at the minimum setting with any tonnage that cannot be crushed simply bypassing the crusher.

Ball Mill Circulating Load
The circulating load (CL) is the amount of material returned to the mill from the classifier expressed as a percentage of the circuit feed. (In many circuits the classifier is a cyclone and because of this will be used as an illustration in the following discussion.) The CL cannot easily be measured directly, but can be inferred from a measurement of cyclone feed rate (density and flow rate measurement) and/or changes in the pump box level or cyclone feed pump speed and pressure (e.g. increasing CL as indicated by increased cyclone pressure, increased pump speed and/or increase pump box level). Cyclone pressure and cyclone feed pump speed control are used to essentially change the capacity of the classifier to accommodate different circulating loads.

There are two independent mechanisms for changing circulating load:
- <u>Changing circuit feed rate</u>. An increase in feed rate causes a coarser product size and also a higher circulating load. The response to feed rate changes is moderately slow. The circuit moves to a new equilibrium with the maximum change occurring once the new equilibrium is reached.
- <u>Changing cyclone feed water addition</u>. An increase in cyclone feed water addition causes an immediate drop in the cut size of the cyclone, giving a finer product in the

short term. This causes a net accumulation of material in the circuit shown by an increase in circulating load. As the circuit reaches steady state, the cyclone product becomes coarser eventually bringing the product fineness down to a level close to where it started. This process can take between 20 minutes and 3 hours depending on the size of mill and the circulating load.

The use of each control mechanism depends on the control objective. Ball mill circuit control is complicated by the interaction between classification and grinding and is an area where carefully executed simulation and plant tests are recommended to determine appropriate control solutions. One solution is to use a circulating load controller to control water addition and let product size control feed tonnage (Wills 1988, Pg. 298).

Balancing Primary and Secondary Circuits

In multiple-stage circuits the balance between the grinding load of the primary and secondary (and tertiary) circuits can be controlled to a moderate degree to increase performance. If a secondary ball mill circuit is at maximum capacity, a supervisory control can determine if it is possible to shift part of the load to the primary circuit. The converse is also possible; that is, if the primary circuit dictates capacity it may be possible to shift part of the load to the secondary circuit. The idea is to balance the circuits such that both are utilized to their maximum capacity.

EXPERT SYSTEMS AND ADVANCED CONTROLLERS

Expert Systems

Expert systems (ES) have become widely used to monitor and control the interacting mechanisms in grinding circuits and are particularly popular in large circuits where the economic benefit of even marginal increases in performance are significant. Herbst, Pate and Oblad (1989), Hales, Colby and Ynchausti (1996), Broussaud and Guyot (1999), Sloan et al. (2001) have documented successful installations with "typical" performance improvements in the range of 4% - 8% increases in throughput. The challenge for a concentrator's management team is to qualify these "small" benefits in light of the various process factors that complicate the calculations. Gritton (2001) and McKay and Broussaud (2001) have provided techniques using random period on/off tests to accurately perform such a "statistically-defensible" benefits analysis. That being said, it is important that the circuit first be characterized to find the appropriate structure and control parameters to ensure that maximum benefit is derived from the ES. That is, the control strategy should be determined first and only then should the implementation strategy be contemplated.

Whilst the term "expert system" is now used to describe a range of advanced control solutions there are important differences between the actual *control* implemented. The term "expert system" strictly refers to systems that use *crisp* or *fuzzy logic* with heuristic "rules", but has become synonymous with a more general range of products that include ES's at their core in support of models, neural nets, etc. Broadly categorizing, the distinguishing features of the systems are:
- Rule-based control (fuzzy or crisp)
- Rule-based control + models (soft sensors)
- Rule-based control + models (soft sensors) + model based optimization

Grinding circuit ES's rely on encapsulating the type of analysis and decision making used by an impeccably trained and experienced operator. This knowledge is written as a set of interrelated rules to derive actions (e.g. setpoint changes) based on prevailing circuit conditions. The rules are applied consistently and continuously and if well written will allow all operating shifts to perform, as a minimum, at the level of the best operating crew. The setpoints are manipulated to achieve the predefined objectives within safe operating limits as set by the technical staff, operations, management, etc. This lets the operator direct the operation of the circuit whilst the expert system deals with interactions and identifies and eliminates any developing problems.

Augmenting field measurements with "soft sensor" estimates may further enhance control. As mentioned previously, models can be used to predict unmeasurable values, such as grindability, or estimate measured values (e.g. between measurements) based on current conditions. Current modeling techniques being employed provide on-line updating of the model parameters, such that the models are continuously tested with new process data and recalibrated when necessary. At the highest level, these models are used to calculate circuit setpoint changes that will maximize grinding performance.

Rule-based control
The structure of a typical rule-based grinding controller can be broken down into sequentially executed components that:
- Check the validity of each measurement used
- Identify any alarm conditions and decide on corrective action
- Analyse the circuit to identify any process problems and decide on corrective action
- Analyse the circuit to identify whether objectives are being reached and decide on action.
- Evaluate the relative importance of the actions and execute the most important.

This cycle is repeated at each control interval; e.g. between 20 seconds and 1 minute, depending on the dynamic response time of the circuit.

Optimizing control: Model based
When a model is available as a soft sensor, it is a natural extension to use this model to perform optimizing control. *Model Predictive Control* (MPC) is typically multivariable in nature and often involves predicting mill behavior for the near future (Hales, Colby and Ynchausti 1996, Herbst, Pate and Oblad 1989, Broussaud and Guyot 1999). One of the keys to a successful MPC implementation is to establish the appropriate optimization criterion for the process variables.

The criterion will often involve weighting certain variables more heavily than others, or penalizing utilization of certain manipulated variables more heavily than others. For example, grinding circuit product size is more important than pump box level and manipulating water should be preferred to manipulating feed tonnage for short-term variations. Though not used extensively in the mineral processing industry the optimization function can also include model predictions. The optimization function can also include constraints (i.e. equipment capacity limitations) either within the grinding circuit itself, or within upstream and/or downstream processes. Ultimately, the optimization function could directly express economic considerations.

Part of the hesitation in using MPC in mineral processing is the need to provide a shell around the main MPC algorithm to ensure a robust application. This shell may consist of an expert system that validates sensors, checks the model predictive capabilities and ensures that the MPC results are sensible. It may also include a component that controls the model parameter adaptation so that the model does not try to adapt to abnormal situations such as a sensor failure.

EXAMPLE CONTROL STRATEGIES
To summarize the discussion let us briefly consider some examples of grinding circuit control. The first example is a circuit under simple regulatory controls, and the second is a circuit including model based controls.

SAG mill / Ball mill circuit basic control
A simple PID control structure, as shown in Figure 6, is comprised of:
- SAG mill bearing pressure controller cascading a setpoint to the mill feed tonnage control
- SAG mill feed water ratioed to mill fresh feed + recycle tonnage
- SAG mill speed controlled by sound

- SAG mill overload detection or high motor power alarm causing a forty percent cut in feed tonnage
- Recycle (pebble) crusher level controlling feed to the crusher
- Cyclone feed density controlling water addition to the sump
- Cyclone feed pressure controlling number of operating cyclones (e.g. opening and closing cyclones on pressure)
- Cyclone feed sump level controlling pump speed
- Particle size analyzer monitoring cyclone overflow size

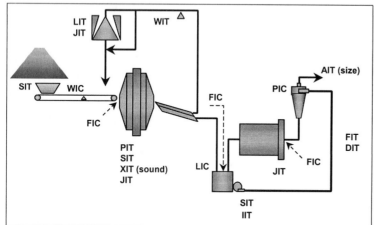

Figure 6 – Simplified Regulatory Control Loops for a SAG-Ball Mill-Crusher Circuit

SAG Mill / Ball Mill Circuit Advanced Control

The same regulatory control loops are employed as in the basic circuit of Figure 6, with the addition of a model-based component. The model-based optimizing control uses a Kalman filter to continuously adapt the parameters of a SAG mill model, ball mill model and cyclone classification model. An optimizer with the objective of maximizing throughput subject to a constraint of 80% finer than 120 microns uses the model inputs and finds the tonnage, SAG feed water and cyclone feed density setpoints that move the circuit toward higher throughput. An expert system verifies these new setpoints are reasonable, that the modeling is valid, and that the physical circuit is capable of accepting the recommended changes. The new setpoints are then applied subject to limits set by the operator (e.g. operator limited tonnage increases because of low stockpile levels, limited flotation capacity, etc.).

CONCLUSIONS

Strategies for the instrumentation and control of grinding circuits have matured over the past few decades and now encompass regulatory through optimizing controls. Where once regulatory PID controllers were the only available tools the practitioner may now select from a number of more advanced techniques. The fundamentals of circuit control, however, remain unchanged. Field instruments and PID controllers must not be ignored in the drive towards more sophisticated applications. Correct instrument installation and maintenance, suitable signal conditioning, good loop tuning and the appropriate controller implementation remain critical to sustain stable operations from a solid foundation of control.

The control hierarchy builds on the regulatory base to include supervisory and optimizing strategies. Regulatory controls provide process stability, supervisory controls ensure setpoint adaptation to changing process conditions, and optimizing controls maximize the fineness or circuit throughput objectives. A strategy based on a set of focused and obtainable objectives, supported by the requisite field instrumentation, will help ensure its success.

The future will see the incorporation of enterprise-wide models to optimize not only the grinding circuit, but also the mine and haulage systems (feeding the circuit), the flotation plant (upgrading the product), the dewatering plant and transportation system, and the smelter / refinery producing the metal. When taken as a whole, the inefficiencies that exist between the various production stages can be minimized and the overall *enterprise* optimized for maximum economic returns.

REFERENCES

Anderson L., R. Perry, A. Neale 1996. Application of Dead-time and Gain Compensation to SAG Feeder Control at P.T. Freeport Indonesia. *Proceedings of the 28th Annual Meeting of the Canadian Mineral Processors, Ottawa, Ont. January 1996. Pg. 360.*

Austin J.W., B.C. Flintoff 1987. Production Improvements Through Computer Control at Brenda Mines Ltd. *Paper presented at the American Mining Congress Meeting in San Francisco, September, 1987.*

Brown C.M. 1982. The Selection of Instrumentation and Control Systems for Semi-Autogenous Grinding Circuits. In *Design and Installation of Comminution Circuits*, ed. A.L. Mular and G.V. Jergensen II, Chapter 41. American Institute of Mining, Metallurgical, and Petroleum Engineers, Inc. Baltimore, MA: Port City Press, Inc.

Broussaud A., O. Guyot, 1999, Factors Influencing the Profitability of Optimizing Control, *Symposium Proceedings, Control & Optimization in Mineral, Metallurgical, and Materials Processing*, ed. Hodouin, Bazin and Desbiens. Quebec City, Canada, CIM, pp. 393-403

Broussaud A., O. Guyot, J. McKay 2001. Advanced Control of SAG and FAG Mills with Comprehensive or Limited Instrumentation. *Proceedings of the International SAG Conference held in Vancouver, B.C., September, 2001. Vol. II.* Vancouver, B.C. Canada: Pacific Advertising Printing & Graphics. Pg. 358.

Carriere K.C. 1982. Computer Control of Grinding Circuits. In *Design and Installation of Comminution Circuits*, ed. A.L. Mular and G.V. Jergensen II, Chapter 38. American Institute of Mining, Metallurgical, and Petroleum Engineers, Inc. Baltimore, MA: Port City Press, Inc.

Chang J.W. 1982. Primary Ball Mill Circuits. In *Design and Installation of Comminution Circuits*, ed. A.L. Mular and G.V. Jergensen II, Chapter 40. American Institute of Mining, Metallurgical, and Petroleum Engineers, Inc. Baltimore, MA: Port City Press, Inc.

Edwards R.P., B.C. Flintoff, R. Perry, 1997. Developments in Distributed Control and Expert Systems for Process Control. In *Proceedings from the SAG '97 Workshop, Vina del Mar, Chile.*

Evans G. 2001. A New Method for Determining Charge Mass in AG/SAG Mills. *Proceedings of the International SAG Conference held in Vancouver, B.C., September, 2001. Vol. II.* Vancouver, B.C. Canada: Pacific Advertising Printing & Graphics. Pg. 331.

Flament F., J. Thibault and D. Hodouin 1990. Potential Applications of Neural Networks for the Control of Mineral Processing Plants. *CIM Conf.*, Hamilton.

Flintoff, B.C., A.L. Mular Eds. 1992, *A Practical Guide to Process Controls in the Minerals Industry*. Chapter 7. MITEC. Published in Vancouver: Gastown Printers Ltd.

Girdner K., J. Handy, J. Kemeny 2001. Improvements in Fragmentation Measurement Software for SAG Mill Process Control. *Proceedings of the International SAG Conference held in*

Vancouver, B.C., September, 2001. Vol. II. Vancouver, B.C. Canada: Pacific Advertising Printing & Graphics. Pg. 270.

Gritton K., 2001, Methods to Document the Benefits of Advanced Control Systems, SME Annual Meeting, Preprint 01-020.

Hales L.B., R.W. Colby, R.A. Ynchausti, 1996. Optimization of AG and SAG Mills Using Intelligent Process Control Software. *Proceedings of the International SAG Conference held in Vancouver, B.C., September, 1996. Vol. II.* Vancouver, B.C. Canada: Pacific Advertising Printing & Graphics. Pg. 632.

Hathaway R.E. 1982. Selection and Sizing of Instrumentation and Control Systems; Size Controlled Grinding Circuits. In *Design and Installation of Comminution Circuits*, ed. A.L. Mular and G.V. Jergensen II, Chapter 39. American Institute of Mining, Metallurgical, and Petroleum Engineers, Inc. Baltimore, MA: Port City Press, Inc.

Herbst J.A., W.T. Pate, A.E. Oblad 1989. Experiences in the Use of Model Based Expert Control Systems in Autogenous and Semi Autogenous Grinding Circuits. *Proceedings of the International SAG Conference held in Vancouver, B.C. September 1989. Vol. II.* Vancouver B.C. Canada: First Folio Printing Co., Ltd. Pg. 669.

Jones R., A. Wright, 2001. Selecting and Configuring Load Cells for AG/SAG Grinding Mill Applications. *Proceedings of the International SAG Conference held in Vancouver, B.C., September, 2001. Vol. II.* Vancouver, B.C. Canada: Pacific Advertising Printing & Graphics. Pg. 227.

Lynch A.J. 1977. *Mineral Crushing and Grinding Circuits; Their Simulation, Optimization, Design and Control.* New York: Elsevier Scientific Publishing Co.

Maerz N.H. 2001. Automated On-Line Optical Sizing Analysis. *Proceedings of the International SAG Conference held in Vancouver, B.C., September, 2001. Vol. II.* Vancouver, B.C. Canada: Pacific Advertising Printing & Graphics. Pg. 250.

Marshall C., 2000. Successful Load Cell Utilisation to Measure AG/SAG Mill Charge Mass, *IIR Crushing & Grinding 2000*, Perth WA, 18-19 May 2000.

McKay J., A. Broussaud, 2001, Benefits Analysis of Expert Control Systems, SME Annual Meeting, Preprint 01-171.

Mular A.L., A. Burkert 1989. Automatic Control of Semiautogenous Grinding (SAG) Circuits. *Proceedings of the International SAG Conference held in Vancouver, B.C. September 1989. Vol. II. Vancouver B.C. Canada: First Folio Printing Co., Ltd. Pg. 651.*

Napier-Munn T.J., S. Morrell, R.D. Morrison, T. Kojovic. 1996. *Mineral Comminution Circuits; Their Operation and Optimization*, Julius Kruttschnitt Mineral Research Centre, The University of Queensland, Australia.

Needham T.M., G.V. Folland, 1994. Grinding Circuit Expansion at Kidston Gold Mine. Annual Meeting of the Society of Mining Engineers, Littleton, CO. Preprint 94-104.

Nguyen D.H., and B. Widrow 1990. Neural Networks for Self-Learning Control Systems. *IEEE Control System Magazine*, Vol. 10, No. 3, pg. 18.

Pax R.A. 2001. Non-Contact Acoustic Measurement of In-Mill Variables of a SAG Mill. *Proceedings of the International SAG Conference held in Vancouver, B.C., September, 2001. Vol. II.* Vancouver, B.C. Canada: Pacific Advertising Printing & Graphics. Pg. 386.

Perry R., L.Anderson 1996. Development of Grinding Circuit Control at P.T. Freeport Indonesia's New SAG Concentrator. *Proceedings of the International SAG Conference held in Vancouver, B.C., September, 1996. Vol. II.* Vancouver, B.C. Canada: Pacific Advertising Printing & Graphics. Pg. 671.

Rogers J.A. 1985. Optimizing Process and Economic Gains. *Chemical Engineering.* Vol. 92, No. 25, pg. 95.

Samsog P.O., P. Soderman, U. Storeng, J. Bjorkman, O. Guyot, A. Broussaud, 1996. Model-Based Control of Autogenous and Pebble Mills at LKAB Kiruna KA2 Concentrator (Sweden). *Proceedings of the International SAG Conference held in Vancouver, B.C., September, 1996. Vol. II.* Vancouver, B.C. Canada: Pacific Advertising Printing & Graphics. Pg. 599.

Sloan R., S. Parker, J. Craven, M. Schaffer 2001. Expert Systems on SAG Circuits: Three Comparative Case Studies. *Proceedings of the International SAG Conference held in Vancouver, B.C., September, 2001. Vol. II.* Vancouver, B.C. Canada: Pacific Advertising Printing & Graphics. Pg. 346.

Stanley G.G. 1987. *The Extractive Metallurgy of Gold in South Africa,* Chapter 3, The South African Institute of Mining and Metallurgy, Monograph Series M7, Johannesburg, RSA.

Stephanopoulos G. 1984. *Chemical Process Control; an introduction to theory and practice.* New Jersey: Prentice-Hall, Inc.

Vien A., R.P. Edwards, R. Perry, B.C. Flintoff 1998. Back to Basics in Process Control. *Proceedings of the 28th Annual Meeting of the Canadian Mineral Processors, Ottawa, Ont. January 1998. Pg. 337.*

Vien A., R.P. Edwards, R. Perry, B.C. Flintoff 2000. Making Regulatory Control a Priority. In *Control 2000,* ed. J.A. Herbst. Society of Mining Engineers. Littleton, CO. Pg. 59-70.

Vien A., J. Palomino, P. Gonzalez, R. Perry 2000, Multiple Feeder Control, Proceedings of the 32nd Annual General. Meeting of the Canadian Mineral Processors, CIM, Ottawa, pp. 295-312.

Will B.A. 1988. *Mineral Processing Technology; An Introduction to the Practical Aspects of Ore Treatment and Mineral Recovery.* Chapter 7. Camborne School of Mines. Printed by Pergamon Press.

Strategies for the Instrumentation and Control of Solid-Solid Separation Processes

Gerald H. Luttrell[1] *and Michael J. Mankosa*[2]

ABSTRACT

Solid-solid separation processes are used to upgrade a variety of materials such as coal, tin, iron ore, and heavy mineral sands. These processes, which include density, magnetic and electrostatic, are often operated without online instrumentation or automatic controls. In many cases, this mode of operation is dictated by complexities associated with the control of large numbers of individual separators. This article discusses the problems associated with the real-time adjustment of process setpoints based on online measurements of concentrate grade and suggests alternative modes of operation that are better suited for plant control and optimization. Case studies that illustrate the potential impact of these different approaches are presented.

INTRODUCTION

Solid-solid separation processes are routinely used to separate minerals based on differences in intrinsic properties such as specific gravity, magnetic susceptibility, or conductivity. Common examples of solid-solid separation processes are listed in Table 1. These unit operations are often incorporated into parallel and/or multistage processing circuits at industrial sites. Parallel circuits are required because of particle size limitations and throughput restrictions associated with these processes. For example, coal preparation plants are typically forced to include three or more parallel circuits as part of their basic flowsheet because no unit operation currently exists that can upgrade a very wide range of particle sizes. Likewise, mineral sands plants are forced to distribute feed slurry to hundreds of spiral separators or dozens of electrostatic separators because of throughput restrictions associated with these processes. Multistage circuits are commonly used in many of these operations to sequentially recover a variety of saleable minerals from the same feed stream. Several stages of cleaning and scavenging in series are also frequently needed to overcome inherent inefficiencies in these processes created by the random misplacement (imperfection) and bypass (short circuiting) of particles.

Table 1. Examples of common solid-solid separation processes.

Separator Type	Coarse Particles	Intermediate Particles	Fine Particles
Density	Dense Media	Spirals	Centrifugal Separators
	Pneumatic Jigs	Shaking Tables	Tilting Frames
Magnetic	Drum Magnet	Drum Magnet	High-Gradient Matrix
	Roll Magnet	Roll Magnet	WHIMS Carousel
High Tension	Electrodynamic Roll	Electrostatic Screen	---
		Electrostatic Plate	
Miscellaneous	Ore Sorters	---	---

[1] Department of Mining & Minerals Engineering, Virginia Polytechnic Institute & State University, Blacksburg, Virginia.
[2] Eriez, Erie, Pennsylvania.

The inherent complexities associated with the use of multistage unit operations and/or parallel circuits makes solid-solid separation processes less amenable to traditional approaches for automatic control. In some cases, this difficulty is created by a lack of reliable and affordable online analyzers that are needed to monitor large numbers of process streams. In other cases, the online manipulation of individual process setpoints to automatically control concentrate quality is simply impractical. For example, the development of a control system that could simultaneously monitor and adjust the setpoints of hundreds of spirals would be a formidable task. Some processes, such as wet drum magnets, are essentially impossible to control since they do not have any user adjustable variables. As a result, industrial instrumentation is typically limited to conventional sensors for monitoring volumetric flow rate, mass flow rate, slurry density, or pulp level. These parameters are commonly maintained at predefined setpoints using simple feedback loops for stabilizing control. Many of the key process variables that impact process performance are visually monitored and manually adjusted by plant operators. Common examples include the selection of specific gravity setpoints for dense media separators or the setting of splitter/cutter positions for spirals, magnetic separators, and electrostatic separators. Control loops are rarely used in industrial plants to optimize performance based on online measurements of concentrate grade or recovery. The installation of online analyzers and controls for the realtime adjustment of individual process variables has been attempted in some cases, but these endeavors have been largely unsuccessful in terms of improving productivity or profitability.

Despite the problems described above, most industrial plants that utilize solid-solid separation processes have the potential to significantly improve metallurgical performance using online instrumentation and automatic controls. This improvement can only be realized, however, by properly implementing a control strategy that incorporates a fundamental optimization principle known as the *incremental grade concept*. This concept points out the shortcomings associated with the realtime manipulation of process setpoints for plant optimization. More importantly, the concept also suggests alternative methods for optimizing plant yield that involve the supervisory control of plant blending practices. This article provides a general review of the concept of incremental grade control and describes industrial control schemes that currently make use of this optimization strategy.

INCREMENTAL GRADE CONCEPT

The optimization of solid-solid separation processes based on the concept of constant incremental grade has long been recognized in the technical literature (Mayer, 1950; Dell; 1956). The basis for this important concept can be best illustrated using the simple circuit diagram shown in Figure 1. In this example, the circuit incorporates two different separators that are configured to upgrade the oversize and undersize streams from a classifier. The combined mass yield (Y) for the overall circuit can be calculated using:

$$Y = S_1 Y_1 + S_2 Y_2 \qquad (EQ\ 1)$$

in which S_1 and S_2 are the respective percentages of feed material reporting to separators 1 and 2 and Y_1 and Y_2 are the respective yields of concentrate generated by separators 1 and 2. Likewise, the grade of the combined concentrate from the circuit can be calculated using:

$$G = (S_1 Y_1 G_1 + S_2 Y_2 G_2)/Y \qquad (EQ\ 2)$$

where G_1 and G_2 are the concentrate grades produced

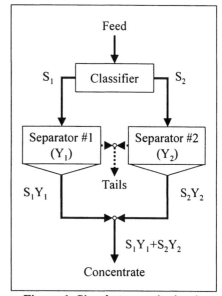

Figure 1. Simple two-unit circuit.

by separators 1 and 2, respectively. The optimization of circuit performance requires that the combined yield (given by Equation 1) be maximized subject to a constraint imposed on concentrate grade (given by Equation 2). For the case of Equation 1, this objective can be achieved mathematically by taking the derivative of Y with respect to Y_1 and setting the result equal to zero. This gives:

$$\frac{\partial Y}{\partial Y_1} = S_1 + S_2 \frac{\partial Y_2}{\partial Y_1} = 0 \tag{EQ 3}$$

$$\frac{\partial Y_2}{\partial Y_1} = -\frac{S_1}{S_2} \tag{EQ 4}$$

Equation 2 can also be easily rearranged to provide a second governing expression for Y. In this case, an expression for maximum yield is obtained by taking the derivative of Y with respect to Y_2. After setting the result equal to zero, this gives:

$$\frac{\partial Y}{\partial Y_2} = \frac{S_1}{G}\left(Y_1 \frac{\partial G_1}{\partial Y_2} + G_1 \frac{\partial Y_1}{\partial Y_2}\right) + \frac{S_2}{G}\left(Y_2 \frac{\partial G_2}{\partial Y_2} + G_2\right) = 0 \tag{EQ 5}$$

$$-\frac{S_1}{S_2}\left(Y_1 \frac{\partial G_1}{\partial Y_2} + G_1 \frac{\partial Y_1}{\partial Y_2}\right) = Y_2 \frac{\partial G_2}{\partial Y_2} + G_2 \tag{EQ 6}$$

Combining Equations 4 and 6 gives:

$$Y_1 \frac{\partial G_1}{\partial Y_1} + G_1 = Y_2 \frac{\partial G_2}{\partial Y_2} + G_2 \tag{EQ 7}$$

The optimization criteria expressed by Equation 7 is not readily apparent. This dilemma can be resolved by considering how a separation process is impacted when the concentrate yield is changed by an infinitesimally small amount (ΔY). In such a case, the resulting yield and grade can be computed using the following mass balance expressions:

$$Y_{new} = Y_{old} + \Delta Y \tag{EQ 8}$$

$$Y_{new}(G_{new}) = Y_{old}(G_{old}) + \Delta Y(G^*) \tag{EQ 9}$$

The term G^* is simply the grade of the last increment of mass added to the concentrate when the yield is increased by an infinitesimal amount. Equation 9 can be rearranged and simplified to show that:

$$G^* = \frac{Y_{new}G_{new} - Y_{old}G_{old}}{\Delta Y} = \frac{(Y_{old} + \Delta Y)(G_{old} + \Delta G) - Y_{old}G_{old}}{\Delta Y} = Y\frac{\Delta G}{\Delta Y} + G \tag{EQ 10}$$

Thus, the left hand and right hand sides of Equation 7 are simply the incremental grades produced by separation processes 1 and 2, respectively. Thus, the maximum yield of concentrate can only be obtained when each separator is operated at an identical incremental grade. This concept is valid for any number of parallel circuits and is independent of the characteristics of the feed streams. It can also be shown that plant profitability is maximized when constant incremental grade is maintained regardless of the operating costs of the different separation processes (Abbott, 1982).

The fundamental basis for the incremental grade concept can be demonstrated using the simple illustrations provided in Figure 2. Each of the two illustrations show two feed streams comprised of various amounts of valuable particles (dark colored material), gangue particles (light colored material), and middlings. The makeup of the first feed stream makes it relatively easy to upgrade since most of the valuable particles and gangue particles are well liberated. In contrast, the second feed stream is difficult to upgrade due to the large percentage of middlings particles.

Figure 2(a) shows the result that is obtained when the separation is conducted to provide a target grade of 75% for each feed stream. For the first feed, all but the pure gangue particles must be recovered to reach the target grade of 75%. This operating point produces a concentrate yield of 75%. For the second feed, a yield of only 53.8% can be realized before the target grade of 75% is exceeded due to the poorer quality of this feed. The concentrate from these two feeds produces a combined yield of 64% at the desired target grade of 75%. At first inspection, this appears to be an acceptable result. However, a closer inspection shows that particles containing up to 75% gangue reported to concentrate when treating the first feed stream, while at the same time many particles containing just 50% gangue were discarded when treating the second feed stream. The only way to prevent this problem is to treat the feed streams at the same incremental grade. As shown in Figure 2(b), operation under this condition increases the concentrate grade for the first feed from 75.0% to 89.3% and decreases the grade for the second from 75.0% to 65.9%. However, when the two streams are blended together, the combined concentrate still meets the required target grade of 75% while providing a concentrate yield of 72%. Thus, the yield obtained by operating at constant incremental grade is significantly higher than that obtained by operating at constant cumulative grade (i.e., 72% versus 64%). In practice, it is easy to be fooled into accepting the lower yield because of the presence of multiple feed streams. The optimum result is intuitive, however, when the feed streams are blended together prior to treatment. In this case, the pure particles would be recovered first, then the 75% pure particles, then the 50% pure particles, and so on until no additional particles could be taken without exceeding the target grade.

The incremental grade concept was originally developed for separations involving mineral systems that included true middlings particles (such as those shown in Figure 2). However, the

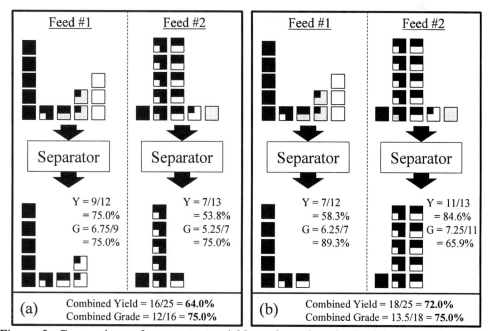

Figure 2. Comparison of concentrate yields and grades obtained by operating at (a) constant cumulative grade and (b) constant incremental grade.

basic concept can be extended to include even those particle systems that are completely liberated. A separation that involves only two pure minerals would still generate products with different "effective" incremental grades because of inefficiencies in the separation process. In cases such as this, the separation is optimized when the effective incremental grade is maintained at the same value in all producing circuits. As such, the incremental quality concept can be universally applied to nearly all types of solid-solid separations.

The question that arises at this point is how can incremental grade be monitored? There is currently no technique for the online measurement of incremental grade. Fortunately, this important parameter can be related to a measurable parameter for many mineral systems. For example, consider the separation of a high-density valuable mineral from a low-density host gangue. For this two-component system, the grade of valuable mineral in an individual particle is directly proportional to the reciprocal of the particle density (ρ) according to the expression:

$$\text{Incremental Grade (\%)} = \left(\frac{100\rho_1\rho_2}{\rho_2 - \rho_1}\right)\frac{1}{\rho} + \left(\frac{100\rho_2}{\rho_2 - \rho_1}\right) \qquad \text{(EQ 11)}$$

where ρ_1 and ρ_2 are the densities of the light and dense components, respectively (Anon., 1966). The incremental grade concept states that maximum plant yield is obtained by operating all circuits at the same incremental grade. Since Equation 11 implies that incremental grade is linearly related to the inverse of specific gravity, this concept can now be modified to show that maximum yield is obtained by operating all circuits at the same specific gravity setpoint. This statement is true regardless of the size distribution or liberation characteristics of the feed, provided that ideal separations are maintained in each circuit. If the separation is less than ideal, minor corrections are necessary to determine the actual specific gravity setpoints required to provide a given incremental grade (Armstrong and Whitmore, 1982; Rong and Lyman, 1985; Clarkson, 1992; Lyman, 1993; King, 1999). Similar types of expressions can be derived for other mineral systems that relate incremental grade to other physical parameters. The importance of this type of relationship is that incremental grade can generally be maintained indirectly in industrial operations by holding the separation parameter of interest at a constant value. Common examples include maintaining specific gravity setpoints for density separations and maintaining field strength and splitter positions for magnetic separations.

PLANT CONTROL STRATEGIES
The incremental grade concept is an important consideration in the control of solid-solid separation processes. As stated earlier, this optimization principle requires that all parallel circuits that contribute to a plant concentrate must be operated at the same incremental grade to achieve maximum yield. What is less obvious is that this optimization concept also requires that the same incremental grade be maintained at all points in time throughout the entire duration of a production cycle. For example, a poorly designed control system may recover individual particles containing a low amount of valuable mineral at one point in time (when the feed contains an abundance of high-grade particles), and then later discard individual particles containing a larger amount of valuable mineral (when the feed contains few high-grade particles). The lower yield from a low-grade production period will never be compensated by the increase in yield realized during a period of high-grade production. As a result, a plant that continuously raises and lowers incremental grade for grade control purposes will always produce less concentrate than a plant that maintains the same incremental grade. Obviously, this realization has tremendous implications in the design of a plant control strategy. The incremental grade concept simply does not support the traditional control strategy that utilizes feedback from online analyzers to make real-time adjustments to circuit setpoints. This strategy can improve the consistency of the concentrate grade, but cannot optimize plant yield. On the other hand, circuits operated under constant incremental quality optimize total plant yield. The only downside is that the concentrate grade may vary considerably throughout the production cycle in response to fluctuations in the feed

quality. For some period of time, these natural variations may cause the concentrate to exceed the target grade and to be unacceptable to downstream customers.

An attractive solution to the variability problem associated with incremental grade control is blending. Blending may be conducted before or after the separation. There are two alternatives for the former scenario. The first involves complete homogenization of the feed ore over extended periods of time (e.g., several days) using stacking and reclaim systems. This type of blending system eliminates short-term variations in feed quality and makes it possible to maintain a stable and constant setpoint for the plant separators. In fact, a control system with a rapid response time would not be needed in this case since any change in the physical properties of the homogenized feed ore would occur very slowly. In many cases, adjustments to setpoints can be made manually in response to data received from standard sampling programs. The turnaround time for the analytical data would simply need to be shorter than any gradual change in overall stockpile quality. This approach would also provide the plant with a relatively constant feed in terms of size and grade. The stable feed would improve plant performance by eliminating overloaded operations and would permit the throughput capacity of the plant to be maximized. Unfortunately, blending systems capable of handling the full feed tonnage treated by most modern processing facilities would require very large stockpile areas and would be expensive to install, operate and maintain.

A second alternative for feed blending is to mix different quality feed streams in appropriate proportions just prior to separation to provide a consistent concentrate grade. Figure 3 provides a flowchart that illustrates the logic behind this type of control strategy. In this particular example, the feed stream to the processing plant is segregated into two stockpiles according to how each responds to upgrading (e.g., difficult and easy). In practice, these two feeds could be segregated based on the fact that they were extracted from different high grade or low grade areas of the same mine or occurred as different geologic splits in the same mined area. In any case, the ore from both stockpiles is fed to the plant while an online analyzer is used to monitor the overall grade of the final concentrate. Based on feedback from the analyzer, the ratio of feed material from the piles is adjusted as required to maintain a constant concentrate grade. This strategy allows the quality of the concentrate to be adjusted online without changing the predetermined setpoints that optimize plant performance. Furthermore, precise determination of the cleaning potential of the different feeds is not required since the control scheme determines the required mix ratios in real time. The feed streams simply need to be sorted into piles with "better" and "worse" qualities. Since complete homogenization of the feed ore is not required in advance, these stockpiles can have relatively small volumes that are easy to manage and more cost effective. Unfortunately, the

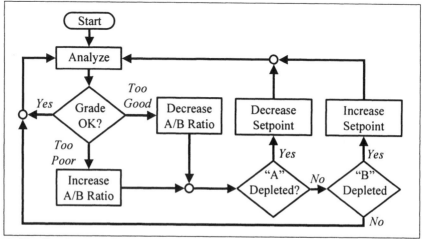

Figure 3. Feedback control strategy designed to manipulate the ratio of two feed streams (A and B) to maintain a constant concentrate grade.

potential exists for one of the stockpiles to be depleted if long-term changes in the feed quality occur. In this case, a supervisory control scheme must be used to adjust the plant setpoint. For example, if the easily treated ore is in danger of being completely consumed, the setpoint may need to be lowered and yield sacrificed to maintain the desired concentrate grade. Alternatively, the system could be designed to issue an alarm to an operator who could choose to maintain the same setpoint and ship lower quality ore to a different market. In any case, the goal of this particular control scheme is to absorb short-term changes in ore quality through variations in stockpile levels without requiring a change in circuit setpoints.

If feed blending is not practical, another scenario for plant yield optimization is to blend products on the concentrate side of the plant. As with feed blending, two alternatives are possible. One option would be to homogenize the entire concentrate from the separation processes over a sufficiently long period (e.g., several days) to ensure that a consistent product is generated. This system has the advantage that the tonnage of concentrate to be blended would be substantially smaller than the tonnage of feed that would need to be homogenized using a feed blending system. Second, the blend system could be configured to combine different proportions of material from different concentrate stockpiles that have been sorted according to grade. This could be conducted automatically using an online analyzer or manually if the turnaround time for sampling and analysis it not too great. The variations in product quality are handled by shifts in the quantities of material stored in the stockpiles. As with the feed based system, a supervisory control loop would be used to monitor the stockpile volumes so that adjustments to setpoints could be made if one or more of the stockpiles starts to be depleted. The major downside to a control system based on concentrate blending is that it does not provide the plant with a consistent feed. Variations in the feed quality can make it very difficult to maintain stable setpoints for many solid-solid separation processes. For example, changes in feed quality that alter the feed tonnage to a spiral circuit can have a dramatic impact on the specific gravity setpoint (Mikhail et al., 1988).

CASE STUDIES
Control of a Coal Preparation Plant

Modern coal preparation plants commonly use density-based separators to remove high-density inorganic matter (rock) from low-density carbonaceous matter (coal). The primary market specification for the concentrate is usually ash content. Unfortunately, the feed coals to preparation plants are typically subject to significant variations in terms of particle size and quality. Factors responsible for these variations include natural fluctuations in the physical properties of the coals and routine changes in production rates from multiple sections or mines. These disturbances make it difficult for plant operations to maintain a consistent coal quality and to maximize clean coal production. Therefore, control systems are needed to help alleviate these difficulties.

Consider the simple case of a 500 tph preparation facility that operates only with heavy media circuits. The plant currently receives run-of-mine feed from two different coal seams. The primary seam, which is mined during three 8 hr shifts, is capable of providing a high yield at the target grade of 7.5% ash. In contrast, the second seam is very difficult to upgrade and, as such, is mined during only one 8 hr production shift. In order to select an appropriate control system for this plant, two sets of partition simulations were conducted over a 24 hr production period. The first set of computations were conducted to simulate the performance of a traditional control strategy involving the realtime adjustment of specific gravity setpoints based on feedback from an online ash analyzer (Figure 4a). As discussed previously, this type of control system does not optimize yield. For comparison, another set of simulations were conducted in which the specific gravity setpoint was held constant and the feed blends were adjusted in response to feedback from the online analyzer to ensure that an acceptable product grade was maintained (Figure 4b). Previous studies have demonstrated that the incremental ash of coal particles can be directly related to the specific gravity of any particular density class (Abbot and Miles, 1990; Luttrell et al., 2000). This relationship was presented earlier in Equation 11. Therefore, constant incremental ash (and

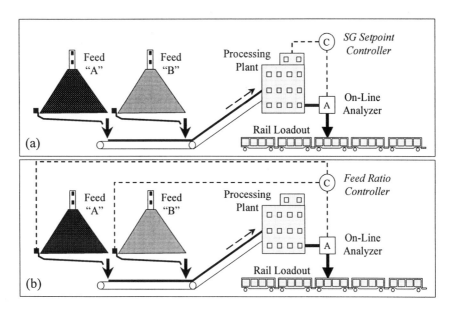

Figure 4. Online control systems for a coal preparation plant based on (a) manipulation of specific gravity setpoints and (b) manipulation of feed rate ratios.

maximum yield) can be maintained by holding specific gravity setpoints in the heavy media circuits at a constant value.

Figure 5 compares the operating data from the two sets of control simulations. When the circuit is configured to automatically adjust the specific gravity setpoints, the ash content produced at any point in time remained relatively constant as a result of the online feedback from the analyzer. However, this requirement forced the specific gravity setpoint to vary greatly over the production period to compensate for (i) natural variations in the quality of the two feed coals and

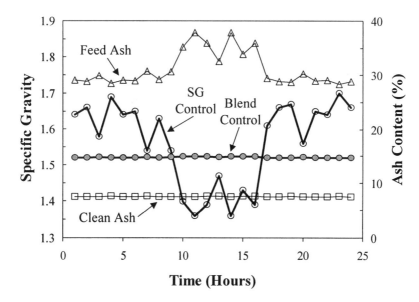

Figure 5. Simulation data comparing two online control strategies (open circles - heavy media setpoint control, filled circles – feed blend ratio control).

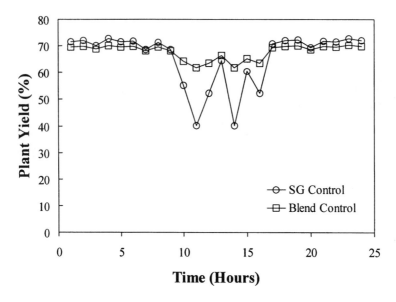

Figure 6. Comparison of plant yields obtained via manipulation of specific gravity setpoints and manipulation of feed blend ratios.

(ii) the large disturbance created by the addition of the poor quality coal during the middle third of the 24 hr production period. In contrast, the control system based on feed blending provided a clean coal of consistent quality while keeping the same specific gravity setpoint. In this case, the large surge of poor quality feed coal that would normally have to be dealt with during one shift is now spread out over the entire duration of the production period by the control system. The fluctuations in quality are handled by increases and decreases in the stockpile volumes of the two coals. More importantly, the blend control is based on the incremental grade concept that maximizes total plant yield. As a result, the blend based control system produced a total of 8133 tons of clean coal over the 24 hr production period, compared to just 7878 tons for the traditional control approach. This improvement provides 255 tons of additional saleable coal from the same tonnage of feed. As shown in Figure 6, the lower production associated with the traditional control strategy is due to sharp drops in yield during brief production periods that can never be made up. In today's market, the production improvement represents an increase in revenue of more than $1.6 million annually for this particular example (i.e., 255 tons/day x $25/ton x 250 days/yr = $1,625,625). Thus, the financial gains afforded by a properly designed control system can be very substantial.

Control of a Magnetic Separation Circuit for Mineral Sands
The incremental grade concept can also be applied to control circuits that involve the blending of concentrates. Consider the simplified magnetic and electrostatic separation circuit shown in Figure 7. This multistage circuit is designed to separate heavy mineral sands, primarily rutile and zircon. Both minerals are nonmagnetic and are separated from other magnetic sands in the first stage of the circuit by drum-type magnetic separators. In the electrostatic circuit, rutile is a conductor (C) and is thrown from the drum of the electrostatic separator, while zircon is a nonconductor (N) and is pinned to the separator drum. Multiple stages of concentrate cleaning are required to generate products of an acceptable grade. As a result, material that reports as middlings (nonconductors from the rutile cleaners or conductors from the zircon cleaners) represents an off-spec product that cannot be sold without further processing. As a standard operating practice, the off-spec material is normally recycled back through the circuit for recleaning through the electrostatic section of the separation circuit

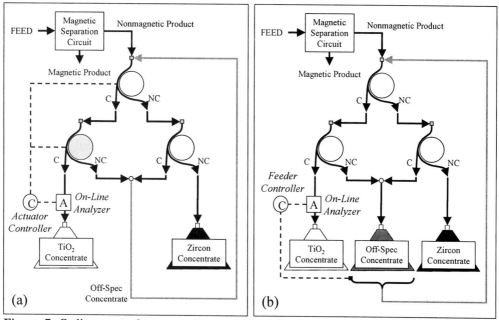

Figure 7. Online control systems for a mineral sands concentration circuit based on (a) adjustment of splitter setpoints and (b) adjustment of middlings back blending.

For control purposes, an online analyzer was installed to monitor the TiO_2 content of the rutile concentrate. As shown in Figure 7a, one possible control strategy is to use the analyzer to provide a feedback signal to pneumatic actuators that in turn vary the position of the product splitters on the electrostatic separators. Several series of mathematical simulations were conducted to evaluate the performance of this control strategy. These simulations were carried out using dynamic partition models based on empirical characterization data. The response of the control system to variations in feed stream quality is summarized in Figure 8. The simulation data show that this

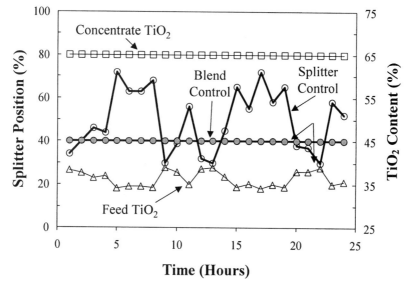

Figure 8. Simulation data comparing two online control strategies (open circles – splitter setpoint control, filled circles – middlings back blending control).

system provides a stable grade by varying the positions of the splitters. However, this configuration does not necessarily optimize circuit performance since the large changes in the splitter setpoints throughout the production period are likely to create large variations in the incremental grade. As discussed earlier, this will not permit the yield of rutile concentrate to be maximized.

A better approach would be to design a control system that makes use of the incremental grade concept to optimize circuit performance. Unlike the previous case study involving coal, a control system based on feed stream blending is difficult in this particular situation due to limitations associated with upstream processes. Fortunately, a control system based on material blending is possible via the recycling of off-spec concentrate. As shown in Figure 7b, a temporary stockpile of off-spec material can be created for blend control purposes. The signal from the analyzer can be used to control the rate at which off-spec concentrate is recycled back through the circuit. Mathematical simulations conducted using this control strategy show that this approach allows a relatively constant concentrate grade to be produced without varying the setpoints of the splitters (Figure 8). Normal variations in the quality of the feed to the circuit are absorbed by changes in the active volume of the off-spec stockpile. The simulation data summarized in Figure 9 shows that the control system based on concentrate blending provided a higher overall yield than the system based on splitter setpoint control. In this case, the average yield was increased from 45.0% to 45.8% with no reduction in concentrate grade. This improvement, which represents a net increase in concentrate tonnage of about 1.76%, would provide over 3,000 tons per year of additional concentrate. Thus, the implementation of this type of control system would be expected to provide a significant financial return.

It is important to note that this particular control strategy does not necessarily ensure that the incremental grade is maintained. This shortcoming is due to the fact that the effective incremental grade does not depend entirely on splitter position for this particular type of separation. For example, it is widely recognized that the effectiveness of electrostatic separators can be influenced by environmental factors such as temperature and humidity. Researchers at the Julius Kruttschnitt Mineral Research Centre in Australia have recently developed a variety of online sensors that may make it possible to address some of these problems. Nevertheless, it is certainly more likely that the effective incremental grade is more consistent when blend control is used than when the splitter positions are varied over large ranges.

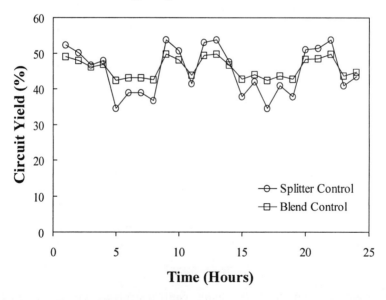

Figure 9. Comparison of plant yields obtained via manipulation of splitter setpoints and manipulation of middlings back blend ratios.

SUMMARY

A theoretical principle known as the incremental grade concept provides useful insight concerning the optimization and control of solid-solid separation circuits. According to this concept, a plant limited by a constraint on concentrate grade will produce maximum total yield when all circuits are operated at the same incremental grade. This concept applies not only for a fixed point in time, but also for the entire duration of a given production cycle. Therefore, the incremental grade concept supports (i) operation at fixed process setpoints and blending of feed ore before separation or concentrates after separation to maintain product consistency and (ii) the online measurement of concentrate grade to adjust feed or concentrate blends. This approach dictates that plant engineers think outside the box (figuratively and literally) since blend-based control systems may influence plant operations as well as mining extraction, materials handling, and sales/marketing activities. In most cases, the incremental grade concept does not support the real-time adjustment of separator setpoints based on online measurements of overall concentrate quality. Data obtained from case studies for two different solid-solid separation systems indicate that large financial returns are possible through the implementation of supervisory control systems that utilize the incremental grade concept to optimize the performance of solid-solid separation processes.

REFERENCES

Abbott, J., 1982. The Optimisation of Process Parameters to Maximise the Profitability from a Three-Component Blend, 1st Australian Coal Preparation Conf., April 6-10, Newcastle, Australia, 87-105.

Abbott, J. and Miles, N.J., 1990. Smoothing and Interpolation of Float-Sink Data for Coals, Inter. Symp. on Gravity Separation, Sept. 12-14, Cornwall, England.

Anonymous, 1966. Plotting Instantaneous Ash Versus Density, *Coal Preparation*, Jan.-Feb., Vol. 2, No. 1, p. 35.

Armstrong, M. and Whitmore, R.L, 1982. The Mathematical Modeling of Coal Washability, 1st Australia Coal Preparation Conf., April 6-10, Newcastle, Australia, 220-239.

Clarkson, C.J., 1992. Optimisation of Coal Production from Mine Face to Customer, 3rd Large Open Pit Mining Conference, Aug. 30 – Sept. 3, Makcay, Australia, 433-440.

Dell, C.C., 1956. The Mayer Curve, *Colliery Guardian*, Vol. 33, pp. 412-414.

King, R.P., 1999. Practical Optimization Strategies for Coal-Washing Plants, *Coal Preparation*, Vol. 20, pp. 13-34.

Luttrell, G.H., Catarious, D.M., Miller, J.D. and Stanley, F.L., 2000. "An Evaluation of Plantwide Control Strategies for Coal Preparation Plants," *Control 2000*, Mineral and Metallurgical Processing, (J.A. Herbst, Ed.), Society for Mining, Metallurgy, and Exploration, Inc., (SME), Littleton, Colorado, pp. 175-184.

Lyman, G.J., 1993. Computational Procedures in Optimization of Beneficiation Circuits Based on Incremental Grade or Ash Content, *Trans. Inst. Mining and Metallurgy*, Section C, 102: C159-C162.

Mayer, F.W., 1950. A New Washing Curve. *Gluckauf*, Vol. 86, pp. 498-509.

Mikhail, M.W., Salama, A.I.A., Parsons, I.S., and Humeniuk, O.E., 1988. "Evaluation and Application of Spirals and Water-Only Cyclones in Cleaning Fine Coal," *Coal Preparation*, Vol. 6, pp. 53-78.

Rong, R.X. and Lyman, G.J., 1985. Computational Techniques for Coal Washery Optimization – Parallel Gravity and Flotation Separation, *Coal Preparation*, 2: 51-67.

Strategies for Instrumentation and Control of Thickeners and Other Solid-Liquid Separation Circuits

Fred Schoenbrunn, Lynn Hales, and Dan Bedell

ABSTRACT

Some of the process variables that are commonly monitored on a thickener are torque, rake height, bed level, bed pressure, feed rate and density, underflow rate and density, settling rate, and overflow turbidity. Many of these are easily measured, while some can be difficult. Combining these signals into a coherent control strategy requires forethought and an understanding of the fundamentals of thickener operation. A wide variety of control strategies have been implemented on thickeners, using various combinations of sensors.

In recent years improved flocculants, higher throughput rates per unit area, and desired higher density underflow concentrations have required the development of better control strategies to successfully operate sedimentation equipment. This has been complicated by plant expansions that have placed increased loads on existing sedimentation equipment. Successful control strategies consider the process goals, plant fluctuations, sensor reliability, and system response times.

A historical review will be discussed followed by discussion of the latest developments in sensors, control equipment, and control strategies.

INTRODUCTION

Thickeners are one of the workhorses for industry and are involved in numerous steps in flowsheet design. They have been instrumental to the mining industry growth and in the processing of minerals. The thickener over the years has received little attention and has generally been treated as a wide spot in the process line with little or no instrumentation. Concentrate thickeners and some clarifiers were the first units to have instruments applied to them due to their impact on production and profitability of a plant. Before the use of polymers thickeners were normally sized with excess capacity so that they could "respond or handle flow and feed variations" and not adversely impact production.

The advent of natural flocculants followed in recent years by synthetic flocculants has increased the efficiency of sedimentation rates allowing increased production through a given size of thickener. This has resulted in the time variable being dramatically shortened. Thickeners may use a "single" flocculant system or a "dual" flocculant system. The single component system may use a cationic, anionic, or non-ionic flocculant to achieve underflow density. The dual flocculant system will use a cationic or coagulant (to obtain acceptable clarity) followed by the addition of an anionic or an-ionic flocculant to achieve good underflow density. The use of flocculants had led to the discovery of a relationship of solids concentration to flocculant addition concentration that gives what is called "Settling Flux" (See Figure 1). It now understood that there is an optimal feed concentration at which the solids will settle at a given flocculant dosage. This discovery has helped advance sedimentation technology but introduced yet another variable into the equation.

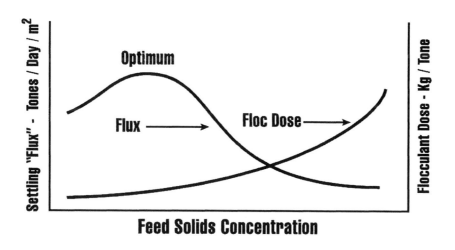

Figure 1 Settling flux

Thickener controls have been generally based upon the actual underflow or overflow solids with some controls based upon polymer addition rates. The biggest challenge for conventional thickeners has been the long lag time between variable changes and results. High rate thickeners have reduced the residence time but are still relatively slow in cause and effect relationships. The emergence of ultra high rate and ultra high density thickeners has further pushed the sedimentation envelope. With each of these developments and advances in sedimentation technology there has been the need for more and better instrumentation and controls. Modern thickeners are becoming better instrumented and industry is discovering that unit operations are more predictable and profitable as a result. The modern generation of thickeners will not successfully function without the aid of quality instrumentation. Because of instrumentation the mining industry is now beginning to control the thickener, inventories, and manage the process flows and improve the overall efficiency of the operations. The following sections will take some of these variables, types of instruments to measure them, and discuss the types of controls used to manage them.

The general objectives for thickeners and clarifiers are to produce clean overflow and maximum solids concentration in the underflow. Flocculants are typically used to agglomerate the solids to increase the settling rate and improve the overflow clarity. Thickeners generally operate continuously and are used in a wide variety of industries, and in numerous applications. The term "thickener" will apply to both thickener and clarifier unless noted otherwise throughout this section.

Thickener control involves a number of complexities and variables such as varying feed characteristics, changes in feed concentration, solids specific gravity, particle size distribution, pH, temperature, and reaction to flocculant which can all contribute to variations in performance. Accurate information about what's happening inside the thickener is often difficult to obtain. In addition to these variables, various phenomena such as "sanding" and "island formation" may occur which can be difficult to predict and interpret from the data.

The two independent variables that are typically used for control of thickeners are underflow rate and flocculant addition rate. A third variable, the feed rate, is generally used only in an emergency to avoid impacting plant production. The dependent variables include underflow density, overflow turbidity, rake torque, solids interface level (bed depth), solids inventory (bed mass), solids settling rate and underflow viscosity.

Historically, most control schemes have used one or two of the dependent variables to control the independent variables. For example, using the underflow density to control the underflow rate, which can be by variable speed pump or if the underflow is by gravity, using a control valve or orifice. Another possible scheme is to use the bed pressure to control the underflow rate and bed level to control the flocculant rate. The range of conditions that it can recognize and to which it can respond will limit any control scheme. None of them so far have been able to resolve all of the possible inputs for specific conditions and react to them. For example, the two control schemes described above do not consider the rake torque and both can have problems from high torque if the feed particle size distribution suddenly becomes coarser.

Various algorithms have been used to control thickeners with varying degrees of success. Rule based expert systems have been developed for use on thickeners since the early days of computerized control systems, but have been cumbersome for implementation, troubleshooting, modification, and tuning. With the recent developments in expert control software, these issues have been greatly simplified.

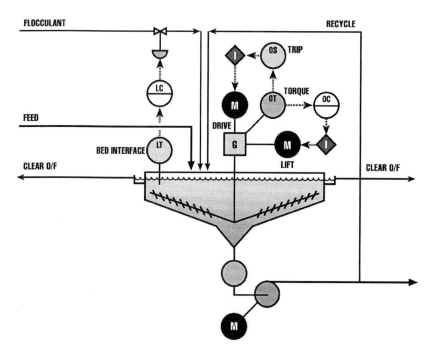

Figure 2 Thickener with minimal instrumentation

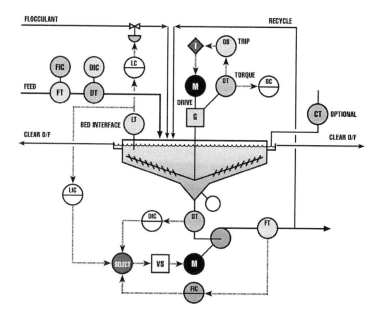

Figure 3 Thickener with additional instrumentation

INSTRUMENTATION

Torque
Rake torque is an indication of the force necessary to rotate the rakes. Higher rake torque is an indication of higher underflow density or viscosity, deeper mud bed, higher fraction of coarse material, island formation, or heavy scale build up on the rake arms.

Rake torque measurement is usually provided by the thickener manufacturer as part of the rake drive mechanism. Typical methods involve load cells, motor power measurement, hydraulic pressure, or mechanical displacement against a spring. They are all generally reliable and reasonably accurate if set up correctly. The type supplied depends on the manufacturer and the type of drive supplied. For example, if a hydraulic drive is used, then hydraulic pressure is the best method to use for torque measurement. Torque measuring devices are designed to produce a signal that may be utilized for alarming or control.

Rake height
Rake lifting devices are frequently used to minimize the torque on the rake arms by lifting them out of the heavy solids and enable the rake to continue running during upset conditions. It is desirable to prevent the rake drives from running extended periods at torques above 50-60%, to prevent accelerated wear on the drive. Lifting the rakes a small distance is usually effective at relieving the pressure on the rakes and thus reducing the torque. Because of this, using the torque indication in a control strategy must also consider the rake height in order to effectively control the thickener.

Rake height indicators are typically supplied by the thickener manufacturer. The two most common methods are ultrasonic or a potentiometer type with a reeling cable. Both are reliable and accurate. In many plants a bar or rod on the end of the rakes serves as a "visual" indicator of height and arm position in the tank. Lifting of the rakes allows a short period of time to make corrections before being forced to shut down the thickener.

Bed level
There are several general types of bed level detection instruments; ultrasonic, nuclear, float and rod, and reeling (with various sensors). Each has advantages and disadvantages, which are discussed below. There is not a standard bed level sensor that is recommended for all applications.

- Ultrasonic bed level sensors work by sending a pulse down from just under the surface, which in theory bounces off the bed surface back to the receiver. Elapsed time is used to calculate the distance. Advantages are non-interfering location, measures over a large span, and relatively inexpensive. The downside is that they do not work on all applications. If the overflow is cloudy, it can interfere with the transmission or causes too much reflection to give a reliable signal. Scaling affects accuracy and can cause drifting or loss of signal. Using them on concentrate thickeners has proved to be particularly troublesome.
- Nuclear bed level sensors work by either sensing background radiation level or attenuation between a source and detector, depending on whether the solids have a natural background radiation level. The sensor is comprised of a long rod that extends down into the bed with radiation detectors spaced along the length. If the ore changes from not having radiation to having it, there will be problems. The advantages are that it is relatively reliable when properly applied. The downside is that it measures over a limited range, may interfere with the rakes (a hinged version that will swing out of the way when the rakes pass by is available), and is relatively expensive.
- Float and rod types work with a ball with a hollow sleeve that slides up and down on a rod that extends down into the bed. The ball weight can be adjusted to float on top of the bed of solids. These are subject to fouling and sticking, and can be installed and measure only in the area above the rakes, however, they are relatively inexpensive.
- Reeling devices work by dropping a sensor down on a cable, and sensing the bed level by optical or conductivity sensors. In theory they are non-fouling and get out of the way of the rakes, but in practice, stories abound of sensors wrapped in the rakes. A number of plants have found them reliable and they can cover a large range. The price is midrange. Freezing wind and cold temperatures can cause icing problems.
- Vibrating or Tuning fork sensor. These are designed to sense a difference in the vibrating frequency in different masses of solids. These are used in Europe and Africa with some success.
- Bubble tube or differential pressure. This is an old but tried and true method of bed level detection. There may be some plugging or fouling of the tube over time.
- External density through sample ports. Slurry samples are taken from nozzles on the side of the tank and pass through a density meter to determine the presence of solids. This system can be set up with automated valves to measure several different sample points, typically sampling once every five or ten minutes. Requires external piping and disposal of the sample stream. Access can be problematic on occasion.

Bed pressure
Because thickeners maintain a constant liquid level, the pressure at the bottom of the thickener is an indication of the overall specific gravity in the tank. If the liquor specific gravity is constant, the overall specific gravity is an indication of the amount of solids in the tank and can be converted into a rough solids inventory. This can be a very effective tool for thickener control. Because of relative height to diameter ratios, it is considered somewhat less useful for very large diameter thickeners.

Differential pressure sensors are used to measure the bed pressure, leaving one leg open to the atmosphere to compensate for barometric pressure variations. Care must be taken in the installation to minimize plugging with solids. This is frequently done by tilting the tank nozzle on which the DP cell is mounted downwards from the sensor so that solids tend to settle away from the sensor surface. A shutoff valve and a water flush tap are recommended to allow easy maintenance.

Flow rate

Flow rates for feed and underflow lines are useful, particularly when combined with density measurements in order to generate solids mass flow rates. Since flocculant is usually dosed on a solids mass basis, knowing the mass flow rate is very useful for flocculant control, providing a fast response system. Flow rate measurement is an absolute necessity for the newer generation of the ultra high rate and ultra high density thickeners.

Since the streams being measured are usually slurries, the flow rate is usually measured by either magnetic flow meters or Doppler type flow meters. As long as these instruments are properly installed in suitable full straight pipe sections, avoiding air if possible, they are accurate and reliable. If the feed stream is in an open launder, flow measurement is more difficult but can be accomplished using ultrasonic devices.

Density

Nuclear gauges are the norm for density measurement. Nuclear density instruments require nuclear handling permits in most countries. It should be noted that there are now some types that use very low level sources that do not need nuclear licensing, reducing the hassle of using these. Density gauges should be recalibrated regularly, roughly every 6 months, as they are subject to drift. As with flow meters, proper installation is important for reliable operation.

Small flow applications may be able to use a coroilis meter to measure both mass flow and percent solids with one instrument.

Settling rate

The settling rate in the feedwell is a good indication of the degree of flocculation, and can be used to maintain consistent flocculation over widely varying feed conditions. A settleometer is a device that automatically pulls a sample from the feedwell and measures the settling rate. The flocculant can then be adjusted to maintain a consistent settling rate. In most applications they require regular maintenance to maintain consistent operation.

Overflow turbidity

Overflow turbidity can be used to control flocculant or coagulant. There may be some significant lag time between the actual flocculation process and when the clarified liquor reaches the overflow discharge point where the sensor is typically positioned. These sensors and meters are generally used as alarms or for trim only. In most applications they require regular maintenance to maintain consistent operation.

CONTROL ARCHITECTURE AND EQUIPMENT

Normally thickeners are part of an integrated control system where the objective is to remotely start and stop the equipment, monitor operating conditions and performance, stabilize operations based on operating and feed conditions and lastly, optimize performance based on economics and/or operational goals. Tools to accomplish these goals include programmable logic controllers (PLC's), distributed control systems (DCS's) and expert control systems.

Programmable Logic Controllers are normally used to perform starting and stopping functions in processing plants. They do have the ability to also implement continuous control loops but this use is only used minimally.

Distributed Control Systems are the workhorse of plant control systems and normally are used to coordinate the monitoring of process data and the subsequent stabilizing control of important process parameters. Specifically, for thickeners, monitored process parameters might include feed flow rate, overflow flow rate, overflow turbidity, underflow density, underflow flow rate, rake position, rake power and torque, flocculent dosage rate, bed level, and bed mass.

Stabilizing control loops might include flocculent dosage rate, underflow density, or underflow flow rate. Bed pressure is sometimes used as an indication of solids inventory. This can be used to help the system to determine whether a high bed level is the result of decreased settling rate or increased solids inventory. In some cases, a drive torque target is used as an indication of acceptable underflow rheology.

A typical P&ID incorporating the instrumentation required by stabilizing control loops previously discussed is shown in Figure 4.

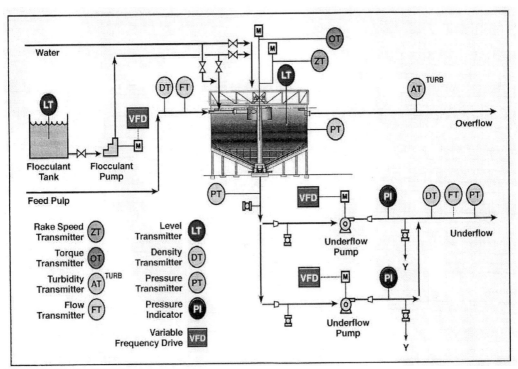

Figure 4. Typical P&ID

CONTROL STRATEGIES

Process goals vary widely throughout the industry and are necessarily aligned to some degree with the basic plant control hardware. For plants with distributed control systems it is common to have a number of stabilizing control loops such as, flocculant dosage rate that is ratioed to the solids mass flow rate of the feed and an underflow density control loop that is maintained by varying the discharge pumping rate.

Where expert control systems exist in grinding and flotation it is natural to extend them to include thickening. The basic concept behind expert control is the continuous monitoring of process conditions along with process and economic objectives and then using advanced *expert* logic and or on-line models to determine what the best process set points should be to optimize the performance of the target unit operation or plant area.

A generic thickener expert control strategy that monitors the performance of the thickener in a tailings dewatering circuit and specifically controls polymer dosage and discharge rates is now described. Table 1 shows a list of process measurements and control actuators employed by the generic thickener expert strategy.

Table 1 - Generic thickener expert strategy measurement and control parameters

Process Parameter	*Process Measurement*	*Control Actuator*
Feed rate	Mag flowmeter	
Feed distribution	Gate valve position	Gate valve position
Underflow discharge rate	Mag flowmeter	Pump speed
Underflow density	Nuclear density meter	
Bed pressure	Pressure transducer	
Flocculant feed rate	Flowmeter	Metering pump
Rake torque	Torque transmitter	
Water overflow rate	Flowmeter	

It is common for expert strategies to run the monitoring and control rules at a fixed time interval. Often times this frequency is a function of the retention time of the unit operation being controlled. For example every five minutes the generic thickener expert strategy determines the appropriate control actions through the following decision tree:

Evaluate for emergency response situations: These are situations that require immediate attention. It is assumed that these conditions, if not addressed, would lead to a process emergency.

1. Check the current drawn by the underflow pump. If the current is too high, decrease the load on the pump by
   - reducing flow through the thickener; or
   - reducing flocculant rate.
2. Check the underflow density. If the density is too high, decrease the polymer dosage.
3. Evaluate the torque on the rake. If the torque is high, increase the pump speed.
4. Check the sludge level in the thickener. If the level is dropping rapidly, decrease the pumping speed and decrease polymer dosage.

Optimize thickener performance: If no *emergency* conditions exist then the generic thickener expert strategy seeks to optimize performance. To obtain optimum performance, two targets are used.

1. An underflow density target is used to ensure optimum solids content in the tailings impoundment, and optimum water reclaim for the mill.
2. A bed level target is used to obtain optimum loading in the thickener without overloading the drive mechanism.

An important detail that the expert system designers have to deal with is how simultaneous emergencies are dealt with. If it is decided that every five minutes only the first emergency determined would be dealt with then a condition where the rake torque is high and the sludge level is dropping quickly, the generic thickener expert strategy will only respond to check the rapidly rising torque by increasing the discharge rate. While the increased pumping rate will also affect the bed level situation, the generic thickener expert strategy described here only responds to the highest priority situation. It is easy to see however that there may be a more *expert* way in taking into account all process conditions simultaneously so that it is possible to deal with multiple emergency conditions while taking into account optimization goals. This additional level of control logic prevents multiplying (or masking) the effects of control actions by responding to multiple emergency conditions with similar or conflicting remedies.

Table 2 shows possible combinations of the underflow density, bed pressure, and bed level process variables and what an expert system might do to optimize the thickener as conditions vary.

Table 2 Optimizing control actions

Underflow Density	Bed Level	Bed Pressure	Polymer Addition	Underflow Rate
above target	above target	Rising	increase	increase
above target	above target	Steady	increase	increase slightly
above target	above target	Falling	no action	increase slightly
above target	on target	Rising	no action	increase
above target	on target	Steady	no action	increase slightly
above target	on target	Falling	no action	no action
above target	below target	Rising	decrease	increase
above target	below target	Steady	decrease	increase slightly
above target	below target	Falling	decrease slightly	no action
on target	above target	Rising	increase	increase
on target	above target	Steady	increase slightly	no action
on target	above target	Falling	increase slightly	no action
on target	on target	Rising	no action	no action
on target	on target	Steady	decrease slightly	no action
on target	on target	Falling	decrease slightly	decrease slightly
on target	below target	Rising	decrease	increase
on target	below target	Steady	decrease	no action
on target	below target	Falling	decrease slightly	no action
below target	above target	Rising	increase	no action
below target	above target	Steady	increase	decrease
below target	above target	Falling	decrease slightly	decrease
below target	on target	Rising	no action	no action
below target	on target	Steady	no action	decrease
below target	on target	Falling	decrease slightly	decrease
below target	below target	Rising	no action	decrease slightly
below target	below target	Steady	decrease slightly	decrease
below target	below target	Falling	decrease	decrease

The amount of the set point changes that are made for the optimization strategy depends upon a number of factors which is beyond the scope of this discussion, however, the very idea of the complications of this issue introduces the concept of the *art of control*.

CONCLUSIONS

Reliable instruments are a prerequisite for a functional control system. The control system must be designed based on the variables that can be consistently measured. Stabilizing control systems can be designed to use these signals to control the operation of the equipment. Expert control systems can be implemented on top of functional stabilizing systems to optimize the operation of the solid/liquid separation equipment. Successful control strategies consider the process goals, plant fluctuations, sensor reliability, and system response times.

REFERENCES

V. Dooley, "Mud Level Gauges – A Comparison of Techniques", Fourth International Alumina Quality Workshop, Darwin, Australia, June 1996.

Nelson M. G., R. P. Klepper, and K. S. Gritton, "Design of an Expert Control System for Thickeners", Presented at SME '97, Denver, Colorado, February 1997.

Allen J. P., M. G. Nelson, and K. S. Gritton, "Automated Thickener Control; Instrumentation and Strategy", Presented at SME '97, Denver, Colorado, February 1997.

Johnson G., S. Jackson, I. Arbuthnot, "Control of a High Rate Thickener on Gold Plant Tailings", The AusIMM Annual Conference, Rotorua, New Zealand, 1990.

Strategies for Instrumentation and Control of Flotation Circuits

Heikki Laurila[1], Jarkko Karesvuori[1] and Otso Tiili[1]

ABSTRACT

Flotation cells have increased in size dramatically over the past ten years, making it more viable to integrate greater levels of instrumentation on each cell, given the value of the material. Advances in instrumentation also have allowed better measurement of specific flotation parameters. Combined, these factors have created an opportunity for the development of more effective strategies for flotation control.

Over the past decade, strategies for flotation control have been delivered using advanced control techniques such as, model based controls, expert systems, and neural networks, all of which have been employed with varying degrees of success. The most recent applications of expert and neural controls look promising and may also lead to more robust model based controls in the future.

The ideas, strategies and instrumentation behind these concepts are discussed in this paper.

INTRODUCTION

Flotation has a long history and nowadays it is one of the most broadly used processes in the mineral separation industry. However, it remains quite an inefficient process. Although a great deal of research and development has been conducted in the field of flotation process, over many years, there are still economic benefits to be achieved through improved operations.

Flotation can be used in most mineral separation applications, as it is suitable for a large variety of minerals and a wide range of particle sizes and densities. Flotation is performed in cells ranging in size from laboratory scale up to 160 m^3 tanks. The newest tanks available on the market have volume of 200 m^3. This allows use of flotation in processes of very large as well as very small capacities.

Flotation is the most difficult stage of ore benefication and the performance of the flotation circuit has a very significant effect on the total performance of the concentration path. Typically, the recovery rates in flotation are between 85 and 95 %, the difficulties in improving performance derive mainly from the complexity and the nonlinearity of the flotation process itself. Earlier the lack of instrumentation and proper technologies for instruments also complicated the achievement of acceptable efficiency. A large number of different control strategies for flotation have been presented in literature and implemented in flotation plants over the past 30 years and development of new instruments continues.

Flotation is facing a new era in terms of automation and process control. There are three main reasons for this. Flotation circuit design is moving away from multiple recycle streams and towards simpler circuits. This removes a degree of safety and stability, as poor operation will send material directly to the tailings rather than recirculating it. The advantage is that the self-compensating nature of the circuit is reduced and it becomes easier to regulate and optimize the process (Henning, Schubert and Atasoy 1998).

The size of the flotation tanks in large-scale flotation plants has been increasing in recent years (Figure 1). While earlier plants had a large number of relatively small tanks in series, future

---

[1] Outokumpu Mintec Oy., Riihitontuntie 7 C, P.O. Box 84, FIN-02201 Espoo, Finland

plants will have a smaller number of larger cells. This trend leads to decreasing number of instruments but along with simple circuit design, the demands of reliability and accuracy for instrumentation have increased.

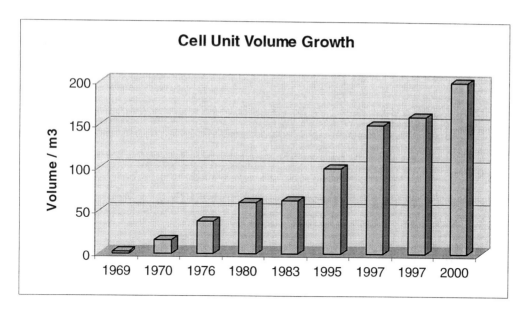

Figure 1 The recent growth of the cell volume

Recent developments in instrumentation have provided new devices such as image analysis based devices for froth characteristics measurement and new digital fieldbus technology has enabled the production of smart instruments. These instruments can provide more information than traditional analog instruments in using self-diagnostics to provide information about the quality of measurements and status of the device.

FLOTATION PROCESS

Separation of valuable mineral from the ore in flotation process is based on the different surface properties of minerals. Some minerals have a tendency to attach stronger to air bubbles than others in water treated with special chemicals. Minerals having this tendency are called hydrophobic, however, most of the minerals are not naturally hydrophobic. Therefore, several chemicals must be introduced in water-ore slurry in order to convert valuable mineral particles into hydrophobic and to make other particles as hydrophilic as possible. Air bubbles formed by airflow to which mineral particles attach, rise to the surface of the slurry, forming the froth, which can be collected.

Flotation is very complex process having a large number of affecting variables, nonlinearities, interaction between variables and randomness. Estimates have been presented that there are about 100 affecting variables in the flotation process (Arbiter and Harris 1962). The raw material and grinding process prior to flotation contribute a relatively large part of the variables. From the viewpoint of process control the most important variables are:

- Slurry properties (density, solids content) and slurry flow rate (retention time)
- Electrochemical potentials (pH, Eh, conductivity)
- Chemical reagents and their addition rate (frothers, collectors, depressants, activators)
- Slurry levels and aeration rates in the cells
- Froth properties (speed, bubble size distribution, stability)
- Particle properties (size distribution, shape, degree of mineral liberation)
- Mineralogical composition of the ore

- Mineral concentrations in feed, concentrate and tailings (recovery, grade)
- Froth wash water rate (specially in flotation columns)

In industrial scale, flotation takes place in interconnected cells, which compose the different sections of flotation circuit. Each section has its function in the total flotation process, as high recovery and high grade cannot be achieved in a single stage of the process. Whilst the design of a flotation circuits vary significantly, the basic operation is similar and some basic rules can be stated.

Each section is composed of flotation cells. Earlier sections could contain up to dozens of relatively small cells but nowadays the total number of cells has reduced as cell sizes have increased. The tailing of each cell becomes the input of the next cell in the circuit. From the last cell of the section, tailing is led to the first cell of other section, except the cells where the tailing is sent to final tailing. Usually concentrates of the single cells in a section are combined to one concentrate flow, which is then directed to next phase of process. Sometimes regrinding or thickening is situated between the flotation sections. Regrinding may be necessary when floating two minerals with very different optimal particle size distribution or when final concentrate grade cannot be achieved because of gangue contamination in non-liberated particles. Thickening may be needed to increase solids content of slurry.

Figure 2 presents the outline of two flotation circuits. The circuits are generalized examples but describe some basic structures and sections of flotation circuits. Both circuits are divided into three sections – rougher, scavenger and cleaner but connections between sections are not similar.

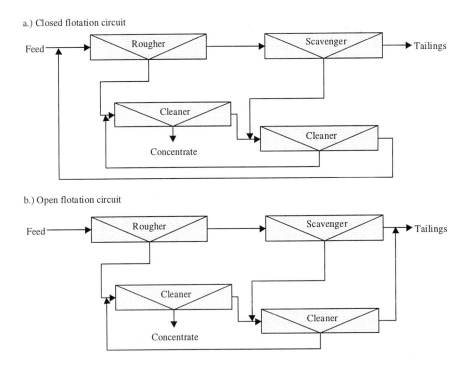

Figure 2 Flotation circuits (a. Closed, b. Open)

Feed coming from the grinding circuit is led to the rougher. Before the actual flotation, slurry may be conditioned in a conditioner, where some of the reagents are added to slurry. Some measurements are often performed in conditioner, pH, for instance. In rougher flotation, most of the fast floating valuable minerals are separated from the slurry directly to the concentrate. In other words, the recovery is held high at the expense of the grade. This is performed in conditions where fairly thin froth beds and high aeration rates are used in the flotation cells.

The tailings of the rougher are introduced to the scavenger, where slowly floating fine and coarse mineral particles are floated in conditions where smaller aeration rate and even thinner froth beds than in rougher section are used. The tailings of the scavenger are sent to the final tailings of the circuit.

The concentrates of the rougher and scavenger are refloated in cleaner section in order to increase the grade of the final concentrate. The froth beds in cells of the cleaner flotation are thicker than in other sections, which leads to the decrease of the water recovery and increased rejection of hydrophilic gangue.

The connection of the tailings of the cleaner flotation makes the difference between the circuits presented in Figure 2. As in the open circuit, the cleaner tailings form the final tailings of the circuit, with scavenger tailings closed circuit, re-circulates the cleaner tailings back to the rougher.

In a closed flotation circuit, the control of circulating loads is essential. Material can accumulate in the circuit as circulating loads increase. This may lead to situations where recovery suddenly drops far below a satisfactory level after being on target and stable for long periods. This occurs when the accumulation has reached certain point where circuit conditions become unstable. The means to correct the situation after it has occurred are minimal. Almost inevitably, the recovery stays low for certain period and may even reduce the average recovery below long-time target values. Figure 3 illustrates the effect of poorly controlled circulating loads on tailings grade.

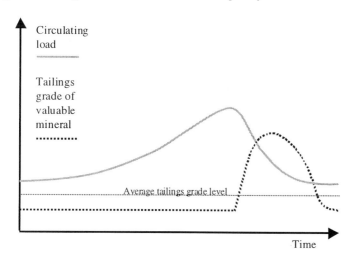

Figure 3 The effect of poorly controlled circulating loads

CONTROL STRATEGIES FOR FLOTATION

Flotation process control is a challenging and important task in the ore benefication chain. The efficiency of the flotation process largely controls the economics of the overall mineral processing plant (Hodouin et al. 2000).

Flotation plants are difficult to operate. Non-linear dynamics, coupling among control loops, large and variable dead times, strong and continuous unmeasured input disturbances, imperfect knowledge of the phenomenology of flotation, impede process control. Frequent lack of appropriate and precise instrumentation makes supervision and control even more difficult. (Osorio, Pérez-Correa and Cipriano 1999).

A universal way to control a flotation plant cannot be given. Each plant has its special features in terms of cell configuration, instrumentation, ore and chemistry, which have led to a large number of different control strategies and methods used and reported in literature. The only undoubtedly common feature between flotation plants is to maximize profit. This objective is pursued in many ways. Some typical aspects and methods are presented in this paper.

Plant-wide flotation control strategy can be divided in layers, which are presented in Figure 4.

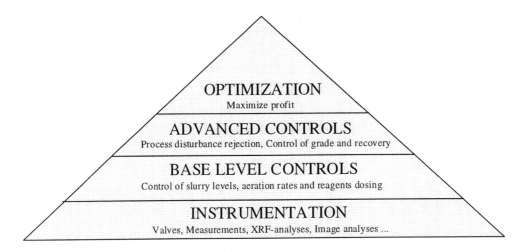

Figure 4 Control system hierarchy of the flotation plant

This starts with instrumentation, which is the basis of all control and must function well. The whole plant optimization is strongly dependent on both well performing and reliable instruments and properly tuned single control loops. For each measurement or control task, a large selection of devices is offered in the market. Attention must be paid in choosing the right ones for the process. As in every instrumentation case, intimate knowledge of the process, for it's special features and demands, is vital.

As a process of many affecting variables and high complexity, flotation has a large number of variables to be measured and to be manipulated. This leads to extensive variety of different instruments used in a flotation plant. In spite of the new digital fieldbus technology, the majority of instruments are still traditional devices, using 4-20 mA analog signal technique. Instruments are connected to plant automation system via I/O-units, where analog signals are converted to digital signals. The digital fieldbus technology will replace the analog technique and provide totally digital communication between field instruments and the automation system.

Base level control consists of traditional PID (Proportional Integral Derivative) controls of slurry levels and aeration rates and ratio control of reagent additions. The derivative term is usually excluded in tuning. The plant-wide flotation control strategies use slurry levels, aeration rates and reagent additions as control variables.

Instrumentation and Base Level Controls
Slurry Flow Measurement. Slurry flow measurement is mainly performed by magnetic flow meter. Measured fluid must be at least weakly conductive (more than 5 microsiemens per centimeter). The measurement is based on Faraday's principle of induction, which states that tension is inducted to conductor when moving in the magnetic field. Magnetic flow meter consists of an electro-magnet wrapped around a length of process pipe, which is lined with an insulating material. Electrodes are installed in the wall of the pipe on opposite sides and these enable an electrical circuit to be formed through the liquid and measuring device.

The material selection for the electrode and linings depends on the properties of slurry. Common lining materials are; hard rubber, Teflon, polyurethane and aluminum oxide. Electrode materials are; stainless steel, platinum and tantalum. The AC-magnetization technique used in earlier times has been replaced by DC-magnetization, with high measurement frequency (more than 30 measurements per minute). Magnetic flow meters do not impose energy loss, as there is no obstruction to flow.

In general, slurry flow measurement is problematic, due to solid particles of slurry and suspended air bubbles, particularly the latter, which will decrease the performance of magnetic flow meter. If slurry contains magnetic material (e.g., magnetite) special demagnetization is needed.

Concentrate flow measurement in open channel can be done by a dam arrangement, using a V-shaped cutout and ultrasonic level transmitter. Schematic illustration of arrangement is presented in Figure 5. Known geometry of the channel and level measurement can be converted to flow rate. This method does not provide flow rate accurately but gives a rough estimate, which can be used in cases where other flow measurements are not possible.

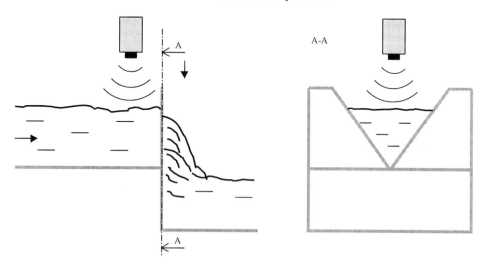

Figure 5 Arrangement for flow measurement in open channel

Slurry flow measurements have an important role in determination of circulating loads. They are also used in circuit mass balance calculations. Although frother addition rate is mostly adjusted according to the tons of ore in feed, some flotation plants have found it reasonable to use feed volume rate instead.

Elemental Assaying. On-stream XRF analyzers are very important instruments in flotation, as they provide elemental assays from the process flows. Manual elemental assays in the laboratory are not as useful for real time process control because of the long delays. On-stream XRF analyzers have decreased the delay to reasonable scale. A primary sampler acquires a primary sample flow from the process stream. It is directed to the secondary sampler, which reduces the size of the sample suitable for the X-ray analysis equipment, then returned to the process. Modern XRF analyzers can report the assay of several elements and solids content from the sample. Usually the analyzer system has many sampling points from different stages of the flotation circuit and one XRF-analyzer can accommodate up to 24 sample lines, measured in sequence. The measurement time for one sample is from fifteen seconds to one minute, the cycle time 5-15 minutes, depending on the number of streams to be assayed. An on-stream XRF analyzer schematic is presented in Figure 6.

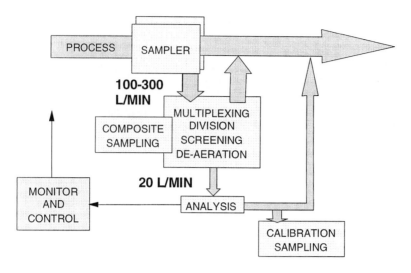

Figure 6 The flowchart of the on-stream XRF analyzer

Elemental assays by on-stream analysis is the only way to obtain on-line information about the performance of flotation in real-time, which enables remedial action for concentrate and tailings grade control as well as recovery control. Based on grades reported by an on-stream analyser, advanced controls adjust the setpoints of base level controllers. Also feedforward action for proper feed type treatment is made possible by real time assays from on-stream analyzers.

Density Measurement. Some on-stream XRF analyzers and particle size analyzers provide density measurement but specific density meters are also commonly used. Nuclear density meters are suitable for slurry density measurement, although suspended air bubbles in the slurry often make measuring impossible. Therefore the location of the instrument must be carefully chosen. A suitable place for density measurement is a process pipe, which in normal conditions is always full of slurry. Nuclear density meters are based on the attenuation of the nuclear radiation by the media. Density measurements, along with slurry flow measurements, enable the calculation of total mass flows, which are required in mass balance calculations.

Slurry Level Measurement. Accurate level measurement in flotation cell is troublesome, due to usually thick froth bed and variations in slurry density, which complicate methods using direct ultrasonic measurement of level or hydrostatic pressure. Also, the concept of level is not definitive, since the transition from slurry with bubbles to froth with slurry, is not sharp. The most typical instruments used for measuring the slurry level in cells are a float with target plate and ultrasonic level transmitter, a float with angle arms and capacitive angle transmitter and reflex radar. Instruments are presented in Figure 7.

Ultrasonic level transmitters emit a series of ultrasonic pulses, which echo from the target plate. The transmitter receives echoed pulses and measures the travel time. In order to separate the real echo from false echoes, acoustic and electronic noise, the signal must be filtered. Similarly, before converting the travel time to distance and output, it must be temperature compensated.

Reflex radar, using a plastic or Teflon plated metal rod, can sense various boundary layers and gives the possibility of measuring both slurry level and froth thickness. Problems associated with this technology are due to severe changes in slurry and froth conductivity.

Froth bed thickness measurement by microwave radar or certain special ultrasonic transmitters has also been under development, although problems occur because echoes are influenced by froth properties.

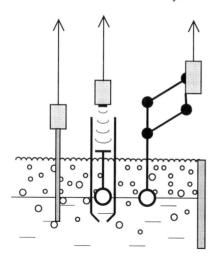

Figure 7 Instruments for slurry level measurement

Slurry Control Valves Process features such as large flowrate capacity changes and abrasive mineral slurries, limit the number of possible valve solutions for slurry flow control. Control valves used in flotation must be robust and durable, specific selection of outflow valve depending on the structure of the cell. Pinch valves and dart valves are suitable solutions for flotation duty. Traditionally, manually operated overflow weirs have been used on tails streams. In older flotation plants, the mechanical design of the cell bank has forced the automation of overflow weirs instead of the installation of control valves.

The pinch valve is simple and economical slurry valve solution. It consists of three main components, valve body, sleeve and actuator. The pinch valve actuator manipulates the cell outflow by flattening the sleeve with jaws. The only component in contact with slurry is the sleeve, which, with the correct material selection for the application, makes the pinch valve relatively easy to maintain. However, sleeve material may lose elasticity over time. This can cause stickiness in the valve operation, so that the sleeve does not follow the jaws correctly. This problem can be eliminated with positive opening tags, which connect the sleeve to the jaws.

The construction of the dart valve is well known. The slurry flow through the valve is manipulated by a vertically moving cone, controlled by an actuator, which varies the area of the cell outlet. The dart valve is located in additional box connected to cell.

Along with choosing the right valve type, the sizing is also important. The control valve should be sized to operate in normal process conditions between 30 and 60% open, optimal for control performance and allows some disturbances to occur before saturating. Dart valves have wider opening range for reasonable control performance than pinch valves. Pinch valves have highly nonlinear characteristic curves near extreme positions of the opening. Depending on the size of the valve, a response time of 10-20 seconds from fully open to fully close is desirable. Particularly large pinch valve actuators require special attention to ensure sufficient actuator speed and in large cells, the use of two valves in parallel is practical, since the control properties of big valves are not adequate. In a dual arrangement, one valve may be operated manually whilst the other is automatically controlled or both valves may be automatically controlled by a slurry level control loop.

Slurry Level Control. Slurry levels are controlled by manipulating the outflow of the cell. The present value of the level is measured and compared to the setpoint and the controller calculates the signal used to manipulate the opening of the control valve in the cell. The majority of flotation plants use the PID-algorithm for this task. In Figure 8 the instrumentation for slurry level control in the flotation cell is shown.

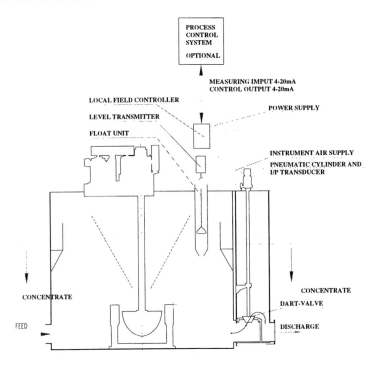

Figure 8 Instrumentation for level control in the flotation cell

In a series of flotation cells, PID-control of slurry levels is troublesome. Feed disturbances travel slowly through the cell bank and as each cell independently compensates for the disturbance separately, they concurrently drive the level of the following cell off the setpoint. Therefore more sophisticated methods have been developed to deal with level control. These strategies consider the problem as a multivariable task, where the whole series of cells is monitored and compensations between cells are calculated.

Jämsä-Jounela et al. (Jämsä-Jounela et al. 2001) presented feedforward control combined with traditional feedback control in order to diminish disturbances caused by the inflow of the cell bank. Stenlund and Medvedev (Stenlund and Medvedev 2000) presented a model-based decoupling control for levels in a series of flotation cells. Single level control loops are strongly interconnected and the aim of the control strategy was to decouple the consecutive level control loops. Perfect decoupling in this context means that, a change in either the slurry level or the control signal will have no influence on the level in the cell before or after the cell where the changes occur and are measured.

A block diagram of feedforward control and decoupling control is presented in Figure 9.

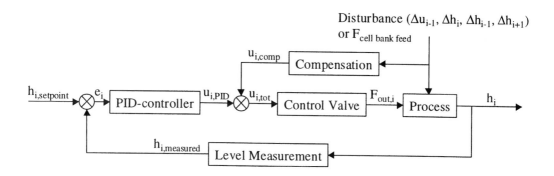

Figure 9 Level control loop with either decoupling or feedforward compensation (i^{th} cell, h cell level, u control signal, e error, F flow)

Airflow Measurement and Control. Airflow can be measured in many ways. Common measuring instruments for flotation airflow are thermal gas mass flow sensor or differential pressure transmitters with venturi tube, Pitot tube or annubar element. Each instrument type has its advantages and limitations.

The functionality of thermal gas mass flow sensors is based on the tendency of the flowing air to cool the sensor. The instruments using the sensor of this type are considered accurate but they are relatively expensive compared to other devices and measured air must be clean. These instruments are also factory calibrated, which complicates any later changes.

Differential pressure flow meters are popular in industry, including flotation, due to their low price, simple principle and fairly low requirement of maintenance. There is a large selection of differential pressure flow meters available and for flotation airflow measurement, the transmitter combined with the venturi tube or Pitot tube is the most common. An orifice plate is not suitable solution due to the significant pressure loss it causes.

A Venturi tube (see Figure 10) consists of converging conical inlet, a cylindrical throat, and a diverging recovery cone. Pressure is measured before the conical inlet and in the center of cylindrical throat. The pressure difference is related to flow and is calculated from the measurements by the transmitter. The Venturi tube is simple, reliable, and accurate if well calibrated and causes tolerable pressure loss. However, it is an expensive device and demands lot of space.

The transmitter with Pitot tube or annubar element (see Figure 10) provides airflow in the pipe by measuring the static and the total pressure. A Pitot tube has only one measuring point, while annubar element has several across the pipe, providing the average velocity in the pipeline. Hence the result provided by annubar element is not as dependent on the velocity profile as by Pitot tube.

Both techniques are quite accurate and the observed pressure drop is small. Problems associated with differential pressure flow meters are installation related, that is, large pipe sizes and flotation plant layouts where it is difficult to locate sufficient straight pipe sections. One solution is to use smaller pipe diameter in the section where instrument is to be installed, since straight pipe section length requirement is a function of the diameter.

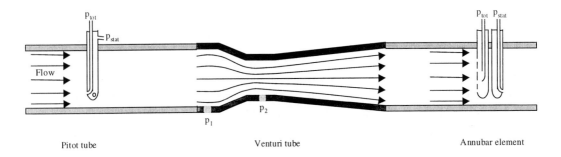

Figure 10 Schematic illustration of Pitot tube, venturi tube and annubar element

Butterfly valves are used for airflow control in the cells, as they are low-priced and their performance is satisfactory for the task.

Airflow control loop contains a measuring device and control valve, where the measuring device is situated before the control valve in the pipe. Airflow is measured and that value is compared to the setpoint. Flotation airflow control is not as problematic as slurry level control; therefore, a properly tuned PID-controller has been determined adequate for this control task. However, correct sizing is important for airflow valves because many problems in airflow control are due to oversized valves and poor airflow control can disturb slurry level control. Unfortunately, oversized airflow valves are very common in flotation plants, often a result of the incorrect assumption that bigger is better. Advanced control strategies and operators use aeration rate as control variable for grade control and circuit balancing. Therefore the setpoints of the airflow control loop are frequently manipulated.

Flotation cells with self-aspirating aeration mechanisms often do not have automatic airflow control. The available range of airflow control is anyhow limited. This problem is pronounced at high altitude.

Reagent Control. Reagent feed rates are often set to worst-case values that contain considerable safety margin under average conditions. In addition to increased chemical costs, this may result in loss of selectivity of the flotation.

A large variety of different instrumentation solutions for reagent dosing are available, due mainly to two special features of this task. Firstly, the reagent flows are quite small and therefore difficult to control and to measure. For instance common term of frother addition is milliliters per minute. Secondly, the reagents have different physical and chemical properties.

A simple but inaccurate way of reagent dosing is the use of on/off valves with estimated liquid flow. Estimates are regularly corrected by manual checks. If accurate reagent addition is needed and flow volumes are large enough for individual control loops, the best results are achieved by using inductive flow meters and control valves. Metering pumps are also suitable, even though they are high-priced and have high maintenance requirements. If flow volumes are too small for earlier mentioned methods or if economy is a priority, the flowing systems are possible solutions.

A dosing system presented in Figure 11 is suitable for flotation chemicals with reasonable conductivity (e.g., Xhantates, Cyanides, H_2SO_4). Slow process response makes the use of this partly On-Off system possible.

Levels of the reservoir vessels are measured by ultrasonic level meters or by flange model pressure transmitters. Instead of actual measurement, simple level switches can be used to indicate high and low level. The idea is to control the main flow with inductive flow meter and control valve or variable speed drive of the pump, then afterwards, split the main flow to adequately small flows for addition points. It is also possible to have a storage tank so high that gravity drives the flow, which is adjusted by a control valve.

Individual reagent dosing control for each addition point is performed by On-Off ball valves with pneumatic actuator and sequence logic.

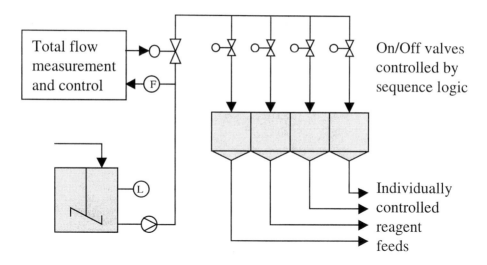

Figure 11 Dosing system for reagents with reasonable conductivity

Reagents having no conductivity (oils, frother) cannot be measured by a magnetic flow meter and a solution could be the use of a Coriolis flow meter. A metering pump is another possibility but often the volume of reagent involved is too low for satisfactory accuracy in operation.

One useful way to control reagents having no conductivity is described in Figure 12. The system consists of constant level tank (overflow) and 3-way solenoid valves having constant volume pipes above and process lines below. As the solenoid opens, it fills the constant volume pipe and when the solenoid closes, it lets the filled liquid to go to process line. The control system takes care of the On-Off pulses to solenoid valves.

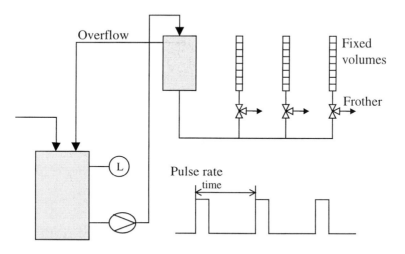

Figure 12 Dosing system for reagents with no conductivity

Most flotation plants use a ratio type control (grams of reagent per ton of ore) for the addition of reagents to the flotation circuit. The feedback action being carried out by plant operators, who adjust the ratio according to measurements provided by the on-stream analyzer (Hodouin et al. 2000).

Figure 13 shows a typical feedforward ratio control of xanthate dosing in copper-lead bulk flotation. In this case collector addition is already made in grinding circuit. On-stream analyzer assays provides lead and copper grades in feed and mill feed is measured with a belt scale. Predicted collector addition is determined from regression calculation, which is multiplied by an operator factor giving operator possibility to make percentage changes to predicted addition in either direction. Regression constants are determined with history data analysis (Jones et al.).

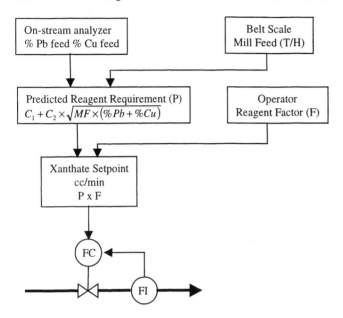

Figure 13 Feedforward ratio control of xanthate (Jones et al.)

pH, Eh and Conductivity Measurements. During the last decades attention has been paid to the effect of electrochemical potentials in slurry on the performance of flotation. Electrochemical measurements can give important information about the surface chemistry of valuable and gangue minerals in the process. Electrochemical potential measurements are almost the only method for detection of the chemical situation directly from the process (Ruonala 1995). pH is the most commonly measured electrochemical potential.

pH measurement is a special case of ion selective measurement, whereby pH measurement is sensitive to the hydrogen ion concentration. By using ion selective electrodes the concentration of certain ion can be measured. Ion selective electrodes are electrochemical cells where potential difference between the membrane of the electrode and the examined solution is developed. The magnitude of the potential difference is proportional to the logarithmic activity of the selected ion in the solution. Unfortunately, the potential difference cannot be reliably measured, directly. In addition, a reference electrode is required to measure the potential of solution. Measuring equipment are electrodes and pH-transmitter.

Sometimes pH measurement can be replaced or complemented by conductivity measurement, which can give approximately the same information about the flotation process and the instruments are less expensive. In general the conductivity measurement may be better than pH measurement in highly alkaline conditions, where reliable and accurate pH measurement is difficult to achieve. Conductivity measurement does not work well if large amount of air in the slurry or ore properties cause variations in the conductivity.

Eh measurement with platinum electrode (redox) can be useful in some special cases, such as Cu-Mo flotation, where additional nitrogen is added to flotation air.

pH measurement is troublesome because the electrodes are easily contaminated by active substances in the slurry. Therefore pH measurement often needs a sampler system or automatic washing system along with regular checks and maintenance.

Slurry pH is controlled by PID control, which adjusts lime or acid addition rate to the slurry. pH control is also problematic due to very slow response of the system to control actions. Therefore, any control interval must be set long; to allow the process to react to the control actions before the controller executes the next action.

Other electrochemical Potential measurements. Recently other electrochemical potential measurements have been under study. The use of minerals as working electrodes makes it possible to detect the oxidation state of different minerals and handle their flotability. The idea and equipment are similar to pH measurement, apart from the electrode materials, which are chosen according to the mineral. Stability of the electrodes has been a significant problem in on-line use although some encouraging results have been achieved using different mineral electrodes for process studies (Ruonala, Heimala and Jämsä-Jounela 1997).

Some minerals float well within certain limits of electrochemical potential. Improved recovery by the control based on electrochemical potential measurements may be possible if the correlations between assays of feed, tails, concentrate, recovery and electrochemical potentials can be found in data analysis.

Ruonala (Ruonala 1995) developed and implemented the control of sulfuric acid by cascade control, where the master loop potential of NiS-electrode gave the setpoint changes to pH of the slave loops in conditioner and flotation cells. Data collection was made before the implementation of the control. In the daily study, statistical and local optimum potentials were found. About 0-2 % increases in nickel recovery was reported, depending on the ore type.

Froth Image Analysis. Operators have always used their eyes to characterize froth. Based on their visual perception they have determined proper control actions for the particular process. However, the weak repeatability over time and the uncertainty due to the varying judgement of each individual have made this information unreliable. In automatic control systems, the use of information obtained by human visual perception has been difficult as well. Therefore new machine vision based instruments have been developed to obtain information about froth characteristics and some commercial products are already available on the market. These instruments build their functionality on the methods of image processing. Images provided by integrated digital camera of the instrument are analyzed with different models and algorithms. There are structurally two types of froth image analysis systems. A centralized type gathers images from several cameras installed on flotation cells, to one central computer, which performs the image analysis for images collected from all units. A distributed type device has camera and computer integrated together, so that all calculations are performed in the instruments. Therefore distributed system is very flexible, consisting of single instruments, that one by one can be connected directly to the plant automation system. Conversely, a centralized system forms a distinct froth image analysis system, which then may be connected to plant automation system. Along with the froth measurements, both types provide a visual image for the operator to view froth in the cells.

In general the difficulty with froth imaging has been relating the information extracted from the images to the flotation performance. Different indices and measurements describing properties of the froth such as bubble size distribution, bubble shape distribution, color, number, density, speed and stability, are provided by these instruments, yet there are only few real applications reported (Cipriano et al. 1998; Sadr-Kazemi and Cilliers 1997). The use of froth speed in control tasks has, however, proven its usefulness (Brown et al. 2001).

To summarize the issues discussed above, an example of a flotation circuit flowsheet with typical instrumentation is presented in Figure 14. Conductivity and pH are measured only in the conditioner. On-stream analyses are taken from feed, tailings and concentrate of the circuit, also from several flows between the flotation sections. Slurry flows, pulp levels and airflows are measured at several critical points. Most of the reagents are already added in grinding circuit, except frother, which is introduced to slurry in the conditioner and the re-addition of sodium cyanide in the cleaner.

Monitoring of the overall process, as well as each section of the process, is enabled by the extensive instrumentation. Level and aeration rate controls perform the fine adjustment of process condition, as flow measurements and on-stream analyses supervise critical circulating loads and performance of various sections. Conductivity measurements provide information about the electrochemical state of the feed and pH-measurement is needed for pH-controlled lime addition.

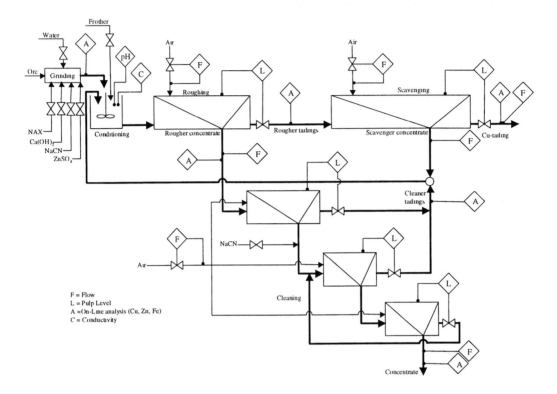

Figure 14 Typical flowsheet of flotation process ,Cu-circuit of Pyhäsalmi concentrator (Koivistoinen and Miettunen 1985)

Advanced Controls

Advanced controls are mainly controlling recovery and concentrate grade. The setpoints for an advanced controller are determined by a plant optimization system or by an operator. Based on setpoints, elemental assays provided by on-stream analyzers and other measurements, the advanced controller determines the control action, which often means the adjustment of the setpoints of base level controllers. Elemental assays from the tailings and the concentrate enable the feedback control of flotation as the performance of the circuit is monitored and corrective actions are made by the advanced controller. Elemental assay from the feed provides the possibility of the feedforward control actions. Reagent additions, aeration rates and slurry levels are adjusted, based on information of the incoming ore. Many advanced controls and particularly the feedforward type controllers, are relying either on process models or rule-based advanced model. Aeration rates, slurry levels or reagent additions rates are increased or decreased by If-Then rules, if recovery or concentrate grade is deviating from its setpoint.

Several deterministic process models to estimate metallurgical responses in flotation have been developed. Niemi, Maijanen and Nihtilä (Niemi, Maijanen and Nihtilä 1974) presented a fairly simple model for flotation, derived from mass balances. The flotation cell is considered as a tank with ideal mixing and first order flotation kinetics are used to describe floating of the mineral. Several affecting variables were ignored.

Bascur (Bascur 1982) chose a population model approach for development of the flotation process model. The idea was to divide mineral particles into several populations depending on their size, mineralogical composition and state in the slurry (e.g., free in slurry, in slurry attached to bubble). Transfer functions between populations are defined and the hydraulic model is included in the population model.

Flotation models have considerable weaknesses due to the complexity of process. Generally all the models suffer from the influence of unmeasured variables on the results of the model. Very complicated models make their use in control purposes difficult, as they demand many measurements and parameters, with inevitable problems in robustness of the model. On the other hand, simple models have problems in accuracy.

Various advanced control techniques such as adaptive control, multivariable control, optimal control, predictive control and fuzzy control, have been developed and explored for flotation control. All methods have generally the same control objective and the same operational principles but the approaches are different. Model-based control strategies are often based on models presented above, however the population model, developed by Bascur, is usually substantially simplified.

The control objective is to drive the recovery and the grade to their setpoints by manipulating the control variables, which are reagent addition, aeration rate, and slurry level. The suitable values for control variables are obtained by advanced control techniques and the values obtained are then used as setpoints for respective base level controls. The tendency is towards the use of aeration rate as a control variable since the slurry level control is more complicated and response of the reagent addition suffers from with considerable delay.

Control objectives are normally combined to one criterion, containing weighted terms for both recovery error (deviation from setpoint) and concentrate grade error. Often control variables are as well included to the criterion, which is then minimized with one of the techniques mentioned above.

Brown et al. (Brown et al. 2001) stated that the speed at which the froth is recovered over the lip of the cell has a very direct and consistent influence on the grade and the recovery of the flotation circuit. Exploitation of this relation has been difficult due to the absence of a suitable device for measuring the froth speed. As mentioned earlier, new machine vision based instruments can now provide this function. Therefore, control of the froth speed is possible to execute by using slurry level, aeration rate and frother dosage as manipulated variables.

A hierarchical control for the froth speed has been developed (Brown et al. 2001). The aeration rate was used for fast and fine froth speed adjustment, while for relatively big step speed adjustments, level and frother changes were applied. The aeration rate was chosen as the primary adjustment variable, as air could be controlled more tightly to a setpoint than the level, level measurement being noisier than air measurements. The froth speed control was successfully extended to control of concentrate grade. It was found that the concentrate grade is more strongly correlated to froth speed than to level or to aeration rate. For this reason, the concentrate grade controller was developed. Based on the grade information obtained from on-stream analyzer, a fuzzy controller adjusted the setpoint of the froth speed. The recovery from the cells under the new control strategy was found to be significantly better than the recovery from the corresponding cells in other bank under manual control. Authors are confident that controls of this type will become common in many flotation plants.

Optimization

When considering the optimization of the flotation process, there are some requirements for the automation system and the knowledge of the process. Appropriate instrumentation and well-tuned base level controls are essential. A quality and extensive data bank with process data including on-stream and laboratory analysis data is also required.

The data bank enables the creation of grade-recovery curves for different ore types. Grade-recovery curves are the basis of most optimization methods as it describes the relationship between the mineral or metal recovery and the grade of the concentrate for a certain ore type. Typical relationship between these factors is presented in Figure 15.

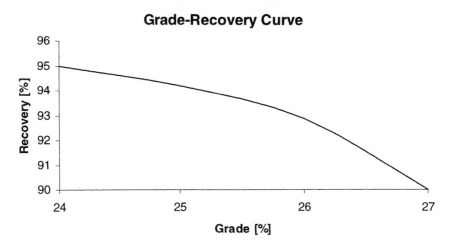

Figure 15 Grade-recovery curve

As seen in Figure 15 an increase in recovery decreases grade and vice versa. Therefore the optimal point on the curve is somewhere between the maximum recovery and maximum grade. There are some ways to determine the optimal point.

Formulas to calculate the value of products and operating costs of production are desirable as some strategies approach the problem from the economical point of view. One idea of how to find the optimal point on certain grade-recovery curve is the principle of isoeconomic contours (Flintoff 1992).

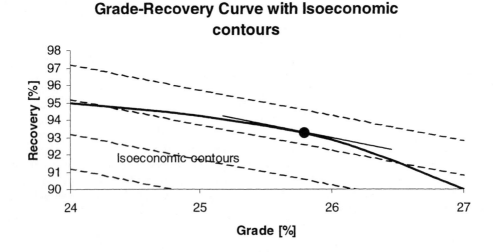

Figure 16 Grade-recovery curve with isoeconomic contours

In Figure 16 dotted lines are describing the equal economical recovery of the process as a function of recovery and concentrate grade. The maximal profit will be achieved by finding the maximum point of grade-recovery curve in proportion to the isoeconomic contours.

Calculated isoeconomic contours are based on net smelter return, which sets the price for produced concentrate. It consists of market price of the produced metal, concentrate grade, refining and smelting fee, quality-based penalties depending on the contents of certain impurities and other costs (freight rate and dewatering costs). The following equations show an example of net smelter return calculation.

$$T_1 = (M_1\% - M_1\%_{GDF})P_{M1} + (M_2\% - M_2\%_{GDF})P_{M2} - C_{SMELTING}$$
$$- (M_1\% - M_1\%_{GDF})C_{REFINING,M1} - (M_2\% - M_2\%_{GDF})C_{REFINING,M2} \quad (1)$$
$$- (IMP\% - IMP\%_{LIMIT})C_{IMPURITY} - (Mo - Mo_{LIMIT})C_{Mo} - C_{OTHER}$$

$$T = \frac{T_1 \text{Recovery}}{M_1\%} \quad (2)$$

where
T_1 is total price for concentrate ton
$M_1\%$ is primary valuable metal grade in concentrate, (e.g., copper)
$M_2\%$ is secondary valuable metal grade in concentrate, (e.g., silver)
$M_i\%_{GDF}$ is deduction factor for valuable metal i (i=1...2) grade in concentrate
P_{Mi} is price factor of valuable metal i (i=1...2), USD/(% x concentrate ton)
$C_{SMELTING}$ is smelting fee, USD/concentrate ton
$C_{REFINING, Mi}$ is valuable metal i (i=1...2) refining fee, USD/(% x concentrate ton)
IMP is impurity grade in concentrate (e.g. zinc in copper flotation)
IMP_{LIMIT} is impurity grade limit
$C_{IMPURITY}$ is impurity fee, USD/(% x concentrate ton)
Mo is concentrate moisture
Mo_{LIMIT} is concentrate moisture limit
C_{Mo} is moisture fee, USD/(% x copper concentrate ton)
C_{OTHER} is fixed fee for dewatering and freight, (USD/concentrate ton)
T is total price for primary valuable metal ton in ore
Recovery is primary valuable metal recovery

In spite of control strategies and expert systems developed, there still are a large number of flotation plants operating purely on the experience of the operators. In these cases, operators examine the curves above, the recent state of process, the available measurements and adjust setpoints of the controllers on the strength of their knowledge. Sometimes cells are even operated totally manually. An experienced operator can often find right corrections to make a given process perform well but inconsistency and lack of continuity provide the need for optimization systems.

New expert systems are concentrating to solve the issue of the feed type classification, which is a challenging and important task. Each feed type should be specially treated, as their grade-recovery curves are feed type-specific as well as finding the optimal reagent dosing.

The performance of the expert system is founded on the success of the feed type classification (Laine 1995). The success of the classification is, in turn, based on the on-line information that the system receives and on the algorithm the system uses. On-line measurements input to the system must include adequate information to distinguish feed types. The algorithm also must be able to classify the information into different feed types. The knowledge on the type of the feed of the process can be utilized in feedforward control, which sets the suitable process conditions for each feed type.

The classification algorithm can be based for instance, on neural networks, principal component analysis or cluster analysis.

In Figure 17 The expert system in Hitura concentrator is described, which uses Kohonen self-organizing maps for classification. As the feed type is classified, it is sent to the knowledge base of the system, which then determines the correct process actions by adjusting the setpoints of the controllers. Information is also provided to operator and the performance is indicated by an economical success index display. The classification module is updated when a new type of feed is encountered (Laine 1995; Jämsä-Jounela, Karesvuori and Laurila 2000).

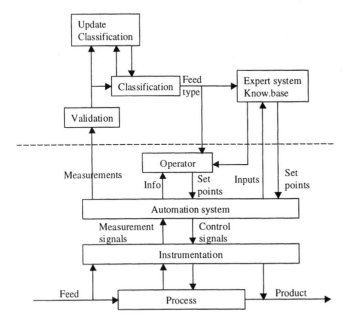

Figure 17 **The structure of the expert system in Hitura concentrator (Jämsä-Jounela, Karesvuori and Laurila 2000)**

The expert system knowledge base consists of rules to which the knowledge of process is converted. Normally If-Then rules use measurements, classified feed type and process models to define optimal control actions. In Figure 18 an example of optimization of reagent dosing is shown (Koivistoinen and Miettunen 1985).

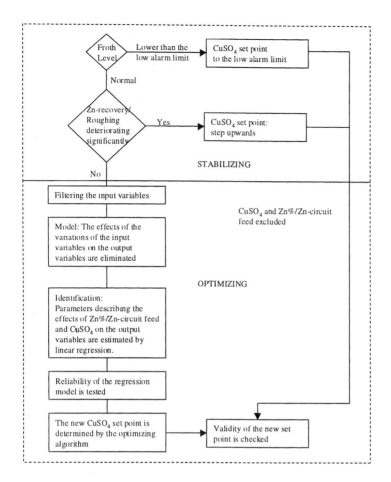

Figure 18 Optimization of CuSO₄ dosing in Pyhäsalmi concentrator (Koivistoinen and Miettunen 1985)

CONCLUSIONS

There has been a veritable plethora of strategies, techniques, technologies, and instrumentation applied over many years by many people, scientists, researchers, suppliers and the industry itself, pursuing the elusive goal of optimizing flotation performance over prolonged periods. The pursuit has met with some encouraging progress in the past few years. A combination of extensive accrued knowledge of the process and control regimes, modern instrumentation and data processing all contribute to the possibility of optimum performance being achievable through plant automation. Nevertheless, the highest and most successful level of optimization will always require that the most basic levels of control are in place, operating well and maintained so.

REFERENCES

Andersen, R.W., Grönli, B., Olsen T.O., Kaggerud I., Romslo K., and K.L. Sandvik 1981. An optimal control system of the rougher flotation at the Folldal Verk concentrator, Norway. *Proceedings of 13th International Mineral Processing Congress vol. 2*, ed. J. Laskowski, Varsova. 1517 – 1537.

Arbiter, N., and C.C. Harris 1962. Flotation kinetics. In *Froth Flotation*, ed. D. W. Fuerstenau, New York, Edwards Brothers Inc. 215-246.

Aumala, O. 1996. *Teollisuusprosessien mittaukset*. Tampere: Pressus.

B ascur, O.A. 1982. *Modelling and computer control of a flotation cell*. Ph.D. thesis, University of Utah, Utah.

Borer J. 1985. *Instrumentation and Control for the Process Industries*. New York: Elsevier.

Brown, N., Bourke, P., Ronkainen, S., and M. van Olst 2001. Improving flotation plant performance at Cadia by controlling and optimizing the rate of froth recovery using Outokumpu Frothmaster, *Proceedings of the 33rd Annual Meeting of the Canadian Mineral Processors*, ed. M. Smith. 25-37.

Cipriano, A., Guarini, M., Vidal, R.,Soto, A., Sepúlveda. C., Mery, D., and H. Briseño 1998. A real time visual sensor for supervision of flotation cells. *Minerals Engineering*. 11(6):489.

Flintoff, B.C. 1992. Measurement issues in quality "control". *Presented at the 1992 Toronto CMP Branch Meeting*.

Herbst, J.A., Pate, W.T., and A.E. Oblad 1992. Model-based control of mineral processing operations. *Powder Technology* 69:21.

Hodouin, D., Bazin, C., Gagnon, E., and F. Flament 2000. Feedforward-feedback predictive control of a simulated flotation bank. *Powder Technology*. 108:173.

Henning, R.G.D., Schubert, J.H., and Y. Atasoy 1998. Improved flotation performance at Fimiston plant through better level control. *Presented at International Symposium on Gold Recovery*, Montreal.

Hulbert, D.G. 1995. Multivariable control of pulp levels in flotation circuits. *Preprints of the 8th IFAC International Symposium on Automation in Mining, Mineral and Metal Processing*, ed. I. J. Baker, Sun City. 71 – 76.

Jones, J.A., Deister II, R.D., Hill, C.W., Barker, D.R. and P.B. Crummie. Process control at the Doe Run company.

Jämsä-Jounela, S.-L., Dietrich, M., Halmevaara, K., and O. Tiili 2001. Control of pulp levels in flotation cells. *Preprints of the 10th IFAC International Symposium on Automation in Mining, Mineral and Metal Processing*, ed. M. Araki, Tokyo, 81-86.

Jämsä-Jounela, S-L., Karesvuori, J. and H. Laurila 2000. Flotation process neural data analysis and on-line monitoring, *Proceedings of the 32nd Annual Operator´s Conference of the Canadian Mineral Processors*, ed. M. Tagami, Ottawa. 441-457.

Koivistoinen, P., and J. Miettunen 1985. Flotation control at Pyhäsalmi. In *Developments in Mineral Processing 6, Flotation of Sulphide Minerals*, ed. K.S.E. Forssberg, Amsterdam: Elsevier. 447-472.

Koivo, H.N., and R. Cojocariu 1977. An optimal control for a flotation circuit. *Automatica* 13:37.

Laine, S. 1995. Ore type based expert system for Hitura concentrator. *Preprints of the 8th IFAC International Symposium on Automation in Mining, Mineral and Metal Processing*, ed. I. J. Barker, Sun City. 321-327.

Niemi, A.J., Maijanen, J.S., and M.T. Nihtilä 1974. Singular optimal feedforward control of flotation. *Preprints of the IFAC/IFORS Symposium on Optimization Methods Applied Aspects*, ed. I. Tomov, I. Popchev, G. Gatev, M. Kitov, N. Naplatanov, Varna. 277-283.

Osorio, D., Pérez-Correa, J.R., and A. Cipriano 1999. Assessment of expert fuzzy controllers for conventional flotation plants. *Minerals Engineering*. 12(11):1327.

Pérez-Correa, R., González, G., Casali, A., Cipriano, A., Barrera, R., and E. Zavala 1998. Dynamic modelling and advanced multivariable control of conventional flotation circuits. *Minerals Engineering*. 11(4):333.

Ruonala, M. 1995. The use of electrochemical mixed potential measurements for the process control and expert system development at the Hitura mine. *Proceedings of Automation in Mining, Mineral and Metal Processing*.

Ruonala, M., Heimala, S., and S. Jounela 1997. Different aspects of using electrochemical potential measurements in mineral processing. *Int. J. Miner. Process.* 51:97.

Sadr-Kazemi, N., and J.J. Cilliers 1997. An image processing algorithm for measurement of flotation froth bubble size and shape distribution. *Mineral Engineering*. 10(10):1075.

Schubert, J.H., Valenta, M., Henning, R.G.D., and I.R. Gebbie 1999. Improved flotation performance at Karee platinum mine through better level control. *Journal of the SAIMM Jan/Feb (1999)*.

Stenlund, B., and A. Medvedev 2000. Level control of cascade coupled flotation tanks. *Preprints of the IFAC workshop on Future Trends in Automation in Mineral and Metal Processing*, ed. S-L. Jämsä-Jounela, Helsinki. 194-199.

Pressure Oxidation Control Strategies

John Cole and John Rust[1]

ABSTRACT

In mineral processing operations there are more items one wants to control than can control. Gold pressure oxidation facilities are the same way. It would be nice to see, touch, taste and feel what is happening inside an autoclave, as with other unit operations. To measure pH online inside the autoclave could be critical to enhancing recovery or lowering costs. The only sense left to an operator is smell, which is very useful to discover a leak in the circuit, but not for process control. This paper will address current and future control strategies utilized in gold pressure oxidation operations.

INTRODUCTION

Newmont Mining Corporation owns and operates the Twin Creeks and Lone Tree Mines in Humboldt County, Nevada. Twin Creeks is located 85 kilometers northeast and Lone Tree 55 kilometers east of Winnemucca, Nevada. Both mines utilize pressure oxidation to treat high-grade refractory ores. Control strategies for these plants and others are similar. This paper will discuss control strategies for gold pressure oxidation circuits for whole ore feed. Control of concentrate pressure oxidation circuits are similar, but not exactly the same. The greatest difference being the lack of slurry pre-heating. The Twin Creeks refractory process facility is typical for all whole ore pressure oxidation plants. The process flowsheets are shown on Figures 1 and 2.

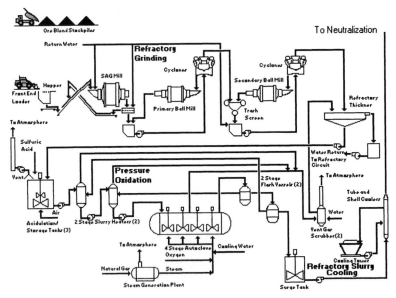

Figure 1

---

[1] John Cole is Chief Process Engineer and John Rust is Senior Metallurgist for Newmont in Nevada

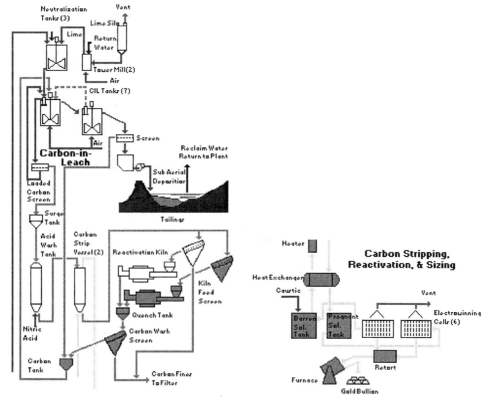

Figure 2

Process Flowsheet

Pressure oxidation is a pre-treatment step to gold recovery. Ore is delivered from the mine to stockpiles. All facilities utilize size reduction to start the ore processing. Specific ore body needs dictates the configuration and size of grinding equipment. Twin Creeks utilizes a SAG mill, primary ball mill and secondary ball mills for size reduction. Target grind size at Twin Creeks is very fine. Post grinding most facilities de-water the slurry to the maximum extent. Excess water consumes too much heat in the autoclaves affecting the circuit heat balance.

Depending on the geochemistry of ore, the thickened slurry may be acidified. Slurry is conditioned with sulfuric acid as required to reduce the carbonate content of the feed. The slurry is then fed to the pressure oxidation circuit. Generally there are several stages of slurry heating utilizing steam from the let down flash vessels. The pre-heating step is critical in whole ore circuits. Heated slurry is pumped into the autoclave via positive displacement diaphragm pumps. The autoclaves are horizontal cylindrical pressure vessels with multiple compartments. There is at least one agitator in each compartment. Oxygen is generated onsite by a cryogenic air separation plant. Oxygen required for the oxidation reaction is added to each compartment. Steam and water may also be added to each compartment to control the temperature. Autoclaves operate anywhere from 160°C to 230°C at pressures ranging from 1,225 kPa to 3,150 kPa. These pressures reflect the saturated vapor pressure plus the non-condensable gas pressure. Vessel retention times are generally 45 to 60 minutes. Oxidized slurry exits the autoclave through a ceramic choke valve. The pressure let down system consists of flash vessels, same number as slurry heaters, which reduce the slurry pressure and subsequently the slurry temperature to atmospheric levels. The slurry is further cooled either through a series of tube and shell heat exchangers or counter-current decantation thickeners. The cooled, oxidized slurry then reports to a neutralization circuit where limestone and/or milk-of-lime is added to neutralize the acid and provide protective alkalinity for the gold leaching and recovery circuit. Neutralized slurry may then be processed through any number of conventional gold recovery means.

PROCESS CONTROL

Process control for pressure oxidation circuits can be categorized as physical and metallurgical controls. It is the same as any mineral processing plant. Physical controls include; level, flow, mass flow, pressure and temperature. Metallurgical controls include; ore blending, slurry density, slurry viscosity, eH/pH, free acid level, level of sulfur oxidation, rate of sulfur oxidation, oxygen utilization, and ferrous iron titration. Without proper physical controls, metallurgical control of the process would be extremely difficult. Physical control of a pressure oxidation circuit is relatively straightforward and not too complex. Though proper physical control is not complex, the outcome of that control may not be cost effective. An example would be over using steam and water for temperature control. Figure 3 shows a simplified pressure oxidation flow sheet with only slurry flow shown.

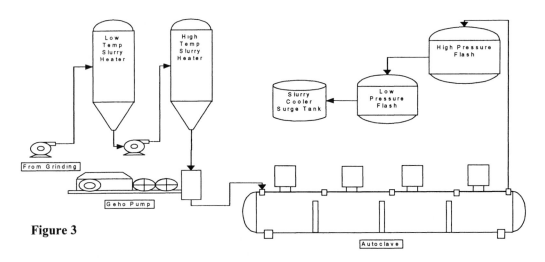

Figure 3

Physical Controls

Level Control. There are many types of instruments to measure level such as ultra-sonic, differential pressure, nuclear devices and even strain gages. Because all of the vessels operate at elevated temperatures and some at elevated pressures, measuring the level in these vessels is difficult. Through years of experience nuclear devices have proved to be the most reliable means of pressure oxidation vessels level measurement. Nuclear sources are used in the heaters, autoclaves, flash vessels and other minor vessels for level measurement. Vessel level is controlled in the slurry heaters and the autoclave, measured only in the other vessels.

In the case of the heaters a nuclear source is mounted on the outside of the heater vessel with a strip detector mounted on the opposite side. Level is measured by the amount of radiation reaching the detector. Level set points are set at a maximum comfortable level, about 80% of range. This provides the maximum suction head on the discharge pump and maximum surge capacity in case of upset conditions. The actual level range monitored is a small part of the vessel. The measured level range starts near the bottom or outlet of the vessel and ends below the flash steam inlet. Inside the vessel above the steam inlet pipe are sets of trays designed to cascade the slurry contacting it with the steam. If the slurry level rises above the steam inlet, the steam will collapse causing very violent shaking of the heaters. To prevent this a separate point source nuclear device is installed as a high level interlock. The heater level controls the speed of it's feed pump, either directly or cascading through flow control. The goal of the control scheme is to have steady levels in the heaters.

At Twin Creeks the first stage heater operates at atmospheric pressures. The second stage heater operates at elevated pressures. There is a pressure control valve on the vent outlet of the second stage heater. Steam to heat the slurry is supplied by the respective first or second stage flash vessels on the autoclave discharge. Level control for two vessels in series utilizing pumps to control the levels is easy. The added complexity of the autoclave circuit is the temperature, steam flow and source of steam. Since the source of steam is the flash vessels any change in throughput or autoclave level will impact the amount of steam reporting to the heaters. In the second stage heater this change in steam flow will change the vessel pressure which in turn will either increase or decrease the head on the feed pump. This will cause a change in flow, which leads to an upset in heater vessel level control. As stated earlier the goal of the control scheme is steady heater level. With the added complexity of varying pressures, a preferred control method utilizes a level signal cascading through flow control. Figure 4 is an example of slurry heater level control.

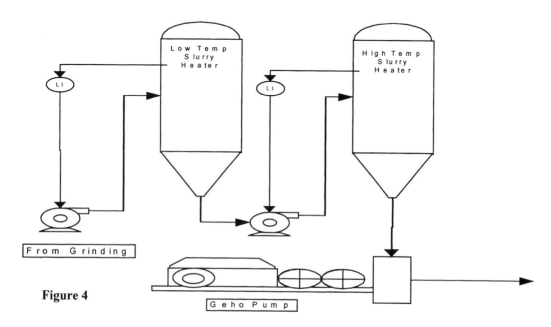

Figure 4

For the autoclave, the nuclear source is mounted inside the autoclave in a double-lined well. The well is located in the vapor zone near the top of the autoclave. A strip detector is mounted on the outside of the autoclave in a position that the slurry level in the autoclave is measured. Actual level measured is generally only at the top of the vessel. The level range is from the horizontal centerline to the top of the last compartment wall in the autoclave. This represents a 0% to 100% level. The autoclave slurry outlet is set near the horizontal centerline of the vessel therefore it is desirable to have the level higher than the outlet. Otherwise both gas and slurry would exit the autoclave through this pipe. A slurry level above the compartment wall could allow short-circuiting. The desired slurry level is fifty millimeters below the last compartment wall. The level detector sends a signal to a right angle choke valve with a modulating plug. As the autoclave level increases the plug opens allowing more flow to exit the autoclave. The reverse is true for decreasing autoclave level. Tuning and proper operation of this valve is critical to autoclave operation. If this valve operates too quickly or is tuned to hold a tight level it can cause several other problems in the circuit. These will be described in detail later. Figure 5 shows the autoclave vessel level control scheme.

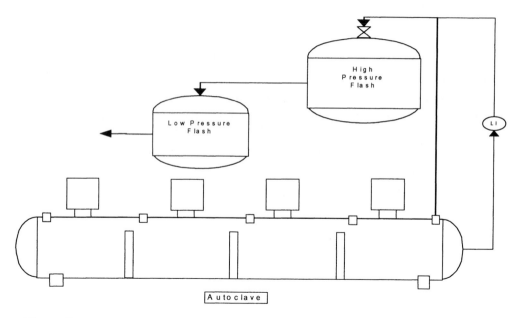

Figure 5

Flow Control. There are two major items where flow is controlled, slurry and oxygen. Both are mass flow, but controlled as volumetric flow. There are other flow controllers such as water or steam but these are utilized for temperature control. Flow set points are set based on throughput requirements.

Slurry flow is measured by a magnetic flow meter. Slurry density is measured by a nuclear device. Combination of flow and density provides mass flow. Slurry flow into the autoclave is controlled by the autoclave feed pump speed. These pumps are positive displacement diaphragm pumps. The slurry flow in these pumps is linear to the speed. The speed controller is a manual set constant. The operator increases or decreases flow based on operating conditions, generally the faster the better. As the operator increases the speed of the feed pump the heater feed pumps will also increase in speed based on the level of the heater vessels. An increase autoclave feed pump speed will lower the level in the second stage heater. The lower level will send a signal to the second stage heater feed pump to increase the speed. This in turn will lower the level of the first stage heater and cause the first stage heater to send a signal to the first stage heater feed pump increasing its speed. Figure 6 shows this control method.

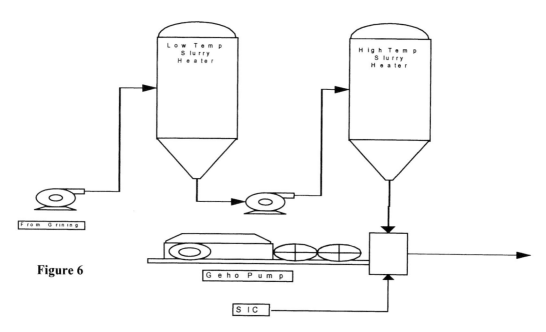

Figure 6

Oxygen flow is measured on the main oxygen header from the oxygen plant by a differential pressure orifice plate. This flow is pressure and temperature compensated to standard conditions. This gives the total oxygen flow to the autoclave. The operator utilizing a manual set constant controls total oxygen flow. There are also individual compartment flow meters. These are vortex shudder types. They are not pressure and temperature compensated. The compartment flow meters are used to distribute oxygen for metallurgical purposes therefore the level of accuracy required is not as high as total oxygen flow. Oxygen flow changes and distribution of the flow to the compartments are made based on the metallurgical performance of the plant. Oxygen flow control is shown on Figure 7.

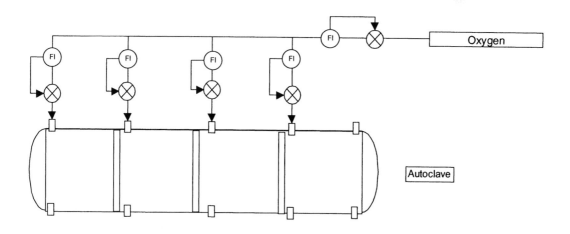

Figure 7

Pressure Control. Pressure control is utilized in the second stage slurry heater, the autoclave and the autoclave agitator seal water system. In all cases the control device is a valve. Set points vary with operating conditions. Autoclave pressure set points are generally 500 to 700 kPa above the boiling point at that temperature. This is known as the non-condensable over-pressure. The majority of the non-condensable gas is oxygen.

A right angle choke valve with a modulating plug controls pressure in the autoclave. The autoclave control valve has to be quick reacting due to the rapid changes in pressure. Autoclave pressure can swing as much as 350 kPa with the loss of a feed pump, oxygen flow or an agitator. Most of the oxygen is added to the first and second compartments in the autoclave. If one of these agitators stops all of the oxygen in those compartments reports to the vapor zone rapidly thereby increasing pressure rapidly. Oxygen is added to the autoclave at a fixed rate based on the tonnes of sulfur fed. Failure in one of the two feed pumps would halve the sulfur feed though the oxygen flow would remain constant. The result would be a rapid increase in autoclave pressure until the pressure control valve compensates for the change. A loss of oxygen flow would cause a rapid decrease in autoclave pressure. Autoclave pressure control is illustrated on Figure 8.

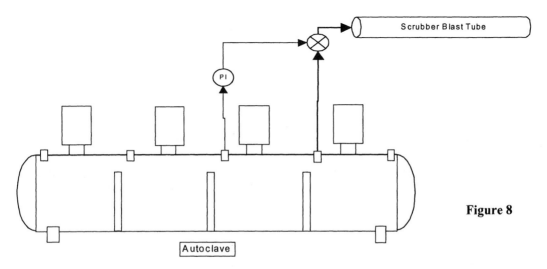

Figure 8

The seal water system is designed to provide a water pressure slightly higher than autoclave pressure for the agitator double mechanical seals. The set point is generally 350 kPa higher than the autoclave. Since the set point pressure is tied to the autoclave pressure, any changes in the autoclave pressure affects the seal water system. Therefore the pressure control valve must be very quick acting. Rapid decrease in autoclave pressure does not cause a problem for the seal water system. A rapid increase in autoclave pressure will encroach on the seal water system pressure causing the system to shut down and go into emergency back up. Once pressure is stabilized the seal water system would have to be started.

Temperature Control. There is a relatively narrow band the temperature in the autoclave is controlled at. For most autoclaves that band is 15°C. Too low of temperature and oxidation is inhibited or ceases. Too high of temperature and the pressure limits on the vessel are approached. Temperature control in the autoclave is simple providing the ore is within design specifications. There is the ability to add steam for heat and water for cooling to all compartments. The goal in operation is not to add either steam or water for temperature control. Proper ore blending and adjusting the incoming slurry temperature and density is the preferred control method. Feed rate is also important in temperature control. Too fast of feed rate causes low temperatures in the first compartments, high temperatures in last compartments. Too slow of feed rate causes high temperatures in the first compartments, low temperatures in last compartments. Temperature control for the autoclave is shown on Figure 9.

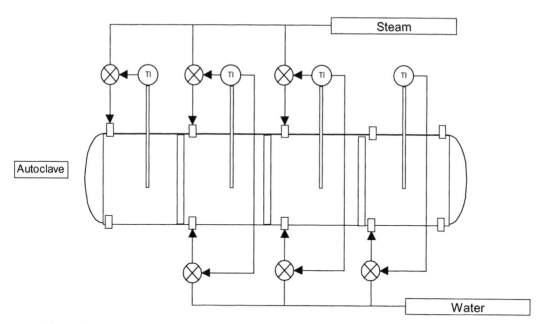

Figure 9

Slurry temperature control in the heaters is essentially pressure control for the second stage. The first stage temperature is not controlled but merely a function of steam flow. Since the first stage heater is operated at atmospheric pressures the highest temperature obtainable is 100°C. Steam from the first stage flash vessel is directed through the second stage heater. The steam comes in direct contact with the slurry. A control valve on the vent outlet regulates the vessel pressure for both the flash and heater vessels. The slurry temperature will approach the steam temperature in steady state operation. Therefore to increase the slurry temperature increase the vessel pressure and decrease pressure to decrease slurry temperature. Changes in operating heater temperatures are made based on autoclave performance. For ores low in sulfur a higher slurry heater temperature would be appropriate. For high sulfur ores lower heater temperatures are would be correct. There is a maximum slurry temperature to autoclave feed pumps can be exposed to, but no minimum. The maximum is dictated by the rubber components in the feed pump. Slurry heater temperature control is shown on Figure 10.

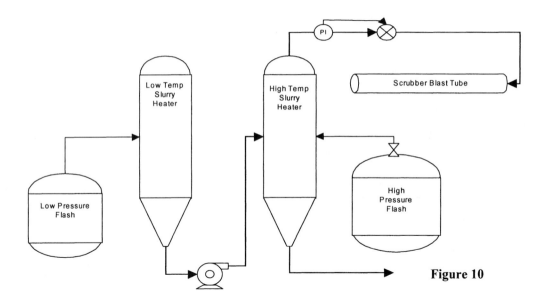

Figure 10

Metallurgical Controls
Without proper control of any of the above-mentioned physical controls, metallurgical performance would suffer. There are metallurgical control aspects of level, pressure, and temperature and flow control. These will be discussed also.

Ore Blending Proper feed to the autoclave is the first and most important aspect of successful operations. An autoclave's function is to pre-treat an ore so that valuable elements may be recovered. The autoclave must burn the sulfur that occludes the gold to allow gold recovery. An autoclave operator is not concerned with the gold grade feeding the plant. Gold will enter and exit the autoclave in the same form. An operator is extremely concerned with the sulfur and carbonate assays, and other ore constituents that oxide or change form. The sulfur is the fuel for the process and the carbonate can be beneficial or detrimental depending on the amount. There is a minimum and maximum amount of sulfur or carbonate an autoclave can process. Too low of sulfur and not enough heat to sustain the reaction, too high of sulfur and not enough oxygen to maintain throughput. Carbonate behaves the same way, too low is usually not an autoclave problem but can be a neutralization problem. Too high of carbonate and the reaction will be quenched due to lack of acid. Other elements such as copper can affect down stream circuit performance with increased cyanide consumption. As with any circuit the autoclave operates best with the least amount of upset. Having a consistent feed is the best way to prevent upsets. An unanticipated high carbonate feed can be very expensive in terms of unwanted down time and the cost acid to destruct the carbonate. An unanticipated low sulfur feed can also be expensive due to the steam required to keep the reaction going. In ore blending absolute assay values are not as critical as anticipating assay values. It does not matter as much if the ore has a 3% sulfur assay or 5% sulfur assay, as long as the operator knows the sulfur value and can adjust the circuit for them.

Slurry Density/Viscosity Slurry density is important for circuit throughput. Slurry viscosity plays a role in oxidation. The higher slurry percent solids, the more circuit throughput at a constant pumping rates. From strictly a production standpoint the thicker the better. Usually, though not always, the more dense the slurry the greater the viscosity. A typical autoclave feed solids target would be 50 percent solids. Too low of density, less than 45 percent solids, throughput is lowered due to pumping limitations and heat balance is affected. Lower density means more water to pump and heat and cool. Water takes five times more energy to heat than solids. Therefore it is critical to have the minimal amount of water in the feed as possible. Even with high sulfur in the feed, a low density feed will require steam addition to maintain proper temperature. Slurry density is controlled upstream of the autoclave in the grinding thickener prior or by water addition to the autoclave feed. The latter works only if the slurry is thicker than desired.

The measure of slurry viscosity is relative to previous viscosity measurements. The control of slurry viscosity is more art than science. There are several ways to increase viscosity and only one cost-effective way to reduce viscosity. Ore types, high clay, generally produce high viscosity slurries. Other operator-induced methods of viscosity enhancement could be over flocculating a thickener or excessive use of acid in the acidulation circuit. The one true viscosity reduced is dilution with water. An operator desires the highest feed percent solids possible thereby maximizing throughput. If the slurry becomes too viscous, sulfur oxidation is ultimately hampered. The autoclave agitators can not mix and distribute the oxygen effectively in too viscous of slurry. The oxygen added to the autoclave must be physically dispersed and intimately contacted with sulfur in order to oxidize it. The more viscous the slurry the smaller area of influence the agitator will have. The oxygen will form large bubbles and vent through the slurry to the vapor zone in the autoclave. Therefore only part of the slurry will be oxidized. Excessively high viscosity can cause similar temperature control problems as too low of density. With limited oxidation in the first compartment due to high viscosity there will not be enough heat generated to maintain the proper temperature. Steam addition would be required to maintain the temperature. The oxidation reaction would move towards the later compartments where typically less oxygen is added. The result would be to starve the slurry for oxygen caused by higher than normal sulfur assays in the later compartments without commensurate levels of oxygen. Though the only control for high viscosity is dilution, knowing when to dilute can be as difficult. Generally if the first compartment temperature is not at anticipated levels and the reason is not apparent, dilute the feed. This will accomplish two things, the solids feed rate will be reduced along with the viscosity. Both of which should help oxidation rate and temperature profile.

Oxidation. To achieve sulfur oxidation a few items are required, heat, acidic environment, and oxygen. To a large extent the amount of each of the items determines the rate of oxidation; more heat, more acid, and excess oxygen equals faster rates of oxidation. On the surface one would argue that complete oxidation at the fastest rate possible is always desired. This is not always true. The extent of oxidation required is ore specific and the minimum required oxidation should be targeted. In the case of the Lone Tree Mine, the original design was for partial oxidation of the sulfides. The desire was to oxidize the entire fine-grained pyrite, which was gold bearing, and none of the coarse pyrite, which was barren. The target was set at 75% overall sulfur oxidation. Though it was possible to achieve more oxidation, it was undesirable from a cost standpoint. The more sulfur oxidation meant more acid generation, which in turn meant more lime consumption. The gold recovery did not improve with increased oxidation.

To start an oxidation circuit the autoclave is filled preferably with acidic slurry. The acidic slurry allows for easier start ups since the slurry is already acidic. When starting without acidic slurry the first addition of oxygen will go to creating an acidic environment. The autoclave is then heated at a slow rate with steam. Slurry flow through the circuit is started when the autoclave is at operating temperatures. As soon as slurry flow is established oxygen flow to each compartment is established. As the oxidation reaction starts the temperature will increase. As the temperature increases in the autoclave the slurry flow is increased. Both oxygen and slurry flows are increased simultaneous until the desired flow rates are achieved. It is easy to overcome the reaction by increasing the slurry feed too fast. The start up and operation of an autoclave is analogous to a campfire. To start a fire one must have kindling and as the fire is burning larger pieces of wood are added. Too much wood too early and the fire will go out. Not enough air and the fire will go out. Not enough wood or too much air and the fire will burn out quickly. The operator must strike a balance with the fuel (sulfur) and air (oxygen). Once the circuit is operating normally decisions must be made as to the targeted throughput, oxygen addition rate, oxygen distribution, temperature, pressure and acid addition to the feed. Most of these decisions are made in advance and not always by the operator. Throughput is dictated by management, oxygen addition is a function of sulfur assay and throughput, temperature and pressure are usually at design levels, leaving acid addition and oxygen distribution for the operator to manipulate.

Ultimate controlling parameters for daily operations vary from facility to facility. Different plants use ferrous iron titrations on the autoclave discharge, free acid titrations on the autoclave discharge, eH of autoclave discharge or sulfur assays. Sulfur assays are the desired control parameter though possible the most difficult to obtain quickly and accurately. Circuit feed rate is adjusted based on the assays. Oxygen addition rate is determined by assays from the autoclave feed. The pounds per hour of oxygen fed to each autoclave is determined by the pound per hour of sulfide sulfur fed and a set oxygen utilization factor. Oxygen utilization is the amount oxygen consumed in the reaction. The design oxygen utilization ranges from 60% for whole ore facilities to more than 80% for concentrate autoclaves. Adjustments to the preset oxygen flow rates are made based on compartment temperatures or other control parameters. Generally if parameters are less than expected oxygen flow is increased. Oxygen distribution is also preset, generally most of the oxygen reports to the first half of the autoclave, since the amount of sulfur is greater \in the first half. The carbonate assays of the feed determine acid addition to the autoclave feed. The amount of acid addition is tempered against the autoclave discharge free acid titration. If free acid titrations are higher than desired, feed acid addition is reduced or stopped. There is a minimum desirable free acid level. This is maintained to insure an acidic, oxidizing environment inside the autoclave. Temperature and pressure usually remain at design levels. A decrease in either temperature or pressure usually reduces the rate of oxidation. Since most if not all facilities operate at near maximum temperature and pressure there is little upside by increasing either one. The use of eH/emf readings in the neutralization circuit has been used as an online measure of oxidation at the Twin Creeks and Lone Tree Mines. The eH/emf varies with ferrous iron levels. It is also an excellent cyanide consumption predictor. As the emf becomes more negative cyanide consumption increases. This is an indication of less than optimal oxidation. Throughput or any of the other controllable parameters are adjusted accordingly.

Figure 11 illustrates the complete control strategy for a pressure oxidation circuit. Individually the control loops are simple. When all of the loops are put together control of the circuit becomes more difficult. As pressures in the heater vessels change so does the flows and heater levels. This affects the temperatures in the heaters and subsequently the autoclave temperature, pressure and level.

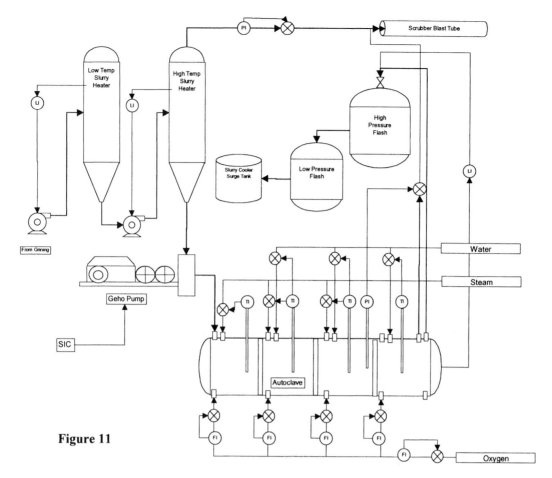

Figure 11

CONTROL IMPROVEMENTS

This list of desired control improvements is short. To improve pressure oxidation control would require the development of a few key instruments. In some cases the instruments exist only are not reliable, such as oxygen analyzers. With such instruments other improvements in pressure oxidation equipment or circuit configuration could be implemented with confidence of success. The single highest operating cost in a pressure oxidation plant is oxygen, followed closely by lime for neutralization. Therefore improvements in the measure and control of oxygen and acidity will lower the operating cost of a pressure oxidation plant.

Oxygen control would involve both online sulfide sulfur analysis on the slurry feed and discharge streams, and gas analysis of the autoclave vent. Online sulfide sulfur analysis would provide the level of oxidation. Precise oxygen flow could then be added to the autoclave based on the oxidation. The vent gas analysis would be used as a trim to the oxygen addition and provide online oxygen utilization. Operating conditions such as individual compartment temperatures and autoclave pressure, along with oxygen distribution could be changed to improve performance. Reliable vent flow measurements would also be required to complete a gas balance.

Online measurement of pH or eH inside the autoclave could be used to measure the rate of oxidation. The goal being minimizing acid generation without sacrificing the level of oxidation. An ore feed change or the addition of a neutralizing agent, such as limestone or trona, to the autoclave feed could be made to control the acidity of the autoclave without fear of over addition and quenching the reaction.

With some or all of these control improvements a supervisory control system could be incorporated. An expert system would be the next step in pressure oxidation control. Such a system could currently be utilized, but could only control portions of the plant. Expert systems could be used control acid addition based on feed carbonate assays and free acid levels on the autoclave discharge. Such a system could minimize the acid used thereby reducing operating costs. An expert system on oxygen addition may be beneficial for the autoclave operation, but would have to also control the oxygen plant to realize full benefit. To simply adjust the oxygen addition rate to the autoclave without adjusting the oxygen production rate would serve no purpose.

CONCLUSIONS
Pressure oxidation control has advanced greatly through time. Though the control can be difficult it is not extraordinarily complex. For the most part all pressure oxidation facilities utilize similar control strategies and equipment. Therefore through the innovation of a few key instruments and control strategies the operation of all pressure oxidation facilities would be easier and less expensive.

ACKNOWLEDGEMENTS
The authors wish to acknowledge all those responsible for the successful operation of the pressure oxidation circuits throughout the world. We also would like to thank the management of Lone Tree Mine, Twin Creeks Mine, and Newmont Mining Corporation for their support and allowing presenting this paper.

19

Engineering, Procurement, Construction, and Management

Section Co-Editors:
Roger M. Nendick and Robert C. Schenk

Development of a Mineral Processing Flowsheet—Case History, Batu Hijau
T. de Mull, S. Saich, K. Sobel 2211

Specification and Purchase of Equipment for Mineral Processing Plants
C. Hunker, S. Maldonado 2223

The Management and Control of Costs of Capital Mineral Processing Plants
D.W. Stewart 2230

Schedule Development and Schedule of Control of Mineral Processing Plants
P. Kumar 2238

The Risks and Rewards Associated with Different Contractual Approaches
P.J. Gard 2245

Success Strategies for Building New Mining Projects
R.J. Hickson 2250

Development of a Mineral Processing Flowsheet - Case History, Batu Hijau

Tom de Mull, Stuart Saich and Karen Sobel

ABSTRACT

The Batu Hijau project was the largest mining project executed up to the time of completion. The concentrator was the largest ever built from "grassroots". Initial ore discovery was made in May 1990 followed by several years of work investigating the geology and mineralogy, which culminated in the preparation of a final feasibility study in July 1996. An EPC contract was awarded to Fluor Daniel for the basic design, detailed engineering and development of the project in August 1996.

During the detailed engineering phase the original design basis was thoroughly scrutinised and updated as a result of further flotation testwork carried out using sea water. This testwork allowed for a second review of the flotation kinetics leading to significant mid stream changes in circuit design which were incorporated into the final installation at a net cost savings. A flotation circuit model was developed using pilot plant data obtained from testwork carried out by AMMTEC in Perth, Australia and subsequently used as the basis for flotation circuit mass balances and equipment sizing. This flotation model together with a well defined process design criteria were used to size flotation circuit equipment and associated hydraulic systems.

The initial ramp up rate of the concentrator achieved the aggressive targets set by the operations personnel. Analysis of early operating data during startup indicated that overall plant performance was well within original design expectations, but that internal circulating loads where greater than expected. Debottlenecking studies were subsequently carried out which confirmed visual observation as to certain equipment modifications that could be made to enhance recoveries. The flotation kinetic model has been updated to reflect actual operating conditions, and to develop a greater understanding of scale up issues for the large flotation circuit equipment involved.

INTRODUCTION

The $1.83 billion Batu Hijau project is located in Sumbawa, Indonesia and is owned and operated by PT Newmont Nusa Tenggara (PTNNT). PTNNT is an Indonesian company owned 80 percent by a partnership between Newmont Mining Corporation (Newmont) and Sumitomo Corporation. A local mining company, PT. Pukuafu Indah, holds the remaining 20 percent.

Fluor Corporation (Fluor) was responsible for engineering, procurement and construction of Batu Hijau, the world's largest greenfield startup mining project ever constructed. The copper concentrator is designed to treat 120,000 tonnes of ore per day. The open-pit mine uses electric shovels and haul truck to transport the ore to the primary crushers. A 5.6-kilometer long overland conveyor carries the ore to the concentrator, which consists of a coarse ore stockpile, two-train SAG and ball mills, primary and scavenger flotation cells, vertical regrind mills, cleaning flotation cells and counter-current decantation thickeners to wash the sea water from the concentrate. The thickened concentrate slurry is stored in two tanks, then pumped to the port site where the concentrate is filtered and stored for shipment.

Tailings produced by the concentrator flow by gravity from the process plant to the ocean, where they are disposed of via submarine tailings placement. The tailings are deposited three kilometers from the coast, at a depth of approximately 108 meters below the surface. The tailings migrate towards the Java Trench and are ultimately deposited at depths of several thousands meters.

Early metallurgical test work on Batu Hijau ore was carried out at Newmont's research facilities. This involved extensive metallurgical testing involving impact crushing, grindability tests, abrasion, flotation, filtration, flocculation etc to enable the definition of initial process configuration. Newmont also enlisted the services of Lakefield Research (Ontario, Canada), and other testing facilities, during this phase to support their effort.

Fluor first became involved in the Batu Hijau project in 1994 to carry out the Optimization Study. The purpose of this work was to optimize major aspects of the project such as plant throughput, process plant location, mode of concentrate transport and location of the port facilities.

In 1995, Newmont and Fluor began work on the feasibility study. Core samples from the ore body were initially sent to Lakefield for preliminary laboratory locked cycle testing. Results of this work were used to define a proposed flotation circuit which was subsequently tested using composite samples of the expected mine plan feed in a batch pilot plant. This work was completed in late 1995, with an addendum issued in July 1996. The main deliverables from this stage of work were the definition of a proposed flotation circuit configuration and preliminary equipment sizing for the final feasibility study report. The feasibility report was issued in February 1996. This paper focuses on development of the process from this point forward.

PILOT PLANT TESTWORK

During the preliminary design phase the need for a source of fresh water, suitable tailings location and subsequent water recovery was investigated. Due to the adverse seismic conditions in the region an alternative source of water (sea water) was recommended, along with the use of sub-sea tailings disposal. The proposal to use sea water in the circuit prompted the need to carry out further pilot plant testwork on suitable composites to confirm flotation kinetics and overall recovery. This was carried out at AMMTEC in Perth, Australia during the early part of 1996. The pilot plant used was set up as a scaled version of the proposed circuit developed during the feasibility study. Results from the revised flotation testwork using sea water instead of fresh water were made available to the design team in late 1996.

Samples of feed, concentrate and tailings material from the AMMTEC sea water pilot plant testwork were sent to the Council for Scientific Research Organisation (CSIRO) in Australia for Scanning Electron Microscopy (SEM) analysis. Results from this work proved invaluable in interpreting the results of the flotation testwork.

RAW DATA ANALYSIS

Several pilot plant testwork campaigns were carried out at AMMTEC, using sea water and further composite samples. This work resulted in the generation of a significant amount of raw data that required further analysis before valid conclusions could be reached. Of the various test campaigns "Trial 12" was run with the desired intent to provide metallurgical information for scale up

Raw data obtained from 'Trial 12' included the following typical metallurgical information.

- Solids flow rates.
- Pulp densities for each stream.
- Particle size distributions for individual streams.
- Metallurgical assays (Copper, Gold, Sulphur and Iron) for individual streams and as a function of particle size within each stream.

To facilitate meaningful conclusions, analysis of the raw data was carried out using the following methodology:

- Mass balancing of raw data using SysCAD/MassBal.
- Visual data smoothing using EXCEL.
- Generation of recovery curves for both particles and individual elements within each stream using EXCEL.

Mass Balancing

As is typical for metallurgical sampling the data obtained from the Trial 12 test runs did not balance, eg. total mass of copper reporting to concentrate and tailings streams did not equal total mass of copper in feed stream. Some form of statistical mass balancing of raw data is required prior to subsequent analysis or modelling.

In order to achieve a suitable mass balance for solids, liquids, particle sizes and individual elements a statistical mass balancing package, SysCAD/Massbal was used. With this tool a reasonable mass balance of particle sizes across multiple unit operations and grades within each size fraction was developed. The raw metallurgical data for each process stream (i.e. solids flow rate, particle size distribution and assays) were used together with typical sample standard deviations to statistically manipulate the raw data into a mathematically balanced data set. This 'massaged stream data', was subsequently used in the data smoothing exercise, detailed below, prior to use in process modeling. The use of unbalanced data, or only using feed and concentrate stream data to develop a process mass balance can easily lead to incorrect conclusions.

Visual Data Smoothing

After mass balancing the raw data it became apparent that a discontinuity existed in the particle size and grade around the 38-micron size fraction. In reviewing the methodology used to analyze particle sizes and prepare sufficient material for subsequent assaying, it was revealed that the samples were wet-sieved at 38 microns and that all material finer than 38 microns was subsequently sized in a "cyclosizer." This equipment uses a series of very small cyclones to separate the minus 38 micron material into size fractions. Because the mineral specific gravity affects the split, any gold or gold bearing material reported to coarser size fractions than their size warranted. This discontinuity was removed via visual data smoothing and further mass balancing using EXCEL.

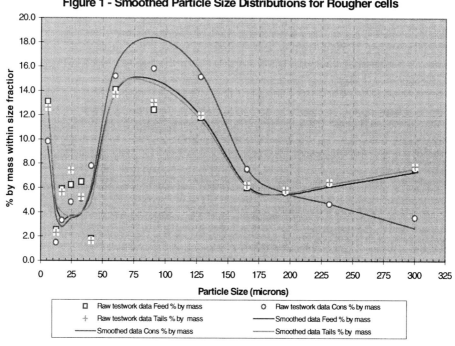

RESULTS OBTAINED FROM RAW DATA ANALYSIS

Analysis of the mass balanced and smoothed data focused on the recovery of material to the concentrate streams as a function of particle size. This analysis was completed both globally across the overall circuit, and also for specific flotation banks within the circuit. By plotting the mass of material within each size fraction for both the feed and concentrate streams, the size range over which preferential flotation occurs could be identified. This method of analysis allowed for an 'easy to visualize' representation of the effect of either 'over' or 'under grinding' of both the primary grinding circuit product and the regrind circuit product. The main objective of the exercise was to minimise (and equalize) flotation losses in either the coarse or fine size fractions through optimisation of grinding design criteria.

In addition to the above analysis an investigation into the optimum rougher concentrate polishing mill retention time was carried out. Results are indicated in the following sections:

Optimum Primary Grind size

Analysis of the material mass recovery to the concentrate stream for the overall flotation circuit was used to identify the optimum primary grind size. Figure 2 indicates the particle size distribution for both the feed and concentrate streams.

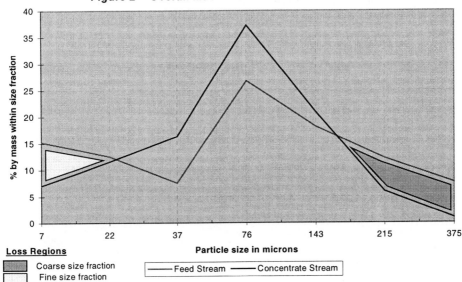

Figure 2 indicates that the flotation circuit does not effectively recover very coarse material (i.e. > 210 microns), and that recovery of particles finer than 38 microns is diminished. This is typical of flotation circuits, but of importance was the easily identifiable size range across which primary grinding should be carried out.

In addition to the mass recovery distribution for individual particle sizes, the recovery of copper as a function of particle size was also examined. Figure 3 presents the copper distribution as a function of particle size for both the feed and concentrate streams.

Figure 3 - Overall Copper distribution in Flotation circuit

The above figure once again highlights a similar trend in recovery as a function of particle size with a slightly narrower recovery range indicated.

Optimum Regrind Particle Size
A similar analysis of particle size distributions for the feed and concentrate streams around the cleaner circuit was also carried out. Figure 4 highlights the results.

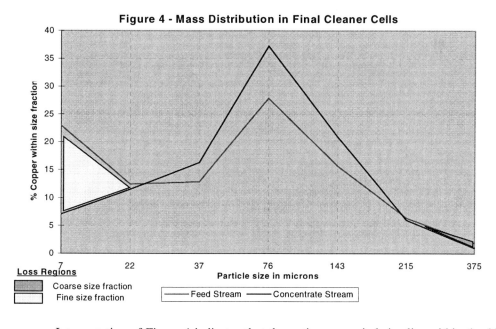

Interpretation of Figure 4 indicates that the optimum regrind size lies within the 38-145 micron size range. The relatively large amount of minus 38 micron material present should be avoided due to the high losses (low recovery) of material within this size fraction. The pilot plant was operated at the proposed regrind specification of 80.0 percent passing 25 microns, which resulted in the generation of a significant quantity of ultra fines material.

One of the significant conclusions reached from the raw data analysis was the requirement to increase the regrind size from 80.0 percent passing 25 microns to 80.0 percent passing 80 microns. Significant cost savings were realized as the design team grinding experts were able to reduce the regrind milling installation to three mills from the original configuration of six. The recommendation to increase the regrind size was subsequently confirmed in further locked cycle flotation tests.

Rougher Concentrate Grinding Optimisation
The concentrate collected from the first cells in the pilot plant ran at a grade just slightly lower than desired final grade. A review of the SEM photographs of this rougher concentrate identified that several large (>180 microns), high grade (>30.0% Cu) particles were present. The recommendation was made that this material be collected separately (split flotation circuit) and passed through a polishing mill to enable some minor particle breakage without over grinding the concentrates. In order to determine the effect of polishing on recovery a concentrate sample was subjected to several batch grind–flotation tests to optimise design basis. Figure 5 indicates the findings.

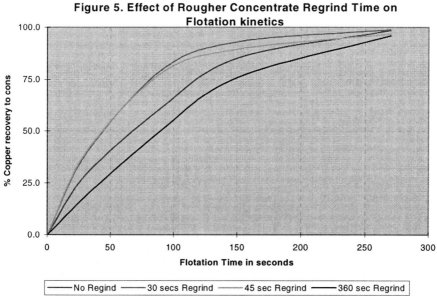

The above figure indicates the improved kinetics as a result of polishing the product for up to 45 seconds but with longer grinding times kinetics drop off and losses occur as a result of overgrinding. The optimum polishing mill product size found was 80 percent passing 82 microns.

FLOTATION CIRCUIT MODELLING
A flotation kinetic model was developed in EXCEL to reflect the proposed circuit design. The kinetic model used the following basic parameters:

- Simple first order rate equation.
- Rate kinetics for each significant copper bearing mineral.
- All non-valuable material lumped together as gangue.
- Mineralogy obtained from Scanning Electron Microscopy (SEM) work.
- Rate kinetics as a function of particle size developed for copper for each unit operation.
- Pulp densities from testwork used.

Rate Equation Used
The form of the rate constant applied within the flotation model was as follows:

$$R_i = R_{i(max)} * (1 - \exp^{(-k_i * t)})$$

Where
- R_i = Recovery of mineral "i" at time t minutes
- $R_{i(max)}$ = Maximum recovery of mineral "i" at time t = infinity
- K_i = Rate constant for mineral "i"
- "i" = Each mineral and gangue modelled

The mass balanced and smoothed data obtained from the sea water pilot plant testwork was then used as the basis to tune relevant rate constants to equate to actual pilot plant performance. For the copper balance, recovery data as a function of particle size for each section of the circuit allowed for the development of flotation rate constants for each size fraction. By developing the rate constants on a size basis the effect of a change in regrind size criteria could be analysed. The rate constants for individual particle sizes were then summated to represent overall copper mineral flotation rate constants. Revised rate constants for the increased regrind size were used in the final model

Rate constants for gangue and gold were tuned to achieve overall grades and recoveries as experienced in the pilot plant.

Minerals used in flotation model
The following minerals or elements were included within the flotation model:

Chalcopyrite
Bornite
Covellite
Gold
Gangue

Results from Scanning Electron Microscope (SEM) analysis were used to define the mineralogy for the original model. The ratio between valuable copper bearing minerals was changed to suit expected mine plans for final modelling but the rate constants for individual minerals assumed to be the same.

Pulp Densities
Results from both the Lakefield and AMMTEC pilot plant testwork indicated that the pulp densities for scavenger concentrate streams were of the order of 5-10% solids by mass. Review of similar operations and experience of project consultants (Dr R Klipmel) indicated that operating pulp densities should be higher at around 10.0 – 15.0 % solids by mass.

Lower pilot plant pulp densities were expected as a result of the objective to maximise recovery at the expense of grade. However when the recommended pulp densities were used within the flotation model, poor copper recoveries as a result of short pulp residence times indicated the need to re-evaluate sizing of the cleaner flotation circuit.

Flotation Modelling Results
The kinetic flotation model as based on the sea water pilot plant testwork was then used to predict performance of the proposed flotation circuit that was, at that time, in the detailed design phase. Conclusions from the flotation modelling were as follows:

- The volume of scavenger concentrate, at the relatively low pulp density, was significantly higher than expected. Pilot plant pulp densities of the order of 2-8% were achieved, whereas the simulated value of 12.0-14.0 % still resulted in excessive volume of scavenger concentrate.
- The original regrind size specified for the scavenger concentrate would result in excessive losses in the fine size fraction.

- Overgrinding of concentrates should be avoided due to high losses.
- An optimum rougher concentrate polishing mill retention was developed.

Based on these findings several modifications to the original flotation circuit configuration were recommended and subsequently tested using the flotation model. These were as follows:

- Install a dewatering circuit on the scavenger concentrate stream.
- Install regrind screens (as used in the iron ore industry) to avoid overgrinding of high density, fine particles that would report to cyclone underflows.
- Resize cleaner flotation circuit to suit revised mass flows.
- Reduce the number of regrind mills from six to three as a result of reduced regrind duty.

PROCESS DESIGN

Process Design Criteria
The basic design criteria for the Batu Hijau project was as follows:

Design Criteria	Value	Units
Nominal throughput	43,800,000	tpa
Nominal throughput	120,000	tpd
Plant availability	92.0	%
Plant surge factor	+/- 15.0	%
Average plant throughput	5,435	tph
Maximum plant throughput	6,250	tph

In addition to the above listed mass throughputs, a range of expected feed grades (from the mine plan) was imposed upon the system to enable calculation of concentrate production rates and grade. The flotation model developed and based on the AMMTEC "Trial 12" data was updated to reflect the proposed dewatering circuit, the proposed circuit configuration and the process design criteria to verify equipment selection.

Process Engineering
At the same time of completion of the flotation modelling development the final vendor bids for flotation equipment were received. This proved opportune as the proposed vendor equipment sizes were not exactly the same as the equipment listed in the feasibility study. Typical issues identified were such as one vendor offering six rows of $100.0 m^3$ Rougher scavenger cells versus a competitor offering five rows of $127.0 m^3$ cells for the same duty.

Once the preferred vendor (commercial and mechanical) was selected the relevant equipment sizes were entered into the flotation model and equipment selection optimised. At this time a detailed review of the proposed flotation circuit configuration and equipment sizing was carried out by owner representatives, engineering company personnel and outside consultants (Dr R Klimpel). During this phase the recommended operating parameters for the dewatering circuit were set within the flotation model such that the regrind circuit would be fed with a constant pulp density independent of rougher scavenger circuit operation. The flotation model was then tested using the ranges specified in the design criteria and overall recovery and grade recorded. Then number and configuration of rougher/scavenger flotation cells was fixed but he configuration and number of cleaner cells varied to optimise operating flexibility and maximise recovery. The following table lists the feasibility study circuit flotation equipment versus final selected equipment.

Description	\multicolumn{4}{c}{Feasibility Study}			
	Rows #	Cells/row	Cell Size m³	Total volume m³
Rougher cells	6	2	100	1200
Scavenger cells	6	8	100	4800
First cleaner cells	1	4	100	400
Cleaner scavenger cells	1	4	100	400
Second cleaner cells	1	5	20	100
Final cleaner cells	1	3	20	60
Description	\multicolumn{4}{c}{Recommended Final Installation}			
Rougher cells	5	1	127	635
Scavenger cells	5	9	127	5715
First cleaner cells	1	4	42.5	170
Cleaner scavenger cells	1	4	42.5	170
Second cleaner cells	1	10	17.0	170
Final cleaner cells*	1	4	14.5	58

* The vendor recommended a final installation of five flotation cells due to insufficient concentrate collection lip length.

Once the final circuit sizing and configuration had been completed the flotation model was run at three design tonnages and associated grades to generate expected mass balances for subsequent slurry material handling equipment. Slurry pump and sump design and sizing were now redone using the new mass balance.

CIRCUIT PERFORMANCE AND DEBOTTLENECKING

Startup of the Batu Hijau copper concentrator and subsequent ramp up to design tonnage went extremely well. Within ten days of starting the second primary grinding circuit the hourly average mill throughput had achieved the design 5,435 t/hr. The design maximum throughput of 6,250 t/hr was achieved four weeks later.

Performance testing was included within the EPC contract. In order to verify the successful completion of each required performance test, operating data from key process points was recorded into daily data sheets and reconciled against production reports. These were then collated over time to verify successful completion of each performance test.

In addition to being a useful method of capturing successful performance, the data was also used to verify the original design basis and update the flotation kinetic model to identify any differences in original design expectations versus actual plant performance. Detailed analysis of the operating data, flotation model updates and equipment sizing has been carried out in the form of de-bottlenecking studies. Three of these have been completed out to date.

The following are some of the major issues highlighted as a result of reviewing operating data and updating the flotation model to actual operating conditions.

- Initial full scale plant flotation kinetics were significantly lower than pilot plant results.
- Rougher and scavenger concentrate pulp densities were significantly higher than expected.
- Circulating load around cleaner circuit higher than expected.

The debottlenecking studies have been used to assist operations personnel in confirming site conclusions and making changes to the circuits to enhance overall performance. Further details as to findings are as follows:

Flotation Kinetics

The initial ore fed to the concentrator consisted of slightly oxidised material, which resulted in a reduced pH of pulp reporting to the rougher-scavenger flotation circuit. Primary lime addition was re-directed from the regrind circuits to the primary grinding circuits to increase pH and improve flotation performance. The types of frother and collector, and addition points were also changed during initial startup to optimize flotation recovery.

The first two cells in each row of rougher-scavenger cells were designed such that concentrate could be collected from both of them, and sent to the polishing mill. The expected operating procedure was that only concentrate from the first cell would normally be sent to the polishing mill. On original startup however, it was found that the collection rate of concentrate from the first cells in each row was significantly lower than expected. The grade was slightly higher than expected but due to the low flow rate, concentrate was collected from the first two cells in each row of cells and sent to the polishing mill.

Discussions with the flotation cell vendor and review of similar operations lead to the conclusion that internal concentrate collection launders were required. Once installed the kinetics improved significantly, and were comparable to original pilot plant testwork. One of the key parameters not investigated within the flotation model was lip loading factors. Back calculation of expected lip loading from the pilot plant testwork indicated a requirement of approximately 1.05 t/m/hr concentrate whereas the physical maximum obtained was 0.55 t/m/hr.

Figure 6 on the following page indicates the original rate constants used for design and typical operating values.

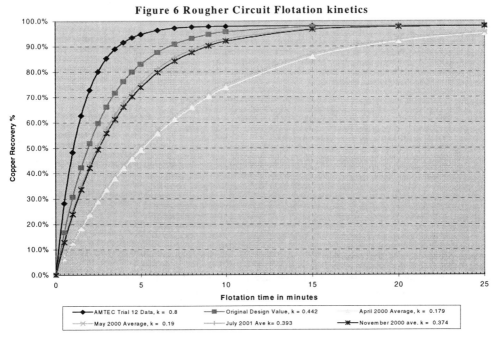

Operation of the rougher-scavenger cells improved back towards original design expectations once the internal launders were installed in late 2000 and early 2002.

Rougher-Scavenger Concentrate Pulp Densities
The initial expectation of the rougher concentrate pulp density as developed from the pilot plant testwork was 20.0 to 25.0 percent solids by mass. Once the plant was started and began to operate at design tonnage it quickly became apparent that the actual pulp densities were significantly higher. These have been found to run as high as 45.0 percent solids by mass and at the expected copper grade. Of note is that the flotation feed pulp density ranges from 32.0 to 38.0 percent solids by mass.

Cleaner Circuit Feed And Circulating Loads
A dewatering cone settler was installed to enable the recycle of excess water back to the scavenger circuit due to the expected low concentrate pulp densities from the scavenger flotation cells. Whilst this primary function has become of lesser importance due to higher than expected

concentrate pulp densities, the cone settler served another critical function. This is in the decoupling of operation of the cleaner circuit from the rougher-scavenger cells as a result of the large residence time of the settler itself. Surges from the primary grinding and scavenger flotation circuits do not affect the downstream processes. The feed rate and pulp density of material to the cleaner circuit was thus very stable and assisted operators in not having a continuously fluctuating feed to this part of the circuit.

Initially the recirculating load around the first cleaner and first cleaner scavenger flotation cells was higher than expected. This was found to be due to the reduced flotation rate within the rougher/scavenger circuit with higher scavenger concentrates reporting to the first cleaner cells. Once the internal launders in the rougher cells were installed and the load shifted back to rougher cells and subsequently the second cleaner cells the circulating load improved significantly.

Regrind Screens and Cyclones

The regrind screens as installed became a significant maintenance effort due to continual blocking and scaling up as a result of the presence of lime. This resulted in a dilute screen product to the regrind mills. The regrind screens were replaced with cyclones (after startup) which have performed well. The required pulp density is now being fed to the regrind mills.

Throughput Studies

Three debottlenecking studies have been completed using hourly average operating data over a minimum of one month operating time frame. As a result of this a valuable database of plant performance has been used to evaluate flotation kinetics as a function of throughput. This was analysed on a shift basis and the model tuned to reflect overall performance. An 'effective cleaning performance' of the rougher cells has been developed using the ratio of rate constants for valuable mineral versus gangue. As the total SAG feed rate increases the grind becomes coarser and the flotation circuit residence time decreases.

By plotting the ratio of rate constants vs SAG throughput an indication of plant performance can be gained. Figure 7 illustrates this.

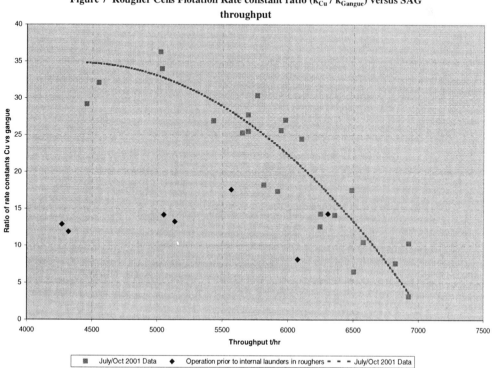

Figure 7 Rougher Cells Flotation Rate constant ratio (k_{Cu} / k_{Gangue}) versus SAG throughput

CONCLUSIONS

The successful startup of the Batu Hijau copper concentrator and subsequent ramp up to design tonnages was the result of extensive metallurgical studies into flotation performance coupled with simple methods of analysis. The evolution of the flotation circuit design in tandem with detailed engineering, and subsequent incorporation of resulting recommendations, resulted in the construction of a flexible and robust installation.

The scale up factor from pilot plant to full scale installation was of the order of 20,000. Notwithstanding this large scale up factor, initial plant recoveries were within expectations and individual streams within 30.0 % of initial mass balances. Subsequent equipment and circuit changes have reduced this to approximately 10.0 %.

By "closing the loop" the original kinetic parameters have been updated using actual plant data within the original flotation model and equipment performance factors defined. This has enabled the flotation model to become more accurate resulting in an excellent de-bottlenecking tool for future plant operations.

ACKNOWLEDGEMENTS

The authors wish to thank the management of PTNNT for their assistance and opportunity to present this paper. In addition a special note of mention of the invaluable input from Dr R Klimpel in guiding the team in the flotation circuit sizing and selection.

Specification and Purchase of Equipment for Mineral Processing Plants

Carey Hunker and Sam Maldonado

ABSTRACT
Effective specification, bidding and purchase of equipment can have a significant impact on the capital cost and quality of new facilities. It is essential for the equipment purchaser to understand the competitive bid process and the types of commercial agreements which govern. The use of industry standards has become a predominant method of defining equipment design and manufacture. In specifying and selecting equipment suppliers from the global marketplace, the role of quality standards such as ISO 9000 is increasingly important.

THE COMPETITIVE BID PROCESS
Considering that suppliers of equipment are frequently willing to provide quotations based on a brief verbal request, why go to the time and expense of preparing detailed specifications? The competitive bid is the reason. There is perhaps no greater tool in the competitive bidding process than a clearly written specification which clearly summarizes the purchaser's requirements using generic, industry standard terminology. Unlike a telephone call, the specification provides sufficient detail to allow each bidder to understand all the criteria against which the bids will be compared. Use of a specification provides identical information to each bidder, promoting a fair and ethical bidding process.

TYPES OF COMMERCIAL AGREEMENTS
Two types of commercial agreements are most common in the purchase of equipment.

- *Purchase Ord*ers are applicable when equipment is installed/erected by the purchaser. Typical examples might include valves and pumps.
- *Contracts* are applicable when equipment is installed/erected by the supplier. Typical examples might include field-erected tanks and conveying systems.

Other commercial arrangements may include rentals and loan arrangements (such as field trials of new prototypes). It is important for the purchaser to understand which commercial approach is appropriate and use appropriate terminology in the specification.

THE SPECIFICATION
Specifications are intended to ultimately form part of a purchase order or contract. The specification contains the technical and engineering details that explain exactly what is being purchased as well as defining the responsibilities of each party to the purchase order (or contract).

Types of Specifications
Two distinct approaches are typically followed in preparation of specifications.

- The *fabrication specification* is applicable when the equipment is designed by the purchaser. Within this type of specification, the purchaser essentially requests a quotation on fabricating the purchaser's design. The specification typically contains specifics on

materials, welding, tolerances and coatings. A fabrication specification is typically accompanied by detailed drawings outlining the purchaser's design. A typical example might include a chute or storage bin.
- The *performance specification* (sometimes referred to as a *duty specification*) is applicable when equipment is designed by the supplier. With this type of specification, the expected performance of the equipment is specified, but the detailed design of the equipment is left to the supplier.

In reality, there are numerous gray areas between these two approaches. For example, when specifying atmospheric tanks, it is common for the purchaser to dimensionally define the tanks, while the determination of wall thickness and location of weld seams is typically left to the tank fabricator. A challenge for the purchaser is therefore to determine what aspects of the equipment design are best performed by the purchaser, and what by the supplier. Luckily, for most common types of equipment, guidelines are typically provided through the use of industry standards.

Industry Standards
Prior to the advent of industry standards, equipment specifications often described every component, nut and bolt of specified equipment in voluminous detail. Suppliers struggled to decide which standard model to offer in response to long, wordy and often unclear specifications. As a response, many industries introduced standards which not only standardized the design of equipment between manufacturers, but also clarified the relationship between the purchaser and the supplier. An industry standard typically classifies equipment into types, groups, classes, categories or services that are easily identified by the purchaser and recognized by the supplier, simplifying and clarifying the bid process significantly. The purchaser therefore need only identify the applicable industry standard, and provide the data recommended by the standard.

An example of an organization preparing industry standards is the Crane Manufacturer's Association of America (CMAA). A purchaser wishing to specify a "top running single girder electric overhead traveling crane" needs only to refer to CMAA Standard 74, which fully defines the design for this equipment. As part of this standard, a brief, three page "Crane Inquiry Data Sheet" is provided which summarizes the data requested by CMAA manufacturers to provide a quotation. Within this standard, four classes of cranes are defined, from Class A (standby or infrequent service) to Class D (heavy service). Use of this standard greatly simplifies the purchaser's job, minimizes the data needed to be communicated while allowing the purchaser to fully define the type of equipment requested in a manner that will be clearly understood by the supplier.

Many industry standards in the United States now fall under the auspices of the American National Standards Institute. Founded in 1918, ANSI is a private, non-profit organization that administers and coordinates the U.S. voluntary standardization and conformity assessment system. Similar organizations in other parts of the world include International Organization for Standardization (ISO) in Europe and the Japanese Industrial Standards Committee (JISC) in Japan. Further information on these organizations may be accessed at http://www.ansi.org (ANSI), http://www.iso.ch (ISO) or http://www.jisc.org (JISC).

Format
While the format of specifications regrettably varies widely, the master specification format defined by the Construction Standards Institute (CSI) has gained widespread acceptance in the United States as a common format for specifications used for construction and procurement. Founded in 1948, the Alexandria, VA, based organization's goal is to enhance communication by providing a common system of organizing and presenting construction information. CSI's members include architects, engineers, constructors, specifiers of construction products, suppliers of construction products, building owners, and facilities managers.

The CSI Manual of Practice, Section II, provides techniques that are useful in preparing specifications. Further information on CSI formats can be accessed at http://www.csinet.org. Some general guidelines for preparing specifications follow:

Tenses: Use the imperative mood, the command form, to explain actions that the reader must perform, or is responsible for. In this style of writing, the subject, "you" is understood, and the verb is at or near the beginning of the sentence or the main idea, e.g., "Grind all welds to 1/8" and clean thoroughly".

Use the indicative mood to define, identify, or describe. For these functions, the verb, shall be, is like the equals sign in an equation, and the verb links the subject of the sentence to the discussion of it, e.g. "The completed weldments shall be free of defects and slag inclusion."

Wording: The purchaser should select and use words carefully. Each should be used in its precise meaning. Once a word and its meaning are selected for use, the same word should be used throughout the specification whenever that particular meaning is intended. Spelling used in specifications should be consistent. Whatever standard is established the most important thing is to be consistent throughout the documents that are being prepared. Some common problem areas:

Insure, Assure, and Ensure.

- To insure is to issue or procure an insurance policy. Assure is to give confidence to or convince a person of something. Ensure is to make certain in a way that eliminates the possibility of error.

Shall and Will.

- Shall is used with reference to the work required to be done by a contractor. Will is used in connection with acts and actions required of the owner or the architect / engineer. The words "must" and "is to" should be avoided.

Install, Furnish, and Provide.

- Install means to place in position for service or use. Furnish means to provide or supply, and provide means to furnish, supply, or make available.

Definite article "the" and indefinite articles "a" and "an" need not be used in most instances.

- Poor: Apply an oil paint with a brush to the wall.
- Correct: Apply oil paint with brush to walls.

All. The use of the word all is usually unnecessary.

- Poor: Store all millwork under shelter.
- Correct: Store millwork under shelter.

Contractor. Avoid using Contractor as the subject of the sentence.

- Poor: Contractor shall lay brick in common bond.
- Correct: Brick shall be laid in common bond.
- Preferred: Lay brick in common bond.

Content

As discussed earlier, specifications are intended to ultimately form part of a purchase order or contract. The typical specification therefore defines both the responsibilities of each party, explains exactly what is being purchased, and provides all relevant technical and engineering details. A typical specification may contain the following:

- Scope of supply: a summary of what is to be supplied, typically clearly defining the purchaser's and supplier's responsibilities.
- References: a summary of applicable industry standards and codes.
- Performance criteria: For performance specs, a description of the specified performance. This is often provided in tabular fashion on data sheets.
- Materials: Materials of construction.
- Tolerances: Tolerances of dimensions or performance.
- Design life: The required life of the equipment, or specific components.
- Dimensional parameters: Physical size of equipment. This is often completely defined in fabrication specification by the purchaser's drawing or sketch.
- Protective Coating: Requirements for painting or protective coatings.
- Noise Levels: Acceptable noise limits.
- Site Conditions: A summary of climate, seismic zone, geographic location and environmental conditions.
- Utilities Available: Utilities may include steam, air, water.
- Power: Voltages and power availability.
- Spare Parts: Requirements for capital and maintenance spares.
- Welding: Specifications for welding, particularly with fabrication specifications.
- Acceptable Sub-Component Manufacturers: The purchaser may only accept sub-components from a specific supplier.
- Quality Requirements: Purchaser's requirements for quality control, non-destructive examination and testing.
- Data Requirements: Purchaser's information requirements, both for bid and after purchase, including timeframes.
- Special requirements for package systems, which may include emissions limits, modularization and requirements for hazardous operation reviews.

LOCATING AND QUALIFYING BIDDERS

Bidder Selection

Mineral processing is one of the most geographically diverse industries in the world. Locating quality suppliers and contractors in some locales can be challenging. In some countries comprehensive trade directories are available to help locate suppliers of specific types of equipment. Examples of these include:

- The Thomas Register (http://www.thomasregister.com/) offers both North American and European Directories.
- The Fraser's Trade Directory (http://www.frasers.com/) provides a listing of Canadian suppliers.
- The South African Bureau of Mines provides a listing of suppliers at (http://www.bullion.org.za/bulza/chaorg/wmdir/wmdsrvc1.htm#equip)

Bidder Qualification

Ensuring all bidders are qualified prior to commencing the bid process will benefit both the purchaser and supplier. For this reason, bidders are usually pre-qualified based on a selection process that may use any of the following criteria as a basis:

- A demonstrable track record, experience and references

- Degree of shop/factory loading
- Stability, financial strength
- Geographic location
- Innovation, technology
- Maintainability (availability of spare parts and service)
- Reliability and complexity
- Safety record/experience modifier
- Quality assurances and/or ISO 9001 compliance
- Pre-manufactured equipment or stockpiled materials
- Qualifications of tradesmen (machining, welding, electrical)

THE BID PROCESS

Ethical Bidding

The need to maintain a legal and ethical bid process needs no justification. Suppliers can expend considerable money and effort in preparing bids. Significant effort should be made to ensure that bidders are afforded an equal opportunity to compete on the same terms as their competitors. In bids involving significant sums of money, sealed bids are often received, and opened only after all are received in a witnessed environment. Some aspects of an ethical bid process:

- Technical discussions, bid clarification responses and meetings, and discussions of alternative proposals conducted in a manner fair to all bidders.
- Bid information considered confidential as it may contain proprietary information. Distribution limited to those performing evaluation.
- No indication given to a supplier regarding the competitive nature of his bid during the evaluation period or negotiations with bidders.

The following practices are not considered part of an ethical bid process

- Bid shopping - playing one bidder against another as a negotiation strategy.
- Requesting quotations without intent to buy, without advising the supplier that the request is for information purposes only and only for the purpose of providing leverage against a competing firm.
- Threatening suppliers with loss of future business unless they comply with unreasonable requests.
- Requesting suppliers to bid on a larger volume of business in order to secure lower pricing for a known lesser quantity.
- Deliberately introducing confusion into the negotiations with unusual issues, terms or false figures.

Evaluation of Bids

A Bid Evaluation typically includes two components, a technical and commercial evaluation. The technical evaluation may include an evaluation and comparison of:

- Conformance/exceptions to specifications
- Performance and efficiency
- Safety and operability
- Utility requirements
- Materials of construction
- Life cycle
- Physical size

- Supplier's quality plan

The commercial evaluation may include and evaluation and comparison of:

- Approval drawing cycle
- Price adjustment (escalation)
- Service representative cost
- Spare parts cost
- Storage Costs
- Return conditions & restocking charges
- Taxes
- Terms of payment, discounts, early payment, progress payments, incentive, and penalty
- Customs clearance costs
- Export packing charge and storage
- Freight (inland, air, ocean, & rail)
- Import duties
- Insurance costs
- Warranty & performance guarantee
- Export credit financing
- Nearest supplier service
- Shipping point
- Liquidated damages acceptance
- Price basis
- Currency and exchange rates
- Schedule
- Supplier's shopload
- Previous experience with supplier
- Governing items and conditions of purchase
- Bid validity period
- Union labels
- Cancellation charges

ENSURING QUALITY

The topic of quality control and quality assurance is too broad to be addressed fully here. However it should be noted that a specification does need to define the purchaser's expectation of the quality of the finished product. In many cases, but not always, industry standards can assist in defining quality requirements. Where this is not the case, other standards may be utilized to define to define one aspect of the manufacture. For example it can be specified that all welds must meet the requirements of a specific standard of the American Welding Society.

The standards of the International Organization for Standardization (ISO), in particular ISO 9001 (Quality management systems - Requirements) are gaining in popularity worldwide. Further information on the ISO 9000 series of quality standards is available at http://www.iso.ch/iso/en/iso9000-14000/iso9000/selection_use/iso9000family.html. Whether based on ISO 9001 or not, it is common practice for the purchaser to review and ensure the supplier's quality plan is adequate, preferably prior to award.

It is also common practice for the purchaser to visit or inspect the purchaser's shop before or during the fabrication of the equipment. The purchaser may wish to impose "hold points" in the fabrication process, where fabrication must cease until the purchaser is satisfied that all quality requirements have been met. If this is the purchaser's intent, it should be identified clearly in the specification so the supplier's schedule can incorporate these work stoppages

OTHER CONSIDERATIONS

Used Equipment

When used equipment is desired, it is important to remember that the supplier's technical resources may not be available to assist in selection. It is not uncommon for used equipment to be sold by clearing houses and agents who may have little technical knowledge of the equipment itself. For this reason, the purchaser may wish to specify a specific size or model number, rather than a performance requirement. Where possible, involving the original manufacturer will likely prove highly valuable in determine if equipment can be re-used and the degree of refurbishment required.

Language/International Considerations. Due to the geographically diverse nature of the mineral processing industry, equipment may be procured from any corner of the globe. When working in any region, the purchaser must be prepared to work in the local language, and in accordance with local laws and customs, which may vary widely.

Impact of the Internet. The advent of the internet has changed, and will no doubt continue to change the manner in which commercial transactions occur throughout the world. At the time of this writing, the mineral processing industry awaits the debut of Quadrem, http:\\www.quadrem.com. Founded by 14 companies (Alcan Aluminium Limited, Alcoa Inc., Anglo American plc, Barrick Gold Corp, The Broken Hill Proprietary Company Limited (BHP), Corporacion Nacional del Cobre de Chile (CODELCO), Companhia Vale do Rio Doce (CVRD), De Beers Consolidated Mines Ltd., Inco Ltd., Newmont Mining Corporation, Noranda Inc., Phelps Dodge Corporation, Rio Tinto, and WMC Limited) the venture intends to create a platform to bring together mining, minerals, and metals producers and suppliers in more than 100 countries utilizing a common catalogue of products in multiple languages. The intent is to allow participants, regardless of size and location, to access and trade with a large pool of suppliers both locally and around the world.

The Management and Control of Costs of Capital Mineral Processing Plants

David W. Stewart

ABSTRACT

The management and control of costs of capital mineral processing plants is dependent on the actions and understanding of all those involved in the development and execution of the project. The following process describes how project costs are monitored against an established budget and how the process varies through the different phases of the project from basic engineering, through detailed engineering, procurement and construction.

INTRODUCTION

The basics of cost management and cost control are straightforward. They are:

- Develop the budget
- Monitor against the budget
- Report deviations early for corrective action
- Revise the budget/forecast (as appropriate)
- Monitor against revised budget/forecast
- Repeat the cycle.

This cost control cycle is illustrated in Figure 1.

Figure 1 – The Cost Control Cycle

For cost management and cost control to be effective this cycle must:

- Occur continuously
- Be diligent
- Involve all project team members
- Result in all project team members being kept informed.

PROJECT SCOPE

A clear and concise definition of the project scope is an essential ingredient in establishing and controlling the costs of a project. All participants in the project should be kept fully informed on the defined project scope. The project owner must make clear what his expectations are.

The physical boundaries of the project should be defined and should address:

- Infrastructure – such as access roads, power supply, communications facilities, housing, port facilities and air strips
- Mine facilities – such as mine development, maintenance and offices
- Process facilities – such as crushing, grinding, flotation, etc.
- Operating materials – spare parts, initial fills of reagents and initial grinding mill ball charge.

The operating requirements of the plant should address:

- Plant throughput
- Plant operating and maintenance schedules
- Plant reliability expectations
- Manual versus automatic control requirements.

Each party with project budget responsibility must have a defined scope on which to base their budget development. Any deviations from that scope would then result in a related change in budget, making it important that the budget based scope definition be as precise as possible to minimize any cost "surprises" in the future.

COST ESTIMATE

After successful completion of a feasibility study, basic engineering typically ensues to refine and confirm the project scope. At the end of basic engineering, a preliminary estimate is prepared. As the project develops, estimates of the total cost of the project are updated. The project scope details improve as the project progresses, and the further the development of the project, the greater should be the accuracy of the cost estimate.

Cost estimates can be categorized by the accuracy of the estimate and may be referred to as order-of-magnitude, preliminary and definitive as illustrated in Figure 2. These terms for classes of estimates are sometimes used during the feasibility study phase also. The order-of-magnitude estimate being very early in the study with minimal engineering completed and the preliminary estimate after the study has been completed.

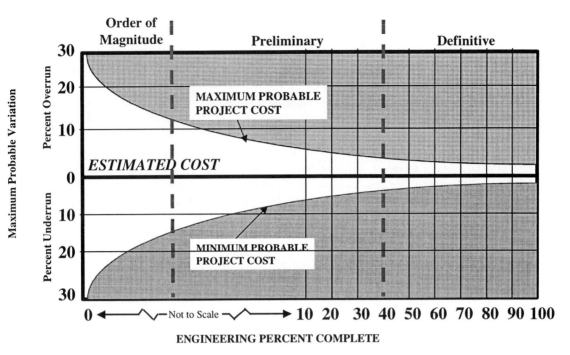

Figure 2 – Classes of Estimates

Cost estimates/forecasts are a combination of direct costs and indirect costs. Direct costs (equipment, material and labor) are generally specifically priced, whereas indirect costs tend to be a function of the project duration and schedule. For this reason and to estimate escalation costs, it is essential that any cost estimate or forecast is based on a carefully prepared project schedule.

Cost estimates are generally assembled by assessing the costs of :

- Process equipment (crushers, mills, pumps, etc.)
- Construction materials (concrete, steel, pipe, cable, etc.)
- Construction labor
- Temporary construction facilities (offices, warehousing, power generation, etc.)
- Temporary construction services (survey, inspection, safety, quality control, etc.)
- Engineering (design), procurement and construction supervision services
- Design and growth allowances
- Contingency and escalation
- Owner's costs (these may include the cost of the owner's organization, land acquisition, interest costs during construction.)

To develop these estimates of costs, each of these components have to be quantified and pricing determined.

Quantifying these components depends largely on progress of the design. Sizes and quantities of pieces of equipment will normally be obtained off the process flow diagrams or the equipment list. Quantities of bulk materials (concrete, steel, cable, etc.) will be extracted from the 3-D design model or taken off detailed drawings as those are developed or, prior to model/drawing development, can be based on historical ratios. Services will generally be based on staffing plans and expected hours to perform those services.

The pricing of materials is obtained by getting quotes from manufacturers for equipment and materials. Labor and services pricing is obtained by getting quotations from contractors who will perform the work or provide the services.

If specific current data is not available, such as during early stage feasibility studies, all pricing data can be developed from historical data. This data would have to be adjusted for escalation, location and any other special circumstances. Even if current data is used, comparison with historical data is helpful in verifying the validity of the data.

A design (quantity) allowance is an amount included in the estimate/forecast which is based on the degree of design completion and a comparison of the designed quantities with historical experience. The design allowance covers quantities we know from experience are not included in the designed quantity at a particular point in the engineering phase. A growth (cost) allowance is an amount included in the estimate/forecast which we know from experience is needed to cover adjustments within the defined scope that cause increases in the price of quoted or awarded materials. As the project progresses and scope and pricing are more refined, these allowances will be reduced.

Contingency is an amount of money included in a budget, estimate or forecast for costs, which based on past experience, are likely to be encountered but are difficult or impossible to identify at the time the estimate or forecast is prepared. Contingency is intended to cover estimate errors and omissions, design developments, pricing variations, and the like. It is not intended to cover changes in scope, force majeure events, labor strikes, etc. There is often a tendency on the owner's side to view contingency as available for his use to cover additional scope and owner cost over-runs. This is not the intent of contingency.

ESTABLISHING A BASELINE BUDGET

Early in the project's development, a baseline budget must be established, if costs are to be controlled effectively. The initial cost estimate for the project becomes the project's initial budget and is the baseline against which all project costs are monitored and reported.

The preparation of the project budget consists of reformatting the approved cost estimate into significant and readily identifiable components, such as purchase orders, contracts, and work packages which can be monitored in accordance with the planned and actual progression of work. In preparing the budget, the estimate details will be expanded or compressed to assure the optimal alignment for monitoring and reporting.

The budget data is organized to conform to project requirements regarding reporting format, code of accounts and the project-specific conditions.

As the project progresses, budget details will be adjusted, as necessary, to conform to changes in the project implementation plan or the code of accounts. This is accomplished through a system of formal budget adjustments. The cost control group maintains a log of budget adjustments, budget transfers between cost accounts, and scope changes. A continuous audit trail of the changes in budget should be maintained at all times.

A series of tabulations and trend lines (curves and graphs) are developed during the budget process. These include quantities, job-hours, and costs for such items as plant equipment, bulk materials, installation labor, contracts, indirect accounts, services and construction equipment, at both summary and detail levels. These will be used during the monitoring process to comparatively depict actual quantities, job-hours, and costs and expenditures against the planned timeline. These tabulations and trend lines are further updated at each formal cost forecast development.

CODE OF ACCOUNTS

Controlling costs requires a constant comparison of the three major categories of project costs. These categories are:

- Committed costs
- Paid costs
- Project budget.

To facilitate this process, the code of accounts for the project is established, so that these three categories can be collected into their respective code of accounts "buckets". In simple terms, monitoring the project costs involves comparing the three categories in each of the related "buckets".

The code of accounts provides the framework for identifying the project's physical facilities and for categorizing the quantities, job-hours and costs for plant equipment and materials, installation and services. It is a standardized system for defining, recording, monitoring, reporting and auditing cost information.

The code normally consists of one code for facility/sub-facility (usually numeric) for direct cost accounts, distributable cost accounts and EPCM services accounts and another code for commodities (often an alpha code) in the direct facility/sub-facility accounts.

For example: 0310.RAD

0310 would indicate the facility, such as Grinding

RAD would indicate 10" to 12" carbon steel pipe – R being pipe, A being carbon steel and D being 10" to 12".

The development and maintenance of the project code of accounts should be the responsibility of the cost control group. The code of accounts must be established and agreed to by the owner and the project team and issued immediately after project award to confirm the defined scope. The code of accounts is applied to labor, material, contracts and services, and is monitored throughout the project life both in the design offices and at the construction site.

COST TRENDING AND FORECASTING

A trend is a deviation from the established baseline - budget or schedule. The trend program is a formalized process for identifying and evaluating deviations as early as possible, such that timely appropriate action can be taken by the project team to alleviate the impact of detrimental trends, or to exploit the benefits of improving trends. This program is most effective during the formative design and procurement stages of the project. The essence of cost trending is timeliness rather than precision, enabling the project team to make decisions early enough that it can influence the direction of project costs. This is illustrated in Figure 3.

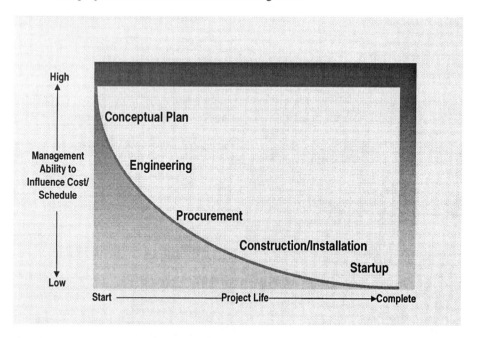

Figure 3 – Cost / Schedule Influence Curve

Trends are generally of the following types:

- Additions or deletions to project scope
- Design development resulting from improved engineering definition
- Price variances resulting from bid quotations
- Changes in installation productivity
- Changes in work execution methodology
- Estimate errors and omissions
- Escalation, re-evaluation, etc.

The trending program is initiated at the inception of the project and should continue throughout the project period.

Each member of the project team should be encouraged to initiate trends. These will then be analyzed for validity. For this reason, it is essential that all project team members have a clear understanding of the project scope and the budget.

Trends are order-of-magnitude estimates prepared in accordance with the code of accounts and include direct, indirect, and engineering/services costs, contingency and schedule impact.

A weekly trend meeting attended by key members of the project team is held for the purpose of:

- Validating identified potential trends
- Discussing pending trends that have been previously identified for status and action
- Reviewing trend estimates prepared during the week
- Reviewing the project status and latest project decisions for identification of new trends
- Discussing owner comments.

On a regular basis (weekly) trend reports and approved trends are issued formally for information and decision-making. The "current forecast" of project costs is continuously updated by adding the approved trends to the previously approved forecast.

At the start of the project a plan should be established for the frequency of formal forecasts of the total project cost. These should occur every 3-6 months throughout the life of the project. A cut-off date is defined for each forecast and on that date all available data is collected, quantified and priced. A forecast establishes the actual costs through the cut-off period and adds to it the expected costs of the remaining "to-go" work. The resulting figures are the expected cost at completion determined at that period in the project's life.

DEFINITIVE ESTIMATE

At a certain stage in the development of the project, the baseline budget should be updated to reflect firm project scope, finalized plant layout, specifications for major plant equipment and materials, layout and process design for the major plant systems (essentially design for major buildings and structures), and a firm construction plan and schedule. This is commonly known as the definitive estimate.

At the time a definitive estimate is prepared, design engineering is ideally at least 40% complete (and no less than 30%), all major design decisions have been made, all major equipment has been committed or firm price quotations have been received, and construction has started, assuring that actual construction experience has been obtained as the basis for the cost of construction.

SCOPE CHANGE CONTROL

The scope change monitoring program establishes a method for the timely assessment of the effect of changes in scope on project cost and schedule from an established base estimate and schedule. This program records and reports the overall effect of scope changes and provides a basis for obtaining owner agreement on contract scope amendments.

Changes in scope include alterations to the project's technical scope and/or to the contractor's scope of services as requested, directed, and/or authorized by the owner.

The importance of control of the scope of the project and identification of deviations from the base defined scope cannot be over-emphasized. The identification of these scope changes and realistic pricing and impact on the project schedule in a timely manner are a key to good project cost control. As soon as a change in scope has been identified, a rough order-of-magnitude estimate is needed for the owner to decide whether or not to proceed with the change.

It is the responsibility of the project team to ensure that scope changes to the project's physical facilities and/or the contractor's scope of services are identified, documented and processed.

For the change control procedure to work effectively, it is important that all members of the project team be encouraged to identify anything that they believe to be a change in scope. This information must be transmitted to the cost control group who will then prepare a cost estimate for the defined change.

The cost control group will prepare a preliminary evaluation of the change at the total project cost level and submit the estimate and schedule impact to the appropriate members of the project team and the project manager. After their review and approval, it will be forwarded to the owner for review and approval to proceed.

After approval, the cost of the change will be included in the current budget (project budget plus approved scope changes).

The status of all changes will be summarized monthly and will be submitted to the project team to keep them fully informed of the changes to the budget.

TRACKING QUANTITIES, HOURS AND COSTS

An essential part of cost containment is tracking the components that have a direct impact on the project's cost. Two key components are the quantities of materials to be installed and the engineering, construction and management service hours involved. To be effective these have to be tracked as soon as they are identified and compared with the budget figures. Ideally, as the detailed design progresses, quantities of bulk materials (concrete, steel, cable, etc.) that the design represents should be compared regularly with the associated quantities in the budget. Any deviations should be noted and discussed immediately with the project team.

Similarly, as materials are installed, they should be quantified and compared with the "trended" budget quantities and, again, deviations from the baseline should be recorded and brought to the attention of the project team.

A further control of quantities ideally would occur at the following stages:

- Quantity budgeted
- Quantity designed
- Quantity purchased
- Quantity shipped
- Quantity received at site
- Quantity distributed to the location for installation
- Quantity installed.

The expenditure of service hours on design, construction and management, should be tracked against the planned weekly or monthly expenditure of the budgeted hours.

Material costs are monitored via a budget allocation system. The engineering department (in the design office or at the construction site) prepares material requisitions (MR's) which provide type, specification, quantity and disposition of material. The cost control group compares MR information to the budget and provides a cost code for each item and the associated budget amount. After bids are received and analyzed, the cost control group compares the pending commitment value with the budget amount, noting deviations. Deviations will be reviewed by the project team to determine if possible action is required to stay within the budget value. The deviations will be identified as "trends" and will be incorporated into the current cost forecast update.

COMMITMENTS AND COSTS

A commitment is a contractual obligation to pay another party for performing services or providing materials. Cost and commitment reporting is a monthly function that keeps the project informed of the commitment and paid cost status of the project. It enables a comparison by cost code of the following items:

- baseline budget
- current budget (original budget plus approved scope changes)
- current forecast
- paid costs this month
- paid costs to-date
- commitments this month
- commitments to-date

Preparation of the cost and commitment report consists of compiling, into a single database, the commitments from the awarded purchase orders and contracts, the paid cost data from the accounting and labor ledgers, updating the current budget, and timely transmittal of the report to the project team.

The current budget data is updated by budget adjustments each month to reflect approved scope changes and budget transfers. The current forecast is revised as trends are approved or when a new project forecast is developed and approved.

Cost and commitment ledgers are maintained in native currencies and therefore separate ledgers are required for each currency. For project reporting, the commitments and costs should then be converted to a common project reporting currency. The reporting currency and the method for conversion from native currencies must be established and agreed to at the start of the project.

HISTORICAL DATA

At the completion of the project, all final actual costs, quantities and hours should be assembled into a project historical report. This report should include a complete definition of what the costs cover and should include flow diagrams, key project drawings, major specifications and data sheets, the project schedule, etc.

This data is then available for use as the basis for studies for plant expansions or similar new plants. In addition, it can be used as a check for estimates and pricing received from third parties when reviewing other potential projects.

CONCLUSION

The key elements to the management and control of costs of capital mineral processing plants are having a clearly defined scope, establishing a realistic baseline budget, monitoring commitments and costs against the baseline budget and identifying deviations early enough to take corrective action and updating the baseline budget as conditions change. This process must be adhered to consistently and frequently.

Schedule Development and Schedule of Control of Mineral Processing Plants

Parmod Kumar

ABSTRACT
The schedule development and schedule control of mineral processing plants is dependent on all those involved in the development and execution of the project. The following process describes how project schedules for engineering, procurement, and construction are developed through the different phases of the project and what steps should be considered in managing the project schedule.

INTRODUCTION
The basic philosophy of schedule development and schedule control is simple. It is:
- plan the work
- work the plan
- monitor the plan
- report deviations as early as possible
- take corrective action
- repeat the cycle

For the schedule to be effective this cycle must be continuous and involve all project team members

The schedules are developed starting with a milestone schedule and then progressing to more levels of detail. The levels of schedule concept is called schedule Hierarchy and this is described in the following section.

LEVEL OF SCHEDULES
The schedule hierarchy is shown below:

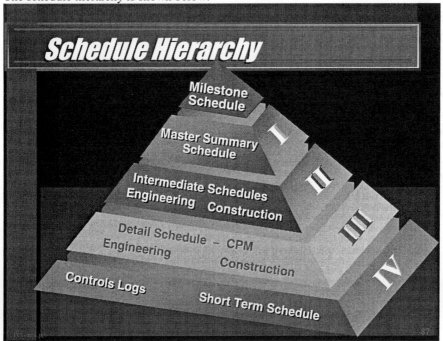

Milestone Schedule

A clear and concise definition of the project scope is essential for developing project schedules. All participants in the project should be kept fully informed as to what the scope of the projects is and the project owner must make clear what his expectations are.

The physical boundaries of the project should define and address:

- infrastructure – such as access roads, power supply, water supply, communications facilities, port facilities and air strips,
- mine facilities – such as mine development, maintenance, offices
- process facilities.
- temporary facilities – camp requirements, temporary shops required for construction, etc.

Master Summary Schedule - Level I

The Master Summary Schedule covers the entire scope of the project and is the basis for client progress reporting. It is used by the project team to evaluate summary progress against established milestones and sets the overall project schedule parameters in the development of more detailed schedules. Client interface activities are also shown on this schedule. A Level 1 schedule identifies the contractual milestones for the project and the key activities to be performed to ensure those milestones are achieved.

When applicable, the planner ensures that "Freeze Design" dates are identified as key milestones for design documents including design criteria, process flow diagrams, general arrangements and the main single line diagram. These milestones will typically be carried down into more detailed lower level schedules.

The schedule is prepared by the planning engineer and developed in a simple bar chart format. The upper section of the schedule identifies the major contractual and/or key project milestones; the lower section shows the major activities grouped by engineering, procurement and construction. The construction activities identify the physical plant facilities. The critical path of the project will be highlighted, so that it is readily visible.

The Master Summary Schedule defines the time commitments between contractor and the client in terms of notice to proceed, receipt of design criteria, major equipment lead-time, and completion dates. The summary checklist (listed below) for schedule development is offered to assist the planner in the development of the project schedule

- client desired milestones: notice to proceed, permits issued, plant ready to accept feed
- major equipment /material lead times
- project scope
- schedule assumptions
- schedule restraints
    - environmental permits
    - major commitment restraints; full funding, etc.
    - site accessibility
    - relocation of utilities
    - availability of construction power/water
    - construction camp
    - climatic conditions – weather, altitude
    - manpower availability
    - sources of permanent power/water
    - availability of client provided items: permits, permanent power, equipment

Based on the above checklist, the applicable information is collected and the Master Summary Schedule is developed. At the minimum it should reflect notice to proceed, start engineering, award and delivery of major equipment, availability of power and water, start construction of major facilities, and plant ready to accept feed.

In developing the schedule, the planner takes into account when engineering can be started for critical areas, what information is required to complete certain parts of engineering so the major equipment can be ordered, and when construction can start. Normally the critical path in the mineral processing plant is installation of the grinding mills and drives. The planner will verify when the engineering has sufficient information so the mills and drives can be procured and delivered. He will also establish when the foundation design can be completed based on mill vendor information available, so the construction can start. Simultaneously, it will be established when the building steel is required and when the design will be complete to support fabricating the building steel. Based on estimated quantities and job hours required to install the concrete and steel the installation durations are established with input from experienced construction personnel. Any other critical areas are also planned based on the similar logic. The planner will also review when the permanent power and water are required to start the plant.

Project Facility Summary (Intermediate) Schedule - Level II

The next level of project scheduling is called the Project Facility Summary Schedule. The work breakdown structure (WBS) provides the framework for identifying the project's physical facilities and for categorizing the type of work. The WBS is established in co-ordination with the cost engineering group so that the code of accounts and WBS are in line with each other. WBS establishes the levels of detail, so that project can be depicted by area, facilities within the area, sub–facilities within major facilities and then by type of work. Type of work is defined by engineering, procurement and construction.

    For Example:
    Level 1 – Project
    Level 2 – Area (infrastructure, mine facilities, process facilities)
    Level 3 – Facilities (crushing, concentrator, port facilities)
    Level 4 – Sub-facilities (grinding, flotation, regrind)
    Level 5 – Type of work (engineering, procurement, construction
    Level 6 – Discipline (civil, architectural, mechanical, electrical)
    Level 7 – Commodity (earthwork, concrete, steel, mechanical equipment)

This schedule identifies the total scope of each facility of the project, its major components and milestones. It establishes the fundamental logic for performing engineering, procurement and construction activities on the project. Dates for major design releases, commitments and deliveries, durations of plant and facility installations, mechanical completion, and commissioning are clearly identified.

The Project Facility Summary Schedule serves as the main link between engineering/procurement activities and construction activities during the development of the detailed engineering/procurement schedules. The construction activities at this level drive the engineering and procurement schedules. This schedule is the basis for estabilshing initial planning dates on the interface table. An Interface Table defines the construction need dates and engineering delivery dates for each facility and by type of work. As subsequent Level III and IV schedules are developed, impractical durations or logic may be identified which result in changes to the Level II schedule to ensure a workable plan.

The Level II schedule is prepared by the planning engineer and developed in a bar chart format. The upper section identifies the major contractual and/or key project milestones. The section below shows the major engineering, procurement, and construction activities within each major project facility. Major equipment/material issue-for-bid, award, and delivery dates are all major construction contract issue for bid, award, and mobilization dates are shown. The critical path of the project will be highlighted, so that it is readily visible.

CPM Schedule - Level III

This schedule covers the entire scope of engineering, procurement, construction, and pre-operational testing activities. It is prepared at a level of detail to effectively plan, schedule and coordinate the project's work activities. The CPM Scheduleprovides a detailed basis to monitor and evaluate progress for the identification of potential schedule impacts and possible development of workarounds. The schedule is comprised of the engineering and procurement logic schedule that is the basis for the milestone dates for each engineering deliverable in an engineering progress and performance reporting (EPPR) system (see Level IV schedules), and of the construction logic schedule which is the basis for the detailed construction schedules (also identified under Level IV schedules). Construction contractor schedules are reviewed against this schedule, and project detailed cash flow projections are based on this schedule. The construction portion is resource loaded with job hours and commodities to develop construction progress curves and commodity curves.

The Level III schedule is prepared by the planning engineer and developed as a CPM schedule using a computerized program such as Primavera. It is prepared by facility and by discipline for engineering and procurement activities and by commodity for construction activities. For the engineering and procurement section, major categories of deliverables (e.g. design criteria, flow diagrams, equipment foundations) are identified for each discipline. Key milestones for each also are identified (e.g, issue for internal co-ordination, issue for approval, issue for construction, incorporation of vendor prints). Development of major material requisitions and contracts are shown and the logic relationships between key vendor prints and design drawings are identified. Engineering Work Packages (EWP) are identified and tied to the relevant contracts, fabrication or construction activities. The EWP is a descrete package of construction activity and includes all the design drawings, specifications, and material lists necessary to allow construction effort to be complete. Key information from outside parties (e.g. topographical and geotechnical surveys) are identified and logically linked.

For the construction section, all major activities are identified by commodity within each facility. The links to the engineering and procurement section will be through the EWPs, and contracts, and equipment and material deliveries.

Look ahead extracts (90/120 day schedules) are run from the Level III schedule as needed and will be used to identify any deviations from the plan. At each update period the reason for any actual or forecast deviation from the target (plan) will be recorded and if related to a trend or change order the related number will also be recorded. This will result in a continuously updated record of schedule impacts on the overall project.

The schedule development sequence and updating process is shown in Figure 1 below:

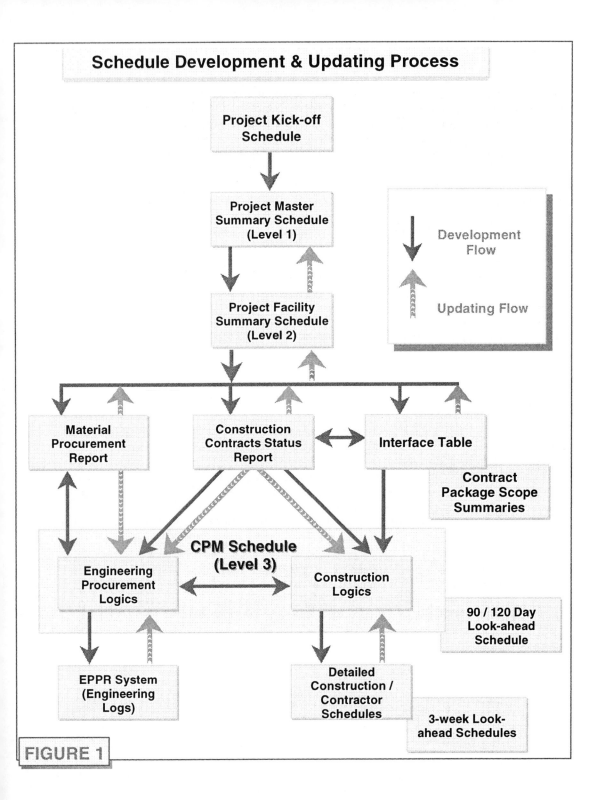

FIGURE 1

Control Logs and Short term Schedules – Level IV

The engineering control logs and shortterm schedulesare referred to as the Level IVschedules. These are used by engineering and construction for planning and performing their work. The engineering control logs show drawings, specifications, material requistions, and related engineering tasks. Each deliverable is identified together with construction issue dates and intermediate milestone dates such as issued for coordination and client approval. These control logs are used for monitoring engineering progress and performance.

PROGRESS AND PERFORMANCE MONITORING

Earned Value (EV) is used to determine performance and progress. This EV system is often overlooked because the term "Earned Job-hours" is used so frequently. Jobhours are, in fact, a "weighting" factor in the EV system. This point is discussed in further detaile below.

Performance

Performance is <u>always</u> measured against budget. For Engineering, it is the number of hours spent to achieve the deliverable milestone versus the hours budgeted. For Construction, it is the number of hours spent to install a unit of work versus the hours budgeted. To develop a composite performance for non-similar tasks, hours are spent and "earned" for each task, then summed and compared. The "earned" hours are calculated from the completed work item and the budgeted hours.

Progress

Progress is <u>always</u> measured against <u>planned</u>, which reflects the latest scope. Physical progress represents the "quantity" of work performed. Schedule or planned progress represents the time frame in which the work is performed. For engineering, scope is generally identified by deliverables. For construction, scope is generally identified by quantities. For any single item, it is a measure of work performed versus work planned. A composite progress for non-similar tasks can be calculated using a job-hour weighting, similar to the performance calculation, but in this case the "earned" hours are calculated from the completed work item and the planned hours.

Some of the progress and performance reporting tools are:

- Progress and performance reports
- Progress and performance curves
- Engineering drawing/deliverable release curves
- Procurement purchase order commitment curves
- Commodity release and installation curves
- Quantity installation curves

SCHEDULE MANAGEMENT

The objective of schedule monitoring and managing the schedule is to assess the current status of the project, identify deviations to the plan and implement any corrective action so that the project is completed in the most timely and cost effective manner. The planning engineer becomes the eyes and ears of the project as far as project schedule is concerned. The planning engineer, with input from the project team, (engineering, procurement, and construction) updates the schedule on a regular basis and compares the status against the approved schedule. Any deviations are reviewed and analyzed and workarounds are developed to ensure that the work is completed within the overall schedule.

Curves for engineering progress and construction progress as well as procurement purchase order commitments are developed to monitor the progress. In addition manpower curves are developed for engineering and construction work. The planning engineer tracks the actual progress versus the planned on a regular basis. Also the number of people planned versus actual are compared. The monitoring process results are discussed with the project team, the project management, and the project owner on a regular basis. This provides the project a tool for corrective actions if and when required.

In order for the schedule to be effective, it is essential that all project team members participate in the schedule planning and monitoring activities for the project.

CONCLUSION

The key elements to the schedule development and schedule control of mineral processing plants are having a clearly defined scope, considering the realistic schedule restraints, client identified milestones, and developing a realistic schedule. The schedule control is simple, by working the plan, monitoring the plan, reporting the deviations and taking the corrective action. This process must be adhered to consistently and frequently.

The Risks and Rewards Associated with Different Contractual Approaches

P.J. (Jeff) Gard[1]

INTRODUCTION

Significant changes have evolved in the mining industry toward the end of the last millennium. Mining companies have become significantly more focused on cost and return on assets. This has led to dramatic changes in the contracting industry with owners expecting contractors to accept significantly more risk. The object of this paper is to examine the risks and rewards, both from an owner and a contractor perspective, associated with differing contractual approaches.

PROJECT GOALS

All too often the goals for a project are stated as delivering a high quality facility, in a safe manner, within the capital budget and ahead of the agreed schedule. These aspects are all very important but the true measure of project success in any industry is normally the same – to maximise shareholder value. Herein lies a serious dichotomy. How can this goal for the mining company be achieved coincident to the identical goal for the contractor? If the correct balance between risks and rewards for both parties can be achieved, then the goals for both parties will be similarly achieved.

RISK SHARING PRINCIPLES

Various authors have identified principles which should govern the allocation of project risks between the various parties involved in the project. To determine these principles the following questions must be answered:

- What is the source of the particular risk?
- Which party can best manage the events that may lead to the occurrence of this risk?
- Which party can best manage this risk, should it eventuate?
- Will the actual cost of the risk or the premiums charged by the party accepting the risk be reasonable and appropriate?
- Will the occurrence of this risk lead to other possible risks for any of the various parties involved?

If at project commencement the allocation of the risks is not understood or is deemed to be inequitable, then the project goal is unlikely to be achieved. The proper identification of the risks involved will lead to the form of contract that will best suit the project.

---

[1] President, Mining and Minerals, Fluor

STANDARD FORM OF CONTRACT

The following standard forms of contract will be considered:

- RCPPF – Reimbursable Cost Plus Percentage Fee where all contractor's "true" costs will be reimbursed plus a fixed percentage of these costs will be paid as a fee.
- RCPFF – Reimbursable Cost Plus Fixed Fee where all contractor's "true" costs will be reimbursed and a fixed fee paid.
- RCPIF – Reimbursable Cost Plus Incentive Fee where all contractor's "true" costs will be reimbursed and an incentive fee paid on the achievement of certain pre-stated goals.
- FP – Fixed Price where the contractor will deliver the facility for a fixed price.
- LSTK – A fixed price contract in which the contractor delivers a facility that is warranted to perform at a specified level.

CONTRACT TYPE VS RISK

The following chart depicts the relative allocation of risk between the owner and contractor for the various contracting scenarios.

Project Definition	Poor		Reasonable		Complete
Amount of Risk	High		Moderate		Low
Financial Uncertainty	High		Moderate		Low
Risk Allocation	High (Owner) → Low (Contractor)				
Typical Contract Form	RCPPF	RCPFF	RCPIF	FP	LSTK

CONTRACT FORMS

Reimbursable Cost Contracts

This form of contract gives minimal risk for the contractor but the highest risk for the owner, since the majority of risks associated with errors and omissions, schedules and quality are borne by the owner.

Most owners agree that the following advantages are given by a RCPFF contract:

- the owner pays for what he wants
- the owner can control who the contractor assigns to project and level of effort to meet schedule and/or cost targets
- the contractors profit margin is virtually transparent
- the project is generally not adversely affected by a loss making contractor

Most contractors agree that the following advantages are given by this form of contract:
- a guaranteed fee or profit margin for the work performed
- limited exposure for errors and omissions

Unfortunately this form of contract has many disadvantages for the high-risk taker – the owner:

- What are the "true" contractor costs to be reimbursed? Disputes all too often ensue in the areas of payroll burdens, overheads and managerial time. Some portions of these items may have to be paid from the fixed fee and therefore the contractor is almost forced to minimise such costs. It is paramount to the success of this form of contract that all "true" contractor costs are reimbursed.
- What is the contractor's commitment to minimise project cost?
- What is the commitment to utilise the best-suited people for the project? This will be dependent on the total workload of the contractor and the mix of contract forms, within this backlog of work.
- What will stop the contractor overstaffing the project in times of low workload?

The most common form of reimbursable contracts are based on a fixed fee, where the contractor has frozen their profit margin irrespective of final cost of services provided. This form of contract does not, however, distribute significant profit risk to the contractor. Other variations of the conventional cost reimbursable contract have evolved to provide a more equitable sharing of risk between the owner and contractor.

Contractor's fees can be put at risk by tying attainment of project goals of cost, schedule, safety or plant performance to payment.

Other variations of the cost reimbursable contract include placing caps on the total services contract value, as is the case in the Guaranteed Maximum Contract or to a lesser extent capping certain components of the services cost such as contractors overheads or expenses related to the work.

Another important issue with the reimbursable cost contract is the selection criteria used to select the contractor. If the owner chooses a competitive bidding process, do they select the contractor on the basis of lowest fee? This will not necessarily guarantee the optimum overall capital cost or full life cycle cost for the project.

INCENTIVE CONTRACTS

Incentive contracts are thought to more equally share the risks between the contractor and the owner. They give a middle of the road position between a RCPFF and a FP contract but are often difficult to reach agreement to give the equal risk sharing.

Incentive contracts must be based on clearly defined and quantified parameters. Care must be taken not to have parameters which could in fact be in conflict with each other. The most common parameters that are used are:

- capital cost
- schedule, possibly including ramp-up time
- safety
- quality

The difficulty in this form of contract often arises from the lack of agreement of the risk sharing by both parties. It is important that differences in risk sharing are recognised when the risks are:

- controllable by the owner
- controllable by the contractor
- not controllable by either

Fixed Price (FP) Contracts

This form of contract gives less risk for the owner but greatly increases the risk for the contractor. With this contract, the owner will pay a fixed price to the contractor, irrespective of the actual cost to the contractor for the performance of the contract.

The contractor is always attempting to reduce costs by improved efficiencies, value engineering and intense negotiation with vendors and subcontractors. The owner's risk in this instance is that there may be a reduction in the quality of the facility where this has not been identified in the contract. It is common knowledge that it is extremely difficult to completely specify, without ambiguity, the requirements in a contract. This often gives a loophole for the contractor with this form of contract, or conversely, it leaves the contractor open to commercial risk by the owner in trying to claim something that the contractor did not envisage.

Another disadvantage for the owner with a fixed price contract is that the contractor may require an excessive premium to fix the price. Even then, the owner has no guarantee that this will be the actual final cost of the facility – it is, in effect, a guaranteed minimum cost.

If the fixed price is too low the entire project may be at risk since the contractor may not be able to meet their contractual commitments. A contract awarded to the lowest fixed price bid through a competitive tender process is not necessarily the best for the project.

Choosing the fixed price contract strategy should be avoided when:

- specifications are not clear and enforceable
- uncertainty is significant
- the reputation and financial security of the contractor are not beyond question

A variation to the fixed price contract format is the inclusion of incentives for attainment of project goals of schedule attainment, process ramp-up, or safety. These incentives may further focus the contractor to the owner's objectives in addition to their own fiscal targets.

There is no doubt that this form of contract should be avoided in the very early stages of any project. At this time there will be poor scope definition, high owner risk and probably financial uncertainty. A more appropriate use of this form of contract is to start the project on a RCPPF basis and convert to a fixed price when the scope is properly defined and there is more financial certainty.

Lump Sum Turn Key

The extreme of passing risk from the owner to the contractor is a Lump Sum Turn Key (LSTK) contract. A LSTK contract passes virtually all of the project risk to the contractor. For an agreed, fixed price (the "lump sum") the contractor undertakes to build a facility and hand it over to the owner when it is a fully operational unit confirming to previously agreed design criteria (the "turn key"). In addition there are usually bonus/penalty agreements on schedule for mechanical completion and the time needed for the plant to reach its design metallurgical performance.

The contract has many advantages for the owner, particularly small and medium sized mining companies who do not have the financial resources of their major competitors. Because the risk is nearly all transferred to the contractor, then the risk for the project to the financial institution that is providing finance for the project is against the contractor rather than the mining company. Providing that the contractor has a strong balance sheet, then it becomes considerably easier for the owner to arrange project financing.

For the contractor, as well, there are also advantages. He becomes master of his own destiny, insofar as the design of the facility is concerned. Provided that the facility conforms to the criteria laid out in the contract, he is free to build it and commission it how he chooses rather than as a compromise to the owner's wishes.

One of the few remaining risks for the owner is that he must provide an ore feed that corresponds to the conditions of the contract. The warranties for a flotation plant, for example, that has been designed to accept an ore feed of 3% copper, 1% zinc and 5% iron may well not be valid if the ore feed on start-up is 1% copper, 3% zinc and 15% iron.

Lastly, because the contractor is taking nearly all of the risk, then his fee premium for this type of contract is the highest of all of the normal contracting methods.

PROJECT ALLIANCES

A project alliance takes the concept of risk sharing to its highest level. In a project alliance the owner, usually multiple contractors, as well as other stakeholders form an alliance to execute the project in which they are collectively incentivised on the successful outcome of the project. As an example, an alliance might comprise:

- The Owner
- An Engineering Contractor
- A Construction Contractor
- A Key Equipment Supplier
- An Environmental Consultant

Targets might be set on capital cost target and a schedule target and bonus/penalty arrangements put in place around those targets. The project might miss its schedule target because of an environmental delay for example, but all of the parties, not just the environmental consultant, would share the penalty.

The object of an alliance is to encourage teamwork in achieving the project goals. In the example used, the environmental delay may have been caused by an engineering error, or perhaps an incident caused by the construction contractor. By incentivizing the group collectively cooperation to avoid delays and overruns is rewarded.

Project alliance have typically been used for large "mega projects" involving multiple contractors.

CONCLUSION

Even with the inherent risks of a fixed price contract it is still a very common form of contract. The RCPFF contracts are not popular with risk-averse owners. Incentive contracts do offer the middle of the road, but they are not used as widely as they could be. This is probably due to the lack of knowledge of this form of contract and their possible motivational effects on the contractor.

Negotiation of a fixed price with a trusted contractor may be the preferred option for an experienced owner.

The setting of realistic goals is necessary for all relationships but is particularly true for incentivized contracts. When a contractor realizes that he cannot achieve any of his incentive, his performance rapidly changes to become a "salvage" exercise and the quality of his performance may suffer.

In any contract situation, there must be trust between both parties and a definite intent to minimise adversarial situations. This will give benefits to both the contractor and the owner and allow information to be shared and decisions to be made, culminating in a successful project.

Success Strategies for Building New Mining Projects

Robin J. Hickson

ABSTRACT

The increasing complexity of issues surrounding new mining projects, whether inside or outside North America, today stretches the technical competence of the selected Engineering & Construction firm, as well as the operational and financial capabilities of the Client Owner. In today's knowledge-driven economy, wealth creation is both embraced and challenged from multiple quarters. A compilation of best practices, i.e. "do's and don'ts" for mining projects is offered to practicing engineers for both Engineering & Construction contractors and for Owners. These strategies are globally drawn from a range of recent mineral undertakings, some successfully completed, some not. While the focus of the paper is toward assisting the young new project engineer, the strategies themselves apply to most mining projects.

INTRODUCTION

Successful building of a new mine project, like any business undertaking involving the interaction of human beings requires not only a clear understanding of the goal by the participants, but also a knowledge of the strategies that are best likely to achieve that goal. Before discussing the key elements of these strategies, it is first important to know what each of the two major players; the Client Owner and the Engineering & Construction contractor (E&C) is seeking when they come together at project initiation. (Note: "Client" and "Owner" terms are used interchangeably herein.)

THE OWNER PERSPECTIVE ON E&C COMPANIES

What is it that mining Clients expect when they select an E&C Company? The answer is simple; competence to successfully execute the project while, at the same time, providing the Owner with real project risk reduction. The bigger the Client, typically the more risk adverse he will be! The components that the Owner looks for within the E&C organization include:

1. A successful track record - of completing similar projects on time and within budget.
2. Engineering competence - instantly available, specific, above-average specialized process and technical skills, i.e. those skills that the Client does not keep in-house.
3. Project management capability:
   - Familiarity with the particular geographic location
   - Experience in building projects of comparable size
   - Knowledge of the selected technical process type.
4. Best team. A core of key players who have successfully worked together previously.
5. Global interacting offices with genuine willingness to provide skills from all offices, not just the signature office. Owners seek an E&C organization with offices that can work together worldwide, and complement each other to provide the Owner with a best global solution that maximizes local in-country experience and knowledge.
6. True understanding of the Request for Proposal (RFP) requirements, not a mere restatement of Client's words in the resultant bid submittal.
7. Proven advanced technology designs, and an ability to timely deliver internal know-how.
8. Up-to-date, reliable, accurate, and comprehensive, in-house cost estimation database.
9. Realistic forecasts - of cost, schedule and process technical performance.

10. The cost accuracy figure provided to the Client should reflect work completed; not just restate the RFP. Any accuracy better than requested will reduce the Owner finance costs.
11. Non-alignment - with a particular technology or subcontractor. Clients mistrust an E&C's own patent recommendation, even when it's truly best.
12. Sourcing prowess - i.e. volume pricing, inspection capability, expediting competence, and logistics skills. The E&C contractor should have appropriate technical and global understanding of the scope of supply, as well as a sufficient network relationship with vendor organizations to both provide procurement areas of opportunity as well as to facilitate resolution of potential problems before they become major project issues.
13. Ability to control both time and cost, and to punctually and accurately report project performance and progress.
14. Competitive rates. (Note: While E&C cost is important, it is rarely the final decider.)
15. Vision to think in terms of total project, not as separate study phases for engineering (E), procurement (P), construction (C) and start-up. Segmentation of project elements can create gaps. It takes a successful execution and meshing of all project phases from initial idea to Client turnover to achieve a successful project outcome. (See Figure 1)
16. A partner that genuinely desires to align with the Client and all other key project players.
17. Exceptional safety performance track record.
18. An executive sponsor with clout and know-how to fix project issues to Client satisfaction.
19. Unswerving commitment to stay in minerals industry to be there for the Client tomorrow!

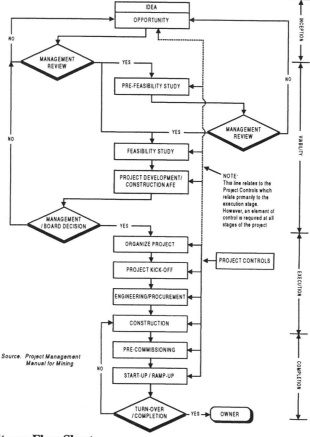

Figure 1 Project Stages Flow Sheet

The Client's selection of its E&C firm should be objective, not subjective. Usage of a 'Project Bidder Selection Evaluation & Ranking' form such as shown in Table 1 is a proven methodology to accomplish this task. The criteria and the criteria weights used on such evaluation form need to be consensually set by the Client evaluation team prior to receipt of bid packages from the E&Cs.

THE E&C COMPANY PERSPECTIVE ON OWNERS

What does an E&C contractor hope for, when it vies for an Owner's business? The answer is more than just "being chosen for the job!" Winning the work, but then being treated unfairly, can be worse than losing. Thus, along with winning the bid, an E&C seeks the following from the Owner:

1. Open and frank communication.
2. A Client willing to do things right. No forcing of the E&C firm to cut critical corners.
3. All parties working to the same key schedule milestone dates, i.e. no hidden agendas.
4. "Total cost" focus, not transaction price. Quality disappears when lowest price is key.
5. A belief in the merit of adequate funding and staffing for excellent project controls.
6. An open-minded willingness to benchmark and incorporate "lessons learned" from others.
7. If advocating "fast tracking" via construction-driven engineering, then a full understanding of the additional risks being incurred is required, along with a willingness to fund early involvement of construction and commissioning personnel into the up-front project design.
8. Owner personnel (with responsibility for design and engineering) located in the same office as E&C contractor. Even in today's cyber world, human interface still is critical.
9. Allowed usage of E&C contractor's own in-house systems, i.e., their 3-D CAD technology, their own document management, materials sourcing, project controls, scheduling systems, etc. It is a detriment to have to work within another party's system.
10. Strong belief in partnering principles, and meaningful alignment with, and respect for all stakeholders. The Client needs to view the E&C contractor as a partner, rather than the enemy!
11. Freezing of scope, process design and process flow drawings (PFDs) in a timely manner; and then sticking to the freeze.
12. Realism in setting Client review times for design criteria, drawings, studies, bid lists, contracts, purchase orders (POs), change orders, etc. Clients who extend their own review times, but still expect overall schedule to stay on track are being unfair.
13. Willingness to issue (and act upon) negative trend notices. Clients that insist on burying negative trends (to avoid bad news getting to their head office) eliminate their best chance of reducing the undesirable consequences.
14. A gainshare system of risk and reward; alignment of Owner, E&C and major suppliers.
15. Concurrence on definition of project completion spelled clearly out at project start.
16. Clients need to be honest and realistic. Too many facts are being spun today. Clients, particularly one-project juniors, often demand that the E&C buy-in to dubious data, as the price of project work.
17. Ability for the E&C to earn a reasonable profit margin.
18. RFP should be a bid document, not a guise for free value engineering. A billion-dollar RFP costs thousands of dollars to prepare. Asking for multiple rebids is unethical.
19. Payment of invoices on time. A number of Clients don't pay their entire bill, and there is a disquieting trend to not pay the final invoice, or the retention. With retention often greater than the E&C profit margin, such non-payment practice can cause E&C doom.

Table 1 Project Bidder Selection - Evaluation and Criteria Ranking Form

Items	Weight Factor	ABC Company		CBA Company		BAC Company	
		%	wt	%	wt	%	wt
PROJECT TEAM STRENGTH & QUALIFICATIONS	(37)						
• Executive Management / Sponsor Support	2						
• Project Manager	9						
• Total Engineering Capability - Civil/Structural/Material Handling/Mechanical/Pipe/Electrical/Instrumentation	6						
• Project Controls Competence:	5						
- Controls Manager							
- Estimator/Scheduler							
• Procurement/Expediting/Logistics Knowledge	3						
• Construction Management Experience & Site Capabilities:							
- Construction Manager	5						
- Field Engineering/QA/QC/Contracts/Safety	4						
• Feasibility Study-Financial Analyst	1						
• Organization Chart - Completeness/Responsiveness	1						
• Key People - Commitment to Keep on the Job	1						
TECHNICAL EXPERTISE	(14)						
• Process Technology Familiarity	3						
• Process Controls Design Capability	1						
• Product Expertise, e.g. Cu; Mo; Au; Ni; Coal	3						
• Familiarity in Building a Like Facility	5						
• Infrastructure Design and Tie-ins Capability	2						
RESPONSIVENESS TO PROJECT RFQ	(15)						
• Technical Understanding of Project	4						
• Design Innovation/Alternatives	3						
• Mine Plan Integration Capability	1						
• Constructability	2						
• Operations Simplicity	1						
• Maintenance Minimization	1						
• Performance Warranty	2						
• Operating & Maintenance Manuals	1						
LOCATION KNOWLEDGE	(9)						
• Familiarity with Client	1						
• Construction Experience at Project Site	2						
• Construction Experience in an On-going Operation	3						
• Special Concerns: : e.g. Host Country/Bonding	3						
PROJECT IMPLEMENTATION	(7)						
• Work Plan Approach	2						
• Ability to Team with Owner - Partnering	2						
• Environmental / Permit Compliance Know-how	2						
• Safety Commitment	1						
PROJECT CONTROLS SYSTEM	3						
SCHEDULE REASONABLENESS	6						
• Ability to Complete by ??							
• Establishment of Key Milestone Dates							
OPERATING COST ESTIMATION EXPERTISE	2						
CAPITAL COST ESTIMATE REASONABLENESS	2						
• Is it Lowballed? Is CF Optimized? Battery Limits?							
COMMERCIAL TERMS	3						
• Responsiveness/Alignment with Client Requirements							
• Is Incentive/Penalty Fee Option Proposed?							
• Clarity/Simplicity							
OFFICE LOCATION	1						
PRESENTATION EFFECTIVENESS	1						
TOTAL SCORE	100						

20. Both Client and E&C should only pursue litigation/arbitration as the final resolution of differences, not a first step. Disagreements between Client and contractor too quickly go to legal resolution rather than to face-to-face settlement by the parties, and it seems that any issue is fair game today, whether or not it was in the original work scope. Lawsuits are being filed against E&C contractors today for project over-production, for under-production, for the E&C not accepting Client data, and for the E&C not challenging the Client data sufficiently up front.

SCOPE DEFINITION

Project success requires candid decision making at the project front-end leading to a clearly defined scope of work. Too often this is all too vague. The task of defining scope and design concepts starts before the contract between the E&C and the Owner is signed.

The proposed E&C Project Manager (PM) and process engineers need to be involved alongside the E&C sales team in proposal evaluation and review of Client's RFP scope. The process engineer understands the functionality of the project scope. He or she may be the sole capable judge for the E&C of how difficult the project will be to design and operate as expected, what the technical risks to costs are, and, ultimately, how the design will (or won't) work. Caution: If internal expertise is lacking, the E&C must acquire appropriate outside specialist consultants to review the pertinent issues of concern.

CAPITAL COST ESTIMATE

The control budget estimate, be it a lump sum project or reimbursable, must be priced realistically; not set arbitrarily to come under a Client target nor forced by an E&C to beat bidder competition. Once a capital estimate is set, the PM and the Project Controls Manager must stay sufficiently familiar with the components to be capable of effectively monitoring the trending program.

Contingency

Capital cost contingency is a specific monetary provision within the capital cost budget that is intended to cover variations in the forecast value of cost or schedule. Contingency is not a provision to cover variations in scope or quality. The contingency allowance is added into a project based on the level of engineering completed and on the deemed risk. A contingency allowance, when correctly calculated, will normally be fully spent during the course of the project.

A project contingency needs to be set properly to cover risk, i.e. at the 90th percentile confidence level. A Monte Carlo simulation of accuracy risk on each line item of the capital budget to achieve a <10% probability of overrun post contingency is one procedure commonly used, and is the procedure recommended. Confidence limits are a quantifiable statistical tool. One must never adjust contingency for front-end marketing purposes.

As a word of caution, the 50 to 85% confidence level range that some firms utilize to assign contingency is insufficient to ensure that the capital budget will not be overrun. The 50 to 85% range is too optimistic; risks are unrealistically assessed. (Note: Using the 90% figure, a major international mining company delivered a commendable "within 1.7% of forecast" budget performance for eight projects, entailing some $3 billion over 6 years, from 1993 to 1999.)

If the Owner wishes to additionally cover the risk of possible changes in scope or quality (such changes more typically being created by the Owner than the E&C) then an allowance, separate to contingency, typically known as a "management reserve" should be created. If such management reserve is utilized, its calculation would be based upon the following:

- Project location
- Degree of definition of project scope
- Level of undefined project risks
- Potential for Owner's scope to occur
- Potential for project design quality change to take place.

This reserve is an unusual component; it is not present in most projects. And if present, unlike normal contingency, it may or may not be spent, dependent upon whether the envisaged scope / quality change actually occurs. The management reserve needs careful administration, typically needing an Owner level above that of Project Director for expenditure authorization. This higher level approval is introduced to avoid the reserve being used as a "slush fund" by the project team.

Estimation Accuracy and Minimum Engineering Level
For a lump sum estimate, the E&C has to establish what constitutes the minimum-adequate level of engineering upon which to safely base a lump sum price? While, theoretically, if contingency is set properly for risk, a lump sum can be established from any engineering level, experience shows mineral projects cannot portray risk well without minimally framing the project with at least some 15% of the engineering effort, i.e. a preliminary or "Type 3" estimate. A "Type 4" definitive estimate, with some 40% engineering complete is the normal minimum for a lump sum estimate however, but a "Type 5" detailed estimate, with >65% engineering, is always preferred.

Capital cost estimates are categorized into Types 1 through 5 based upon the level of effort behind the quantity take-offs and the pricing knowledge. The types can be summarized as follows:

- Type 1 Magnitude Estimate, with 0-2% engineering complete. Used for scoping studies. Accuracy ±30% or worse.
- Type 2 Conceptual Estimate, with some 5-10% engineering complete. Used for prefeasibility studies. Accuracy ±20-25%.
- Type 3 Preliminary Estimate, with some 15-30% engineering complete. Used for feasibility studies and baseline budgets. Accuracy ±15%.
- Type 4 Definitive Estimate, with some 40-60% engineering complete. Used for project control and simple lump sum bids. Accuracy ±10%.
- Type 5 Detailed Estimate, with > 65% engineering complete. Used for project control and more complex lump sum bids. Accuracy ±5% or better.

Bulk material quantities, particularly steel, cabling and piping, are perennial problems for estimating engineers whenever engineering definition is lacking. Material pricing and unit cost seem to be less of an issue.

Capital cost estimates rely heavily on the E&C in-house estimation database. This database thus needs continuous updating to maintain relevance. Real life variances discerned in the Home Office (HO) and field must be routinely fed back to the E&C's estimating department.

To achieve project success within the engineering phase, prior to engineering start the discipline and process engineers must buy-in to the manhour budget for their areas. If not, successful adherence to budget and schedule is unlikely. They then must be held accountable!

SCHEDULE
A realistic schedule with logic tie-in milestones to procurement cycles and to subcontract activities is a necessity. Initially, a project master schedule, using a Gantt bar chart that shows overall time frames of each major project phase and the significant milestones will suffice. However, to actually execute the engineering or the construction phases a master integrated schedule with logic tie-ins for all the individual work elements (typically Primavera Level 3 or beyond) is a requisite.

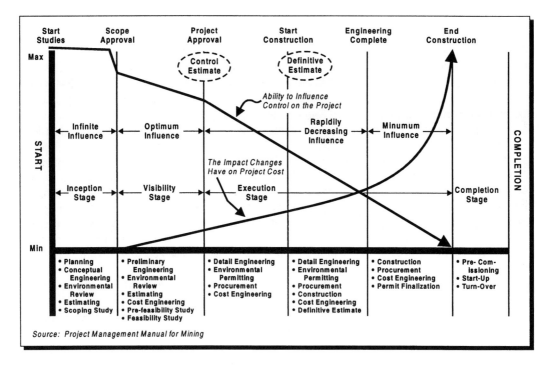

Figure 2 Profile of Project Management Control Influence

PROJECT EXECUTION - OWNER ISSUES

Today's Owner approaches a new project with a myriad of concerns. Avoidance of the all too common historical mining project execution issues is paramount:

- Capital overruns
- Completion dates missed
- Forecast production output achieved late
- Operating cash costs frequently never achieved
- Environmental standards compromised, particularly in the less developed countries
- Insufficient regard for safety.

Project overstatement could also be added to the above. Compulsive optimism seems to drive certain mining project advocates. Insufficient benchmarks are taken to check projects against reality, or if taken, they are ignored or explained away. But these pitfalls are preventable. Consistent project success lies in a rigorous execution of basic project management tenets.

Project Characterization

The Owner must be willing to spend sufficiently up front, where the optimal influence over project outcome resides. (See Figure 2.) This is the time to decide to build or not. Cheap estimates up-front lead to bad outcomes. Proper initial project characterization requires a formal process:

1. **Multi-department, integrated initial evaluation (scoping) assessment process.** This is used by Owner in the initial idea/opportunity phase, when initial "go/no go" business strategy is executed. This initial scoping assessment process requires timely participation by the Owner's pertinent support disciplines, (sales, human resources, tax, legal, treasury, etc). These disciplines are brought into the decision-making process early, to introduce a quick, objective, collective fatal flaw assessment of front-end project risk.

2. **A Project Management "Best Practices" program.** A well-defined Project Management process provides the originator of an "idea" with the tools to develop that idea into a winning project that will meet corporate Owner objectives. A formal project management program, following a regimen of proven best practices should kick in post the scoping phase, i.e. from pre-feasibility through start-up. By following a set outline of best practices, management is assured that project viability is adequately challenged, correct questions are posed, the project is properly characterized, necessary personnel are placed, and that an appropriate process is implemented. The goal is to set an optimum execution strategy to deliver the best life-cycle performance for the project.
3. **Technical underpin to design criteria.** Each project needs a focused set of site-specific technical plans and designs. And, as a complement, rigorous benchmarking against comparable project history is also necessary, to honestly assess the potential for success and to fully highlight the risks.
4. **Fatal flaw review.** Periodic reviews should be scheduled to look for fatal flaws and better technical solutions. These should be conducted with non-project personnel, experienced with the concepts involved. Project and/or process engineers are not always able to stand back at sufficient distance to take a look at the total picture. Opinions from outside the project team should thus be routinely sought and never refused consideration.
5. **Feasibility Study completed to "bankable quality."** Even if external monies are not sought, a feasibility study should be of a quality that could merit bank project financing. When a project cannot persuade outsiders to put up their own money without demanding that the corporate Owner underpin the investment, it is a strong indication that the project should not be pursued!

Project Independence

To achieve an unbiased project characterization, the Owner's project group needs independence from the Owner's operations and exploration departments. Unless the Owner's project department is separated from an Owner's operation's "stay-in-business" concerns, and from Exploration's "pet project" influences, mischaracterization by project advocates will occur. However, the large central project group, the norm of 30 years ago, is no longer tenable for this task. The solution is to keep a small central, specialized Owner's project group, ideally less than six persons, that has the knowledge to hire the appropriate outsource entities to execute any project. The small size of this group ensures that the carry cost for the Owner is perceived as an opportunity, not a burden. A low carrying cost also minimizes the job security concerns of Owner personnel, thus avoiding any project creation tendency. Outsourcing can provide the same, or better, cadre of skills as an internal central department, but without the overhead.

Project Success

To arrive at a successful outcome, everyone has to have the same goal. Thus a project must adopt a Project Creed - define project success before starting out. From Day One everyone must understand the exact success goals, and the metrics that they will be measured by. Final project success should mean achievement of all six of the following parameters:

- Project handed over to Operations on schedule
- Project completed within budget
- Plant's name-plate production met after ramp-up
- Predicted unit cash costs achieved by the date specified
- Zero Lost Time Accidents
- Full environmental compliance.

It is vital that the first four criteria are all given equal weight. If an Owner elects to emphasize only one, two, or three of these first four success criteria, then the project team will utilize the non-selected criteria as their "safety valve." This is bad. Ideally, the metrics for these success parameters should be set initially by the Owner and the E&C together to create the highest net present value (NPV) for the project, preferably using a gainshare program for the potential shared benefit of Owner, E&C contractor and suppliers.

And, as the final success ingredient; one has to have the right corporate culture to have any chance of project success. The requisite elements of this culture are as follows:

- **Single Point Accountability.** One cannot segment out project portions to different corporate departments and expect a successful project outcome. All areas: initial characterization, engineering and construction, owner costs, project commissioning, project controls etc., need to be under one entity, the Owner's Project Director. He/she needs overall control of project. These project elements are inter-linked; a modification to any one element affects all the others.
- **Honest Characterizations**. The project group must operate within a corporate mandate in which it is more important to characterize a project correctly than to achieve a specified internal rate of return (IRR). If job security hinges on coming up with a positive NPV, human frailty will ensure mischaracterization ensues.
- **One Team.** Successful outsource utilization requires that external personnel be treated as insiders. Second class treatment of external people will result in a second-class outcome.

PROJECT EXECUTION - E&C ISSUES
Constant reinforcement of proper project management procedures by the E&C firm is a fundamental necessity, to ensure that a success mentality pervades. Simply handing out a project manual and hiring a world class E&C firm cannot alone ensure a successful project outcome.

Project Team
Senior E&C management has to ensure that appropriate human resources are supplied at project inception, and that the key project members are compatible, within the project team and with the Client. Further, the project team should be based together, in one place, wherever the project focus is. A PM cannot be effective during the field construction phase when based in the comfort of the home office. The PM has to be constantly engaged, wherever the action is. Figures 3 and 4 illustrate typical project organization charts for the home office and for the field respectively.

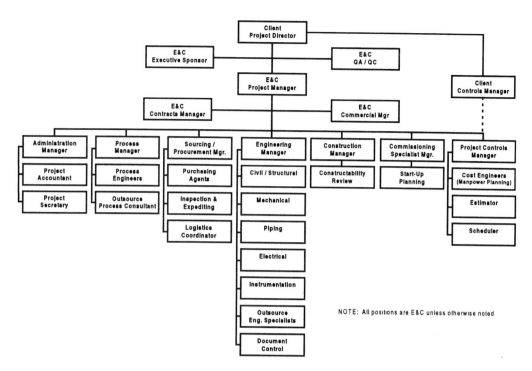

Figure 3 Project Organization Chart: Home Office

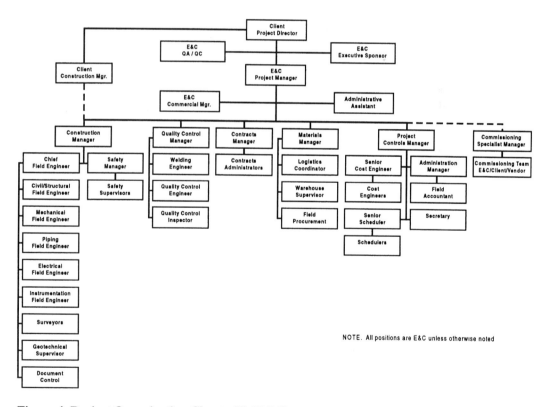

Figure 4 Project Organization Chart: Field Office

Cost control and planning manpower needs are, unfortunately, often under-budgeted, particularly in the field. As a generality projects need to budget, then assign, significantly more cost and control engineers than have historically been utilized; i.e. sufficient personnel capable of timely and accurately capturing project trends.

A key to smooth project execution in the field is integration of construction and commissioning personnel into the project team at the front end of engineering. Proper attention to construction and start-up planning during early engineering will positively affect many aspects of project outcome.

A Contracts Administrator is needed to handle the subcontractors on projects of any significant complexity or size (>$25 million); the PM cannot effectively handle these additional duties alone. For major lump sum projects (>$50 million) a Commercial Manager to interface with the Client should probably also be installed reporting to the PM, as shown in Figures 3 and 4.

Partnering

Project orientation needs to include formal team building. In larger projects, a formal team building exercise (partnering), led by a third party, should be included at project kick-off. Partnering builds upon the merits of aligning Owner stakeholders and outsource project support resources together to empower team success. It is important to preach and employ an integrated team approach throughout the project life.

- Partnering builds consensus and eliminates the "us" versus "them" mentality
- Partnering creates a trusting environment where issues are discussed openly and frankly
- Partnering allows a "best man for the job" philosophy to exist, and removes turf battles.

Project Procedures

Existing generic project procedures have to be modified to best facilitate the setting of 'project specific' requirements, such as mobilization, logistics, start-up, etc.

Projects need policing for quality. Efforts must be extended toward uniform application of best practices during all phases of a project. E&C management and/or quality audit teams should target at least one multi-day site review per quarter. Further, in the first three months of project initiation, it is useful to introduce an external reviewer to ensure proper procedures are established.

Insurance coverage (deductibles, caps, coverage limits) must be fully understood at project start. Insurance limitations should be noted as a reminder in the Project Manager's monthly report.

Formal dispute resolution tactics need to be established early in the project. This will not avoid Client and subcontractor confrontations, but it will facilitate a civil path through the issues.

Project Sponsor

The E&C project executive sponsor has to take his/her role seriously, and needs to be willing to commit time, particularly in cultivating Client senior management contacts and relations. While the sponsor role traditionally is mostly focused on Client concerns, it is suggested here that the E&C project sponsor also play a more active internal E&C role. The suggestion is that regular, though not necessarily frequent, internal meetings led by the PM take place, to include both the E&C sponsor and the senior project staff. If internal conflicts are present, this process can provide a mechanism to let those issues rise in a timely manner to the sponsor's attention.

Draw on Experience

The smart Client recognizes the merit of bringing into the project the wisdom of others who have "trod the path before" to:

- Benchmark critical components against external operating installations
- Incorporate constructability and commissioning reviews during engineering
- Conduct an external fatal flaw analysis

- Insert seasoned operations personnel into all project phases through to commissioning
- Remember, there are no old mistakes that can't be repeated.

Project Site
Site familiarity is critical. The engineering leads and the Construction Manager (CM) should walk the site together, prior to engineering start. Site conditions, equipment availability, plant tie-ins, Client preferences, labor rates, critical equipment deliveries, and construction package preferences all need to be established and understood by the team members, up front.

Project Information
Field construction disciplines (Controls, Procurement, Accounting etc. as shown in Figure 4) must retain a report link to their HO lead. While project disciplines always report to the PM via a line structure, each must also retain a reporting link to their functional HO department head.

Electronic transmittal of HO drawings to the field is a necessity today. The computers of all project team members on site need e-mail linkage and connection by LAN (Local Area Network) as well as WAN (Wide Area Network) to the E&C contractor's global Intranet.

CONTRACT
The discussion that follows is focused on the contract between the Owner and the E&C contractor. However, most of the points would also apply to a contract between the E&C and subcontractors.

Judicial Usage of Lump Sums
Lump sum bids should be used only when the project scope is well defined. It is a misconception that a "hard money" lump sum contract will shift the cost overrun risk from the Owner on to the contractor. Lump sum is merely an agreement for the contractor to provide, for the lump sum price, all services necessary to satisfy the contract battery limits. If limits are not precisely defined, all one gets from a lump sum, are change orders to the contractor's benefit.

Because Owner and Contractor interests are not identical, no contract is truly ideal. Various combinations of the lump sum (L) and reimbursable (R) elements are shown in Table 2.

Banks and junior mining companies prefer lump sum, in a mistaken belief that their financial exposure is capped, and that their risk is transferred to the E&C. Major mining companies are more pragmatic about the evolving nature of pertinent data within a grass-roots project; thus they tend to prefer cost reimbursable, at least through detail engineering.

Lump sums are not generally appropriate until after the feasibility study is complete. And, if value engineering is contemplated, a Client should ask that this be undertaken prior to fixing the lump sum amount.

The contract for lump sum work has to include sufficient detail such that it reads, to all parties, as an absolute fixed scope of work. Thus any "allowances" included within a lump sum submittal must be fully explained, defined, and agreed-to by the Client prior to project kick-off.

Table 2 Types of Contract

COST ELEMENT	CONTRACT TYPE									
	1	2	3	4	5	6	7	8	9	10
Design Services	L	L	L	R	R	R	R	R	R	R
Engineering Services	L	L	R	L	R	R	R	R	R	R
Management Services*	L	L	L	L	L	L	L	R	R	R
Equipment Supply	L	R	R	R	L	R	R	R	R	R
Material Supply	L	R	R	R	L	R	R	R	R	R
Construction Management	L	L	L	L	L	L	L	L	R	R
Construction	L	L	L	L	L	L	R	R	R	R
Indirect Costs	L	L	L	L	L	L	R	R	R	R
Profit and Overheads	L	L	L	L	L	L	L	L	L	R

* Management Services include project management, scheduling, procurement, expediting, logistics, cost control, and financial administration

L stands for Lump Sum; R stands for Cost Reimbursable

Type 1:	Represents the Turnkey Lump Sum contract
Type 2 through 6:	Represent Lump Sum contracts, Type 5 being the most common lump sum contract within the minerals industry.
Type 8:	Represents the "Fee Plus" contract.
Type 10:	Represents the "Cost Plus" contract where all elements are fully reimbursable. Profit and overheads would be expressed as a percentage.

Source: Project Management Manual for Mining

Use of Bonuses / Penalties

Bonuses for achievement of successful project completion are worthwhile. The bonus needs to be set up-front, when recipients have the maximum chance of influence over outcome. The bonus should not be set, however, until after project scope is set in an approved Feasibility Study. An incentive/penalty scheme should include both key E&C and Owner staff, tied to the following:

- Key milestone dates, including project completion
- Capital cost
- Safety
- Constructed facility performance

Incentives and penalties preferably should be symmetrical (equal amounts around the set target). A gainshare system of risk and reward for the Owner and E&C contractor/suppliers is best - where all participants have the potential to share the commercial benefit of a successful outcome and, equally, all have the potential to share the pain of failure. i.e. everyone wins, or everyone loses. (See Figure 5, Examples of Project Gainshare, courtesy of M. Entwistle, Kvaerner E&C 2001 - see footnotes, Reference 7.)

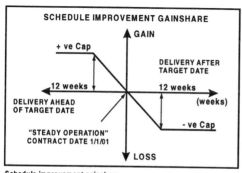 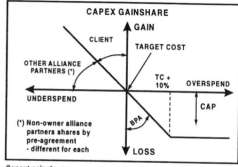

Schedule improvement gainshare   Capext gainshare

Figure 5 Examples of Project Gainshare Bonus / Penalty

A project incentive (bonus) program, if established, has to be fully understood by all parties at project start; and if the bonus is to act as a true incentive for performance improvement, then the bonus monies need to be paid to the participating parties immediately after project completion.

PROJECT MANAGER

An obvious critical component of project success is the assignment of appropriate project leadership. Odds favor a best candidate for the Owner's Project Director coming from outside the Owner company. A project leader needs:

- Familiarity with the geographic location
- Knowledge of the process type
- Experience in handling a project of this size
- Fluency with the "local" language.

Overcoming Client Preferences / Demands

Prior similar experience for the E&C Project Manager is also good. But, while a PM's similar experience is always helpful to others on the project team, it is not an absolute requisite (except, perhaps at a remote location site). At some level within the E&C team though, relevant experience does become valuable and even necessary. Otherwise, the Client probably would never have selected this particular E&C to perform the work. The E&C PM must therefore proactively embrace the relevant experiences of subordinates on his team.

The E&C organization must ensure that its PM is someone who can serve the project's needs. A good PM is one who cares about project success to the point of being passionate about it, and is willing to accept ownership of the effort. Moreover, a good PM is committed to the right approach and an honest and forthright manner of applying the skills necessary for success. A Client will always state that he prefers a manager with prior experience in the type of project being constructed. While "nice to have" this is not always warranted. More importantly, it may not necessarily satisfy the E&C contractor's need for adequate project management and control.

Project Manager Responsibility / Accountability

Project Managers (and similarly, Owner Project Directors) need to be given clear project responsibility then held accountable for project outcome, as much for properly informing corporate management of project issues, status and performance each month, as for delivering the desired result. PM non-performance, certainly repeated non-performance, needs to result in removal of the PM. The identification of problems within a project is not, however, synonymous with negative performance. By addressing problems, solutions are sought and implemented. A constant vigil toward problem areas and their management is the hallmark of a good PM.

If a PM is to be burdened with overall project accountability (which he should be!), then he must be similarly accorded overall project responsibility. Thus the PM, not the corporate office, is the person who should be accorded jurisdiction over who can be removed from the project team, and when. Personnel reassignment during a project life must be the exception, rather than the rule.

The PM also needs to be sole entity drawing on project contingency and escalation. Outside monetary draws on a project will kill PM accountability.

Project Manager Performance
A PM must have the all-around skill-sets to be capable of managing all aspects of a project, including procurement, controls, construction, start-up, and project closeout, i.e. not just the more frequently encountered front-end studies and engineering.

The PM needs to be fully involved with establishing, negotiating and closing the Client contract. The PM and other key personal must fully understand the contract, particularly scope definition, performance guarantees, and penalty clauses. The PM and the lead process engineer should review the contract before it is signed. Advice needs to be timely sought from all relevant quarters to avoid contract provisions that may inject undesirable implications.

A quality oversight process to measure the fulfillment of the PM's responsibilities needs to be visibly in place. Regular project oversight review helps prevent problems with project performance that could remain hidden to management until too late in the schedule to correct.

Pitfalls of Project Management
Many things can derail a project. As W. J. Tinsley noted in his 1999 IMM Presidential address, PM success must avoid focusing on apparent shortcomings; but rather it requires an addressing of the actual failings themselves. (See footnotes, Reference 8.) These pitfalls of project management have historically included:

- Insufficient planning for project execution
- Loss of emphasis on overall project goals
- Inattention to safety and quality
- Lack of focus on value
- Poorly defined scope, budget, and/or schedule
- Inadequate project control system
- Improper management of change
- Lack of understanding and acknowledgement of the effects of change
- Poor communication between parties / lack of trust.

ENGINEERING
Clear, full definition of technical concepts must be established prior to the start of detail engineering. This should be "a given", but all too frequently is not! Recognize that starting on detail engineering before process concept (i.e. PFD's, mass balance) is complete will result in rework and inefficiency.

To further minimize any possible rework, the "failure mode analysis" (Hazop) review, which is typically conducted at the 70% engineering stage, should additionally be undertaken at an earlier stage of drawings, i.e. prior to drawing number Rev. 0.

Engineering Checks
Budget and schedule must allow the appropriate manhours and time to conduct a proper check of drawings by engineering such that complete documents can be issued in accordance with the agreed schedule. Drawings must not be sent out to meet a preset schedule knowing that the drawings are incomplete, not properly checked, or that corrections will be needed. Accountability

for quality of engineering documents resides with the E&C department lead manager. These discipline leads have to enforce the protocols for checking as outlined in their own procedures.

Cost and schedule constraints may lead E&Cs to sometimes try and cut corners, to do insufficient internal checking. The PM must be constantly alert to not let any such quality compromises creep into the project. Design reviews must always be held; and such reviews need to be conducted with all affected disciplines. The results of design reviews must then not be allowed to be suppressed simply because they might cause delay or cause conflict with the Client.

Freezing Design
The design freeze must be a hard freeze, not a soft freeze, and have the commitment of the Client, project management and design engineering. One must avoid the endless desire to continue engineering to achieve "perfection". An inability to freeze design concept leads to engineering overruns and then to "knock-on" delays. A viable freeze starts with communicating expectations, and then consensually establishing appropriate budgets for each discipline at project start.

A freeze will not hold if unilaterally imposed. A freeze must be based upon sound technical judgment and not be arbitrarily undertaken to merely accomplish a shortsighted goal of satisfaction of say, a schedule milestone. If sincere doubts persist, then design must be revised.

Certified Vendor Drawings
Detail engineering should ideally proceed on the basis of approved vendor drawings. However, the Client must recognize that, today, critical vendor certified drawings are not truly final until after fabrication is released, in spite of the certified connotation from the vendor. Without this release, certified drawings can change. Thus it is a fatal flaw to claim 100% engineering prior to award of fabrication. Vendors do not release their own subs to fabricate prior to the E&C drawing release.

3-D CAD Models
Use of 3-D CAD models should be utilized to the maximum practical extent. While there is little, if any, engineering cost or time reduction from utilization of 3-D over 2-D, there is a real field construction savings from the better interference recognition of 3-D, and the better visualization of the ultimate facility layout. Exceptions need to be made, however, where there are limited 3-D CAD model skills in joint venture partners, to accommodate third party engineers, and for certain small projects. It must also be recognized that 3-D models are not always usable by a segment of the foreign fabrication shops or by all constructors in the field. Thus the E&C needs to be prepared to supply 2-D drawings when required, and not blindly insist on supplying 3-D models in all instances.

Engineering Leadership / Skill Base
Engineering skills serving the minerals industry are constantly aging; capability and expertise can erode over time. A project must not be allowed to suffer from a lack of guidance by experienced project engineers. While training programs and adding younger engineers into the project engineering talent pool are the best long term answers for the E&C, outsourcing often has to be embraced for the short term project good, particularly where internal expertise is lacking.

The PM needs to keep aware that the E&C contractor may not proactively accept the outsourcing of project work elements. Many E&C companies struggle to recognize that they do not contain all the specialized skills for every job, or that their mature engineers may not all be usefully experienced for the particular project in hand.

Remove the "Maximization of Billable Hours (cost reimbursable?) Mindset"
This is more than just recognizing the difference between lump sum and cost reimbursable at project outset. Cost reimbursable contracts, by themselves, are not open doors to an endless engineering effort. Schedule constraints exist on all projects; the best method of reducing overall project cost is a reduction in schedule along with the proper management of construction labor.

Timely delivery of engineering documents is key. Attempts to maximize billable manhours flow contrary to this objective and, in the eyes of many Clients, are constant points of contention. The number of manhours that a project requires is simply a tool for use in project planning, not an objective to be met.

With respect to lump sum contracts, all participants must understand the philosophy of "minimum adequate" billable manhours. What isn't spent, is additive to profit, a portion of which can be returned to the project team either through an incentive plan established at project start or through a performance reward plan distributed at project end.

Client Representative
Client personnel in the E&C office are to be encouraged, but must be managed. Clients must not be allowed to go to individual engineers and change engineering concepts without going through the E&C contractor's trend and change order process.

Selection of the control DCS/PLC system and configuration should always involve the Client.

PROCUREMENT
Procurement normally encompasses sourcing, purchasing, expediting, and then shipping and receiving. However these last two items have been separated out into a following separate Logistics section within this paper - to highlight their importance.

Sourcing
The first procurement step on most major minerals projects is to agree on a clear procurement plan, then identify all critical, long lead-time equipment and prioritize the sourcing of those items.

The next step is to prepare a bidder list for early Client approval. Most major E&C contractors maintain a database for pre-qualifying vendors/suppliers. This database needs to be referenced before the project team (PM, project engineer, process, procurement, Client) nominates their vendor list. This ensures optimal vendor selection. For schedule efficiency, the Client needs to place a person in the E&C office with authority to sign purchase orders.

Specific requirements of the financing institutions (e.g. Exim, EDC) that relate to tax credits, financing assistance, etc., must be identified prior to order placement, and then actively tracked.

Keep in mind when placing orders the need to maximize the standardization of components, i.e. define "smart spares." This needs to be addressed in both engineering and in procurement.

Purchasing
For domestic projects it is generally more efficient to purchase goods "FOB jobsite" as ownership of purchased goods then stays with the vendor until goods are received on site. The vendor is thus responsible for packing, preparation and accuracy of packing lists, inland freight, and delivery.

The terms of purchase must be determined as a strategy during the early stages of the project to account for overall insurance coverage.

Expediting
For all critical goods, the project must appoint project expeditors and prepare a shop expediting/surveillance plan with early warning status reporting of potential slippage.

Before critical goods are released for shipment from vendors, the material must be inspected, packing inspected and packing lists verified. Pre-shipment inspection at a supplier's facility does not typically include verifying packing lists because the goods are open for inspection and not yet packed. A second visit would thus be required for designated "critical" goods after packing is complete. If export packing is to be performed by others, then a new packing list will be needed.

"1 Lot as per Attached List" terminology is to be avoided as a line item on purchase orders. The use of "1 Lot" may be unavoidable, however, in cases of complex equipment as the buyer and seller may not know component breakdown until after the goods are produced.

Field Issues

The E&C contractor will normally furnish the large-quantity imported bulks (pipe, fittings, cable, etc.) to subcontractors as the E&C can generally do this at considerable cost and time savings for the project. Subcontractors (subs) should generally purchase the small bulk items, especially on overseas projects. Subs typically have better local knowledge; therefore shorts, damaged and small bulk items should remain their responsibility.

Procurement continuity must be maintained from home office engineering into the field. While it is generally beneficial for the responsible procurement person to move with the project to site, it may not always be appropriate. For example:

- If engineering continues in home office and it is necessary that purchasing remain close
- Where local agents can better perform the small value, rapid response field transactions
- For work overseas, where it is essential to employ a field agent with language fluency.

LOGISTICS

Offshore projects as well as large, logistically complex projects will require a logistics study. Such study usually results in a requirement for the services of a freight forwarder and a customs broker. To properly handle logistics, first one must understand the transit cycle. This probably depends upon the season and the state of the roads. Three days from port site may stretch to three weeks, or more, in the wet season.

Freight Forwarder

If a freight forwarder is utilized, one should award his contract early. A freight forwarder is chosen on basis of real (not claimed) experience and competency. Principal issues for selection are strength of representation in the country where the project is located, strength of representation where goods are being shipped from, and past successful performance.

The freight forwarder's scope of responsibility typically includes collection, inland transportation to port of export, seaworthy packing, export crating, booking of vessels, customs clearance, and inland transportation to site, as well as producing his own packing list based on actual packed contents. (Such list needs to be as accurate as the vendor's list.)

If the freight forwarder is tasked with subcontracting trucking services from vessel to jobsite, then the freight forwarder assumes responsibility for proper equipment use and rigging.

Receiving Port

It is vital to appoint a good, local transit agent at project outset. More important, make sure "your guy" knows how to work with government officials to move goods in/out of port. Develop procedures for seamless customs clearance.

If building one's own port, hire experienced port facility personnel early (harbor master, warehouse personnel etc.) Review and improve, as necessary, existing port infrastructure. Establish a heavy equipment off-loading facility as well as a "bonded yard" for "in bond" shipments at the port of entry.

To save transportation costs, try to use dedicated ships with consolidated cargoes. This will require establishment of consolidation marshalling yards, both at the port of entry and at the staging area.

Materials Tracking

The E&C procurement group typically provides a bulk material management procedure (usually a proven computer software module today) to track materials from MTO (materials take-off) estimation, procurement, shipping, warehousing through issuance of materials to contractors. The E&C logistics plan should provide specific transport routing instructions for each PO package, for each geographic area of supply.

Consolidated packing lists are needed on site for each container. The freight forwarder (or vendor) must provide a consolidated packing list per container. Site uses these packing lists to receive all pieces, for customs clearance and for tax clarification.

Be vigilant in ensuring that the vendor interprets PO specifications accurately. The logistics lifeline to a remote site can be four months or more; receiving the wrong part can ruin a schedule.

SPECIAL NEEDS OF THE REMOTE LOCATION PROJECT
While the project execution strategies outlined to this point could pretty well all apply to both domestic North American as well as to remote foreign locations, it is worthwhile looking at the unique special needs of the remote location, international project.

Local Knowledge
Application of knowledge gained from similar undertakings in the same region is paramount. Lack of familiarity with local conditions will negatively impact progress and construction productivity.

Understand the weather seasons. When are the rainy months, the droughts, the freeze-ups, and the thaws? In many locations, land (or river) transportation can be impossible in certain weeks of the year. Seasonal effects must be taken into account in the schedule.

Experienced Personnel
The key is to assign a resourceful, innovative, experienced site project team, preferably a team with relevant international construction experience, that has successfully worked together previously in a remote environment, i.e., one with a winning track record.

Ensure that there is a commitment of these selected personnel to relocate. If they don't want to be there, they won't succeed!

Procedures Set Early
A project execution plan must be in place prior to mobilization. Establish the work breakdown structure and control budget in the initial 90 days. The control schedule (up to Primavera Level 3) and overall manpower loading also has to be established early, with a build-in of job factors (labor skills) for local conditions. It is imperative to remain schedule-driven through all project phases.

Set the task force center in a civilized location, as close to site as possible, equipped with a computer and CAD network.

Get agreement on design standards (e.g. US, Australian, French, Russian); then make sure that everyone is cognizant of that agreement.

Pre mobilization, understand all project-permitting requirements through operations start-up.

Environmental Issues
Embrace Western environmental standards. Lending agencies require these anyway, so why fight them? As the 2001 Wellcome Trust Survey on public attitudes to science guided (see footnotes, Reference 9), for acceptance by society, projects need to respond positively to the public concerns of the non governmental organizations (NGOs.)

- Select designs (and fuels and reagents) that are environmentally friendly
- Don't make environmental compliance an engineering encumbrance
- Involve the environmental manager from Day One
- Force the alignment of the E&C contractor with the Owner's commitment to the local community and environment.

An open, pro-active involvement with the key stakeholders (including local community, state and federal agencies) does work. Such a strategy facilitated the successful execution of the recent

Alaska-Juneau Gold Mine closure project by the E&C firm Kvaerner, and Echo Bay Mines, along with receipt of the Year 2000 Alaska Governor's "Reclamation Project of the Year" Award.

Establishment of Initial Site

Undertake a familiarization visit to the project site by the core project team within the first four weeks. Establish a temporary facility and camp as close to the permanent site as possible. Adequate mobile equipment and transportation mechanisms need to be available from Day One.

The initial focus is not on the facility that the Owner wants built; rather it is the rapid establishment of the site for habitation. Thus, appoint a knowledgeable camp manager immediately - to get the initial, basic catering and accommodation facilities in place. A realistic understanding of the true availability of existing housing is crucial.

Data Gathering

Assess local road and beaching capabilities. Survey the site for available materials and resources (aggregates, sand, concrete, craneage, fab shops). Gather geotechnical and survey data.

Early Works

Upgrade site access roads at the project front-end. Time and effort expended on ensuring access suitability for equipment will eventually pay large dividends in maintaining schedule. Know where your earth-moving equipment spare parts are sourced. Don't wait for the first breakdown!

Place satellite communications as soon as possible. Land-based telephone lines in remote locations are generally unreliable. (It is OK to use local landlines as back up, however). Today, a project web site connecting site, Client, task force center, and support offices is becoming a must.

SAFETY AND HEALTH

The project safety, health and environmental plans need to incorporate the Client, local and E&C contractor standards. Safety can never be an afterthought.

Raise the Safety Profile: Have a Commitment to Zero

Every project needs a senior management commitment to a safe project and working environment. Statistics from the Construction Industry Institute of the USA show that when the company president and senior management regularly review construction safety performance, accident incidents reduce a staggering 86%. (See footnotes, Reference 10.)

It is also a good practice to make safety performance part of contractor remuneration. Mine construction safety performance is worse, on average, than mine operations safety performance. This does not have to be so; 11% of mining related construction projects completed over the past four years had zero lost time accidents (LTAs), e.g. Kvaerner's Solvay Soda Ash Project in Wyoming (1.2 M hours), Fluor Daniel's Cerro Verde Copper Project in Peru (3.2 M hours), Henderson Molybdenum's 2000 Project in Colorado (1.4 M hours). Every project's goal should be zero LTAs.

"Best practices" for achieving zero accidents are as follows:

- Demonstrated senior management commitment
- Proper staffing for safety (<50 site workers per safety professional)
- Site specific pre-project / pre-task safety planning
- Formal safety orientation, training and education (> 4 hours per month)
- Worker (and family) involvement (including safety perception surveys)
- Formal recognition and rewards (distributed at least bi-weekly for 0 LTAs)
- Subcontractor management (requirement for site specific plans)
- Accident / incident reporting and investigation (with sanctions for non-compliance)
- Drug and alcohol testing.

Health & Medical
A reliable supply of water is crucial, both construction and potable, particularly on the remote site. Know where the water is coming from before personnel mobilize. Problems with potable water supply will ultimately result in the outbreak of illness.

Have medical evacuation facilities in-place before the site team is established. Establish medical facilities onsite; at the very least, a first aid center - and, if in a remote location, preferably with a nurse and/or doctor.

TRAINING
A training needs survey is required very early in a project; i.e. a local job skills assessment coupled with workforce requirements (for both operations and construction crafts).

CONSTRUCTION
Set the construction-contracting plan with its engineering deliverables early, then establish the work packages, and structure the schedule (and cost estimate) around the work packages.

E&Cs typically furnish project licenses; they have to ascertain that all engineering, construction management and construction licenses necessary for the jurisdiction of the project are in hand.

Materials & Subcontractor Resources
In addition to standard vendor assessment criteria, construction management has to verify sources of raw materials, financial strength of suppliers and local subcontractors, as well as local fabrication shop capabilities. Local subcontractors are generally willing, but not always sufficiently competent.

- Maximize usage of local and regional resources. Work packages should reflect this resource base.
- Maximize offsite fabrication, plant packaging and/or modularization (e.g. substations, rebar cages, pipe racks). Match to port and site craneage, as well as to assembly yard facilities.
- Set up a bulk materials management entity, with multi-discipline support.
- On-site power needs to be sufficiently under project management's own control to be reliable.

Fast Track
Approach fast-track methodology of project execution with extreme caution. While a construction-driven project is normal, and makes sense (the majority of money is expensed in this project phase), fast-track has, in multiple hind sights, been used more as an excuse to delete candid front-end risk evaluation, and to justify insufficient planning and/or poor project controls by proponents of less-than-robust projects, than for real project execution improvement reasons. Lessons learned from fast-track projects indicate that:

- For "construction driven" fast-track projects to succeed, project management procedures need bolstering, not finessing. This is not the time to miss out any crucial project execution steps.
- Undertake constructability reviews during both the design and procurement phases. Take construction related activities into full account prior to site mobilization.
- Move a cadre of key experienced design engineers from the engineering phase to the field for the duration of the construction phase, to quickly clarify design interpretation queries.
- Quality assurance (QA) and quality control (QC) plans are always required. Remoteness of site cannot justify substandard quality.

COMMISSIONING

Install a separate start-up commissioning team. A separate commissioning team best assures that the construction contractor has truly achieved "practical" completion, and that the plant is really ready for the Owner to take "care, custody, and control."

- Commissioning team leader needs to be a "start-up" specialist, inserted for the task.
- A start-up and commissioning plan is needed by sequence and by task through both cold commissioning (each plant system initiated in isolation without any process product) and hot commissioning (integrated plant run-in with process product) on into production ramp-up. Such plan should identify all necessary spares, tooling and first fills.
- The commissioning team needs to integrate multiple participants (Owner plant operations and maintenance personnel, vendors, construction leads, and experienced start-up staff) within a defined strategy and hands-on roles.
- One should avoid lump sum for all the post mechanical completion services, such as start-up and commissioning activities. Scope, in these stages, is too susceptible to change.

PROJECT COMPLETION

Define "completion" at project start. To facilitate a smooth and painless demobilization, the E&C needs to establish as tight a definition of project completion as possible within the contract (particularly for mechanical completion). To achieve this goal, all parties must agree upon the completion criteria (and language) Many Clients are not cognizant of the differences between mechanical completion, commissioning completion, substantial completion, practical completion, first-product completion, ramp-up completion, performance test completion, etc.

It is best to set specific dates for release of design and field construction staff, to try and prevent key personnel leaving early for the next job's paycheck. Consider offering demobilization bonuses to key team members to encourage them to stay through completion release date.

Capital Savings

Capital under-runs go back to the Owner's treasury. These project savings aren't a free "slush fund" to be spent on non-project items. The Owner's operations team needs to be clearly told this at project outset.

Lessons Learned

Every project needs an honest appraisal of project outcome, i.e. a lessons-learned close-out report. This should be a written audit, conducted three to six months after the formal project turnover. The appraisal findings need to be fully circulated. Burying mistakes does no one any good.

Use the circulated close-out report as a catalyst to stop repeating the same errors again in the future. Incorporate lessons learned from prior projects into future projects.

PROJECT CONTROLS

Project control is a vital component of the project management system, and a big part of what Clients expect when they hire an E&C firm; i.e. the ability to organize and control a concentrated expenditure of money. Every project needs a timely, clear report of true project status from the project site each month and a clear forecast characterization of where it is trending. If an E&C company cannot do this, it cannot effectively manage projects.

Project Controls System

Project success requires adherence to a full suite of project controls throughout the life of the project. Controls have to be implemented at project outset; not part way through the project. The controls function is poorly understood by most Clients. Clients like to believe a "good" project manager alone makes the difference. Intelligent Clients know that good controls create good project managers!

A project control system is a complete and comprehensive process, whereby all aspects of project execution are monitored and reported against the originally approved scope of work, budgeted costs and project schedule. The foundation of the project control system starts with the "project execution plan". It then flows through the project procedures manual and project control reporting documents to the conducting of formal review meetings and the regular issuance of required management reports.

The project control system provides the timely data that gives the PM the basis for decisions regarding the activities of the project. Effective project control requires:

- A clearly defined scope of work at the front end of the project - and agreed to by all!
- An understanding of what is controllable
- Constant attention to the details
- Assignment of a controls manager
- Freeze of the scope of work, post project definition
- Circulation of regular weekly and monthly reports to all key stakeholders
- Full, open reporting of problems and risk issues
- Mandate of strict adherence to the change order procedure
- One person, the Project Manager, to approve change orders.

The goal in every project is "no surprises." This requires that a control schedule and a base cost estimate (control budget) be set early (in the first 90 days), followed by constant analysis and forecast trending.

Controls Issues between Client and E&C
A project is controlled by the E&C contractor to a set scope, not to a budget! This is frequently a very difficult concept for the Owner to grasp - Clients are used to solely thinking in terms of budgets.

The degree of Client approval needed in the various project execution steps must be established in writing and communicated to the E&C Project Manager and his staff.

It is best to establish the progress-reporting philosophy for both Owner and E&C contractor prior to project start. Set the types and frequency of report, agree on a trending procedure; and how "percent complete" is to be measured and combined between activities.

Project Changes
The field cost engineer's role is one of proper forward trending and planning, not of historical record reporting. Commitment reporting is mandatory and must be maintained current at all times. Procurement and contracts managers must route their data through project controls. Project control is a communications issue. It is imperative that cost engineers and accountants be kept apprised of project changes. An individual needs to be assigned to handle project controls in the field with the single responsibility of accurately getting the trends into the project reports on a timely basis.

All projects need a formal change notification and change control procedure established up front, set in the procedures manual, and then proactively adhered to. The trend notification form and the change log, itemizing those changes are the primary control documents. Project staff need to properly and fully utilize the trending mechanism; the PM has to have the backbone to formally write and then timely submit all resultant change orders immediately to the Client. Late or non-notification from field E&C personnel on the basis that a "delicate" or "friendly" Client relationship will somehow better facilitate future change submittals is simply unacceptable. Typically lump sum contracts require Client change notification within a specified, short elapsed time from the event. Delayed notification is most times tantamount to the E&C conceding any recovery for the change.

Contingency Management

A contingency / escalation drawdown management procedure must be formalized for all projects. Tracking of contingency is a project controls tool. Drawdown logs must be instigated on all projects. Tracking contingency (including escalation) against "remaining to commit" is a key indicator toward forecasting the health of a project.

Transparent contingency management is requisite. Non-contingency slush funds (escalation, growth etc.) need eliminating. All non-specific funds (escalation, growth, risk allowance etc.) need to be on one budget line, within contingency.

Audits

Routine audits should be conducted throughout the project life cycle. A project audit provides formal oversight that evaluates adherence to project management "best practices" and gives independent confirmatory evidence of project progress.

REPORTS

Regular weekly and monthly reports must circulate in a timely manner to all key stakeholders. These reports are an integral part of the controls function.

Monthly Report

For North American corporations, the "executive summary" and the "issues and concerns" section at the front of the monthly report has to be in English, as well as the host language. Each report needs an overall project status curve and table comparing "overall actual" against target each month, not just individual element progress. Each report also needs a contingency drawdown log.

Overall project progress is preferably reported on an earned work value basis, but if this isn't available, then manhours are generally the next best substitute, particularly for the front-end engineering phase. It is imperative to agree up-front on the relative weighting of the project elements, i.e. for the Engineering (E), Procurement (P), and Construction (C) activities utilized in reporting overall project progress.

Monthly reports have to be readable, accurate, short and concise. Summary costs, a schedule with a table of milestones, charts, graphs and photos are preferred over repetitious text.

Variance reporting within the monthly report must be timely. Shortfalls being buried as long as possible do the project no good. Honest variance reporting needs to be enforced. Negative issues have to be dealt with openly. Monthly reports are not external political documents published to give the Client a "warm and fuzzy" feeling. The importance of publishing honest reports needs to be established with the Client up front; particularly the pricing of cost trends, the delays in schedule trends, and the timing of when these are reported.

The monthly report due date needs to be a set (preferably the same) date each month. Accommodation, however, has to be given for incorporation of the E&C's accounting system into the monthly report. Client thus needs to agree up-front on each month's cut-off date. Due dates have to be met, even if gaps show in the monthly report.

Flash Reports

Single-page flash reports showing project KPIs (Key Performance Indicators) should be issued by the end of the first business week of each month.

Project Closure Reports

Close-out reports should be requisite on all major projects. Manhours should be allocated for this at project initiation. These reports are a vital road map to project performance improvement.

CONCLUSION
Careful attention to the following:

- Defining a clear scope prior to project initiation
- Setting of a control budget with appropriate contingency
- Establishing a milestoned schedule within a detailed plan of project execution
- Adherence to proper control and management of change over project life,

are all vital, fundamental ingredients to project success. With these elements firmly in place, then by following the strategies outlined within this paper, having the right team of E&C and Client individuals, and working under a partnered framework wherein all stakeholders are aligned to a common set of goals, any mineral project can be successfully completed anywhere on this globe.

ACKNOWLEDGMENTS
Certain strategies outlined within this manuscript are borrowed from prior unpublished works of the author and Terry Owen - VP Capital Projects, Inco Limited. Most of the strategies themselves were developed from real life Client experiences gained by Owen and Hickson while working on mineral projects around the globe with Kerr McGee, Freeport McMoRan, Cyprus Amax and Phelps Dodge. The author is grateful to the management and employees of all these organizations, as well as to prior employers, New Jersey Zinc, Asarco and Gold Fields, and to his present employer, the international E&C firm Kvaerner, for the education that they all so generously endowed over the past four decades. The opinions expressed within this paper, however, are solely those of the author.

REFERENCES
1. Hawley, R. February 14, 2001. *Making the best of Valuable Talent.* Hawley Group Report to the Engineering & Technology Board. London, UK.
2. Editorial, February 2001. *Leading the Field.* Engineering First Magazine - Issue #15. UK.
3. Hickson, R. J. July 31, 2000. *E&C Companies; the Mining Client Perspective.* Unpublished report to the Kvaerner Board, San Ramon, CA.
4. Hickson, R. J. May 18, 2001. *Mining Project "Do's and Don'ts" for E&C Firms and Owners.* Proceedings of 51st Annual MPD Meeting of the SME. Colorado Springs.
5. Hickson, R. J. March 1, 2000. *Project Management for Dummies; How to improve your project success ratio in the new millennium.* Preprint No. 00-133 in Proceedings of 2000 Annual SME Meeting. Salt Lake City. Published in SME Transactions 2000, Volume 308.
6. Owen, T. L., Hickson, R. J. September 1997. *Project Management Manual for Mining.* Unpublished manuscript.
7. Entwistle, M. Spring 2001. *Delivering Value - Clean Fuels Alliance in Australia.* Examples of Project Gainshare. Kvaerner E&C Bulletin #6, page 13.
8. Tinsley, W. J. October 1, 1999. *Change & Challenge.* IMM Presidential Address to Yorkshire Branch of Institution of Mining & Metallurgy. Pontefract, UK.
9. Wellcome Trust Survey, February 2001. *Science and the Public: A review of Science Communication and Public attitudes to Science.* Office for Science & Technology, UK.
10. Charles, R. January 24, 2002. *Best Practices Zero Accidents Program; Construction Industry Institute of USA.* Kvaerner Environmental, Health & Safety Champions Meeting. London, UK.

20

Start-Up, Commissioning, and Training

Section Co-Editors:
Ken Major and Mike Mular

Pre-Commissioning, Commissioning, and Training *T. Watson*	2277
Plant Ramp Up and Performance Testing *R.M. Nendick*	2285
Preparation of Effective Operating Manuals to Support Operator Training for Metallurgical Plant Start-Ups *S.R. Brown*	2290
Planning and Staffing for a Successful Project Start-Up *K.A. Brunk, L.J. Buter, K.M. Levier*	2299
Maintenance Scheduling, Management, and Training at Start-Up: A Case Study *P. Vujic*	2315
Operator Training *A. Vien*	2328
Safety and Health Considerations and Procedures During Plant Start-Up *L.A. Schack*	2337

Pre-Commissioning, Commissioning and Training

Timothy Watson, P.Eng., AMEC

ABSTRACT

The preparation for the start-up and commissioning of a new facility needs to begin at the time the project is conceptualized so that adequate time can be built into the project master schedule to allow for the ramp-up to full production. Significant planning and preparation are required by both the owner and the engineer throughout the various phases of the project to ensure the smooth and successful start-up of the facility. It is not enough to begin the preparation for start-up a few months prior to the introduction of fresh feed if the owner expects to go into the start-up with a well trained and confident operations and maintenance team. The overall schedule for the project needs to developed from back to front; how the facility will be started up must drive the sequence of construction which will in turn set the priorities for the engineering and procurement of the new facility.

The start-up and commissioning of a new facility is preceded by the following stages:

- Owner's education program
- Mechanical completion
- Pre-commissioning.

It is essential that the owner's education program start long before the new facility is mechanically complete and it should continue throughout the plant checkout and pre-commissioning phases.

Mechanical completion of a new facility involves the construction contractors and the owner's project management team with assistance from the pre-commissioning team. As each system reaches mechanical completion the pre-commissioning team will be responsible for the final checkout without the introduction of fresh feed. The introduction of fresh feed is usually the responsibility of the owner with assistance from the pre-commissioning team.

This paper will provide an overview to the procedures and the responsibilities of the various participants involved in the mechanical completion, pre-commissioning and commissioning of a new facility.

INTRODUCTION

The start-up of a new facility can be significantly enhanced by the development and implementation of a commissioning plan of approach. Some of the elements a commissioning plan of approach should address are:

- Commissioning sequence
- Mechanical completion, pre-commissioning and commissioning requirements
- Performance tests
- Roles and responsibilities during construction, pre-commissioning and commissioning
- Owner's education and training program
- Safety.

The development of the commissioning plan of approach during the initial phase of detail design will allow the requirements of plant commissioning and start-up to be built into the project master schedule. When the master schedule is developed, the sequence in which the plant will be commissioned should be the basis of the schedule.

This paper will discuss the various elements that are key in planning for the successful start-up of a new facility.

DISCUSSION

Commissioning Sequence

The development of a commissioning plan should begin during the initial phase of detail design and should continue throughout the detail design and construction phases of the project. The plan as to how the facility is going to be commissioned should play an integral role in the development of the project master schedule. The development of the master schedule should be based on the sequence in which the facility is going to be commissioned. The commissioning sequence will then drive the construction completion requirements, which will then set the priorities for the procurement and engineering activities.

The transfer of care, custody and control (TCCC) of a new facility to the owner will be facilitated by the use of a construction testing, pre-commissioning and commissioning plan. The basis of the commissioning plan is the identification of the systems, both process and non-process, into which the project can be sub-divided, that will assist in the orderly completion of construction and pre-commissioning ahead of plant commissioning. The identification of the systems should consider plant layout, process requirements, mechanical, electrical and control systems requirements and any specific contract and/or sub-contract requirements.

Depending upon the simplicity or complexity of the plant, the size of the plant and/or the requirements of the owner, the definition of the process and non-process systems can be presented in different ways. For a small plant with a relatively simple process, marked up flowsheets and single line diagrams can define the process and non-process systems. For a larger, more complex plant, the definition of the process and non-process systems can be accomplished by assigning each piece of mechanical equipment, each pipeline, instrument and electrical item to a process or non-process system. The presentation of this data should be supplemented using a narrative system description, color-coded marked-up process and instrument diagrams (P&ID's) and electrical single line diagrams.

To assist with the timely completion of construction and the transition into pre-commissioning and commissioning as soon as realistically possible, system-based scheduling should be used to drive the completion of construction. For the pre-commissioning and commissioning plan to be effective, the system definitions and a systems-based schedule needs to be presented to construction in a timely manner so that construction can complete their work on a system basis rather than an area basis.

Mechanical Completion, Pre-Commissioning and Commissioning Requirements

In the engineering and construction business, there are a number of different terms that have meaning in relationship to project completion and responsibility for the assets of the new facility. The terms may vary from industry to industry and country to country, but what is important is to ensure there is no misunderstanding with respect to who is responsible for the facility and the necessary testing requirements. For the purposes of this paper, the following terms will be used and defined: mechanical completion, pre-commissioning and commissioning.

Mechanical Completion. Mechanical completion is generally defined as the installation of the facility in accordance with the contract, drawings, specifications and vendor documentation. There are a number of activities that fall within mechanical completion, some of which are:

- Soils and compaction testing, particle distribution testing of backfill materials

- Torquing of structural steel bolts
- Hi-pot testing of power cables
- Ground resistance testing
- Die penetrant or x-ray testing of welds
- Cold alignment of rotating equipment
- Paint thickness testing
- Hydro-testing, flushing and re-installment of piping systems
- Chemical or mechanical cleaning of specialty piping systems
- Removal of rust inhibitors and installation of lubricant, greases and fluids
- Instrument calibration where possible
- Functional checkout of electrical equipment and instruments both locally and through the distributed control system (DCS) and/or programmable logic controllers (PLC)
- Motor run-in following the functional checkout of the motor control circuit
- Splicing and alignment of belt conveyors, belt feeders, apron feed feeders, drag conveyors, bucket elevators, etc.
- Final alignment of rotating equipment following motor run-in
- Completion of touch-up painting, pipe and equipment thermal insulation, pipe and equipment labeling
- Completion and turnover of all quality assurance and quality control (QA/QC) documentation.

Pre-Commissioning. Pre-commissioning is the period when all of the process systems are run on air and/or water without introducing fresh feed into the plant or facility. All the equipment is run through the PLC and/or the DCS in order to verify all personnel safety, equipment safety and process interlocks are functioning as intended.

In some instances it is not possible or desirable to run some pieces of equipment or entire process trains on air and/or water. In these instances it is still possible to verify a portion or all of the control system through the use of a process simulator or simulating the individual control loops utilizing signal generators.

The pre-commissioning period is a good time to introduce the Owner's maintenance and operating personnel into the checkout team. Maintenance personnel can assist in collecting data while the equipment is running on air and/or water. The data collected during this period can form the basis of the ongoing plant condition monitoring program. The control room operators can begin to get hands-on operating experience on a system-by-system basis without the pressure of running with feed.

Pre-commissioning is the time when equipment vendor representatives assist with the checkout of individual pieces of equipment or entire process trains. The use of vendor representatives ensures the equipment is checked out and started up in accordance with the manufacturer's requirements and also ensures the equipment warranties are not violated.

Some of the activities that occur during pre-commissioning include:

- Verification of all personnel safety, equipment safety and process interlocks
- Drop tests can be performed to verify the accuracy of level transmitters, flow transmitters and pump capacities on water
- Conveyors should be run empty, the belts trained under no load conditions and the weigh scales should be calibrated
- Nuclear density transmitters should be zeroed on water

- Motor power and/or current draw and equipment baseline vibrations should be recorded while the equipment is running on air or water
- Preliminary loop tuning should be carried out where possible. It is important to realize that the majority of the loop tuning can only occur when the facility is running on actual process fluids.

Some of the other activities that occur during pre-commissioning to get the plant ready for the introduction of fresh feed include:

- Dry out of refractory lined fireboxes for burners or dryers
- Dry out and heat up of refractory lined furnaces and reactor vessels
- Chemical curing of brick-lined hydrometallurgical vessels
- Chemical cleaning of boilers
- Cleaning of steam piping.

Generally, the completion of pre-commissioning activities signifies the transfer of care, custody and control (TCCC) from either the engineer or the contractor to the owner. The TCCC from the engineer or contractor to the owner is usually accompanied by a number of process system turnover packages. The turnover packages should contain the QA/QC documentation that was prepared by the contractor as part of their mechanical completion requirements, the pre-commissioning test reports, any reports prepared by specialty vendors and a copy of the contractually agreed upon as-built drawings.

Commissioning. The start of the commissioning period of a new facility is usually signified by the introduction of fresh feed and extends until the new facility has reached full design production or a percentage of design production. The owner is usually responsible for both the facility and the necessary operations and maintenance activities during the commissioning period. Many times a new facility will ramp up to full production quickly without experiencing significant difficulties. At other times, the time to get a new facility to design production may be extended due to a number of different factors or a combination of factors, for example:

- The bulk sample taken for the testwork program may not be representative of the entire ore body
- The friction factor of the flowing fine ore or concentrate may not have been incorporated into the design of fine ore or concentrate mass flow bins
- Screen harmonics may not have been dampened properly when supporting one or more vibrating screen within a steel structure
- Pumps that require their motors or impellers changed due to the solution or slurry specific gravity or froth factors being different than those used for the design
- Atmospheric or climatic data may not have been incorporated into the design correctly.

Performance Tests
Some contracts may have specific performance test provisions that dictate the facility is run at nameplate capacity, or a percentage thereof, for a period of time. The performance test is usually run after the facility has been commissioned and started up and is conducted by the owner with assistance from the engineer or process licensor. During the performance test it will be necessary to record confirming data. To ensure the accuracy of the data critical instruments should be re-calibrated prior to the commencement of the performance test. If the method of calculating and evaluating the results are not defined in the contract, they should be agreed to before the initiation of the performance test.

Roles and Responsibilities During Construction, Pre-Commissioning and Commissioning

The roles and responsibilities of the owner, engineer and contractor during construction, pre-commissioning and commissioning should be contractually defined and understood by all.

Construction. Through to mechanical completion the contractor and/or contractors have prime responsibility for the assets of the new facility. The individual contractors should be expected to use the process commissioning sequence and scheduling information in planning their activities and work with other contractors to complete large multi-contractor systems. The project quality assurance/quality control (QA/QC) documentation should be utilized to document the various tests or measurements conducted throughout the construction period.

The main interface with the contractors is the project construction management team. As construction progresses, members of the pre-commissioning team should participate in the functional check-out of all equipment and control loops by providing assistance to the contractor through the operation of the DCS and/or PLC. Having members of the pre-commissioning team assisting the contractors with the functional checkout can reduce the overall duration of the schedule by eliminating the independent verification of the plant electrical and control system wiring.

Pre-Commissioning. The completion of the last 15 to 20% of construction should occur on a system basis rather than an area basis.

This approach allows pre-commissioning activities to commence as early as possible on the project because the remaining construction activities will be completed on a prioritized system basis.

Prior to the pre-commissioning of a system all the pieces of equipment and components that make up the system should be visually inspected for completeness and assurances should be provided that all the required QA/QC testing and documentation has been completed.

Utmost care needs to be exercised during the pre-commissioning period as new equipment is put into service or heated up. Many of the construction workers may still be on the job completing the last of the systems or punch list items and the operation of equipment may present many new hazards to the construction workers.

The pre-commissioning of a system should continue for a period of time necessary to prove out the system. The system should be considered complete when all the mechanical, piping, electrical, instrumentation and control systems have been operated, checked and calibrated, and are in a state that commissioning and/or start-up can commence.

During the pre-commissioning period, the project construction manager is usually still responsible for the site, but there is significant interface between the construction management team and the pre-commissioning team to ensure all tests are properly scheduled, coordinated and conducted in a safe manner. The pre-commissioning period is a good time to begin to get some of the Owner's operation and maintenance personnel involved as part of their education program prior to start-up.

Commissioning. The commissioning phase of a new facility usually commences with the introduction of fresh feed into the facility. During the commissioning phase the owner is usually responsible for both the operation and maintenance of the new facility. It is not unusual for members of the pre-commissioning team, construction management team and selected members from contractors to be requested by the owner to provide support during the commissioning phase. The additional manpower during the commissioning phase of a new facility is required to assist with those activities that are not usually encountered during the normal operation of an existing facility, for example;

- PLC or DCS support may be required to assist with the final loop tuning once fresh feed has been introduced.
- For those facilities containing horizontal grinding mills, millwright support may be required to re-torque the head and shell bolts after the grinding mills have operated for a defined period of time.

- For plants containing equipment driven by V-belts, the V-belts will need to be re-tensioned a number of times as the belts go through the initial stretching.
- PLC and/or DCS support may be required to assist with troubleshooting configuration problems that were not identified during the pre-commissioning period.
- Variable speed drives may need to be re-ranged once actual operating conditions have been encountered during the initial commissioning period.

Over and above the engineering and craft labor support that is required to deal with the additional work load that normally occurs during the commissioning period, problems may be encountered that also require additional support. Examples of the types of problems that may arise during the commissioning period are:

- Replacement of screen decks on sizing screens with improperly sized apertures
- Replacement of pump impellers or motors that were improperly sized or that require re-sizing due to the actual conditions being different than those allowed for in the initial system
- Pipes or pipelines that were improperly sized
- Premature replacement of wear components because the ore or slurry is more abrasive than initially anticipated
- Premature replacement of refractory lining systems because they were over stressed or cycled during the initial commissioning period
- Replacement of instruments, electrical or mechanical components that failed prematurely during plant commissioning.

Despite the various problems that may be encountered during the commissioning period, a well-trained operations and maintenance staff provided with some additional support can deal with most situations quickly and efficiently.

Owner's Education and Training Program
The start-up of a new facility that has been designed and constructed properly will be significantly enhanced by an operator team that has been well trained and possesses the skills and confidence to operate and maintain the new facility. To accomplish this goal the owner's education and training program needs to start long before the facility is mechanically complete and should continue throughout the pre-commissioning and commissioning phases and throughout the life of the operation. Depending upon the size of the operation, a full-time training coordinator should be an integral part of the owner's operating team. To be effective, a comprehensive training program should consist of the following elements; classroom training, on-site practical field training, and possibly off-site training.

As part of the overall training program, the training opportunities that exist during mechanical completion and pre-commissioning should not be overlooked. Examples of the many different training opportunities that exist as part of mechanical completion or pre-commissioning are summarized below:

- During the functional checkout or bumping and running in of motors the control room operators can participate in the checkout by calling up the appropriate PLC and/or DCS screen to initiate the start or stop signal for the motor. As the plant checkout progresses from functional checkout and the bumping and running of motors to the pre-commissioning activities with entire systems or sub-systems, the control room operators can obtain significant hands-on experience prior to plant commissioning or start-up.

- More and more in the design of new facilities the owner's people responsible for the plant control system are participating in the initial configuration of the PLC and/or DCS. Their participation may include the development of the operator interface displays, controller configuration or possibly the configuration of the entire control system. Whether their involvement in the initial configuration is minor or significant, they should be expected to participate in the verification of the plant control system during the functional checkout and pre-commissioning phases. Having the owner's personnel participate in PLC and/or DCS verification will provide them with the confidence that the plant control system was configured as intended and the various input and outputs (I/O) are reporting to the correct locations within the control system.
- During pre-commissioning many equipment vendors are brought to site to assist with the checkout and initial operation of individual pieces of equipment or entire processes. When the vendors are onsite, it is an ideal time for both the operations and maintenance staff to receive additional classroom training or field training for that specific piece of equipment directly from the factory representative.
- For electricians and instrument mechanics there are a number of different pre-commissioning activities they can participate in that will enhance their training ahead of commissioning, for example;
  - Field calibration or calibration verification of instruments
  - Trouble shooting of field devices that may not be functioning properly
  - Assist with recording data during the initial equipment run-in
  - Assist with trouble shooting protection relays for medium or high voltage switchgear or large horsepower motors.
- Similarly for mechanics and millwrights hands-on training opportunities exist during pre-commissioning, for example;
  - Assist vendor representatives with checkout and start-up of specialty equipment
  - Assist in collecting and recording baseline vibration data that will be used as part of the ongoing condition monitoring
  - Many pieces of equipment need to be re-greased or have the oil changed after the first few days or few hundred hours of operation. Assisting with this exercise can provide hands-on experience with many different pieces of new equipment
  - During the initial run-in of conveyors the belts will need to be trained.

In summary, the functional checkout period of mechanical completion and the various activities during pre-commissioning can provide excellent hands-on training opportunities for the owner's operation and maintenance staff. Coordinating these activities with the owner requires a considerable amount of effort as many other activities will be ongoing simultaneously as the owner is preparing his people for start-up.

Safety
The conditions that exist during the construction phase of a project are well understood by the construction and safety professionals within the industry and the necessary work processes and procedures are implemented on construction sites to assist with worker safety. The hazards that arise during the pre-commissioning and commissioning phases of a new facility are different from those encountered during construction and special precautions should be taken to safeguard the remaining construction workers and operating personnel.

On most sites, a lockout tag-out procedure is used to assist with the identification of responsibility for an individual piece of equipment. The same procedure also identifies the potential hazards associated with a piece of equipment during the testing phase. For example purposes, a three tag lockout tag-out system utilizing the colors red, orange, and green will be reviewed.

As construction progresses and the construction testing phase is approaching, red tags should be attached to all pieces of equipment – mechanical or electrical, all pipeline isolation valves, and all instruments. The red tag indicates the piece of equipment is under construction and under no circumstances can the piece of equipment or pipeline be energized or put into service. As construction progresses and the piece of equipment goes into the testing phase, the red tag is replaced by an orange tag.

The orange tag indicates the piece of equipment – mechanical or electrical, valves, or instruments – can be energized or put into service at anytime. The orange tag is a visual indication to the construction workers that they are near a live or potentially live piece of equipment. Should they be required to work on that piece of equipment, a work permit is required. As an additional safeguard during the testing phase, precautions such as barrier tape should be used when running in a motor or performing a load test on a piece of equipment. Upon the completion of the testing phase and the individual piece of equipment being turned over to the owner, the orange tag is replaced with a green tag.

The green tag indicates that the piece of equipment – mechanical or electrical, pipeline or pipeline isolation valve or instrument – has been turned over to the owner. Should the contractor be required to work on a piece of green tagged equipment, a work permit should be obtained from the operations group. Whenever a work permit is required to work on either an orange or green tagged piece of equipment, it is important to follow the appropriate lockout procedure.

The potential hazards associated with the initial commissioning phase of a new facility may be far greater than a facility that has been in operation for an extended period of time. The increased risk may be associated with lack of familiarity of the new facility by the operations and maintenance staff, or increased congestion on-site due to the additional personnel retained to assist with plant commissioning. During this initial commissioning period, it is essential that the owner's safety personnel be fully trained in all aspects of fire fighting, rescue and evacuation, utility isolation, and the health hazards associated with various reagents that may be used and stored on site.

Plant Ramp Up and Performance Testing

Roger M. Nendick [1]

ABSTRACT

The economic success of a mining project can be significantly impacted by the rapidity with which the plant gets to its full operating performance, both in terms of production and cost. The time taken for the plant to get to full production from the time ore is first introduced into the facility is referred to as the ramp up period. In extremes this time can range from days to years.

The financial institutions providing the capital for new mining projects recognize the importance of this ramp up period to the overall project economics and usually insist on a performance test to demonstrate that the facility is operating as per design. There is usually a significant financial incentive for the owner to pass this test and so, in turn, the owner normally has a performance test as part of his contract with the E&C Contractor.

The organization of the Ramp Up, and the execution of these performance tests becomes a mini project in itself.

INTRODUCTION

A mining company makes a decision to proceed with a new mining project. Funding for the project has been arranged with a consortium of financial institutions who, in an attempt to place some assurance on receiving their money back, have imposed a performance test on the mining company to ensure that the property performs according to its financial parameters.

As an incentive for the mining company to pass that performance test, there is usually a financial incentive in the form of a reduction in interest rate, or changing of the loan status from recourse to non-recourse once the test has been completed satisfactorily.

The mining company then negotiates a contract for the construction of the facility with an engineering and construction company (contractor). In order to reduce its risk, it passes on as much as the performance test as it can to the engineering and construction contractor; how much it is able to pass onto the contractor is all part of the contract negotiation.

With the contract signed, the project proceeds to construction stage. The sequence of events during the course of construction are as depicted in the bar chart shown below.

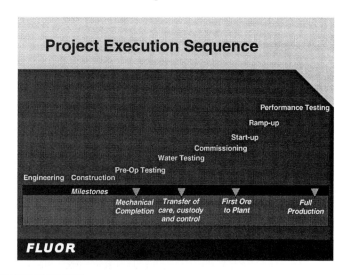

---

[1] Vice President, Consulting & Studies, Fluor Mining & Minerals

The final months of construction are typically a frenzy of activity on both sides:

- The contractor is busy trying to finish the physical construction and to do the pre-commissioning of the various units of the plant in order to hand them over to the owner;
- The owner in turn is trying to recruit the necessary labor force and to order in the spares and operating supplies for the operation of the plant.

Often, but not always, insufficient attention is given to the Commissioning of the plant, the ramping up of the plant and the execution of the performance testing. As a result the commissioning and ramp up take place with much anguish and acrimony, and the performance test is forgotten about until some one turns a page in the contract and realizes it has to be carried out.

The above describes the nightmare, but one that is all too frequent in the history of new mining projects.

For the rest of this paper I would like to deal with how the agony of this kind of start up can be avoided.

THE CONTRACT

The ideal situation would be that the owner has identified his metallurgical superintendent and that the contractor has identified his commissioning leader and that both of these people are involved in the contract negotiation, particularly with regard to the performance test.

If this is not done, then frequently the terms of the performance test will render it extremely difficult and costly to perform. The practicality of executing the performance test needs to be considered at the time of writing the contract. Lawyers are ignorant of plant operation. The test needs to be designed to minimize the disruption of normal plant performance. As an example, tests that require samples to be taken from the SAG mill feed belt should consider that stopping, locking out, sampling and restarting the belt can take as long as 30 minutes, during which time there will be no feed to the mill. This alone is equivalent to over 6% downtime at only one sample per eight-hour shift.

Test Duration

Typically Performance Tests are specified for two time frames:

- A long duration test typically for 60 days but sometimes 90 days or more, which requires the plant to operate at an average of 90% of its rated capacity during this time;
- A short duration test where the plant is expected to perform at 100% or even 110% or 120% of its average rated capacity.

The purpose of the latter test is to ensure that the plant has some catch-up capacity to make up for occasions when it has performed below the average rate.

Definitions

It is absolutely vital that the expectations of plant performance are clearly spelled out in the Contract.

Throughput must be specified in terms of average values and at peak design values, and the derivation of the plant availability should be clearly spelled out. The performance test expectations must clearly state which values are being measured for each test.

Where a recovery and product quality performance test is also to be provided, the ore grades and the expected impurity levels, as well as the physical and lithological characteristics of the ore must be clearly specified. It is prudent for the owner at this time to make sure that the mine can supply ore with the required physical and chemical characteristics at the time that the performance test will be conducted. If the mine plan does not permit the mining of this typical ore when required, then further discussion and debate will be required at the time of the test in order to agree upon acceptable feed material.

As well as specifying the main throughput parameters for the entire plant, the contract should also spell out the expectation for the various unit processes within the plant. In a typical flotation concentrator, for example, the capacity of the concentrate filtration plant may well be significantly different from the design average, in order to account for varying ore grades during the life of the mine.

Responsibilities

The contract must clearly spell-out who is in-charge of each stage of the project. Typically the contractor has responsibility for the plant up until the end of pre-commissioning. When the pre-commissioning packages are signed over to the owner together with care custody and control, they become the owner's responsibility. When all of the pre-commissioning packages have been handed over the owner generally takes control of the complete facility along with responsibility for the commissioning. An exception to this is if the plant has been built on a lump sum turn key basis, in which case the contractor may have responsibility up until the completion of the performance test.

Whatever the case, it is important that both parties understand exactly who is responsible so that responsibility for equipment damage, along with insurance responsibilities, is clearly understood.

ENGINEERING

Once the contract is signed and the legal details of the performance test agreed, it is important that it is not forgotten during the engineering process.

Frequent changes may be made to the plant design, or to the scope of the facilities, that can impact on the performance of the test. The owner and contractor should schedule meetings approximately monthly to discuss progress on how the test will be executed, as well as any engineering matters that have arisen that impact on the test. Planning for how the test will be executed can be started at this early stage.

At a point midway through the engineering, the metallurgical superintendent on the owner's side may wish to appoint a performance test manager whose sole responsibility is the execution and passing of the performance test. Similarly on the contractor's side the commissioning manager for the contractor may also wish to appoint an individual to specifically look after the performance test. The planning and execution of the test is a mini project in itself. As an example of this, the performance testing for a major concentrator in South America which was undertaken recently, necessitated the hiring of 18 additional temporary staff on the owners side in order to handle the sampling, sample preparation and analysis of the samples required for the test. The requirement for sample storage in the event of dispute over the test result required the purchase of 1800 steel drums for bulk samples, and a shipping container for pulverized samples. The tests themselves required individual tests at locations at the concentrator, the filtration plant and the port shipping facilities. These facilities were geographically spread over a 1000km distance.

COMMISSIONING PLANS

It is never too early to start the commissioning plans for a project. Although the commissioning plan is typically the responsibility of the owner, it is as well that he also involves the contractor to get his buy in to this important phase of the project.

It is often claimed that commissioning plans are futile because "we don't know what is going to happen once we push the button". The inference of this is that it is impossible to plan for the unknown. There are many advantages to the preparation of a commissioning plan:

- It focuses the attention of all parties on the task that lays ahead;
- It enables plans to be made for the recording of the many items of data that need to be collected about the performance of individual pieces of equipment;
- It provides the basis for a platform from which to build in the eventuality that something does go wrong;

- It provides a document that will become a written record of all the events that took place during commissioning.

Items that need to be considered in a commissioning plan, include:

- If there are is more than one module in the plant, which module will be started first (obviously this will require liaison with the construction and pre-commissioning teams);
- Will the plant be started at its full tonnage for each module or will it be started at a reduced tonnage;
- Are there any special ore feed requirements that need to be addressed, for example special ore requirement for bedding in of stock piles or bedding in of thickeners etc;
- What will be the water supplies for start up of the plant? Frequently there is no return dam water available during the initial operation of the plant and this requires additional fresh water supply to be available.

RAMP UP

The period from the first introduction of ore into the plant through until the plant reaching its ultimate design capacity, both in terms of quantity and quality of product, it is referred to as the ramp up period.

Depending on the complexity of the design and the degree of preparation and planning that has gone into the start up, the time necessary for this to be achieved can require anything from a few days to 2 or 3 years to reach its full capacity. There are no hard and fast rules for defining this period, but an excellent paper by McNulty[2] provides good guidelines. A graph from that paper, reproduced below, shows the ramp up times derived from studying the case histories of various types of plant. Series one through series five shows increasing plant complexity from simple flotation concentrators to integrated hydrometallurgical facilities.

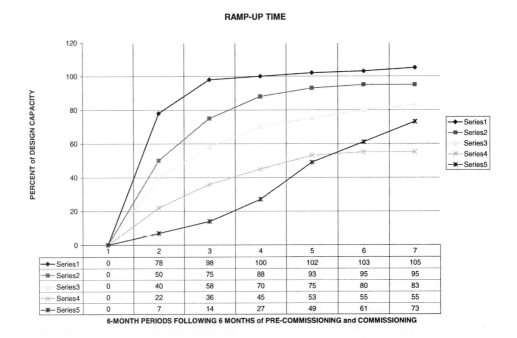

---

[2] "Developing Innovative Technology" by Terry McNulty, Mining Engineering, October 1998.

As the tonnage is gradually being increased, bottlenecks will appear in various areas of the plant. Below is a listing of some of the more frequent bottlenecks to production, but each plant ramp up will have its own particular array of problems. General items include:

- Wrongly sized pumps (This is not always that pumps are too small. Frequently pumps that are too big can be as big a barrier to production);
- Undersized motors for equipment.
- Poorly designed transfer chutes that either plug or wear out too quickly;
- Pipelines that sand out or wear out preventing reliable material transfer;
- Excessive spillage and inadequate sump pump arrangements (Note: this is difficult to design for, as some pumps that are adequate for handling routine spillage when a plant is in normal operation may be inadequate for the excessive spillage that often occurs during the start up. Although this can be an irritation during a plant start up it is as well not to overreact and install massive pumps that become a problem when the plant reaches routine operation).

RECORD KEEPING

It is extremely important that both the owner and contractor keep good and accurate records of events during commissioning. The owner should invite the contractor's representative to attend the daily production meeting for the operation and accurate minutes should be kept at these meetings. A further meeting should be held on a regular basis between the owner and the contractors commissioning representatives to set the priorities for repair, rectification and punch-listed items. Demands on maintenance crews during the project ramp up are often excessive. It is prudent for the owner to arrange for a temporary SWAT Team of maintenance people to attend to project work arising from the ramp up.

In this way the regular maintenance requirements of the plant do not get behind such that the long-term reliability of the equipment suffers.

THE PERFORMANCE TEST

By the time the performance test is ready to be conducted, the plant should be operating in a reasonably stable manner. Notification of the date of the test needs to be agreed a few weeks in advance, so that the necessary arrangements can be made for the test. Also the Independent Engineer for the financial institutions needs to be given time to get to the property to hold some preliminary meetings. The independent engineer becomes a third member of the team conducting the performance test and will need to be given full sets of records for each day's production.

Frequently the performance test also requires the plant to perform within certain cost parameters and this will necessitate the Independent Engineer being given data on reagent consumptions, power consumptions and other components of the operating cost. Hopefully sufficient trust is built between the three members of the team, such that all individuals do not consider it necessary to be present on site throughout the entire test, i.e. 60 days or more.

At the end of the test period it becomes a major exercise to gather together all of the information and produce a performance test report. Explanation of any unusual events that have occurred during the performance test may be required if the requirements have not been exactly met during the 60-day period. A well documented and written report may obviate the need for re-performance of the test with its attendant cost.

CONCLUSION

The success of both plant ramp up and performance tests can be reduced to two 'P's:

People and Planning

The Planning needs to start before the contract is signed and continue right through the execution.

The People can cause failure or create success, despite everything that has gone before. It is close to impossible to successfully ramp up a plant and conduct a successful performance test unless there is close cooperation between all parties.

Preparation of Effective Operating Manuals to Support Operator Training for Metallurgical Plant Start-Ups

Stephen R. Brown

ABSTRACT

Effective plant operating manuals used in a formal training program can make the difference between a successful start-up and a failure. Once the plant process design and control strategies have been fixed, equipment has been ordered, and the plant is under construction, the only major variable affecting success is the capability of plant operating personnel. It is essential that the myriad details concerning plant operation are documented in comprehensive operating manuals suitable for training the non-technical personnel who will operate the plant. This paper describes the best approach for producing the operating manuals and conducting operator training.

WHAT OPERATORS NEED TO KNOW

Performance Associates has been in the business of assisting mining companies to successfully start up new mineral processing plants for the past 18 years. Based on this experience, it is obvious that certain, key information must be known—and applied—by the operators and front-line supervisors during the start-up. Failure to impart this information, and to apply this knowledge during the commissioning phase, will likely result in either outright failure or a long, agonizing, and protracted effort to achieve design capacity—if design capacity can be achieved at all. This key information consists of several elements.

Process Unit Operations

First, it is essential that the operators have a conceptual understanding of the process and the principle of operation of each major unit operation in their area of responsibility. Conceptual knowledge allows for more effective reasoning when process upset conditions occur. Rather than attempting to provide a recipe covering any conceivable upset, the operator's conceptual knowledge will allow for drawing the appropriate conclusions based on the situation at hand. Specifically, the following elements concerning the functioning of individual unit operations should be documented and thoroughly understood by front-line supervisors and operators.

Objective. Describes the purpose of the unit operation. For example, the objective of a ball mill circuit is to reduce the size distribution of feed material to allow for liberating minerals locked in the host rock.

Basic Theory. Describes the chemical, mechanical, electrical, etc., methodology (e.g., magnetism, chemical reaction, mechanical action, differential density) used by the unit operation to effect the objective without reference to the physical layout of the equipment.

Principle of Operation. Describes the physical layout and how the basic theory is applied by the actual equipment being described. The inclusion of diagrams, photos, etc., as necessary, to illustrate the important principles of operation is required.

Critical Variables. In any unit operation, the output quality is a function of certain critical variables. For example, cyclone feed density and pressure drop affect how effective a cyclone is at making a size differentiation of the feed slurry. This element identifies the important variables associated with each unit operation described.

Safe Job Procedures

To ensure each employee works safely, information on the correct methods for performing potentially hazardous jobs must be learned. Each new plant operation will contain potential hazards that must be understood by every employee. These hazards include working with various reagents, such as sodium cyanide, caustic, sulfuric acid, etc. They also include working around various types of moving equipment.

Process Control

Each operator must also fully understand each control loop in his area of responsibility. This understanding includes the variable being controlled, the instruments and control strategy employed, how to recognize when control problems occur, the backup options available, and when it is appropriate to exercise those options. Distinguishing between process problems and process control problems is also important. This understanding has become more difficult over the recent past since control strategies have become increasingly complex with the advent of more and more powerful process control software.

Interlocks

In addition to the control loops, all interlocks must be thoroughly understood, including how interlock logic is affected by various operating parameters, such as remote operation, local operation, maintenance operation, etc. We have seen a very significant increase in the complexity of plant interlocks being designed into new plants.

Alarms

Once the process and its critical variables are understood, along with the controls and interlocks, the operator must then learn the fault, cause, and remedy associated with each alarm. This learning can be a tall order since many of today's new plants have literally hundreds of programmed alarms in each plant area.

Start-Up and Shutdown Procedures

Each operator must also learn the correct steps to start up and shut down the plant under various conditions. These conditions normally include: start-up from complete shutdown, start-up from standby shutdown, start-up from power failure, and start-up from emergency shutdown. Additionally, each operator must know how to manipulate the distributed control system (DCS) to determine what is happening in the process, to take control of a particular controller, to adjust set points, etc. Operators must also know how to effect control using any local control panels in the plant. These panels are typically used for packaged boilers, samplers, and solution heaters. The new complexity of metallurgical plants make all of these procedures much more involved than they used to be.

Operator Tasks

Finally, each operator must learn other operating procedures associated with optimizing the plant. These procedures include such tasks as checking pulp density, optimizing flotation cell performance, tapping a furnace, taking solution samples, conducting routine inspections, and ensuring that the plant operates within permit requirements.

Conclusion

Operators having any limitations in the above described knowledge will cost the company during the start-up and subsequent initial operation—the more the knowledge gap, the more it will cost.

In plant start-up situations where very little of this knowledge has been transmitted to operators, the start-up can be little short of disastrous. In a typical start-up scenario, personnel react inappropriately to process upset conditions, causing further upsets. These upsets result in a new series of inappropriate reactions, sometimes including physical plant changes. Many times a never-ending cycle of operator reactions causing problems—resulting in different reactions—

causing more and more serious problems, occurs. Once this cycle has started, it can quickly get out of hand. Just getting back to the base plant condition can be virtually impossible.

To add to these problems, once the plant actually starts and problems develop, there is no time left to train the operators. The problems begin to multiply; since everyone is working extra hours to deal with the problems, there is no chance to catch up. Operators are left to absorb the necessary information by trial and error while dealing with the start-up problems. In some cases, more complex plants never do successfully start up. Simpler plants may eventually operate at production capacities approaching design, but only after long, arduous start-up periods.

Generally, operators of plants started under these conditions all have their own pet methods for controlling the operation. We have observed many plants where critical variables are controlled with entirely different home-grown strategies on each shift. In some cases, even the target values are different.

AN ALTERNATIVE APPROACH
Introduction
Few people would dispute the necessity of transmitting literally thousands of critical pieces of information about the new plant to the operator. In fact, there is really only one way to actually do it. We have found that writing a series of custom plant operating manuals, specifically designed for an education level of the target employee pool, is the correct approach. These manuals are then used in a formal classroom training program, complete with graphic support, workbooks, and tests. The training must occur prior to mechanical completion. Ideally, the trained operators complete the class and field training and then assist with the final stages of preoperational testing. Only then are they ready to introduce feed and perform their normal operating functions.

For our plant operating manuals, we have found that the following contents work well, both for training, and as a continuing reference.

Operating Manuals Contents
Introduction. This section describes the purpose of the manuals and identifies those volumes in the set. It also illustrates the scope of the particular manual volume.

Use of the Manual. This section describes how pages are numbered and how to find information. It is important that the manual contents are well organized and it is easy for an operator to find the information needed.

Safe Job Procedures. This section provides formal written procedures, including any special equipment required, to be followed by the operator when performing potentially hazardous job functions.

Process Design. This section provides a written description, complete with necessary schematic diagrams and illustrations, describing the process. It also includes principles of operation for all of the major process unit operations. Refer to Figure 1 for a typical graphical illustration. Unit operations such as vacuum pumps, flotation cells, furnaces, dryers, filters, etc., are described. It is essential that the operator is provided with the necessary information so he or she can describe the key operating principles. This section also provides an equipment list and color flowsheets. Each process flow stream is illustrated with a different distinctive color.

Process Control. This section provides a table identifying each critical process variable, such as temperatures, flows, pressures, densities, etc. It also summarizes their target values, methods of control, and their impact on the process. Following the process variable table, each control loop is described using text, a simple block diagram, and a simple loop diagram extracted from the piping and instrument diagram (P&ID). Each method of control such as automatic remote set point, automatic local set point, and manual, is discussed, as applicable.

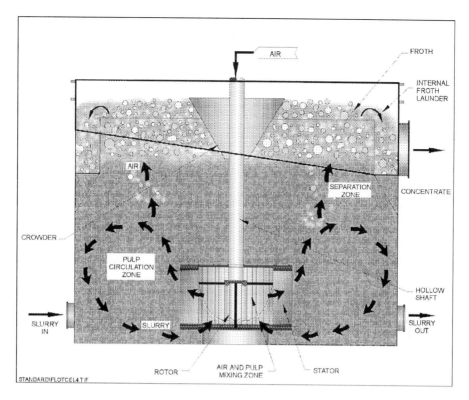

Figure 1 Typical Illustration in the Process Description—Flotation Cell

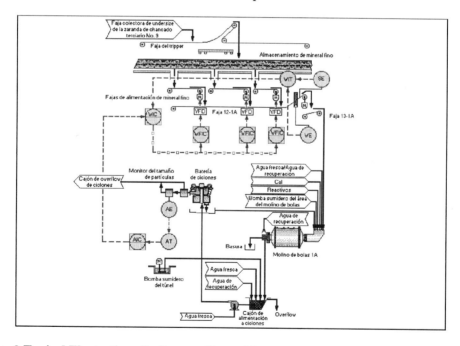

Figure 2 Typical Illustration of a Process Control Loop

Interlocks. This section provides tables identifying all interlocks and permissives for each motor and affected instrument, along with cause-and-effect diagrams. The diagrams cover the same information as the tables and are used to complement the tables. The interlock tables and diagrams are organized by logical process system.

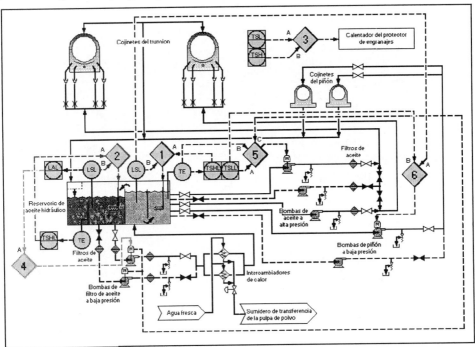

Figure 3 Typical Illustration of an Interlock

Alarms. This section illustrates each alarm in the process sorted by tag number. It is in a tabular format and includes the affected equipment, the fault, the potential causes, and the steps to take to remedy the alarm.

Operating Procedures. This section is divided into three sections: Start-Up, Shutdown, and Operator Tasks. Start-up describes the detailed procedures necessary to start up the plant from a complete shutdown, from a standby shutdown, from a power failure, and from an emergency shutdown. Shutdown describes the procedures necessary for a complete shutdown, a standby shutdown, and an emergency shutdown. It also describes the effect of a power failure and any specific procedures the operator should perform. Operator tasks describe additional procedures required to operate the plant. These procedures always include preoperational inspections necessary to set up the plant for start-up. They also include procedures necessary for the operator to perform his job function. They may also include any steps the operator must take to manually control key process variables. Typical operator tasks for various job functions are:

- Manual boiler blowdown.
- Measuring pH.
- Preparing a batch of flocculant.
- Furnace tapping.
- Shift inspection.

An operator task procedure is important whenever consistency is critical.

DEVELOPING THE MANUALS
Writing each manual is a tedious and involved process. The following source material is needed:

- Process flow diagrams.
- Piping and instrument diagrams.
- Equipment operating and maintenance instructions.
- Functional descriptions.
- Motor control schematics.
- Control valve specification sheets.
- Design criteria.
- Equipment list.
- Alarm list.

This material is used to develop each of the sections previously described. Manual writers must be experienced in plant operation and should be good writers; this is often a difficult combination. Typically, several months must be dedicated to the manual writing process. In many cases, engineering changes are still occurring as the manuals are being prepared; this adds to the complexity.

Once the manuals are completed, we suggest preparing an accompanying training module for each manual. The training module optimizes use of each of the manuals in a formal classroom instruction setting.

TRAINING MODULE
Learning Objectives and Module Outline
This section provides a list of the objectives that the trainee should be able to accomplish once the training is over. The module outline provides the instructor with an outline of the manual and a suggested time duration for training on each section.

Overhead Transparencies or Computer Projector
All graphics in the operating manual are made into overhead training aids for use during classroom instruction.

Workbook
The workbook is a learning device comprising a series of fill-in-the-blank-type questions which the trainee answers while referring to the operating manual. It is used to reinforce learning after the material is covered in a traditional lecture. The instructor is provided with an answer sheet.

Knowledge Assessment Test (Theory Assessment)
The knowledge assessment test is a validation device designed to determine how much of the material was learned by the trainee. It comprises multiple choice and true-false questions and is given after the module's classroom instruction is completed. Results can be used to determine if remedial training is required; they can also be used to determine where individual operators are ultimately placed in the operation.

Qualification Checklist (Practical Assessment)
The qualification checklist is designed to validate that the trainee can apply the theory learned in the classroom on the job. It is completed by the trainees' immediate supervisor during the initial stages of operation. It can be used in combination with a probationary period during which the operator proves he or she can accomplish the job functions required.

Once the modules have been completed, the next step is to conduct classroom training.

CLASSROOM AND IN-PLANT TRAINING
It is important to use credible personnel with previous experience in plant operations for training instruction. Ideally, the personnel who have prepared the manuals and modules should carry out the classroom instruction. In many cases, we use our personnel to conduct train-the-trainer

classroom sessions for the client's trainers, then they, in turn, train their plant operators. We have found this approach very successful.

The formal training consists of three components:

- Classroom lecture.
- Trainee completion of workbooks.
- Site visits to observe the plant equipment and instrumentation.

We have found that the lecture, workbook, and site visits work best when they are distributed throughout the training day. Too much time in the classroom can dull the learning process.

During the classroom phase, it is important to get the trainees involved. Trainee participation results in better retention and makes for a more interesting experience. Near the end of each module, simulation drills can be held. These drills require that the group is split into teams. Each team then attempts to determine the cause of hypothetical process upsets postulated by other teams or by the instructor. The simulation drills require knowledge of the full breadth of information contained in each manual.

Once the formal classroom training sessions are completed, additional time can be spent in the field tracing pipelines, identifying every control valve and instrumentation element, and generally marking up P&IDs as instruments, equipment, and pipelines are identified.

The final phase of training is trainee participation in preoperational testing prior to introduction of feed. Operators, having completed training, are extremely knowledgeable about the new plant. They make ideal personnel to walk the plant and prepare punch lists of discrepancies.

When functional testing of completed plant systems occurs, operators can also participate in that testing. Ideally, the new operators can use the distributed control system or local PLC controls to operate the equipment necessary under the direction of appropriately qualified engineering personnel. As problems are identified during the testing phase, the new operators and supervisors can participate in problem-solving teams investigating the problems.

We have found that there is no substitute for highly trained operating personnel during the testing and start-up phase of any new plant. The training program described above will provide those highly trained operators and supervisors. We know of no other satisfactory method for ensuring that your personnel are ready to operate the new plant.

COMPUTER-BASED TRAINING

The current state of computer technology allows for taking the plant operating manuals just-described to the next step—either a web browser-based interface allowing for hyperlinks to navigate the manuals on a company intranet or a *full-blown*, interactive multimedia interface.

Web Browser Interface

Using a web browser version of the manuals, the user can click on the manual desired to obtain the detailed manual table of contents. The user then clicks on the hyperlinked table of contents item to access that section of the manual. Individual manual subsections are accessed in the same way by clicking on hyperlinks. The text and graphics in the hard-copy manual are the same as those accessed by the computer.

Hyperlinks can also be added to any references. In other words, a hyperlink could be provided whenever another section of the manual is referenced such as a safe job procedure. Therefore, the user can simply click on the hyperlink reference to go directly to the referenced section. The user can then return to the previous section by using the *Back* button on the web browser.

The web browser version is essentially the same as the hard-copy version; the difference is that it is electronic and all volumes, sections, subsections, etc., are accessed by hyperlinks.

Interactive Multimedia Version

The multimedia version contains the same information contained in the hard-copy manuals, but it is in a Microsoft Windows environment. Specifically, the multimedia interface is designed as follows.

Safe Job Procedures. Each safe job procedure is selected from a drop-down menu selected from the main bar menu for each area. The safe job procedures are in text format, with hyperlinks to graphics, if appropriate.

Process Description. In the multimedia application, the process components of each site-specific area, such as grinding, flotation, etc., are divided into process systems. A text box containing the process description for each system can be scrolled down the left side of the computer screen. The center of the screen is used to display graphics associated with the process description. The graphics box displays color flow diagrams from the operating manual, along with any other appropriate illustrations including schematic diagrams and principles of operation. In addition, hypertext links are accessible and provided at pertinent points in the process description text to allow the user to link to glossary definitions, relevant principles of operation and their associated graphics, full-motion video, digital photographs, and other useful illustrations. These hyperlinked objects appear in the graphics box.

This version can also be provided with a voice-over narration of the process description. The script for the voice-over appears in a text box on screen so the user can follow along. As the voice-over proceeds, the graphics in the graphics box automatically change to illustrate what is being discussed in the narration.

Process Control. Process variables can be selected from a menu box that appears on the main screen for each site-specific system. Each variable is shown in a list box, and text boxes provide information regarding the target range, control method, and impact on the process for the variable selected.

Control loops in the system are selected from a menu box. The loops each include text and diagrams. As in the process description, the user can scroll down the text on the left side of the screen. The graphic box illustrates the loop diagram or a simple block diagram depending on the user's selection. For automatic sequence controls, if applicable, animations of the sequence can be provided.

Interlocks. Each site-specific interlock can be selected from the menu box. When an interlock is selected, the text box on the left side of the screen shows each required logical input. The graphics box shows the corresponding interlock diagram.

Alarms. Alarms can also be selected from the menu box. For each site-specific system, a list box containing the relevant groups of alarms (for example, grinding lubrication alarms) appears. A graphic showing all of the alarms in that group on a flow diagram background is displayed. Once a group has been selected, each individual alarm within that group is listed in a second list box. As the user selects an alarm from the second list box, the selected alarm is highlighted on the graphic diagram. Text boxes then illustrate the fault, cause, and remedy for the selected alarm. Alternatively, the user can click directly on any alarm on the diagram to link to text boxes containing the specific fault, cause, and remedy associated with it.

Start-Up/Shutdown. From the main site-specific area bar menu, the user selects operating procedures and then start-up. The different types of start-up are listed. The user selects the kind of start-up—for example, start-up from complete shutdown—and can then move through the procedures using the mouse. Each step is displayed, as are any observations, cautions, warnings, or notes associated with that step.

Operator Tasks. Operator tasks are also selected from operating procedures on the main bar menu for each site-specific area. Once selected from a drop-down menu, they appear in scroll-down text boxes and contain hyperlinks to appropriate graphics.

Workbook. Workbook questions from the hard-copy module are provided in a separate window below the main multimedia window. The user can progress through the workbook questions while finding the answers in the main window above. The workbook questions lead the trainee through the technical information to be learned.

Tests. Knowledge assessment tests are provided to validate that each trainee has learned the material. The tests are composed of multiple choice and true-false questions. Once each test is started, the trainee must finish it without reference to the reference material.

Training Curriculum and Data Tracking System. A database tracking system can be integrated with the multimedia system. This component allows for establishing a custom curriculum for each defined job position. Personnel are then assigned to jobs resulting in each person's learning hierarchy. Test results and qualification checklist results are tracked in the database for each individual trainee.

Videos and Digital Photographs. Full-motion video and photos can also be included with appropriate hyperlinks. Videos are best used for illustrating operating equipment and for illustrating the correct performance of procedures.

Interactive Simulations. Interactive simulations of unit operations are also possible. These simulations provide the operator with the opportunity to change set points or operating conditions and observe the effect on the process.

CONCLUSIONS

There are literally thousands of specifics that must be learned by each operator involved in a new plant. In addition to facts concerning the new plant, operators must also learn principles and theory associated with the new equipment, controls, and methods. No matter how carefully the plant has been designed, or how well the new equipment works, the start-up will not be a success until the operator has completed this learning.

There are only two ways for operators to learn the material necessary. They can learn it in a controlled classroom environment as has been discussed in this paper. Alternatively, they can learn it as they are attempting to operate the plant by trial and error. The cost of the former approach, while certainly not inexpensive, is very low compared to the lost production and damage usually associated with the latter approach.

Planning and Staffing for a Successful Project Start-up

Kenneth A. Brunk, Larry J. Buter and K. Marc Levier

ABSTRACT
A successful project start-up can be defined as the culmination of a series of tasks that result in an operation that produces the saleable product at designed tonnage rates, to design specifications and at designed recoveries from design feed at designed costs to meet schedule.

Successful project start-ups are the result of vision, cooperation, commitment, planning, and execution involving every area of corporate responsibility. Such successful start-ups have their roots in the executive offices where the philosophies and standards for ethics, health and safety, environmental and cultural responsibilities, training, design, and closure are set. The effective communication of these philosophies to the balance of the corporation and the project team, with the expectation that they are complied with, forms the basis for successful project execution.

This chapter will identify the teams of people, the project organizations, and tasks that are needed to accomplish the project start-up. Additionally, the chapter will discuss the integration of the various engineering and corporate disciplines required to design, construct and start-up the project. Schedules for key events such as metallurgical treatment, final design, mine design, pre-stripping, project engineering and design, site pioneering, site construction and staff hiring, training manual development, training, procurement of project consumables and start-up spares, pre-commissioning, commissioning and finally start-up itself will be presented and explained.

Definition of a Successful Start-up
What is the definition of a successful project start-up? It can be defined as the culmination of a series of tasks that result in an operation that produces the salable product at designed production rates, to design specifications and at designed recoveries on schedule and within budget. The importance of a successful start-up is plain and simple economics. However, the morale of the project team and the new operations team is equally important in creating continued operating success.

Successful project start-ups are the result of vision, cooperation, commitment, planning, and execution involving every area of corporate responsibility. Such successful start-ups have their roots in the executive offices where the philosophies and standards for ethics, health and safety, environmental and cultural responsibilities, training, project design, and operations closure are set. The effective communication of these philosophies to the balance of the corporation and the project team forms the basis for successful project execution with the understanding and performance expectation that these philosophies will be followed.

The start of success for a project begins with the net-present-value (NPV) and cash flow calculations which must meet or exceed corporate minimum standards before the corporate board of directors will approve the project for construction. These economic numbers are important to the corporation because personnel and monetary resources available are limited and the corporation must assure that these resources are utilized on projects that will provide the best return on the investment. Every project submitted to the headquarters office for consideration is in direct competition for funding approval. The funding for a project may come from the corporate funds, standard loans from outside sources or loans where money is borrowed from the sale of future production. The success or failure of a project start-up has a major and critical impact on meeting these expectations. Many projects have failed to meet the required financial criteria and this has had a direct impact on the ability of the parent corporation to remain in the mining business.

An article published 1984[1] displays impact on the NPV and DCF on delayed start-ups, Although it is now dated, it still is a good example for today's projects. An excerpt from the article is shown in the next several paragraphs.

"Based on recent experience, it is prudent to consider the following for estimating cash flow in project feasibility analysis:

1) New mines may be expected to have average annual production equal to 50-70% of the designed capacity during the first year from start-up, 80-100% of designed capacity during the second year, and to be near or at designed capacity after the third year.

2) New beneficiation plants may expect to have average annual production of 40-60% of designed capacity during the first year from start-up, 80-100% of designed capacity in the second and third year from start-up, and to be near or at designed capacity after the fourth year.

3) New processing plants may expect an annual production of 40-60% of designed capacity in the first and second years from start up, and 80-90% of the designed rate during their third and fourth years.

The CRA study for the World Bank clearly indicates that firms do not realistically estimate the time necessary to bring a project to full capacity. Based on the record of recent projects—in the Americas, Europe, Asia, Oceania, and the Mid-East—investors should be prepared for long delays. On the average, most new mining and smelting operations have not produced positive cash flow in the year following start-up. In fact, many projects not only take about two years to achieve design capacity but, in the interim, increase the cash flow exposure of the sponsor.

As an illustration of the effect of start-up delays on project profitability and cash flow exposures, consider a copper mine/smelter complex scheduled to produce 100,000 st/yr starting seven years from today. Considering all costs in constant dollars, the investment required is roughly $700 million. Assume that the project was justified on the basis of a 20% discounted cash flow rate of return (DCF-ROR) and the net present value (NPV) of $760 million at a 10% real discount rate, for example.

As shown in Fig. 4, with no additional costs, a one-year delay decreases the DCF-ROR to about 19%, while an extreme five-year delay reduces it to about 15%. When annual costs equal to 15% of total investment are added to the delays, the DCF-ROR will drop to 17% with a one-year delay and to about 12% with a five-year delay.

The impact of start-up delays will also be disastrous, because of the following cash drains: 1) high operating costs, due to curtailed or no production; 2) low revenues from lower-than-forecast production; 3) low revenues from off-specification production during prolonged start-up; 4) special unplanned start-up crew expenses, 5) fix-up costs; and 6) increasing debt due to inability to service the debt.

---

[1] Agarwal, J. C., Brown, S.R., Katrak, S.E., "Taking the Sting Out of Project Start-up Problems", E&MJ, September 1984, pp. 62-66.

The foregoing cash draw-downs are all cumulative. In the copper project mentioned, the cash flow implications of a delayed start-up may be as follows:

Original capital expenditure	$700 million
Interest charges for 3-year delay at 10%	220 million
Fix-up expenses	100 million
Operating costs (net of any revenues) for 3 years at 40¢/lb of copper at an average of 50% capacity	120 million
Total	**$1,140 million**

What had started out as a $700 million project with a 20% DCF-ROR will now have roughly an 11% DCF-ROR (Fig.5). More importantly, the total cash exposure for the project has increased from $700 million to $1,140 million. Therefore, even a modest improvement in decreasing start-up delays can have impacts of hundred millions of dollars.

If corporate executives use similar start-up data for production scenarios as a guide for timing future cash flows, they stand a better chance of meeting ROR targets. A baseline schedule reflecting the reality of start-up delays justifies a comprehensive analysis and training program that: 1) examines the various pitfalls in scale-up design and allows for adequate safeguards; 2) reviews the start-up procedures; and 3) trains the operating and supervisory personnel to meet the challenge of a preplanned start-up schedule."

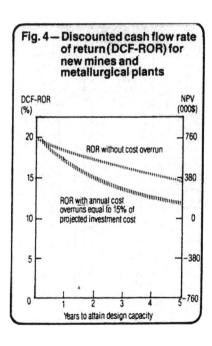

Fig. 4 — Discounted cash flow rate of return (DCF-ROR) for new mines and metallurgical plants

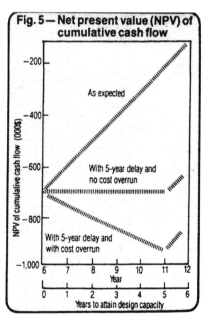

Fig. 5 — Net present value (NPV) of cumulative cash flow

The co-authors of this paper have been involved in start-ups that have met designed production goals in less then a day and in others that have met the above stated "normal" industry standards. The difference in the two time lines, is that in the one-day cases, the start-up was given a very high level of priority with a great attention to detail and forward planning.

CORPORATE STRUCTURE AND PHILOSOPHY FOR PROJECT EXECUTION
Corporate Standards

Large or small companies have some of the same issues in common that must be addressed at the senior level of management. The first of these is the company's or corporation's standards. These include the philosophies and standards for ethical conduct, health and safety, environmental and cultural responsibilities, training, design, and closure. The effective communication of these philosophies to the balance of the corporation and the project team with the expectation that they are complied with forms the basis for successful project execution. These items are very important to projects since each can have a major impact on the project economics as well as staying in operation. Major violations of most of these items can lead to the project being delayed or temporarily closed until brought into compliance with the standards. The company executives must set in place a system that audits the compliance of their standards to assure that all of the project objectives will be achieved. A method of correcting noncompliance issues must be in place before the project is developed to assure that start-up will not be delayed.

Management must also provide the support, both financially and in commitment, to allow their standards to be met. If they are stating that the standards will be met but do not provide the resources of management and financial support, failure to some degree will result.

An example of this would be to state that every employee will be properly task trained before start-up. This is a requirement for all operations in the United States by OSHA and MSHA law but may not be required by law in foreign countries. However management will often proceed to cut funds needed to develop a good training program and have the proper personnel hired early enough for the training to be completed if expenditures are higher than originally projected. Such conduct serves only to undermine the goals of the corporation and the project.

Management must also define the reporting structure of the project team to senior management. This can range from the team having total autonomy to total control of the project by senior management, or more normally, somewhere between the two styles. When the company management establishes a clear line of responsibility and authority for the project team, the project is given an increased probability of success. This includes the level of involvement that the senior management will have as the project develops. Without clear lines being established before the project starts, confusion will result around the project direction (Who is in charge?) and will destroy the team morale. This will only cause frustration for both parties. The project will also have an increased probability of being over budget and have schedule delays. Both of these will greatly increase the project costs and reduce the ability of the project to meet the project economics.

Projects can be located near the company's home office, another company operation or in a remote part of the world. The location of a project can contribute to confusion within the company as to how the project will be developed. Projects close to the home office or another company operation can be the easiest to develop since the company's standards, ethics and culture can be more easily imposed on the new project since these are a part of the local culture.

If the project is in a remote location or in another part of the world, the local customs, business ethics, and worker productivity must be addressed when calculating the project economics. A standard of operation for all of the company's facilities should be the goal, but local customs or decrees may require the standards to be modified to meet them. The problem then becomes that not all of the company's units are being held to the same standard and it is easy for a unit to argue that other standards should also be modified to meet what they feel is needed for success. Caution must be exercised when corporate standards are modified!

An example of this is the cost of providing benefits for employees. In some areas of the world, it is required that the company provide a meal for the employees on every shift. This is a cost not normally expected if the company only has operations in the United States; but it must be taken into account in other locations of the world. In some instances this may amount to thousands of meals per day which increases the food supply requirements for procurement and logistical systems that may already be challenged to provide the necessities for the project team and construction workers.

Productivity

Another often overlooked factor is worker productivity as compared to U.S. Gulf Coast standards. In dealing with a project in a former Soviet Union country, the government officials gave guarantees of their excellent construction capabilities and worker productivity. On advice of the engineering company "experienced" in these matters, all estimates of labor man-hours were increased by a factor of 2.5. In fact after the project concluded, the correct factor was more like 5.5. This in itself led to significant budget overruns and further start-up delays.

Project Management Approach

Each corporation has a culture and a method for developing projects. This may vary from small companies, with no development resources to large corporations that have dedicated teams that move from project to project. Small companies may need to depend totally on outside resources to develop a project. This requires extensive evaluation of the outside project team to assure the company's economic objectives are met, since the control of the project is now outside the company's direct control.

Large corporations that have dedicated project teams may have more control of their success but this is not always the case nor is it guaranteed. They too must undertake extensive evaluation of all outside resources to be used on the project to assure project success. Large corporations must also safeguard that the project team is thorough and acts in the best interest of the company. The major reason for concern here is that the project team must make projects viable in order to stay employed. This can sometimes lead to over optimism and bias on the part of the team leaders when only the upside conditions are used to produce acceptable economics for the project. This can lead to failure of the project since it is unrealistic to assume that everything will meet the most optimistic conditions. The project leaders and team members must be incentivized regardless of company size to maintain objectivity in their evaluations and recommendations.

For example, if a project that has been optimistically estimated is approved by management and later it is learned the costs escalated then the project may become uneconomic. Often this happens when the project is well under way. At this late stage of the project life, most of the money for engineering, design, equipment procurement and construction has been spent or committed and only the owner's cost portion of the project cost can be targeted for spending reductions. This can lead to project management taking short term cost cutting measures in the area of hiring, training, and procurement of operating and maintenance supplies. This can also contribute to redesign of the plant to incorporate "cheaper" equipment that appears to save money but in fact is being misapplied in its application. This is false economy, since this will cause a successful start-up to be almost impossible to achieve. The simple advent of equipment improperly designed and misapplied can result in millions of dollars in loss, for the most simple problem, when you add up the time and effort that will go into analyzing the problem, attempting to make the original design work, the evaluation of a correct fix, the re-engineering of the fix, and the installation and start-up of the fix. Additional cost consideration must also be added in for the loss in production and plant availability. Of course at this point the operator is stuck to solve the problem while the engineering team has already departed for the next job and spent their bonus! (The engineering team is available to come fix the problem they created for a price!!!) It is up to management to assure such optimistically engineered projects do not make it off the drawing board and into the field.

PROJECT TEAM OVERVIEW
Team Makeup

Projects can be developed using a fixed cost or cost plus basis for engineering, design, procurement, and construction management (EPCM). Depending on the company's philosophy of project development, <u>one of these methods</u> will be chosen. Either requires an individual or a team of individuals to interact with the EPCM firm in order to track the cost and schedule of the project.

To develop and start-up a project successfully, a group of individuals must be chosen to form a project team. This team must have a leader commonly referred to as the project manager. This person must have experience in dealing with the size and type of project that the company intends to develop. This individual may already exist in corporations that have dedicated project teams or, as the case with a small company, an individual may need to be selected from outside the company. In either case, for the project to be completed on time and within budget and start up successfully, the manager must have experience with similar projects. It seldom works to allow an individual that has only managed a $50 million project to manage a $500 million project or vice versa, since the level of activity for the projects are very different. In small projects, normally the project manager has a small number of individuals to assist in the design and construction activities. These people must have multidiscipline or varied backgrounds in order to review and assist the engineering company to provide a project that will start-up successfully.

In a very large project, the project manager can have an individual with specialized expertise in each area, since the level of activity for each area is much greater and requires a full time person to maintain the project schedule. Management must make it clear that the project manager will oversee the project under the direction of the general manager who will be responsible for the project once it is started. Both managers need to be dedicated to the overall objective of the project, i.e. a successful start-up, and be willing to cooperate with each other to fulfill this common goal. Neither the project manager or the engineering company can be allowed to believe or act as if they have supreme control over the total project because their actions can have a very large detrimental impact on the long term operation of the project.

For example, when the project manager and engineering company are allowed to manage all affairs in the community and make promises of long term employment, building hospitals, churches, recreational facilities, or providing free transportation to the project site, the GM and operations people are left holding the bag and must deliver. The general manager will be the person ultimately held accountable for the projects cost and operating performance and will be held accountable by the local people if the promises made during development are not met.

Once the project manager has been assigned to the project, the remainder of the team can be selected. These individuals may be from within company operations, or they can be selected from outside of the company. If they are selected from within the company, they should be relieved of their current position obligations in order to concentrate on the new project. An individual who is an excellent employee in an operating situation may not be successful in the role of a project team member since the level of control for each position is quite different. If the project is being developed in a country foreign to the company, the project manager and general manager must at least have been issued a passport and traveled out of the country as part of the job qualifications. Individuals that have zero experience outside of their country, will find it a challenge to succeed. They may have been put in a position to fail by their company.

If members are selected from within the company for the project team, they should be assigned to the start-up and operating group after start-up. This allows them to have ownership of their decisions during the development process. This also leads to a better start-up since they are very familiar with the project and the requirements that need to be met in order for it to be successful. They must have an adequate level of expertise to allow them to succeed in their assignment. A means of retaining them for the duration of the development and for at least a year after start-up needs to be addressed by management before the individuals are selected. If incentives are granted to the team, they must be fair, clearly understood by both parties and must be in writing to prevent problems as the project progresses.

If contract individuals are hired outside the company, their contract should clearly state the duration of their contract is through start-up and for a fixed time thereafter. An incentive needs to be in place in the contract to make sure they complete the assignment. These people will be critical to having a successful start-up and an incentive is one way to prevent disruption of the continuity of the project. With a large turnover of either in-house or contract people, the project continuity is lost.

The location of the project team is a decision that must be made early in the development process. Facilities should be obtained in the engineering house for the project team. These facilities need to be close to the engineering team's location but yet separate enough for the project team to have some privacy so discussions can be held without disrupting the engineering team. The goal of the total project team is to have a relationship that is open and beneficial to the successful start-up of the project. A clear method of communicating between the owner's team and the engineering team is needed to prevent confusion and added cost. All change orders need to be approved by several members of the project team including the project manager to allow tracking of costs associated with design changes. At some time in the process, all changes must be stopped or the project will never be constructed on time. If the practice of change is allowed to continue, the project can never be developed at the estimated cost.

After the completion of the flowsheet, P&ID's, G.A's, and equipment specifications the operating teams involvement is changed from one of providing design to insure the criteria are included into the project. This phase of effort needs to be carefully managed and controlled to provide oversight while not being disruptive to the overall project design effort. On every project design issues will come up that need to be thought through and evaluated on a very rapid basis to contain costs and meet schedules. An adept project manager needs to manage the situation with the goal of providing a process plant that works. At the same time an observer from the operating group needs to understand cost, schedule and be able to evaluate alternative means of accomplishing the task.

As project development is nearing completion, the method for transition of members from the project team to the start-up and operating teams located in the field needs to be decided. The timing of this transition is determined on a case by case basis. Close communication between the general manager (who should be part of the project team from the start) and the project manager is a must for this to happen smoothly.

As important as having facilities in the engineering office, are having facilities for the team in the field. Construction of the permanent site offices needs to be a high priority. This allows the key operating people to be hired early in the process and allows them to become familiar with the facility they will be held accountable to operate. Temporary facilities will also be required for the owner's project team that will be onsite earlier than the permanent office completion. If the project is remote, housing arrangements need to be in place or contract man camps hired to house both operating and construction overseers. If family housing is part of the project, these must be constructed very early in order to allow families to become part of the project and to reduce turnover and extensive travel costs. This is not normally considered the best use of project cash but in the end can contribute significantly to a committed team and allow a successful project completion. There is no worse situation in the world than contractor's project personnel, and operations personnel fighting over bunks, office space or food!

Frozen Flowsheet
Successful projects result from the integration of the geology, mining, and metallurgy to produce a mine design and process flowsheet that work together. Ample test work needs to be done in the mine design and process plant design to develop process criteria and equipment specifications that can be translated into engineering drawings and specifications needed to construct the facilities.

Successful projects also begin with a flowsheet that is completely fixed or "frozen". This means all metallurgical and process related test work has been completed prior to commencing process plant design engineering. From this test work all process parameters such as flow rates, temperatures, pressures, retention times, pipe line velocities, slurry densities, rheology, bulk densities, etc. have been determined to enable material of construction selection and equipment selection that is compatible with those characteristics.

Coupled with the frozen flowsheet is the development of a plant operating philosophy that fully describes how the plant is to be operated and controlled. For example, a detailed description of a process circuit is needed to enable the design engineers to include the correct instrumentation and control package necessary to control the process.

Good project practice should include the plant operating personnel in the development of the general arrangement drawings, the flowsheet, the process instrumentations diagrams, the project equipment specifications (P&ID's) and spare part selection.

By including the operating team in the above, the project manager has tapped the most talented people in those disciplines to define the project details. This inclusion will ensure that issues related to personnel safety, traffic flow, operator access, maintenance access, flow sampling, and so on have been addressed.

ENGINEERING COMPANY SELECTION
Overview
After the flowsheet has been frozen, the first order of business is for the project team to select an engineering company. Engineering companies with demonstrated technical capabilities for the type of project and project size and complexity should be placed on a bid list. Each company should be researched on the projects of the anticipated size that they have completed in the location of the project. This is very important if the project is located in a remote location because logistics of manpower and supplies will be critical to the project success. Companies without foreign or remote project expertise should be removed from bid lists for foreign/remote projects. A list of specific criteria that will be evaluated for each engineering company must be developed and weighted so the evaluation of engineering firms can be done objectively. Higher weighting should be given to critical areas of experience which will make or break the project's success. When several companies have been selected, the project manager and several of the company's senior staff must visit the engineering companies and personally evaluate this group of firms. This decision as well as the selection of the project manager is critical to the project success.

Design Phase Criteria
Corporate objectives for the project must be clearly understood by all parties before design is started. It must also be decided if the capital cost will be the important driving force or will operating cost be emphasized. Obviously, management wants both the low capital and operating cost but this is normally not possible. Normally capital cost can be increased slightly to slightly lower the operating cost. An example of this is in the general arrangement of equipment. If operating, cleanup and maintenance accessibility is included in the design process, it will require more floor space than if only equipment installation is considered. If not included, it will increase the operating costs will increase due to downtime and excess manpower required to perform normal operating and maintenance functions. These decisions must be justified through trade off studies and properly documented.

Accounting procedures must be determined by both the owner and engineering firm in order to track the cost and progress of the project. The procedures need to be adequate to allow management of the project but not so complicated that a great deal of cost is incurred to manage a very small amount of money.

The same concept is true for the security of the site during construction and operation. The amount of money that is allocated for security should be evaluated against the potential cost of loss. In locations where there is civil unrest, the safety of employees must be included in the evaluation. In this case evacuation planning and cost must be included. Losses of expensive key components during construction will delay start-up and increase costs. Loss of simple items such as zerk fittings or stainless steel valves and piping can cause a major delay in start-up in remote areas where these items are not available locally. A good rule of thumb in these locations is that the item will disappear if it is not bolted to the ground, and even then they sometimes disappear too!

Corporate, national and local safety standards must be determined and the standard for the design must be established. In some cases, the corporate standard may be different than what the local or national law requires. For North American companies doing project internationally the corporate standards are generally more stringent. The standard that provides the proper protection should be used as long as it meets the law.

If the project is located in an area that does not speak the same language as the engineering firm, then the documents must be provided in each of the major languages. The interpretation of the documents in an accurate and timely manner is necessary for the local people to be involved in the project review and design. Failure to recognize the importance of this will occur in the training of the employees who must operate the plant. An example of this is the labeling of equipment. If the local language is different than the label it can be confusing and dangerous. Dual labeling may be required for mixed language locations.

Projects located in remote areas may require that mancamps be established as the project is being constructed. If this is the case, then the company must decide how the camp will be operated. Included are the security, health and safety of all camp employees. Most camps should be operated where alcoholic beverages are not allowed. This keeps the in-camp trouble minimized. Other forms of recreation must then be included in the camp setting to occupy the employee's free time. Control of outside visitors must be done to limit the visitation of unauthorized people. Camp facilities during construction and start-up must be adequate to prevent people turnover and to prevent sickness. Both of these conditions will cause labor unrest and may delay the project.

Transportation of employees during and after construction must be decided before and during construction. Of importance here is if buses or cars will be used to transport people. If cars are used, then an established parking area must be located to allow construction and still provide reasonable access and protection of employee's vehicles. If buses are used, then established routes and stops must be determined. Parking at pickup points must be developed at the pickup points if it is not available.

ACTIVITIES PARALLEL TO THE DESIGN AND CONSTRUCTION PHASE
Overview

The project will have many activities that are being done concurrently with the construction. Some of these include in-fill drilling, pre-production mine development, background environmental sampling and monitoring, staffing of operating personnel, and procurement of the project's supplies and equipment. These activities need to be coordinated with the engineering and construction contractor to prevent delays to start-up. The site manager and his early key people must coordinate and communicate with the construction personnel to prevent delays and hazardous conditions. An example is the blasting required for pre-production stripping of an open pit. Since the mine is normally located near the process facility to prevent extra cost of long ore hauls, blasting during construction must be coordinated with the construction firm. If possible, blasting should occur when construction crews are changing shifts or at lunchtime when work has stopped. The production people must be adaptable and accommodate the contractors to cause limited delays of construction.

The owner and his people, with the help of the engineering firm, develop a plan for the training of employees and a start-up plan. As the plant is completed in sections, a clear plan must be developed to transfer the sections from the contractor to the operator. Operations will want to commission sections and systems within the plant before the project is completed. These areas, however, must be signed off by the contractor and owner to prevent disagreement if damage occurs to the section after turnover.

Since no plant is totally without items that need correcting at the end of construction, a list of unfinished or unacceptable items must be developed (punch list). A crew from the contractor then must complete corrections to this list, usually while the plant is running. The operators must coordinate with the construction crew to prevent hazardous conditions for the employees. An example of this is the use of cutting equipment by the contractor in modifying conveyor transfer chutes. This can cause fires and do major damage to plant equipment causing start-up delays.

The site general manager is responsible for coordinating the activities described above as well as many other business related activities. For a grass roots operation the efforts needed are to identify, organize, design and implement all the items necessary to conduct the operations of the business. These include items such as:

- Accounting
- Purchasing
- Payroll
- Employee benefits
- Communications
- Computer and data handling systems
- Permits
- Licenses
- Human relations
- Warehousing
- Maintenance systems
- Mine development and operations
- Taxes
- Safety or loss control
- Insurance
- Shipping and traffic
- Off take agreements
- Loan compliance documents
- Inspections
- Training, salary, operations, maintenance
- Community relations
- Management reporting
- Operations reporting
- Process plant operations
- Analytical support

While this list is not complete it serves to point out that a successful start up and operation require the coordination and input of many people. None of these areas can be neglected or the operation will suffer. Also, these operations need to be coordinated with the project manger to minimize expenditures and obtain the most efficient use of manpower as possible.

Below is a sample organization chart for a large project. It may not include all the key positions needed for every project as well as it may contain additional positions above what is required for a smaller project.

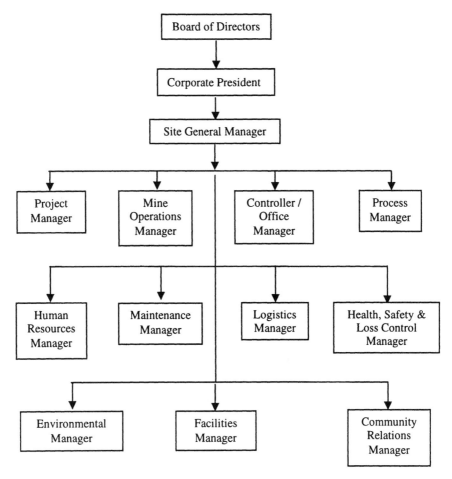

The table below is a summary of the anticipated schedule to hire people for the project. This schedule allows for people to be allocated to the engineering effort as well as to allow the site to have employees to become familiar with the project. Employees must be on the site early to allow policies, procedures and training efforts to be completed before start-up.

When Project has Approval
Site General Manager
Project Manager

Selection of Engineering Company for Final Design
Controller
Human Resources
Environmental Manager
Process Manager
Maintenance Manager

Start of Detailed Design
Mine operations Manager
Community Relations Manager
Health, Safety & Loss Control Manager
Computer & Information Manager

Completion of Detailed Design
Mine Planning Engineers
Logistics Manager

6 months before Start of Mine Development
Mine General Foremen
Planning and Scheduling General Foreman-Mine
Accounting Personnel
Warehouse General Foreman
Environmental Engineers
Health, Safety and Loss Control Personnel
Analytical Department General Foreman
Mine Operation Foremen

3 months before Start of Mine Development
Analytical Personnel
Warehouse Foremen
Maintenance Foremen-Mine
Planning and Scheduling Foremen-Mine
Mine Operators
Mine Mechanics
Mine Electricians

9 Months before Process Mechanical Completion
Maintenance General Foreman-Process
Planning and Scheduling General Foreman-Process
Maintenance General Foreman-Facilities
Planning and Scheduling Foremen-Process
Planning and Scheduling Foremen-Facilities

6 Months before Process Mechanical Completion
Process General Foreman
Planning and Scheduling General Foreman-Facilities
Metallurgical Engineers
Maintenance Foremen-Process
Maintenance Foremen-Facilities

3 Months before Process Mechanical Completion
Process Operating Foremen
Process Operators
Process Mechanics
Process Electricians
Facilities Electricians

GENERAL DESCRIPTION OF FUNCTIONAL RESPONSIBILITIES
Mine Development Team
Schedules for mine development, including equipment delivery, crew training, pre-stripping, underground development, de-watering and the like must be meshed with the facilities

construction schedules. Such coordination will ensure a source of proper plant feed for start-up and operation.

Generally the owner's team carries out the mine development. This team is specialized in mining techniques and mine planning. However, the team must be very diligent and keep the project manager informed of progress, problems, and performance against the overall schedule. Also, often times much needed earthworks or building materials are supplied from the early mine development. Therefore, adherence to schedule and quality standards for the materials are a must.

Accounting

The areas of accounting, payroll, data handling, taxes, insurance, management reporting and operations reporting can often be placed under the direction of the controller or chief accountant. This person must be articulate and knowledgeable in the software systems necessary to meet the daily requirements of a mine site while still complying with the corporate reporting requirements and standards. The individual will need to hire and train a staff composed of as many people local to the area as possible.

Purchasing and Warehousing

Purchasing and warehousing are two areas of an operation that are also suited to report to one individual. This person must be honest, understand the expectations of performance of the corporation and knowledgeable in the general business of mining.

Furthermore, the individual must be able to analyze situations, able to handle conflict, and create a can-do group of employees in the department. It will be important for the individual to understand lead times, shipping routes and cost and contract types and terms. Purchasing and warehousing are service centers on a mine and they need to operate as such.

This group needs to have a place at the start-up to store the necessary spare parts for the plant and mine equipment. These items must be arranged in an orderly manner and stored in a place that prevents contamination by the environment they are in. Also, if the first fill reagents are taken as part of the owners responsibility and cost, they must be ordered and delivered in time to prevent start-up delays.

Human Relations—Employee Benefits

Human relations and employee benefits are two areas that historically are placed together. This area needs to be headed up by a person with a can-do attitude who is knowledgeable in labor law and local labor issues and practices. Additionally this person must be capable of building and leading a team to staff this mine facility and provide for new hires when vacancies occur. Criteria for hiring must be established through interfacing with the various department heads. Also, training programs and training critique should be coordinated by this group. This is especially true in regions where skills training is a determinant in pay scales for individuals. Record keeping is also a must for this aspect of the company as good records can be an asset in workers compensation cases, etc.

Employee benefits must be designed to meet the criteria of the corporation while recognizing local customs. Where necessary dual standards of compensation and living may prevail.

The hiring of a competent human relations professional is one of the first orders of business for today's operations manager.

Prior to full scale hiring of the employee contingent for a mine site it is necessary to have in place the wage ranges, benefits plans, and employee qualification guidelines to enable successful hiring and team building characteristics. Such planning will help ensure the success of the operation and mining future employee related issues.

Maintenance and Maintenance Systems

For today's large and complex operations it is generally necessary to hire an experienced maintenance professional early in the project life. Ideally this individual would lead the effort of establishing design specifications for the mine and process plant equipment. The individual is also invaluable for overseeing the design of the maintenance shops, workshops and selection of spare parts and tools.

This individual is also responsible to specify, evaluate and select the maintenance planning software to be utilized to ensure the equipment reaches and maintains design availability. The person must be able to critically evaluate problems, and make decisions. The Maintenance Superintendent will have in the crew individuals who are organized, who can evaluate problems and who have a can-do attitude. In addition to knowing maintenance skills they must have a working knowledge of the processes in the plant to enable good communications with the operators.

Mine Operations and Development
The mine superintendent must be on the project team almost from inception of the project. Mine production for today's complex ores and environmental constraints requires that mine development work occur often before plant construction.

To achieve the mine production rates and ore blends necessary to meet the project requirements, mine development needs to be evaluated and planned very carefully. Indeed, process plant location should be such to minimize haulage of material, ore and waste, from the mine. This decision can only result from high quality mine plans.

As alluded to earlier, mine operations are often integrated into the plant construction effort to provide fill material, construct tails dams, fresh water ponds, access roads, and to provide major cut and fill operations as required. This means that mine engineers, mine geologists, equipment operators and maintenance people must be on site early in the project life. These people need accommodations, training, equipment, facilities, and so forth to function.

The mine superintendent must be knowledgeable in the mining method to be used, be appreciative of the construction requirements for the project, be cooperative with the project manger and be accountable to the general manager.

The individual needs to be innovative to be able to direct crews in often harsh environments with minimal facilities to work with, especially in early stages of the project.

Computer Systems and Communications
Mine sites require significant computing power for the technical support systems, mine design systems, business systems and accounting systems. Also, associated with this computing power is the necessity of fast, reliable communications within this site and to the outside world.

Proper system sizing and design requires the identification of the various software packages and systems that will be used on site as well as those necessary to communicate management information off site. Approximate data and communications requirements need be developed and systems sized accordingly.

Following that exercise the requirements for system support and training need to be determined. Also, the philosophy of how the training will be accomplished and by whom needs to be addressed.

Therefore it is necessary to place on the payroll a technical systems administrator who understands the needs of the business and the uses of the programs. The various systems would ideally be integrated to eliminate multiple data entry and efficient reporting. This individual needs to be on the team in time to organize the soft and hardware acquisition, and to establish training programs for the users.

The systems person should also specify the communications system to be used on and off site and enable their use during construction to eliminate cost duplication.

Environmental Control
Environmental permitting and compliance are cornerstones of today's mining operations regardless of what country in the world the mine is located. Permits must be obtained for everything from operating the actual mine and mill to establishing the landfill in which to place garbage. The permit effort must be timed such that when the project construction is complete the operation can commence start-up and construction.

Permitting effort must start as soon as the project appears to be viable, even at the conceptual stage. Studies to obtain background information to provide site characterization and base line information must take place even during the late stage exploration on the property. Also, the road map of effort and timing of the studies that will be needed to enable the permitting of the project must be established and evaluated from a risk point of view before the commitment of major dollars to develop the project.

Therefore the placement of a high caliber individual whose expertise is the permitting of projects needs to occur early on in the project life cycle. This person should be available to see the project permitting through to completion and be available for consultation at least for a year after start-up.

The individual will need to develop environmental compliance and monitoring plans to assure the process design takes into account the environmental constraints and conditions of the permits. Items such as the allowable treatment rates, discharge water quality, pounds per hour of emissions from stacks, and so forth have direct impact on plant design and cost.

Additionally the individual must have hired and trained a team of professionals on-site to monitor the operation and assure compliance with the permits and environmental laws. These people will need to be skilled to work with the operators at site and have first hand knowledge of the operation. They will need to interface with the local officials and community to establish and maintain workable relationships over the life of the project and its ultimate closure.

Health, Safety and Loss Control
Industrial health and safety responsibilities for an operation must be addressed early in the project to assure the process plant and mining operation meet the minimum standards established by local governments, the insurance companies, the lenders and the corporation. Identifying problem areas early in the project allows for the design to address the issues. Items such as chemical storage, ventilation, equipment access, fire protection, fire fighting systems, ambulance requirements, medical evacuation requirements, personnel evacuation and security requirements, employee education and training and the like need to be designed into the project and available upon start-up.

The crews and management need to have been trained in the operation of the facilities and emergency response systems prior to start-up. Fire response teams, and first aid response teams need to be in place and functional at the start-up stage of the project. Also, the safety officer needs to have established disaster plans and coordinate the plans with the local medical facilities, should they exist.

Therefore the person to head this aspect of the operation needs to be on board and functional well prior to start-up. The individual must be involved in the employee training at all levels to assure common implementation of the corporate expectations and laws of the region. The person needs to be articulate, able to think on his feet and able to manage in emergency situations.

Analytical Support
Today's ore bodies require a high degree of rather sophisticated analytical characterization on a daily basis to assure the delivery of ore that meets the requirements of the processing facilities. To accomplish this requirement requires that a laboratory capable of the minimum analyses within rapid turn around time be located on or near the mine site. The laboratory needs to functional prior to the commencement of the actual operation of the process plant. This lab may also serve to monitor various flow streams to maintain permit requirements, even if an outside lab is used to obtain the results that are officially reported to the watchdog agencies.

The chemist and the staff will need to be on board and functional at start-up to provide the analytical information and sample turn around that is needed at this time of the project.

Plant Operations Superintendent
The plant operations superintendent will be a very key individual in the life cycle of the project, and needs to be on board, and involved immediately after the project is deemed ready for flowsheet design. This person will provide valuable operating input to the design team in all

aspects of the site layout and process plant design. The superintendent will provide guidance in areas such as operator and maintenance access, ergonomics and traffic flow, operating philosophy, equipment type, selection, equipment configuration, and process control.

While the plant is in detail design, the individual will monitor the drawings and equipment selection process to ensure the layout and equipment is in compliance with the concepts agreed with. He is there to ensure that the engineer provides an operable plant.

The superintendent is also involved in the interviewing and hiring of his staff of foremen, general foreman, metallurgists, clerks, etc. The individual is responsible to have in place a functional crew of plant operators, and technicians capable of running the operation during the start-up exercise.

Additionally the reporting systems need to have been designed and tested prior to start-up to provide management with the production they need.

Transition from Construction to Operations
As construction of the project nears completion, some construction employees will desire to become part of the operating team. This can be good for both the contractor and the operation if done properly. Operating staff can evaluate the employee's work performance during construction and decide if it matches the expectations of the operating staff. For the contractor, it can be positive since a good employee can be given a permanent position rather than being laid off at the end of the project. It can also be a negative for the contractor if the transition timing is not agreed to in advance, and the contractor is left without key employees to finish construction. The answer is to have very open communication between all parties and a well thought out plan before the construction is completed.

CONCLUSION
Successful plant start-up does not just happen! It occurs because a tremendous amount of planning and cooperation of many different factions has happened due to excellent resource planning. Communications between all people involved is required to prevent misunderstandings and delays. The project must be a team effort by all members if the project is to start-up successfully.

Maintenance Scheduling, Management and Training at Start-up: A Case Study

Peter Vujic[1]

ABSTRACT

We all subscribe to the philosophy of thinking about maintenance before the start-up of any new plant. To those of us in the maintenance fraternity the resultant benefits are obvious. Activities include: having maintenance input at the design stage; ensuring maintainability in design; ensuring quality information is received from vendors; optimizing, selecting and purchasing start-up and one-year spares; developing maintenance plans; and having selected and trained (educated) maintenance personnel prior to start-up.

Questions remain of how to go about doing all this. How many people are needed to make this happen? When is the best time to introduce a maintenance development team to the project and at what cost? Is there a typical plan and time frame for all these activities? What problems will be confronted and what are some of the solutions?

This paper, by illustration of case studies within the BHP/Billiton group, will cover the above issues and provide a better insight into what is meant by ´Maintenance Scheduling, Management and Training at Start-up´.

INTRODUCTION

The last few years, there has been a change in the thinking and commitment to new business ventures. The traditional approach to projects has been: Equipment Design and Delivery, Construction, Commissioning and Handover. The maintaining aspect of the project has often been neglected. Today BHPBilliton no longer considers these projects as "projects" but "business ventures" and include as part of the venture, the maintenance function. The objective is *to reduce overall venture development costs whilst maximizing start-up effectiveness and subsequent whole of life venture costs and the key theme in using this approach is do the right things, at the right time, in the right sequence and to the right level of detail*[2]. HATCH, in conjunction with the Global Maintenance Network of BHPBilliton, has developed the ´Venture Maintainability Guidelines´ in order to realize the overall objectives.

The development of the Maintenance Function is a project in itself and with each new venture there is a trend to appoint a Maintenance Development Manager to deliver the maintenance function. Some examples of such appointments have been at Escondida Oxide plant in Chile in 1998, and the Oxide plant at Tintaya in Peru and the Phase 4 Expansion at Escondida in Chile in 2000. The venture maintainability guidelines become the road map to achieve the set objectives. Figure 1 below identifies the tasks ahead and also shows where the EPCM approach interacts with the Maintenance Function.

This paper follows the activities as they occurred during the process of developing the maintenance function at BHPBilliton Tintaya Oxide project.

---

[1] Maintenance Manager – HATCH, Peru.
[2] From ´Venture Maintainability Guideline´, May 1999, HATCH & BHPBilliton GMN.

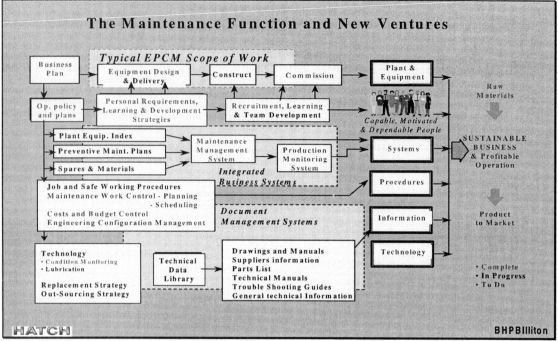

Figure 1: Venture Maintainability Process Flow

This paper is a "worked example" of using the Venture Maintainability Guidelines at the Oxide Leach Plant venture at Tintaya in Peru. The paper looks at how the Venture Maintainability Guidelines were applied, the issues confronting those responsible for setting up the Maintenance Function for a new project, some helpful hints and ideas, and opportunities to improve for the next project.

What is Venture Maintainability?
Simply put, it is about minimizing the total venture development and early life operating costs, whilst maximizing start-up effectiveness and subsequent whole-of-life venture performane. We use *Venture* Maintainability and not *Project* because *Venture* implies the development of a concept through financial justification, Board approval, design, construction, set-up and then sustainable and profitable business operation. Whereas *Project* has traditionally implied a more specific focus on the design, procurement, construction and commissioning of the plant and equipment associated with a venture. In the past, the relationship between "the Project" and "Operations" was at best poorly integrated, and could even be adversarial. A Venture approach aims to pull all these players together.

In practical terms, Venture Maintainability aims to ensure plant and equipment "operability", "maintainability" and a logistics start-up plan are considered and incorporated into capital ventures from the time of conception; not left until after commissioning. It is not necessarily about doing something new, nor is it about increasing project costs – rather it is designed to leverage what we know now and do it better. The underlying premise is to carry out the venture development by doing the right things, at the right time, in the right sequence, and to the right level of detail.

In the context of the Venture Maintainability process, the concept of *maintainability* can be defined by the desired outcomes of the maintainability approach:

- The provision of capable, motivated and dependable people at the right time to be fully prepared to take ownership and responsibility for carrying out the work required.
- The provision of an appropriate organizational learning environment within which those people can develop and work effectively.
- The provision of appropriate information and knowledge in the right form, at the right time.

- The provision of appropriate business systems and support facilities to enable those people to work together effectively.
- The provision of plant and equipment which is accessible and safe to operate and maintain.
- The provision of appropriate operating and maintenance plans for effective management of equipment condition and performance.
- The provision of the appropriate tools and facilities to support safe and effective operations and maintenance.
- The provision of an appropriate spares support strategy.

THE TINTAYA OXIDE PROJECT

The Oxide Leach plant was built along side an existing Copper Concentrate plant at BHPBilliton Tintaya in the hills (4,200 meters above sea level) of the Andes in southern Peru. Cathode copper production rate is planned at 34,000 tpa in the first two years and rising to 40,000 tpa for the next five years. The plant economic life is seven years. The Oxide Process Flowsheet is shown below:

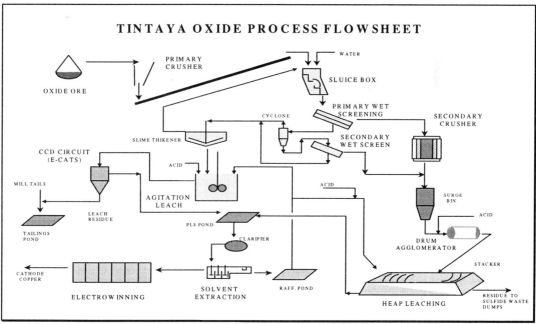

Figure 2: Tintaya Oxide Process Flow diagram.

The overall workforce, operators and maintenance personnel number eighty. The maintenance team is made up of one Maintenance Superintendent, two planners, one reliability engineer, five electrical and six mechanical trades staff.

DEVELOPING THE MAINTENANCE FUNCTION FOR THE OXIDE PROJECT

The paper is titled *Maintenance Scheduling and Training at Start-up,* however as you will see, there is much more involved than the title implies. For this reason, The term *"Maintenance Function"* has been chosen to explain all the maintenance type activities that are necessary before start-up.

The Maintenance Development Manager and his team

The project had a number of stop/starts (due to poor copper prices) but finally was given the green light to proceed in February 2000. The project was resurrected at 35% design completion. The Maintenance Development Manager was appointed to the design team almost three months after the restart of the project and at 40% design completion.

The overall initial objective was to have all aspects of Maintenance Function (MF) in place before start-up. Activities to support this objective were to:

- Develop an overall Maintenance Function Development Plan (MFDP) of how to plan to develop the MF and clearly define what was required.
- Develop a Capital Budget to realize the Maintenance Function that was submitted to management for approval. This budget also included the cost of other engineers required to assist the Maintenance Development Manager achieve the objectives.
- Identify manning requirements for maintenance.
- Carry out a Maintenance Criticality Assessment on all equipment in order to obtain an agreed list of critical equipment. This agreed list of equipment will be used initially, to prioritise maintenance development efforts and then prioritise ongoing maintenance and operational activities during production.

An additional objective was to try and optimize/minimize the overall maintenance development costs by taking advantage of work that had already been carried out by the Maintenance Development Manager for the BHPBilliton Oxide plant at Escondida in Chile.

Results:

- Maintenance Function Development Plan (MFDP) was completed within the first two months and was the blue print for the work ahead. Advantage was taken from the MFDP developed for Escondida Phase 4 project. The MFDP document incorporated "best practice" ideas from BHPBilliton Global Maintenance Network (GMN)[1], PHASE IV PROJECT, moreCLASS[2] ideals and HATCH´s Venture Maintainability. This document is available to the next Maintenance Development Manager and can be easily configured to suit the requirement.
- Maintenance Capital Budget – once the MFD Plan was completed the preparation of the capital budget was relatively straight forward. Details included:
    - Cost of two Maintenance Development Engineers for spares, special tools, vendor manual review and specific equipment data collection.
    - Development of equipment codes and preventive/predictive plans – the maintenance strategy for the equipment.
    - Development of an overall Condition Monitoring Strategy.
    - Development of first year and critical corrective procedures.
    - Development of instrument calibration procedures.
    - Development of lifting procedures
    - Development of isolation procedures and risk analysis review.
    - Development of an electronic technical library.
    - Translational service costs.
    - An allowance for special development.
    - Training activities.
    - First year operating and insurance spares.

- An equipment criticality assessment was conducted using a maintenance criticality assessment system called MCAS[3]. The assessment was conducted with the operations manager and selected process and equipment experts. An example of the end result of the criticality assessment review is seen in Figure 3.

---

[1] **GMN** – A dedicated group/network of maintenance practitioners within BHPBilliton.

[2] **moreCLASS** – **M**inerals **O**perating Excellence Capability Assurance Strategy - name given to the strategic direction for maintenance within BHPBilliton.

[3] **MCAS** – a HATCH proprietary decision support software.

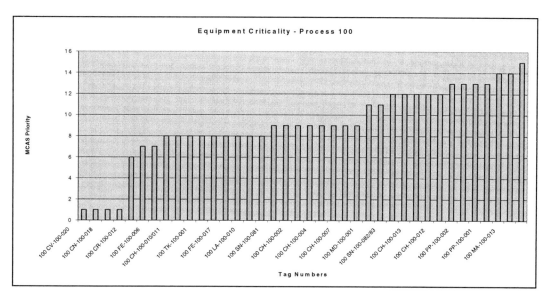

Figure 3: Graphical output from MCAS showing final criticality priorities of equipment

Issues/tips
MFDP

- **Overall Timeline Plan** – a timeline model (developed at Phase 4) was used, see figure 4 below. No detailed, project type plan was developed. The bulk of the project was managed from this initial timeline, however a detailed plan was developed for the last three months of the project. Some people however may feel more comfortable with a detailed plan and that is entirely up to the individual.

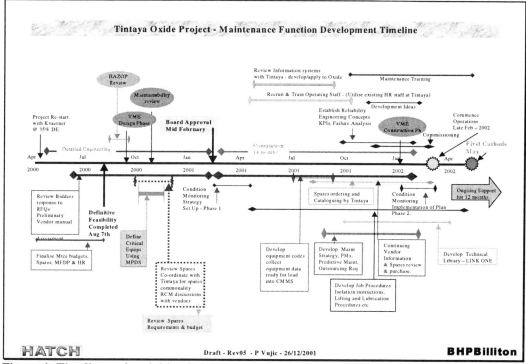

Figure 4: Timeline and activities developing the Maintenance Function.

- **Determining development costs of projects within the plan** – Previous experience in this area proved very valuable and close to the mark. Using local companies in some areas of the maintenance development work proved successful from a cost saving point of view in particular and the work was also of a high standard. (Some money was given to back to the project manager).
- **Determining the initial spares budget** – ideally, the "first year operating spares" (The term "first year operating spares" has been used for convenience only, however it should be noted that the term "first year spares" implies initial purchases of spares. Spares may be insurance type or on consignment or required for the first year of operation and are determined after discussions with vendors on equipment function and failure modes) budget should be created from a detailed study in conjunction with the development of the equipment maintenance plans, however this is not realistic. The detailed maintenance plan cannot be developed until:
    o all the vendor manuals including data specifications are available,
    o after discussions with vendors about their equipment, related to reliability, past experiences, failures and failure modes. (this is a time consuming exercise and needs to be carefully managed, most vendors reside in the states). Face to face contact is ideal however telephone conference calls were also used and proved reasonably successful.
    o other sites with similar equipment have been canvassed, (again a time consuming exercise but very valuable)
    o final equipment design and selection. This was a problem as there were ongoing design changes on some equipment well into the project.

The spares budget was developed by reviewing the equipment list and making a considered judgement on what spares would be required and by consulting experienced people from within the organizations. There is a rule of thumb and percentages from past projects but we opted not to use them. There was sufficient doubt, due to lack of historical evidence, to substantiate these percentages.

One of the objectives was to optimize on spares purchase. With dedicated people and well defined maintenance strategy, spares should be a minimum. Spares requirements and costs could be a function of the equipment expenditure and not the overall project value. The initial spares recommendation was accepted but was also criticized based on previous projects that it was too light. Advice given was found to be closer to the real value of spares. The spares cost was approximately 7% of the cost of the equipment and for future projects a suggested percentage of 7.5% is recommended for the initial spares budget. The spares issue is further discussed later in the paper.

- **Determination of the number on the maintenance team and the operating budget**
    There are a number of schools of thought here. One is to use benchmarking and industry standards. The danger here is the competitive nature of being the best, each business trying to get the lowest ratio of employees per output unit. Are we sure we have enough people to do the job efficiently and effectively or are we just trying to compete? Do we really understand the demands on the maintenance team?

The approach used was to work from first principles, a simpleapproach. It is the number of equipment and components that will determine the workload for the maintenance team. The maintenance objective is 100% reliability of equipment – zero failures (during the required operating period). The only way to achieve this is to be able to understand or know equipment condition at any given time. To understand equipment condition we have to understand the failure behaviour of each component because the ´equipment´ doesn't fail as such, it is the individual components that make up the equipment that fail.

If we appreciate the number of components and each component has an average of three failure modes then there is an inspection activity associated with each failure mode – we can soon calculate the required time and people required for inspection tasks alone.

In addition to the inspection tasks are shutdowns and repairs, which again can be calculated by understanding the failure modes. Inspections make up 30% of the workload with another 60% of the workload as planned corrective maintenance. (These percentages are consistent with studies conducted by BHPBilliton Global Maintenance team). Consideration should also be given to a commitment of approximately four weeks per year for ongoing training of personnel (implies a working year of approx 1800 hrs).

- **Timing of the start of the Maintenance Development Team.**
  - **Maintenance Development Manager**: Initial thoughts on the appointment of a Maintenance Development Manager were for him/her to be appointed at the beginning of the project full time.
    - That appointment should be made a the beginning of the project and involvement on a part time basis early in the project, to be involved in feasibility studies, involvement in the bidders requirement documents (to ensure that maintenance requirements are met). A fulltime involvement at no later than 30% engineering completions is recommended.
    - Alternatively, for the BHPBilliton group (possibly via GMN) to create a specialist "Maintenance Feasibility Development" role that who is involved fulltime in the feasibility phase and initial bidders requirements to ensure the Maintenance side of the business venture is understood.

- **Maintenance Development Engineer(s)**: One engineer (as a minimum) commenced approximately four months after bids were received. His main role was to review spares (start-up and first year spares) recommendations and centralize all equipment maintenance related data, i.e. folders were created against Purchase Order numbers and relevant information was copied and included in each folder. This exercise proved worthwhile and saved valuable time later in the project. This engineer continued to work full time on spares review and ordering (via the existing Tintaya stores group). Spares review will be considered separately later in the paper.
- **Maintenance Development Engineer(s)**: A second engineer commenced once the project was formally approved. His role was document control and management and included review of vendor manuals; collection of lubrication data; management of the contracts for corrective procedures and the development of the technical library.

Spares Requirements for the first year of operation

This by far was the most interesting part of the whole project and also presents an area of opportunity to save time and money by getting the vendors to change their ways. The overall objective was to optimize on the number (and cost) of the first year spares requirements.

It was decided that the purchase of the first year spares would be organized by the owners as apposed to the project contractors. Note that the start-up spares were the responsibility of the project contractors.

The first task of the Maintenance Development Manager was to issue to the project contractors a Maintenance Bid Requirement Document [4] as an addendum to the purchase orders. This document spelled out in detail what maintenance type information was required with the bid and included, format and presentation of spares details, format and presentation of lubrication data, standard of vendors manuals and requirements of maintenance recommendations. Unfortunately only a handful of vendors complied which made the task of spares review a very lengthy exercise.

---

[4] **Maintenance Bid Requirement Document** – developed by the Escondida Phase 4 Maintenance Development Manager.

- **Reviewing the data submitted by vendors** An introduction to world of spares. Spares details were submitted in various formats which included, faxed copies of outdated recommended spares, word documents, different Excel type spread sheets, ink jet copies with hand written changes, other copies with the description truncated (printer set incorrectly) and all fell well short of the requirements. This format was supposed to save time and money during review and it is still believe it will provided that vendors comply. As can be expected the next step was to contact each vendor in turn to seek clarification of details and cost of all spares. This continued for the length of the project (into 2002).
- **Reviewing the spares recommendations with the vendors.** The objectives here were threefold. First, to take advantage of the equipment vendor´s knowledge and experience with his equipment to gain valuable information of the possible failure modes and past experiences that they may have in order to develop the best maintenance strategy possible for the equipment. A modified Reliability Centred Maintenance (RCM) approach was used and well appreciated by the vendors. Second, to review the recommended spares submitted by the vendors in light of the discussions and come to a mutual agreement on spares requirements. It was interesting to note that many vendors willingly agreed to change the number of spares required after the review. Third, to clarify all spares details. These tasks took time, organizing discussions and countless email communications. To date the benefits of the review have been:

    o A "saving" of almost US$200,000 on vendor recommended first year spares.
    o Identification of critical spares overlooked by vendors.
    o Correction of details that may have resulted in the wrong spares being ordered.
    o Having all information available so that spares can be catalogued and ordered with a minimum delay by Tintaya´s purchasing department.
    o Agreements on ´on consignment´ and vendor held stock – mainly related to known wear items.

- **Purchase of Spares by Tintaya.** Due to the detailed analysis conducted, the ordering of spares was relatively straight forward. The detailed review gave us confidence that the correct spares will be available when required, eliminating one of the possible failure modes of equipment and often cause for prolonged equipment delays – the wrong spare!
  In addition, in order to save time and money and being conscious of equipment warranty, Oxide took the decision to buy ´first year operating spares´ direct from the original vendors. This policy saved many hassles and enquiries from local agents who were trying to sell us equivalent spares, references and requests from ´friends of friends´ and enquiries from our own stores personnel. A review of the equipment under operating conditions is the responsibility of our maintenance team and is part of the continuous improvement cycle. They will decide (cost and reliability point of view) later if other vendors will be invited to submit alternative spares.
- The information collected will also be valuable when the contractor hands over their start-up spares. Purchases by the contractor were based on the initial information included with the purchase orders this was shown to be lacking and on handover to Oxide the maintenance development team had to go through the same process in getting detailed information required to catalogue the spares in the Stores.

Condition Monitoring Strategy
The focus today is on predictive maintenance, also known as Condition Monitoring. HATCH was commissioned to develop an initial strategy and then to assist in its implementation (assisting the nominated reliability engineer) just before, during and after start-up. The development of the strategy was made easier as the maintenance development team had all relevant equipment data available in separate folders.

There was a strong focus on "clean oil" and monitoring of same. During the development phase of the condition monitoring strategy it was found that many gearboxes lacked oil sampling points and/or sight glasses. In hind sight, it would have been more appropriate if this level of detail was carried out during the equipment selection phase to ensure when equipment arrived on sight they had the appropriate facilities to monitor oils easily.

Development of Corrective Procedures.
This activity is a time consuming one and a local Peruvian engineering company was commissioned to develop these. Corrective procedures take an average of two to three days to complete and considering that some three hundred initial procedures were required, the estimated time for a team of three people was one hundred days.

The timing of this activity was approximately four months after the project approval when some 50% of the vendor manuals were available. Tintaya Oxide has a preferred format for standard procedures and the contractor's task was to extract the information from the vendor manuals and develop the Tintaya standard.

There were two stages to the development of these procedures. Approximately 80% of the corrective procedure details were obtained from the vendor manuals, the other 20% was a sanity check of the procedure once the equipment was physically on site.

Development of the equipment maintenance strategy. (preventive and predictive plans).
Previously developed guidelines for the development of equipment maintenance strategies and these guidelines were used on the Oxide project. The approach is one based on the RCM methodology and was supported by the software system called MPDS[5] (Maintenance Plan Development System). A team of engineers was contracted to develop the maintenance strategies. An overview of the process is seen in figure 5.

Concepts in the Maintenance Strategy Development Process

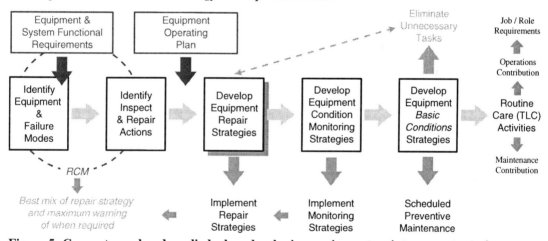

Figure 5: Concepts used and applied when developing equipment maintenance strategies.

- **Issues – Ownership of the Maintenance Strategy**

Ideally, this process is carried out with the maintenance team so that they have ownership of the strategies developed. A compromise was required at Oxide as there was insufficient time for the maintenance team to be involved in all the analysis due to process and specific equipment training programs. The compromise reached included that the initial equipment strategy reviews were conducted with the maintenance team with the objective for them to appreciate the review, input into the strategies and to gain confidence not only in the approach but the strategy development

---

[5] **MPDS** – maintenance decision software support tool provided by HATCH was developed by BHP/BHPE.

team who would continue with the project. The data collected from the vendor spares reviews proved very valuable during the development of the equipment strategies.

Development of Calibration Procedures.

Instrumentation has been, for a long time, neglected in terms of having appropriate maintenance strategies. Today, with automation of our plants, instruments must be given the attention they deserve. They are controlling devices, protection devices, safety devices, environmental protection devices and quality devices. The reliability and accuracy of these devices is vital to reliable production and its quality.

Regular calibrations of instruments is an important part of the maintenance strategy for the plant. HATCH were commissioned to undertake this exercise as they have had past experience in this type of work (in particular at ESCONDIDA Oxide in Chile) and have built up a library of calibration procedures for a large number of instrument types.

Development of Lifting Procedures.

Any lifting activity should be clearly defined by a procedure because of the inherent safety risks with the activity of lifting equipment. Vendor manuals were surprisingly lacking in detail and clarity of lifts in this area. Without lifting procedures, the risk of accidents increases, as people may try anything to carry out a lift. Each corrective procedure was reviewed and any activity that required lifting was developed into a procedure. The procedure included:

- calculations to determine the correct wire rope and lifting angles.
- weight calculations and nominated hoist or crane to carry out the lift safely.
- clear photos or sketches to identify the lifting lugs or connection points.
- analysis of risks during the lift.

Development of an Electronic Library.

It is important to get standardization of manuals from the vendors because at the end of the day you want anyone to be able to search and FIND information quickly, in particular the planners. Simply storing electronic copies of vendor manuals is not the answer to an electronic technical library. Each manual differs in format, each manual tends to use different maintenance terminology, each manual has maintenance information in different locations throughout the manuals and are not conducive to finding specific information quickly.

Where electronic manuals were not available from the vendors, the manuals can be digitized (Oxide project utilized a company called *metanoia*[6] *in Chile). The digitized manuals were then edited to a standardized format. A standard contents page was then developed to facilitate the finding of information as required.*

This exercise may appear to be costly but the opportunity exists for substantial savings for future projects especially if we can get vendors to work towards standardizing their manuals.

Special Development Projects.

It is a good idea to reserve some funds for special development work. ´Special Development´ means; bright ideas, enhancement of procedures, special guests, training and leaving the door open for the ability to fund improvements and pursue those bright ideas.

For the Oxide project there were three special development projects and they consisted of:

- Development of Posters The posters were designed to reinforce the maintenance philosophy and maintenance message. Motivational posters such as: TLC[7] – Tighten Lubricate and Clean, a very basic but extremely important maintenance philosophy.

---

[6] *metanoia* – Metanoia Ltda of Santiago in Chile.
[7] **TLC** – The origin of TLC comes from Tender Loving Care and an analogy is drawn with the need to look after equipment in a similar manner.

- Taking photos and mounting them next to the new equipment. The challenge to the team is to keep the equipment in the as new condition. The photos are a motivator and challenge for the total operating crew and reinforces the TLC philosophy.
- I NEED ATTENTION (INA) tags. Tags that had two parts. The top part is detached and using a tie, is secured on or close to the equipment that needs attention. The detached part has a unique number, the equipment tag number is identified, date and signature is included. The bottom portion also has the same number and tag number but also room for text to further describe the attention required. Main benefits include:
    o All equipment that needs attention has an identifying tag (no equipment is forgotten)
    o Reduced safety risk of working on the wrong job – the work order for repair references the INA tag number.
    o Provides a visual view on the condition of the plant and valuable for the Maintenance Supt on his walk around inspections.

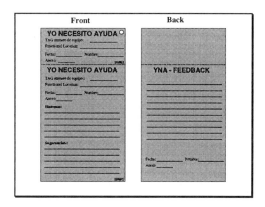

Equipment and Component Tagging.
The major equipment had contractor tag numbers installed (part of the contract) however there was much equipment (valves, instruments, miscellaneous pumps etc) that did not have specific equipment tag numbers. Unique equipment/component identification is important, first and foremost from a safety point of view and then from a point of view of equipment history and costing.

Equipment tagging would be carried out once the 'Functional Location' (term used in the maintenance module of the SAP business management system used at Tintaya) for equipment had been completed. Each component would have a tag prepared with a dual Functional Location and Equipment Tag No. These would be placed adjacent to the components.

People Selection and Training Activities.
Capable people are an important ingredient in ensuring equipment and process reliability. It has been said that the root cause of any failure can be ultimately traced to 'people'.

- **People Selection**
  Selecting the "right" people is half the battle in ensuring there are capable people. At Tintaya, a DDI[8] selection process was used. The process proved successful. It was the first time Tintaya had used the process and the opportunity was taken to review the selection process itself. DDI were given feedback and agreed with the recommendations. Some recommendations were to streamline the process as it is a time consuming task.
  Oxide has one dedicated person responsible for -coordinating selection and training activities. The job was very demanding but well supported not only by all of us on the management team at Oxide but by training personnel from Escondida Phase 4.

---

[8] **DDI** - Development Dimensions International - Addison, Texas, USA.

- **Training Activities**

  The Escondida Phase 4 project currently underway has a dedicated team of "training" specialists. Their focus has been to ensure that any learning activities undertaken will be prepared, delivered and assessed in a way to maximize benefits to the participants. They developed learning guidelines which were sent to selected vendors of equipment who were invited to offer training.

  Oxide also made use of these guidelines when inviting vendors. The response to the guidelines was very encouraging and some vendors commented that the guidelines really made them think again on how they were traditionally delivering this particular service.

  We also made extensive use of the local technical college, TECSUP with refresher courses on basic mechanics, electrics, hydraulics, pneumatics and instrumentation.

  Vendor training was also carried on targeted critical equipment. Further equipment specific training is planned after start-up of operations.

Managing Maintenance.

The responsibility of ´managing´ maintenance falls on the shoulders of the planners. Everyone has a part to play of course but the planners are the central figure. The overall process is seen below in figure 6.

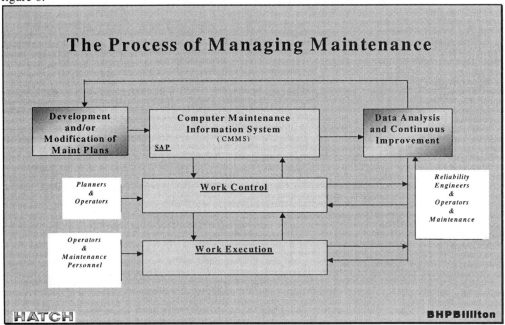

Figure 6: The process of managing maintenance and the interrelationships with other members of the team.

The computerized maintenance management system used at Oxide will be SAP. A "pipeline process" was developed to re-educate planners in the process of managing maintenance with a strong emphasis on correct planning and scheduling. Another strong theme and objective for Oxide is that ´all work will be planned´ and it was imperative that the planners understood the function of planning and scheduling. During the process KPI´s (Key Performance Indicators) were identified and set. The exercise proved valuable and is recommended for all planners.

Figure 7 below demonstrates the ´pipeline process´ for planning.

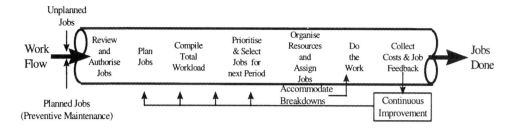

Figure 7: The 'Pipeline Process' for planning and scheduling.

Project Audits and Reviews.
Three audits or reviews were planned and conducted on the Oxide Maintenance Function Development project. They were:

- Venture Maintainability Review early in the Project
- Equipment Maintainability Review at about 80% design phase
- Mid term Venture Maintainability Review

The reason for the reviews are: to demonstrate support to the project manager by the maintenance fraternity in BHPBilliton, to identify issues within the project so that corrective action/s can take place, to ensure a consistent approach to developing the maintenance function and to identify opportunities to improve the overall process of developing the maintenance function. GMN/HATCH developed what is called the VME, Venture Maintainability Evaluation and looks at all aspects and interrelationships within the project. This evaluation can be conducted as a series of interviews with project personnel or as a self evaluation.

The Equipment Maintainability Review process looks specifically at equipment maintainability and considers such things as, accessing equipment for both predictive and corrective maintenance, standardisation of equipment, facilities to repair equipment . The timing of this review is recommended at approximately 80% design. If the review is conducted too early there is usually insufficient information to give justice to the review. If it is done too late, example design completion, all information is available however it is virtually impossible to influence any changes.

CONCLUSION
For the uninitiated, developing the maintenance function or preparing 'maintenance' before start-up can be a daunting exercise. This paper has shown that there are a multitude of issues to consider but with proper support in particular the standards and guidelines, the task is no longer daunting. It is still complex and involved as all projects are but very manageable. The paper also shows that there is opportunity for improvement and potential to reduce the overall capital investment of developing the maintenance function and for the venture itself.

ACKNOWLEDGEMENTS

1. BHPBilliton for granting permission for me use the Oxide project as my worked example.
2. Alan Pangbourne (Oxide Project Manager) for his support and encouragement.
3. GMN (Global Maintenance Network) for their support and input.
4. Mike Duggan (Maintenance Development Manager for the Phase 4 project) of whom I hold high regard in this area of developing the maintenance function. Many of the ideas and methodologies developed by Mike and his team I have used to best advantage in the Oxide project.
5. The whole Tintaya Oxide team and their commitment and support to what we a trying to achieve with developing the 'maintenance function'.

Operator Training

André Vien[1]

ABSTRACT

Training, taken in the broader context, actually involves two separate functions: education, which teaches knowledge based on principles; and training, which teaches skills based on procedures. Both are essential to shorten the learning curve during the start-up of a new piece of equipment, and indeed of an entirely new plant. Shortening this learning curve has a significant impact on attaining, and possibly exceeding, design production targets sooner. Combine this with the requirements to have a significant portion of indigenous workforce, which may not have had any exposure to mining, the necessity to undertake an educational program is heightened.

INTRODUCTION

When we buy a piece of process equipment it is critical that the mechanical installation be done properly to ensure that the device will operate at peak performance on start-up. Once the equipment is running, some form of preventive maintenance is required to ensure continued efficient operation. The same philosophy should be applied to operator training, as they are also elements of the entire processing system. That is, operator training should be considered an integral part to circuit operation. Using the analogy of a piece of machinery, operators need proper installation (initial training) and preventive maintenance (continuous training - or refreshers - after start-up).

In the mineral processing industry, operator training has been quite variable. Their knowledge and skills are very much a function of training and experience, which are clearly corporate and site specific. In fact, it can be crew specific. A quantitative way to assess this is to conduct a statistical analysis of historical production data (Vien et al. 1994a). If one operator or crew is statistically superior to the others, then the performance difference is a measure of the benefits of better training, i.e. bringing everyone to at least the same level as the best operator/crew. Variation of performance between operators is an indication of training deficiencies. The cost of these training deficiencies can be very significant, and in some cases in of the order of $US500,000 per year (Vien et al. 1994a). Potential benefits of this magnitude seem to be the rule, and not the exception.

Another telltale sign of training deficiency is that in many plants the operating crews employ quite different strategies, all in pursuit of a common operating goal. This is typified by the changes made to set points just after a shift change. Although the different strategies may produce similar metallurgical results, typically, one will be more economical than the others. Additionally, changing the set points will create a disturbance and hence, for a period of time, circuit operation will not be optimal. The existence of different operating strategies is therefore a clear indication of the need for better training.

As the trend for newer mines is to be located in regions where mining is foreign to a significant portion of the indigenous workforce, an educational program to teach the fundamentals of mining and mineral processing is now essential to most new mine sites. This educational program must start at the very beginning by describing the series of transformations required to go from a rock to a metal product. This provides a foundation to provide context and relevance to the training on operating procedures and operator duties.

---

[1] Metso Minerals, Business Line Mineral Processing – Process Technology, Kelowna, B.C., Canada

Under the current economic pressures, a good operator must possess good knowledge and good skills in order for the company not only to perform, but indeed to survive. Knowledge and skills are taken to be equivalent to theoretical and practical abilities, respectively. To fully develop in these areas an operator requires a good grounding in both procedures and principles, as well as the opportunity to apply and hone his/her knowledge and skill sets. As Figure 1 indicates (the text appearing to the far right is simply the dictionary definition of the corresponding term on the left.), knowledge is imparted through education whereas skills are imparted through training. This difference is explored in more detail in the next section. Since the reader may not be familiar with educational material aimed at the mineral processing plant operator, selected material of such a course is presented.

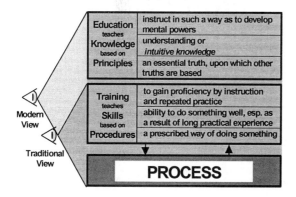

Figure 1: A perspective on education and training in the mineral processing industry (Vien *et al.* 1994b)

As a separate topic from the education vs training issue is the question of mode of delivery. Should the material be delivered in a classroom setting or through computer based training or a combination of both? The section on computer based training addresses this issue and the potential benefits and pitfalls of such a system.

Finally, an overview of the contents of a training manual (procedures) is presented. The purpose of this section is to show that there is little overlap in the scope of material in a training manual compared to that of an educational program. They really serve two different purposes and both are essential.

EDUCATION VS TRAINING

As mentioned in the introduction, a distinction is made between education, which teaches knowledge based on principles; and training, which teaches skills based on procedures, as depicted in Figure 1. This is an important distinction. A trained operator will try to minimize upsets after they occur (applying cause and effect reasoning) whilst an educated operator can anticipate and mitigate upsets before they occur (applying conceptual models faithful to the physics and chemistry of the process).

Traditionally, the mining industry has been involved in operator training, and it is only recently that educational programs have emerged as an important aspect of operator development. However, and probably for historical reasons, the educational aspects are still combined under the general umbrella of training. While this conventional terminology is acknowledged, it is important to understand the roles of these two distinctly different activities. A truly effective training program will include the proper balance for the needs of the operation, and these two components should be properly integrated to support learning.

Training is aimed at issues such as safety, regulations and operating procedures (startup, shutdown, etc.). Education provides an understanding of the principles of operation behind the process, and of the principles of process control. Management's expectations of plant performance have been continually increasing in response to an increasingly competitive world. In order to meet these expectations, operators must actively endeavor to optimize the process, rather than to just get through the shift without a disruption. The level of knowledge required of the operator to meet today's performance expectations can only be achieved with a more thorough understanding of the process. This understanding also gives the educated operator the knowledge necessary to troubleshoot unforeseen situations.

One of the major differences between training and education is that training is explicit, e.g. a startup procedure, whereas education is implicit, e.g. reasoning about the current situation. This reasoning is only possible by providing the trainee with the fundamental blocks upon which our current understanding of the principles of operation is based (Vien et al. 1994a). To understand grinding circuit operation, one needs to first understand unit operations, such as a grinding mill. To understand unit operation, one needs to first understand fundamental concepts, such as particle size measurement, rate of breakage, fragmentation, liberation, etc. An educational program must therefore start by covering fundamentals, which include basic physico-chemical properties and engineering concepts (e.g. work index, liberation spectrum, rate of breakage, etc.). Only then can the design, operating and control characteristics of unit operations (e.g. grinding mills, cyclones, pumps, etc.) be added to the curriculum. Finally, the knowledge gained from the previous sections is used to perform reasoning on the dynamic impact of interactions between unit operations in the circuit, both for changes in process variables and for diagnostic reasoning. As can be seen, each level builds on the previously acquired knowledge to form a complete understanding of the principles of operation of a circuit.

Since the principles of operation are generic in nature, the knowledge imparted to the trainee is transferable to another plant or flowsheet. It can also be more easily transferred to entirely different situations. For example, a knowledge of the property of viscosity, as introduced in the context of milling, is useful in the understanding and diagnostics of other processes such as flotation, pumping, dewatering, etc. In contrast, the training on the site-specific procedures is not directly transferable to another flowsheet. Indeed, being skilled in the startup and shutdown of a grinding circuit does not help for the startup and shutdown of a flotation bank.

A beneficial side effect of implementing an educational system is that it provides a platform for unification of knowledge standards, development of a common jargon and more harmony in the conceptual understanding of milling processes between operators, management and process engineers. This, in turn, fosters more and better communication and interaction between all levels of personnel.

Illustration of Educational (Knowledge-Based) Content (after Rybinski et al. 2001)
For the purposes of illustrating knowledge-based content, the author has elected to describe the Metso CBT software package. The origins (Vien *et al.* 1994a), prototype evaluation test work (Vien *et al.* 1994b) and a more detailed illustration of content (Vien and Grondin 1996) have been described elsewhere. In this section, only a brief description and example[2] is presented.

As a background, it is interesting to note that the educational system was originally conceived as something that would enhance existing training programs in North American operations, i.e. where basic training programs already existed. The objective was to help operators extend their understanding of the principles behind their circuit operation to enable them to operate the circuits closer to optimum. However, where an educational system has perhaps been even more helpful is in new mining operations where the operators are recruited from local communities, and they have essentially no training or experience in the industry. In these cases, feedback from users indicates that the educational system is probably best used first, to provide a basis of understanding (context) for the normal skills training efforts.

The development approach for a given module consists of utilizing focus groups comprising: (a) academics and technical practitioners to set the curriculum and develop explanations, and (b)

---

[2] The example is from a course that is designed for an interactive multimedia system and thus it can only be roughly approximated here.

operators and subject matter experts to test relevance and refine explanations. The magnitude of the effort for development of educational material should not be underestimated. The grinding and flotation modules offered by Metso contain over 400 and 800 pages, respectively, and took several man-years to develop.

The development philosophy was designed to:

- educate operators – i.e. faithfully teach engineering principles without any mathematics, ensuring the explanations were based on sound science
- promote communication among operators and between operators and technical staff, by introducing a common conceptual understanding of the science and technology, and establishing a common vocabulary.

To try to illustrate the content accuracy and depth, as well as the level of the presentation, an example is drawn from the Metso CBT Flotation Module. This particular example, starting in the Fundamentals section, combines the processes of grinding and flotation using the liberation spectrum – as would be the case in most university courses on this subject. (The reader will have to accept that terms such as grade, recovery, etc. have already been explained elsewhere in the software.)

Figure 2 includes two of the numerous images from a narrated animation on a page in the Mineralogy topic area, introducing the idea of a Liberation Spectrum. The illustration shows that one can take broken particles and sort them on the basis of their grade (in this case expressed in terms of the copper mineral, chalcopyrite). It also shows that with finer grinds one tends to get better liberation, i.e. more material ends up in the two extreme regions, and less in the middle area. The resulting distribution of particles by grade is the liberation spectrum, and it is a handy practical (and theoretical) way to think about the role and results of comminution in mineral processing.

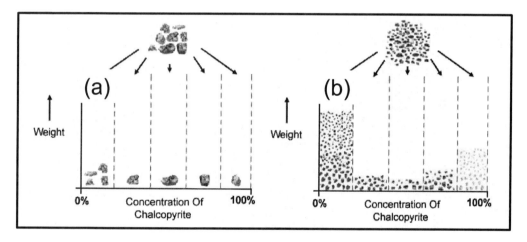

Figure 2: Introduction of the notion of liberation spectrum for grinding a copper ore

Having introduced the liberation spectrum, it now provides a basis for other explanations, including an illustration of how a perfect flotation process would operate; an explanation which is conceptually consistent with those provided for other separation devices, such as the hydrocyclone. Figure 3 is again a series of images taken from a narrated animation on a page dealing with the ultimate grade-recovery curve.

The graphic in Figure 3a simply illustrates the notion of perfect separation by drawing a vertical line, which represents the division of the feed material into a tailings and a concentrate product. Figure 3b expands on this idea by showing that different separations can be made by drawing the vertical line in different spots. The accompanying narration would point out that, for example, in the case of point B, the separation would produce a fairly high grade concentrate lots of pure and near-pure chalcopyrite, but the recovery would be low, as quite a lot of the locked chalcopyrite would be lost to tailings. The third graphic (Figure 3c) shows the various combinations of grade-recovery associated with the four points in Figure 3b, thereby linking grade-recovery to both the grinding (liberation) and separation performance. The final figure is simply an observation that because of inefficiencies in separation, the real grade-recovery curve lies below the ultimate grade-recovery.

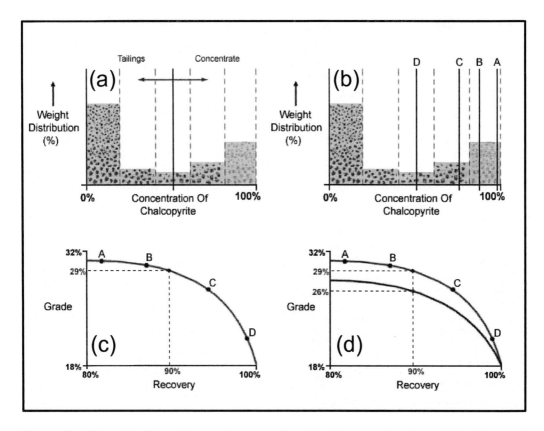

Figure 3: Using the liberation spectrum and a simple separator to explain the grade-recovery curve

In the section on unit operations, one of the effects explained relates to changing air flow rate. (Elsewhere the ideas of collision, attachment, detachment and entrainment are explained.) Increasing air results in increased recovery (see Figure 4), and because of the grade - recovery relationship, this means lower concentrate grade. The link between air flow rate and grade (recovery) control is thus established.

Finally, in the section on circuits, the impact of changing air flow rate on the entire cleaner circuit is illustrated (see Figure 5). Increasing air flow rate will increase cleaner concentrate recovery (point 2) at the expense of grade (the explanation would summarize and refer to the detailed description in the unit operation section described above). The explanation would continue by indicating that the cleaner tails grade and mass flow would drop (point 3). The consequence is that the cleaner scavenger concentrate grade and mass flow would drop (point 4), as would those of the cleaner scavenger tails (point 5). The end results is that the overall recovery is increased, but at a lower grade. The link between air flow rate and the overall cleaner circuit recovery is thus established, based on the descriptions in the unit operation section as well as the fundamental engineering concepts such as the grade-recovery curve and liberation spectrum.

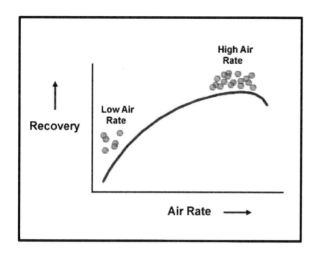

Figure 4: Illustrating the effect of air flow rate on copper recovery (and grade)

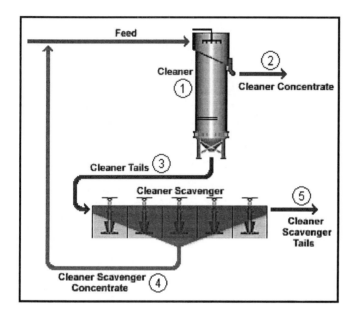

Figure 5: Illustrating the effect of air flow rate on recovery and grade in the cleaner circuit

Hopefully this example achieved its objectives. While it was short, the description conveys a reasonable picture of what operator education should entail. Perhaps this will also help to explain why it is that such systems are useful to bring operators and engineers together as a technical team, to some extent eliminating the traditional hierarchical relationship.

It is important to reiterate here that such educational systems are entirely focussed on these kinds of knowledge concepts, without regard for the basic safety and operations issues. These should be handled by a complementary (skills) training system, as described in the training manual section.

There can be little doubt that the reader would have had a more enjoyable time working though this example on a computer based system, rather than reading it here. Computer based training is the subject of the next section.

COMPUTER BASED TRAINING

To achieve this goal of increased process knowledge, principles of operation must be added to the list of operator training educational requirements. Since the principles of operation are generic in nature, the educational material can be standardized throughout the industry. This is desirable because it saves time and money by avoiding duplication of effort, and it provides the best material possible to the trainee. This can also apply to training on industry-wide standardized procedures such as first aid, lockout and rigging. The best way to disseminate this information widely is through computer based training.

Two major benefits of computer based training are the use of multi-media presentations and the ability to use simulation. Rather than have the trainee learn on-the-job with all the risks that entails, simulation can teach the trainee how to address a specific situation. It also allows the trainee to experience situations that are uncommon, without having to upset the plant or wait perhaps years before a certain problem occurs. Computer Aided Instruction (CAI) allows the trainees to see, hear and do!

Computer based instruction is the only method that provides a high level of retention of the material. This is principally due to the engaging and interactive nature of such a package. It keeps the interest level of the trainee high throughout the learning experience as compared to a classroom setting where the mind of the trainee may wander at times. As a result, the amount of time required for training is shorter. Different people learn at different rates. Computer aided instruction provides an individualized pace which is better than the common pace of a classroom setting, keeping the trainee alert throughout the learning process.

However, a computer based training system cannot anticipate all the questions and cannot provide interaction outside the preset system configuration. Discussions with a trainer are invaluable in this regard. In fact, a computer based training system, meant for individual use, has been successfully used in a facilitated group discussion mode (Rybinski et al. 2001). This ingenious use of computer based training assisted the trainer by providing the best of both worlds: a software package providing top-notch educational material and interaction with the trainee to foster better communication, deeper understanding and a greater sense of team work. The end result was a faster assimilation of the knowledge required for the operation of new equipment involving technology not previously used at the site.

Of course, skills practice, which can only be performed by manipulating the actual equipment, can only be done in the field and not with a computer based (or paper based) training system. Undoubtedly, preparation for this skills practice through computer based training makes the practice more effective and less time consuming.

In the introduction it was stated that training should be provided on a continual basis to maintain a high level of competency. A computer based training system can be made available twenty-four hours a day. Why provide training only during eight hours a day on week days when the operators are there twenty-four hours a day, seven days a week? Additionally, night shifts tend to be less busy than day shift. Operators could take advantage of the slack time for training. This also means that training can be provided in manageable chunks that are smaller than the saturation level of the learner (when the learner can no longer absorb the material being presented). Additionally, timely accessibility to training means that the material can be reviewed as required. It provides an on-going source of technical support to operators for diagnostic and decision support purposes.

Automated instructional software is readily available to create the course content. However, trainers should heed to the following warning:

> We are here to tell you that an authoring package does not a multimedia developer make. Please do not subject your trainees to a program developed with no more skills than word processing requires. This is cruel and unusual punishment. (Salopek 1998)

In other words, it is more important to focus on the contents and the judicious choice of media to enhance the learning experience as opposed to a glitzy presentation that is no more than a flash in the pan. "There's a social behavior of staying in the classroom. There's no such compunction against leaving a bad screen ... The tolerance for mediocrity is extra low." (Salopek 1998) To avoid this possibility the creator of the material should have a good grounding in instructional system design and understand the principles of adult learning e.g. when to use animations, when to use videos, what length should they be, how much information to put on a page, where on the page should it be, etc. There are several books on the topic, including that of Lee and Mamone 1995.

One of the keys in the design of a successful computer based training system is to recognize the various learning styles:

> A visual learner learns information best by seeing it, an auditory learner by hearing it, a kinesthetic learner by feeling or experiencing it. Effectively designed multimedia environments accommodate all three styles. The use of graphics, animation, video and audio appeal to people who learn by seeing or hearing. Getting kinesthetic learners to "feel it" is trickier, although good results can be obtained with simulation techniques. (Cohen and Rustad 1998)

In the end, it may be more effective to buy commercially available packages and/or contract out the work. In this role, the trainer becomes a manager of the training system rather than a developer. This benefits both the trainer and the trainee as it provides the trainer with (arguably) higher quality material and more time to devote to assessment and management of training needs.

TRAINING MANUALS

The purpose of training manuals, whether in printed or electronic form, is two-fold: to provide the initial skills training i.e. learning the safety, operational and emergency procedures; and to be used as a reference for refreshing and maintaining those skills. Unfortunately, the latter seldom occurs. This is partly due to difficulty in accessing the manuals at the time of need (locked up in an office as opposed to an open shelf, or available on a computer, in the control room) and partly due to the manuals being outdated (e.g. flowsheet or procedural changes). Training manuals must be kept current at all times so that they remain relevant to plant operators.

The training manuals should contain the following information (Wilmot, Sass and VanDeBeuken 1986):
- Cover letter (statement of objective and management's commitment)
- Table of content (to facilitate access to specific topics)
- Overview (process area covered, summary description, flowsheet)
- Safe job procedures (operational, maintenance, emergency, hazards, first aid)
- Process description (design basis, flowsheet, general arrangements, equipment description, brief explanation of how equipment works)
- Process control (loop narratives, interlocks, alarms, P&IDs, control system interface, instrument location)
- Operating procedures (start-up from long shutdown, start-up from short shutdown, start-up from emergency shutdown, short term shutdown, long term shutdown, emergency shutdown, changing circuit configuration)
- Operator duties and responsibilities
- Trouble-shooting guide.

Although the initial manuals cannot be completed until detailed engineering is performed, the basic operating guide can, and should, be created at the time of the P&ID review (Wilmot, Sass and VanDeBeuken 1986). Not only will it make the P&ID review easier but, perhaps more importantly, it can be used as a tool to make the final design more operation friendly. This is where extensive plant operating experience on the part of the design team is essential. No amount of operator training can compensate for a flawed design – they can only learn to cope with it.

CONCLUSIONS

Companies spend very large sums of money building mineral processing plant. To make the most of this investment it is necessary that it be operated well and safely. Both operator education (which teaches knowledge based on principles) and operator training (which teaches skills based on procedures) are essential components required to meet this objective.

Both knowledge and skills are vital to shorten the learning curve during the start-up of a new piece of equipment, and indeed of an entirely new plant. Shortening this learning curve has a significant impact on attaining, and possibly exceeding, design production targets sooner.

Training should be undertaken before, during and after plant construction. Even before detailed engineering is completed, but after the flowsheet has been selected, the educational portion of training can commence since the knowledge imparted is generic in nature. This provides the necessary foundation to minimize the time required to learn the site-specific skills, which can only be undertaken once detailed engineering has been completed (and based on current practices this means that the construction phase is well underway). After start-up, the knowledge and skills can be honed by providing continual access to education and training.

With fewer operators on the plant floor, each decision made has a greater economical impact since there is a lower chance for correction or review by other frontline personnel. Combined with the requirements to have a significant portion of indigenous workforce, which may not have had any exposure to mining, the necessity to undertake an educational program, possibly involving computer based training, is heightened. The material taught should include the fundamental principles of operation to provide a deeper understanding of the process. This will allow the operator to not only stabilize the process, but to optimize it as well.

Studies have shown that a higher retention of the material is obtained when a computer based instruction system is used. Computer based training can therefore be more effective. However, this is an aid and not a replacement for trainers as interaction with the trainee is essential to complete the learning process.

A computer based instruction system can be made available in one of several plant locations twenty-four hours a day. This access would allow operators to train at times other than during the normal staff working hours.

It is imperative that training manuals be kept up to date to remain relevant to the plant operator.

REFERENCES

Cohen, S.L. and J.M Rustad. 1998. High-Tech High Time? *Training & Development*. December 1998:31.

Lee, W.W., and R.A. Mamone. 1995. *The Computer Based Training Handbook: Assessment, Design, Development, Evaluation*. Englewood Cliffs: Educational Technology Publications.

Rybinski, E., R. Zunich, M. Grondin and B. Filntoff. 2001. Operator Education: An Important Element of the Corporate Knowledge Management Effort. *SME Annual Meeting*, Preprint 01-156, Denver, February

Salopek, J.J. 1998. Coolness is a State of Mind. *Training & Development*. November 1998:22

Vien, A., D. Willett, B.C. Flintoff and D.H. Hendriks. 1994a. Mill Operator Training – Where Do We Go?. *Proc. 26th Canadian Mineral Processors Mtg.*, Paper No. 16, Ottawa, January.

Vien, A., D. Willett, B.C. Flintoff and D.H. Hendriks. 1994b. Mill Operator Training: Prototype Evaluation. *Paper pres. 16th CIM District 6 Mtg.*. Vancouver, October.

Vien, A. and M. Grondin. 1996. Conceptual Model of AG/SAG Operation for Operator Training. *Proc. International Symposium on Autogenous and Semi-Autogenous Grinding Technology*. Eds. A.L. Mular, D.J. Barratt and D.A. Knight, 2:729.

Wimot, C.I., A. Sass and J. VanDeBeuken. 1986. Preparation of Operating Manuals. In *Design and Installation of Concentration and Dewatering Circuits*. eds. A.L. Mular and M.A. Anderson, Chapter 50. SME

Safety and Health Considerations and Procedures During Plant Start-up

Louis A. Schack

ABSTRACT

The safety and health of employees and contractors is of primary importance before and during any plant start-up. This is especially true with regard to the start-up of complex processing plants staffed by large numbers of employees with differing skills and levels of experience. The early assignment of employees to plant operations crews provides for detailed review and adaptation of start-up manuals and the development of safe practices well before actual start-up. Teams of employees must thoroughly examine and trace the production circuit as construction proceeds. Only well-trained, knowledgeable and safe operators can achieve or exceed the production targets envisioned by plant designers.

INTRODUCTION

Newmont Mining Corporation operates both surface and underground gold mines on Nevada's Carlin Trend. This geologic feature is at the center of most productive gold mining district in North America. The modern practice of mining low-grade deposits of disseminated gold began at Newmont's Carlin pit in 1965. Today, several surface and underground mines remain in production on the Carlin Trend, yielding more than three million ounces of gold annually.

Historically, on-site ore processing methods include cyanide leaching of low grade oxide ores, traditional milling (semi-autogenous grinding and ball mills) followed by gold recovery through carbon-in-leach (CIL) and carbon-in-column (CIC) circuits.

As mining in several surface pits expanded throughout the 1980s, exploration drilling revealed substantial deposits of refractory ores in which gold is associated with sulfide and carbonate mineralization. Such ores are not generally amenable to the traditional processing methods applied to oxide ores. These refractory deposits included ores accessible from existing surface pits such as Gold Quarry (1984) and ores of higher grade that presented opportunities for the development of major underground mines such as Carlin East (1993) and the multi-deposit Leeville Mine, now under construction.

Gold recovery from refractory deposits requires the application of complex milling and chemical processes including high-pressure oxidation (autoclave process) and fine grinding/high-temperature oxidation (roasting process). Newmont's 1997 acquisition of Santa Fe Pacific Gold Corporation included the purchase of two additional refractory processing facilities in Nevada: the Lone Tree Autoclave (near Valmy, Nev.) and the Sage Autoclaves (Twin Creeks Mine). Both facilities are located west of the Carlin Operations in Humboldt County.

Prior to the Santa Fe merger and in coordination with sequential mining plans for a minimum of 20 years of refractory ore production, Newmont Mining Corporation designed and built a Refractory Ore Treatment Plant (ROTP) adjacent to existing oxide milling facilities at the company's Carlin South Area Operations.

Construction of the ROTP began in 1993. The ROTP consists of a secondary crushing circuit, twin roasting circuits (north and south) fed by a dry grinding, double rotating mill with a nominal capacity of 352 st/hr. (2.8 Mt/yr.). The plant also includes air and gas preheating circuits, gas recovery, cleaning and cooling circuits and an acid recovery and storage circuit. The chemical constituents of many refractory ores create large volumes of recoverable sulfuric acid during the roasting process.

The roasters are of a Circulating Fluid Bed (CFB) design. Commissioning was completed and refractory ore processing began in 1994. At the time of its commissioning, the ROTP was the largest plant of its kind in the gold mining industry. The designs of many of its circuits and related components are unique to this particular mill. Initial design and construction capital approached $400 million.

The immense size, complexity, and intrinsic operational hazards of the ROTP and its ancillary facilities required a long term, comprehensive approach to employee training, operation and maintenance procedure development and coordination of activities between departments, vendors and construction contractors.

Of particular importance was the integration of safety considerations and situational response protocols into all operations and maintenance procedures. Beginning in mid-1993, a team of four experienced operations, electrical and maintenance supervisors were charged with establishing a training program for all employees involved in the start-up of this complex mill. To most involved, the ROTP and its accompanying processes presented many new and unknown challenges.

This team, dubbed the "Vector Group," eventually created more than 20 training modules covering all areas of operation and maintenance. Many others contributed to this core group's efforts during the months leading to start-up in late 1994.

STRATEGIC APPROACH

The acronym ROTP took on a secondary meaning as start-up preparations progressed - Reliable On-target Training Program. The process began with a comprehensive study of construction drawings, circuit schematics, flow charts and available manuals. In many cases, specific systems were broken down for detailed study of individual components. As each component was analyzed relative to its function in the larger combination of systems, the team developed a broad understanding of the plant's complex operation. Throughout this early phase, team members attempted to identify and assess potential hazards, establishing safe procedures and emergency response protocols.

The Vector Group also traveled to similar operations at various U.S. locations to study safety practices and to identify application opportunities for the ROTP. These visits included tours of acid plants in Idaho and Tennessee and a cement plant in Iowa. Special attention was paid to personal protective equipment and engineering controls.

Rather than developing procedures that support only availability and production goals, the teams were charged with building safety considerations into all operations and maintenance practices. Any procedure in which safety was not the primary concern was rejected. Throughout this process – up to and beyond start-up - vendors provided technical staff to assist Newmont teams in deciphering drawings and documents, identifying hazards and developing preventive measures and response plans.

The inclusion of those who would eventually manage the plant in the early development of all operations and maintenance procedures created a sense of ownership toward the final procedures manuals and a healthy respect for the process and its potential hazards. These advantages proved valuable as former start-up team leaders assembled crews and trained employees using materials they had personally developed and tested.

Several of the larger vendors, including Lurgi - supplier of the roasters, gas cleaning components and acid plant - and Krupp - supplier of the grinding circuit components - provided commissioning groups to participate directly in Newmont's start-up preparations. These vendors contributed to detailed investigation of all processes and shared anecdotal information from previous installations of similar processing systems. The role of these vendors' representatives proved vital to safe start-up of the plant. Two of the initial four members of the Vector Group were designated liaisons to Lurgi and Krupp, sharing information on a daily basis.

Newmont's internal safety department assigned a safety representative to review and contribute to the development of all safety practices. This representative also assisted in the testing, selection and purchase of all safety supplies and personal protective equipment.

OBSERVATION AND ADAPTATION

While initial efforts focused on studying technical materials and building a common understanding of the process, components and controls, it was recognized that technical knowledge is no substitute for practical, visual study of the complex process under construction.

Throughout the construction phase, team representatives toured the project twice weekly. As equipment was installed, photographic records created a permanent reference for many components that are enclosed or inaccessible during plant operation and routine maintenance. Where needed, drawings and flow charts were modified to reflect slight variations in component construction, circuit flows and integrated systems. Controls, sensors and other support equipment installations were verified and evaluated as well.

In most cases, operating and troubleshooting controls and software were provided by contracted suppliers. All control software was adapted and tested for applicability and reliability by Newmont Information Systems programmers. Software adaptation is a continuous process that allows for improvements to safe practices, systems integration and operational stability while minimizing downtime and lost production. Such improvements will continue throughout the operating life of the ROTP.

TRAINING MANUALS

A training manual covering operation, safety considerations, start-up and shutdown procedures and other activities was produced for each of the many processes and circuits that make up the ROTP. A typical manual includes 14 sections, the first of which includes this disclaimer:

> *"The instructions and procedures represent decisions based on limited knowledge of equipment and design criteria. As the construction evolves and more precise information becomes available, the procedures should be reviewed for correctness and applicability. This is especially critical for personnel safety, as changes may impact previous safety considerations."*

The training manual for employees in the Fine Ore Handling circuit begins with an *Organization and Application* section that details the manual's purpose - "to provide the information that operators need to know to operate the equipment properly, efficiently and safely." Also included in this section is a discussion of hazard recognition and the safe work practices each employee is expected to follow in each of seven operational conditions.

Section 2 includes a general description of the circuit and the function of each piece of equipment, including dust collection and air filtration. Design capacity and operation limits are detailed as well.

Section 3 provides general safety procedures, with descriptions of personal protective equipment and basic procedural guidelines.

Section 4 describes hazardous tasks and begins with lockout procedures for all equipment within the circuit. The section covers safety considerations for working with compressed air, working inside an ore storage bin, using hand tools in confined spaces, cleaning plugged transfer points and other situations.

Sections 5 through 11 explain in detail the steps for safe operation of the circuit in each of seven operational conditions:

1. start-up from complete shutdown
2. start-up from standby shutdown
3. start-up from emergency shutdown
4. complete shutdown
5. standby shutdown
6. emergency shutdown
7. normal operation

Each section includes cross-references to applicable drawings and flowcharts.

Section 12 details the effect failures of or upset conditions in one piece of equipment or area can have on the circuit as a whole. Several tables guide the trainee through a troubleshooting process that includes the alarms and control features that are activated under each condition.

Section 13 provides appropriate procedures for inspection and preventive maintenance of major equipment before start-up and during normal operation.

Section 14 is a series of drawings of each piece of equipment and flow charts of the Fine Ore handling circuit.

As stated above in the disclaimer, several years of experience in operating the plant have required varying levels of modification to the training materials for each ROTP work area and process circuit.

EMPLOYEE HEALTH AND SAFETY

The ROTP presents a great number of health and safety concerns for employees and contracted workers. These include various corrosive acids, chemical by-products, gases and reagents. Many of these materials are contained at high pressures and temperatures well over 1,000° F. Other important hazards include burning fuels, high voltage electrical sources, rotating equipment, fall hazards and confined spaces. Maintenance employees undergo extensive training in confined space and "hot work" safety.

Members of the start-up teams developed specific safety guidelines and personal protective equipment (PPE) requirements that reflect the hazards present in each area of the plant. PPE requirements range from respirators for all employees and plant visitors to acid-resistant suits, boots and face shields for employees in acid handling circuits.

Consultation with operators of similar plants and various equipment vendors were invaluable in providing the appropriate safety equipment to ROTP employees. Examples include:

- A fiber metal hard hat in combination with a specific face shield and visor for employees exposed to acids
- Goretex™ suits for daily work in the acid plant
- Heavyweight acid-resistant suits for close maintenance work in the acid plant
- Safety harnesses rather than belts and lanyards for fall protection

START-UP

As construction proceeded, ROTP operations personnel were divided into four crews in early 1994. At this time, a bidding process began which filled the remaining operations and maintenance crews, drawing many experienced employees from other mills on the mine site. Their training began with classroom instruction and plant tours during the final phases of plant construction. These employees observed plant components and modified training materials as they became familiar with their new work areas.

The operations crews performed small-scale start-up simulations in preparation for coordinated simulations involving other ROTP functions. Many of these simulations included tracing and identifying equipment and control systems. Initial mill start-up began with oxide ore feed through the grinding circuit as construction of the roasters was completed. Refractory ore processing began soon thereafter.

CHALLENGES AFTER START-UP

Start-up and operation of the complex plant led quickly to the formation of a team assembled to identify, prioritize and develop solutions to a variety of operations and maintenance challenges. The *ROTP Phase I Maintenance HAZOP Review* (L. Davies, et al. 1995) was completed over a period of three weeks in June 1995. Increased plant availability, higher gold production and improved safety were the stated goals of the HAZOP review.

Using a methodical approach, 13 areas ranging from the ROTP's acid pumps to its propane system were analyzed. Of the problems identified, three levels of priority were established. Priority 1 items included all safety or environmental concerns and those that had a significant impact on availability and production. Priorities 2 and 3 included items of less acute concern. Phase 1 addressed only Priority 1 problems. These were further identified as High probability and High severity (HH), High probability and Medium severity (HM) or Medium probability and High severity (MH).

The Phase 1 review led to 140 solutions that marked the beginning of a continuous improvement process that remains a driving principle in maintaining and operating the ROTP.

RESULTS/BENEFITS

Despite numerous challenges and concerns encountered early in the ROTP's operational life, a history of effective troubleshooting and solution implementation has proven its worth as plant availability, ore throughput, gold recovery and safety performance continue to improve.

Scheduled major maintenance outages that originally consumed - at minimum - one month of production each year are now shortened to a single 3-week outage. Unscheduled maintenance has also been dramatically reduced. Ore throughput averages well above 3 million tons annually and gold recovery consistently exceeds 90%. The ROTP produces about 700,000 ounces of gold each year.

Beyond the production improvements, the safety records accomplished by the ROTP's crews are the most significant result of a comprehensive and effective training program. As of May 2002, the mill's maintenance crews marked 6 years without a lost time accident while the operations crews achieved 5 years versus the same measure. Efficient and safe performance also support environmental performance. The ROTP boasts a commendable performance record with regard to its air quality and operations permits.

From a broader perspective, the successful start-up and operation of the ROTP minimizes potential obstacles to the financing and permitting of projects of similar scale at other Newmont operations. The ROTP provides hundreds of Newmont employees with invaluable experience in the safe and efficient processing of refractory ores. This produces a commodity more valuable than gold itself - a steady supply of expertise that supports the growth of a minerals company that now leads the world in gold production.

ACKNOWLEDGEMENTS

I am grateful to Tony Gunter, former Process General Foreman at Newmont's Refractory Ore Treatment Plant, and to Rick Folkmire, Process Operator, for providing much of the information contained in this manuscript.

REFERENCES

L. Davies, et al., H.A. Simons, Ltd. 1995. Newmont Gold Company Refractory Ore Treatment Plant Phase I Maintenance HAZOP Review Report.

21
Case Studies

Section Co-Editors:
Dr. Martin C. Kuhn and Donald C. Gale

Sunrise Dam Gold Mine—Concept to Production
W.R. Lethlean, P.J. Banovich **2345**

A Case Study in SAG Concentrator Design and Operations at P.T. Freeport Indonesia
R. Coleman, A. Neale, P. Staples **2367**

High Pressure Grinding Roll Utilization at the Empire Mine
D.J. Rose, P.A. Korpi, E.C. Dowling, R.E. McIvor **2380**

The Raglan Concentrator—Technology Development in the Arctic
J. Holmes, D. Hyma, P. Langlois **2394**

SUNRISE DAM GOLD MINE – CONCEPT TO PRODUCTION

WR Lethlean[1] and PJ Banovich[2]

ABSTRACT

The Cleo deposit in the Eastern Goldfields region of Western Australia was discovered by the Shell Company of Australia in 1993. In 1994, Acacia Resources was floated from Shell, taking with it the 500,000-ounce Cleo resource under the name of Sunrise Dam Gold Mine (SDGM). This paper describes the metallurgical testwork, plant design and initial development that led to pouring of first gold in March 1997 and the subsequent Stage 1 plant upgrade that occurred in 1999.

INTRODUCTION/PROJECT HISTORY

The Sunrise Dam Gold Mine is located approximately 55 km south of Laverton and 730 km north west of Perth in the Eastern Goldfields region of Western Australia. The Placer-Granny Smith Joint Venture has exploited the Sunrise resource located on the adjacent mining lease to the Cleo resource via their Granny Smith plant located approximately 30 km to the north of SDGM (Figure 1).

Initial tenements in the Sunrise Dam area were acquired by Shell in 1987. Assessment of preliminary geophysical and drilling data throughout the tenement area during the period 1988 to 1991 resulted in the identification of the Golden Delicious, Pink Lady, Red Delicious and Cleo Prospects. Considerable difficulties were experienced in penetrating the lake sediments which blanket the western half of the tenement area to a depth of 60 to 90 m.

Exploration activities focused on the Golden Delicious Prospect, where, by late 1990, a large, predominantly low-grade mineralised system was established.

Minimal exploration activity was conducted outside Golden Delicious until early 1993 when additional aircore drilling was conducted in the Cleo area. This program identified the need to extend drillholes into fresh rock, and a number of anomalous intercepts were obtained.

Follow-up drilling was conducted during 1993 at Cleo, with a number of encouraging intercepts, including 55 m at 10 g/t gold. A program of RC drillholes, with seven diamond tails, was completed by the end of 1993, resulting in the discovery of the Cleo Resource.

Extensive RC and diamond drilling programs were conducted at Cleo during the following three years, resulting in a reserve of 600,000 recoverable ounces being considered in the May 1996 Feasibility Study.

---

[1] Chief Metallurgical Engineer, AngloGold Australia Ltd
[2] Project Manager, Sunrise Dam Gold Mine

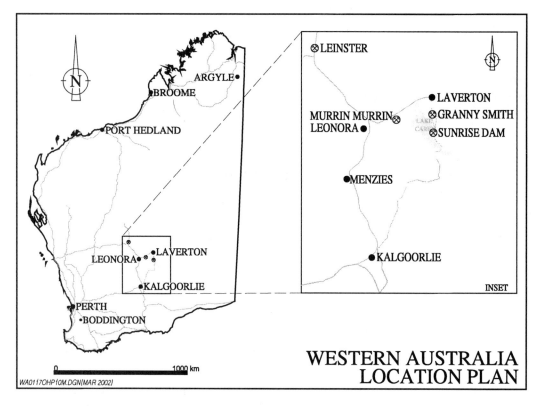

Figure 1 Sunrise Dam Gold Mine location plan

GEOLOGY

The Sunrise tenements are located within the southern portion of the Archaean age Laverton Tectonic Zone, within the Yilgarn Block of Western Australia. The geology of the region is poorly exposed, deeply weathered, and extensively covered by surficial sediments and deep soils. The region consists of a north-trending greenstone package bounded by undifferentiated granitoids to the east and west. The central portion of the greenstone package, in which the Sunrise tenements are located, consists of acid to intermediate volcanics and sedimentary rocks, sandwiched between two corridors of predominantly mafic and ultramafic extrusive and intrusive rocks. Late stage intrusives of loosely granitoid composition occur throughout the area.

Cleo Geology

There is no outcropping geology or surface expression of the mineralisation and the geological understanding of Cleo is based entirely on drilling information. The area is covered by 20 to 90 m of lake sediments and aeolian sand dunes. Beneath the lake sediments is a sequence of interbedded Banded Iron Formation (BIF), Volcanoclastic (VC) and intermediate Volcanics, which have been intruded by narrow quartz felspar porphyries.

Primary gold mineralisation is associated with shallowly and steeply north-westerly dipping structures and is hosted by all major rock-types.

Shallow dipping structures are characterised by intense pervasive carbonate-sericite-chlorite alteration, pyrite and quartz carbonate veining. These zones exhibit intense shearing over narrow zones. Higher gold grades often occur where these zones intersect BIF structures.

Steeper dipping zones are characterised by zones of quartz carbonate veining and breccias with pyrite and arsenopyrite. Wallrock alteration is generally less intense and pervasive relative to the shallow dipping zones.

Gold occurs as coarse visible gold within veins and breccias and in association with pyrite, arsenopyrite and arsenious pyrite. The gold associated arsenic minerals lead to some refractory behaviour.

The Cleo deposit is divided into three general ore types depending on degree of weathering, namely oxide, transition and fresh. Mining would commence with oxide ore for a period of two to three years, followed by treatment of the transition and fresh material.

FEASIBILITY STUDY

Geological and metallurgical investigations were conducted between 1994 and 1996, culminating in a recommendation to develop the SDGM in May 1996. The key parameters contained in the Feasibility Study recommendation are presented in Table 1.

Table 1 Key feasibility study parameters

Mineable Reserve	Mt	5.8
Annual Throughput	Mtpa	1.0
Mine Life	Years	5.8
Head Grade	Au g/t	3.8
Gold Recovered	Oz	592,000
Total Operating Cost	$A/t	31.20
Cash Operating Cost	$A/Oz	307.03
Initial Capital Cost	$AM	65.2

Approval was granted in May 1996 and pre-stripping of the 60 to 90 m of overburden commenced at the end of the third quarter. First gold was poured in March 1997.

METALLURGICAL TESTWORK

The SDGM metallurgical testwork program was carried out in three basic phases as outlined below.

Resource Definition Phase

During this phase the geological crew were in the process of confirming a resource that was considered to have the potential for further development. The metallurgical support consisted of preliminary leach tests to determine approximate gold recovery, cyanide and lime consumption and a determination if weak acid dissociable cyanide (CN_{wad}) would be formed.

Specifically, the aim of the testwork was to determine if any of the three generalised ore types, oxide, transition and fresh, were refractory or not. The information assisted in deciding the size of the resource required before proceeding to the pre-feasibility phase of the project.

Prefeasibility Phase

The prefeasibility phase is the critical stage that determines the fate of the project. The key metallurgical work centres on determining gold recoveries for all identified ore types, developing data to feed into the design criteria, generating operating costs for all ore types and identifying environmental issues to be considered.

Sample selection was achieved by reviewing the drilling logs with the geologists, and agreeing on the intercept composites to be selected to represent the ore types throughout the known ore resource. In addition the locations of the PQ diamond core samples required for comminution testwork were determined.

The principal areas of the testwork program carried out for the SDGM project were:

- Gravity/leach testwork program carried out on numerous intercept composites to determine the gold recovery percentages by gravity and cyanide leach, cyanide and lime consumption, oxygen requirements, leach time required and any CN_{wad} generated. The testwork program was designed in-house with the testwork being carried out at AMMTEC, a commercial metallurgical laboratory located in Perth Western Australia. The data generated in this program, as summarised in Table 2 below, was used in financial analysis and design criteria.

Table 2 Summary of gravity/leach testwork

	Bench G/L Test Fresh Ore $P_{80} = 53\mu m$ Tests in site water		Bench G/L Test Transition $P_{80} = 75\mu m$ Tests in site water		Bench G/L Test Oxide Ore $P_{80} = 75\mu m$ Tests in site water	
	g/t	%	g/t	%	g/t	%
HEAD GRADE						
assay	6.00		3.94		3.57	
calculated	5.21		4.39		3.73	
RESIDUE	1.23		0.77		0.3	
EXTRACTION	3.97	76.3	3.62	82.4	3.43	92
GOLD DISTRIBUTION						
Gravity	1.49	28.6	0.91	20.7	1.05	28.2
Leach	2.48	47.7	2.71	61.6	2.38	63.9
Residue	1.23	23.7	0.77	17.6	0.3	8
Calculated head	5.21	100	4.39	100	3.73	100
REAGENT CONSUMPTION						
NaCN (kg/t)	0.80		0.70		0.70	
CaO (kg/t)	2.20		4.80		4.00	

- Comminution testwork carried out on numerous intercept composites to determine the traditional Bond Work Index (Wi) values for rod mill, ball mill and abrasion (Ai), JK Tech data for SAG mill simulations, Unconfined Compressive Strength (UCS) and crusher work index by size (not shown). A summary of results is shown in the Table 3 below.

Table 3 Comminution summary

	Ball Wi kWh/t	Rod Wi kWh/t	Ai kWh/t	UCS MPa
Oxide ORE SAMPLES				
Average BIF	14.54	19.67	0.284	nd
Average VC	7.30	7.30	0.011	nd
Design Oxide	10.92	13.48	0.148	nd
TRANSITION SAMPLES				
Average BIF	16.97	nd[3]	nd	nd
Average VC	11.63	nd	nd	nd
Design Transition	14.30	nd	nd	nd
FRESH ORE SAMPLES				
Average BIF	17.93	24.42	0.491	74
Average VC	18.63	26.08	0.168	57
Design Fresh	18.28	25.25	0.330	142

- Tailings characterisation testwork to generate data for designing tailings disposal, both physical containment and any potential acid mine drainage (AMD) issues.
- Geochemical characterisation testwork on both ore and waste material to identify potential AMD, either within the waste dump or within the tailing storage facility.

Feasibility Phase

Limited leach testwork was carried out during this stage and was restricted to drill core intercepts not included in the previous program if it was considered to add to the database. Other minor testwork programs carried out included characterisation of the impact of hypersaline water on reagent consumption and the fate of cyanide in the tailings system, both important issues to be considered in the feasibility study.

A program to determine a suitable and cost effective refractory process route to lift primary ore recovery from 76% to 95% was also commissioned. Because of the degree of difficulty of this task, the program remained active over the next four years. The key obstacle to progress was the poor quality of process water. The water available for this project contained more than 250,000 ppm total dissolved solids.

PLANT DESIGN AND LAYOUT

The plant design and layout was carried out in two distinct stages. The first stage was the prefeasibility study where design concepts and criteria were developed and costed to a ± 25% accuracy. The prefeasibility information was used in the initial project financial analysis to determine project viability. The second stage was the feasibility study stage, where advanced engineering drawings were developed and costed to ± 10% to confirm the accuracy of the previous capital cost. Generally very few new concepts were added to the project during this stage.

Capacity

The key design criteria of the SDGM process plant capacity was decided by the then owners, Acacia Resources. Their corporate strategy had gold production targets, for which SDGM was the major building block. Prior to project approval, the two objectives were to produce 100,000 ounces per year, and establish a minimum life of seven years. The project development approval

---

[3] not determined

was given in May 1996, at completion of the feasibility study and confirmation of the corporate objectives.

DESIGN CONSIDERATIONS

Mine Development
The SDGM project was based on being developed as an open pit mine. The mine design dictated that ore processing would commence on 100% oxide material for approximately two years followed by a period of blends of oxide, transition and fresh until the mill feed would revert to 100% fresh ore in year four.

Comminution
The decision was made early in the prefeasibility phase to copy the successful comminution circuit concept installed at Acacia Resources' Union Reefs Gold Mine (URGM) and install a two-stage crushing plant followed by single stage ball mill for the oxide ore processing. Equipment was sized to achieve an annual throughput of 0.75 Mtpa of oxide ore. The throughput was subsequently upgraded to 1.00 Mtpa during the feasibility study.

In making this decision, due cognisance was taken of the potential problems that can occur with wet clay based ores and secondary cone crushers and screens. The design considerations supporting the decision to proceed with the two stage crushing circuit were:

- The experience gained at URGM
- The lithology of the deposit that showed the oxide ore zones were generally composed of thick widths of competent BIF material associated with the soft "clayey" volcanoclastic ore, hence competent material was available to assist in keeping the clayey ore moving through the secondary crusher
- The secondary crusher would be operated with a wider gap than normal
- The product screen aperture would be set at a relatively coarse 16 mm
- Acceptance that the ball mill would scat the coarser material and that the scats would be recycled when processing primary ore.

At the time that the decision was made to proceed with the oxide comminution circuit, it was recognised that major changes/upgrade of the circuit would be required to enable the competent primary ore to be processed. The two options considered were three-stage crushing followed by two-stage ball milling or SAG and ball milling with scats crushing (SABC). The plant layout allowed for either option to be installed. The reason to not proceed further with the design was the strong possibility that the ore reserve would increase sufficiently to allow throughput to increase substantially above the start-up 1.00 Mtpa rate.

The decision not to lock in on throughput capacity, other than for the oxide ore, which had a defined tonnage, was proven correct as outlined later.

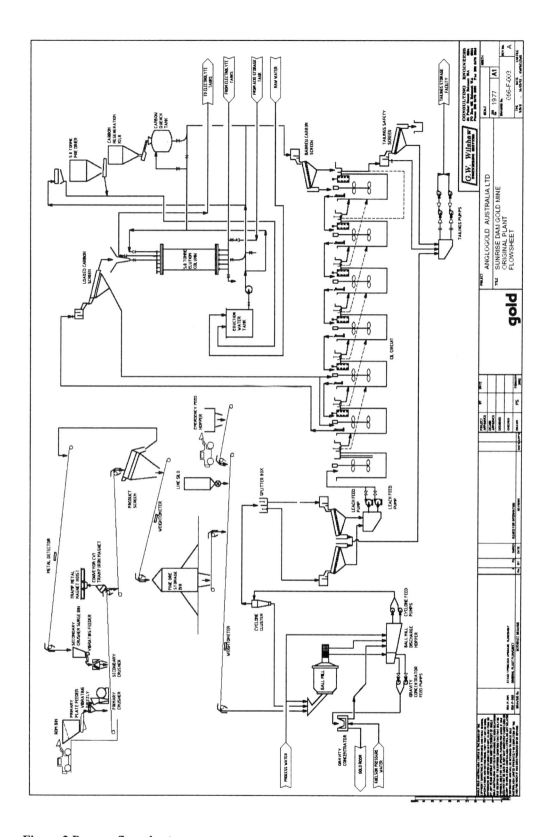

Figure 2 Process flow sheet

Gravity Recovery

The testwork indicated that approximately 20% of the gold in all ore types could be recovered in a gravity circuit. Diamond drill core consistently contained free visible gold in ore intercepts. The decision was made to copy the URGM flowsheet with the use of Knelson Concentrators as primary concentrating units and a Gemini shaking table as the cleaning unit.

The circuit design was based on taking approximately 20% of the ball mill discharge from the front of the ball mill discharge trommel screen protected by a section of the trommel screen at 5 mm aperture. A Warman pump would deliver the required pulp flow rate to the 2 mm aperture vibrating screen ahead of two 30 inch automatic discharge Knelson Concentrators. The concentrate would discharge at set intervals to a storage hopper above the Gemini table located in the Gold Room. Knelson Concentrator tail, at low pulp density, would be discharged into the ball mill discharge hopper. Gemini table tail would report to the ball mill discharge hopper via the Gold Room sump pump.

Historically, the gravity circuit has consistently recovered 40% of the gold.

Leach

The testwork indicated that a 24-hour leach time was required at cyanide concentrations of 200 ppm in the feed tank. A hybrid CIL circuit consisting of one leach tank followed by six CIL adsorption tanks was chosen. Cyclone overflow at an average 44% solids pulp density would be screened at 0.8 mm for trash removal prior to being fed to the leach tanks.

The carbon retention screens were the standard rotary swept vertical screen.

Gold Recovery

Gravity-recovered gold would be produced by direct smelting the Gemini table concentrate after calcining the concentrate at 700 °C for 16 hours. The calcining stage was required to oxidise iron from the balls and sulphides (pyrite and arsenopyrite) recovered into the Gemini table concentrate.

A refinement to be installed later was to process the Knelson Concentrator concentrate through the inhouse-developed ACACIA Reactor and recover the gravity gold via an electrowinning cell. The major benefits achieved by the ACACIA Reactor were:

- Improved security by designing an automated hands-free operation
- Improved safety by eliminating arsenic fumes generated during calcining and smelting
- Improved gold recovery by recovering the fine free gold circulated back to the grinding circuit in the Gemini table tail
- Improved gravity gold accounting.

The elution circuit design was based on the split Anglo American system principally for conserving precious potable water. The column was sized to enable the expected batches of high grade feed material to be treated without the need to reduce tonnage or sacrifice recovery. The eluted gold would be recovered in electrowinning cells with the cathodes calcined at 700 °C for 16 hours prior to smelting in a gas-fired pot furnace.

Tailing Storage Facility

A paddock style tailing storage facility was chosen for the oxide ore treatment phase. The structure would be built to water retaining specifications with underdrainage to ensure consolidation of the settled solids.

As described later, due to process plant capacity increases, a Central Thickened Discharge tailing storage facility was installed after the Stage 1 upgrade.

Metallurgical Balance

The requirements of the metallurgical balance were considered during the design. The mill feed conveyor weightometer would record tonnage and the tail sample would be collected via an automatic sampler. The issue of sampling the mill feed was debated at length with the main consideration being the influence of coarse gold on sample size required. In the end, it was decided to place an automatic sampler on the leach feed stream and back-calculate a mill feed grade by including the gravity-recovered gold into the calculation.

Reagents

The key considerations in the reagent design were safety and environment protection. The four major reagents to be used in the plant were cyanide, quicklime, sodium hydroxide and hydrochloric acid. The characteristics of each reagent dictated individual treatment from delivery to storage through to distribution within the process plant.

Cyanide, as solid briquettes, was initially designed to be delivered to site in one tonne boxes and the two-month consumption equivalent would be stored within a fenced and locked compound. The cyanide mixing and liquid cyanide storage tank was located within the plant structure, but isolated from the daily human traffic routes. The standard mixing procedure was to add water, sodium hydroxide and the solid sodium cyanide briquettes to the mixing tank. The procedure was designed to prevent hydrogen cyanide gas generation. An additional safety feature is to add a dye colour to the mixed cyanide solution so that any failure that results in cyanide solution spillage is easily identified and appropriate safe clean-up procedures can be applied.

A change to liquid cyanide delivery occurred after three years operation. The liquid is delivered at 30% concentration and offloaded into a storage tank with a one-week capacity.

Sodium hydroxide would be delivered to site as a 50% sodium hydroxide liquid and stored adjacent to the cyanide-mixing unit.

Quicklime would be unloaded and stored in a 300-tonne silo located adjacent to the ball mill feed conveyor such that the quicklime could be added to the circuit via this conveyor, utilising the ball mill as a slaker. Quicklime storage was equivalent to one month's consumption.

Hydrochloric acid would be delivered as a 35% HCl solution. The storage tank was located remote from both the cyanide and caustic storage tanks.

In designing all liquid reagent storage tanks, the principal consideration was that they were sized to hold the equivalent of one and a half truck container loads. The reasons behind this concept were that

- The re-order level was such that the operation did not run precariously low while waiting for deliveries
- On delivery, it was guaranteed to hold the full truck container load without overflowing the storage tank.

In addition, each storage tank was located within a concrete bunded area designed to contain 110% of the largest tank within the bund and capture any spillage from a rupture in any tank based on the requirements of the Dangerous Goods Act.

In designing the reagent solution distribution systems, it was decided to use high-density polyethylene pipes colour coded on the outside to identify the reagent being distributed within the pipe. The use of tags was considered to be ineffective within the pipe trace and only effective at feed and discharge ends. The concept has proven very effective at the mine site.

Infrastructure - Process Plant

The remoteness of the SDGM site meant that the facilities were needed to provide power, water, communications and an access road surface suitable for equipment and supply transport.

A power requirement of 4 MW meant that six 1-MW generators were required. Due to the climatic conditions at site, the capacity of each generation set was down-rated to approximately 0.8 MW continuous supply. SDGM purchased the generation sets from StateWest Power Supply and then contracted that company to operate the station.

Telephone communications were non-existent at the site until the development of the project. An arrangement was entered into with the national telephone company to provide voice and data communications to the site via a microwave system.

The development of the water supply was not an easy task due to the lack of good quality ground water or suitable terrain to build water storage catchments within a practical distance of the mine site location. Consequently, process water was sourced from the pit dewatering bores which produced water at 250,000 ppm TDS. Water for sprays, hose-down and reagent mixing was sourced from a small limited bore field producing water at 30,000 ppm TDS. Potable water was generated on-site via a reverse osmosis (RO) plant fed with water at less than 4,000 ppm TDS from a small borefield approximately 12 km from site.

Access to site from the nearest bitumen road during geological exploration was via a deteriorated haul road. The priority job after project approval was the upgrading of this road to allow truck access to bring in the mining and construction equipment.

The other major consideration was to protect the plant during the heavy torrential rain events that occur in the area. Because the landscape was generally flat with a shallow fall from the east to the west, the mine site and associated facilities would be subject to sheet flooding. Shallow but wide protection drains and bunds were constructed around each of the facilities. Hindsight has proven this to be a wise decision.

Hazan and Hazop

Once the design was confirmed and the piping and instrumentation and final plant layout drawings were approximately 90% completed, a detailed HAZAN/HAZOP review was carried out. The review identified a number of issues that had been either overlooked or had not been given sufficient consideration during design. The timing of the analysis was such that the corrections required could be carried out without affecting the project schedules.

The concept of Hazan/Hazop analysis has been used consistently in the subsequent mill upgrades to assist in the management of change.

Infrastructure - Administration

A major consideration during design was that the mine site was to operate under a fly in/fly out (FIFO) arrangement for the staff. This meant that an airstrip capable of accommodating commuter aircraft was to be constructed along with a 156-person accommodation village. The location of these two facilities presented interesting challenges as they had to be located on the mine lease, in an area that was not prospective for gold mineralisation and that catered for the special requirements of the facilities. Specifically, the airstrip required a "protection" zone around it, the size being dependent on the largest aircraft likely to use the strip. The accommodation village was to be located at a distance from the mine operation such that industrial noise would not be an issue. The village also increased load on the power and potable water facilities.

The most difficult design task in the whole operation was the main administration office. Everybody within the organisation had a view on what was required, but no concept of the costs required to fulfil their dreams. Common sense, logic and cost control prevailed!

Table 4 Design criteria

Material Characteristics		
Ore type		Oxide
Grade	g/t	4.0
Mineralisation		
- BIF	%	43
- Volcanoclastic	%	57
Rod mill work index	kWh/t	13.5
Ball mill work index	kWh/t	10.9
Abrasion index	kWh/t	0.148
Specific gravity		2.81
Bulk density	t/m^3	1.65
Production		
Throughput	tpa	1,000,000
Gold recovery		
- Gravity	%	20.0
- Leach	%	72.5
- Total	%	92.5
Gold production	oz/a	100,000
Crushing rate	tph	200
Crushing plant utilisation	%	55
Milling rate	tph	109
Milling plant utilisation	%	94.3
CIL circuit configuration		1 leach + 6 CIL
Total CIL residence time	h	24
CIL feed pulp density	% solids	44
Consumables		
Cyanide consumption	kg/t	0.94
Lime consumption	kg/t	4.00
Grinding media consumption	kg/t	1.15
Total power consumption	kWh/t	25.9
Process water salinity	ppm	250,000

CAPITAL COSTS

Capital costs for both the prefeasibility and feasibility stages were generated using the following methods:

- Obtaining three quotations for major equipment estimated to cost in excess of A$5,000 with all other equipment costs from the engineering company's records
- Material quantities taken from engineering drawings and quotations from suppliers
- Construction costs from budget quotations from construction companies based on the quality of drawings available at time of quotation
- Project engineering and construction management costs from the engineering company
- Owner's cost estimates supplied by the owner.

The cost comparison between the prefeasibility study and the feasibility study is set out in Table 5 below.

Table 5 Project cost estimates

Description	Feasibility Study A$	Prefeasibility Study A$	Difference A$
Plant construction	15,842,760	12,748,949	3,093,811
Water	1,559,887	2,712,496	-1,152,609
Tailings	4,066,706	2,558,424	1,508,282
Services	5,921,031	7,523,685	-1,602,654
Support facilities	7,584,730	4,540,037	3,044,693
Construction indirect	5,271,551	4,882,000	389,551
Total	**40,246,665**	**34,965,591**	**5,281,074**

The reason for the increase between the pre-feasibility study and the feasibility study costs were

- Increase in tonnage from 0.75 Mtpa to 1.00 Mtpa
- Increase in tailing storage facility area to account for the reduced consolidation affect due to the use of hypersaline water
- Village accommodation increased from 80 rooms to 156 rooms due to the operation agreeing to provide mining contractor's accommodation
- Increased accuracy of the estimate.

The above prefeasibility costs were developed on the EPCM concept. The feasibility study tender documents requested that the engineering companies submit both an EPCM bid and a Lump Sum bid. The Lump Sum bid by JR Engineering Services (JRES) was too attractive to ignore, and following clarification meetings, the JRES Lump Sum tender was accepted. The final capital cost for the process plant and associated facilities was A$38,967,431 which included a sum of A$1,684,538 for additional water bores required because of the poor production performance of the original bores.

In addition to the above costs, A$3,585,000 was spent to purchase the power generation equipment rather than purchase power under contract at an elevated rate that included the capital component as well as the operating cost.

OPERATING COSTS

The prefeasibility and feasibility operating costs were developed generally from first principles using data generated in the testwork and information from the engineering design. Quotations from suppliers were obtained for all materials and reagents. The feasibility operating costs are set out below in Table 6.

Table 6 Feasibility study oxide ore operating cost

Description	Annual costs A$	Unit costs A$/t
Manpower	2,115,456	2.12
Consumables	3,201,000	3.20
Power	3,412,167	3.41
Liners and Media	1,685,889	1.69
Maintenance Contract	1,002,222	1.00
Maintenance materials	991,556	0.99
Outside Services	289,444	0.29
ROM Pad Ore Rehandle	600,000	0.60
Total	13,297,733	13.30

OPERATING PHILOSOPHY AND STAFFING

The decision was made early in the plant design process to install a high level of process control and automation and utilise a relatively small number of highly trained staff to operate the plant. The control platform chosen was based on Allen-Bradley PLC's and the CiTect SCADA system. It was not deemed appropriate to utilise expert control for the relatively simple flowsheet, however, wherever possible, routine plant control functions were provided by the process control facilities. This was aimed at providing the process technicians with sufficient time during each shift to focus on process optimisation and variance analysis, with the ultimate aim of improved production and safety.

The staffing philosophy was centred around using 'green' process technicians, based on the success of this approach at URGM in 1995. A longer-term vision of self managed work teams also led to an extremely flat organisational structure as outlined in Figure 3. Plant maintenance was to be carried out by a contract workforce supervised by company coordinators.

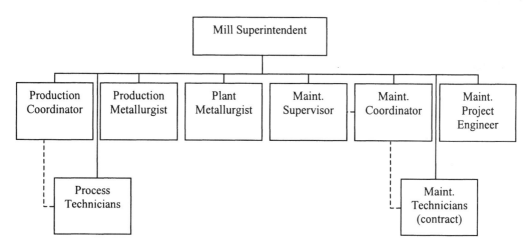

The dotted lines represent functional reporting relationships.

Figure 3 Organisation chart – SDGM processing department

SDGM works on a FIFO roster system where employees commute to the mine from their place of residence (generally Perth) for a continuous work cycle then commute back to their place of residence for an extended leave cycle. In the case of technical staff, such as metallurgists and coordinators, employees work nine days followed by five days off. Production and maintenance technicians work 14 consecutive 12-hour shifts followed by seven days off. The 14-shift block was divided into seven dayshifts and seven nightshifts for the process technicians.

The FIFO roster system results in three shift crews of process technicians. At any time, one crew is on dayshift, one on nightshift and one is out on leave. A total of 18 process technicians were recruited to operate the plant, resulting in three shifts of 6 people each, allocated as follows:

Crushing Technician	1	(working seven dayshifts and seven nightshifts roster)
Control room technician	1	(working seven dayshifts and seven nightshifts roster)
Wet plant technician	1	(working seven dayshifts and seven nightshifts roster)
Gold room technician	1	(working fourteen dayshifts roster)
Laboratory Technician	1	(working fourteen dayshifts roster)
Services Technician	1	(working fourteen dayshifts roster)

No foreman or shift leader was appointed, with the control room technician assuming responsibility for the performance of the shift. Individuals are rotated through the positions on a cycle determined by skill levels and training requirements. Technical staff are on 24-hour call to render assistance on nightshift if needed.

Of the 18 process technicians recruited, 3 were transferred to SDGM from URGM as experienced operators. These individuals were responsible for leading the shifts through the commissioning period and assisting in hands-on training of the fresh recruits.

Process Technician Training Program

The decision to recruit 'green' process technicians was based on the opportunity such a strategy provides to train the individuals in company procedures without having to 'undo' previous training that may not be compatible with the company's objectives. In addition to this benefit, the employees exhibited a large degree of enthusiasm for the job given their exposure to a new industry and technologies. A significant proportion of the technicians recruited were transferred from the company's exploration division, bringing with them a considerable amount of project development history.

In order to prepare the technicians for the commissioning of the plant an intensive training program was developed. An analysis of training needs and skills was conducted and the requisite skills of a competent process technician were divided into five levels of increasing complexity. These levels were then linked to remuneration in accordance with the local employee relations regulations. The skills and training needs were combined into a training matrix outlining theory training requirements and competency requirements as shown in Figure 4.

The matrix was designed around normal operation where technicians would be recruited in small numbers to cope with staff turnover. In general, a technician would be recruited as a Level 1 Process Technician and work their way through the matrix as they received training and developed skills and competencies. For the initial influx, however, all technicians were required to receive all of the training and were then classified into levels and assigned roles according to demonstrated skills and competencies. This required an intensive six-week off-site training program, coordinated and delivered by the Production Coordinator and Production Metallurgist in Perth.

Much of the training material was available in a general format from URGM. The two plants shared many common features. However, in order to customise the training material to SDGM, the services of Normet Pty Ltd, a metallurgical and training consultant, were employed. Normet assisted in producing training modules based on the SDGM plant design, largely interpreted from design criteria and engineering drawings. This information was compiled into modules incorporating metallurgical and plant operating theory, equipment descriptions and operating procedures. Extensive use of schematic representation was made to assist in the learning process.

Because the training program was carried out in Perth where there is little opportunity to view real equipment, a video was made of the URGM operations outlining each unit process and detailing equipment operation and care and operating practices. Original equipment manufacturer technical or promotional videos were used when available.

The training program was carried out between January 6 and February 20 1997, after which the teams were mobilised to site in preparation for commissioning of the crushing plant. One of the unforseen benefits of the program was the strong team building that resulted from the team spending 6 weeks together in an intensive learning environment. As is natural in any group exposed to such an environment, leaders emerged and individuals took on responsibility for their team mates to ensure that no one lagged behind in skills acquisition. As this phenomenon was identified and other members of the processing department were recruited (i.e. maintenance technicians, metallurgists), these employees were required to attend certain of the training classes to gain exposure to their new colleagues and enhance the team building aspect of the program. Once the workgroup was divided into shifts, this early contact with the other members of the group paid dividends in terms of inter-shift and inter-discipline cooperation.

The approach detailed above was not without its problems. The major downside associated with commissioning a new plant with new operators was the workload imposed on the metallurgists and experienced technicians. These employees were called on to ensure the plant ran smoothly and provide hands-on training to their work groups, which was a demanding role. The overall result, however, was extremely positive and would likely be repeated at any future company projects.

COMMISSIONING

In order to capitalise on the learning opportunities presented by plant commissioning, the arrangements for plant start-up were focussed on allowing the process technicians to assume as much control of the plant as possible. To this end, SDGM operations and metallurgical staff were responsible for ore commissioning of the plant, whilst the engineering contractor was responsible for pre-commissioning activities and technical support during ore commissioning.

The crushing plant was commissioned in February 1997. It was decided that a quantity of stemming material would be crushed for use in drill and blast activities in the mine before crushing any ore, thus providing an opportunity to run the crusher on dayshift for a period of two weeks. During this period, as many of the technicians as possible were rotated through the circuit. Had this exercise not been undertaken, the crusher would only have run for a few days before the crushed ore stockpiles were full, resulting in the requirement to crush on specification once the milling circuit was commissioned, with very little training time.

No major operational problems were encountered during crusher commissioning. Design throughput and availability were achieved immediately, paying testimony to the level of design and the quality of construction.

The remainder of the plant was commissioned in early March (grinding, leach, gold recovery). A parcel of low-grade ore was treated initially to fill the CIL circuit and allow reagent levels to be established prior to introducing ROM grade ore. A design fault with the mill feed arrangement resulted in overdesign tonnages being treated from the start. However, the ore from the early stages of the pit was considerably softer than design, so the higher throughput did not lead to any operational problems.

Level 1	Level 2	Level 3	Level 4	Level 5
Competencies	**Competencies**	**Competencies**	**Competencies**	**Competencies**
▲ Communicate in the workplace ▲ Work safely ▲ Apply local risk control process ▲ Operate gantry crane ▲ Operate load shifting equipment ▲ Handle reagents ▲ Manage water services	▲ Conduct conveying operations ▲ Manage crushing process ▲ Operate Rockbreaker ▲ Manage isolation process – crusher only ▲ Respond to unplanned shutdown – crusher only ▲ Perform control room operations – crusher only ▲ plus 2 electives **Or** ▲ *Manage laboratory* **Or** ▲ *Manage Goldroom and Conduct ACACIA Reactor process* *plus 3 electives*	▲ Conduct leaching process ▲ Conduct elution process ▲ Conduct electrowinning process ▲ Manage isolation process – leaching/elution only ▲ Respond to unplanned shutdown – leaching/elution only ▲ Perform control room operations – leaching/elution only ▲ Conduct thickening process plus 2 electives	▲ Conduct milling process ▲ Perform control room operations ▲ Conduct wet gravity separation ▲ Conduct gold room operations ▲ Manage unplanned shutdowns – all areas ▲ Manage isolations – all areas ▲ Leadership (Frontline Management Standards)	▲ Defined on an individual basis
Theory	**Theory**	**Theory**	**Theory**	**Theory**
▲ General terminology ▲ Intro to Mineral Processing ▲ Safe manual handling ▲ Skid steer loader – skills training ▲ Forklift – skills training ▲ Reagents module ▲ Water systems module ▲ 4WD ▲ Intro to Citect ▲ Slinging & lifting and Introductory to O/H cranes	▲ Introduction to conveying module ▲ Crushing module ▲ Isolations module ▲ Rockbreaker module **Or** ▲ *Laboratory module* **Or** ▲ *Goldroom Module*	▲ Leaching / adsorption module ▲ Elution & Carbon regeneration module ▲ Thickening module ▲ Oxygen module ▲ Isolations module ▲ Instrumentation ▲ Introduction to process control	▲ Grinding module ▲ Isolations module ▲ Citect ▲ Introduction to continuous improvement ▲ Gravity separation module ▲ Front line supervisors course	▲ Data analysis ▲ Project optimisation (mentor) ▲ Met. Technician ▲ Statistical Process Control ▲ Workplace trainer & Assessor ▲ Report Writing ▲ Presentation
	Electives	**Electives**	**Electives**	**Electives**
	▲ Maintain Cyclones + Cyclone module ▲ Conduct pump operations + Introduction to pumping ▲ Conduct valve operations + Valves module	▲ Maintain Cyclones + Cyclone module ▲ Conduct pump operations + Introduction to pumping ▲ Conduct valve operations + Valves module		

Figure 4 Training matrix

The most disappointing aspect of the startup was the significant extent of problems that were encountered with the process control system. The engineering contractor had not allowed sufficient time in the development schedule to complete the programming prior to startup and many changes were required on the run. This led to much frustration and a less than ideal final product. The decision was taken some 6 months after startup to completely reconfigure the software to provide greater operability and maintainability of the system.

First gold was poured on March 27, 1997, one month ahead of schedule. Practical completion and final hand-over of the plant took place on April 19, 1997, following achievement of the hand-over criteria listed below, and applying to a continuous 28-day period:

- 90% utilisation of the mill
- Average throughput of 90% of design
- 100% utilisation for 72 hours
- 90% of design gold recovery.

The effectiveness of the training program was evident immediately as the inexperienced technicians rapidly gained the knowledge to operate the plant proficiently.

Table 7 Actual performance versus design

		Design	Month 1[4] April 1997	Year 1 Apr-97 – Mar-98
Production				
Throughput	tpa	1,000,000	1,012,572	1,178,046
Gold recovery				
- Gravity	%	20.0	42.0	45.1
- Leach	%	72.5	53.5	50.5
- Total	%	92.5	96.5	95.6
Gold production	oz/a	100,000	149,844	205,956
Crushing rate	tph	200	220	313
Crushing plant utilisation	%	55.0	53.2	47.7
Milling rate	tph	109	124	139
Milling plant utilisation	%	94.3	94.4	97.4
Consumables				
Cyanide consumption	kg/t	0.940	0.743	0.478
Lime consumption	kg/t	4.000	4.430	3.563
Grinding media consumption	kg/t	1.150	0.880	0.740
Total power consumption	kWh/t	25.900	23.430	22.001

[4] Annual equivalents

STAGE 1 UPGRADE

By mid-1998 the resource had grown from eight million tonnes to 22 million tonnes. In addition, the oxide reserves were scheduled to be largely depleted by mid-1999 and increased milling capacity was required in order to maintain production rates with an increasing proportion of transition and fresh ore content in the mill feed. A study into upgrade options for the comminution circuit commenced in May 1998.

The primary objective of the study was to design a flowsheet that provided a throughput of up to 2 Mtpa on blended ores and 1.25 Mtpa on fresh ore, without prejudicing future expansion options in the event that a further increase in throughput was warranted. At the time, it was considered that the plant would ultimately end up as a SAG/Ball/Crush (SABC) circuit, treating in excess of 2 Mtpa of fresh ore. However tertiary crushing and ball milling could not be ruled out, so the Stage 1 upgrade had to provide a circuit that would fit both scenarios in future.

Studies into comminution circuit options were carried out by JRES. In-house studies using various specialty consultants (Orway Mineral Consultants, JK Tech) were also carried out to cross check the engineer's conclusions and further explore available options.

The final flowsheet was established in September 1998 in a joint review process between JRES, JK Tech and SDGM. Installation of a second ball mill, identical to the existing ball mill, in series configuration was chosen. The benefits of this choice included ease of adaptation into the plant layout, commonality of spares and vendors and perfect fit into a future SABC circuit if installed. Modelling of the new circuit indicated a capacity of greater than 2 Mtpa on oxide ore and 1.45 Mtpa on fresh ore, with blends falling between the two extremes.

Considerable difficulty in finalising the crushing flowsheet was experienced due to the fact that the simulation and modelling work all suggested an upgrade was required, but historical plant performance indicated that the existing circuit would handle the new duty. A decision was finally taken to delay any upgrade of the crushing circuit until some operating experience on fresh ore had been gained.

Given the excellent leaching kinetics displayed by the oxide ore to date, it was not considered necessary to upgrade the CIL circuit at the time. Leaching was typically completed halfway through the circuit so even though the throughput was effectively doubling, no loss of recovery was expected.

Stage 1 upgrade design criteria are presented in table 8.

A lump sum contract to construct the upgraded facilities was entered into with JRES in September 1998, for a total value of $A10.1 M. The new facilities were due to be commissioned in June 1999 with the overall schedule being driven by the ball mill delivery. The scope of works covered by the lump sum contract included:

- Installation of a second 1.9 MW (4.27 m x 6.4 m) ball mill in series with the original mill
- Upgrade of the classification circuit
- Upgrade of the gravity recovery circuit by installation of a single 48 in. Knelson Concentrator
- Upgrade of pumps, pipelines and services to match the new duty
- Installation of three 1-MW generators.

The expanded flowsheet is presented in Figure 5.

In choosing a Lump Sum contractor to carry out the design engineering and construction, it was considered that the original contractor, JRES, would provide a clear advantage due to their familiarity with the plant and association with the organisation and its staff. The project was not bid competitively but rather conducted under an open book arrangement with the contractor providing access to all cost and profit figures. SDGM were responsible for final equipment selection and specification as in the original project.

Table 8 Stage 1 upgrade design criteria

Material Characteristics		
Ore type		Oxide/Fresh Blend
Grade	g/t	4.0
Blend Characteristics		
- Oxide	%	10
- Fresh	%	90
Rod mill work index	kWh/t	25
Ball mill work index	kWh/t	19.5
Abrasion index	kWh/t	0.300
Specific gravity		2.80
Bulk density	t/m^3	1.75
Production		
Throughput	tpa	2.0
Gold recovery		
- Gravity	%	30.0
- Leach	%	65.0
- Total	%	95.0
Gold production	oz/a	>200,000
Crushing rate	tph	335
Crushing plant utilisation	%	75
Milling rate	tph	250
Milling plant utilisation	%	94.3
CIL circuit configuration		1 leach + 6 CIL
Total CIL residence time	h	11.7
CIL feed density	% solids	40.8
Consumables		
Cyanide consumption	kg/t	0.98
Lime consumption	kg/t	4.00
Grinding media consumption	kg/t	1.216
Total power consumption	kWh/t	29.401
Process water salinity	ppm	250,000

Because the upgrade had to be built in and around an operating plant, cooperation between construction and operating teams was of paramount importance, as was minimising any downtime associated with tying in new equipment. This interface was very well managed by both parties and the project was completed with an excellent safety record and a minimum of interruption to production. Good cooperation between the plant maintenance group and the contractor allowed the majority of the tie-ins to be carried out during scheduled plant outages, with a final 24-hour shutdown being the only additional downtime required.

The expanded processing plant was commissioned in June 1999. Official hand-over took place on June 30, 1999. Following the success of the initial commissioning, the same approach was employed for the upgrade, with SDGM personnel being responsible for ore commissioning activities. Once again the plant came on-line extremely quickly and design throughput was achieved without any fuss. Process control configuration was carried out by the contractor that had reconfigured the original system, providing continuity and standardisation. This ensured that the system was fully operational prior to startup and the operations personnel had had ample opportunity to gain familiarity with the new features and equipment.

The increase in throughput also required an expansion of the on-site diesel power generating facilities and the tailings storage facilities.

Table 9 Actual performance versus design

		Design	Month 1[5] July 1999	Year 1 Jul-99 – Jun-00
Production				
Throughput	tpa	2,000,000	2,023,512	1,792,154
Gold recovery				
- Gravity	%	30.0	37.9	46.9
- Leach	%	65.0	50.1	41.8
- Total	%	95.0	88.0	88.7
Gold production	oz/a	>200,000	115,380	223,465
Crushing rate	tph	335	335	313
Crushing plant utilisation	%	75.0	62.2	64.1
Milling rate	tph	250	236	212
Milling plant utilisation	%	94.3	96.0	96.9
Consumables				
Cyanide consumption	kg/t	0.980	0.587	0.519
Lime consumption	kg/t	4.000	1.936	3.008
Grinding media consumption	kg/t	1.216	0.895	0.978
Total power consumption	kWh/t	29.401	21.948	25.129

---

[5] Annual equivalents

Tailings Storage

The original paddock style tailing storage facility (TSF) was designed to accept 1 Mtpa of residue for a period of 5 years, with staged raising of the TSF embankments. When it was recognised that the plant throughput was to exceed 2 Mtpa and that the annual rate of rise of the stored tailings surface would exceed what was considered to be the limit of good practice, options for long term storage were investigated. Australian Tailings Consultants (formerly MPA Williams and Associates) were commissioned to evaluate a number of storage options including multiple paddocks, in-pit disposal, in-dump disposal and central thickened discharge (CTD) disposal. Following a desktop review of the available options, central thickened discharge disposal was chosen for its low cost, low energy and excellent environmental characteristics. The CTD storage facility and tailing thickener were commissioned in December 1999.

The testwork carried out during the prefeasibility and feasibility studies showed that AMD and CN_{wad} were not issues that would occur during tailing disposal. The operation of the original paddock style tailing storage facility confirmed the testwork results, with CN_{wad} in tailings return water at less than 10 ppm. The operation of the CTD storage facility has continued to confirm the low CN_{wad} content of the return water.

Crushing Circuit

During the second half of 1999 it became apparent that the crushing circuit was not capable of handling the increased throughput and ore hardness. In particular, maintenance requirements of the secondary crusher increased dramatically and a real risk of extended downtime was present. Because of the capacity constraints, operating hours were increased which in turn reduced the time available for maintenance activities and the overall quality of the operation was jeopardised. The safety and health risks associated with the condition of the plant and the pressures on the operating and maintenance staff were also recognised.

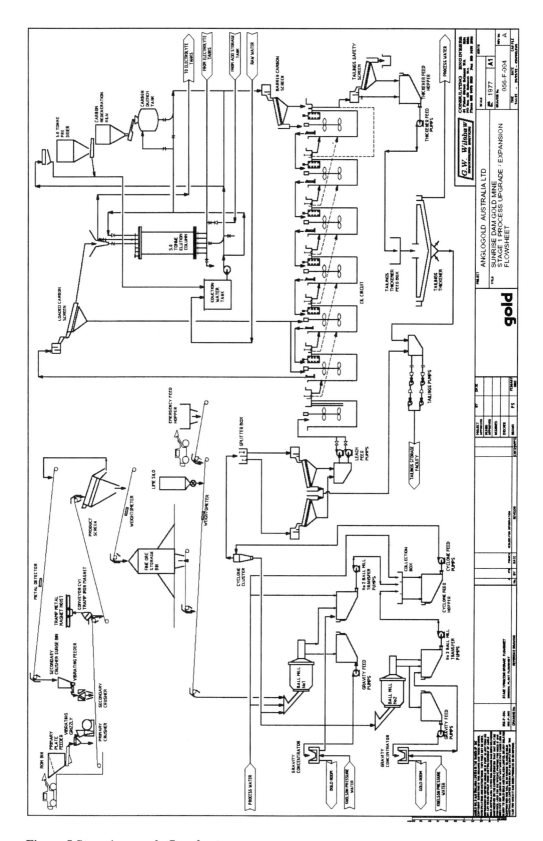

Figure 5 Stage 1 upgrade flowsheet

Surveys were carried out on fresh ore to allow sufficient data to be gathered to accurately model the circuit and the ore. By the end of 1999 sufficient information on future expansions was available so that the requirements of any such expansions could be factored into any short term upgrades of the crushing circuit. Approval to upgrade the secondary crusher from the Metso Omnicone 1560 to a Metso HP500 was granted in January 2000 and the crusher was commissioned in April 2000. JRES were again used to carry out all EPC activities.

CONCLUSIONS

The Sunrise Dam Gold Mine commenced its life as a 1Mtpa oxide operation with a mine life of 6 years. Between its inception in 1996 and mid 1999 the operation grew to more than 2 Mtpa via a major plant upgrade and ongoing optimisation. Further development of the operation to over 3 Mtpa occurred in 2001. However the second upgrade is beyond the scope of this paper.

From the very start, a quality approach was taken to plant design and every effort was made to ensure that the right equipment was installed and the right level of operability and maintainability were achieved. This did not necessarily mean taking a 'Rolls Royce' approach, but rather involving operating and maintenance staff at every stage to ensure their concerns were addressed and an appropriate level of 'buy-in' and ownership was achieved. The benefits of this approach are apparent in the performance statistics and the condition of the plant, as well as in the calibre and attitudes of the people that operate and maintain it.

The model used to manage engineering activities has also proved to be extremely successful, and the relationship developed between SDGM and the principal engineering contractor, JR Engineering Services, has contributed significantly to the success of the operation.

ACKNOWLEDGEMENTS

The success of the SDGM project is a credit to the dedication, hard work and vision of many people. The authors would therefore like to thank all staff and contractors, past and present, for their contribution. The management of AngloGold Australia are thanked for their support and permission to publish this paper.

A Case Study in SAG Concentrator Design and Operations AT P.T. Freeport Indonesia

Rick Coleman[1], Andrew Neale[2] and Paul Staples[3]

ABSTRACT

P.T. Freeport Indonesia (PTFI) currently operate a 760,000 t/d mining and 245,000 t/d milling operation in the remote highlands of Papua, the easternmost province of the Republic of Indonesia. The operations stretch from the dewatering and concentrate shipping facility at the Port of Amamapare on the Arafura Sea to the Grasberg open pit mine located more than 114 km away at 4,000 m above sea level. The four concentrators (C1, C2, C3 & C4) are located an elevation of 2,800 m above sea level. The isolated location and generally difficult working conditions make for challenges not encountered in most North American operations. In spite of this, the mill design, technical and operating groups have been able to develop and maintain an effective level of plant wide productivity through the appropriate selection of available technologies and the implementation of innovative management practices. The history of the mining operation in general and the expansion of the milling operations in particular have been well documented by McCulloch (1991), Russel and Kieffer (1994), and Coleman and Veloo (1996). Van Nort et al (1991) have described the geology of the Grasberg deposit in detail.

In 1995 C3, consisting of a single (10.4 m) 34ft SAG mill, two ball mills, and rougher and cleaner flotation circuits, was commissioned. Three years later C4, incorporating a single (11.6 m) 38ft SAG mill, four ball mills, and rougher flotation, was started-up. The C3 cleaner circuit was expanded and is common to both concentrators. Today the two SAG mills process approximately 175,000 t/day, with the remainder (70,000 t/day) provided by the North/South concentrators (C1 and C2), which are conventional two stages crushing, with single stage ball milling. This case study details SAG mill design considerations and operating experience since start-up in 1995.

DESIGN OVERVIEW

Before starting the C3 design, Freeport engineers, operators and maintenance personnel visited several large-scale SAG milling operations looking for best practice techniques to incorporate into the Freeport facilities. The importance of this exercise cannot be over estimated as Freeport learned from the successes and failures of other operations. The open sharing of ideas and concepts between mining operations is a testament to the spirit of cooperation within the industry.

The key design features incorporated into both the C3 and C4 concentrators were as follows:

1. Variable speed SAG mills.
2. Pebble recycle systems with conventional as opposed to high lift conveyors.
3. No recycling of slurry to the SAG mills.
4. Vibrating SAG discharge screens with built in redundancy (i.e. no SAG trommel screens).
5. A high level of process automation and process control.
6. Rougher flotation utilizing the largest cells available at the time of selection.
7. Flotation columns for the cleaning circuits.

---

[1] Senior Vice President Mine Operation, PT Freeport Indonesia
[2] Vice President Technical Services, PT Freeport Indonesia
[3] General Superintendent, Concentrator Planning and Operations, PT Freeport Indonesia

Key factors impacting design considerations for both C3 and C4 were:

1. Variable ore hardness and feed size distribution indicated highly variable mill throughput rates.
2. Unplanned downtime would have a significant impact on overall operating economics therefore equipment redundancy was essential.
3. Importance of designing for ease of operation and maintenance.
4. The layout of C3 incorporated anticipated the future C4 expansion.

In mid-1995, less than 6 months after commissioning C3, a site-based design team was formed to evaluate opportunities for the C4 expansion. By this time a better understanding of the Grasberg deposit concluded that the C4 expansion would not be a simple duplication of the C3 expansion. During the early days of the C4 expansion project, several additional key design criteria were established:

1. Maximize component consistency with C3 to minimize inventory costs.
2. Project cost considerations would be decided on a Net Present Value (NPV) basis, rather than on a capital allocation basis.
3. Competitive bidding would not be required on all processing equipment.

The last point was critical. Sole sourcing can expedite a project and allows the vendor to work as a project partner to ensure that the process equipment meets the overall project criteria. Because Freeport had recently completed a similar expansion, the project team was familiar with current financial and commercial terms, so there was little concern that any particular vendor would try to take advantage of the lack of competitive bidding.

Table 1 summarizes key installed equipment, while the following sections summarize the specific C3/C4 flow sheets, with further details being provided by Coleman and Veloo (1996) and Coleman and Napitupulu (1997).

Table 1 – Summary of Key Process Equipment

AREA EQUIPMENT	C 3 Number	C 3 Size	C 3 Installed kW	C 3 Capacity	C 4 Number	C 4 Size	C 4 Installed H.P.	C 4 Capacity
Feed Conveyors	3	183cm (72")	N/A	3,500 tph	2	213cm (84")	N/A	6,000 tph
SAG Mill	1	10.36 m Dia	10,600 kW	3,300 tph	1	11.6 m dia	19,500 kW	5,500 tph
Vibrating Screens	2	3.0 x 7.3 m	90 kW N/A	3,000 tph	3	3.0 x 7.3 m	90 kW N/A	3,000 tph
Pebble Crusher	1	2.1	350 kW	400 ph	1	MP1000	750 kW	650 tph
Ball Mills	2	6.1 x 9.3 m	6,375 kW ea	1,500 tph	4	7.3 m x 9.3 m	10,500 kW ea	1,500 tph
Cyclones	14 / mill	66cm (26")	1.1 mw Pump	1,500 tph	14 / mill	66cm (26")	1.1 mw pump	1,500 tph
Rougher Flotation Cells	12 x 3 banks	85 m3	N/A	3,300 tph	9 x 4 banks	127 m3	N/A	5,500 tph
Cleaner Flotation Cells (combined for C3 / C4)	8	3.6 m Dia x 15 m high	N/A	3,000 tpd	12	3.6 m Dia x 15 m high	N/A	6,600 tpd
Regrind Mills	1	3.96 x 7.6 m	1,875 kW	150 tph	1	1	3.96 x 7.6 m	215 tph

Ore Delivery

Ore from the Grasberg orebody is crushed to minus 20cm (8") in three primary crushers located at the 3,400m (11,150ft) level. Crushed ore is conveyed by two parallel conveyor systems, a 213cm (72") system that transfers 7,500 t/hr and an 183cm (84") system that transports 10,000 t/hr, to a series of four vertical ore passes. The 3m diameter (10ft) ore passes deliver Grasberg ore to the mill at the 2,800m (9,200ft) elevation with two in operation, and two on standby. Ore is reclaimed at the bottom of the ore passes and directed to a common C3/C4 stockpile and a separate North/South (C1/C2) stockpile via redundant conveying systems.

Ore Reclaim

Four variable speed hydraulic 213cm (84") reclaim apron feeders are installed in the C3 reclaim tunnel, while three similar units are installed in the C4 reclaim tunnel. Each of the C3 units has a maximum output of 2,600 t/hr, and, in general two of the four are operated to provide a blend of fine and coarse ore while the other two are on standby.

Both C3 and C4 share a single coarse ore stockpile. The C4 ore reclaim tunnel was designed and constructed as part of the C3 project to ensure that the C4 construction would not interrupt C3 production. However, due to the overall plant orientation, and restrictions on the reclaim tunnel length, only three feeders could be installed. To meet production requirements, the bed depth was increased to allow a unit capacity of 3,500 t/hr, with two of the three feeders running during normal operation.

Grinding

Both grinding circuits were designed and commissioned as conventional SABC circuits. SAG mill product is screened at 10-11 mm, the oversize routed to a pebble crusher, with crusher product returned to the SAG mill feed conveyor. The SAG screen undersize flows by gravity to a reverse closed circuit ball mill circuit.

The C3 SAG mill is a 10.4m$\varnothing$ x 5.2m (34ft$\varnothing$ x 17ft) mill with a 10,600 kW gearless variable speed drive, also known as a wrap-around motor. The C4 SAG mill is an 11.6m$\varnothing$ x 5.8m (38ft$\varnothing$ x 19ft) mill with a 19,500 kW gearless drive. The SAG variable speed drives have provided operational flexibility for processing variable hardness ores, ease of starting, grinding-out and stopping the SAG mills, and excellent availability. The mills and bearings were structurally designed to facilitate a 20% ball charge in C3, and a 21% ball charge in C4.

Separate vibrating screens are installed behind each SAG mill. The higher capital costs relative to a SAG mill trommel screen is more than offset by the increase in SAG mill running time from decoupling the grinding and screening processes. The original two parallel 3.0m x 7.3m (10ft x 24ft) double deck vibrating screens in C3 (one operating, one on standby except during periods of high circulating loads) have been replaced with larger 3.7m x 7.3m (12ft x 24ft) and more robust screens. Three similar screens are installed in C4, with two in operation, and one on standby.

The C3 grinding circuit incorporates two 6.1m x 9.3m (20ft x 30ft) ball mills with single pinion 6,400 kW (8,500 hp) drives behind the SAG mill operating at 78% of critical speed. In C4 there are four 7.3 m x 9.3 m (24ft x 30ft) ball mills with 5,250 kW (7,500 hp) dual pinion drives installed. The C3 SAG screen undersize is gravity fed to two of three cyclone feed pump boxes each equipped with a 1,120 kW (1,500 hp) 51cm x 61cm (20in x 24in) variable speed centrifugal pump feeding a cluster of fourteen Krebs D26B cyclones. The third system is a standby unit and the cyclone underflow can be directed to either of the two ball mills. The C4 SAG screen undersize is gravity fed to a four-way splitter feeding four cyclone feed pump boxes with the same pumps and cyclones as C3. There is a standby pump installed on each pump box. While a standby cyclone feed pump has immediate capital and design implications, the benefit in terms of overall plant availability more than compensates. Freeport's position is that a high capital item such as a ball mill should not be taken out of service due to the lack of redundancy of a relatively low cost cyclone feed pump. Operating experience to date has validated this decision.

Flotation

The C3 rougher flotation circuit consists of three parallel banks of 12 Wemco 85m^3 (3,000ft^3) cells providing 26 minutes residence time. The C4 circuit has four parallel banks of 9 Wemco 127m^3 (4,500ft^3) flotation cells providing approximately 21 minutes of residence time. The rougher tails is final tails, the rougher concentrates are pumped to dedicated regrind circuits comprised of a 3.96m x 7.62m (13ft x 25ft) regrind mill in closed circuit with a cluster of Krebs D20B cyclones. Regrind cyclone overflow is pumped to a common cleaner circuit comprised of 3.66m$\varnothing$ x 15.24m (12ft$\varnothing$ x 50ft) cleaner and cleaner-scavenger columns. Both sets of columns produce final concentrate. The cleaner-scavenger column tails are directed to a bank of 12 Wemco 85 m^3 (3,000ft^3) mechanical cells. The concentrate from this bank is pumped back to the cleaner-scavenger columns, the tails to final tails. Cleaner circuit copper recovery averages 95-97% while concentrate grades average 30% Cu.

Dewatering

Final concentrate flows to one of two 2.2mØ x 13.7m (7.25ftØ x 45ft) vertical tower mills to reduce the top size to 80% passing 45 microns in order to minimize abrasive wear in the concentrate pipelines which deliver the concentrate to the port site dewatering plant. Concentrate is thickened to approximately 65% solids by weight, prior to pumping to port site. Dewatering at port site is done in conventional vacuum disk filters and rotary dryers to reduce moisture content to about 9% by weight prior to loading onto ocean-going ships.

Tailings are thickened in a 75m (245ft) center-drive and a 122m (400ft) tractor-drive tailings thickener. The Freeport milling complex is located at about 2,590 meters (8,500ft) above sea level, and water recovery is key to the operation. Reclaim water provides about half of the mill water requirements. Thickened tailings are discharged into the tailings river transport system, which carries the tailings to the tailings deposition system located in the lowlands.

Process Control

Both C3 and C4 are controlled from a single manned control room, supported by satellite control booths located throughout the milling complex. The Foxboro DCS/Allen-Bradley PLC control system currently has in excess of 1,700 analog inputs, 500 analog outputs, 10,000 digital inputs and 5,000 digital outputs. The PLC network is comprised of ten (10) fully redundant Allen-Bradley (AB) PLC5's communicating between processors over a redundant AB Data Highway Plus (DH+) fiber optic data highway, and with the DCS via appropriate foreign device gateways. There are two Siemens PLC's controlling the two SAG mill drives and four GE Fanuc PLC's controlling the dual pinion drives of the four C4 ball mils. The major components of the multi-node DCS include the operator stations, engineering work stations, control processors and foreign device gateways are all connected by a fault tolerant Carrier Band Local Area Network (LAN). By mineral processing standards, this is probably one of the larger control systems in the world. Process control was an integral part of the concentrator design, and significant efforts have been made to incorporate the appropriate level of advanced control throughout the operation as described by Neale and Veloo (1996), Perry et al (1996), and Anderson and Perry (1996).

CONCENTRATOR 3 START-UP

C3 Ball Mill Start-Up

The C3 ball mills and flotation circuit were ready for commissioning in early February 1995, before the SAG mill was ready. In response to this, crusher slurry from the wet screening plant in the North/South crushing plant was temporarily directed to the C3 cyclone feed pump boxes. A graded ball charge of 250 tonnes of 25mm (1") to 65mm (2.5") was used to start operation of the ball mills. The 25 mm balls were rejected within the first week of operation and replaced with the standard 65mm balls. No attempt was made to start the C4 ball mills with a graded charge.

Many of the grinding and flotation control loops were commissioned in manual as the operators became familiar with the control system. However, by the end of the first week of operation, most loops were running in automatic, a testament to the efforts of the pre-start-up training team. Cyclone feed pump box level, cyclone feed density and cyclone pressure control required significant tuning efforts, and some sanding of the cyclone underflow tubs, flotation cells and cyclone feed lines were experienced during the start-up period.

A significant problem during the ball mill start-up was sanding of the cyclone feed pump boxes and cyclone feed lines. This was due to highball mill operating densities "floating" balls out of the mills into the pump boxes and tripping the pumps. The ball mills had a large-diameter trunnion liner, which made it impossible to increase ball charge as part of the effort to maximize mill power draw. This was resolved by installing a ball-retaining ring made from steel screen meshing at the interface between the trunnion liner and the trommel, and also at the end of the trommel. Higher flights were also installed on the trommel to improve the separation of slurry and coarse rejects and ball mill scats, and preventing short-circuiting of slurry directly to the reject scoop.

C3 SAG Mill Start-Up

The SAG mill was commissioned autogenously in late February. The SAG mill start-up was quite smooth with only minor adjustments required to electrical and mechanical components. Part of the reason was that the operators had almost three weeks experience with the ball mill and flotation circuit, so could focus on the SAG mill start-up, once it was ready for commissioning. Figure 1 below illustrates the changes made to the grinding circuit that resulted in throughput increases during the first eight months of operation.

Figure 1: SAG Mill Average Tonnes per Hour

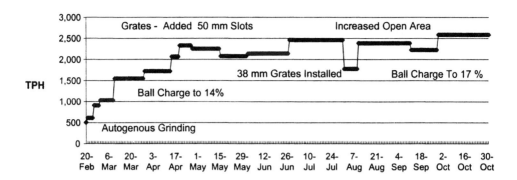

Initially it was intended to operate the SAG mill autogenously for 2-3 weeks to allow the operators to gain some experience with the new equipment. However, a combination of lower than expected throughput rates, the demonstrated abilities of the operators, and the fact that the ore was otherep and had to be milled, led to the decision to start charging grinding balls after two days of SAG mill operation. Initially 75mm (3p) balls were charged, but within a few days 105mm (4p) balls were introduced, and charge levels gradually raised to approximately 14% within four weeks.

Grate Design

It was recognized during the C3 design stage that the SAG mill feed contained a high levels of fines, as much as 50% -25mm (1p). As a result, the original grates were designed with 25mm (1p) slots. However, it was soon realized that the grates were actually restricting throughput to 1,500 t/hr at a ball charge of 11% by volume, compounded by the fact that the grates started to peen over on the outer section, reducing the slot width to 20mm (0.75p). Because alternative discharge grates were not available on site it was decided to cut 50mm (2p) pebble ports in half of the installed grates to increase the open area from 9% to approximately 11%. This resulted in an increase in throughput to 1,700 t/h. A further increase in ball charge to 14% ball charge increased throughput to 2,000 t/h.

A set of 38mm (1.5") grates was designed, ordered and expedited to site in late May and installed in June. The open area of the 25mm (1") grates had been increased to approximately 12% by cutting more of the inner and outer grates. While this allowed for some very crude experimenting to find the ideal open area, the cutting activities eventually led to the failure of the grates. The new 38mm grates provided an equivalent open area of 12.4%. Throughput rates at this time were approximately 2,200 t/h. Larger slot width grates were considered, and a set of 50mm (2") grates were designed but never ordered.

The original 38mm grates were designed with minimal structural support between each slot in an effort to maximize open area. Although these castings did provide the desired mill throughput the early breakage due to excessive steel-on-steel contact, reduced SAG mill availability, and overall mill throughput. A revision was made to the grate pattern to add the structural strength back into the casting, which had the undesirable effect of reducing the open area to 10.3%. The center discharge plates were then redesigned as a half-plate/half-grate to restore the open area to 12.2%.

The original grate design incorporated a 35cm (14p) high lifter bar. It soon became apparent that this high lifter bar was creating a low-pressure zone on the trailing side of the grate, particularly at mill operating speeds of 80% of critical. This resulted in less material passing through the trailing half of the grate than through the leading side. To counteract this effect the new 38mm grates were designed with a lift section reduced to 20cm (8p).

Liner Design
The original SAG mill liner concept was to maximize mill availability by designing very large, long-life lifters incorporating a "High-Low" configuration, and near vertical lifter angles. This liner profile provides for wear protection of the low lifter by the high lifter, and allows for the replacement of only half the shell liners at any one time thus reducing downtime hours for a particular liner change.

Unfortunately, the 35cm (14") high lifters with no relief angle, overthrew the grinding media, missing the impact zone and impacting the shell above charge. This resulted in minimal breakage of the critical size material, and accelerated liner breakage. The solution was to remove all the high shell liners and install the spare on-site set of low liners. A new set of shell liners was designed with a greater angle of relief on the leading faces to ensure the grinding media released and impacted where required.

SAG Charge Level
The SAG mill was structurally designed to accommodate a 20% ball charge since the mill was expected to act more like a very large ball mill rather than a conventional SAG mill to accommodate the fine feed size distribution. The results of increasing the charge level from 14% to 20% were somewhat inconclusive given all the other adjustments being made. However, the end result was that the ball charge was maintained at approximately 18% with SAG mill throughput averaging approximately 2,600 t/hr. At this throughput the ball mills started to become the constraint therefore the focus was placed on ball mill improvements and optimizing SAG mill product size distribution by adjusting SAG screen apertures and crusher cavity profiles.

SAG 1 Discharge Screens and Pebble Crusher
The discharge screens were originally fitted with parallel slots of 9 mm aperture bottom deck panels. The panels very quickly blinded resulting in significant quantities of minus 9 mm material circulating to the pebble crusher. This in turn caused the crusher bowl and mantle to build up with mud and resulted in "ring bounce". The solution to this problem turned out to be replacement of the screen panels with a zigzag design, which was significantly more efficient.

The pebble crusher was capable of processing only approximately 300-400 t/h even though at times feed rates were in excess of 500 tph resulting in a fraction being bypassed back to the SAG mill.

C3 Ball Mill Throughput Limitations
Ball mill circuit limitations were initially due to high circulating loads resulting from low cyclone feed density, undercharged ball mills and inefficient use of installed ball mill power. Increasing cyclone feed density significantly improved ball mill capacity and increasing the ball charge level to 39% by volume with the aid of the ball retaining screens reduced the circulating load.

The original ball mill liners were a very thick single wave shell liners designed for longevity. The single wave liners were redesigned as double wave liners with reduced thickness. This resulted in an increase in mill volume when installed and increased mill power draw at the same ball charge, due to the increased lever arm, i.e. the distance from the mill axis to the inside of the liner.

The success of the expansion project is summarized in Table 2 below.

Table 2: Comparison of Plant Performance with Design					
SAG MILL	Units	Design Normal Ore Ave	Design Soft Ore Ave	Best or Optimum Day	Best or Optimum Month *
Tonnes / Day	t/day	52,000	62,500	80,878	69,181
Tonnes/ Hour	t/hr	2,407	2,894	3,429	2,923
Availability	%	90	90	100	98
Mill Feed Size 80% Passing	mm	35 - 45	30	15	16
SAG Mill Power Req'd	kW	8,860	8,910	8,000	10,300
SAG Specific Power	kWhrs/t	3.68	3.08	2.74	3.95
SAG Mill Operating Wi	kWhrs/t	17.3	14.5	13.0	13.0
Ball Charge	%	10 - 13	10 - 13	20	18
Pebble Prod.	t/hr	600	430	350	350
BALL MILLS					
New Feed Per Ball Mill	t/hr	1,204	1,447	1,715	1,462
Flot Feed 80% Passing	Microns	150	150	190	210
Ball Charge	%	30 - 35	30 - 35	39	39
Power Req'd / Utilized	kW	5,990	6,014	6,400	6,200
No. of Cyclones		26	24	28	26
Cyclone Feed % Solids	%	56.5	56.5	65	67
Copper Rec.	%	90.5	90.5	92	88.8

* Best month data not all from one month.

The preceding table suggests the C3 expansion was a success. SAG mill throughput exceeded expectations while above forecast operating time is indicative of a concentrator which was designed and built with top priority given to operational and maintenance considerations.

CONCENTRATOR 4 EXPANSION PROJECT

By mid-1995, less than 6 months after the C3 start-up, a site-based design team was formed to evaluate the opportunities for a further concentrator expansion. In January 1996, a decision was made to proceed with the project, and by early 1998 site commissioning of the C4 concentrator commenced. Some of the key issues of the expansion project are reviewed below, with further details of the start-up provided by Coleman et al (2001).

The first phase of the project involved expanding the existing C3 cleaner flotation circuit into a combined C3-C4 cleaner circuit, including:

1. Relocating two Vertimills (tower mills).
2. Installing six column flotation cells.
3. Modifying eight existing column flotation cells.
4. Replacing twelve 42.5m^3 (1,500ft^3) flotation cells with twelve 85m^3 (3000ft^3) cells.

During the 40 days required to accomplish this work the cleaner circuit tailings were directed to the rougher feed with little impact on existing operations.

The first of four 7.3 m (24ft) diameter x 9.3 m (30ft) long ball mills and a section of rougher flotation circuit were commissioned in December 1997 utilizing crusher slurry from the North-South crusher wet screening plant. Initial start-up problems were experienced with the clutch control on the dual pinion "Quadramatic" drives for the ball mills. Alignment issues, which are often a problem with dual pinion drives were experienced, but limited.

The 11.6 m (38 ft) diameter x 5.8 m EGL SAG was commissioned autogenously in January 1998 with ball charging following within days. This was a mere three years after the start-up of C3, and only 17 months after breaking ground for C4. The SAG mill was commissioned with 38mm (1.5") grates and a high-high shell liner configuration. Several days were required to balance the ball charge in the SAG and single ball mill.

By early February, after commissioning the second of the four ball mills, the SAG mill was capable of processing between 1,700 and 3,300 t/hr depending on ore characteristics. The third ball mill was commissioned in May followed by the fourth ball mill in June. Shortly after the start-up of the third ball mill C4 achieved a throughput of 100,000 tonnes in one day.

The operating objective at this point was to continuously maximize SAG mill throughput with the following being evaluated during the months subsequent to start-up.

- **Grate Apertures**: The 38 mm grate openings initially peened over to 32 mm apertures and restricted throughput. However, as the grates wore the aperture size increased resulting in additional throughput. Grates with 50 mm openings were ordered and installed.

- **Pulp Dischargers**: During the C4 construction phase it was recognized that the original set of pulp dischargers were undersized and would restrict slurry flow through the mill. A larger set of pulp dischargers was ordered and installed in May 1998.

- **Pebble Crusher**: The MP1000 pebble crusher was started up once tonnage exceeded 400 t/hr. When choke fed, the crusher ran well but required regular cleaning to prevent build-up below the crusher due to the excessive water carryover due to uneven loading on the SAG discharge screens. A number of modifications to the SAG discharge box were required to address this problem.

- **Mill Power**: During the design phase the motor power rating was increased above the original specifications of 18–19 MW to ensure adequate power was available if required. Mill power typically averages 16–18 MW and it is unlikely full power will be utilized.

- **Mill Speed and Sound**: Experience with C3 suggested mill throughput was optimized while operating at a mill speed of approx 76% critical. The C4 SAG mill was designed to operate at 76% critical under normal conditions up to a maximum of 80% critical. Optimum mill sound levels were determined by field measurements with a hand held sound meter and comparison with the local sound meters installed near the mill shell. Set points were then adjusted accordingly to optimize throughput.

- **Shell Liners**: The original shell liners were designed with 3 inches of plate section and 10 inches of lift with a face angle of 12 degrees. The mill contained 69 rows of liners therefore spacing was generous for a mill this size. Experimentation with shell liners designs began in 1999.

SAG MILL OPTIMIZATION

By late 1998, the C4 concentrator had achieved near design throughput on a consistent basis. PTFI concentrator personnel identified areas in each of the major unit processes (SAG mill, ball mills and flotation) that were production bottlenecks. The following describes the most significant opportunities for process improvements identified in each area; see Staples et al (2001).

Shell Liners

The steep face (12°) lifter angle resulted in a ball trajectory that impacted the shell above the charge toe at normal operating speeds. This risked liner damage. Consistent with practice at several other operations, and with output from a study using the MillSoft modeling software, PTFI decided to test shallower face angles. In July 1999 a set of 18-degree lifters were installed and in May 2000, a set of 25-degree lifters were installed.

Data analysis suggests that only a small gain in throughput can be directly attributed to face angle changes. However, due to the decrease in shell impacts, improved resistance to packing and steadier operation, PTFI continues to use face angles between 18 and 25 degrees.

Due to the inherent difficulty of interpreting noisy production data over long periods of time (shell liners typically have a life of 9 months in the C4 SAG mill), PTFI has been working closely with suppliers and technology groups to investigate new off-line techniques for optimizing shell liner design. Discrete Element Modeling (DEM) simulations indicate that PTFI is running the correct range of face angle. The final consideration is lifter spacing and simulations indicate that increased spacing will provide some benefit. A 46-row set has been designed and is under consideration.

Discharge End

Under fine feed conditions the C4 SAG mill can treat sustained tonnages up to 5,800 t/hr plus 1,500 t/hr recycle. The ability of the mill to discharge this very large volume of material is unique to PTFI. Thus, a great deal of focus and energy has been placed on optimization of discharge end design.

At start-up, 38mm grates were installed with an open area of approximately 11%. Shortly thereafter, 50mm grates were trialed and by mid 1999 had become the standard. Further grate design changes were made over the following 18 months to first optimize the slot layout and then to increase open area. In addition to these considerations were issues of structural integrity and liner life. With the move to very large double sized grates, designed to reduce re-line time, came casting and metal flow challenges for liner suppliers. It is the experience at PTFI that good grate design must consider several factors:

1. Grate slot aperture size – Apertures must be sized correctly to discharge critical size material but also must match the recycle systems capacity. Increases in slot aperture size have resulted in significantly increased throughput and further increases are expected as testing continues.
2. Open area – Open area must be balanced with aperture size to match recycle system capacity and obtain the desired wear life. Higher open areas have generally been beneficial at PTFI. However, though open area is important, placement of that area is even more critical. PTFI has tested open areas as high as 15% that wore to nearly 20% after 4 months. Recent modifications discussed below have reduced this slightly.
3. Slot design – A crucial aspect of grate design lies in the positioning of the grate slots. Again, there is a balance with wear life that must be made. Other critical factors that should be addressed are ensuring adequate channel depth for flow in the pulp lifter chamber and positioning of the slots to minimize flow-back into the mill.
4. Relief angle – The design must also ensure adequate relief angle to prevent blinding of the slots with grinding media. PTFI experience indicates that 3-degree relief is inadequate and current designs utilize a 5-degree angle.

By examining wear patterns in the pulp lifter chambers and on the front and backside of the grates, the decision was made to remove three slots at the inner most radius of the inner grate. It appeared these slots were contributing more to flow-back than to transport of material into the pulp chamber. Figure 2 shows the three inner slots that were removed from the inner grate. Figure 3 shows the most recent design in C4 without these slots. The reader may also note that, at the expense of some open area, there is more steel between the slots to reduce incidence of premature breakage. Tests of the new design are currently underway and initial results are very encouraging.

Pulp Lifters

The other significant change tested was to increase the pulp lifter depth by 4 inches at the cone and taper the increase down as you move to the outer radius of the mill. It was felt that pumping capacity of the pulp chamber was marginal and that there was significant restriction at the pulp discharger. This made it possible to increase the volume of the pulp discharger without the loss of mill volume at the shell, where most of the power is drawn. This change was made in March 2001 and along with changes to the grate design, indications are that pumping capacity has improved with C4 achieving record throughput levels.

Again, PTFI is working closely with suppliers and technology groups to evaluate options to further increase throughput. It is hoped that DEM will provide solid information for options such as converting to a vortex/curved discharge end. Because discharge end pumping capacity is such a major concern with both the PTFI SAG mills, moving to a curved end would likely have significant benefits. However, the risks and downtime associated with such a drastic change are daunting and more confidence is required before proceeding. Development of an entire discharge end model is underway and it is hoped that the necessary information can be gained using this technique.

Figure 2 - Previous High Open Area C4 Grate Design

Figure 3 - Current Test C4 Grate Design

SAG Bypass Project
A recent development is the C4 SAG Bypass. First simulated in 1998, this flow sheet change introduced two additional recycle conveyors to transport pebble crusher discharge directly back to the SAG discharge screens, thus bypassing the SAG mill. Simulation studies estimated that for every 4 tonnes of crusher discharge that bypassed the SAG mill, an additional tonne of new feed could be processed. The new system was commissioned in June 2001 and is credited with increasing C4 SAG mill throughput by 5,000 t/day, which is remarkably consistent with

the simulator predictions. This is a particularly exciting development in that it may well define the new flow sheet standard for SABC circuits with separate SAG product screening.

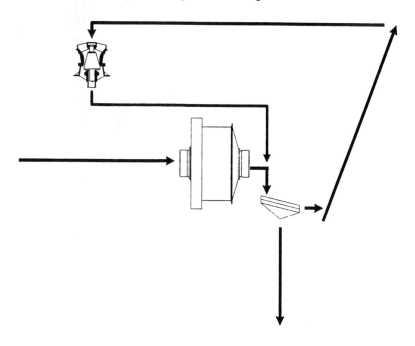

Figure 4 - SAG Pebble Crusher Discharge Bypass Flow sheet

C3 To C4 Slurry Transfer
C3 has 13 MW of installed ball mill power, which at a typical feed rate of 2,800 t/hr, translates to 4.6 kW installed per tonne of fresh feed. Recognizing that C3 SAG mill throughput was often restricted by ball mill capacity, C4 was designed with 42 MW of installed ball mill power, a significantly higher 8.4 kW installed per tonne at 5,000 t/hr. Typically, C3 grinds averaged 23% +212 micron while C4 averaged 15% +212 micron (28 mesh).

A 35cm slurry pipeline was installed and connected to the spare cyclone feed pump in order to divert a portion of the C3 SAG screen undersize (SAG circuit product) to the C4 ball mill circuit. Pumping this material to C4 utilized the additional ball mill capacity in the C4 circuit and removed a bottleneck that often restricted C3 SAG mill throughput. This improved overall grinding power utilization, resulting in finer grinds and increased metal recovery. The current system is capable of pumping 600 to 1500 t/hr of slurry to C4 and has resulted in a throughput increase of 3,000 t/day in C3. In addition, the overall C3/C4 combined grind has improved by 1% +212 micron, and copper recovery has increased by 0.3%.

SUMMARY OF LESSONS LEARNED
As usual with projects the size of C3 & C4, a number of valuable lessons were learned. Following is a brief summary of some of the more memorable lessons from the grinding circuits.

- The C4 SAG mill discharge box and the distribution of feed to the vibrating screens was one of the more significant challenges. After various modifications and the replacement of the original vibrating screens most of the problems were solved. The importance of getting this right in an SABC circuit cannot be overemphasized as the penalties include excessive downtime, and poor dewatering of the SAG discharge slurry resulting in poor pebble crusher performance.

- Conveyor speed and CEMA loadings are important features to consider in the design of a recycle system for an SABC circuit. Magnet efficiency is significantly reduced when attempting to remove steel from a fast moving highly loaded belt. Tramp metal recovery was improved by adjusting the idler configuration to flatten the belt and spread the load under the magnet.
- The pebble crusher circuits were designed without a surge bin. Although it is possible to keep the crushers fed through a combination of improved magnet operation and higher circulating loads, the SAG circuit stability suffers when the crusher feed is inconsistent.
- The design of the SAG mill discharge end liner configuration is important. Not only the size of the grate apertures and amount of open area but also the volume behind the grates and the cross sectional area at the narrowest section of the pulp dischargers. Due to the very high SAG mill volumetric throughputs the pulp chamber depths were increased during the design phase. Recent changes have further increased the volume of both the pulp chamber and pulp dischargers resulting in additional throughput increases.
- During start-up it is prudent to have two or more sets of varying aperture grates on the ground in anticipation of throughput issues.
- Ball mill lining requires the removal of the feed end trunnion liner prior to installation of the lining machine, adding hours onto each lining job. Either lining from the discharge end through the large opening or a redesigned feed trunnion would significantly reduce lining time.
- Wrap around variable speed motors were originally considered for all four ball mills. However due to the relatively small power grid it was felt that the harmonics could potentially disrupt power supply to all process equipment. As such the twin pinion drives were selected. The soft start capabilities of the wrap around motors would have significantly reduced mill downtime and likely improved grinding circuit performance for all conditions experienced.
- Flotation cleaner circuits bixes were all designed within large sumps with pumps placed at a low elevation within the sumps. The pumps were susceptible to flood conditions causing the V belts to operate in slurry. It was necessary to raise all the pump one meter to keep them high and dry.
- The slope of the underflow launder for the 400 ft thickener was set at 1/16 inch per foot. Upon start-up of the thickener, the underflow launder overflowed when transporting higher density slurry (+ 60% solids) containing coarser material (+30% 65 Mesh). As C4 ramped up in tonnage the launder problem diminished through a combination of higher launder velocities and better slurry rheology. On reflection, a slope of 1/8 or 3/16 of an inch per foot may have been more appropriate.
- The 400 foot thickener center well had a number of design flaws including feed pipe arrangement, lack of rubber lining in the center well and insufficient size, all of which contributed to high wear due to extreme slurry velocity. This in turn resulted in excessive downtime to repair or replace the feed well. The solution is a larger rubber lined center well combined with lowering the feed pipes thus reducing the incoming slurry velocity.
- The dewatering plant expansion was accomplished using disk filters and a thermal dryer rather than more recent technology (filter presses, ceramic filters etc) without fully investigating the advantages or disadvantages of the latter. Secondly, auxiliary equipment design criteria were duplicated without addressing current operational and maintenance issues. Vacuum pump capacity was one such example while CEMA loading on conveyors was another. Although the expansion equipment continues to operate, Freeport has since decided to replace older, high cost dewatering equipment with a pressure filter and upgrade most of the conveyors.

CONCLUSIONS
The start-up of the expansion SAG Plant at P.T. Freeport Indonesia's Papuan operations has, by all standards, been a tremendous success. This can be attributed to a combination of a good overall plant design and layout, unique mill feed characteristics, SAG mill liner and circuit design, a practical approach to plant wide process control, and the emphasis placed on training prior to the SAG Plant start-up. The success of the expansion can also be attributed to a teamwork philosophy from all those involved in these projects.

ACKNOWLEDGEMENTS
The authors wish to thank the management of P.T. Freeport Indonesia for their support and permission to write and present this paper.

REFERENCES

Anderson, L., Perry, R, & Neale, A.J., (1996), **Application of dead-time and gain compensation to SAG feeder control at P.T. Freeport Indonesia**, Proceedings, 28th Annual Meeting of the Canadian Mineral Processors, Ottawa, Ontario.

Coleman, R.E., Nugroho, S., Neale A.J., (2001). **Design and Start-up of the PT Freeport Indonesia No. 4 Concentrator**, SAG 2001, Vancouver, B.C. October 2001

Coleman, R.E and Napitupulu, P., (1997). **Freeport's Fourth Concentrator – A large Step Towards the 21st Century**, AUSIMM Sixth Mill Operators Conference, Madang PNG, Oct 6 – 8, 1997.

Coleman, R.E. and Veloo, C., (1996). **P.T. Freeport Indonesia Concentrator Expansion**, SME Annual Meeting, Phoenix, Arizona, March 11 – 14, 1996.

McCulloch, Jr., W.E. (1991). **Mill Expansions at Freeport Indonesia - four-fold production increase in a decade**, Copper 91-Cobre 91, Volume II, pp. 3-17.

Neale, A.J. and Veloo, C., (1996). **Process Control at P.T. Freeport Indonesia's Milling Operations**, CIM Annual Meeting, Ottawa, Ontario, January 1996.

Perry, R., and Anderson, L., (1996). **Development of Grinding Circuit Control at PT Freeport Indonesia's New SAG Concentrator**, Proceedings of SAG 96 Conference, Vancouver, British Columbia.

Russell, R.L. & Kieffer, L.D. (1994). **Mill Expansions at P.T. Freeport Indonesia**, paper presented at the 1994 SME Annual Meeting, Albuquerque, New Mexico.

Staples, P., Siewert, H., Stuffco, T., and Mular, M., (2001). **SAG Concentrator Improvements at PT Freeport Indonesia**, SAG 2001, Vancouver, B.C. October 2001

Van Nort, S.D., Atwood, G.W., Collinson, T.B., Flint, D.C. & Potter, D.R., (1991). **Geology and Mineralization of the Grasberg Porphyry Copper-Gold Deposit, Irian Jaya, Indonesia**, Mining Engineering, March 1991.

HIGH PRESSURE GRINDING ROLL UTILIZATION AT THE EMPIRE MINE

David J. Rose, Metallurgical Engineer, Empire Iron Mining Partnership
Paul A. Korpi, General Manager, Empire Iron Mining Partnership
Edward C. Dowling, Senior Vice President-Operations, Cleveland Cliffs Inc.
Robert E. McIvor, General Manager, Cliff's Technology Center

Abstract

The Empire Mine is currently using High Pressure Grinding Rolls (HPGR) to process magnetite ore at its facility in Palmer, MI. The evolution of the flowsheet with emphasis on the needs for and advantages of the High Pressure Grinding Rolls will be outlined and discussed. Plant performance and circuit improvements will be reviewed.

Empire Mine History

Empire Mine is an integrated open pit mine, concentrating plant and pelletizing facility that is currently capable of producing 8.0 million LT of pellets annually. Opened in 1963, Empire was capable of producing 1.6 million LTPY of pellets. Expansions in 1966, 1975 and 1980 have increased the capacity to its current level. Empire Mine is owned by a joint venture of Ispat Inland Steel Co. and Cleveland Cliffs Incorporated.

Flowsheet Design

Ore at Empire Mine is crushed to minus 9 inches in a Primary Crusher (Traylor 60 x 89). Crushed ore is transported to one of two crude ore buildings where it provides feed to 23 Primary Grinding Lines. Fully Autogenous Primary Milling is used at Empire with double deck screens separating cobber feed size material (-1mm) from the bottom deck oversize (+1mm to –½ in.). Bottom deck oversize material is returned to the Primary mill along with excess top deck oversize material (+½ in. to – 2½ in.), which helps provide grinding media to the secondary (pebble) milling circuit. Figure 1 shows the original circuit, before excess pebble crushers were added to some lines. This flowsheet is still in place for ten grinding lines.

Magnetic separators (cobbers) treat the Primary Mill product, rejecting approximately half of the crude ore from the plant. Cobber concentrate is then pumped to hydrocyclones to separate the minus 25 micron fraction from the coarse fraction, which is further ground in pebble mills. The pebble mill discharge is in closed circuit with the cyclones. Cyclone overflow material flows through a siphonsizer / thickener sizer to deslime the slurry and then through a second magnetic separation stage (finishers) to produce flotation feed grade material.

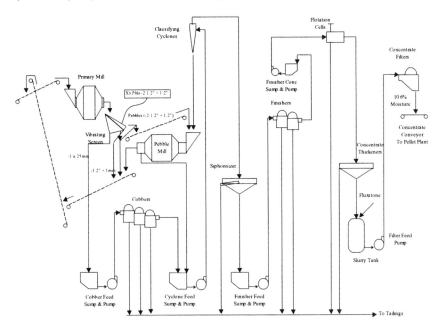

Figure 1   Empire Original Flowsheet (No Pebble Crushing)

The flotation plant at Empire is a reverse float with the gangue (silica) being floated off the top using an amine reagent. Flotation concentrate is thickened in conventional concentrate thickeners and filtered on disc filters to produce balling plant feed for the pellet plant.

Figure 2 shows the modified flowsheet for the Empire in which excess pebble crushers have been installed on thirteen grinding lines. Excess pebbles are defined as those produced by the primary mill not required to maintain horsepower setpoint in the pebble (secondary) mill. A variable speed feeder provides all the pebbles required in the mill and the excess enters the pebble crusher circuit by means of an overflow from the pebble hopper. The crushed pebbles are returned to the primary mill just as the uncrushed excess pebbles would be in the lines without pebble crushing.

The line 1 Primary Mill was converted to a ball mill grinding fluxstone for the pre-fluxed pellets produced in the pellet plant.

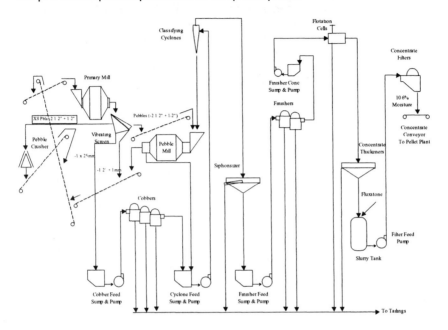

Figure 2   Empire Flowsheet With Pebble Crushing

Pelletizing at Empire is done using eighteen balling lines feeding four rotary kilns. Three pellet product grades are produced annually at the plant, which ships these pellets to several different blast furnace operators around the Great Lakes area.

Plant Equipment

Three expansions in the years since initial startup have left Empire with four unique plant sections, each utilizing different equipment, in terms of manufacturer, design and size.

Empire I was originally comprised of six Hardinge-Cascade 24-ft by 8-ft diameter primary mills and six Nordberg pebble mills 25 ½ feet long by 12½ ft in diameter. Eighteen banks of three drum cobber magnetic separators (6 ft long by 3 ft diameter) provided the primary concentrating step. These were followed by 18 banks of six 10-in. Krebs cyclones, six 38-ft diameter siphonsizers and 18 banks of two drum magnetic finishers (6 ft long by 3 ft diameter). Final concentrate upgrade was achieved in one bank of eight 500 ft^3 Wemco flotation cells. The magnetic separators in Empire I are a mix of Dings and Jeffrey's units.

Empire II started with 10 Allis-Chalmers Rockcyl primary mills, 24 ft by 12½ ft diameter that are twinducer driven. The pebble mills are the same Nordberg 25½ ft by 12½ ft units as installed in Empire I followed by two 3-unit 8 ft by 3 ft diameter cobber magnetic separator drums from Dings. Thirty banks of six Krebs 10-in. diameter cyclones are followed by ten 38-ft diameter siphonsizers and 30 Dings finisher magnetic separators (6 ft by 3 ft diameter). The flotation section in Empire II is a 5-cell unit comprised of Wemco 1000 ft^3 cells.

Empire III is a section of five lines using Allis-Chalmers Rockcyl mills 24 ft by 12½ ft in diameter. These are standard pinion drive units. The pebble mills are Allis-Chalmers Rockcyl units 25½ ft by 15½ ft in diameter, followed by three Dings three-drum magnetic separators (8 ft by 3 ft diameter). Five banks of twelve 15 in. Krebs cyclones are used with five 46 ft diameter siphonsizers and Dings magnetic finishers to provide flotation circuit feed. The magnetic finishers are comprised of eight 6 ft by 3 ft units and nine 8 ft by 3 ft units. Empire III flotation is achieved in a bank of five Wemco 500 ft^3 cells.

Empire IV is the newest and largest section of the Empire plant. It utilizes 3 Koppers 32-ft diameter by 16½ ft primary mills and six Nordberg 32 ft by 15½ ft diameter pebble mills. There are 18 banks of three drum (10 ft by 3 ft diameter) cobber separators and 21 banks of two drum (10 ft by 3 ft diameter) finisher magnetic separators. Siphonsizers are replaced in the flowsheet by three 85 ft diameter thickener sizers. The original flowsheet included one bank of ten 500 ft^3 flotation cells.

Each line at Empire is independent of the others. There are grinding units dedicated to a certain section of concentrating equipment. The changes to this came about with the addition of distribution systems, which effectively "disconnected" the primary and secondary halves of the grinding circuit.

Process improvements have been realized over the years by adding excess pebble crushing (Lines 11 – 24), cobber concentrate distribution (Lines 2 – 6, 12 – 16 and 22 – 24), additional flotation capacity (Lines 22 – 24) and the HPGR (Lines 22 – 24).

Flowsheet Evolution

From the time Empire first produced pellets, the plant flowsheet has maintained the same basic design. Primary Autogenous milling followed by magnetic separation followed by secondary grinding in pebble mills and then finishing the concentrating process in magnetic separators and flotation cells.

Ore hardness and lower iron content in the crude started to make it necessary to find ways to increase the throughput of the grinding section in the Plant in order to produce enough concentrate for the owners. The first addition to the flowsheet was the Excess Pebble Crusher in Empire IV (the last expansion section), where a 7 ft Symons cone crusher was installed to crush the excess pebbles (+½ in.) material and return the crushed pebbles to the Primary Mills. The smaller material that is returned to the Primary mill is easier to grind to target size range and therefore horsepower consumption is reduced and the throughputs can be increased.

More crushers were added to the older section of the plant in the mid 1990's as the orebody was expected to get harder. Four Nordberg HP200 crushers were added to the grinding lines from 11 through 21 in differing combinations. Pebble Crushing proved to improve circuit throughputs by approximately 20% for the times when the crushers were running. These improvements allowed Empire to produce over 8.5 million LT of concentrate from 1995 through 1998.

In the mid 1990's, mine planning projections were still indicating a higher work index crude ore material coming into the plant. Another option to be investigated was the addition of another way to release the load in the Primary Mills and increase throughputs. Using the theory of the crushers, high pressure grinding rolls promised to provide better comminution at lower costs and higher productivity.

High Pressure Grinding Roll Design

High Pressure Grinding Rolls have been developed over the last 15 to 20 years and have primarily been used in the aggregate industry. The units that are in use in cement have proven to be powerful workhorses and are capable of running at high availabilities and tonnage rates. In the mining industry, they are widely used in diamond operations throughout the world, such as a recent installation in a diamond mine in the Northwest Territories. More recently High Pressure Grinding Rolls are being installed in applications for gold and phosphate minerals. Also, a wide range of iron ore applications successfully utilize HPGR's for both coarse ore and pellet feed grinding, although Empire's HPGR installation continues to be the only iron ore application in North America.

The HPGR design is comprised of a fixed and a moveable roll that run counter rotation to one another at the same speed. The moveable roll is positioned at such a distance from the fixed roll to exert a specific amount of pressure on the material passing between the two surfaces. The feed material is introduced at the top of the rolls in the area where the two surfaces begin to meet. The movement of the rolls and the static pressure from the feed above provides for a positive downward force and ensures an effective nipping of the feed into the gap between the rolls. The material between the rolls forms a particle bed and the force from the rolls causes the grinding action that reduces the material size. The grinding force acts throughout the particle bed, and the main size reduction mechanism is defined as interparticle crushing. This is considered effectively a grinding action between the ore particles as opposed to the roll-particle-roll contact crushing as in conventional rolls crushers. The manufacturer of the machine has stressed that the units are grinding machines, not crushers so the top size feed material must be limited based on the roll size.

The roll surfaces are lined with hard metal alloy studs. These studs are designed to create high and low spots in the surface where the feed material collects. This makes the effective grinding surface smooth and provides a so-called autogenous lining. The second effect is that the wear on both the studs and base metal surface is minimized as the amount of metal area exposed to contact with the ore is reduced.

The mechanical characteristics of HPGR include electro-hydraulic control systems for the roll positioning and pressure control. A separate lubrication system is included for the rolls and bearings. The entire unit is housed in an enclosed frame with dust shields to limit the amount of fugitive dust that escapes. On the ends of the rolls there are cheek plates that keep the material between the roll faces from getting squeezed out and bypassing the grinding action of the HPGR. These cheek plates are a wear area and therefore, are comprised of a high wear alloy. These need to be adjusted for position and pressure occasionally.

Getting feed to a HPGR is relatively simple, once you have prepped the feed material to reduce the top size. A gravity feed chute can be employed to maintain a choked feed chute and allow for proper distribution into the HPGR. The drives for the rolls are variable feed, which can be used to maintain a level in the feed chute and keep the rolls choke fed. Access doors at the top and sides of the HPGR frame are used for minor maintenance and inspections of the rolls, cheek plates and feed chute.

Empire Options

For many years, Empire had been expecting the harder ores and looking for ways to increase tonnages (or at least maintain current levels) to produce high enough concentrate tonnages. In the late 1980's and early 1990's, Empire IV, with its excess pebble crusher was significantly more efficient than the other lines at treating harder ores. For this reason, a split ore blend program was used to maximize concentrator productivity. In this plan, the harder ores were directed towards the Empire IV section. The logistics associated with trying to maintain this schedule proved to be difficult.

Other options for plant productivity included removal of the original DSM screens, which were a secondary screening step between the vibrating screens and the magnetic separators, the use of smaller pebbles in the pebble mills and changing the cobber feed top size to 1 mm from 2 mm. These changes have been kept in place with all the other changes that have been made.

In the mid 1990's, the orebody was still getting harder and leaner, requiring the plant to treat more ore to maintain concentrate levels. With the success of the pebble crushing in Empire IV, excess pebble crushers were planned and added to lines 11 though 21 in Empire II and Empire III. These installations were complete in late 1995 and helped Empire to maintain production at 8.5 million LT of concentrate for the next three years.

Mine planning in 1995 indicated that the silicate content of crude ores would be steadily increasing over the next few years from a historical 30 to 40% of the blend to 40 to 60% in the next few years to greater than 85% in later years. Harder ores reduce grinding rates and excess pebble crushing had helped to offset the shortfall up until then, but the future was going to consist of harder ores, with lower iron content. The only way to maintain high enough production levels was going to be increasing throughput to offset the reduction in recovery that was inevitable.

The Empire Solution

HPGR technology had been pilot plant tested on the Empire ores and had proven in the controlled environment to be successful in treating high tonnages of ore and relieving the load in the primary mill, which increase throughput rates. The HPGR installed at Empire was originally intended to be a test for the technology and its application at Empire.

Options investigated include a second stage of pebble crushing, but the technology available at the time would not produce the product size distribution that the HPGR was capable of and the energy costs would be higher.

A requisition for funding was approved to purchase a HPGR for Empire IV, with performance guarantees in place to ensure the improvements Empire was looking for would be attained. KHD from Germany was the supplier of the HPGR based on the testing program they had completed in conjunction with Empire and Cliff's technical personnel at the Coleraine Minerals Research Laboratory (Duluth University of Minnesota).

The HPGR installed at Empire is a KHD model RPSR 7.0 – 140/80, the rolls being 1400 mm (55 in.) diameter by 800 mm (31.5 in.) wide, and providing a maximum specific pressing force of 6.25 N/mm^2 (906.5 psi). It is powered by two 670 kW (900 HP) variable speed controlled motors to provide a roller surface speed of 0.9 to 1.8 m/s (2.95 to 5.91 ft/s) through planetary reducers. The ROLVIS control package oversees pressure, gap width and speed, as well as monitoring power, oil pressure, the lubrication system, and other operational features.

Rollers are carried in a machine frame in four-row cylindrical roller bearings as well as two self-aligning roller thrust bearings. One roll is moveable and the other fixed. The moveable roller moves closer or farther away from the fixed roller to meet the pre-set force as required at the surface where the rock is pressed. The system consists of 2 hydraulic cylinders with spherical pistons and accumulators and associated other equipment. Cheek plates are used on the edges of the rolls to keep material from bypassing the roll surfaces. These are adjusted to keep the spacing between the edge of the roll at a minimum.

The HPGR was installed in the Empire flowsheet to follow the Symons pebble crusher in Empire IV treating crushed product and returning the fine product to the primary mills, as shown in Figure 3.

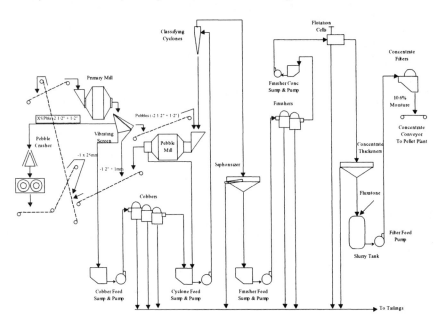

Figure 3   Empire Flowsheet With HPGR Installation

HPGR Performance

The HPGR supplied to Empire was guaranteed to meet certain criteria regarding unit availability, tonnage throughputs, product size distribution, liner life and specific power consumption.

The HPGR supplied to Empire was initially installed with segmented studded liners that could be replaced without having to remove the rolls from the frame. These segments (six per roll) were held in place by two rows of six bolts that were covered by alloy pucks over the bolt head. The one thing that became an issue with the segmented liners was that the bolt covers would come loose and fall out or cracks would develop and they would break into pieces and fall out requiring maintenance for the replacement of and repair to the bolt covers and holes.

The unit availability guarantees from KHD specified that the unit would be ready to run at least 95% of the time. The frequency of the cover changes was getting to be such that the unit availability was suffering. The tire style liners were provided to Empire at a prorated cost based on the hours that the segmented liners lasted. Along with some other scheduled maintenance, the liners were replaced, with help from KHD and the HPGR was put back into operation.

The unit installed at Empire was guaranteed to operate at 400 LTPH by KHD. The feed rates in 2001 averaged 323 LTPH. This is a result of the process not having enough material to consistently feed 400 LTPH to the HPGR. The tonnage of excess pebbles generated by the primary mills is not enough to maintain this feed rate to the rolls. This problem is being addressed by some material handling changes in the circuit allowing for recirculating loads around the HPGR.

Product size distribution from the HPGR was guaranteed to be 50% passing 2.5mm. Performance testing in the circuit showed that the product was running approximately 47% passing 2.5mm. The HPGR product, as with any grinding operation, is dependent upon the feed size of material in the process. At the time of the performance testing, the feed size was coarser than the specified range from the pebble crusher and the reduction ratio was actually a little better than projected by KHD. Based on these results, the Empire testing was considered to be a success.

The liner life for the segmented liners was guaranteed for 12,000 operating hours from KHD. Liner life on HPGR rolls is determined by the height of the studs on the studded surface. Measurements of the stud height are made on a regular basis to determine the wear rates of the studs and project liner life. Despite the difficulties with the bolt covers, the liner life was projected to meet the guarantee from KHD, so this was not the issue with the segments.

Figure 4   Original Segmented Liners Showing Bolt Covers

Specific power consumption in the HPGR was guaranteed to be 2.5 kWh/LT (3.35 HP/LT). Actual power consumption has declined steadily since the HPGR was commissioned and in 2001 averaged less than half of the target (1.2 kWh/LT).

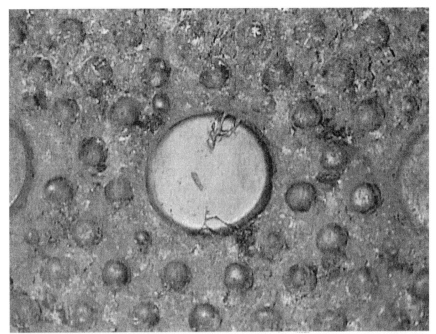
Figure 5   Damaged Bolt Cover on Original Liners

A significant change in technology allowed the manufacturer to construct a seamless tire style liner that did not require the use of bolts or bolt covers. This eliminated the need for replacing and repairing the bolt covers and bolts as we had seen with the original liners.

The original liners have been replaced with the new seamless liners as part of the optimization of the HPGR circuit at Empire. HPGR performance has not been affected by the new liners, but availability has been much improved. The only downtime taken now is essentially scheduled downtime to perform routine maintenance tasks.

Figure 6 New "Tire Style" One Piece Liner

Empire HPGR Plant Results

Table 1 HPGR Results 1998 – 2001

	HPGR Performance Data					
	RP-3 Hrs	RP-3 LT	HP/LT	% Op Time (with feed)	Feed Rate (LTPH)	% PC Product Roll Pressed
1998	2,375.3	645,229	2.25	27.1	271.6	81.8
1999 *	1,971.5	570,051	2.02	22.5	289.1	66.3
2000	4,198.8	1,305,918	1.96	47.8	311.0	90.9
2001 *	3,943.9	1,273,460	1.61	45.0	322.9	92.3

* 1999 and 2001 were interrupted by shutdowns affecting performance data.

Table 1 shows the HPGR run data for the past four years. Operating time (defined as with feed) is still low, but the key is the percent of potential feed that is actually fed to the HPGR. This has increased to over 92% indicating that the unit will handle the tonnage Empire can currently provide. The feedrates have been higher than the 400 LTPH target at times, but this cannot be sustained because of feed material limitations.

Table 2 Effect of HPGR on Primary Mill Performance

	Productivity Without Roll Press Operating		
	PRIMARY MILL	PRIMARY MILL	TOTAL
	LTPH	HP/LT	HP/LT
2000	356.9	23.9	37.7
2001	366.6	22.7	35.0
	Productivity With Roll Press Operating		
	PRIMARY MILL	PRIMARY MILL	TOTAL
	LTPH	HP/LT	HP/LT
2000	417.3	20.1	32.3
2001	429.3	19.1	30.7
	Primary Mill Improvements		
	PRIMARY MILL	PRIMARY MILL	TOTAL
	LTPH	PR HP/LT	HP/LT
	17.3%	-16.2%	-13.9%

Table 2 shows the effect on primary and total milling that the HPGR product has. Without HPGR product going back to the primary mills, the feed rates are lower and the power consumption rates are higher. The feed rates are a little more than 17% better with the HPGR than without and the specific power consumption rate drops by almost the same amount.

Summary

Empire Mine has successfully integrated high pressure grinding roll technology into an autogenous grinding circuit treating high tonnages of magnetite iron ores. Being the first application of its kind, the cooperation and support of the manufacturer has enabled Empire and KHD to learn more about the capabilities and characteristics of these grinding units.

The Raglan Concentrator – Technology Development in the Arctic

J. Holmes[1], D. Hyma[2], and P. Langlois[3]

Abstract

The pre-operational period of the Raglan mine/mill complex was characterised by two significant exploration and engineering study programs, the first during the mid 1970's and the second in the early 1990's. During both periods, the challenges associated with the remote arctic site led to the requirement for novel construction methods together with the adaptation of technology. This paper focuses on three specific metallurgical studies conducted in 1991-1992, each aimed at defining the most feasible option for the project. The fully autogenous grinding circuit, the concentrate dewatering, storage and transportation system and the tailing dewatering and deposition system are presented in detail. Technology development is then examined by comparing each initial design to the current operation, including a summary of the lessons learned.

PROJECT INTRODUCTION

Falconbridge's Raglan Concentrator, located at the northern limit of Quebec's Nunavik Region, was piloted in 1991. Detailed engineering started in January 1995. The plant was constructed in modules in Quebec City, where virtually all of the process equipment was installed. During the summer of 1997, the modules were shipped via barges to Deception Bay, and transported 96 km by land to the concentrator site in Katinniq. The concentrator modules were assembled and connections completed between August and November 1997, with ore first added in December. Located north of the 62nd parallel, the immediate region is typified by shallow topography, little or no vegetation, and permafrost. All electrical energy is generated on-site with diesel generators. There are no road or rail connections to the south, and all of the workers at the site must be transported by air. Workers from the south are flown from Rouyn-Noranda in a company-owned and operated jet, and workers from northern communities in Nunavik are flown to Raglan by a regional airline. The ore is mined from underground and open pit operations, and contains significant quantities of copper, cobalt, and PGM's. The ore occurs in discrete lenses, varying in size from 250,000 tonnes to 1,000,000 tonnes. Characteristics of the ore can vary greatly. Originally designed to process 800,000 tonnes per year of high-grade nickel ore, plant capacity is currently at 1,000,000 tonnes per year.

---

1 Senior Metallurgical Engineer – Falconbridge Chile S.A. (formerly Raglan Concentrator Superintendent)
2 Engineering Manager – Koniambo Project, Falconbridge (Australia) Pty Ltd (formerly Raglan Project Metallurgist)
3 Strathcona Mill Senior Process Engineer – Falconbridge Sudbury Operations (formerly Raglan Mill Metallurgist)

STUDY #1 – GRINDING CIRCUIT

Flowsheet Development

Early bench scale tests revealed that Raglan ores were very hard, with typical work indices ranging between 18 and 23. At the same time, flotation tests revealed the need to grind to a P_{80} of 65 μm to achieve acceptable metal recovery. Both observations were a direct result of the finely disseminated occurrence of sulphide minerals throughout the hard ultramafic host rock peridotite. During a 1991 pilot plant program, the following objectives were set:

- Design a circuit to produce a particle size of 80% passing 65 μm to ensure sufficient sulphide mineral liberation for flotation
- Design a circuit that provides high certainty of reliability and performance
- Design a circuit with efficient power utilization characteristics since power would be generated on-site
- Design a circuit that minimized operating costs, primarily by reducing the need for grinding steel

A number of design parameters were investigated during the campaign, including semi versus fully autogenous grinding, the impact of pebble crushing, the opening size on the primary mill discharge screen, and pebble versus ball milling for the secondary mill. For the primary mill, the main focus was on using autogenous grinding to eliminate the cost of shipping SAG mill grinding steel to the high arctic. Conceptually, the design would be typical of the circuits used for milling hard Taconite iron ores in the U.S. The main focus for the secondary mill was achieving the targeted product size while also coping with tonnage swings from the primary circuit.

Ore samples for pilot plant grinding testing were taken from one underground zone, which was accessed in the predevelopment period. No single zone could be described as representative of all of the ore at Raglan, however this zone was accessible, and could be described as "average" according to the knowledge of the local mineralogy at that time. Lab testing was used to determine metallurgical differences between ore types.

The main observations from the pilot plant were the following:

- The operating work index of the various ore samples was consistent with previous test data. Values ranged from 19 to 23 with an average value of 20
- The net power consumption from the two stages of milling ranged from 23 to 28 kWh/tonne
- The autogenous circuit was marginally less power efficient than the SAG circuit
- Pebble crushing was shown to be very important with this ore. Crushing pebbles increased throughput by approximately 35% and reduced power consumption by approximately 30%
- SAG milling with a 6% steel charge increased throughput by as much as 50%
- Secondary pebble milling consumed 7 kWh/tonne less power on average at a pebble wear rate of 1.5%. Pebble milling, however, was less tolerant to variations on feed rate.

Based upon the pilot plant test data, an autogenous mill was selected for the primary mill including a mill discharge screen and recycle pebble crusher. For future increases in throughput, the mill would be equipped with a variable speed drive and structurally designed to receive an 8% steel charge. A standard ball mill was selected for the secondary mill. The ball mill was equipped with an oversized 3000 hp motor that was standardised with the primary mill motor.

Selected Flowsheet Description

Ore from open pit and underground mining operations is introduced to the underground ore handling system via underground truck dumps. Two underground bins with a combined capacity of 4000 tonnes provide roughly a day of live capacity for ore storage. Ore from the two bins is conveyed to a 150 tonne surge bin immediately ahead of the grinding circuit in the concentrator.

The Raglan grinding circuit has a conventional ABC configuration (Figure 1) with a design capacity of 100 mtph (800,000 tonnes per year). The primary 24 ft x 8.5 ft (EGL) autogenous mill is powered by a 3000 hp motor with variable speed drive. Primary mill discharge is sized using an 8 ft x 16 ft single deck vibrating screen, with screen oversize re-circulating to a 5½ ft short head cone crusher (400 hp) before returning to the primary mill. Screen undersize is combined with secondary ball mill discharge to feed up to five 15 in. diameter Krebs hydrocyclones. The ball mill is a 14 ft x 21 ft overflow mill driven by a 3000 hp motor at 73% critical speed. Target P_{80} for the grinding circuit product was 65 μm.

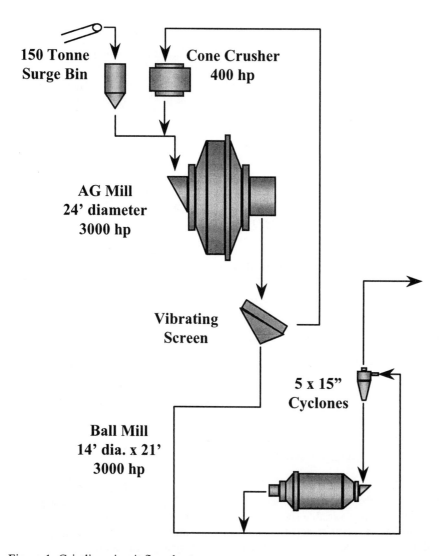

Figure 1: Grinding circuit flowsheet

Current Circuit Operation

The Raglan grinding circuit was commissioned in December 1997. Initial problems with low grinding circuit throughput were overcome by adding many more pebble ports in the AG mill discharge than were originally configured. The target throughput of 100 tph was achieved by the summer of 1998.

Table 1 shows some of the key data used in the design of the circuit compared with the actual plant operating data. It can be seen that overall, the plant performance is quite similar to the design data. With the correct pebble port configuration, the range of plant throughput very closely matches design. And at greater than 100%, the crusher recirculating load is very high for AG/SAG circuits, as provided for in the design. Most importantly, the circuit produces the design product P80 of 65μm, even at throughputs well over 100 tph. And while the AG mill design had provided for the option to convert to SAG milling if necessary, the AG circuit has demonstrated its cost-effectiveness through very low liner wear and reasonable mechanical availability.

While the grinding circuit has met its overall objectives, there are a few notable differences between design and actual plant performance that are worthwhile highlighting. Firstly, the overall specific power consumption is somewhat higher in the plant (31.5 kWh/t vs. 27.2 kWh/t design). This increased energy requirement is the result of inefficient grinding of the coarsest fraction of the circuit feed (+150 mm). Whereas the top-size tested in the pilot plant was 100 mm, the full-scale mill sees feeds with F_{80} of 130 to 180 mm. The +100 mm rocks have a low grinding rate and tend to build-up in the mill, forming a critical size which limits throughput. The disproportionately-high energy requirement for the feed top-size was not evident at the smaller scale of the original pilot plant. The grinding rate of the coarse rocks is discussed further below under Critical Size and Pebble Crushing.

The second important difference between pilot plant and full-scale results is the screen transfer size (T_{80}), which is rather coarser than design. In order to compensate for the slow grinding rate of the coarse particles, it became necessary in the plant to use a larger screen size than originally intended (9 mm vs. 3.4 mm design) to get the fines out of the AG mill. It was also necessary to run at a much lower density in the AG mill (~55-60% vs. 75% design) to keep the mill flushed of fines as much as possible. Both of these operating parameters helped to reduce the mill charge level and increase the energy input to the coarse particles, but resulted in a coarser transfer size to the ball mill.

A final key difference is the ball mill percent recirculating load, which is much lower than design (~200% vs. 300% design). This is interesting because the pilot plant very closely predicted the fines content of the AG mill screen undersize (AG mills typically produce a significant amount of fines). Table 1 shows the percent passing 37 μm in the cyclone O/F, design and plant. With similar fines content in the feed as estimated in the design, the ball mill is clearly producing more fines than predicted in the piloting, resulting in a reduced recirculating load with the given cyclones.

Variation in Ore Hardness

Net energy consumption for the AG mill can range from 17 to 26 kWh/t. This variation in power requirement is indicative of the significant variation in ore hardness seen in the plant, and results in throughputs ranging from 80 to 120 tph. Short-term changes in ore hardness frequently cause rapid changes in AG mill charge level and recirculating load, forcing the operator to make quick tonnage changes to compensate.

Table 1. Design vs. Operating Data

	Circuit Tonnage	New feed F80	Net AG Power	Pebble Crusher % C.L.	Screen U/S T80	Net Ball Mill Power
	(mtph)	(mm)	(kWh/t)	(%)	(mm)	kWh/t (net)
Design	75-125	200	13.5	150	0.95	13.7
Actual	80-120	127-180	17 - 26	75 - 130	1.7 - 2.6	14
	Bond Wi	Ball Mill Operating Work Index	Ball Mill % C.L.	Cyclone O/F P80	Cyclone O/F % -37 um	Total Specific Power Consumption
	(kWh/t)	(kWh/t)	(%)	(mm)	(%)	(kWh/t)
Design	23	15	300	65	47	27.2
Actual	19.4 - 24.2	14.2	175 - 225	49 - 76	60 - 70	31.5

The variation in ore hardness was quantified in 1999 through grinding circuit benchmarking and extensive drill core sampling for MinnovEx SPI testing (SAG Power Index). The MinnovEx SPI test is designed to measure ore hardness and infer power requirement under the prevailing ore breakage mechanisms of SAG milling, and is analogous to the Bond work index for ball milling[2]. Figure 2 shows 190 SPI test results in a plot of cumulative distribution of required energy for AG milling with crusher to produce a transfer size of 1,200 microns (standard correction factors used)[3]. The median energy value is 17.6 kWh/t corresponding to a median SPI time of 321 minutes, indicating that the ore is very hard relative to MinnovEx's database of results from other grinding circuits[2]. Furthermore, 20% of the material has a value greater than 20.8 kWh/t, nearly a 20% increase in energy requirement, whereas the softer 20% of the feed requires less than 14.2 kWh/t. This range of ore hardness reflects the variation in plant throughput over time. It is noteworthy that 5% of the samples showed extremely high energy requirements. These samples correspond to waste fractions. The results point to the strong influence of ore dilution on the energy consumption in the grinding circuit, and are corroborated by circuit behaviour.

Critical Size and Pebble Crushing
As mentioned earlier, the AG mill was observed to fill with large rocks (4 to 8 in.) to the point that throughput was limited. The presence of a very coarse critical size is illustrated using the grinding rate curves derived from JKSimMet modelling. The grinding rate curve represents the breakage rates required for each size fraction to satisfy a steady-state mass balance around the mill (perfect mixing mill model)[2], and is derived through model fitting.

Figure 3 depicts grinding rate (units of hr^{-1}) as a function of particle size on log-log scales. The Raglan AG mill curve is shown in contrast with the default SAG grinding rate curve, which represents the average grinding rate distribution for the database on which the JK SAG model is based. The difference in shape of these two curves is telling. Whereas the default curve shows increasing grinding rates at particle sizes greater than about 25 mm, the Raglan AG mill curve shows decreasing rates above that size, with the lowest grinding rates in the coarsest size fractions. This feature of the Raglan AG mill model points to the coarse particles as representing a critical size for the mill which serves to limit overall circuit throughput.

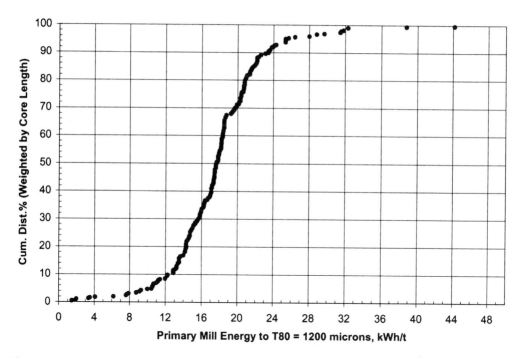

Figure 2: Cumulative AG Mill Energy Distribution for 1,200 μm Transfer Size[3]

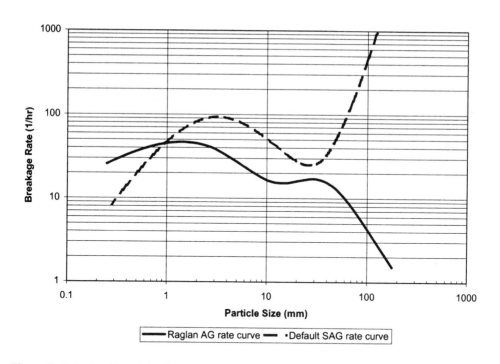

Figure 3: Grinding Rate Distribution Curves

Increasing Grinding Circuit Throughput – 1 MTPY Project
There was, and continues to be, strong economic pressure to push circuit throughput at Raglan, as there is at many milling operations. As the benchmarking and modelling data show, efforts to maximise throughput must focus on the breakage of the coarse particles, and must take into account the variation in ore hardness seen in the process.

In 1999 and 2000, a number of process changes were made in an attempt to increase the rate of breakage of the coarse particles. The intent was to push the AG option to its limits before considering the more expensive conversion to SAG milling. The process changes included:

- Higher mill speed to increase impact frequency in the mill
- Larger 4 in. x 5 in. pebble ports to enable discharge of some of the coarser rocks
- New liner configuration on the pebble crusher to allow tight closed side settings.

These changes had limited impact on circuit throughput. Not much effort has been put into AG mill lifter re-design. This might be an area of focus in the future, given the successes seen elsewhere with SAG operations.

In the spring of 2000, two changes were made which together produced a significant increase in throughput. Pebble port size was increased to 5 in. x 5 in., replacing the 4 in. x 5 in. ports, and this allowed significantly larger rocks out of the mill. This change initially produced poor results, as the crusher was unable to adequately reduce the coarser particles. The effect was to build an excessive circulating load of fines and to deplete the mill of coarse media, thereby shifting the circuit to a much more inefficient mode of operation. A grizzly with 4 in. bar-spacing was then added in front of the pebble crusher to bypass some of the largest rocks directly back to the AG mill. The impact of this second change was dramatic, with crusher performance returning to normal and average plant throughput increasing from 115 tph to 130 tph. Effectively, the combination of very large pebble ports coupled with the recirculation of +4 in. material allowed the plant to achieve the 1 MTPY target.

While the critical importance of effective pebble crushing in the AG circuit has been demonstrated, the reason for the significantly improved throughput with the large ports and grizzly combination is not obvious. The negative impact of recirculating uncrushed coarse material to the AG feed would seem to be outweighed by the higher rate of production of fines due to improved crusher operation. One possible explanation for this favorable trade-off is that with the bypass of large pebbles back to the mill, the charge is kept much coarser (i.e. medium-sized particles are removed and crushed while the big rocks go back). The presence of more grinding media may result in more efficient grinding of the fine fractions (-25 mm), thereby increasing the rate of production of minus screen size material. Another possibility is that mass transport through the mill has been affected. It has always been observed during inspection of the charge that coarse particles (plus pebble port size) build up in the mill charge next to the discharge grates. There is a very clear size gradient along the axis of the mill, with fines toward the feed end and coarse particles at the discharge. The presence of coarse material at the discharge may present an added restriction to the flow of medium-sized particles through to the grates and out of the mill. By removing more of the large rocks and sending them back to the feed end, this restriction may be diminished, allowing slightly faster discharge to the crusher and lowering the charge level. Though the specific cause of improved performance is unclear, it is evident that a more optimum balance between crushing and grinding has been struck with the addition of large ports and the bypass grizzly.

STUDY #2 - CONCENTRATE DEWATERING, STORAGE AND TRANSPORTATION

Flowsheet Development
Under specific conditions, some sulphide concentrates that contain pyrrhotite are known to exhibit self-heating properties that can lead to a significant hazard for both production personnel and property. Not surprisingly, the majority of pyrrhotite-containing concentrate is therefore handled in slurry form to eliminate this risk. The high arctic location of Raglan eliminated slurry transport as a feasible option due to the prohibitive cost of shipping the water contained in concentrate, and

also due to the problems associated with freezing. All transport options, therefore, needed to minimize the amount water contained in concentrate.

Standard testing methods applied to Raglan concentrate samples obtained from the pilot plant program confirmed this self-heating behavior and helped to quantify its severity. The phenomenon of self-heating is discussed in detail in the literature, most recently by Rosenblum et al[5]. The result of self-heating is the release of heat, the formation of sulphates that bind the concentrate particles together to form solid agglomerates, the smell of sulphur dioxide fumes and the eventual physical transformation of the concentrate filter cake into a sintered mass. Extensive testing at Lakefield Research and the Noranda Technology Centre concluded that to avoid spontaneous heating, Raglan concentrate must be dried to a moisture level less than 0.3% and also cooled to a temperature less than 30°C following drying. The cooling requirement was discovered as a secondary conclusion of the work at Noranda Technology Centre. Using the information described above, several drying and material handling unit processes were examined.

Four types of drying unit operations were investigated, most of which required some form of upstream filtration:

- Fluid bed dryer
- Flash dryer
- Steam coil (Myren) dryer
- Spray dryer.

The fluid bed dryer was selected on the basis of proven operating history on sulphide concentrates, a tolerance to variations in feed moisture content in the event of upstream pressure filtration problems, and the ability to easily utilize the exhaust gas from the diesel generators located in the power station. The major disadvantage with this unit was the requirement for a 700 HP blower to boost the pressure of the diesel generator off-gas.

The other dryer options were rejected for a variety of technical reasons. The flash dryer was of interest since the diesel generator off-gas could be used directly for drying without a boost in pressure. However, the unit lacked proven operating history and was known to be intolerant to variations in feed moisture content. The steam coil dryer also lacked proven operating history and was not suited to using the diesel generator exhaust in place of steam. The spray dryer was of interest since it combined filtration and drying into one operation. However, it required more heat for drying than was available from the diesel exhaust flow. The cost of installing a separate combustion chamber and burning additional arctic diesel fuel to generate dryer hot-gas more than offset the capital and operating cost of a filtration operation.

Following drying, three dry concentrate handling alternatives were considered, as follows:

- Bulk handling using conventional conveying and loading equipment, similar to many base metal operations
- Flexible bulk bags of 1 to 2 tonne capacity, similar to those used for matte and other high value products
- Pneumatic handling using equipment typical to what most producers of bulk cement use.

A simple qualitative comparison was made between the three alternatives. While the bulk handling option presented very low technical risk, the likelihood of exposure to rain and/or snow would surely trigger spontaneous heating. (i.e. high technical risk) In addition, high product loss and environmental contamination through the system would be expected due to the fine particle size of the bone-dry concentrate. (P_{80} = 20 μm). The use of flexible bags may have addressed the above-mentioned problems, however operating costs were expected to be high as a result of transferring

13,000 bags to the cargo ship, five to six times during a nine month shipping season. The final option, pneumatic handling, was selected for Raglan based on cement industry experience with clean storage and high rate loading of dry, fine bulk cement products, often very near highly populated areas.

Selected Flowsheet Description
Concentrate is thickened in a 13-m diameter Supaflo high-rate thickener, and pumped via peristaltic pumps to a filter feed tank. Two Svedala VPA 1530 filters treat the slurry, from which the filter cake is conveyed at 10 to 12% moisture to a dryer feed bin using twin screw-type feeders and a rotary air lock. The 2.4 m diameter by 7 m high Fuller fluid bed dryer is supplied with hot exhaust gas from the power plant generators via a 700 HP blower. The fluid bed media is currently generated from the mill grinding circuit. The inlet gas, tempered by ambient outside air, varies according to the feed rate to a maximum of 330 °C. The outlet temperature is maintained between 100 to 110 °C. Outlet gas and dry concentrate pass through a vertical duct to two parallel pulse jet dust collectors. Clean gas is expelled to atmosphere via a fan and stack, while hot concentrate passes along air slides to a cooler. The cooler consists of a 1.8-m by 1.6-m column, 10-m high, with an extensive network of cooling pipes through which cold water passes. Cool concentrate is discharged into an airlift, which transports the material to a distribution air slide feeding any of three storage silos, with a combined capacity of 4,000 tonnes. The silos discharge into air slides which feed 50-tonne tanker trucks. Trucks travel 96 km to Deception Bay, and are pneumatically discharged into a 50-m diameter storage dome, with 50,000-tonne capacity. A mechanical arm in the dome reclaims concentrate to a central bottom discharge point followed by conveying to pneumatic pods. The pods transfer the concentrate into the MV Arctic for ocean voyage to Quebec City. The ship transports 27,000 tonnes per voyage, five or six times per year, with a cessation of shipping between March 1 and June 1 due to local agreements.

Current Circuit Operation
The concentrate thickener is operating without any significant problems. There is a persistent froth accumulation on the surface, however a downstream clarifier on the overflow stream has made this problem less important.

The concentrate pressure filters are operating at satisfactory levels of availability, throughput and product quality. A number of minor mechanical and programming modifications have been implemented in order to facilitate operation and maintenance. As with any filter system, proper maintenance is critical to preserve performance of the filters.

In order to increase plant throughput and peak filter capacity, the concentrate filters were increased from 8 to 12 chambers, with the relatively simple addition of more plates. It is highly recommended that this flexibility be preserved when sizing filters, especially for new plants. There are several reasons for this:

- The plant throughput will inevitably increase within a few years of start-up. Potential for added chambers means potential for "instant" additional capacity.
- Flowsheet modifications such as additional regrinding may change the size distribution of the concentrate, increasing the filtration requirements.
- The ore used in pilot plants and/or filter selection testing is often not completely representative. An expandable filter is prudent.

The provision to add more plates means having a slightly longer frame, and a few other provisions. The additional cost should not be great, provided that the filter length is reasonable.

Filter cloth selection can play a significant role in the filter performance, as well as cost. At Raglan, the use of a monofilament fabric has been very successful in the concentrate filtering application. However, the same fabric has not yet proven completely successful in the tailings application. A conservative well-proven filter cloth fabric should be used when sizing filters and selecting initial cloths. The client should insist that the filter supplier not use "high-tech" cloths in order to obtain optimistic filtering rates and hence smaller filter areas. Expanding an existing plant with proven experience with such cloths might be an exception to this recommendation. A solid working relationship with a cloth supplier was very important during the first few years of operation at Raglan.

A fluid-bed dryer is not a piece of equipment often installed in concentrators, and therefore there was a fairly steep learning curve. Early in the operation, there was also a catastrophic failure of the fluidizing fan, leading to massive equipment damage and some production losses. This event reinforces the need to conduct independent design and fabrication audits on equipment which are of unique design or application, and where the results of failure can be severe.

The material handling equipment feeding the filter cake had some difficulties with abrasion and concentrate build-up, despite careful design. A number of modifications were required in this area in order for the dryer sector to operate at full capacity. The dryer itself has undergone only one minor modification. Being shorter than ideal due to building constraints, there is an increased probability of media being picked up by high velocity gas localised at the dryer exit. Therefore, a deflector was installed inside the dryer to reduce the carry-over of bed media into the exhaust duct. The media size was also increased slightly to compensate for this problem. The dryer capacity meets current needs, however, there are opportunities for further increases, if required.

One problem that persisted over the first year and a half of operations was the occurrence of concentrate fires. Principally occurring in the dust collectors, the fires were a result of exposing concentrate to the conditions determined in laboratory self-heating test work to be dangerous. The conditions include warm temperatures, some small amount of moisture, and a supply of oxygen. When the dust collector exit became blocked, concentrate would accumulate, containing a small amount of moisture, and the exhaust gas/fresh air mixture supplied oxygen and heat. In a surprisingly short period of time, the concentrate would start to oxidize. In order to eliminate the problem, the discharge air slides were modified to have a steeper slope and larger exit, allowing for more efficient transport of concentrate from the dust collectors. This has almost eliminated the potential for build-up of concentrate and greatly reduced the risk of fire. More sensitive level alarms were installed which immediately alert the operator of any build-up.

A number of other upgrades have been incorporated in the two main dryer dust collectors. The original inlet design created excessive turbulence, leading to excess dust loading in the air surrounding the bags. This led to pulses in the dust collectors being too frequent, causing excessive bag replacement and limiting the dust collector capacity. A new inlet design was modelled to determine velocity vectors and was able to greatly reduce inlet velocity and turbulence, hence increasing bag life and capacity. The tube sheets in the dust collectors were also changed, partly due to fire damage, and partly to provide a different style of bag that was easier to install and remained in-place more securely. All of the air slides transporting concentrate were also replaced with high-top units, allowing less dust capture, and reducing subsequent plugging of the air slide dust collection system. Plugging of this dust collection system has also led to poor flow conditions in the air slides. Some additional modifications are necessary to reduce dusting while trucks are being loaded.

STUDY #3 - TAILING DEWATERING AND DEPOSITION

Flowsheet Development

Site conditions under which Raglan tailing are discharged generally are as follows:

- Below freezing temperatures for roughly nine months
- Permanently frozen ground (permafrost) with a temperature of $-7\ °C$
- An active surface layer of broken rock approximately 1 to 2 metres thick
- Average wind speeds of 30 km/h from the northwest
- Virtually no vegetation, and little topographical relief
- Limited aggregate supply for dam construction
- Annual precipitation of 650 mm (net) and no nearby lakes.

The challenge was to design an environmentally safe and cost effective tailing system taking into account a number of regulatory trends, such as:

- Zero effluent discharge
- Perpetual monitoring
- Submission of closure plan and posting of insurance bond prior to start-up
- Requirements for progressive restoration during the operating life of the mine.

The above conditions and challenges resulted in a decision to evaluate tailing discharge design "first principles". After considering many options, three were selected for more comprehensive study, each being characterized by the pulp density of the water/solid mixture. Each option was required to safely contain 680,000 tonnes of solids annually for a period of 25 years. The end result of the study was the selection, design and commissioning of a tailing system that worked "hand-in-hand" with the natural permafrost ground conditions of the Arctic.

Conventional Option: 55% Solids

At 55% solids, the Raglan tailing is typical of most slurry streams from a mineral processing plant. Well proven, simple technology could be used to thicken, pump and contain the slurry. Operating costs could easily be benchmarked and were relatively low. However, due to the characteristic flat terrain and limited supply of construction aggregates at this site, the capital cost to construct tailing dam walls was estimated to be very high. This option would also result in a large tailing site "footprint" and a significant recycle water management system. Advantages and disadvantages of this option are summarised in Table 2.

Table 2- Conventional Tailings Option

Advantage	Disadvantage
Proven, simple technology	Recycle water management in extreme arctic conditions
Known operation/maintenance	Large surface area to manage environmentally
Low operating cost - $1.34/tonne ore	No potential for continuous reclamation
	Negative visual impact on landscape
	Only suitable site > 5 km from the concentrator
	High capital cost - $45/annual tonne installed capacity
	High closure cost - $12/annual tonne installed capacity

A net present value cost estimate was prepared, taking into account the initial capital outlay, annual operating costs and post-production phase closure costs. A value of Cdn$53 million (1992 dollars) was estimated for this base option. It was concluded that the technically riskier higher density options should be thoroughly researched in an attempt to mitigate the problems of water management, large open surface area, and high capital plus closure cost.

Paste Option: 70% Solids
When the pulp density of the Raglan tailings reaches 70% solids, a typical thixotropic paste is formed. Conceptually, the idea involved utilizing equipment and operating techniques employed by aluminum producers that dewater and contain "red mud" as their tailing. High density deep bed thickeners followed by positive displacement pumps would be used to thicken and transport the tailing paste to a disposal area. At the disposal area, the tailing paste would be placed in layers that gradually would freeze into permafrost. The greatest perceived advantage of this approach was the positive impact on the size of the tailing containment facility. With only 70% of the original slurry volume in the form of a paste, the size and cost of a containment facility would be significantly reduced vs. a conventional design. Also, the high potential for the majority of the tailing water to freeze in-situ would result in minimal recycle and a much simpler, smaller, lower cost water management system. Against these advantages was the overwhelming concern of using thickening technology that was unproven for a sulphide tailing together with difficult placement logistics in an arctic situation where flexibility during upsets is vital to achieving target production rates. Added to this was the unknown long-term behaviour of residual water and percentage of "ice lensing" within the tailings. Laboratory testwork predicted a stable material. However, extrapolation of results to "real world" was considered risky in this situation. Advantages and disadvantages for this option are summarised in Table 3.

Table 3- High Density Option

Advantage	Disadvantage
Simplified, lower cost recycle water management system	Less proven, more complex technology
Potential for continuous reclamation	Less flexibility in upset conditions
Smaller surface area to manage environmentally	Less known operation/maintenance
Suitable site < 1 km from the concentrator	Higher operating cost - $1.75/tonne ore
Lower capital cost - $21/annual tonne installed capacity	
Lower closure cost - $8/annual tonne installed capacity	

The net present value cost calculation for this option was Cdn$31 million (1992 dollars). Significantly lower estimated capital and closure costs more than offset the increase in estimated annual operating cost. However, due a much higher perceived technology risk, a further option was considered that would hopefully capture the advantages of these first two options.

Filtering Option: 85% Solids
At 85% solids, the Raglan tailings moves from an unstable paste state to a stable solid state in the form of a filter cake. The concept involved the use of pressure filters to produce a cake that could be easily conveyed, stored in silos, transferred to trucks and subsequently moved to a containment area for stockpiling using standard mobile equipment. Many of the previously stated advantages would be achieved but at the expense of higher long term operating costs to maintain pressure filters and to handle tailing filter cake with mobile equipment. Project management decided to accept this higher cost in an effort to minimize technology risk, maximize operational flexibility at a harsh arctic site, maximize environmental protection by utilizing continuous reclamation

practices, minimize water recycle and snowmelt run-off problems and minimize both the initial capital and final closure cost. Advantages and disadvantages for this option are summarised in Table 4.

Table 4- Filtered Option

Advantage	Disadvantage
Proven technology	Highest operating cost - $3.15/tonne ore
Flexibility in upset conditions	Potential dusting of active surfaces
Known operation/maintenance	
Simplified, lower cost recycle water management system	
Potential for continuous reclamation	
Smaller surface area to manage environmentally	
Suitable site < 3 km from the concentrator	
Second lowest capital cost - $24/annual tonne installed capacity	
Low closure cost - $8/annual tonne installed capacity	

The net present value cost estimate for this option was Cdn$40 million (1992 dollars), representing a value midway between the previous two options.

Selected Flowsheet Description
Tailings are thickened in a 30.5 m diameter Supaflo High Rate thickener, and pumped via SRL pumps to a filter feed tank. Three Svedala VPA 2040 filters treat the slurry, with completely independent operation. Three Elliot 1000 HP centrifugal compressors are installed to provide air to the filters and other plant requirements. Filter cake is discharged via individual belt feeders onto a single common conveyor discharging into a 150 tonne silo. The silo is equipped with an air-actuated clam gate, operated to fill two Volvo 35 ton articulated trucks. Tailings filter cake is hauled 3 km one-way to the disposal site, where it is dumped, spread, and compacted. Haulage is based on a 24-hour operation, and spreading and compacting is done on dayshift only. Run-off water from the tailings pad is collected in an excavated pond and pumped into the mill recycle water system.

Filter operation
Operation of the tailings filters has been successful overall, producing good quality filter cake. As with any filter, a rigorous maintenance program is essential in order to maintain availability. Some modifications to controls and mechanical elements have been implemented. The availability of membrane air at the proper pressure and filter cloth condition are both very important factors in the filter performance. The filters have been expanded from the original design of 42 chambers to 46 in order to accommodate increased throughput. In retrospect, it would have been desirable to include the provision for a few additional chambers beyond 46. The tailings filters are occasionally subject to significant swings in performance with different ore types. Some of the more highly altered ore zones have very poor filtering characteristics, sometimes to the point of slowing down plant throughput. These types of problems are difficult to foresee in pilot the pre-production testing and design stages. Constant efforts are made to reduce cycle times for the filters, in order to increase the peak capacity, and lower overall filter utilization. The centrifugal compressors have performed satisfactorily, and their operation has been optimized significantly in order to realize substantial power savings.

Tailing Placement Operation

The most important issue related to the tailings operation is that of freezing. So far the tailings have frozen very successfully, and seasonal thawing is limited to minor depths of uncovered tailings. This is despite an accumulation of salts in the process water due to the use of brine for drilling underground.

During the first several years' tailings deposition, there were a number of operational issues to be resolved. Rainfall patterns were not entirely as expected, with occasional heavy rain for extended periods, particularly in late summer. During these periods, it was almost impossible to drive on the tailings to dump loads. As long as bare terrain within the tailings impoundment area was available, dumping could be done. This option was not possible as the occupied area expanded, and a system of ramps constructed from waste rock had to be prepared in advance of the rainy season. The tailings stack is constantly growing higher, therefore the ramps need to be replaced regularly.

Due to freezing conditions during most of the year, large quantities of tailings filter cake built up in the truck boxes, affecting net payloads. Heating the boxes with exhaust did not prevent the problem. Eventually, bolted box liners of PTFE were found to be the best solution. The cold temperatures also mean that tailings freeze fairly quickly after dumping, and therefore the piles must be bulldozed and compacted before this happens. Accumulated snow must be cleared in before applying tailings to an area, adding to the work during the nine or ten months of winter.

Another issue relates to dusting of the tails. Dust from the pad is a winter issue, due to sublimation of frozen water in the tailings, essentially desiccating the tailings. Extremely low temperature and low air humidity are experienced during winter at Raglan. If the tailings are compacted with a roller compactor before the water freezes or is sublimated, the dusting is reduced greatly, but not eliminated. Several methods of dust control have been attempted, with varying degrees of success. Water spray is not effective due to rapid sublimation of the ice cover. Trials with a chemical binder have not been effective either. Snow is an effective dust suppressant, however maintaining a snow cover on the pad has proven to be difficult due to persistent and sometimes extreme wind conditions on the exposed top surface. Snow fences were tested, but could not withstand the wind.

Trucked and compacted snow remains in place much longer. The most effective solution at this time is a thin cover of gravel, which is about to be applied in "dormant" areas, minimising exposed surface areas while leaving the "active" working surface for tailings deposition. Use of a roller compactor in active working areas is the best defence against dust.

Costs of the tailings disposal method are higher than expected. A summary of costs (from 2000) is presented in Table 5, based on one million tonnes per year operation. Even with some improvements in productivity, vehicle selection, and specific methodology at the pad, the costs are likely to remain high. The conventional disposal of tailings in an abandoned asbestos open pit mine was examined briefly in 2000/2001, however high capital costs and higher than expected operating costs were determined in a preliminary study, and the idea was abandoned.

Table 5- Filtered Tailings Costs

Operating/Maintenance Costs	$/yr	$/tonne(ore)
Filtering (includes feed system, filters, compressor allowance, and delivery to bin)	1,800,000	1.80
Trucking/spreading/compacting (includes costs at load out)	1,700,000	1.70
Capital Costs	**$/yr**	**$/tonne(ore)**
Ditching and cover cost	2,000,000	2.00
Mobile fleet replacement	600,000	0.60
Total		6.10

In the summer of 1999, after 18 months of operation in a demonstration mode, a project was initiated to complete the final detailed design for the water containment ditches and cover, which is based on the "walk-away" approach. AGRA (Amec) was selected from several consultants to perform the engineering. Several other firms are involved for their particular expertise in geochemistry, thermal modelling, and civil engineering. Starting with the original preliminary designs, and with the practical operating experience gained, the designs are now completed and the first phases of construction have begun. Currently a progressive reclaim approach is proposed, and therefore the construction will be completed in phases. This will allow for evaluation of technical and economic factors in the design at each stage of the project. Now that the detailed design of the cover is complete, it is clear that the cover costs are higher than originally thought. It is not the intention of this paper to discuss in detail the design and construction, however a publication focussing on the tailings may be made in the future.

OVERALL CONCLUSIONS AND LESSONS LEARNED

The grinding, concentrate dewatering, and tailings disposal sectors of the Raglan concentrator have all performed at or above the original design throughput. The plant is currently operating at 1,000,000 tonnes per year, compared with an original target of 800,000 tonnes per year. The flotation circuit, while not discussed in this paper, is also producing recoveries better than in the plant design. Therefore, while the Raglan concentrator must be judged a success overall, there are some lessons to be learned.

The lessons learned from the grinding circuit design are not dramatic, but they are important. While the circuit has met its objectives, the extreme variability of the ore hardness and its impact on throughput was perhaps not fully appreciated. Better tools are widely accepted today that perhaps allow less reliance on pilot plant data as the key design factor. These tools include the SPI approach developed by Starkey and MinnovEx, and simulators such as JKSimMet. While the pilot plant is certainly a key part of any greenfield project, one must always ask whether or not the feed stock for the pilot plant is representative enough to yield a sufficiently robust grinding circuit design. The answer in most cases will probably be no, and therefore the best available tools must be employed to look at the variations, and account for them in the design.

The concentrate dewatering circuit design has demonstrated that the concerns and precautions about concentrate self-heating were indeed well founded. Concentrate oxidation does take place under conditions more or less predicted with the lab testing. Careful design of the process can avoid those conditions and prevent fires on a consistent basis. While a number of the details were not exactly right initially, a good understanding of the situation helped to remedy the problem by identifying the required design modifications.

With respect to tailings disposal, the main lessons learned are essentially about cost. The day-to-day operating cost of the disposal system is somewhat higher than predicted, and the cover costs are very much higher. With respect to the cover cost, both the quantities and unit construction rates were underestimated. Quantities would have been better predicted by completing a more detailed design of the cover much earlier. These kinds of details are most often determined later in the operating life of a mine. At Raglan, the progressive reclamation aspect of the tailings deposition plan meant that a detailed cover design was required very early in the mine life. If the design had been detailed in parallel with the plant design, then this cost could have been more accurately forecast and used in the decision making process. Having said this, it is quite likely that the choice of design would have remained the same, due to the other factors involved in the decision.

There are a couple of other aspects peculiar to Raglan, which are not discussed in the previous sections, but are worth mentioning briefly. The first of these is layout. There probably does not exist an operator who thinks his plant is big enough. In any plant design, there is a struggle between efficiency (i.e. keep it smaller and cheaper) and spaciousness. In the case of a modular plant, the drive for efficient use of space is very strong. After a few years of operation at Raglan, it is probably fair to say that the layout is perhaps too efficient, to the point where maintenance of equipment is more time-consuming and costly. In a few areas, such as the dryer module, building restrictions were such that equipment design was compromised, which has led to costly modifications. There is no easy solution to this eternal struggle, except for a constant quest for balance between the two needs.

The second aspect was the heat recovery system associated with the power plant. The diesel generators at Raglan were designed to be extremely energy efficient. Waste heat is captured from the engine cooling systems and hot exhaust gas, and is then transferred to a glycol system, which provides most of the heat for the buildings, and also provides all the hot water. In addition, the exhaust gas is used to dry the concentrate, saving enormous amounts of energy. All of these systems have been very successful, due to a very careful design.

ACKNOWLEDGEMENTS

The authors would like to acknowledge all of the people who have contributed to the development and operation of the Raglan concentrator over the past 10 years. Their tremendous efforts and enthusiasm has resulted in a very successful operation in a very inhospitable part of the world. Finally, Falconbridge Limited and SMRQ are thanked for supporting this work and allowing it to be published.

REFERENCES

1. Hyma, D. and Williams, S., "Falconbridge's Raglan Project: A development Update and Description of the Concentrator Circuit Design", Proceedings of the 1993 Canadian Mineral Processors Conference, Ottawa, Ontario, Canada.

2. Starkey, J. and Dobby, G.S., Application of the MinnovEx SAG Power Index at Five Canadian SAG Plants; International Autogenous and Semiautogenous Grinding Technology, Volume 1 of 3, pp.345-360, 1996.

3. MinnovEx Technologies Inc., "SPI Testing and Grinding Circuit Expansion Study: Preliminary Report"; Report to SMRQ, 1999.

4. Napier-Munn, T.J., Morrell, S., Morrison, R.D., and Kojovic, T., Mineral Comminution Circuits, Their Operation and Optimisation, Indooroopilly:JKMRC, 1999.

5. Rosenblum, F., Nesset, J., and Spira, P., "Evaluation and Control of Self-Heating in Sulphide Concentrates", Proceedings of the 2001 Canadian Mineral Processors Conference, Ottawa, Ontario, Canada

Author Index

A
Abulnaga, B., 1403
Alderman, J.K., 978
Alexander, D., 63
Aliasso Jr., R.E., 1962
Altman, K.A., 1631, 1652
Anderson, C.G., 1709, 1760, 1778
Arvidson, B., 1033
Ashley, K.J., 25

B
Baczek, F., 1295
Banovich, P.J., 2345
Barfoot, G., 1446
Barratt, D., 99, 539, 755
Bascur, O.A., 507
Bedell, D., 2164
Beerkircher, G., 621
Bennett, D., 1446
Bigg, T., 123
Blois, M., 1358
Bootle, M.J., 1373
Bothwell, M.A., 894
Botz, M.M., 1866
Bourgeois, F., 479
Boyd, K., 606, 669
Braden, T.F., 2020
Brierley, C.L., 1540
Briggs, A.P., 1540
Brochot, S., 479
Brodie, M.N., 1973
Brown, B.S., 1809
Brown, S.R., 2290
Brunk, K.A., 2299
Bunk, S., 1493
Burchardt, E., 636
Burt, R.O., 947
Buter, L.J., 2299

C
Callow, M.I., 698, 801
Carson, J., 1478
Coats, G., 1911
Col, M., 1446
Cole, A., 1493, 1530
Cole, J., 2196
Coleman, R., 2367
Cook, R.E., 2077
Cox, C., 1342
Crandall, T.E., 1962
Craven, J.W., 2114

D
Danecki, C.D., 819
Davey, G., 783
Davies, M.P., 1828
de Mull, T., 2211
Dobby, G., 1239
Dowling, E.C., 2380
Durance, M.V., 479

E
Easton, J.H., 1255
Edwards, R., 2130
Erickson, M., 1358

F
Filas, B.A., 1902
Fleming, C.A., 1644
Flintoff, B., 383, 2051

G
Gale, C.O., 1747
Gale-Lee, M., 1847
Gard, P.J., 2245
Ginsberg, D.W., 528
Giralico, M.A., 1709
Graber, G., 1463
Grandy, G.A., 819
Gu, Y., 270
Guillaneau, J.C., 479
Gulyas, J., 1510

H
Halbe, D., 326
Hales, L., 2164
Halupka, R., 371
Hampton, A.P., 1663
Hanks, J., 99
Harris, M.C., 461
Hearn, S., 929
Hedvall, P., 421
Herbst, J.A., 383, 495
Hickson, R.J., 2250
Holmes, J., 2394
Holmes, T., 1478
Hosford, P., 1694
Hunker, C., 2223
Hyma, D., 2394

J
Johnson, N.W., 1097
Johnston, B., 281

K
Kaja, D.M., 404
Kappes, D.W., 1606
Karesvuori, J., 2174
Kennedy, J.P., 507
Kerr, A., 1142
Klymowsky, R., 636
Knecht, J., 636
Kongas, M., 2020
Korpi, P.A., 2380
Kram, T., 207
Kumar, P., 2238

L
Lamb, K., 1510
Langlois, P., 2394
Laplante, A.R., 160, 995
Laros, T., 1295, 1331
Laurila, H., 2174
Lelinski, D., 1179
Lethlean, W.R., 2345
Leung, W., 1262
Levier, K.M., 2299
Lichter, J.K.H., 783
Lighthall, P.C., 1828
Lim, K., 621, 628
Luttrell, G.H., 2152

M
Major, K., 566, 1403
Malbon, S., 1011
Maldonado, S., 2223
Malone, D.M., 1962
Mankosa, M.J., 176, 1069, 2152
Martin, T.E., 1828
McClelland, G.E., 251
McCloskey, J., 1847
McCuaig, G.W., 1953
McIvor, R.E., 2380
McMullen, J., 211
McNulty, T.P., 119
McPartland, J.S., 251
McTavish, S., 1631, 1652
Meadows, D.G., 801
Medower, R.A., 2077
Merks, J.W., 37
Miller, J., 1159
Moon, A.G., 698
Morrison, R.D., 442, 461
Mosher, J., 63, 123
Mudder, T.I., 1866

Mular, A.L., 310, 383, 894, 1049
Munro, P.D., 1097

N
Neale, A., 2367
Nelson, M.G., 1179
Nendick, R.M., 2285
Newman, L.C., 1760
Nordell, L.K., 495
Nordin, M., 421
Norrgran, D.A., 176, 1069

O
O'Bryan, K., 621, 628
Olson, T.J., 880
Ounpuu, M.O., 145

P
Parker, S.J., 2114
Parnow, C.C., 1902
Parsons, S.D., 2114
Patzelt, N., 636
Perry, R., 2130
Pitard, F.F., 77
Pocock, B.K., 201
Post, T.A., 1709
Prokesch, M.E., 1463
Pruett, R., 1159

R
Rajamani, R.K., 383
Reed, W.M., 867
Reeves, R.A., 962
Reisinger, R.W., 1902
Rice, S., 1828
Richardson, J.M., 442
Ricks, B.L., 1422
Robinson, T.G., 1709

Rose, D.J., 2380
Roset, G.K., 1760
Rowland Jr., C.A., 710
Runge, K.C., 461
Rust, J., 2196

S
Saich, S., 2211
Saloheimo, K., 2020
Sarbutt, K.W., 145
Schack, L.A., 2337
Schaeffer, L.E., 1932
Schaffner, M., 1631
Scheffel, R.E., 1571
Schmiermund, R.L., 1886
Schnarr, P., 1932
Schoenbrunn, F., 1331, 2164
Scott, J.W., 3, 281
Semple, P.G., 1680
Sherman, M., 539, 755
Sienkiewicz, J.R., 2104
Silverblatt, C.E., 1255
Sloan, R.P., 2114
Slottee, S., 1295
Smith, C.B., 201, 1313
Smith, L.D., 346
Smolik, T.J., 326
Sobel, K., 2211
Spiller, D.E., 160
Staples, P., 2367
Stewart, D.W., 2230
Storms, R.J., 1962
Stuffco, T., 2066
Sunna, K., 2066
Sutherland, D., 270
Svalbonas, V., 840
Symonds, D.F., 1011

T
Thomas, K.G., 211, 1530
Thomas, P.F., 819, 1983
Thompson, D., 264
Thompson, P., 136
Tiili, O., 2174
Tinkler, O.S., 1709
Tippin, B., 1159
Townsend, I.G., 1313
Traczyk, F.P., 1179, 1342
Turner, P.A., 880
Twidwell, L., 1847

U
Utley, R.W., 584, 606

V
Valine, S.B., 917
Vien, A., 2130, 2328
Villeneuve, J., 479
von Beckmann, J., 1680
Vujic, P., 2315

W
Warnica, D., 1493
Watson, T., 2277
Weinand, H.J., 1932
Welch, G.D., 201, 1289
Weldon, T.A., 1747
Wells, J., 1694
Wells, P.F., 2011
Wells, P.J., 1403
Wennen, J.E., 917
Whiten, W.J., 461
Williams, R.A., 1530
Williams, S.R., 145
Winckers, A., 1124
Wood, K.R., 1204

Subject Index

A

Acid rock drainage   1886–1894, 1899–1901
   prediction techniques   1894–1899
Acid wash circuits   1680–1683
Activated carbon adsorption in gold recovery
   264–266
AG mills
   bench-scale and pilot plant testing for circuit
      design   123–134
   motor and drive system selection   1983–2010
   selection and sizing   755–782
Agitated tank leaching
   compared with heap leaching   1634–1635
   and counter current decantation (CCD)
      thickeners   1631
   history   1632–1633
   and Merrill-Crowe zinc precipitation recovery
      process   1631
   process selection and design   1631–1643
Air jigs   990–991
Air tables   989–990
Aluminum   1924–1925
AMMTEC   2211–2212
Anglo American Research (AARL) elution
   process   1694–1697
Arctic grinding and dewatering studies
   (Falconbridge Raglan Concentrator)
   2394–2410
Arsenic removal from wastewater   1847–1853,
   1857–1865
Assay
   fire assaying   1778–1805
   laboratories   2011–2019
Australian Minerals Industry Research
   Association   462
Autoclaves
   Barrick Goldstrike acidic autoclave
      1530–1534, 1538–1539
   pressure oxidation and pressure leaching
      autoclaves   1510–1529
Autogenous grinding mills. *See* AG mills
Automated mineralogical analysis   270–277

B

BacTech Environment Corporation   1547,
   1548, 1550
Ball mills
   characteristics   698–709
   fine and ultrafine grinding   783–800
   following SAG mills   801–818
   motor and drive system selection   1983–2010
   selection factors   721–731, 745–748
Barrick Goldstrike Mines   1530
   acidic autoclave   1530–1534, 1538–1539
   oxygenated roaster   1535–1539
   two-stage roaster   1493–1509
Batu Hijau case study   2211–2222
Belt conveyor systems
   calculations   1449–1453
   components   1447–1449
   types and selection criteria   1446–1447
Bench-scale and pilot plant testing
   automated mineralogical analysis   270–277
   in cyanide leach circuit design   251–263
   in comminution circuit design   123–134
   in filtration circuit design   207–210
   in flotation circuit design   145–159
   in gold- and copper-recovery circuit design
      264–269
   in gravity concentration circuit design
      160–175
   locked cycle testing   150–153
   in magnetic concentration circuit design
      176–200
   overview   119–122
   in selection of flotation reagents   136–144
   in selection of pre-oxidation process to
      enhance gold recovery   211–250
   in thickening and clarification circuit design
      201–206
BHPBilliton   2315
Bidding   2223–2229
Bin selection   1478–1483
Biooxidation   1540–1541, 1565–1568
   aerated, stirred-tank   1547–1561
   in heap leaching of sulfidic-refractory gold ore
      1562–1564
   principles   1541–1547
BIOX   1547, 1548, 1550
Bond Work Index   544
BRGM   1547, 1548, 1550
BRUNO Crushing Plant Simulator   404–420
Bubble-surface-area flux   1187–1188
Bullion
   determination by spectroscopy and other
      instrumental methods   1797–1802
   fire assaying   1778–1805
   platinum group metals   1760–1777
   production and refining   1747–1759

C

Cadmium removal in bullion production   1755
Carbon reactivation   1680–1693
Carbon-in-columns processing. *See* CIC, CIL,
   and CIP processing
Carbon-in-leach processing. *See* CIC, CIL, and
   CIP processing

Carbon-in-pulp processing. *See* CIC, CIL, and CIP processing
Case studies
    Batu Hijau 2211–2222
    coal preparation plant control 2158–2160
    Empire Mine high pressure grinding roll 2380–2393
    Falconbridge Raglan Concentrator arctic grinding and dewatering studies 2394–2410
    PT Freeport Indonesia SAG mill 2367–2379
    solid–solid separation process control 2158–2163
    Sunrise Dam Gold Mine start-up 2345–2366
Centrifugal separation
    applications 1275–1288
    calcium carbonate 1285–1286
    mills 783–800
    separators 953–954
    terminology 1266–1267
    theory of 1261–1266
    types of centrifuges 1267–1274
CIC, CIL, and CIP processing
    circuit design and equipment selection 1652–1659
    compared with agitated tank leaching 1631, 1635
    selecting among CIP, CIL, and CIC processing 1644–1651
Circuit surveys 63–73
Clarification 201–206
Coal
    centrifugal separation 1275–1278
    flotation separation 1171–1172
    and heavy-media cyclones 962–977
    heavy-media separator sizing and selection 1011–1032
    preparation plant control case study 2158–2160
Comminution
    autogenous and semi-autogenous grinding mill selection and sizing 755–782
    bench-scale and pilot plant testing for circuit design 123–134
    Bond Work Index 544
    circuit flowsheets 579–583
    circuit sampling 63–73
    crusher types 566–583
    crushing plant design and layout 669–697
    factors in circuit selection 539–565
    grinding equipment types and circuit flowsheets 698–709
    grinding mill design 840–864
    grinding mill drive selection and evaluation 819–839
    grinding mill selection 710–754
    grinding plant design and layout 801–818
    high pressure grinding roll selection and sizing 636–668
    in-pit crushing design and layout 606–620
    JKSimMet comminution circuit simulator 442–460
    pebble crusher selection and sizing 628–635
    primary crusher selection and sizing 584–605
    sampling for process design 99–116
    secondary and tertiary cone crusher selection and sizing 621–627
    ultrafine and stirred grinding mill selection and sizing 783–800
Compartment mills 731–733
Computational fluid dynamics 495–506, 1730–1731
    in design of mechanical flotation equipment 1189
Computer Integrated Manufacturing model 2052
Concentrate
    Falconbridge Raglan Concentrator dewatering, storage, and transportation studies 2400–2403
    handling 1474–1476
Cone separators 987–988
Construction materials
    aluminum 1924–1925
    copper and copper alloys 1925–1926
    elastomers 1932–1952
    irons 1913–1914
    masonry and membrane linings for process vessels 1962–1969
    metallic 1911, 1927–1931
    nickel alloys 1923–1924
    plastics 1953–1961
    stainless steels 1914–1923
    steels 1912–1913
    titanium 1926–1927
Construction projects 2250–2274
Contracts 2245, 2249
    bonuses and penalties 2262–2263
    Fixed Price 2246, 2248
    lump sum 2261
    Lump Sum Turn Key 2246, 2248–2249
    and performance tests 2286
    and ramp up 2286–2287
    Reimbursable Cost Plus Fixed Fee 2246–2247
    Reimbursable Cost Plus Incentive Fee 2246–2247
    Reimbursable Cost Plus Percentage Fee 2246–2247
    types 2262
Copper
    as construction material 1925–1926
    flotation separation of copper gold ores 1124–1141
    heap leach design and practice 1571–1605
    recovery by solvent extraction and electrowinning 267–268, 1709–1744
Corrosion control 1962–1969

Cost estimations
   capital costs   314–324, 2254–2255
   Class I (order of magnitude)   310, 315, 327
   Class II (preliminary)   315
   Class III (definitive)   315
   cost indexes   310–314
   equipment costs   310–314
   O'Hara method   317–320
   process operating costs   326–345
   terminology   314
Costs
   heap leaching   1627–1629
   management and control   2230–2237
   weighted average cost of capital   351
Counter current decantation (CCD) thickeners   1631
Crowe, T.B.   1665
Crushers and crushing   566–567, 578–583
   control strategies   2114–2115, 2122–2129
   crusher types   571–578, 588–595
   factors in crusher selection   567–571
   in-pit crushing design and layout   606–620
   instrumentation strategies   2114–2122
   plant design and layout   669–697
   selection and sizing of pebble crushers   628–635
   selection and sizing of primary crushers   584–605
   selection and sizing of secondary and tertiary cone crushers   621–627
Crystal Ball software   341–344
Cyanide leaching
   bench-scale and pilot plant testing in circuit design   251–263
   bench-scale and pilot plant testing in gold recovery from cyanide solutions   264–266
   and Merrill-Crowe process   1663, 1666–1671
   process chemistry   1666–1667
   solution characteristics   1671
Cyanide removal from solutions and slurries   1866–1885

D

Davis–Ritchie equations   1889
Davy BB   1716
Demonstration plants   121
Design criteria   3–12
   format   14–16
   requirements   9–11
   sample excerpt   18–21
   sample index   7–9
Discounted cash flow method   346–370
Discrete element methods   495–506
Discrete grain breakage   495–506
Drill mud   1285
Dry separation   954, 989–991
Drying equipment
   selection criteria   1463–1466
   storage   1476–1477
   types   1466–1474
Dual laminate   1953–1961
Dutch State Mines   963
Dynawhirlpool   962–963

E

Elastomers   1932–1934, 1947–1952
   defined   1932
   types   1934–1940
   uses   1941–1947
Electric power distribution systems   1973–1982
Electrical separation methods   1049–1068
Electrowinning
   in copper heap leaching   1571–1605
   gold   1699–1708
   plant sizing and selection for copper EW   1734–1739
   and refining   1748, 1755–1756
   and solvent extraction in copper recovery   267–268, 1709–1744
Elution
   Anglo American Research (AARL) process   1694–1697
   in carbon reactivation   1683
   gold   1694–1699
   Zadra process   1697–1699
Empire Mine high pressure grinding roll case study   2380–2393
Engineering
   company selection   2306–2307
   engineering and construction (E&C) firms   2250–2274
Enhanced gravity separation   988–989
Enterprise dynamic simulation models   528–535
Enterprise resource management systems   507–527
Environmental factors in facility siting   1902–1908
Equipment specification and bidding   2223–2229

F

Facilities and support systems. *See also* Construction materials
   electric power distribution systems   1973–1982
   environmental and social factors in siting   1902–1908
   metallurgical and assay laboratory equipment, design, and layout   2011–2019
   motor and drive system selection for comminution circuits   1983–2010
   on-line composition analysis of mineral slurries   2020–2047
   research and development laboratories   121–122

Falconbridge Raglan Concentrator arctic grinding
and dewatering studies   2394–2410
Feasibility studies   281–282
   final   295–309
   and prefeasibility studies in operating cost
      estimates   328–336
   preliminary (intermediate economic
      evaluations)   288–295
   preliminary evaluations   282–288
Feeder selection   1486–1490
Filtration
   bench-scale and pilot plant testing in circuit
      design   207–210
   and centrifugal separation   1262–1288
   equipment testing, sizing, and specifying
      1313–1330
   equipment types and design features
      1342–1357
   filter media   1355–1356
   pressure   1348–1354
   principles and theory for equipment
      characterization   1289–1294
   theory   1314–1315
   vacuum   1343–1348
Finance. *See also* Contracts
   discounted cash flow method   346–370
   project financing   371–379
   risk-reward balance   371–372
Fine particle separators   952–953
Fire assaying   1778–1805
Fisher's F-test   54
Fixed Price contracts   2246, 2248
Flocculation   201–206
Flotation   2174–2177
   Batu Hijau kinetic modeling   2211–2222
   bench and pilot plant programs in circuit
      design   145–159
   circuit sampling   63–73
   column flotation   1239–1252
   complex sulphide ores   1097–1123
   control strategies   2177–2195
   copper gold ores   1124–1141
   design of mechanical flotation equipment
      1179–1203
   equipment selection   1204–1218
   nickel ores   1142–1158
   nonsulfide minerals   1159–1178
   plant layout   1218–1238
   reagent selection via batch flotation tests
      136–144
Fluor Corporation   2211–2212
FOXCOM protocol   2086

G

Gates   1489
GENCOR S.A. Ltd.   1547

Gold
   Barrick Goldstrike acidic autoclave
      1530–1534, 1538–1539
   Barrick Goldstrike oxygenated roaster
      1535–1539
   Barrick Goldstrike two-stage roaster
      1493–1509
   bench-scale and pilot plant testing in
      selection of pre-oxidation process
      211–250
   biooxidation chemistry and tests   238–245
   biooxidation in heap leaching of
      sulfidic-refractory gold ore   1562–1564
   bullion production and refining   1747–1759
   determination by spectroscopy and other
      instrumental methods   1797–1802
   electrowinning   1699–1708
   elution   1694–1699
   fire assaying   1778–1805
   flotation separation of copper gold ores
      1124–1141
   gravity recovery   995–1010
   heap leach design and practice   1606–1630
   pressure oxidation chemistry and tests
      231–238
   pressure oxidation control strategies
      2196–2208
   recovery by activated carbon adsorption
      (bench-scale and pilot plant testing)
      264–266
   recovery by zinc dust cementation
      (bench-scale and pilot plant testing)   266
   roasting chemistry and tests   219–231
   sampling theory and practice   77–98
   selecting among CIP, CIL, and CIC processing
      1644–1651
Gravity separation
   air jigs   990–991
   air tables   989–990
   applications   955–958
   basic technology   947–950
   bench-scale and pilot plant testing in circuit
      design   160–175
   centrifugal separators   953–954
   cone separators   987–988
   density separators   932–936
   dry separation   954, 989–991
   enhanced   988–989
   fine particle separators   952–953
   in gold recovery   995–1010
   jigs   983–985
   non-heavy media   978–994
   rising current washers   986–987
   selection of gravity classifiers   867–879
   shaking tables   952, 985
   sluices   951–952
   spiral concentrators   867–879, 929–931, 952,
      985–986
   wet separation   983–989

Grinding 2130–2131
   control objectives and elements 2131–2140
   control strategies 2147–2151
   Empire Mine high pressure grinding roll case study 2380–2393
   equipment types and circuit flowsheets 698–709
   expert systems and advanced controllers 2146–2147
   Falconbridge Raglan Concentrator studies 2395–2400
   high pressure grinding roll selection and sizing 636–668
   mill design 840–864
   plant design and layout 801–818
   regulatory control loops 2140–2142
   selection and evaluation of mill drives 819–839
   selection and sizing of autogenous and semi-autogenous mills 755–782
   selection and sizing of ultrafine and stirred grinding mills 783–800
   supervisory control 2143–2146
Gy, Pierre 77
   sampling constant 47–48
   Sampling Theory 77–98

H

HART protocol 2085–2086
Heap leaching
   with biooxidation for sulfidic-refractory gold ore 1562–1564
   compared with agitated tank leaching 1634–1635
   copper heap leach design and practice 1571–1605
   costs 1627–1629
   defined and described 1607–1608
   engineering design 1584–1598
   history 1571–1572, 1606–1607
   irrigation (sprinkling) systems 1596–1597, 1614–1615, 1618–1619
   and ore types 1610–1611
   pads and liners 1585–1591, 1619–1620
   precious metal heap leach design and practice 1606–1630
   preliminary evaluations and testing 1572–1584, 1612
   reasons for using 1609–1610
   stacking and reclaiming 1459, 1623–1625
Heavy metal removal from wastewater 1847–1848, 1857–1865
Heavy-media cyclones 962–977
Heavy-media separators 962–977
   sizing and selection of 1011–1032
Heterogeneity Test 77–98
High fidelity simulation 397–400, 495–506

High pressure grinding rolls
   Empire Mine case study 2380–2393
   selection and sizing of 636–668
High-gradient magnetic separator (HGMS) 1087, 1091–1093
Hindered settlers 929–943
HIOX 1550
Holmes & Narver pumper 1716
Homestake Mine 1644
Hopper outlet sizing 1483–1486
HPGR. *See* High pressure grinding rolls
Hydraulic classifiers. *See* Hindered settlers
Hydrocyclones 880–893

I

In-pit crushing and conveying 606–620
INCO Canada 1760, 1767–1769, 1770–1772, 1868–1869
Instrumentation. *See also* Process control
   control devices 2109–2111
   crusher instrumentation strategies 2114–2122
   measurement devices 2106–2108
Internal rate of return 348–350
IPCC. *See* In-pit crushing and conveying
Iron Ore Company (Canada) 939–943
Irons 1913–1914

J

Jet mills 783–800
Jigs 983–985
JKSimFloat flotation simulator 461–478
JKSimMet comminution circuit simulator 442–460
Julius Krutschnitt Mineral Research Centre 442–443, 462

K

Kaolin
   centrifugal separation 1281–1285
   flotation separation 1168–1171

L

Laboratories
   metallurgical and assay (equipment, design, and layout) 2011–2019
   research and development 121–122
LARCODEMS. *See* Large Coal Heavy-media Separator
Large Coal Heavy-media Separator 962–963
Leaching
   agitated tank leaching selection and design 1631–1643
   bench-scale and pilot plant testing in cyanide leach circuit design 251–263
   biooxidation in heap leaching of sulfidic-refractory gold ore 1562–1564

copper heap leach design and practice 1571–1605
heap leach stacking and reclaiming 1459, 1623–1625
precious metal heap leach design and practice 1606–1630
pressure leaching autoclaves 1510–1529
Lead 1097–1123
LIGHTNIN hydrofoil 1716–1718
LKAB Iron Ore Mine (Sweden) 936–939
Lonrho Refinery 1774–1776
Lump Sum Turn Key contracts 2246, 2248–2249

M

Magnetic separation 1069–1071, 1093
  bench-scale and pilot plant testing in circuit design 176–200
  high-gradient magnetic separator (HGMS) 1087, 1091–1093
  integral 1074–1093
  tramp metal removal 1071–1074
  wet high-intensity magnetic separator (WHIMS) 1087–1091
Maintenance
  predictive 2080
  process control 2073–2075, 2104–2113
  and start-up 2315–2327
Masonry and membrane linings for process vessels 1962–1969
Matthey Rustenburg Refinery 1772–1774
MEMS technology 2080–2081
Mercury retorting 1750–1755
Merrill, C.W. 1665
Merrill Crowe process 1663–1679
  and agitated tank leaching 1631
  compared with CIP, CIL, and CIC processing 1644, 1646–1647, 1665–1666
  in gold recovery 266
  history 1663–1665
  and refining 1748–1750, 1756–1757
Metso CBT software 2330–2333
Metso Minerals 404
Mica 1162–1164
MicroElectro Mechanical systems. *See* MEMS technology
Mine Environment Neutral Drainage 1887, 1889
MINEWALL 2.0 1889
Mining-influenced waters (MIWs) 1886–1894, 1899–1901. *See also* Wastewater
  prediction techniques 1894–1899
Model Predictive Control 2147
Modeling and simulation
  Batu Hijau flotation model 2211–2222
  BRUNO Crushing Plant Simulator 404–420
  computational fluid dynamics 495–506

Computer Integrated Manufacturing model 2052
  in design of mechanical flotation equipment 1179–1203
  discrete element methods 495–506
  discrete grain breakage 495–506
  empirical 383, 389–393
  enterprise dynamic simulation models 528–535
  enterprise resource management systems 507–527
  fundamental 383
  hierarchy 383–384
  high fidelity simulation 397–400, 495–506
  history of mineral processing simulation 384–389
  JKSimFloat flotation simulator 461–478
  JKSimMet comminution circuit simulator 442–460
  mining-influenced waters 1888–1889
  Model Predictive Control 2147
  overview 383–403
  phenomenological 383
  PlantDesigner crushing and screening program 421–441
  population balance models 393–397, 495–506
  USIM PAC 3.0 simulator 479–494
ModL programming language 423–434
Mount Isa Inlier (Australia), 1097

N

National Institute of Metallurgy (South Africa) 1774
NEMA 4 and 4X enclosures 2105
Net present value 348–350
.NET technology 2080, 2081
Newmont Mining Corporation 2196, 2211–2212, 2337–2341
Nickel 1142–1158
Nickel alloys 1923–1924
Nonsulfide minerals 1159–1178

O

OCA. *See* On-line composition analysis of mineral slurries
OCA-XRF. *See* X-ray fluorescence
On-line composition analysis of mineral slurries 2020–2021, 2046–2047
  methods 2021–2023
Operating manuals 2292–2295
Operator training 2290–2298, 2328–2336
  computer based 2334–2335
Overflow 868

P

Pebble mills 733–735, 748

Performance testing   2286, 2289
PGMs.   *See* Platinum group metals
Phosphate   1172–1175
Photometric ore sorting   1033–1048
Pilot plant testing.   *See* Bench-scale and pilot-plant testing
PlantDesigner crushing and screening modeling program   421–441
Plastic construction materials   1953–1961
Platinum group metals
   bullion production and refining   1760–1777
   defined   1760
   determination by spectroscopy and other instrumental methods   1797–1802
   fire assaying   1778–1805
Population balance models   393–397, 495–506
Potash
   centrifugal separation   1278–1279
   flotation separation   1175–1176
Pre-commissioning   2277–2284
Pre-oxidation
   Barrick Goldstrike acidic autoclave   1530–1534, 1538–1539
   Barrick Goldstrike oxygenated roaster   1535–1539
   Barrick Goldstrike two-stage gold ore roaster   1493–1509
   biooxidation (bioleaching) equipment and circuits   1540–1568
   pressure oxidation and pressure leaching autoclaves   1510–1529
Precious metals.   *See* Gold, Platinum group metals, Silver
Pressure leaching   1510–1529
Pressure oxidation
   autoclaves   1510–1529
   chemistry and tests   231–238
   process control strategies   2196–2208
Process control   2051–2056, 2063–2065, 2078–2080.   *See also* Instrumentation
   asset management   2080
   control devices   2109–2111
   crusher control strategies   2114–2115, 2122–2129
   DCS   2090–2091, 2093–2096
   design   2066–2070, 2075–2076
   flotation instrumentation and control strategies   2174–2195
   FOXCOM protocol   2086
   grinding instrumentation and control strategies   2130–2151
   hardware   2060
   hardware selection   2077–2103
   HART protocol   2085–2086
   hybrid   2090–2091, 2096–2097
   instrumentation   2057–2059
   level 1 communications   2084–2087
   level 2 communications   2087–2088
   level 3 and 4 communications   2088–2089
   maintenance   2073–2075, 2104–2113
   measurement devices   2106–2108
   MEMS technology   2080–2081
   Model Predictive Control   2147
   NEMA 4 and 4X enclosures   2105
   .NET technology   2080, 2081
   PLC   2090–2093
   predictive maintenance   2080
   pressure oxidation control strategies   2196–2208
   process   2056–2057, 2067
   process management   2053
   production management   2053
   SEVA technology   2080, 2081
   solid–solid separation instrumentation and control strategies   2152–2163
   strategies   2060–2062
   system selection   2070–2072
   system types   2090–2097
   thickening instrumentation and control strategies   2164–2173
   users   2063
PT Freeport Indonesia SAG mill case study   2367–2379
PT Newmont Nusa Tenggara   2211
PT Pukuafu Indah   2211
PYROX   1889

Q

QEM*SEM   28
Qualitative Evaluation of Materials by Scanning Electron Microscopy.   *See* QEM*SEM
Quartz/feldspar   1164–1168
Quebec Cartier Mine   929

R

Raglan Concentrator arctic grinding and dewatering studies   2394–2410
Ramp up   2285–2289
Regrind mills   735, 748–751
Reimbursable Cost Plus Fixed Fee contracts   2246–2247
Reimbursable Cost Plus Incentive Fee contracts   2246–2247
Reimbursable Cost Plus Percentage Fee contracts   2246–2247
Research and development laboratories   121–122
Rising current washers   986–987
Risk analysis
   conventional sensitivity analysis   339–340
   in feasibility studies   288, 306–307
   Monte Carlo technique (Crystal Ball software)   341–344
   risk components in mineral project   351–352
   risk factors in project financing   373–376
   risk–reward balance   371–372

Risk and contracts 2245–2249
Rod mills 712–721, 743–744

S

SAG mills
   bench-scale and pilot plant testing for circuit design 123–134
   equipment types and characteristics 698–709
   followed by ball mills 801–818
   motor and drive system selection 1983–2010
   PT Freeport Indonesia case study 2367–2379
   selection and sizing 755–782
Sampling
   central values 40–41
   for comminution processes 99–116
   for feasibility studies 25–36
   Fisher's F-test 54
   gold 77–98
   Gy's sampling constant 47–48
   Heterogeneity Test 77–98
   interleaving sampling protocol 56–57
   large comminution and flotation circuits 63–73
   in mineral processing 37–55
   for mineral separation processes 99–116
   presentation of results 35–36
   spatial dependence 43–45
   Student's t-test 50–53, 61
   terminology 38–39
   types of 26–27
   variance 41–43, 45–46, 58–59
Sand
   centrifugal separation of tar sand 1281
   process control for magnetic separation of mineral sands 2160–2162
Sandvik Rock Processing AB 421
Schedule development and control 2238–2244
Screening
   coarse 894–916
   fine 917–928
Sedimentation
   bench-scale and pilot plant testing 201–206
   centrifugal 1262–1288
   equipment testing, sizing, and specifying 1295–1312
   equipment types and design features 1331–1341
   theory 1295–1296
Selenium
   removal in bullion production 1755
   removal from wastewater 1847–1848, 1853–1856, 1857–1865
Semi-autogenous grinding mills. *See* SAG mills
Separation processes 99–116
Settling rate 868
SEVA technology 2080, 2081
Shaking tables 952, 985

Silver
   bullion production and refining 1747–1759
   determination by spectroscopy and other instrumental methods 1797–1802
   fire assaying 1778–1805
   flotation separation of zinc–silver–lead ores 1097–1123
   heap leach design and practice 1606–1630
Simulation. *See* Modeling and simulation
Size separation
   coarse screening 894–916
   fine screening 917–928
   gravity classifiers 867–879, 929–943
   hydrocyclones 880–893
   overflow 868
   settling rate 868
   spiral concentrators 867–879, 929–931, 952, 985–986
   underflow 868
Sluices 951–952
Slurry
   Bingham plastic fluids 1392–1393, 1399, 1427
   classifications 1392–1399, 1427
   conventional 1430
   conventional tailings 1431
   cyanide removal 1866–1885
   heterogeneous 1392–1395
   launders 1409–1415
   Newtonian viscous 1392–1393, 1395–1396, 1427
   non-conventional 1431
   non-conventional steam coal slurries 1431
   on-line composition analysis 2020–2047
   pipeline systems 1422–1445
   piping system sizing and design 1403–1409, 1417–1418
   properties 1426–1427
   pumpboxes 1416–1417
   thickened tailings disposal 1431
   upcomers 1415–1416
   valve selection 1418–1419
Slurry pumps
   applications 1399–1401
   bearing assembly 1375
   casings 1379–1380
   construction materials 1380–1383
   hydraulics 1385–1388
   impellers 1375–1377
   pumpboxes 1416–1417
   selection and sizing 1373–1402
   settling velocity 1388–1391
   specific speed 1378–1379
   wear and cavitation 1383–1385
Social factors in facility siting 1902–1908
Soda ash 1279–1280
Solid–liquid separation
   available equipment designs 1256

centrifugal sedimentation and filtration 1262–1288
control architecture and equipment 2169–2170
filtration equipment characterization 1289–1294
filtration equipment testing, sizing, and specifying 1313–1330
filtration equipment types and design features 1342–1357
instrumentation 2167–2169
instrumentation and control strategies 2164–2173
plant design, layout, and economic considerations 1358–1369
process characterization 1255–1261
sedimentation equipment testing, sizing, and specifying 1295–1312
sedimentation equipment types and design features 1331–1341
Solid–solid separation 2152–2153
electrical methods 1049–1068
gravity recovery of gold 995–1010
gravity separation 947–961
heavy-media cyclone, 962–977
heavy-media separator sizing and selection 1011–1032
incremental grade concept 2153–2156
magnetic methods 1069–1093
non-heavy media 978–994
photometric ore sorting 1033–1048
plant control strategies 2156–2158
process control case studies 2158–2163
Solvent extraction
in copper heap leaching 1571–1605
and electrowinning in copper recovery 267–268, 1709–1744
McCabe-Thiele diagrams 1709–1714
mixing concepts 1716–1726
settler concepts 1726–1734
Specifications 2223–2226
Spiral concentrators 867–879, 929–931, 952, 985–986
Stacking systems
cascading conveyor ("grasshopper") system 1456–1457
fixed stackers 1455
heap leach stacking and reclaiming 1459, 1623–1625
overhead tripper or shuttle 1455
radial stacker 1456
silo storage 1460
stockpile calculations 1461–1462
stockpile types and selection criteria 1454
traveling stackers (stacker/reclaimers) 1457–1458
Start-up and commissioning
maintenance scheduling, management, and training 2315–2327

operator training and manuals 2290–2298, 2328–2336
and performance testing 2286, 2289
planning and staffing 2299–2314
and pre-commissioning 2277–2284
and ramp up 2285–2289
safety and health considerations 2337–2341
successful start-up defined 2299–2302
Sunrise Dam Gold Mine case study 2345–2366
Steel 1912–1913
stainless 1914–1923
Stillwater Mining 1761–1767
Stirred media mills 783–800
Student's t-test 50–53, 61
Sumitomo Corporation 2211
Sunrise Dam Gold Mine start-up case study 2345–2366

T

Tailings
co-disposal with other wastes 1817
conventional 1431, 1815
dams and impoundments 1828–1846
dewatering 1837–1838
dry, 1817
dry cake disposal 1838
Falconbridge Raglan Concentrator dewatering and deposition studies 2404–2408
land disposal (site development and operations) 1809–1827
new developments in disposal 1833–1840
paste, 1816, 1839
subaqueous disposal 1835
thickened tailings disposal 1431, 1815, 1839–1840
Tar sand 1281
Thermal reactivation 1683–1692
Thermoplastic 1953–1961
Thermoset 1953–1961
Thickened tailings disposal 1431, 1815, 1839–1840
Thickening
bench-scale and pilot plant testing in circuit design 201–206
control architecture and equipment 2169–2170
counter current decantation (CCD) 1631
instrumentation 2167–2169
instrumentation and control strategies 2164–2173
Titanium 1926–1927

U

Underflow 868
USIM PAC 3.0 simulator 479–494

W

Wasp, Edward J. 1422
Wastewater. *See also* Mining-influenced
    waters (MIWs)
  arsenic removal 1847–1853, 1857–1865
  heavy metal removal 1847–1848, 1857–1865
  selenium removal 1847–1848, 1853–1856,
    1857–1865
Weighted average cost of capital 351
Wet high-intensity magnetic separator (WHIMS)
    1087–1091
Wet separation 983–989

X

X-ray fluorescence 2023–2045

Z

Zadra elution process 1697–1699
Zinc
  cementation or precipitation. *See* Merrill
    Crowe process
  flotation separation of zinc–silver–lead ores
    1097–1123